LIFE

The Science of Biology

EIGHTH EDITION

Sinauer Associates, Inc.

W. H. Freeman and Company

EIGHTH EDITION

LIFE

The Science of Biology

DAVID
SADAVA
The Claremont Colleges
Claremont, California

H. CRAIG
HELLER
Stanford University
Stanford, California

GORDON H.
ORIANS
Emeritus, University of Washington
Seattle, Washington

WILLIAM K.
PURVES
Emeritus, Harvey Mudd College
Claremont, California

DAVID M.
HILLIS
University of Texas
Austin, Texas

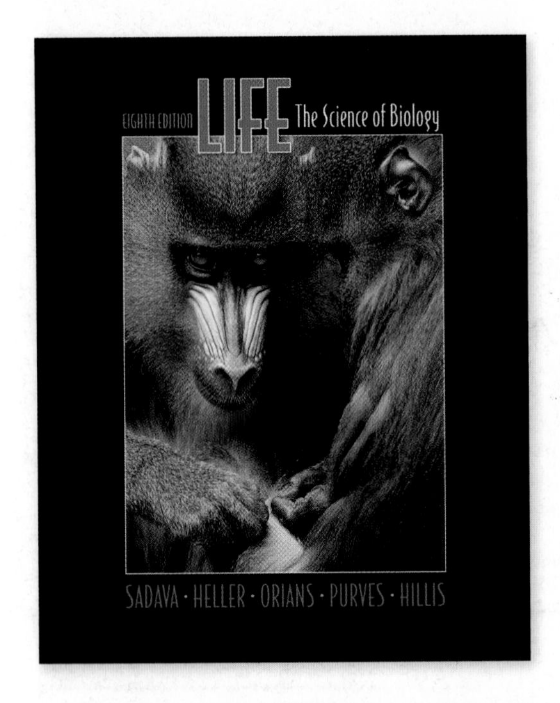

Cover Photograph

"Amor de Madre," photograph of mandrill (*Mandrillus sphinx*) mother and young.

Frontispiece

Grizzly bears (*Ursus arctos horribilis*) hunt salmon in an Alaskan river.

LIFE: The Science of Biology, Eighth Edition

Address editorial correspondence to:

Sinauer Associates Inc., 23 Plumtree Road, Sunderland, MA 01375 U.S.A.
www.sinauer.com
publish@sinauer.com

Address orders to:

VHPS/W.H. Freeman & Co., Order Dpt., 16365 James Madison Highway,
U.S. Route 15, Gordonsville, VA 22942 U.S.A.
www.whfreeman.com

Examination copy information: 1-800-446-8923
Orders: 1-888-330-8477

Library of Congress Cataloging-in-Publication Data

Life: the science of biology / David Sadava ... [et al.]. — 8th ed.
p. cm.
Includes index.
ISBN-13: 978-0-7167-7671-0 (hardcover) – ISBN 978-0-7167-7673-4 (Volume I) –
ISBN 978-0-7167-7674-1 (Volume 2) – ISBN 978-0-7167-7675-8 (Volume 3)
1. Biology. I. Sadava, David E.
QH308.2.L565 2007
570—dc22 2006031320

Printed in U.S.A.
First Printing December 2006
The Courier Companies, Inc.

To our students, especially the more than 30,000
we have collectively instructed in introductory biology over the years.

The Authors

Craig Heller — Gordon Orians — Bill Purves — David Sadava — David Hillis

David Sadava is the Pritzker Family Foundation Professor of Biology at the Keck Science Center of Claremont McKenna, Pitzer, and Scripps, three of The Claremont Colleges. Twice winner of the Huntoon Award for superior teaching, Dr. Sadava has taught courses on introductory biology, biotechnology, biochemistry, cell biology, molecular biology, plant biology, and cancer biology. He is a visiting scientist in medical oncology at the City of Hope Medical Center. He is the author or coauthor of five books on cell biology and on plants, genes, and crop biotechnology. His research has resulted in over 50 papers, many coauthored with undergraduates, on topics ranging from plant biochemistry to pharmacology of narcotic analgesics to human genetic diseases. For the past 15 years, he and his collaborators have investigated multi-drug resistance in human small-cell lung carcinoma cells with a view to understanding and overcoming this clinical challenge. Their current work focuses on new anti-cancer agents from plants.

Craig Heller is the Lorry I. Lokey/Business Wire Professor in Biological Sciences and Human Biology at Stanford University. He earned his Ph.D. from the Department of Biology at Yale University in 1970. Dr. Heller has taught in the core biology courses at Stanford since 1972 and served as Director of the Program in Human Biology, Chairman of the Biological Sciences Department, and Associate Dean of Research. Dr. Heller is a fellow of the American Association for the Advancement of Science and a recipient of the Walter J. Gores Award for excellence in teaching. His research is on the neurobiology of sleep and circadian rhythms, mammalian hibernation, the regulation of body temperature, and the physiology of human performance. Dr. Heller has done research on sleeping kangaroo rats, diving seals, hibernating bears, and exercising athletes. Some of his recent work on the effects of temperature on human performance is featured in the opener to Chapter 40.

Gordon Orians is Professor Emeritus of Biology at the University of Washington. He received his Ph.D. from the University of California, Berkeley in 1960 under Frank Pitelka. Dr. Orians has been elected to the National Academy of Sciences and the American Academy of Arts and Sciences, and is a Foreign Fellow of the Royal Netherlands Academy of Arts and Sciences. He was President of the Organization for Tropical Studies, 1988–1994, and President of the Ecological Society of America, 1995–1996. He is a recipient of the Distinguished Service Award of the American Institute of Biological Sciences. Dr. Orians is a leading authority in ecology, conservation biology, and evolution. His research on behavioral ecology, plant–herbivore interactions, community structure, and environmental policy has taken him to six continents. He now devotes full time to writing and to helping apply scientific information to environmental decision-making.

Bill Purves is Professor Emeritus of Biology as well as founder and former Chair of the Department of Biology at Harvey Mudd College in Claremont, California. He received his Ph.D. from Yale University in 1959 under Arthur Galston. A fellow of the American Association for the Advancement of Science, Dr. Purves has served as head of the Life Sciences Group at the University of Connecticut, Storrs, and as Chair of the Department of Biological Sciences, University of California, Santa Barbara, where he won the Harold J. Plous Award for teaching excellence. His research interests focused on the hormonal regulation of plant growth. Dr. Purves elected early retirement in 1995, after teaching introductory biology for 34 consecutive years, in order to concentrate entirely on research directed at learning and science education. He is currently participating in the development of a virtual technical high school, with responsibility for curriculum design in scientific reasoning and health science.

David Hillis is the Alfred W. Roark Centennial Professor in Integrative Biology and the Director of the Center for Computational Biology and Bioinformatics at the University of Texas at Austin, where he also has directed the School of Biological Sciences. Dr. Hillis has taught courses in introductory biology, genetics, evolution, systematics, and biodiversity. He has been elected into the membership of the American Academy of Arts and Sciences, awarded a John D. and Catherine T. MacArthur Fellowship, and has served as President of the Society for the Study of Evolution and of the Society of Systematic Biologists. His research interests span much of evolutionary biology, including experimental studies of evolving viruses, empirical studies of natural molecular evolution, applications of phylogenetics, analyses of biodiversity, and evolutionary modeling. He is particularly interested in teaching and research about the practical applications of evolutionary biology.

Preface

As active scientists working in a wide variety of both basic and applied biology, we are fortunate to be part of a field that is not only fascinating but also changes rapidly. It is apparent not just in the time span since we started our careers—we see it every day when we open a newspaper or a scientific journal. As educators of both introductory and advanced-level students, we desire to convey our excitement about biology's dynamic nature.

This new edition of *Life* looks, and is, quite different from its predecessors. In planning the Eighth Edition, we focused on three fundamental goals. The first was to maintain and enhance what has worked well in the past—an emphasis on not just what we know but how we came to know it; the incorporation of exciting new discoveries; an art program distinguished by its beauty and clarity; plus a unifying theme. As should be the case in any biology textbook, that theme is evolution by natural selection, a 150-year-old idea that more than ever ties together the living world. We have been greatly helped in this endeavor by the addition of a new author, David Hillis. His knowledge and insights have been invaluable in developing our chapters on evolution, phylogeny, and diversity, and they permeate the rest of the book as well.

Our second goal has been to make *Life* more pedagogically accessible. From the bold new design to the inclusion of numerous learning aids throughout each chapter (see New Pedagogical Features), we have worked to make our writing consistently easy to follow as well as engaging.

Third, between editions we asked seven distinguished ecologists—all of whom teach introductory biology—to provide detailed critiques of the Ecology unit. As a result of their extensive suggestions, Part Nine, Ecology, has a fresh organization (see The Nine Parts). And one of the seven, May Berenbaum, has agreed to join the *Life* author team for the Ninth Edition. The other six stalwarts are thanked in the "Reviewers of the Eighth Edition" section.

Enduring Features

As stated above, we are committed to a blending of a presentation of the core ideas of biology with an emphasis on introducing our readers to the process of scientific inquiry. Having pioneered the idea of depicting seminal experiments in specially designed figures, we continue to develop this here, with 96 EXPERIMENT figures (28 percent more than in the Seventh Edition). Each follows the structure: Hypothesis, Method, Result, and Conclusion. Many now include "Further Research," which asks students to conceive an experiment that explores a related question.

A related feature is the RESEARCH METHOD figures, depicting many laboratory and field methods used to do this research. All the Experiment and Research Method figures are listed in the endpapers at the back of the book.

Another much-praised feature—which we pioneered ten years ago in *Life*'s Fifth Edition—is the BALLOON CAPTIONS used in our figures. We know that many students are visual learners. The balloon captions bring explanations of intricate, complex processes directly into the illustration, allowing the reader to integrate the information without repeatedly going back and forth between the figure and its legend.

Life is the only introductory biology book for science majors that begins each chapter with a story. These OPENING STORIES, most of which are new to this edition, are meant to intrigue students while helping them see how the chapter's biological subject relates to the world around them.

New Pedagogical Features

There are several new elements in the Eighth Edition chapters. Each has been designed as a study tool to aid the student in mastering the material. In the opening page spread, IN THIS CHAPTER previews the chapter's content, and the CHAPTER OUTLINE gives the major section headings, all numbered and all framed as questions to emphasize the inquiry basis of science.

Each main section of a chapter now ends with a RECAP. This key element briefly summarizes the important concepts in the section, then provides two or three questions to stimulate immediate review. Each question includes reference to pertinent text or a figure or both.

The CHAPTER SUMMARY boldfaces key terms introduced and defined in the chapter. We have kept the highlighted references to key figures and to the Web tutorials and activities that support a topic in the chapter.

Another new element, BIOBITS, is not strictly a learning aid, but offers intriguing (occasionally amusing) supplemental information. BioBits, like the opening stories, are intended to help students appreciate the interface between biology and other aspects of life.

Redesigned WEB ICONS alert the reader to the tutorials and activities on *Life*'s companion website (www.thelifewire.com). Each of these study and review resources, many of which are new for the Eighth Edition, has been created specifically for *Life*. A full list, by chapter, is found in the front endpapers of the book.

The Nine Parts

We have reorganized the book into nine parts. Part One sets the stage for the entire book, with the opening chapter on biology as an exciting science, starting with a student project, and how evolution unites the living world. This is followed by chapters on the basic chemical building blocks that underlie life. We have tried to tie this material together by relating it to theories on the origin of life, with new discoveries of water in our solar system as an impetus.

In Part Two, Cells and Energy, we present an integrated view of the structure and biochemical functions of cells. The discussions of biochemistry are often challenging for students; thus we have reworked both the text and illustrations for greater clarity. These discussions are presented in the context of the latest discoveries on the origin of life and evolution of cells.

Part Three, Heredity and the Genome, begins with continuity at the cellular level, and then outlines the principles of genetics and the identification of DNA as the genetic material. New examples, such as the genetics of coat color in dogs, enliven these chapters. This is followed by chapters on gene expression and on the prokaryotic and eukaryotic genomes. Many new discoveries have been made in this new field of genomics, ranging from tracking down the bird flu virus to the genomes of wild cats, such as the cheetah.

Part Four, Molecular Biology: The Genome in Action, reinforces the basic principles of classical and molecular genetics by applying them to such diverse topics as cell signaling, biotechnology, and medicine. We use many new experiments and examples from applied biology to illustrate these concepts. These include the latest information on the human genome and the emerging field of systems biology. The chapter on natural defenses now includes a discussion of allergy.

Part Five, The Patterns and Processes of Evolution, has been updated in several important ways. We have emphasized the importance of evolutionary biology as a basis for comparing and understanding all aspects of biology, and have described numerous practical applications of evolutionary biology that will be familiar and relevant to the everyday lives of most students.

Recent experimental studies of evolution are described and explained, to help students understand that evolution is an ongoing, observable process. The chapters on phylogenetics and molecular evolution have been completely rewritten to reflect recent advances in those fields. Other changes reflect our growing knowledge of the history of life on Earth and the mechanisms of evolution that have given rise to all of biodiversity.

Part Six, The Evolution of Diversity, reflects the latest views on phylogeny. It continues to emphasize groups united by evolutionary history over classically defined taxa. This emphasis is now supported by an appendix on the Tree of Life that clearly maps out and describes all groups discussed in the text, so that students can quickly look up unfamiliar names and see how they fit into the larger context of life. We now discuss aspects of phylogeny that are still under study or debate (among major groups of eukaryotes, plants, and animals, for instance).

In Part Seven, Flowering Plants: Form and Function, we report on several exciting new discoveries. These include the receptors for auxin, gibberellins, and brassinosteroids as well as great progress on the florigen problem. We have updated our treatment of signal transduction pathways and of circadian rhythms in plants. The already strong treatment of environmental challenges to plants has been augmented by new Experiment figures on plant defenses against herbivores, one confirming that nicotine does help tobacco plants resist certain insects.

Part Eight, Animals: Form and Function, is about how animals work. Although we give major attention to human physiology, we embed it in a background of comparative animal physiology. Our focus is systems physiology but we also introduce the underlying cellular and molecular mechanisms. For example, our explanations of nervous system phenomena—whether they be action potentials, sensation, learning, or sleep—are discussed in terms of the properties of ion channels. The actions of hormones are explained in terms of the molecular mechanisms. Maximum athletic performance is explained in terms of the underlying cellular energy systems. Throughout Part Eight we try to help the student make the connections across all levels of biology, from molecular to behavioral, and to see the relevance of physiology to issues of health and disease. Of central importance in each chapter is mechanisms of control and regulation.

Part Nine, Ecology, begins with a new chapter that describes the scope of ecological research and discusses recent advances in our understanding of the broad patterns in the distribution of life on Earth. The next chapter, also new, combines Behavior and Behavioral Ecology. It shows how the decisions that organisms make during their lives influence both their survival and reproductive success, and also the dynamics of populations and the structure of ecological communities. The chapter on Population Ecology has new material that explains how ecologists are able to mark and follow individual organisms in the wild to determine their survival and reproductive success. Following a chapter on Community Ecology, another new chapter, Ecosystems and Global Ecology, shows how ecologists are expanding the scope of their studies to encompass the functioning of the global ecosystem. This discussion leads naturally to the final chapter in the book, Conservation Biology, which describes how ecologists and conservation biologists work to reduce the rate at which species are becoming extinct as a result of human activities.

Full Books, Paperbacks, or Loose-Leaf

We again provide *Life* both as the full book and as a cluster of paperbacks. Thus, instructors who want to use less than the whole book, or who want their students to have more portable units, can choose from these split volumes:

Volume I, The Cell and Heredity, includes: Part One, The Science and Building Blocks of Life (Chapters 1–3); Part Two, Cells and Energy (Chapters 4–8); Part Three, Heredity and the Genome (Chapters 9–14); and Part Four, Molecular Biology: The Genome in Action (Chapters 15–20).

Volume II, Evolution, Diversity, and Ecology, includes: Chapter 1, Studying Life; Part Five, The Patterns and Processes of Evolution (Chapters 21–25); Part Six, The Evolution of Diversity (Chapters 26–33); and Part Nine, Ecology (Chapters 52–57).

Volume III, Plants and Animals, includes: Chapter 1, Studying Life; Part Seven, Flowering Plants: Form and Function (Chapters 34–39); and Part Eight, Animals: Form and Function (Chapters 40–51).

Note that each volume also includes the book's front matter, Appendixes, Glossary, and Index.

Life is also available in a loose-leaf version. This shrink-wrapped, unbound, 3-hole punched version is designed to fit into a 3-ring binder. Students take only what they need to class and can easily integrate any instructor handouts or other resources.

Media and Supplements for the Eighth Edition

The media and supplements for *Life*, Eighth Edition have been assembled with two main goals in mind: (1) to provide students with a collection of tools that helps them effectively master the vast amount of new information that is being presented to them in the introductory biology course; and (2) to provide instructors with the richest possible collection of resources to aid in teaching the course—preparing, presenting the lecture, providing course materials online, and assessing student comprehension.

All of the *Life* media and supplemental resources have been developed specifically for this textbook. This gives the student the greatest degree of consistency when studying across different media. For example, the animated tutorials and activities found on the Companion Website were built using textbook art, so that the manner in which structures are illustrated, the colors used to identify objects, and the terms and abbreviations used are all consistent.

The rich collection of visual resources in the Instructor's Media Library provides instructors with a wide range of options for enhancing lectures, course websites, and assignments. Highlights include: layered art PowerPoint® presentations that break down complex figures into detailed, step-by-step presentations; a collection of approximately 200 video segments that can help capture the attention and imagination of students; and the new set of PowerPoint® slides of textbook art with editable labels and leaders that allow easy customization of the figures.

For a detailed description of all the media and supplements available to accompany the Eighth Edition, please turn to "Life's Media and Supplements package" on page xiii.

Many People to Thank

One of the wisest pieces of advice ever given to a textbook author is to "be passionate about your subject, but don't put your ego on the page." Considering all the people who looked over our shoulders throughout the process of creating this book, this advice could not be more apt. We are indebted to many people, who gave invaluable help to make this book what it is. First and foremost are our colleagues, biologists from over 100 institutions. Some were users of the previous edition, who suggested many improvements. Others reviewed our chapter drafts in detail, including advice on how to improve the illustrations. Still others acted as accuracy reviewers when the book was almost completed. Our publishers created an advisory group of introductory course coordinators. They advised us on a variety of issues, ranging from book content and design to elements of the print and media supplements. All of these biologists are listed in the Reviewer credits.

We needed a fresh editorial eye for this edition, and we were fortunate to work with Carol Pritchard-Martinez as development editor. With a level head that comes from years of experience, she was a major presence as we wrote and revised. Elizabeth Morales, our artist, was on her second edition with us. This time, she extensively revised almost all of the prior art and translated our crude sketches into beautiful new art. We hope you agree that our art program remains superbly clear and elegant. Once again, we were lucky to have Norma Roche as the copy editor. Her firm hand and encyclopedic recall of our book's many chapters made our prose sharper and more accurate. For this edition, Norma was joined by the capable and affable Maggie Brown. Susan McGlew coordinated the hundreds of reviews that we described above. David McIntyre was a truly proactive photo editor. Not only did he find over 500 new photographs, including many new ones of his own, that enrich the book's content and visual statement, but he set up, performed, and photographed the experiment shown in Figure 36.1. The elegant new interior design is the creation of Jeff Johnson. He also coordinated the book's layout and designed the cover. Carol Wigg, for the eighth time in eight editions, oversaw the editorial process. Her influence pervades the entire book—she created many BioBits, shaped and improved the chapter-opening stories, interacted with David McIntyre in conceiving many photo subjects, and kept an eagle eye on every detail of text, art, and photographs.

W. H. Freeman continues to bring *Life* to a wider audience. Associate Director of Marketing Debbie Clare, the Regional Specialists, Regional Managers, and experienced sales force are effective ambassadors and skillful transmitters of the features and unique strengths of our book. We depend on their expertise and energy to keep us in touch with how *Life* is perceived by its users.

Finally, we are indebted to Andy Sinauer. Like ours, his name is on the cover of the book, and he truly cares deeply about what goes into it.

DAVID SADAVA

CRAIG HELLER

GORDON ORIANS

BILL PURVES

DAVID HILLIS

Reviewers for the Eighth Edition

Between-Edition Reviewers (Ecology and Animal Parts)

May Berenbaum, University of Illinois, Urbana-Champaign
Carol Boggs, Stanford University
Judie Bronstein, University of Arizona
F. Lynn Carpenter, University of California, Irvine
Dan Doak, University of California, Santa Cruz
Jessica Gurevitch, SUNY, Stony Brook
Margaret Palmer, University of Maryland
Marty Shankland, University of Texas, Austin

Advisory Board Members

Heather Addy, University of Calgary
Art Buikema, Virginia Polytechnic Institute and State University
Jung Choi, Georgia Technical University
Rolf Christoffersen, University of California, Santa Barbara
Alison Cleveland, Florida Southern University
Mark Decker, University of Minnesota
Ernie Dubrul, University of Toledo
Richard Hallick, University of Arizona
John Merrill, Michigan State University
Melissa Michael, University of Illinois
Deb Pires, University of California, Los Angeles
Sharon Rogers, University of Nevada, Las Vegas
Marty Shankland, University of Texas, Austin

Manuscript Reviewers

John Alcock, Arizona State University
Charles Baer, University of Florida
Amy Baird, University of Texas, Austin
Patrice Boily, University of New Orleans
Thomas Boyle, University of Massachusetts, Amherst
Mirjana Brockett, Georgia Institute of Technology
Arthur Buikema, Virginia Polytechnic Institute and State University
Hilary Callahan, Barnard College
David Champlin, University of Southern Maine
Chris Chanway, University of British Columbia
Mike Chao, California State University, San Bernardino
Rhonda Clark, University of Calgary
Elizabeth Connor, University of Massachusetts, Amherst
Deborah A. Cook, Clark Atlanta University
Elizabeth A. Cowles, Eastern Connecticut State University
Joseph R. Cowles, Virginia Polytechnic Institute and State University
William L. Crepet, Cornell University
Martin Crozier, Wayne State University
Donald Dearborn, Bucknell University
Mark Decker, University of Minnesota
Michael Denbow, Virginia Polytechnic Institute and State University
Jean DeSaix, University of North Carolina, Chapel Hill
William Eldred, Boston University
Andy Ellington, University of Texas, Austin
Gordon L. Fain, University of California, Los Angeles
Kevin M. Folta, University of Florida
Miriam Goldbert, College of the Canyons
Kenneth M. Halanych, Auburn University
Susan Han, University of Massachusetts, Amherst
Tracy Heath, University of Texas, Austin
Shannon Hedtke, University of Texas, Austin
Mark Hens, University of North Carolina, Greensboro
Albert Herrera, University of Southern California
Barbara Hetrich, University of Northern Iowa
Erec Hillis, University of California, Berkeley
Jonathan Hillis, Austin, Texas
Hopi Hoekstra, University of California, San Diego
Kelly Hogan, University of North Carolina, Chapel Hill
Carl Hopkins, Cornell University
Andrew Jarosz, Michigan State University
Norman Johnson, University of Massachusetts, Amherst
Walter Judd, University of Florida
David Julian, University of Florida
Laura Katz, Smith College
Melissa Kosinski-Collins, Massachusetts Institute of Technology
William Kroll, Loyola University of Chicago
Marc Kubasak, University of California, Los Angeles
Josephine Kurdziel, University of Michigan
John Latto, University of California, Berkeley
Brian Leander, University of British Columbia
Jennifer Leavey, Georgia Institute of Technology
Arne Lekven, Texas A&M University
Don Levin, University of Texas, Austin
Rachel Levin, Amherst College
Thomas Lonergan, University of New Orleans
Blase Maffia, University of Miami
Meredith Mahoney, University of Texas, Austin
Charles Mallery, University of Miami
Ron Markle, Northern Arizona University
Mike Meighan, University of California, Berkeley
Melissa Michael, University of Illinois, Urbana-Champaign
Jill Miller, Amherst College
Subhash Minocha, University of New Hampshire
Thomas W. Moon, University of Ottawa
Richard Moore, Miami University of Ohio
John Morrissey, Hofstra University
Leonie Moyle, University of Indiana
Mary Anne Nelson, University of New Mexico
Dennis O'Connor, University of Maryland, College Park
Robert Osuna, SUNY, Albany
Cynthia Paszkowski, University of Alberta
Diane Pataki, University of California, Irvine
Ron Patterson, Michigan State University
Craig Peebles, University of Pittsburgh
Debra Pires, University of California, Los Angeles
Greg Podgorski, Utah State University
Chuck Polson, Florida Institute of Technology

Donald Potts, University of California, Santa Cruz
Jill Raymond, Rock Valley College
Ken Robinson, Purdue University
Sharon L. Rogers, University of Nevada, Las Vegas
Laura Romano, Denison University
Pete Ruben, Utah State University, Logan
Albert Ruesink, Indiana University
Walter Sakai, Santa Monica College
Mary Alice Schaeffer, Virginia Polytechnic Institute and State University
Daniel Scheirer, Northeastern University
Stylianos Scordilis, Smith College
Kevin Scott, University of Calgary
Jim Shinkle, Trinity University
Denise Signorelli, Community College of Southern Nevada
Thomas Silva, Cornell University
Jeffrey Tamplin, University of Northern Iowa
Steve Theg, University of California, Davis
Sharon Thoma, University of Wisconsin, Madison
Jeff Thomas, University of California, Los Angeles
Christopher Todd, University of Saskatchewan
John True, SUNY, Stony Brook
Mary Tyler, University of Maine
Fred Wasserman, Boston University
John Weishampel, University of Central Florida
Elizabeth Willott, University of Arizona
David Wilson, University of Miami
Heather Wilson-Ashworth, Utah Valley State College

Accuracy Reviewers

John Alcock, Arizona State University
John Anderson, University of Minnesota
Brian Bagatto, University of Akron
Lisa Baird, University of San Diego
May Berenbaum, University of Illinois, Urbana-Champaign
Gerald Bergtrom, University of Wisconsin, Milwaukee
Stewart Berlocher, University of Illinois, Urbana-Champaign
Mary Bisson, SUNY, Buffalo
Arnold Bloom, University of California, Davis
Judie Bronstein, University of Arizona
Jorge Busciglio, University of California, Irvine
Steve Carr, Memorial University of Newfoundland
Thomas Chen, Santa Monica College
Randy Cohen, California State University, Northridge
Reid Compton, University of Maryland, College Park
James Courtright, Marquette University
Jerry Coyne, University of Chicago
Joel Cracraft, American Museum of Natural History
Joseph Crivello, University of Connecticut, Storrs
Gerrit De Boer, University of Kansas, Lawrence
Arturo DeLozanne, University of Texas, Austin
Stephen Devoto, Wesleyan University
Laura DiCaprio, Ohio University
John Dighton, Rutgers Pinelands Field Station
Jocelyne DiRuggiero, University of Maryland, College Park
W. Ford Doolittle, Dalhousie University
Emanuel Epstein, University of California, Davis
Gordon L. Fain, University of California, Los Angeles
Lewis J. Feldman, University of California, Berkeley
James Ferraro, Southern Illinois University
Cole Gilbert, Cornell University
Elizabeth Godrick, Boston University
Martha Groom, University of Washington
Kenneth M. Halanych, Auburn University
Mike Hasegawa, Purdue University
Mark Hens, University of North Carolina, Greensboro
Richard Hill, Michigan State University
Franz Hoffman, University of California, Irvine
Sara Hoot, University of Wisconsin, Milwaukee
Carl Hopkins, Cornell University
Alfredo Huerta, Miami University
Michael Ibba, The Ohio State University
Walter Judd, University of Florida
Laura Katz, Smith College
Manfred D. Laubichler, Arizona State University
Brian Leander, University of British Columbia
Mark V. Lomolino, SUNY College of Environmental Science and Forestry
Jim Lorenzen, University of Idaho
Denis Maxwell, University of Western Ontario
Brad Mehrtens, University of Illinois, Urbana-Champaign
John Merrill, Michigan State University
Allison Miller, Saint Louis University
Clara Moore, Franklin and Marshall College
Julie Noor, Duke University
Mohamed Noor, Duke University
Theresa O'Halloran, University of Texas, Austin
Norman R. Pace, University of Colorado
Randall Packer, George Washington University
Walt Ream, Oregon State University
Eric Richards, Washington University
Steve Rissing, The Ohio State University
R. Michael Roberts, University of Missouri, Columbia
Pete Ruben, Simon Fraser University
David A. Sanders, Purdue University
Mike Sanderson, University of California, Davis
Marty Shankland, University of Texas, Austin
Jeff Silberman, University of Arkansas
Margaret Silliker, DePaul University
Dee Silverthorn, University of Texas, Austin
M. Suzanne Simon-Westendorf, Ohio University
Alastair G.B. Simpson, Dalhousie University
John Skillman, California State University, San Bernardino
Frederick W. Spiegel, University of Arkansas
John J. Stachowicz, University of California, Davis
Heven Sze, University of Maryland
E.G. Robert Turgeon, Cornell University
Mary Tyler, University of Maine
Mike Wade, Indiana University
Leslie Winemiller, Texas A&M University
Mimi Zolan, Indiana University

Tree of Life Appendix Reviewers

John Abbott, University of Texas, Austin
Joseph Bischoff, National Center for Biotechnology Information
Ruth Buskirk, University of Texas, Austin
David Cannatella, University of Texas
Joel Cracraft, American Museum of Natural History
Scott Federhen, National Center for Biotechnology Information
Carol Hotton, National Center for Biotechnology Information
Robert Jansen, University of Texas, Austin
Brian Leander, University of British Columbia
Detlef Leipe, National Center for Biotechnology Information
Beryl Simpson, University of Texas, Austin
Richard Sternberg, National Center for Biotechnology Information
Edward Theriot, University of Texas
Sean Turner, National Center for Biotechnology Information

Supplements Authors

Dany Adams, The Forsyth Institute
Erica Bergquist, Holyoke Community College
Ian Craine, University of Toronto
Ernest Dubrul, University of Toledo
Edward Dzialowski, University of North Texas
Donna Francis, University of Massachusetts, Amherst
Jon Glase, Cornell University
Lindsay Goodloe, Cornell University
Celine Muis Griffin, Queen's University
Nancy Guild, University of Colorado at Boulder
Norman Johnson, University of Massachusetts, Amherst
James Knapp, Holyoke Community College
Jennifer Knight, University of Colorado, Boulder
David Kurjiaka, University of Arizona
Richard McCarty, Johns Hopkins University
Betty McGuire, Cornell University
Nancy Murray, Evergreen State College
Deb Pires, University of California, Los Angeles
Catherine Ueckert, Northern Arizona University
Jerry Waldvogel, Clemson University

LIFE's Media and Supplements Package

For the Student

Companion Website www.thelifewire.com

(Also available as a CD, optionally packaged with the book)

The *Life*, Eighth Edition Companion Website is available free of charge to all students (no access code required). The site features a variety of study and review resources designed to help students master the wide range of material presented in the introductory biology course. Features of the site include:

- *Interactive Summaries.* These summaries combine a review of important concepts with links to all the key figures from the chapter as well as all of the relevant animated tutorials and activities.
- *Animated Tutorials.* Over 100 in-depth animated tutorials present complex topics in a clear, easy-to-follow format that combines a detailed animation with an introduction, conclusion, and quiz.
- *Activities.* Over 120 interactive activities help the student learn important facts and concepts through a wide range of activities, such as labeling steps in processes or parts of structures, building diagrams, and identifying different types of organisms.
- *Flashcards.* For each chapter of the book, there is a set of flashcards that allows the student to review all the key terminology from the chapter. Students can review the terms in the study mode, and then quiz themselves on a list of terms.
- *New! Experiment Links.* New for the Eighth Edition, each experiment featured in the textbook has a corresponding treatment on the companion website that links to further information about the experiment, further research that followed, and applications derived from the research.
- *Interactive Quizzes.* Every question includes an image taken from the textbook, thorough feedback on both right and wrong answer choices, references to textbook pages, and links to electronic versions of book pages, where the related material is highlighted.
- *Online Quizzes.* These quizzes test the student's comprehension of the chapter material, and the results are stored in the online gradebook.
- *Key Terms.* The key terminology introduced in each chapter is listed, with definitions and audio pronunciations from the Glossary.
- *Suggested Readings.* For each chapter of the book, a list of suggested readings is provided as a resource for further study.
- *Glossary.* The language of biology is often difficult for students taking introductory biology, so we have created a full glossary with audio pronunciations.
- *Math for Life* (Dany Adams, *The Forsyth Institute*). *Math for Life* is a collection of mathematical shortcuts and references to help students with the quantitative skills they need in the laboratory.
- *Survival Skills* (Jerry Waldvogel, *Clemson University*). *Survival Skills* is a guide to more effective study habits. Topics include time management, note-taking, effective highlighting, and exam preparation.

Study Guide (ISBN 978-0-7167-7893-6)

Edward M. Dzialowski, *University of North Texas*; Jon Glase, *Cornell University*; Lindsay Goodloe, *Cornell University*; Nancy Guild, *University of Colorado*; and Betty McGuire, *Smith College*

For each chapter of the textbook, the *Life* Study Guide offers a variety of study and review tools. The contents of each chapter are broken down into both a detailed review of the Important Concepts covered and a boiled-down Big Picture snapshot. In addition, Common Problem Areas and Study Strategies are highlighted. A set of study questions (both multiple-choice and short-answer) allows students to test their comprehension. All questions include answers and explanations.

Lecture Notebook (ISBN 978-0-7167-7894-3)

This invaluable printed resource consists of all the artwork from the textbook (more than 1,000 images with labels) presented in the order in which they appear in the text, with ample space for note-taking. Because the Notebook has already done the drawing, students can focus more of their attention on the concepts. They will absorb the material more efficiently during class, and their notes will be clearer, more accurate, and more useful when they study from them later.

MCAT® Practice Test (ISBN 0-7167-5907-1)

A complete printed MCAT exam, with answers, allows students to test their knowledge of the full range of introductory biology content as they prepare for the medical school entrance exams.

CatchUp Math & Stats

Michael Harris, Gordon Taylor, and Jacquelyn Taylor

This primer will help your students quickly brush up on the quantitative skills they need to succeed in biology. Presented in brief, accessible units, the book covers topics such as working with powers, logarithms, using and understanding graphs, calculating standard deviation, preparing a dilution series, choosing the right statistical test, analyzing enzyme kinetics, and many more.

Student Handbook for Writing in Biology, Second Edition

Karen Knisely (ISBN 0-7167-6709-0)

This book provides practical advice to students who are learning to write according to the conventions in biology. Using the standards of journal publication as a model, the author provides, in a user-friendly format, specific instructions on: using biology databases to locate references; paraphrasing for improved comprehension; preparing lab reports, scientific papers, posters; preparing oral presentations in PowerPoint®, and more.

Bioethics and the New Embryology: Springboards for Debate

Scott F. Gilbert, Anna Tyler, and Emily Zackin (ISBN 0-7167-7345-7)

Our ability to alter the course of human development ranks among the most significant changes in modern science and has brought embryology into the public domain. The question that must be asked is: Even if we *can* do such things, *should* we do such things?

BioStats Basics: A Student Handbook

James L. Gould and Grant F. Gould (ISBN 0-7167-3416-8)

BioStats Basics provides introductory-level biology students with a practical, accessible introduction to statistical research. Engaging and informal, the book avoids excessive theoretical and mathematical detail to focus on how core statistical methods are put to work in biology.

Laboratory Manuals

W. H. Freeman publishes a range of high-quality biology lab texts, all of which are available for bundling with *Life,* Eighth Edition. Our laboratory texts are available as complete paperback texts, or as Freeman Laboratory Separates.

- *Biology in the Laboratory,* Third Edition
  Doris R. Helms, Carl W. Helms, Robert J. Kosinski, and John C. Cummings (ISBN 0-7167-3146-0)
- *Laboratory Outlines in Biology-VI*
  Peter Abramoff and Robert G. Thomson (ISBN 0-7167-2633-5)
- *Anatomy and Dissection of the Frog,* Second Edition
  Warren F. Walker, Jr. (ISBN 0-7167-2636-X)
- *Anatomy and Dissection of the Rat,* Third Edition
  Warren F. Walker, Jr. and Dominique Homberger (ISBN 0-7167-2635-1)
- *Anatomy and Dissection of the Fetal Pig,* Fifth Edition
  Warren F. Walker, Jr. and Dominique Homberger (ISBN 0-7167-2637-8)
- *Atlas and Dissection Guide for Comparative Anatomy,* Sixth Edition
  Saul Wischnitzer and Edith Wischnitzer (ISBN 0-7167-6959-X)

Custom Publishing for Laboratory Manuals

http://custompub.whfreeman.com

Instructors can build and order customized lab manuals in just minutes, choosing material from Freeman's biology laboratory manuals, as well as their own material.

For the Instructor

Instructor's Media Library

In order to give you the widest possible range of resources to help engage students and better communicate the material, we have assembled an unparalleled collection of media resources. The Eighth Edition of *Life* features an expanded Instructor's Media Library (available on a set of CDs and DVDs) that includes:

- *Textbook Figures and Tables.* Every image from the textbook is provided in both JPEG (high- and low-resolution) and PDF formats.
- *Unlabeled Figures.* Every figure in the textbook is provided in an unlabeled format. These are useful for student quizzing and custom presentation development.
- *Supplemental Photos.* The supplemental photograph collection contains over 1,500 photographs (all in addition to those in the text), forming a rich resource of visual imagery.
- *Animations.* A collection of over 100 in-depth animations, all of which were created from the textbook's art program, and which can be viewed in either narrated or step-through mode.
- *Videos.* This collection of approximately 200 video segments covering topics across the entire textbook helps demonstrate the complexity and beauty of life.
- *PowerPoint® Resources.* For each chapter of the textbook, we have created several different types of PowerPoint® presentations. These give instructors the flexibility to build a presentation in the manner that best suits their needs. Included are:
  - Figures and Tables
  - Lecture Presentation
  - New! Editable Labels
  - Layered Art
  - Supplemental Photos
  - Videos, Animations
- *Clicker Questions.* A set of questions written specifically to be used with classroom personal response systems ("clickers") is provided for each chapter. These questions are designed to reinforce concepts, gauge student comprehension, and provide an outlet for active participation.
- *Chapter Outlines, Lecture Notes,* and the complete *Test File* are all available in Microsoft Word® format for easy use in lecture and exam preparation.
- An intuitive *Browser Interface* provides a quick and easy way to preview all of the content on the Instructor's Media Library.
- *Computerized Test Bank.* The entire printed Test File, plus the textbook end-of-chapter Self-Quizzes, the Companion Website Online Quizzes, and the Study Guide questions are all included in Brownstone's easy-to-use Diploma® software.
- *Instructor's Website.* A wealth of instructor's media, as well as electronic versions of other instructor supplements, are available online for instant access anytime.
- *Online Quizzes.* The Companion Website includes an Online Quiz for each chapter of the textbook. Instructors can choose to use these quizzes as assignments, and can view the results in the online gradebook.

- *Course Management System Support.* As a service for adopters using WebCT, Blackboard, or ANGEL for their courses, full electronic course packs are available.

Instructor's Resource Kit

The *Life,* Eighth Edition Instructor's Resource Kit includes a wealth of information to help instructors in the planning and teaching of their course. The Kit includes:

- *Instructor's Manual,* featuring:
  - A "What's New" guide to the Eighth Edition
  - A brief chapter overview
  - A key terms section with all the boldface terms from the text
  - Chapter outlines
- *Lecture Notes*—detailed notes for each chapter that can serve as the basis for lectures, including references to figures and media resources.
- *Media Guide*—A visual guide to the extensive media resources available with the Eighth Edition of *Life.* The guide includes thumbnails and descriptions of every video, animation, PowerPoint®, and supplemental photo in the Media Library, all organized by chapter.
- *Lab manual* and custom lab manual information.

Overhead Transparencies

This set includes over 1,000 transparencies—all the four-color line art and all the tables from the text—along with convenient binders. Balloon captions have been removed and colors have been enhanced for clear projection in a wide range of conditions. Labels and images have been resized for improved readability.

Test File

Ernest Dubrul, *University of Toledo;* Jon Glase, *Cornell University;* Norman Johnson, *University of Massachusetts;* Catherine Ueckert, *Northern Arizona University*

The test file offers more than 5,000 questions, including fill-in-the-blank and multiple-choice test questions. The electronic version of the Test File also includes all of the textbook end-of-chapter Self-Quiz questions, all of the Student Website Online Quiz questions, and all of the Study Guide questions.

iclicker

Developed for educators by educators, iclicker is a hassle-free radio-frequency classroom response system that makes it easy for instructors to ask questions, record responses, take attendance, and direct students through lectures as active participants. For more information, visit www.iclicker.com.

The *Life* eBook

The *Life,* Eighth Edition eBook is a complete online version of the textbook that can be purchased online, or packaged with the printed textbook. This online version of *Life* is a substantially less expensive alternative that gives your students an efficient and rich learning experience by integrating all of the resources from the companion website directly into the eBook text. In addition, the eBook offers instructors unique opportunities to customize the text with their own content. Key features of the eBook include:

- *For Students:*
  - Integration of all website **activities** and **animated tutorials**
  - In-text **self quiz questions**
  - Interactive **summary exercises**
  - Custom text **highlighting**
  - A **notes** feature that allows students to annotate the text
  - Complete **glossary**, **index**, and **full-text search** features
- *For Instructors:*
  - A powerful notes feature that can incorporate text, Web links, and documents directly into the text
  - Easy integration of instructor media resources, including videos and supplemental photographs
  - Ability to link directly to any eBook page from other sites

BIOPORTAL

New for the Eighth Edition, BioPortal is the digital gateway to all of the teaching and learning resources that are available with *Life.* BioPortal integrates the *Life* textbook, all of the student and instructor media, extensive assessment resources, and course planning resources all into a powerful and easy-to-use learning management system. All of this means that your students get easy access to learning resources, presented in the proper context and at the proper time, and you get a complete learning management system, ready to use, without hours of prep work. Features of BioPortal include:

- *eBook*
  - Completely integrated with all media resources
  - Customizable with notes, sections, images, and more
- *Student Resources*
  - Animated Tutorials and Activities
  - Assessment: Online Quizzes and Interactive Quizzes
- *Instructor Resources*
  - Complete Test Bank
  - All quizzes
  - Media resources: Videos, PowerPoints, Supplemental Photos, and more
- *Assignments*
  - Quizzes and exams
  - Assignable textbook sections
  - Assignable animations and activities
  - Custom assignments
- *Easy-to-Use Course Management*
  - Complete course customization
  - Custom resources/Document posting
  - Announcements/Calendar/Course Email/Discussions
  - Robust Gradebook

Contents in Brief

Contents

Part One ■ The Science and Building Blocks of Life

Part Two ■ Cells and Energy

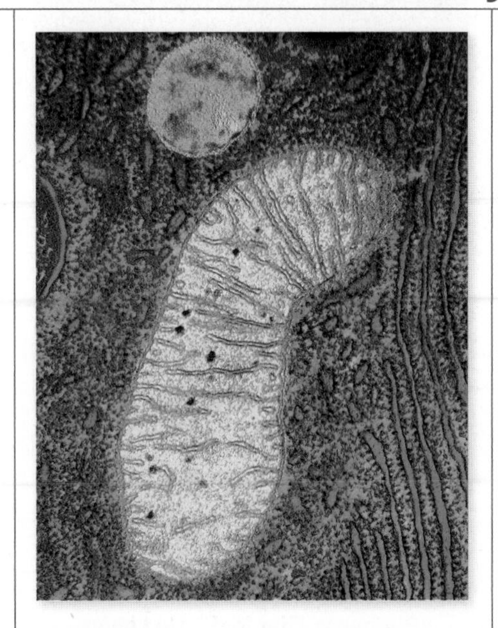

Part Three ■ Heredity and the Genome

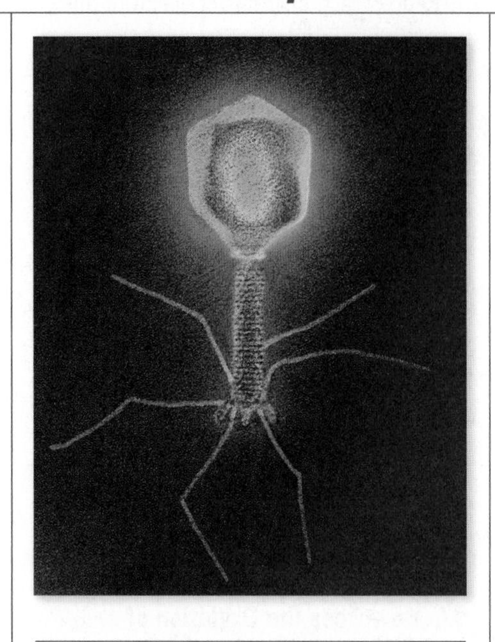

Part Four ■ Molecular Biology: The Genome in Action

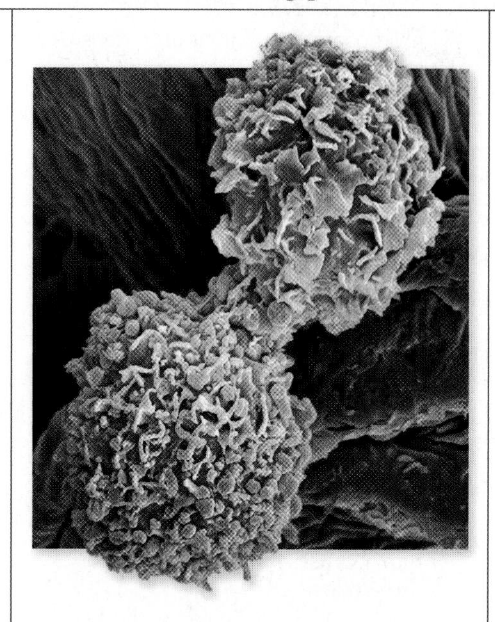

Part Five ■ The Patterns and Processes of Evolution

Part Six ■ The Evolution of Diversity

Part Seven ■ Flowering Plants: Form and Function

Part Eight ■ Animals: Form and Function

Part Nine ■ Ecology

PART ONE

The Science and Building Blocks of Life

CHAPTER 1 Studying Life

Why are frogs croaking?

In August of 1995, a group of Minnesota middle school students on a field trip were hiking through some local wetlands when they discovered a horde of young frogs, most of them with deformed, missing, or extra legs. The students' find made the national news and focused public attention on amphibian population declines, an issue already being studied by many scientists.

There are a number of possible reasons for the problems amphibians are facing. Water pollution is an obvious possibility, since these animals breed and spend their early lives in ponds and streams. Acidic rain resulting from air pollution could also affect their watery homes. Could ultraviolet radiation be causing "mutant" frogs? Is global warming adversely affecting amphibians? Is some disease attacking them? Evidence exists to support each of these possibilities, and there is no a single answer. In one case, a college undergraduate came up with an answer, and in the process gave scientists a whole new perspective on the question.

In 1996, Stanford University sophomore Pieter Johnson was shown a collection of Pacific tree frogs with extra legs growing out of their bodies. He decided to focus his honors research project on finding out what caused these deformities. The frogs came from a pond in an agricultural region near abandoned mercury mines; thus two possible causes of the deformities were agricultural chemicals, and heavy metals from the old mines.

Pieter applied the *scientific method*. Based on what he knew and on his library research, he proposed a logical explanation for the monster frogs—environmental water pollution—and designed an experiment to test his idea. His experiment compared ponds where there were deformed frogs with ponds where the frogs were normal and tested for the presence or absence of pollutants. As frequently happens in science, his proposed explanation, or *hypothesis*, was disproved by his experiment. But his field work led to a new hypothesis: that the deformities are caused by a parasite. Pieter conducted laboratory experiments, the results of which supported the conclusion that a certain type of parasite is present in some ponds. These parasites burrow into newly hatched tadpoles and disrupt the development of the adult

Frogs Are Having Serious Problems These preserved Pacific tree frogs (*Hyla regilla*) exhibit multiple deformities of the hind legs. Similar deformities have been found in frogs from different regions of the world.

A Biologist at Work As a college sophomore, Pieter Johnson studied numerous ponds that were home to Pacific tree frogs, trying to discover why some ponds had so many deformed frogs.

frog's hind legs. Pieter's research did not explain the global decline in amphibians, but it did illuminate one type of problem amphibians encounter. Science usually progresses in such small but solid steps.

Biologists use the scientific method to investigate the processes of life at all levels, from molecules to ecosystems. Some of these processes happen in millionths of seconds and others cover millions of years. The goals of biologists are to understand how organisms and groups of organisms function, and sometimes they use that knowledge in practical and beneficial ways.

IN THIS CHAPTER we examine the most common features of living organisms and put those features into the context of the major principles that underlie all biology. Next we offer a brief outline of how life evolved and how the different organisms on Earth are related. We then turn to the subjects of biological inquiry and the scientific method.

CHAPTER OUTLINE

1.1 What Is Biology?

Biology is the scientific study of living things. Biologists define "living things" as all the diverse organisms descended from a single-celled ancestor that appeared almost 4 billion years ago. Because of their common ancestry, living organisms share many characteristics that are not found in the nonliving world. Most living organisms:

- consist of one or more cells
- contain genetic information
- use genetic information to reproduce themselves
- are genetically related and have evolved
- can convert molecules obtained from their environment into new biological molecules
- can extract energy from the environment and use it to do biological work
- can regulate their internal environment

This list can serve as a rough guide to the major themes and unifying principles of biology that you will encounter in this book. A simple list, however, belies the incredible complexity and diversity of life. Some forms of life may not display all of these characteristics all of the time. For example, the seed of a desert plant may go for many years without extracting energy from the environment, converting molecules, regulating its internal environment, or reproducing; yet the seed is alive.

And what about viruses? Although they do not consist of cells, viruses probably evolved from cellular organisms, and many biologists consider them to be living organisms. Viruses cannot carry out physiological functions on their own, but must parasitize the machinery of host cells to do those jobs for them—including reproduction. Yet viruses contain genetic information, and they certainly evolve (as we know because evolving flu viruses require annual changes in the vaccines we create to combat them). Are viruses alive? What do you think?

This book explores the characteristics of life, how these characteristics vary among organisms, how they evolved, and how they work together to enable organisms to survive and reproduce. *Evolution* is a central theme in biology, and therefore in this book. Through differential survival and reproduction, living systems evolve and become adapted to Earth's many environments. The processes of evolution have generated the enormous diversity that we see today as life on Earth (**Figure 1.1**).

Living organisms consist of cells

Cells and the chemical processes within them are the topic of Part Two of this book. Some organisms are *unicellular,* consisting of a single cell that carries out all the functions of life, while others are *multicellular,* made up of a number of cells that are specialized for different functions.

The discovery of cells was made possible by the invention of the microscope in the 1590s by father and son Dutch spectacle makers Zaccharias and Hans Janssen. The first biologists to improve on their technology and use it to study living organisms were Antony van Leeuwenhoek (Dutch) and Robert Hooke (English) in the middle to late 1600s. It was van Leeuwenhoek who discovered that drops of pond water teemed with single-celled organisms, and he made many other discoveries as he progressively improved his microscopes over a long lifetime of research. Hooke carried out similar studies. From observations of plant tissues (cork, to be specific) he concluded that the tissues were made up of repeated units he called *cells* (**Figure 1.2**).

1.1 The Many Faces of Life The processes of evolution have led to the millions of diverse organisms living on Earth today. Archaea (A) and bacteria (B) are single-celled organisms. The three different bacteria in (B) represent the three shapes (rods, or bacilli; helices; and spheres, or cocci) often seen in these organisms, which are described in Chapter 26. (C) Organisms whose single cells display more complexity are commonly known as protists. Protists are the subjects of Chapter 27. (D) Chapters 28 and 29 cover the multicellular green plants. The other broad groups of multicellular organisms are (E) the fungi, discussed in Chapter 30, and (F,G) the animals, covered in Chapters 31–33.

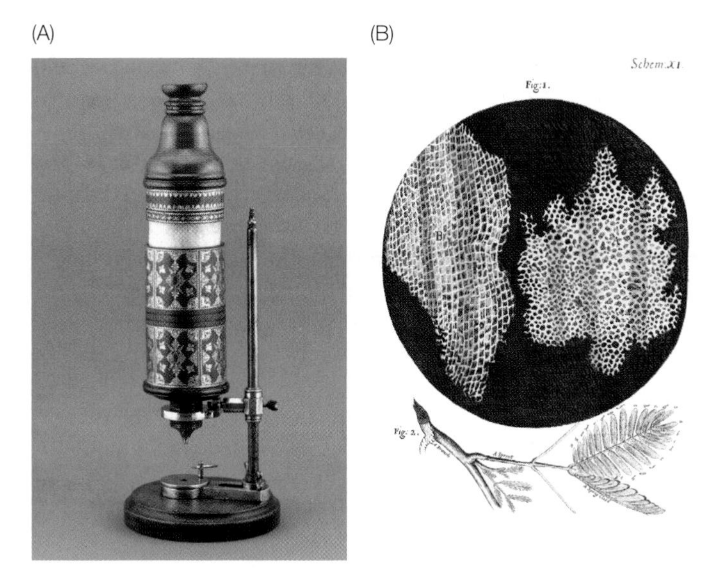

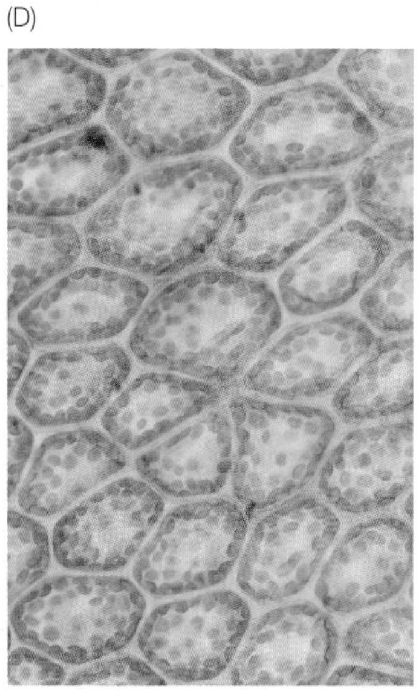

1.2 All Life Consists of Cells (A) The development of microscopes, such as this instrument of Robert Hooke's, revealed the microbial world to seventeenth-century scientists. (B) Hooke was the first to propose the concept of cells, based on his observations of thin slices of plant tissue (cork) under his microscope. (C) A modern version of the optical, or "light" microscope. (D) A modern light micrograph reveals the intricacies of cells in a leaf.

In 1676, Hooke wrote that van Leeuwenhoek had observed "a vast number of small animals in his Excrements which were most abounding when he was troubled with a Loosenesse and very few or none when he was well." This simple observation represents the discovery of bacteria.

More than a hundred years passed before studies of cells advanced significantly. In 1838, Matthias Schleiden, a German biologist, and Theodor Schwann, from Belgium, were having dinner together and discussing their work on plant and animal tissues, respectively. They were struck by the similarities in their observations and came to the conclusion that the structural elements of plants and animals were essentially the same. They formulated their conclusion as the **cell theory**, which states that:

- Cells are the basic structural and physiological units of all living organisms.
- Cells are both distinct entities and building blocks of more complex organisms.

But Schleiden and Schwann did not understand the origin of cells. They thought cells emerged by the self-assembly of nonliving materials, much as crystals form in a solution of salt. This conclusion was in accordance with the prevailing view of the day that life arises from nonlife by spontaneous generation—mice from dirty clothes, maggots from dead meat, insects from mixtures of straw and pond water. The debate over whether or not life could arise from nonlife continued until 1859, when the French Academy of Sciences sponsored a contest for the best experiment to prove or disprove spontaneous generation. The prize was won by the great French scientist Louis Pasteur; his experiment proving that life must be present in order for life to be generated is described in Figure 3.30. Pasteur's vision of microorganisms led him to propose the germ theory of disease and explain the role of single-celled organisms in the fermentation of beer and wine. He also designed a method of preserving milk by heating it to kill microorganisms, a process we now know as pasteurization.

Today we readily accept the fact that all cells come from preexisting cells. In addition, we understand that the functional properties of organisms derive from the properties of their cells. We also understand that cells of all kinds share many essential mechanisms because they share a common ancestry that goes back billions of years. We therefore add a few more elements to the cell theory:

- All cells come from preexisting cells.
- All cells are similar in chemical composition.
- Most of the chemical reactions of life occur within cells.
- Complete sets of genetic information are replicated and passed on during cell division.

At the same time Schleiden and Schwann were building the foundation for the cell theory, Charles Darwin was beginning to understand how organisms undergo evolutionary change.

The diversity of life is due to evolution by natural selection

Evolution by **natural selection**, as proposed by Charles Darwin, is perhaps the major unifying principle of biology and is the topic of Part Five of this book.

Darwin proposed that living organisms are descended from common ancestors and are therefore related to one another. He did not have the advantage of understanding the mechanisms of genetic inheritance that you will learn about in Part Three, but even so he surmised that such mechanisms existed because offspring resembled their parents in so many different ways. That simple fact is the basis for the concept of a **species**. Although the precise definition of a species is complicated, in its most widespread usage it refers to a group of organisms that look alike ("are morphologically similar") and can breed successfully with one another.

But offspring also differ from their parents. Any population of a plant or animal species displays variation, and if you select breeding pairs on the basis of some particular trait, that trait is more likely to be present in their offspring than in the general population. Darwin himself bred pigeons, and was well aware of how pigeon fanciers selected for unusual feather patterns, beak shapes, and body sizes. He realized that if humans could select for specific traits, the same process could operate in nature; hence the term *natural selection*.

How would selection function in nature? Darwin postulated that different probabilities of survival and reproductive success would do the job. He reasoned that the reproductive capacity of plants and animals, if unchecked, would result in unlimited growth of populations, but we do not observe such unlimited population growth in nature; therefore only a small percentage of offspring must survive to reproduce. So, any trait that confers even a small increase in the probability that its possessor will survive and reproduce would be strongly favored and would spread in the population. Darwin called this phenomenon *natural selection*.

Because organisms with certain traits survive and reproduce best under specific sets of conditions, natural selection leads to **adaptations**: structural, physiological, or behavioral traits that enhance an organism's chances of survival and reproduction in its environment (**Figure 1.3**). The many different environments and ecological communities organisms have adapted to over evolutionary history have led to a remarkable amount of diversity, which we will survey in Part Six of this book.

If all cells come from preexisting cells, and if all the diverse species of organisms on Earth are related by descent with modification from a common ancestor, then what is the source of information that is passed from parent to daughter cells and from parental organisms to their offspring?

1.3 Adaptations to the Environment The leaves of all plants are specialized for photosynthesis—the sunlight-powered transformation of water and carbon dioxide into larger structural molecules called carbohydrates. The leaves of different plants, however, display many different adaptations to their individual environments.

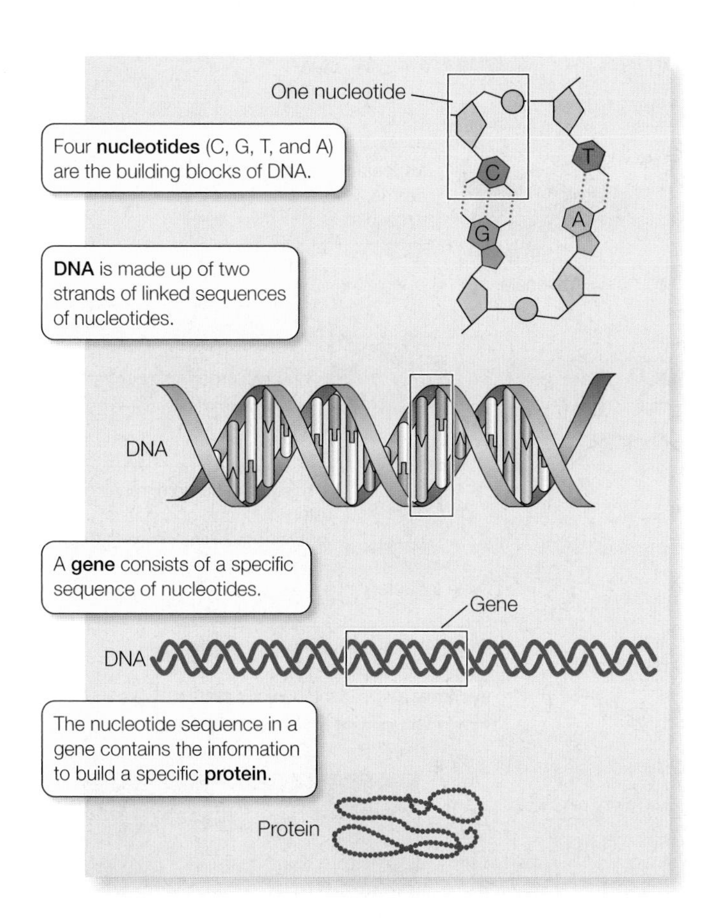

1.4 The Genetic Code Is Life's Blueprint The instructions for life are contained in the sequences of nucleotides in DNA molecules. Specific DNA sequences comprise genes, and the information in each gene provides the cell with the information it needs to manufacture a specific protein. The average length of a single human gene is 16,000 nucleotides.

Biological information is contained in a genetic language common to all organisms

A cell's instructions—or "blueprints" for existence—are contained in its **genome**, which is the sum total of all the DNA molecules in the cell. **DNA** (deoxyribonucleic acid) molecules are long sequences of four different subunits called **nucleotides**. The sequence of the nucleotides contains genetic information. Specific segments of DNA called **genes** contain the information the cell uses to make proteins (**Figure 1.4**). **Proteins** make up much of an organism's structure and are the molecules that govern the chemical reactions within cells. By analogy with a book, the nucleotides of DNA are like the letters of an alphabet. Protein molecules are the sentences they spell. Combinations of proteins that constitute structures and control biochemical processes are the paragraphs. The structures and processes that are organized into different systems with specific tasks (such as digestion or transport) are the chapters of the book, and the complete book is the organism. Natural selection is the author and editor of all the books in the library of life.

If you were to write out your own genome using four letters to represent the four nucleotides, you would have to write a total of more than 3 billion letters. If you used the same size type as you are reading in this book, your genome would fill about a thousand books the size of this one.

All the cells of a multicellular organism contain the same genome, yet different cells have different functions and form different structures. Therefore, different types of cells in an organism must express different parts of their genome. How the control of gene expression enables a complex organism to develop and function is a major focus of current biological research.

The genome of an organism consists of thousands of genes. If the nucleotide sequence of a gene is altered, it is likely that the protein that gene encodes will be altered. Alterations of genes are called *mutations*. Mutations occur spontaneously; they can also be induced by many outside factors, including chemicals and radiation. Most mutations are deleterious, but occasionally a change in the properties of a protein alters its function in a way that improves the functioning of the organism under the environmental conditions it encounters. Such beneficial mutations are the raw material of evolution.

Cells use nutrients to supply energy and to build new structures

Living organisms acquire substances called *nutrients* from the environment. Nutrients supply the organism with energy and raw materials for building biological structures. Cells take in nutrient molecules and break them down into smaller chemical units. In doing so, they can capture the energy contained in the chemical bonds of the nutrient molecules and use that energy to do different kinds of work. One kind of cellular work is the building, or *synthesis*, of new complex molecules and structures from smaller chemical units. For example, we are all familiar with the fact that carbohydrates eaten today may be deposited in the body as fat tomorrow (**Figure 1.5A**). Another kind of work cells do is mechanical work—for example, moving molecules from one cellular location to another, or even moving whole cells or tissues, as in the case of muscles (**Figure 1.5B**).

Living organisms control their internal environment

Life depends on thousands of biochemical reactions that occur inside cells. These reactions require materials to be moved into and out of cells in a controlled manner. Within the cell, those reactions are linked in that the products of one are the raw materials of the next. If this complex network of reactions is to be properly integrated, reaction rates within a cell must be precisely controlled. A large proportion of the activities of cells are directed toward the regulation of the multiple chemical reactions continuously in progress inside the cell.

Organisms that are made up of more than one cell have an *internal environment* that is not cellular. That is, their individual cells are bathed in extracellular fluids, from which they receive nutrients and into which they excrete wastes. The cells of multicellular organisms are specialized to contribute in some way to the maintenance of that internal environment. However, with the evolution of specialized

1.5 Energy from Nutrients Can Be Stored or Used Immediately (A) The cells of this Arctic ground squirrel have broken down the complex carbohydrates in plants and converted their molecules into fats, which are stored in the animal's body to provide an energy supply for the cold months. (B) The cells of this kangaroo are breaking down food molecules and using the energy in their chemical bonds to do mechanical work—in this case, to jump.

functions, these cells lost many of the functions carried out by single-celled organisms, and therefore depend on the internal environment for essential services. The interdependence of the different kinds of cells in a multicellular organism can by expressed in the famous motto of the Three Musketeers: "One for all, and all for one!"

To accomplish their specialized tasks, assemblages of similar cells are organized into *tissues*. For example, a single muscle cell cannot generate much force, but when many of these cells combine to form the tissue of a working muscle, considerable force and movement can be generated (see Figure 1.5B). Different tissue types are organized to form *organs* that accomplish specific functions. Familiar organs include the heart, brain, and stomach. Organs whose functions are interrelated can be grouped into *organ systems*. The functions of cells, tissues, organs, and organ systems are all integral to the multicellular *organism* (**Figure 1.6**). The biology of organisms is the subject of Parts Seven and Eight of this book.

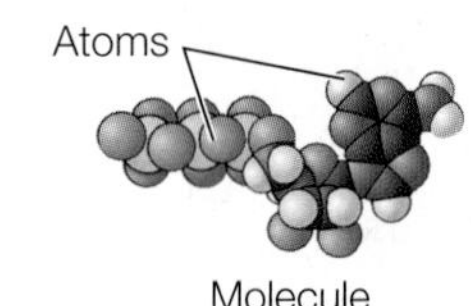

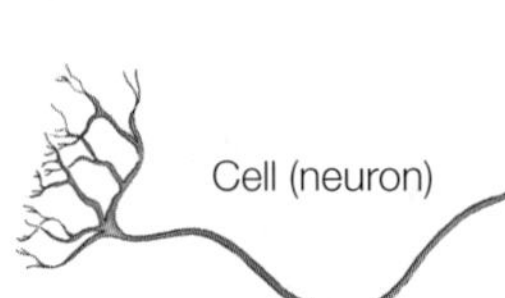

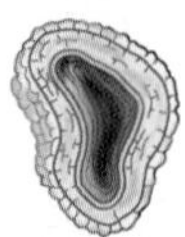

Organ (brain)

Organs combine several tissues that function together. Organs form **systems**, such as the nervous system.

1.6 Biology Is Studied at Many Levels of Organization Life's properties emerge when DNA and other molecules are organized in cells. Energy flows through all the biological levels shown here.

(A)

(B)

1.7 Conflict and Cooperation Organisms of the same species interact with one another in various ways. (A) Territorial elephant seal bulls defend stretches of beach from other males. The single male who controls a stretch of beach is able to mate with the many females (seen in the background) that live there. (B) Members of a meerkat colony are usually related to one another. Meerkats cooperate in many ways, such as watching out for predators and giving a warning bark if one appears.

As Figure 1.6 shows, individual organisms do not live in isolation, and the hierarchy of biology continues well beyond that level.

Living organisms interact with one another

Organisms interact with their external environments as well as the internal environment. Individual organisms are part of *populations* that interact among themselves and with populations of different organisms.

Organisms interact in many different ways. For example, some animals are *territorial* and will try to prevent other individuals of their species from exploiting the resource they are defending, whether it be food, nesting sites, or mates (**Figure 1.7A**). Animals may also *cooperate* with members of their species, forming social units such as a termite colony, a school of fish, or a meerkat colony (**Figure 1.7B**). Such interactions among individuals have resulted in the evolution of social behaviors such as communication.

The interaction of populations of many different species forms a *community*, and interactions between different species are a major evolutionary force. Adaptations that give an individual of one species an advantage in obtaining members of another species as food (and the converse, adaptations that lessen an individual's chances of becoming food) are paramount in evolutionary history. Organisms of different species may compete for the same resources, resulting in natural selection for specialized adaptations that allow some individuals to exploit those resources more efficiently than others can.

In any given geographic locality, the interacting communities form *ecosystems*. Organisms in the ecosystem can modify the environment in ways that affect other organisms. For example, in most terrestrial (land) environments, the dominant plants greatly modify the environmental conditions in which animals and other plants must live. The ways in which species interact with one another and with their environment is the subject of *ecology*, the topic of Part Nine of this book.

Discoveries in biology can be generalized

Because all life is related by descent from a common ancestor, shares a genetic code, and consists of similar building blocks—cells—knowledge gained from investigations of one type of organism can, with care, be generalized to other organisms. Therefore, biologists can use **model systems** for research, knowing that they can extend their findings to other organisms and to humans. For example, our basic understanding of the chemical reactions in cells came from research on bacteria, but is applicable to all cells, including those of humans. Similarly, the biochemistry of photosynthesis—the process by which plants use sunlight to produce biological molecules—was largely worked out from experiments on *Chlorella* (a type of pond scum; see Figure 8.12). We learned much of what we know about the genes that control plant development from work on a single species of plant (see Chapter 19). Knowledge about how animals develop has come from work on sea urchins, frogs, chickens, roundworms, and fruit flies. And recently, the discovery of a major gene controlling human skin color came from work on zebrafish. Being able to generalize from model systems is a powerful tool.

1.1 RECAP

Living organisms are made of cells, evolve by natural selection, contain genetic information, extract energy from their environment and use it to do biological work, control their internal environment, and interact with one another.

- Can you describe the relationship between evolution by natural selection and the genetic code? See pp. 5–7
- Do you understand why results of biological research on one species can be generalized to very different species? See p. 9

Now that we have an overview of the major features of life that will be explored in depth in this book, we can ask how and when life first emerged. In the next section we will describe the history of life from the earliest simple life forms to the complex and diverse organisms that inhabit our planet today.

1.2 How Is All Life on Earth Related?

What do biologists mean when they say that all organisms are *genetically related*? They mean that species on Earth share a *common ancestor*. If two species are similar, as dogs and wolves are, then they probably have a common ancestor in the fairly recent past. The common ancestor of two species that are more different—say, a dog and a deer—probably lived in the more distant past. And if two organisms are very different—such as a dog and a clam—then we must go back to the *very* distant past to find their common ancestor. How can we tell how far back in time the common ancestor of any two organisms lived? In other words, how do we discover the evolutionary relationships among organisms?

For many years, biologists investigated the history of life by studying the *fossil record*—the preserved remains of organisms that lived in the distant past (**Figure 1.8**). Geologists supplied knowledge about the ages of fossils and the nature of the environments in which they lived. Biologists then inferred the evolutionary relationships among living and fossil organisms by comparing their anatomical similarities and differences. The development of modern molecular methods for comparing genomes, described in Chapter 24, has enabled biologists to more accurately establish the degrees of relationship between living organisms and use that information to help us interpret the fossil record.

In general, the greater the differences between the genomes of two species, the more distant was their common ancestor. Using molecular techniques, biologists are exploring fundamental questions about the history of life on Earth. What were the earliest forms of life? How did those simple organisms give rise to the great diversity of organisms alive today? Can we reconstruct the family tree of all life?

1.8 Fossils Give Us a View of Past Life The most prominent of the many fossilized organisms in this rock sample are ammonoids, an extinct mollusk group whose living relatives include squids and octopus. Ammonoids flourished between 200 million and 60 million years ago; this particular group of fossils is about 185 million years old.

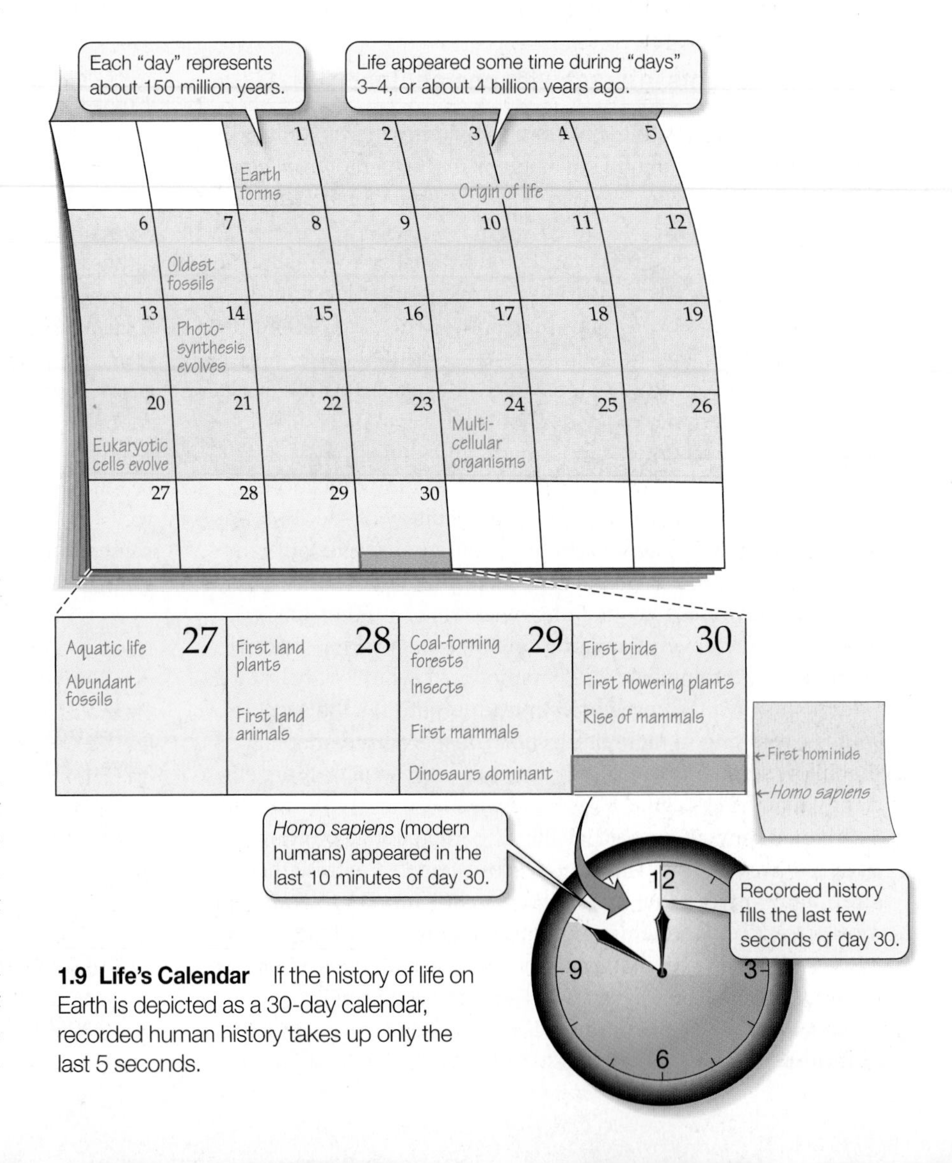

1.9 Life's Calendar If the history of life on Earth is depicted as a 30-day calendar, recorded human history takes up only the last 5 seconds.

Life arose from nonlife via chemical evolution

Geologists estimate that Earth is about 5 billion years old. For the first billion years, it was not a very hospitable place for life, and there was no life. If we picture the history of Earth as a 30-day calendar, life probably arose around the middle of the first week, or about 4 billion years ago (**Figure 1.9**).

When we consider how life might have arisen from nonliving matter, we must take into account the properties of the young Earth's atmosphere, oceans, and climate, all of which were very different from today's. Biologists postulate that complex biological molecules first arose through the random physical association of chemicals in that environment. Experiments that simulate the conditions on early Earth have confirmed that the generation of complex molecules under such conditions is possible, even probable. The critical step, however, for the evolution of life had to be the appearance of molecules that could reproduce themselves and also serve as templates for the synthesis of large molecules with complex but stable shapes. The variation of the shapes of these large, stable molecules (described in Chapter 3) enabled them to participate in increasing numbers and kinds of interactions with other molecules: chemical reactions.

Biological evolution began when cells formed

The second critical step in the origin of life was the enclosure of complex biological molecules in *membranes*, which kept them close together and increased the frequency with which they interacted. Fatlike molecules were the critical ingredient: because these molecules are not soluble in water, they form membrane-like films. These films tend to form spherical *vesicles*, which could have enveloped assemblages of other biological molecules. Scientists postulate that about 3.8 billion years ago, this natural process of membrane formation resulted in the first cells with the ability to replicate themselves—an event that marked the beginning of biological evolution.

For 2 billion years after cells originated, all organisms consisted of only one cell. These first unicellular organisms were (and are, as multitudes of their descendants exist in similar form today) **prokaryotes**. Prokaryotic cell structure consists of DNA and other biochemicals enclosed in a membrane.

These early prokaryotes were confined to the oceans, where there was an abundance of complex molecules they could use as raw materials and sources of energy. The ocean shielded them from the damaging effects of ultraviolet light, which was intense at that time because there was no oxygen in the atmosphere, and hence no protective ozone layer.

Photosynthesis changed the course of evolution

The sum total of all the chemical reactions that go on inside a cell constitutes the cell's **metabolism**. To fuel their metabolism, the earliest prokaryotes took in molecules directly from their environment, breaking these small molecules down to release the energy contained in their chemical bonds. Many modern species of prokaryotes still function this way, and very successfully.

An extremely important step that would change the nature of life on Earth occurred about 2.5 billion years ago with the evolution of **photosynthesis**. The chemical reactions of photosynthesis (which are explained in Chapter 8) transform the energy of sunlight into a form of energy that can power the synthesis of large biological molecules. These large molecules become the building blocks of cells; they can also be broken down to provide metabolic energy. Because its energy-capturing processes provide food for other organisms, photosynthesis is the basis of much of life on Earth today.

Early photosynthetic cells were probably similar to present-day prokaryotes called *cyanobacteria* (**Figure 1.10**). Over time, photosynthetic prokaryotes became so abundant that vast quantities of oxygen gas—O_2, which is a by-product of photosynthesis—began slowly to accumulate in the atmosphere. O_2 was poisonous to many of the prokaryotes that lived at that time. However, those organisms that tolerated O_2 were able to proliferate as the presence of oxygen opened up vast new avenues of evolution. Metabolism based on the use of O_2, called *aerobic metabolism*, is more efficient than the *anaerobic* (non-oxygen-using) *metabolism* that characterized earlier organisms. Aerobic metabolism allowed cells to grow larger, and today it is used by the majority of Earth's organisms.

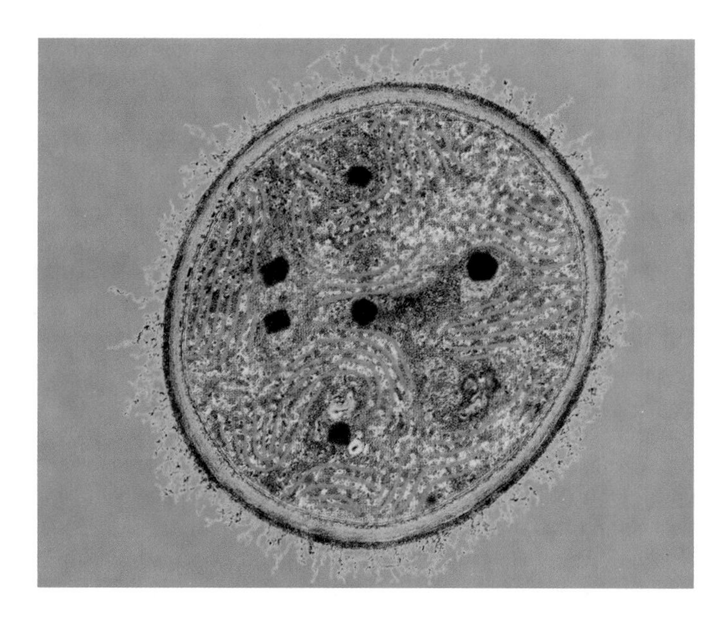

1.10 Photosynthetic Organisms Changed Earth's Atmosphere This modern cyanobacterium may be very similar to the early photosynthetic prokaryotes that introduced oxygen into Earth's atmosphere.

Over millions of years, the vast quantities of oxygen released by photosynthesis formed a layer of ozone (O_3) in the upper atmosphere. As the ozone layer thickened, it intercepted more and more of the sun's deadly ultraviolet radiation. Only in the last 800 million years has the presence of a dense ozone layer allowed organisms to leave the protection of the ocean and live on land.

Eukaryotic cells evolved from prokaryotes

Another important step in the history of life was the evolution of cells with discrete intracellular compartments, called **organelles**, that were capable of taking on specialized cellular functions. This event happened about 3 weeks into our calendar of Earth's history (see Figure 1.9). One of these organelles, the *nucleus*, came to contain the cell's genetic information. The nucleus has the appearance of a dense kernel, giving these cells their name: **eukaryotes** (from the Greek *eu*, "true," and *karyon*, "kernel"), as distinguished from the cells of prokaryotes, which lack internal compartments (*pro*, "before").

Some organelles are hypothesized to have originated when cells ingested smaller cells (see Figure 4.26). For example, the organelle specialized to conduct photosynthesis, the *chloroplast*, could have originated as a photosynthetic prokaryote that was ingested by a larger eukaryote. If the larger cell failed to break down this intended food object, a partnership could have evolved in which the ingested prokaryote provided the products of photosynthesis and the host cell provided a good environment for its smaller partner.

Multicellularity arose and cells became specialized

Until slightly more than 1 billion years ago, all the organisms that existed—whether prokaryotic or eukaryotic—were unicellular. Yet another important evolutionary step occurred when some eukaryotes failed to separate after cell division, remaining attached to each other. The permanent association of cells made it possible for some cells to specialize in certain functions, such as reproduction, while other cells specialized in other functions, such as absorbing nutrients and distributing them to neighboring cells. This **cellular specialization** enabled multicellular eukaryotes to increase in size and become more efficient at gathering resources and adapting to specific environments.

Biologists can trace the evolutionary Tree of Life

If all the species of organisms on Earth today are the descendants of a single kind of unicellular organism that lived almost 4 billion years ago, how have they become so different? And why are there so many species?

As long as individuals within a population mate at random, structural and functional changes may evolve within that population, but the population will remain one species. However, if some event isolates some members of a population from the others, structural and functional differences between them may accumulate over time. In short, the evolutionary paths of the two groups may diverge to the point where their members can no longer reproduce with each other. They have evolved into different species. This evolutionary process, called *speciation,* is detailed in Chapters 22 and 23.

Biologists give each species a distinct scientific name formed from two Latinized names (a *binomial*). The first name identifies the species' *genus*—a group of species that share a recent common ancestor. The second is the name of the species. For example, the scientific name of the human species is *Homo sapiens*: *Homo* is our genus and *sapiens* is our species. Scientific names usually refer to some characteristic of the species. *Homo* is derived from the Latin word for "man," and *sapiens* is derived from the Latin word for "wise" or "rational."

As many as 30 million species of organisms may exist on Earth today. Many times that number lived in the past but are now extinct. Many millions of speciation events created this vast diversity, and the unfolding of these events can be diagrammed as an evolutionary "tree" showing the order in which populations split and eventually evolved into new species. An evolutionary tree traces the descendants of ancestors that lived at different times in the past. The organisms on any one branch share a common ancestor at the base of that branch. The most closely related groups are placed together on the same branch; more distantly related organisms are on different branches. In this book, we adopt the convention that time flows from left to right, so the tree in Figure 1.11 (and other trees in this book) lies on its side, with its root—the ancestor of all life—at the left. Although many details remain to be clarified, the broad outlines of the Tree of Life have been determined. Its branching patterns are based on a rich array of evidence from fossils, structures, metabolic processes, behavior, and molecular analyses of genomes.

No fossils exist to help us determine the earliest divisions in the lineage of life because those unicellular organisms had no parts that could be preserved as fossils. However, molecular evidence has been used to separate all living organisms into three major **domains**: Archaea, Bacteria, and Eukarya (**Figure 1.11**). The organisms of each domain have been evolving separately from organisms in the other domains for more than a billion years.

Organisms in the domains **Archaea** and **Bacteria** are all prokaryotes. Archaea and Bacteria differ so fundamentally from each other in their metabolic processes that they are believed to have separated into distinct evolutionary lineages very early.

Members of the third domain—**Eukarya**—have eukaryotic cells. Three major groups of multicellular eukaryotes—plants, fungi, and animals—all evolved from unicellular *microbial eukaryotes*, more generally referred to as *protists*. The photosynthetic protist that gave rise to plants was completely distinct from the protist that was ancestral to both animals and fungi, as can be seen from the branching pattern of Figure 1.11.

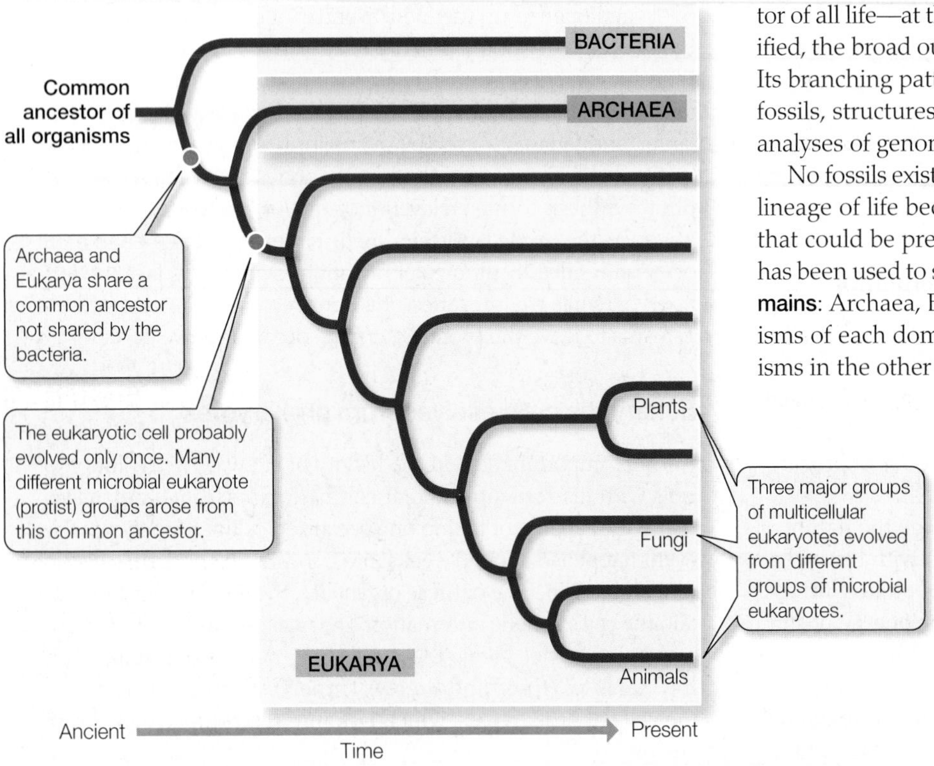

1.11 The Tree of Life The classification system used in this book divides Earth's organisms into three domains: Bacteria, Archaea, and Eukarya. The unlabeled blue branches within the Eukarya represent various groups of microbial eukaryotes, more commonly known as "protists."

Some bacteria, some archaea, some protists, and most plants are capable of photosynthesis. These organisms are called *autotrophs* ("self-feeders"). The biological molecules they produce are the primary food for nearly all other living organisms.

Fungi include molds, mushrooms, yeasts, and other similar organisms, all of which are *heterotrophs* ("other-feeders")—that is, they require a source of molecules synthesized by other organisms, which they then break down to obtain energy for their own metabolic processes. Fungi break down energy-rich food molecules in their environment and then absorb the breakdown products into their cells. Some fungi are important as decomposers of the waste products and dead bodies of other organisms.

Like fungi, animals are heterotrophs, but unlike fungi they ingest their food source, then break down the food in a digestive tract. Animals eat other forms of life, including plants, fungi, and other animals. Their cells absorb the breakdown products and obtain energy from them.

1.2 RECAP

The first cellular life on Earth was prokaryotic and arose about 4 billion years ago. The complexity of the organisms that exist today is the result of several important evolutionary events, including the evolution of photosynthesis, eukaryotic cells, and multicellularity. The genetic relationships of all organisms can be shown as a branching Tree of Life.

- Can you explain the evolutionary significance of photosynthesis? See p. 11
- What do the domains of life represent? What are the major groups of eukaryotes? See p. 12 and Figure 1.11

In February of 1676, Robert Hooke received a letter from the physicist Sir Isaac Newton. In this letter Newton famously remarked to Hooke, "If I have seen a little further, it is by standing on the shoulders of giants." We all stand on the shoulders of giants, building on the research of earlier scientists. By the end of this course, you will know more about evolution than Darwin ever could have, and you will know infinitely more about cells than Schleiden and Schwann did. Let's look at the methods biologists use to expand our knowledge of life.

1.3 How Do Biologists Investigate Life?

Biologists use many tools and methods in their research, but regardless of the methods they use, biologists take two basic approaches to their investigations of life: they observe and they conduct experiments.

Observation is an important skill

Biologists have always observed the world around them, but today their abilities to observe are greatly enhanced by many sophisticated technologies, such as electron microscopes, DNA chips, magnetic resonance imaging, and global positioning satellites. Advances in technology have been responsible for most major advances in biology. For example, not too long ago it was extremely difficult and time-consuming to decipher the nucleotide sequence that makes up a single gene. New technologies enabled biologists to sequence the entire human genome in only 13 years (1990–2003). Scientists now use these methods routinely, sequencing the genomes of organisms (including organisms that cause serious diseases) in only days. We will explore some of these technologies and what we have learned from them in Part Four of this book.

Our ability to observe the distributions of organisms, such as fish in the world's oceans, has also improved dramatically. A short time ago, researchers could put physical tags on fish and then only hope that someday a fisherman would catch one and send back the tag, which would at least reveal where the fish ended up. Today, electronic recording devices attached to fish can continuously record not only where the fish is, but also how deep it swims at different times of day and the temperature and salinity of the water around it (**Figure 1.12**). At set intervals, these tags download their information to a satellite, which relays it back to researchers. Suddenly we are acquiring a great deal of new knowledge about the distribution of life in the oceans.

The scientific method combines observation and logic

Observations lead to questions, and scientists make additional observations and do experiments to answer those questions. The conceptual approach that underlies the design and conduct of most modern scientific investigations is called the **scientific method**. This

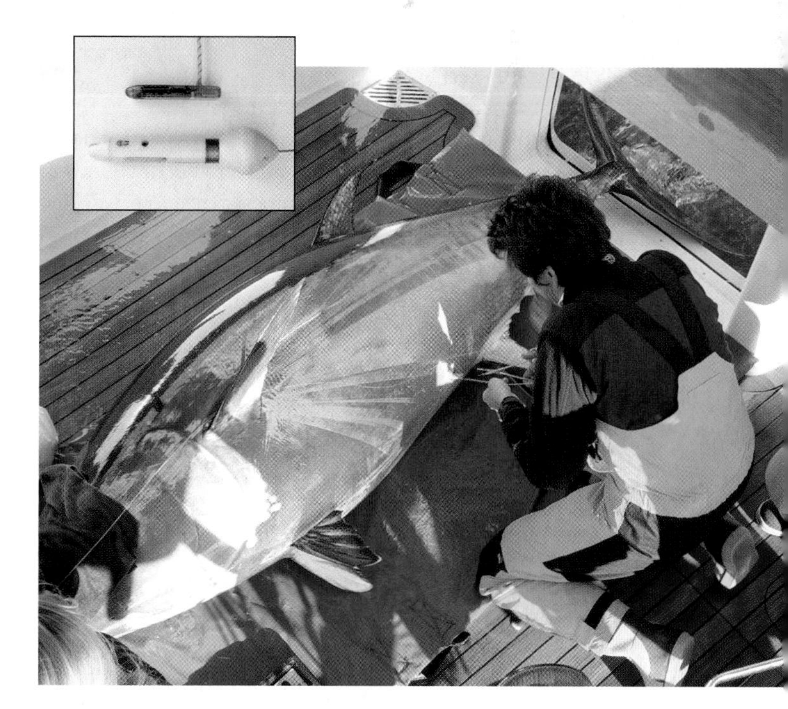

1.12 Tuna Tracking Marine biologist Barbara Block attaches a computerized data recording tag (inset) to a bluefin tuna. The use of such tags makes it possible to track individual tuna wherever they travel in the world's oceans.

powerful tool, also called the *hypothesis–prediction (H–P) method,* is a conceptual approach that provides a strong foundation for making advances in biological knowledge. The scientific method has five steps:

- Making *observations*
- Asking *questions*
- Forming *hypotheses*, or tentative answers to the questions
- Making *predictions* based on the hypotheses
- *Testing* the predictions by making additional observations or conducting experiments

Once a question has been posed, a scientist uses *inductive logic* to propose a tentative answer to the question. That tentative answer is called a **hypothesis**. For example, at the opening of this chapter, you learned that Pieter Johnson was shown abnormal frogs gathered in certain ponds. The first question stimulated by this observation was, is there something in these ponds that caused frogs to develop such extreme anatomical abnormalities?

In formulating a hypothesis, scientists put together the facts they already know to formulate one or more possible answers to the question. Pieter knew that there were likely to be contaminants in the ponds where deformed frogs were found because agricultural pesticides were used heavily in the region. In addition, mercury had once been mined nearby, and the abandoned mines could be a source of heavy metals in the water. He also knew that there were nearby ponds in which the frogs were normal. His first hypothesis, therefore, was that contaminants in the water caused mutations in the frog eggs.

HYPOTHESIS: Something in the environment is causing developmental limb abnormalities in Pacific tree frogs (*Hyla regilla*).

METHOD

1. Identify a test area of small ponds in an area where abnormal tree frogs have been found (agricultural land in Santa Clara County, California).
2. Collect and analyze water samples from the ponds.
3. Census the organisms in the ponds.
4. Look for correlations between the presence of frog abnormalities and the characteristics of the ponds.

Deformed hind leg

Santa Clara County, CA

RESULTS

Pacific tree frogs were found in 13 of 35 ponds. Frogs with limb abnormalities were found in 4 of these 13 ponds. Water and census analyses of the 13 ponds containing frogs revealed no difference in water pollution, but did reveal the presence of snails infested with parasitic flatworms of the genus ***Ribeiroia*** in the 4 ponds with abnormal frogs.

	Pesticide residues in water?	Heavy metals in water?	Industrial chemicals in water?	Snails in water?	*Ribeiroia* in water?	*Ribeiroia* larvae in frogs?
Ponds with normal frogs	No	No	No	No	No	No
Ponds with abnormal frogs	No	No	No	Yes	Yes	Yes

CONCLUSION: Infection by parasitic *Ribeiroia* may cause abnormalities in the limb development of Pacific tree frogs.

1.13 Comparative Experiments Look for Differences between Groups Pieter Johnson analyzed the differences between ponds in which deformed frogs were present versus nearby ponds in which there were no deformed frogs. Such comparisons can result in valuable insights.

The next step in the scientific method is to apply a different form of logic—*deductive logic*—to make predictions based on the hypothesis. Based on his hypothesis, Pieter predicted (1) that he would find contaminants in the ponds with the abnormal frogs, and (2) that eggs from those ponds would produce abnormal frogs when they were hatched in the laboratory.

Good experiments have the potential of falsifying hypotheses

Once predictions are made from a hypothesis, **experiments** can be designed to test those predictions. The most informative experiments are those that have the ability to show that the prediction is wrong. If the prediction is wrong, the hypothesis must be questioned, modified, or rejected.

Both of Pieter Johnson's initial predictions proved to be wrong. He counted frogs and other organisms in 35 ponds in the region where the deformed frogs had been found and measured chemicals in the water. Thirteen of the ponds were home to Pacific tree frogs, but he found deformed frogs in only four ponds. To Pieter's surprise, analysis of the water samples failed to reveal higher amounts of pesticides, industrial chemicals, or heavy metals in the ponds with deformed frogs. Also surprisingly, when he collected eggs from those ponds and hatched them in the laboratory, he always got normal frogs. The original hypothesis that contaminants caused mutations in the frog eggs had to be rejected. A new hypothesis had to be formulated and new experiments had to be conducted.

There are two general types of experiments and Pieter used both:

- In a **comparative experiment**, we predict that there will be a difference between samples or groups based on our hypothesis. We then test whether or not the predicted difference exists.
- In a **controlled experiment**, we also compare samples or groups, but in this case we start the experiment with groups that are as similar as possible. We predict on the basis of our hypothesis that some factor, or *variable*, plays a role in the phenomenon we are investigating. We then use some method to manipulate that variable in an "experimental" group while leaving the "control" group unaltered. We then

test to see if the manipulation created the predicted difference between the experimental and control groups.

COMPARATIVE EXPERIMENTS Comparative experiments are valuable when we do not know or cannot control the critical variables. Pieter Johnson performed a comparative experiment when he tested the water in the ponds (**Figure 1.13**). His challenge was to find some variable that differed between the ponds with normal and abnormal frogs. Finding no differences in the water chemistry of the two types of ponds, he had to reject his hypothesis that environmental contaminants were causing mutations in the frogs. So he compared the two types of ponds to see what variables *were* different between them.

Pieter found that a species of freshwater snail was present in the ponds with abnormal frogs, but absent from the ponds with normal ones. Freshwater snails are hosts for many parasites. His new hypothesis was that a parasite infecting the snail was in some way responsible for the frogs' deformities. To test that hypothesis, he performed controlled experiments.

CONTROLLED EXPERIMENTS In controlled experiments, one variable is manipulated while others are held constant. The variable that is manipulated is called the *independent variable* and the response that is measured is the *dependent variable*. A good controlled experiment is not easy to design because biological variables are so interrelated that it is difficult to alter just one.

Many parasites go through complex life cycles with several stages, each of which requires a specific host animal. Pieter focused on the possibility that some parasite that used freshwater snails as one of its hosts was infecting the frogs and causing their deformities. Pieter found a candidate parasite with this type of life cycle: a small flatworm called *Ribeiroia*, which was present in the ponds where the deformed frogs were found.

In Pieter's controlled experiment, the independent variable was the presence or absence of the snail and the parasite (**Figure 1.14**). He controlled all other variables by collecting frog eggs from ponds where there were no snails or flatworm parasites and hatching them in the laboratory. He divided the resulting tadpoles into two groups and placed them in separate tanks. He introduced snails and parasites into half of the tanks (the experimental group) and left the other tanks (the control group) free of snails and parasites. His dependent variable was the frequency of abnormalities in the frogs that developed under the different sets of conditions. He found that 85 percent of the frogs in the experimental tanks with *Ribeiroia* alone, but none of the frogs in the other tanks, developed abnormalities. Thus Pieter's results supported his hypothesis, and he could go on to investigate how the parasites caused abnormalities in the developing frogs.

Ribeiroia uses three hosts in California ponds: snails, frogs, and predatory birds such as herons. For the parasite to complete its life cycle and reproduce, it must be able to move from a frog to a bird. The limb deformities *Ribeiroia* causes may actually make infected frogs easier for predatory birds to capture and eat.

EXPERIMENT

HYPOTHESIS: Infection of Pacific tree frog tadpoles by the parasite *Ribeiroia* causes developmental limb abnormalities.

METHOD

1. Collect *Hyla regilla* eggs from a site with no record of abnormal frogs.
2. Allow eggs to hatch in laboratory aquaria. Randomly divide equal numbers of the resulting tadpoles into control and experimental groups.
3. Allow the control group to develop normally. Subject the experimental groups to infection with *Ribeiroia*, a different parasite (*Alaria*), and a combination of both parasites.
4. Follow tadpole development. Count and assess the resulting adult frogs.

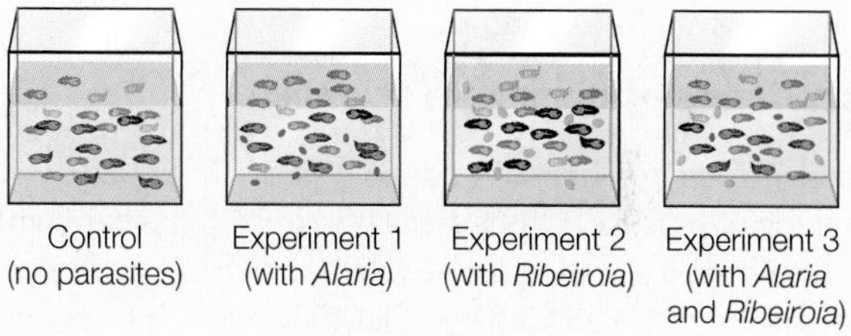

Control (no parasites)
Experiment 1 (with *Alaria*)
Experiment 2 (with *Ribeiroia*)
Experiment 3 (with *Alaria* and *Ribeiroia*)

RESULTS

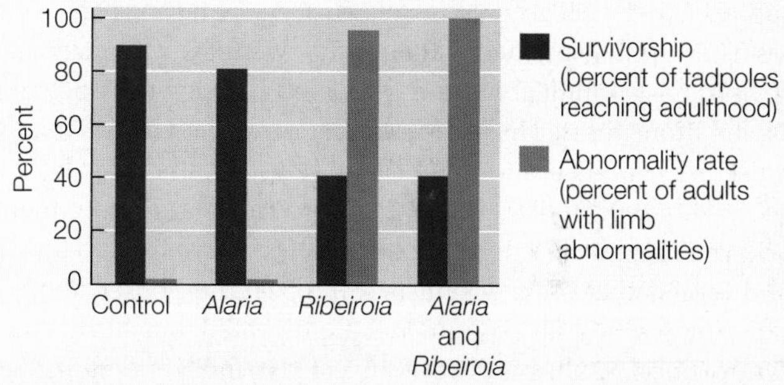

Percent
100
80
60
40
20
0
Control
Alaria
Ribeiroia
Alaria and *Ribeiroia*
Survivorship (percent of tadpoles reaching adulthood)
Abnormality rate (percent of adults with limb abnormalities)

CONCLUSION: *Ribeiroia* causes developmental limb abnormalities in Pacific tree frogs.

1.14 Controlled Experiments Manipulate a Variable The variable Johnson manipulated was the presence or absence of two species of parasitic flatworm. Other conditions of the experiment remained constant.

Statistical methods are essential scientific tools

Whether we are doing comparative or controlled experiments, at the end we have to decide whether there is a difference between the samples, individuals, groups, or populations in the study. How do we decide whether a measured difference is enough to support or falsify a hypothesis? In other words, how do we decide in an unbiased, objective way that the measured difference is significant?

Significance can be measured with statistical methods. Scientists use statistics because they recognize that variation is ubiquitous. Statistical tests analyze that variation and calculate the probability that the differences observed could be due to random variation. The results of statistical tests are therefore probabilities. A statistical test starts with a **null hypothesis**—the premise that no difference exists. When quantified observations, or **data**, are collected, statistical methods are applied to those data to calculate the likelihood that the null hypothesis is correct.

More specifically, statistical methods tell us the probability of obtaining the same results by chance even if the null hypothesis were true. Put another way, we need to eliminate insofar as possible the chance that any differences showing up in the data are merely the result of random variation in the samples tested. Scientists generally conclude that the differences they measure are significant if the statistical tests show that the *probability of error* (the probability that the results can be explained by chance) is 5 percent or lower. In particularly critical experiments, such as tests of the safety of a new drug, scientists require much lower probabilities of error, such as 1 percent or even 0.1 percent.

Not all forms of inquiry are scientific

Science is a unique human endeavor that is bounded by certain standards of practice. Other areas of scholarship share with science the practice of making observations and asking questions, but scientists are distinguished by what they do with their observations and how they answer their questions. Data, subjected to appropriate statistical analysis, are critical in the testing of hypotheses. The scientific method is the most powerful way humans have devised for learning about the world and how it works. Scientific explanations for natural processes are objective and reliable because the hypotheses proposed *must be testable* and *must have the potential of being rejected* by direct observations and experiments. Scientists clearly describe the methods they have used to test hypotheses so that other scientists can repeat their observations or experiments. Not all experiments are repeated, but surprising or controversial results are always subjected to independent verification. All scientists worldwide share this built-in process of testing and rejecting hypotheses, so they all contribute to a common body of scientific knowledge.

If you understand the methods of science, you can distinguish science from non-science. Art, music, and literature are activities that contribute to the quality of human life, but they are not science. They do not use the scientific method to establish what is fact. Religion is not science, although religions have historically purported to explain natural events ranging from unusual weather patterns to crop failures to human diseases and mental afflictions. Many such phenomena that at one time were mysterious are now explicable in terms of scientific principles.

The power of science derives from the uncompromising objectivity and absolute dependence on evidence that comes from *reproducible and quantifiable observations*. A religious or spiritual explanation of a natural phenomenon may be coherent and satisfying for the person or group holding that view, but it is not testable, and therefore it is not science. To invoke a supernatural explanation (such as an "intelligent designer" with no known bounds) is to depart from the world of science.

Science describes the facts about how the world works, not how it "ought to be." Many of the recent scientific advances that have contributed so much to human welfare also raise major ethical issues. Developments in genetics and developmental biology, for example, now enable us to select the sex of our children, to use stem cells to repair our bodies, and to modify the human genome. Although scientific knowledge allows us to do these things, science cannot tell us whether or not we should do them, or if we choose to do so, how we should regulate them.

Making wise decisions about such issues requires a clear understanding of the implications of available scientific information. Success in surgery depends on an accurate diagnosis. So does success in environmental management. However, to make wise decisions about public policy, we also need to employ the best possible ethical reasoning in deciding which outcomes we should strive for. For a bright future, society needs both good science and good ethics, as well as an educated public that understands the importance of both and the critical differences between them.

1.3 RECAP

The scientific method of inquiry starts with the formulation of hypotheses based on observations and data. Comparative and controlled experiments are carried out to test hypotheses.

- Can you explain the relationship between a hypothesis and an experiment? See p. 14
- What features characterize questions that can be answered only by using a comparative approach? See p. 14 and Figure 1.13
- What is controlled in a controlled experiment? See p. 15 and Figure 1.14
- Do you understand why arguments must be supported by quantifiable and reproducible data in order to be considered scientific? See p. 16

The vast amount of scientific knowledge accumulated over centuries of human civilization allows us to understand and manipulate aspects of the natural world in ways that no other species can. These abilities present us with challenges, opportunities, and, above all, responsibilities. Let's look at how knowledge of biology can affect the formulation of public policy.

1.4 How Does Biology Influence Public Policy?

The study of biology has long had major implications for human life. Agriculture and medicine are two important human activities that depend on biological knowledge. Our ancestors unknowingly applied the principles of evolutionary biology when they domesticated plants and animals. People have also been speculating about the causes of diseases and searching for methods of combating them since ancient times. Long before the causes of diseases were known, people recognized that diseases could be passed from one person to another. Isolation of infected persons has been practiced as long as written records have been available, but most so-called cures were not effective until scientists found out what caused diseases.

Today, thanks to the deciphering of genomes and the ability to manipulate them, vast new possibilities exist for improvements in the control of human diseases and agricultural productivity. At the same time, these capabilities have raised important ethical and policy issues. How much and in what ways should we tinker with the genetics of humans and other species? Does it matter whether our crops and domesticated animals are changed by traditional breeding experiments or by gene transfers? What rules should govern the release of genetically modified organisms into the environment? Science alone cannot provide answers to those questions, but wise policy decisions must be based on accurate scientific information.

Another reason for studying biology is to understand the effects of the vastly increased human population on its environment. Our use of natural resources is putting stress on the ability of Earth's ecosystems to continue to produce the goods and services on which our society depends. Human activities are changing global climates, causing the extinctions of a large number of species, and spreading new diseases while facilitating the resurgence of old ones. The rapid spread of the SARS and West Nile viruses, for example, was facilitated by modern modes of transportation, and the recent resurgence of tuberculosis is the result of the evolution of bacteria that are resistant to antibiotics. Biological knowledge is vital for determining the causes of these changes and for devising wise policies to deal with them. An understanding of biology also helps people appreciate the marvelous diversity of living organisms that provides goods and services for humankind and also enriches our lives aesthetically and spiritually.

Biologists are increasingly called on to advise government agencies concerning the laws, rules, and regulations by which society deals with the increasing number of problems and challenges that have at least a partial biological basis. As an example of the value of scientific knowledge for the assessment and formulation of public policy, let's return to the tracking study of bluefin tuna introduced in Section 1.3. Prior to this study, both scientists and fishermen knew that bluefins had a western Atlantic breeding ground in the Gulf of Mexico and an eastern Atlantic breeding ground in the Mediterranean Sea. Overfishing was endangering the western breeding population. Everyone assumed that the fish from the two breeding populations had geographically separate feeding grounds as well as separate breeding grounds, so an international commission drew a line down the middle of the Atlantic Ocean and established stricter fishing quotas on the western side of the line. The intent was to allow the western population to recover. However, new data revealed that in fact the eastern and western bluefin populations mix freely on the feeding (and hence fishing) grounds across the entire North Atlantic (**Figure 1.15**). Thus a fish caught on the eastern side of the line could be from the western breeding population, so the established policy was not appropriate for achieving its intended goal.

Throughout this book we will share with you the excitement of studying living things and illustrate the rich array of methods that biologists use to determine why the world of living things looks and functions as it does. The most important motivator of most biologists is curiosity. People are fascinated by the richness and diversity of life and want to learn more about organisms and how they interact with one another. The trait of human curiosity

1.15 Bluefin Tuna Do Not Recognize the Lines Drawn on Maps by International Commissions Because it was assumed that western (red dots) and eastern (gold dots) breeding populations of bluefin tuna also fed on their respective sides of the Atlantic Ocean, separate fishing quotas were established to either side of 45°W longitude (dashed line). It was believed this would allow the endangered western population to recover. However, tracking data showed that the two populations mix freely, especially in the heavily fished waters of the northernmost Atlantic (blue circle); so in fact the established policy does not protect the western population.

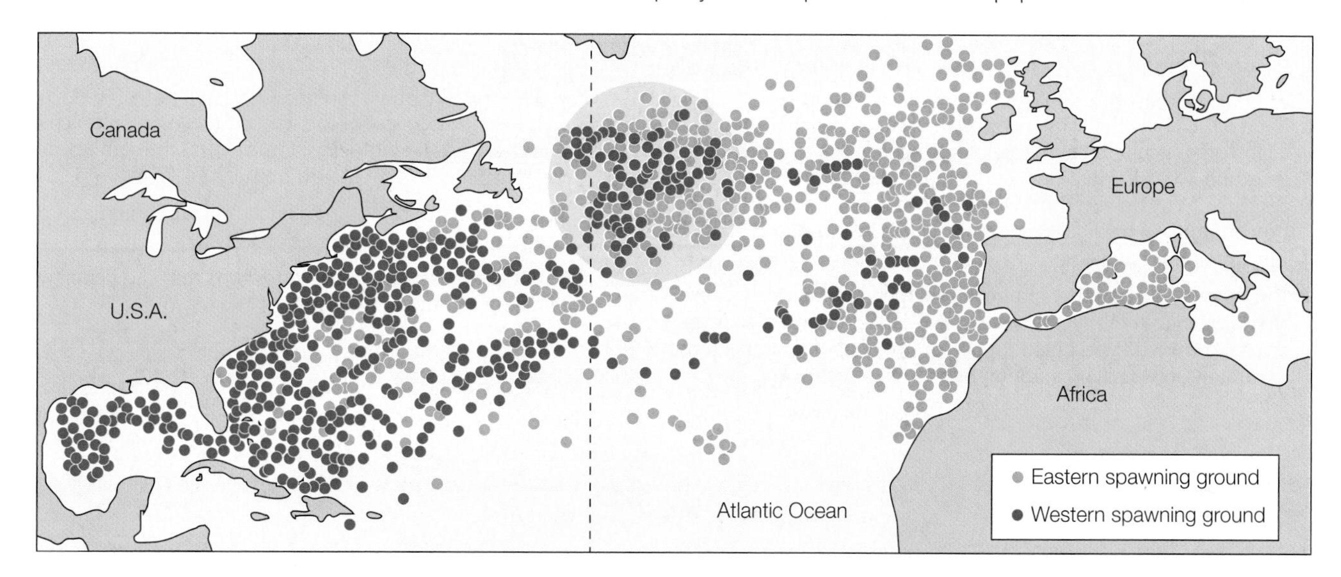

might even be seen as adaptive, and could have been selected for if individuals who were motivated to learn about their surroundings were likely to have survived and reproduced better, on average, than their less curious relatives!

There are vast numbers of questions for which we do not yet have answers, and new discoveries usually engender questions no one thought to ask before. Perhaps you will eventually pose and answer one or more of those questions.

CHAPTER SUMMARY

1.1 What is biology?

Biology is the study of life at all levels of organization, ranging from molecules to the biosphere.

The **cell theory** states that all life consists of cells, and all cells come from preexisting cells.

All living organisms are related to one another through descent with modification. **Evolution** by **natural selection** is responsible for the diversity of **adaptations** found in living organisms.

The instructions for a cell are contained in its **genome**, which consists of DNA molecules made up of sequences of **nucleotides**. Specific segments of DNA called **genes** contain the information the cell uses to make **proteins**. Review Figure 1.4

Cells are the basic structural and physiological units of life. Most of the chemical reactions of life take place in cells. Living organisms control their internal environment. They also interact with other organisms of the same and different species. Biologists study life at all these levels of organization. Review Figure 1.6, Web/CD Activity 1.1

Biological knowledge obtained from a **model system** may be generalized to other species.

1.2 How is all life on Earth related?

Biologists use fossils, anatomical similarities and differences, and molecular comparisons of genomes to reconstruct the history of life. Review Figure 1.9

Life first arose by chemical evolution. Biological evolution began with the formation of cells.

Photosynthesis was an important evolutionary step because it changed Earth's atmosphere and provided a means of capturing energy from sunlight.

The earliest organisms were **prokaryotes**; organisms with more complex cells, called **eukaryotes**, arose later. Eukaryotic cells have discrete intracellular compartments, called **organelles**, including a **nucleus** that contains the cell's genetic material.

The genetic relationships of **species** can be represented as an evolutionary tree. Species are grouped into three **domains: Archaea, Bacteria**, and **Eukarya**. The domains Archaea and Bacteria consist of unicellular prokaryotes. The domain Eukarya contains the microbial eukaryotes (protists), plants, fungi, and animals. Review Figure 1.11, Web/CD Activity 1.2

1.3 How do biologists investigate life?

The **scientific method** used in most biological investigations involves five steps: making observations, asking questions, forming hypotheses, making predictions, and testing those predictions.

Hypotheses are tentative answers to questions. Predictions made on the basis of a hypothesis are tested with additional observations and two kinds of **experiments: comparative** and **controlled experiments**. Review Figures 1.13 and 1.14

Statistical methods are applied to **data** to establish whether or not the differences observed are significant or whether they could be expected by chance. These methods start with the **null hypothesis** that there are no differences.

Science can tell us how the world works, but it cannot tell us what we should or should not do.

1.4 How does biology influence public policy?

Wise public policy decisions must be based on accurate scientific information. Biologists are often called on to advise governmental agencies on the solution of important problems that have a biological component.

FOR DISCUSSION

1. Even if we knew the sequences of all of the genes of a single-celled organism and could cause those genes to be expressed in a test tube, we still could not create one of those organisms in the test tube. Why do you think this is so? In light of this fact, what do you think of the statement that the genome contains all of the information for a species?
2. If someone told you that giraffes developed long necks because they stretched their necks to reach leaves higher and higher on trees, how would you help that person think about giraffes more accurately, in terms of evolution by natural selection?
3. In a recent discovery of the genes that control skin color in zebrafish, why did the biologists assume that the same genes might be responsible for skin color in humans?
4. Why is it so important in science that we design and perform tests capable of falsifying a hypothesis?
5. What features characterize questions that can be answered only by using a comparative approach?

FOR INVESTIGATION

1. The abnormalities of frogs in Pieter Johnsons's study were associated with the presence of a parasite. How would you investigate how the parasites induced the formation of monster frogs? Hint: When tadpoles were exposed to the parasites *after* they began to develop legs, they did not show abnormalities.

2. Just as all cells come from preexisting cells, mitochondria—the cell organelles that convert energy in food to a form of energy that can do biological work—all come from preexisting mitochondria. Cells do not synthesize mitochondria from the genetic information in their nuclei. What investigations would you carry out to understand the nature of mitochondria?

PART SEVEN
Flowering Plants: Form and Function

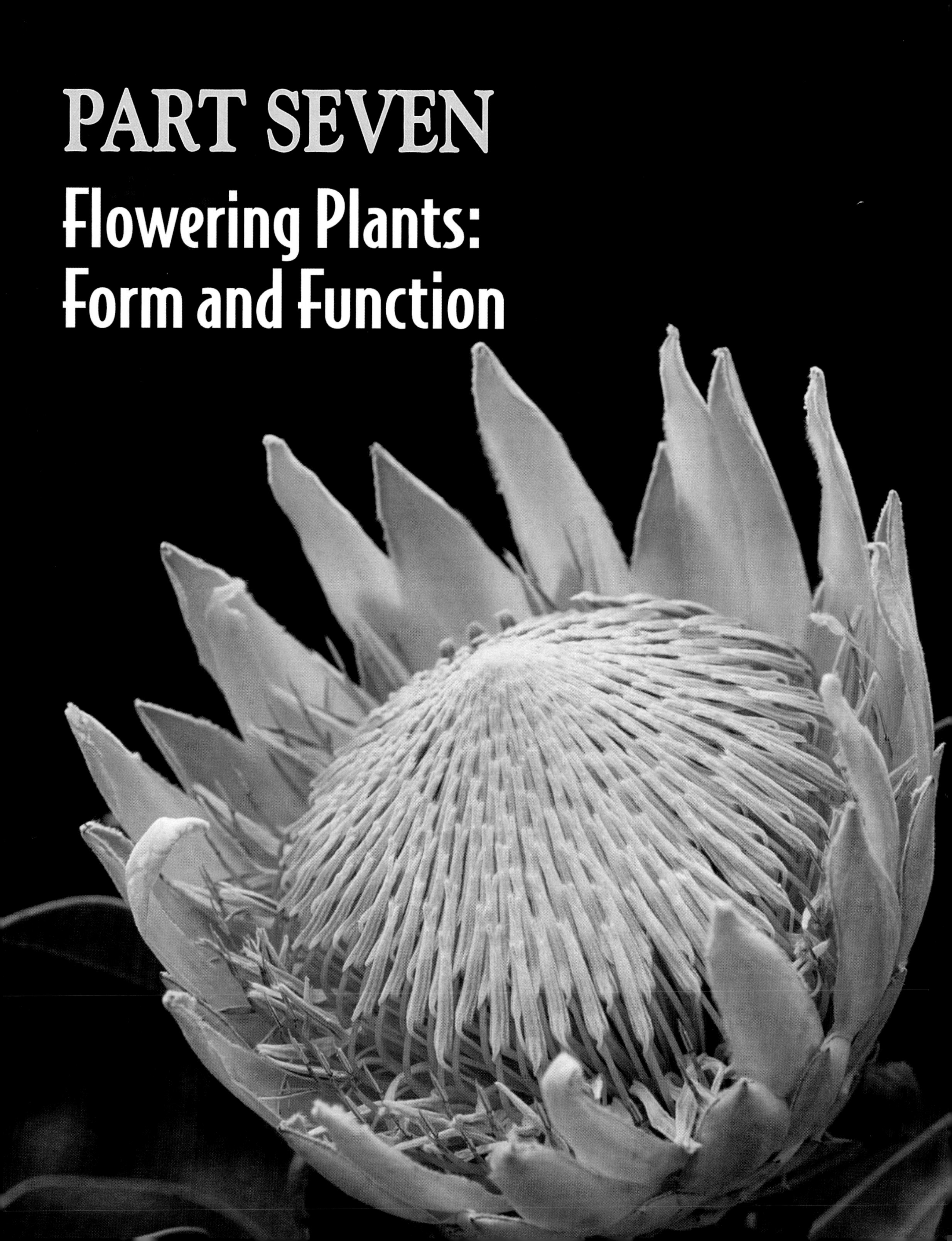

CHAPTER 34 The Plant Body

CSI: Wood anatomy convicts a killer

In the evening of May 21, 1927 Charles Lindbergh landed his plane, *Spirit of St. Louis*, in Paris, becoming the first person to fly nonstop across the Atlantic Ocean. He also became a national hero in the United States, but five years later, "Lucky Lindy" suffered a terrible tragedy when his 20-month-old son was kidnapped. Clues were few: the handwriting on twelve ransom notes, footprints (which were never measured!), a carpenter's chisel, indentations in the ground below the baby's window, and, about 60 feet away in some bushes, a crude wooden ladder in three parts.

The case was assigned to the superintendent of the New Jersey State Police, H. Norman Schwarzkopf (father of General Norman Schwarzkopf, of Gulf War fame), who decided that the ladder was his best clue. He hired Arthur Koehler from the U.S. Forest Products Laboratory in Madison, Wisconsin, to examine the ladder for evidence. What could wood anatomy tell Koehler about the kidnapping?

Microscopic examination of the ladder revealed its cell structure, enabling Koehler to learn that some of the rungs were made from Douglas fir and others from Ponderosa pine, rails were made from Douglas fir and North Carolina pine, and dowels for the railing from birch. Clearly, the ladder was assembled from scraps. Koehler began to extend his examination to other features of the wood.

In the meantime, thousands of leads poured in. The baby's body was discovered, intensifying the search. Gold certificates from the ransom began to appear, eventually leading to the arrest of Bruno Richard Hauptmann. Hauptmann resembled a description provided by the man who had passed the ransom in the dark of night, and his handwriting strongly resembled that on the ransom notes. This and other evidence was circumstantial, however, and prosecutors sought further evidence that would convince a jury that Hauptmann was the kidnapper.

Koehler worked painstakingly to amass evidence based on the wood in the ladder. His most exhausting task was canvassing some 1,600 lumber mills that processed North Carolina pine. He was seeking a particular machine planer that could have left the distinc-

WANTED

INFORMATION AS TO THE WHEREABOUTS OF

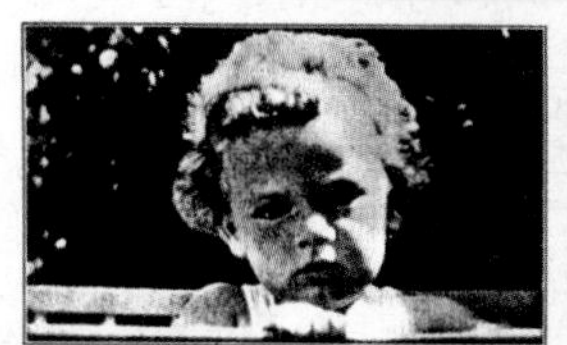

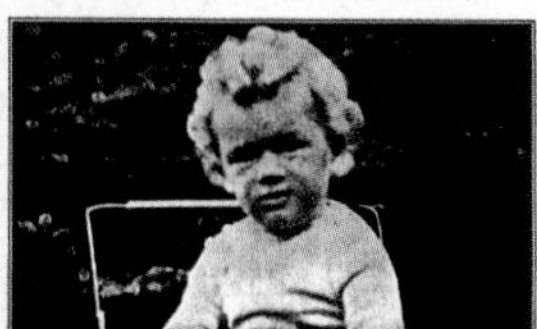

CHAS. A. LINDBERGH, JR.

OF HOPEWELL, N. J.

SON OF COL. CHAS. A. LINDBERGH

World-Famous Aviator

This child was kidnaped from his home in Hopewell, N. J., between 8 and 10 p. m. on Tuesday, March 1, 1932.

DESCRIPTION:

Age, 20 months — Hair, blond, curly
Weight, 27 to 30 lbs. — Eyes, dark blue
Height, 29 inches — Complexion, light

Deep dimple in center of chin
Dressed in one-piece coverall night suit

ADDRESS ALL COMMUNICATIONS TO
COL. H. N. SCHWARZKOPF, TRENTON, N. J., or
COL. CHAS. A. LINDBERGH, HOPEWELL, N. J.

ALL COMMUNICATIONS WILL BE TREATED IN CONFIDENCE

A Desperate Search In the spring of 1932, people across the United States were glued to newspaper and radio updates on the kidnapping of the son of national hero Charles Lindbergh. Posters like this one resulted in thousands of tips from concerned citizens but, sadly, what the police eventually discovered was the child's murdered body.

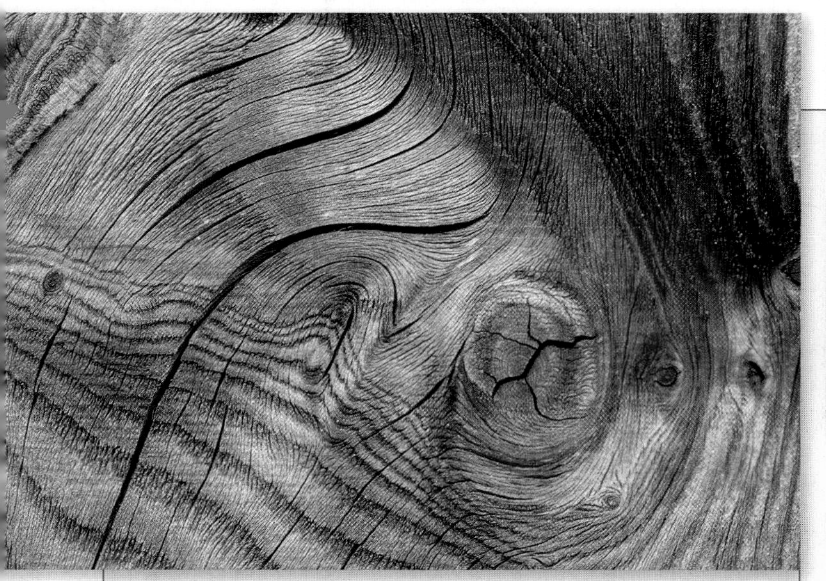

Fingerprints in Wood Wood's unique grain is affected by year-to-year variations in moisture and temperature, by branching (knots), uneven shading of the leaf canopy, whether the wood came from the tree's trunk or a branch, and many other factors. Forensic analysis of wood is one more tool in the arsenal of crime-fighters.

tive pattern of grooves found on the rails. He found it at a mill in South Carolina and traced a shipment from that mill to a lumber company in the Bronx. The Bronx store had employed Hauptmann and sold him lumber.

In court, Koehler used a hand plane from Hauptmann's own toolbox to show that it produced the same pattern of ridges found on one of the rails. He matched three nail holes with protruding nails in a joist in Hauptmann's floor.

Most convincing of all was Koehler's courtroom comparison of the wood grain in a floorboard from Hauptmann's attic—missing a section—and a section of a ladder rail. In spite of a gap in the rail, Koehler showed clearly that the grain of the two pieces matched. That is, the "rings" in the wood of the two sections coincided perfectly, proving that wood from the attic was used to construct the ladder. A jury found Hauptmann guilty, and he was executed on April 3, 1936.

IN THIS CHAPTER we will examine plant structure at the levels of organs, cells, tissues, and tissue systems. We will see how organized groups of dividing cells, called meristems, contribute to the growth of the plant body. The chapter concludes with a consideration of how leaf structure supports photosynthesis.

CHAPTER OUTLINE

34.1 How Is the Plant Body Organized?

The lumber Hauptmann used to build his ladder was the product of the plant body—specifically, the woody stems of trees. All vascular plants have essentially the same structural organization. This chapter describes the basic architecture of the angiosperm body plan.

Most angiosperms (flowering plants) belong to one of two major clades. **Monocots** are generally narrow-leaved flowering plants such as grasses, lilies, orchids, and palms. **Eudicots** are broad-leaved flowering plants such as soybeans, roses, sunflowers, and maples. These two clades, which account for 97 percent of flowering plant species, differ in several important basic characteristics (**Figure 34.1**). Most of the remaining species (including water lilies and magnoliids, discussed in Section 29.4) are structurally similar to the eudicots.

As Chapter 29 described, angiosperms are vascular plants characterized by double fertilization, a triploid endosperm, and seeds enclosed in modified leaves called carpels. These are all reproductive characteristics; flowers, which are the plant's devices for sexual reproduction, consist of modified leaves and stems and will be considered in detail in Chapter 38. But flowering plants also possess three kinds of *vegetative* (nonreproductive) organs: roots, stems, and leaves. In both monocots and eudicots, all the organs are organized in two systems: the **shoot system** and the **root system**. The basic body plans of a generalized monocot and a generalized eudicot are shown in **Figure 34.2**.

- The *shoot system* of a plant consists of the stems, leaves, and flowers. Broadly speaking, the **leaves** are the chief organs of photosynthesis. The **stems** hold and display the leaves to the sun and provide connections for the transport of materials between roots and leaves. The **nodes** are the points of attachment of leaf to stem, and the stem regions between successive nodes are **internodes**.
- The *root system* anchors the plant in place and provides nutrition. The extreme branching of plant roots and their high surface area-to-volume ratio allow them to absorb water and mineral nutrients from the soil.

Each of the vegetative organs can be understood in terms of its structure. By *structure* we mean both its overall form, called its *morphology*, and its component cells and tissues and their

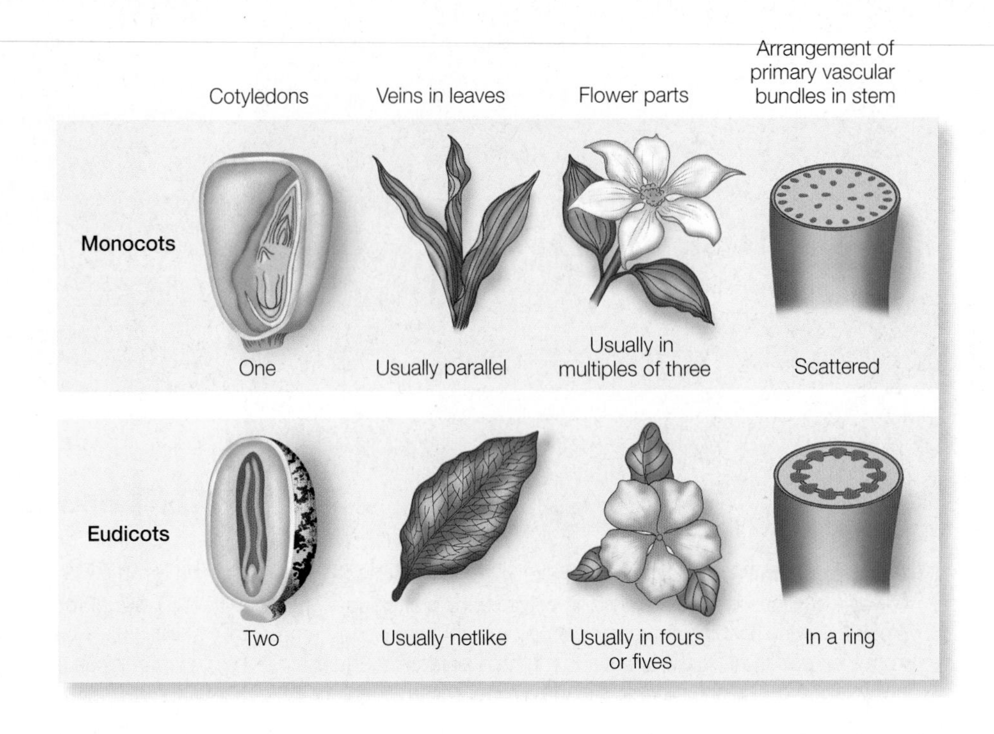

34.1 Monocots versus Eudicots The possession of a single cotyledon clearly distinguishes the monocots from the other angiosperms. Several other anatomical characteristics also differ between the monocots and the eudicots. Most angiosperms that do not belong to either clade resemble eudicots in the characteristics shown here.

arrangement, called its *anatomy.* Let's first consider the overall forms of roots, stems, and leaves.

Roots anchor the plant and take up water and minerals

In most plants, water and minerals enter the plant through the root system, which lies in the soil, where light does not penetrate. Roots typically lack the capacity for photosynthesis even when removed from the soil and placed in light.

There are two principal types of root systems. Many eudicots have a *taproot system*: a single, large, deep-growing primary root accompanied by less prominent lateral roots. The taproot itself often functions as a nutrient storage organ, as in carrots (**Figure 34.3A**).

By contrast, monocots and some eudicots have a *fibrous root system,* which is composed of numerous thin roots that are all roughly equal in diameter (**Figure 34.3B**). Many fibrous root systems have a large surface area for the absorption of water and minerals. A fibrous root system clings to soil very well. The fibrous root systems of grasses, for example, may protect steep hillsides where runoff from rain would otherwise cause erosion.

Some plants have *adventitious roots*. These roots arise above ground from points along the stem; some even arise from the leaves. In many species, adventitious roots can form when a piece of shoot is cut or broken from the plant and placed in water or soil. Adventitious rooting enables the cutting to establish itself in the soil as a new plant. Such a cutting is a form of asexual reproduction, also called *vegetative reproduction* in plants, which we will discuss in Chapter 38. Some plants—corn, banyan trees, and some palms, for example—use adventitious roots as props to help support the shoot.

Stems bear buds, leaves, and flowers

The central function of the **stem** is to elevate and support the reproductive organs (flowers) and the photosynthetic organs (leaves).

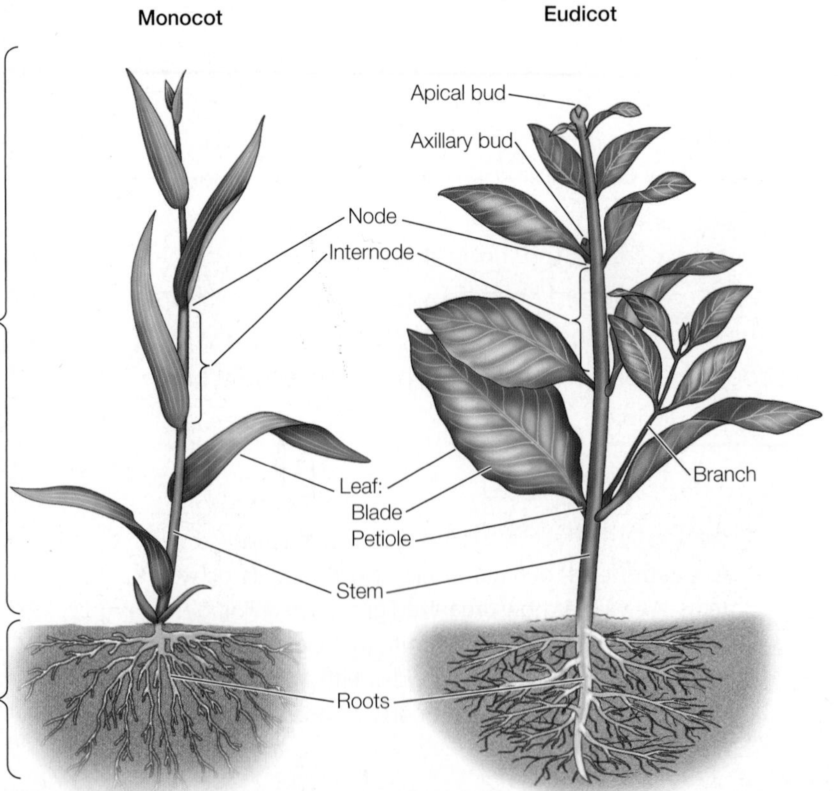

34.2 Vegetative Organs and Systems The basic plant body plan and the principal vegetative organs are similar in monocots and eudicots.

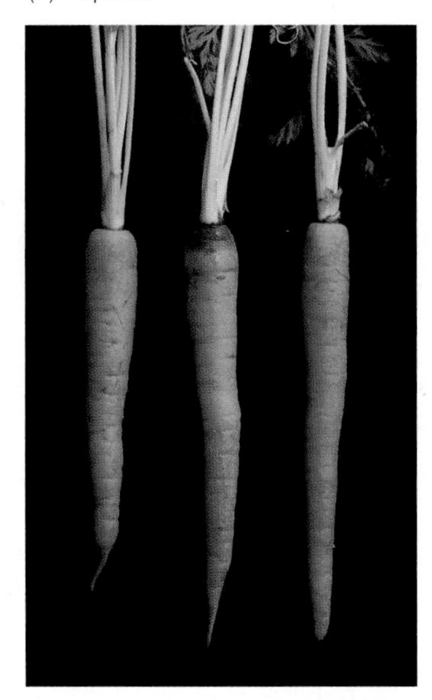

34.3 Root Systems (A) The taproot system of a carrot contrasts with (B) the fibrous root system of a leek.

Unlike roots, stems bear buds of various types. A *bud* is an embryonic shoot. A stem bears leaves at its nodes, and in the angle (axil) where each leaf meets the stem there is an **axillary bud** (see Figure 34.2). If it becomes active, the axillary bud can develop into a new *branch*, or extension of the shoot system. The branching patterns of plants are highly variable, depending on the species, environmental conditions, and a gardener's pruning activities.

At the tip of each stem or branch is an **apical bud**, which produces the cells for the upward and outward growth and development of that shoot. Under appropriate conditions, other buds form that develop into flowers.

Some stems are highly modified. The *tuber* of a potato, for example—the part of the plant eaten by humans—is an underground stem rather than a root. Its "eyes" are depressions containing axillary buds; thus, a sprouting potato is just a branching stem (**Figure 34.4A**). Many desert plants have enlarged, water-retaining stems (**Figure 34.4B**). The *runners* of strawberry plants and Bermuda grass are horizontal stems from which roots grow at frequent intervals (**Figure 34.4C**). If the links between the rooted portions are broken, independent plants can develop on each side of the break—a form of vegetative reproduction.

Although young stems are usually green and capable of photosynthesis, they usually are not the principal sites of photosynthesis. Most photosynthesis takes place in leaves.

Leaves are the primary sites of photosynthesis

In gymnosperms and most flowering plants, the leaves are responsible for most of the plant's photosynthesis, producing energy-rich organic molecules and releasing oxygen gas (see Section 8.1). In certain plants, the leaves are highly modified for more specialized functions, as we will see below.

As photosynthetic organs, leaves are marvelously adapted for gathering light. Typically, the **blade** of a leaf is a thin, flat structure attached to the stem by a stalk called a **petiole** (see Figure 34.2). In many plants, the leaf blade is held by its petiole at an angle almost perpendicular to the rays of the sun. This orientation, with the leaf surface facing the sun, maximizes the amount of light available for photosynthesis. Some leaves track the sun over the course of the day, moving so that they constantly face it.

34.4 Modified Stems (A) A potato is a modified stem called a tuber; the sprouts that grow from its eyes are shoots, not roots. (B) The stem of this barrel cactus is enlarged to store water. Its highly modified leaves serve as thorny spines. (C) The runners of beach strawberry are horizontal stems that produce roots at intervals. Runners provide a local water supply and allow rooted portions of the plant to live independently if the runner is cut.

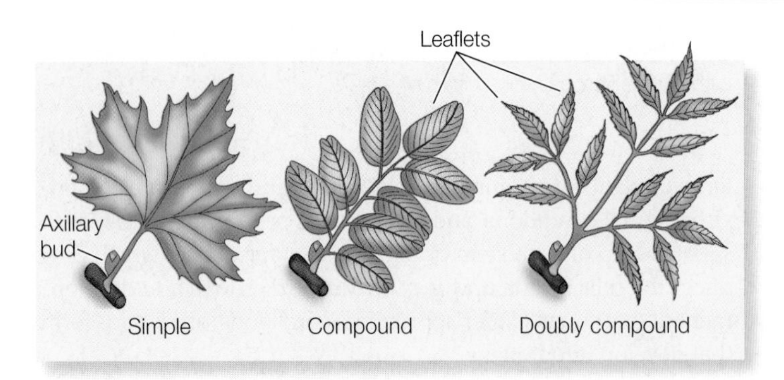

34.5 Simple and Compound Leaves Simple leaves are those with a single blade. Some compound leaves consist of leaflets arranged along a central axis. Further division of the axis results in a doubly compound leaf.

The leaves at different sites on a single plant may have quite different shapes. These shapes result from a combination of genetic, environmental, and developmental influences. Most species, however, bear similar, if not identical, leaves of a particular broadly defined type. A leaf may be *simple*, consisting of a single blade, or *compound*, with multiple blades called *leaflets* arranged along an axis or radiating from a central point (**Figure 34.5**). In a simple leaf, or in a leaflet of a compound leaf, the veins may be parallel to one another, as in most monocots, or in a netlike arrangement, as in eudicots. Differential growth of the leaf veins and the tissue between the veins determines the overall shape of a blade or leaflet.

During development in some plant species, leaves are highly modified for special functions. For example, modified leaves serve as storage depots for energy-rich molecules, as in the bulbs of onions. In other species, the leaves store water, as in succulents. The spines of cacti are modified leaves (see Figure 34.4B). Many plants, such as peas, have modified portions of leaves called *tendrils* that support the plant by wrapping around other structures or plants.

What do all the vegetative organs of flowering plants have in common? Roots, stems, and leaves are composed of three tissue systems. Let's see what they are and what they do.

The tissue systems support the plant's activities

A **tissue** is an organized group of cells that have features in common and that work together as a structural and functional unit. Tissues, in turn, are grouped into **tissue systems**. Three tissue systems extend throughout the body of the vascular plant: *vascular, dermal,* and *ground* tissues in a concentric arrangement (**Figure 34.6**).

The **vascular tissue system** is the plant's plumbing or transport system. Its two constituent tissues, the xylem and phloem, distribute materials throughout the plant. The **xylem** distributes water and mineral ions taken up by the roots to all the cells of the stem and leaves. As a result of its cellular complexity, xylem can perform a variety of functions, including transport, support, and storage. All the living cells of the plant body require a source of energy and chemical building blocks. The **phloem** meets these needs by transporting carbohydrates from sites of production (called *sources*, primarily leaves) to sites of utilization or storage (called *sinks*, such as growing tissue, storage tubers, and developing flowers).

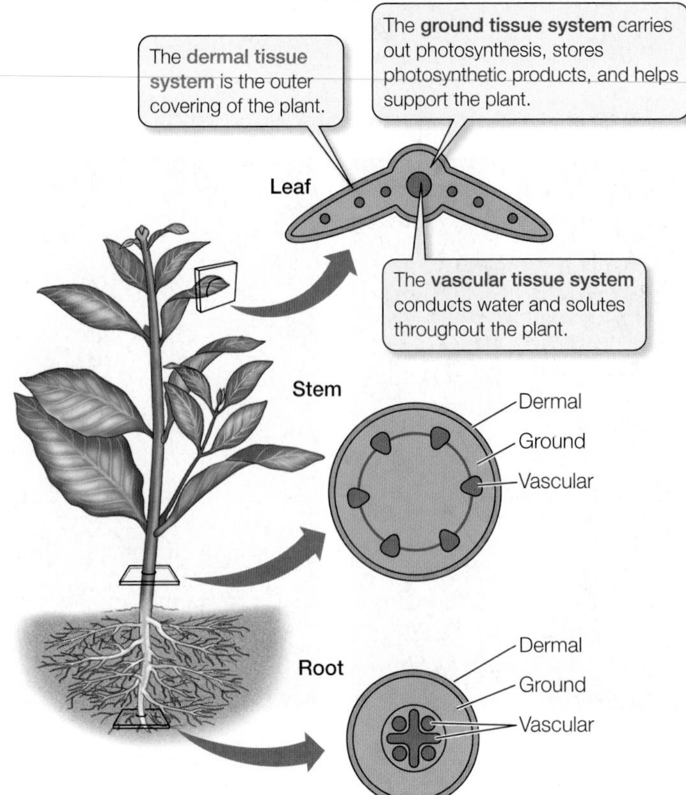

34.6 Three Tissue Systems Extend Throughout the Plant Body The arrangement shown here is typical of eudicots, but the three tissue systems are continuous in the bodies of all vascular plants.

The **dermal tissue system** is the outer covering of the plant. All parts of the young plant body are covered by an **epidermis**, which may be a single layer of cells or several layers. The epidermis is a complex tissue that may include specialized cell types, such as the *guard cells* that form stomata (pores) for gas exchange in leaves. The shoot epidermis secretes a layer of wax-covered *cutin*, the **cuticle**, that helps retard water loss from stems and leaves. The stems and roots of woody plants have a dermal tissue system called the *periderm*, a protective covering that will be discussed later in this chapter.

The **ground tissue system** makes up the rest of the plant. Ground tissue functions primarily in storage, support, photosynthesis, and the production of defensive and attractive substances.

34.1 RECAP

The basic body plan of both monocots and eudicots consists of a root system and a shoot system. Stems and leaves are part of the shoot system. A bud is an embryonic shoot. Plant tissues form three tissue systems: vascular, dermal, and ground.

- How would you distinguish between a piece of stem and a piece of root? See pp. 746–747
- Can you distinguish among the three tissue systems in terms of their location and function? See p. 748 and Figure 34.6

All plant organs are composed of tissues and tissue systems. Let's now consider the basic structural and functional units of plant tissues: their cells.

34.2 How Are Plant Cells Unique?

Plant cells have all the essential organelles common to eukaryotes (see Figure 4.7), but certain additional structures and organelles distinguish them from many other eukaryotes:

- They contain *chloroplasts* or other plastids.
- They contain *vacuoles*.
- They possess cellulose-containing *cell walls*.

Plant cells are alive when they divide and grow, but certain cells function only after their living parts have died and disintegrated. Other plant cells develop specialized metabolic capabilities; for example, some can perform photosynthesis, and others produce and secrete waterproofing materials. A plant has several different types of cells that differ dramatically in the composition and structure of their cell walls. The walls of each cell type have a composition and structure that correspond to its special functions.

Cell walls may be complex in structure

The cytokinesis of a plant cell is completed when the two daughter cells are separated by a cell plate (see Figure 9.12B). The daughter cells then deposit a gluelike substance within the cell plate; this substance constitutes the **middle lamella**. Next, each daughter cell secretes cellulose and other polysaccharides to form a **primary wall**. This deposition and secretion continue as the cell expands to its final size (**Figure 34.7**).

Once cell expansion stops, a plant cell may deposit one or more additional cellulosic layers to form a **secondary wall** internal to the primary wall (see Figure 34.7). Secondary walls are often impregnated with unique substances that give them special properties. Those impregnated with the polymer *lignin* become strong, as in wood cells. Walls to which the complex lipid *suberin* is added become waterproof.

Although it lies outside the plasma membrane, the cell wall is not a chemically inactive region. In addition to cellulose and other polysaccharides, the cell wall contains proteins, some of which are enzymes. Chemical reactions in the wall play important roles in cell expansion and in defense against invading organisms. Cell walls may thicken or be sculpted or perforated as cells differentiate into specialized cell types. The genome of the tiny plant *Arabidopsis thaliana* contains more than a thousand genes related to cell wall biosynthesis and function. Except where the secondary wall is waterproofed, the cell wall is permeable to water and mineral ions and allows molecules other than large macromolecules to reach the plasma membrane.

Localized modifications in the walls of adjacent cells allow water and dissolved materials to move easily from cell to cell. The primary wall usually has regions where it becomes quite thin. In these regions, cytoplasm-filled canals called **plasmodesmata** (singular, *plasmodesma*) pass through the primary wall, allowing direct communication between plant cells. A plasmodesma is traversed by a strand of endoplasmic reticulum (**Figure 34.8**). Under certain circumstances, a plasmodesma can enlarge dramatically, allowing even macromolecules and viruses to pass directly between cells. Macromolecules passing through enlarged plasmodesmata include transcription factors and RNAs. Substances can move from cell to cell through plasmodesmata without having to cross a plasma membrane.

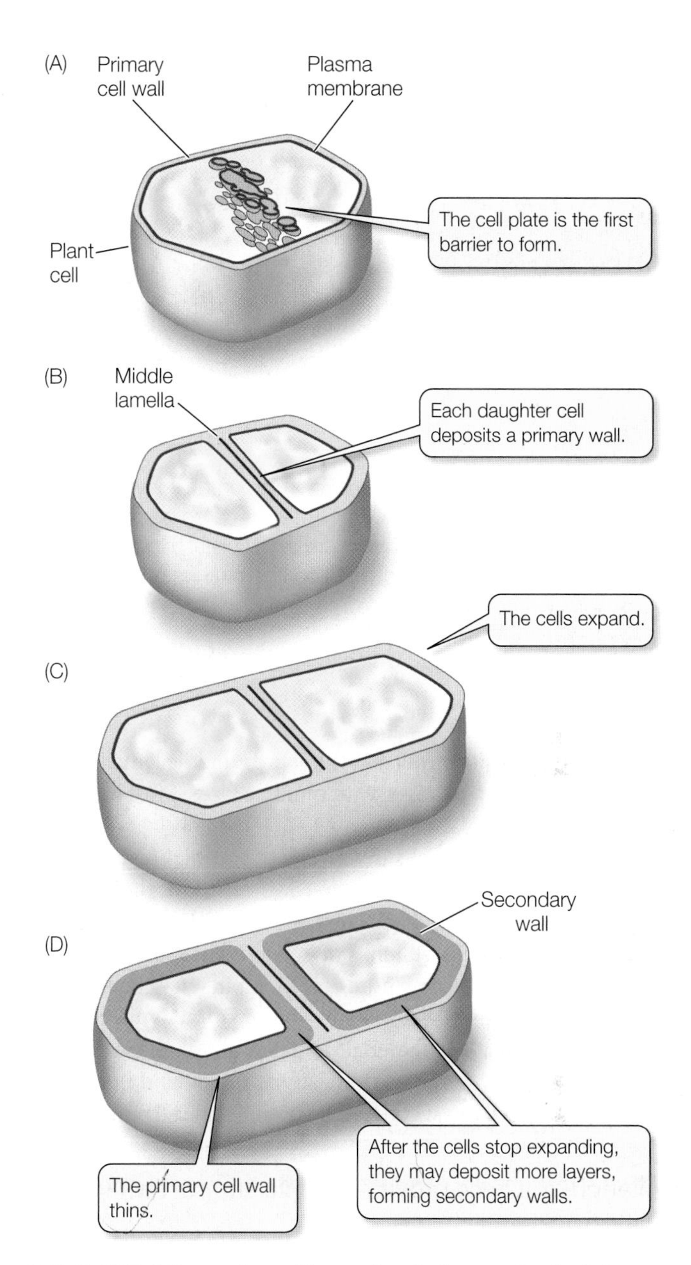

34.7 Cell Wall Formation Plant cell walls form as the final step in cell division.

Parenchyma cells are alive when they perform their functions

The most numerous cell type in young plants is the **parenchyma cell** (**Figure 34.9A**). Parenchyma cells usually have thin walls, con-

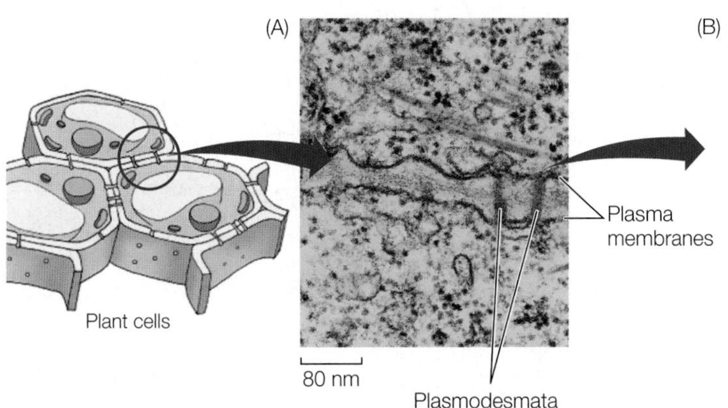

34.8 Plasmodesmata (A) An electron micrograph shows that cell walls are traversed by strandlike structures called plasmodesmata (dark stain). The green objects are cytoskeletal microtubules (see Section 4.3). (B) Plasmodesmata contain strands of endoplasmic reticulum.

sisting only of a primary wall and the shared middle lamella. Many parenchyma cells have shapes with multiple faces. Most have large central vacuoles.

The photosynthetic cells in leaves are parenchyma cells that contain numerous chloroplasts. Some nonphotosynthetic parenchyma cells store substances such as starch or lipids. In the cytoplasm of these cells, starch is often stored in specialized plastids called *leucoplasts* (see Figure 4.16B). Lipids may be stored as oil droplets, also in the cytoplasm. Some parenchyma cells appear to serve as "packing material" and play a vital role in supporting the stem. Many retain the capacity to divide and hence may give rise to new cells, as when a wound results in cell proliferation.

Collenchyma cells provide flexible support while alive

Collenchyma cells are supporting cells. Their primary walls are characteristically thick at the corners of the cells (**Figure 34.9B**). Collenchyma cells are generally elongated. In these cells, the primary wall thickens, but no secondary wall forms. Collenchyma provides support to leaf petioles, nonwoody stems, and growing organs. Tissue made of collenchyma cells is flexible, permitting stems and petioles to sway in the wind without snapping. The familiar "strings" in celery consist primarily of collenchyma cells.

Sclerenchyma cells provide rigid support

In contrast to collenchyma cells, **sclerenchyma** cells have thickened secondary walls that perform their major function: support. Many sclerenchyma cells die after laying down their cell walls and thus perform their supporting function when dead. There are two types of sclerenchyma cells: elongated **fibers** and variously shaped **sclereids**. Fibers provide relatively rigid support in wood and other parts of the plant, where they are often organized into bundles (**Figure 34.9C**). The bark of trees owes much of its mechanical strength to long fibers. Sclereids may pack together densely, as in a nut's shell or in some seed coats (**Figure 34.9D**). Isolated clumps of sclereids, called *stone cells*, in pears and some other fruits give them their characteristic gritty texture.

Fibers from the stems of hemp, *Cannabis sativus*, gave strength to sails, ropes, and rigging of sailing vessels and durability to the Gutenberg Bible, the Declaration of Independence, and the Constitution of the United States. The original Levi's jeans were made of hemp cloth.

Cells of the xylem transport water and minerals from roots to stems and leaves

Xylem contains conducting cells called **tracheary elements**, which undergo programmed cell death (apoptosis; see Section 9.6) before they assume their function of transporting water and dissolved minerals. There are two types of tracheary elements. The evolutionarily more ancient tracheary elements, found in gymnosperms and other vascular plants, are spindle-shaped cells called **tracheids** (**Figure 34.9E**). When the cell contents—nucleus and cytoplasm—disintegrate upon cell death, water and minerals can move with little resistance from one tracheid to its neighbors by way of *pits*, interruptions in the secondary wall that leave the primary wall unobstructed.

Flowering plants evolved a water-conducting system made up of *vessels*. The individual cells that form vessels, called **vessel elements**, must also die and become empty before they can transport water. Vessel elements have pits in their cell walls as do tracheids, but are generally larger in diameter than tracheids. Vessel elements secrete lignin into their secondary cell walls, then partially break down their end walls, and finally die and disintegrate, resulting in a hollow tube. They are laid down end-to-end, so that each ves-

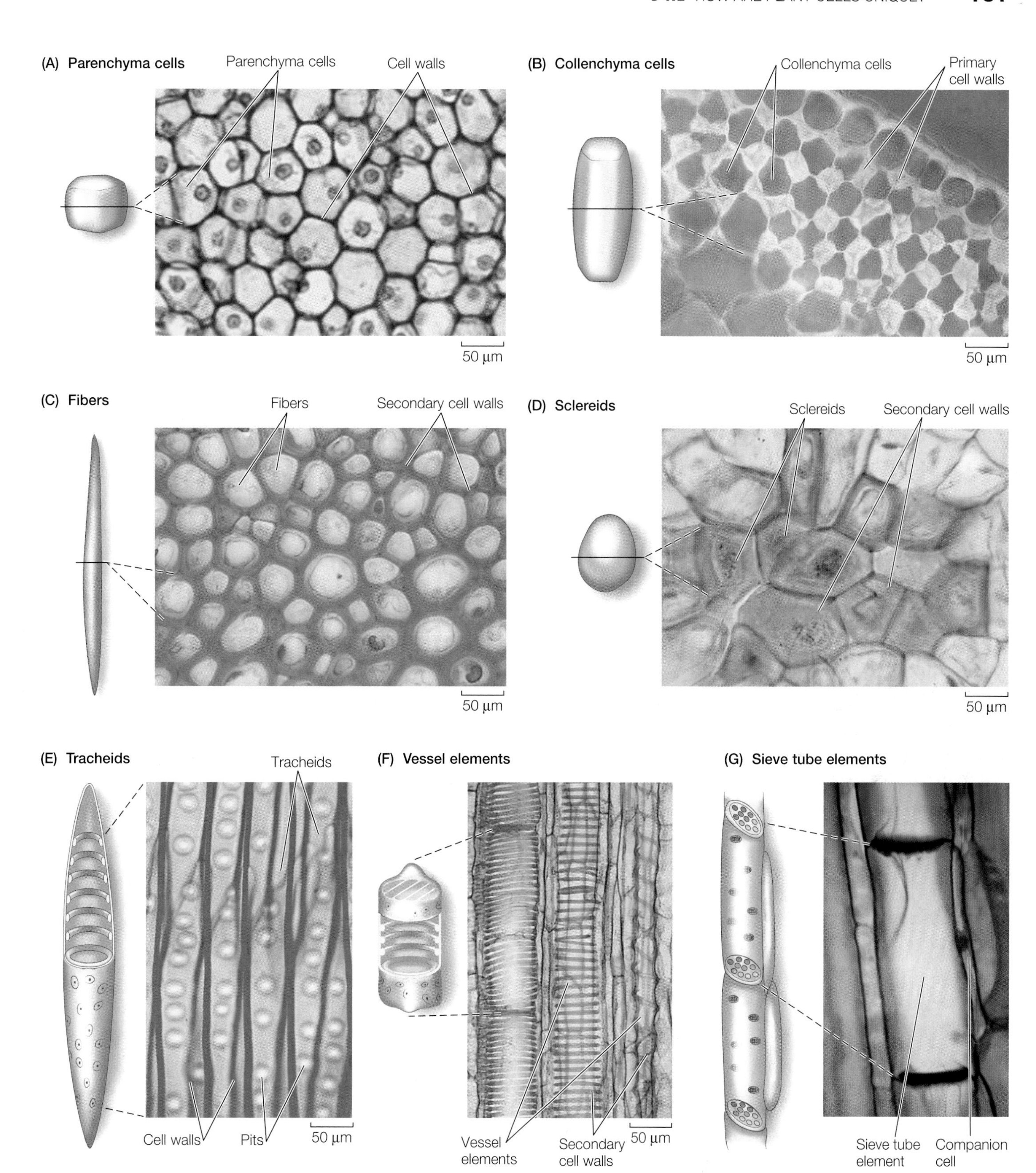

34.9 Plant Cell Types (A) Parenchyma cells in the petiole of *Coleus*. Note the thin, uniform cell walls. (B) Collenchyma cells make up the five outer cell layers of this spinach leaf vein. Their cell walls are thick at the corners of the cells and thin elsewhere. (C) Sclerenchyma: Fibers in a sunflower plant (*Helianthus*). The thick secondary walls are stained red. (D) Sclerenchyma: Sclereids. The extremely thick secondary walls of sclereids are laid down in layers. They provide support and a hard texture to structures such as nuts and seeds. (E) Tracheary elements: Water-conducting tracheids in pine wood. The thick cell walls are stained dark red. (F) Tracheary elements: Vessel elements in the stem of a squash. The secondary walls are stained red; note the different patterns of thickening, including rings and spirals. (G) Food-conducting sieve tube elements and companion cells in the stem of a cucumber.

sel is a continuous hollow tube consisting of many vessel elements, providing an open pipeline for water conduction (**Figure 34.9F**). In the course of angiosperm evolution, vessel elements have become shorter, and their end walls have become less and less obliquely oriented and less obstructed, presumably increasing the efficiency of water transport through them. The xylem of many angiosperms also includes tracheids.

Cells of the phloem translocate carbohydrates and other nutrients

The transport cells of the phloem, unlike those of the mature xylem, are living cells. In flowering plants, the characteristic cells of the phloem are **sieve tube elements** (**Figure 34.9G**). Like vessel elements, these cells meet end-to-end. They form long *sieve tubes*, which transport carbohydrates and many other materials from their sources to tissues that consume or store them. In plants with mature leaves, for example, products of photosynthesis move from leaves to root tissues.

Unlike vessel elements, which break down their end walls as they mature, sieve tube elements contain plasmodesmata in the end walls that enlarge to form pores, enhancing the connection between neighboring cells. The result is end walls that look like sieves, called **sieve plates** (**Figure 34.10**). As the holes in the sieve plates expand, the membrane that encloses the central vacuole, called the *tonoplast*, disappears. The nucleus and some cytoplasmic components also break down, and thus do not clog the pores of the sieve.

At functional maturity, a sieve tube element is filled with **phloem sap**, consisting of water, dissolved sugars, and other solutes. This solution moves from cell to cell along the sieve tube. The moving sap solution is distinct from the layer of cytoplasm at the periphery of a sieve tube element, next to the cell wall. This stationary layer of cytoplasm contains the organelles remaining in the sieve tube element.

Each sieve tube element has one or more **companion cells** (see Figure 34.10), produced as a daughter cell along with the sieve tube element when a parent cell divides. Numerous plasmodesmata link a companion cell with its sieve tube element. Companion cells retain all their organelles and, through the activities of their nuclei, they may be thought of as the "life support systems" of the sieve tube elements.

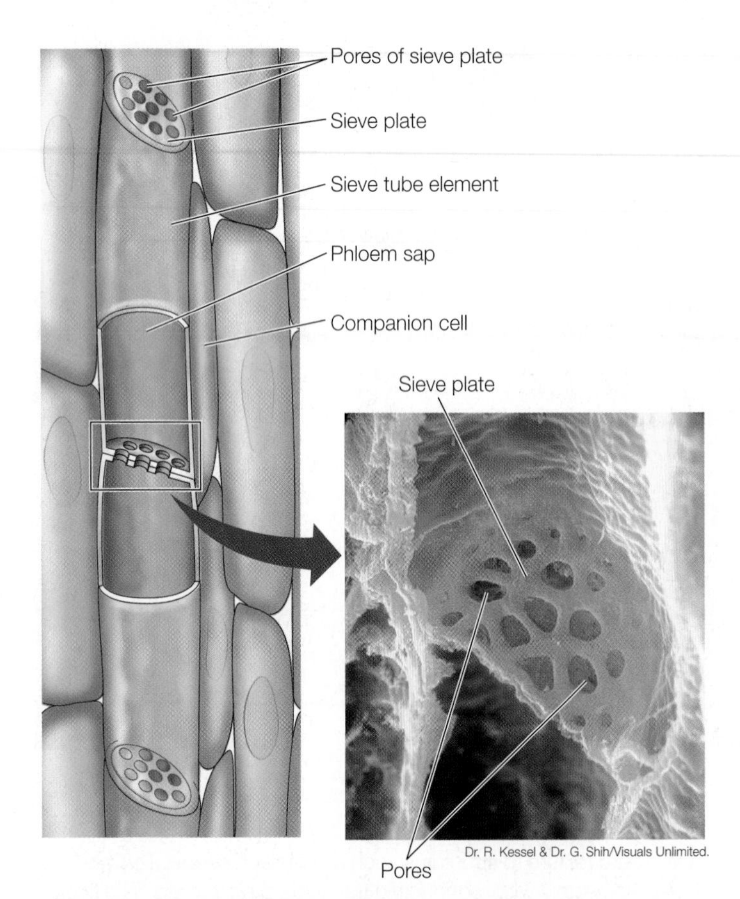

34.10 Sieve Tubes Individual sieve tube elements join together to form long tubes that transport carbohydrates and other nutrient molecules throughout the plant body in the phloem. Sieve plates form at the ends of each sieve tube element, and phloem sap passes through the pores in the sieve plate.

34.2 RECAP

There are several types of plant cells, differing in structure and function. Certain cells specialized for support or transport do not assume their function until they have died, but most plant cells function only when alive.

- Do you understand the structure and role of plasmodesmata? See p. 750 and Figure 34.8
- What structural differences make tissues made of collenchyma cells more flexible than those consisting primarily of sclerenchyma? See p. 750
- How are the transport cells of the phloem different from those of the xylem? What are their respective functions? See pp. 750, 752 and Figure 34.10

In the discussions that follow, we will examine how the cells and tissue systems are organized in the different organs of a flowering plant. Let's begin by seeing how this organization develops as the plant grows.

34.3 How Do Meristems Build the Plant Body?

In its early embryonic stages, a plant establishes the basic body plan for its mature form. Two patterns contribute to the plant body plan:

- The arrangement of cells and tissues along the main axis from root to shoot
- The concentric arrangement of the tissue systems

Both patterns arise through orderly development and are best understood in developmental terms.

Plants and animals grow differently

As the plant body grows, it may lose parts, and it forms new parts that may grow at different rates. The growing stem consists of modules, or units, laid down one after another. Each module consists of a node with its attached leaf or leaves, the internode below that node, and the axillary bud or buds at the base of that internode (see Figure 34.2). New modules are formed as long as the stem continues to grow.

- Each *branch* of a plant may be thought of as a module that is in some ways independent of the other branches. A branch does not bear the same relationship to the remainder of the plant body as a limb does to the remainder of an animal body. Among other things, branches form one after another (unlike limbs, which form simultaneously during embryonic development). Also, branches often differ from one another in their number of leaves and in the degree to which they themselves branch. Branches, like stems, are long-lived, lasting from years to centuries.
- *Leaves* are modules of another sort. They are usually short-lived, lasting weeks to a few years.
- *Root systems* are also branching structures, and lateral roots are semi-independent units. As the root system grows, penetrating and exploring the soil environment, many roots die and are replaced by new ones.

Many seed plants may be thought of as having two units of yet another sort: the primary plant body and the secondary plant body. All seed plants have a **primary plant body**, which consists of *all the non-woody parts* of the plant. Many plants—monocots in particular—consist entirely of primary plant body. Trees and shrubs, however, also have a **secondary plant body** consisting of *wood and bark*. The tissues of the secondary plant body are laid down as the stems and roots thicken; the primary plant body includes leaves, flowers, and all parts of the body that were laid down before thickening began. The secondary plant body continues to grow and thicken throughout the life of the plant. The primary plant body also continues to grow, *lengthening* the shoot and root systems and forming new leaves.

The localized regions of cell division in plants are called **meristems** (**Figure 34.11**). Meristems are forever young, retaining the ability to produce new cells indefinitely. The cells that perpetuate the meristems, called *initials*, are comparable to the stem cells found in animals (discussed in Section 19.2). When an initial divides, one daughter cell develops into another meristem cell the size of its parent, while the other daughter cell differentiates into a more specialized cell.

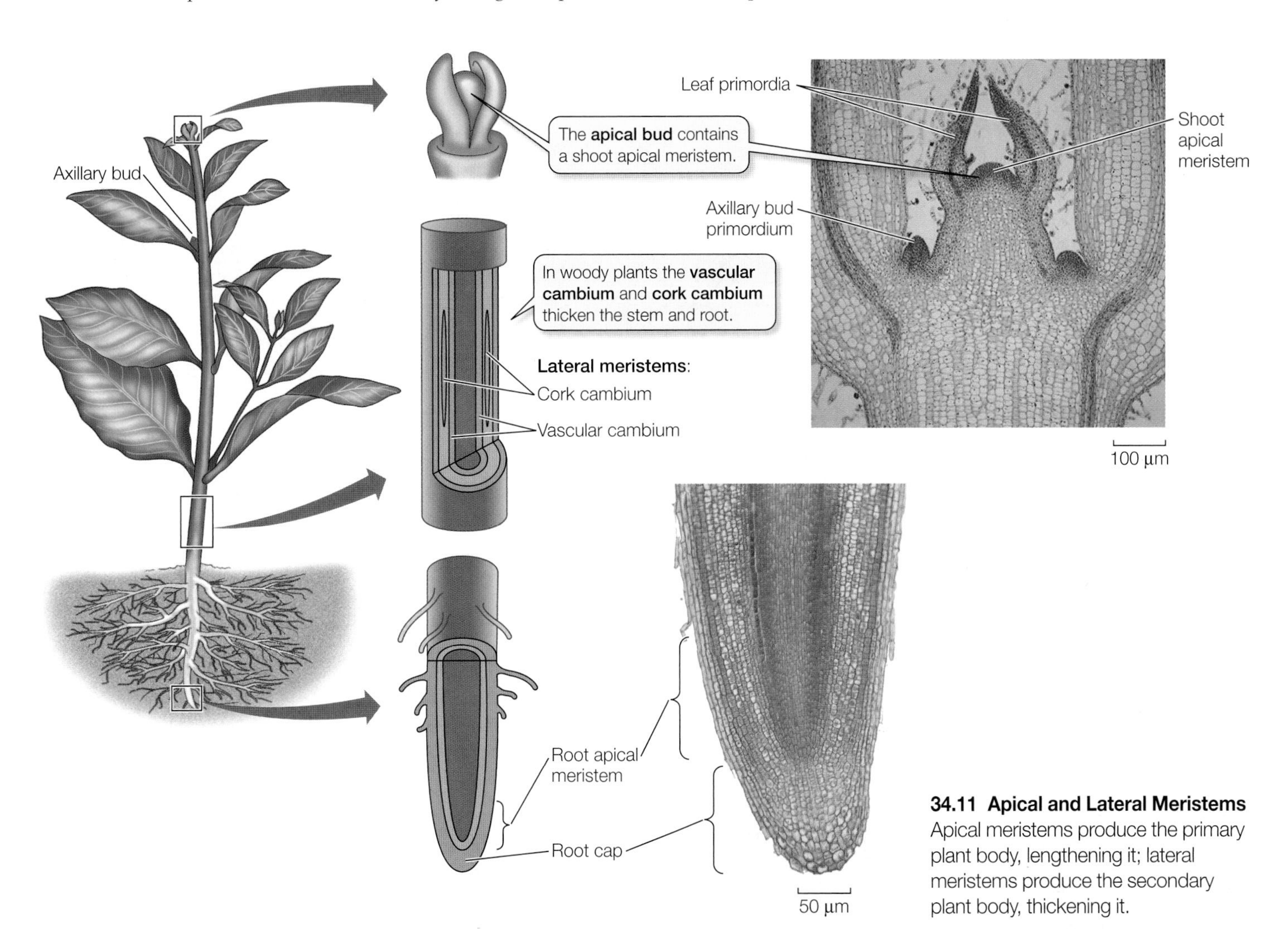

34.11 Apical and Lateral Meristems Apical meristems produce the primary plant body, lengthening it; lateral meristems produce the secondary plant body, thickening it.

Although all parts of the animal body grow as an individual develops from embryo to adult, in most animals, this growth is *determinate*. That is, the growth of the individual and all its parts ceases when the adult state is reached. Determinate growth is also characteristic of some plant parts, such as leaves, flowers, and fruits. The growth of stems and roots, by contrast, is *indeterminate*, and it is generated from specific regions of active cell division and cell expansion.

A hierarchy of meristems generates a plant's body

Two types of meristems contribute to the growth and development of the plant:

- **Apical meristems** give rise to the primary plant body.
- **Lateral meristems** give rise to the secondary plant body.

APICAL MERISTEMS Apical meristems are located at the tips of roots and stems and in buds and are responsible for **primary growth**, which leads to elongation of shoots and roots and formation of organs (see Figure 34.11). All plant organs arise ultimately from cell divisions in the apical meristems, followed by cell expansion and differentiation. Primary growth gives rise to the primary plant body, which is the entire body of many plants.

- *Shoot apical meristems* supply the cells that extend stems and branches, allowing more leaves to form and photosynthesize.
- *Root apical meristems* supply the cells that extend roots, enabling the plant to "forage" for water and minerals.

Apical meristems in both the shoot and the root give rise to a set of cylindrical *primary meristems* that produce the primary tissues of the plant body.

From the outside to the inside of the root or shoot, which are both cylindrical organs, the primary meristems are the **protoderm**, the **ground meristem**, and the **procambium**. These in turn give rise to the three tissue systems:

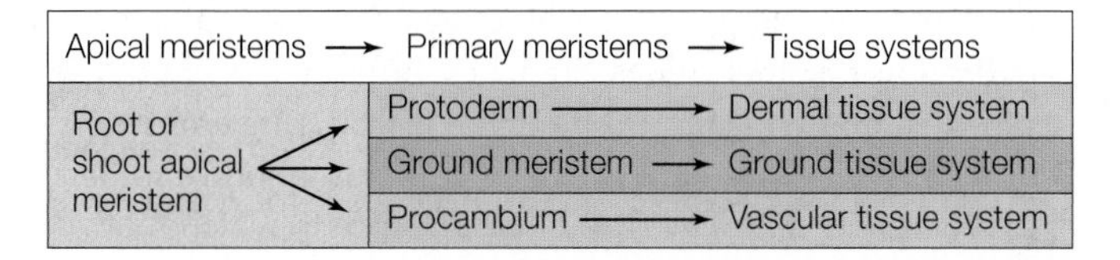

Because meristems can continue to produce new organs throughout the lifetime of the plant, the plant body is much more variable in form than the animal body, whose organs are produced only once.

LATERAL MERISTEMS: VASCULAR AND CORK CAMBIA Some roots and stems develop a secondary plant body, the tissues of which we commonly refer to as *wood* and *bark*. These complex tissues are derived by **secondary growth** from two lateral meristems:

- The **vascular cambium** is a cylindrical tissue consisting predominantly of vertically elongated cells that divide frequently. It supplies the cells of the secondary xylem and phloem, which in trees eventually become wood and bark.
- The **cork cambium** produces mainly waxy-walled *cork cells*. It supplies some of the cells that become bark.

Wood is secondary xylem. **Bark** is everything external to the vascular cambium (periderm plus secondary phloem). Toward the inside of the stem or root, the dividing cells of the vascular cambium form new xylem, the *secondary xylem*, and toward the outside they form new phloem, the *secondary phloem*.

Each year, deciduous trees lose their leaves, leaving bare branches and twigs in winter. These twigs illustrate both primary and secondary growth (**Figure 34.12**). The apical meristems of the twigs and their branches are enclosed in buds protected by bud scales. When the buds begin to grow in the spring, the scales fall away, leaving scars that show us where the bud was and identifying each year's growth. The dormant twig shown in Figure 34.12 is the product of pri-

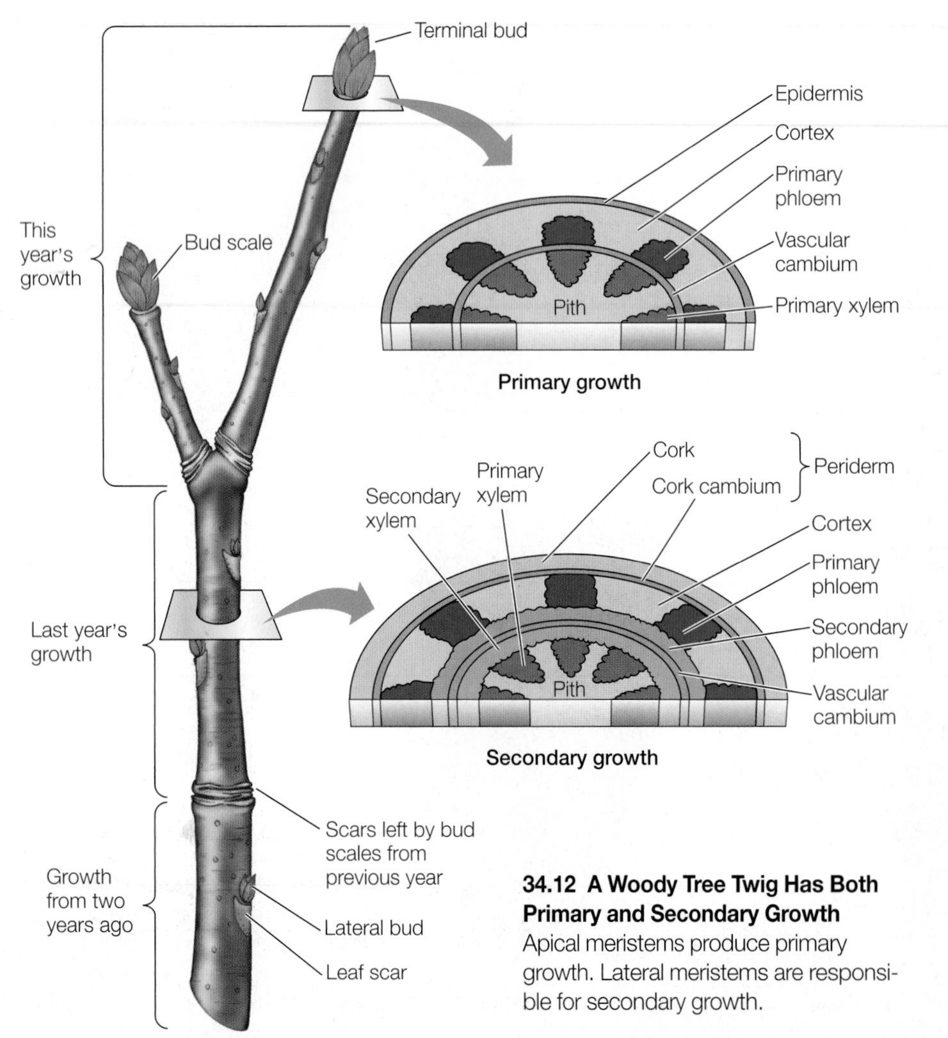

34.12 A Woody Tree Twig Has Both Primary and Secondary Growth Apical meristems produce primary growth. Lateral meristems are responsible for secondary growth.

34.13 An Ancient Individual Bristlecone pines (*Pinus longaeva*) can live for centuries. The oldest known living organism is a bristlecone pine that has been alive for almost 5,000 years—almost as long as recorded human history.

mary and secondary growth. Only the buds consist entirely of primary tissues.

As a tree trunk grows in diameter, the outermost layers of the stem, including the epidermis, crack and fall off. Without the activity of the cork cambium, this sloughing off of tissues would expose the tree to potential damage, including excessive water loss or invasion by microorganisms. The cork cambium produces new protective cells, primarily in the outward direction. The walls of these cork cells become impregnated with suberin. The mass of waterproofed cells produced by the cork cambium is called the **periderm**.

In some plants, meristems may remain active for years or even centuries. The oldest known individual plant is a bristlecone pine that has lived for more than 4,900 years—almost 50 centuries (**Figure 34.13**). In contrast, it is doubtful that any animal has ever lived much longer than 2 centuries. Plants such as the bristlecone pine grow in size, or at least in diameter, throughout their lives. In the sections that follow, we will examine how the various meristems give rise to the plant body.

The root apical meristem gives rise to the root cap and the root primary meristems

The root apical meristem produces all the cells that contribute to growth in the length of the root (**Figure 34.14A**). Some of the daughter cells from the apical (tip) end of the root apical meristem contribute to a **root cap**, which protects the delicate growing region of the root as it pushes through the soil. The cells of the root cap are often damaged or scraped away and must therefore be replaced constantly. The root cap is also the structure that detects the pull of gravity and thus controls the downward growth of roots.

In the middle of the root apical meristem is a *quiescent center*, in which cell divisions are rare. The quiescent center can become more active when needed—following injury, for example.

The daughter cells produced above the quiescent center (that is, away from the root cap) elongate and lengthen the root. After they elongate, these cells differentiate, giving rise to the various tissues of the mature root. The growing region farther above the apical meristem comprises the three cylindrical primary meristems: the protoderm, the ground meristem, and the procambium (see page 756). These primary meristems give rise to the three tissue systems of the root.

The apical and primary meristems constitute the **zone of cell division**, the source of all the cells of the root's primary tissues. Just above this zone is the **zone of cell elongation**, where the newly formed cells are elongating and thus pushing the root farther into

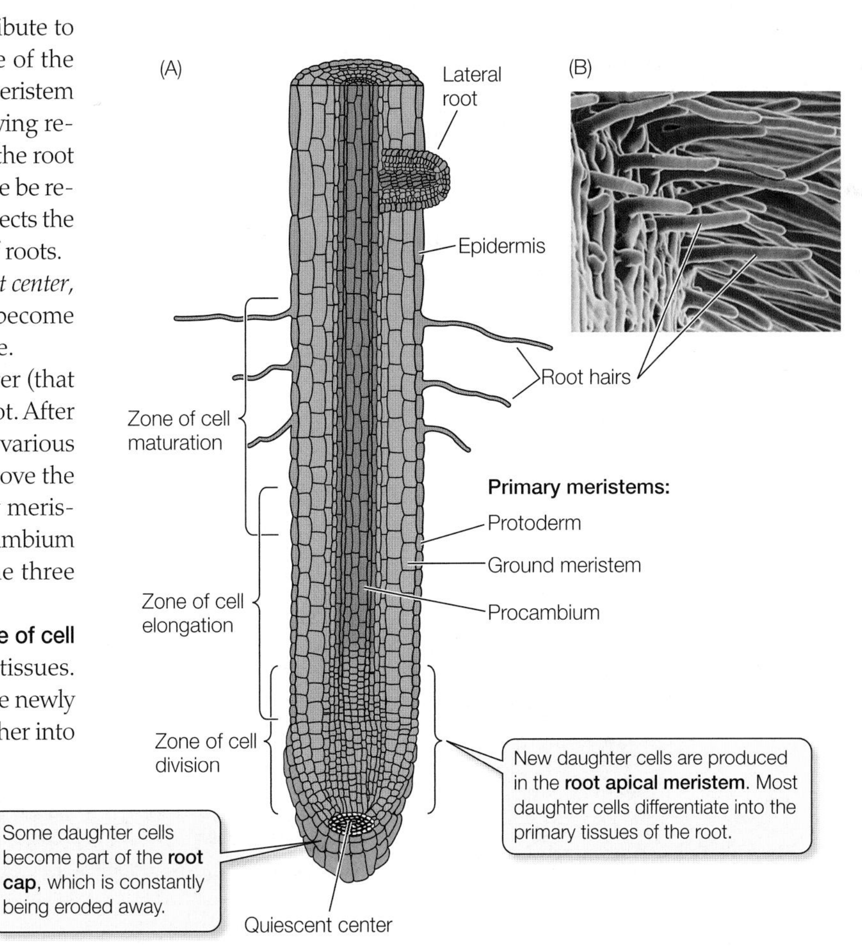

34.14 Tissues and Regions of the Root Tip (A) Extensive cell division creates the complex structure of the root. (B) Root hairs, seen with a scanning electron microscope.

the soil. Above this is the **zone of maturation**, where the cells are differentiating, taking on specialized forms and functions such as water transport or mineral uptake. These three zones grade imperceptibly into one another; there is no abrupt line of demarcation. In the zone of maturation, many of the epidermal cells produce amazingly long, delicate **root hairs**, which vastly increase the surface area of the root (**Figure 34.14B**). Root hairs grow out among the soil particles, probing nooks and crannies and taking up water and minerals.

It has been estimated that the root system of a single mature rye plant has a total absorptive surface of more than 600 square meters (almost half the area of a basketball court).

In the great majority of plants, and especially in trees, a fungus is closely associated with the root tips (see Figure 30.10). Such roots have poorly developed or no root hairs. This association, called a *mycorrhiza*, increases the plant's absorption of minerals and water and in fact these plants cannot survive without the mycorrhizae.

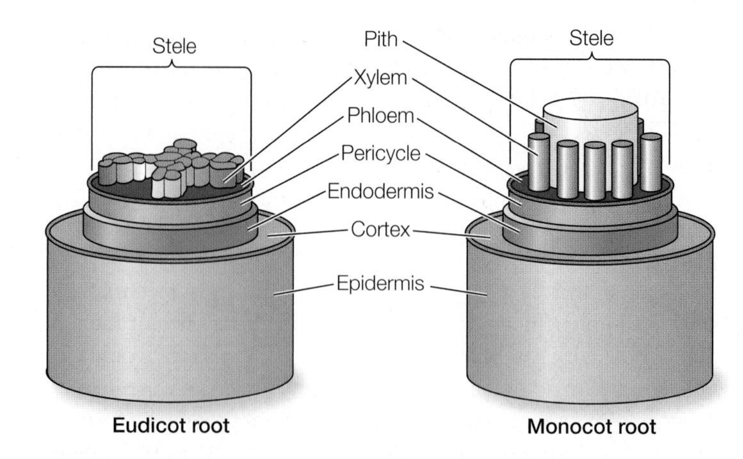

34.15 Products of the Root's Primary Meristems The protoderm gives rise to the outermost layer (epidermis). The ground meristem produces the cortex, the innermost layer of which is the endodermis. The primary vascular tissues of the root are found in the stele, which is the product of the procambium. The arrangement of tissues in the stele differs in the roots of eudicots and monocots.

The products of the root's primary meristems become root tissues

The products of the three primary meristems are shown in **Figure 34.15**:

- The **protoderm** gives rise to the outer layer of cells—the **epidermis**—which is adapted for protection of the root and absorption of mineral ions and water.
- Internal to the epidermis, the *ground meristem* gives rise to a region of ground tissue that is many cells thick, called the **cortex**. The innermost layer of the cortex is the **endodermis** of the root.
- Moving inward past the endodermis, we enter the vascular cylinder, or **stele**, produced by the *procambium*.

The cells of the *cortex* are relatively unspecialized and often function in nutrient storage. Unlike those of other cortical cells, the cell walls of the endodermal cells contain suberin. The placement of this waterproofing substance in only certain parts of the cell wall enables the cylindrical ring of endodermal cells to control the access of water and dissolved ions to the vascular tissues.

The *stele* consists of three tissues: pericycle, xylem, and phloem (see Figure 34.15). The **pericycle** consists of one or more layers of relatively undifferentiated cells. It has three important functions:

- It is the tissue within which lateral roots arise (**Figure 34.16A**).
- It can contribute to secondary growth by giving rise to lateral meristems that thicken the root.
- Its cells contain membrane transport proteins that export nutrient ions into the cells of the xylem.

34.16 Root Anatomy (A) Cross section through the tip of a lateral root in a willow tree. Cells in the pericycle divide and the products differentiate, forming the tissues of a lateral root. (B, C) Cross sections showing the primary root tissues of (B) a eudicot (the buttercup, *Ranunculus*) and (C) a monocot (corn, *Zea mays*).

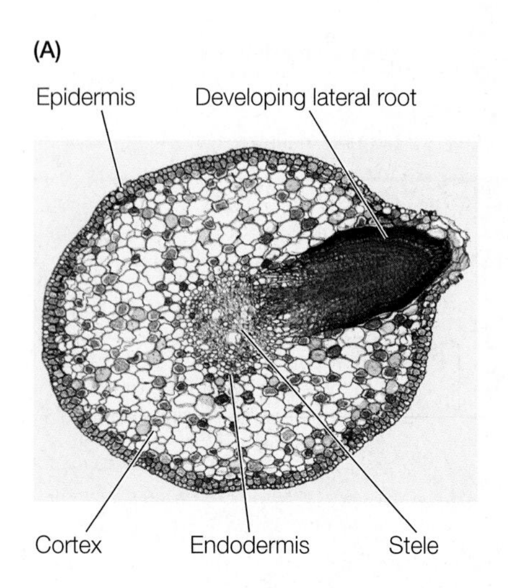

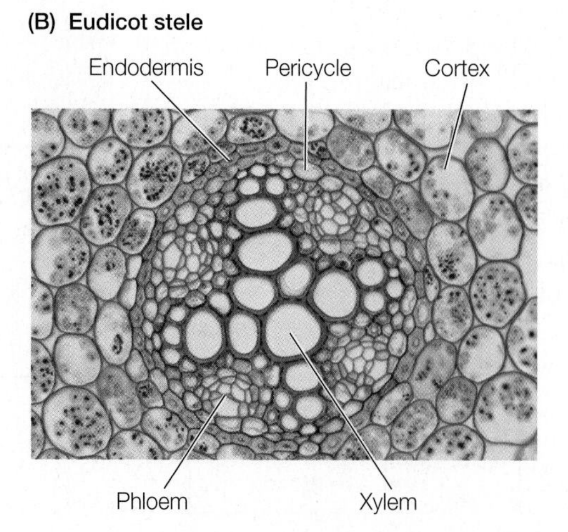

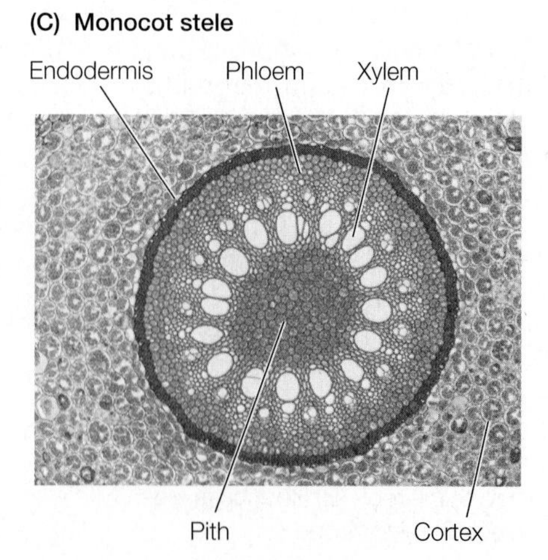

At the very center of the root of a eudicot lies the xylem—seen in cross section in the shape of a star with a variable number of points (**Figure 34.16B**). Between the points are bundles of phloem. In monocots, a region of parenchyma cells, called the **pith**, lies in the center of the root (**Figure 34.16C**). The pith often stores carbohydrate reserves and is also found in the stems of both eudicots and monocots.

The products of the stem's primary meristems become stem tissues

The shoot apical meristem, like the root apical meristem, forms three primary meristems: the protoderm, the ground meristem, and the procambium. These primary meristems, in turn, give rise to the three tissue systems. The shoot apical meristem also repetitively lays down the beginnings of leaves and axillary buds. Leaves arise from bulges called **leaf primordia**, which form as cells divide on the sides of shoot apical meristems (see Figure 34.11). **Bud primordia** form at the bases of the leaf primordia. The growing stem has no protective structure analogous to the root cap, but the leaf primordia can act as a protective covering for the shoot apical meristem.

The plumbing of angiosperm stems differs from that of roots. In a root, the vascular tissue lies deep in the interior, with the xylem at or near the very center (see Figure 34.16B,C). The vascular tissue of a young stem, however, is divided into discrete **vascular bundles** (**Figure 34.17**). Each vascular bundle contains both xylem and phloem. In eudicots, the vascular bundles generally form a cylinder, but in monocots, they are seemingly scattered throughout the stem.

In addition to the vascular tissues, the stem contains other important storage and supportive tissues. Internal to the ring of vascular bundles in eudicots is a storage tissue, the pith, and to the outside lies a similar storage tissue, the cortex. The cortex may contain supportive collenchyma cells with thickened walls. The pith, the cortex, and the regions between the vascular bundles in eudicots—called *pith rays*—constitute the ground tissue system of the stem. The outermost cell layer of the young stem is the epidermis, the primary function of which is to minimize the loss of water from the tissues within.

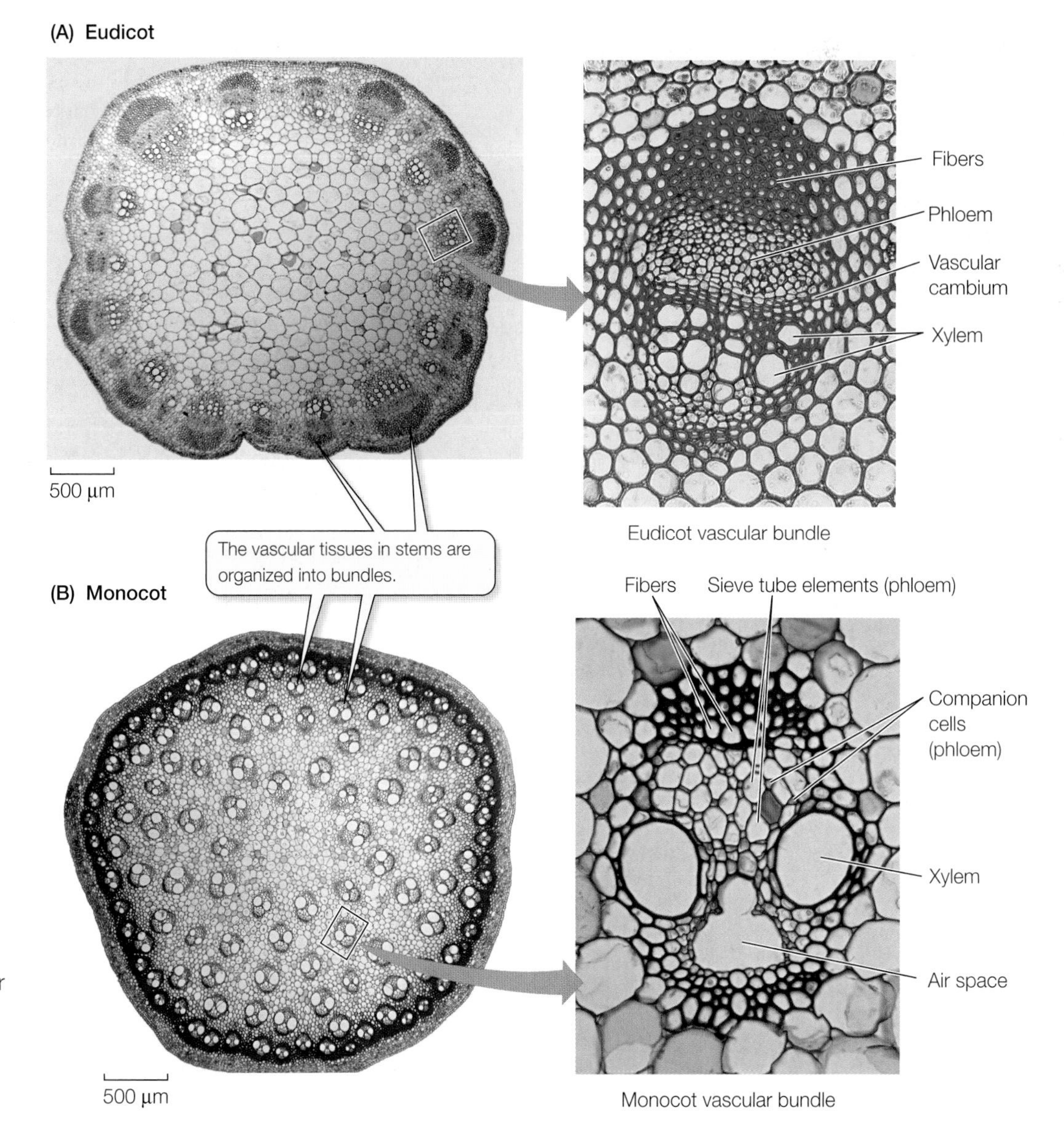

34.17 Vascular Bundles in Stems (A) In eudicot stems, the vascular bundles are arranged in a cylinder, with the pith in the center and the cortex outside the cylinder. (B) A scattered arrangement of vascular bundles is typical of monocot stems.

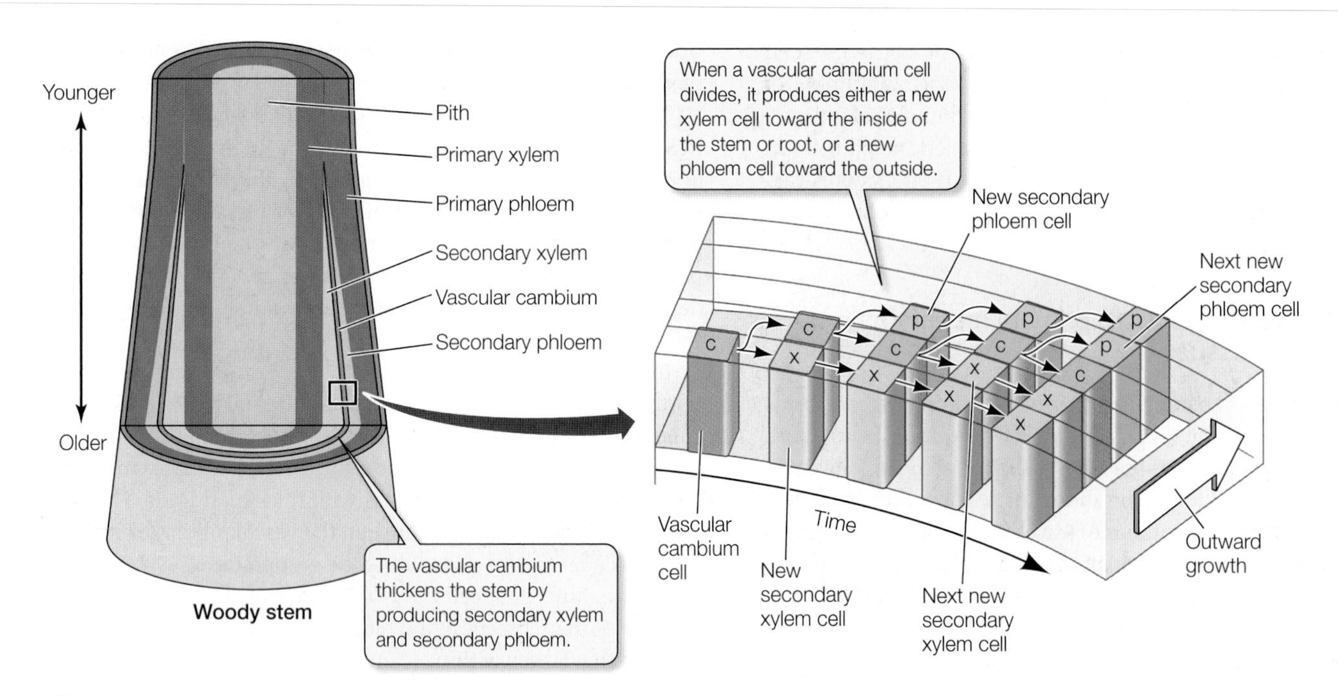

34.18 Vascular Cambium Thickens Stems and Roots Stems and roots grow thicker because a thin layer of cells, the vascular cambium, remains meristematic. These highly diagrammatic images emphasize the pattern of deposition of secondary xylem and phloem by the vascular cambium.

Many eudicot stems and roots undergo secondary growth

Some stems and roots remain slender and show little or no secondary growth. However, in many eudicots, secondary growth thickens stems and roots considerably. This process gives rise to wood and bark, and it makes the support of tall trees possible. As described earlier in the chapter, secondary growth results from the activity of the two lateral meristems, the vascular cambium and cork cambium.

The vascular cambium is initially a single layer of cells lying between the primary xylem and the primary phloem (see Figure 34.17A). The root or stem increases in diameter when the cells of the vascular cambium divide, producing secondary xylem cells toward the inside of the root or stem and producing secondary phloem cells toward the outside (**Figure 34.18**). In the stems of woody plants, cells in the pith rays between the vascular bundles also divide, forming a continuous cylinder of vascular cambium running the length of the stem. This cylinder, in turn, gives rise to complete cylinders of secondary xylem (wood) and secondary phloem, which contributes to the bark.

As the vascular cambium produces secondary xylem and phloem, its principal cell products are vessel elements, supportive fibers, and parenchyma cells in the xylem and sieve tube elements, companion cells, fibers, and parenchyma cells in the phloem. The parenchyma cells in the xylem and phloem store carbohydrate reserves in the stem and root.

Living tissues such as this storage parenchyma must be connected to the sieve tubes of the phloem, or they will starve to death. These connections are provided by **vascular rays**, which are composed of cells derived from the vascular cambium. These rays, laid down progressively as the cambium divides, are rows of living parenchyma cells that run perpendicular to the xylem vessels and phloem sieve tubes (**Figure 34.19**). As the root or stem con-

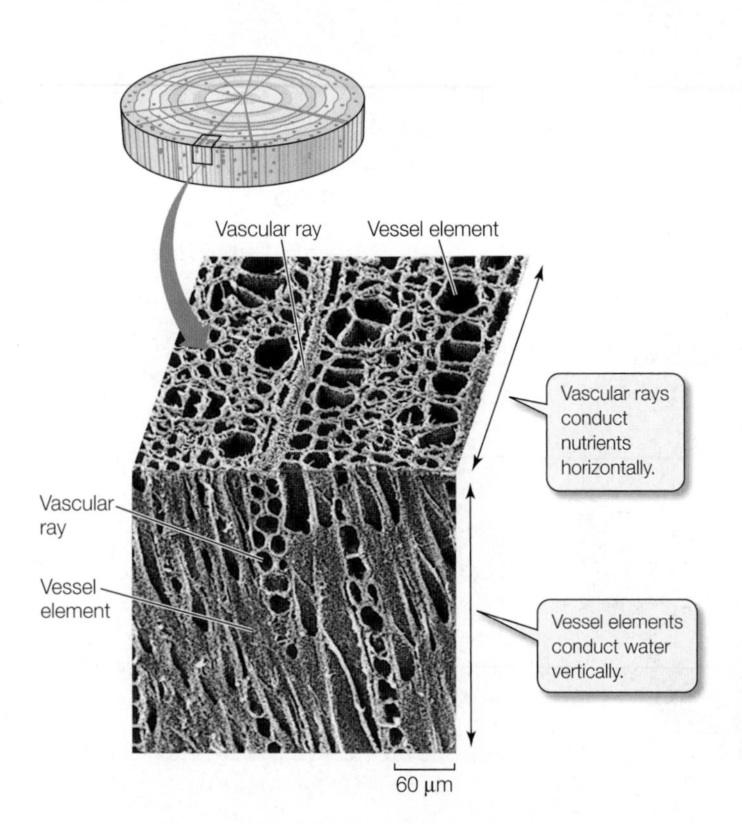

34.19 Vascular Rays and Vessel Elements In this sample of wood from the tulip poplar, the orientation of vascular rays is perpendicular to that of the vessel elements. The vascular rays transport phloem sap horizontally from the phloem to storage parenchyma cells.

tinues to increase in diameter, new vascular rays are initiated so that this storage and transport tissue continues to meet the needs of both the bark and the living cells in the xylem.

The vascular cambium itself increases in circumference with the growth of the root or stem. To do this, some of its cells divide in a plane at right angles to the plane that gives rise to secondary xylem and phloem. The products of each of these divisions lie within the vascular cambium itself and increase its circumference.

Only eudicots and other non-monocot angiosperms have a vascular cambium and a cork cambium and thus undergo secondary growth. The few monocots that form thickened stems—palm trees, for example—do so without using vascular cambium or cork cambium. Palm trees have a very wide apical meristem that produces a wide stem, and dead leaf bases also add to the diameter of the stem. Basically, monocots grow in the same way as do other angiosperms that lack secondary growth.

Wood and bark, consisting of secondary phloem, are unique to plants showing secondary growth. These tissues have their own patterns of organization and development.

WOOD Cross sections of most tree trunks (mature stems) in temperate-zone forests show *annual rings* (**Figure 34.20**), which result from seasonal environmental conditions. In spring, when water is relatively plentiful, the tracheids or vessel elements produced by the vascular cambium tend to be large in diameter and thin-walled. Such wood is well adapted for transporting water and minerals. As water becomes less available during the summer, narrower cells with thicker walls are produced, making this summer wood darker and perhaps more dense than the wood formed in spring. Thus each growing season is usually recorded in a tree trunk by a clearly visible annual ring. Trees in the moist tropics do not undergo seasonal growth, so they do not lay down such obvious regular rings. Variations in temperature or water supply can lead to the formation of more than one "annual" ring in a single year, but commonly each year brings a new annual ring and a new batch of leaves.

How do annual rings and leaves relate to each other? For example, do the needles on a pine tree connect with the current year's annual ring (xylem) or with a previous year's ring (**Figure 34.21**)? As it turns out, the answer varies from species to species.

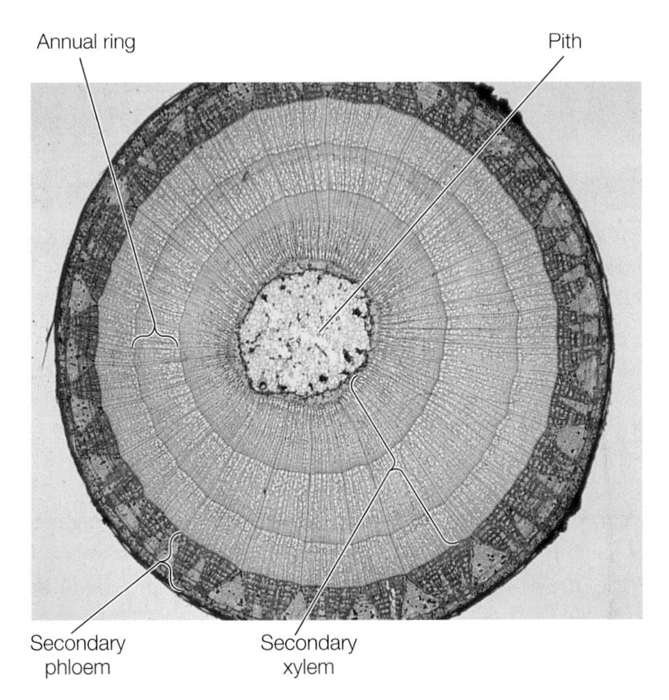

34.20 Annual Rings Rings of secondary xylem are the most noticeable feature of this cross section from a 3-year-old basswood stem.

EXPERIMENT

HYPOTHESIS: A needle of a ponderosa pine tree connects with the xylem laid down in the year the needle was formed.

METHOD

1. Immerse the basal ends of 2-cm-long segments of young branches in a dye solution.
2. Clip the tip of one needle in the segment, and apply vacuum to the cut needle.
3. After 5 minutes of the vacuum treatment, cut the segment several mm above the base. Observe which annual ring(s) contain the dye.

RESULTS

When 1-year-old needles were tested, the dye was always found in the annual ring formed 1 year before the experiment (y–1). When 2-year-old needles were tested, the dye was always found in the annual ring formed 2 years before the experiment (y–2), and sometimes also in the ring formed the following year.

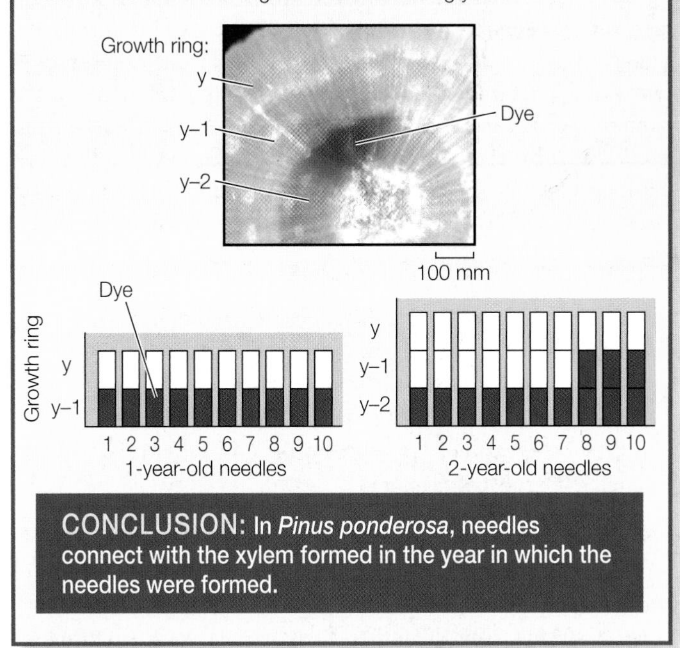

CONCLUSION: In *Pinus ponderosa*, needles connect with the xylem formed in the year in which the needles were formed.

34.21 How Do Leaves Relate to Annual Rings? Clarice Maton and Barbara L. Gartner determined which annual rings supply water to which needles of various gymnosperms. FURTHER RESEARCH: Is this a general phenomenon? That is, does it appear in other woody plants? Suggest how you could perform this experiment on a twig of an angiosperm tree such as a maple.

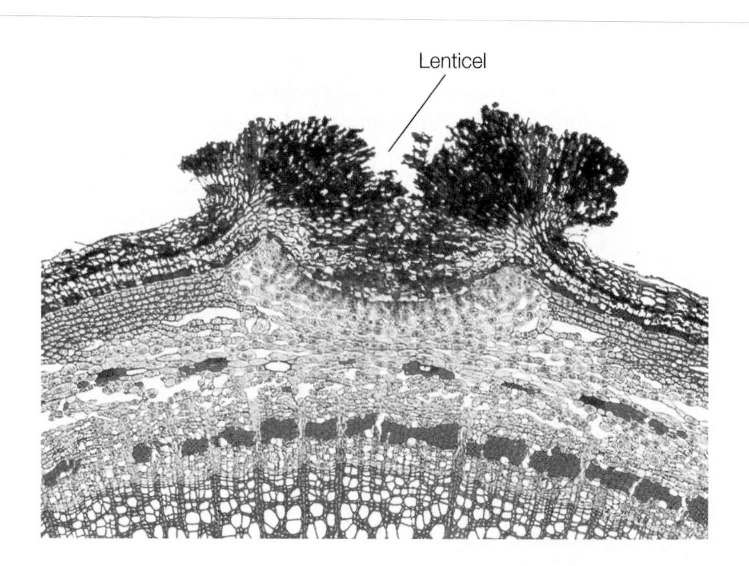

34.22 Lenticels Allow Gas Exchange through the Periderm The region of periderm that appears broken open is a lenticel in a year-old elderberry (*Sambucus*) twig; note the spongy tissue that constitutes the lenticel.

The difference between old and new regions of wood also contributes to its appearance. As a tree grows in diameter, the xylem toward the center becomes clogged with water-insoluble substances and ceases to conduct water and minerals; this *heartwood* appears darker in color. The portion of the xylem that actively conducts water and minerals throughout the tree is called *sapwood* and is lighter in color and more porous than heartwood.

The knots that we find attractive in knotty pine but regard as a defect in structural timbers are cross sections of branches. As a trunk grows, the bases of branches become buried in the trunk's new wood and appear as knots when the trunk is cut lengthwise.

The overall appearance of wood, resulting from annual rings, vascular rays, and unevenness from sawing, knots, and so forth, is referred to as the wood's *figure*. Recall that one of the most convincing pieces of evidence in the "Lindbergh trial" discussed at the beginning of this chapter was a comparison of the figure of two pieces of sawn wood. Recall, too, that wood from no fewer than four tree species was used to construct a single ladder.

BARK As secondary growth of stems or roots continues, the expanding vascular tissue stretches and breaks the epidermis and cortex, which ultimately flake away. Tissue derived from the secondary phloem then becomes the outermost part of the stem. Before the dermal tissues are broken away, cells lying near the surface of the secondary phloem begin to divide and produce layers of **cork**, a tissue composed of cells with thick walls waterproofed with suberin. The cork soon becomes the outermost tissue of the stem or root (see Figure 34.12). The dividing cells, derived from the secondary phloem, form a cork cambium. Sometimes the cork cambium produces cells to the inside as well as to the outside; these cells constitute what is known as the **phelloderm**.

> The international wine market once used more than 15 billion cork bottle stoppers every year. For centuries, people around the western Mediterranean have harvested great sheets of this plant tissue from cork oak trees. The use of plastic stoppers and screw top bottles threatens to undermine the cork oak economy.

Cork, cork cambium, and phelloderm make up the periderm of the secondary plant body. As the vascular cambium continues to produce secondary vascular tissue, the corky layers are in turn lost, but the continuous formation of new cork cambia in the underlying phloem gives rise to new corky layers.

When periderm forms on stems and roots, the underlying tissues still need to release carbon dioxide and take up oxygen for cellular respiration. **Lenticels** are spongy regions in the periderm of stems and roots that allow such gas exchange (**Figure 34.22**).

34.3 RECAP

Apical and lateral meristems are responsible for producing the primary and secondary plant bodies. The primary meristems are the protoderm, the ground meristem, and the procambium. The two lateral meristems, from which wood and bark arise, are the vascular cambium and the cork cambium.

- Can you explain the differences among the three types of primary meristems in both roots and shoots? See p. 754
- Can you describe the cells derived from the root apical meristem and the general process of root growth? See pp. 755–756 and Figure 34.14
- Do you understand how the vascular cambium gives rise to thicker stems and roots? See Figure 34.18

Of the three types of vegetative organs, only roots and stems may undergo secondary growth. Leaves do not. The primary function of leaves is one that is essential not only to the life of the plant but to all other life on the planet: photosynthesis.

34.4 How Does Leaf Anatomy Support Photosynthesis?

We can think of roots and stems as important supporting actors that sustain the activities of the real stars of the plant body, the leaves—the organs of photosynthesis. Leaf anatomy is beautifully adapted to carry out photosynthesis and to support it by exchanging the gases O_2 and CO_2 with the environment, limiting evaporative water loss, and exporting the products of photosynthesis to the rest of the plant. **Figure 34.23A** shows a section of a typical eudicot leaf in three dimensions.

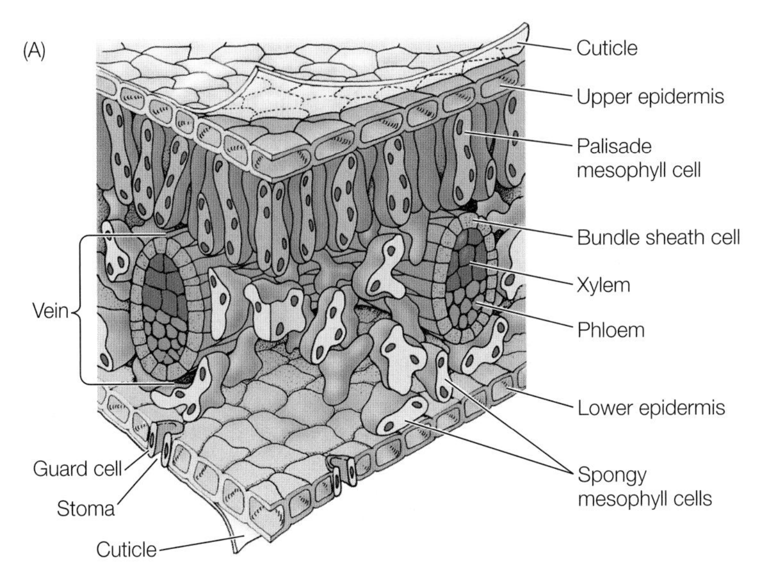

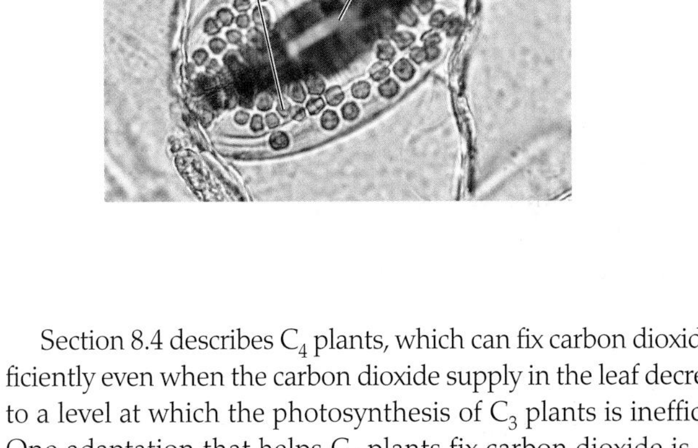

34.23 The Eudicot Leaf (A) This three-dimensional diagram shows a section of a eudicot leaf. (B) The network of fine veins in this maple leaf carries water to the mesophyll cells and carries photosynthetic products away from them. (C) Carbon dioxide enters the leaf through stomata such as this one on the epidermis of a eudicot leaf.

Most eudicot leaves have two zones of photosynthetic parenchyma tissue referred to as **mesophyll**, which means "middle of the leaf." The upper layer or layers of mesophyll consist of elongated cells; this zone is referred to as *palisade mesophyll*. The lower layer or layers consist of irregularly shaped cells; this zone is called *spongy mesophyll*. Within the mesophyll is a great deal of air space through which carbon dioxide can diffuse to and be absorbed by photosynthesizing cells.

Vascular tissue branches extensively throughout the leaf, forming a network of **veins** (**Figure 34.23B**). Veins extend to within a few cell diameters of all the cells of the leaf, ensuring that the mesophyll cells are well supplied with water and minerals. The products of photosynthesis are loaded into the phloem of the veins for export to the rest of the plant.

Covering the entire leaf on both its upper and lower surfaces is a layer of nonphotosynthetic cells, the epidermis. The epidermal cells have an overlying waxy cuticle that is highly impermeable to water. Although this impermeability prevents excessive water loss, it also poses a problem: While keeping water in the leaf, the epidermis also keeps carbon dioxide—the other raw material of photosynthesis—out.

The problem of balancing water retention and carbon dioxide availability is solved by an elegant regulatory system that will be discussed in more detail in the next chapter. *Guard cells* are modified epidermal cells that change their shape, thereby opening or closing pores called *stomata* (singular, *stoma*), which serve as passageways between the environment and the leaf's interior (**Figure 34.23C**). When the stomata are open, carbon dioxide can enter and oxygen can leave, but water vapor can also be lost.

Section 8.4 describes C_4 plants, which can fix carbon dioxide efficiently even when the carbon dioxide supply in the leaf decreases to a level at which the photosynthesis of C_3 plants is inefficient. One adaptation that helps C_4 plants fix carbon dioxide is their modified leaf anatomy (see Figure 8.17). The photosynthetic cells in the C_4 leaf are grouped around the veins in concentric layers, forming an outer mesophyll layer and an inner *bundle sheath*. These layers each contain different types of chloroplasts, leading to the biochemical division of labor illustrated in Figure 8.18.

34.4 RECAP

Leaves are the organs of photosynthesis. They exchange gases with the atmosphere, obtain water and nutrients from the roots, and export the products of photosynthesis by way of the phloem.

- How is a leaf adapted to carry out photosynthesis? See Figure 34.23
- Do you understand how stomata serve the needs of a leaf? See p.19

Leaves receive water and mineral nutrients from the roots by way of the stems. In return, the leaves export products of photosynthesis, providing a supply of chemical energy to the rest of the plant body. And, as we have just seen, leaves exchange gases, including water vapor, with the environment by way of the stomata. All three of these processes will be considered in detail in the next chapter.

CHAPTER SUMMARY

34.1 How is the plant body organized?

Most flowering plants belong to one of two major clades: **monocots** and **eudicots**. Monocots differ from eudicots in a number of structural respects. Review Figure 34.1

The vegetative organs of flowering plants are roots, which form a **root system**, and stems and leaves, which form a **shoot system**.

Stems bear embryonic shoots called buds. **Axillary buds** can develop into branches. **Apical buds** found at the tips of stems and branches produce cells for the elongating shoots. **Leaves** are the primary sites of photosynthesis. The leaf **blade** is attached to the stem by a **petiole**. Review Figure 34.2

Three tissue systems extend throughout the plant body: **vascular tissue**, **dermal tissue**, and **ground tissue**. Review Figure 34.6

The plant vascular tissue system includes the **xylem**, which conducts water and minerals absorbed by the roots, and the **phloem**, which conducts the products of photosynthesis throughout the plant body.

The dermal tissue system protects the plant body surface. In plants without secondary growth, it consists of the **epidermis**.

The ground tissue system produces and stores nutrients and other substances and provides mechanical support.

34.2 How are plant cells unique?

Plant cells are different from other eukaryotic cells in having chloroplasts or other plastids, vacuoles, and cellulose-containing **cell walls.**

The walls of individual cells are separated by a **middle lamella**; each cell also has its own **primary wall**, and some produce a thick **secondary wall**. Review Figure 34.7

Plasmodesmata connect adjacent plant cells. Review Figure 34.8

Many **parenchyma** cells store starch or lipids; some carry out photosynthesis. **Collenchyma** cells provide flexible support. **Sclerenchyma** cells include **fibers** and **sclereids** that provide strength and often do not function until they die. Review Figure 34.9

Tracheary elements include **tracheids** and **vessel elements**, which are conducting cells of the xylem. **Sieve tube elements** are the conducting cells of the phloem; their activities are often controlled by **companion cells. Phloem sap** passes from cell to cell through sieve plates. Review Figures 34.9 and 34.10

34.3 How do meristems build the plant body?

All seed plants possess a **primary plant body** consisting of nonwoody tissues. Shrubs and trees also possess a **secondary plant body** consisting of wood and bark.

A hierarchy of **meristems** (localized regions of cell division) generates the plant body. **Apical meristems** at the tips of stems and roots give rise to three primary meristems (**protoderm**, **ground meristem**, and **procambium**) that in turn produce the three tissue systems of those organs. Apical meristems are responsible for **primary growth** (growth in length). Review Figure 34.11

The **root apical meristem** gives rise to the **root cap** and to the three primary meristems. Root tips have overlapping **zones of cell division**, **elongation**, and **maturation**. Review Figure 34.14

The vascular tissue in young roots is contained within the **stele.** Review Figures 34.15 and 34.16, Web/CD Activities 34.1 and 34.2

The **shoot apical meristem** also gives rise to the three primary meristems. **Leaf primordia** on the sides of the apical meristem develop into leaves.

The vascular tissue in young stems is divided into **vascular bundles**, each containing both xylem and phloem. Review Figure 34.17, Web/CD Activities 34.3 and 34.4

Two **lateral meristems**, the **vascular cambium** and **cork cambium**, are responsible for **secondary growth** (growth in width) when it occurs. Review Figure 34.12

In stems and roots with secondary growth, lateral meristems give rise to **wood** (secondary xylem) and **bark** (secondary phloem plus **cork**). Review Figure 34.18, Web/CD Tutorial 34.1

The **periderm** consists of cork, cork cambium, and phelloderm, all pierced at intervals by lenticels that allow gas exchange. Review Figure 34.22

34.4 How does leaf anatomy support photosynthesis?

Veins bring water and minerals to the **mesophyll** (the photosynthetic tissue) and carry the products of photosynthesis to other parts of the plant body. Review Figure 34.23

A waxy cuticle retards water loss from the leaf and is impermeable to carbon dioxide. Guard cells control openings (stomata) in the leaf that allow CO_2 to enter, but also allow some water to escape. Review Figure 34.23, Web/CD Activity 34.5

SELF-QUIZ

1. Which of the following is *not* a difference between monocots and eudicots?
   a. Eudicots more frequently have broad leaves.
   b. Monocots commonly have flower parts in multiples of three.
   c. Monocot stems do not generally undergo secondary thickening.
   d. The vascular bundles of monocot stems are commonly arranged as a cylinder.
   e. Eudicot embryos commonly have two cotyledons.
2. Roots
   a. always form a fibrous root system that holds the soil.
   b. possess a root cap at their tip.
   c. form branches from axillary buds.
   d. are commonly photosynthetic.
   e. do not show secondary growth.
3. The plant cell wall
   a. lies immediately inside the plasma membrane.
   b. is an impermeable barrier between cells.
   c. is always waterproofed with either lignin or suberin.
   d. always consists of a primary wall and a secondary wall, separated by a middle lamella.
   e. contains cellulose and other polysaccharides.
4. Which statement about parenchyma cells is *not* true?
   a. They are alive when they perform their functions.
   b. They typically lack a secondary wall.
   c. They often function as storage depots.
   d. They are the most numerous cells in the young plant body.
   e. They are found only in stems and roots.

5. Tracheids and vessel elements
   a. must die to become functional.
   b. are important constituents of all seed plants.
   c. have walls consisting of middle lamella and a primary wall.
   d. are always accompanied by companion cells.
   e. are found only in the secondary plant body.
6. Which statement about sieve tube elements is *not* true?
   a. Their end walls are called sieve plates.
   b. They must die to become functional.
   c. They link end-to-end, forming sieve tubes.
   d. They form the system for translocation of organic nutrients.
   e. They lose the membrane that surrounds their central vacuole.
7. The pericycle
   a. is the innermost layer of the cortex.
   b. is the tissue within which branch roots arise.
   c. consists of highly differentiated cells.
   d. forms a star-shaped structure at the very center of the root.
   e. is waterproofed by suberin.
8. Secondary growth of stems and roots
   a. is brought about by the apical meristems.
   b. is common in both monocots and eudicots.
   c. is brought about by vascular and cork cambia.
   d. produces only xylem and phloem.
   e. is brought about by vascular rays.
9. Periderm
   a. contains lenticels that allow for gas exchange.
   b. is produced during primary growth.
   c. is permanent; it lasts as long as the plant does.
   d. is the innermost part of the plant.
   e. contains vascular bundles.
10. Which statement about leaf anatomy is *not* true?
   a. Stomata are controlled by paired guard cells.
   b. The cuticle is secreted by the epidermis.
   c. The veins contain xylem and phloem.
   d. The cells of the mesophyll are packed together, minimizing air space.
   e. C_3 and C_4 plants differ in leaf anatomy.

FOR DISCUSSION

1. When a young oak was 5 m tall, a thoughtless person carved his initials in its trunk at a height of 1.5 m above the ground. Today that tree is 10 m tall. How high above the ground are those initials? Explain your answer in terms of the manner of plant growth.
2. Consider a newly formed sieve tube element in the secondary phloem of an oak tree. What kind of cell divided to produce the sieve tube element? What kind of cell divided to produce that parent cell? Keep tracing back until you arrive at a cell in the apical meristem.
3. Distinguish between sclerenchyma cells and collenchyma cells in terms of structure and function.
4. Distinguish between primary and secondary growth. Do all angiosperms undergo secondary growth? Explain.
5. What anatomical features make it possible for a plant to retain water as it grows? Describe the plant tissues and how and when they form.

FOR INVESTIGATION

Maton and Gartner found that a few other species behaved like ponderosa pine, but that different relationships between needles and annual rings characterized other species. For example, in some species the needles, regardless of age, are served by the current year's xylem. How might you try to make sense of such differences?

CHAPTER 35 Transport in Plants

The curious curate

The curate of Teddington had a lively curiosity and a strong practical bent. As a clergyman he believed that God had "observed the most exact proportions of number, weight and measure in the make of all things." He also believed that we should understand nature by observation and experimentation, and thus he spent much of his life observing and experimenting, and then trusting what his investigations told him. He was several things at once: a clergyman, a physiologist, a chemist, and an inventor. He firmly established experimentation with appropriate controls as an indispensable tool.

The Reverend Stephen Hales (1677–1761) is considered the father of the discipline of plant physiology and is one of the great figures in the history of animal physiology. The leading scientists of his day presumed that the sap of trees circulates like the blood in our bodies. They did not question Aristotle's pronouncement in the fourth century B.C.E. that plants can be understood by analogy to animals.

Hales recognized, however, that experimentation was more likely than Aristotelian dogma to reveal how sap moves in plants. He approached the question by measuring both the uptake of water by a plant (the amount he added) and the amount of water lost by the leaves (measured by change in weight). A single sunflower plant, he found, takes up and releases 17 times as much water in a day as does a human, weight for weight. He also showed that the movement of the sap is always upward and never downward—that is, the sap does *not* circulate.

Trying to understand how "nature wonderfully contrived ... most powerfully to raise and keep in motion the sap," he experimented on what we now call root pressure. He determined how root pressure varies by time of day.

Hales also investigated the relationship between leaves and the atmosphere. He was the first to show that plants take up "food" of some sort from the atmosphere, and that this process requires light. His studies of gases, done before anyone had conceived of such things as oxygen or carbon dioxide, were early landmarks in the history of chemistry. He developed methods for collecting gases over water that were later used by Priestley and Lavoisier in the discovery of oxygen and other gases.

He studied the movement of fluid in animals as well as in plants. He was the first to measure blood pressure and the rate of blood flow in the capillaries.

Sunflowers Sweat More Than You Do Stephen Hales measured the amount of water "imbibed" and the amount "perspired" by sunflower plants.

Water Has to Be Transported a Long Way It took more than 200 years for plant biologists to extend Stephen Hales's work and satisfactorily explain transport of water in plants. From sunflowers to the tallest gymnosperms (the coast redwood, *Sequoia sempervirens*), the leaves of plants pull the sap up through the xylem.

Among his many practical inventions was a ventilator, made from the bellows of a church organ, to refresh the air in prisons, hospitals, and the holds of ships.

Hales did not work in obscurity. In his time he received many honors for his contributions. Now every two years the American Society of Plant Biologists gives its prestigious Stephen Hales Prize to a person who has served the science of plant biology in some noteworthy manner. Over the years, the prize has gone to many of the people responsible for the progress described in the chapters of Part Seven.

IN THIS CHAPTER we will consider the uptake of water and minerals from the soil, the transport of these materials up the plant in the xylem, the control of evaporative water loss from leaves, and the translocation (movement from one location to another) of dissolved substances in the phloem.

CHAPTER OUTLINE

35.1 How Do Plants Take Up Water and Solutes?

Terrestrial plants must obtain both water and mineral nutrients from the soil, usually by way of their roots. The roots, in turn, obtain carbohydrates and other important materials from the leaves (**Figure 35.1**). We learned in Section 8.1 that water is one of the ingredients required for carbohydrate production by photosynthesis in the leaves. Water is also essential for transporting solutes both upward and downward, for cooling the plant, and for developing the internal pressure that supports the plant body. As Stephen Hales showed, plants lose large quantities of water to evaporation, and this water must be continually replaced.

How do leaves high in a tree obtain water from the soil? What are the mechanisms by which water and mineral ions enter the plant body through the roots and ascend as sap in the xylem? Because neither water nor minerals can move through the plant into the xylem without crossing at least one plasma membrane, we will first focus on osmosis. Then we will examine the active uptake of mineral ions by the plant and follow the pathway by which both water and minerals move through the root to gain entry to the xylem.

Water moves through a membrane by osmosis

Osmosis, the movement of water through a membrane in accordance with the laws of diffusion, is described in Section 5.3. The **solute potential** (also called the *osmotic potential*) of a solution is a measure of the effect of dissolved solutes on the osmotic behavior of the solution. The following statement presents an opportunity for confusion, so study it carefully:

- The greater the solute concentration of a solution, the more negative its solute potential, and the greater the tendency of water to move into it from another solution of lower solute concentration (and less negative solute potential).

For osmosis to occur, the two solutions must be separated by a *selectively permeable* membrane (i.e., a membrane that is permeable to water but relatively impermeable to the solute). Recall from Figure 5.9 that if pure water (less negative solute potential) is separated from a solution with a high salt concen-

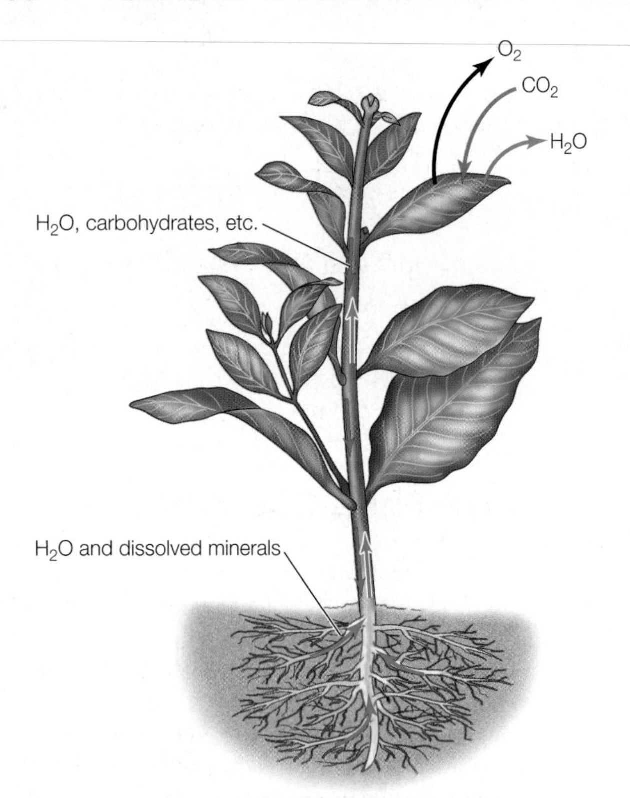

35.1 The Pathways of Water and Solutes in the Plant Water travels from the soil to the atmosphere, and it circulates within the plant, carrying important solutes with it.

tration (more negative solute potential) by a selectively permeable membrane, water molecules will travel across the membrane from the pure water side to the high-salt side. Recall, too, that osmosis is a *passive* process—no direct input of energy is required.

Unlike animal cells, plant cells are surrounded by a relatively rigid cell wall. As water enters a plant cell due to its negative solute potential, the entry of more water is increasingly resisted by an opposing **pressure potential** (called *turgor pressure* in plants). (Pressure potential is a hydraulic pressure analogous to the air pressure in an automobile tire; it is a mechanical pressure that can be measured with a pressure gauge.) As more and more water enters, the pressure potential becomes greater and greater.

Owing to the rigidity of the cell wall, plant cells do not burst the way animals cells do when placed in pure water; instead, water enters plant cells by osmosis until the pressure potential exactly balances the solute potential. At this point, the cell is *turgid;* that is, it has a significant positive pressure potential.

The overall tendency of a solution to take up water from pure water, across a membrane, is called its **water potential** and is represented as ψ, the Greek letter psi (pronounced "sigh") (**Figure 35.2**). The water potential of a solution is simply the sum of its (negative) solute potential (ψ_s) and its (usually positive) pressure potential (ψ_p):

$$\psi = \psi_s + \psi_p$$

For pure water open to the atmosphere and therefore under no applied pressure, all three of these parameters are defined as zero.

Whenever water moves by osmosis, the following important rule applies:

- Water always moves across a selectively permeable membrane toward the region of lower (more negative) water potential.

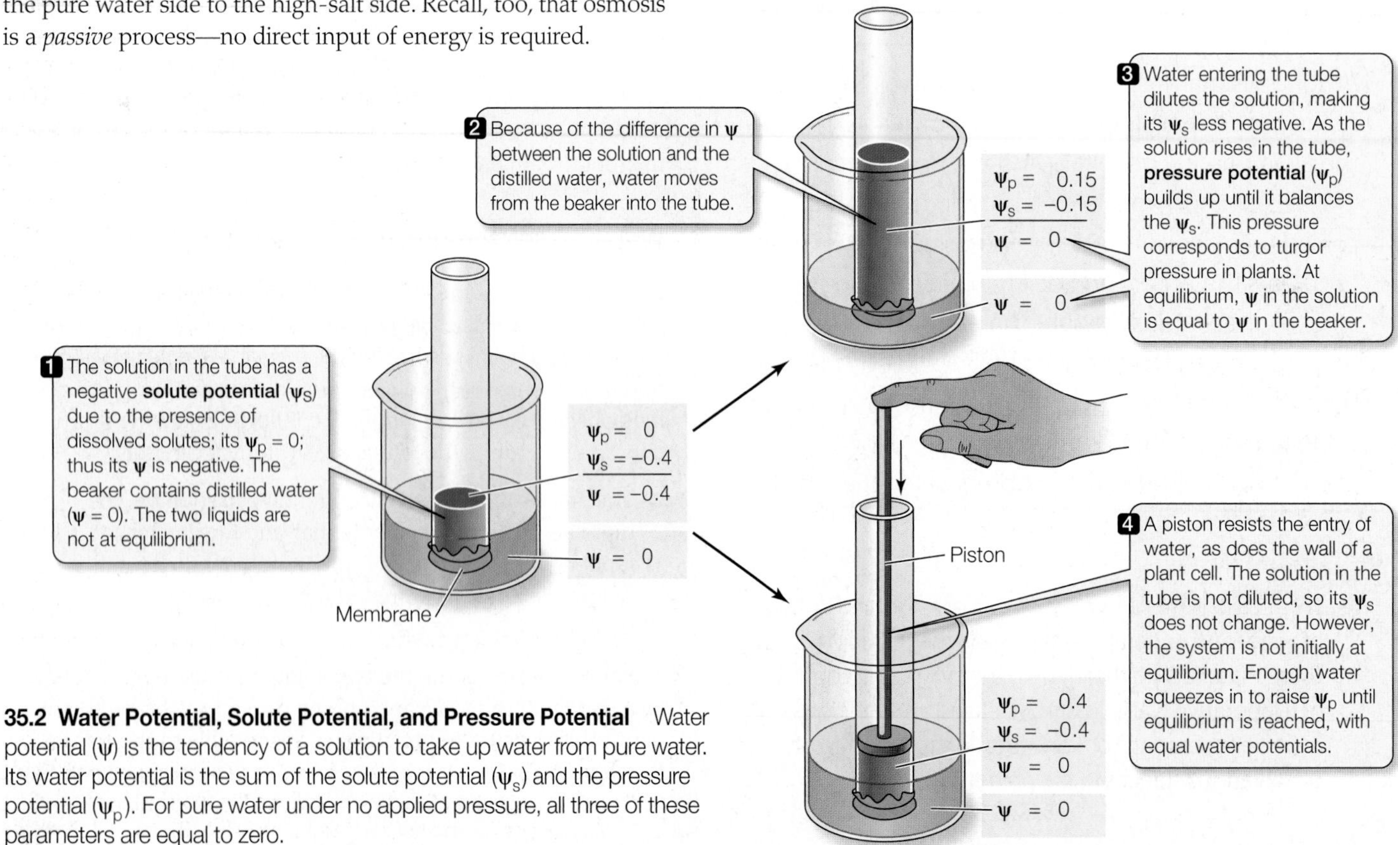

35.2 Water Potential, Solute Potential, and Pressure Potential Water potential (ψ) is the tendency of a solution to take up water from pure water. Its water potential is the sum of the solute potential (ψ_s) and the pressure potential (ψ_p). For pure water under no applied pressure, all three of these parameters are equal to zero.

We can measure solute potential, pressure potential, and water potential in *megapascals* (MPa), a unit of pressure. Atmospheric pressure, "one atmosphere," is about 0.1 MPa, or 14.7 pounds per square inch; typical pressure in an automobile tire is about 0.2 MPa.

Osmosis is of great importance to plants. The physical structure of many plants is maintained by the (positive) pressure potential of their cells; if the pressure potential is lost, the plant *wilts*. Within living tissues, the movement of water from cell to cell follows a gradient of water potential.

Over longer distances, in unobstructed tubes such as xylem vessels and phloem sieve tubes, the flow of water and dissolved solutes is driven by a *gradient of pressure potential*, not a gradient of water potential. The movement of a solution due to a difference in pressure potential between two parts of a plant is called **bulk flow**. We'll see that bulk flow in the xylem is between regions of differing *negative* pressure potential (tension) while bulk flow in the phloem is between regions of differing *positive* pressure potential (turgidity).

Aquaporins facilitate the movement of water across membranes

Water moves readily through biological membranes. How can this be? **Aquaporins** are membrane channel proteins through which water can move without interacting with the hydrophobic environment of the membrane's phospholipid bilayer (see Section 5.3). These proteins, important in both plants and animals, allow water to move rapidly from environment to cell and from cell to cell. Their abundance in the plasma membrane and tonoplast (vacuolar membrane) varies with environmental conditions, depending on a cell's need to obtain and retain water. The permeability of some aquaporins can be regulated, changing the *rate* of osmosis across the membrane. However, water movement through aquaporins is always passive, so the *direction* of water movement is unchanged by alterations in aquaporin permeability.

Uptake of mineral ions requires membrane transport proteins

Mineral ions, which carry electric charges, generally cannot move across a membrane unless they are aided by transport proteins, including ion channels and carrier proteins (see Section 5.3). The ions would otherwise be blocked by the hydrophobic interior of the membrane, and they are too large to pass through aquaporins.

We have just seen that water moves through a water-permeable membrane in response to a gradient of water molecules. Other molecules and ions follow their concentration gradients as permitted by the characteristics of the membrane. When the concentration of these charged ions in the soil is greater than that in the plant, transport proteins can move them into the plant by facilitated diffusion, which is a passive process. The concentrations of most ions in the soil solution, however, are lower than those required inside the plant. Thus the plant must actively take up ions *against* their concentration gradients—a process that requires energy.

Electric charge differences also play a role in the uptake of mineral ions. Movement of a negatively charged ion into a negatively charged region is movement against an electrical gradient and therefore requires energy. The combination of concentration and electrical gradients is called an *electrochemical gradient*. Uptake against an electrochemical gradient involves *active transport*, which is fueled by ATP generated by cellular respiration. Active transport requires specific transport proteins.

Unlike animals, plants do not have a sodium–potassium pump (see Section 5.4) for active transport. Rather, plants have a **proton pump**, which uses energy obtained from ATP to move protons out of the cell against a proton concentration gradient (**Figure 35.3, step 1**). Because protons (H^+) are positively charged, their accumulation outside the cell has two results:

- An electrical gradient is created such that the region outside the cell becomes positively charged with respect to the region inside.
- A proton concentration gradient develops, with more protons outside the cell than inside.

Each of these results has consequences for the movement of other ions. Because the inside of the cell is now more negative than the outside, cations (positively charged ions) such as potassium (K^+) move into the cell by facilitated diffusion through their specific membrane

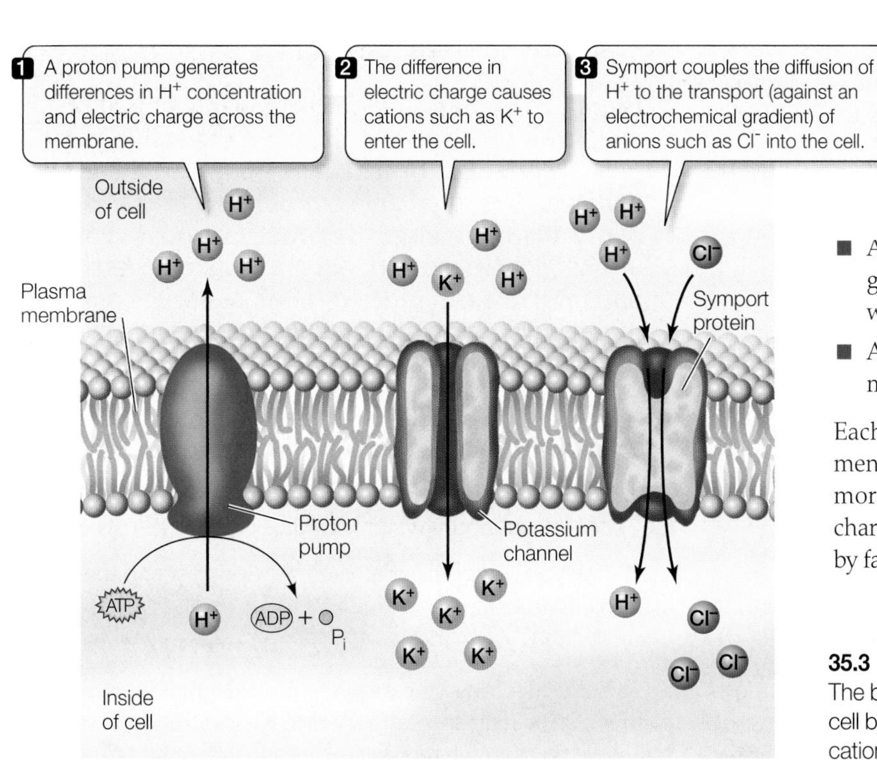

35.3 The Proton Pump in Active Transport of K^+ and Cl^- The buildup of hydrogen ions (H^+) transported outside the cell by the proton pump (1) drives the movement of both cations (2) and anions (3) into the cell.

channels (**Figure 35.3, step 2**). In addition, the proton concentration gradient can be harnessed to drive secondary active transport, in which anions (negatively charged ions) such as chloride (Cl^-) are moved into the cell against an electrochemical gradient by a symport protein that couples their movement with that of H^+ (**Figure 35.3, step 3**). In sum, there is a vigorous traffic of ions across plant cell membranes, involving specific membrane transport proteins and both active and passive processes.

The proton pump and the coordinated activities of other membrane transport proteins cause the interior of a plant cell to be very negatively charged with respect to the exterior; that is, they build up a significant *membrane potential*. Biologists can measure the membrane potential of a plant cell with microelectrodes, just as they can measure similar charge differences in nerve cells and other animal cells (see Section 44.2). Most plant cells maintain a membrane potential of at least –120 millivolts (mV).

Water and ions pass to the xylem by way of the apoplast and symplast

Mineral ions enter and move through plants in various ways. Where water is moving by bulk flow, dissolved minerals are carried along in the stream. Both water and minerals also move by diffusion. At certain sites, where plasma membranes are being crossed, some mineral ions are moved by active transport. One such site is the surface of a root hair, where mineral ions first enter the cells of the plant. Later, within the stele, the ions must cross another plasma membrane before entering the nonliving vessels and tracheids of the xylem.

The movement of ions across membranes can also result in the movement of water. Water moves into a root because the root has a more negative water potential than does the soil solution. Water moves from the cortex of the root into the stele (which is where the vascular tissues are located) because the stele has a more negative water potential than does the cortex.

Water and minerals from the soil can pass through the dermal and ground tissues to the stele via two pathways, the *apoplast* and the *symplast* (**Figure 35.4**):

- The **apoplast** (Greek *apo*, "away from"; *plast*, "living material") consists of the cell walls, which lie outside the plasma membranes, and the intercellular spaces (spaces between cells) that are common to many tissues. The apoplast is a continuous meshwork through which water and dissolved substances can flow or diffuse without ever having to cross a membrane. Movement of materials through the apoplast is thus unregulated—until it reaches the *Casparian strips* of the endodermis.
- The **symplast** (Greek, *sym*, "together with") passes through the continuous cytoplasm of the living cells connected by plasmodesmata. The selectively permeable plasma membranes of the root hair cells control access to the symplast, so movement of water and dissolved substances into the symplast is tightly regulated.

Water and minerals that pass from the soil solution through the apoplast can travel as far as the endodermis, the innermost layer of the root cortex. The endodermis is distinguished from the rest of the ground tissue by the presence of **Casparian strips**. These waxy, suberin-impregnated regions of the endodermal cell wall form a water-repelling (hydrophobic) belt around each endodermal cell where it is in contact with other endodermal cells. Casparian strips act as a seal that prevents water and ions from moving between the cells (see Figure 35.4).

35.4 Apoplast and Symplast Plant cell walls and intercellular spaces constitute the apoplast. The symplast comprises the living cells, which are connected by plasmodesmata. To enter the symplast, water and solutes must pass through a plasma membrane. No such selective barrier limits movement through the apoplast. Casparian strips in the endodermis of the cortex are impregnated with the water-repelling substance suberin and separate apoplast in the cortex from apoplast in the stele.

The Casparian strips of the endodermis completely separate the apoplast of the cortex from the apoplast of the stele. However, they do not obstruct the outer or inner faces of the endodermal cells. Accordingly, water and ions can enter the stele only by way of the symplast—that is, by entering and passing through the cytoplasm of the endodermal cells. Thus transport proteins in the plasma membranes of these cells determine which mineral ions pass into the stele, and at what rates.

Once they have passed the endodermal barrier, water and minerals leave the symplast and enter the apoplast of the stele. Parenchyma cells in the pericycle or xylem can aid this process.

As mineral ions move into the solution in the cell walls, the water potential in the apoplast becomes more negative; thus water moves out of the cells and into the apoplast by osmosis. In other words, active transport of ions moves the ions directly, and water follows passively. The end result is that water and minerals end up in the xylem, where they constitute the *xylem sap*.

35.1 RECAP

Osmotic mechanisms govern the movement of water from the soil into the plant stele; this is a passive process. Uptake of minerals from the soil occurs against an electrochemical gradient and is therefore an active process requiring energy and membrane transport proteins. Water and minerals can move through either the apoplast or the symplast, but must enter and leave the symplast to reach the xylem.

- Can you distinguish among water potential, solute potential, and pressure potential? See pp. 765–766
- Can you explain what happens when you drop a piece of stem into pure water? See p. 766 and Figure 35.2
- Can you distinguish between the apoplast and the symplast? See p. 767 and Figure 35.4

So far we've described the movement of water and minerals into plant roots and their entry into the root xylem. How does the xylem sap move on from the root system into the plant body?

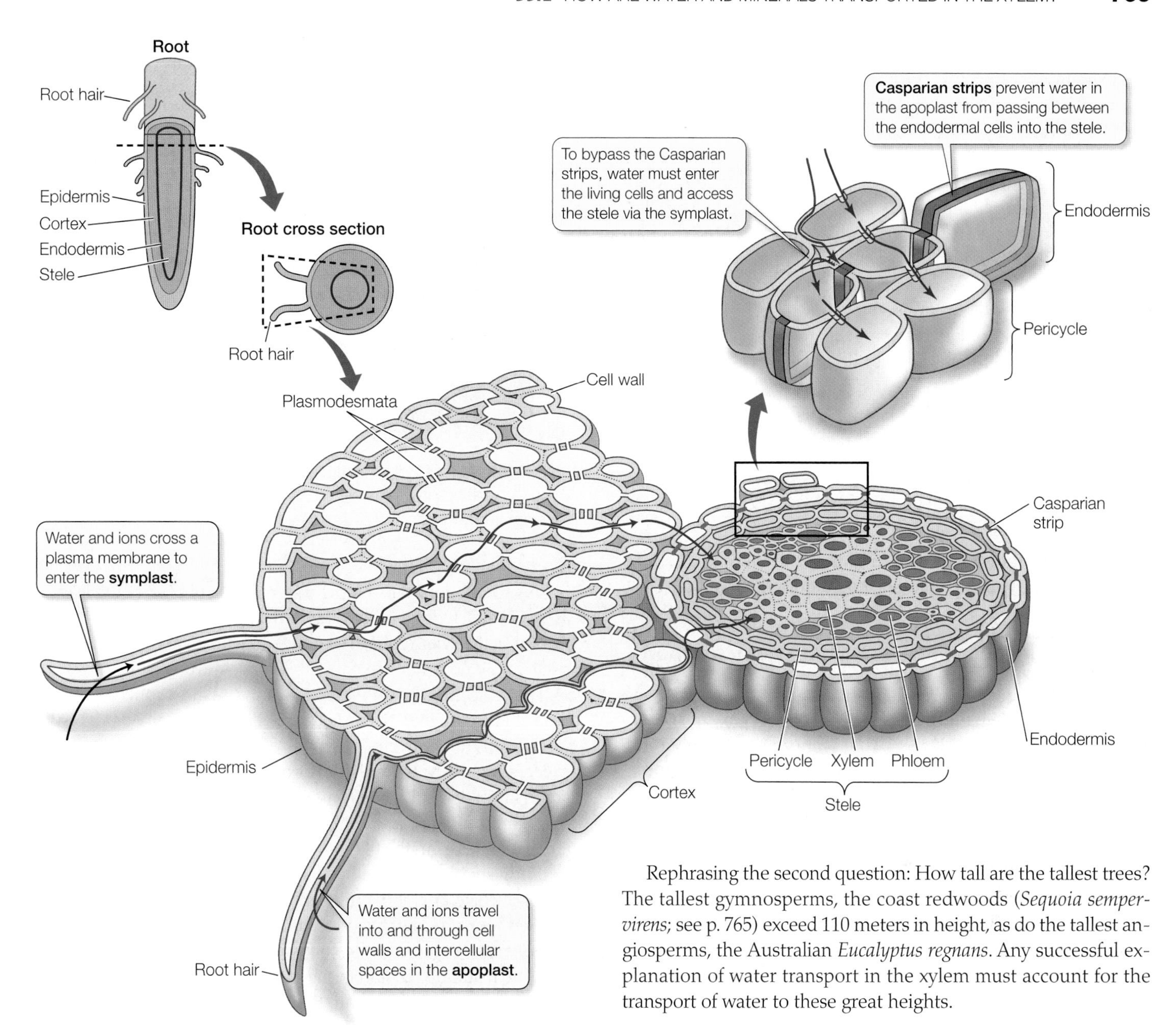

Rephrasing the second question: How tall are the tallest trees? The tallest gymnosperms, the coast redwoods (*Sequoia sempervirens*; see p. 765) exceed 110 meters in height, as do the tallest angiosperms, the Australian *Eucalyptus regnans*. Any successful explanation of water transport in the xylem must account for the transport of water to these great heights.

Movies featuring Tarzan show him swinging through the jungle canopy on lianas—twining jungle vines. And when his exertions leave him thirsty, he just chops open a vine and drinks the water from its stem. Neither activity is particularly realistic, although, like all plants, lianas do move water through the xylem in their stems.

35.2 How Are Water and Minerals Transported in the Xylem?

Before we consider the underlying mechanism of xylem transport, let's pause to reflect on the magnitude of what it accomplishes. Consider two questions: How much water is transported? And how high can water be transported?

The following example illustrates how much water an individual plant can transport. A single maple tree 15 meters tall has been estimated to have some 177,000 leaves, with a total leaf surface area of 675 square meters—half again the area of a basketball court. During a summer day, that tree loses 220 liters of water *per hour* to the atmosphere by evaporation from the leaves. To prevent wilting, the xylem needs to transport 220 liters of water from the roots to the leaves every hour. (By comparison, a 50-gallon drum holds 189 liters.)

Various hypotheses to explain the ascent of xylem sap have been offered over the years. We begin by reviewing some illuminating experiments that ruled out early models, and then turn to evidence in support of the current model.

Experiments ruled out xylem transport by the pumping action of living cells

Some of the earliest attempts to explain the rise of sap in the xylem were based on the hypothesis that a pumping action by living cells in the stem might push the sap upward. However, experiments conducted and published in 1893 by the German botanist Eduard Strasburger definitively ruled out such models.

Strasburger worked with trees about 20 meters tall. He sawed through the trunk of each tree at its base and plunged the cut end into a solution of a poison, such as copper sulfate. The solution rose through the trunk, as was readily evident from the progressive death of the bark higher and higher up. When the solution reached the leaves, the leaves died, too, at which point the movement of the solution stopped (as shown by the liquid level in the bucket, which stopped dropping).

This simple experiment established three important points:

- Living, "pumping" cells were not responsible for the upward movement of the solution, because the solution itself killed all living cells with which it came in contact.
- The leaves played a crucial role in transport. As long as they were alive, the solution continued to move upward; when the leaves died, movement ceased.
- The movement was not caused by the roots, because the trunk had been completely separated from the roots.

35.5 Guttation Root pressure is responsible for forcing water through openings in the margins of this leaf on a lady's mantle plant (*Alchemilla*).

Root pressure does not account for xylem transport

In spite of Strasburger's observations, some plant physiologists hypothesized that xylem transport was based on **root pressure**—pressure exerted by the root tissues that would force liquid up the xylem. The basis for root pressure is a higher solute concentration, and accordingly a more negative water potential, in the xylem sap than in the soil solution. This water potential draws water into the stele; once there, the water has nowhere to go but up, so it rises in the vessels and tracheids.

There is good evidence that root pressure exists—for example, the phenomenon of *guttation*, in which liquid water is forced out through openings at the margins of leaves (**Figure 35.5**). Guttation occurs only under conditions of high atmospheric humidity and plentiful water in the soil, which occur most commonly at night. Root pressure is also the source of the sap that oozes from the cut stumps of some plants when their tops are cut off. Stephen Hales was the first to study root pressure quantitatively.

Root pressure, however, cannot account for the ascent of sap in trees. Root pressure seldom exceeds 0.1–0.2 MPa (1–2 atmospheres). If root pressure were driving sap up the xylem, we would observe a positive pressure potential in the xylem at all times. In fact, as we are about to see, the xylem sap in most trees is under *tension*—has a negative pressure potential—when it is ascending. Furthermore, as Strasburger had already shown, materials can be transported upward in the xylem even when the roots have been removed. If the roots are not pushing the xylem sap upward, what causes it to rise?

The transpiration–cohesion–tension mechanism accounts for xylem transport

An alternative to pushing is pulling: The leaves pull the xylem sap upward, which was first demonstrated by Stephen Hales. The evaporative loss of water from the leaves indirectly generates a pulling force—**tension**—on the water in the apoplast of the leaves, as we'll see. Hydrogen bonding between water molecules makes the sap in the xylem cohesive enough to withstand the tension and rise by bulk flow. Let's see how this process works.

The concentration of water vapor in the atmosphere is lower than that in the leaf. Because of this difference, water vapor diffuses from the intercellular spaces of the leaf, through openings called stomata, to the outside air in the process called **transpiration**. Within the leaf blade, water evaporates from the moist walls of the mesophyll cells and enters the intercellular spaces. As water evaporates from the film coating each cell, the film shrinks back into tiny spaces in the cell walls, increasing the curvature of the water surface and thus increasing its surface tension. This increased tension (negative pressure potential) in the surface film draws more water into the cell walls, replacing that which was lost. The resulting tension in the mesophyll draws water from the xylem of the nearest vein into the apoplast surrounding the mesophyll cells. The removal of water from the veins, in turn, establishes tension on the entire column of water contained within the xylem, so that the column is drawn upward all the way from the roots (**Figure 35.6**).

The ability of water to be pulled upward through tiny tubes results from the remarkable **cohesion** of water—the tendency of water molecules to stick to one another through hydrogen bonding (see Section 2.4). The narrower the tube, the greater the tension the water column can withstand without breaking. The integrity of the column is also maintained by the *adhesion* of water to the xylem walls. The cohesion of water in the xylem is great enough to withstand even the tensions developed there.

In summary, the key elements of water transport in the xylem are

- *Transpiration*, the evaporation of water from the leaves
- *Tension* in the xylem sap resulting from transpiration
- *Cohesion* in the xylem sap from the leaves to the roots

This **transpiration–cohesion–tension mechanism** requires no work (that is, no expenditure of energy) on the part of the plant. At each step between soil and atmosphere, water moves passively toward a region with a more negative water potential. Dry air has the most negative water potential (–95 MPa at 50% relative humidity), and the soil solution has the least negative water potential (between –0.01 and –3 MPa). Xylem sap has a water potential more negative than that of cells in the cortex of the root, but less negative than that of mesophyll cells in the leaf.

In the tallest trees, such as a 110-meter *Sequoia*, the difference in pressure potential between the top and the bottom of the column may be as great as 3 MPa. Recall that the pressure in a typical automobile tire is 0.2 MPa.

Mineral ions contained in the xylem sap rise passively with water as it ascends from root to leaf. In this way the nutritional needs of the shoot are met. Some of the mineral elements brought to the leaves are subsequently redistributed to other parts of the plant by way of the phloem, but the initial delivery from the roots is through the xylem.

In addition to promoting the transport of minerals, transpiration contributes to temperature regulation. As water evaporates from mesophyll cells, heat is taken up from the cells, and the leaf temperature drops. This cooling effect of evaporation (so evident in the cooling of our skin when we sweat) is important in enabling plants to live in hot environments. A farmer can hold a leaf between thumb and forefinger to estimate its temperature; if the leaf doesn't feel cool, that means that transpiration is not occurring, so it must be time to water.

A pressure chamber measures tension in the xylem sap

The transpiration–cohesion–tension model holds true only if the column of sap in the xylem is under tension (has a negative pressure potential). The most elegant demonstrations of this tension, and of its adequacy to account for the ascent of xylem sap in tall trees, were performed by the biologist Per Scholander, who measured tension in stems with an instrument called a **pressure chamber** (**Figure 35.7**).

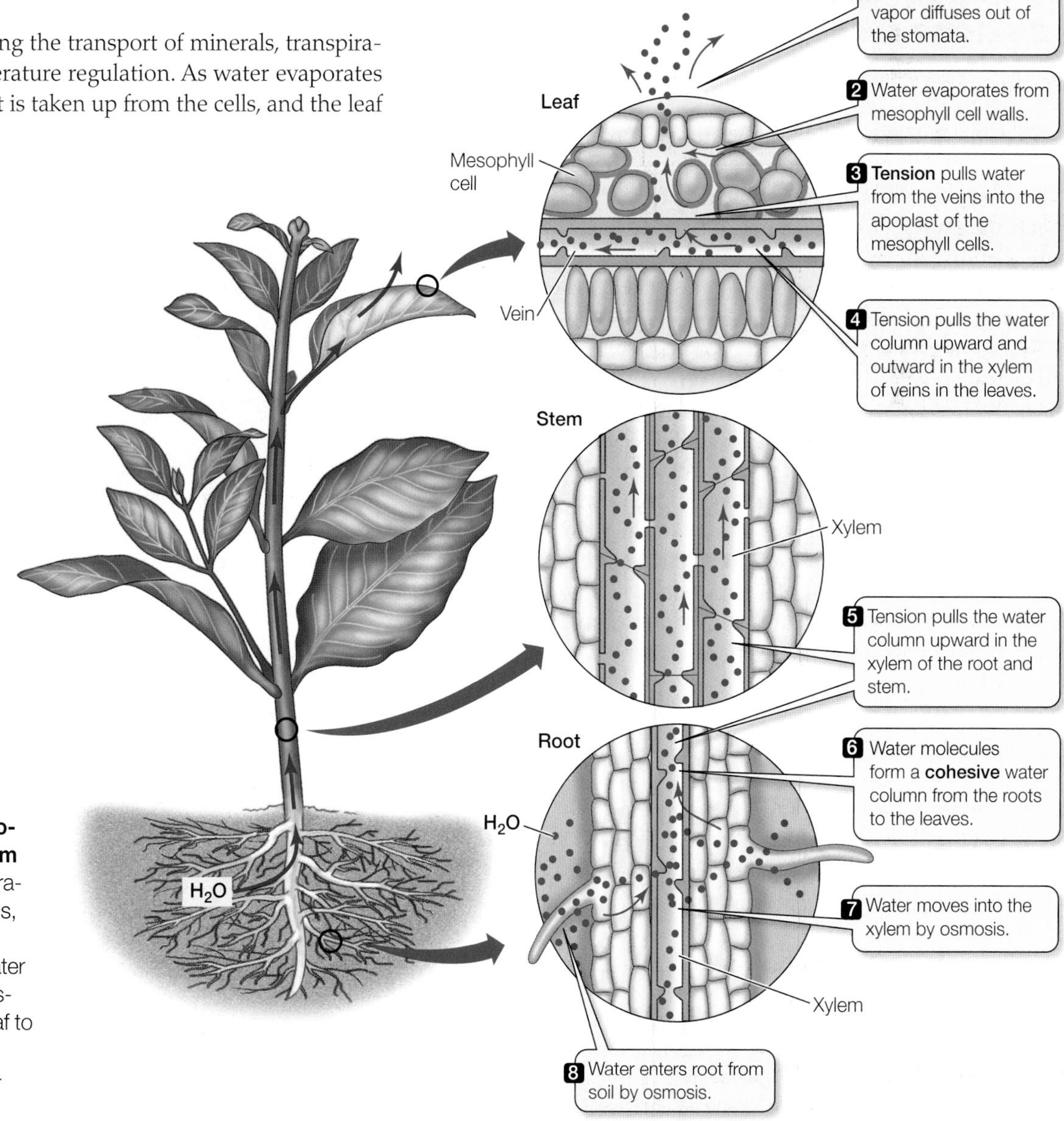

35.6 The Transpiration–Cohesion–Tension Mechanism Transpiration causes evaporation from mesophyll cell walls, generating tension on the xylem. Cohesion among water molecules in the xylem transmits the tension from the leaf to the root, causing water to move from the soil to the atmosphere.

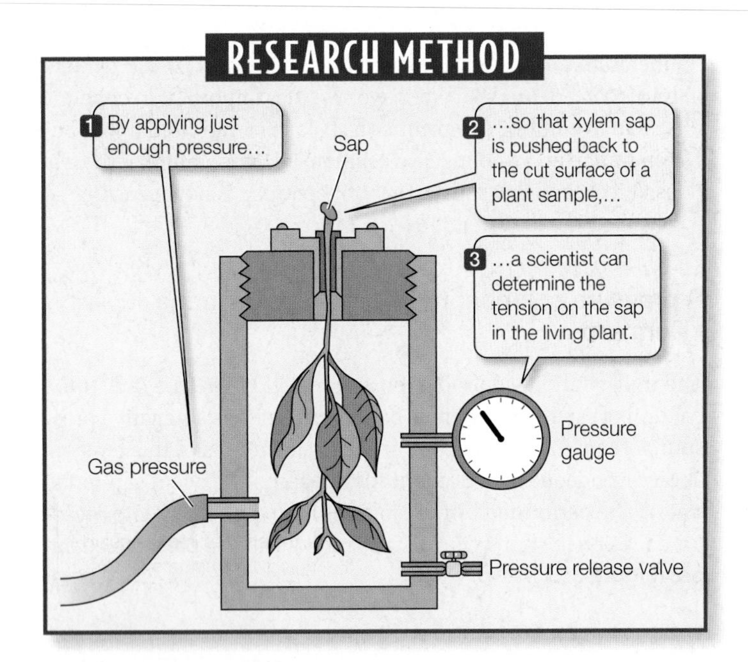

35.7 A Pressure Chamber The amount of tension on the sap in different types of plants can be measured with this device.

Consider a stem in which the xylem sap is under tension. If the stem is cut, the sap pulls away from the cut, into the stem. This behavior indicates that the pressure in the intact xylem is lower than that of the atmosphere. Now the stem is quickly placed in the pressure chamber, in which the pressure may be raised. The cut surface remains outside the chamber. As gas pressure is applied to the plant parts within the chamber, the xylem sap is pushed back to the cut surface. When the sap first becomes visible again at the cut surface, the pressure in the chamber is recorded. This pressure is equal in magnitude but opposite in sign to the tension (negative pressure potential) originally present in the xylem.

Scholander used the pressure chamber to study dozens of plant species, from diverse habitats, growing under a variety of conditions. Whenever xylem sap was ascending, he found that it was under tension. He also noticed that the tension disappeared in some of the plants at night, when transpiration ceased. In developing vines, the xylem sap was under no tension until leaves formed. Once leaves developed, transport in the xylem began and so did the tensions.

Per Scholander used surveying instruments to locate twigs at various heights on tall trees, then had a sharpshooter shoot the twigs down with a rifle. Testing these twigs in the pressure chamber, Scholander confirmed that tension differences at different heights were indeed great enough to keep the sap ascending.

The rate at which xylem sap ascends is not the same at all times. No flow of xylem sap takes place at night, when there is little or no transpiration. By day, when the sap is ascending, the rate of ascent depends on several factors. These include temperature, light intensity, and wind velocity, all of which affect the transpiration rate, and hence the rate of sap flow. Another factor may be the concentration of K^+ in the sap: In greenhouse experiments utilizing tobacco plants, the rate of flow increases as the K^+ concentration increases (**Figure 35.8**). K^+ appears to affect a cell wall component in the membrane of pits between vessel elements, changing the size of the pit. This effect may increase the rate of water flow through a vessel when a neighboring vessel is blocked, as by an air bubble.

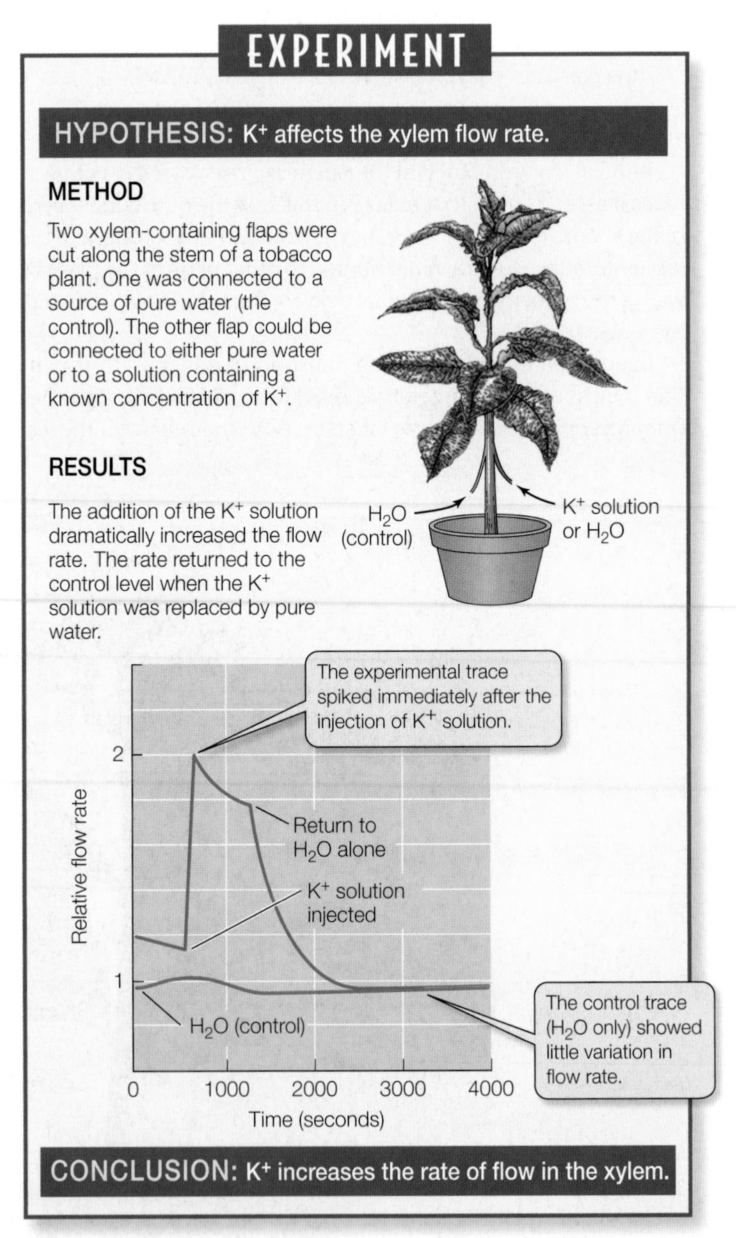

35.8 Potassium Ions Speed Transport in the Xylem This experiment by Maciej Zwieniecki and colleagues showed that the rate of fluid ascending through the xylem spiked when a solution with a known concentration of potassium ions was injected. Repeating the experiment with solutions of different concentrations of K^+ showed that the higher the K^+ concentration, the greater the flow rate. FURTHER RESEARCH: What experiments would you perform to determine whether this effect is specific to K^+?

35.2 RECAP

The transpiration-cohesion-tension mechanism explains the ascent of xylem sap. Transpiration draws water out of leaves, resulting in tension that pulls water from the xylem. Because of cohesion between water molecules, water is pulled passively through the xylem vessels in continuous columns, always toward a region with a more negative water potential.

- Do you understand the roles of transpiration, cohesion, and tension in xylem transport? See pp. 770–771 and Figure 35.6
- What is measured by the pressure chamber technique? See p. 772 and Figure 35.7

Although transpiration provides the impetus for the transport of water and minerals in the xylem, it also results in the loss of tremendous quantities of water from the plant. How plants control this loss is the subject of the next section.

35.3 How Do Stomata Control the Loss of Water and the Uptake of CO_2?

The epidermis of leaves and stems minimizes transpirational water loss by secreting a waxy cuticle, which is impermeable to water. However, the cuticle is also impermeable to carbon dioxide. This poses a problem: How can the leaf balance its need to retain water with its need to obtain CO_2 for photosynthesis?

Plants have evolved an elegant compromise in the form of **stomata** (singular *stoma*), or pores, in the epidermis of their leaves. A pair of specialized epidermal cells, called **guard cells**, controls the opening and closing of each stoma (**Figure 35.9A**). When the stomata are open, CO_2 can enter the leaf by diffusion—but water vapor is lost in the same way. Closed stomata prevent water loss, but also exclude CO_2 from the leaf.

Most plants open their stomata only when the light intensity is sufficient to maintain a moderate rate of photosynthesis. At night, when darkness precludes photosynthesis, their stomata remain closed; no CO_2 is needed at this time, and water is conserved. Even during the day, the stomata close if water is being lost at too great a rate.

The stoma and guard cells seen in Figure 35.9A are typical of eudicots. Monocots typically have specialized epidermal cells associated with their guard cells. The principle of operation, however, is the same for both monocot and eudicot stomata. In what follows, we describe the regulation and mechanism of stomatal opening and the normal cycle of opening and closing.

(A)

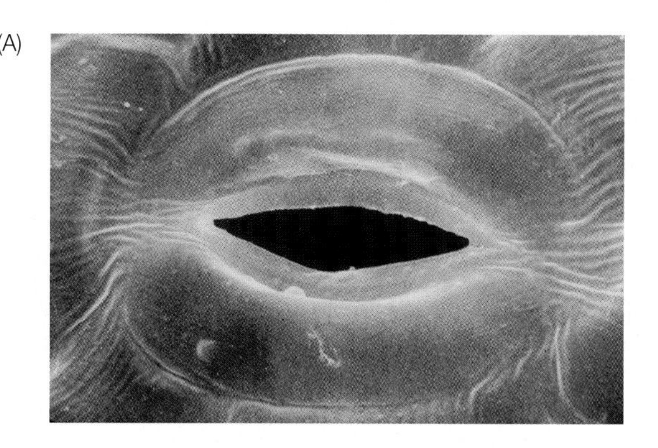

(B)

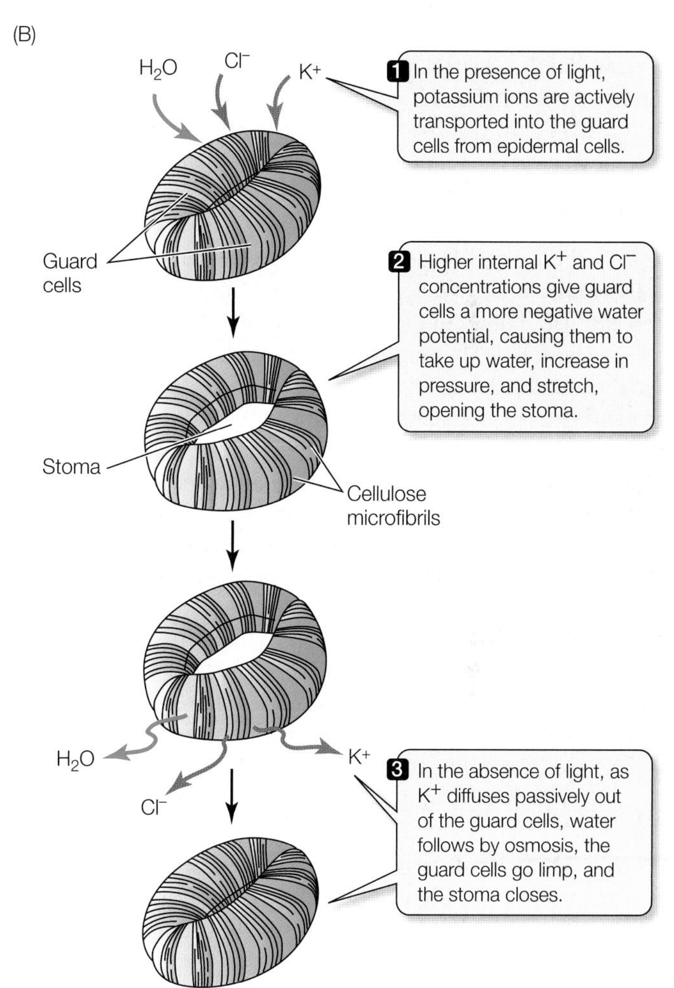

35.9 Stomata (A) Scanning electron micrograph of an open stoma formed by two sausage-shaped guard cells. (B) Potassium ion concentrations affect the water potential of the guard cells, controlling the opening and closing of stomata. Negatively charged ions that accompany K^+ maintain electrical balance and contribute to the changes in water potential that open and close the stomata.

The guard cells control the size of the stomatal opening

Light causes the stomata of most plants to open, admitting CO_2 for photosynthesis. Another cue for stomatal opening is the level of CO_2 in the intercellular spaces inside the leaf. A low level favors opening of the stomata, thus allowing the uptake of more CO_2.

Water stress is a common problem for plants, especially on hot, sunny, windy days. Plants have a protective response to these conditions, which uses the water potential of the mesophyll cells as a cue. Even when the CO_2 level is low and the sun is shining, if the mesophyll is too dehydrated—that is, if the water potential of the mesophyll is too negative—the mesophyll cells release a plant hormone called *abscisic acid*. Abscisic acid acts on the guard cells, causing them to close the stomata and prevent further drying of the leaf. This response reduces the rate of photosynthesis, but it protects the plant.

The opening and closing of stomata is regulated by control of the K^+ concentration in the guard cells (review Figure 35.3). Blue light, absorbed by a pigment in the guard cell plasma membrane, activates a proton pump, which actively transports H^+ out of the guard cells and into the apoplast of the surrounding epidermis. The resulting proton gradient drives K^+ into the guard cell, where it accumulates (**Figure 35.9B**). The increasing internal concentration of K^+ makes the water potential of the guard cells more negative. Water enters the guard cells by osmosis, increasing their pressure potential. The arrangement of the cellulose microfibrils in their cell walls causes the guard cells to respond to this increase by changing their shapes so that a gap—the stoma—appears between them.

The stoma closes by the reverse process when active transport ceases in response to the absence of blue light or the presence of abscisic acid. Potassium ions diffuse passively out of the guard cells, water follows by osmosis, the pressure potential decreases, and the guard cells sag together and seal off the stoma. Negatively charged chloride ions and organic ions also move into and out of the guard cells along with the potassium ions, maintaining electrical balance and contributing to the change in the solute potential of the guard cells.

Showing how much potassium moves into the guard cells to open a stoma was a difficult feat. A typical guard cell has a total volume of less than 0.03 nanoliters when the stoma is closed and almost 0.05 nanoliters when it is open. The scientists who solved the problem used an *electron probe microanalyzer*, an instrument normally used by metallurgists to study the fine structure of alloys (**Figure 35.10**).

EXPERIMENT

HYPOTHESIS: Guard cells of open stomata contain more potassium ions than do those of closed stomata.

METHOD

1. Peel strips of epidermis from leaves of broad beans in the dark (closed stomata) and in the light (open stomata).
2. Examine the strips to locate stomata.
3. Scan across guard cells with the electron probe microanalyzer set to measure K^+ concentration.

RESULTS

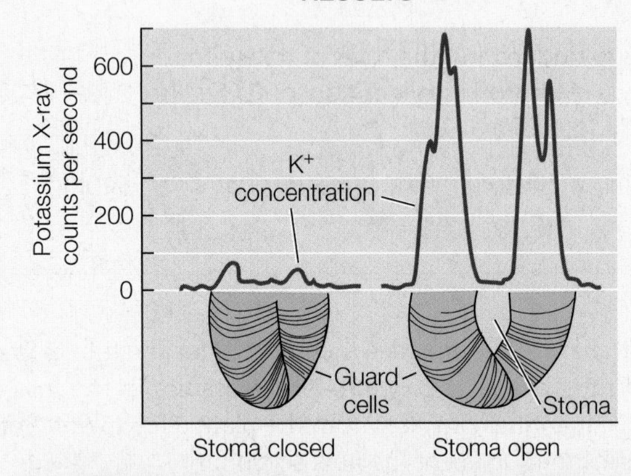

CONCLUSION: K^+ concentration within the guard cells surrounding an open stoma was much greater than that in the guard cells surrounding a closed stoma.

35.10 Measuring Potassium Ion Concentration in Guard Cells G. D. Humble and Klaus Raschke used the electron probe microanalyzer to examine individual stomata of the broad bean. They determined that K^+ concentration in each guard cell increased more than twentyfold as the stomatal aperture increased sixfold, and the guard cell volume nearly doubled. FURTHER RESEARCH: What other ion or ions would you study in order to further explore the mechanism of stomatal opening?

Transpiration from crops can be decreased

Stomata are the "referees" mediating the admission of CO_2 for photosynthesis and the exit of water by transpiration. Farmers would like their crops to transpire less, thus reducing the need for irrigation. Similarly, nurseries and gardeners would like to be able to reduce the amount of water lost by plants that are to be transplanted, because transplanting often damages the roots, causing the plant to wilt or die. What they need is a good **antitranspirant**: a compound that can be applied to plants to reduce water loss from the stomata without excessively limiting CO_2 uptake.

Abscisic acid and its commercial chemical analogs have been found to work as antitranspirants in small-scale tests, but their high cost has precluded commercial use. So research has turned to the question of whether plants can be made more sensitive to their own abscisic acid. The guard cells of transgenic plants with a mutant allele of the *era* gene are highly sensitive to abscisic acid and hence resistant to wilting during drought stress. This recent discovery might lead to an agricultural application.

A totally different type of antitranspirant temporarily seals off the leaves from the atmosphere. Growers use a variety of compounds, most of which form polymer films around leaves, to form a barrier to evaporation by sealing the stomata. These compounds cause undesirable side effects, however, and can be used only for short periods of time. Their most common use is in the transplanting of nursery stock.

35.3 RECAP

Leaf pores called stomata admit the CO_2 needed for photosynthesis but also permit the exit of water by transpiration. Stomata can be opened or closed by guard cells to regulate water loss.

- Can you explain the role of K^+ ions in the functioning of guard cells? See p. 774 and Figure 35.9
- Under what circumstances do the mesophyll cells release abscisic acid, and what is its effect? See p. 774

In the absence of antitranspirants, stomata are normally open during daylight hours, allowing CO_2 to be fixed and converted to the products of photosynthesis. In the next section we'll see how these products are delivered to other parts of the plant, supporting plant growth.

35.4 How Are Substances Translocated in the Phloem?

Photosynthesis takes place in the mesophyll cells and, in C_4 plants, in the bundle sheath cells of the leaf (see Figure 8.18). The products of photosynthesis (primarily carbohydrates) diffuse to the nearest small vein, where they are actively transported into sieve tube elements. The movement of carbohydrates and other solutes through the plant in the phloem is called **translocation.**

Substances in the phloem are translocated from sources to sinks. A **source** is an organ (such as a mature leaf or a storage root) that *produces* (by photosynthesis or by digestion of stored reserves) more sugars than it requires. A **sink** is an organ (such as a root, a flower, a developing fruit or tuber, or an immature leaf) that *consumes* sugars for its own growth and storage needs. Sugars (primarily sucrose), amino acids, some minerals, and a variety of other solutes are translocated between sources and sinks in the phloem.

How do we know that such organic solutes are translocated in the phloem, rather than in the xylem? Just over 300 years ago, the Italian scientist Marcello Malpighi performed a classic experiment in which he removed a ring of bark (containing the phloem) from the trunk of a tree—that is, he *girdled* the tree (**Figure 35.11**). Over time, the bark in the region above the girdle swelled. We now know that the swelling resulted from the accumulation of organic solutes that came from higher up the tree and could no longer continue downward because of the disruption of the phloem. Later, the bark below the girdle died because it no longer received sugars from the leaves. Eventually the roots, and then the entire tree, died.

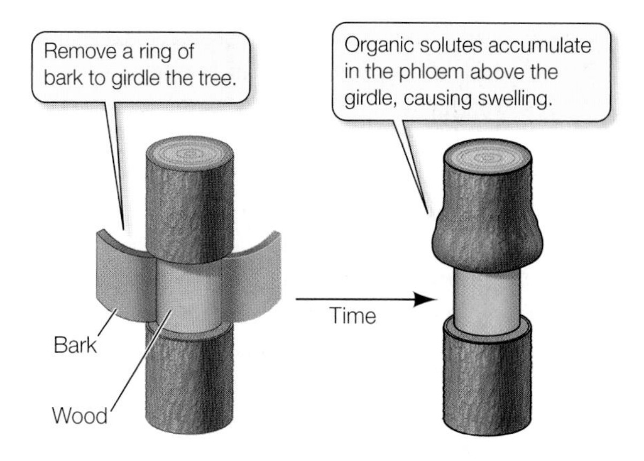

35.11 Girdling Blocks Translocation in the Phloem By girdling—removing a ring of bark containing the phloem—Malpighi blocked the translocation of organic solutes in a tree. Bark below the girdle died because it no longer received nutrients; eventually the entire tree died.

Any explanation of the translocation of organic solutes must account for a few important observations:

- Translocation stops if the phloem tissue is killed by heating or other methods; thus the mechanism must be different from that of transport in the xylem.
- Translocation often proceeds in both directions simultaneously—up the stem and down the stem—depending on the location of sources and sinks.
- Translocation is inhibited by compounds that inhibit respiration and thus limit the ATP supply in the source.

To investigate translocation, plant physiologists needed to obtain samples of pure sieve tube sap from individual sieve tube elements. This difficult task was simplified when it was discovered that a common garden pest, the aphid, feeds on plants by drilling into a sieve tube. An aphid inserts its specialized feeding organ, called a *stylet,* into a stem until the stylet enters a sieve tube. The pressure within the sieve tube is greater than that in the surrounding plant tissues or outside the plant, so the nutritious sieve tube sap is forced through the stylet and into the aphid's digestive tract. So great is the pressure that sugary liquid is forced through the insect's body and out the anus (**Figure 35.12**). This works because the phloem sap is under strongly positive pressure, unlike the negative pressure potential in the xylem.

Plant physiologists use aphids to collect phloem sap. When liquid appears on the aphid's abdomen, indicating that the insect has connected with a sieve tube, the physiologist quickly freezes the aphid and cuts its body away from the stylet, which remains in the sieve tube element. For hours, phloem sap continues to exude from the cut stylet, where it may be collected for analysis. Chemical analysis of phloem sap collected in this manner reveals the contents of a single sieve tube element over time. Physiologists can also infer the rates at which different substances are translocated by

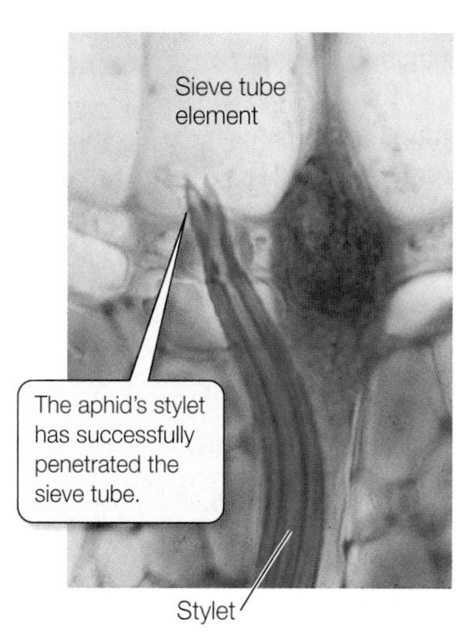

35.12 Aphids Collect Sap Aphids feed on sap drawn from a sieve tube, which they penetrate with a modified feeding organ, the stylet. Pressure inside the sieve tube forces sap through the aphid's digestive tract, from which it can be harvested.

measuring how long it takes for radioactive tracers administered to a leaf to appear at stylets at different distances from the leaf.

These methods have allowed us to understand how, at times, different substances might move in opposite directions in the phloem of a stem. Experiments with aphid stylets have shown that all the contents of any given sieve tube element move in the same direction. Thus, bidirectional translocation can be understood in terms of different sieve tubes conducting sap in opposite directions. These and other experiments led to the general adoption of the *pressure flow model* as an explanation for translocation in the phloem.

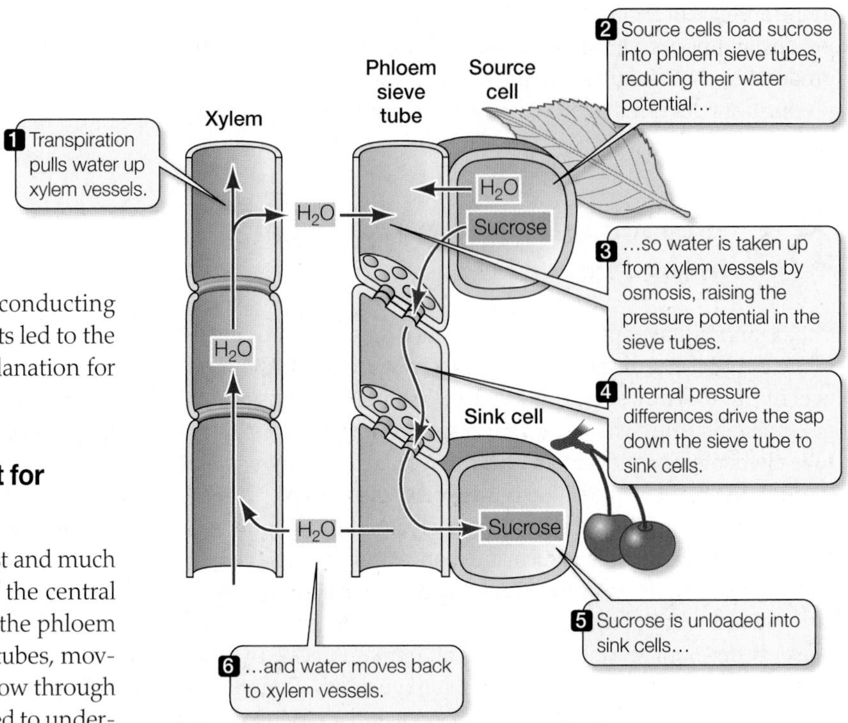

35.13 The Pressure Flow Model Combined pressure potential and water potential differences drive the bulk flow of phloem sap from a source to a sink.

The pressure flow model appears to account for translocation in the phloem

During sieve tube element development, the tonoplast and much of the cytosol breaks down, allowing the contents of the central vacuole to combine with much of the cytosol to form the phloem sap. The sap flows under pressure through the sieve tubes, moving from one sieve tube element to the next by bulk flow through the sieve plates, without crossing a membrane. We need to understand how this pressure is generated in order to understand translocation in the phloem.

Two steps in translocation require metabolic energy:

- Transport of sucrose and other solutes from sources into the sieve tubes, called *loading*
- Removal of the solutes from the sieve tubes into sinks, called *unloading*

According to the **pressure flow model** of translocation in the phloem, sucrose is actively transported into sieve tube elements at a source, giving those cells a greater sucrose concentration than the surrounding cells. Water therefore enters the sieve tube elements by osmosis. The entry of this water causes a greater pressure potential at the source end of the sieve tube, so that the entire fluid content of the sieve tube is pushed toward the sink end of the tube—in other words, the sap moves by bulk flow in response to a pressure gradient (**Figure 35.13**). In the sink, the sucrose is unloaded by active transport, maintaining the gradients of solute potential and water potential needed for movement.

The pressure flow model of translocation in the phloem is contrasted with the transpiration–cohesion–tension model of xylem transport in **Table 35.1**.

The pressure flow model has been experimentally tested

The pressure flow model was first proposed more than half a century ago, but some of its features are still being debated. Other mechanisms have been proposed to account for translocation in sieve tubes, but some have been disproved, and no other theory has as much support as the pressure flow model.

Two essential requirements must be met in order for the pressure flow model to be valid:

- The sieve plates must be unobstructed, so that bulk flow from one sieve tube element to the next is possible.
- There must be an effective method for loading sucrose and other solutes into the phloem in source tissues and removing them in sink tissues.

Let's see whether these requirements are met.

TABLE 35.1

Mechanisms of Sap Flow in Plant Vascular Tissues

	XYLEM	PHLOEM
Driving force for bulk flow	Transpiration from leaves	Active transport of sucrose at source
Site of bulk flow	Non-living vessel elements and tracheids (cohesion)	Living sieve tube elements
Pressure potential in sap	Negative (pull from top; tension)	Positive (push from source; pressure)

ARE THE SIEVE PLATES OPEN? Early electron microscopic studies of phloem samples cut from plants seemed to contradict the pressure flow model. The pores in the sieve plates always appeared to be plugged with masses of a fibrous protein, suggesting that sieve tube sap could not flow freely. The function of that fibrous protein was unclear, however. What purpose would it serve at the sieve plates?

Researchers contemplated the possibility that this protein functions to block the sieve plates when sieve tube elements have been damaged. Perhaps the protein is usually distributed more or less at random throughout the sieve tube elements but damage propels a sudden surge of sap toward the cut surface, carrying the protein into the pores, blocking them and preventing the loss of valuable nutrients. How might this hypothesis be tested? How might phloem for microscopic observation be obtained without causing the sap to surge to the cut surface?

One way to prevent the surge of the sap is to freeze plant tissue rapidly before cutting it. Another way is to let the tissue wilt so that no pressure exists in the phloem before cutting. If these methods are utilized, the sieve plates are not clogged by the protein, indicating that sieve tubes are unobstructed unless damaged. Thus, the first condition of the pressure flow model is met.

HOW ARE SIEVE TUBE ELEMENTS LOADED AND UNLOADED? If the pressure flow model is correct, there must be mechanisms for loading sugars and other solutes into the phloem in source regions and for unloading them in sink regions.

Sugars and other solutes produced in the mesophyll pass from cell to cell in the leaf and eventually enter the sieve tubes of the phloem. In some plants these substances leave the mesophyll cells and enter the apoplast. Then specific sugars and amino acids are actively transported into cells of the phloem, thus reentering the symplast. This passage through the apoplast and back into the symplast allows the selection of substances to be translocated by forcing them to pass through selectively permeable membranes. In many plants, solutes reenter the symplast at the sieve tube elements' companion cells, which then transfer the solutes to adjacent sieve tube elements.

A form of secondary active transport (see Section 5.4) loads sucrose into the companion cells and sieve tubes. Sucrose is carried across the plasma membrane from apoplast to symplast by sucrose–proton symport; thus the entry of sucrose and protons is strictly coupled. For this symport to work, the apoplast must have a high concentration of protons; these protons are supplied by a primary active transport system, the proton pump. The protons then diffuse back into the cell through the symport protein, bringing sucrose with them.

In sink regions, the solutes are actively transported *out* of the sieve tube elements and into the surrounding tissues. This unloading serves two purposes: It helps to maintain the gradient of solute potential and hence of pressure potential in the sieve tubes, and it promotes the buildup of sugars and starch to high concentrations in storage regions, such as developing fruits and seeds. Thus the second requirement of the pressure flow model is met, and the model is supported.

Plasmodesmata allow the transfer of material between cells

Many substances move from cell to cell within the symplast by way of plasmodesmata (see Figures 34.8 and 35.4). Among their other roles, plasmodesmata participate in the loading and unloading of sieve tube elements. The mechanisms vary among plant species, but the story in tobacco plants is a common one. In tobacco, sugars and other solutes in source tissues enter companion cells by active transport from the apoplast and move on to the sieve tube elements through plasmodesmata. In sink tissues, plasmodesmata connect sieve tube elements, companion cells, and the cells that will receive and use the transported compounds.

Plasmodesmata undergo developmental changes as an immature sink leaf becomes a mature source leaf. Plasmodesmata in sink tissues favor rapid unloading: They are more abundant, and they allow the passage of larger molecules. Plasmodesmata in source tissues are fewer in number.

It was long thought that only substances with molecular weights of less than 1 kilodalton could fit through a plasmodesma. Then biologists discovered that cells infected with tobacco mosaic virus (TMV) could allow molecules with molecular weights as great as 20 kDa to exit. We now know that TMV encodes a "movement protein" that produces this change in the permeability of the plasmodesmata—and that the plants themselves normally produce at least one such movement protein. Even large molecules such as proteins and RNAs, with molecular weights up to at least 50,000, can move between living plant cells. We will see some consequences of this movement of macromolecules through plasmodesmata in later chapters. Biologists are exploring possible ways to regulate the permeability, number, and form of plasmodesmata as a means of modifying traffic in the plant. Such modifications might, for example, allow the diversion of more of a grain crop's photosynthetic products into the seeds, increasing the crop yield.

35.4 RECAP

Carbohydrates produced by photosynthesis are translocated from source to sink via the phloem by a pressure flow mechanism.

- Can you distinguish between a source and a sink? See p. 775
- How does loading of sucrose at the source result in bulk flow toward the sink according to the pressure flow model? See p. 776 and Figure 35.13

CHAPTER SUMMARY

35.1 How do plant cells take up water and solutes?

Water moves through biological membranes by osmosis, always moving toward cells with a more negative water potential. The **water potential** (ψ) of a cell or solution is the sum of the **solute potential** and the **pressure potential**. Review Figure 35.2

The movement of a solution due to a difference in pressure potential between two parts of a plant is called **bulk flow**.

Aquaporins allow water molecules to pass through biological membranes without interacting with the hydrophobic interior of the membrane.

Mineral uptake requires transport proteins. Some minerals enter the plant passively by facilitated diffusion; others enter by active transport. A **proton pump** provides energy for the active transport of many mineral ions across membranes in plants. Review Figure 35.3

Water and minerals pass from the soil to the xylem by way of the **apoplast** and **symplast**. In the root, water and minerals can move from the cortex into the stele only by way of the symplast because **Casparian strips** in the endodermis block their movement through the apoplast. Review Figure 35.4, Web/CD Activity 35.1

35.2 How are water and minerals transported in the xylem?

Root pressure is responsible for **guttation** and for the oozing of sap from cut stumps, but it cannot account for the ascent of xylem sap in trees.

Water transport in the xylem results from the combined effects of **transpiration**, **cohesion**, and **tension**. Evaporation from the leaf produces tension in the mesophyll cells, which pulls a column of water—held together by cohesion—up through the xylem from the root. Dissolved minerals are carried passively in the water. Review Figure 35.6

Transport in the xylem is by bulk flow. It does not require the expenditure of energy.

35.3 How do stomata control the loss of water and the uptake of CO_2?

Stomata allow a compromise between water retention and carbon dioxide uptake.

A pair of **guard cells** controls the size of the stomatal opening. A proton pump, activated by blue light, pumps protons out of the guard cells to the walls of surrounding epidermal cells, setting up a proton gradient that drives the active transport of potassium ions into the cells. Water follows osmotically, swelling the cells and opening the stomata. Review Figure 35.9

When threatened by dehydration, mesophyll cells release abscisic acid, which causes guard cells to close the stomata.

Antitranspirants may be used to reduce transpiration rates in some applications but are not used for crop plants.

35.4 How are substances translocated in the phloem?

Products of photosynthesis, as well as some minerals, are translocated through sieve tubes in the phloem by way of living sieve tube elements.

Translocation in the phloem can proceed in both directions in the stem, although in a single sieve tube it goes only one way. Translocation requires a supply of ATP.

Translocation in the phloem is explained by the **pressure flow model**: The difference in solute concentration between **sources** and **sinks** creates a difference in (positive) pressure potential along the sieve tubes, resulting in bulk flow. Review Figure 35.13 and Table 35.1, Web/CD Tutorial 35.1

SELF-QUIZ

1. Osmosis
   a. requires ATP.
   b. results in the bursting of plant cells placed in pure water.
   c. can cause a cell to become turgid.
   d. is independent of solute concentrations.
   e. continues until the pressure potential equals the water potential.
2. Water potential
   a. is the difference between the solute potential and the pressure potential.
   b. is analogous to the air pressure in an automobile tire.
   c. is the movement of water through a membrane.
   d. determines the direction of water movement between cells.
   e. is defined as 1.0 MPa for pure water under no applied pressure.
3. Which statement about aquaporins is *not* true?
   a. They are membrane transport proteins.
   b. Water movement through aquaporins is always active.
   c. The permeability of some aquaporins is subject to regulation.
   d. They are found in both animals and plants.
   e. They enable water to pass through the phospholipid bilayer without encountering a hydrophobic environment.
4. Which statement about proton pumping across the plasma membrane of plants is *not* true?
   a. It requires ATP.
   b. The region inside the membrane becomes positively charged with respect to the region outside.
   c. It enhances the movement of K^+ ions into the cell.
   d. It pushes protons out of the cell against a proton concentration gradient.
   e. It can drive the secondary active transport of negatively charged ions.
5. Which statement is *not* true?
   a. The symplast is a meshwork consisting of the (connected) living cells.
   b. Water can enter the stele without entering the symplast.
   c. The Casparian strips prevent water from moving between endodermal cells.
   d. The endodermis is a cell layer in the cortex.
   e. Water can move freely in the apoplast without entering cells.
6. In the xylem,
   a. the products of photosynthesis travel down the stem.
   b. living, pumping cells push the sap upward.
   c. the driving force is in the roots.
   d. the sap is often under tension.
   e. the sap must pass through sieve plates.

7. Which of the following is *not* part of the transpiration–cohesion–tension mechanism?
   a. Water evaporates from the walls of mesophyll cells.
   b. Removal of water from the xylem exerts a pull on the water column.
   c. Water is remarkably cohesive.
   d. The wider the tube, the greater the tension its water column can withstand.
   e. At each step, water moves to a region with a more strongly negative water potential.
8. Stomata
   a. control the opening of guard cells.
   b. release less water to the environment than do other parts of the epidermis.
   c. are usually most abundant on the upper epidermis of a leaf.
   d. are covered by a waxy cuticle.
   e. close when water is being lost at too great a rate.
9. Which statement about phloem transport is *not* true?
   a. It takes place in sieve tubes.
   b. It depends on mechanisms for loading solutes into the phloem in sources.
   c. It stops if the phloem is killed by heat.
   d. A high pressure potential is maintained in the sieve tubes.
   e. In sinks, solutes are actively transported into sieve tube elements.
10. The fibrous protein in sieve tube elements
    a. may plug leaks when a plant is damaged.
    b. clogs the sieve plates at all times.
    c. never clogs the sieve plates.
    d. serves no known function.
    e. provides the driving force for transport in the phloem.

FOR DISCUSSION

1. Epidermal cells protect against excess water loss. How do they perform this function? What differences might you expect to find in the structure of the epidermis in stems, roots, and leaves?
2. Phloem transports material from sources to sinks. Give examples of each. How might the distribution of sources and sinks change in the course of a year?
3. What is the minimum number of plasma membranes a water molecule would have to cross in order to get from the soil solution to the atmosphere by way of the stele? To get from the soil solution to a mesophyll cell in a leaf?
4. Transpiration exerts a powerful pulling force on the water column in the xylem. When would you expect transpiration to proceed most rapidly? Why? Describe the source of the pulling force.

FOR INVESTIGATION

The experiments by Zwieniecki and colleagues showed that the flow rate in the xylem of the tobacco stem increased with increasing K^+ content. It has been argued that this effect, while interesting, may not be relevant to the xylem of plants in the field. How might you try to resolve this uncertainty?

CHAPTER 36 Plant Nutrition

When the land blew away

The world has known many natural disasters: plagues, famines, wildfires, tsunamis, hurricanes, floods, droughts, and earthquakes. One of the greatest disasters of the twentieth century occurred in North America during the 1930s when a prolonged drought, combined with a culturally modified landscape, turned the central plains of the continent into the Dust Bowl.

The native vegetation in the Plains States in the nineteenth century was grass—long grass in the east and short grass in the west. Cattlemen moved in to where their herds could graze their fill on a seemingly endless supply of food. But in fact, the supply wasn't endless. As one area was overgrazed, the cattle were moved on to new areas, leaving damaged soil in their wake. Settlers followed, "busted the sod" with plows, planted crops, and disrupted vegetation cycles that had gone on for centuries.

In the modern world, events in one location can have consequences far away. In 1914, the first year of World War I, the Turkish navy blocked the supply of Russian wheat to the rest of the world. The resulting shortages prompted farmers in the Plains States to seek a profit in wheat. They plowed both good and marginal soils and planted wheat, and still more wheat. But wheat requires more water than did the native grasses, and rainfall was irregular throughout the 1920s. In 1932 rainfall failed almost completely, not to return in good supply until 1939.

Without water, crops failed—if they even started to grow. The Plains States are windy states, and without plant roots there was nothing to hold the soil in place when the winds blew. The farms literally blew away. Farmers spent their last money to buy seed, hoping for a green year; but dry year followed dry year. The stone-broke farmers migrated westward, along with others whose livelihoods had depended on the farmers. Recall also that these events took place during the period of world economic history known as the Great Depression.

In the meantime, massive dust storms blew the desiccated topsoil eastward. One such storm in May 1934 blew soil all the way to the decks of ships in the Atlantic Ocean, depositing dust on desks in Washington, D.C. and darkening the midday sun

Dreams Disappeared in a Cloud of Dust This photograph of a family displaced by the Dust Bowl was taken by Dorothea Lange in the winter of 1936. The family had traveled to northern California looking for work as migrant farm labor. In Lange's words, "I saw and approached the hungry and desperate mother…I did not ask her name or her history. …She said that they had been living on frozen vegetables from the surrounding fields, and birds that the children killed. She had just sold the tires from her car to buy food."

The Problem Has Not Gone Away This photograph was taken near Fort Benton, Montana, in early May 2002. Drifting topsoil has rendered the plow useless on a field that in previous years would have been ready for spring planting. Unusually dry conditions combined with high, persistent winds turned this agricultural region into a present-day dust bowl.

in New York City. The worst of the many Dust Bowl storms occurred on "Black Sunday," April 14, 1935, creating a spectacular "black blizzard."

What did these black blizzards mean? They didn't just knock down fences and bury homes in the Dust Bowl while damaging crops to the east. They removed the mineral nutrients that once sustained the grasses of the prairies. The loss of nutrient-containing topsoil in the Dust Bowl left the region scarred for years in terms of its ability to sustain profitable agriculture.

This kind of disaster continues to strike all over the world. Today, parts of sub-Saharan Africa's farmland are losing their topsoil as a result of poor land management, swelling populations, and a challenging climate. Crop failures, starvation, and large-scale human displacements are inevitable consequences.

IN THIS CHAPTER we consider the conditions that foster healthy and sustained plant growth. We identify nutrients that are essential to plants and how they acquire them. Because most plant nutrients come from the soil, we discuss the formation of soils and the effects of plants on soils. We devote a section to nitrogen metabolism in plants, and we conclude with a look at carnivorous and parasitic plants.

CHAPTER OUTLINE

36.1 How Do Plants Acquire Nutrients?

Every living thing must obtain raw materials from its environment. These **nutrients** include the major ingredients of macromolecules: carbon, hydrogen, oxygen, and nitrogen. Carbon enters the living world as atmospheric carbon dioxide through the carbon-fixing reactions of photosynthesis. Hydrogen and oxygen enter plants mainly as water, so these elements are in plentiful supply.

Later in this chapter, we will focus on nitrogen, which is in relatively short supply for plants. The movement of nitrogen from the atmosphere into organisms begins with processing by some highly specialized bacteria living in the soil. Some of these bacteria act on nitrogen gas, converting it into a form usable by plants. The plants, in turn, provide organic nitrogen (and carbon) to animals, fungi, and many microorganisms.

Living organisms need certain **mineral nutrients** in addition to nitrogen, of course. The proteins of organisms contain sulfur (S), and their nucleic acids contain phosphorus (P). Chlorophyll contains magnesium (Mg), and many important compounds, such as the cytochromes, contain iron (Fe). Within the soil, these and other minerals dissolve in water, forming a solution—called the **soil solution**—that contacts the roots of plants. Plants take up most of these mineral nutrients from the soil solution in ionic form.

Autotrophs make their own organic compounds

Autotrophs are organisms that make their own *organic* (carbon-containing) compounds from simple inorganic nutrients—carbon dioxide, water, nitrogen-containing ions, and many other soluble mineral nutrients. Plants, some protists, and some bacteria are autotrophs. Plants provide carbon, oxygen, hydrogen, nitrogen, and sulfur to most of the rest of the living world. *Heterotrophs* are organisms, such as animals and fungi, that require pre-formed organic compounds as food. Heterotrophs depend directly or indirectly on autotrophs as their source of nutrition.

Most autotrophs are *photosynthesizers*—that is, they use light as their source of energy for synthesizing organic compounds from inorganic raw materials. Some autotrophs, however, are

chemolithotrophs, deriving their energy not from light but from reduced inorganic substances, such as hydrogen sulfide (H_2S), in their environment. All chemolithotrophs are bacteria. As we'll see below, some chemolithotrophic bacteria in the soil contribute to the nutrition of plants by increasing the availability of nitrogen and sulfur.

How does a stationary organism find nutrients?

Many heterotrophs can move from place to place to find the nutrients they need. An organism that cannot move, termed a *sessile* organism, must obtain nutrients and energy from sources that are somehow brought to it. Most sessile animals, such as clams and barnacles, depend primarily on the movement of water to bring them raw materials and energy in the form of food, but a plant's supply of energy arrives at the speed of light from the sun. However, with the exception of carbon and oxygen in CO_2, a plant's supply of nutrients is strictly local, and the plant may use up the water and mineral nutrients in its local environment as it develops. How does a plant cope with the problem of scarce nutrient supplies?

One way is to extend itself by growing in search of new resources. Growth is a plant's version of movement. Among plant organs, the roots obtain most of the mineral nutrients needed for growth. By growing through the soil, roots mine it for new sources of mineral nutrients and water that help leaves and stems grow. The growth of leaves helps a plant secure light and carbon dioxide, which in turn allows the roots to continue their growth in the soil. A plant may compete with other plants for light by outgrowing and shading them.

As it grows, a plant—or even a single root—must deal with a variable environment. Animal droppings create high local concentrations of nitrogen. A particle of calcium carbonate in the soil may make a tiny area alkaline, while dead organic matter may make a nearby area acidic. Such microenvironments encourage or discourage the proliferation of a root system and help direct its growth.

36.1 RECAP

Plants are autotrophs that obtain carbon by photosynthesis, and mineral nutrients and water from the soil.

- Can you distinguish between autotrophs and heterotrophs? See p. 781
- Can you explain how plants, being sessile, seek out nutrients? See p. 782

We know that plants need nutrients to support their growth. Let's look in more detail at the specific mineral nutrients they need.

36.2 What Mineral Nutrients Do Plants Require?

As roots grow through the soil, what important mineral nutrients do plants take up from their environment, and what are the roles of those nutrients? **Table 36.1** lists the mineral nutrients that have been determined to be essential for plants. Some plants require additional mineral nutrients. Except for nitrogen, they all derive from rock. All of them are usually taken up from the soil solution.

The criterion for calling something an **essential element** is that it be required for the plant to complete its life cycle. An essential element cannot be replaced by another element. In this section,

TABLE 36.1

Mineral Elements Required by Plants

ELEMENT	ABSORBED FORM	MAJOR FUNCTIONS
MACRONUTRIENTS		
Nitrogen (N)	NO_3^- and NH_4^+	In proteins, nucleic acids, etc.
Phosphorus (P)	$H_2PO_4^-$ and HPO_4^{2-}	In nucleic acids, ATP, phospholipids, etc.
Potassium (K)	K^+	Enzyme activation; water balance; ion balance; stomatal opening
Sulfur (S)	SO_4^{2-}	In proteins and coenzymes
Calcium (Ca)	Ca^{2+}	Affects the cytoskeleton, membranes, and many enzymes; second messenger
Magnesium (Mg)	Mg^{2+}	In chlorophyll; required by many enzymes; stabilizes ribosomes
MICRONUTRIENTS		
Iron (Fe)	Fe^{2+} and Fe^{3+}	In active site of many redox enzymes and electron carriers; chlorophyll synthesis
Chlorine (Cl)	Cl^-	Photosynthesis; ion balance
Manganese (Mn)	Mn^{2+}	Activation of many enzymes
Boron (B)	$B(OH)_3$	Possibly carbohydrate transport (poorly understood)
Zinc (Zn)	Zn^{2+}	Enzyme activation; auxin synthesis
Copper (Cu)	Cu^{2+}	In active site of many redox enzymes and electron carriers
Nickel (Ni)	Ni^{2+}	Activation of one enzyme
Molybdenum (Mo)	MoO_4^{2-}	Nitrate reduction

we'll consider the symptoms of particular mineral deficiencies, the roles of some of the mineral nutrients, and the technique by which the essential elements for plants were identified.

Essential elements fall roughly into two categories: *macronutrients* and *micronutrients* (see Table 36.1).

- Plants need **macronutrients** in concentrations of at least 1 gram per kilogram of their dry matter.*
- Plants need **micronutrients** in concentrations of less than 100 milligrams per kilogram of their dry matter.

These two categories differ only with regard to the amounts required by plants. Did you notice the gap between those numbers? The macronutrients are needed in much greater concentrations, but both the macronutrients and the micronutrients are essential for the plant to complete its life cycle from seed to seed. How do we know if a plant is getting enough of a particular nutrient?

Deficiency symptoms reveal inadequate nutrition

Before a plant that is deficient in an essential element dies, it usually displays characteristic **deficiency symptoms**, such as discoloration or deformation of its leaves. **Table 36.2** lists the symptoms of some common mineral deficiencies. Such symptoms help horticulturists diagnose mineral nutrient deficiencies in plants. With proper diagnosis, appropriate treatment can be applied in the form of a **fertilizer** (an added source of mineral nutrients).

Nitrogen deficiency is the most common mineral deficiency in both natural and agricultural environments. Plants in natural environments are almost always limited by nitrogen, but they seldom display deficiency symptoms. Instead, their growth slows to match the available supply of nitrogen. Crop plants, on the other hand, show deficiency symptoms if a formerly abundant supply of nitrogen becomes limiting. The visible symptoms of nitrogen deficiency include uniform yellowing of older leaves (**Figure 36.1, right**).

Dry matter, or *dry weight*, is what remains after all the water has been removed from a plant tissue sample.

TABLE 36.2

Some Mineral Deficiencies in Plants

DEFICIENCY	SYMPTOMS
Calcium	Growing points die back; young leaves are yellow and crinkly
Iron	Young leaves are white or yellow
Magnesium	Older leaves have yellow in stripes between veins
Manganese	Younger leaves are pale with green veins
Nitrogen	Oldest leaves turn yellow and die prematurely; plant is stunted
Phosphorus	Plant is dark green with purple veins and is stunted
Potassium	Older leaves have dead edges
Sulfur	Young leaves are yellow to white with yellow veins
Zinc	Young leaves are abnormally small; older leaves have many dead spots

Chlorophyll, which is responsible for the green color of leaves, contains nitrogen. Without nitrogen there is no chlorophyll, and without chlorophyll, the yellow carotenoid pigments in the leaves become visible.

Nitrogen deficiency is not the only cause of yellowing. Inadequate iron in the soil can also cause yellowing because iron, although it is not contained in the chlorophyll molecule, is required for chlorophyll synthesis. However, iron deficiency commonly causes yellowing of the *youngest* leaves (**Figure 36.1, center**). Nitrogen is readily translocated in the plant and can be redistributed from older tissues to younger tissues to favor their growth. Iron, on the other hand, cannot be readily redistributed. Younger tissues that are actively growing and synthesizing compounds needed for their growth show iron deficiency before older leaves, which have already completed their growth.

36.1 Mineral Deficiency Symptoms The plants on the left were grown with a full complement of essential nutrients. The center plants were deprived of iron, while the plants at the right are deficient in essential nitrogen.

Several essential elements fulfill multiple roles

Essential elements may play several different roles in plant cells—some structural, others catalytic. Magnesium, as we have mentioned, is a constituent of the chlorophyll molecule and hence is essential to photosynthesis. It is also required as a cofactor by numerous enzymes involved in cellular respiration and other metabolic pathways.

Phosphorus, usually in phosphate functional groups, is found in many organic compounds, particularly in nucleic acids and in the intermediates of the energy-harvesting pathways of photosynthesis and glycolysis. The transfer of phosphate groups occurs in many energy-storing and energy-releasing reactions, notably those that use or produce ATP (see Chapter 7). Phosphorylation—the addition or removal of phosphate groups—is also used to activate or inactivate enzymes (see Chapter 15).

Calcium plays many roles in plants. Its function in the processing of hormonal and environmental cues is a subject of great biological interest, as we'll see in the next chapter. Calcium also affects membranes and cytoskeletal activity, participates in spindle formation for mitosis and meiosis, and is a constituent of the middle lamella of cell walls. Other elements, such as iron and potassium, also play multiple roles in plants.

We know that *all* of these elements are essential to the life of all plants; how did biologists discover which elements are essential?

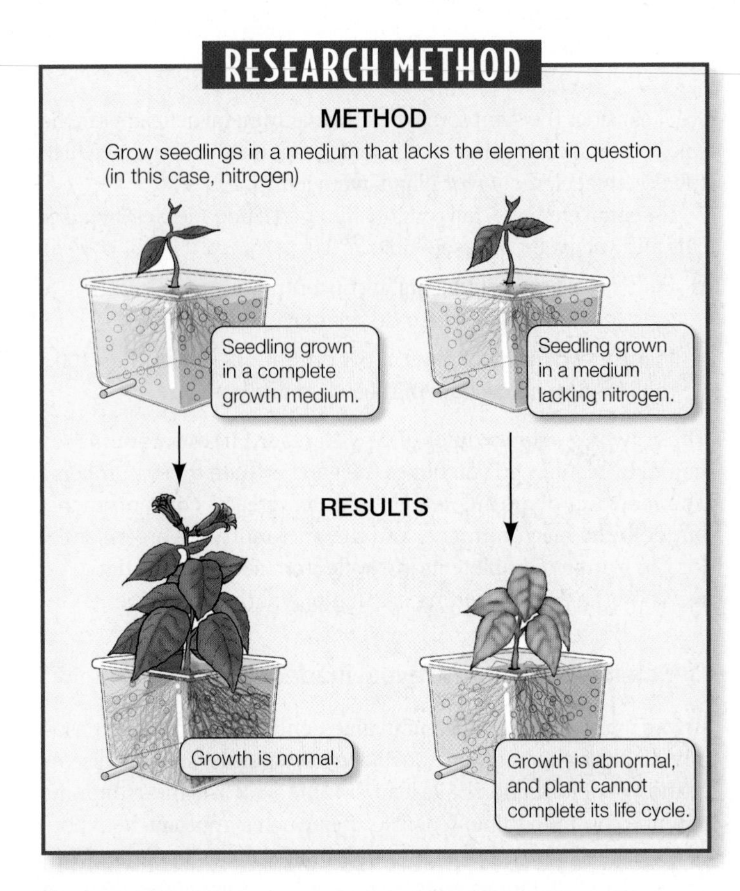

36.2 Identifying Essential Elements for Plants This figure illustrates the classic procedure for identifying nutrients essential to plants, using nitrogen as an example.

Experiments were designed to identify essential elements

An element is considered essential to plants if a plant fails to complete its life cycle or grows abnormally when that element is absent or insufficient. The essential elements for plants were identified by growing plants **hydroponically**—that is, with their roots suspended in nutrient solutions without soil (**Figure 36.2**). In the first successful experiments of this type, performed a century and a half ago, plants grew seemingly normally in solutions containing only calcium nitrate [$Ca(NO_3)_2$], magnesium sulfate [$MgSO_4$], and potassium phosphate [KH_2PO_4]. Omission of any of these compounds resulted in a solution that was incapable of supporting normal growth. Tests with other compounds including these elements soon established six macronutrients—calcium, nitrogen, magnesium, sulfur, potassium, and phosphorus—as essential elements.

Identifying essential *micronutrients* by this experimental approach proved to be more difficult. Nineteenth-century experiments on plant nutrition sometimes used chemicals so impure that they provided micronutrients that investigators thought they had excluded. Furthermore, some micronutrients are required in such tiny amounts that there may be enough of the substance in a seed to supply the embryo and the resultant second-generation plant throughout its lifetime and leave enough in the next seed to get the third generation well started. Such difficulties make it necessary to perform nutrition experiments in tightly controlled laboratories with special air filters that exclude microscopic salt particles in the air, and to use only chemicals that have been purified to the highest degree attainable by modern chemistry.

Even simply touching a plant may give it a significant supply of chlorine, in the form of chloride ions from human sweat.

Iron was the first micronutrient to be clearly established as essential, in the 1840s. The last micronutrient to be listed as essential was nickel, in 1983 (the experiment is described in **Figure 36.3**). Only rarely are new essential elements reported now. Either the list is complete or, perhaps, we will need more sophisticated techniques to add to it.

36.2 RECAP

Mineral nutrients required by plants are classified as macronutrients and micronutrients, depending on the amount needed. Micronutrients are often needed in such minute amounts that only sophisticated chemical experiments can determine their essentiality.

- Can you describe some specific deficiency symptoms seen in plants? See p. 783 and Table 36.2
- Can you outline an experimental method for determining whether an element is essential to a plant? See Figures 36.2 and 36.3

EXPERIMENT

HYPOTHESIS: Nickel is an essential element for a plant to complete its life cycle.

METHOD

1. Grow barley plants for 3 generations in nutrient solutions containing 0, 0.6, and 1.0 μM $NiSO_4$.
2. Harvest seeds from 5–6 third-generation plants in each of the groups.
3. Determine the nickel concentration in seeds from each plant.
4. Germinate other seeds from the same plants and plot the success of germination against nickel concentration.

RESULTS

There was a positive correlation between seed germination and seed nickel concentration. There was no germination at the lowest nickel concentration.

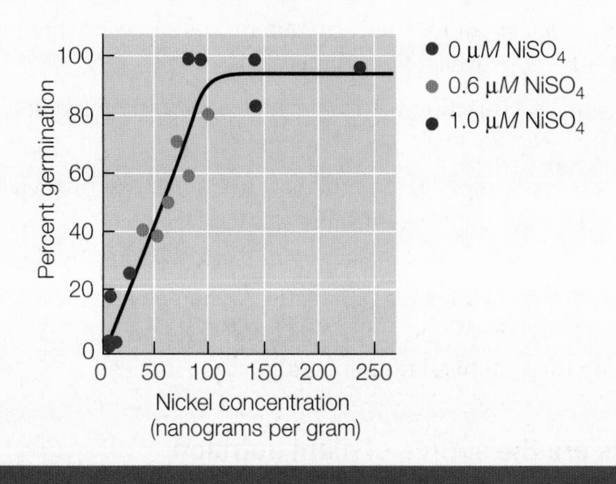

CONCLUSION: Barley seeds from nickel-free plants require nickel in order to germinate, and thereby complete the life cycle.

36.3 Is Nickel an Essential Element for Plant Growth? Using highly purified salts in the growth media, Patrick Brown and his colleagues tested whether barley can complete its life cycle in the absence of nickel. Other investigators showed that no other element could substitute for nickel.

When humans are not adding fertilizer or otherwise "feeding" a plant its mineral nutrients, where does its mineral nutrition come from? The answer lies in the minerals found in the soil, but nutrition is not the only role the soil plays.

36.3 What Are the Roles of Soil?

Most terrestrial plants live their lives anchored to the soil. Soils give growing plants mechanical support, of course, but soils and plants interact in many other noteworthy ways. Plants obtain their mineral nutrients from the soil solution. Water for terrestrial plants also comes from the soil, as does the supply of oxygen for the roots. Soils harbor bacteria, some of which are beneficial to plant life; they may also contain organisms that are harmful to plants, and some contain toxic levels of metal ions such as cadmium, chromium, and lead (see Chapter 39).

In this section, we examine the composition, structure, and formation of soils. We will consider their role in plant nutrition, their care and supplementation in agriculture, and how they are modified in turn by the plants that grow in them.

Soils are complex in structure

Soils are complex systems of living and nonliving components (**Figure 36.4**). The living components include plant roots as well as populations of bacteria, fungi, protists, and animals such as earthworms and insects. The nonliving portion of the soil includes rock fragments ranging in size from large rocks and pebbles through *sand* and *silt* and finally to tiny particles of *clay* that are 2 μm or less in diameter. Soil also contains water and dissolved mineral nutrients, air spaces, and dead organic matter. The air spaces are crucial sources of oxygen (in the form of O_2) for plant roots.

The characteristics of soils are not static. Soils change constantly through the effects of natural phenomena such as rain, extremes of temperature, and the activities of plants and animals, as well as the practices of humans—agriculture in particular.

The structure of many soils changes with depth, revealing a *soil profile*. Although soils differ greatly, almost all soils consist of sev-

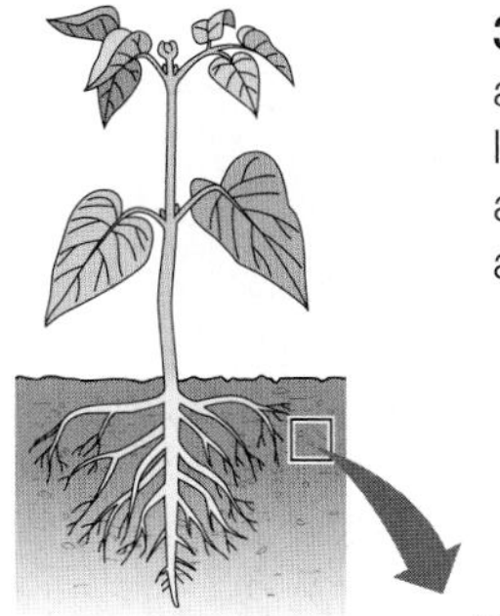

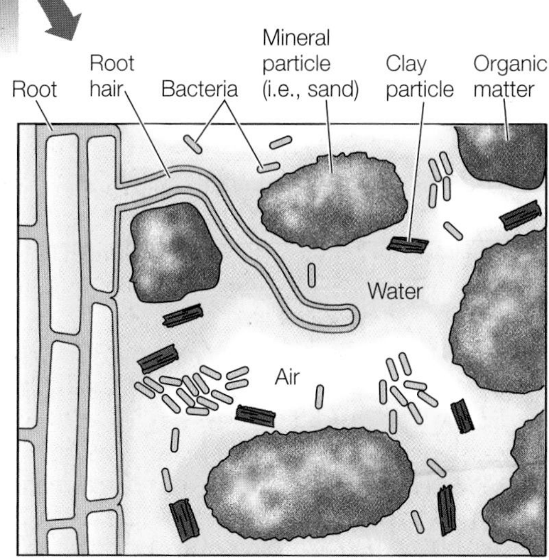

36.4 The Complexity of Soil Soils favorable for plant growth contain both clay and larger mineral particles as well as water, air, and organic matter. Other organisms are also present.

eral recognizable horizontal layers, called **horizons**, lying on top of one another. Mineral nutrients tend to be **leached** from the upper horizons—dissolved in rain or irrigation water and carried to deeper horizons, where they are unavailable to plant roots.

Soil scientists recognize three major *horizons*—termed A, B, and C—in the profile of a typical soil (**Figure 36.5**). **Topsoil** is the **A horizon**, from which mineral nutrients may be depleted by leaching. Most of the soil's dead and decaying organic matter is located in the A horizon, as are most plant roots, earthworms, insects, nematodes, and microorganisms. Successful agriculture depends on the presence of a suitable A horizon; the A horizon is what blew away from the U.S. plains states during the Dust Bowl.

Topsoils are composed of different proportions of sand, silt, and clay. Pure sand contains abundant air spaces between the relatively large particles, but binds little water and does not make mineral nutrients easily available to plant roots. Clay binds more water than sand does, but the tiny clay particles pack tightly together, leaving little space to trap air. As we will see, the charged surfaces of clay particles bind mineral element ions and make them available to plants. A little bit of clay goes a long way in affecting soil properties. A **loam** is a soil that has significant amounts of sand, silt, and clay, and thus has sufficient levels of air, water, and available nutrients for plants. Loams also contain organic matter. Most of the best topsoils for agriculture are loams.

Below the A horizon is the **B horizon**, or **subsoil**, which is the zone of infiltration and accumulation of materials leached from above. Farther down, the **C horizon** is the **parent rock**, also called bedrock, that is breaking down to form soil. Some deep-growing roots extend into the B horizon to obtain water and nutrients, but roots rarely enter the C horizon.

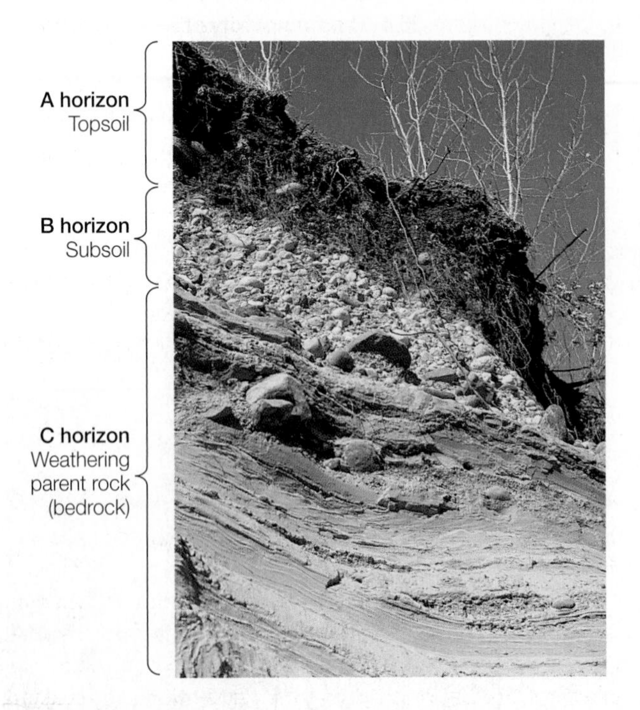

36.5 A Soil Profile The A, B, and C horizons can sometimes be seen in road cuts such as this one in Australia. The dark upper layer (the A horizon) is home to most of the living organisms in the soil.

Soils form through the weathering of rock

The type of soil in a given area depends on many factors, including the type of parent rock from which it formed, the climate, the landscape features, the organisms living there, and the length of time that soil-forming processes have been acting (sometimes millions of years). Both the physical and chemical properties of soils depend to a considerable extent on the amounts and kinds of **clay** particles they contain. These tiny particles, which bind mineral nutrients and aggregate into larger particles (see Figure 36.4), are extremely important to plant growth.

Rocks are broken down into soil particles in part by *mechanical weathering*, which is the physical breakdown—without any accompanying chemical changes—of materials by wetting, drying, and freezing. The most important parts of soil formation, however, include *chemical weathering*, the chemical alteration of at least some of the materials in the rocks. Several types of chemical weathering are required:

- *Oxidation* by atmospheric oxygen makes some essential elements more available to plants.
- *Hydrolysis* (reaction with water) releases some mineral nutrients.
- *Acids* (carbonic acid in particular) free some essential elements from their parent salts.

These chemical weathering reactions leave the surface of clay particles with an abundance of negatively charged chemical groups, to which certain mineral nutrients bind. Let's see how plant roots take up these mineral nutrients from clay particles.

Soils are the source of plant nutrition

The availability of mineral nutrients to plant roots depends on the presence of clay particles in the soil. The negatively charged clay particles bind the positively charged ions (cations) of many minerals that are important for plant nutrition, such as potassium (K^+), magnesium (Mg^{2+}), and calcium (Ca^{2+}). Ammonium ions (NH_4^+), a major form of nitrogen, are also bound by clay. To become available to plants, these cations must be detached from the clay particles.

Plants acquire important cations through reactions generated by protons (hydrogen ions, H^+). These protons are released into the soil by roots, which also release CO_2 through cellular respiration. The CO_2 dissolves in the soil water and reacts with it to form carbonic acid, which then ionizes to form bicarbonate and free protons:

$$CO_2 + H_2O \rightleftharpoons H_2CO_3 \rightleftharpoons H^+ + HCO_3^-$$

These protons bind more strongly to the clay particles than do the mineral cations; in essence, they trade places with the cations in a process called **ion exchange** (**Figure 36.6**). Ion exchange puts important cations back into the soil solution, from which they are taken up by the roots. The capacity of a soil to support plant growth, called *soil fertility*, is determined in part by its ability to provide nutrients in this manner.

Clay particles effectively hold and exchange cations, and cations tend to be retained in the A horizon. However, there is no compa-

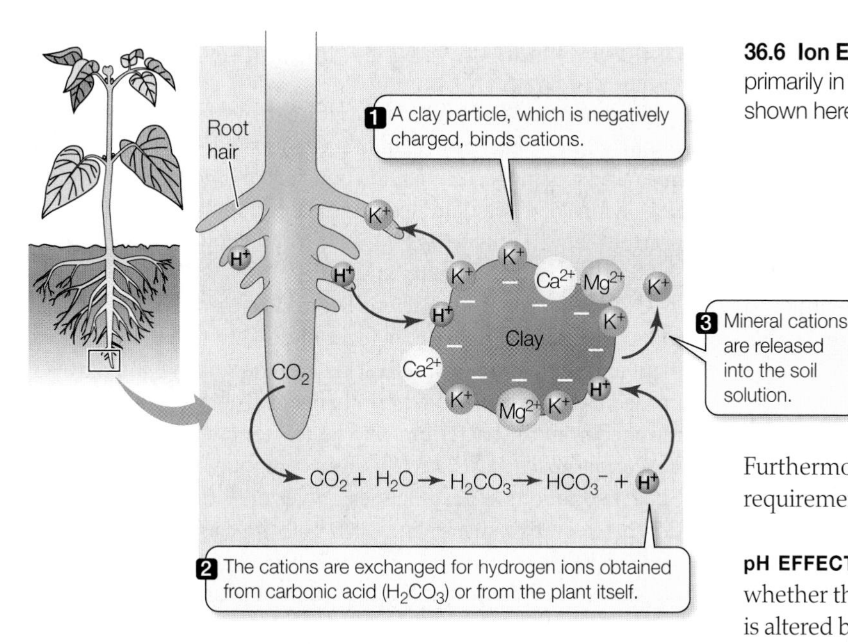

36.6 Ion Exchange Plants obtain mineral nutrients from the soil primarily in the form of positive ions; potassium (K^+) is the example shown here.

rable mechanism for exchanging anions, the negatively charged ions. As a result, important anions such as nitrate (NO_3^-) and sulfate (SO_4^{2-})—direct sources of nitrogen and sulfur, respectively—may leach rapidly from the A horizon. As a consequence of such leaching, the primary soil reservoir of nitrogen is not always in the form of nitrate ions. Most of the nitrogen in the A horizon is found in the organic matter in the soil, which slowly decomposes to release nitrogen as ammonium ions, a form that can be absorbed and used by plants.

Fertilizers and lime are used in agriculture

Agricultural soils often require fertilizers because irrigation and rainwater leach mineral nutrients from the soil and because the harvesting of crops removes the nutrients that the plants took up from the soil during their growth. Crop yields decrease if any essential element is depleted. Mineral nutrients may be replaced by organic fertilizers, such as rotted manure, or by inorganic fertilizers of various types.

ORGANIC AND INORGANIC FERTILIZERS The three elements most commonly added to agricultural soils are nitrogen (N), phosphorus (P), and potassium (K). Commercial fertilizers are characterized by their "N-P-K" percentages. A 5-10-10 fertilizer, for example, contains 5 percent nitrogen, 10 percent phosphate (P_2O_5), and 10 percent potash (K_2O) by weight.* Sulfur, in the form of ammonium sulfate, is also occasionally added to soils.

Either organic or inorganic fertilizers can provide the necessary mineral nutrients for plants. Organic fertilizers such as manure or crop residues release nutrients slowly, which results in less leaching than occurs with a one-time application of an inorganic fertilizer. However, the nutrients from organic fertilizers are not immediately available to plants. Organic fertilizers also contain residues of plant or animal materials that improve the structure of the soil, providing spaces for air movement, root growth, and drainage. Inorganic fertilizers, on the other hand, provide a supply of soil nutrients that is almost immediately available for absorption. Furthermore, inorganic fertilizers can be formulated to meet the requirements of a particular soil and a particular crop.

PH EFFECTS ON NUTRIENTS The availability of nutrient ions, whether they are naturally present in the soil or added as fertilizer, is altered by changes in soil pH. The optimal soil pH for most crops is about 6.5, but so-called acid-loving crops such as blueberries prefer a pH closer to 4. Rainfall and the decomposition of organic substances lower the pH of the soil. Such acidification can be reversed by **liming**—the application of compounds commonly known as *lime*, such as calcium carbonate, calcium hydroxide, or magnesium carbonate. The addition of these compounds leads to the removal of H^+ ions from the soil. Liming also increases the availability of calcium to plants.

Sometimes, on the other hand, a soil is not acidic enough for a crop. In this case, sulfur can be added in the form of elemental sulfur, which soil bacteria convert to sulfuric acid. Iron and some other elements are more available to plants at a slightly acidic pH. Soil pH testing is useful for home gardens and lawns as well as for agriculture. The test results indicate what amendments should be made to the soil.

SPRAY APPLICATION OF NUTRIENTS Spraying leaves with a nutrient solution is another effective way to deliver some essential elements to growing plants. Plants take up more copper, iron, and manganese when these elements are applied as foliar (leaf) sprays than when they are added to the soil. Such foliar application of micronutrients is sometimes used in wheat production, but fertilizers are still delivered most commonly by way of the soil.

The relationship between plants and soils is not a one-way affair—soils affect plants, but plants also affect soils.

Plants affect soil fertility and pH

The soil that forms in a particular place depends on the types of plants growing there as well as on mechanical weathering, the underlying parent rock, and other factors. Plant litter, such as dead roots and fallen leaves, is the major source of the carbon-rich materials that break down to form **humus**—dark-colored organic material, each particle of which is too small to be recognizable with the naked eye. Soil bacteria and fungi produce humus by breaking down plant litter, animal feces, dead organisms, and other organic material. Humus is rich in mineral nutrients, especially nitrogen that was excreted by animals. Humus favors plant growth by trap-

*The analysis is by weight of the nutrient-containing compound and not as weights of the elements N, P, and K. A 5-10-10 fertilizer actually does contain 5 percent nitrogen, but only 4.3 percent phosphorus and 8.3 percent potassium on an elemental basis.

ping supplies of water and oxygen for absorption by roots. Looking at the big picture, we see that successful plant growth can help create conditions that support further plant growth.

Plants also affect the pH of the soil in which they grow. Roots maintain a balance of electric charges. If they absorb more cations than anions, they excrete H^+ ions, thus lowering the soil pH. If they absorb more anions than cations, they excrete OH^- ions or HCO_3^- ions, raising the soil pH. Roots can also actively change the pH in their immediate vicinity by exuding organic acids such as citric and malic acids that acidify the soil, making it easier to take up certain ions such as Fe^{3+}.

36.3 RECAP

Land plants live anchored in the soil and obtain water and mineral nutrients from it. Plants and soil interact in many ways, and plants affect the soils in which they grow.

- Can you name types of mechanical and chemical weathering that form soil from rock? See p. 786
- Can you explain how soil fertility is enhanced by the process of ion exchange? See p. 786 and Figure 36.6

The essential mineral nutrient most commonly in short supply, in both natural and agricultural situations, is nitrogen, even though elemental nitrogen makes up almost four-fifths of Earth's atmosphere. Why is nitrogen so scarce in soil, and how do plants acquire it?

36.4 How Does Nitrogen Get from Air to Plant Cells?

The Earth's atmosphere is a vast reservoir of nitrogen in the form of nitrogen gas (N_2). However, plants cannot use N_2 directly as a nutrient. The triple bond linking the two nitrogen atoms is extremely stable, and a great deal of energy is required to break it; thus N_2 is a highly unreactive substance. How, then, do plants obtain usable nitrogen for the synthesis of proteins and nucleic acids?

A few species of bacteria have an enzyme that enables them to convert N_2 into a more reactive and biologically useful form by a process called **nitrogen fixation**. These prokaryotic organisms—*nitrogen fixers*—convert N_2 to ammonia (NH_3). Although there are relatively few species of nitrogen fixers, and their biomass is small compared to that of other organisms that depend on them, these talented prokaryotes are essential to the biosphere as we know it.

Nitrogen fixers make all other life possible

By far the greatest share of total world nitrogen fixation is performed biologically by **nitrogen-fixing bacteria**, which fix approximately 170 million metric tons of nitrogen per year. About 80 million metric tons is fixed industrially by humans. Smaller amounts of nitrogen, about 20 million metric tons per year, are fixed in the atmosphere by nonbiological means such as lightning, volcanic eruptions, and forest fires. Rain brings these atmospherically formed products to the ground.

Several groups of bacteria fix nitrogen. In the oceans, various photosynthetic bacteria, including cyanobacteria, fix nitrogen. In fresh water, cyanobacteria are the principal nitrogen fixers. On land, free-living soil bacteria make some contribution to nitrogen fixation, but they fix only what they need for their own use and release the fixed nitrogen only when they die.

Other nitrogen-fixing bacteria live in close association with plant roots. They release up to 90 percent of the nitrogen they fix to the plant and excrete some amino acids into the soil, making nitrogen immediately available to other organisms. The plant obtains fixed nitrogen from the bacterium, and the bacterium obtains the products of photosynthesis, high-energy compounds, from the plant. Such associations are excellent examples of *mutualism*, an interaction between two species in which both species benefit. They are also examples of *symbiosis*, in which two different species live in physical contact for a significant portion of their life cycles.

Bacteria of the genus ***Rhizobium*** fix nitrogen in close, mutualistic association with the roots of plants in the legume family. The legumes include peas, soybeans, clover, alfalfa, and many tropical shrubs and trees. The bacteria infect the plant's roots, and the roots develop nodules in response to their presence (**Figure 36.7**). (How nodules develop is further described below.) The various species of *Rhizobium* show a high specificity for the species of legume they infect. Farmers and gardeners coat legume seeds with *Rhizobium* to make sure the bacteria are present. Some farmers alternate their crops, planting clover or alfalfa occasionally to increase the available nitrogen content of the soil.

The legume–*Rhizobium* association is not the only bacterial association that fixes nitrogen. Some cyanobacteria fix nitrogen in association with fungi in lichens or with ferns, cycads, or nonvascular plants. Rice farmers can increase crop yields by growing the water fern *Azolla*, with its symbiotic nitrogen-fixing cyanobacterium, in the flooded fields where rice is grown. Another group of bacteria, the filamentous actinobacteria, fix nitrogen in associ-

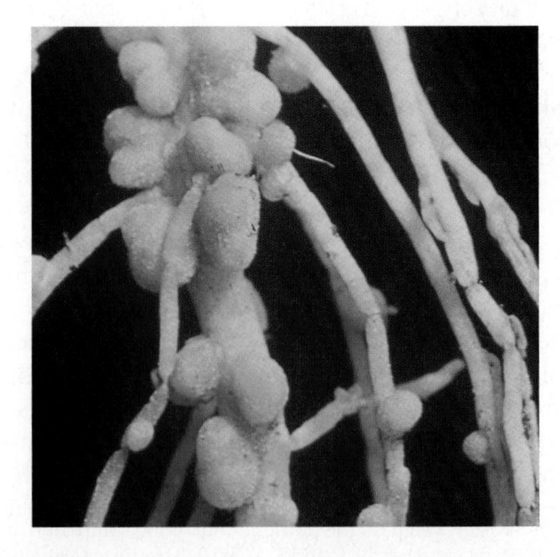

36.7 Root Nodules Large, round nodules are visible in the root system of a pea plant. These nodules house nitrogen-fixing bacteria.

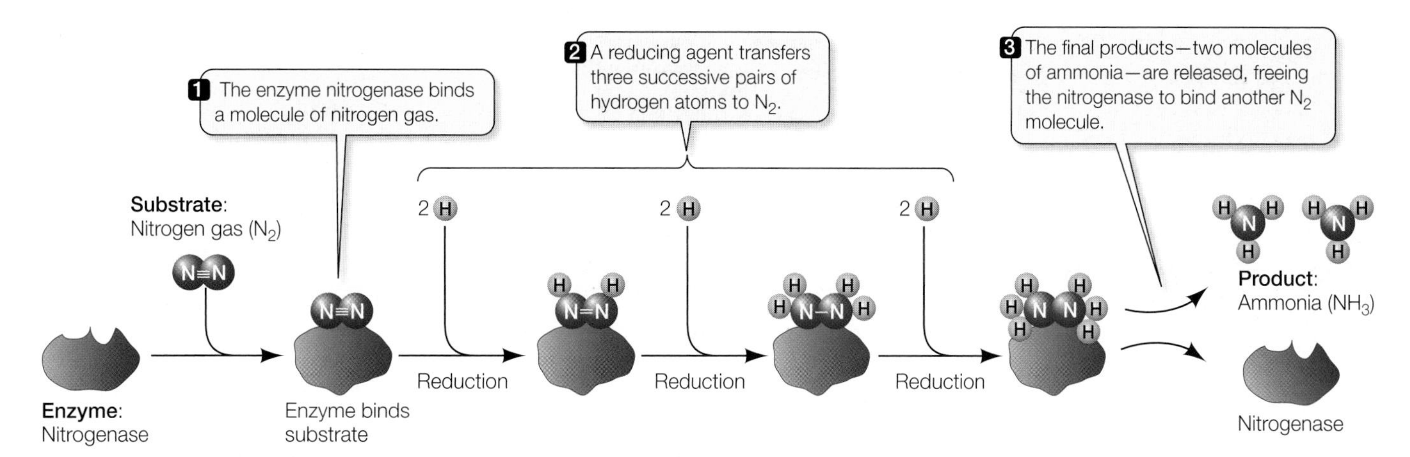

36.8 Nitrogenase Fixes Nitrogen Throughout the chemical reactions of nitrogen fixation, the reactants are bound to the enzyme nitrogenase. A reducing agent transfers hydrogen atoms to nitrogen, and eventually the final product—ammonia—is released.

ation with root nodules on woody species such as alder and mountain lilacs.

How does biological nitrogen fixation work? In the four sections that follow, we'll consider the role of the enzyme *nitrogenase*, the mutualistic collaboration of plant and bacterial cells in **root nodules**, the need to supplement biological nitrogen fixation in agriculture, and the contributions of plants and bacteria to the global nitrogen cycle.

Nitrogenase catalyzes nitrogen fixation

Nitrogen fixation is the *reduction* of nitrogen gas (see Section 7.1). It proceeds by the stepwise addition of three pairs of hydrogen atoms to N_2 (**Figure 36.8**). In addition to N_2, these reactions require three things:

- A strong reducing agent to transfer hydrogen atoms to N_2 and to the intermediate products of the reaction
- A great deal of energy, which is supplied by ATP
- The enzyme **nitrogenase**, which catalyzes the reaction

Depending on the species of nitrogen fixer, either respiration or photosynthesis may provide both the necessary reducing agent and ATP.

Nitrogenase is so strongly inhibited by oxygen that its presence in biochemical extracts was obscured and its discovery delayed because investigators had not thought to seek it under anaerobic conditions. It is therefore not surprising that many nitrogen fixers are anaerobes and live in environments with little or no O_2. Because this crucial enzyme is so inhibited by O_2, it was at first surprising that the reaction occurs in legumes, which respire aerobically, as do *Rhizobium*. Investigation of the root nodules where nitrogenase is found revealed how the enzyme could operate there.

Within a root nodule, O_2 is maintained at a low level sufficient to support respiration, but not so high as to inactivate nitrogenase. The plant makes this possible by producing the protein **leghemoglobin** in the cytoplasm of the nodule cells. Leghemoglobin is a close relative of hemoglobin, the red, oxygen-carrying pigment of animals. Some plant nodules contain enough of it to be bright pink when viewed in cross section. Leghemoglobin, with its iron-containing heme groups, transports enough oxygen to the *bacteroids* to support their respiration.

Some plants and bacteria work together to fix nitrogen

Neither free-living *Rhizobium* species nor uninfected legumes can fix nitrogen. Only when the two are closely associated in root nodules does the reaction take place. The establishment of this symbiosis between *Rhizobium* and a legume requires a complex series of steps, with active contributions by both the bacteria and the plant root (**Figure 36.9**). First the root releases flavonoids and other chemical signals that attract soil-living *Rhizobium* to the vicinity of the root. Flavonoids trigger the transcription of bacterial *nod* genes, which encode Nod (nodulation) factors. These factors, secreted by the bacteria, cause cells in the root cortex to divide, leading to the formation of a primary nodule meristem. This meristem gives rise to the plant tissue that constitutes the nodule.

Among the products of the nodule meristem is a layer of cells that excludes O_2 from the interior of the nodule. The function of leghemoglobin is to carry O_2 across this barrier. Within a nodule, the bacteria take the form of **bacteroids** within membranous vesicles. Bacteroids are swollen, deformed bacteria that can fix nitrogen—in effect, nitrogen-fixing structures.

The partnership between bacterium and plant in nitrogen-fixing nodules is not the only case in which plants depend on other organisms for assistance with their nutrition. Another example is that of **mycorrhizae**, root–fungus associations in which the fungus greatly increases the absorption of water and minerals (especially phosphorus) by the plant (see Figure 30.12). A growing body of evidence suggests that nodule formation depends on some of the same genes and mechanisms that allow mycorrhizae to develop.

Biological nitrogen fixation does not always meet agricultural needs

Bacterial nitrogen fixation is not sufficient to support the needs of agriculture. Traditional farmers used to plant dead fish along with corn; the decaying fish released fixed nitrogen that the developing corn could use. Industrial nitrogen fixation is becoming ever more important to world agriculture because of the need to feed a rapidly expanding population.

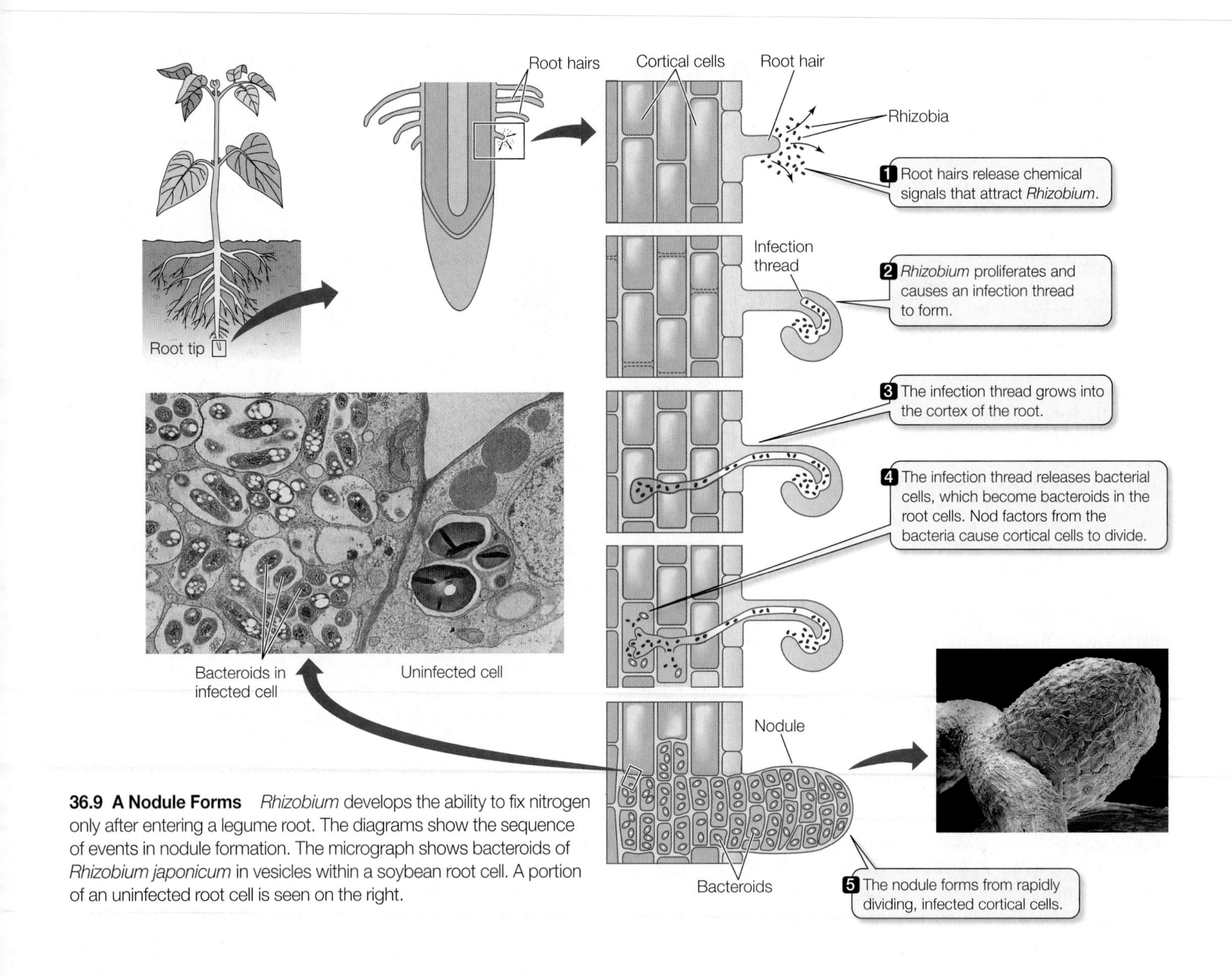

36.9 A Nodule Forms *Rhizobium* develops the ability to fix nitrogen only after entering a legume root. The diagrams show the sequence of events in nodule formation. The micrograph shows bacteroids of *Rhizobium japonicum* in vesicles within a soybean root cell. A portion of an uninfected root cell is seen on the right.

North Korean agriculture was able to feed that country's population until soon after the 1989 collapse of the Soviet Union, which had provided North Korea with chemicals and petroleum. This loss of support was followed by three years of drought, hailstorms, and floods. Today, North Korea cannot produce the fertilizer it needs and is a starving country with a failed farming system.

Most industrial nitrogen fixation is done by a chemical process called the *Haber process*, which requires a great deal of energy. An alternative is urgently needed because of the rising cost of fossil fuel-generated energy. At present, in the United States, the manufacture of nitrogen-containing fertilizer takes more energy than does any other aspect of crop production. The primary energy sources for industrial production of fertilizer are natural gas and hydroelectric power. In biological systems, nitrogen fixation requires a great deal of ATP—about 16 to 20 ATP per N fixed.

Research on biological nitrogen fixation is being pursued, with commercial applications in mind. One line of investigation centers on recombinant DNA technology as a means of engineering new plant–bacterium associations that produce their own nitrogenase. Currently attempts are underway to transfer genes from *Rhizobium* into bacteria that already live in the roots of important cereal plants such as rice. So far the attempts have been unsuccessful.

Plants and bacteria participate in the global nitrogen cycle

Essential nitrogen cycles through the biosphere in a complex **global nitrogen cycle** that we outline in **Figure 36.10**. The nitrogen released into the soil by nitrogen fixers is primarily in the form of ammonia (NH_3) and ammonium ions (NH_4^+). Although ammonia can be toxic to plants if it accumulates in tissues, ammonium ions can be taken up safely at low concentrations. Soil bacteria called **nitrifiers** oxidize ammonia to nitrate ions (NO_3^-)—another

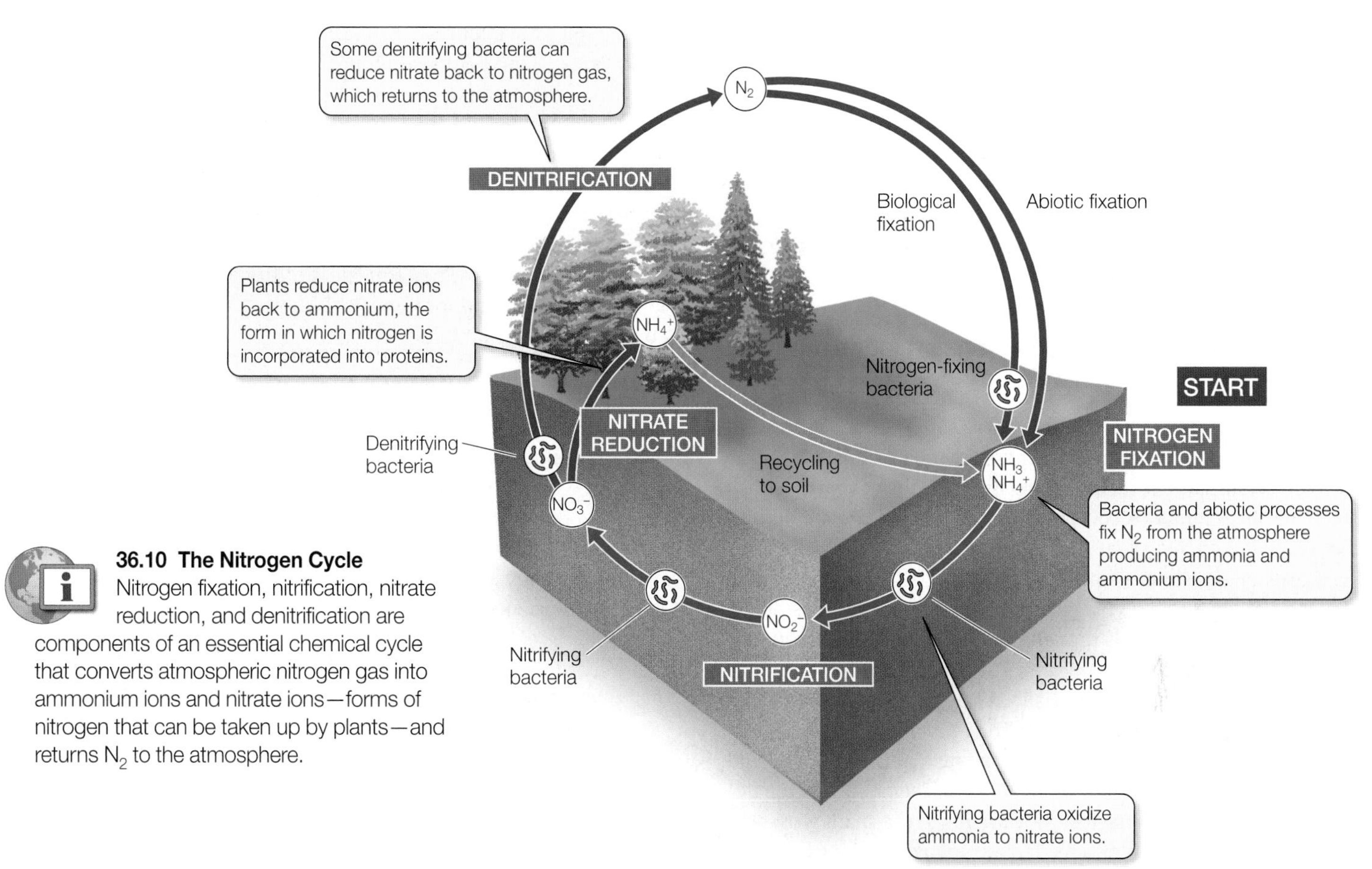

36.10 The Nitrogen Cycle Nitrogen fixation, nitrification, nitrate reduction, and denitrification are components of an essential chemical cycle that converts atmospheric nitrogen gas into ammonium ions and nitrate ions—forms of nitrogen that can be taken up by plants—and returns N_2 to the atmosphere.

form that plants can take up—by the process of **nitrification**. Soil pH affects the uptake of nitrogen: Nitrate ions are taken up preferentially under more acidic conditions, ammonium ions under more basic ones.

These two initial processes are carried out by bacteria: they *reduce* N_2 to ammonia in nitrogen fixation (see Figure 36.8) and *oxidize* ammonia to nitrate in nitrification. The next steps are carried out by plants, which reduce the nitrate they have taken up all the way back to ammonia. All the reactions of **nitrate reduction** are carried out by the plant's own enzymes. The later steps, from nitrite (NO_2^-) to ammonia, take place in the chloroplasts. The plant uses the ammonia thus formed to manufacture amino acids, from which the plant's proteins and all its other nitrogen-containing compounds are formed. Animals cannot reduce nitrogen, and they depend on plants to supply them with reduced nitrogenous compounds.

Bacteria called **denitrifiers** return nitrogen from soil nitrate to the atmosphere as N_2. This process is called **denitrification**. In combination with leaching and the removal of crops, denitrification keeps the level of available nitrogen in soils low.

Thus, the cycle of nitrogen through the biosphere includes four key steps:

1. *fixation* of atmospheric N_2 to NH_3 and NH_4^+ by bacteria and by abiotic processes;
2. *nitrification* of these molecules to nitrate by bacteria;
3. *nitrate reduction* by plants;
4. *denitrification* of nitrate by bacteria back to N_2, which is then released to the atmosphere to begin another cycle.

The nitrogen cycle is essential for life on Earth: nitrogen-containing compounds constitute 5–30 percent of a plant's total dry weight. The nitrogen content of animals is even higher, and all the nitrogen in the animal world arrives there by way of the plant kingdom. Other elements, sulfur for example, also undergo biogeochemical cycling.

36.4 RECAP

Bacteria in soils and root nodules convert nitrogen from an inert gas into forms that plants can use. Nitrogen-fixing bacteria are essential to the biosphere because all of the nitrogen that animals need comes from plants. Denitrification returns nitrogen from dead organisms and animal waste back to the atmosphere, continuing the global nitrogen cycle.

- To reduce nitrogen gas to a form plants can use requires the enzyme nitrogenase and what else? See p. 789 and Figure 36.8
- Can you explain how a root nodule forms on a legume? See p. 789 and Figure 36.9

Let's conclude by looking at some unusual plant species that have evolved special ways to obtain the nutrients they need.

36.5 Do Soil, Air, and Sunlight Meet the Needs of All Plants?

Most plants obtain their mineral nutrients from the soil solution, which is also their source of water. Atmospheric CO_2 is their source of carbon atoms, and sunlight meets their energy needs. Some plants must look to other sources for some or all of these needed commodities. Carnivorous and parasitic plants are examples of plants with such special needs.

Carnivorous plants supplement their mineral nutrition

Some plants that are found primarily in nitrogen-deficient soils augment their nitrogen and phosphorus supply by capturing and digesting flies and other insects. There are about 450 of these **carnivorous plant** species, the best-known of which are Venus flytraps (genus *Dionaea*; **Figure 36.11A**), sundews (genus *Drosera*; **Figure 36.11B**), and pitcher plants (genus *Sarracenia*).

Carnivorous plants are normally found in boggy regions where the soils are extremely nutrient deficient. The carnivorous plants have evolved adaptations that allow them to augment their supply of nitrogen by capturing animals and digesting their proteins.

The Venus flytraps have specialized leaves with two halves that fold together. When an insect trips trigger hairs on a leaf, its two halves quickly come together, their spiny margins interlocking and imprisoning the insect before it can escape. The closing of the flytrap's leaf is one of the fastest movements in the plant world—it requires only one tenth of a second, during which the leaf halves reverse their curvature. The leaf then secretes enzymes that digest its prey. The leaf absorbs the products of digestion, especially amino acids, and uses them as a nutritional supplement.

Pitcher plants produce pitcher-shaped leaves that collect small amounts of rainwater. Insects are attracted into the pitchers either by bright colors or by scent and are prevented from getting out again by stiff, downward-pointing hairs. The insects eventually die and are digested by a combination of enzymes and bacteria in the water. Even rats have been found in large pitcher plants.

Sundews have leaves covered with hairs that secrete a clear, sticky, sugary liquid. An insect touching one of these hairs becomes stuck, and more hairs curve over the insect and stick to it as well. The plant secretes enzymes to digest the insect and eventually absorbs the carbon- and nitrogen-containing products of digestion.

None of the carnivorous plants *must* feed on insects; they can grow quite adequately without insects, but in their natural habitats they grow faster and are a darker green when they succeed in capturing insects. They use the additional nitrogen from the insects to make more proteins, chlorophyll, and other nitrogen-containing compounds.

Parasitic plants take advantage of other plants

Some **parasitic plants** derive their mineral nutrients from the living bodies of other plants. Most of these parasites are autotrophs, but a few plant species have, in the course of their evolution, lost the ability to sustain themselves by photosynthesis. To meet their needs for energy and carbon, these heterotrophs parasitize other, photosynthesizing plants.

Albino mutant plants cannot produce chlorophyll or photosynthesize, and normally die at an early seedling stage. Some 65 years ago, H. A. Spohr grew albino corn seedlings by "feeding" them sucrose solution. The plants remained white but increased in dry weight, produced the same number of leaves as normal plants, and even flowered—they just couldn't produce their own sugar.

Perhaps the most familiar parasitic plants are the several genera of mistletoes and dodders (**Figure 36.12**). Mistletoes are green and carry on some photosynthesis, but they parasitize other plants for water and mineral nutrients and may derive photosynthetic products from them as well. Mistletoes and dodders extract nutrients from the vascular tissues of their hosts by forming absorptive organs called *haustoria*, which invade the host plant's tissues.

Another parasitic plant, the Indian pipe (*Monotropa uniflora*), once was thought to obtain its nutrients from dead organic matter. It is now known to get its nutrients, with the help of fungi, from nearby actively photosynthesizing plants. Hence it, too, is a parasite.

Dwarf mistletoe (*Arceuthobium americanum*) is a serious parasite in forests of the western United States, destroying more than

(A) *Dionaea muscipula*

(B) *Drosera rotundifolia*

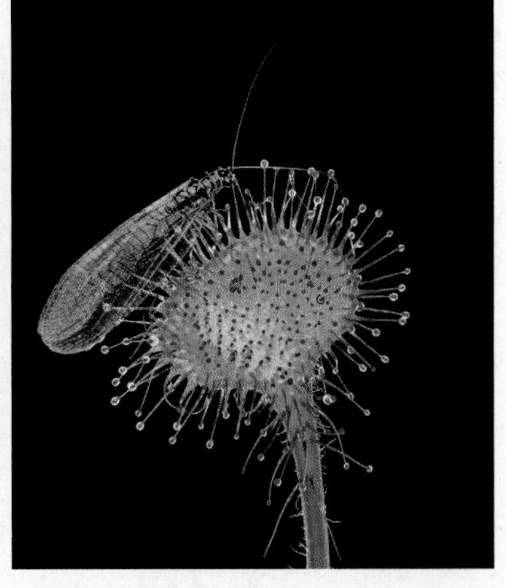

36.11 Carnivorous Plants Some plants have adapted to nitrogen-poor environments by becoming carnivorous. (A) The Venus flytrap obtains nitrogen and phosphorus from the bodies of insects trapped inside the plant when its hinges snap shut. (B) Sundews trap insects on sticky hairs. Secreted enzymes will digest the carcass externally.

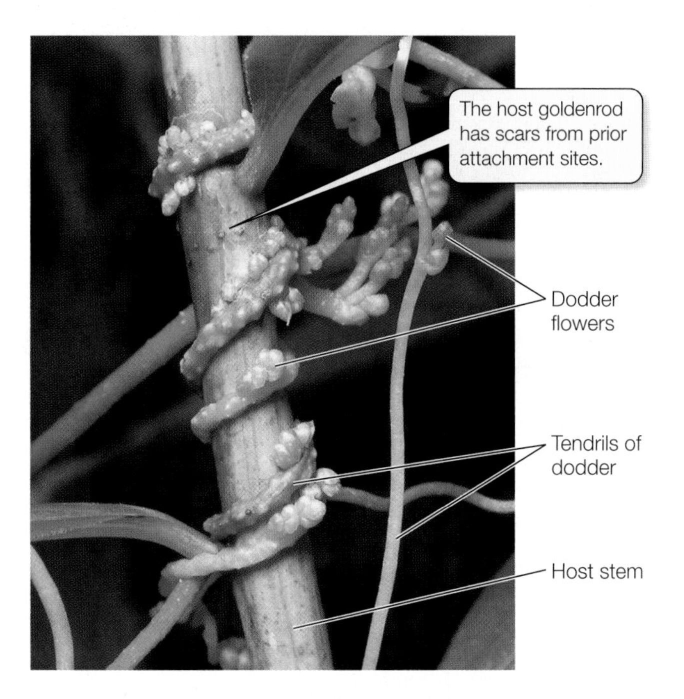

36.12 A Parasitic Plant Tendrils of dodder (genus *Cuscuta*) wrap around a goldenrod (genus *Solidago*). The parasitic dodder obtains water, sugars, and other nutrients through tiny, rootlike protuberances that penetrate the surface of the host plant.

3 billion board feet of lumber per year. Parasitic plants are a much more urgent problem in developing countries. Witchweed (*Striga*) imperils more than 300 million sub-Saharan Africans by attacking their cereal and legume crops. In the Middle East and North Africa, broomrape (*Orobanche*) ravages many crops, especially vegetables and sunflowers.

36.5 RECAP

Carnivorous and parasitic plants supplement their nutrition by extracting materials from animals or other plants.

- How do the needs of heterotrophic parasitic plants differ from those of carnivorous plants? See p. 792

Bad land management led to the Dust Bowl and to current problems in Africa and elsewhere. Invasive parasitic plants can wreak havoc even if the soil contains adequate nutrients. The world's peoples depend on a sufficient nutrient supply to sustain the crops that feed humankind.

CHAPTER SUMMARY

36.1 How do plants acquire nutrients?

Plants are photosynthetic **autotrophs** that can produce all their organic molecules from carbon dioxide, water, and minerals, including a nitrogen source.

Mineral nutrients are obtained from the **soil solution**.

Chemolithotrophic bacteria in the soil increase the availability of nitrogen and sulfur to plants.

Root growth allows plants, which are sessile, to search for mineral resources.

Microenvironments within the soil such as acidic or alkaline areas affect the direction of root growth.

36.2 What mineral nutrients do plants require?

Plants require 14 **essential mineral elements**. Of these, six are **macronutrients** and eight are **micronutrients**. **Deficiency symptoms** suggest what essential element a plant lacks. Review Table 36.1

The requirement for each essential element was discovered by growing plants on **hydroponic solutions** lacking that element. Review Figure 36.2, Web/CD Tutorial 36.1

36.3 What are the roles of soil?

Soils contain water, air, and inorganic and organic substances. Soils have living and nonliving components. Review Figure 36.4

A soil typically consists of two or three horizontal zones called **horizons**. **Topsoil** forms the uppermost or A horizon. Topsoil tends to lose mineral nutrients through **leaching**. **Loams** are excellent agricultural topsoils, with a good balance of sand, silt, clay, and organic matter.

Soils form by mechanical and chemical **weathering** of rock. Chemical weathering imparts mineral nutrients to **clay** particles. Plant litter decomposes to form **humus**. Plants obtain some mineral nutrients through ion exchange between the soil solution and the surface of clay particles. Review Figure 36.6

Farmers use **fertilizers** to make up for deficiencies in soil mineral nutrient content. **Liming** can reverse acidification.

36.4 How does nitrogen get from air to plant cells?

Some **nitrogen-fixing bacteria** live free in the soil; others live symbiotically as **bacteroids** within plant roots. In **nitrogen fixation**, nitrogen gas (N_2) is reduced to ammonia (NH_3) or ammonium ions (NH_4^+) in a reaction catalyzed by **nitrogenase**. Review Figure 36.8

Nitrogenase requires anaerobic conditions, but the bacteroids in **root nodules** require oxygen, which is maintained at the proper level by **leghemoglobin**.

The formation of a root nodule requires interaction between the root system of a legume and ***Rhizobium***. Review Figure 36.9

Mycorrhizae are root-fungus associations that greatly increase a plant's absorption of water and minerals.

Plants and bacteria interact in the **global nitrogen cycle**, which involves series of reductions and oxidations of nitrogen-containing molecules. Review Figure 36.10, Web/CD Activity 36.1

Nitrification by bacteria converts ammonia to nitrate ions in the soil. **Nitrate reduction** is carried out by the plant's own enzymes, enabling plants to form their own nitrogen compounds. **Denitrification** returns nitrogen from animal wastes and dead organisms to the atmosphere.

36.5 Do soil, air, and sunlight meet the needs of all plants?

Carnivorous plants are autotrophs that supplement a low nitrogen supply by feeding on insects. **Parasitic plants** draw on other plants to meet their needs, which may include minerals, water, or the products of photosynthesis.

SELF-QUIZ

1. Macronutrients
   a. are so called because they are more essential than micronutrients.
   b. include manganese, boron, and zinc, among others.
   c. function as catalysts.
   d. are required in concentrations of at least 1 gram per kilogram of plant dry matter.
   e. are obtained by the process of photosynthesis.
2. Which of the following is *not* an essential mineral element for plants?
   a. Potassium
   b. Magnesium
   c. Calcium
   d. Lead
   e. Phosphorus
3. Fertilizers
   a. are often characterized by their N-P-O percentages.
   b. are not required if crops are removed frequently enough.
   c. restore needed mineral nutrients to the soil.
   d. are needed to provide carbon, hydrogen, and oxygen to plants.
   e. are needed to destroy soil pests.
4. In a typical soil,
   a. the topsoil tends to lose mineral nutrients by leaching.
   b. there are four or more horizons.
   c. the C horizon consists primarily of loam.
   d. the dead and decaying organic matter gathers in the B horizon.
   e. more clay means more air space and thus more oxygen for roots.
5. Which of the following is *not* an important step in soil formation?
   a. Removal of bacteria
   b. Mechanical weathering
   c. Chemical weathering
   d. Clay formation
   e. Hydrolysis of soil minerals
6. Nitrogen fixation is
   a. performed only by plants.
   b. the oxidation of nitrogen gas.
   c. catalyzed by the enzyme nitrogenase.
   d. a single-step chemical reaction.
   e. possible because N_2 is a highly reactive substance.
7. Nitrification is
   a. performed only by plants.
   b. the reduction of ammonium ions to nitrate ions.
   c. the reduction of nitrate ions to nitrogen gas.
   d. catalyzed by the enzyme nitrogenase.
   e. performed by certain bacteria in the soil.
8. Nitrate reduction
   a. is performed by plants.
   b. takes place in mitochondria.
   c. is catalyzed by the enzyme nitrogenase.
   d. includes the reduction of nitrite ions to nitrate ions.
   e. is known as the Haber process.
9. Which of the following is a parasite?
   a. Venus flytrap
   b. Pitcher plant
   c. Sundew
   d. Dodder
   e. Tobacco
10. All carnivorous plants
   a. are parasites.
   b. depend on animals as a source of carbon.
   c. are incapable of photosynthesis.
   d. depend on animals as their sole source of phosphorus.
   e. obtain supplemental nitrogen from animals.

FOR DISCUSSION

1. Methods for determining whether a particular element is essential have been known for more than a century. Since these methods are so well established, why was the essentiality of some elements discovered only recently?
2. If a Venus flytrap were deprived of soil sulfates and hence made unable to synthesize the amino acids cysteine and methionine, would it die from lack of protein? Explain.
3. Soils are dynamic systems. What changes might result when land is subjected to heavy irrigation for agriculture after being relatively dry for many years? What changes in the soil might result when a virgin deciduous forest is cut down and replaced by crops that are harvested each year?
4. We mentioned that important positively charged ions are held in the soil by clay particles, but other, equally important, negatively charged ions are leached deeper into the soil's B horizon. Why doesn't leaching cause an electrical imbalance in the soil? (Hint: Think of the ionization of water.)
5. The biosphere of Earth as we know it depends on the existence of a few species of nitrogen-fixing prokaryotes. What do you think might happen if one of these species were to become extinct? If all of them were to disappear?

FOR INVESTIGATION

Brown and coworkers established nickel as an essential element for plant growth, although nickel deficiency has never been observed in nature. What sort of evidence could lead you to suspect that some other element, such as rubidium, was an essential element? (It isn't, by the way.) What sorts of experiments might you perform to test your suspicion?

CHAPTER 37 Regulation of Plant Growth

More rubber, please

Within weeks of attacking Pearl Harbor on December 7, 1941, Japanese forces invaded Malaysia and the Dutch East Indies, thus taking control of most of the world's rubber supply. The governments of Canada and the United States reacted swiftly to what was truly a crisis, since rubber was crucial both to the war effort and to the domestic economy. Rubber tires were the first commodity to be rationed in the United States. The main purpose of gas rationing and a national 35-miles-per-hour speed limit was to conserve tires.

Although rubber trees (*Hevea brasiliensis*) are native to South America, British colonialists established plantations of these magnificent trees in Southeast Asia during the late nineteenth century, and that region continues to be the source of 90 percent of the world supply of natural rubber. A North American shrub, guayule (*Parthenium argentatum*), was used commercially to a limited extent before World War II, but its rubber yield was far from sufficient to ease the shortage.

During the course of the war, scientists developed synthetic rubber, which helped address the problem. Taking a different approach, James Bonner and other plant biologists worked to increase the supply of natural rubber. Bonner and his colleagues launched an urgent study of guayule, but determined that this plant could not become a major source of rubber. Bonner turned his attention back to *H. brasiliensis*, but could do little work with that species until the war ended and access to the great rubber plantations was again possible.

Eventually Bonner became chairman of the Agricultural Science and Biology Subcommittee of the Malaysian Rubber Research and Development Board. Among his many accomplishments was the discovery of a method to speed the collection of rubber-containing latex from *H. brasiliensis*.

To collect latex, workers make a V-shaped cut in the bark of the rubber tree and insert a spout at the bottom of the V. Latex flows from the cut surface into a cup attached to the tree. However, the latex at the cut surfaces coagulates within 1–3 hours and the flow stops, making it necessary to cut frequently. Bonner and his colleagues discovered that coagulation slowed dramatically, and far more latex could be collected, if the region of the cut was exposed to the gas

Rubber Balls Make the World Go Around *Hevea brasiliensis* trees are native to Brazil, where it was discovered that boiling and otherwise processing the milky liquid—latex—that flowed from their cut bark resulted in resilient rubber. Today the worldwide demand for rubber is at an all-time high.

More of a Valuable Substance Latex from *H. brasiliensis* is the source of natural rubber. Workers make V-shaped slashes in the trees from which latex flows; treating the slashes with ethylene keeps latex flowing longer.

ethylene, a natural plant product with dramatic effects on physiology.

Treating cut bark with ethylene, along with improved fertilizers and better control of rubber-tree diseases, ultimately led to a doubling of the world's supply of natural rubber. Even though synthetic substitutes surpassed the production of natural rubber by the late 1950s, consumption of rubber products is so great that today the need for natural rubber is at an all-time high.

In the meantime, biologists have determined that ethylene is an important naturally occurring plant growth hormone. It is but one of the many factors described in this chapter that collaborate to regulate a plant's growth throughout its life cycle, from seed to senescence.

IN THIS CHAPTER we will give a brief overview of the life of a flowering plant and its developmental stages. We will explore the nature of the environmental cues, photoreceptors, and hormones (including ethylene) that regulate plant growth and development. We will also consider the multiple roles and interactions of these different elements.

CHAPTER OUTLINE

37.1 How Does Plant Development Proceed?

The *development* of a plant—the series of progressive changes that take place throughout its life—is regulated in complex ways. Four factors are involved in regulating plant growth:

- *Environmental cues* to which a plant responds
- *Receptors* that allow a plant to sense environmental cues, such as *photoreceptors,* molecules that absorb light
- *Hormones,* chemical signals that mediate the effects of the environmental cues including those sensed by receptors
- The plant's *genome,* which encodes enzymes that catalyze the biochemical reactions of development

Many recent advances in understanding plant growth and development have come from work with *Arabidopsis thaliana,* a weed in the mustard family. This plant is used as a *model* organism by researchers because its body and seeds are tiny, its genome is unusually small for a flowering plant, and it flowers and forms many seeds soon after growth begins. Its genome is fully sequenced, so researchers have an accounting of all genes in the plant. Genes can be inserted or deleted. *Arabidopsis* mutants with altered developmental patterns provide evidence for the existence of hormones and for the mechanisms of hormone and photoreceptor action.

Several hormones and photoreceptors play roles in plant growth regulation

Hormones are regulatory compounds that act at very low concentrations at sites often distant from where they are produced. Unlike animals, which produce each hormone in a specific part of the body, plants produce hormones in many cell types. Each plant hormone plays multiple regulatory roles, affecting several different aspects of plant development (**Table 37.1**). Interactions among these hormones are often complex.

Like hormones, **photoreceptors** are involved in many developmental processes in plants. Unlike plant hormones, which are small molecules, plant photoreceptors are *pigments* (molecules that absorb light) associated with proteins. Light (an en-

TABLE 37.1

Plant Growth Hormones

HORMONE	TYPICAL ACTIVITIES
Abscisic acid	Maintains seed dormancy and winter dormancy; closes stomata
Auxins	Promote stem elongation, adventitious root initiation, and fruit growth; inhibit axillary bud outgrowth and leaf abscission
Brassinosteroids	Promote stem and pollen tube elongation; promote vascular tissue differentiation
Cytokinins	Inhibit leaf senescence; promote cell division and axillary bud outgrowth; affect root growth
Ethylene	Promotes fruit ripening and leaf abscission; inhibits stem elongation and gravitropism
Gibberellins	Promote seed germination, stem growth, and fruit development; break winter dormancy; mobilize nutrient reserves in grass seeds

vironmental cue) acts directly on photoreceptors, which in turn regulate the processes of development, such as the many changes accompanying the growth of a young seedling emerging from the soil and into the light.

No matter what cues regulate development, ultimately the plant's genome determines the limits of plant development. The genome encodes the master plan, but its interpretation depends on conditions in the environment. It is also the target for some hormone actions. For several decades hormones and photoreceptors were the focus of most work on plant development, but recent advances in molecular genetics have allowed us to focus on the underlying processes, such as signal transduction pathways.

Signal transduction pathways are involved in all stages of plant development

Plants, like other organisms, make extensive use of *signal transduction pathways*, sequences of biochemical reactions by which a cell generates a response to a stimulus (Chapter 15). Cell signaling in plant development involves a receptor (for a hormone or for light) and a signal transduction pathway, and concludes with a cellular response. Protein kinase cascades amplify responses to signals in plants, as they do in other organisms (see Figure 15.10).

The details of plant cell signaling are best understood in the context of the general pattern of plant development. The factors just described affect plants through their entire developmental history by acting on three fundamental processes: cell division, cell expansion, and cell differentiation.

The seed germinates and forms a growing seedling

If all developmental activity is suspended in a seed, even when conditions appear to be suitable for its growth, the seed is said to be **dormant**. Cells in dormant seeds do not divide, expand, or differentiate. For the embryo to begin developing, seed dormancy must be broken by one of the mechanisms discussed later in this section.

As the seed begins to **germinate**—to develop into a seedling—it takes up water. The growing embryo then obtains chemical building blocks—monomers—for its development by digesting the polysaccharides, fats, and proteins stored in the seed. The embryos of some plant species secrete hormones that direct the mobilization of these reserves. Germination is completed when the **radicle** (embryonic root) emerges from the seed coat. The plant is then called a **seedling**.

If the seed germinates underground, the new seedling must elongate rapidly (in the right direction!) and cope with a period of life in darkness or dim light. A series of photoreceptors directs this stage of development and prepares the seedling for growth in the light environment.

Early shoot development varies among the flowering plants. **Figure 37.1** shows the shoot development patterns of monocots and eudicots.

Plant growth from seedling to adult is regulated by several hormones. Other hormones are involved in the plant's defenses against herbivores and microorganisms (discussed in Chapter 39).

The plant flowers and sets fruit

Flowering—the formation of reproductive organs—may be initiated when the plant reaches an appropriate age or size. Some plant species, however, flower at particular times of the year, meaning that the plant must be capable of distinguishing different times of the year. In these plants, the leaves measure the length of the night (shorter in summer, longer in winter) with great precision. Light absorption by photoreceptors is the first step in this time-measuring process.

Once the leaves have determined that it is time for the plant to flower, that information must be transmitted as a signal to the places where flowers will form. It was proposed more than 70 years ago that this signal is transmitted from the leaf to the site of flower formation in the form of a "flowering hormone," then termed *florigen*, common to many plants. In spite of intensive research, it was not until 2005 that biologists identified specific compounds that influenced the transition to flowering (see Section 38.2).

After flowers form, hormones play further roles in reproduction. Hormones and other substances control the growth of the pollen tube that brings sperm and egg together (see Figure 38.1). Following fertilization, a fruit develops and ripens under hormonal control.

The plant senesces and dies

Some plants, such as iris and elm, are **perennials**: they continue to grow year after year. **Annuals**, such as petunia and marigold, complete their life cycle in a single year, then **senesce** (deteriorate as a result of aging) and die.

The death of the entire plant, which may be triggered by signals from the environment, follows senescent changes that are controlled by hormones such as ethylene. This life history pattern appears to be an adaptation for producing more offspring by shifting nutrients from nonreproductive tissues into the seeds; in so doing, the parent plant essentially starves itself to death, ensuring that sufficient nutrients are available for seed maturation.

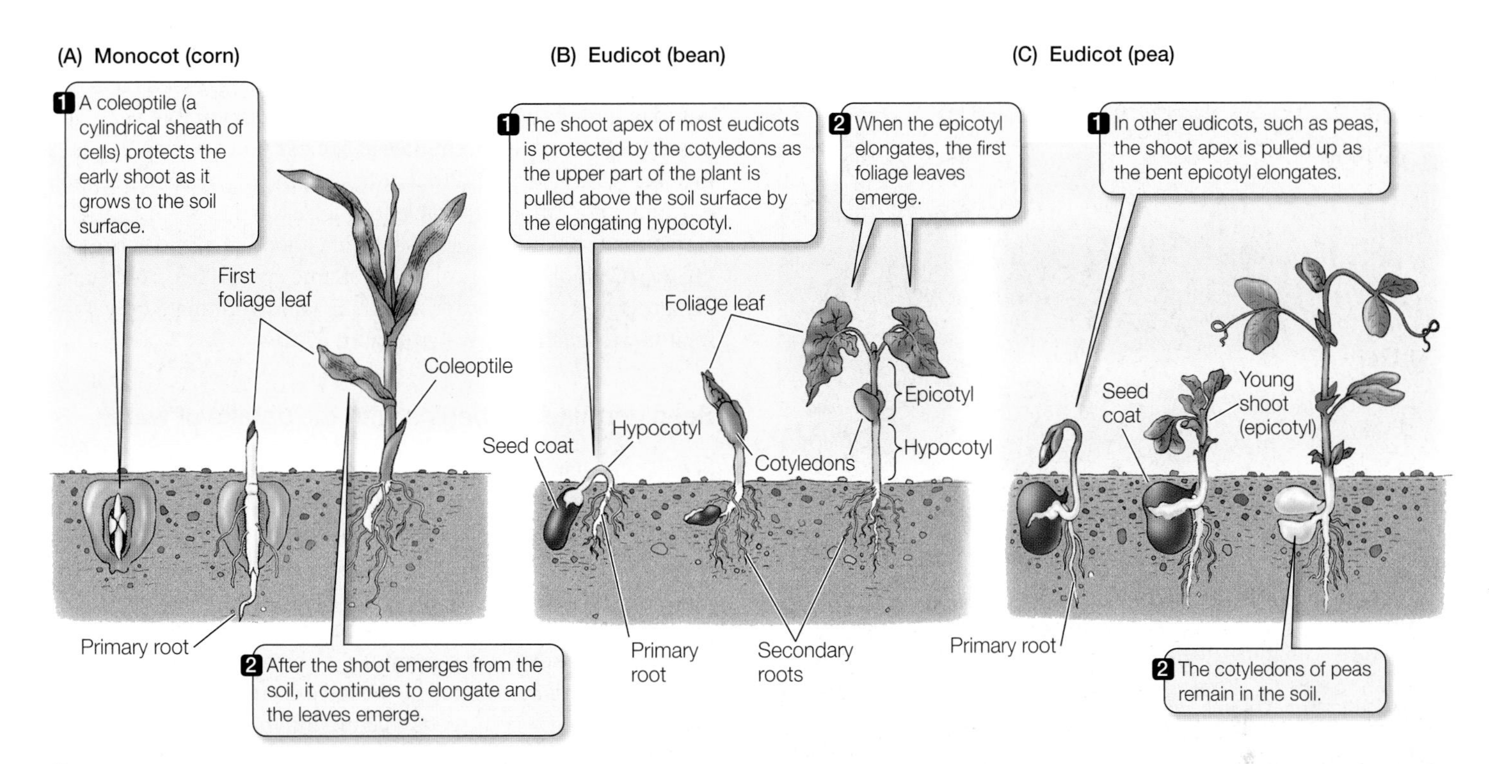

37.1 Patterns of Early Shoot Development (A) In grasses and some other monocots, growing shoots are protected by a coleoptile until they reach the soil surface. (B) In most eudicots, the growing point of the shoot is protected by the cotyledons. (C) In some other eudicots, the cotyledons remain in the soil, and the growing point is protected by the first true leaves.

In many perennials, leaves senesce and fall at the end of the growing season, shortly before the onset of winter. Leaf fall is regulated by an interplay of the hormones ethylene and auxin.

Having described the steps in a plant's life history, let's examine how these various steps are regulated. We'll begin at the start of the life history, with the seed and its germination.

Not all seeds germinate without cues

The seeds of some plant species are capable of germinating as soon as they have matured. All they need for germination is water. But the seeds of many species are dormant at maturity. Seed dormancy may last for weeks, months, years, or even centuries. The mechanisms that maintain seed dormancy are numerous and diverse, but three principal strategies dominate:

- Exclusion of water or oxygen from the embryo by means of an impermeable seed coat
- Mechanical restraint of the embryo by means of a tough seed coat
- Chemical inhibition of embryonic development

Seed dormancy must be broken before germination can begin. The dormancy of seeds with impermeable coats can be broken if the seed coat is abraded as the seed tumbles across the ground or through a creek bed or passes through the digestive tract of an animal. Cycles of freezing and thawing can also aid in making the seed coat permeable. Soil microorganisms probably play a major role in softening seed coats. Fire is a major force for ending seed dormancy (**Figure 37.2**). Fire can melt waterproofing wax in seed coats, allowing water to reach the embryo, and it can release mechanical restraint, cracking the seed coat. Fire can also break down chemical inhibitors of germination. *Leaching*—the dissolving out of water-soluble chemical inhibitors by prolonged exposure to water—is another way in which dormancy can be broken.

Seed dormancy affords adaptive advantages

What are the potential advantages of seed dormancy? For many plant species, dormancy ensures survival through unfavorable conditions and results in germination when conditions are more favorable for growth. To avoid germination in the dry days of late summer, for example, some seeds require exposure to a long cold period before they will germinate. Other seeds will not germinate until a certain amount of time has passed, regardless of how they are treated. This strategy prevents germination while the seeds are still attached to the parent plant. Seeds that must be scorched by fire in order to germinate avoid competition with other plants by germinating only where an area has been cleared by fire. Dormancy also helps seeds to survive long-distance dispersal, allowing plants to colonize new territory.

The dormancy of some seeds is broken by exposure to light. These seeds, which germinate only at or near the surface of the soil, are generally tiny seeds with few food reserves. Such seeds would be incapable of surviving if germination occurred while they were buried deeply. Conversely, the germination of some other seeds is inhibited by light; these seeds germinate only when buried and thus kept in darkness. Light-inhibited seeds are usually large and well stocked with nutrients.

37.2 Fire and Seed Germination This fireweed germinated and flourished after a great fire along the Alaska Highway.

Seed dormancy helps annual plants cope with year-to-year variation in the amount and frequency of rainfall. The seeds of some annuals remain dormant throughout an unfavorable year. The seeds of other species germinate at specific times during the year, increasing the likelihood that at least some of the seedlings will encounter conditions favorable for their growth.

Dormancy may also increase the likelihood of a seed germinating in a favorable ecological setting. Some cypress trees, for example, grow in standing water, and their seeds germinate only if inhibitors are leached by water (**Figure 37.3**).

Seed germination begins with the uptake of water

Seeds can begin to germinate after dormancy is broken and environmental conditions are satisfactory. The first step in germination is the uptake of water, called **imbibition** (from *imbibe*, "to drink"). Typically, only 5 to 15 percent of a seed's weight is water, whereas most other plant parts contain 80 to 95 percent water. A seed's water potential is very negative (see Section 35.1), and water will be taken up if the seed coat is permeable. The magnitude of this water potential is demonstrated by the force exerted by seeds expanding in water. Cocklebur seeds that are imbibing can exert a pressure of up to 1,000 atmospheres (about 100 megapascals) against a restraining force.

As a seed takes up water, it undergoes metabolic changes: Enzymes are activated upon hydration, RNA and then proteins are synthesized, the rate of cellular respiration increases, and other metabolic pathways are activated. In many seeds, no DNA synthesis and no cell division occur during the early stages of germination. Initially, growth results solely from the expansion of small, preformed cells. DNA is synthesized only after the radicle begins to grow and ruptures the seed coat.

The embryo must mobilize its reserves

To fuel these metabolic activities, the embryo must use the reserves of energy and raw materials stored in the seed. Until the young plant is able to photosynthesize, it depends on these reserves, which are stored in the **cotyledons** (see Figure 37.1) or in the **endosperm** (the specialized nutritive tissue) of the seed. The principal reserve of energy and carbon in many seeds is starch. Other seeds store fats or oils. Usually, the endosperm holds amino acid reserves in the form of proteins, rather than as free amino acids.

The giant molecules of starch, lipids, and proteins must be broken down by enzymes into monomers that can enter the cells of the embryo.

37.3 Leaching of Germination Inhibitors The seeds of bald cypress, a tree adapted to moist or wet environments, germinate only after being leached by water, which increases the chances that they will germinate in a location suitable for their growth.

The polymer starch yields glucose for energy metabolism and for the synthesis of cellulose and other cell wall constituents. The digestion of stored proteins provides the amino acids the embryo needs to synthesize its own proteins. Lipids are broken down into glycerol and fatty acids, both of which can be metabolized for energy. Glycerol and fatty acids can also be converted to glucose, which permits fat-storing plants to make all the building blocks they need for growth.

37.1 RECAP

Environmental cues, receptors, hormones, and the genome interact to regulate all stages of plant growth and development. Stages of development include dormancy, germination, seedling growth, flowering, senescence, and death.

- Can you name some circumstances under which seed dormancy is advantageous? See p. 799
- On what can a young plant embryo rely for energy and nutrients before it is able to commence photosynthesis? See pp. 800–801

What signal initiates the breakdown of stored reserves in the seed, releasing energy and building blocks for growth of the new seedling?

37.2 What Do Gibberellins Do?

In germinating barley and other cereal seeds, the embryo secretes **gibberellins**, one of several classes of plant growth hormones. Gibberellins diffuse through the endosperm to a surrounding tissue called the **aleurone layer**, which lies underneath the seed coat. The gibberellins trigger a cascade of events in the aleurone layer, causing it to synthesize and secrete enzymes that digest proteins and starch stored in the endosperm (**Figure 37.4**).

Commercially, gibberellins are used in the brewing industry to enhance the "malting" (germination) of barley and the breakdown of its endosperm, producing sugar that is fermented to alcohol.

Gibberellins play a variety of roles in plant development in addition to triggering the mobilization of seed reserves. We'll begin our discussion of plant growth hormones by describing the discovery of the gibberellins, as well as their many effects.

"Foolish seedling" disease led to the discovery of the gibberellins

The gibberellins are a large family of closely related compounds. Although not steroids themselves, the gibberellins belong to the same broad class of compounds—the *terpenes*—as do rubber, the steroids, and another plant hormone we'll meet later in this chapter. Some gibberellins are found in plants and others in a pathogenic (disease-causing) fungus in which they were first discovered.

In 1809, the study of the gibberellins began indirectly with observations of the *bakanae*, or "foolish seedling," disease of rice. Seedlings affected by this disease grow more rapidly than their healthy neighbors, but this rapid growth gives rise to tall, spindly plants that die before producing seeds (the rice grains used for food). The disease has had a significant impact on rice yields in several parts of the world. It is caused by the ascomycete fungus *Gibberella fujikuroi*.

In 1925, the Japanese biologist Eiichi Kurosawa grew *G. fujikuroi* on a liquid medium and then separated the fungus from the medium by filtration. He heated the filtered medium to kill any re-

37.4 Embryos Mobilize Their Reserves During seed germination in cereal grasses, gibberellins trigger a cascade of events that results in the conversion of starch and protein reserves into monomers that can be used by the developing embryo.

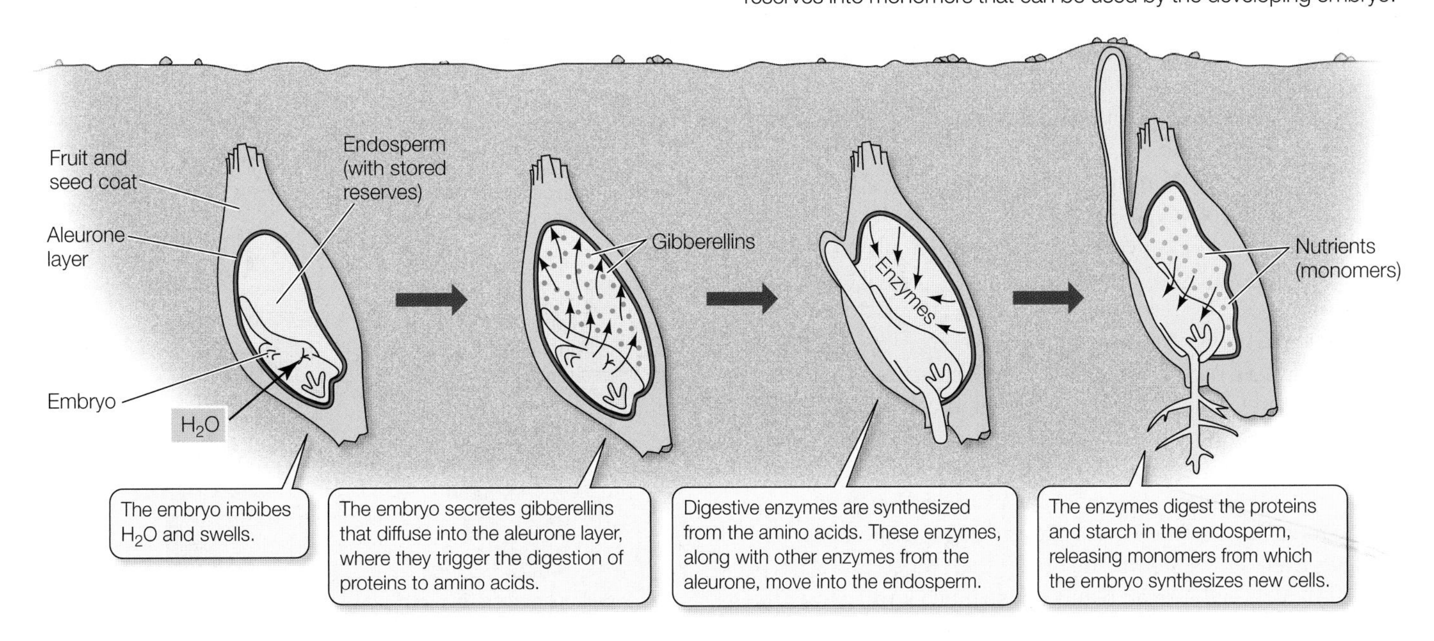

37.5 The Effect of Gibberellins on Dwarf Plants Both of the dwarf tomato plants in this photograph were the same size when the one on the right was treated with gibberellins.

maining fungus, but found that the resulting heat-treated filtrate was still capable of inducing rapid growth in rice seedlings. Medium that had never contained the fungus did not have this effect. This experiment established that *G. fujikuroi* produces a growth-promoting chemical substance, which Kurosawa called a gibberellin.

Were the gibberellins simply exotic products of an obscure fungus, or did they play a more general role in plant growth? Bernard O. Phinney of the University of California, Los Angeles, answered this question in part in 1956, when he reported the spectacular growth-promoting effect of gibberellins on dwarf corn seedlings. He used plants that were known to be genetic dwarfs, in which a particular recessive allele (say, *d1*) was present in the homozygous condition (*d1d1*). Gibberellins applied to nondwarf—wild-type—corn seedlings had almost no effect, whereas dwarf seedlings treated with gibberellins grew as tall as their normal relatives. (A comparable effect of gibberellins applied to a dwarf tomato plant is shown in **Figure 37.5**.)

Phinney drew two conclusions from the results of this experiment: first, that gibberellins are normal constituents of corn, and perhaps of all plants, and second, that some dwarf plants are short because they produce insufficient amounts of gibberellins. According to Phinney's hypothesis, nondwarf plants manufacture enough gibberellins to promote their full growth, whereas dwarf plants do not. Extracts from nondwarf plants of numerous species were found to promote growth in dwarf corn. These findings provided direct evidence that plants that are not genetic dwarfs contain gibberellin-like substances. Phinney's work set the stage for today's use of mutant plants to investigate the control of plant development.

The roots, leaves, and flowers of dwarf corn plants appear normal, but their stems are much shorter than those of wild-type plants. All parts of the dwarf plant contain much lower concentrations of gibberellins than do the organs of a wild-type plant. We can infer, then, that normal stem elongation *requires* gibberellins or the products of gibberellin action. We can further infer that gibberellins play a less essential role in the development of roots, leaves, and flowers.

The gibberellins have many effects on plant growth and development

Gibberellins and other hormones regulate the growth of fruits. It has long been known that grapevines that produce seedless grapes develop smaller fruit than varieties that produce seed-bearing grapes. Experimental removal of seeds from immature seeded grapes prevented normal fruit growth, suggesting that the seeds are sources of a growth regulator. It was then shown that spraying young seedless grapes with a gibberellin solution caused them to grow as large as seeded ones. It is now standard commercial procedure to spray seedless grapes with gibberellins. Biochemical studies showed that the developing seeds produce gibberellins, which diffuse out into the immature fruit tissue.

A different type of gibberellin effect is seen in some **biennials**—plants that grow vegetatively in their first year and flower in their second year and then die. Some biennial plants respond dramatically to an increase in the level of gibberellins. In their second year, the apical meristems of these biennials respond to environmen-

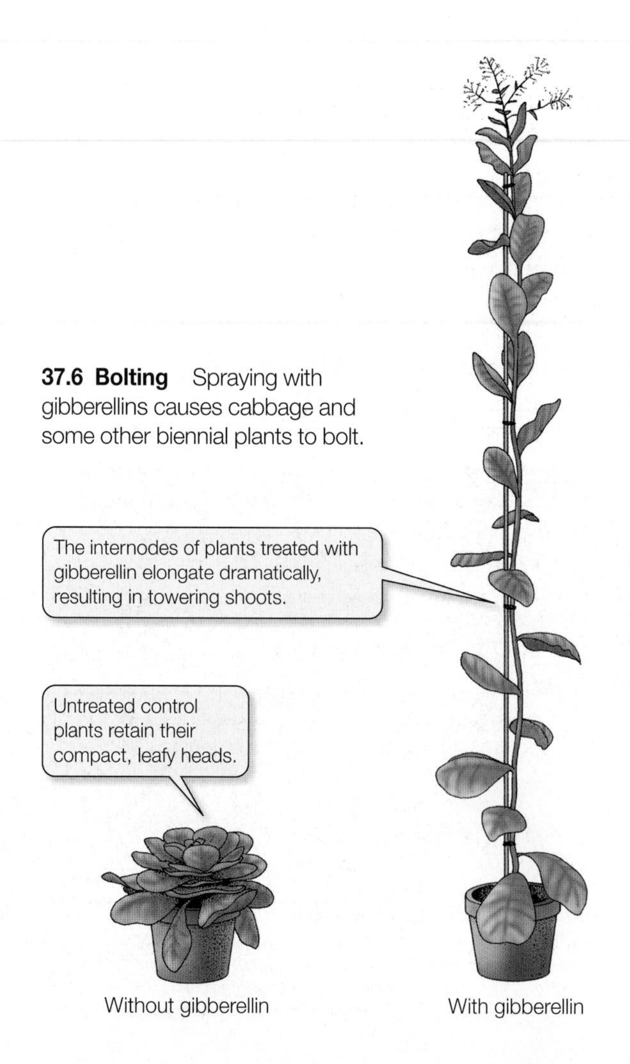

37.6 Bolting Spraying with gibberellins causes cabbage and some other biennial plants to bolt.

tal cues by producing elongated shoots, which eventually bear flowers. This rapid shoot elongation is called **bolting**. When the plant senses the appropriate environmental cue—longer days or a sufficient winter chilling—it produces more gibberellins, raising the gibberellin concentration to a level that causes the shoot to bolt. Some biennial species will bolt when sprayed with a gibberellin solution without exposure to any environmental cue (**Figure 37.6**).

Gibberellins have other important effects. They also cause fruit to grow from unfertilized flowers, promote seed germination, and help bring spring buds out of winter dormancy.

37.2 RECAP

Gibberellins are plant hormones that affect stem growth, fruit size, seed germination, and many other aspects of plant development; the effects vary from species to species.

- Can you explain how gibberellins contribute to the germination of barley seeds? See Figure 37.4
- Why is it believed that gibberellins are more important to the growth of stems than of roots and leaves? See p. 802

Most other hormones, like the gibberellins, have multiple effects within the plant, and they often interact with one another to regulate developmental processes. In controlling stem elongation, for example, gibberellins interact with another hormone, auxin.

37.3 What Does Auxin Do?

If you pinch off the apical bud at the top of a bean plant, inactive axillary buds become active and develop into lateral branches. Similarly, pruning a shrub stimulates the formation of new branches. If you cut off the blade of a leaf but leave its petiole (stalk) attached to the plant, the petiole drops off sooner than it would have if the leaf were intact. If a plant is kept indoors, its shoots will grow toward a window. These diverse responses of shoots are all mediated by plant hormones called **auxins**, of which the most important is *indoleacetic acid* (IAA), a close chemical relative of the amino acid tryptophan.

In this section we will look at the discovery of auxin, its transport within the plant, and its role as a mediator of the effects of light and gravity on plant growth. We'll discover its many effects on vegetative growth and on fruit development. Then we'll examine its mechanism of action.

Phototropism led to the discovery of auxin

The discovery of auxin and its numerous physiological effects on plants can be traced back to work done in the 1880s by Charles Darwin and his son Francis. The Darwins were interested in plant movements. One type of growth movement they studied was **phototropism**, the growth of plant organs toward light (as in most shoots) or away from it (as in roots). They asked, What part of the plant senses the light?

To answer this question, the Darwins worked with canary grass (*Phalaris canariensis*) seedlings grown in the dark. A young grass seedling has a **coleoptile**—a cylindrical sheath a few cells thick that protects the delicate shoot as it pushes through the soil (see Figure 37.1A). When the seedling breaks through the soil surface, the coleoptile soon stops growing, and the shoot emerges unharmed. The coleoptiles of grasses are phototropic—they grow toward the light.

To find the light-receptive region of the coleoptile, the Darwins tried "blindfolding" the coleoptiles of dark-grown canary grass seedlings in various places, then illuminating them from one side (**Figure 37.7**). The coleoptile grew toward the light whenever its tip was exposed. If the top millimeter or more of the coleoptile was covered, however, it showed no phototropic response. Thus, the Darwins were able to conclude that the tip contains the photoreceptor that responds to light. The actual bending toward the light, however, takes place in a growing region a few millimeters below the tip. Therefore, the Darwins reasoned, some type of signal must travel from the tip of the coleoptile to the growing region. Later, others demonstrated that this signal is a chemical substance by showing that it can move through certain permeable materials, such as gelatin, but not through impermeable materials, such as a metal sheet.

Further experiments showed that the tip of the coleoptile produces a hormone that moves down the coleoptile to the growing region, and that this hormone causes cells to grow faster. First, if the tip is removed, the growth of the coleoptile is sharply inhibited. If the tip is carefully replaced, growth resumes—even if the tip and base are separated by a thin layer of gelatin. Furthermore, the hormone moves down from the tip, but it does not move from one side of the coleoptile to the other. If the tip is cut off and moved so that it rests on only one side of the cut end of the coleoptile, the coleoptile curves as the cells on the side below the replaced tip grow more rapidly than those on the other side.

In the 1920s, the Dutch botanist Frits W. Went followed up on the Darwins' experiment. He removed coleoptile tips and placed their cut surfaces on a block of agar. Then he placed pieces of that agar on decapitated coleoptiles—positioned to cover only one side, just as coleoptile tips had been placed in earlier experiments (**Figure 37.8**). As they grew, the coleoptiles curved away from the side with the agar. This curvature demonstrated that a hormone had indeed diffused into the agar block from the isolated coleoptile tips. Went had at last isolated a hormone from a plant. Later chemical analysis showed that this hormone, named auxin, was indoleacetic acid.

Auxin transport is polar and requires carrier proteins

Early experiments showed that the movement of auxin through certain plant tissues is strictly *polar*—that is, it is unidirectional along a line from apex to base. By inverting plants and plant parts, scientists determined that the apex-to-base direction of auxin movement has nothing to do with gravity; the polarity of this movement is a totally biological phenomenon.

Auxin transport is completely or partially polar in many plant parts. In most leaf petioles, for example, auxin moves only from

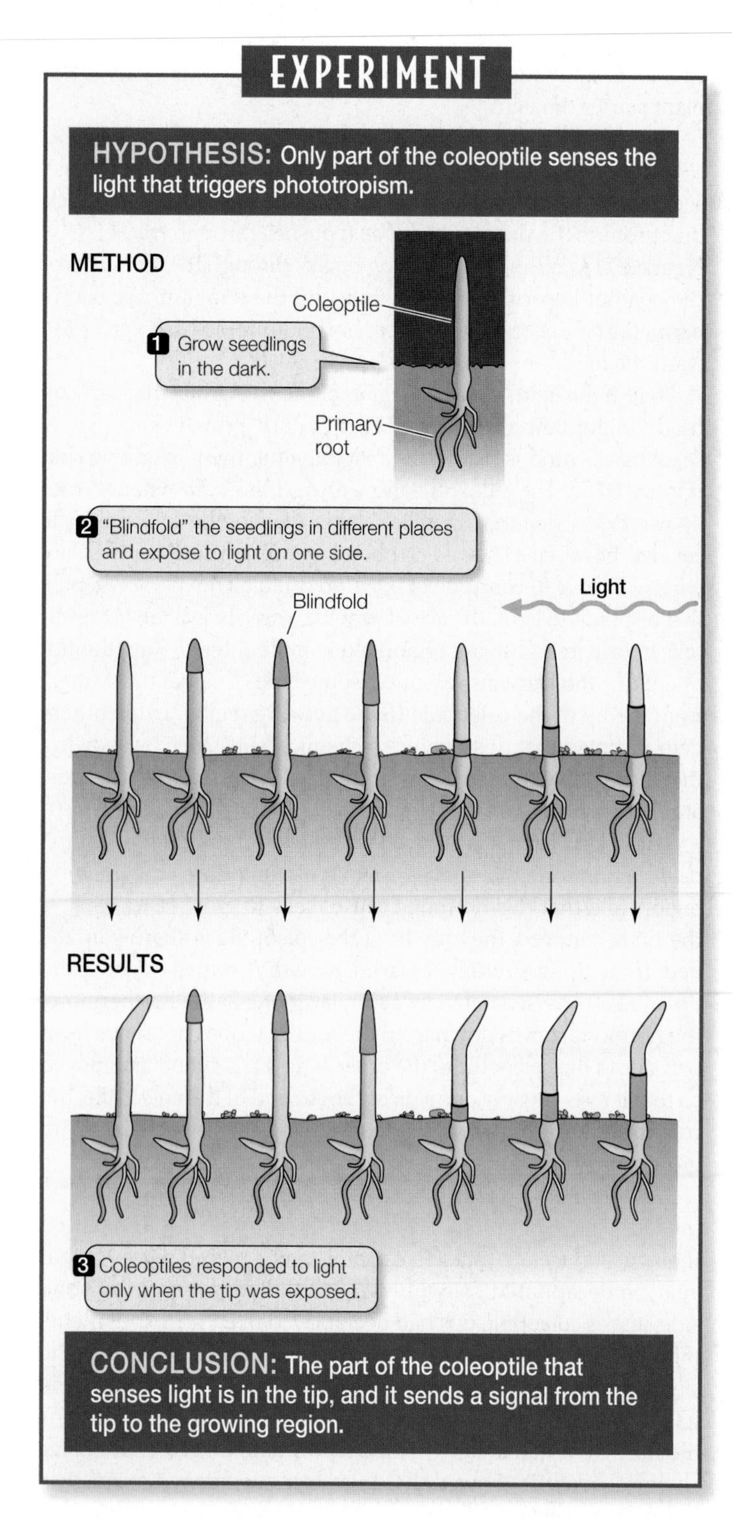

37.7 The Darwins' Phototropism Experiment The series of drawings at the center show some of the ways in which seedlings grown in the dark were "blindfolded"; the drawings below them show the results the Darwins observed in each case. Their observations led them to hypothesize the existence of a growth-promoting signal produced by the coleoptile.

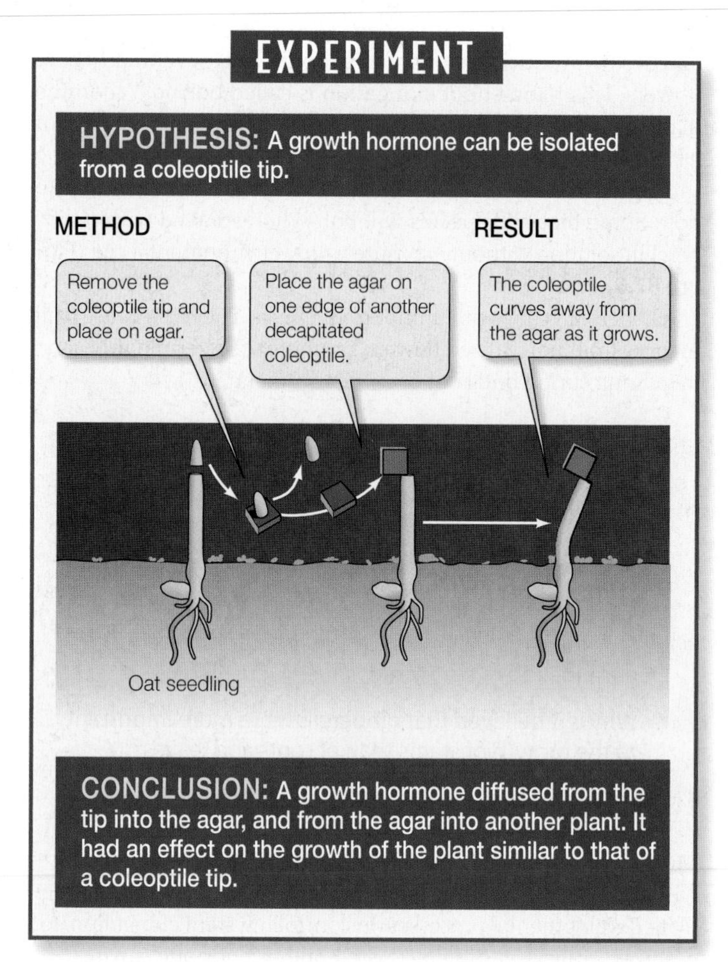

37.8 Went's Experiment By placing coleoptile tips on blocks of agar, Went isolated the growth-promoting hormone whose existence the Darwins had hypothesized.

the blade end toward the stem end. In roots, however, auxin moves toward the root tip, in the phloem. What regulates these movements of auxin?

The polar transport of auxin depends on the location of **auxin anion efflux carriers**, membrane proteins that are confined to the basal ends of cells (the ends closer to the base of the plant than to the shoot apices). Plant cell cytoplasm has a nearly neutral pH, and at this pH auxin exists as an anion—it is negatively charged. Auxin anions can leave the cells only by way of the basally located auxin anion efflux carriers.

Proton pumps in the plasma membrane pump hydrogen ions (H^+) out of the cells, rendering the cell walls acidic. At this lower pH, auxin is present both as an anion and as a free acid. Either form of auxin can enter the cell from any direction. About half the auxin entry into cells is by passive diffusion of the free acid and the other half is by active transport (symport) of the anions along with H^+. However, once auxin molecules enter the cell they become anions, which can depart only from the base (**Figure 37.9**). In this way, auxin anion efflux carriers contribute to the establishment of auxin gradients in the plant. Because it forms a gradient, auxin can act as a *morphogen* (see Section 19.5), instructing cells as to their orientation within the plant and determining how they differentiate.

Other auxin carrier proteins are specific to certain tissues and cells and participate in specific auxin responses. Such auxin carrier proteins are involved in plant responses to light and gravity.

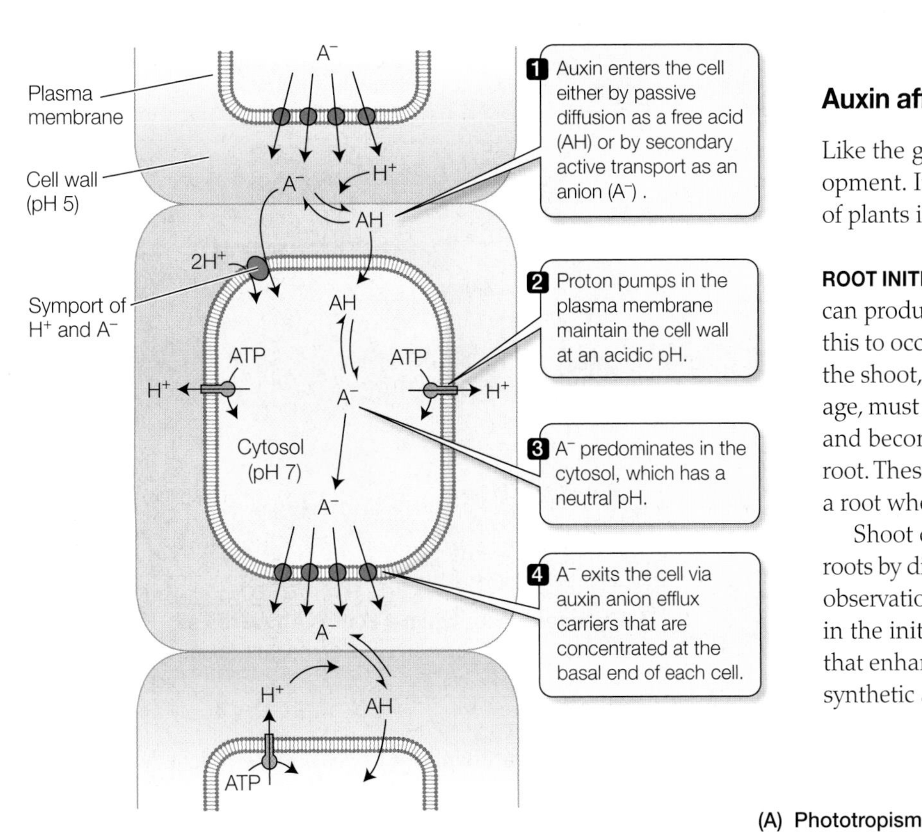

37.9 Polar Transport of Auxin Proton pumps and the basally placed auxin anion efflux carriers lead to a net movement of auxin in a basal direction.

Auxin affects plant growth in several ways

Like the gibberellins, auxin has many roles in plant development. It affects the vegetative and reproductive growth of plants in a number of ways.

ROOT INITIATION Cuttings from the shoots of some plants can produce roots and develop into entire new plants. For this to occur, certain undifferentiated cells in the interior of the shoot, originally destined to function only in food storage, must set off on a new mission: They must differentiate and become organized into the apical meristem of a new root. These changes are similar to those in the pericycle of a root when a lateral root forms (see Section 34.3).

Shoot cuttings of many species can be made to develop roots by dipping the cut surfaces into an auxin solution; this observation suggests that the plant's own auxin plays a role in the initiation of lateral roots. Commercial preparations that enhance the rooting of plant cuttings typically contain synthetic auxins.

Light and gravity affect the direction of plant growth

While polar auxin transport establishes the orientation of growth, *lateral* (side-to-side) redistribution of auxin is responsible for plant movements. This redistribution is carried out by an auxin carrier protein that moves to one side of the cell (as opposed to the base) and thus allows auxin to exit the cell only from that side.

When light strikes a grass coleoptile on one side, auxin at the tip moves laterally toward the shaded side. The imbalance thus established is maintained down the coleoptile, so that in the growing region below, the auxin concentration is highest on the shaded side. Cell growth is thus speeded up on that side, causing the coleoptile to bend toward the light (**phototropism**; **Figure 37.10A**). If you have noticed a houseplant bending toward a window, you have observed phototropism.

Even in the dark, auxin moves to the lower side of a shoot that has been tipped over, causing more rapid growth in the lower side and, hence, an upward bending of the shoot. Such growth in a direction determined by gravity is called **gravitropism** (**Figure 37.10B**). The upward gravitropic response of shoots is defined as *negative gravitropism;* that of roots, which bend downward, is *positive gravitropism.*

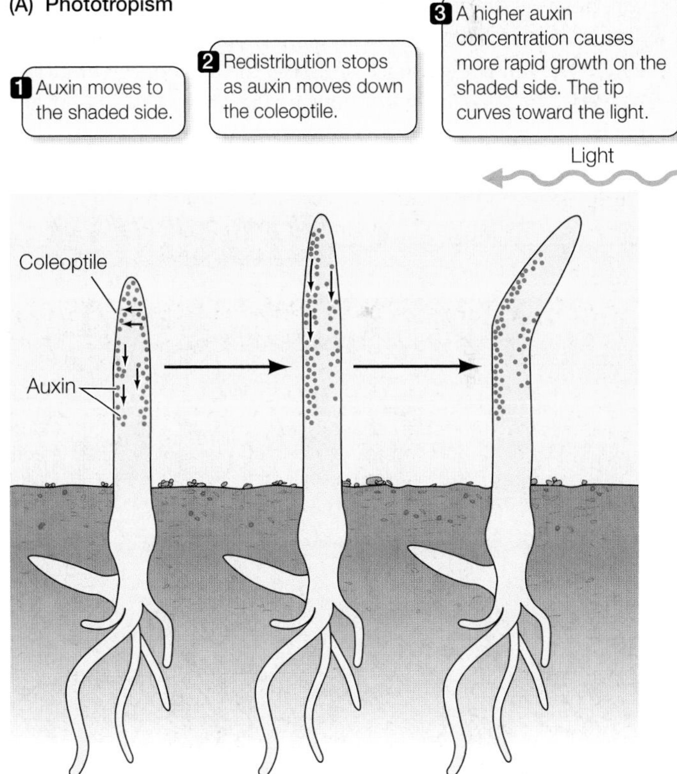

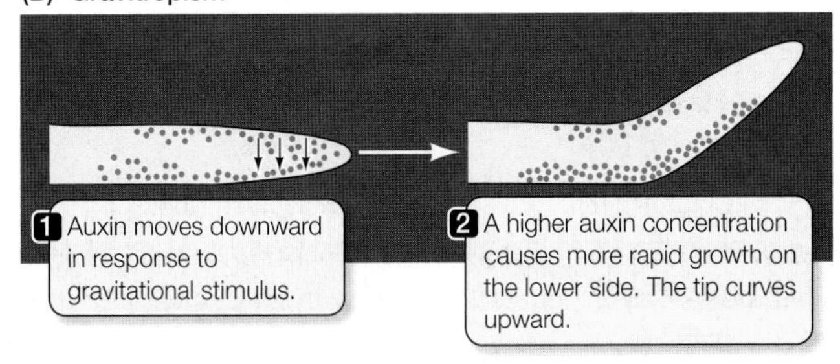

37.10 Plants Respond to Light and Gravity (A) Phototropism and (B) gravitropism occur in shoot apices in response to a redistribution of auxin.

LEAF ABSCISSION In contrast to its stimulatory effect on root initiation, auxin inhibits the detachment of old leaves from stems. This detachment process, called **abscission**, is the cause of autumn leaf fall. Most leaves consist of a blade and a petiole that attaches the blade to the stem. Abscission results from the breakdown of a specific part of the petiole, the *abscission zone* (**Figure 37.11**). If the blade of a leaf is cut off, the petiole falls from the plant more rapidly than if the leaf had remained intact. If the cut surface is treated with an auxin solution, however, the petiole remains attached to the plant, often longer than an intact leaf would have. The timing of leaf abscission in nature appears to be determined in part by a decrease in the movement of auxin, produced in the blade, through the petiole.

APICAL DOMINANCE Auxin helps maintain **apical dominance**, a phenomenon in which apical buds inhibit the growth of axillary buds, resulting in the growth of a single main stem with minimal branching. This phenomenon can be demonstrated by an experiment with young seedlings. If the plant remains intact, the stem elongates and the axillary buds remain inactive. Removal of the apical bud—the major site of auxin production—results in growth of the axillary buds. If the cut surface of the stem is treated with auxin, however, the axillary buds do not grow (**Figure 37.12**). The apical buds of branches also exert apical dominance: The axillary buds on the branch are inactive unless the apex of the branch is removed. That is why gardeners prune shrubs to encourage branching.

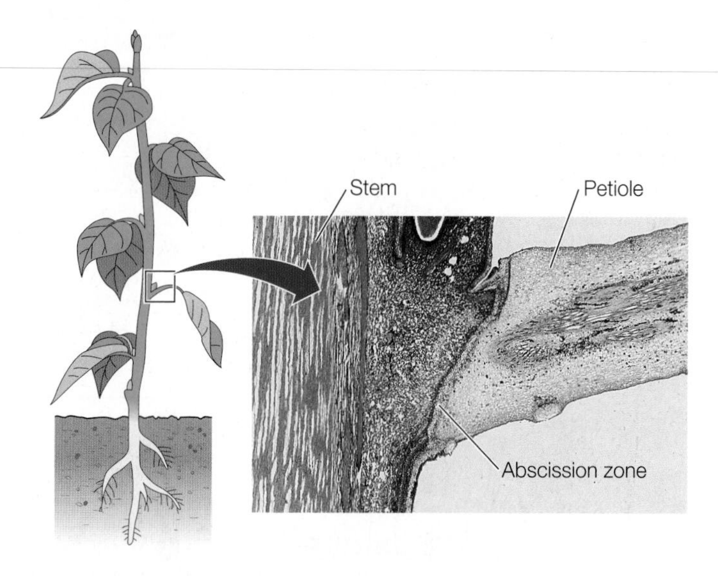

37.11 Changes Occur when a Leaf Is About to Fall The breakdown of cells in the abscission zone of the petiole causes the leaf to fall.

In the two experiments on leaves and stems just discussed, removal of a particular part of the plant elicits a response—abscission or loss of apical dominance—and that response is prevented by treatment with auxin. These results are consistent with other data showing that the excised part of the leaf or stem is an auxin source and that auxin in the intact plant helps maintain apical dominance and delays the abscission of leaves.

STEM AND ROOT ELONGATION Auxin promotes stem elongation but inhibits root elongation. The question of why different organs respond in opposite ways to the same growth hormone remains unanswered, but is a subject of current research.

FRUIT DEVELOPMENT Fruit development normally depends on prior fertilization of the egg, but in many species treatment of an unfertilized ovary with auxin or gibberellins causes **parthenocarpy**—fruit formation without fertilization. Parthenocarpic fruits form spontaneously in some cultivated varieties of plants, including seedless grapes, bananas, and some cucumbers.

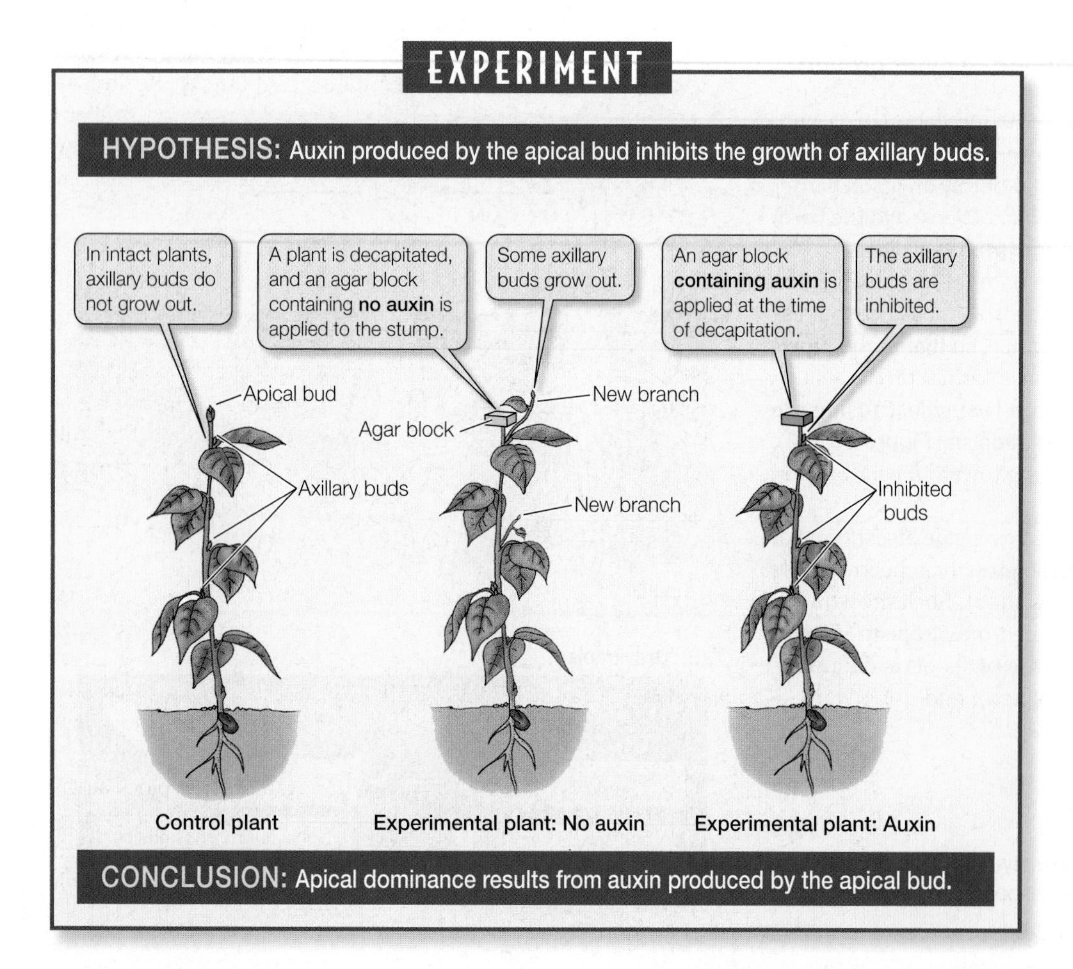

37.12 Auxin and Apical Dominance Auxin produced by the apical bud maintains apical dominance—the growth of a single main stem with minimal branching.

Auxin analogs as herbicides

Auxin is absolutely essential for plant survival; no mutants lacking auxin have ever been found. Many synthetic auxins—chemical analogs of indoleacetic acid—have been produced and studied. One of them, 2,4-dichlorophenoxyacetic acid (2,4-D), has the striking property of being lethal to eudicots at concentrations that are harmless to monocots. Because 2,4-D cannot be broken down by eudicots, it accumulates and causes the plant to "grow itself to death." This property makes 2,4-D an effective *selective herbicide* that can be sprayed on a lawn or a cereal crop to kill weeds that are eudicots.

All of the activities discussed so far illustrate the great diversity of important roles that auxin plays in plant growth and development. Now let's see *how* auxin plays one of its roles—promoting stem elongation through effects on the cell wall.

Auxin promotes growth by acting on cell walls

The expansion of plant cells is what causes plant growth. Thus the cell wall plays key roles in controlling the rate and direction of growth of a plant cell. Auxin acts on cell walls to regulate this process.

CELL EXPANSION The expansion of a plant cell is driven primarily by the uptake of water, which enters the cytoplasm of the cell and accumulates in its central vacuole (see Section 35.1). As the vacuole expands, the cell grows rapidly, with the vacuole often making up more than 90 percent of the volume of a mature cell. The vacuole presses the cytoplasm against the cell wall as it expands, and the wall resists this force.

The principal strengthening component of the plant cell wall is *cellulose*, a large polymer of glucose. In the wall, string-like cellulose molecules tend to associate in parallel with one another. Bundles of approximately 250 cellulose molecules make up *microfibrils* that are visible under an electron microscope (**Figure 37.13**). What makes the cell wall rigid is a network of cellulose microfibrils connected by bridges of other, smaller polysaccharides. The orientation of the majority of cellulose microfibrils determines the direction of cell expansion (**Figure 37.14**).

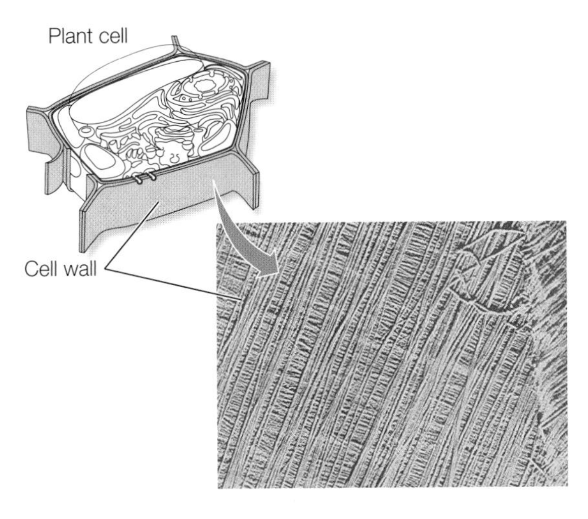

37.13 Cellulose in the Cell Wall The plant cell wall is a network of cellulose microfibrils linked by other polysaccharides. The crisscross pattern results from the deposition of successive layers within each of which the microfibrils are parallel.

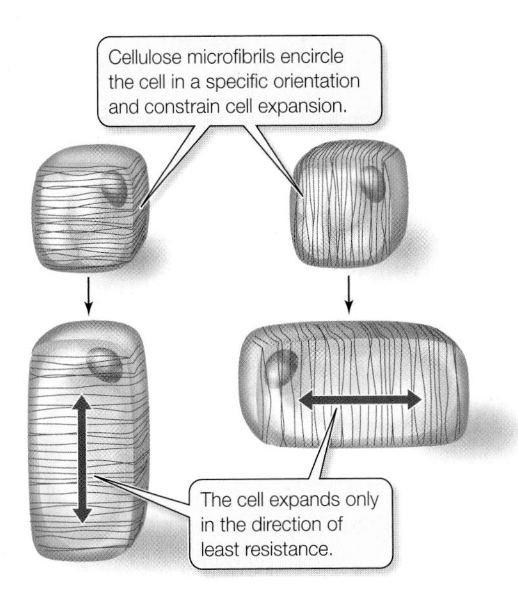

37.14 Plant Cells Expand The orientation of cellulose microfibrils in the plant's cell wall determines the direction of cell expansion.

For the cell to expand, its wall must loosen and be stretched. If the wall were only stretched, however, it would become thinner. Cell expansion involves more than stretching. New polysaccharides are deposited throughout the wall, and new cellulose microfibrils are deposited at the inner surface of the wall, maintaining its thickness. As a consequence of this pattern of cellulose deposition, the microfibrils in the outermost part of the wall are the oldest, and those in the innermost part the youngest. How do these properties of cell walls relate to the action of auxin on plant cell expansion?

CELL WALL LOOSENING Experiments with segments of oat coleoptiles have shown that plant cell walls recover incompletely from being stretched (**Figure 37.15**). Reversible stretching is called *elasticity*, and irreversible stretching is called **plasticity**. Treating the coleoptile segments with auxin before they were stretched significantly increased their plasticity; in other words, it loosened the cell walls. This result suggested that auxin-induced cell expansion might result from just such a loosening effect.

Auxin acts by causing the release of a "wall-loosening factor" from the cytoplasm. Studies in the 1970s indicated that the wall-loosening factor was sometimes simply hydrogen ions (protons, H^+). Acidifying the growth medium (that is, adding H^+) caused segments of stems or coleoptiles to grow as rapidly as segments treated with auxin. Furthermore, treating coleoptile segments with auxin caused acidification of the growth medium. Auxin increases the activity of proton pumps in the plasma membrane, increasing the H^+ concentration in the cell wall. Treatments that block acidification by auxin also block auxin-induced growth. These experimental results led to the hypothesis that hydrogen ions secreted into the cell wall activate one or more cell wall proteins. So, a search began for candidate proteins.

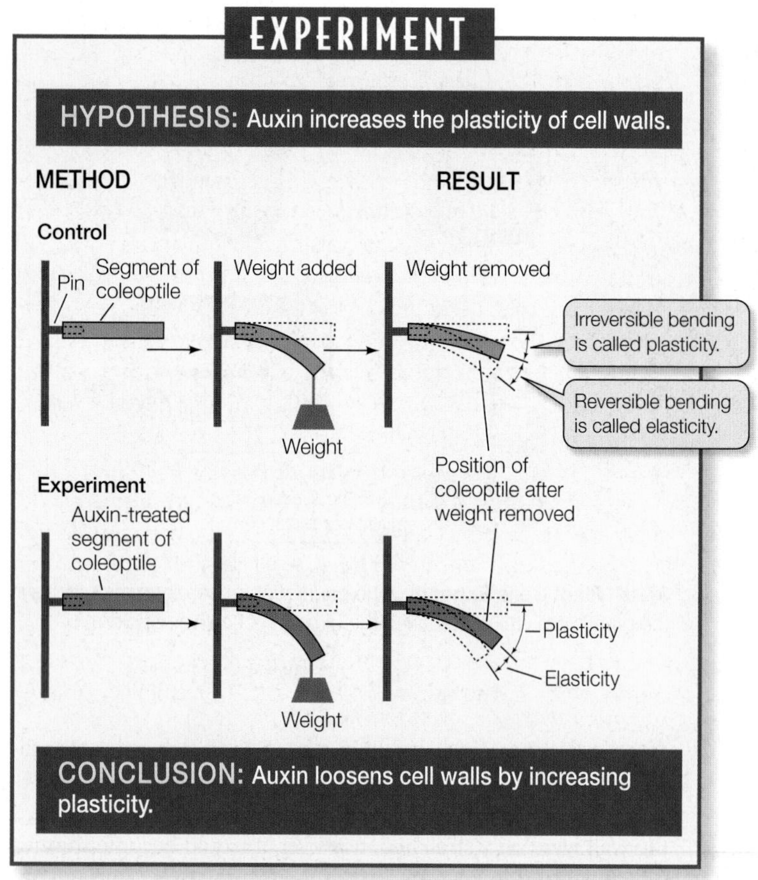

37.15 Auxin Acts on Cell Walls Auxin increases the plasticity, but not the elasticity, of cell walls. FURTHER RESEARCH: Design an experiment to determine whether gibberellins affect either wall plasticity or wall elasticity.

Proteins called *expansins* were isolated and purified from plant cell walls in the 1990s. When the purified proteins were added to isolated cell walls of several plant species, the walls expanded. Expansins are widespread among terrestrial plants. Furthermore, these proteins are activated by hydrogen ions. Expansins modify the pattern of hydrogen bonding between the polysaccharides in the plant cell wall. These modifications may allow polysaccharide macromolecules to slip past each other, so that the wall stretches and the cell expands.

Before auxin can initiate a chain of events such as the one just described, it must first be recognized as a signal by the cell. How does this recognition occur?

Auxin and gibberellins are recognized by similar mechanisms

The initial step in the action of any plant hormone is the binding of that hormone by a specific receptor protein. There are several proteins that can bind various plant hormones, but some of this binding may be nonspecific. To demonstrate that an auxin-binding protein is an auxin receptor, it must be shown that the protein actually mediates one or more of the effects of auxin, whether by promoting the expression of genes, by activating an enzyme, or by activating a proton pump. Different effects of auxin may involve different receptor proteins.

In 2005, more than 20 years after molecular biologists first identified an auxin-binding protein, workers at the University of Indiana and at the University of York (in the United Kingdom) showed conclusively that another protein binds auxin *and* then triggers a normal auxin response. Thus, it is a true auxin receptor. There may be others. When this receptor protein binds an auxin molecule the result is the destruction of proteins that directly inhibit genes involved in certain auxin responses. Thus, auxin indirectly promotes gene expression (**Figure 37.16**).

Almost exactly four months later, workers in Japan reported a strikingly similar finding—a gibberellin receptor protein that binds a gibberellin molecule, resulting in the destruction of proteins that directly inhibit transcription of certain genes. In other words, the mechanism governing the gibberellin receptor protein is exactly

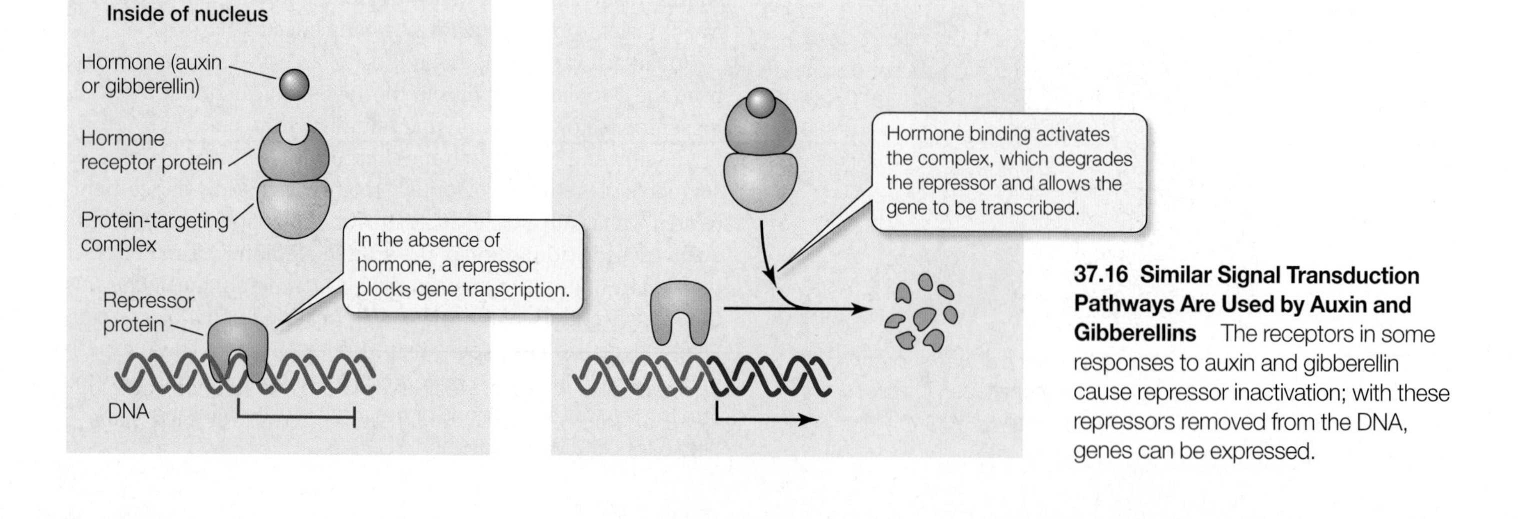

37.16 Similar Signal Transduction Pathways Are Used by Auxin and Gibberellins The receptors in some responses to auxin and gibberellin cause repressor inactivation; with these repressors removed from the DNA, genes can be expressed.

analogous to that for the auxin receptor protein. Both receptors reside in the nucleus, but they belong to different families of proteins.

37.3 RECAP

Auxin regulates stem elongation and mediates phototropism and gravitropism; it also plays roles in apical dominance, leaf abscission, root initiation, and other phenomena.

- What are auxin anion efflux carriers and how do they contribute to the polar transport of auxin? See p. 804 and Figure 37.9
- Can you explain why, even though auxin moves *away* from the lighted side of a coleoptile tip, the coleoptile bends *toward* the light? See p. 805 and Figure 37.10
- Can you describe how the signal transduction pathways for auxin and gibberellin are similar? See p. 808 and Figure 37.16

How can a single hormone, such as auxin or a gibberellin, have so many effects? As we have seen, a hormone may have more than one receptor and it may trigger more than one signal transduction pathway, and a single transduction pathway may affect more than one gene or membrane protein. We learn about other important plant hormones in the next section, and they, too, have multiple effects.

37.4 What Do Cytokinins, Ethylene, Abscisic Acid, and Brassinosteroids Do?

What signals tell different types of plant cells and organs to form? That is, what triggers differentiation of plant tissues? Much of the research on this question has been performed using plant tissues that have been removed from the plant body and grown in culture on an artificial medium. One easily grown tissue is pith—the spongy innermost tissue of a stem. Pith cells proliferate rapidly in culture, but show no differentiation. All the cells are similar and unspecialized; they grow into a lump of pith tissue on the surface of the culture medium.

Cutting a notch in the cultured pith tissue and inserting a stem tip into the notch causes the pith cells below the inserted tip to differentiate. Some of them differentiate to form water-conducting xylem cells. Differentiation of pith cells can also be initiated by adding a mixture of auxin and coconut milk (a product of the coconut endosperm and a rich source of plant hormones) to the notch.

It was Dutch plant physiologist Johannes van Overbeek who discovered in 1941 that adding coconut milk to culture medium caused a dramatic increase in the growth of plant embryos. Purification and analysis of coconut milk was instrumental in research that led to the identification of the cytokinins and other plant growth hormones.

A similar effect can be observed in intact plants. If notches are cut in the stems of coleus plants, interrupting some of the strands of vascular tissue, the strands gradually regenerate from the upper side of the cut to the lower side (recall that auxin moves from the apex to the base of a stem). If the leaves above the cut are removed, regeneration is slowed. However, when the missing leaves are replaced with an auxin solution, vascular tissue regenerates normally. These results show that auxin and other plant hormones signal the formation of specific cell types.

Experiments with cultured plant tissues have helped clarify which hormones control organ formation. The eventual fate of an undifferentiated cell can be adjusted depending on hormone signals. Undifferentiated cultures of tobacco pith form roots when treated with an appropriate concentration of auxin. Another group of plant growth hormones—the **cytokinins**—causes buds and then shoots to form in such cultures. The pattern of organ formation depends on the ratio of auxin to cytokinin in the medium. A high proportion of auxin favors roots, and a high proportion of cytokinin favors buds, but both processes are most active when both hormones are present. These results show that other hormones can modify the effects of auxin, reminding us that plant growth is regulated more by hormone interactions than by a single hormone.

Cytokinins are active from seed to senescence

In studies of plant cell division, botanists discovered that the cytokinins powerfully stimulate cell division in tissue cultures. Cytokinins consist of adenine (the nitrogenous base "A" in DNA and RNA) with an attached group. Two closely related cytokinins, called *zeatin* and *isopentenyl adenine*, occur naturally in plants. A third, *kinetin*, may be considered a synthetic cytokinin because it has never been isolated from plant tissue.

No mutants lacking cytokinins have ever been found. Thus, like auxin, cytokinins seem to be required throughout the life of a plant. Cytokinins are synthesized primarily in the roots and move to other parts of the plant. They have a number of different effects:

- Adding an appropriate combination of auxin and cytokinins to a growth medium induces rapid cell proliferation in cultured plant tissues.
- Cytokinins can cause certain light-requiring seeds to germinate even when kept in constant darkness.
- Cytokinins usually inhibit the elongation of stems, but they cause lateral swelling of stems and roots (the fleshy roots of radishes are an extreme example).
- Cytokinins stimulate axillary buds to grow into branches; thus the balance between auxin and cytokinin levels controls the extent of branching (bushiness) of a plant.
- Cytokinins increase the expansion of cut pieces of leaf tissue in culture and may regulate normal leaf expansion.
- Cytokinins delay the senescence of leaves. If leaf blades are detached from a plant and floated on water or a nutrient solution, they quickly turn yellow and show other signs of senescence. If instead they are floated on a solution containing a cytokinin, they remain green and senesce much more slowly.

One effect of cytokinins is to support the production of seeds by rice. Plant biologists in Japan developed a rice variety that combines enhanced seed production and reduced plant height. By reducing the expression of a gene that encodes cytokinin oxidase, an enzyme that degrades cytokinins, they increased cytokinin levels in the developing heads, greatly enhancing grain yield. Why reduce the height? Rice is subject to *lodging*—excessive bending after wind or rain storms—which makes harvesting difficult. Shorter plants bear the increased load and produce a harvestable crop.

Ethylene is a gaseous hormone that hastens leaf senescence and fruit ripening

Whereas the cytokinins delay senescence, another plant hormone promotes it: the gas **ethylene** ($H_2C{=}CH_2$), which is sometimes called the senescence hormone. Ethylene can be produced by all parts of the plant, and like all plant hormones it has several effects.

Back when streets were lit by gas rather than by electricity, leaves on trees near street lamps dropped earlier than those on trees farther from the lamps. We now know that ethylene, a combustion product of the illuminating gas, is what caused the early abscission. Auxin delays leaf abscission, but ethylene strongly promotes it; thus a balance of auxin and ethylene controls abscission.

FRUIT RIPENING By promoting senescence, ethylene also speeds the ripening of fruit. As the fruit ripens, it loses chlorophyll and its cell walls break down; ethylene promotes both of these processes. Ethylene also causes an increase in its own production. Thus, once ripening begins, more and more ethylene forms, and because it is a gas, it diffuses readily throughout the fruit and even to neighboring fruits on the same or other plants.

The old saying "one rotten apple spoils the barrel" is true. That rotten apple is a rich source of ethylene, which speeds the ripening and subsequent rotting of the other fruit in a barrel or other confined space.

Farmers in ancient times slashed developing figs to hasten their ripening. We now know that wounding causes an increase in ethylene production by the fruit and that the raised ethylene level promotes ripening. Today commercial shippers and storers of fruit hasten ripening by adding ethylene to storage chambers. This use of ethylene is the single most important use of a natural plant hormone in agriculture and commerce—even more important than its use in rubber plantations to prevent the latex from coagulating, as described at the beginning of the chapter. Ripening can also be delayed by the use of "scrubbers" and adsorbents that remove ethylene from the atmosphere in fruit storage chambers.

As flowers senesce, their petals may abscise, to the detriment of the cut-flower industry. Growers and florists often immerse the cut stems of ethylene-sensitive flowers in dilute solutions of silver thiosulfate before sale. Silver salts inhibit ethylene action, probably by interacting directly with the ethylene receptor, and thus delay senescence—enabling florists and consumers to keep their cut flowers from dropping their petals prematurely.

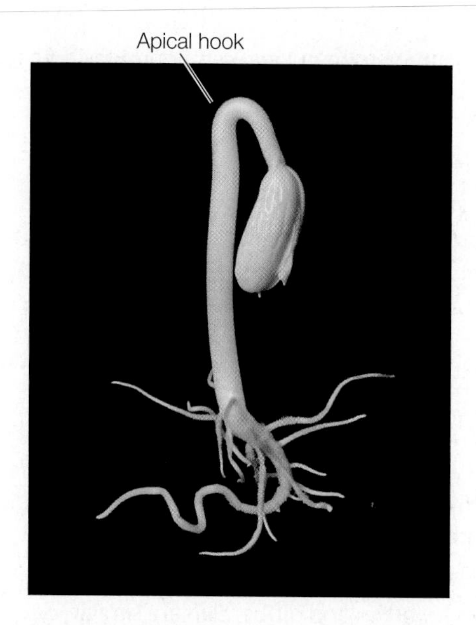

37.17 The Apical Hook of a Eudicot Asymmetrical production of ethylene is responsible for the apical hook of this bean seedling. The ethylene concentration was highest on the right side, so more rapid growth on the left caused and maintained the hook.

STEM GROWTH Although it is associated primarily with senescence, ethylene is active at other stages of plant development as well. The stems of many eudicot seedlings form an **apical hook** that protects the delicate shoot apex while the stem grows through the soil (**Figure 37.17**). The apical hook is maintained through an asymmetrical production of ethylene gas, which inhibits the elongation of cells on the inner surface of the hook. Once the seedling breaks through the soil surface and is exposed to light, ethylene synthesis stops, and the cells of the inner surface are no longer inhibited. These cells now elongate, and the hook unfolds, raising the shoot apex and the expanding leaves into the sun.

Ethylene also inhibits stem elongation in general, promotes lateral swelling of stems (as do the cytokinins), and causes stems to lose their sensitivity to gravitropic stimulation. Together, these three phenomena constitute the *triple response*, a well-characterized stunted growth habit observed when plants are treated with ethylene.

THE ETHYLENE SIGNAL TRANSDUCTION PATHWAY Analysis of *Arabidopsis* mutants has revealed the mechanism of ethylene action. Some of these mutants do not respond to applied ethylene, and others act as if they have been exposed to ethylene even though they have not. Studies of the mutant genes and their protein products, coupled with comparisons of their amino acid sequences with those of other known proteins, have revealed some of the details of the signal transduction pathway through which ethylene produces its effects (**Figure 37.18**). The pathway includes two membrane proteins in the endoplasmic reticulum. The first is an ethylene receptor (labeled A in the figure) and the second is a channel (C). In the absence of ethylene, another protein (B) keeps C inactive. When A binds ethylene it inactivates B. Without B to inactivate it, C acts through a second messenger to activate a transcription factor (D). The transcription factor turns on the genes that produce ethylene's effects in the cell. Ethylene was the first plant hormone to have its mechanism of action elucidated in this way.

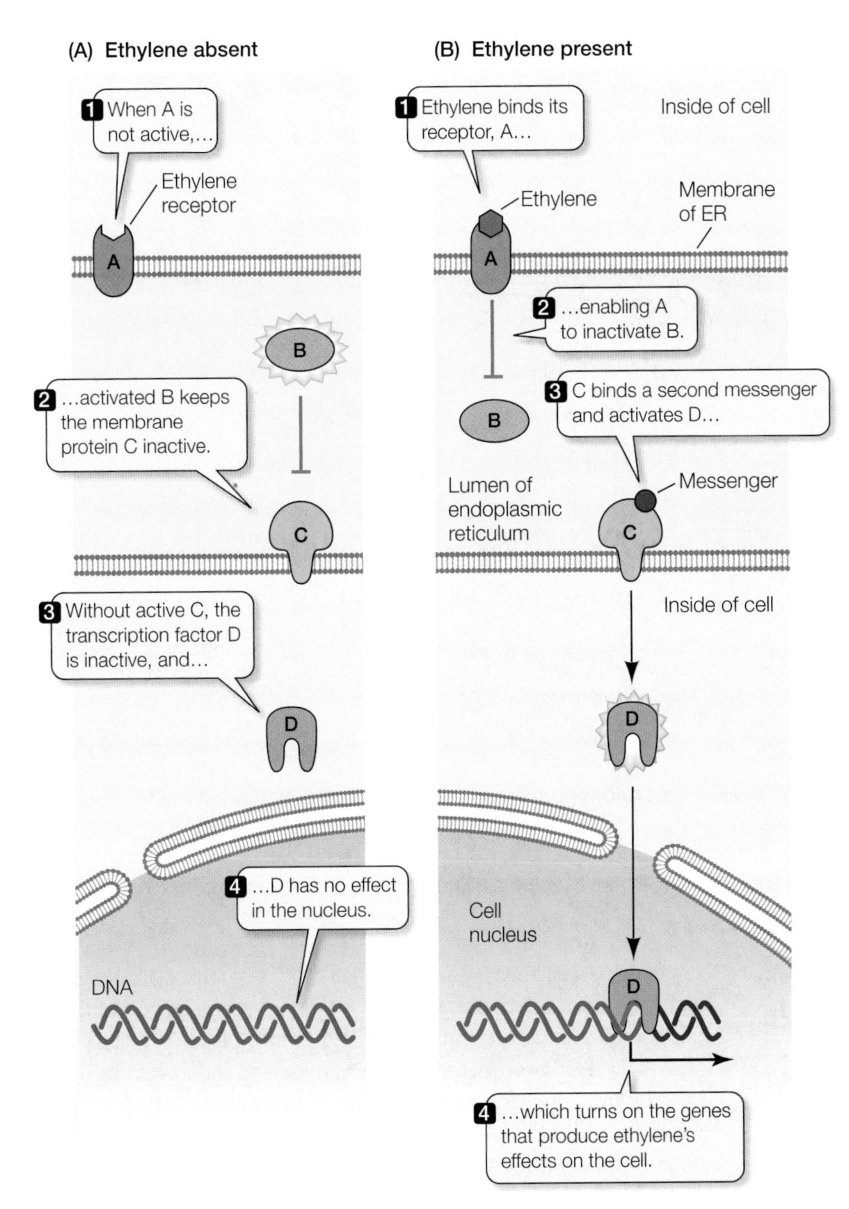

37.18 The Signal Transduction Pathway for Ethylene This diagram shows the roles of four proteins (A, B, C, and D) in the signal transduction pathway through which ethylene exerts its many effects.

Abscisic acid is the "stress hormone"

Abscisic acid is another hormone with multiple effects in the living plant. During seed formation, abscisic acid promotes the accumulation of storage proteins by allowing the expression of the genes that encode those proteins. It is generally present in high concentrations in dormant buds and some dormant seeds, and it is probably the most common of the chemical inhibitors that initiate and maintain dormancy in mature seeds. Abscisic acid also inhibits stem elongation. It is sometimes referred to as the stress hormone of plants, because it accumulates when plants are deprived of water and it may play a role in maintaining the dormancy of buds in winter. Like the gibberellins, abscisic acid belongs to the class of compounds called terpenes.

Sometimes seed dormancy ends prematurely. Some mutant corn plants, called *vp* mutants, have seeds that germinate while still attached to the cob on the parent plant—a condition called **vivipary** (Latin *vivus*,"alive"; *parere*,"give birth"). Several *vp* mutants are naturally deficient in abscisic acid. Applying abscisic acid to these mutants reduces their tendency to show vivipary. Another type of *vp* mutant fails to respond in any way to applied abscisic acid. These results indicate that abscisic acid is the inhibitor that normally prevents seeds from germinating while still attached to the parent plant. The first type of mutant cannot make enough abscisic acid; the second type of mutant is viviparous because it cannot respond to abscisic acid—its own or any applied to it.

Abscisic acid also regulates gas and water vapor exchange between leaves and the atmosphere through its effects on the guard cells of the leaf stomata. Abscisic acid causes stomata to close, and it also prevents the opening of stomata normally caused by light. Both of these effects involve ion channels in the plasma membrane of the guard cells. The first response of a guard cell to abscisic acid is the opening of calcium channels and the entry of calcium into the cell. This calcium causes the cell's vacuole to release calcium, too. The increased concentration of calcium in the cytoplasm leads to a chain of events that result in the opening of potassium channels, the loss of K^+ and water from the cytoplasm, and the closing of the stoma as the guard cells sag together.

Brassinosteroids are hormones that mediate effects of light

More than 20 years ago, biologists isolated an interesting steroid from the pollen of rape, a member of the Brassicaceae, or mustard family. When applied to various plant tissues, this **brassinosteroid** stimulated cell elongation, pollen tube elongation, and vascular tissue differentiation, but it inhibited root elongation. Since then, dozens of chemically related and growth-affecting brassinosteroids have been found in plants. Treatment with as little as a few nanograms of brassinosteroid per plant is enough to promote growth.

The properties of an *Arabidopsis* mutant called *det2* made it clear that brassinosteroids are naturally occurring plant hormones. When grown in darkness, seedlings homozygous for the *det2* allele differ dramatically from wild-type seedlings grown under the same conditions: In many respects, they look like wild-type seedlings grown in the light. Treatment of dark-grown *det2* mutant seedlings with brassinosteroids causes them to grow normally—that is, like wild-type plants grown in the dark. These results, supported by chemical analysis, showed that *det2* plants are unable to synthesize their own brassinosteroids, and that lack of the hormone results in abnormal growth. The gene product DET2 is a link between hormones and photoreceptors throughout the life cycle of *Arabidopsis*.

The receptor protein and signal transduction pathway for brassinosteroids differ sharply from those for steroid hormones in animals. Receptors for animal steroid hormones exist in the cytoplasm of animal cells and move to the nucleus when bound by steroids. In contrast, the receptor for brassinosteroids is an integral protein in the plasma membrane. Binding of a brassinosteroid by the receptor inactivates the protein that turns certain genes on and others off (**Figure 37.19**).

Some of the effects of light on plant development result from effects on the signal transduction pathway for brassinosteroids. Others may result from alterations in brassinosteroid levels in the plant.

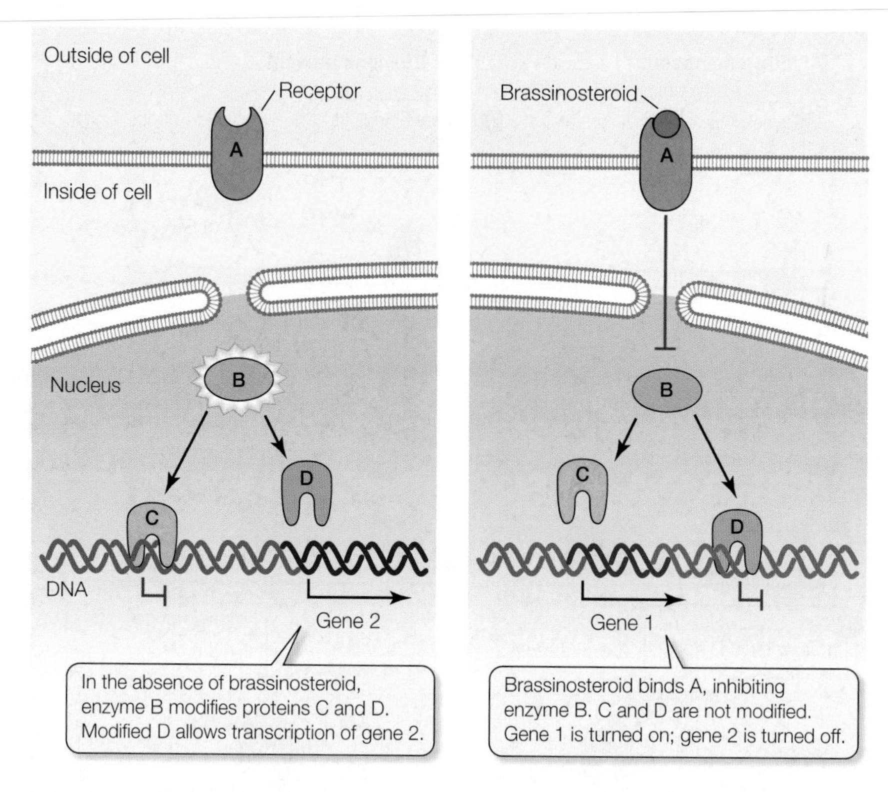

37.19 The Brassinosteroid Signal Transduction Pathway Begins at the Plasma Membrane Unlike the receptors for animal steroid hormones, the brassinosteroid receptor is a membrane protein. The signal transduction pathway concludes by activating certain genes and inactivating others.

37.4 RECAP

Cytokinins, abscisic acid, ethylene, and brassinosteroids work in concert with auxin and gibberellins to mediate plant development. Plant growth is regulated more by hormone interactions than by a single hormone.

- How do cytokinins interact with auxin to regulate a plant's development? See p. 809
- Can you explain how the experiments with *vp* mutants revealed the role of abscisic acid in preventing premature seed germination? See p. 811

Plant response to light—the energy source for photosynthesis—is crucial. We saw how their pioneering investigations of phototropism led the Darwins to the discovery of auxin. Let's now look more closely at how plants respond to light.

37.5 How Do Photoreceptors Participate in Plant Growth Regulation?

The length of the night determines the onset of winter dormancy in many plant species. As summer progresses, the days become shorter (that is, the nights become longer). Leaves have a mechanism for measuring the length of the night, as we will see in the next chapter. Measuring night length is an accurate way to determine the season of the year. If a plant determined the season only by the temperature, it might be fooled by a winter warm spell or by unseasonably cold weather in the summer. The length of the night, on the other hand, is determined by Earth's rotation around the sun and, for a given latitude, does not vary randomly. Plants use the environmental cue of night length to time several aspects of their growth and development.

Night length is only one of several environmental cues detected by plants. Environmental conditions are also signaled by light's presence or absence, its intensity, and its spectral properties (specific wavelengths). Even though temperature can vary from its seasonal norms, it nevertheless also provides important environmental cues, both by its value at any particular time and by the distribution of warmer and colder stretches over a period of time. (Temperature cues are considered further in Chapter 38.) The plant senses these environmental cues and then responds, often by stepping up or decreasing its production of hormones. In this section, we focus on light cues: how certain photoreceptors sense light, its duration, and its wavelength distribution.

Light regulates many aspects of plant development in addition to phototropism. Light affects seed germination, shoot elongation, the initiation of flowering, and many other important aspects of plant development. Several photoreceptors take part in these processes. Five **phytochromes** mediate the effects of red and dim blue light. Three or more types of **blue-light receptors**, discovered more recently, mediate the effects of higher-intensity blue light.

Phytochromes mediate the effects of red and far-red light

Some seeds will not germinate in darkness, but do so readily after even a brief exposure to dim light. Blue and red light are highly effective in promoting germination.

Of particular importance to plants is the fact that far-red light *reverses* the effect of a prior exposure to red light. Far-red light is a very deep red, bordering on the limit of human vision and centered on a wavelength of 730 nm; red wavelengths are around 660 nm. If exposed to brief, alternating periods of red and far-red light in close succession, lettuce seeds respond only to the final exposure: If it is red, they germinate; if it is far-red, they remain dor-

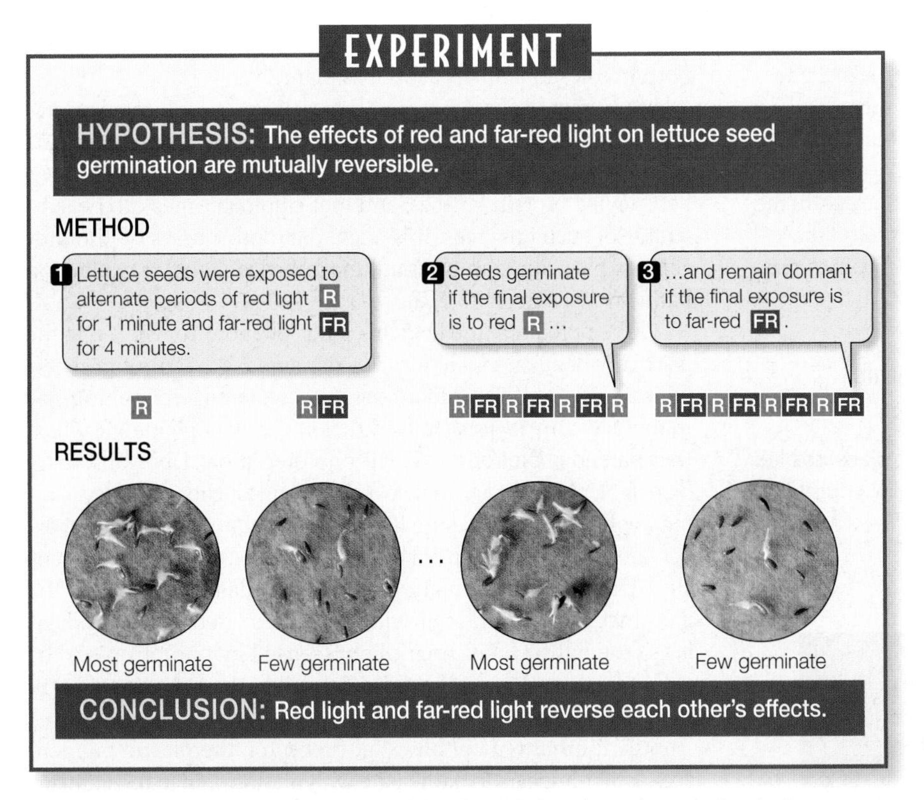

37.20 Sensitivity of Seeds to Red and Far-Red Light In each case, the final exposure determines the seed's germination response.

mant (**Figure 37.20**). This reversibility of the effects of red and far-red light regulates many other aspects of plant development, including flowering and seedling growth.

The basis for these effects of red and far-red light resides in certain bluish photoreceptor proteins called phytochromes. In the cytosol of plants are two interconvertible forms of phytochromes. Light drives the interconversion of the two forms. The form that absorbs principally red light is called $\mathbf{P_r}$. Upon absorption of a photon of red light, a molecule of P_r is converted into $\mathbf{P_{fr}}$. The P_{fr} form absorbs far-red light; when it does so, it is converted to P_r.

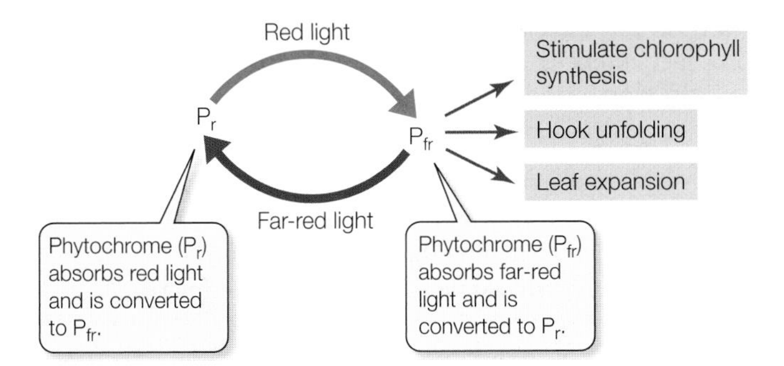

P_{fr} has some important biological effects. As we have just seen, one of them is to initiate germination in certain seeds, such as lettuce.

Phytochromes have many effects on plant growth and development

Phytochromes help to regulate a seedling's early growth. When seeds germinate in the dark below the soil surface, a pale and spindly seedling forms, with undeveloped leaves. Such an **etiolated** seedling cannot carry out photosynthesis. The seedling shoot must reach the soil surface and begin photosynthesis before its nutrient reserves are expended and it starves.

> Etiolated plants can be gourmet fare. "Bean sprouts"—a delicacy in China for more than 3,000 years— are mung bean seedlings that are allowed to germinate for several days in either dim light or total darkness. Sprouts grown in total darkness are the whitest, and also the crispest. Similarly, white asparagus spears are grown under a sand cover that keeps out light and allows a distinctive flavor to develop.

Plants have evolved a variety of ways to cope with the problem of germinating underground. Etiolated flowering plants, for example, do not form chlorophyll. Only when they are exposed to light do they synthesize chlorophyll and turn green; this delay conserves the resources needed to make chlorophyll, which would be useless in the dark. An etiolated shoot uses its stored resources to elongate rapidly and hasten its arrival at the soil surface, where photosynthesis quickly begins. To break through the soil yet protect its delicate, underdeveloped leaves, the shoot of an etiolated eudicot seedling forms an apical hook (see Figure 37.17).

All of these etiolation-associated phenomena (lack of chlorophyll, rapid shoot elongation, production of an apical hook, delayed leaf expansion) are regulated by the phytochromes. In a seedling that has never been exposed to light, all the phytochrome is in the P_r (red-absorbing) form. Exposure to light converts P_r to the P_{fr} (far-red-absorbing) form. The P_{fr} initiates a reversal of the etiolation phenomena: Chlorophyll synthesis begins, shoot elongation slows, the apical hook unfolds, and the leaves begin to expand. These changes constitute *photomorphogenesis*.

Multiple phytochromes have different developmental roles

For years, plant biologists had difficulty accounting for some aspects of phytochrome action. A solution to these problems may lie in the discovery of multiple forms of phytochromes and other photoreceptors. *Arabidopsis* has five genes that encode different phytochromes: such diversity has been found throughout the green plants.

The several phytochromes play differing roles during plant development. Some of them even play off each other to fine-tune plant growth during the day. Consider, for example, the light spectrum available to a seedling that is growing in the shade of other plants. Because chlorophyll in the leaves above it absorbs the light first, the shaded seedling receives a spectrum that is relatively rich in far-red wavelengths (and poor in red wavelengths); the ratio of far-red to red is increased as much as 10 to 20 times in the shade. In some shade-intolerant species, the interplay among signal transduction pathways initiated by the different phytochromes leads to an increased rate of stem elongation, moving the shaded leaves to higher light intensities.

Phytochromes act in both the cytoplasm and the nucleus. In the cytoplasm, phytochromes probably function as protein kinases. When they are transferred to the nucleus, phytochromes promote the transcription of many genes.

Cryptochromes, phototropins, and zeaxanthin are blue-light receptors

Cryptochromes are yellow photoreceptor pigments that absorb blue and ultraviolet light. They affect some of the same developmental processes, including seedling development and flowering, that phytochromes do. Unlike phytochromes, cryptochromes play important roles in animals as well as plants.

In contrast to phytochromes, cryptochromes are located primarily in the plant cell nucleus. The exact mechanism of cryptochrome action is not yet known. It may be significant that phytochromes behave like protein kinases, and that cryptochromes can be substrates of such enzymes. It is likely that both classes of photoreceptors participate in protein kinase-based signal transduction pathways (see Section 15.3).

We've noted that the Darwins' study of phototropism led to the discovery of auxin. But a question remained: What photoreceptor initiates phototropism? Plant scientists working with phototropic mutants of *Arabidopsis* showed that two yellow pigments, which they named **phototropins**, are the photoreceptors. Upon absorbing blue light, phototropins initiate a signal transduction pathway leading to phototropic curvature (**Figure 37.21**). Phototropins also participate in the relocation of chloroplasts in response to light intensity. Under low intensity light, chloroplasts gather on the illuminated side of the cell, with their faces oriented perpendicular to the light. Under bright light (sensed by phototropins) the chloroplasts gather on the sides parallel to the light and present their edges, thus absorbing a minimum of light.

Still another type of blue-light receptor, the plastid pigment **zeaxanthin**, appears to participate with phototropin in the light-induced opening of stomata. Zeaxanthin itself is formed in the guard cells in response to light, but it is light absorbed by the zeaxanthin and phototropin that causes the stomata to open.

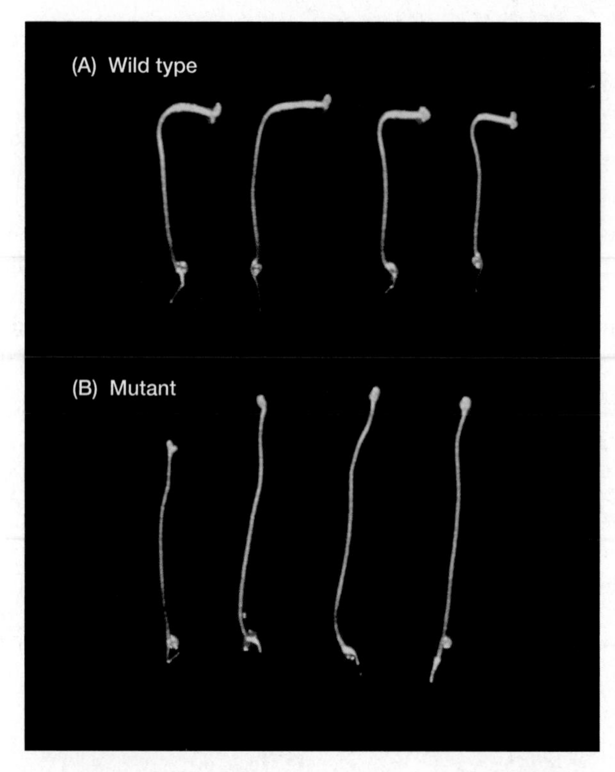

37.21 A Nonphototropic Mutant (A) The four etiolated wild-type *Arabidopsis* seedlings in the top row are demonstrating normal phototropism. (B) These four mutant seedlings cannot produce phototropin. Their lack of phototropic response indicates that phototropin is the photoreceptor that signals the plant to curve toward light.

37.5 RECAP

Phytochromes and blue-light receptors mediate the effects of light on plant growth and development. Several phytochromes exist, each in two forms that are interconvertible by red and far-red light.

- Do you understand why red light affects seed germination differently from far-red light? See p. 813 and Figure 37.20
- Can you explain how the behavior of an etiolated seedling helps it reach the light it needs in order to develop further? See p. 813

Photoreceptors also play a regulatory role in flowering. In addition to light, another environmental cue—temperature—regulates flowering. We will examine these topics and others in the next chapter, which focuses on reproduction in flowering plants.

CHAPTER SUMMARY

37.1 How does plant development proceed?

The environment, photoreceptors, hormones, and the plant's genome all play roles in the regulation of plant development.

Each of several plant hormones plays multiple regulatory roles, affecting several different aspects of development. Interactions among these hormones are often complex. Review Table 37.1

Seed dormancy, which offers adaptive advantages, may be caused by many mechanisms. In nature, it is broken by mechanisms such as abrasion, fire, leaching, and low temperatures. When **dormancy** ends and the seed **imbibes** water, it **germinates** and develops into a **seedling**. Review Figure 37.1, Web/CD Activities 37.1 and 37.2

Hormones and photoreceptors act through signal transduction pathways to regulate seedling development. Before the germinating embryo can begin photosynthesis, it relies on energy reserves in the **cotyledons** and the **endosperm** to grow.

37.2 What do gibberellins do?

The embryos of cereal seeds secrete **gibberellins**, which cause the **aleurone layer** to synthesize and secrete digestive enzymes that break down the large molecules stored in the endosperm. Review Figure 37.4, Web/CD Activity 37.3

Dozens of gibberellins exist. These hormones regulate the growth of stems and of some fruits and cause **bolting** in some **biennial** plants.

37.3 What does auxin do?

See Web/CD Tutorial 37.1

Auxin moves from the tip to the growing region of the **coleoptile**. Review Figures 37.7, 37.8, Web/CD Tutorial 37.2

Auxin transport is polar. **Auxin anion efflux carriers**—membrane proteins confined to the basal ends of cells—cause auxin to move from the tip to the base of the shoot. Review Figure 37.9

Lateral movement of auxin, mediated by auxin carrier proteins, is responsible for **phototropism** and **gravitropism**. Review Figure 37.10

Auxin plays roles in root formation, leaf **abscission**, **apical dominance**, and **parthenocarpic fruit development**. Certain synthetic auxins are used as selective herbicides. Review Figures 37.11 and 37.12

Auxin increases the **plasticity** of the cell wall, promoting cell expansion. It does so by increasing the pumping of protons from the cytoplasm into the cell wall, where the lowered pH activates proteins called expansins. Review Figure 37.15, Web/CD Tutorial 37.3

Receptors for auxin and for gibberellins residing in the cell nucleus initiate similar signal transduction pathways. Review Figure 37.16

37.4 What do cytokinins, ethylene, abscisic acid, and brassinosteroids do?

Cytokinins are adenine derivatives. They promote plant cell division, promote seed germination in some species, inhibit stem elongation, promote lateral swelling of stems and roots, stimulate the growth of axillary buds, promote the expansion of leaf tissue, and delay leaf **senescence**.

A balance between auxin and **ethylene** controls leaf abscission. Ethylene promotes senescence and fruit ripening. It causes the formation of a protective **apical hook** in eudicot seedlings. In stems, it inhibits elongation, promotes lateral swelling, and causes a loss of gravitropic sensitivity. Review Figure 37.18

Abscisic acid maintains winter dormancy in buds, prevents seeds from germinating while still attached to the parent plant, and inhibits stem elongation. It also affects stomatal opening.

Dozens of different **brassinosteroids** affect cell elongation, pollen tube elongation, vascular tissue differentiation, and root elongation. Some effects of light are mediated by changes in the action and levels of brassinosteroids. Review Figure 37.19

37.5 How do photoreceptors participate in plant growth regulation?

Phytochromes are bluish pigments found in the cytosol. They exist in two forms, $\mathbf{P_r}$ and $\mathbf{P_{fr}}$, that are interconvertible by light. They affect seedling growth, flowering, and etiolation. Review Figure 37.20

The five known phytochromes mediate the effects of red, far-red, and dim blue light. They may play different roles in plant development, and their signal transduction pathways may interact to mediate the effects of light environments of differing spectral distribution.

The **blue-light receptors** are yellow pigments that absorb blue and ultraviolet light. **Cryptochromes**, which mediate the effects of high-energy light, interact with phytochromes in controlling seedling development and floral initiation. The signaling pathways for phytochromes and cryptochromes are based on protein kinases.

Other blue-light receptors are **phototropins**, the photoreceptors for phototropism and chloroplast movements, and **zeaxanthin**, which with the phototropins mediates the light-induced opening of stomata.

SELF-QUIZ

1. Which of the following is *not* an advantage of seed dormancy?
   a. It makes the seed more likely to be digested by birds that disperse it.
   b. It counters the effects of year-to-year variations in the environment.
   c. It increases the likelihood that a seed will germinate in the right place.
   d. It favors dispersal of the seed.
   e. It may result in germination at a favorable time of year.
2. Which of the following does/do *not* participate in seed germination?
   a. Imbibition of water
   b. Metabolic changes
   c. Growth of the radicle
   d. Mobilization of nutrient reserves
   e. Extensive mitotic divisions
3. To mobilize its nutrient reserves, a germinating barley seed
   a. becomes dormant.
   b. undergoes senescence.
   c. secretes gibberellins into its endosperm.
   d. converts glycerol and fatty acids into lipids.
   e. takes up proteins from the endosperm.

4. The gibberellins
   a. are responsible for phototropism and gravitropism.
   b. are gases at room temperature.
   c. are produced only by fungi.
   d. cause bolting in some biennial plants.
   e. inhibit the synthesis of digestive enzymes by barley seeds.
5. In coleoptile tissue, auxin
   a. is transported from base to tip.
   b. is transported from tip to base.
   c. can be transported toward either the tip or the base, depending on the orientation of the coleoptile with respect to gravity.
   d. is transported by simple diffusion, with no preferred direction.
   e. is not transported, because auxin is used where it is made.
6. Which process is *not* directly affected by auxin?
   a. Apical dominance
   b. Leaf abscission
   c. Synthesis of digestive enzymes by barley seeds
   d. Root initiation
   e. Parthenocarpic fruit development
7. Plant cell walls
   a. are strengthened primarily by proteins.
   b. often make up more than 90 percent of the total volume of an expanded cell.
   c. can be loosened by an increase in pH.
   d. become thinner and thinner as the cell grows longer and longer.
   e. are made more plastic by treatment with auxin.
8. Which statement about cytokinins is *not* true?
   a. They promote bud formation in tissue cultures.
   b. They delay the senescence of leaves.
   c. They usually promote the elongation of stems.
   d. They cause certain light-requiring seeds to germinate in the dark.
   e. They stimulate the development of branches from axillary buds.
9. Ethylene
   a. is antagonized by silver salts such as silver thiosulfate.
   b. is liquid at room temperature.
   c. delays the ripening of fruits.
   d. generally promotes stem elongation.
   e. inhibits the swelling of stems, in opposition to cytokinins' effects.
10. Phytochrome
   a. is the only photoreceptor pigment in plants.
   b. exists in two forms interconvertible by light.
   c. is a pigment that is colored red or far-red.
   d. is a green-light receptor.
   e. is the photoreceptor for phototropism.

FOR DISCUSSION

1. How may it be advantageous for some species to have seeds whose dormancy is broken by fire?
2. Cocklebur fruits contain two seeds each that are kept dormant by two different mechanisms. How might this use of two mechanisms of dormancy be advantageous to cockleburs?
3. Whereas relatively low concentrations of auxin promote the elongation of segments cut from young plant stems, higher concentrations generally inhibit their growth, as shown:

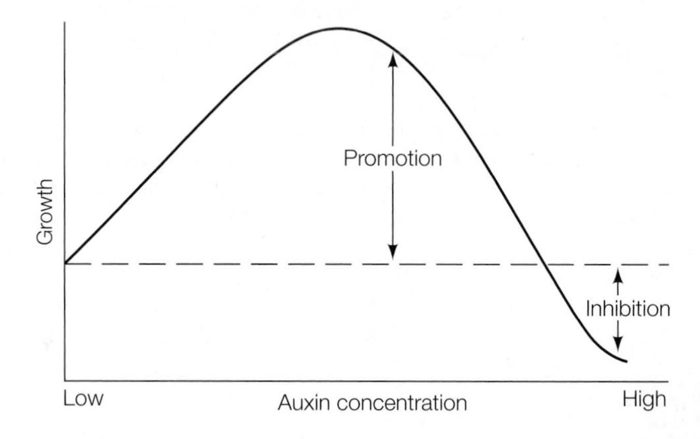

   In some plants, the inhibitory effects of high auxin concentrations appear to be secondary: High auxin concentrations cause the synthesis of ethylene, which is what causes the growth inhibition. Silver thiosulfate inhibits ethylene action. How do you think the addition of silver thiosulfate to the solutions in which the stem segments grew would affect the appearance of the graph at the left?
4. Corn stunt virus causes a great reduction in the growth rate of infected corn plants, so the diseased plants take on a dwarfed form. Since their appearance is reminiscent of the genetically dwarfed corn studied by Phinney, you suspect that the virus may inhibit the synthesis of gibberellins by the corn plants. Describe two experiments you might conduct to test this hypothesis, only one of which should require chemical measurement.
5. Some etiolated seedlings develop hairs on their epidermis when exposed to dim light. Describe an experiment to test the hypothesis that a phytochrome is the photoreceptor for this effect.

FOR INVESTIGATION

A. N. J. Heyn investigated the effects of auxin on the plasticity and elasticity of cell walls by bending segments of oat coleoptiles after treating them with or without auxin. Suggest other ways by which one might measure the plasticity and elasticity of cell walls with and without treatment with auxin or exposure to low pH.

CHAPTER 38 Reproduction in Flowering Plants

Big smelly flowers of Sumatra

The British trading colony of Singapore was founded in 1819 by Sir Thomas Stamford Raffles. Today Singapore is a dynamic, independent city-country whose Raffles Hotel is famous as the home of the tropical gin drink known as the Singapore Sling.

Sir Thomas, however, achieved greater things. He abhorred slavery, and during his tenure as the British governor of Sumatra was able to curtail the then-thriving Dutch slave trade. He was also a noteworthy naturalist with a keen interest in the magnificent flora and fauna of the East Indies. The natural history collection he built is the basis of today's Raffles Museum of Biodiversity Research of the National University of Singapore.

In 1818, Sir Thomas traveled into the Sumatran rainforest with an English friend, the botanist Dr. Joseph Arnold. A native guide led the men to an enormous red flower growing in the seeming absence of any leaves or stems. The flower was found to be that of a parasitic plant, with filamentous roots growing inside climbing vines. The parasite gives no visible sign of its presence until a bud appears and grows into a gigantic flower that lasts about a week.

This plant, dubbed *Rafflesia arnoldii*, holds the world record for flower weight—more than 10 kilograms. But it is not the only giant flower found in Sumatra. *Amorphophallus titanum*, also known as titan arum, "voodoo lily," or "corpse flower," is a relative of skunk cabbage and calla lilies. While *Rafflesia* bears a true flower, the "flower" of *A. titanum* is actually an inflorescence, or cluster of many flowers. The towering central structure, called the *spadix*, reaches 2 meters in length; the surrounding petal-like *spathe* can be a meter wide.

Rafflesia, titan arum, and skunk cabbage all smell truly awful ("like a rotting possum under your porch," as one arboretum curator put it). At the time of pollination, these plants all release the protein breakdown products putrescine and cadaverine, which smell just as their names suggest. These malodorous chemicals attract flies and other carrion-feeding insects, which lay eggs and depart, unwittingly carrying pollen. (The insect larvae that hatch, unfortunately for them, find no dead meat to eat, and thus die.)

As in all angiosperms, the flowers of *Rafflesia* and *Amorphophallus* are reproductive equipment. Flowers produce the gametophytes, female and male, that produce the gametes that give rise to the next sporophyte generation. The male gametophytes take the form of pollen grains, and pollination is crucial to angiosperm reproduction.

The World's Largest Flower *Rafflesia arnoldii* produces the world's largest flower; its blossom attains a diameter of nearly a meter and can weigh up to 11 kg. It is rarely seen, as the buds take many months to develop and the blossom lasts for just a few days.

The World's Largest Inflorescence This massive structure is borne by *Amorphophallus titanum*, which came by its name for obvious reasons. Its individual flowers are arranged on a spadix.

Will *Rafflesia's* specialized pollination system be adequate to preserve the species? Its host plant, *Tetrastigma tuberculatum*, grows only in undisturbed rainforests in Southeast Asia, and that habitat is disappearing under human pressures. *Rafflesia* flowers only rarely, and individuals are dioecious. What are the odds of a female *Rafflesia* plant blooming within a very few days of a male plant, and close enough to allow an insect to deliver pollen from the male to the female? The plant does not appear to be common, and its long-term future is doubtful.

IN THIS CHAPTER we contrast sexual and asexual reproduction, focusing on the details of sexual reproduction. We will consider angiosperm gametophytes, pollination, double fertilization, embryo development, and the roles of fruits in seed dispersal. We will examine the transition from the vegetative state to the flowering state, a key event in angiosperm development. We conclude by considering the role of asexual reproduction in nature and in agriculture.

CHAPTER OUTLINE

38.1 How Do Angiosperms Reproduce Sexually?

Angiosperms have many ways of reproducing—and humans have developed even more ways of reproducing them. Flowers contain the sex organs; thus it is no surprise that almost all angiosperms reproduce sexually. But many reproduce asexually as well; some even reproduce asexually most of the time. What are the advantages and disadvantages of these two kinds of reproduction? The answers to this question involve genetic recombination. As we have seen, sexual reproduction produces new combinations of genes and diverse phenotypes. Asexual reproduction, in contrast, produces a clone of genetically identical individuals.

Both sexual and asexual reproduction are important in agriculture. Many important annual crops are grown from seeds, which are the product of sexual reproduction. Seed-grown crops include the great grain crops, all of which are grasses—wheat, rice, corn, sorghum, and millet—as well as plants in other families, such as soybean and safflower. Other crops, such as strawberry, potato, and banana, are usually produced asexually.

Orange trees, which have been under cultivation for centuries, can be grown from seed—except for the navel orange, which has no seeds. Early in the nineteenth century, on a plantation on the Brazilian coast, a single orange seed gave rise to one tree that had aberrant flowers. Parts of the flowers aborted, and seedless fruits formed. Asexual reproduction is the only way of propagating this plant, and every navel orange in the world comes from a tree derived asexually from that original Brazilian navel orange tree.

Unlike navel oranges, strawberries are capable of forming seeds and need not be propagated asexually. Nonetheless, asexual propagation of strawberries is common because individual plants are highly heterozygous and do not breed true to type; thus asexual propagation of a particularly desirable genotype ensures product uniformity and high quality. We will treat asexual reproduction in greater detail at the end of this chapter.

Our concern for now is sexual reproduction. In sexual reproduction, meiosis and subsequent mating between individuals of different genotypes shuffle genes into new combinations, resulting in a diversity of genotypes in each generation, some of which may be superior to those of the parental generation. This genetic diversity may serve the population well if the environment changes or as the population expands into new environments. The adaptability resulting from genetic diversity is the major advantage of sexual reproduction over asexual reproduction, although sexual reproduction can also break up well-adapted combinations of alleles through the same process of recombination.

38.1 Development of Gametophytes and Nuclear Fusion The embryo sac is the female gametophyte; the pollen grain is the male gametophyte. The male and female nuclei meet and fuse within the embryo sac. Most angiosperms have double fertilization, in which a zygote and an endosperm nucleus form from separate fusion events—the zygote from one sperm and the egg, and the endosperm from the other sperm and two polar nuclei.

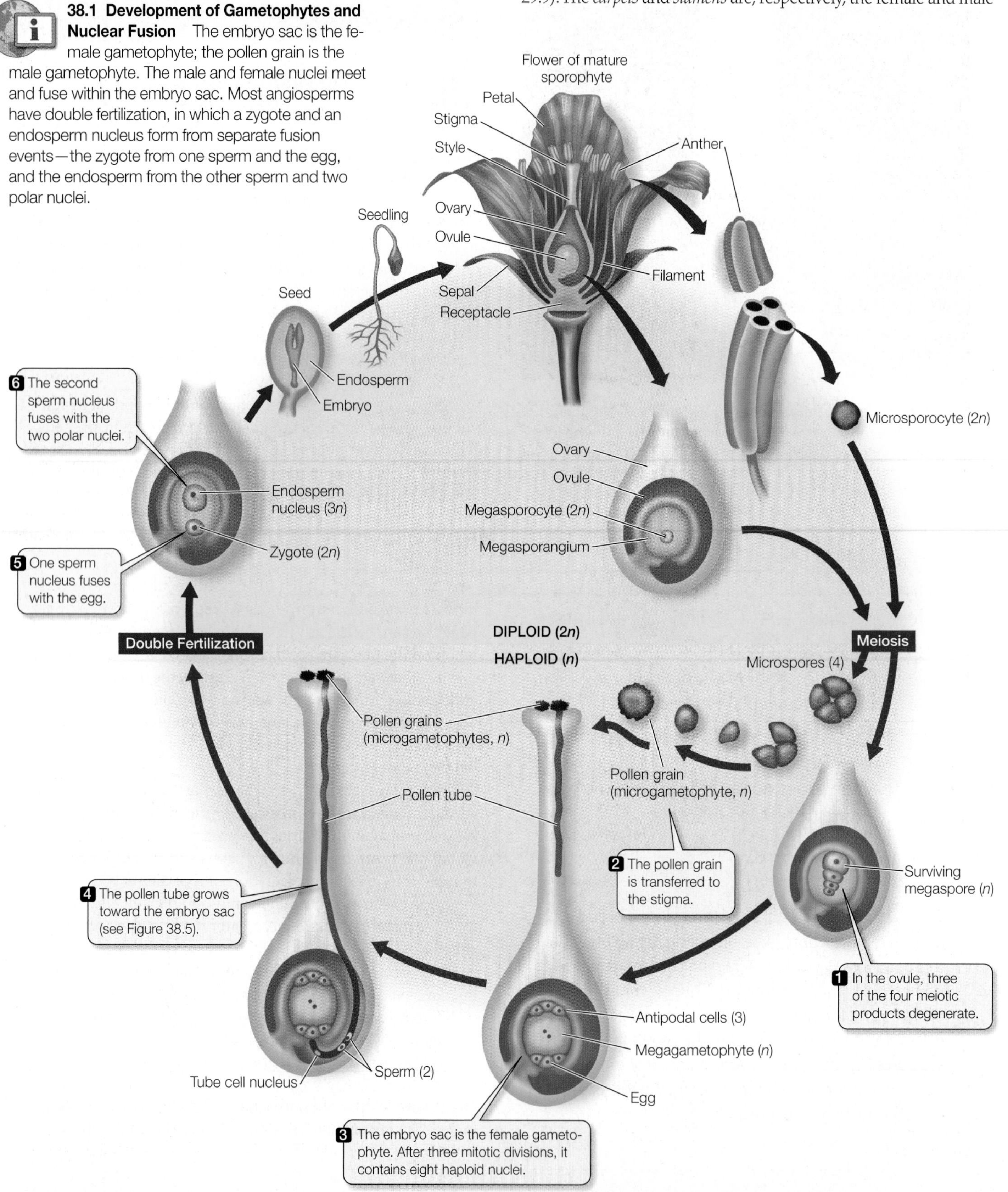

The flower is an angiosperm's structure for sexual reproduction

A complete flower consists of four groups of organs that are modified leaves: the carpels, stamens, petals, and sepals (see Figure 29.9). The *carpels* and *stamens* are, respectively, the female and male sex organs. A *pistil* is a structure composed of one or more carpels. The base of the pistil, called the *ovary*, contains one or more *ovules*, each of which contains a *megasporangium*, within which a female gametophyte may develop. The stalk of the pistil is the *style*, and the end of that stalk is the *stigma*. Each stamen is composed of a *filament* bearing a two-lobed *anther*, which consists of four *microsporangia* fused together. Male *gametophytes* begin their development within the microsporangia.

The *petals* and *sepals* of many flowers are arranged in whorls (circles) or spirals around the carpels and stamens. Together, the petals constitute the *corolla*. Below them, the sepals constitute the *calyx*. The petals are often colored, attracting pollinating animals; the sepals are often green and photosynthetic. All the parts of the flower are borne on a stem tip, the *receptacle*. Flower parts are very diverse in form, in contrast to the microscopic gametophytes that develop within them.

Flowering plants have microscopic gametophytes

Central to understanding plant reproduction is the concept of *alternation of generations*, in which a multicellular diploid ($2n$) generation alternates with a multicellular haploid (n) generation (see Section 28.2).

In angiosperms, the diploid sporophyte generation is the larger and more conspicuous one. The sporophyte generation produces flowers. The flowers produce spores, which develop into tiny gametophytes that begin and, in the case of the megagametophyte, end their development enclosed by sporophyte tissue.

The haploid gametophytes—the gamete-producing generation—of flowering plants develop from haploid spores in sporangia within the flower (**Figure 38.1**):

- Female gametophytes (megagametophytes), which are called **embryo sacs**, develop in megasporangia.
- Male gametophytes (microgametophytes), which are called **pollen grains**, develop in microsporangia.

Within the ovule, a *megasporocyte*—a cell within the megasporangium—divides meiotically to produce four haploid *megaspores*. In most flowering plants, all but one of these megaspores then degenerate. The surviving megaspore usually undergoes three mitotic divisions, producing eight haploid nuclei, all initially contained within a single cell—three nuclei at one end, three at the other, and two in the middle. Subsequent cell wall formation leads to an elliptical, seven-celled megagametophyte with a total of eight nuclei:

- At one end of the elliptical megagametophyte are three tiny cells: the egg and two cells called *synergids*. The egg is the female gamete, and the synergids participate in fertilization by attracting the pollen tube and receiving the sperm nuclei prior to their movement to the egg and central cell.
- At the opposite end of the megagametophyte are three *antipodal cells*, which eventually degenerate.
- In the large central cell are two **polar nuclei**, which together combine with a sperm nucleus.

The embryo sac (megagametophyte) is the entire seven-cell, eight-nucleus structure. You can review the development of the embryo sac in Figure 38.1.

The pollen grain (microgametophyte) consists of fewer cells and nuclei than the embryo sac (**Figure 38.2**). The development of a pollen grain begins when a *microsporocyte* within the anther divides meiotically. Each resulting haploid *microspore* develops a spore wall, within which it normally undergoes one mitotic division before the anthers open and release these two-celled pollen grains. The two cells are the *tube cell* and the *generative cell*. Further development of the pollen grain, which we will describe shortly, is delayed until the pollen arrives at a stigma. In angiosperms, the transfer of pollen from the anther to the stigma is referred to as **pollination**.

Pollination enables fertilization in the absence of water

Gymnosperms and angiosperms do not require external water as a medium for gamete travel and fertilization—a freedom not shared by other plant groups. The male gametes of gymnosperms and angiosperms are contained within pollen grains. But how do these pollen grains travel from an anther to a stigma?

Many different mechanisms have evolved for pollen transport. In some plants, such as peas and their relatives, self-pollination is accomplished before the flower bud opens. Pollen is transferred by the direct contact of anther and stigma within the same flower, resulting in *self-fertilization*.

Wind is the vehicle for pollen transport in many species. Wind-pollinated flowers have sticky or featherlike stigmas, and they produce pollen grains in great numbers. Some aquatic angiosperms are pollinated by water carrying pollen grains from plant to plant. Animals, including insects, birds, and bats, carry pollen among the

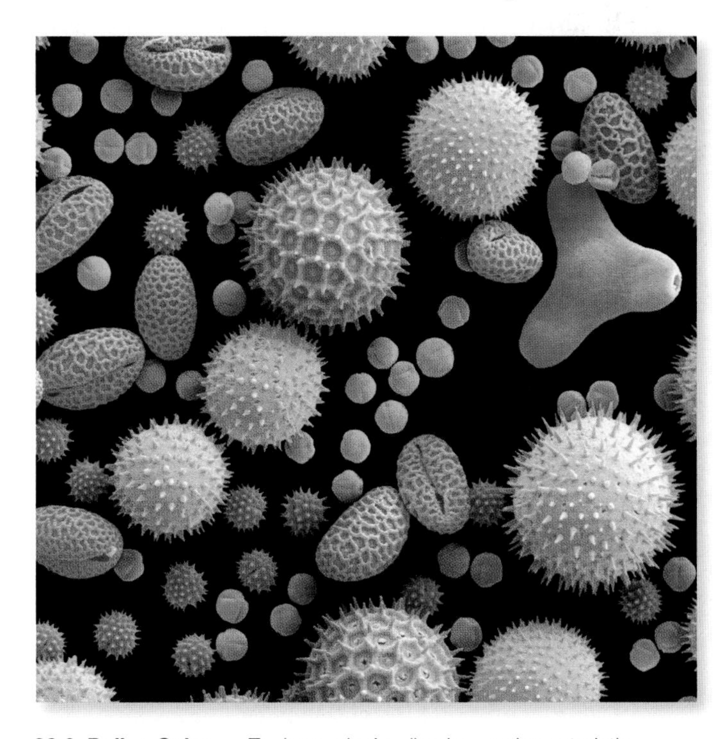

38.2 Pollen Galore Each species' pollen has a characteristic size, shape, and cell wall structure. These grains are the male gametophytes. Color has been artificially added to the micrograph.

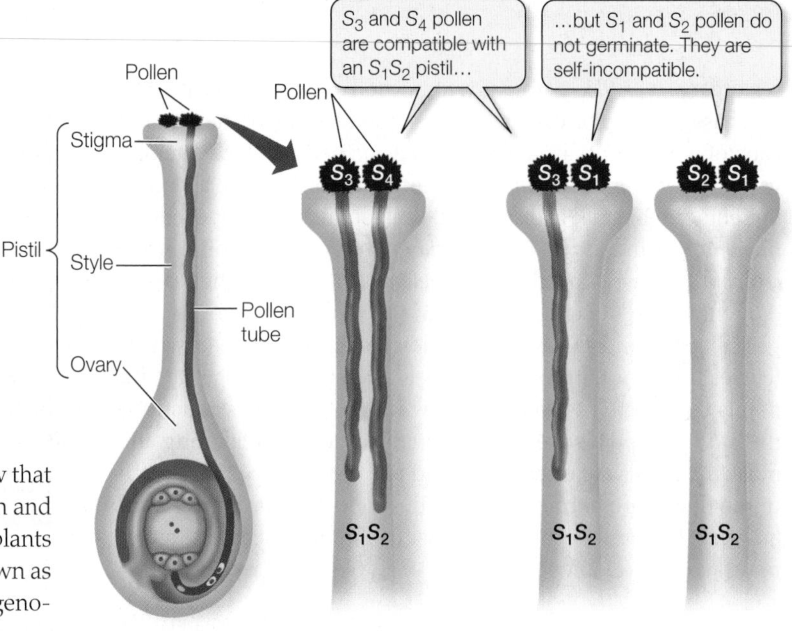

38.3 Self-Incompatibility Pollen grains do not germinate normally if their *S* allele matches one of the *S* alleles of the stigma. Thus, the egg cannot be fertilized by a sperm from the same plant.

flowers of many plants. Pollination among the flowers of different individuals of the same species is called *cross-pollination*.

Some flowering plants practice "mate selection"

In our discussion of Mendel's work (see Section 10.1), we saw that some plants can reproduce sexually by both cross-pollination and self-pollination. But not all plants have this flexibility. Many plants reject pollen from their own flowers. This phenomenon, known as **self-incompatibility**, promotes outcrossing between different genotypes.

A single gene, the *S* gene, is responsible for self-incompatibility in most plants. The *S* gene has dozens of alleles. A pollen grain is haploid and possesses a single *S* allele; the recipient pistil is diploid. In self-incompatible plants, pollen fails to germinate, or the pollen tube fails to traverse the style, if the *S* allele of the pollen matches one of the two *S* alleles in the pistil (**Figure 38.3**).

The stigma plays an important role in "mate selection" by flowering plants. The stigmas of most plants are exposed to the pollen of many other species as well as their own. Pollen from the same species binds strongly to the stigma by cell–cell signaling between the stigma and the cell walls of the pollen grains. In contrast, foreign pollen falls off readily or fails to germinate—or the pollen tubes are incapable of penetrating the stigma.

A pollen tube delivers sperm cells to the embryo sac

When a functional pollen grain lands on the stigma of a compatible pistil, it germinates. Germination for a pollen grain is the development of a **pollen tube** (**Figure 38.4**). The pollen tube either traverses the spongy tissue of the style or, if the style is hollow, grows downward on the inner surface of the style until it reaches an ovule. The pollen tube may grow millimeters or even centimeters in the process.

The downward growth of the pollen tube is guided in part by a chemical signal, a small protein produced by the synergids within the ovule. If one synergid is destroyed, the ovule still attracts pollen tubes, but destruction of both synergids renders the ovule unable to attract pollen tubes, and fertilization does not occur.

Angiosperms perform double fertilization

In most angiosperm species, the mature pollen grain consists of two cells, the tube cell and the generative cell. The larger tube cell encloses the much smaller generative cell. Guided by the tube cell nucleus, the pollen tube eventually grows through the megasporangial tissue and reaches the embryo sac. The generative cell, meanwhile, has undergone one mitotic division and cytokinesis to produce two haploid **sperm cells**.

Both of the sperm cells enter the embryo sac, where they are released into the cytoplasm of one of the synergids. This synergid degenerates, releasing the sperm cells (**Figure 38.5**). Each sperm cell then fuses with a different cell of the embryo sac. One sperm cell fuses with the egg cell, producing the diploid zygote. The other fuses with the two polar nuclei in the central cell, forming a **triploid (3*n*) nucleus**. While the zygote nucleus begins mitotic division to form the new sporophyte embryo, the triploid nucleus undergoes rapid mitosis to form a specialized nutritive tissue, the **endosperm**. The endosperm will later be digested by the developing embryo, as discussed in Chapter 37. The antipodal cells and the remaining synergid eventually degenerate, as does the pollen tube nucleus.

This process is known as **double fertilization** because it involves two nuclear fusion events:

- One sperm cell fuses with the egg cell.
- The other sperm cell fuses with the two polar nuclei.

The fusion of a sperm cell nucleus with the two polar nuclei to form endosperm takes place only in angiosperms. The fusion of these

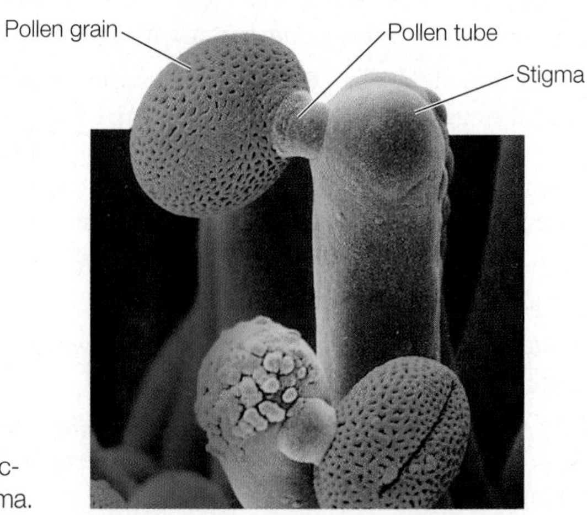

38.4 Pollen Tubes Begin to Grow These pollen grains have landed on fingerlike structures on the stigma of an *Arabidopsis* flower, and pollen tubes have penetrated the stigma.

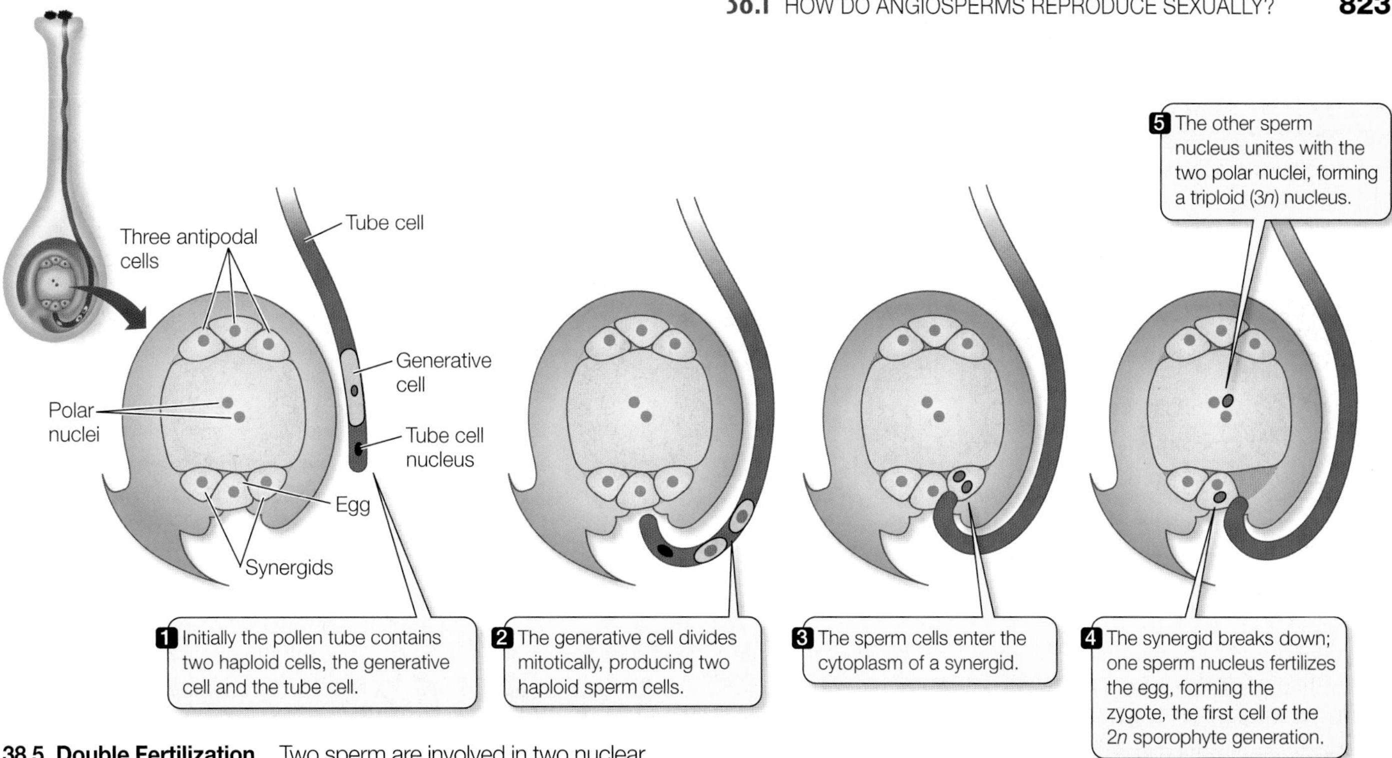

38.5 Double Fertilization Two sperm are involved in two nuclear fusion events, hence the term "double fertilization." One sperm is involved in the formation of the diploid zygote and the other results in the formation of the triploid endosperm. Double fertilization is a characteristic feature of angiosperm reproduction.

three nuclei, the possession of flowers, and the formation of fruit are the three most definitive characteristics of angiosperms.

The fusion of a sperm cell with the two polar nuclei produces not a zygote but a nutritionally supportive tissue, the endosperm. You consume endosperm in many foods—including popcorn, which is really popped endosperm.

Embryos develop within seeds

Shortly after fertilization, highly coordinated growth and development of embryo, endosperm, integuments, and carpel ensues. The *integuments*—tissue layers immediately surrounding the megasporangium—develop into the seed coat, and the carpel ultimately becomes the wall of the fruit that encloses the seed.

The first step in the formation of the embryo is a mitotic division of the zygote that gives rise to two daughter cells. These two cells face different fates. An asymmetrical (uneven) distribution of cytoplasm within the zygote causes one daughter cell to produce the embryo proper and the other daughter cell to produce a supporting structure, the **suspensor** (**Figure 38.6**). The suspensor pushes

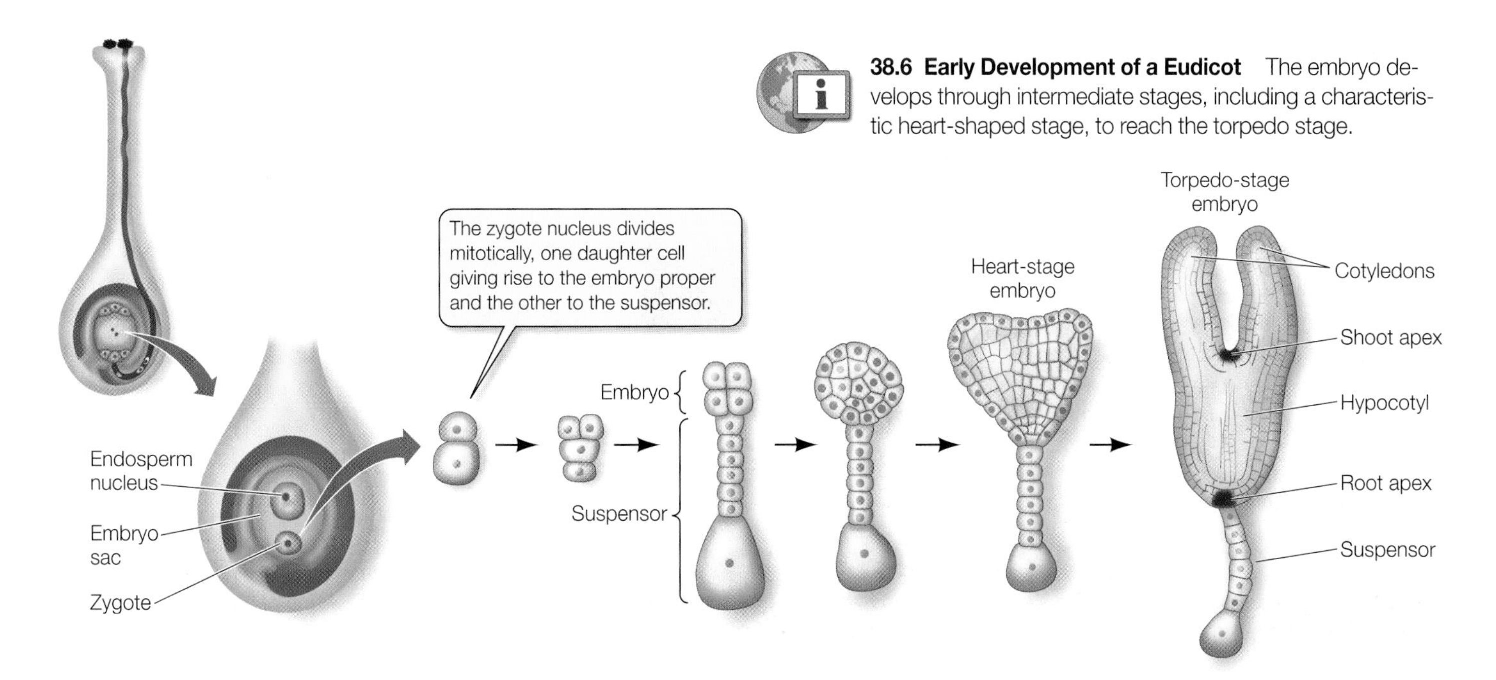

38.6 Early Development of a Eudicot The embryo develops through intermediate stages, including a characteristic heart-shaped stage, to reach the torpedo stage.

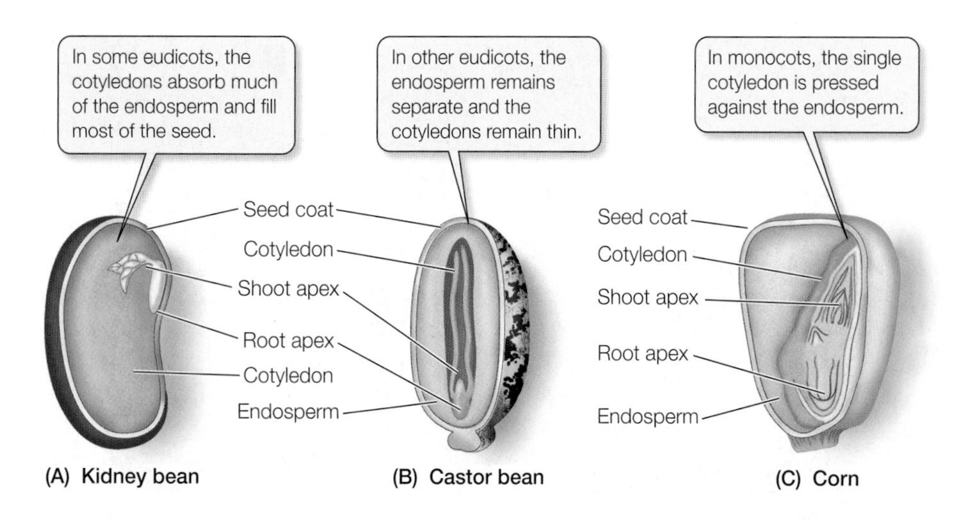

38.7 Variety in Angiosperm Seeds In some seeds, such as kidney beans (A), the nutrient reserves of the endosperm are absorbed by the cotyledons. In others, such as castor beans (B) and corn (C), the reserves in the endosperm will be drawn upon after germination.

the embryo against or into the endosperm and provides one route by which nutrients pass from the endosperm into the embryo.

The asymmetrical division of the zygote establishes polarity as well as the longitudinal axis of the new plant. A long, thin suspensor and a more spherical or globular embryo are distinguishable after just four mitotic divisions. The suspensor soon ceases to elongate. However, cell divisions continue, the primary meristems form, and the first organs begin to form within the embryo.

In eudicots, the initially globular embryo develops into the characteristic *heart stage* as the cotyledons ("seed leaves") start to grow. Further elongation of the cotyledons and of the main axis of the embryo gives rise to the *torpedo stage*, during which some of the internal tissues begin to differentiate (see Figure 38.6). Between the cotyledons is the *shoot apex*; at the other end of the axis is the *root apex* (**Figure 38.7**). Each of the apical regions contains the meristematic cells that continue to divide to give rise to the organs developing over the life of the plant.

During seed development, large amounts of nutrients are moved in from other parts of the parent plant, and the endosperm accumulates starch, lipids, and proteins. In many species, the cotyledons absorb the nutrient reserves from the surrounding endosperm and grow very large in relation to the rest of the embryo (Figure 38.7A). In others, the cotyledons remain thin (Figure 38.7B) and draw on the reserves in the endosperm as needed when the seed germinates.

In the late stages of embryonic development, the seed loses water—sometimes as much as 95 percent of its original water content. In this desiccated state, the embryo is incapable of further development; it remains quiescent until internal and external conditions are right for germination. (As mentioned in Section 37.1, a necessary early step in seed germination is the massive imbibition of water.) As the embryo and endosperm develop, the structures of the ovule and ovary are also undergoing developmental changes to form a seed and fruit, respectively.

Some fruits assist in seed dispersal

In angiosperms the ovary wall—together with its seeds—develops into a fruit after fertilization has occurred. A **fruit** may consist of only the mature ovary and seeds, or it may include other parts of the flower or structures that are closely related to it. In some species, this process produces fleshy, edible fruits such as peaches and tomatoes, while in other species the fruits are dry or inedible. Some major variations on this theme are illustrated in Figure 29.15, which shows only fleshy, edible fruits. Whatever its form, the fruit serves to promote seed dispersal.

Some fruits help disperse seeds over substantial distances, increasing the probability that at least a few of the many seeds produced by a plant will find suitable conditions for germination and growth to sexual maturity. Various plants, including milkweed and dandelion, produce a fruit with a "parachute" that may be blown some distance from the parent plant by the wind (**Figure 38.8A**). Water disperses some fruits; coconuts have been known to travel thousands of miles between islands. Still other fruits move by hitching rides with animals—either on, as with burrs on your hiking socks (**Figure 38.8B**), or inside them, as with berries in birds. Seeds swallowed whole along with fruits such as berries travel through the animal's digestive tract and are deposited some distance from the parent plant. In some species, seeds must pass through an animal in order to break dormancy.

(A) *Asclepias syriaca*

(B) *Arctium* sp.

38.8 Dispersing Fruit (A) A milkweed seed pod. Silky filaments catch the wind currents and carry the brown seeds with them. (B) Animals who rub up against the "hook and loop" surface of burdock fruit walk away with it attached to their fur, thus making them unwitting agents of dispersal. This feature of the fruit is said to have inspired the invention of modern Velcro™.

38.1 RECAP

Sexual reproduction in angiosperms is accomplished by flowers. After fertilization, the flower develops seed(s) and fruit.

- Explain the relationship between an ovule and an ovary, and between a fruit and a seed. See pp. 821 and 824 and Figure 38.1
- What function does self-incompatibility serve, and how do plants engage in "mate selection"? See p. 822 and Figure 38.3
- Explain the roles of the two sperm nuclei in double fertilization. See p. 822 and Figure 38.5

We have now traced the sexual life cycle of angiosperms from the flower, to the fruit, to the dispersal of seeds. Seed germination and the vegetative development of the seedling are presented in Chapter 37. The next section covers the rest of the angiosperm life cycle—the transition from the vegetative to the flowering state—and how this transition is regulated.

38.2 What Determines the Transition from the Vegetative to the Flowering State?

If we view an angiosperm as something produced by a seed for the function of bearing more seeds, then the act of flowering is one of the supreme events in the plant's life. The transition to the flowering state marks the end of vegetative growth for some plants. In other plants, vegetative growth may accompany flowering or resume after flowering is completed. But whatever the specific pattern, flowering always entails major developmental changes.

Apical meristems can become inflorescence meristems

The first visible sign of a transition to the flowering state may be a change in one or more apical meristems in the shoot system. During vegetative growth, an apical meristem continually produces leaves, axillary buds, and stem (**Figure 38.9A**) in a kind of unrestricted growth called *indeterminate growth* (see Section 34.3).

Flowers may appear singly or in an orderly cluster that constitutes an *inflorescence*. If a vegetative apical meristem becomes an **inflorescence meristem**, it ceases production of leaves and axillary buds and produces other structures: smaller leafy structures called *bracts*, as well as new meristems in the angles between the bracts and the stem (**Figure 38.9B**). These new meristems may also be inflorescence meristems, or they may be **floral meristems**, each of which gives rise to a flower.

Each floral meristem typically produces four consecutive whorls or spirals of organs—the sepals, petals, stamens, and carpels discussed earlier in the chapter—separated by very short internodes, keeping the flower compact (**Figure 38.9C**). In contrast to vegetative apical meristems and some inflorescence meristems, floral meristems are responsible for *determinate* growth—growth of limited duration, like that of leaves.

A cascade of gene expression leads to flowering

How do apical meristems become floral meristems or inflorescence meristems, and how do inflorescence meristems give rise to floral meristems? How does a floral meristem give rise, in short order, to four different floral organs (sepals, petals, stamens, and carpels)? How does each flower come to have the correct number of each of the floral organs? Numerous genes are expressed and interact to produce these results. We'll refer here to some of the genes whose actions have been most thoroughly studied in *Arabidopsis* and snapdragons (*Antirrhinum*).

- Expression of a group of **meristem identity genes** initiates a cascade of further gene expression.
- This cascade begins with **cadastral genes**, which participate in pattern formation—the spatial organization of the whorls of organs.
- Cadastral genes trigger the expression of **floral organ identity genes**, which work in concert to specify the successive whorls (see Figure 19.15).

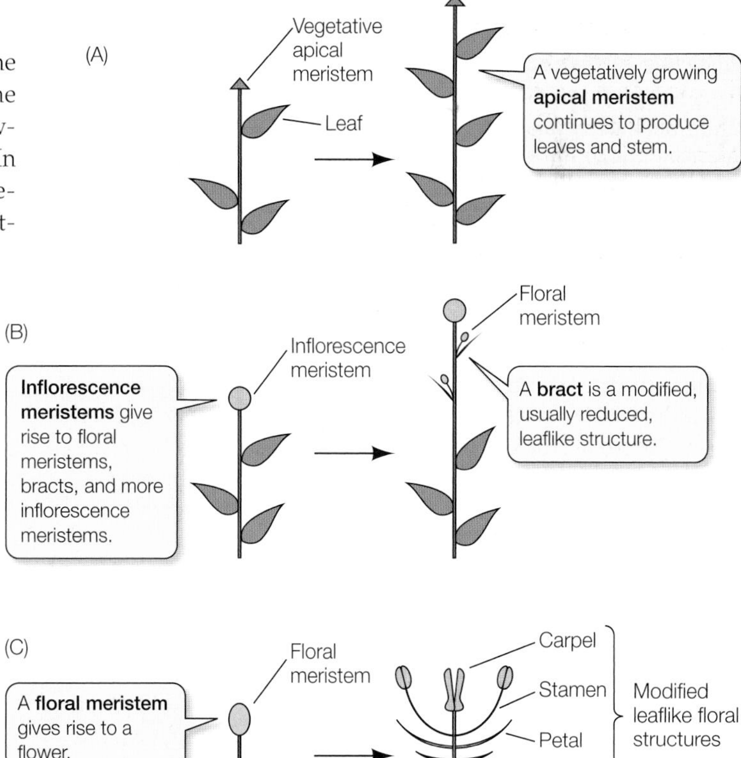

38.9 Flowering and the Apical Meristem A vegetative apical meristem (A) grows without producing flowers. Once the transition to the flowering state is made, inflorescence meristems (B) give rise to bracts and to floral meristems (C), which become the flowers.

Floral organ identity genes are homeotic genes (see Section 19.5), and their products control the transcription of other genes and assign an identity to cells in the floral meristems to control their developmental patterns.

Let's now consider how the transition from the vegetative to the flowering state is initiated.

Photoperiodic cues can initiate flowering

Environmental cues trigger the transition to the flowering state in many cases, depending on the genetic makeup of the species. The life cycles of flowering plants fall into three categories: annual, biennial, and perennial. **Annuals**, such as many food crops, complete their life cycle in one growing season. **Biennials**, such as carrots and sweet William, grow vegetatively for all or part of one growing season, then flower, form seeds, and die in the second growing season. **Perennials**, such as oak trees, live for a few to many growing seasons, during which both vegetative growth and flowering occur once the plants have reached sexual maturity. What control systems give rise to these and other differences in flowering behavior?

In 1920, W. W. Garner and H. A. Allard of the U.S. Department of Agriculture studied the behavior of a newly discovered mutant tobacco plant. The mutant, named 'Maryland Mammoth,' had large leaves and exceptional height. When the other plants in the field flowered, 'Maryland Mammoth' plants continued to grow and remained vegetative. Garner and Allard took cuttings of 'Maryland Mammoth' into their greenhouse, and the plants that grew from those cuttings finally flowered in December.

Garner and Allard guessed that this flowering pattern had something to do with the mutant's response to some environmental cue. They tested several likely environmental variables, such as temperature, but the key variable proved to be day length. By moving plants between light and dark rooms at different times to vary the day length artificially, they were able to establish a direct link between flowering and day length. As we will discuss below, we now know that the key variable is not day length but the length of the *night*; however, Garner and Allard did not make this distinction.

'Maryland Mammoth' plants did not flower if the light period they were exposed to was longer than 14 hours per day, but flowering commenced once the days became shorter than 14 hours. Thus, the **critical day length** for 'Maryland Mammoth' tobacco is 14 hours (**Figure 38.10**). Control of flowering and several other plant responses by the length of day or night is called **photoperiodism**.

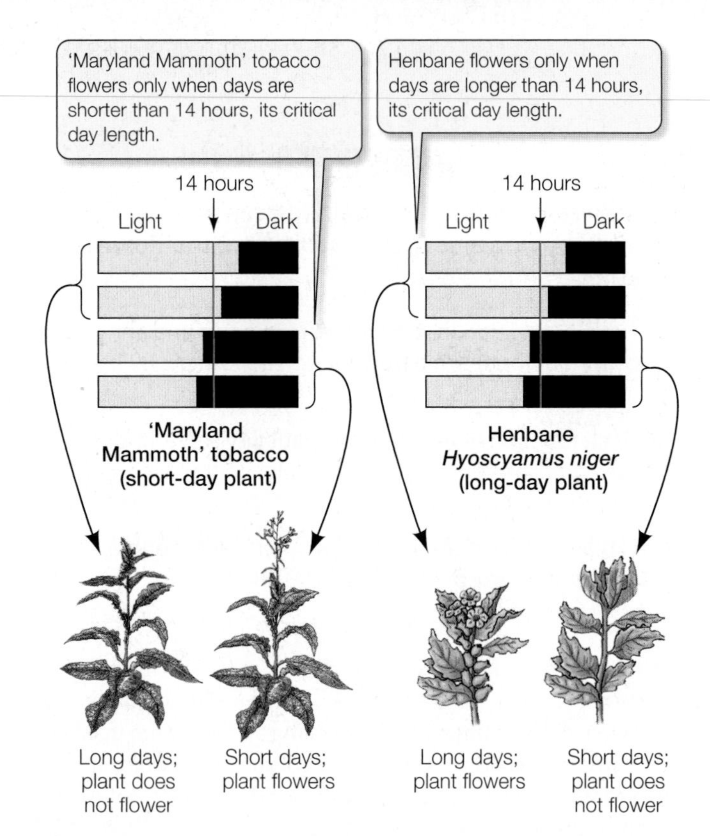

38.10 Day Length and Flowering By artificially varying the day length in a 24-hour period, Garner and Allard showed that the flowering of 'Maryland Mammoth' tobacco is initiated when the days become shorter than a critical length. 'Maryland Mammoth' tobacco is thus called a short-day plant. Henbane, a long-day plant, shows an inverse pattern of flowering.

Plants vary in their responses to different photoperiodic cues

Plants that flower in response to photoperiodic stimuli fall into several classes. Poinsettias, chrysanthemums, and 'Maryland Mammoth' tobacco are **short-day plants** (SDPs), which flower only when the day is *shorter* than a critical *maximum*. Thus, for example, we see chrysanthemums in nurseries in the fall, and poinsettias in winter. Spinach and clover are examples of **long-day plants** (LDPs), which flower only when the day is *longer* than a critical *minimum*. Spinach in the garden tends to flower and become bitter in the summer, and is thus normally planted in early spring. Generally, LDPs are triggered to flower in midsummer and SDPs in late summer, fall, or sometimes in the spring. Because short days occur both before and after midsummer, there is a degree of ambiguity in this signal. Could there be a more precise way for plants to regulate flowering?

Some plants require photoperiodic signals that are more complex than just short or long days. One group, the *short-long-day plants*, must experience first short days and then long ones in order to flower. Accordingly, white clover and other short-long-day plants flower during the long days before midsummer. Another group, the *long-short-day plants*, cannot flower until the long days of summer have been followed by shorter ones, so they bloom only in the fall. Kalanchoe (see Figure 38.18B) is a long-short-day plant.

What do these four patterns of photoperiodic response have in common? All of them serve the point of photoperiodism, which is to synchronize the flowering of plants of the same species in a local population to promote cross-pollination and successful reproduction.

Near the Equator, the annual variation in day length is only about 2 minutes. How do angiosperms growing in these regions manage to flower at around the same time? Recent work suggests that their flowering cues on the time of sunrise and/or sunset, which varies predictably by about 30 minutes over the course of a year.

The flowering of some angiosperms, such as corn, roses, and tomatoes, is not photoperiodic. In fact, there are more of these **day-**

neutral plants than there are short-day and long-day plants. Some plants are photoperiodically sensitive only when young and become day-neutral as they grow older. Others require specific combinations of day length and other factors—especially temperature—to flower.

Other processes besides flowering are also under photoperiodic control. We have learned, for example, that short days trigger the onset of winter dormancy in plants. They trigger the formation of tubers in some begonias and storage roots in dahlias. (Animals, too, show a variety of photoperiodic behaviors, as discussed in Chapter 53.)

The length of the night is the key photoperiodic cue determining flowering

The terms "short-day plant" and "long-day plant" became entrenched before scientists determined that photoperiodically sensitive plants actually measure the length of the *night*, or of a period of darkness, rather than the length of the day. This fact was demonstrated by Karl Hamner of the University of California at Los Angeles and James Bonner of the California Institute of Technology (**Figure 38.11**).

Working with cocklebur, an SDP, Hamner and Bonner ran a series of experiments using two sets of conditions:

- For one group of plants, the light period was kept constant—either shorter or longer than the critical day length—and the dark period was varied.
- For another group of plants, the dark period was kept constant and the light period was varied.

The plants flowered under all treatments in which the dark period exceeded 9 hours, regardless of the length of the light period. Thus, Hamner and Bonner concluded that it is the length of the *night* that matters; for cocklebur, the **critical night length** is about 9 hours. Thus, it would be more accurate to call cocklebur a "long-night plant" than a short-day plant.

In cocklebur, a single long night is sufficient to trigger full flowering some days later, even if the intervening nights are short ones. Most plants are less sensitive than cocklebur and require from two to several nights of appropriate length to induce flowering. For some plants, a single shorter night in a series of long ones, even one day before flowering would have commenced, inhibits flowering.

Through other experiments Hamner and Bonner gained some insight into how plants measure night length. They grew SDPs and LDPs under a variety of light/dark conditions. In some experiments, the dark period was interrupted by a brief exposure to light; in others, the light period was interrupted briefly by darkness. Interruptions of the light period by darkness had no effect on the flowering of either short-day or long-day plants. Even a brief interruption of the dark period by light, however, completely nullified the effect of a long night (experiment A in **Figure 38.12**). An SDP flowered only if the long nights were uninterrupted. An LDP experiencing long nights flowered if those nights were interrupted by exposure to light. Thus, the investigators concluded, these plants must have a timing mechanism that measures the length of a continuous dark period.

The nature of this timing mechanism has been partially revealed, beginning with the determination of the effective wavelengths of light and the identity of the photoreceptors. In the interrupted-night experiments, the most effective wavelengths of light were in the red range (experiment B in **Figure 38.12**), and the effect of a red-light interruption of the night could be fully reversed by a subsequent exposure to far-red light, indicating that a phytochrome is the photoreceptor. Phytochromes and blue-light receptors, which affect several aspects of plant development (Section 37.5), also participate in the photoperiodic timing mechanism. How might such a mechanism operate?

Experiments showed that when a plant is subjected to a dark period several days in duration, the plant's sensitivity to a light flash during the long night varies on a roughly 24-hour cycle. This suggests that phytochrome functions as a photoreceptor only, and that something else plays the timekeeping role. A biological clock is

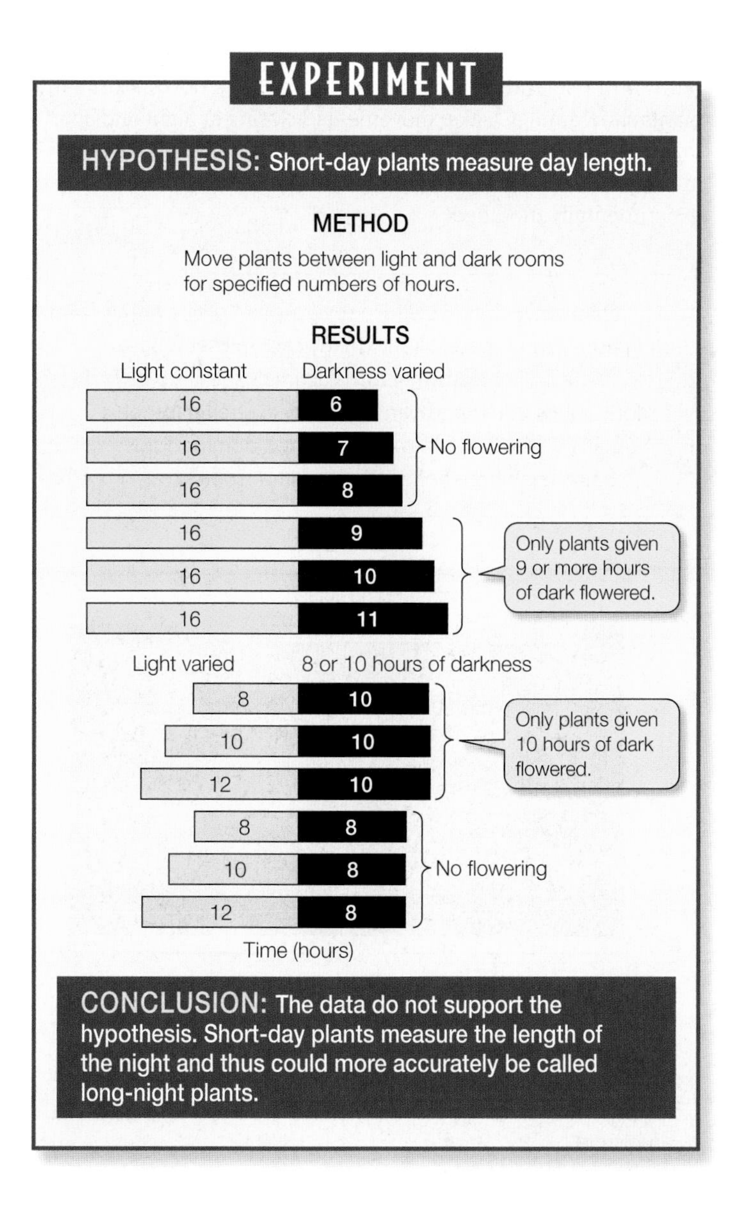

38.11 Night Length and Flowering The length of the dark period, not the length of the light period, determines flowering.

EXPERIMENT A

HYPOTHESIS: Photoperiodically sensitive plants measure the length of the night.

Short-day plants	Light/dark combinations	Long-day plants
No flowering		Flowering
No flowering		Flowering
Flowering		No flowering
No flowering		Flowering

CONCLUSION: Photoperiodic plants measure the length of the night, not the day.

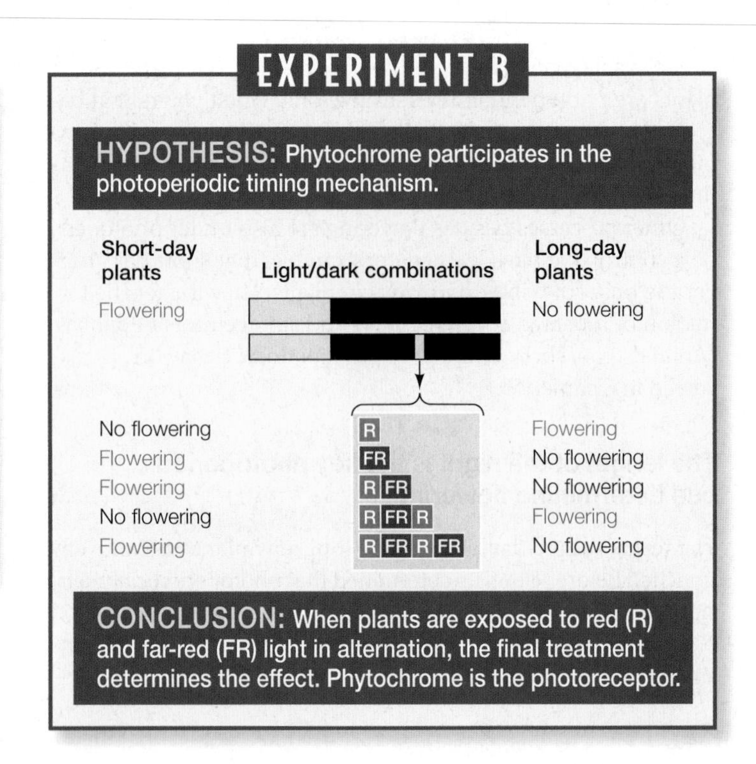

38.12 The Effect of Interrupted Days and Nights (A) Experiments suggest that plants are able to measure the length of a continuous dark period and use this information to trigger flowering. (B) Phytochromes seem to be involved in the photoperiodic timing mechanism.

linked to the phytochrome (which sets the clock) and to the production of flowers.

Circadian rhythms are maintained by a biological clock

It is clear that organisms have some way of measuring time, and that they are well adapted to the 24-hour day–night cycle of our planet. A "biological clock"—an oscillator that "ticks" back and forth between two states at roughly 12-hour intervals—resides within the cells of all eukaryotes and some prokaryotes. The major outward manifestations of this clock are known as **circadian rhythms** (Latin *circa*, "about," and *dies*, "day").

We can characterize circadian rhythms, as well as other regular biological cycles, in two ways: The **period** is the length of one cycle, and the **amplitude** is the magnitude of the change over the course of a cycle (**Figure 38.13**).

The circadian rhythms of cyanobacteria, protists, animals, fungi, and plants share some important characteristics:

- The period is remarkably insensitive to *temperature*, although lowering the temperature may drastically reduce the amplitude of the rhythmic effect.
- Circadian rhythms are *highly persistent;* they may continue for days even in an environment in which there are no environmental cues, such as light–dark periods.
- Circadian rhythms can be *entrained*, within limits, by light–dark cycles that differ from 24 hours. That is, the period an organism expresses can be made to coincide (within limits) with that of the light–dark cycle to which it is exposed.
- A brief exposure to light can shift the peak of the cycle—it can cause a *phase shift*.

Plants provide innumerable examples of circadian rhythms. The leaflets of plants such as clover normally hang down and fold at night and rise and unfold during the day. The flowers of many plants show similar "sleep movements," closing at night and opening during the day. They continue to open and close on an approximately 24-hour cycle even when the light and dark periods are experimentally modified.

Night flowering has evolved in several plant groups, ranging from certain water lilies to many cacti. Most of these plants are pollinated by nocturnal animals such as bats and moths, but some Himalayan bumblebees shelter for the night inside the wooly flowers of night-blooming *Saussurea* species, whose pollen they then transport.

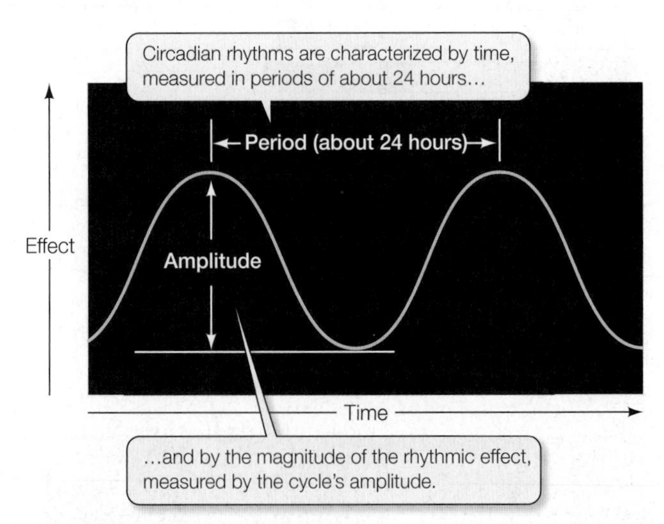

38.13 Features of Circadian Rhythms Circadian rhythms, like all biological rhythms, can be characterized in two ways: by period and by amplitude.

The period of circadian rhythms in nature is approximately 24 hours. If a clover plant, for example, were to be placed in light on a day–night cycle totaling exactly 24 hours, it would express a rhythm with a period of exactly 24 hours. However, if an experimenter used a day–night cycle of, say, 22 hours, then over time the rhythm would change—it would be **entrained** to a 22-hour period.

If an organism is maintained under constant darkness, it will express a circadian rhythm with an approximately 24-hour period. However, a brief exposure to light under these circumstances can cause a **phase shift**—that is, it can make the next peak of activity appear either later or earlier than expected, depending on when the exposure is given. Moreover, the organism does not then return to its old schedule if it remains in darkness. If the first peak is delayed by 6 hours, the subsequent peaks are all 6 hours late. Such phase shifts are permanent—until the organism receives more exposures to light.

EXPERIMENT

HYPOTHESIS: Plants fix more carbon photosynthetically when their circadian clock matches the environment's light-dark cycle.

METHOD

1. Select two mutant strains of *Arabidopsis thaliana*: one (*ztl-1*) with a long-cycle rhythm, and the other (*toc1-1*) with a short-cycle rhythm.
2. Grow plants of both mutant strains under either a 28-hour light-dark cycle or a 20-hour cycle instead of the normal 24-hour cycle.
3. Determine photosynthetic C fixation for each of the four groups.

RESULTS

Long-cycle mutants fixed more carbon per hour when grown on a longer-than-24-hour cycle, while short-cycle mutants performed better on a shorter-than-24-hour cycle.

CONCLUSION: Each mutant performed best under the cycle that corresponded to its genetically determined circadian rhythm.

38.14 Does the Circadian Clock Help a Plant Interact with Its Environment? Anthony Dodd and his coworkers studied the relationship between the plants' natural circadian cycles and light-dark cycles by observing the responses of mutant plants to experimentally imposed light-dark cycles. The clock does help the plants perform more efficiently. FURTHER EXPERIMENT: Design experiments to test other measures of plant performance in wild-type plants and the mutants studied by Dodd et al.

The general occurrence of circadian rhythms in plants suggests that they confer a selective advantage, but what might that advantage be? Recent experimental evidence indicates that the advantage consists of a coordination of the expression of certain genes with the phases of the daily light-dark cycles (**Figure 38.14**). The coordination results in increased efficiency—chlorophyll is not produced in the dark, when it is not needed, for example.

Photoreceptors set the biological clock

Phytochromes and blue-light receptors are known to affect the period of the biological clock, with the different pigments "reporting" on different wavelengths and intensities of light. This diversity of photoreceptors could be an adaptation to the changes in the light environment that a plant experiences in the course of a day or a season. How do these photoreceptors interact with a plant's biological clock?

The biological clock of *Arabidopsis* is based on the activities of at least three "clock" genes. The clock genes encode regulatory proteins that interact to produce a circadian oscillation. How does this oscillating clock interact with photoreceptors and the environment?

Arabidopsis is an LDP. Its clock controls the activity of *CONSTANS* (a gene that is *not* part of the clock mechanism) in such a way that the *CONSTANS* product, CO protein, accumulates in one phase of the clock's cycle—the phase in which night falls. Under long nights (short days), CO protein accumulates at night. Under short nights (long days), CO is also relatively abundant at dawn and dusk. When CO protein levels are high, light absorbed by phytochrome A and the blue-light receptor cryptochrome 2 leads to flowering (**Figure 38.15**). Thus *Arabidopsis* flowering results from the coincidence of light (detected by the two photoreceptors) with a sufficient amount of CO protein (which varies according to a circadian rhythm). Where is this coincidence-based photoperiodic mechanism located in relation to where flowering occurs?

The flowering stimulus originates in a leaf

Is the timing mechanism for flowering located in a particular plant tissue or organ, or are all parts able to sense the length of the night? This question was resolved by "blindfold" experiments in which it quickly became apparent that each leaf is capable of timing the night (**Figure 38.16**).

If a cocklebur plant—an SDP that needs long nights to flower—is kept under a regime of short nights and long days, but a leaf is covered so as to give it the needed long nights, the plant will flower (experiment A in Figure 38.16). If one leaf is given a photoperiodic treatment conducive to flowering—called an *inductive* treatment—other leaves kept under noninductive conditions will tend to inhibit flowering. (An inductive treatment works best if only one leaf is left on the plant.)

Although it is the leaves that sense an inductive photoperiod, the flowers form elsewhere on the plant. Thus, some kind of signal must be sent from the leaf to the site of flower formation. Three lines of evidence suggest that this signal is a chemical substance—a flowering hormone.

38.15 Photoreceptors and the Biological Clock Interact in Photoperiodic Plants One of the genes regulated by the circadian clock in *Arabidopsis* encodes the CO protein. Flowering depends on a combination of CO protein and light: enough CO must be present when photoreceptors have light available to them.

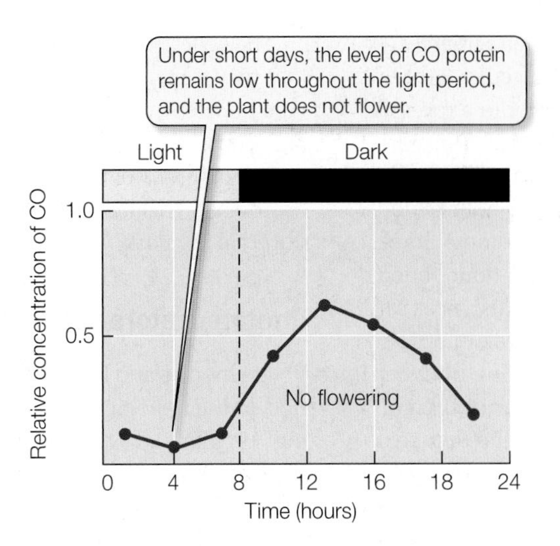

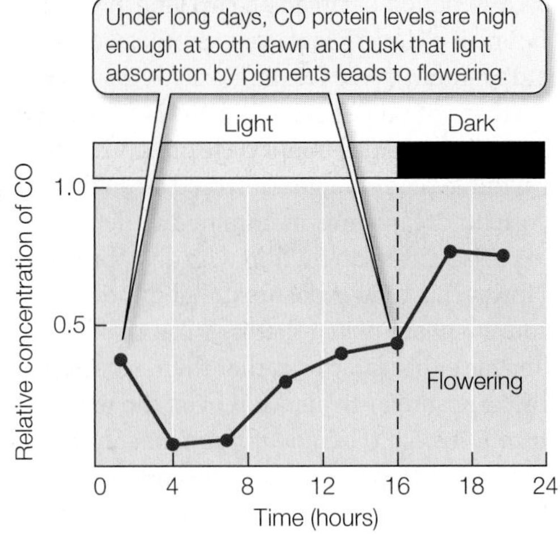

- If a photoperiodically induced leaf is immediately removed from a plant after the inductive dark period, the plant does not flower. If, however, the induced leaf remains attached to the plant for several hours, the plant will flower. This result suggests that something is synthesized in the leaf in response to the inductive dark period, then it moves out of the leaf to induce flowering.
- If two or more cocklebur plants are grafted together, and if one plant is exposed to inductive long nights and its graft partners are exposed to noninductive short nights, all the plants flower (experiment B in Figure 38.16).
- In at least one species, if an induced leaf from one plant is grafted onto another, noninduced plant, the host plant flowers.

38.16 Evidence for a Flowering Hormone If even a single leaf of cocklebur is exposed to inductive conditions, a signal travels to the entire plant (and even to other plants, in grafting experiments), inducing it to flower.

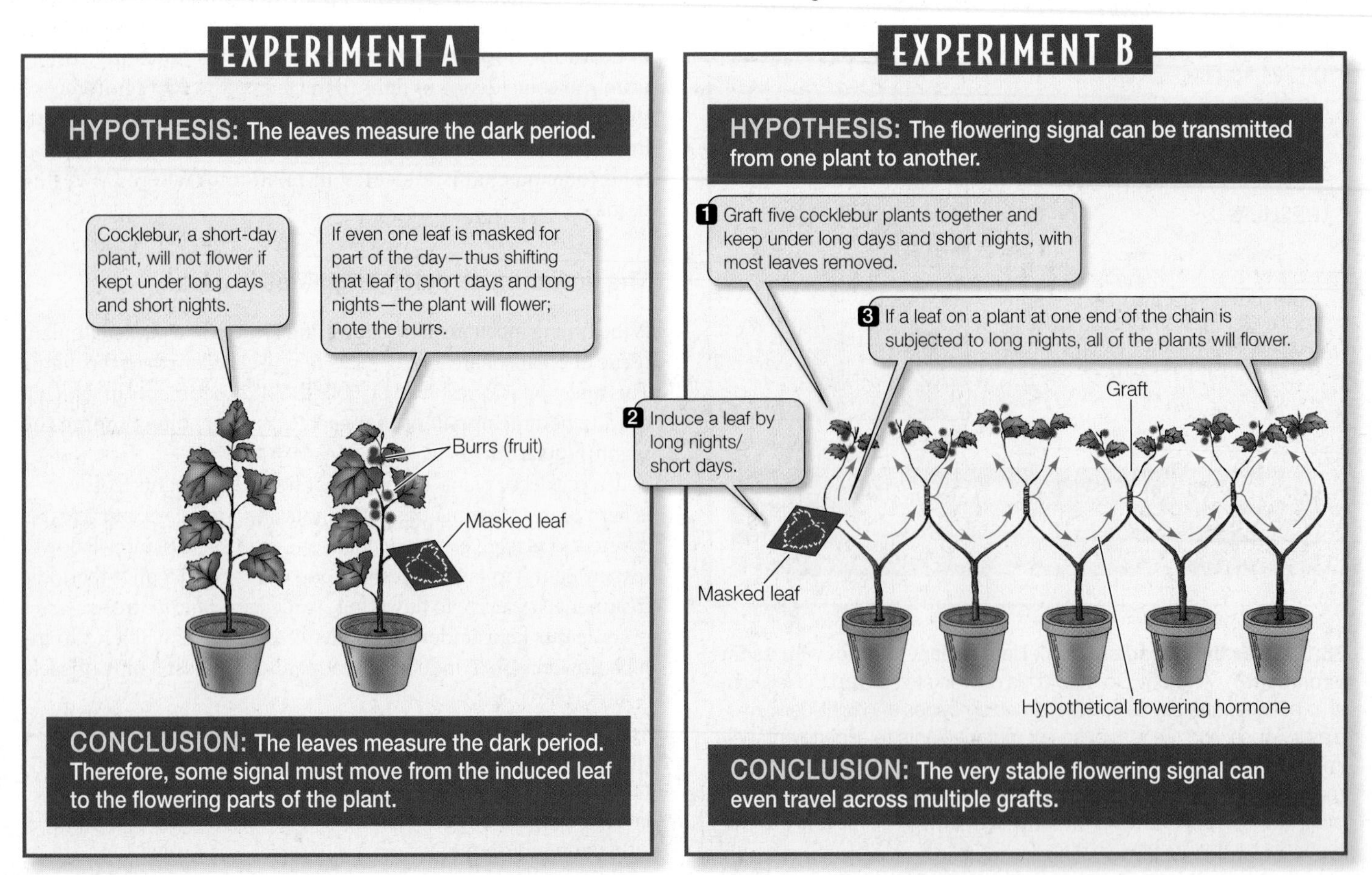

Some 50 years ago, Jan A. D. Zeevaart, a plant physiologist at Michigan State University, performed this last experiment. He exposed a single leaf of the SDP *Perilla* to a short-day/long-night regime, inducing the plant to flower. Then he detached this leaf and grafted it onto another, noninduced, *Perilla* plant—which responded by flowering. The same leaf grafted onto successive hosts caused each of them to flower in turn. As long as 3 months after the leaf was exposed to the short-day/long-night regime, it could still cause plants to flower.

Experiments such as Zeevaart's led to the conclusion that the photoperiodic induction of a leaf causes a more or less permanent change in the leaf, causing it to start and continue producing a flowering hormone that is transported to other parts of the plant, where the hormone initiates the development of reproductive structures. Biologists named this hypothetical hormone **florigen**, although decades passed without its being isolated or characterized.

An elegant experiment suggested that the florigen of SDPs is identical to that of LDPs, even though SDPs produce it only under long nights and LDPs only under short nights. An SDP and an LDP were grafted together, and both flowered, as long as the photoperiodic conditions were inductive for one of the partners. Either the SDP or the LDP could be the one induced, but both would always flower. These results suggest that a flowering hormone—the elusive florigen—was being transferred from one plant to the other.

The direct demonstration of florigen activity did not occur until 2005. It had long been thought that florigen could be neither a protein nor an RNA because those molecules were too large to pass from one living plant cell to another. However, we now know that such macromolecules can be transferred by way of plasmodesmata. A group in Sweden has shown that exposure of a leaf of *Arabidopsis* (an LDP) to an inductive light regime leads to the production in the leaf of mRNA by the *FLOWERING LOCUS T* (*FT*) gene, and the mRNA travels to the shoot apex. There, other genes are activated, leading to flowering. The *FT* mRNA is, then, at least part of the mobile flower stimulus—florigen.

We have considered the photoperiodic regulation of flowering, from photoreceptors in a leaf to the biological clock to the signal that travels from the induced leaf to the sites of flower formation. Light is not the only environmental variable that affects flowering, however. In some plants, low temperatures are a cue that eventually triggers flowering.

In some plants flowering requires a period of low temperature

Certain cereal grains serve as classic examples of the control of flowering by temperature. In both wheat and rye, we distinguish two categories of flowering behavior: annual and biennial. Spring wheat, for example, is an annual plant: it is sown in the spring and flowers in the same year. Winter wheat is biennial: it must be sown in the fall, and it flowers the following summer (**Figure 38.17**). If winter wheat is not exposed to cold in its first year, it will not flower normally the next year.

The implications of this finding were of great agricultural interest in the newly communist Soviet Union of the 1920s. Winter wheat yields more grain than spring wheat, but it cannot be grown in some parts of Russia because the winters are too cold for its survival. The Soviet agronomist Trofim Lysenko demonstrated that if seeds of winter wheat were premoistened and prechilled, they could be sown in the spring and would develop and flower normally the same year. Thus, high-yielding winter wheat could be grown even in previously hostile regions.*

This induction of flowering by low temperatures is called **vernalization** (Latin *vernum*, "spring"). Vernalization may require as many as 50 days of low temperatures (from about –2° to +12°C). The low-temperature treatment inhibits the expression of a gene whose protein product represses other genes that contribute to flower development. Some plant species require both vernalization and long days to flower. There is a long wait from the cold, short days of winter to the warm, long days of summer, but because the vernalized state easily lasts at least 200 days, these plants do flower when they experience the appropriate night length.

*Lysenko went on to claim that this vernalized state, once imposed by cold treatment, was inherited by the untreated progeny of the treated plants. There was no genetic evidence for his outlandish statement, which repudiated the chromosome model of inheritance and ignored all advances in evolutionary thinking. His argument was well received by the Soviet establishment, however, because it meshed with a Marxist doctrine of inheritance of acquired social and political characteristics. Lysenko was lionized and put in charge of the Soviet Academy of Agricultural Sciences. His reign there, which lasted into the 1960s, destroyed the lives and careers of a generation of Russian geneticists and left Soviet biology in a backward state from which it took years to recover.

38.17 Vernalization The seeds of winter wheat must be exposed to cold temperatures for some amount of time in order to flower, a phenomenon known as vernalization.

38.2 RECAP

Flowering of some angiosperms is controlled by night length, a phenomenon called photoperiodism. Exposure to low temperatures—vernalization—triggers flowering in others, and some plants require low temperatures followed by an appropriate night length to flower.

- Can you explain the differences between apical meristems, inflorescence meristems, and floral meristems? See p. 825 and Figure 38.9
- Why is "short-day plant" a misleading term? See p. 827
- Do you understand how a biological clock in *Arabidopsis* controls the release of CO protein, and how photoreceptors are responsible for determining when this plant should flower? See p. 829 and Figure 38.15

We have seen how environmental factors interact with genes to control flowering in angiosperms. The entire function of flowers is sexual reproduction, which maintains beneficial genetic variation in a population. Many angiosperms, however, also benefit from being able to reproduce asexually.

38.3 How Do Angiosperms Reproduce Asexually?

Although sexual reproduction takes up most of the space in this chapter, asexual reproduction is responsible for many of the new plant individuals appearing on Earth. This fact suggests that in some circumstances, asexual reproduction must be advantageous.

At the beginning of this chapter, we saw that one of the advantages of sexual reproduction is genetic recombination. Self-fertilization is a form of sexual reproduction, but when a plant self-fertilizes, fewer opportunities for genetic recombination exist than with cross-fertilization. A diploid, self-fertilizing plant that is heterozygous for a certain locus can produce both kinds of homozygotes for that locus plus the heterozygote among its progeny, but it cannot produce any progeny carrying alleles that it does not itself possess. Yet many plant species are capable of self-fertilization and produce viable and vigorous offspring.

Asexual reproduction eliminates genetic recombination altogether. When a plant reproduces asexually, it produces a clone of progeny genetically identical to the parent. If a plant is well adapted to its environment, asexual reproduction may spread its superior genotype throughout that environment. This ability to exploit a particular environment is an advantage of asexual reproduction.

Many forms of asexual reproduction exist

We call stems, leaves, and roots *vegetative organs* to distinguish them from flowers, the reproductive parts of the plant. The modification of a vegetative organ is what makes **vegetative reproduction** in plants—asexual reproduction—possible. In many cases, the stem is the organ that is modified. Strawberries and some grasses, for example, produce horizontal stems, called *stolons* or *runners*, that grow along the soil surface, form roots at intervals, and establish potentially independent plants (see Figure 34.4C). *Tip layers* are upright branches whose tips sag to the ground and develop roots, as in blackberry and forsythia.

Some plants, such as potatoes, form enlarged fleshy tips of underground stems, called *tubers* (see Figure 34.4A). *Rhizomes* are horizontal underground stems that can give rise to new shoots. Bamboo is a striking example of a plant that reproduces vegetatively by means of rhizomes. A single bamboo plant can give rise to a stand—even a forest—of plants constituting a single, physically connected entity.

Whereas stolons and rhizomes are horizontal stems, bulbs and corms are short, vertical, underground stems. Lilies and onions form *bulbs* (**Figure 38.18A**), short stems with many fleshy, highly modified leaves that store nutrients. These storage leaves make up most of the bulb. Bulbs are thus large underground buds. They can give rise to new plants by dividing or by producing new bulbs from axillary buds. Crocuses, gladioli, and many other plants produce *corms*, underground stems that function very much as bulbs do. Corms are disclike and consist primarily of stem tissue; they lack the fleshy modified leaves that are characteristic of bulbs.

Around 1600, the first tulip bulbs were imported into Holland from central Asia. During the 1630s, the flower's popularity blossomed into a bizarre "tulipomania"—at one point a single bulb sold for the equivalent of $76,000! Six weeks later, the price fell to less than a dollar, resulting in a debacle for several Dutch banks.

Not all vegetative organs modified for reproduction are stems. Leaves may also be the source of new plantlets, as in some succulent plants of the genus *Kalanchoe* (**Figure 38.18B**). Many kinds of angiosperms, ranging from grasses to trees such as aspens and poplars, form interconnected, genetically homogeneous populations by means of *suckers*—shoots produced by roots. What appears to be a whole stand of aspen trees, for example, may be a clone derived from a single tree by suckers.

Plants that reproduce vegetatively often grow in physically unstable environments such as eroding hillsides. Plants with stolons or rhizomes, such as beach grasses, rushes, and sand verbena, are common pioneers on coastal sand dunes. Rapid vegetative reproduction enables these plants, once introduced, not only to multiply but also to survive burial by the shifting sand; in addition, the dunes are stabilized by the extensive network of rhizomes or stolons that develops. Vegetative reproduction is also common in some deserts, where the environment is often not suitable for seed germination and the establishment of seedlings.

Dandelions, citrus trees, and some other plants reproduce by the asexual production of seeds, called **apomixis**. As we have seen, meiosis reduces the number of chromosomes in gametes by half, and fertilization restores the sporophytic number of chromosomes in the zygote. Some plants can skip over *both* meiosis and fertilization and still produce seeds. Apomixis produces seeds within

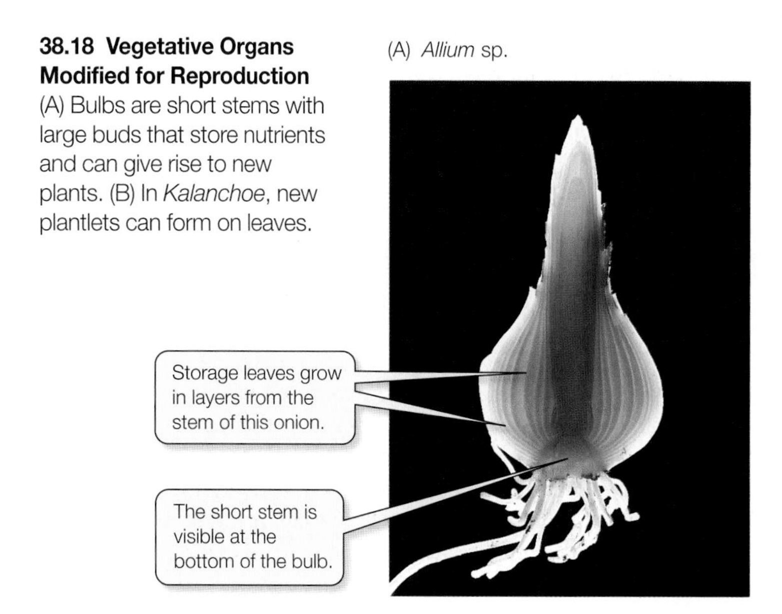

38.18 Vegetative Organs Modified for Reproduction (A) Bulbs are short stems with large buds that store nutrients and can give rise to new plants. (B) In *Kalanchoe*, new plantlets can form on leaves.

the ovary without the mingling and segregation of chromosomes and without the union of gametes. Certain cells within the ovule simply develop into seeds, and the ovary wall develops into a fruit. An apomictic embryo has the sporophytic number ($2n$) of chromosomes. The result of apomixis is a fruit with seeds that are genetically identical to the parent plant.

Apomixis sometimes requires pollination. In some apomictic species, a sperm nucleus must combine with the polar nuclei for the endosperm to form. In other apomictic species, the pollen provides the signals for embryo and endosperm formation, although neither sperm nucleus participates in fertilization. This observation emphasizes that pollination and fertilization are not the same thing.

Vegetative reproduction has a disadvantage

Vegetative reproduction is highly efficient in a stable environment. A change in the environment, however, can leave an asexually reproducing species at a disadvantage.

The English elm, *Ulmus procera*, provides a striking example. It was apparently introduced into England as a clone by the ancient Romans. This tree reproduces asexually by suckers and is incapable of sexual reproduction. In 1967, Dutch elm disease first struck the English elms. After 2 millennia of clonal growth the population lacked genetic diversity and was unable to meet the challenge. Today the English elm is all but gone from England.

Vegetative reproduction is important in agriculture

Farmers and gardeners take advantage of some natural forms of vegetative reproduction. They have also developed new types of asexual reproduction by manipulating plants. One of the oldest methods of vegetative reproduction used in agriculture consists of simply making cuttings of stems, inserting them in soil, and waiting for them to form roots and thus become autonomous plants. The cuttings are usually encouraged to root by treatment with a plant hormone, auxin, as described in Section 37.3.

Horticulturists reproduce many woody plants by **grafting**—attaching a bud or a piece of stem from one plant to the root or root-bearing stem of another plant. The part of the resulting plant that comes from the root-bearing "host" is called the **stock**; the part grafted on is the **scion** (**Figure 38.19**).

For a graft to succeed, the vascular cambium of the scion must associate with that of the stock. By cell division, both cambia form masses of wound tissue. If the two masses meet and connect, the resulting continuous cambium can produce xylem and phloem, allowing transport of water and minerals to the scion and of photosynthate to the stock. Grafts are most often successful when the stock and scion belong to the same or closely related species. Much fruit grown for market in the United States is produced on grafted trees.

Scientists in universities and commercial laboratories have been developing new ways to produce useful plants via tissue culture. Because many plant cells are *totipotent*, cultures of undifferentiated tissue can give rise to entire plants, as can small pieces of tissue cut directly from a parent plant (see Figure 19.3). Tissue cultures sometimes are used commercially to produce new plants.

Culturing tiny bits of apical meristem can produce plants free of viruses. Because apical meristems lack developed vascular tis-

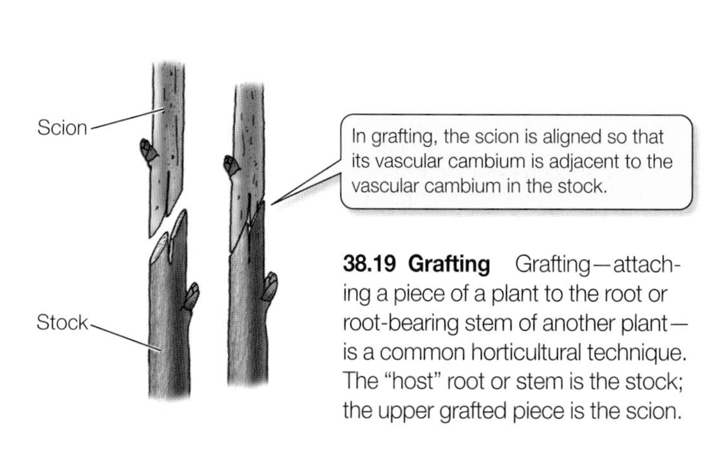

38.19 Grafting Grafting—attaching a piece of a plant to the root or root-bearing stem of another plant—is a common horticultural technique. The "host" root or stem is the stock; the upper grafted piece is the scion.

sues, viruses tend not to enter them. Treatment with hormones causes a single apical meristem to give rise to 20 or more shoots; thus, a single plant can give rise to millions of genetically identical plants within a year by repeated meristem culturing. Using this approach, strawberry and potato producers are able to start each year's crop from virus-free plants.

Recombinant DNA techniques applied to tissue cultures can provide plants with increased resistance to pests or increased nutritive value to humans. There is also interest in making certain valuable, sexually reproducing plants capable of apomixis. By causing cells from sexually incompatible species to fuse, one can obtain plants with exciting new combinations of properties.

38.3 RECAP

Angiosperms may reproduce asexually by means of modified stems, roots, or leaves, or by apomixis.

- Do you see how apomixis differs from sexual reproduction? See pp. 832–833
- Explain how vegetative reproduction of plants can be advantageous to humans. See pp. 833–834

We have seen how angiosperms reproduce sexually and asexually. A disadvantage of asexual reproduction is that its genetic inflexibility may leave a population unable to cope with a changing environment. In the next chapter we see that even unchanging environments, both biological and physical, present challenges to plants—and that plants have evolved ways of coping with them.

CHAPTER SUMMARY

38.1 How do angiosperms reproduce sexually?

Sexual reproduction promotes genetic diversity in a population. The flower is an angiosperm's structure for sexual reproduction.

Flowering plants have microscopic gametophytes. The megagametophyte is the **embryo sac**, which typically contains eight nuclei in a total of seven cells. The microgametophyte is the **pollen grain**, which usually contains two cells. Review Figure 38.1, Web/CD Tutorial 38.1

Following **pollination**, the pollen grain delivers sperm cells to the embryo sac by means of a **pollen tube**. Review Figure 38.3

Most angiosperms exhibit **double fertilization**: One sperm nucleus fertilizes the egg, forming a zygote, and the other sperm nucleus unites with the two **polar nuclei** to form a **triploid endosperm**. Review Figure 38.5

The zygote develops into an embryo (with an attached **suspensor**), which remains quiescent in the seed until conditions are right for germination. Review Figure 38.6, Web/CD Activity 38.1

Ovules develop into seeds, and the ovary wall and the enclosed seeds develop into a **fruit**.

38.2 What determines the transition from the vegetative to the flowering state?

For a vegetatively growing plant to flower, an apical meristem in the shoot system must become an **inflorescence meristem**, which in turn must give rise to one or more **floral meristems**. Review Figure 38.9

Angiosperms are classified as **annuals**, **biennials**, or **perennials**, depending upon the length of their life cycle.

Flowering results from a cascade of gene expression. **Floral organ identity genes** are expressed in floral meristems that give rise to sepals, petals, stamens, and carpels.

Short-day plants flower when the nights are longer than a critical night length specific to each species; **long-day plants** flower when the nights are shorter than a critical night length. Some angiosperms have more complex photoperiodic requirements, but most are **day-neutral**. Review Figure 38.10

Plants exhibit **circadian rhythms**, which are characterized by both their **period** and their **amplitude**. Circadian rhythms can be **entrained** and can be induced to experience a **phase shift**. Review Figure 38.13

The mechanism of photoperiodic control involves phytochromes and a biological clock. Review Figures 38.12 and 38.15, Web/CD Tutorial 38.2

A flowering hormone, called **florigen**, is formed in a photoperiodically induced leaf and is translocated to the sites where flowers will form. Review Figure 38.16

In some angiosperm species, exposure to low temperatures—**vernalization**—is required for flowering.

38.3 How do angiosperms reproduce asexually?

Asexual reproduction allows rapid multiplication of organisms that are well suited to their environment.

Vegetative reproduction involves the modification of a vegetative organ—usually the stem—for reproduction.

Some plant species produce seeds asexually by **apomixis**.

Horticulturists often **graft** different plants together to take advantage of favorable properties of both **stock** and **scion**. Review Figure 38.19

Agriculturalists use natural and artificial techniques of asexual reproduction to reproduce particularly desirable plants.

SELF-QUIZ

1. Sexual reproduction in angiosperms
   a. is by way of apomixis.
   b. requires the presence of petals.
   c. can be accomplished by grafting.
   d. gives rise to genetically diverse offspring.
   e. cannot result from self-pollination.
2. The typical angiosperm female gametophyte
   a. is called a megaspore.
   b. has eight nuclei.
   c. has eight cells.
   d. is called a pollen grain.
   e. is carried to the male gametophyte by wind or animals.
3. Pollination in angiosperms
   a. never requires external water.
   b. never occurs within a single flower.
   c. always requires help by animal pollinators.
   d. is also called fertilization.
   e. makes most angiosperms independent of external water for reproduction.
4. Which statement about double fertilization is *not* true?
   a. It is found in most angiosperms.
   b. It takes place in the microsporangium.
   c. One of its products is a triploid nucleus.
   d. One sperm nucleus fuses with the egg nucleus.
   e. One sperm nucleus fuses with two polar nuclei.
5. The suspensor
   a. gives rise to the embryo.
   b. is heart-shaped in eudicots.
   c. separates the two cotyledons of eudicots.
   d. ceases to elongate early in embryonic development.
   e. is larger than the embryo.
6. Which statement about photoperiodism is *not* true?
   a. It is related to the biological clock.
   b. A phytochrome plays a role in the timing process.
   c. It is based on measurement of the length of the night.
   d. Most plant species are day-neutral.
   e. It is limited to plants.
7. Before florigen was isolated, we thought it exists because
   a. night length is measured in the leaves, but flowering occurs elsewhere.
   b. it is produced in the roots and transported to the shoot system.
   c. it is produced in the coleoptile tip and transported to the base.
   d. we think that gibberellin and florigen are the same compound.
   e. it may be activated by prolonged (more than a month) chilling.
8. Which statement about vernalization is *not* true?
   a. It may require more than a month of low temperatures.
   b. The vernalized state generally lasts for about a week.
   c. Vernalization makes it possible to have a winter wheat crop each year.
   d. It is accomplished by subjecting moistened seeds to chilling.
   e. It was of interest to Russian scientists because of their native climate.
9. Which of the following does *not* participate in asexual reproduction?
   a. Stolon
   b. Rhizome
   c. Zygote
   d. Tuber
   e. Corm
10. Apomixis involves
    a. sexual reproduction.
    b. meiosis.
    c. fertilization.
    d. a diploid embryo.
    e. no production of a seed.

FOR DISCUSSION

1. Which method of reproduction might a farmer prefer for a crop plant that reproduces both sexually and asexually? Why?
2. Thompson Seedless grapes are produced by vines that are triploid. Think about the consequences of this chromosomal condition for meiosis in the flowers. Why are these grapes seedless? Describe the role played by the flower in fruit formation when no seeds are being formed. How do you suppose Thompson Seedless grapes are propagated?
3. Poinsettias are popular ornamental plants that typically bloom just before Christmas. Their flowering is photoperiodically controlled. Are they long-day or short-day plants? Explain.
4. You plan to induce the flowering of a crop of long-day plants in the field by using artificial light. Is it necessary to keep the lights on continuously from sundown until the point at which the critical day length is reached? Why or why not?
5. Dodd and coworkers pointed out the need for crop breeders to be aware if genes for the phase and period of circadian rhythms are closely linked to genes being studied for enhanced crop yields. If such a tight linkage occurs, how might it affect a research program?

FOR INVESTIGATION

The space program may need to deal with methods for crop production on another planet or during transit through space. Outline an experimental program to address this problem in terms of the findings on circadian rhythms by Dodd and coworkers (see Figure 38.14).

CHAPTER 39 Plant Responses to Environmental Challenges

Salt in the delta threatens a hungry nation

Bangladesh is an impoverished country with a high population density and a weak economy. Its 150 million citizens live on an area of about 144,000 square kilometers. (For comparison, imagine half the people in the United States—or more than 4 times the total population of Canada—all living in the single state of Iowa.) Bangladesh is far from being able to feed its own population and must rely heavily on international food aid. It is crucial that the country maximize its ability to feed its citizens, even in the face of ecological challenges.

Aside from the fact of overpopulation, the people of Bangladesh face agricultural problems arising from the country's almost unique geography. Much of its total land area lies at or below sea level in the Ganges River delta. A massive network of 230 rivers threads through the country, carrying runoff from melting Himalayan snow through agricultural land to the Indian Ocean. About one-third of Bangladesh is flooded annually by the torrential rains of the monsoon season (June through October). From November through May, however, a very different problem arises: salinization, or too much salt in the soil.

During the monsoon season, rainfall and rivers provide fresh water that is low in salt content. Things change during the dry season. When the flow of fresh water is sharply reduced, saline (salty) water from the delta's estuaries near the ocean penetrates inland, sometimes as far as 300 kilometers. This results in progressive salinization of the soil. Furthermore, as the soil dries out from November onward, ground water rises and moves laterally, bringing with it dissolved salts from deeper in the soil. With further drying, the soil becomes coated with salt.

Humans have contributed to salinization as well. Some years ago, India diverted part of the flow of the Ganges River toward the city of Calcutta—and away from Bangladesh. The subsequent reduction of the river's flow into the delta meant further intrusion of sea water. Bangladesh has contributed to its own problems by failing to regulate shrimp farming in the delta, which is carried out by allowing brackish water to spread over the soil. Inappropriate agricultural practices continue to favor salinization rather than checking it.

Too much salt in the soil is toxic and can kill a plant outright. A bit less salt can kill a plant osmotically by making it difficult to take up water from the soil. Still lower soil salt concentrations simply reduce

Monsoon Season Each year the monsoon rains of Southeast Asia force many people to retreat from flooded homes.

Salty Soils Much of Bangladesh's agricultural land is in the Ganges River delta, at or below sea level. This cropland suffers from increasing amounts of salt in the soil and diminishing agricultural returns.

CHAPTER OUTLINE

plant growth and crop yields. Worldwide, about 20,000 square kilometers of agricultural soil are lost each year to salinization. Even in the United States, almost one-fourth of all irrigated lands experience salinization, resulting in significant crop losses.

Agriculture in many parts of the world is challenged by salty, arid, or waterlogged soils. Some plants have adaptations that allow them to thrive or at least survive in such environments; others simply don't grow in them. In discussing these and other environmental challenges to plant growth, we'll touch on the possibility of engineering plants to become salt-tolerant without becoming unpalatable.

IN THIS CHAPTER we begin by examining interactions between plants and plant pathogens. We then consider interactions between plants and herbivores. Next we will discuss some of the adaptations plants exhibit to cope with temperature extremes, and, finally, we describe how certain plants can survive in physical environments that are either extremely dry or water-saturated, that are dangerously salty, or that contain high concentrations of toxic substances.

39.1 How Do Plants Deal with Pathogens?

The environment teems with organisms that cause plant diseases. We know of more than a hundred diseases that can kill a tomato plant, each of them caused by a different pathogen, including bacteria, fungi, protists, and viruses. Like animals, plants have a variety of defenses against pathogens. Like the defenses of our own bodies, these mechanisms are not perfect, but they generally keep the plant world in competitive balance with its pathogens.

Plants and pathogens have evolved together in a continuing "arms race." Pathogens have evolved mechanisms with which to attack plants, and plants have evolved mechanisms for defending themselves against pathogens. Each set of mechanisms uses information from the other. For example, the pathogen's enzymes may break down the plant's cell walls, and the breakdown products may signal to the plant that it is under attack. In turn, the plant's defenses alert the pathogen that it is under attack.

What determines the outcome of a battle between a plant and a pathogen? The key to success for the plant is to respond to the information from the pathogen quickly and massively. Plants use both mechanical and chemical defenses in this effort.

Plants seal off infected parts to limit damage

Tissues such as epidermis or cork protect the outer surfaces of plants, and these tissues are generally covered by cutin, suberin, or waxes. This protection is comparable to the nonspecific immune defenses of animals (see Section 18.2). When pathogens pass these barriers, other nonspecific plant defenses are activated.

The defense systems of plants and animals differ. Animals generally repair tissues that have been damaged by pathogens, but plants do not. Instead, they seal off and sacrifice the damaged tissues so that the rest of the plant does not become infected. This approach works because most plants, unlike most animals, are modular and can replace damaged parts by growing new ones.

One of a plant cell's first defensive responses is the rapid deposition of additional polysaccharides on the inside of the cell wall, reinforcing this barrier to invasion by the pathogen (**Figure 39.1**). These polysaccharides block the plasmodesmata, limiting the ability of viral pathogens to move from cell to cell.

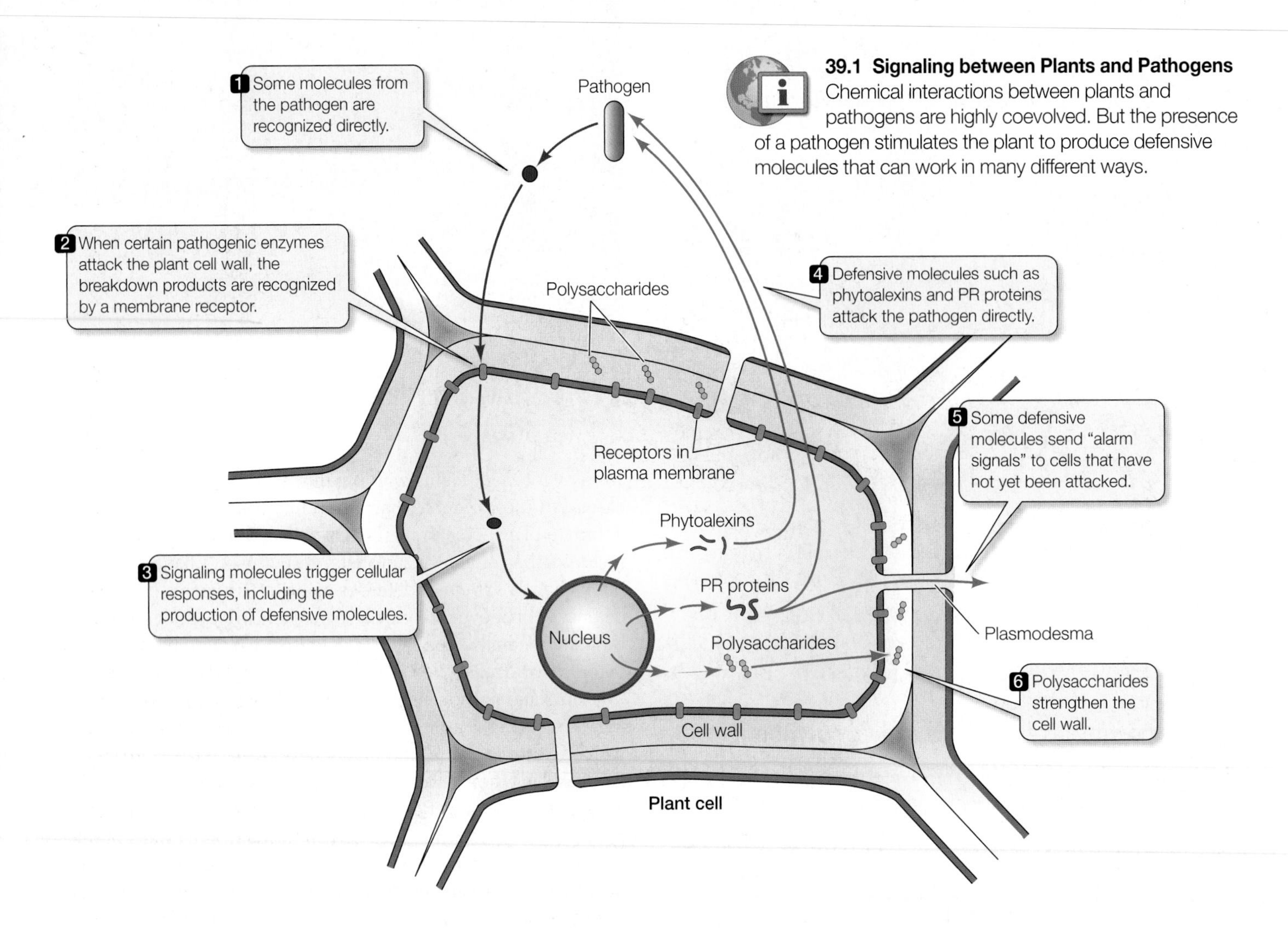

39.1 Signaling between Plants and Pathogens Chemical interactions between plants and pathogens are highly coevolved. But the presence of a pathogen stimulates the plant to produce defensive molecules that can work in many different ways.

They also serve as a base upon which lignin may be laid down. Lignin enhances the mechanical barrier, and the toxicity of lignin precursor chemicals makes the cell inhospitable to some pathogens. These lignin building blocks are only one example of the toxic substances that plants use as chemical defenses.

Some plants have potent chemical defenses against pathogens

When infected by certain fungi and bacteria, plants produce a variety of defensive compounds. Two important kinds of defensive compounds are small molecules called *phytoalexins* and larger proteins called *pathogenesis-related proteins* (see Figure 39.1).

Phytoalexins are toxic to many fungi and bacteria. Most are phenolics or terpenes, compounds that protect plants against herbivores as well as pathogens (**Table 39.1**). They are produced by infected cells and their immediate neighbors within hours of the onset of infection. Enzymes from a pathogenic fungus can cause plant cell walls to release signaling molecules called *oligosaccharins*, which trigger phytoalexin production. Because their antimicrobial activity is nonspecific, phytoalexins can destroy many species of fungi and bacteria in addition to the one that originally triggered their production. Physical injuries, viral infections, and chemical compounds produced in response to damage by herbivores can also induce the production of phytoalexins.

Plants also produce several types of **pathogenesis-related proteins**, or **PR proteins**. Some are enzymes that break down the cell walls of pathogens. These enzymes destroy some of the invading cells, and in some cases the breakdown products of the pathogen's cell walls serve as chemical signals that trigger further defensive responses. Other PR proteins may serve as alarm signals to plant cells that have not yet been attacked. In general, PR proteins appear not to be rapid-response weapons; rather, they act more slowly, perhaps after other mechanisms have blunted the pathogen's attack.

PR proteins and phytoalexins do not act alone. Rather, they are tools used in complex defensive responses, such as the *hypersensitive response* and *systemic acquired resistance*.

The hypersensitive response is a localized containment strategy

Plants that are resistant to fungal, bacterial, or viral diseases generally owe this resistance to the **hypersensitive response**. Cells around the site of infection die, preventing the spread of the pathogen by depriving it of nutrients. Some of the cells produce phytoalexins and other chemicals before they die. The dead tissue, called a *necrotic lesion*, contains and isolates what is left of the microbial invasion (**Figure 39.2**). The rest of the plant remains free of the infecting microbe.

TABLE 39.1

Secondary Plant Metabolites Used in Defense

CLASS	TYPE	ROLE	EXAMPLE
Nitrogen-containing	Alkaloids	Affect herbivore nervous system	Nicotine in tobacco
	Glycosides	Release cyanide or sulfur compounds	Dhurrin in sorghum
	Nonprotein amino acids	Disrupt herbivore protein structure	Canavanine in jack bean
Phenolics	Flavonoids	Phytoalexins	Capsidol in peppers
	Quinones	Inhibit competing plants	Juglone in walnut
	Tannins	Deter herbivores and microbes	Many woods, such as oak
Terpenes	Monoterpenes	Insecticides	Pyrethroids in chrysanthemums
	Sesquiterpenes	Phytoalexins; deter herbivores	Gossypol in cotton
	Steroids	Mimic insect hormones and disrupt insect life cycles	α-Ecdysone in ferns
	Polyterpenes	Feeding deterrent?	Latex in rubber tree

One of the defensive chemicals produced during the hypersensitive response is a close relative of aspirin. Since ancient times, people in Asia, Europe, and the Americas have used willow (*Salix*) leaves and bark to relieve pain and fever. The active ingredient in willow is *salicylic acid*, the substance from which aspirin is derived.

It now appears that all plants contain at least some salicylic acid. This compound often evokes a second complex defensive response, which we will examine next.

Systemic acquired resistance is a form of long-term "immunity"

Systemic acquired resistance is a general increase in the resistance of the entire plant to a wide range of pathogenic species. It is not limited to the pathogen that originally triggered it or to the site of the original infection, and it may have a long-lasting effect.

39.2 The Aftermath of a Hypersensitive Response These necrotic spots on the leaves of a broad bean plant are a response to "chocolate spot" fungus, *Botrytis fabae*.

Systemic acquired resistance is accompanied by the synthesis of PR proteins. Treatment of plants with salicylic acid or aspirin leads to the production of PR proteins and to a resistance to pathogens. Salicylic acid treatment provides substantial protection against tobacco mosaic virus (a well-studied plant pathogen) and some other viruses. In some cases salicylic acid inhibits virus replication and in others it interferes with the movement of viruses out of the infected area.

Salicylic acid also serves as a plant hormone. In some cases, microbial infection in one part of a plant leads to the export of salicylic acid to other parts of the plant, where it causes the production of PR proteins before the infection can spread. The PR proteins then limit the extent of the infection. Infected plant parts also produce the closely related compound *methyl salicylate* (also known as oil of wintergreen). This volatile substance travels to other plant parts through the air, and may trigger the production of PR proteins in neighboring plants that have not yet been infected.

How does a plant know when it should activate the hypersensitive response and systemic acquired resistance? An interaction between plant and pathogen initiates these responses.

Some plant genes match up with pathogen genes

Many plants use the hypersensitive response and systemic acquired resistance as nonspecific defenses against various pathogens. However, the triggering of these responses resides in a highly specific mechanism, called **gene-for-gene resistance**. The ability of a plant to defend itself against a specific strain of a pathogen depends on the presence a particular allele of a gene in the plant that corresponds to a particular allele of a gene in the pathogen (**Figure 39.3**). Let's see how this matching works.

Plants have a large number of ***R* genes** (resistance genes), and many pathogens have sets of ***Avr* genes** (avirulence genes). Dominant *R* alleles favor resistance, and dominant *Avr* alleles make a pathogen less effective. If a particular plant has the dominant allele of one *R* gene and a pathogen strain infecting it has the dominant allele of the corresponding *Avr* gene, the plant will be resistant to that strain. This is true even when none of the other *R-Avr* pairs fea-

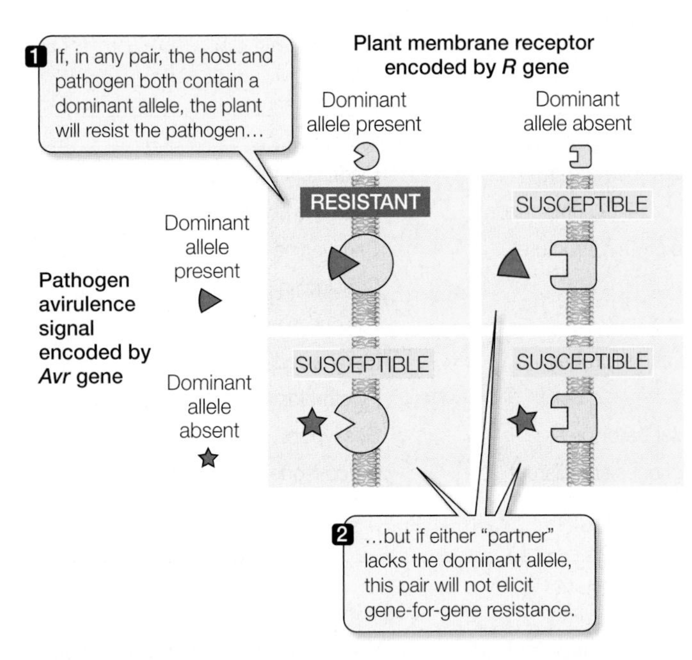

39.3 Gene-for-Gene Resistance A single pair of corresponding dominant alleles promotes resistance even if all the other pairs are mismatches.

tures corresponding dominant alleles. (This effect of one *R-Avr* pair overruling the others is an example of *epistasis*; see Section 10.3.)

The mechanism of gene-for-gene resistance is not completely understood. There are thousands of specific *R* genes among the plants, and their products have different functions. The *Avr* genes in pathogens are simply the genes that cause the pathogen to produce a substance, often toxic, that elicits a defensive response in the plant. Most gene-for-gene interactions trigger the hypersensitive response.

Before we leave the topic of plant defenses against pathogens, let's consider a recently discovered specific defense mechanism directed against RNA viruses (viruses that have RNA instead of DNA as their hereditary material).

Plants develop specific immunity to RNA viruses

Plants respond to attack by RNA viruses by mounting a specific immune response. The plant uses its own enzymes to convert some of the single-stranded RNA of the invading virus into *double-stranded RNA* (dsRNA) and to chop that dsRNA into small pieces called *small interfering RNAs* (siRNAs). Some of the viral RNA is transcribed, forming mRNAs that advance the viral infection. However, the siRNAs interact with another cellular component to degrade those mRNAs, blocking viral replication. This phenomenon is an example of *RNA interference* (*RNAi*), or *posttranscriptional gene silencing* (see p. 365). Molecular biologists are exploring applications of RNAi in plant biotechnology.

The immunity conferred by RNAi spreads quickly throughout the entire plant, by mechanisms not yet fully understood. However, the establishment of immunity depends on the extent of the original infection and the speed of the plant's response. Plant viruses do fight back: most have evolved mechanisms to confound RNA interference. Natural selection favors both improved attack mechanisms for the pathogens and improved defense mechanisms for the plants.

39.1 RECAP

In the hypersensitive response to pathogens, plants seal off infected areas and produce chemical defenses. Systemic acquired resistance, providing a longer-lasting, more general immunity may follow. Gene-for-gene resistance triggers some of these phenomena.

- What are some of the defensive compounds produced by plant cells when infected by bacteria or fungi? See pp. 838–839, Figure 39.1, and Table 39.1
- Can you explain how *R* and *Avr* genes determine which pathogens a plant may be unable to resist? See pp. 839–840 and Figure 39.3

Not all biological threats to plants come from microorganisms and viruses that cause diseases. Another threat comes from the many animals, from inchworms to elephants, that eat plants.

39.2 How Do Plants Deal with Herbivores?

Herbivores—animals that eat plants—depend on plants for energy and nutrients, and they often spread disease. Plants have many defense mechanisms that protect them against herbivores, as we will see. First let's consider how herbivores can have a *positive* effect on some of the plants they eat.

Grazing increases the productivity of some plants

Herbivores are predators that prey on plants, but rarely do they kill their prey. In **grazing**, a herbivore eats part of a plant, such as the leaves, without killing the plant, which then has the potential to grow back.

What are the consequences of grazing? How detrimental is grazing to plants? How well have plants adapted to their place in the food chain? Certain plants and their predators have evolved together, each acting as the agent of natural selection on the other. Coevolution has favored increased photosynthetic production in some grazed-upon plant species.

Removing some leaves from a plant can increase the rate of photosynthesis in the remaining leaves, for several reasons. First, nitrogen obtained from the soil by the roots no longer needs to be divided among as many leaves. Second, the export of sugars and other photosynthetic products from the leaves may be enhanced because the demand for those products in the roots is undiminished, while the sources for those products—leaves—have been decreased. That is, the remaining leaves may compensate by photosynthesizing more rapidly.

A third and particularly significant factor increasing photosynthesis, especially in grasses, is an increase in the availability of light to the younger, more active leaves or leaf parts. The removal of older or dead leaves by a grazer decreases the shading of younger leaves. Unlike most other plants, which grow from their shoot and leaf tips, grasses grow from the base of the shoot and leaf, so their growth is not cut short by grazing.

Mule deer and elk graze many plants, including one called scarlet gilia (*Ipomopsis aggregata*). Although grazing removes about 95 percent of the aboveground plant, the scarlet gilia quickly regrows not one but four replacement stems (**Figure 39.4**). Grazed plants produce three times as many fruits by the end of the growing season as do ungrazed plants.

Some grazed trees and shrubs continue to grow until much later in the season than do ungrazed but otherwise similar plants. The growing season is extended in part by the grazers' removal of apical buds, which stimulates axillary buds to become active and produces a more heavily branched plant. Leaves on ungrazed plants may also die earlier in the growing season than leaves on grazed plants.

A plant also may benefit from moderate herbivory by attracting animals that spread its pollen or that eat its fruit and thus disperse its seeds—in such cases, the benefits to the plant outweigh the costs. Nevertheless, resisting attack by herbivores is often advantageous to a plant.

Some plants produce chemical defenses against herbivores

Although a plant cannot flee its herbivorous enemies, it may be able to defend itself chemically. Many plants attract, resist, and inhibit other organisms by producing special chemicals known as *secondary metabolites*. *Primary metabolites* are substances, such as proteins, nucleic acids, carbohydrates, and lipids, that are produced and used by all living things. **Secondary metabolites** are substances that are not used for basic cellular metabolism. Although all organisms use the same kinds of primary metabolites, plants can differ as radically in their secondary metabolites as they do in their external appearance.

The more than 10,000 known secondary plant metabolites range in molecular weight from about 70 to more than 390,000 daltons, but most have a low molecular weight. Some are produced by only a single species, while others are characteristic of entire genera or even families. The effects of defensive secondary metabolites on animals are diverse. Some secondary metabolites act on the nervous systems of herbivorous insects, mollusks, or mammals. Others mimic the natural hormones of insects, causing some larvae to fail to develop into adults. Still others damage the digestive tracts of herbivores. Some secondary metabolites are toxic to fungal pests. Humans make commercial use of many secondary plant metabolites as fungicides, insecticides, rodenticides, and pharmaceuticals.

The secondary metabolite nicotine was one of the first insecticides to be used by farmers and gardeners. Yet tobacco and related plants that produce nicotine are still attacked by pests such as the tobacco hornworm. The question of whether nicotine deters pests was conclusively investigated by biologists in 2004. The study used tobacco plants in which an enzyme in nicotine biosynthesis had been silenced, lowering the nicotine concentration by more than 95 percent. The low-nicotine plants suffered much more damage than normal plants (**Figure 39.5**).

While many secondary metabolites have protective functions, others are essential as attractants for pollinators and seed dispersers. Table 39.1 lists the major classes of defensive secondary plant metabolites and their biological roles.

Let's look at a specific example of an insecticidal secondary metabolite, canavanine.

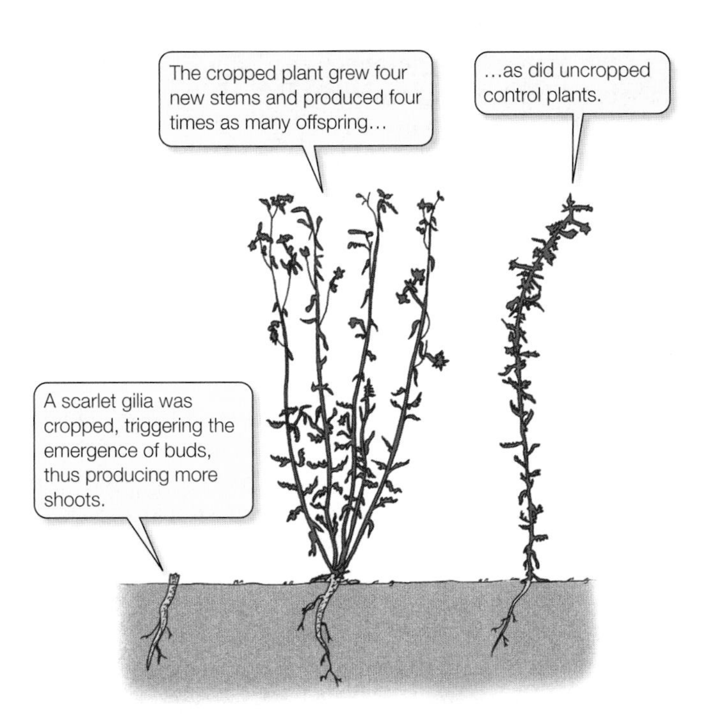

39.4 Overcompensation for Being Eaten Experiments confirm that some plants benefit from grazing.

Some secondary metabolites play multiple roles

Canavanine is an amino acid that is not found in proteins, but is similar to the amino acid arginine, which is found in almost all proteins. Canavanine has two important roles in plants that produce it in significant quantities. The first is as a nitrogen-storing compound in seeds. The second role is defensive, and is based on the similarity of canavanine to arginine:

A seemingly slight chemical difference... ...produces an inactive protein.

Arginine	Canavanine
NH_2	NH_2
C=NH	C=NH
N—H	N—H
H—C—H	O
H—C—H	H—C—H
H—C—H	H—C—H
H_2N—C—H	H_2N—C—H
C—OH	C—OH
‖ O	‖ O

EXPERIMENT

HYPOTHESIS: Nicotine helps protect tobacco plants against insects.

METHOD

1. Create a low-nicotine line of plants by modifying a gene that encodes a key enzyme in the biosynthetic pathway to nicotine.
2. Transplant both low-nicotine and wild-type tobacco plants into a field plantation where they are accessible to insects.
3. Assess the extent of leaf damage by insects at 2-day intervals.

RESULTS

The low-nicotine plants suffered more than twice as much leaf damage as did the wild-type controls.

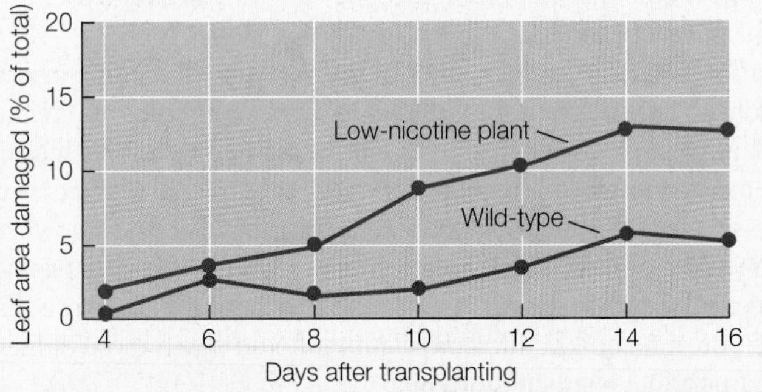

Leaf area damaged (% of total)
20
15
10
5
0
Low-nicotine plant
Wild-type
4 6 8 10 12 14 16
Days after transplanting

CONCLUSION: Nicotine provides tobacco plants with at least some protection against insects.

39.5 Some Plants Use Nicotine to Reduce Insect Attacks Anke Steppuhn and her coworkers showed that wild-type tobacco plants lost less leaf area to pests than did transgenic plants that contained less nicotine. FURTHER RESEARCH: Treatment of tobacco plants with jasmonate (a hormone) elicits the production of nicotine and other compounds. How would you modify this experiment to determine whether nicotine is the only insecticidal compound produced by tobacco?

When an insect larva consumes canavanine-containing plant tissue, the canavanine is incorporated into the insect's proteins in some of the places where the DNA has coded for arginine, because the enzyme that charges the tRNA specific for arginine fails to discriminate accurately between the two amino acids (see Section 12.4). The structure of canavanine, however, is different enough from that of arginine that some of the resulting proteins end up with a modified tertiary structure and hence reduced biological activity. These defects in protein structure and function lead to developmental abnormalities that kill the insect.

A few insect larvae are able to eat canavanine-containing plant tissue and still develop normally. Why? In these larvae, the enzyme that charges the arginine tRNA discriminates correctly between arginine and canavanine. The canavanine they ingest is thus not incorporated into the proteins they form, and the larvae are not harmed.

In plants that produce it, canavanine is present regardless of whether the plant is under attack. Other chemical defenses come into play only when a predator strikes.

Some plants call for help

If you are attacked, it makes sense to call for help. Some plants do this, too. When caterpillars begin to chew on the leaves of corn, cotton, or some other plant species, the leaves synthesize chemical signals that attract other insects that feed on the caterpillars.

Although most such calls for help are generated in the leaves, in 2005 biologists found that some plant roots attacked by beetle larvae respond by releasing an attractant for a tiny nematode that attacks the larvae (**Figure 39.6**). Perhaps it will soon be possible to enhance the production of such attractants and increase crop yields.

Many defenses depend on extensive signaling

Many plant defenses are activated by a series of signals. Insects feeding on tomato leaves damage the cells, leading to a chain of events that includes the formation of hormones and ends with the production of an insecticide. The signaling steps in the production of one defensive compound, shown in **Figure 39.7**, involve two hormones. **Systemin**, which is formed in response to an insect attack, is a polypeptide hormone—the first polypeptide hormone to be discovered in plants. **Jasmonates**, whose production is initiated by systemin, are formed from the unsaturated fatty acid linolenic acid. The final step in the chain is the synthesis of a protease inhibitor. The inhibitor, once in an insect's gut, interferes with the digestion of proteins and thus stunts the insect's growth.

Jasmonates also take part in the "call for help" caterpillars attack a corn plant. A volatile substance released from the leaves by chewing caterpillars is the first signal, leading to the formation of jasmonates by the plant. The jasmonates, in turn, trigger the formation of the volatile compounds that attract insects that prey on the caterpillars.

Although plants have many effective natural defenses, agricultural researchers are attempting to provide crop plants with even more effective ones, including transfer of natural, effective mechanisms between species.

Recombinant DNA technology may confer resistance to insects

Wild and domesticated common beans (*Phaseolus vulgaris*) differ in their resistance to two species of bean weevils. Some wild bean seeds are highly resistant to these insects, but no cultivated bean seeds show such resistance. Scientists discovered that all weevil-resistant bean seeds contain a specific seed protein, *arcelin*. This protein has never been found in cultivated bean seeds. Therefore, the scientists hypothesized that arcelin is responsible for the resistance of some seeds to predation by the weevils.

To test the hypothesis that arcelin is the compound that confers resistance in wild species, the scientists performed two series of experiments. In one series, they crossed cultivated and wild bean plants. All of the progeny seeds of such crosses that contained arcelin showed resistance to weevils. In the other series of experiments, the scientists removed the seed coats of domesticated beans and ground the remainder of the seeds into flour. They added different concentrations of arcelin to different batches and molded the flour into artificial seeds. They then let weevils attack the artificial seeds. The more arcelin the artificial seeds contained, the more resistant they were to weevils. These data support the hypothesis that arcelin is the active component that confers weevil resistance.

Now that the data indicate that arcelin is the active component in pest resistance, it was of interest to assess its toxicity, if any, in animal systems. Preliminary tests showed that arcelin in cooked beans was not harmful to rats—a first step toward determining whether arcelin is safe in food for humans. Agricultural scientists must sometimes choose between crop protection and appeal to humans. A plant with sturdy chemical defenses may taste bad, make us sick, or even kill us.

Scientists are now seeking to introduce genes for arcelin and other resistance-conferring proteins into agriculturally important crops such as beans. The development of crop plants that produce

EXPERIMENT

HYPOTHESIS: Corn roots attacked by beetle larvae attract nematodes that will attack the larvae.

METHOD

1. Construct a test system with 6 arms radiating from a central chamber. Add soil, connecting all parts of the system.
2. Three of the control chambers contain only soil. The other three chambers each contain a single corn plant: One healthy plant, one with roots damaged by beetle larvae, and one with roots damaged by stabbing with a metal tool.
3. After three days, add ~2,000 nematodes to the central chamber.
4. After 24 hours count the nematodes in each connecting arm.

RESULTS

Nematodes moved into each of the arms, but by far the most moved into the arm leading to the larvae-damaged plant.

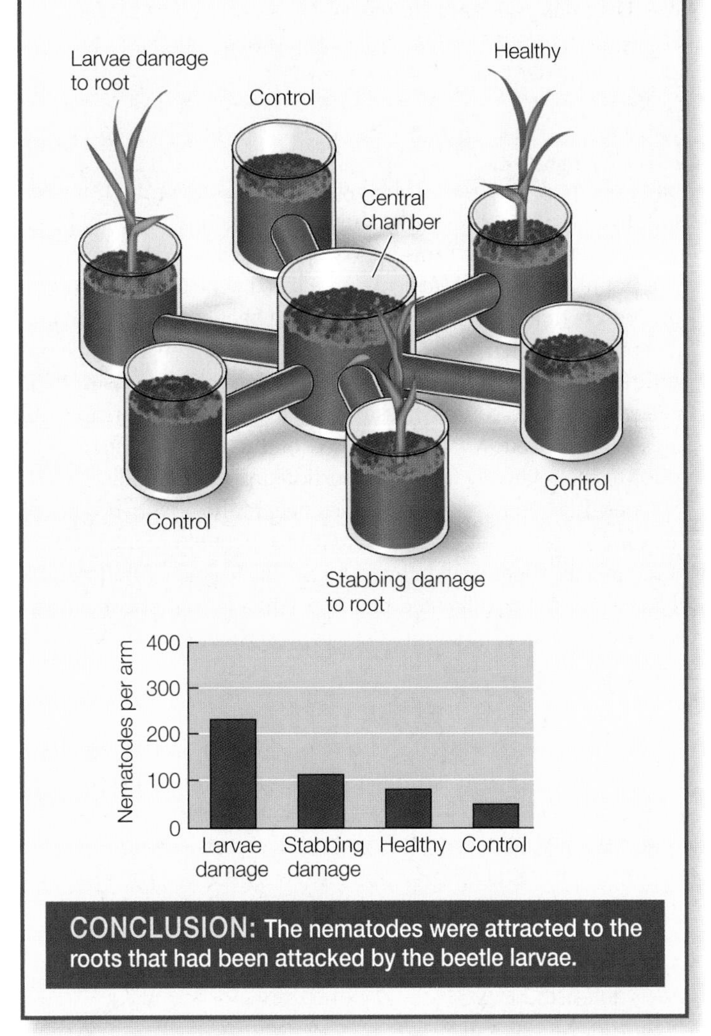

39.6 Roots May Recruit Nematodes as Defenders When attacked by beetle larvae, corn roots generate a chemical signal that attracts nematodes—natural enemies of the larva. FURTHER RESEARCH: The biologists who did this study also showed that roots attacked by the beetle larvae released a secondary metabolite called (E)-β-caryophyllene. How would you test the hypothesis that this compound is the attractant for the nematodes?

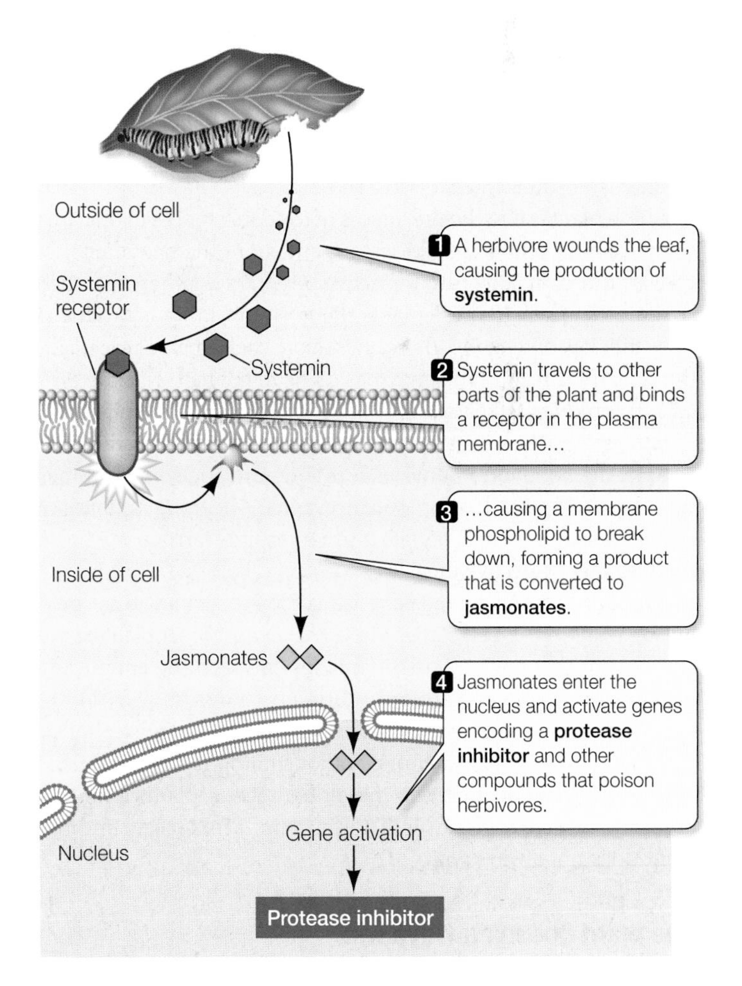

39.7 A Signaling Pathway for Synthesis of a Defensive Secondary Metabolite The chain of events initiated by an insect's attack leads to the production of a defensive chemical and can consist of many steps. These steps may include the synthesis of one or two hormones, binding of receptors, gene activation, and, finally, synthesis of insecticides.

their own pesticides is an active area of research in agricultural biotechnology. One of the most widely applied approaches has been the engineering of several crops, such as tomatoes, corn, and cotton, to express the toxin genes from *Bacillus thuringiensis*. The toxin kills insect pest larvae, enhancing crop yields, as discussed in Section 16.6.

Why don't plants poison themselves?

Why don't the chemicals that are so toxic to herbivores and microbes kill the plants that produce them? Plants that produce toxic secondary metabolites generally use one of the following measures to protect themselves:

- The toxic material is isolated in a special compartment.
- The toxic substance is produced only after the plant's cells have already been damaged.
- The plant uses modified enzymes or modified receptors that do not recognize the toxic substance.

The first method is the most common. Plants using this method store their poisons in vacuoles if they are water-soluble. If they are hydrophobic, the poisons are stored in **laticifers** (tubes containing a white, rubbery latex) or dissolved in waxes on the epidermal surface. This compartmentalized storage keeps the toxic substance away from the mitochondria, chloroplasts, and other parts of the plant's own metabolic machinery. One example of a plant with this mechanism is milkweed, discussed below.

Some plants store the precursors of toxic substances in one area of the plant, such as the epidermis, and store the enzymes that convert those precursors to the active poison in another area, such as the mesophyll. In these plants the toxic substance is produced only after it is damaged. When a herbivore chews part of the plant, the cells rupture, and the enzymes come in contact with the precursors, producing the toxic product. The only part of the plant that is damaged by the toxic substance is that which was already damaged by the herbivore. Plants such as sorghum and some legumes that respond to attack by producing cyanide—a strong inhibitor of cellular respiration in all organisms that respire—are among those that use this protective measure.

The third protective measure is used by the canavanine-producing plants described earlier. They, unlike most other plants, have an enzyme that correctly distinguishes between the chemically similar canavanine and arginine during protein synthesis. However, as we have seen, some herbivores can evade poisoning by canavanine in a similar manner, demonstrating that no plant defense is perfect. Like plants and their pathogens, plants and their predators evolve together in a continuing "arms race," and the plant does not always win.

The plant doesn't always win

Milkweeds such as *Asclepias syriaca* are latex-producing (laticiferous) plants. When damaged, a milkweed releases copious amounts of toxic latex from its laticifers, which run alongside the veins in its leaves. Latex, a milky liquid, has long been suspected to deter insects from eating the plant because some insects that feed on neighboring plants of other species do not attack laticiferous plants. This observed behavior is consistent with, but does not prove, the hypothesis that the latex keeps the insects at bay.

39.8 Disarming a Plant's Defenses This beetle is inactivating a milkweed's defense system by cutting its laticifer supply lines.

Stronger support for this hypothesis was obtained by studying field populations of *Labidomera clivicollis*, a beetle that is one of the few insects that feed on *A. syriaca*. These beetles show a remarkable prefeeding behavior: They cut a few veins in the leaves before settling down to dine (**Figure 39.8**). Cutting the veins, with their adjacent laticifers, causes massive latex leakage and interrupts the latex supply to a downstream portion of the leaf. The beetles then move to the relatively latex-free portion and eat their fill.

Does this behavior of the beetles negate the adaptive value of latex protection? Not entirely. Great numbers of potential insect pests are still effectively deterred by the latex. And evolution proceeds. Over time, milkweed plants producing higher concentrations of toxins may be selected by virtue of their ability to kill even the beetles that cut their laticifers.

39.2 RECAP

Plants must defend themselves against attack by herbivores and disease despite the fact that they are immobile. Not all herbivory is detrimental, but where it is, many plants use secondary metabolites as chemical defenses.

- What roles do systemin and jasmonates play in plant defense? See p. 843 and Figure 39.7
- Why don't a plant's toxic secondary metabolites poison the plant itself? See p. 844

A plant's survival depends not only on successful defense against predators and pathogens. The physical environment can be hostile to a plant as well. Next we consider how plants adapt to climate-imposed threats.

39.9 Desert Annuals Evade Drought Seeds of desert plants often lie dormant for long periods awaiting conditions appropriate for germination. When they do germinate, they grow and reproduce rapidly before the short wet season ends. During the long dry spells, only dormant seeds remain alive.

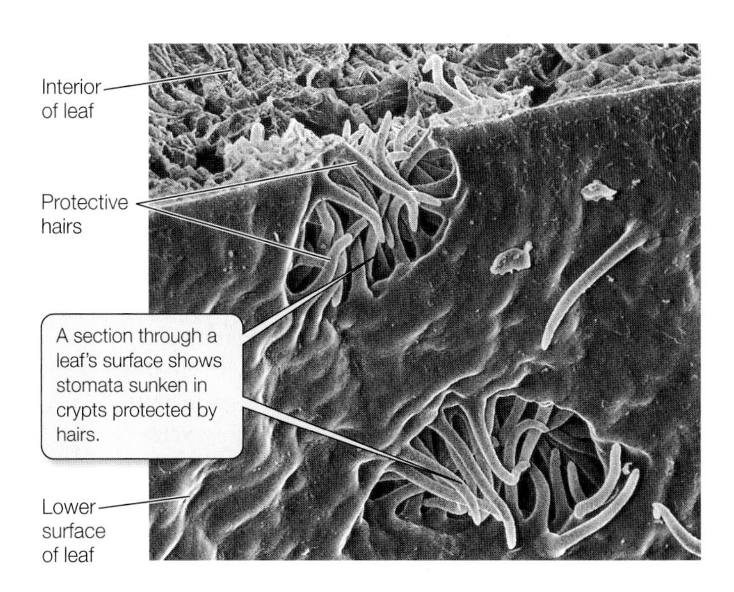

39.10 Stomatal Crypts Stomata in the leaves of some xerophytes are located in sunken cavities called stomatal crypts. The hairs covering these crypts trap moist air.

39.3 How Do Plants Deal with Climate Extremes?

Plants are threatened by many aspects of the physical environment, such as excess dryness, waterlogged soils, and extremes of temperature, both high and low. Water is often in short supply in the terrestrial environment. Extreme terrestrial habitats such as deserts intensify this challenge. Some desert plants have no special *structural* adaptations for water conservation other than those found in almost all flowering plants. Instead, they have an alternative *strategy*. These desert annuals simply evade the periods of drought. They carry out their entire life cycle—from seed to seed—during a brief period in which rainfall has made the surrounding desert soil sufficiently moist (**Figure 39.9**).

Many plants that inhabit particularly dry areas, however, have one or more adaptations that allow them to conserve water. Plants adapted to dry environments are called **xerophytes**.

Xeros is the Greek word for "dry." Xerox, one of the most recognized brand names in commercial history, made its reputation (and fortune) from its patent on the first "dry" office copying process, based on powdered toner. Up until that time, office copiers such as mimeograph machines required messy liquids to "duplicate" a master copy.

Some leaves have special adaptations to dry environments

Plants that remain active during dry periods must have structural adaptations that enable them to survive. The secretion of a thick cuticle over the leaf epidermis helps retard water loss in dry environments. An even more common adaptation is a dense covering of epidermal hairs. Some species have stomata only in sunken cavities below the leaf surface, which reduces the drying effects of air currents; often these **stomatal crypts** contain hairs as well (**Figure 39.10**). The hairs slow the air currents around the stomata.

Succulence—the possession of fleshy, water-storing leaves or stems—is an adaptation to dry environments. Ice plants and their relatives have fleshy leaves in which water may be stored. Other xerophytes, such as ocotillo, produce leaves only when water is abundant, shedding them as the soil dries out (**Figure 39.11**). Cacti

39.11 Opportune Leaf Production The ocotillo, a xerophyte that lives in the lower deserts of the southwestern United States and northern Mexico, produces leaves only when there is sufficient water for photosynthesis.

and similar plants have spines rather than typical leaves, and photosynthesis is confined to the fleshy stems. The spines may reflect incident radiation, or they may dissipate heat. Corn and some related grasses have leaves that roll up during dry periods, thus reducing the leaf surface area through which water is lost. Some trees, such as eucalyptus, that grow in arid regions have leaves that hang vertically at all times, thus evading direct rays of the midday sun.

These xerophytic adaptations of leaves minimize water loss by the plant. However, such adaptations simultaneously minimize the uptake of carbon dioxide and thus limit photosynthesis. In consequence, most xerophytes grow slowly, but they utilize water more efficiently than do other plants—that is, they fix more grams of carbon by photosynthesis per gram of water lost to transpiration than other plants do.

39.12 Mining Water with Deep Taproots In Death Valley, California, this mesquite must reach far down into the sand dunes for its water supply.

Plants have other adaptations to a limited water supply

Roots may also be adapted to dry environments. Mesquite trees (genus *Prosopis*; **Figure 39.12**) obtain water through taproots that grow to great depths, reaching water supplies far underground, as well as from condensation on their leaves. The Atacama Desert in northern Chile often goes several years without measurable rainfall. The landscape there is almost barren of plant life, save for many surprisingly large mesquite trees.

A more common adaptation of desert plants is a root system that grows rapidly during rainy seasons but dies back during dry periods. Cacti have shallow but extensive fibrous root systems that effectively intercept water at the surface of the soil following even light rains.

Xerophytes and other plants that receive inadequate water may accumulate the amino acid proline or other solutes to substantial concentrations in their vacuoles. As a consequence, the solute potential and water potential of their cells become more negative; thus these plants tend to extract more water from the soil than do plants that lack this adaptation. Plants living in salty environments share this and several other adaptations with xerophytes, as we will see.

In water-saturated soils, oxygen is scarce

For some plants, the environmental challenge is the opposite of that faced by xerophytes: too much water. Some plants live in environments so wet that the diffusion of oxygen to their roots is severely limited. Since most plant roots require oxygen to support respiration and ATP production, most plants cannot tolerate saturated soil conditions for long.

Some species, however, are adapted to life in a water-saturated habitat. Their roots grow slowly and hence do not penetrate deeply. Because the oxygen level is too low to support aerobic respiration, the roots carry on alcoholic fermentation (an anaerobic process; see Chapter 7), which provides ATP for the activities of the root system. This adaptation explains why their growth is slow—fermentation is much less efficient in producing ATP than aerobic respiration.

The root systems of some plants adapted to swampy environments have **pneumatophores**, which are extensions that grow out of the water and up into the air (**Figure 39.13**). Pneumatophores have lenticels and contain spongy tissues that allow oxygen to diffuse through them, aerating the submerged parts of the root system. Cypresses and some mangroves are examples of plants with pneumatophores.

Submerged or partly submerged aquatic plants often have large air spaces in the leaf parenchyma and in the petioles. Tissue con-

39.13 Coming Up for Air The roots of the mangroves in this tidal swamp obtain oxygen through pneumatophores.

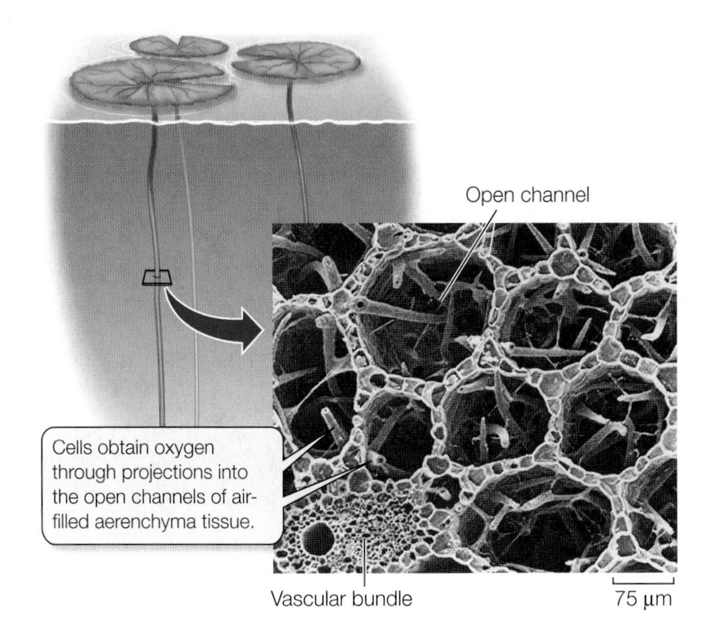

39.14 Aerenchyma Lets Oxygen Reach Submerged Tissues The scanning electron micrograph, a cross section of a petiole of the yellow water lily, shows the air-filled channels of aerenchyma tissue. The cells that line these channels obtain oxygen by extending projections into these channels.

taining such air spaces is called **aerenchyma** (**Figure 39.14**). Aerenchyma stores oxygen produced by photosynthesis and permits its ready diffusion to parts of the plant where it is needed for cellular respiration. Aerenchyma also imparts buoyancy. Furthermore, because it contains far fewer cells than most other plant tissue, respiratory metabolism in aerenchyma proceeds at a lower rate, and the need for oxygen is much reduced.

We have seen that plants are threatened by shortages or excesses of water in their physical environment. We now examine two more threats in the physical environment: high and low temperatures.

Plants have ways of coping with temperature extremes

Temperatures that are too high or too low can stress plants and even kill them. Plants differ in their sensitivity to heat and cold, but all plants have their limits. Any temperature extreme can damage cellular membranes.

- High temperatures destabilize membranes and denature many proteins, especially some of the enzymes of photosynthesis.
- Low temperatures cause membranes to lose their fluidity and alter their permeabilities to solutes.
- Freezing temperatures may cause ice crystals to form, damaging cellular membranes.

Transpiration (loss of water by plants through evaporation) can cool a plant, but it also increases the plant's need for water. Therefore, it is not surprising that many plants living in hot environments have adaptations similar to those of xerophytes. These adaptations include epidermal hairs and spines that radiate heat, leaf displays that intercept less direct sunlight, and an alternative form of metabolism that allows plants to perform some metabolic processes in the cool of night—crassulacean acid metabolism (CAM; see Section 8.4).

Plants respond within minutes to high temperatures by producing several kinds of **heat shock proteins**. Among these proteins are chaperonins (see Figure 3.12), which help other proteins maintain their structures and avoid denaturation. Threshold temperatures for the production of heat shock proteins vary, but 39°C is sufficient to induce them in most plants. We have much to learn about the dozens of heat shock proteins, but we do know that some other types of stress also induce their formation. Among these stresses are chilling and freezing.

Low temperatures above the freezing point injure many plants, including important crops such as rice, corn, and cotton as well as tropical plants such as bananas. This is referred to as *chilling injury*. Many plant species can be modified to resist the effects of cold spells by a process called **cold-hardening**, which involves repeated exposure to cool, but not injurious, temperatures. The hardening process, however, requires many days. A key change that occurs during the hardening process is an increase in the relative amount of unsaturated fatty acids in membranes. Unsaturated fatty acids solidify at lower temperatures than do saturated ones. Thus, the membranes retain their fluidity and function normally at cooler temperatures.

Low temperatures induce the formation of certain heat shock proteins that protect against chilling injury. There are also cases of "cross-protection" by heat shock proteins that are induced by one type of stress and that protect against other stresses. Tomatoes stressed by 2 days of high temperatures, for example, formed heat shock proteins and became resistant to chilling injury for the next 3 weeks.

If ice crystals form within plant cells, they can kill the cells by puncturing organelles and plasma membranes. More importantly, the growth of ice crystals outside the cells can draw water from the cells and dehydrate them, causing them to plasmolyze. Freezing-tolerant plants have a variety of adaptations to cope with these problems. A common one is the production of **antifreeze proteins** that slow the growth of ice crystals.

39.3 RECAP

Xerophytes are plants that cope with dry environments through structural or behavioral adaptations. Other plants are adapted to life in a water-saturated environment. Many plants are resistant to high and low temperatures. Heat shock proteins play roles in adapting to both extremes.

- Describe the tradeoff between water conservation and photosynthesis in xerophytes. See p. 846
- What conditions induce the formation of heat shock proteins, and what functions do they serve? See p. 847

Just as extremes of climate can limit plant growth, the presence of certain substances can make an environment inhospitable to plant growth. These substances include salt and heavy metals.

39.4 How Do Plants Deal with Salt and Heavy Metals?

Worldwide, no toxic substance restricts angiosperm growth more than salt (sodium chloride) does. *Saline*—salty—habitats support, at best, sparse vegetation. Saline habitats themselves are diverse, ranging from hot, dry, salty deserts to moist, cool, salty marshes. Along the seashore saline environments are created by ocean spray. The ocean itself is a saline environment, as are river estuaries, where fresh and salt water meet and mingle. The salinization of agricultural land is an increasing global problem, not just in Bangladesh, which we described at the beginning of this chapter. Even where crops are irrigated with fresh water, sodium ions from the water accumulate in the soil to ever greater concentrations as the water evaporates.

Saline environments pose an osmotic problem for plants. Because of its high salt concentration, a saline environment has an unusually negative water potential. To obtain water from such an environment, a plant must have an even more negative water potential than that of a plant in a nonsaline environment; otherwise, it would lose water, wilt, and die. A second problem is the potential toxicity of high concentrations of certain ions, notably sodium and chloride.

The **halophytes**—plants adapted to saline habitats—belong to a wide variety of flowering plant groups. How do these plants cope with a saline environment?

Most halophytes accumulate salt

Most halophytes share one adaptation: They accumulate sodium and, usually, chloride ions and transport those ions to the leaves. The accumulated ions are stored in the central vacuoles of leaf cells, away from more sensitive parts of the cells. Nonhalophytes accumulate relatively little sodium, even when placed in a saline environment; of the sodium that is absorbed by their roots, very little is transported to the shoot. The increased salt concentration in the tissues of halophytes makes their water potential more negative, so they can take up water more easily from the saline environment.

> Scientists have succeeded in causing the overexpression of a gene that enables tomato plants to take up salt into the vacuoles of cells. This increased sodium uptake converts the tomato to a halophyte, enabling it to grow in saline environments. Crops with such gene overexpression can be watered with diluted seawater.

Some halophytes have other adaptations to life in saline environments. Some, for example, have **salt glands** in their leaves. These glands excrete salt, which collects on the leaf surface until it is removed by rain or wind (**Figure 39.15**). This adaptation, which reduces the danger of poisoning by accumulated salt, is found both in some desert plants, such as tamarisk, and in some mangroves growing in seawater.

39.15 Excreting Salt This salty mangrove has special salt glands that excrete salt, which appears here as crystals on the leaves.

Salt glands can play multiple roles, as in the desert shrub *Atriplex halimus*. This shrub has glands that secrete salt into small bladders on the leaves, where, by increasing the gradient in water potential, the salt helps the leaves obtain water from the roots. At the same time, by making the water potential of the leaves more negative, the salt reduces the transpirational loss of water to the atmosphere.

The adaptations we have just discussed are specific to halophytes. Several other adaptations are shared by halophytes and xerophytes.

Halophytes and xerophytes have some similar adaptations

Many halophytes, like some xerophytes, accumulate the amino acid proline in their cell vacuoles, making the water potential of their tissues more negative. Unlike sodium, proline is relatively nontoxic.

Succulence is another adaptation that halophytes and xerophytes have in common, as might be expected, since saline environments, like dry ones, make water uptake difficult. Succulence characterizes many halophytes that occupy salt marshes. There the salt concentration in the soil solution may change throughout the day. When the tide is out, evaporation increases the salt concentration. Succulence may offer a reserve of water for the plant during the period of maximum salinity; when the salinity drops as the tide comes in, the leaf's store of water is replenished.

Many succulents—both xerophytes and halophytes—use crassulacean acid metabolism which allows them to take up and store CO_2 as carboxyl groups at night and release the CO_2 for use in photosynthesis during the day. They have reversed stomatal cycles that enable them to conserve water by closing their stomata in the daytime (**Figure 39.16**). Other general adaptations to a saline environment include high root-to-shoot ratios, sunken stomata, reduced leaf areas, and thick cuticles.

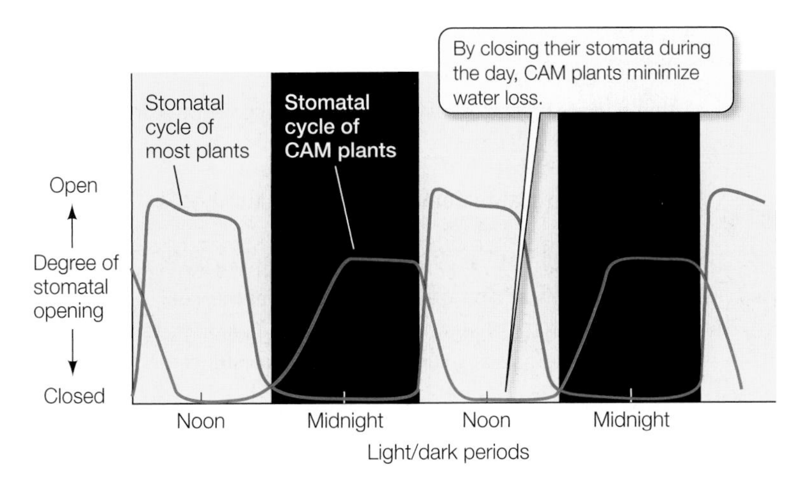

39.16 Stomatal Cycles Most plants open their stomata during the day. CAM plants reverse this stomatal cycle; their stomata open during the night.

Some habitats are laden with heavy metals

Salt is not the only toxic solute found in soils. High concentrations of some heavy metal ions, such as chromium, mercury, lead, and cadmium, poison most plants; these ions often are more toxic than sodium at equivalent concentrations.

Some geographic sites are naturally rich in heavy metals as a result of normal geological processes. In other places, acid rain leads to the release of toxic aluminum ions in the soil. Other human activities, notably the mining of metallic ores, leave localized areas—known as *tailings*—with substantial concentrations of heavy metals and low concentrations of nutrients. Such sites are hostile to most plants, and seeds falling on them generally do not produce adult plants.

39.17 Life after Strip Mining Although high concentrations of heavy metals kill most plants, some plants possess mechanisms to survive the stress and can populate a vacant niche. Here grass colonizes this eroded strip mine in North Park, Colorado.

Mine tailings rich in heavy metals, however, generally are not completely barren (**Figure 39.17**). They may support healthy plant populations that differ genetically from populations of the same species on the surrounding normal soils. How can these plants survive?

Initially, some plants were thought to tolerate heavy metals by excluding them: By not taking up the metal ions, it was believed, the plants avoided being poisoned. However, measurements have shown that tolerant plants growing on mine tailings do take up heavy metals, accumulating them to concentrations that would kill most plants. Thus these plants must have a mechanism for dealing with the heavy metals they take up. Such tolerant plants may be useful agents for *bioremediation*, a decontamination process by which the heavy metal content of some contaminated soils is decreased by living organisms.

We know the mechanism of at least one case of tolerance to a different toxic metal. When the roots of a buckwheat grown in China are exposed to aluminum concentrations high enough to inhibit root growth in other plants, they secrete oxalic acid. Oxalic acid combines with aluminum ions, forming a complex that does not inhibit growth.

From mine to mine, the heavy metals in the soil differ. In Wales and Scotland, bent grass (*Agrostis*) grows near many mines. Samples of bent grass from several such sites were tested for their ability to grow in various solutions, each containing only one heavy metal. In general, the plants tolerated particular heavy metals—the ones most abundant in their habitat—but were sensitive to others. That is, they tolerated only one or two heavy metals, rather than heavy metals as a group.

Tolerant plant populations can evolve and colonize an area surprisingly rapidly. The bent grass population around a particular copper mine in Wales is resistant to copper and is relatively abundant, although the copper-rich soil dates from mining done only a century ago.

39.4 RECAP

Halophytes are adapted to saline habitats by various means, some of which are the same as those employed by xerophytes to cope with dry habitats. A few species are adapted to heavy-metal-rich habitats that are toxic to most other plants.

- What are some roles of salt glands in halophyte leaves? See p. 848
- Can you name some adaptations that are useful to both halophytes and xerophytes? See p. 848 and Figure 39.16
- Can you explain why some heavy metal-tolerant plants may be useful agents for bioremediation? See p. 849

CHAPTER SUMMARY

39.1 How do plants deal with pathogens?

Plants and pathogens evolve together. Review Figure 39.1, Web/CD Tutorial 39.1

Plants can strengthen their cell walls when attacked. Additional lignin and polysaccharides limit the ability of viral pathogens to move from cell to cell.

PR proteins may break down cell walls of pathogens and trigger other defensive responses.

In the **hypersensitive response**, cells produce **phytoalexins** and then die, trapping the pathogens in dead tissue.

The hypersensitive response is often followed by **systemic acquired resistance**, in which salicylic acid activates further synthesis of defensive compounds.

Gene-for-gene resistance matches up alleles in a plant's resistance genes (*R* genes) and a pathogen's **avirulence genes** (***Avr* genes**); when these match, the plant mounts a more vigorous defense. Review Figure 39.3

Plants use **RNA silencing** to develop immunity to invading RNA viruses.

39.2 How do plants deal with herbivores?

Grazing by **herbivores** increases the productivity of some plants. Review Figure 39.4

Some plants produce **secondary metabolites** as defenses against herbivores. Review Figures 39.5, 39.6 and Table 39.1

Hormones including **systemin** and **jasmonates** participate in pathways leading to the production of defensive chemicals. Review Figure 39.7

Toxic chemicals produced by plants are often isolated in plant compartments such as vacuoles and **laticifers**. In some plants enzymes that activate precursors of toxic substances come into contact only when a plant is damaged.

39.3 How do plants deal with climate extremes?

Xerophytes are adapted to dry environments. Some adaptations are behavioral: evasion of drought by the timing of germination by desert annuals, and by the shedding of leaves by ocotillo.

Other xerophytic adaptations are structural, including thickened cuticle, epidermal hairs, sunken stomata, **succulence**, and long taproots.

Adaptations to water-saturated habitats include **pneumatophores**, which allow oxygen uptake from the air, and **aerenchyma**, in which oxygen can diffuse and be stored.

Membranes and proteins can be damaged at high or low temperatures. Plants respond to high or low temperatures by producing **heat shock proteins**.

Some plants undergo **cold-hardening**, which includes changes in membrane lipids and production of heat shock proteins.

Some plants resist freezing by producing **antifreeze proteins**.

39.4 How do plants deal with salt and heavy metals?

Most **halophytes** accumulate salt. Some have **salt glands** that excrete the salt to the leaf surface.

Halophytes and xerophytes have some adaptations in common, such as succulence and the ability to make the water potential of their tissues more negative.

Some plants take up and decontaminate heavy metals; others are tolerant to specific heavy metals.

See Web/CD Activity 39.1 for a concept review of this chapter.

SELF-QUIZ

1. Which of the following is *not* a common defense against bacteria, fungi, and viruses?
   a. Lignin formation
   b. Phytoalexins
   c. A waxy covering
   d. The hypersensitive response
   e. Mycorrhizae
2. Plants sometimes protect themselves from their own toxic secondary metabolites by
   a. producing special enzymes that destroy the toxic substances.
   b. storing precursors of the toxic substances in one compartment and the enzymes that convert those precursors to toxic products in another compartment.
   c. storing the toxic substances in mitochondria or chloroplasts.
   d. distributing the toxic substances to all cells of the plant.
   e. performing crassulacean acid metabolism.
3. Herbivory
   a. is predation by plants on animals.
   b. always reduces plant growth.
   c. usually increases the rate of photosynthesis in the remaining leaves.
   d. reduces the rate of transport of photosynthetic products from the remaining leaves.
   e. is always lethal to the grazed plant.
4. Which statement about secondary plant metabolites is *not* true?
   a. Some attract pollinators.
   b. Some are poisonous to herbivores.
   c. Most are proteins or nucleic acids.
   d. Most are stored in vacuoles.
   e. Some mimic the hormones of animals.
5. Which statement about latex is *not* true?
   a. It is sometimes contained in laticifers.
   b. It is typically white.
   c. It is often toxic to insects.
   d. It is a rubbery solid.
   e. Milkweeds produce it.
6. Which of the following is *not* an adaptation to dry environments?
   a. A less negative solute potential in the vacuoles
   b. Hairy leaves
   c. A heavier cuticle over the leaf epidermis
   d. Sunken stomata
   e. A root system that grows each rainy season and dies back when it is dry

7. Some plants adapted to swampy environments meet the oxygen needs of their roots by means of a specialized tissue called
   a. parenchyma.
   b. aerenchyma.
   c. collenchyma.
   d. sclerenchyma.
   e. chlorenchyma.
8. Halophytes
   a. all accumulate proline in their vacuoles.
   b. have solute potentials that are less negative than those of other plants.
   c. are often succulent.
   d. have low root-to-shoot ratios.
   e. rarely accumulate sodium.
9. Which of the following is *not* a commonly toxic heavy metal?
   a. Chromium
   b. Cadmium
   c. Lead
   d. Potassium
   e. Mercury
10. Plants that tolerate heavy metals commonly
   a. differ genetically from other members of their species.
   b. do not take up the heavy metals.
   c. are tolerant to all heavy metals.
   d. are slow to colonize an area rich in heavy metals.
   e. weigh more than plants that are sensitive to heavy metals.

FOR DISCUSSION

1. How might plant adaptations affect the evolution of herbivores? How might the adaptations of herbivores affect plant evolution?
2. The stomata of the common oleander, *Nerium oleander*, are located in sunken crypts in its leaves. Whether or not you know what an oleander is, you should be able to describe an important feature of its natural habitat. What is that feature?
3. Explain why halophytes often use the same mechanisms for coping with their challenging environments as xerophytes do for coping with theirs.
4. In ancient times, people used less sophisticated methods for mining than we use today. Thus ancient mines often yield substantial profits to modern-day miners who find and work them. On the basis of the information in this chapter, how might you try to locate the site of an ancient mine?

FOR INVESTIGATION

The tobacco hornworm *Manduca sexta* is adapted to feeding on nicotine-producing plants. Using the genetically modified tobacco developed by Steppuhn and coworkers (see Figure 39.5), how might you test the hypothesis that dietary nicotine protects the tobacco hornworm against its parasite *Cotesia congregata*?

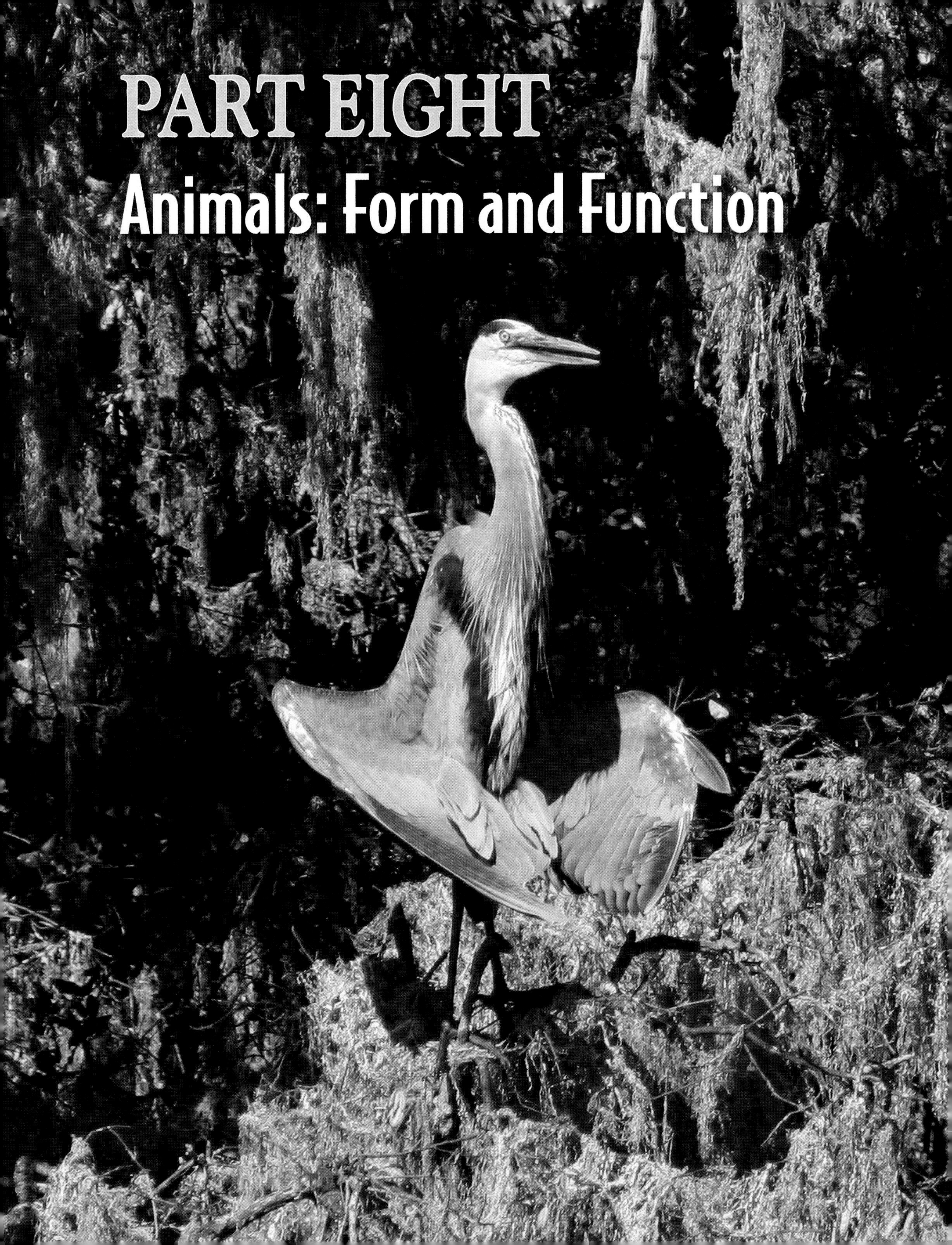
PART EIGHT
Animals: Form and Function

CHAPTER 40 Physiology, Homeostasis, and Temperature Regulation

Cool it!

"A new world record!" These words convey the thrill of world-class athletic competition. But as records are broken by mere centimeters or by fractions of a second, are we reaching absolute limits to human performance? We can assess many physiological limits to extreme performance—maximum breathing rate, for example, or the maximum rate at which the heart can supply blood to the muscles. One less obvious physiological limit is temperature.

The 2004 Olympic women's marathon was held on a hot, humid day in Athens. World record holder Paula Radcliffe was the favorite in the 42-kilometer race. But, overcome by heat stress, Radcliffe collapsed 6 kilometers from the finish line. It could have been much worse. Heat stroke, which can occur when internal body temperature exceeds 41°C, causes failure of major organs such as the heart and brain and results in death in over 20 percent of cases. Soldiers fighting in the desert are at extreme risk of heat-related illnesses, as are firefighters. Agricultural, industrial, and construction workers are also subject to the adverse affects of heat. Biologists at Stanford University developed a technology to cool individuals in such situations, and in the process discovered a way to enhance athletic performance.

Working muscles produce heat. This heat is carried by the blood to skin surfaces, where it is lost to the environment. Not all skin surfaces are equally good at dissipating heat, however. Because fur impedes heat loss, mammals evolved efficient bare skin heat-loss portals such as the nose, tongue, footpads, and parts of the face. These areas have specialized blood vessels that can act like radiators to disperse heat or close down to conserve heat. Although humans are not furred, we retain these specialized blood vessels in our hands, feet, and face (which is why we blush). The Stanford biologists designed a device to amplify heat extraction from these areas.

The heat extractor is a chamber that encases the hand and is sealed at the wrist. The hand is in contact with a cooled surface, but the critical component is a mild vacuum produced inside the chamber. The vacuum pulls more blood into the hand, allowing the cool surface to extract more heat. With this device, an active individual's body temperature rises more slowly, and it cools more rapidly when they rest. An additional unexpected benefit is that cooling reduces fatigue and greatly increases exercise capacity. "Cooled" athletes work out harder and longer, and

Heat Limits Performance Heat stress forced Paula Radcliffe to drop out of the 2004 Olympic marathon. When the body's internal temperature is stressed by extreme heat, its homeostatic mechanisms may fail.

A Cooling Glove The heat extractor increases heat loss and allows the body to perform at a higher level in severe conditions.

their physical conditioning increases substantially. In one study, college freshmen improved their pushup performance at a rate of 5 pushups a day without cooling, but 9 pushups a day with cooling. Some men and women in the study achieved more than 800 pushups in a workout session.

Human beings can survive in almost any terrestrial environment, from steamy tropical forests to frigid polar regions. Some of the ways we adapt to temperature are behavioral; our ability to control fire undoubtedly affected the course of our evolution. And, like all other mammals, we are adapted to maintain our internal temperature when confronted with heat or cold—an example of the need for living organisms to maintain a "steady state," or *homeostasis*.

IN THIS CHAPTER we explore the internal environment that provides for the needs of all of the body's cells. We survey the cell and tissue types that make up physiological systems and discuss how these systems maintain the internal environment within certain physiological limits, a condition called homeostasis. Homeostasis is explained in detail using one important example, the regulation of body temperature.

CHAPTER OUTLINE

40.1 Why Must Animals Regulate Their Internal Environments?

All animals need nutrients and oxygen and must eliminate carbon dioxide and other waste products of metabolism. Single-celled organisms meet all these needs by direct exchanges with the external environment. Even some simple multicellular animals meet the needs of their cells in this way. Such animals are common in the sea; they tend to be small and flat or, as in sponges, perforated with channels through which seawater can flow. In such an animal, no cell is far from direct contact with seawater, which contains nutrients, absorbs waste, and provides a relatively unchanging physical environment. In larger animals, however, most cells do not have direct contact with the external environment.

An internal environment makes complex multicellular animals possible

The cells of multicellular animals exist within an **internal environment** of extracellular fluid that bathes every cell of the body (**Figure 40.1**). Individual cells get their nutrients from this extracellular fluid and dump their waste products into it. As long as the conditions in the internal environment are held within certain limits, the cells are protected from changes or harsh conditions in the external environment. Thus, a stable internal environment makes it possible for an animal to occupy habitats that would kill its cells if they were exposed to it directly. How is the internal environment kept constant?

As multicellular organisms evolved, cells became specialized for maintaining specific aspects of the internal environment. In turn, the development of an internal environment enabled these specializations, since each cell did not have to be a generalist and provide for all of its own needs. Some cells evolved to be the interface between the internal and the external environments and to provide the necessary transport functions to get nutrients in, move wastes out, and maintain appropriate ion concentrations in the internal environment. Other cells became specialized to provide internal functions such as circulation of the extracellular fluids, energy storage, movement, and information processing. The evolution of physiological systems to maintain different aspects of the internal environment made it possible for multicellular animals to become

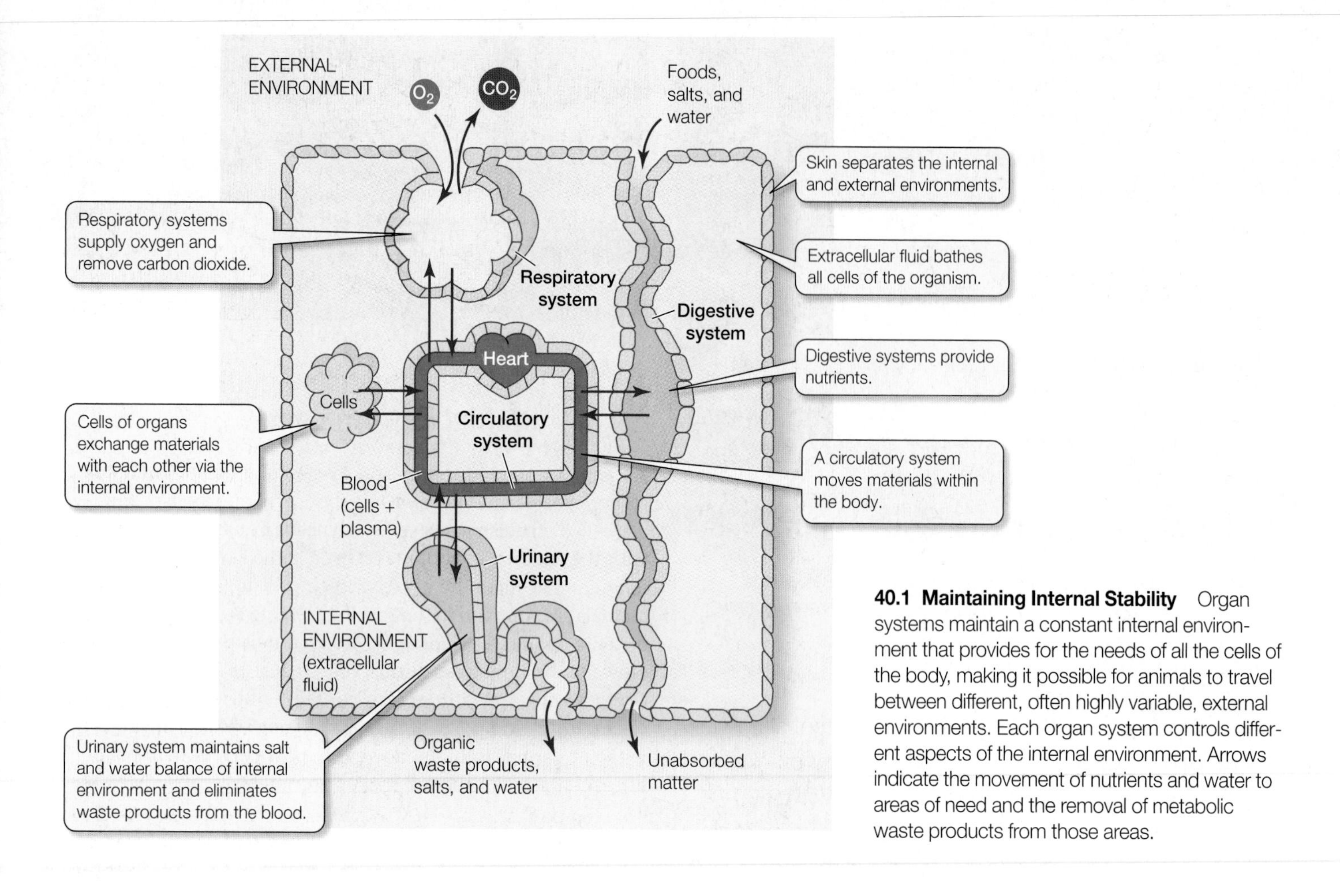

40.1 Maintaining Internal Stability Organ systems maintain a constant internal environment that provides for the needs of all the cells of the body, making it possible for animals to travel between different, often highly variable, external environments. Each organ system controls different aspects of the internal environment. Arrows indicate the movement of nutrients and water to areas of need and the removal of metabolic waste products from those areas.

larger, thicker, more complex, and more adaptable to external environments that are very different from the internal environment.

The composition of the internal environment is constantly being challenged by the external environment and by the metabolic activity of the cells of the body. The maintenance of stable conditions (within a narrow range) in the internal environment is called **homeostasis**. Homeostasis is an essential feature of complex animals. If a physiological system fails to function properly, homeostasis is compromised, and as a result cells are damaged and can die. To avoid loss of homeostasis, physiological systems must be controlled and regulated in response to changes in both the external and internal environments.

You are 60 percent water. One-third of that water is outside of your cells. About 20 percent of that extracellular fluid is circulating in your blood vessels; the rest (some 11 liters) bathes the cells of the body.

Homeostasis requires physiological regulation

The activities of all physiological systems are controlled—speeded up or slowed down—by actions of the nervous and endocrine systems. But to *regulate* the internal environment, information is required.

Think of it this way. You *control* the speed of your car with the accelerator and the brakes, but when you use the accelerator and brakes to *regulate* the speed of your car, you have to know both how fast you are going and how fast you want to go. The desired speed is a **set point**, or reference point, and the reading on your speedometer is **feedback information**. When the set point and the feedback information are compared, any difference between them is an **error signal**. Error signals suggest corrective actions, which you make by using the accelerator or brake (**Figure 40.2**).

Some components of physiological systems are called **effectors** because they *effect* changes in the internal environment. Effectors are **controlled systems** because their activities are controlled by commands from regulatory systems. **Regulatory systems**, in contrast, obtain, process, and integrate information, then issue commands to controlled systems. An important component of any regulatory system is a **sensor**, which provides the feedback information that is compared to the internal set point.

A fundamental way to study a regulatory system is to identify the information it uses. What are its sensors? How is the information from the sensors used? **Negative feedback** is the most common use of sensory information in regulatory systems. The word "negative" indicates that this feedback information causes the effectors to reduce or reverse the process or counteract the influence that created an error signal. In our car analogy, the recognition that you are going too fast is negative feedback if it causes you to slow down. Negative feedback is a stabilizing influence in physiological systems; it tends to return a variable of the internal environment to the set point from which it deviated.

Although not as common as negative feedback, **positive feedback** is also seen in some physiological systems. Rather than returning a system to a set point, positive feedback amplifies a response (i.e., increases the deviation from the set point). Examples of regulatory systems that use positive feedback are the responses that empty body cavities, such as urination, defecation, sneezing, and vomiting. Another example is sexual behavior, in which a little stimulation causes more behavior, which causes more stimulation, and so on.

Feedforward information is another feature of regulatory systems. The function of **feedforward information** is to change the set point. Seeing a deer ahead on the road when you are driving is an example of feedforward information (see Figure 40.2); this information takes precedence over the posted speed limit, and you change your set point to a slower speed. Before the start of a race, the "on your marks, get set" commands are feedforward information that raises your heart rate before you begin to run. Feedforward information anticipates change in the internal environment before that change occurs.

These principles of control and regulation help organize our thinking about physiological systems. Once we understand how a system works, we can then ask how it is regulated. The example we will explore in this chapter is the regulation of body temperature. But before we explore this first example of what will be our recurrent theme—the function, evolution, control, and regulation of each physiological system—we need to get acquainted with the important structural features that all physiological systems have in common.

Physiological systems are made up of cells, tissues, and organs

Each physiological *system* is composed of discrete *organs*, such as the liver, heart, lungs, and kidneys, that serve specific functions in the body. These organs are made up of tissues. A **tissue** is an assemblage of cells and, although there are many, many types of cells, there are only four kinds of tissue: *epithelial*, *connective*, *muscle*, and *nervous*. The word "tissue" is often used in a general way to refer to a piece of an organ, such as "lung tissue" or "kidney tissue"; however, as we will see, a piece of an organ will almost always consist of more than one of the four kinds of tissues.

EPITHELIAL TISSUES **Epithelial tissues** are sheets of densely packed, tightly connected *epithelial cells* that cover inner and outer body surfaces (**Figure 40.3A**). They act as barriers and provide transport across those barriers. Epithelial tissues form the skin and line the hollow organs of the body, such as the gut.

Epithelial cells have many roles in the body. Some secrete hormones, milk, mucus, digestive enzymes, or sweat. Others have cilia that move substances over surfaces or through tubes. Epithelial cells can also provide information to the nervous system. Smell and taste receptors, for example, are epithelial cells that detect specific chemicals. Epithelial cells create boundaries between the inside and the outside of the body and between body compartments; they line the blood vessels and make up various ducts and tubules (**Figure 40.3B**). Filtration and transport are important functions of epithelial cells. They control what molecules and ions can leave the blood to enter the internal environment or the urine. They can selectively transport ions and molecules from one side of an epithelial mem-

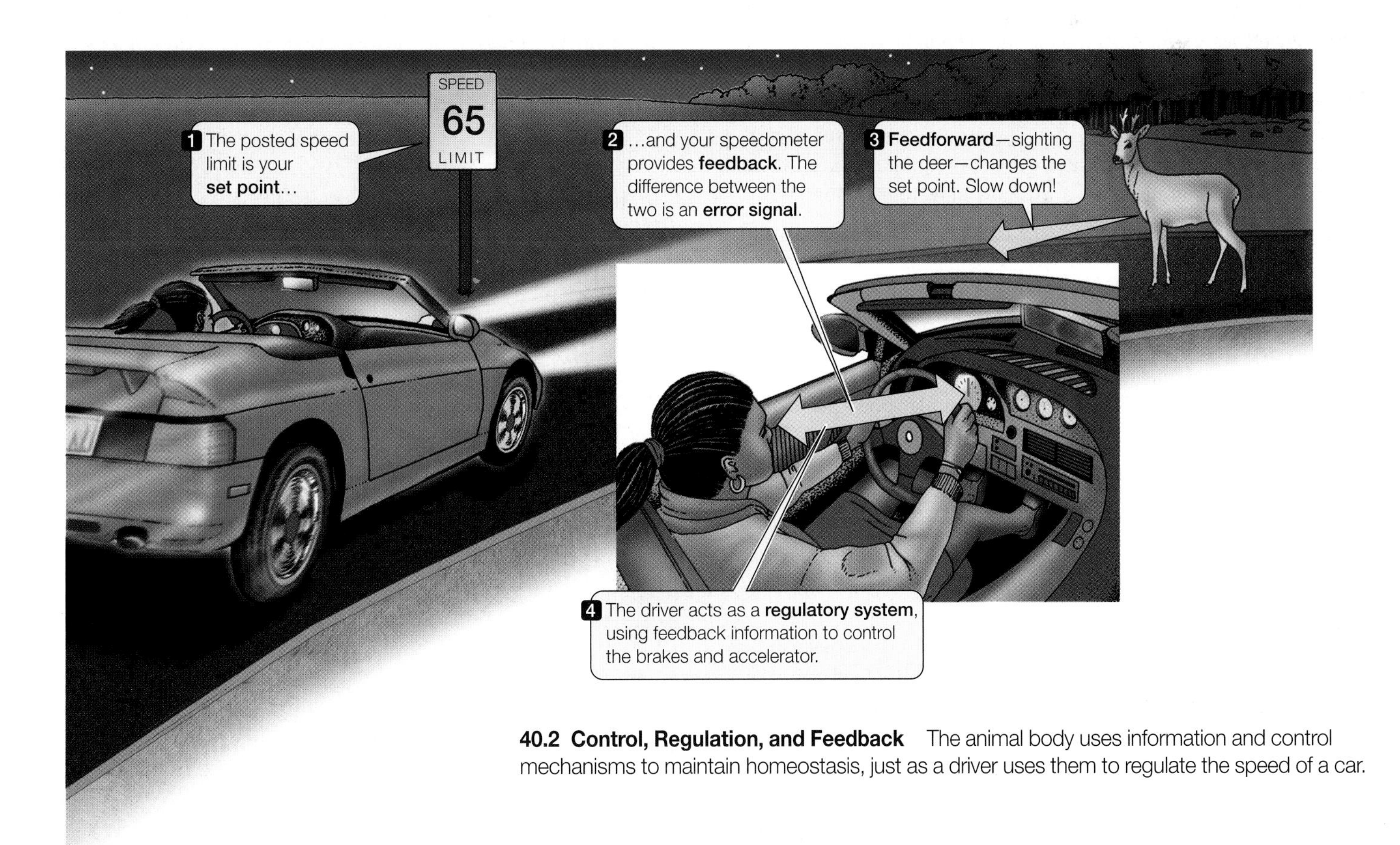

40.2 Control, Regulation, and Feedback The animal body uses information and control mechanisms to maintain homeostasis, just as a driver uses them to regulate the speed of a car.

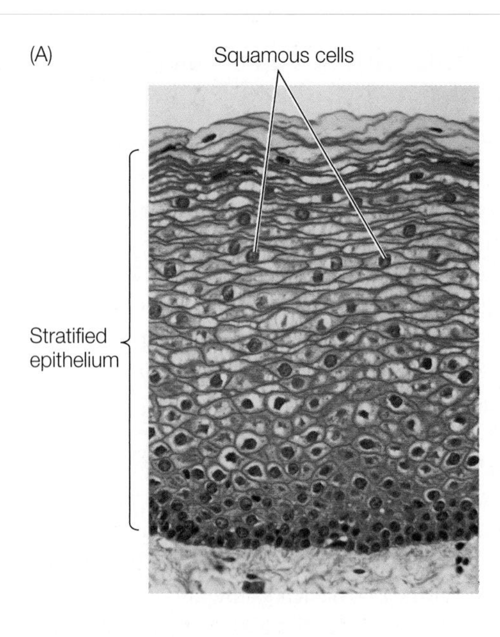

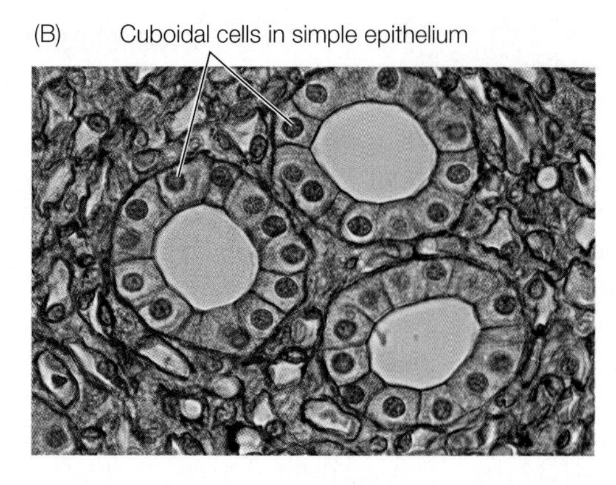

40.3 Epithelial Tissue (A) Epithelial cells make up the outer layers of skin. (B) A single layer of epithelial cells forms a tubule in the kidney. These cells have many transport functions.

brane to the other. Examples are the absorption of nutrient molecules from your gut and the secretion of acid into your stomach.

The skin and the lining of the gut are examples of epithelial tissues that receive much wear and tear. Accordingly, cells in these tissues have a high rate of cell division to replace cells that die and are shed. Dandruff is discarded skin cells.

Parts of you are younger than you think. The epithelium of your digestive tract is renewed about every 5 days, the surface layer of your skin every 2 weeks. Your blood is renewed every 4 months. The turnover of your liver cells is 1–2 years.

MUSCLE TISSUES **Muscle tissues** consist of elongated cells that can contract to generate forces and cause movement. Muscle tissues are the most abundant tissues in the body, and when animals are active, muscles use most of the energy produced in the body. The three types of muscle tissues—**skeletal**, **cardiac**, and **smooth**—are discussed in Section 47.1. Skeletal muscles (so named because they mostly attach to bones) are responsible for locomotion and other body movements such as facial expressions, shivering, and breathing (**Figure 40.4**). These muscles are under both conscious and unconscious control. Cardiac muscle makes up the heart and is responsible for the heart beat and the pumping of blood. Smooth muscle is responsible for movement and generation of forces in many hollow internal organs such as the gut, bladder, and blood vessels. Cardiac and smooth muscle are not under conscious control, but are controlled by physiological regulatory systems.

CONNECTIVE TISSUES In contrast to densely packed epithelial and muscle tissues, **connective tissues** are generally dispersed populations of cells embedded in an *extracellular matrix* that they secrete. The composition and properties of the matrix differ among types of connective tissues.

Protein fibers are an important component of the extracellular matrix of connective tissue cells. The dominant protein in the extracellular matrix is *collagen* (see Figure 4.25). Collagen is the most abundant protein in the human body, representing 25 percent of total body protein. Collagen fibers are strong and resistant to stretch, giving strength to the skin and the connections between bones and between bones and muscles. The fibers provide a net-like framework for organs, giving them shape and structural strength. Connective tissue that fills spaces between organs has a low density of collagen fibers.

Elastin is another type of protein fiber in the extracellular matrix of connective tissues. It is so named because it can be stretched to several times its resting length and then recoil. Fibers composed of elastin are most abundant in tissues that are regularly stretched, such as the walls of the lungs and the large arteries.

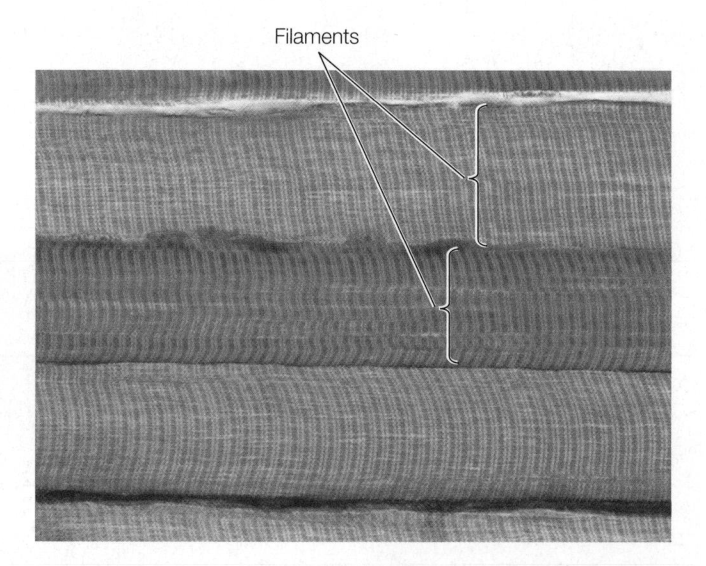

40.4 Filaments in Skeletal Muscle Cells A skeletal muscle cell is packed with protein filaments that interact to cause contraction. The regular arrangement of the filaments, which are made up of two different proteins, results in the striated (striped) appearance.

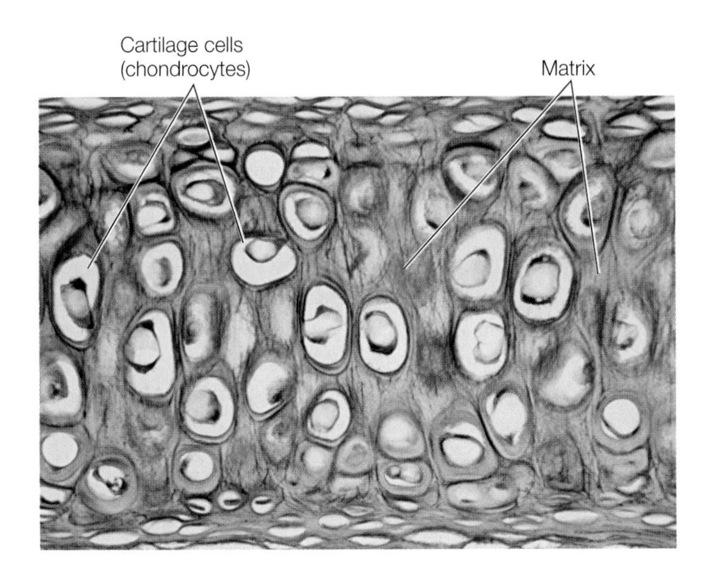

40.5 Cartilage Cartilage makes structures such as the ear stiff but flexible. Cartilage cells, or chondrocytes, secrete an extracellular matrix rich in collagen fibers and elastin fibers. In this micrograph, the elastin fibers are stained dark blue.

Cartilage and bone are connective tissues that provide rigid structural support. In *cartilage,* a network of collagen fibers is embedded in a flexible matrix consisting of a protein–carbohydrate complex, along with a specific type of cell called a *chondrocyte* (**Figure 40.5**). Cartilage, which lines the joints of vertebrates, is resistant to compressive forces. Since it is flexible, it provides structural support for flexible structures such as external ears and noses. The extracellular matrix in *bone* also contains many collagen fibers, but it is hardened by the deposition of the mineral calcium phosphate. Cartilage and bone are discussed in greater detail in Section 47.3.

Adipose tissue is a form of loose connective tissue that includes adipose cells, which form and store droplets of lipids. Adipose tissue, or "fat," is a major source of stored energy. It also serves to cushion organs, and layers of adipose tissue under the skin can provide a barrier to heat loss.

Blood is a connective tissue consisting of cells dispersed in an extensive extracellular matrix, the blood *plasma.* The blood plasma is much more liquid than the extracellular matrices of the other connective tissues, but it too contains an abundance of proteins. Many of the proteins and cellular elements of blood were presented in Section 18.1, and blood will discussed again in Section 49.4.

NERVOUS TISSUES Two basic cell types in **nervous tissues** are *neurons* and *glial cells.* Neurons come in many shapes and sizes, and encode information as electrical signals called nerve impulses. Nerve impulses travel over long extensions of the neurons, called *axons,* to reach other neurons, muscle cells, or secretory cells (**Figure 40.6A**). Where the axon is in close proximity to a target cell, nerve impulses trigger the release of chemical signals that bind to receptors on the target cell and stimulate a response. Neurons are involved in controlling the activities of most organ systems.

Glial cells do not generate or conduct electrochemical signals, but they provide a variety of supporting functions for neurons (**Figure 40.6B**). There are more glial cells than neurons in our nervous system. The properties of nervous tissues are detailed in Chapters 44, 45, and 46.

Organs consist of multiple tissues

Organs include more than one kind of tissue, and most organs include all four. The wall of the stomach is a good example (**Figure 40.7**). The inner surface of the stomach is lined with a sheet of epithelial cells. Different epithelial cells secrete mucus, enzymes, or stomach acid. Beneath the epithelial lining is connective tissue. Within this connective tissue are blood vessels, neurons, and glands (clusters of secretory epithelial cells). Concentric layers of smooth muscle tissue enable the stomach to contract to mix food with digestive juices. A network of neurons between the muscle layers controls these movements and influences the secretions of the stomach. Surrounding the stomach is a sheath of connective tissue.

An individual organ is usually part of an **organ system**—a group of organs that function together to achieve a particular physiological function or set of functions (see Figure 40.1). The stomach is part of the digestive system.

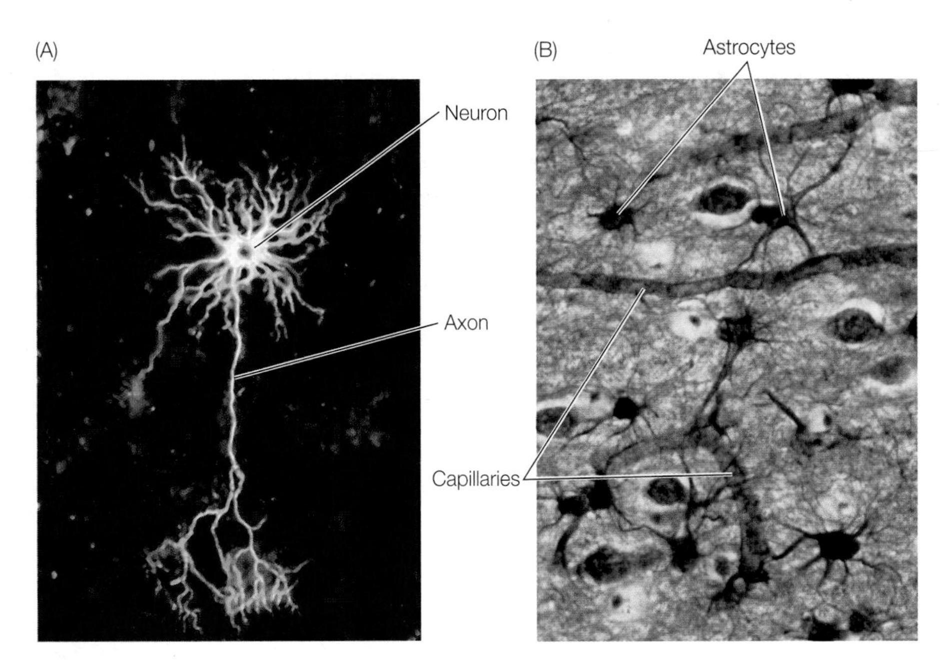

40.6 Nervous Tissue Includes Neurons and Glia (A) This human neuron consists of a cell body, a number of processes that receive input from other neurons, and one long axon that sends information to other cells. (B) A section through human brain tissue shows astrocytes, a type of glia. Glial cells provide support and protection for neurons including creating a barrier that protects the brain from many chemicals circulating in the blood.

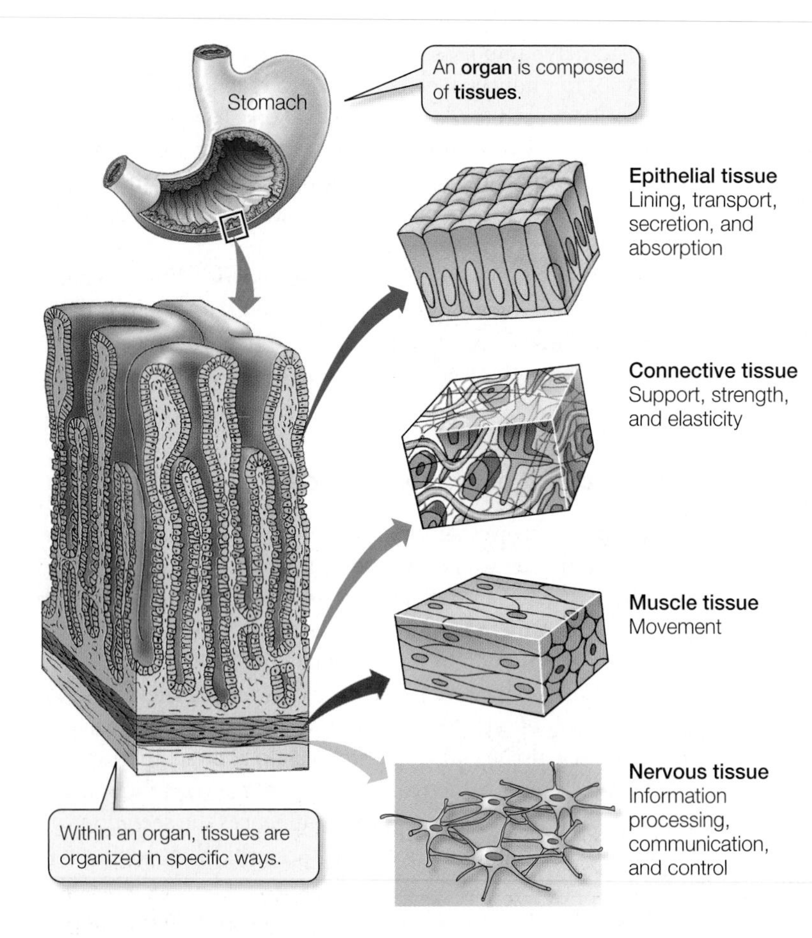

40.7 Tissues Form Organs Most organs contain more than one of the four different kinds of tissue. An organ such as the stomach is made up of all four.

40.1 RECAP

In all but the smallest multicellular animals, the internal environment provides for the needs of all of the cells. Organs and organ systems provide physiological regulation to maintain stability or homeostasis of the internal environment.

- Do you understand how negative feedback regulation tends to return a physiological variable to its set point? See p. 856 and Figure 40.2
- Can you describe a key function of each of the four kinds of tissue found in animals? See pp. 857–859 and Figure 40.7

Subsequent chapters in this unit will describe each of the organ systems mentioned above in much greater detail. The remainder of this chapter will focus on the mechanisms of homeostasis, using one important variable of the internal environment—its temperature—as our example.

40.2 How Does Temperature Affect Living Systems?

Temperatures vary enormously over the face of Earth, from the boiling hot springs of Yellowstone National Park to the interior of Antarctica, where the temperature can fall below –80°C. Living cells, however, can function over only a narrow range of temperatures. If cells cool below 0°C, ice crystals form and damage their structures. Some animals have adaptations, such as antifreeze molecules in their blood, that help them resist freezing; others can survive freezing. Generally, however, cells must remain above 0°C to stay alive.

The upper temperature limit for survival in most cells is about 45°C (although some specialized algae can grow in hot springs at 70°C, and some archaea live at near 100°C). In general, proteins begin to denature and lose their function as temperatures rise above 40°C. Therefore, most cellular functions are limited to the range between 0°C and 40°C, which approximates the thermal limits for life. A particular species, however, usually has much narrower limits.

Q_{10} is a measure of temperature sensitivity

Even between 0°C and 45°C, changes in tissue temperature create problems for animals. Most physiological processes, like the biochemical reactions that constitute them, are temperature-sensitive, going faster at higher temperatures (see Figure 6.22). The temperature sensitivity of a reaction or process can be described in terms of $\mathbf{Q_{10}}$, a quotient calculated by dividing the rate of a process or reaction at a certain temperature, R_T, by the rate of that process or reaction at a temperature 10°C lower, R_{T-10}:

$$Q_{10} = \frac{R_T}{R_{T-10}}$$

Q_{10} can be measured for a simple enzymatic reaction or for a complex physiological process, such as rate of oxygen consumption. If a reaction or process is not temperature-sensitive, it has a Q_{10} of 1. Most biological Q_{10} values are between 2 and 3. A Q_{10} of 2 means that the reaction rate doubles as temperature increases by 10°C, and a Q_{10} of 3 indicates a tripling of the rate (**Figure 40.8**).

Changes in tissue temperature can disrupt an animal's physiology because not all of the component reactions that constitute the metabolism of the animal have the same Q_{10}. Biochemical reactions are linked together in complex networks. If different reactions have different Q_{10} values, changes in tissue temperature will shift the rates of some reactions more than others. Thus, a change in tissue temperature can disrupt the balance and integration of the reactions that constitute a physiological process. To maintain homeostasis, organisms must be able to compensate for or prevent changes in body temperature.

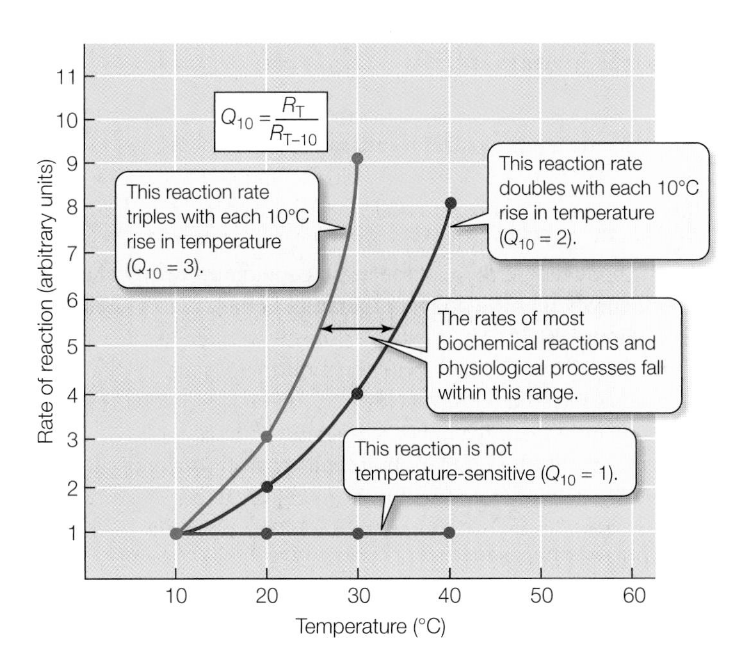

40.8 Q_{10} and Reaction Rate The larger the Q_{10} of a reaction or process, the faster its rate rises in response to an increase in temperature.

Animals can acclimatize to a seasonal temperature change

The body temperature of some animals (especially aquatic animals) is tightly coupled to environmental temperature. The body temperature of a fish in a pond, for example, will always be the same as the water temperature, which might range from 4°C in midwinter to 24°C in midsummer. How do such differences in water temperature affect the fish? We would predict that because of Q_{10} relationships, all of the fish's physiological functions would be much slower in the winter than in the summer. In support of our prediction, we could expose summer fish to a range of temperatures in the laboratory and show that whatever physiological function we measure, it goes slower and slower as we decrease the temperature of the water. However, if we study the same fish in the winter, we see that its physiological functions at low temperatures are not as slow as they were in our experiment with summer fish. The reason is that the fish's physiology has *acclimatized* to the seasonal change in water temperature.

What might account for acclimatization to seasonal changes in temperature in this fish? As discussed in Section 6.5, an organism may call upon one of several different temperature-optimized isozymes to catalyze a given reaction. If the fish can express a number of isozymes that operate at different optimal temperatures, it can catalyze reactions with one set of enzymes in summer and another set in winter. The result is that metabolic functions of the fish are much less sensitive to long-term changes in temperature than they are to short-term changes. Most animals cannot compensate completely for a seasonal change in body temperature. Partial compensation is most commonly observed.

40.2 RECAP

Life can exist in only a narrow range of temperatures, but even changes within that range can be disruptive because different physiological processes have different temperature sensitivities.

- Plot a Q_{10} curve for a physiological process. See p. 860 and Figure 40.8
- Do you understand why a change in body temperature can disrupt physiological processes? See p. 860
- Can you explain how aquatic organisms can acclimatize to seasonal temperatures? See p. 861

We have seen how animals are affected by the temperature of their environment. Now let's take a look at the adaptations of animals that allow them to control and regulate their body temperatures.

40.3 How Do Animals Alter Their Heat Exchange with the Environment?

Many of us learned to think of animals as being either "cold-blooded" or "warm-blooded," which implies a comparison with our own body temperature and sets mammals and birds apart from other animals. This simple classification breaks down when we realize that mammals that hibernate become cold, and that many reptiles and insects can be quite warm when they are active. Physiologists sometimes classify animals according to whether they have a constant body temperature (homeotherms) or a variable body temperature (poikilotherms). But a deep-sea fish has a constant body temperature. Should it be classified with mammals?

A thermal classification system that avoids such irrational results is one based on the source of heat that predominantly determines the temperature of the animal. **Ectotherms** are animals whose body temperatures are determined primarily by external sources of heat. **Endotherms** can regulate their body temperature by producing heat metabolically or by using active mechanisms of heat loss.

Most mammals and birds are endotherms; other animals are ectotherms most of the time. This statement suggests that, like the homeotherm/poikilotherm classification, this scheme is not perfect, and therefore we need a third category. A **heterotherm** is an animal that behaves sometimes as an endotherm and other times as an ectotherm. For example, a mammal that hibernates is a perfect endotherm over the summer, but during the winter it has bouts of hibernation during which its internal heat production falls and it behaves much like an ectotherm.

How do endotherms produce so much heat?

Section 6.1 describes how transfers of energy in biological systems are always inefficient. With every transfer of energy—from food molecules to ATP, from ATP to biological work—some of the energy is lost as heat. This is true for both ectotherms and endotherms, so why do endotherms produce more heat? The answer is that the cells of endotherms are less efficient at using energy than those in ectotherms.

In a resting endotherm, most of the energy expended goes into pumping ions across membranes. K^+ is the dominant positive ion inside cells and Na^+ is the dominant positive ion outside cells. To the extent that cell membranes permit, these ions diffuse down their concentration gradients. To maintain their proper concentrations inside and outside cells, the ions must be transported back "uphill," which requires an expenditure of energy. While this is true for both ectotherms and endotherms, the cells of endotherms tend to be more "leaky" to ions than those of ectotherms. Thus endotherms must expend more energy (and thus release more heat) than do ectotherms to maintain ion concentration gradients. This is akin to running on a treadmill: the faster the treadmill goes (analogous to leaking ions) the faster you have to run (analogous to pumping ions) to remain in the same position.

We can speculate that a mutation allowing seemingly faulty or leaky ion channels may have led to the evolution of endothermy. Such a mutation in a small ectotherm may have promoted sufficient heat production to allow this ectotherm to remain active longer after the sun went down. Thus, for the first endotherms a whole new nocturnal world of ecological opportunities opened, one in which there was less competition from similarly sized ectotherms. Major differences between endotherms and ectotherms are their resting metabolic rates, the sum total of all energy expenditures in their bodies when at rest, and their responses to changes in environmental temperature.

Ectotherms and endotherms respond differently to changes in temperature

Let's compare how two similar-sized animals, a lizard (an ectotherm) and a mouse (an endotherm) respond to changes in temperature. We put each animal in a closed chamber and measure its body temperature and metabolic rate as we change the temperature of the chamber from 37°C to 0°C.

The body temperature of the lizard equilibrates with that of the chamber, whereas the body temperature of the mouse remains at 37°C (**Figure 40.9A**). The metabolic rate of the lizard (already much lower than the metabolic rate of the mouse) decreases as the temperature is lowered (**Figure 40.9B**). In contrast, the mouse's metabolic rate increases as chamber temperature is lowered below 27°C. The increase in the mouse's metabolism produces enough heat to prevent its body temperature from falling. In other words, the mouse regulates its body temperature by increasing its metabolic rate; the lizard does not.

We can test our laboratory observation—that the lizard does not regulate its body temperature—by measuring its temperature after we return it to its desert habitat. In nature, in contrast to in the laboratory, the lizard's body temperature is sometimes considerably different from the environmental temperature, which in the desert can fluctuate by 40°C in a few hours. The lizard achieves this difference by using behavior to alter its heat exchange with the environment (**Figure 40.10A**). Its behavioral strategies include spend-

In the early spring, the hooded flower of the skunk cabbage becomes "endothermic" for two weeks and actually melts its way up through overlying ice and snow. During this time the flower's temperature can be 20°C above that of the environment, and its metabolic rate is about the same as a mammal of similar size.

40.9 Ectotherms and Endotherms React Differently to Environmental Temperatures (A) The body temperatures of a lizard and a mouse of the same body size are different when equilibrated to different environmental temperatures. (B) The metabolic rates of the lizard and mouse react in opposite manners to cooler temperatures. The mouse curve turns upward again at higher temperatures, however, because it takes metabolic energy to sweat or pant to dissipate heat.

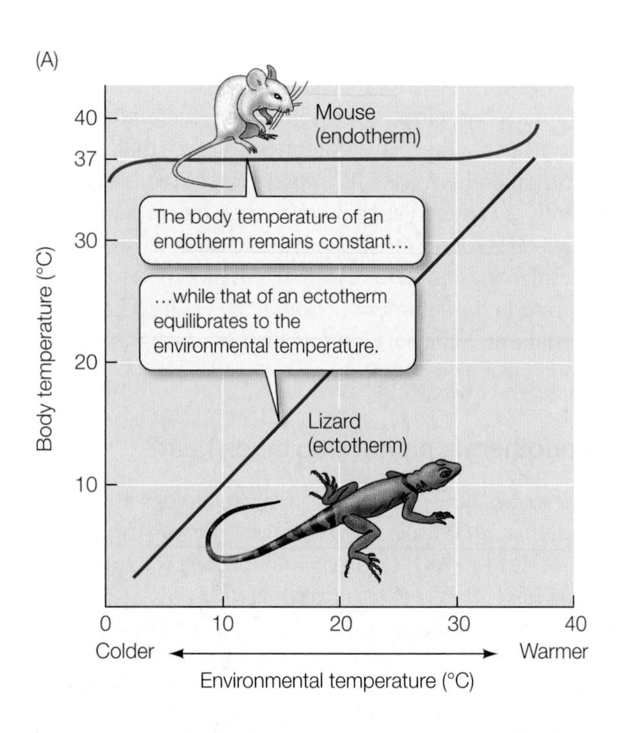

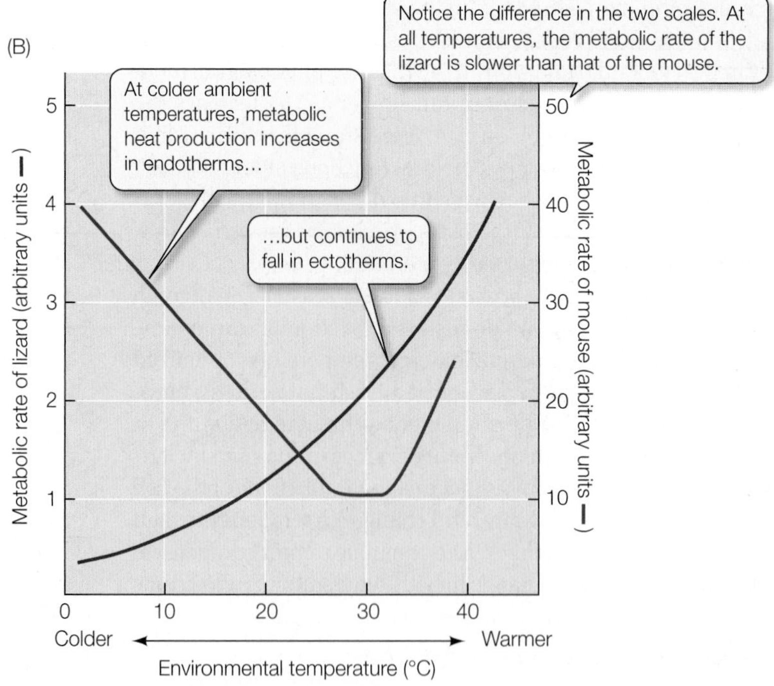

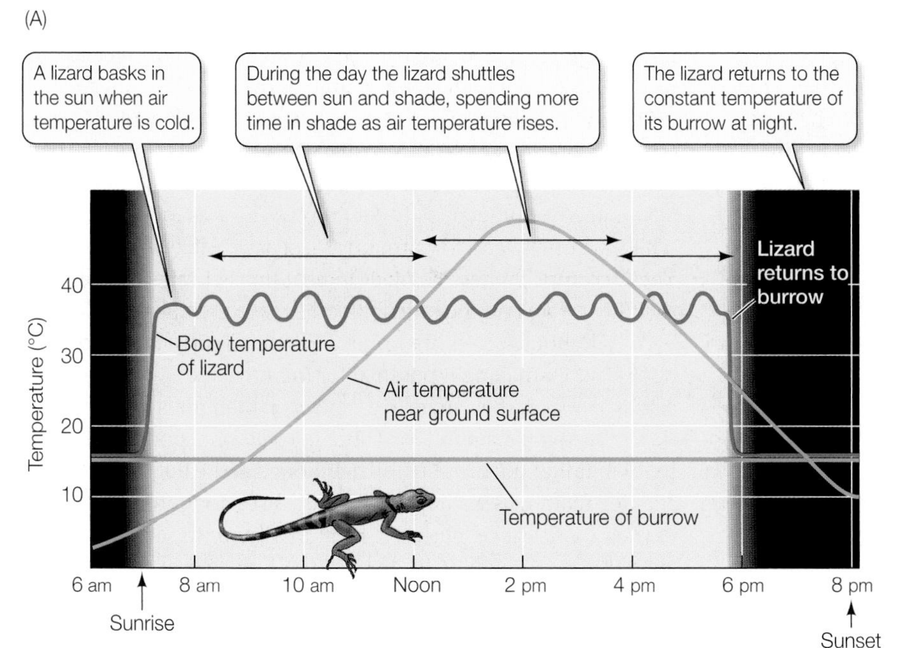

(B) *Loxodonta africana*

40.10 Ectotherms and Endotherms Use Behavior to Regulate Body Temperature (A) The lizard's body temperature is dependent on environmental heat, but it can regulate its temperature by moving from place to place within its environment. (B) When air temperatures on the African savanna soar, an elephant may thermoregulate by showering itself with water.

ing time in a burrow, basking in the sun, seeking shade, climbing vegetation, and changing its orientation with respect to the sun.

While the lizard can regulate its body temperature quite well, it does so by behavioral mechanisms rather than by altering its internal metabolic heat production. In our laboratory experiment, the lizard could not use its thermoregulatory behavior, but in its natural environment it could move from place to place to alter the heat exchange between its internal and external environments.

Behavioral thermoregulation is not the exclusive domain of ectotherms (**Figure 40.10B**). Endotherms usually select the most comfortable thermal environment possible. They may change their posture, orient to the sun, move between sun and shade, and move between still air and moving air, the same as the ectotherm in our field experiment. Examples of more complex thermoregulatory behavior include nest construction and social behavior such as huddling. Humans put on or remove clothing; we also burn fossil fuels to generate the energy to heat or cool buildings.

Energy budgets reflect adaptations for regulating body temperature

Both ectotherms and endotherms can influence their body temperatures by altering four avenues of heat exchange between their bodies and the environment (**Figure 40.11**):

- **Radiation**: Heat transfers from warmer objects to cooler ones via the exchange of infrared radiation (what you feel when you stand in front of a fire).
- **Conduction**: Heat transfers directly when objects of two different temperatures come into contact (think of putting an icepack on a sprained ankle).
- **Convection**: Heat transfers to a surrounding medium such as air or water as that medium flows over a surface (the wind chill factor).
- **Evaporation**: Heat transfers away from a surface when water evaporates on that surface (the effect of sweating).

40.11 Animals Exchange Heat with the Environment An animal's body temperature is determined by the balance between internal heat production and four avenues of heat exchange with the environment: radiation, conduction, convection, and evaporation.

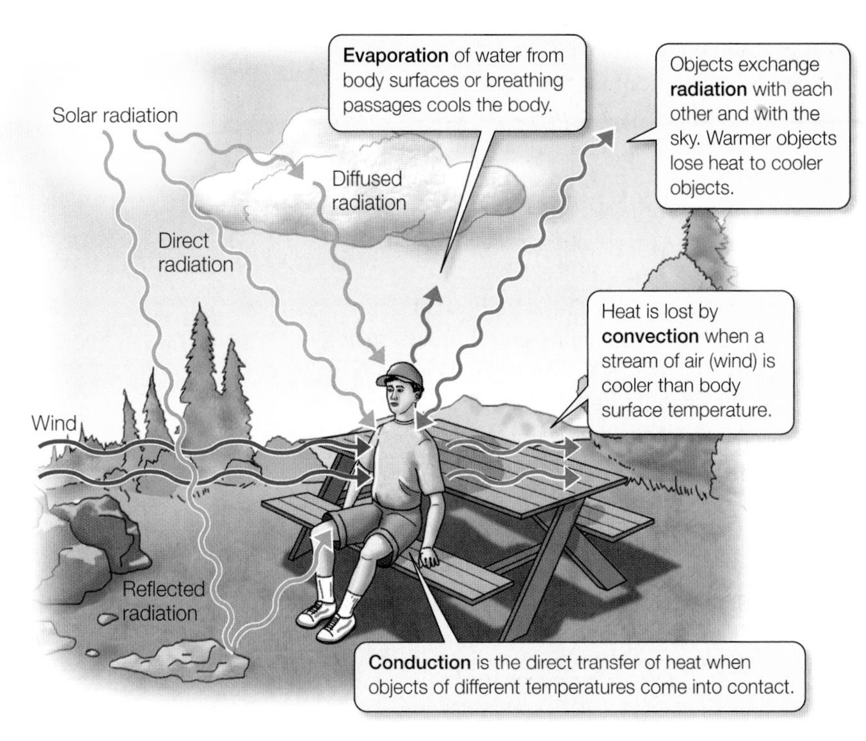

The total balance of heat production and heat exchange can be expressed as an **energy budget**, based on the simple fact that if the body temperature of an animal is to remain constant, the heat entering the animal must equal the heat leaving it. The heat coming in usually comes from metabolism and solar radiation (R_{abs}, for radiation absorbed). Heat leaves the body via the four mechanisms listed above: radiation emitted (R_{out}), convection, conduction, and evaporative heat loss. The energy budget takes the following form:

$$\overbrace{\text{metabolism} + R_{abs}}^{\text{heat}_{in}} = \overbrace{R_{out} + \text{convection} + \text{conduction} + \text{evaporation}}^{\text{heat}_{out}}$$

Anyone who has experienced a very hot environment will realize that heat can also *enter* the body through convection (e.g., the hot desert wind) and conduction (e.g., a hot car seat—ouch!). In that case, the values of those factors will become negative in the energy budget equation.

The energy budget is a useful concept because any adaptation that influences the ability of an animal to deal with its thermal environment must affect one or more components of the budget. So the energy budget gives us the ability to quantify and compare the thermal adaptations of animals. One interesting observation is that all of the components on the right side of the energy budget equation—the heat loss side—depend on the surface temperature of the animal. One way surface temperature can be controlled is by altering the flow of blood to the skin.

Both ectotherms and endotherms control blood flow to the skin

Heat exchange between the internal environment and the skin occurs largely through blood flow. As described at the beginning of this chapter, when body temperature rises due to exercise, blood flow to the skin increases, and the skin surface becomes warm. The heat brought from the body core to the skin by the blood is lost to the environment through the four avenues listed above, and this heat loss helps to bring the body temperature back to normal. In contrast, when body temperature is too low or the environment is too cold, the blood vessels supplying the skin constrict, reducing heat loss to the environment.

The control of blood flow to the skin can be an important adaptation for an ectotherm such as the marine iguana (a reptile) of the Galápagos archipelago (**Figure 40.12**). The Galápagos are volcanic islands that lie on the Equator but are bathed by cold ocean currents. The iguanas bask on hot black lava rocks on shore, then enter the cold ocean water to feed on seaweed. When the iguanas are feeding, they cool to the temperature of the sea. This cooling lowers their metabolism, making them slower, more vulnerable to predators, and incapable of efficient digestion. They therefore alternate between feeding in the cold sea and basking on the hot rocks. It is advantageous for iguanas to retain body heat as long as possible while swimming and to warm up as fast as possible when basking. They accomplish these changes in heat transfer rates by changing their heart rate and the rate of blood flow to their skin.

What about furred mammals? Fur acts as *insulation* to keep body heat in, making it possible for mammals to live in very cold climates. When they are active, however, mammals still must get rid of excess heat, and it does little good to transport that heat to the skin under the fur. Thus, as mentioned at the beginning of this chapter, mammals have special blood vessels for transporting heat to their hairless skin surfaces. Heat loss from these areas of skin is tightly controlled by the opening and closing of these special blood vessels. When you are cold, the blood flow to your hands and feet decreases and they can feel very cold, but when you exercise, these same surfaces can get very hot quickly.

Some fish elevate body temperature by conserving metabolic heat

Active fish can produce substantial amounts of metabolic heat, but they have difficulty in retaining any of that heat. Blood pumped from the heart goes directly to the gills, where it comes very close to the surrounding water to exchange respiratory gases. So any

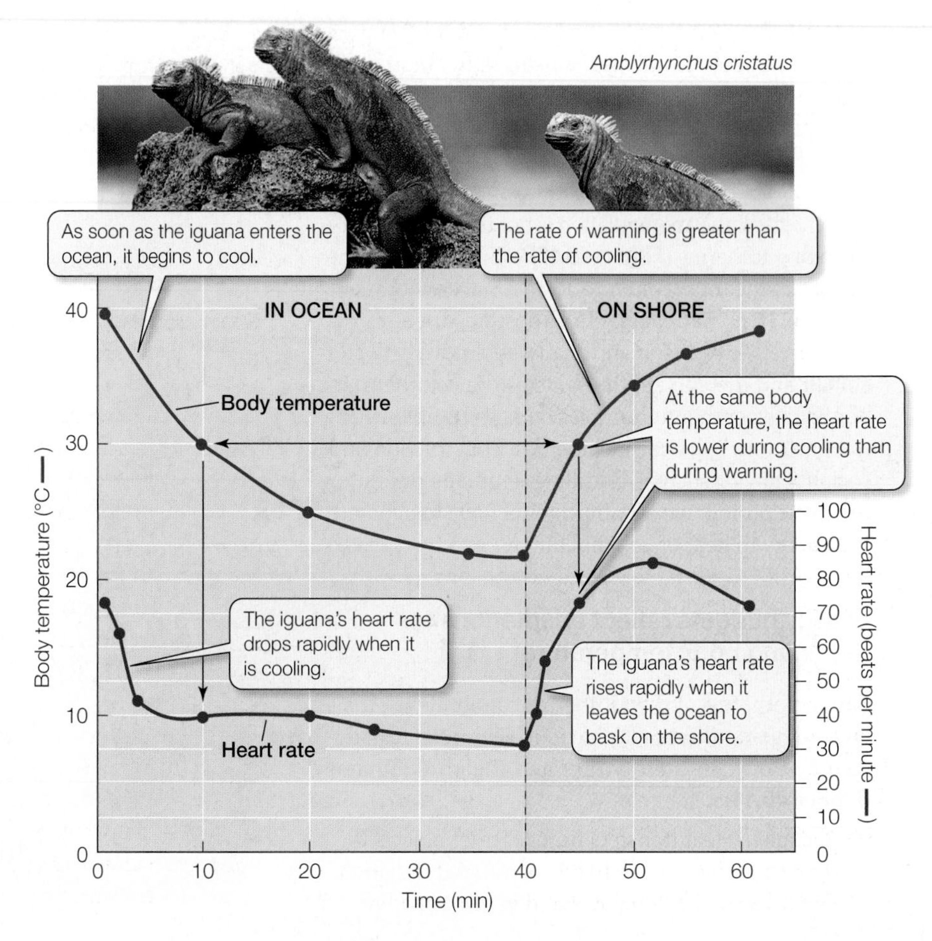

40.12 Some Ectotherms Regulate Blood Flow to the Skin Galápagos marine iguanas control blood flow to the skin to alter their heating and cooling rates.

(A) "Cold" fish

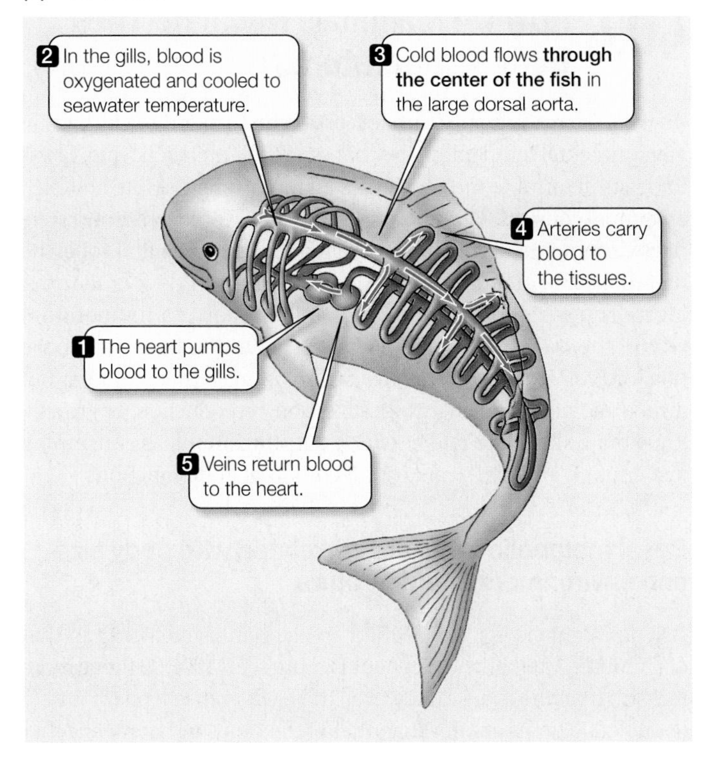

(B) "Hot" fish

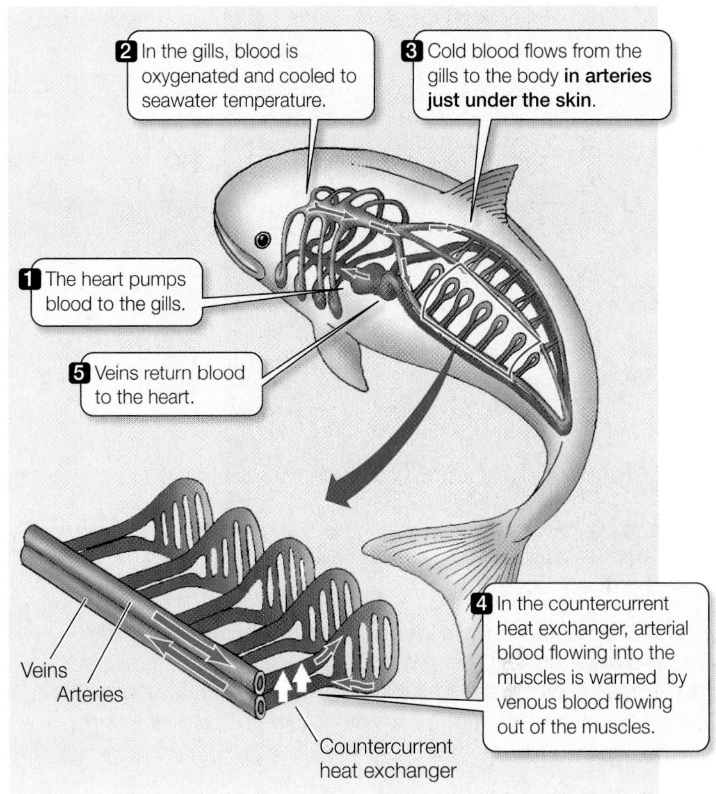

40.13 "Cold" and "Hot" Fish (A) Most fish species are "cold" fish. Their circulatory systems conduct cool, oxygenated blood from the gills through a large dorsal aorta to the rest of the body. (B) The anatomy of "hot" fish species includes a mechanism—the countercurrent heat exchanger—that allows heat to pass into cold arterial blood from venous blood that has been warmed by the metabolism of the muscles.

heat that the blood picks up from metabolically active muscles is lost to the surrounding water as it flows through the gills. It is thus surprising that some large, rapidly swimming fishes, such as bluefin tuna and great white sharks, can maintain temperature differences as great as 10°–15°C between their bodies and the surrounding water. The heat comes from their powerful swimming muscles, and the ability of these "hot" fish to conserve that heat is based on the remarkable arrangements of their blood vessels.

In the usual ("cold") fish circulatory system, oxygenated blood from the gills collects in a large dorsal vessel, the aorta, which travels through the center of the fish, distributing blood to all organs and muscles (**Figure 40.13A**). "Hot" fish have a smaller central dorsal aorta, and most of their oxygenated blood is transported in large vessels just under the skin (**Figure 40.13B**). The cold blood from the gills is thus kept close to the surface of the fish. Smaller vessels transporting this cold blood into the muscle mass run parallel to vessels transporting warm blood from the muscle mass back toward the heart. Since the vessels carrying the cold blood into the muscle are in close contact with the vessels carrying warm blood away, heat flows from the warm to the cold blood and is therefore retained in the muscle mass.

Because heat is exchanged between blood vessels carrying blood in opposite directions, this adaptation is called a **countercurrent heat exchanger**. It keeps the heat within the muscle mass, enabling these fish to have an internal body temperature considerably above the water temperature. Why is it advantageous for the fish to be warm? Each 10°C rise in muscle temperature increases the fish's sustainable power output almost threefold!

Some ectotherms regulate heat production

Some ectotherms raise their body temperature by producing heat. For example, the powerful flight muscles of many insects must reach 35°–40°C before the insects can fly, and they must maintain these high temperatures during flight. Such insects produce the required heat by contracting their flight muscles in a manner analogous to shivering in mammals. The heat-producing ability of insects can be quite remarkable. Probably the most impressive case is a species of scarab beetle that lives mostly underground in mountains north of Los Angeles, California. To mate, these beetles come above ground, and males fly in search of females. They undertake this mating ritual at night, in winter, and only during snowstorms.

Honeybees regulate temperature as a group. They live in large colonies consisting mostly of female worker bees that maintain the hive and rear the larval offspring of the single queen bee. During winter, honeybee workers cluster around the brood of larvae. They adjust their individual metabolic heat production and density of clustering so that the brood temperature remains remarkably constant, at about 34°C, even as the outside air temperature drops below freezing (**Figure 40.14**).

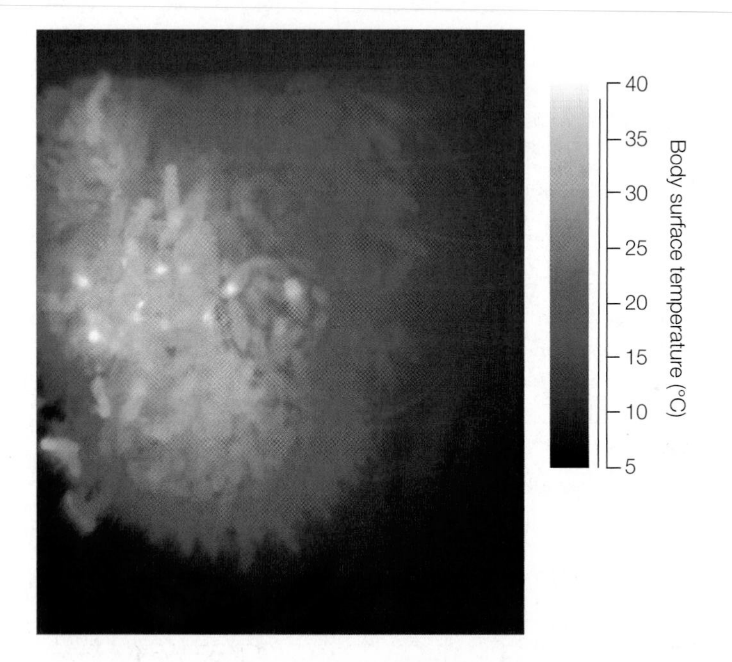

40.14 Bees Keep Warm in Winter Honeybee colonies survive winter cold because workers generate metabolic heat. In this infrared photograph of the center of an overwintering hive, individual bees are discernible by the heat their bodies produce as they cluster around their queen.

40.3 RECAP

Animals that metabolically produce their own heat are called endotherms. Those that depend on environmental sources of heat are called ectotherms. Heat exchange between an animal and its environment occurs via radiation, convection, conduction, and evaporation.

- In terms of the energy budget relationship, why is the control of blood flow to the skin so important for thermoregulation? See p. 864
- Can you explain how countercurrent heat exchange makes it possible for some fish to have a body temperature above that of the surrounding water? See pp. 864–865 and Figure 40.13

Endotherms must keep their body temperatures within a critical physiological range. Let's look more closely at the evolutionary adaptations that enable endothermic mammals to maintain this optimal temperature range.

40.4 How Do Mammals Regulate Their Body Temperatures?

As we saw in Figure 40.9, endotherms can respond to changes in environmental temperature by changing their metabolic rate. Physiologists determine metabolic rate by measuring the rate at which an animal consumes O_2 and produces CO_2. Within a narrow range of environmental temperatures, called the **thermoneutral zone**, the metabolic rate of endotherms is low and independent of temperature. The metabolic rate of a resting animal at a temperature within the thermoneutral zone is known as the **basal metabolic rate**, or **BMR**. It is usually measured in animals that are quiet but awake and not using energy for digestion, reproduction, or growth. Thus the BMR is the rate at which a resting animal is consuming just enough energy to carry out its minimal body functions.

Basal metabolic rates are correlated with body size and environmental temperature

As you might expect, the BMR of an elephant is greater than that of a mouse. After all, the elephant is more than 100,000 times more massive than the mouse. However, the BMR of the elephant is only about 7,000 times greater than that of the mouse. That means that a gram of mouse tissue uses energy at a rate 20 times greater than a gram of elephant tissue (**Figure 40.15**). Across all of the endotherms, BMR per gram of tissue increases as animals get smaller.

Why should this disproportionate difference exist? We don't know for sure. As animals get bigger, they have a smaller ratio of surface area to volume (see Figure 4.2). Since heat production is related to the volume, or mass, of the animal, but its capacity to dissipate heat is related to its surface area, it was once reasoned that larger animals evolved lower metabolic rates to avoid overheating. This explanation is insufficient for several reasons, one being that the relationship between body mass and metabolic rate holds for even very small organisms and for ectotherms, in which overheating is not a problem.

In an endotherm, the basal metabolic rate versus ambient temperature curve represents the integrated response of all of the animal's thermoregulatory adaptations (**Figure 40.16**). The thermoneutral zone is bounded by a *lower* and an *upper critical temper-*

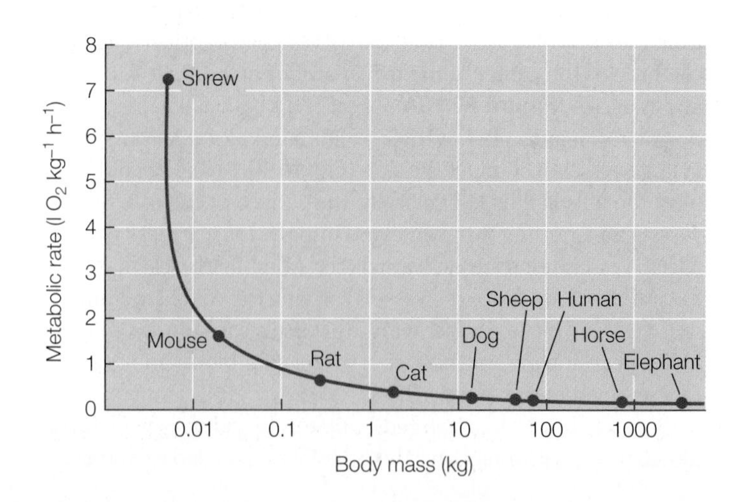

40.15 The Mouse-to-Elephant Curve On a weight-specific basis, the metabolic rate of small endotherms is much greater than that of larger endotherms. This graph plots O_2 consumption per kg of body weight (a measure of the metabolic rate) against a logarithmic plot of body weight.

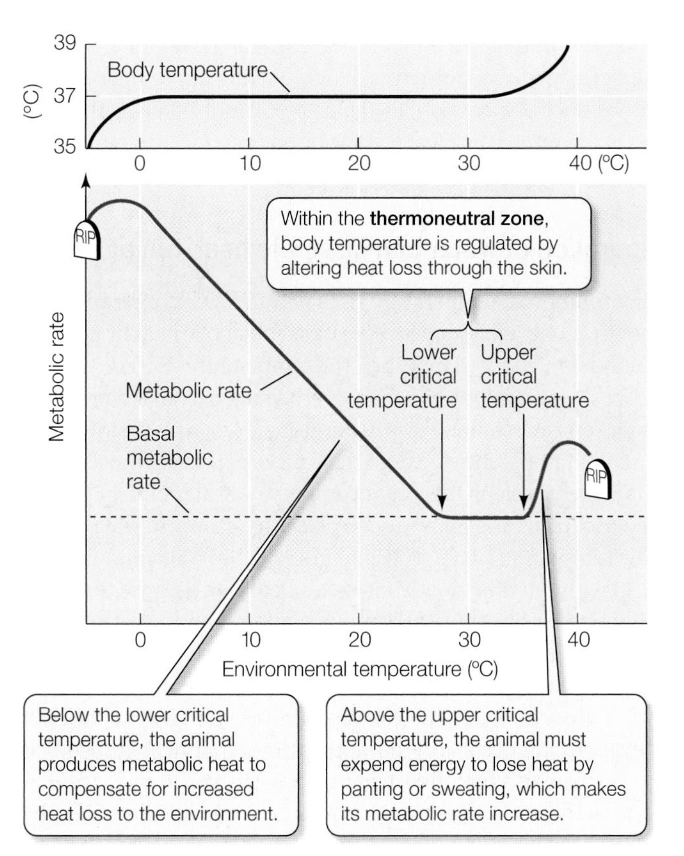

40.16 Environmental Temperature and Mammalian Metabolic Rates Outside the thermoneutral zone, maintaining a constant body temperature requires the expenditure of energy. Outside extreme limits (0°C and 40°C in this instance), the animal cannot maintain its body temperature and dies.

ature. Within its thermoneutral zone, an endotherm's thermoregulatory adaptations do not require much energy and could be considered passive; such adaptations include changing posture, fluffing fur, and controlling blood flow to the skin. Outside its thermoneutral zone, however, thermoregulatory responses are active and require considerable metabolic energy.

One interesting aspect of the thermoneutral zone is that many endotherms can survive at a far greater range of temperatures *below* the lower critical temperature than *above* it (see Figure 40.16). The coldest habitats on Earth are the Arctic, the Antarctic, and the high mountains. Many birds and mammals, but almost no reptiles or amphibians, live in these places. Adaptations to life in the cold have been a big part of the evolutionary success of these animals.

Endotherms respond to cold by producing heat and reducing heat loss

When environmental temperatures fall below the lower critical temperature, endotherms must produce heat to compensate for the heat they lose to the environment. Mammals can accomplish this in two ways: shivering and nonshivering heat production. Birds use only shivering heat production.

Shivering uses the contractile machinery of skeletal muscles to consume ATP without causing any visible behavior. Shivering muscles pull against each other so that little movement other than a tremor results. The energy from the conversion of ATP to ADP in this process is released as heat. Shivering heat production is perhaps too narrow a term, however; increased muscle tone and increased body movements also contribute to increased heat production in cold environments.

Most *nonshivering heat production* occurs in a specialized adipose tissue called **brown fat** (**Figure 40.17**). This tissue looks brown because of its abundant mitochondria and rich blood supply. In brown fat cells, a protein called *thermogenin* uncouples proton movement from ATP production, allowing protons to leak across the inner mitochondrial membrane rather than having to pass through the ATP synthase and generate ATP (review the discussion of the chemiosmotic mechanism in Section 7.4). As a result, metabolic fuels are consumed without producing ATP, but heat is still released. Brown fat is especially abundant in newborn infants of many mammalian species, including humans, in some adult mammals that are small and acclimatized to cold, and in mammals that hibernate.

The most important adaptations of endotherms to cold environments are those that reduce heat loss. Since most heat is lost from the body surface, many cold-climate species have a smaller surface area than their warm-climate cousins, even when their body masses are the same. Rounder body shapes and shorter appendages reduce the surface area-to-volume ratios of some cold-climate species (**Figure 40.18**).

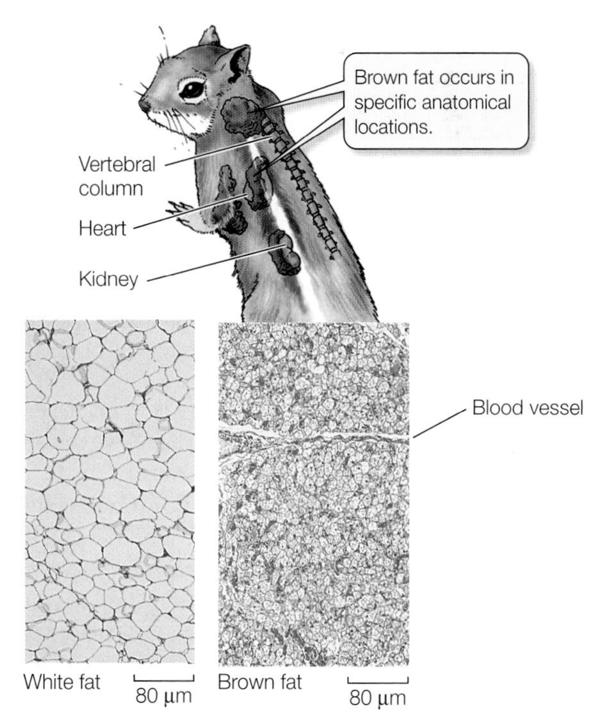

40.17 Brown Fat In many mammals, specialized brown fat tissue produces heat. When viewed through a microscope at similar magnifications, we see that white fat cells (left) contain large droplets of lipid but have few organelles and limited blood supply, while brown fat cells (right) are packed with mitochondria and richly supplied with blood.

40.18 Adaptating to Hot and Cold Climates (A) The antelope jackrabbit is found in the Sonoran Desert of Arizona. Its large ears serve as heat exchangers, passing heat from the animal's blood to the surrounding air. (B) The thick fur of the arctic hare provides insulation in the frigid winter. Its ears and extremities are relatively smaller than those of the jackrabbit.

Because each strand is hollow, a polar bear's fur is an even better insulator than normal hair. It also acts like optical fibers, transmitting solar radiation to the skin. Beneath its white fur coat the bear's skin is black and readily absorbs the solar radiation that reaches it.

Another means of decreasing heat loss is to increase thermal insulation. Animals adapted to cold climates have much thicker layers of fur, feathers, or fat than do their warm-climate relatives. The fur of an arctic fox or a northern sled dog provides such good thermal insulation that these animals don't even begin to shiver until the air temperature drops considerably below freezing. Fur and feathers are good insulators because they trap a layer of still, warm air close to the skin surface. If that air is displaced by water, insulation is drastically reduced. In many species, oil secretions spread through fur or feathers by grooming are critical for resisting wetting and maintaining a high level of insulation.

Decreasing blood flow to the skin is an important thermoregulatory adaptation in the cold. Constriction of blood vessels in the skin, and especially in the appendages, greatly improves the ability of an animal to conserve heat. Countercurrent heat exchange like we saw in the "hot" fish is also an important adaptation in the appendages of endotherms. Blood flowing out to the paw of a wolf, the hoof of a caribou, or the foot of a bird parallels the flow of the blood returning. Heat is transferred from the outgoing to the returning blood, thus retaining heat in the animal's core.

Evaporation of water can dissipate heat, but at a cost

As environmental temperature rises within an endotherm's thermoneutral zone, it dissipates more of its metabolic heat by increasing blood flow to the skin. When the temperature exceeds the upper critical temperature, however, overheating becomes a problem. For an exercising animal, overheating can occur at even low environmental temperatures. Large mammals, especially those in hot habitats such as elephants, rhinoceroses, and water buffaloes, have little or no insulating fur and seek places to wallow in water when the air temperature is high (see Figure 40.10B). Having water in contact with the skin greatly increases heat loss because the heat-absorbing capacity of water is much greater than that of air.

Evaporation of water from external or internal body surfaces through sweating or panting can also cool an endotherm. A gram of water absorbs about 580 calories of heat when it evaporates. If this evaporation occurs on the skin, most of that heat is absorbed from the skin and the underlying blood. Any sweat or saliva that falls off of the body, however, provides no cooling. Thus when the need for heat loss is greatest, a lot of water from the internal environment can be squandered with no cooling benefit. Since water is heavy, animals do not carry an excess supply of it, and many hot environments are also arid. In habitats that are both hot and dry, sweating and panting are cooling adaptations of last resort.

Sweating and panting are *active* processes that require the expenditure of metabolic energy. That is why the metabolic rate increases when the upper critical temperature is exceeded (the rising curve Figure 40.16). A sweating or panting animal is generating heat in the process of dissipating heat, which can be a losing battle.

The vertebrate thermostat uses feedback information

The thermoregulatory mechanisms and adaptations we have just discussed work through a regulatory system that integrates information from environmental and physiological sources and then issues commands that control body temperature. Such a regulatory system is based on feedback information, and can be thought of as a *thermostat*. We focus now on the vertebrate thermostat, particularly in the mammal.

Where is the vertebrate thermostat? Its major integrative center is at the bottom of the brain in a structure called the **hypothalamus**. If you slide your tongue back as far as possible along the roof of your mouth, it will be just a few centimeters below your hypothalamus. The hypothalamus is a key part of many regulatory systems, so we refer to it again in chapters to come. If the hypothalamus of a mammal's brain is damaged, the animal loses (among other things) its ability to regulate its body temperature, which then rises in warm environments and falls in cold ones.

In many species, the temperature of the hypothalamus itself is the major source of feedback information to the thermostat. Cooling the hypothalamus causes fish and reptiles to seek a warmer

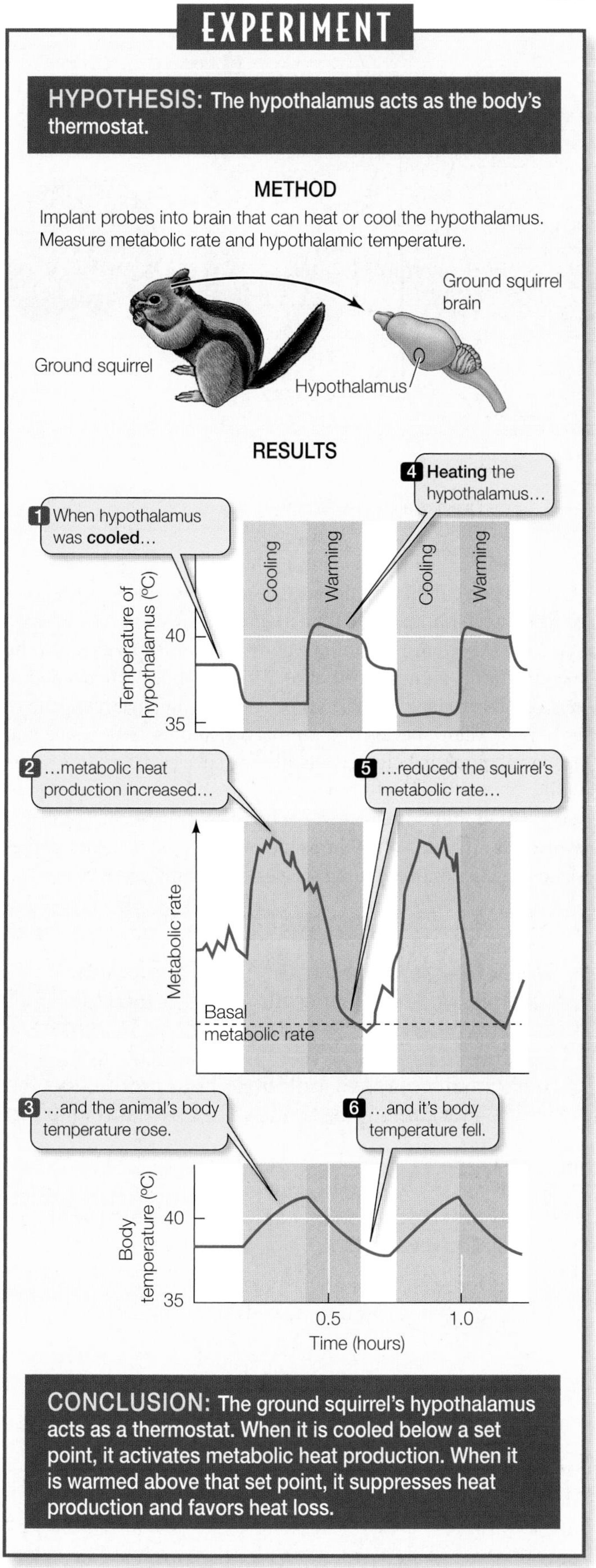

EXPERIMENT

HYPOTHESIS: The hypothalamus acts as the body's thermostat.

METHOD

Implant probes into brain that can heat or cool the hypothalamus. Measure metabolic rate and hypothalamic temperature.

Ground squirrel
Ground squirrel brain
Hypothalamus

RESULTS

1 When hypothalamus was **cooled**...
2 ...metabolic heat production increased...
3 ...and the animal's body temperature rose.
4 **Heating** the hypothalamus...
5 ...reduced the squirrel's metabolic rate...
6 ...and it's body temperature fell.

Cooling
Warming
Cooling
Warming
Temperature of hypothalamus (°C)
40
35
Metabolic rate
Basal metabolic rate
Body temperature (°C)
40
35
0.5
1.0
Time (hours)

CONCLUSION: The ground squirrel's hypothalamus acts as a thermostat. When it is cooled below a set point, it activates metabolic heat production. When it is warmed above that set point, it suppresses heat production and favors heat loss.

40.19 The Hypothalamus Regulates Body Temperature The observation that damage to the hypothalamus disrupts thermoregulation led to the finding that the hypothalamus acts as a thermostat in the vertebrate body.

environment, and warming the hypothalamus causes them to seek a cooler environment. In mammals, cooling the hypothalamus can stimulate constriction of the blood vessels supplying the skin and increase metabolic heat production. Because it activates these thermoregulatory responses, cooling the hypothalamus causes the body temperature to rise. Conversely, mild warming of the hypothalamus stimulates dilation of the blood vessels supplying the skin, while stronger hypothalamic heating stimulates sweating or panting. Consequently heating the hypothalamus can actually cause the overall body temperature to fall (**Figure 40.19**).

The hypothalamus generates a set point like a setting on the thermostat of a house. When the temperature of the hypothalamus exceeds or drops below that set point, thermoregulatory responses (the controlled system) are activated to reverse the direction of temperature change. Hence, hypothalamic temperature is a negative feedback signal.

Experimental warming and cooling of the hypothalamus show that mammals have separate set points for activating different thermoregulatory responses. If the hypothalamus of a mammal is cooled, the vessels supplying blood to the skin constrict at a specific hypothalamic temperature. A slightly lower hypothalamic temperature initiates shivering. If the hypothalamic temperature is then raised, shivering ceases; then blood vessels supplying the skin dilate. At still higher hypothalamic temperatures, sweating or panting starts.

The vertebrate thermoregulatory system has adjustable set points and integrates sources of information in addition to hypothalamic temperature. For example, temperature sensors in the skin register environmental temperature; change in skin temperature is feedforward information that shifts the hypothalamic set point for thermoregulatory responses. The set point for metabolic heat production is higher when skin is cold and lower when skin is warm.

Other factors can shift hypothalamic set points for thermoregulatory responses. Set points are higher during wakefulness than during sleep, and they are higher during the active part of the daily cycle than during the inactive part, even if the animal is awake at both times. Even when an endotherm is kept under constant environmental conditions, its body temperature displays a daily cycle of changes in set point. This kind of cycle is controlled by an internal *circadian rhythm;* these endogenous bodily rhythms are discussed further in Chapter 53.

Fever helps the body fight infections

Growing evidence suggests that fever is an adaptive response that helps the body fight pathogens. A fever is a rise in body temperature in response to substances called **pyrogens**. *Exogenous pyrogens* come from foreign substances such as bacteria or viruses that invade the body. *Endogenous pyrogens* are produced by cells of the immune system in response to infection.

The presence of a pyrogen in the body causes a rise in the hypothalamic set point for the metabolic heat production response. As a result, you shiver, put on a sweater, or crawl under a blanket, and your body temperature rises until it matches the new set point. At the higher body temperature you no longer feel cold, and you may not feel hot, but someone touching your forehead will say that you are "burning up." Taking aspirin lowers your set point to normal.

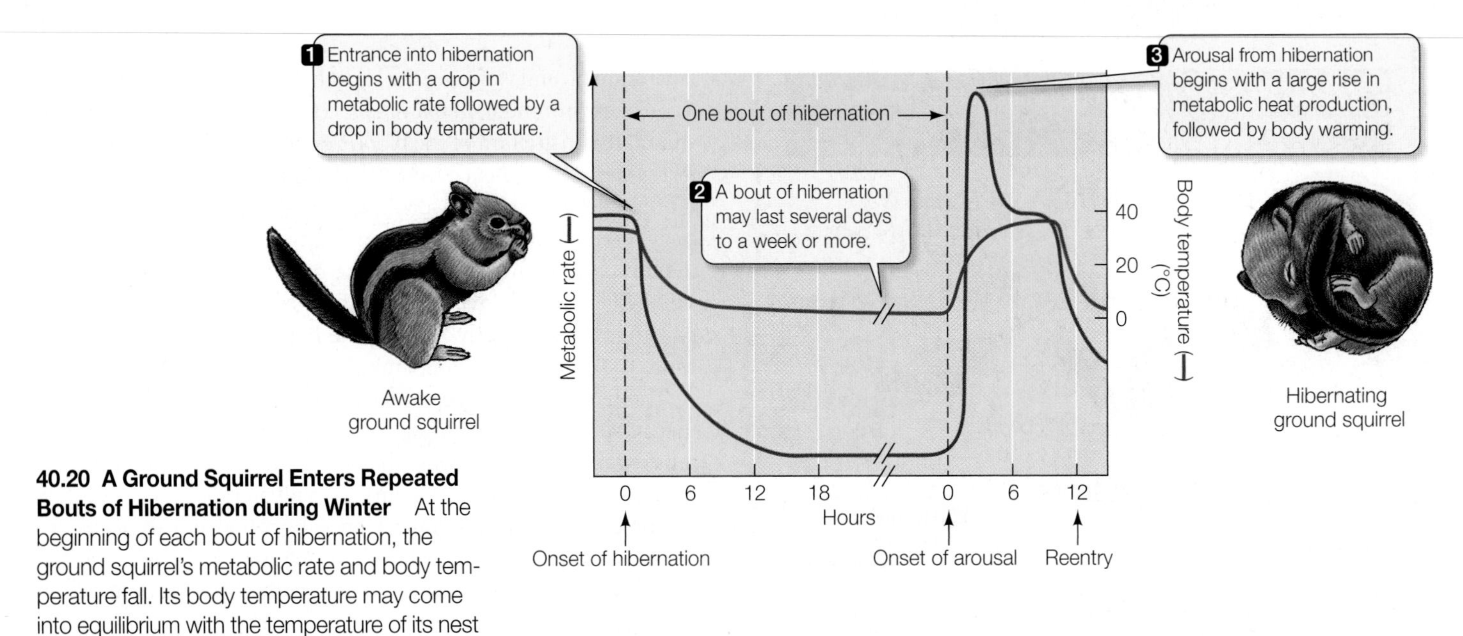

40.20 A Ground Squirrel Enters Repeated Bouts of Hibernation during Winter At the beginning of each bout of hibernation, the ground squirrel's metabolic rate and body temperature fall. Its body temperature may come into equilibrium with the temperature of its nest and stay at that level for days. The bout is ended by a rise in metabolic heat production that returns body temperature to a normal level.

Now you feel hot, take off clothes, and even sweat until your elevated body temperature returns to normal. Although modest fevers help the body fight infections, extreme fevers can be dangerous and must be controlled, usually with fever-reducing drugs.

Turning down the thermostat

Hypothermia refers to a state of below-normal body temperature. It can result from a natural turning down of the thermostat or from traumatic events such as starvation (lack of metabolic fuel), exposure to extreme cold, serious illness, or anesthesia. Many species of birds and mammals use *regulated hypothermia* as a means of surviving periods of cold and food scarcity. Some even become hypothermic daily.

Hummingbirds, for example, are very small endotherms with a high metabolic rate. Just getting through a single day without food could exhaust their metabolic reserves. Hummingbirds and other small endotherms can extend the period over which they can survive without food by dropping their body temperature during the portion of day or night when they would normally be inactive. This adaptive hypothermia is called **daily torpor**. Body temperature can drop 10°–20°C during daily torpor—an enormous savings of metabolic energy.

Regulated hypothermia that lasts for days or even weeks, with drops to very low body temperatures, is called **hibernation** (**Figure 40.20**). During the deep sleep of hibernation, the body's thermostat is turned extremely low to maximize energy conservation. Arousal from hibernation occurs when the hypothalamic set point returns to the normal level for a mammal.

Many hibernators maintain body temperatures close to the freezing point during hibernation. The metabolic rate needed to sustain a hibernating animal may be only one-fiftieth its basal metabolic rate. Many species of mammals, including bats, bears, and ground squirrels, hibernate, but only one species of bird (the poorwill) has been shown to hibernate. The ability of hibernators to reduce their thermoregulatory set point so dramatically probably evolved as an extension of the set point decrease that accompanies sleep even in nonhibernating species of mammals and birds.

40.3 RECAP

Within the thermoneutral zone, an endotherm controls its body temperature by altering insulation and blood flow to the skin. When the temperature drops below the thermoneutral zone, the animal increases metabolic heat production. Above this zone, it dissipates heat by panting or sweating.

- Can you describe how endotherms produce heat, and how heat production changes with body size? See p. 866 and Figure 40.15
- Why is dependence on evaporative water loss a dangerous strategy for dealing with hot environments? See p. 868
- What is the nature of negative feedback information and feedforward information used by the mammalian thermostat? See pp. 868–869 and Figure 40.19

CHAPTER SUMMARY

40.1 Why must animals regulate their internal environments?

Multicellular animals provide for the needs of all of their cells by maintaining a stable internal environment, which consists of the **extracellular fluid**.

Each cell, tissue, and organ contributes to **homeostasis** of the internal environment. Review Figure 40.1

The regulation of physiological systems is mostly through **negative feedback regulation. Feedforward information** functions to change **set points**. Review Figure 40.2

Tissues are assemblages of different cells. **Organs** are made up of tissues, and most organs contain all four kinds of tissue. Organs are grouped into **organ systems**. Review Figure 40.7

Epithelial tissues provide barriers and have secretory and transport functions.

Muscle tissues contract. The three types of muscle tissue are skeletal, cardiac, and smooth muscle.

Connective tissues, in which cells are embedded in an extracellular matrix, provide support. Cartilage, bone, adipose tissue, and blood are types of connective tissue.

Nervous tissues process and communicate information. They contain two cell types, neurons and glial cells.

40.2 How does temperature affect living systems?

Life is sustained within a narrow range of environmental temperatures. Q_{10} is a measure of the sensitivity of a life process to temperature. A Q_{10} of 2 means that the reaction rate doubles as temperature increases by 10°C. Review Figure 40.8

Metabolic rate is a measure of the energy turnover of an animal's cells and is often measured as the rate of O_2 consumption.

40.3 How do animals alter their heat exchange with the environment?

Endotherms can produce considerable metabolic heat to compensate for heat loss to the environment. **Ectotherms** generally do not. Review Figure 40.9

Energy budgets describe all pathways for heat exchange between an organism and its environment. The four avenues of heat exchange are **radiation**, **conduction**, **convection**, and **evaporation**. Review Figure 40.11

Skin temperature is an important variable and it can be influenced by blood flow. Circulatory system adaptations such as countercurrent heat exchange can conserve metabolic heat. Review Figures 40.12 and 40.13

40.4 How do mammals regulate their body temperatures?

Within the **thermoneutral zone**, mammals have a minimal or **basal metabolic rate**. BMR scales with body size. Review Figures 40.15 and 40.16, Web/CD Activity 40.1

In vertebrates, the control of thermoregulatory effectors relies on commands from a regulatory center in the **hypothalamus.** This thermostat uses its own temperature as a major negative feedback signal and skin temperature as a feedforward signal. Review Figure 40.19, Web/CD Tutorial 40.1

Fever is a regulated increase in body temperature and hibernation is a regulated decrease. Review Figure 40.20

SELF-QUIZ

1. Which of the following characterizes the protein elastin?
   a. It functions predominantly in muscle tissue to resist excess stretching.
   b. It is found predominantly in epithelial tissue.
   c. It is found in the extracellular matrix of connective tissue.
   d. It is the most abundant protein in the body.
   e. It is responsible for the elasticity of the long extensions of neurons.
2. If the Q_{10} of the metabolic rate of an animal is 2, then
   a. the animal is better acclimatized to a cold environment than if its Q_{10} is 3.
   b. the animal is an ectotherm.
   c. the animal consumes half as much oxygen per hour at 20°C as it does at 30°C.
   d. the animal's metabolic rate is not at basal levels.
   e. the animal produces twice as much heat at 20°C as it does at 30°C.
3. Which statement about brown fat is true?
   a. It produces heat without producing ATP.
   b. It insulates animals acclimatized to cold.
   c. It is a major source of heat production for birds.
   d. It is found only in hibernators.
   e. It provides fuel for muscle cells.
4. Which of the following is the most important and most general characteristic of endotherms adapted to cold climates compared to those adapted to warm climates?
   a. Higher basal metabolic rates
   b. Higher Q_{10} values
   c. Brown fat
   d. Greater insulation
   e. Ability to hibernate
5. Which of the following would cause a decrease in the hypothalamic temperature set point for metabolic heat production?
   a. Entering a cold environment
   b. Taking an aspirin when you have a fever
   c. Arousing from hibernation
   d. Getting an infection that causes a fever
   e. Cooling the hypothalamus
6. Mammalian hibernation
   a. occurs when animals run out of metabolic fuel.
   b. is a regulated decrease in body temperature.
   c. is less common than hibernation in birds.
   d. can occur at any time of year.
   e. lasts for several months, during which body temperature remains close to environmental temperature.
7. Which of the following is an important difference between an ectotherm and an endotherm of similar body size?
   a. The ectotherm has higher Q_{10} values.
   b. Only the ectotherm uses behavioral thermoregulation.
   c. Only the endotherm can constrict and dilate the blood vessels to the skin to alter heat flow.
   d. Only the endotherm can have a fever.
   e. At a body temperature of 37°C, the ectotherm has a lower metabolic rate than the endotherm.

8. How would you describe the role of skin temperature in the human thermoregulatory system?
   a. It provides feedforward information.
   b. It acts as a set point for metabolic heat production.
   c. It provides positive feedback information.
   d. It provides an error signal.
   e. It provides negative feedback information.
9. What is the biggest difference between a "cold" fish such as a trout and a "hot" fish such as a tuna?
   a. The temperature of the blood leaving the heart.
   b. The temperature of the blood entering the gills.
   c. The arrangement of blood vessels in the gills.
   d. The temperature of the brain.
   e. The volume of blood flowing in lateral arteries just under the skin.
10. Which of the following statements about the thermoneutral zone is true?
   a. Metabolic heat production is variable.
   b. Skin blood flow is variable.
   c. The environmental temperature equals body temperature.
   d. Its lower boundary (lower critical temperature) is lower for small than for large endotherms.
   e. It is the range of hypothalamic temperatures that do not alter metabolic heat production.

FOR DISCUSSION

1. What is the advantage of feedforward information for homeostasis? Can you suggest what some sources of feedforward information could be for regulation of breathing, blood pressure, secretion of digestive juices, and elimination of wastes?
2. In some epithelial tissues there are "tight junctions" between the individual cells that prevent anything from passing between them (see Figure 5.7), and in other cases the junctions between epithelial cells are quite loose. What are the possible advantages in different organs of loose versus tight junctions between epithelial cells? Give some examples in which these differences would be important.
3. Newton's law of cooling describes how a physical object comes into thermal equilibrium with its environment. The law is expressed

$$HL = K(T_o - T_a)$$

   HL is the rate of heat loss, K is the thermal conductance constant (how easily an object loses heat), T_o is the temperature of the object and T_a is the ambient temperature. Compare this expression with the metabolic rate/temperature curve for endotherms. In Newton's law of cooling, K is a constant reflecting the properties of the object. What would K represent for an endotherm? Using a version of Newton's law that replaces T_o with T_b (body temperature), explain why the metabolic rate curve projects to zero at an ambient temperature that equals body temperature.
4. The range of temperatures compatible with life is about 0°C to the low 40s. Endotherms have regulated body temperatures much closer to the upper limit of this range than to the lower. What are the advantages of living so close to the upper limit?
5. Discuss what it means when we say that the metabolic rate of mammals scales to the 3/4 power of body mass. In contrast, heart size of mammals scales according to the first power of body mass. What does this difference imply for the functions of the hearts of mammals of different sizes?

FOR INVESTIGATION

1. The text described the drop of body temperature of a hibernator as regulated hypothermia—a turning down of the thermostat. Yet we also saw that if we put an ectotherm in a cold environment, its body temperature will also fall. What experiment could you do to prove that the mammalian hibernator did not just simply turn off or inactivate its thermoregulatory system in order to behave like an ectotherm in the cold?
2. The observations on the Galápagos marine iguana showed that its body temperature rose faster in air than it fell in water. The inference was that the iguana was influencing its gain or loss of heat by altering the blood flow to its skin. However, the thermal properties of air and water are different, and in the case described in the text, the animal was breathing when in air but not when diving in the water. What experiment could you do to strengthen the argument that the iguana was actively altering the flow of heat across its skin?

CHAPTER 41 Animal Hormones

Testosterone abuse

The use of performance-enhancing steroid drugs—generally known as *anabolic steroids*—has become a scandal in athletic competition. Olympic champions have lost medals, professional athletes have been suspended, coaches have lost their jobs, suppliers have been arrested, and record-breaking performances have been expunged from the books. The problem may be epitomized by the prestigious and venerable international bicycle race, the Tour de France. In recent years this grueling competition, a month-long trek over flatlands and mountains, has been undermined by accusations and revelations of illegal anabolic steroid use by many top competitors, including the presumed winner of the 2006 race. But just what is the problem?

Shortly before puberty, the male reproductive system increases its production of an important chemical signal—testosterone, the male sex hormone. Testosterone enters cells, where it binds to certain receptors and alters gene expression. The cells that have these receptors are those involved in the development of male secondary sex characteristics such as deep voice, facial and body hair, and increased muscle mass. Supplemental anabolic steroids are used therapeutically to treat conditions such as delayed puberty, some types of impotence, and the loss of muscles that occurs with certain diseases. But hormone signals in the body usually have many different actions, and testosterone is no exception.

When a muscle is exercised, an interaction between the exercise and the sex steroids results in growth of the muscle—something that is very obvious among body builders. Body builders who abuse anabolic steroids typically use them in doses 10–100 times greater than the therapeutic doses. The resulting extreme growth of skeletal muscle mass extends to women as well, because a female's cells have steroid receptors, although females normally have only low levels of testosterone. When women body builders use anabolic steroids they develop male muscle patterns. They also develop deep voices and body and facial hair, and because these steroids generate negative feedback in the control of the female reproductive system, their breast tissue diminishes, they stop menstruating, and they become infertile. Similar negative feedback in males causes their sex organs to shrink and become nonfunctional. Even more serious (for both men and women) is a greatly increased risk of cancer and of heart, liver, and kidney disease.

A Grueling Race The prestigious Tour de France is one example of an international sporting event marred by revelations about the illegal use of anabolic steroids among some competitors.

Anabolic Steroids Build Big Muscles Anabolic steroids greatly enhance the development of skeletal muscle in response to exercise. Steroids have this effect on women as well as men.

Despite the risks, steroid drugs are consistently taken by those seeking an unfair advantage; athletic governing organizations thus seek to detect their use. Anabolic steroids circulate in the blood and are broken down in the liver; some residual byproducts are excreted in the urine. Illicit drug makers constantly seek to design new anabolic steroids that produce the desired physical results but do not produce detectable byproducts.

IN THIS CHAPTER we will examine how hormones—chemical signals such as the sex steroids—produce and coordinate anatomical, physiological, and behavioral changes in animals. First we examine the hormonal control of invertebrate life cycles. Building on that example, we discuss the general characteristics of hormones and their receptors. We then describe the functions, control, and mechanisms of action of mammalian hormones, paying particular attention to the extensive interactions between the neural and hormonal information systems in mammals.

CHAPTER OUTLINE

41.1 What Are Hormones and How Do They Work?

Control and regulation require information. In multicellular animals, most of this information is transmitted as electric signals and as chemical signals. The electric signals are impulses generated by the nervous system, conducted along cell processes of nerve cells to their targets on specific cells. The chemical signals are **hormones**, secreted by cells of the **endocrine system** into the extracellular fluid.

To compare the two informational systems of the body—the nervous and endocrine systems—think of neuronal communication (which will be detailed in Chapters 44–46) as a telephone system that sends specific messages to specific receivers. Hormonal communication, in contrast, is like a radio or TV network that sends out a broadcast message that can be picked up by whoever has an appropriate receiver that is turned on and tuned in.

Hormones can act locally or at a distance

The cells that secrete hormones are called **endocrine cells**, and the cells that have receptors for those hormones are called **target cells**. Hormones secreted into the extracellular fluid can diffuse into the blood, which distributes them throughout the body so they can activate target cells far from the site of release (**Figure 41.1A**). Such hormones are called **circulating hormones**; testosterone is an example of a circulating hormone.

Some hormones are released in such tiny quantities, or are so rapidly inactivated by enzymes, or are taken up so efficiently by local cells that they never diffuse into the blood in sufficient amounts to act on distant cells. Because these hormones affect only target cells near their release site, they are called **paracrine hormones** (**Figure 41.1B**). An example of a paracrine hormone is histamine, one of the mediators of inflammation (see Figure 18.4). The most local action a hormone can have is when there are receptors on the same cell that released it. When a hormone influences the cell that released it, it is said to have an **autocrine** function. Autocrine functions can provide negative feedback to control rates of secretion.

Some endocrine cells exist as single cells within a tissue. Hormones of the digestive tract, for example, are secreted by isolated endocrine cells in the wall of the stomach and small intestine. Many hormones, however, are secreted by aggregations of endocrine cells forming secretory organs called **en-**

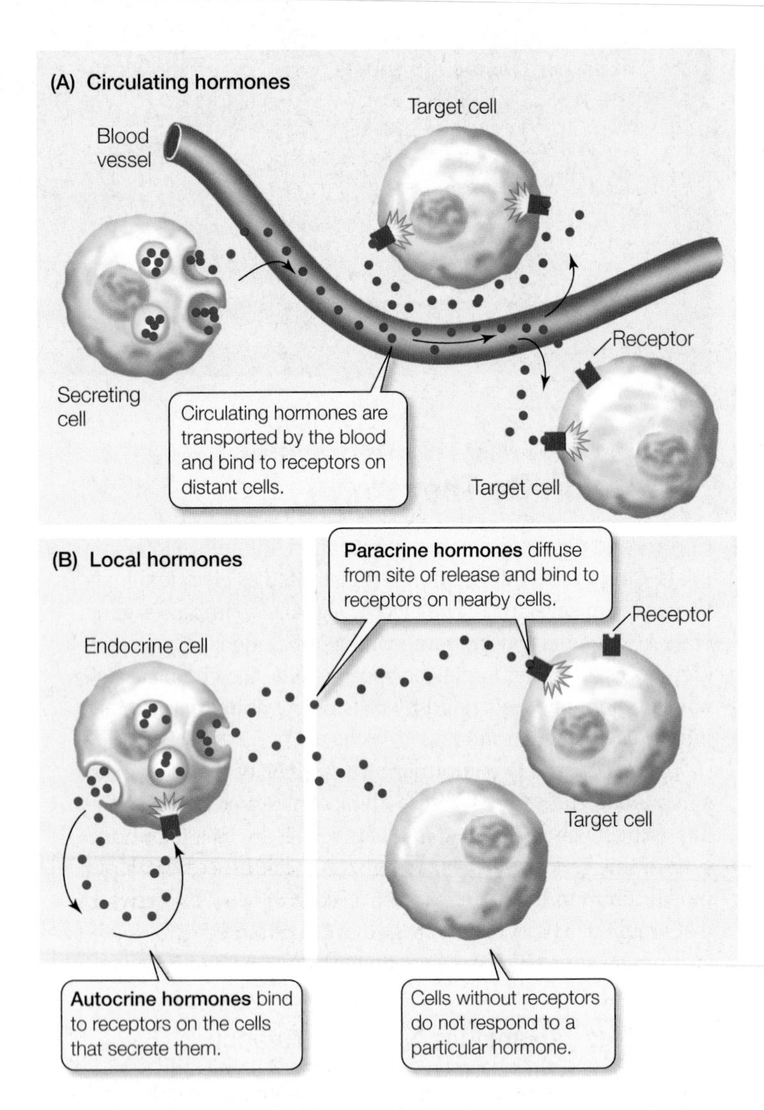

41.1 Chemical Signaling Systems (A) Most hormones are distributed throughout the body by the circulatory system. (B) An autocrine hormone influences the cell that releases it; a paracrine hormone diffuses to nearby cells.

docrine glands. The name "endocrine" reflects the fact that these glands do not have ducts that lead to the outside of the body; rather, they secrete their products directly into the extracellular fluid. In contrast, **exocrine glands** have ducts that carry their products to the surface of the skin (for example, sweat glands) or to the surface of a body passageway that leads to the outside of the body (salivary glands). A single endocrine gland may secrete several different hormones.

Hormonal communication arose early in evolution

Plants do not have nervous systems, but they do have hormones. The most primitive of the multicellular animals, the sponges, also do not have a nervous system, but they do have chemical communication. Even a protist, the social amoeba, which produces multicellular fruiting bodies by the aggregation of individuals, coordinates the aggregation with a chemical signal, cAMP (see Figure 27.32). This chapter provides only a few examples of invertebrate hormone action, but there are many. Here we will explore the hormonal control of molting and metamorphosis—two important events in the lives of arthropods, the most numerous animal group on Earth (see Chapter 32). The hormones involved represent an ancient system of hormonal communication that may be related to the anabolic steroid system discussed in the chapter's opening paragraphs.

Hormones from the head control molting in insects

The largest group of arthropods are the insects, and like all arthropods they have rigid exoskeletons. Therefore, their growth is episodic, punctuated with *molts* (shedding of the exoskeleton). Each growth stage between two molts is called an *instar*.

The British physiologist Sir Vincent Wigglesworth was a pioneer in the study of the hormonal control of growth and development in insects. Wigglesworth conducted experiments on the blood-sucking bug *Rhodnius prolixus*. Upon hatching, *Rhodnius* looks like a miniature version of an adult, lacking some adult features. The juvenile molts five times before developing into a mature adult; a blood meal triggers each episode of molting and growth.

Rhodnius is a hardy experimental animal; it can live a long time even after it is decapitated. If decapitated within an hour after having a blood meal, *Rhodnius* may live up to a year, but it does not molt. If decapitated a week after its blood meal, it does molt (**Figure 41.2, Experiment 1**). These observations led Wigglesworth to the hypothesis that something diffusing slowly from the head controls molting.

Wigglesworth tested his hypothesis with a clever experiment. He decapitated two *Rhodnius*: one shortly after its blood meal and another that had its blood meal a week earlier. The two decapitated bodies were connected with a short piece of glass tubing that allowed body fluid transfer between them. They both molted (**Figure 41.2, Experiment 2**). Thus one or more substances from the bug fed a week earlier must have crossed through the glass tube and stimulated molting in the other bug.

We now know that two hormones working in sequence regulate molting: prothoracicotropic hormone (PTTH) and ecdysone. Cells in the brain produce PTTH, which is why it has also been called brain hormone. PTTH is transported to and stored in a pair of structures called the *corpora cardiaca* attached to the brain. After appropriate stimulation (which for *Rhodnius* is a blood meal), the PTTH is released from these structures and it diffuses in the extracellular fluid to an endocrine gland, the prothoracic gland. PTTH stimulates the prothoracic gland to release the hormone ecdysone. Ecdysone diffuses to target tissues and stimulates molting.

Ecdysone is a lipid-soluble steroid molecule that readily enters its target cells (mostly cells of the epidermis). In the target cells, ecdysone binds to a receptor that is probably ancestral to the vertebrate testosterone receptor. The hormone–receptor complex acts as a transcription factor and induces expression of the genes for enzymes involved in digesting the old cuticle and secreting a new one.

The control of molting by PTTH and ecdysone is a general arthropod hormonal control mechanism and is an example of how a hormonal system works with the nervous system to integrate diverse information and induce a long-term effect. The nervous system receives various types of information (such as day length, temperature, crowding, and nutrition) that help determine the optimal

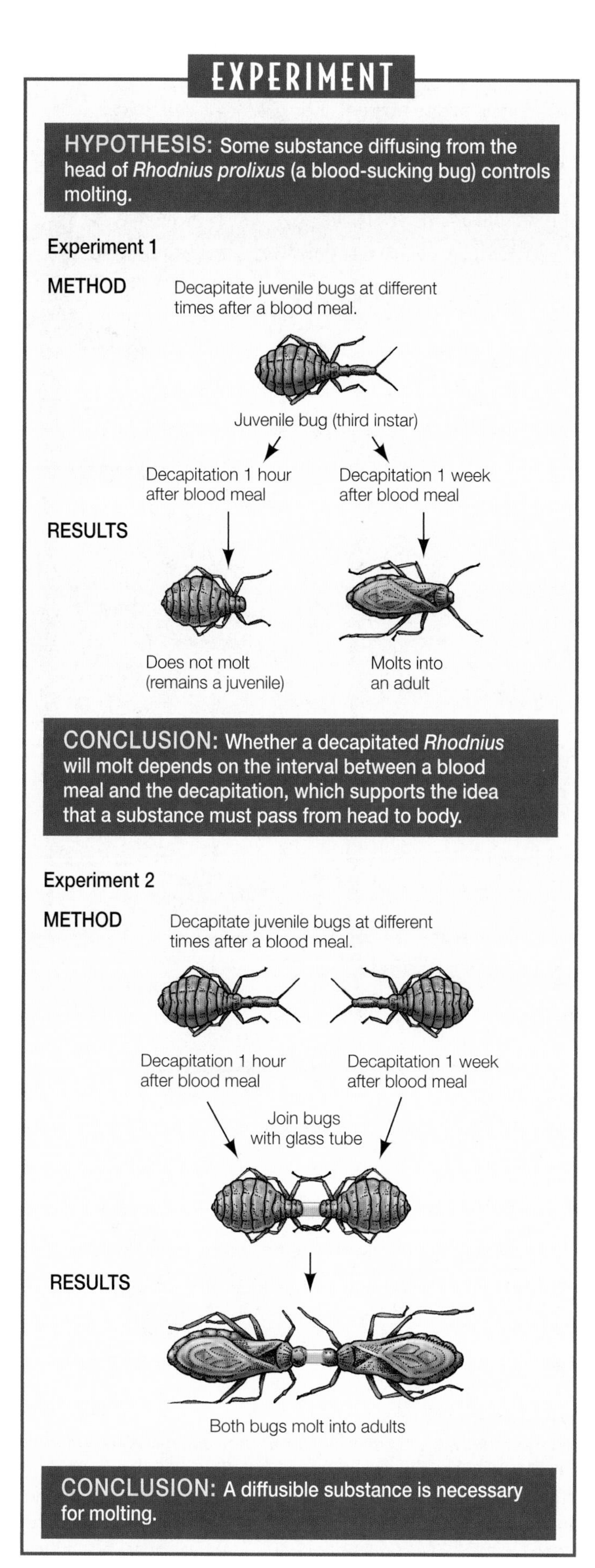

41.2 Diffusible Substance Triggers Molting The effect of time since the last blood meal on *Rhodnius* molting (experiment 1) led to the hypothesis that some substance diffusing slowly through the insect's body stimulated molting. Experiment 2 showed that molting is indeed controlled by a substance—a hormone—diffusing from the head.

timing for growth and development. The nervous system (the brain) then controls the endocrine gland (the prothoracic gland) producing the hormone (ecdysone) that orchestrates the physiological processes involved in development and molting. Later in this chapter we will see similar links between the nervous system and endocrine glands in vertebrates.

Juvenile hormone controls development in insects

The *Rhodnius* decapitation experiments yielded a curious result. Regardless of the instar used, the decapitated bug always molted directly into an adult form. Additional experiments by Wigglesworth demonstrated that a hormone other than those responsible for molting determines whether a bug molts into another juvenile instar or into an adult.

Because the head of *Rhodnius* is long, it is possible to remove just the front part of the head, which contains the brain, while leaving the rear part intact. That rear part contains the structures that release the PTTH. When fourth-instar bugs that had been fed a week earlier were partly decapitated, leaving these structures intact, they molted into fifth instars, not into adults.

This experiment was followed by more experiments using glass tubes to connect individual bugs. When an unfed, completely decapitated fifth-instar bug was connected to a fed, partly decapitated fourth-instar bug (with only the front part of its head removed), both bugs molted into juvenile forms. A substance from the rear part of the head of the fourth-instar bug prevented both bugs from molting into adults.

The substance responsible for preventing maturation is **juvenile hormone**, which is released continuously from the same structures that release PTTH in response to feeding. As long as juvenile hormone is present, *Rhodnius* molts into another juvenile instar. Normally *Rhodnius* stops producing juvenile hormone during the fifth instar, and then it molts into an adult.

The control of development by juvenile hormone is more complex in insects, such as butterflies, that undergo complete metamorphosis. These animals undergo dramatic developmental changes in their life cycles. The fertilized egg hatches into a *larva*, which feeds and molts several times, becoming bigger each time. After a fixed number of molts it enters an inactive stage called a *pupa*. The pupa undergoes major body reorganization and finally emerges as an adult.

An excellent example of complete metamorphosis is provided by the silkworm moth, *Hyalophora cecropia* (**Figure 41.3**). As long as juvenile hormone is present in high concentrations, larvae molt into larvae. When the level of juvenile hormone falls, larvae spin cocoons and molt into pupae. Because no juvenile hormone is produced in pupae, they molt into adults.

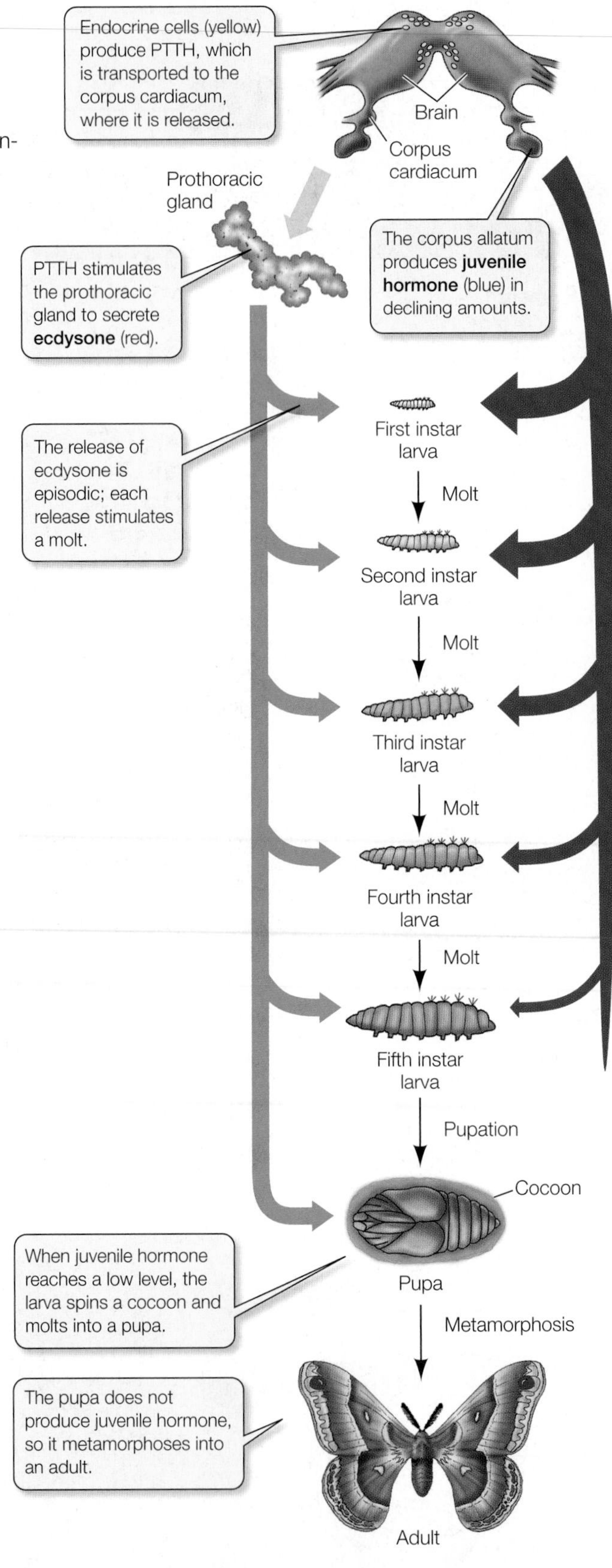

41.3 Complete Metamorphosis Butterflies and moths undergo complete metamorphosis in which the feeding larvae (caterpillars) bear no resemblance to the reproductive adult. Three hormones control molting and metamorphosis in the silkworm moth *Hyalophora cecropia*.

The existence and function of insect hormones was experimentally demonstrated many years before the hormones were identified chemically. That is not surprising when you consider the tiny amounts of certain hormones that exist in an organism. In one of the earliest studies of ecdysone, biochemists produced only 250 milligrams of pure ecdysone (about one-fourth the weight of an apple seed) from 4 tons of silkworms!

Silk is produced from the fibers of the cocoon that shelters the pupa of the silkworm moth. Silk producers spray the larvae with juvenile hormone to delay pupation. The larvae grow bigger, and bigger larvae make bigger cocoons that yield more silk.

Hormones can be divided into three chemical groups

We have seen examples of the roles hormones can play in long-term physiological and developmental processes in humans and insects. Now we can step back and ask some general questions about hormones. What kinds of hormones exist? What is their chemical nature, and how do they act upon organisms? There is enormous diversity in the chemical structure of hormones, but most of them can be divided into three groups:

- The majority of hormones are *peptides* or *polypeptides* (proteins). These hormones, of which insulin is an example, are water-soluble and are thus easily transported in the blood, but they cannot pass readily through lipid-rich cell membranes. Therefore, peptide and protein hormones are packaged in vesicles in the cells that make them and are released by exocytosis.
- *Steroid hormones* such as testosterone and estrogen are derivatives of the steroid cholesterol. They are lipid-soluble and easily dissolve in and pass through cell membranes. Steroid hormones diffuse out of the cells that make them as they are synthesized. Because steroid hormones are not soluble in blood, however, they must be bound to carrier proteins in order to be transported to their target cells.
- *Amine hormones* are mostly derivatives of the amino acid tyrosine (thyroxine is one example). Some amine hormones are water-soluble and others are lipid-soluble; their modes of release differ accordingly.

Hormone receptors are found on the cell surface or in the cell interior

The chemical structure of hormones is related to the location of their receptors. Lipid-soluble hormones can diffuse through plasma membranes, and therefore their receptors are inside the cell, in either the cytoplasm or the nucleus. In most cases, the complex formed by the lipid-soluble hormone and its receptor acts by altering gene expression in the cell (see Figure 15.8).

Water-soluble hormones cannot readily pass through plasma membranes, so their receptors are on the cell surface. These recep-

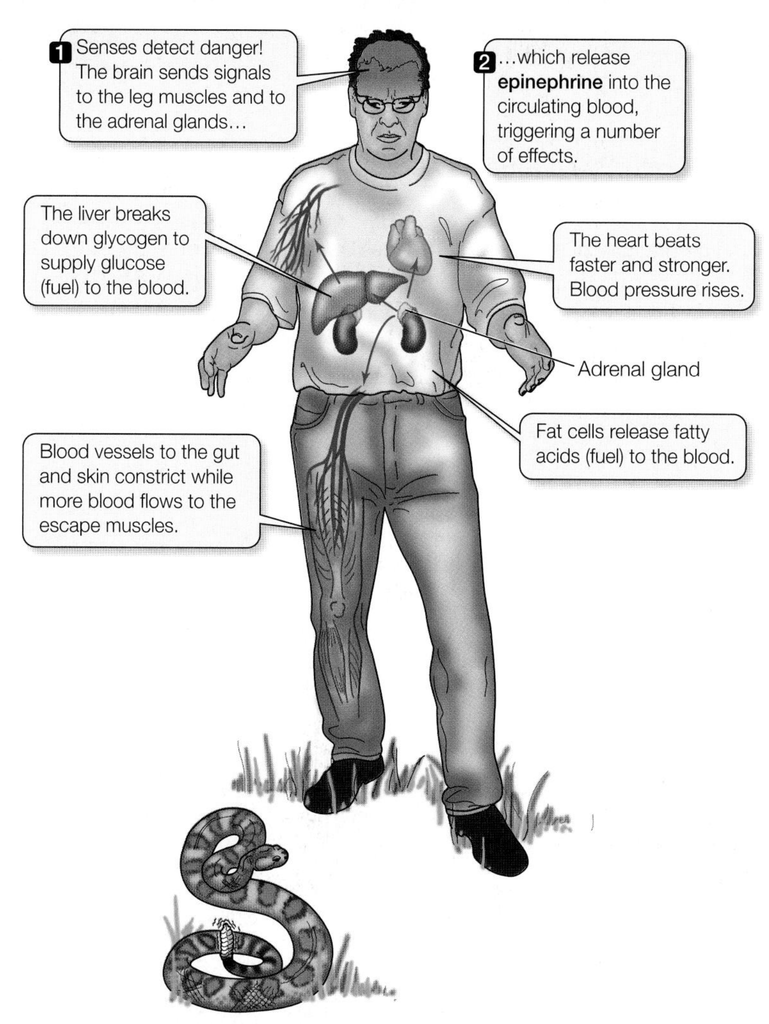

41.4 Epinephrine Stimulates "Fight or Flight" Responses When surprised by a threat, the brain sends nerve impulses to the adrenal medulla (see Figure 41.11), where epinephrine is released almost instantaneously. Epinephrine circulates around the body and induces different responses in different tissues to help you get out of danger fast.

tors are large glycoprotein complexes with three domains: a **binding domain** that projects outside the plasma membrane, a **transmembrane domain** that anchors the receptor in the membrane, and a **cytoplasmic domain** that extends into the cytoplasm of the cell. The cytoplasmic domain initiates the target cell's response by activating protein kinases or protein phosphatases (see Figures 15.6 and 15.7). In most cases these protein kinases and phosphatases activate or inactivate enzymes in the cytoplasm, which leads to the cell's response, but the signaling cascade initiated by the membrane receptors can also generate chemical signals that enter the nucleus and alter gene expression (see Figure 15.10).

Hormone action depends on the nature of the target cell and its receptors

Most hormones diffuse through the extracellular fluid and are picked up by the blood, which distributes them throughout the body. Wherever such a hormone encounters a cell with a receptor to which it can bind, it triggers a response. The nature of the response depends on the responding cell and its receptors. The same hormone can cause different responses in different types of cells.

Consider the hormone *epinephrine*. Suppose you are walking in the forest and almost trip over a rattlesnake. You jump back. Your heart starts to thump and a whole set of protective actions are set in motion. The jump and the initial heart thumping are driven by your nervous system, which reacts very quickly. Simultaneously with these muscular responses, however, your nervous system stimulates endocrine cells in the adrenal gland just above your kidneys to secrete epinephrine. Within seconds, epinephrine is diffusing into your blood and circulating around your body to activate the many components of the **fight-or-flight response** (**Figure 41.4**).

Epinephrine (an amine) binds to receptors in the heart and blood vessels, sustaining the higher heart rate and causing the heart to beat more strongly. Your heart is now pumping more blood, because your muscles need that blood to fuel your escape. Epinephrine causes more of your circulating blood to flow to the muscles by causing blood vessels in your digestive tract to constrict (digesting lunch can wait!). Similarly, it decreases blood flow to the skin and to the kidneys and suppresses some of the functions of the immune system.

Epinephrine also binds to cells in the liver and to receptors on fat cells. In the liver, epinephrine stimulates the breakdown of glycogen into glucose for a quick energy supply. In fatty tissue, it stimulates the breakdown of fats to yield fatty acids—another source of energy. These are just some of the many actions triggered by one hormone. They all contribute to increasing your chances of escaping from a dangerous situation.

Because it stimulates the heart, epinephrine is a frontline treatment for cardiac arrest (heart attack). Several of its effects, including dilation of the airways, make it a lifesaving drug for people who suffer from asthma or from potentially fatal allergies to things like peanuts and bee stings.

41.1 RECAP

Hormones are chemical signals released by endocrine cells into the extracellular environment, where they diffuse to nearby cells or into the blood. The receptors for water-soluble hormones are on the surface of target cells; receptors for lipid-soluble hormones are in the cytoplasm.

- How does juvenile hormone help effect metamorphosis? See p. 877 and Figures 41.2 and 41.3
- Can you describe the different methods by which water-soluble and lipid-soluble hormones reach their receptors? See p. 879
- Do you understand why a single hormone can have diverse effects in the body?

Since the nervous system and the endocrine system are the two major informational systems of the body, we might expect their activities to be coordinated—and, indeed, they are.

41.2 How Do the Nervous and Endocrine Systems Interact?

The list of hormones known to exist among the vertebrates is long and growing longer. To make the subject manageable, we will focus primarily on the hormones found in mammals (**Figure 41.5**). We will begin our survey by considering the hormones involved in the integration of nervous system and endocrine system functions.

The pituitary connects nervous and endocrine functions

The **pituitary gland** sits in a depression at the bottom of the skull just over the back of the roof of the mouth (**Figure 41.6A**). It is attached by a stalk to the **hypothalamus**, which is involved in many physiological regulatory systems. In turn, the pituitary is involved in the hormonal control of many physiological processes. The nervous system is involved in pituitary functions in two ways. First, two

Hypothalamus (see Figure 41.6)
Release and release-inhibiting hormones control the anterior pituitary
ADH and *oxytocin* are transported to and released from the posterior pituitary

Anterior pituitary (see Figure 41.7)
Thyroid stimulating hormone (TSH): activates the thyroid gland
Follicle stimulating hormone (FSH): in females, stimulates maturation of ovarian follicles; in males, stimulates spermatogenesis
Luteinizing hormone (LH): in females, triggers ovulation and ovarian production of estrogens and progesterone; in males, stimulates production of testosterone
Adrenocorticotropic hormone (ACTH): stimulates adrenal cortex to secrete cortisol
Growth hormone: stimulates protein synthesis and growth
Prolactin: stimulates milk production

Posterior pituitary (see Figure 41.6)
Receives and releases two hypothalamic hormones:
Oxytocin: stimulates contraction of uterus, stimulates flow of milk.
Antidiuretic hormone (ADH): promotes water conservation by kidneys

Thymus gland (disappears in adults)
Thymosin: activates immune system T cells

Pancreas (islets of Langerhans)
Insulin: stimulates cells to take up and use glucose
Glucagon: stimulates liver to release glucose
Somatostatin: slows digestive tract functions including release of insulin and glucagon

Pineal gland
Melatonin: helps to regulate circadian rhythms

Thyroid gland (see Figures 41.9 and 41.10)
Thyroxine (T_3 and T_4): stimulates cell metabolism
Calcitonin: stimulates incorporation of calcium into bone

Parathyroid glands (on posterior surface of thyroid; see Figure 41.10)
Parathormone (PTH): stimulates release of calcium from bone

Adrenal gland (see Figure 41.11)
Cortex
Cortisol: mediates metabolic responses to stress
Aldosterone: involved in salt and water balance

Medulla
Epinephrine (adrenaline) and *norepinephrine* (noradrenaline): stimulate immediate fight or flight reactions

Gonads (see Chapter 42)
Ovaries (female)
Estrogens: development and maintenance of female sexual characteristics
Progesterone: supports pregnancy

Testes (male)
Testosterone: development and maintenance of male sexual characteristics

Other organs include cells that produce and secrete hormones

Organ	Hormone
Adipose tissue	Leptin
Heart	Atrial natriuretic peptide
Kidney	Erythropoeitin
Stomach	Gastrin
Intestine	Secretin, cholecystokinin
Skin	Vitamin D (cholecalciferol)

41.5 The Endocrine System of Humans The cells that produce and secrete hormones may be organized into discrete endocrine glands, or they may be embedded in the tissues of other organs such as the digestive tract or kidneys. The hypothalamus is part of the brain, but it includes cells that secrete neurohormones into the extracellular fluids.

hormones produced by nerve cells in the hypothalamus are transported to the pituitary via long extensions of those nerve cells and released there. Second, many of the hormones produced by the pituitary are controlled by other hypothalamic hormones that reach the pituitary through the blood.

The two parts of the pituitary gland have different functions and separate developmental origins. The *anterior pituitary* originates as an outpocketing of the roof of the embryonic mouth cavity, and the *posterior pituitary* originates as an outpocketing of the floor of the developing brain. Thus the anterior pituitary originates from gut epithelial tissue and the posterior pituitary originates from neural tissue.

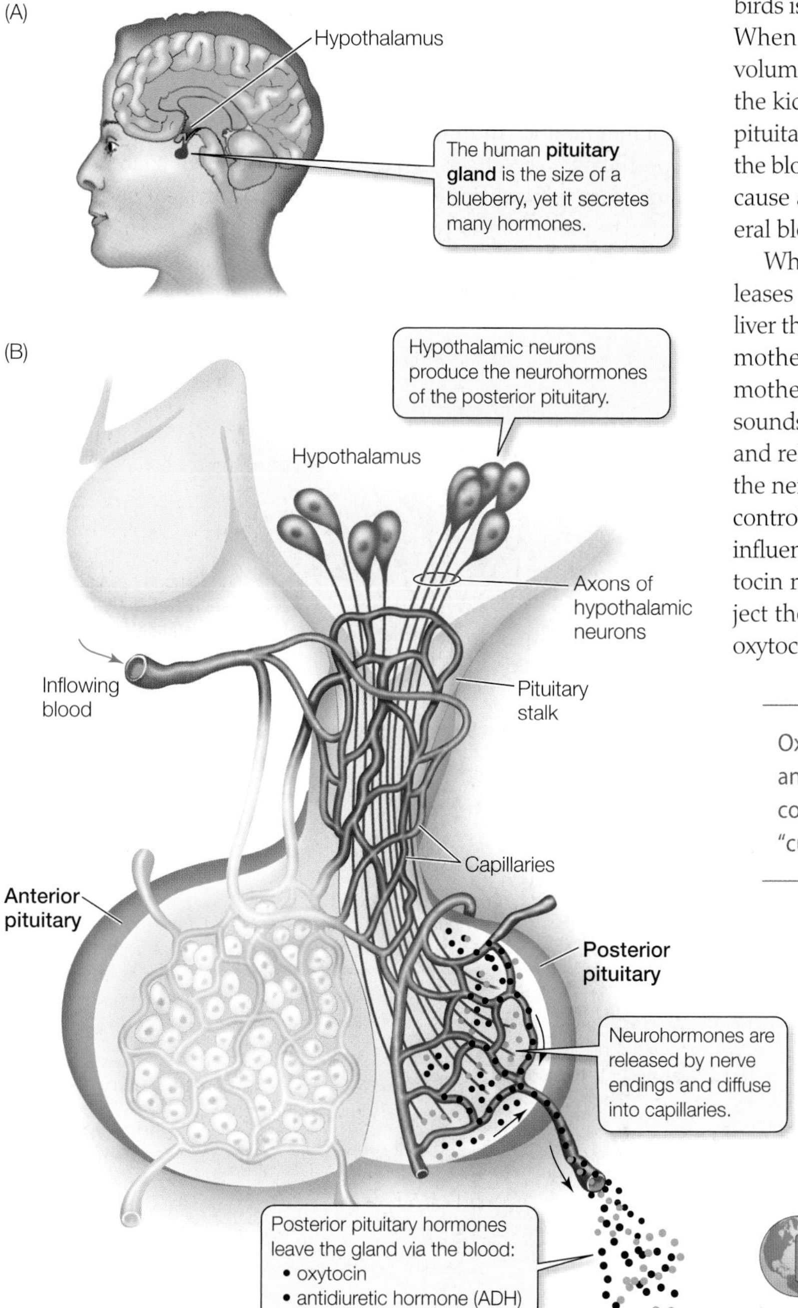

41.6 The Posterior Pituitary Releases Neurohormones The two hormones stored and released by the posterior pituitary are peptide neurohormones produced in the hypothalamus.

THE POSTERIOR PITUITARY The **posterior pituitary** releases two peptide hormones, antidiuretic hormone (also called vasopressin) and oxytocin. Because these hormones are synthesized in neurons in the hypothalamus, they are called **neurohormones**. The insect hormone PTTH discussed earlier is also a neurohormone (produced in neural tissue). As antidiuretic hormone and oxytocin are produced by cells, they are packaged in vesicles. These vesicles are then transported down long extensions (axons) of the neurons that run from the hypothalamus through the pituitary stalk and terminate in the posterior pituitary. The vesicles are stored until a nerve impulse stimulates their release (**Figure 41.6B**). How do the vesicles move down the axons? Proteins called *kinesins* grab onto the vesicles and, powered by ATP, "walk" step by step down microtubules in the axons.

The main action of **antidiuretic hormone** (**ADH**) in mammals and birds is to increase the amount of water conserved by the kidneys. When ADH secretion is high, the kidneys produce only a small volume of highly concentrated urine. When ADH secretion is low, the kidneys produce a large volume of dilute urine. The posterior pituitary increases its release of ADH when blood pressure falls or the blood becomes too salty. ADH is also known as *vasopressin* because at high concentrations it causes the constriction of peripheral blood vessels as a means of elevating blood pressure.

When a woman is about to give birth, her posterior pituitary releases **oxytocin**, which stimulates the uterine contractions that deliver the baby. Oxytocin also brings about the flow of milk from the mother's breasts. The baby's suckling stimulates neurons in the mother that cause the secretion of oxytocin. Even the sight and sounds of her baby can cause a nursing mother to secrete oxytocin and release milk from her breasts. This is a good example of how the nervous system integrates information and contributes to the control of hormonally mediated processes. In turn, hormones can influence the nervous system. Oxytocin promotes bonding. If oxytocin release is blocked, new mothers from rats to sheep will reject their newborn offspring, but if a virgin rat is given a dose of oxytocin, she will adopt strange pups as if they were her own.

Oxytocin promotes pair bonding and trust in a variety of animals. In humans, its secretion rises with intimate sexual contact. Not surprisingly, oxytocin has been nicknamed the "cuddle hormone."

THE ANTERIOR PITUITARY Four peptide and protein hormones released by the **anterior pituitary**—*thyrotropin, corticotropin* (also known as *adrenocorticotropic hormone*, or *ACTH*), *luteinizing hormone*, and *follicle-stimulating hormone*—are **tropic hormones**, meaning they control the activities of other endocrine glands (**Figure 41.7**).

Each tropic hormone is produced by a different type of pituitary cell. We will say more about the tropic hormones when we describe their target glands (thyroid, adrenal cortex, testes, and ovaries) later in this chapter and in Chapter 42.

Other peptide and protein hormones produced by the anterior pituitary are *growth hormone, prolactin, melanocyte-stimulating hormone, enkephalins,* and *endorphins.*

Growth hormone (GH) acts on a wide variety of tissues to promote growth. One of its important effects is to stimulate cells to take up amino acids. Growth hormone promotes growth also by stimulating the liver to produce chemical messages called *somatomedins* or *insulin-like growth factors* (*IGFs*), which stimulate the growth of bone and cartilage. Thus, growth hormone can also be considered in part a tropic hormone because it stimulates liver cells to produce and release hormones.

Overproduction of growth hormone in children causes *gigantism* (individuals may grow to nearly 8 feet tall). Underproduction causes *pituitary dwarfism,* in which individuals fail to reach normal adult height. Beginning in the late 1950s, children with serious growth hormone deficiencies were treated with growth hormone extracted from pituitaries of human cadavers. The treatment was successful in stimulating substantial growth, but a year's supply of the hormone for one individual required up to *50* cadaver pituitaries! In the mid-1980s, scientists using genetic engineering technology isolated the gene for human growth hormone and introduced it into bacteria that could be grown in large quantities, making it possible to purify enough of the hormone to make it widely available.

Prolactin stimulates breast development and the production and secretion of milk in female mammals. In some mammals, prolactin also functions as an important hormone during pregnancy. In human males, prolactin plays a role in controlling the endocrine function of the testes.

Endorphins and **enkephalins** are the body's natural opiates. In the brain, these molecules act as neurotransmitters in pathways that control pain. Their production in the anterior pituitary is normally quite small and probably not significant. They are a byproduct of the production of two other pituitary hormones. One gene encodes a large parent molecule called *pro-opiomelanocortin.* POMC is cleaved to produce several peptides. corticotropin, melanocyte-stimulating hormone, endorphins, and enkephalins all result from the cleavage of POMC.

The anterior pituitary is controlled by hypothalamic hormones

The secretion of hormones by the anterior pituitary is under the control of neurohormones from the hypothalamus. The hypothalamus receives information about conditions in the body and in the external environment through both neuronal signals and hormones that reach it through the circulation. If the connection between the hypothalamus and the pituitary is experimentally cut, pituitary hormones are no longer released in response to changes in the internal or external environment. In experiments in which pituitary cells were maintained in culture, extracts of hypothalamic tissue stimulated some of those cells to release their hormones into the culture medium. Therefore, scientists hypothesized that secretions of the hypothalamic cells control the activities of anterior pituitary cells.

Although hypothalamic neurons do not extend into the anterior pituitary as they do into the posterior pituitary, a special set of **portal blood vessels** connects the hypothalamus and the anterior pituitary (see Figure 41.7). It was thus proposed that secretions from neurons in the hypothalamus enter the blood and are conducted down the portal vessels to the anterior pituitary, where they stimulate the release of anterior pituitary hormones.

In the 1960s, two large teams of scientists, led by Roger Guillemin and Andrew Schally, initiated the search for these hypothalamic secretions. Because the amounts of such neurohormones in any individual mammal would be tiny, massive numbers of hypothalami from pigs and sheep were collected from slaughterhouses and shipped to laboratories in refrigerated trucks. One extraction effort began with the hypothalami from 270,000 sheep

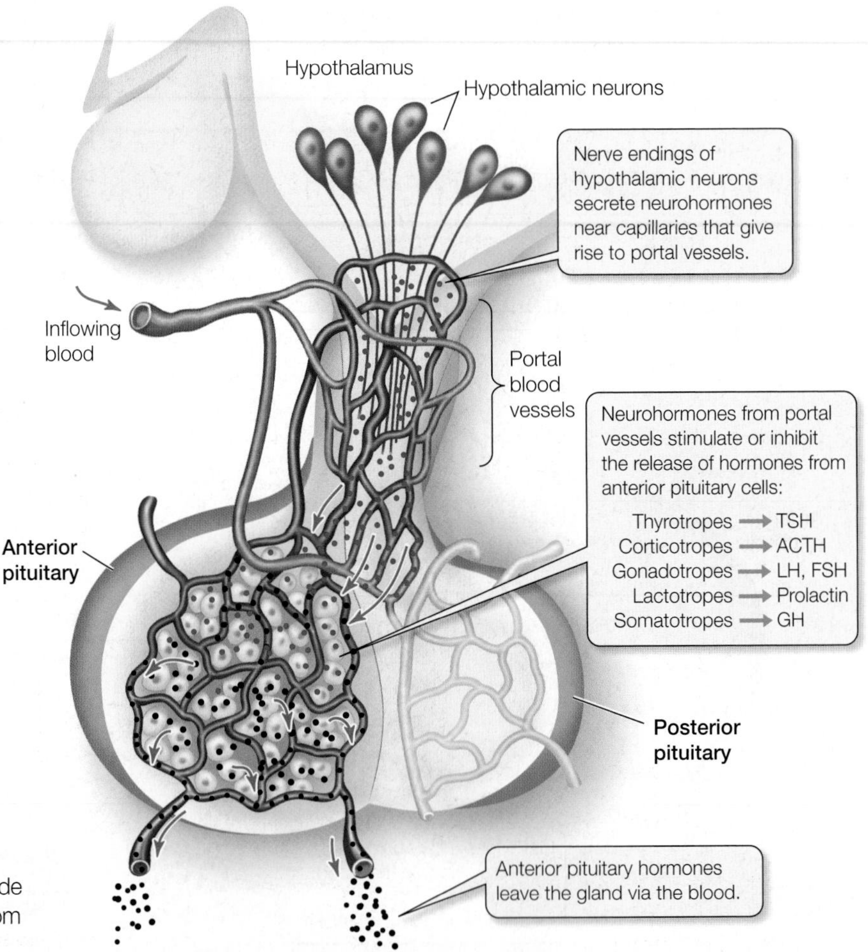

41.7 The Anterior Pituitary Produces Many Hormones
Cells of the anterior pituitary produce four tropic hormones that control other endocrine glands, and several other peptide hormones. These cells are controlled by neurohormones from the hypothalamus delivered through portal blood vessels.

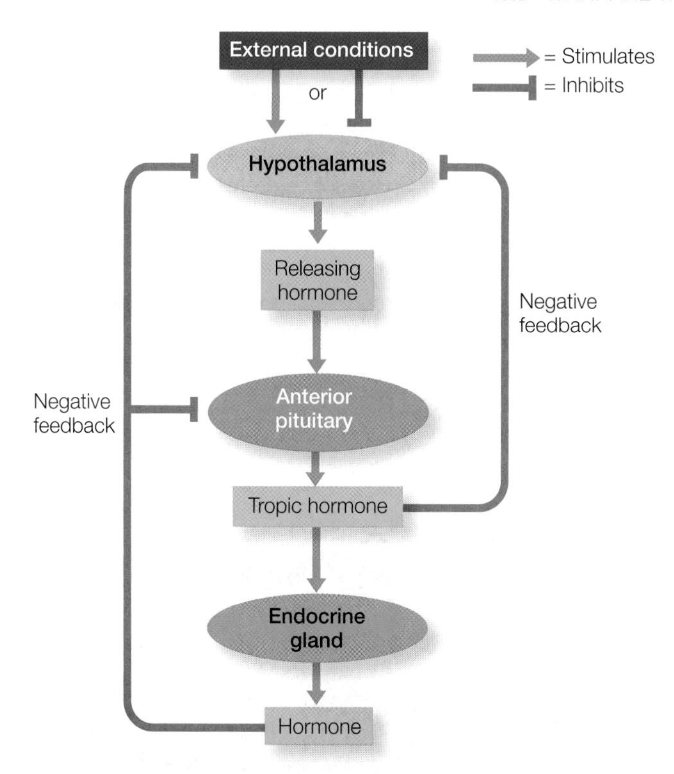

41.8 Multiple Feedback Loops Control Hormone Secretion
Multiple negative feedback loops regulate the chain of command from hypothalamus to anterior pituitary to endocrine glands.

and yielded only 1 milligram of purified **thyrotropin-releasing hormone (TRH)**. Biochemical analysis of this pure sample revealed that TRH is a simple tripeptide consisting of glutamine, histidine, and proline. TRH was the first hypothalamic *releasing hormone* (that is, release-stimulating hormone) to be isolated and characterized. It causes certain anterior pituitary cells to release the tropic hormone thyrotropin, which in turn stimulates the activity of the thyroid gland.

Soon after discovering thyrotropin-releasing hormone, Guillemin's and Schally's teams identified **gonadotropin-releasing hormone (GnRH)**, which stimulates certain anterior pituitary cells to release the tropic hormones that control the activity of the gonads (the ovaries and the testes). For these discoveries, Guillemin and Schally received the 1977 Nobel prize in medicine. Many other hypothalamic neurohormones, including both releasing hormones and release-inhibiting hormones, are now known, including:

- Prolactin-releasing and release-inhibiting hormones
- Growth hormone-releasing hormone
- Growth hormone release-inhibiting hormone (somatostatin)
- Adrenocorticotropin-releasing hormone
- Melanocyte-stimulating hormone and release-inhibiting hormone

Negative feedback loops control hormone secretion

As well as being controlled by hypothalamic releasing and release-inhibiting hormones, the endocrine cells of the anterior pituitary are also under direct and indirect negative feedback control by the hormones of the target glands they stimulate (**Figure 41.8**). For example, the hormone cortisol, produced by the adrenal gland in response to corticotropin secreted by the anterior pituitary, returns to the pituitary in the circulating blood and inhibits further release of that tropic hormone. Cortisol also acts as a negative feedback signal to the hypothalamus, inhibiting the release of corticotropin-releasing hormone. In some cases a tropic hormone also exerts negative feedback control on the hypothalamic cells producing the corresponding releasing hormone.

41.2 RECAP

Many interactions between the nervous system and endocrine systems are controlled by the pituitary. The posterior pituitary releases two neurohormones, and the anterior pituitary, under the control of other neurohormones, releases hormones that control other endocrine glands in the body.

- Describe the anatomical and functional relationships between the brain and the two parts of the pituitary. See pp. 881–882 and Figures 41.6 and 41.7
- What are the tropic hormones of the anterior pituitary, and how do they influence endocrine mechanisms? See p. 882 and Figure 41.7

Now that we know some of the mechanisms by which endocrine systems are controlled, we will take a more detailed look at the functions of the major endocrine glands of the body.

41.3 What Are the Major Mammalian Endocrine Glands and Hormones?

Hormones help regulate functions in all physiological systems. In this section we will examine a few major examples of hormonal action in physiological processes, but will add many more in the chapters that follow.

Thyroxine controls cell metabolism

The **thyroid gland** wraps around the front of the windpipe (*trachea*) and expands into a lobe on either side (see Figure 41.5). Two cell types in the thyroid gland produce the hormones thyroxine and calcitonin. Thyroxine is produced, stored, and released by many round structures called *follicles* (**Figure 41.9A,B**). Cells in the spaces between the follicles produce calcitonin.

Thyroxine, which begins as the glycoprotein *thyroglobulin* in the follicle lumen, is built out of two molecules of tyrosine bound to four atoms of iodine.

Thyroxine (T_4)

Thus, thyroxine is called T_4.

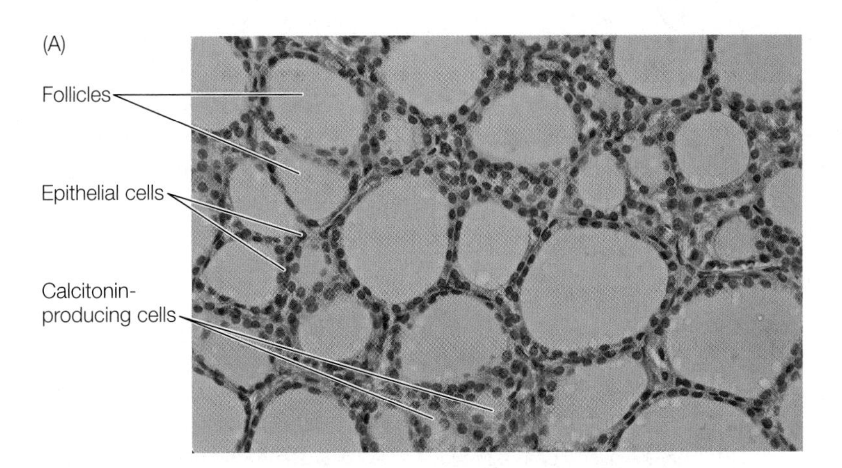

41.9 The Thyroid Gland Consists of Many Follicles (A) Cross section through a thyroid gland showing numerous follicles bounded by epithelial cells. Calcitonin-secreting cells are found in the spaces between the follicles. (B) Epithelial cells synthesize and iodinate thyroglobulin. Iodinated thyroglobulin is stored in the follicles until it is processed by epithelial cells to generate T_3 and T_4. (C) Iodine deficiency can result in hypothyroid goiter. In this condition, a lack of functional thyroxine results in oversynthesis of thyroglobulin and subsequent enlarged follicles.

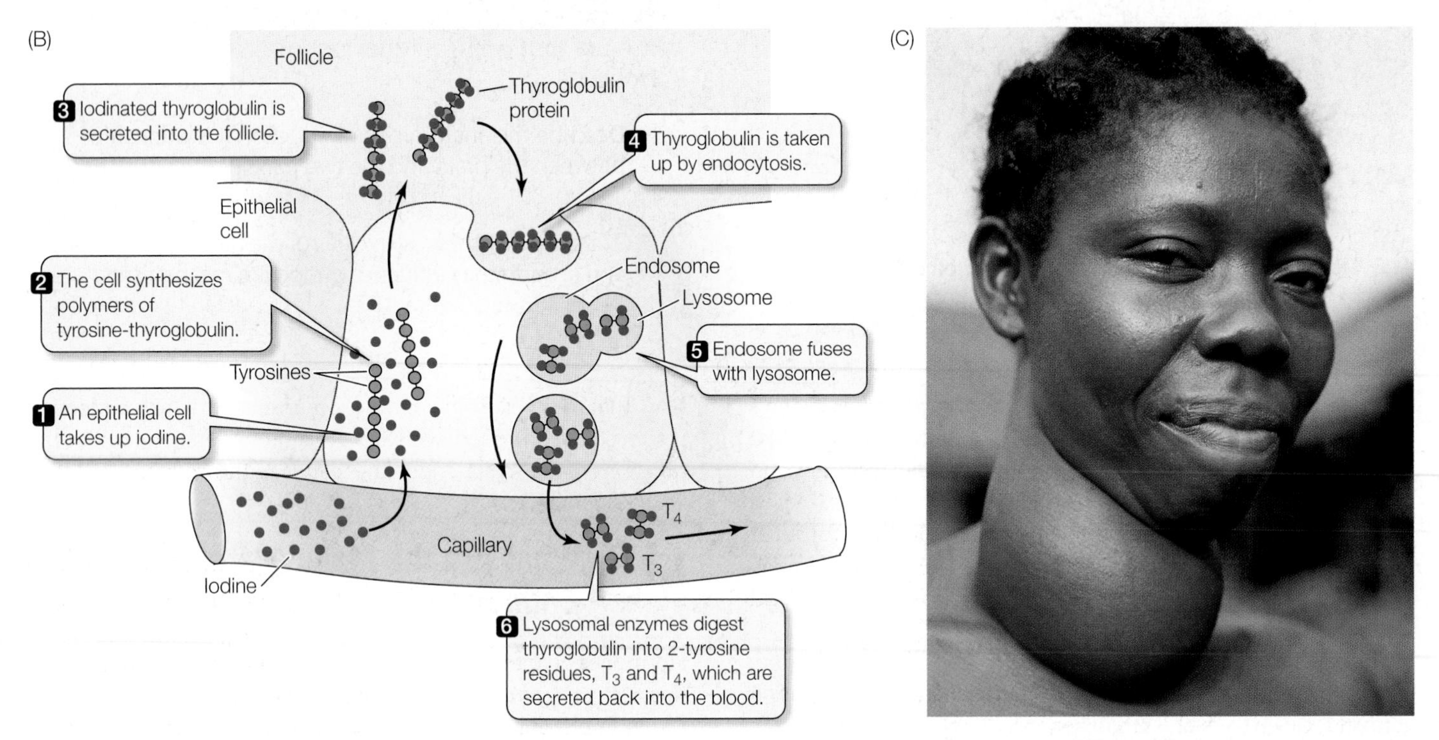

The follicles also produce and release *triiodothyronine*, a version of thyroxine that has only three atoms of iodine and is called T_3:

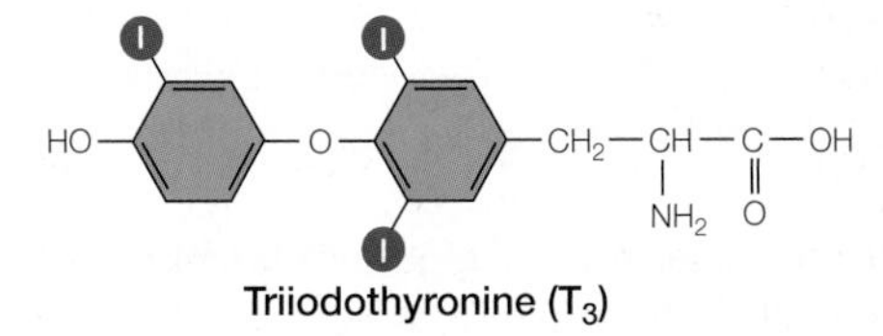

Triiodothyronine (T_3)

The thyroid usually releases about four times as much T_4 as T_3. T_3 is the more active hormone in the cells of the body, but when T_4 is in circulation, it can be converted to T_3 by an enzyme within target cells. Each target cell can thus set its own sensitivity to thyroid hormones by controlling the conversion of T_4 to the more active T_3. When you read about thyroxine, keep in mind that the actions discussed are primarily those of T_3.

Thyroxine in mammals plays many roles in regulating cell metabolism. Thyroxine is lipid-soluble, so it enters cells readily where it binds to receptors. The thyroxine–receptor complex acts in the nucleus as a transcription factor that affects the activity of a large number of genes. These genes encode enzymes in energy pathways, transport proteins, and structural proteins. As a result, thyroxine elevates the metabolic rates of most cells and tissues. Exposure to cold for several days leads to an increased release of thyroxine, an increased conversion of T_4 to T_3, and therefore an increased basal metabolic rate (see Section 40.4). Thyroxine is especially crucial during development and growth, as it promotes amino acid uptake and protein synthesis by cells. Insufficient thyroxine in a human fetus or growing child greatly retards physical and mental growth, resulting in a condition known as *cretinism*.

The tropic hormone **thyrotropin**, or **thyroid-stimulating hormone (TSH)**, produced by the anterior pituitary, activates the thyroxine-producing follicle cells in the thyroid. Thyrotropin-releasing hormone (TRH; discussed earlier in this chapter), produced in the hypothalamus and transported to the anterior pituitary through the portal blood vessels, activates the TSH-producing pituitary cells. The hypothalamus uses environmental information, such as temperature or day length, to determine whether to increase or decrease the secretion of TRH. This sequence of steps is also regulated by a negative feedback loop: circulating thyroxine inhibits the response of pituitary cells to TRH, so less TSH is released when thyroxine levels are high, and more TSH is released when thyroxine levels are low. Circulating thyroxine also exerts negative feedback on the production and release of TRH by the hypothalamus.

Thyroid dysfunction causes goiter

A *goiter* is an enlarged thyroid gland that causes a pronounced bulge on the front and sides of the neck. Goiter results from either **hyperthyroidism** (thyroxine excess) or **hypothyroidism** (thyroxine deficiency). The negative feedback loop whereby thyroxine controls TSH release helps explain how two very different conditions can result in the same symptom, but it is also necessary to understand how the thyroid makes, stores, and releases thyroxine (**Figure 41.9C**).

Each thyroid follicle consists of a layer of epithelial cells surrounding a mass of glycoprotein called **thyroglobulin**. The epithelial cells make the thyroglobulin by creating long polymers of tyrosine residues. As the thyroglobulin is secreted into the lumen of the follicle, iodine is added to the tyrosine residues. When thyroxine is needed, the same epithelial cells that made the thyroglobulin take it back through endocytosis and digest it to the smaller thyroxine molecules.

If there was enough iodine available when the thyroglobulin was made, its digestion releases molecules of T_3 and T_4. If there was not enough iodine available when the thyroglobulin was made, many of the residues released will not be T_3 or T_4 and will not bind to receptors on target cells.

Goiter occurs when the production of thyroglobulin is far above normal and the follicles become greatly enlarged. *Hyperthyroid* goiter results when the negative feedback mechanism fails to turn off the follicle cells even with high blood levels of thyroxine. The most common cause of hyperthyroidism is an autoimmune disease in which an antibody to the TSH receptor is produced. This antibody can bind to the TSH receptor on the follicle cells, causing them to produce and release thyroxine. Even though blood levels of TSH may be quite low because of the negative feedback from high levels of thyroxine, the thyroid remains maximally stimulated and it grows bigger. Hyperthyroid patients have high metabolic rates, are jumpy and nervous, usually feel hot, and may develop a buildup of fat behind the eyeballs, which in turn causes their eyes to bulge.

Hypothyroid goiter results when there is not enough circulating thyroxine to turn off TSH production. The most common cause of this condition is a deficiency of dietary iodide, without which the follicle cells cannot make thyroxine (**Figure 41.9C**). Without sufficient thyroxine, TSH levels remain high, and the thyroid continues to produce large amounts of thyroglobulin. But because insufficient iodide is available, few of the tyrosine residues in the thyroglobulin are iodinated. When this iodine-poor thyroglobulin is digested by the follicle cells, it does not yield functional thyroxine. Without functional thyroxine, the TSH levels remain high and stimulate more and more synthesis of thyroglobulin, and the follicles get bigger. The symptoms of hypothyroidism are low metabolism, intolerance of cold, and general physical and mental sluggishness.

Goiter affects about 5 percent of the world's population. The addition of iodide to table salt has greatly reduced the incidence of hypothyroid goiter in industrialized nations, but the condition is still common in the other parts of the world.

Calcitonin reduces blood calcium

The regulation of calcium levels in the blood is a crucial and difficult task. It is crucial because changes in blood calcium levels can cause serious problems. When blood calcium falls more than 30 percent below normal, the nervous system becomes overly excited, resulting in muscle spasms and even seizures. When blood calcium rises above normal, the nervous system becomes depressed and muscles—including the heart—weaken. Regulation of blood calcium is difficult because only about 0.1 percent of the calcium in the body is located in the extracellular fluids. About 1 percent is within cells, and almost 99 percent is in the bones. Therefore, the body must regulate a tiny pool of calcium in the blood, and that tiny pool can be influenced greatly by relatively small shifts in the much larger pools of calcium in the cells and bones.

There are multiple mechanisms for changing blood calcium levels, including:

- deposition and absorption of bone
- excretion of calcium by the kidneys
- absorption of calcium from the digestive tract

These mechanisms are controlled by the hormones calcitonin, parathyroid hormone, and vitamin D.

Calcitonin lowers the concentration of calcium in the blood (**Figure 41.10**). Bone is continually remodeled through resorption of old bone and synthesis of new bone, as we will see in Section 47.3. Cells called *osteoclasts* break down bone and release calcium; *osteoblasts* take up circulating calcium and deposit new bone. Calcitonin decreases the activity of osteoclasts and thereby shifts the balance of bone turnover to favor removal of calcium from the blood. Because the turnover of bone in adult humans is not very high, calcitonin does not play a major role in calcium homeostasis in adults. It is probably more important in young, growing individuals, but overall calcium levels are more influenced by parathyroid hormone than by calcitonin.

Parathyroid hormone elevates blood calcium

The **parathyroid glands** are four tiny structures embedded in the posterior surface of the thyroid gland. Their single hormone product, **parathyroid hormone** (also called **PTH** or parathormone), is the critical hormone in the regulation of blood calcium levels. Levels of calcium in the blood are sensed by receptors in the plasma membrane of the parathyroid cells. When these receptors are activated, they inhibit the synthesis and release of PTH. A fall in blood calcium removes this inhibition and triggers the synthesis and release of PTH.

PTH raises blood calcium levels in several ways. Its actions provide a good example of the complexity of physiological regulation. One major action of PTH is to stimulate bone turnover and remod-

41.10 Hormonal Regulation of Calcium Calcitonin, parathyroid hormone, and vitamin D help regulate calcium levels in the blood.

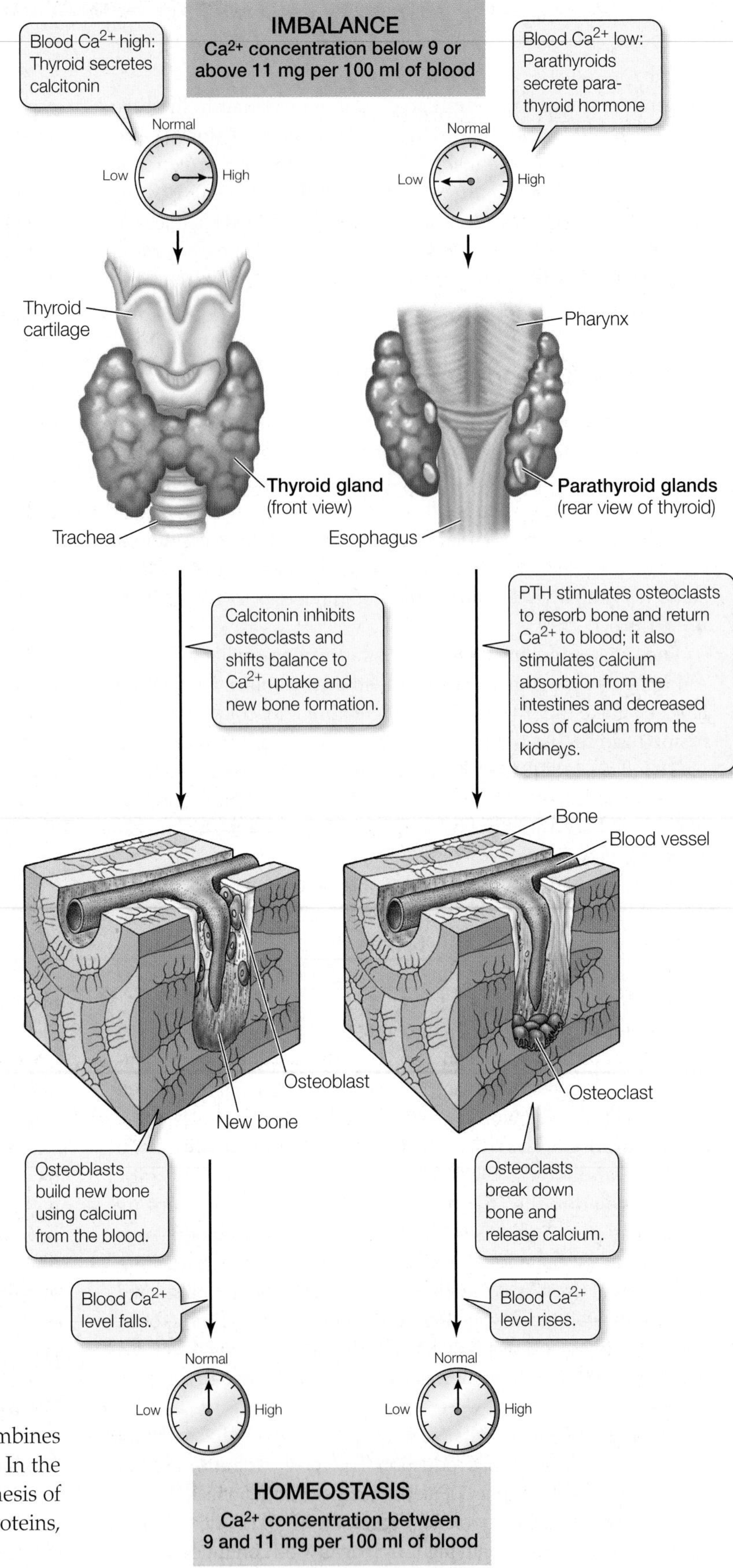

eling. This is a dynamic process that involves both resorption of old bone and laying down of new bone. Osteoblasts have PTH receptors and are stimulated by the hormone to remove calcium from the blood and form new bone. However, stimulated osteoblasts also release paracrine signals—cytokines—that stimulate osteoclasts. This process of bone turnover entails a net loss of calcium from bone. It was previously thought that this indirect path was the only way that PTH influenced bone resorption, but recently investigators have shown that osteoclasts also have PTH receptors and are stimulated by PTH directly (see Figure 41.10). PTH also conserves calcium by stimulating the kidneys to reabsorb it rather than excrete it in the urine. Increased secretion of PTH causes the digestive tract to absorb more calcium from food. This is an indirect effect, however, dependent on vitamin D. PTH activates vitamin D, which in turn causes the digestive tract to enhance absorption of dietary calcium.

Vitamin D is really a hormone

A *vitamin* is a substance that the body needs in small quantities, but cannot synthesize and therefore must obtain from the diet. By this definition, **vitamin D** is not a vitamin, because the body can and does synthesize it.

It had long been known that fragile bones were common among people living at high latitudes where winter days are short and the winter diet often lacks meat, fish, dairy products, and fresh vegetables. Since the condition could be reversed taking cod-liver oil (which, as it turns out, contains large amounts of vitamin D), it was assumed that a dietary vitamin was involved. We now know that vitamin D is synthesized in skin cells, where cholesterol is converted into vitamin D (also called calciferol) by ultraviolet light. Vitamin D circulates in the blood, and acts on distant cells; thus it is actually a hormone.

The vitamin D produced in the skin is not very active, but as it passes through the liver it receives one hydroxyl (—OH) group, and in the kidneys it receives another to become (1,25)-dihydroxyvitamin D, the most active form. PTH stimulates this final step in the kidneys. Active vitamin D is lipid-soluble, so readily enters cells. In cells it combines with a cytoplasmic receptor to form a transcription factor. In the digestive tract, this transcription factor increases the synthesis of calcium pumps, calcium channels, and calcium-binding proteins, all of which promote the uptake of calcium.

In the kidneys, vitamin D acts synergistically with PTH to decrease calcium loss in the urine. In bone, vitamin D, like PTH, stimulates bone turnover and liberates calcium—which seems the opposite of what would be expected. However, through all of its actions, vitamin D raises blood calcium levels, and that is essential to promote bone deposition. Vitamin D also acts on parathyroid cells to inhibit the transcription of the PTH gene, thus forming a negative feedback loop for the regulation of PTH.

PTH lowers blood phosphate levels

Bone minerals are a combination of calcium and phosphate. Thus, when PTH stimulates the release of calcium from bone, it also causes the release of phosphate. Increases in blood levels of both calcium and phosphate can be dangerous. The normal levels of calcium and phosphate in the blood approach the concentration at which they would precipitate out of solution as calcium phosphate salts, leading to maladies such as kidney stones and calcium deposits in the arteries (hardening of the arteries). To reduce this problem, PTH acts on the kidneys to increase the elimination of phosphate via the urine.

Insulin and glucagon regulate blood glucose levels

Before the 1920s, *diabetes mellitus* was a fatal disease, characterized by weakness, lethargy, and a dramatic loss of body mass. The disease was known to be connected somehow with the **pancreas**, a gland located just below the stomach (see Figure 41.5), and with abnormal glucose metabolism, but the link was not clear.

Today we know that diabetes mellitus is caused by a lack of the protein hormone **insulin** (in type I or juvenile-onset diabetes) or by a lack of insulin receptors on the target tissues (in type II or adult-onset diabetes). For patients in which the hormone is lacking, insulin replacement therapy is extremely successful. At present, more than 1.5 million people with diabetes in the United States lead almost normal lives by using manufactured insulin.

Insulin binds to a receptor on the plasma membrane of a target cell, and this insulin–receptor complex allows glucose to enter the cell (see Figure 15.6). In the absence of insulin or insulin receptors, glucose entry into cells is impaired and glucose accumulates in the blood until it is lost in the urine. High levels of blood glucose cause water to move from cells into the blood by osmosis, and the kidneys increase urine output to remove this excess fluid volume from the blood. Because glucose uptake by most cells is impaired without insulin, those cells must use fat and protein for fuel instead of glucose. As a result, the body of the untreated diabetic wastes away, and critical tissues and organs are damaged.

For centuries the prospects for diabetics were bleak. A change came almost overnight in 1921, when the physician Frederick Banting and a medical student, Charles Best of the University of Toronto, discovered that they could reduce the symptoms of diabetes by injecting an extract prepared from pancreatic tissue. The active component of this extract was found to be a small protein hormone—insulin—consisting of just 51 amino acids.

Banting was awarded a Nobel prize for the discovery of insulin, but his co-worker, Charles Best, was not nominated. The rules of the Nobel Committee precluded awards to medical students. Banting protested and shared both the award and the credit with Best.

Insulin is produced in clusters of endocrine cells in the pancreas. These clusters are called **islets of Langerhans** after the German medical student who discovered them. There are three types of cells in the islets:

- Beta (β) cells produce and secrete insulin.
- Alpha (α) cells produce and secrete the hormone glucagon, which has effects opposite from those of insulin.
- Delta (δ) cells produce the hormone somatostatin.

The rest of the pancreas is an exocrine gland producing enzymes and other secretions that travel through ducts to the intestine to participate in digestion.

After a meal, the concentration of glucose in the blood rises as glucose is absorbed from the food in the gut. This increase stimulates the β cells of the pancreas to release insulin. Insulin stimulates cells to use glucose as fuel and to convert it into storage products, such as glycogen and fat. When the gut contains no more food, the glucose concentration in the blood falls, and the pancreas stops releasing insulin. As a result, most cells of the body shift to using glycogen and fat, rather than glucose, for fuel. If the concentration of glucose in the blood falls substantially below normal, the islet α cells release **glucagon**, which stimulates the liver to convert glycogen back to glucose to resupply the blood. These actions will be discussed in greater detail in Section 50.4.

Somatostatin is a hormone of the brain and the gut

Somatostatin is released from the cells of the pancreas in response to rapid rises of glucose and amino acids in the blood. This hormone has paracrine functions within the islets: it inhibits the release of both insulin and glucagon. Its actions outside the pancreas slow the digestive activities of the gut. Pancreatic somatostatin extends the period of time during which nutrients are absorbed from the gut. Somatostatin is also produced in very small amounts by cells in the hypothalamus. Acting as a neurohormone, hypothalamic somatostatin is transported in the portal vessels to the anterior pituitary, where it inhibits the release of growth hormone and thyrotropin.

The adrenal gland is two glands in one

An **adrenal gland** sits above each kidney, just below the middle of your back. Functionally and anatomically, each adrenal gland consists of a gland within a gland (**Figure 41.11**). The core, called the **adrenal medulla**, produces the hormone **epinephrine** (also known as *adrenaline*) and, to a lesser degree, **norepinephrine** (or *noradrenaline*), which also acts as a neurotransmitter in the nervous system. Surrounding the medulla is the **adrenal cortex**, which produces steroid hormones. The medulla develops from nervous tissue and is under the control of the nervous system; the cortex is under hormonal control, largely by **corticotropin** (**ACTH**) from the anterior pituitary.

THE ADRENAL MEDULLA The adrenal medulla produces epinephrine and norepinephrine in response to stressful situations, arousing the body to action. As we saw earlier in this chapter, epinephrine increases heart rate and blood pressure and diverts blood flow to active muscles and away from the gut.

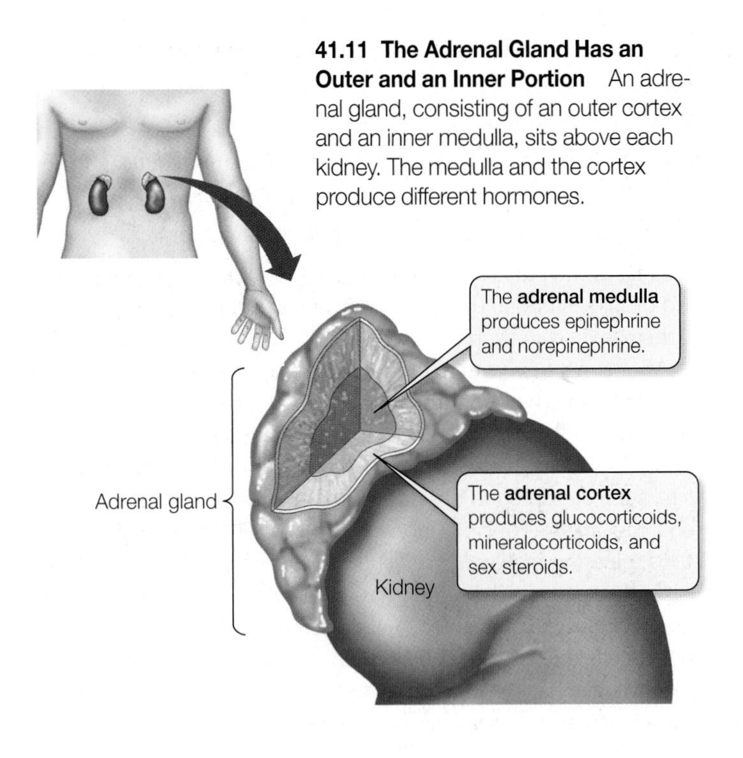

41.11 The Adrenal Gland Has an Outer and an Inner Portion An adrenal gland, consisting of an outer cortex and an inner medulla, sits above each kidney. The medulla and the cortex produce different hormones.

Epinephrine and norepinephrine are both amine hormones—derivatives of the amino acid tyrosine. They are water-soluble, and both bind to the same receptors on the surfaces of target cells. These receptors can be grouped into two general types, *α-adrenergic* and *β-adrenergic receptors,* which stimulate different actions within cells (see Figure 41.17 later in this chapter). Epinephrine acts equally on both types, but norepinephrine acts mostly on α-adrenergic receptors. Therefore, drugs called *beta blockers,* so named because they selectively block β-adrenergic receptors, can reduce the fight-or-flight responses to epinephrine without disrupting the physiological regulatory functions of norepinephrine. Beta blockers are commonly prescribed to reduce the symptoms of anxiety, such as dry mouth and elevated heart rate (heart palpitations).

THE ADRENAL CORTEX The cells of the adrenal cortex use cholesterol to produce three classes of steroid hormones, collectively called **corticosteroids**:

- The *glucocorticoids* influence blood glucose concentrations as well as other aspects of fat, protein, and carbohydrate metabolism.
- The *mineralocorticoids* influence the ionic balance of extracellular fluids.
- The *sex steroids* play roles in sexual development, sex drive, and anabolism.

The adult adrenal cortex secretes sex steroids in only negligible amounts. The major producers of sex steroids are the gonads, as we will see in the following section.

Aldosterone, the main mineralocorticoid (**Figure 41.12A**), stimulates the kidneys to conserve sodium and to excrete potassium. If the adrenal glands are removed from an animal, sodium must be added to its diet, or its sodium will be depleted and it will die.

The main glucocorticoid in humans is **cortisol** (**Figure 41.12B**), which is critical for mediating the body's response to stress. Within minutes of a stressful stimulus (one provoking fear or anger, for example), your blood cortisol level rises. Cells not critical for your action (fight or flight) are stimulated by cortisol to decrease their use of blood glucose and shift instead to utilizing fats and proteins for energy. This is no time to feel sick, have allergic reactions, or heal wounds, so cortisol also blocks immune system reactions. That is why cortisol or drugs that mimic cortisol action are useful for reducing inflammation and allergies.

Cortisol release is controlled by corticotropin from the anterior pituitary, which in turn is controlled by **corticotropin-releasing hormone** from the hypothalamus. Because the cortisol response to a stressor has this chain of steps, each involving secretion, diffusion, circulation, and cell activation, it is much slower than the epinephrine response. Also, many of the actions of cortisol involve changes in gene expression, and that takes time.

Turning off the cortisol response is as important as turning it on. A study of stress in rats showed that old rats could turn on their stress responses as effectively as young rats, but they had lost the ability to turn them off as rapidly. As a result, they suffered from the well-known consequences of stress seen in humans: ulcers, cardiovascular problems, strokes, impaired immune system function, and increased susceptibility to cancers and other diseases. Further research showed that the stress responses are turned off

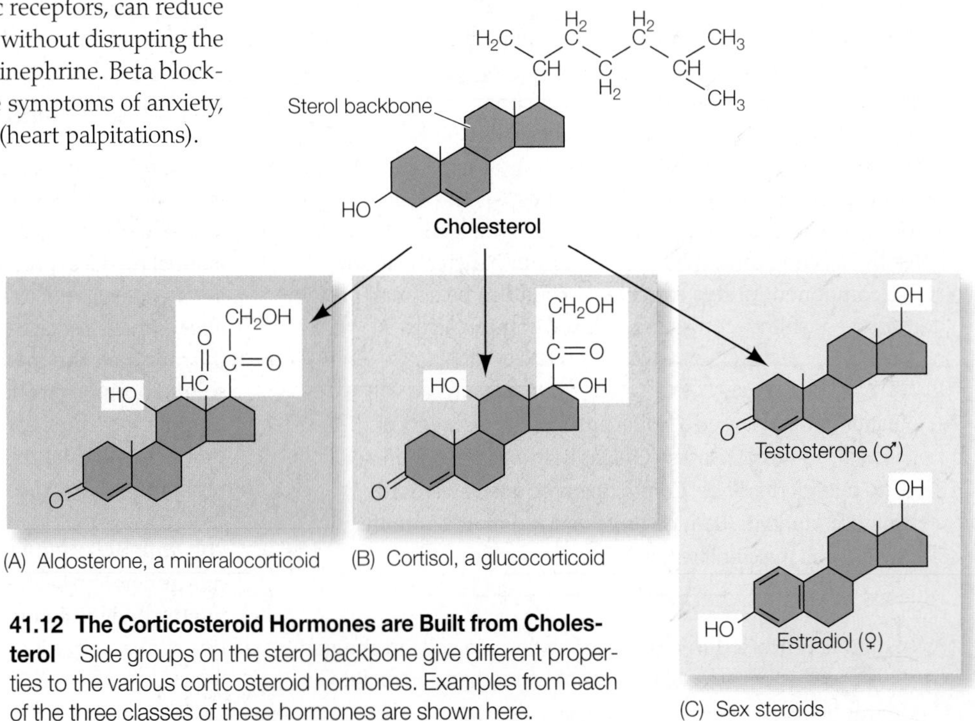

41.12 The Corticosteroid Hormones are Built from Cholesterol Side groups on the sterol backbone give different properties to the various corticosteroid hormones. Examples from each of the three classes of these hormones are shown here.

by the negative feedback action of cortisol on cells in the brain, which causes a decrease in the release of corticotropin-releasing hormone (see Figure 41.8). Repeated activation of this negative feedback mechanism, either through repeated stress or through prolonged medical use of cortisol, leads to a gradual loss of cortisol-sensitive cells in the brain, and therefore to a decreased ability to terminate stress responses.

The sex steroids are produced by the gonads

The **gonads**—the testes of the male and the ovaries of the female—produce hormones as well as sperm and ova. The male steroids are collectively called **androgens**, and the dominant one is *testosterone*. The female steroids are **estrogens** and **progesterone**. The dominant estrogen is *estradiol*, which is made from testosterone. Thus, males and females both synthesize testosterone, but females have an enzyme (aromatase) that converts testosterone to estradiol (**Figure 41.12C**).

The sex steroids have important developmental effects; they determine whether a fetus develops into a female or a male. (A *fetus* is the latter stage of a mammalian embryo; a human embryo is called a fetus from the eighth week of pregnancy to the moment of birth.) After birth, the sex steroids control the maturation of the reproductive organs and the development and maintenance of secondary sexual characteristics, such as breasts and facial hair.

The sex steroids begin to exert their effects in the human embryo in the seventh week of development. Until that time, the embryo has the potential to develop into either sex. In mammals and birds, the instructions for sex determination reside in the genes. In mammals, individuals that receive two X chromosomes normally become females, and individuals that receive an X and a Y chromosome normally become males (see Section 10.4).

These genetic instructions are carried out through the production and action of the sex steroids. In humans, the presence of a Y chromosome normally causes the undifferentiated embryonic gonads to begin producing androgens in the seventh week. In response to the androgens, the reproductive system develops into that of a male. If androgens are not produced at that time, female reproductive structures develop (**Figure 41.13**). In other words, androgens are required to trigger male development in humans, and the default condition is female. The opposite situation exists in birds; male characteristics develop unless estrogens are present to trigger female development.

Changes in control of sex steroid production initiate puberty

Sex steroids have dramatic effects at **puberty**—the time of sexual maturation in humans. Sex steroids are produced at low levels by the juvenile gonads, but their production increases rapidly at the beginning of puberty—around the age of 12 to 13 years. Why does this sudden increase occur?

In the juvenile, as in the adult, the production of sex steroids by the ovaries and testes is controlled by the anterior pituitary tropic hormones **luteinizing hormone** (**LH**) and **follicle-stimulating hormone** (**FSH**), which together are called the **gonadotropins**. The production of these tropic hormones is under the control of the hypothalamic gonadotropin-releasing hormone (GnRH). Before puberty, the gonads can respond to gonadotropins and the pituitary can re-

41.13 The Development of Human Sex Organs The sex organs of early human embryos are similar. Male sex steroids (androgens) promote the development of male sex organs. Without androgen action, female sex organs form, even in genetic males.

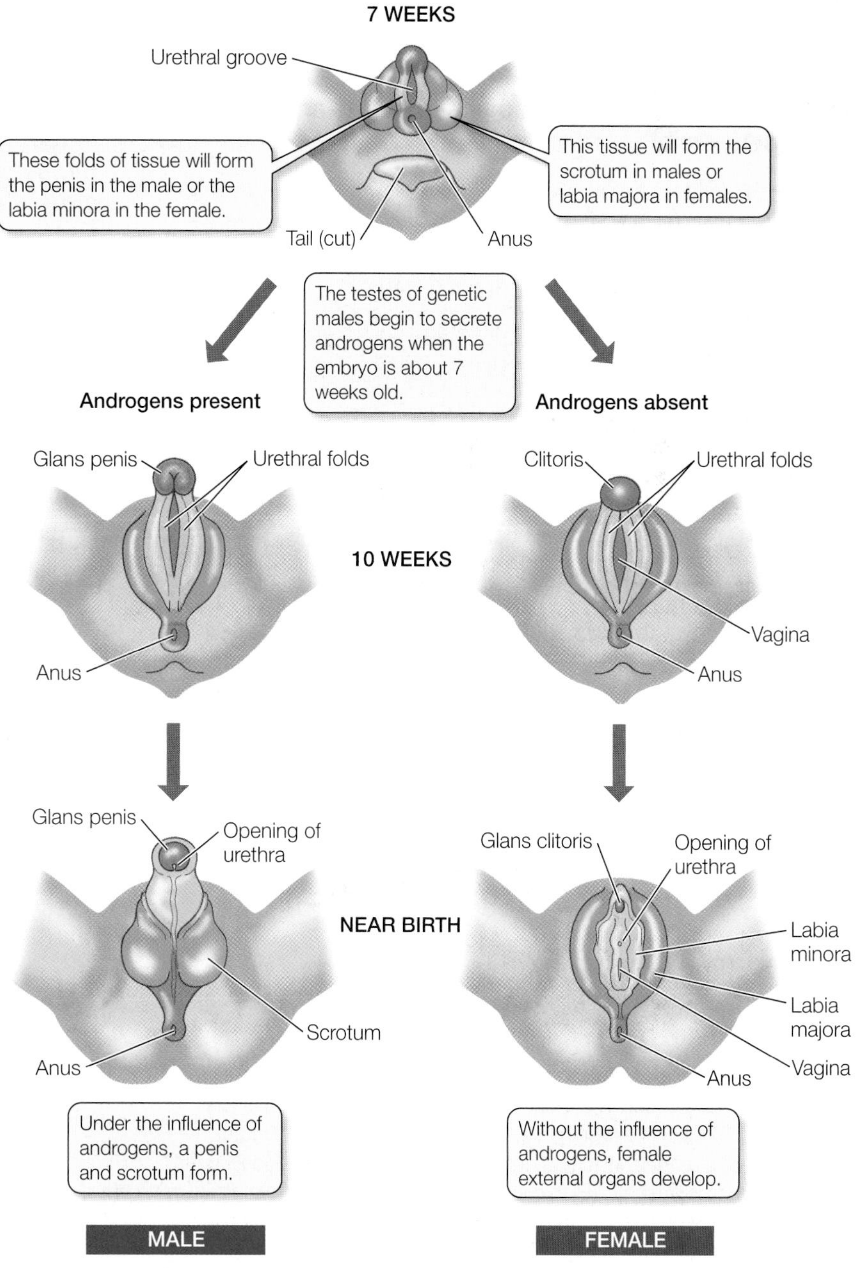

spond to GnRH, but the hypothalamus produces only very low levels of GnRH. Puberty is initiated by a reduction in the sensitivity of hypothalamic GnRH-producing cells to negative feedback from sex steroids and from gonadotropins. As a result, GnRH release increases, stimulating increased production of gonadotropins and hence increased production of sex steroids.

In females, increasing levels of LH and FSH at puberty stimulate the ovaries to begin producing the female sex hormones. The increased circulating levels of these hormones initiate the development of the traits of a sexually mature woman: enlarged breasts, vagina, and uterus; broadened hips; increased subcutaneous fat; growth of pubic hair; and the initiation of the menstrual cycle.

In males, an increasing level of LH stimulates groups of cells in the testes to synthesize testosterone, which in turn initiates the physiological, anatomical, and psychological changes associated with adolescence. The voice deepens, hair begins to grow on the face and body, and the testes and penis grow. As we saw at the beginning of the chapter, androgens also help bones and skeletal muscles grow.

Melatonin is involved in biological rhythms and photoperiodicity

The **pineal gland** is situated between the two hemispheres of the brain and is connected to the brain by a little stalk. It produces the amine hormone **melatonin** from the amino acid tryptophan.

(A)

Plasma melatonin (=)
Light
Dark
Light
Winter (long nights)
Winter
Summer
Light
Dark
Light
Summer (short nights)
Time of day

(B)

41.14 The Release of Melatonin Regulates Seasonal Changes (A) Melatonin is released in the dark and is inhibited by light exposure. The duration of daily melatonin release thus changes as day length (photoperiod) changes, inducing dramatic seasonal physiological changes in some animals. (B) In winter, these Siberian hamsters are white and do not reproduce. In summer, they are mottled brown and breed.

The release of melatonin by the pineal gland occurs in the dark and therefore marks the length of the night. Exposure to light inhibits the release of melatonin.

In vertebrates, melatonin is involved in biological rhythms, including **photoperiodicity**—the phenomenon whereby seasonal changes in day length cause physiological changes in animals. Many species, for example, come into reproductive condition when the days begin to lengthen (**Figure 41.14**). Humans are not strongly photoperiodic, but melatonin in humans may play a role in entraining daily biological rhythms to the daily cycle of light and dark.

The list of hormones is long

We have discussed the major endocrine glands and their hormones in this chapter, but many more hormones exist. As we discuss the organ systems of the body in the chapters that follow, we will frequently describe hormones that their tissues produce as well as hormones that control their functions.

41.3 RECAP

In mammals, the major endocrine glands include the hypothalamus, pituitary, thyroid, parathyroid, pancreas, adrenals, gonads, and pineal gland. Each of these glands secrete, and respond to, hormones that play crucial roles in physiology and development.

- Can you describe how thyroxine is produced and how its production and release are controlled? See p. 884 and Figure 41.9
- How is blood calcium level regulated? See pp. 885–886 and Figure 41.10
- What are the hormonal bases for the two forms of diabetes? See p. 887
- Can you describe the changes in the feedback control of sex steroids that result in puberty? See pp. 889–890

Studies of endocrine systems are not easy. Many hormones are released in very small quantities, and some disappear from the extracellular fluids rapidly. A hormone's receptors may be found on diverse cells around the body, and those different cells can respond in different ways to the same hormone. How have we overcome these difficulties to learning how hormones work?

41.4 How Do We Study Mechanisms of Hormone Action?

We can break the study of hormone actions into different sets of problems. First, we must be able to detect, identify, and measure hormones. Second, we must be able to identify and characterize the receptors for the hormones. Third, we must understand the signal transduction pathways activated by hormones in different tissues.

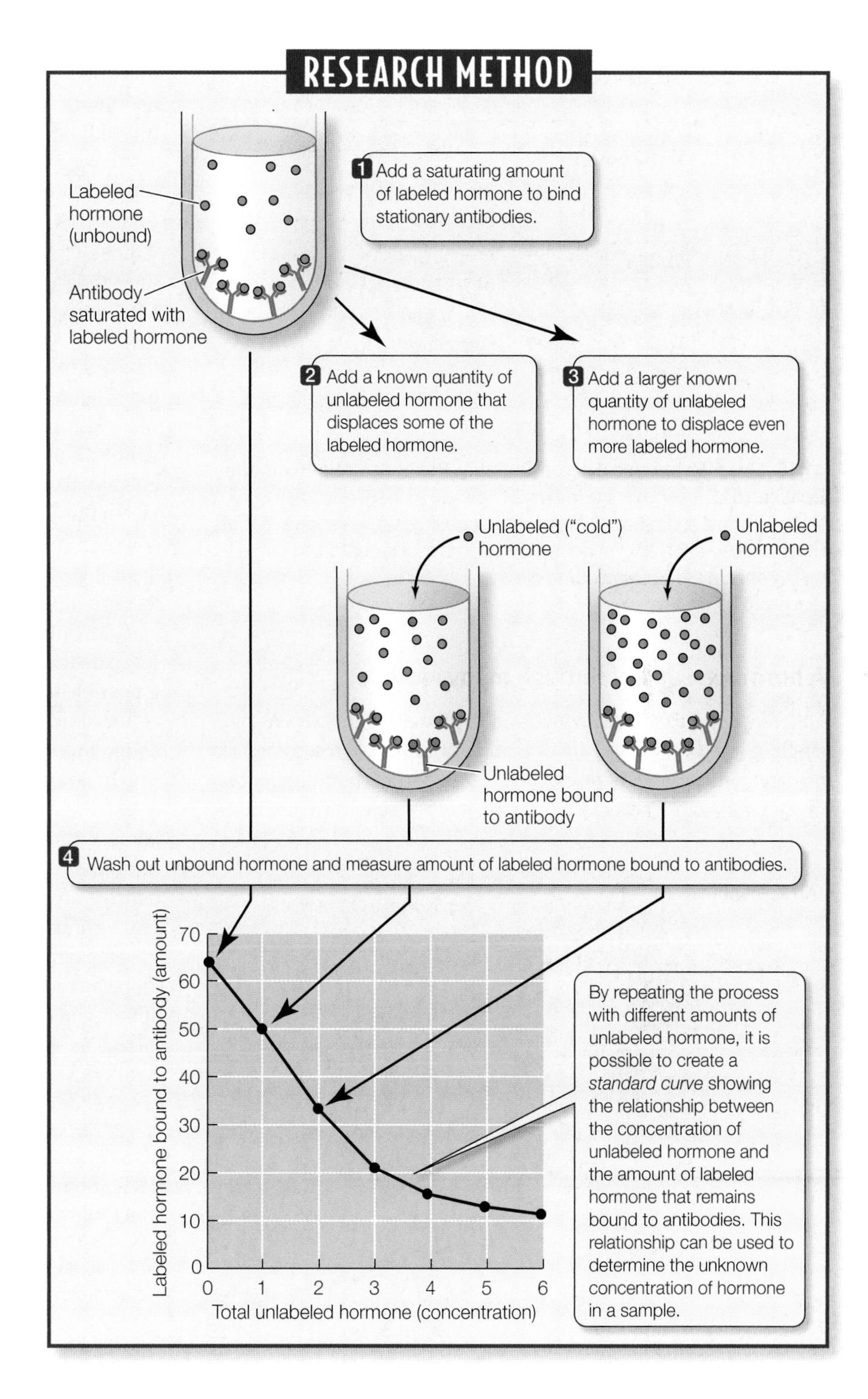

41.15 An Immunoassay Measures Hormone Concentration To develop an immunoassay, the immune response of an animal is used to produce antibodies to the hormone to be measured. A purified sample of the hormone is labeled in some way. A known amount of the antibody is mixed with enough labeled hormone to saturate all of the antibody-binding sites (step 1), and the excess is washed away. When a sample of unlabeled ("cold") hormone is added to the mix, it displaces some of the labeled hormone. The decrease in the amount of label is a measure of the amount of cold hormone added (step 2). By repeating the process using larger known amounts of cold hormone (step 3), a standard curve is created that can be used to determine the amount of hormone in an unknown sample (step 4).

Hormones can be detected and measured with immunoassays

As we have seen, testosterone has many dramatic and diverse effects, yet its concentration in the blood of adult human males is only about 30 to 100×10^{-9} g/ml; that is 30 to 100 billionths of a gram per milliliter. To measure hypothalamic releasing hormones requires calibrations in the range of *trillionths* of a gram per milliliter.

The ability to detect and measure infinitesimal quantities of hormone was an important breakthrough Rosalyn Yalow developed a method called *radioimmunoassay* because it used radioactive labels (she used radioactive isotopes of iodine) to follow interactions between antigens (the hormone to be measured) and antibodies made to that antigen. Today we are more likely to use nonradioactive labels, so we the technique is called simply **immunoassay** (**Figure 41.15**). Being able to measure hormones in the blood made it possible to study many important hormonal mechanisms, and in 1977 Yalow shared the Nobel prize with Guillemin and Schally, who used her technique in their work on hypothalamic neurohormones (see p. 883).

Hormones are not simple on–off switches, and an important characteristic of a hormone is the time course over which it acts. This time course can be measured by the hormone's *half-life* in the blood (defined as the length of time it takes for one-half of the hormone molecules to be depleted. Soon after endocrine cells are stimulated to secrete their hormone, the hormone reaches its maximum concentration in the blood. By taking subsequent blood samples, researchers can determine how long it takes for the circulating hormone to drop to half of that maximum concentration. The fight-or-flight response to epinephrine, for example, is relatively quick in its onset and termination; the half-life of epinephrine in the blood is only 1–3 minutes. The effects of other hormones, such as cortisol and thyroxine, are expressed over much longer periods, and their half-lives are on the order of days or weeks.

Immunoassays have also facilitated the measurement of dose-response curves. Whether one is studying a drug or a natural hormone for therapeutic use, it is critical to know the sensitivity of the body to the drug or hormone. Drugs may stimulate or block release of hormones. Being able to measure the concentrations of the hormones or other molecules in the blood makes it possible to construct dose–response curves that help physicians adjust dosages appropriately. (**Figure 41.16**).

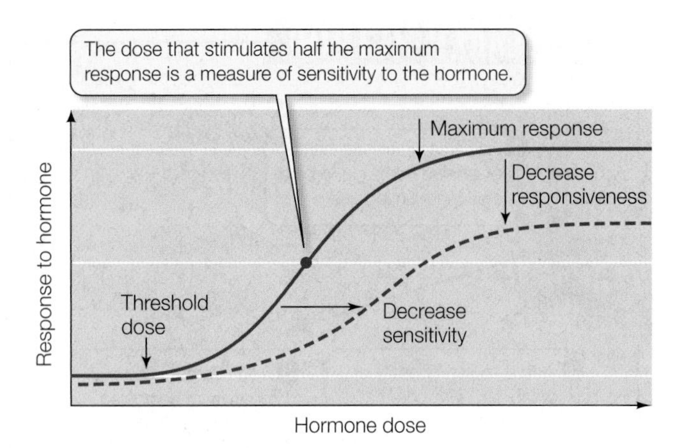

41.16 Dose–Response Curves Quantify Response to a Hormone Between the threshold and maximum values, a dose–response curve frequently has an S shape. Anything that changes the responsiveness of a system—such as a change in the number of receptors in target cells—affects the position of the curve.

A hormone can act through many receptors

Different receptors may be involved in mediating the actions of a single hormone, thus the investigation of hormone receptors has become a major area of research. Because of the slight differences in receptors for a particular hormone, it is possible to create drugs that are very selective in blocking or stimulating specific responses. Receptors have mostly been identified, isolated, and purified through biochemical separation techniques. For example, a hormone can be bound to a substrate such as resin beads packed into a glass column. When an extract of cells suspected of containing receptors to that hormone is run through the column, the receptors bind to the hormone molecules on the beads. The hormone–receptor complexes can be washed off of the beads and the receptors isolated. This technique is called **affinity chromatography**.

As more receptors are isolated and characterized, researchers discover that they frequently exist in families with common structural features. The common features result from common nucleotide sequences in their genes. Genomic analyses have led to the discovery of many receptors. Investigators "scan" the genome for sequences that bear homologies to known receptor gene sequences. When they get a "hit," they have found a candidate gene for a new receptor. They can then identify the molecule to which it binds (its ligand), describe its localization within the body, and characterize its physiological effects.

Knowing the molecular identity of receptors and being able to measure their concentration with immunoassay procedures makes it possible to study their regulation. We saw above that the release of hormones can be under negative feedback control. Similarly, the abundance of receptors for a hormone can be under feedback control. In some cases, continuous high levels of a hormone can decrease the number of its receptors, a process known as **downregulation**. **Upregulation** of receptors can occur when the levels of hormone secretion are suppressed. The regulation of receptor abundance is an important mechanism controlling sensitivity of the system to hormonal signaling.

An example of downregulation of a receptor occurs in type II diabetes mellitus, which is characterized by elevated levels of circulating insulin but a loss of insulin receptors. Although genetic factors are likely involved, a possible immediate cause of the disease is an overstimulation of pancreatic release of insulin by excessive carbohydrate intake, which leads to downregulation of the

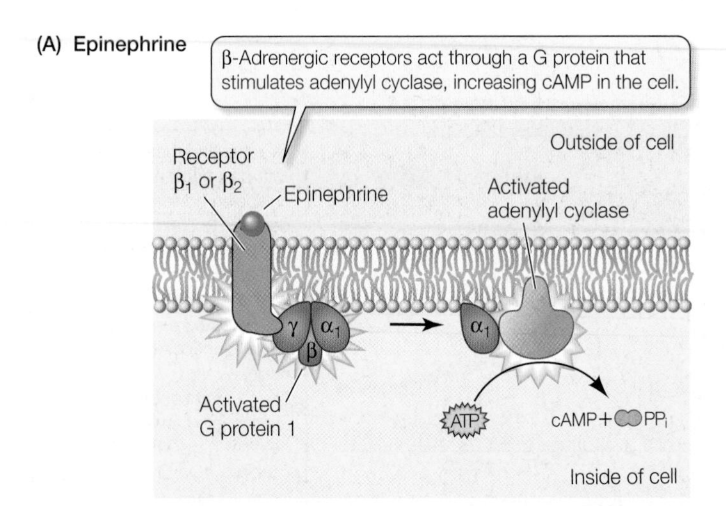

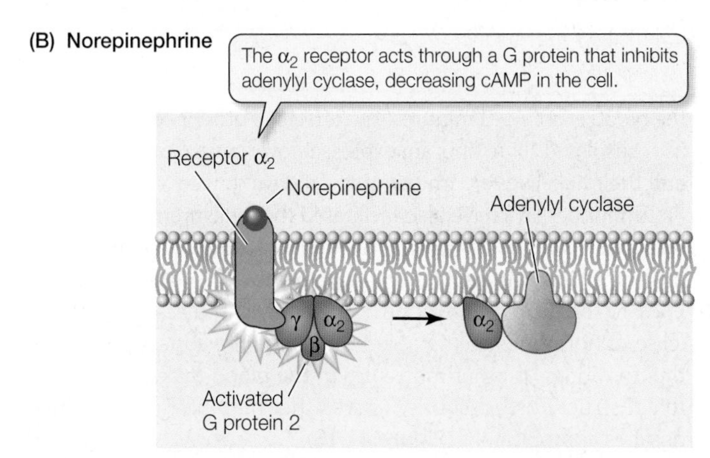

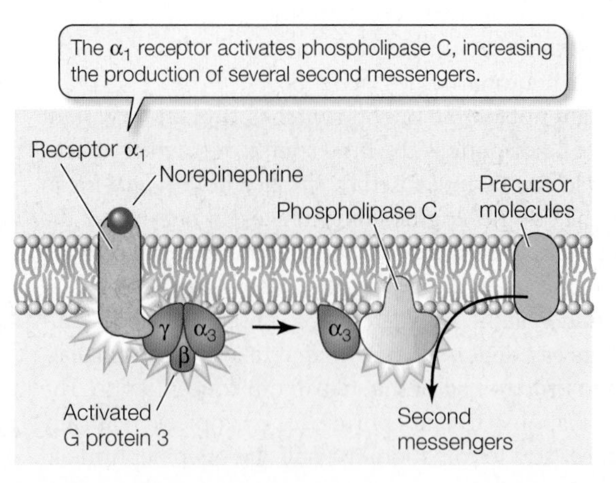

41.17 Some Hormones Can Activate a Variety of Signal Transduction Pathways Epinephrine and norepinephrine bind to G protein-linked adrenergic receptors that act through different signal transduction pathways. Epinephrine acts equally on both α- and β-adrenergic receptors; norepinephrine acts mostly on α-adrenergic receptors

insulin receptors. An example of upregulation occurs when someone has been on a regular dose of beta blockers (see page 888). As the activity of the beta receptors is blocked over time, more receptors are produced. If the person goes off the medication suddenly, the beta receptor effects are amplified, resulting in heightened anxiety. Thus dosage changes in the long-term use of such medications are usually gradual and carefully supervised.

A hormone can act through different signal transduction pathways

A hormone may affect only certain cells, and its effects can differ greatly between different cell types. The cell- or tissue-specific nature of hormone action is due to the fact that only cells with appropriate receptors can respond to the hormone. The diversity of actions of a hormone in different cells, however, arises because receptors can be linked to different signal transduction pathways. For example, the two types of receptors mentioned above for epinephrine and norepinephrine (the α-adrenergic and the β-adrenergic receptors) are both cell-surface G protein-linked receptors, but they connect with different signal transduction pathways within cells (**Figure 41.17**). Therefore, the two hormones can elicit different responses, even in the same cells.

Signal transduction pathways also partly explain why hormones secreted in such tiny quantities can have huge physiological effects. Many of these pathways involve signaling "cascades" in which each step amplifies the signal. For example, the binding of the hormone to its receptor might activate an enzyme (such as adenylyl cyclase), which produces multiple second messenger molecules (cAMP). Each of those may in turn activate another enzyme—a kinase—which can phosphorylate and thereby activate a large number of molecules of another enzyme, each of which in turn will catalyze the production of a final product molecule. Thus, a single event of a hormone binding to its receptor might result in a million or more molecules of a final product. An example is the response of liver cells to epinephrine (see Figure 15.18).

41.4 RECAP

Studies of mechanisms of hormone action involve being able to measure hormone concentrations, identify and characterize hormone receptors, and investigate signal transduction pathways.

- Can you describe what an immunoassay is? See p. 891 and Figure 41.15
- How can you identify receptors to a particular hormone? See p. 892
- Explain how the same hormone can induce different responses in different tissues. See p. 892 and Figure 41.17

CHAPTER SUMMARY

41.1 What are hormones and how do they work?

Endocrine cells secrete chemical messages called **hormones**, which bind to receptors on or in **target cells**. In some cases endocrine cells are aggregated into endocrine glands.

Paracrine hormones diffuse to targets near the site of secretion. **Autocrine** hormones influence the cell that secretes them. Most hormones are delivered to target cells by the circulatory system. Review Figure 41.1

Two diffusible substances, prothoracicotropic hormone and ecdysone, control molting in insects. A third hormone, juvenile hormone, prevents maturation. When an insect stops producing juvenile hormone, it molts into an adult. Review Figures 41.2 and 41.3, Web/CD Tutorial 41.1

Most hormones are either peptides, proteins, steroids, or amines. Peptide and protein hormones and some amines are water-soluble; steroids and some amines are lipid-soluble.

Receptors for water-soluble hormones are on the cell surface. Receptors for lipid-soluble hormones are inside the cell.

Hormones cause different responses in different target cells. Review Figure 41.4

41.2 How do the nervous and endocrine systems interact?

Some vertebrate hormones are released by discrete endocrine glands. Many other hormones are produced and released by endocrine cells incorporated into organs of the body. Review Figure 41.5, Web/CD Activity 41.1

The **pituitary** gland is an interface between the brain and endocrine system. The anterior pituitary develops from embryonic mouth tissue; the posterior pituitary develops from the brain.

The posterior pituitary secretes two **neurohormones**: **antidiuretic hormone** and **oxytocin**. The anterior pituitary secretes **tropic hormones** (**thyrotropin**, **corticotropin**, **luteinizing hormone**, and **follicle-stimulating hormone**) as well as **growth hormone**, **prolactin**, **melanocyte-stimulating hormone**, **endorphins**, and **enkephalins**.

The anterior pituitary is controlled by neurohormones produced by cells in the **hypothalamus** and transported through **portal blood vessels** to the anterior pituitary. Review Figures 41.6 and 41.7, Web/CD Tutorial 41.2

Hormone release in the hypothalamus–pituitary–endocrine gland system is controlled by negative feedback. Review Figure 41.8

41.3 What are the major mammalian endocrine glands and hormones?

The thyroid gland is controlled by **thyrotropin** and secretes **thyroxine**, which controls cell metabolism. Review Figure 41.9

The level of calcium in the blood is regulated by three hormones. **Calcitonin** from the thyroid lowers blood calcium by promoting bone deposition. **Parathyroid hormone** raises blood calcium by promoting bone turnover and decreased calcium excretion. **Vitamin D** promotes calcium absorption from the digestive tract. Review Figure 41.10, Web/CD Tutorial 41.3

CHAPTER SUMMARY

The pancreas secretes three hormones. **Insulin** stimulates glucose uptake by cells and lowers blood glucose, **glucagon** raises blood glucose, and **somatostatin** slows the rate of nutrient processing.

The **adrenal gland** has two portions, one within the other. The hormones of the adrenal medulla, epinephrine and norepinephrine, cause fight-or-flight responses such as stimulating the liver to supply glucose to the blood. Review Figure 41.11

The adrenal cortex produces three classes of corticosteroids: glucocorticoids, mineralocorticoids, and small amounts of sex steroids. Review Figure 41.12

Aldosterone is a mineralocorticoid that stimulates the kidney to conserve sodium and excrete potassium. **Cortisol** is a glucocorticoid that decreases glucose utilization by most cells.

Sex hormones (androgens in males, estrogens and progesterone in females) are produced by the gonads in response to tropic hormones. Sex hormones control sexual development, secondary sexual characteristics, and reproductive functions. Review Figure 41.13

The pineal hormone **melatonin** is involved in controlling biological rhythms and photoperiodism. Review Figure 41.14

41.4 How do we study mechanisms of hormone action?

Immunoassays are used to measure concentrations of hormones and receptors. Review Figure 41.15

The sensitivity of a cell to hormones can be altered by **up- or down-regulation** of the receptors in that cell.

The response of a cell to a hormone depends on its receptors and the signal transduction pathways those receptors activate. Review Figure 41.17

See Web/CD Activity 41.2 for a concept review of this chapter.

SELF-QUIZ

1. Before puberty
   a. the pituitary secretes luteinizing hormone and follicle-stimulating hormone, but the gonads are unresponsive.
   b. the hypothalamus does not secrete much gonadotropin-releasing hormone.
   c. males can stimulate massive muscle development through a vigorous training program.
   d. testosterone plays no role in development of the male sex organs.
   e. genetic females will develop male genitals unless estrogen is present.
2. Both epinephrine and cortisol are secreted in response to stress. Which of the following statements is also true for *both* of these hormones?
   a. They act to increase blood glucose availability.
   b. Their receptors are on the surfaces of target cells.
   c. They are secreted by the adrenal cortex.
   d. Their secretion is stimulated by corticotropin.
   e. They are secreted into the blood within seconds of the onset of stress.
3. Growth hormone
   a. can cause adults to grow taller.
   b. stimulates protein synthesis.
   c. is released by the hypothalamus.
   d. can be obtained only from cadavers.
   e. is a steroid.
4. PTH
   a. stimulates osteoblasts to lay down new bone.
   b. reduces blood calcium levels.
   c. stimulates calcitonin release.
   d. is produced by the thyroid gland.
   e. is released when blood calcium levels fall.
5. Steroid hormones
   a. are produced only by the adrenal cortex.
   b. have only cell surface receptors.
   c. are water-soluble.
   d. act by altering the activity of proteins in the target cell.
   e. act by altering gene expression in the target cell.
6. The hormone ecdysone
   a. is released from the posterior pituitary.
   b. stimulates molting in insects.
   c. maintains an insect in larval stages unless PTTH is present.
   d. stimulates the secretion of juvenile hormone from the prothoracic glands.
   e. keeps the insect exoskeleton flexible to permit growth.
7. The posterior pituitary
   a. synthesizes oxytocin.
   b. is under the control of hypothalamic releasing neurohormones.
   c. secretes tropic hormones.
   d. secretes neurohormones.
   e. is under feedback control by thyroxine.
8. Which of the following contributes to the development of goiter?
   a. Inadequate iodine in the diet
   b. Autoimmune antibodies that stimulate the TSH receptor
   c. Lack of feedback from circulating T_3 and T_4
   d. Overproduction of thyroglobulin
   e. All of the above
9. Which of the following is a likely cause of diabetes?
   a. Overproduction of insulin by beta cells of the pancreas
   b. Loss of alpha cells of the pancreas
   c. Loss of insulin receptors
   d. Overproduction of glucagon
   e. Loss of receptors for somatostatin
10. Which statement is true of all hormones?
    a. They are secreted by glands.
    b. They have receptors on cell surfaces.
    c. They may stimulate different responses in different cells.
    d. They target cells that are distant from their site of release.
    e. When the same hormone occurs in different species, it has the same action.

FOR DISCUSSION

1. Explain how both hyperthyroidism and hypothyroidism can cause goiter. Refer to the roles of the hypothalamus and the pituitary in your answer.
2. There are several apparently enigmatic aspects of the role of PTH in regulating blood calcium. First, PTH raises blood calcium levels, yet osteoclasts may not have PTH receptors. Second, parathyroid cells have calcium-sensing receptors on their plasma membranes, and these receptors are activated by rising levels of blood calcium. Third, bone consists of calcium and phosphate salts, yet PTH causes blood phosphate to decline. Explain these various aspects of PTH actions.
3. Various side effects of anabolic steroid use were mentioned in this chapter. Some of these effects are due to the direct action of the steroid, but others are due to the negative feedback action of the steroid. Discuss an example of each and explain possible mechanisms.
4. Compare the characteristics you would expect of a hormone signaling system that controls a short-term process, such as digestion, with the characteristics you would expect of a hormone signaling system that controls a long-term process, such as development.
5. In the perpetual war between agriculturists and insects, a new weapon has been developed: a chemical similar to juvenile hormone. Explain how this chemical could be used as an insecticide. What factors do you think should be considered before it is released indiscriminately into the environment?

FOR INVESTIGATION

Each spring male deer grow antlers that they use in male–male competition for mates. Each fall they shed those antlers. Although females of most deer species do not grow antlers, the caribou are an exception. Female caribou grow antlers in the spring but do not shed them in the fall. (Over the winter, they use their antlers to defend patches of food from males.) In addition, newborn caribou begin to grow antlers after birth. Assuming that antler growth is controlled by hormones, how would you investigate the control of antler growth in caribou? What hormonal assays might you want to use? What hypotheses would you test with respect to the relationship between time of year and antler growth?

CHAPTER 42 Animal Reproduction

Explosive sex

Producers of valuable honey and pollinators of many crucial plants, the common honeybee, *Apis mellifera*, has been the subject of study and fascination for humans throughout recorded history. The unique sex life of these social insects is among their most intriguing aspects.

Over 99 percent of female honeybees do not reproduce. They exist to help one female—the queen—reproduce. At any given time, there is only one queen in a bee hive. She lays all of the eggs, which is a full-time job. She also determines whether an egg will be fertilized or not. Fertilized eggs develop into females; unfertilized eggs develop into males. Wait, you might think: it's the *male* that fertilizes the egg! Yes, it is the male's sperm that fertilizes the egg, but in this case a female—the queen bee—controls sperm delivery.

Eventually, every hive must have a new queen. Sometimes the old queen dies. Other times, in a phenomenon known as *swarming*, the queen and a retinue of workers leave their current hive to start a new one. Whether the queen dies or leaves, a new queen must be produced. The remaining worker bees enlarge a few cells in the honeycomb that contain fertilized eggs laid by the old queen. The larvae that hatch from those eggs are fed special food that stimulates their growth and development into prospective queens.

The first queen to pupate and emerge from her royal chamber kills any other aspiring queens. She then leaves the hive for her mating flight. Males from all around get the message that a virgin queen is available and congregate around her. While in flight, she will mate with 15–20 males, and each coupling is an event. After a male manages to insert his penis into her vagina, he literally explodes, leaving not only his sperm but also his sex organs (which drop out later) in the female. The process is repeated with other males, all of whom die as a result. The queen returns to the hive with a lifetime supply of sperm and sets about laying eggs. She will live for about 2 years and lay as many as 3000 eggs each day. If she releases sperm, the eggs will be fertilized and develop into females who will devote their lives to feeding her, maintaining the hive, foraging for nectar and pollen, and raising their sisters. An unfertilized egg develops into a male, who will hang around the hive doing nothing useful until he takes off to woo a virgin queen—an unlikely

A Unique Reproductive Strategy Among honeybees and some other insects, the only reproductive female is the queen, seen here in the center as she deposits eggs in the comb. The female workers who attend her are sterile.

Swarming Will Mean a New Queen Bee When a honeybee colony swarms, as seen here, the queen leaves the hive and takes a retinue of workers with her. A new queen will emerge to take over the old hive and a few males will get to perform their one brief function—fertilizing a virgin queen—before they die.

event. It is thus in the queen's best interest to limit the number of males she produces.

Natural selection has resulted in some amazing adaptations, none more so than those involved in reproduction. Sexual or asexual, bizarre or otherwise, the practices necessary for the continuation of a species must evolve to be advantageous for that species.

IN THIS CHAPTER we examine the diverse ways in which animals produce offspring. We first examine asexual mechanisms of reproduction, in which only a single parent is involved. We then turn to sexual reproduction, in which an egg and a sperm unite to create a new diploid individual. Next we focus on the anatomy, function, and endocrine control of the human reproductive system, as well as the technologies used to limit or to enhance human fertility. We end this chapter with a discussion of human sexual health and sexually transmitted diseases.

CHAPTER OUTLINE

42.1 How Do Animals Reproduce Without Sex?

Sexual reproduction is a nearly universal trait in animals, although many species can also reproduce asexually and some reproduce only asexually. Offspring produced asexually are genetically identical to one another and to their parents. Asexual reproduction is efficient because no mating is required. Furthermore, asexual populations can use resources efficiently because all individuals in the population can convert resources into offspring. However, asexual reproduction does not generate the kind of genetic diversity produced by sexual reproduction, as described in Chapter 9. Genetic diversity is the raw material that enables natural selection to shape adaptations in response to environmental change, and a lack of genetic diversity can be disadvantageous to species in changing environments.

A variety of animals, mostly invertebrates, reproduce asexually. They tend to be species that are sessile and cannot search for mates or that live in sparse populations and rarely encounter potential mates. Asexually reproducing species are likely to be found in relatively constant environments where genetic diversity is less important for species success. Three common modes of asexual reproduction are *budding, regeneration,* and *parthenogenesis.*

Budding and regeneration produce new individuals by mitosis

Many simple multicellular animals produce offspring by **budding**. New individuals form as outgrowths or buds from the bodies of older animals. A bud grows by mitotic cell division, and the cells differentiate before the bud breaks away from the parent (**Figure 42.1A**). The bud is genetically identical to the parent, and it may grow as large as the parent before it becomes independent.

Regeneration is usually thought of as the replacement of damaged tissues or lost limbs, but in some cases pieces of an organism can regenerate complete individuals. Echinoderms, for example, have remarkable abilities to regenerate. If sea stars are cut into pieces, each piece that includes a portion of the central disc grows into a new animal (**Figure 42.1B**). In the early 1900s oyster fishermen in Narragansett Bay tried to eliminate the sea stars (starfish) that were preying on their oysters. Whenever they encountered a sea star, they chopped it up with knives

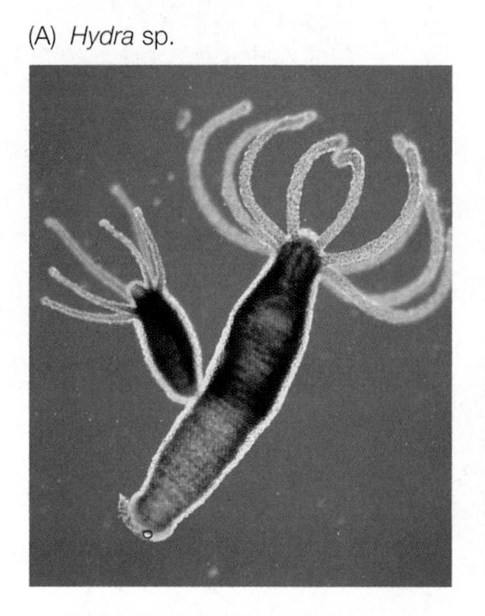

42.1 Asexual Reproduction in Animals (A) Budding: A new individual forms as an outgrowth from an adult hydra. (B) Regeneration: The single severed arm of a mature sea star is regenerating an entire animal.

and threw it back into the water. As a result, the sea star population increased explosively.

Regeneration frequently results when an animal is broken by an outside force. A storm, for example, can cause heavy surf that breaks colonial cnidarians such as corals. Pieces broken off the colony can regenerate into new colonies. In some species, breakage occurs in the absence of external forces. Some species of segmented marine worms develop segments with rudimentary heads bearing sensory organs, then break apart. Each fragmented segment forms a new worm.

Parthenogenesis is the development of unfertilized eggs

Not all eggs must be fertilized to develop. A common mode of asexual reproduction in arthropods is the development of offspring from unfertilized eggs. This phenomenon, called **parthenogenesis**, also occurs in some species of fish, amphibians, and reptiles. Most species that reproduce parthenogenetically also engage in sexual reproduction or sexual behavior at other times.

In some species, parthenogenesis is part of the mechanism that determines sex. As we saw at the beginning of this chapter, in honey bees (as well as in most ants and wasps), males develop from unfertilized eggs and are haploid. Females develop from fertilized eggs and are diploid. Most females are sterile workers, but a select few become fertile queens. After a queen mates, she has a supply of sperm that she controls, enabling her to produce either fertilized or unfertilized eggs. Thus the queen determines when and how much of the colony resources are expended on males.

Parthenogenetic reproduction in some species requires sexual behavior even though sperm are not delivered to the female reproductive tract and eggs are not fertilized. One case that has been investigated extensively by David Crews and his students at the University of Texas is parthenogenetic reproduction in a species of whiptail lizard. This species has no males, but females can act as males, engaging in all aspects of courtship display and mating, although no sperm are produced or transferred (**Figure 42.2**). Whether a specific female acts as a female or as a male depends on cyclical hormonal states. When estrogen levels are high, she acts as a female. When her progesterone level peaks, she acts as a male. The stimulation resulting from the sexual activity triggers the release of eggs from the ovary.

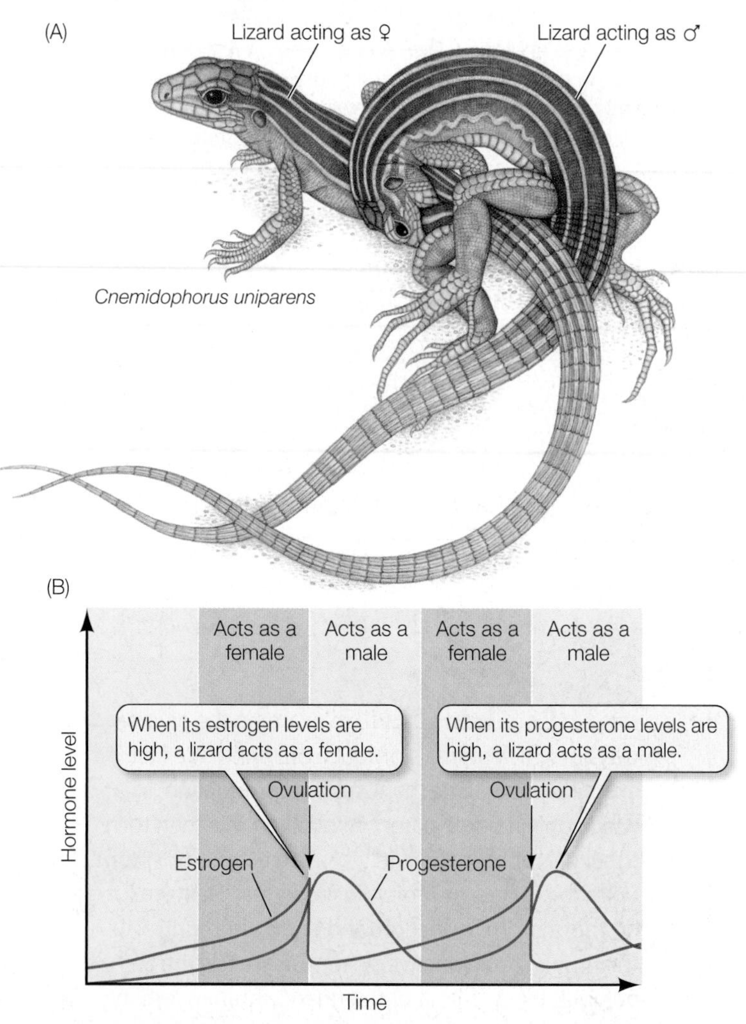

42.2 Sexual Behavior May Be Required for Asexual Reproduction (A) Parthenogenetic whiptail lizards are all female, but take turns acting the male role in reproductive behavior. (B) The stage of the ovarian cycle determines the role an individual whiptail plays.

42.1 RECAP

Most animals reproduce sexually, but many can also or can only reproduce asexually through budding, regeneration, or parthenogenesis.

- Can you explain why asexual reproduction might be disadvantageous for an animal living in a changing environment? See p. 897
- Do you understand how parthenogenesis can be related to sex determination? See p. 898

Asexual reproduction is an efficient way to use resources. However, the fact that sexual reproduction produces genetic diversity must be a tremendous advantage, because most species reproduce sexually.

42.2 How Do Animals Reproduce Sexually?

A large portion of the time and energy budgets of sexually reproducing animals goes into mating, which exposes them to predation, can result in physical damage, and detracts from other useful activities, such as feeding and caring for existing offspring. Furthermore, mating requires that resources be used to maintain a large population of males that do not bear offspring. Despite all these disadvantages, the production of genetic diversity is an overwhelming evolutionary advantage resulting from sexual reproduction.

Sexual reproduction requires the joining of two haploid sex cells to form a diploid individual. These haploid cells, or *gametes*, are produced through **gametogenesis**, a process that involves meiotic cell divisions. Two events in meiosis contribute to genetic diversity: *crossing over* between homologous chromosomes and the *independent assortment* of chromosomes (see Sections 9.6 and 10.1). Sexual reproduction itself also contributes to genetic diversity. The genetic variation among the gametes of a single individual and the genetic variation between any two parents produce an enormous potential for genetic variation between any two offspring of a sexually reproducing pair of individuals.

Sexual reproduction in animals consists of three fundamental steps:

- Gametogenesis (making gametes)
- Mating (getting gametes together)
- Fertilization (fusing gametes)

The process of gametogenesis is very similar across animal species. Processes of fertilization are also rather similar in widely different species. Therefore, while our discussion of gametogenesis will focus generally on mammals, and our discussion of fertilization will feature sea urchins, the facts would not be dramatically different were we to consider different groups of animals. Adaptations for mating, in contrast, show incredible anatomical, physiological, and behavioral diversity across species. Thus, we will take our topics out of order by considering gametogenesis and fertilization first before turning to the more diverse topic of mating and finally discussing human reproduction in detail.

Gametogenesis produces eggs and sperm

Gametogenesis occurs in the **gonads**, which are **testes** (singular *testis*) in males and **ovaries** in females. The tiny gametes of males, called **sperm**, move by beating their flagella. The larger gametes of females, called eggs or **ova** (singular *ovum*), are nonmotile.

Gametes are produced from **germ cells**, which have their origin in the earliest cell divisions of the embryo and remain distinct from the rest of the body. All other cells of the embryo are called *somatic cells*. Germ cells are sequestered in the body of the embryo until its gonads begin to form. The germ cells then migrate to the developing gonads, where they take up residence and proliferate by mitosis, producing **spermatogonia** (singular *spermatogonium*) in males and **oogonia** (singular *oogonium*) in females. Spermatogonia and oogonia, which are diploid, multiply by mitosis, eventually producing **primary spermatocytes** and **primary oocytes**.

Meiosis, the next step in gametogenesis, reduces the chromosomes to the haploid number, and the resulting haploid cells eventually mature into sperm and ova. (You may want to review the discussion of meiosis in Section 9.5 before reading further.) Although the steps of meiosis are similar in males and females, gametogenesis differs between the sexes.

SPERMATOGENESIS The initial proliferation of male germ cells into spermatogonia proceeds by mitosis in the embryo. As illustrated in **Figure 42.3A**, primary spermatocytes then undergo the first meiotic division to form **secondary spermatocytes**. The second meiotic division produces four haploid **spermatids** for each primary spermatocyte that enters meiosis. In mammals, the progeny of primary spermatocytes remain connected by *cytoplasmic bridges* after each division.

One reason that mammalian spermatocytes remain in cytoplasmic contact throughout their development is the asymmetry of sex chromosomes in males. Half the secondary spermatocytes receive an X chromosome, the other half a Y chromosome. The Y chromosome contains fewer genes than the X chromosome, and some of the products of genes found only on the X chromosome are essential for spermatocyte development. By remaining in cytoplasmic contact, all four spermatocytes can share the gene products of the X chromosomes, although only half of them have an X chromosome.

A spermatid bears little resemblance to a mature sperm. Through further differentiation, however, the spermatid becomes compact, streamlined, and motile. We will look at the differentiation of human sperm in more detail below.

OOGENESIS Oogonia, like spermatogonia, proliferate through mitosis (**Figure 42.3B**). The resulting primary oocytes immediately enter prophase of the first meiotic division. In many species, including humans, the oocyte experiences developmental arrest at this point and may remain so for days, months, or years. In the human female, this period of arrest is at least 10 years (i.e., until puberty) and some primary oocytes may remain in prophase I for up to 50 years (i.e., until menopause)! In contrast, male gametogenesis continues steadily to completion once the primary spermatocyte has differentiated.

During this prolonged prophase I, or shortly before it ends, the primary oocyte grows larger through increased production of

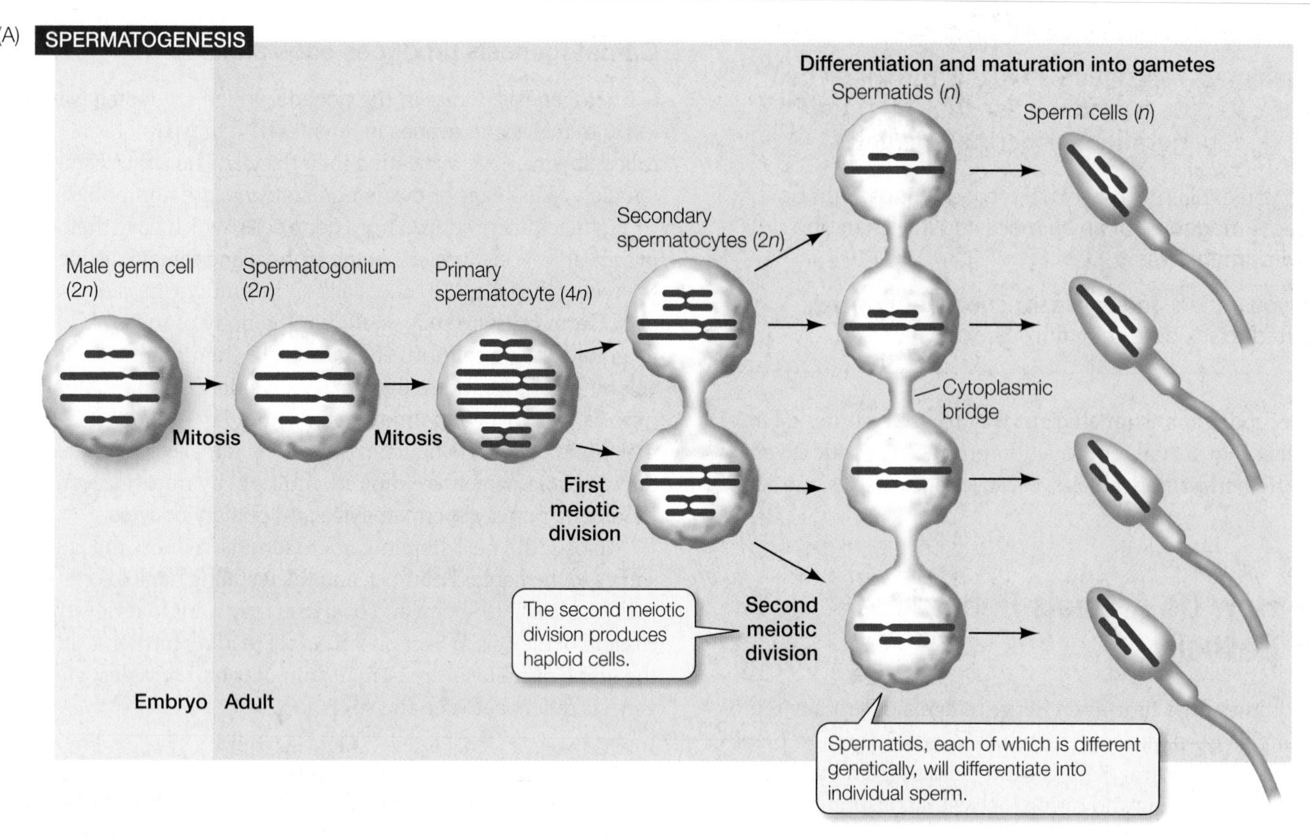

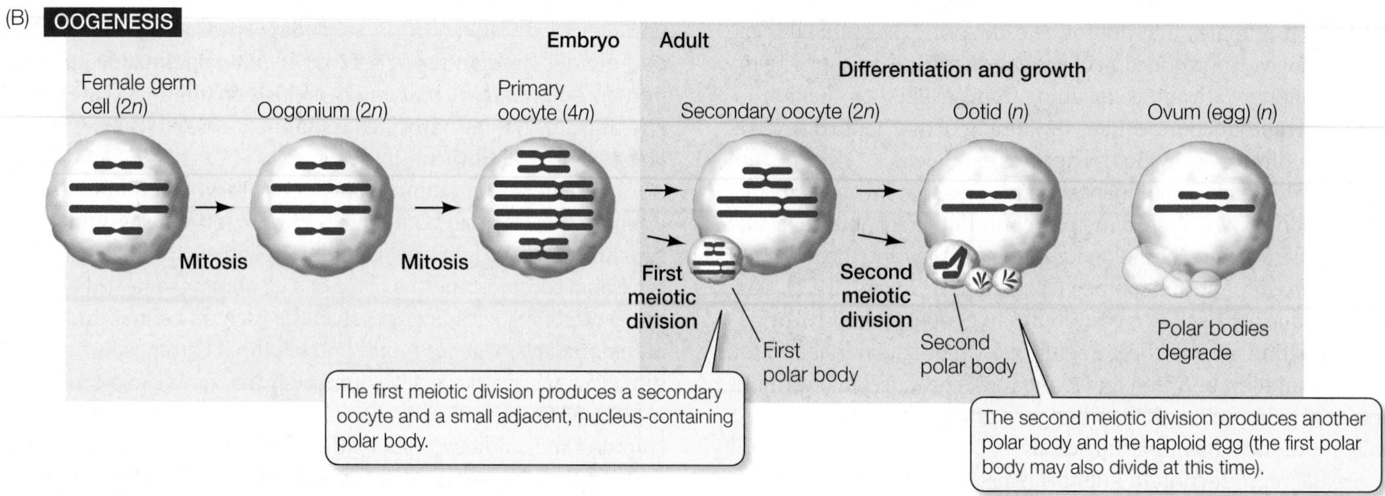

42.3 Gametogenesis Male and female germ cells proliferate by mitosis and produce diploid spermatogonia and oogonia that mature into primary spermatocytes and oocytes before entering meiosis. (A) Spermatogonia continue to divide by mitosis in adults, producing a steady supply of spermatocytes that divide meiotically to produce haploid spermatids, which differentiate into sperm. In many species the progeny of spermatocytes remain in contact through cytoplasmic bridges until the sperm mature. (B) In mammals, oogonia cease division in the embryo, and primary oocytes remain arrested in prophase I of meiosis until they are ovulated and fertilized. Each oocyte will produce one ootid which matures into an ovum.

ribosomes, RNA, cytoplasmic organelles, and energy stores. At this point, the primary oocyte acquires all the energy, raw materials, and RNA that the ovum will need to survive its first cell divisions after fertilization. In fact, the nutrients in the egg must maintain the embryo until it is either nourished by the maternal circulatory system or can feed on its own.

When a primary oocyte resumes meiosis, its nucleus completes the first meiotic division near the surface of the cell. The daughter cells of this division receive grossly unequal shares of cytoplasm. This asymmetry represents another major difference from spermatogenesis, in which cytoplasm is apportioned equally. The daughter cell that receives almost all the cytoplasm becomes the **secondary oocyte**, and the one that receives almost none forms the **first polar body** (see Figure 42.3B).

The second meiotic division of the large secondary oocyte is also accompanied by an asymmetrical division of the cytoplasm. One daughter cell forms the large, haploid **ootid**, which eventually differentiates into a mature ovum, and the other forms the **second polar body**. Polar bodies degenerate, so the end result of oogenesis is only one mature egg for each primary oocyte that entered meiosis. However, that egg is a very large, well-provisioned cell.

A second period of arrested development occurs after the first meiotic division forms the secondary oocyte. The egg may be expelled from the ovary in this condition; however, to simplify discussion, we will call the female gamete an egg once it leaves the ovary. In many species, including humans, the second meiotic division is not completed until the egg is fertilized by a sperm.

Fertilization is the union of sperm and egg

The union of the haploid sperm and the haploid egg in **fertilization** creates a single diploid cell, called a **zygote**, which will develop into an embryo. Fertilization does more, however, than just restore the full genetic complement of the animal. The processes associated with fertilization help eggs and sperm get together, prevent the union of sperm and eggs of different species, and guarantee that only one sperm will enter and activate the egg metabolically. Fertilization involves a complex series of events:

- The sperm and the egg recognize each other.
- The sperm is *activated*, enabling it to gain access to the plasma membrane of the egg.
- The plasma membranes of the sperm and the egg fuse.
- The egg blocks entry of additional sperm.
- The egg is metabolically activated and stimulated to start development.
- The egg and sperm nuclei fuse to create the diploid nucleus of the zygote.

SPECIFICITY IN SPERM–EGG INTERACTIONS Specific recognition molecules mediate interactions between sperm and eggs. These molecules ensure that the activities of sperm are directed toward eggs and not other cells, and they help prevent eggs from being fertilized by sperm from the wrong species. The latter function is particularly important in aquatic species that release eggs and sperm into the surrounding water. The sea urchin is such a species, and its mechanisms of fertilization have been well studied.

The eggs of sea urchins and various other marine invertebrates release chemical attractants that increase the motility of sperm and cause them to swim toward the egg. These chemical attractants are species-specific. For example, eggs of one species of sea urchin release a specific peptide consisting of 14 amino acids. As this peptide diffuses from the egg, it binds to receptors on the sperm of the same species. The sperm respond by increasing their mitochondrial respiration and motility. Before exposure to the peptide, the sperm swim in tight little circles, but after binding the peptide, they swim energetically up the concentration gradient of the peptide until they reach the egg that is releasing it.

When sperm reach an egg, they must get through two protective layers before they can fuse with the egg plasma membrane. The eggs of sea urchins are covered with a **jelly coat**, which surrounds a proteinaceous **vitelline envelope** (**Figure 42.4**). The sperm's assault on these protective layers depends on a membrane-enclosed structure called an acrosome.

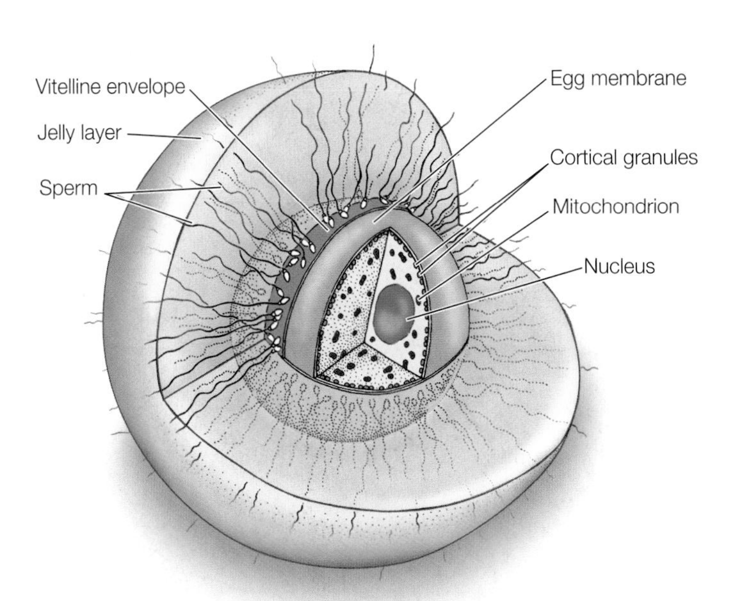

42.4 The Sea Urchin Egg Sea urchin eggs are released into seawater. They are protected by a jelly layer and a proteinaceous vitelline envelope. Sperm must penetrate both to reach the egg plasma membrane. Many sperm attach to the vitelline envelope, but only one will penetrate the egg cell membrane and achieve fertilization.

The **acrosome**, which contains enzymes and other proteins, is located at the front of the sperm head, where it forms a cap over the sperm nucleus. When the sperm makes contact with an egg of its own species, substances in the jelly coat trigger an *acrosomal reaction* which begins with the breakdown of the plasma membrane covering the sperm head and the underlying acrosomal membrane (**Figure 42.5**). The acrosomal enzymes are released, and they digest a hole through the jelly coat.

As a result of the polymerization of actin triggered by the acrosomal reaction, a structure called the *acrosomal process* extends out of the head of the sperm. The acrosomal process is coated with species-specific recognition molecules called *bindin*, and there are bindin receptors on the vitelline envelope of the egg. The interaction of these two molecules enable the sperm to contact the egg plasma membrane. That contact results in fusion of the sperm and egg plasma membranes and the formation of a *fertilization cone* that engulfs the sperm head, bringing it into the egg cytoplasm.

In animals that practice internal fertilization, mating behaviors help guarantee species specificity, but egg–sperm recognition mechanisms still exist. The mammalian egg is surrounded by a thick layer called the **cumulus**, which consists of a loose assemblage of maternal cells in a gelatinous matrix (**Figure 42.6**). Beneath the cumulus is a glycoprotein envelope called the **zona pellucida**, which is functionally similar to the vitelline envelope of sea urchin eggs. When mammalian sperm are deposited in the female reproductive tract, they are metabolically activated and made capable of an acrosomal reaction if they should meet an egg. An activated sperm can penetrate the cumulus and interact with the zona pellucida.

Unlike the jelly coat of sea urchin eggs, the cumulus of mammalian eggs does not trigger the acrosomal reaction. When sperm make contact with the zona pellucida, a species-specific glyco-

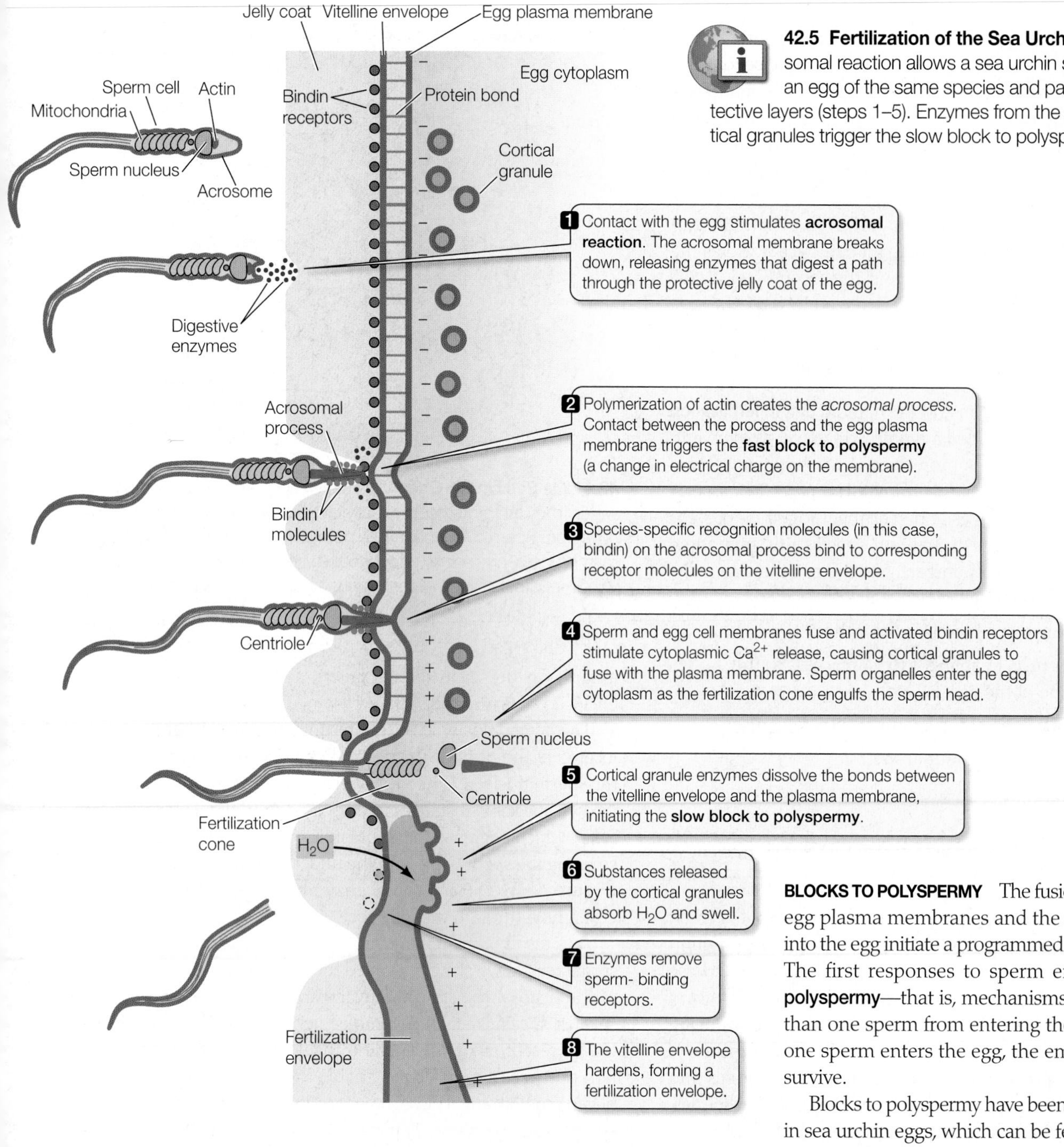

42.5 Fertilization of the Sea Urchin Egg The acrosomal reaction allows a sea urchin sperm to recognize an egg of the same species and pass through its protective layers (steps 1–5). Enzymes from the sea urchin egg's cortical granules trigger the slow block to polyspermy (steps 6–8).

1. Contact with the egg stimulates **acrosomal reaction**. The acrosomal membrane breaks down, releasing enzymes that digest a path through the protective jelly coat of the egg.
2. Polymerization of actin creates the *acrosomal process.* Contact between the process and the egg plasma membrane triggers the **fast block to polyspermy** (a change in electrical charge on the membrane).
3. Species-specific recognition molecules (in this case, bindin) on the acrosomal process bind to corresponding receptor molecules on the vitelline envelope.
4. Sperm and egg cell membranes fuse and activated bindin receptors stimulate cytoplasmic Ca^{2+} release, causing cortical granules to fuse with the plasma membrane. Sperm organelles enter the egg cytoplasm as the fertilization cone engulfs the sperm head.
5. Cortical granule enzymes dissolve the bonds between the vitelline envelope and the plasma membrane, initiating the **slow block to polyspermy**.
6. Substances released by the cortical granules absorb H_2O and swell.
7. Enzymes remove sperm- binding receptors.
8. The vitelline envelope hardens, forming a fertilization envelope.

protein binds to recognition molecules on the head of the sperm. This binding triggers the acrosomal reaction, releasing acrosomal enzymes that digest a path through the zona pellucida. When the sperm head reaches the egg plasma membrane, other proteins facilitate its adhesion to and fusion with the egg plasma membrane.

The importance of the zona pellucida and its sperm-binding molecules as a species-specific recognition mechanism was revealed in experiments on mammalian eggs and sperm in culture dishes. When the zona was stripped from human eggs and they were exposed to hamster sperm, fertilization took place, resulting in a hamster–human hybrid zygote. The hybrid zygote did not survive its first cell division, but the experiment demonstrated that the recognition mechanism in mammalian species resides in the zona pellucida.

BLOCKS TO POLYSPERMY The fusion of the sperm and egg plasma membranes and the entry of the sperm into the egg initiate a programmed sequence of events. The first responses to sperm entry are **blocks to polyspermy**—that is, mechanisms that prevent more than one sperm from entering the egg. If more than one sperm enters the egg, the embryo is unlikely to survive.

Blocks to polyspermy have been studied extensively in sea urchin eggs, which can be fertilized in a dish of seawater. Within seconds after the sperm membrane contacts the egg membrane, an influx of sodium ions occurs, which changes the electric charge difference across the egg's plasma membrane. This *fast block to polyspermy* prevents the fusion of other sperm with the egg plasma membrane, but it is transient. The change in membrane electrical charge lasts only about a minute, but that is enough time to allow a slower block to sperm entry to develop.

Before fertilization, the vitelline envelope is bonded to the egg plasma membrane. Just under the plasma membrane are vesicles called *cortical granules,* which contain enzymes and other proteins that dissolve the bonds between the vitelline envelope and the egg plasma membrane. The sea urchin egg, like all animal cells, contains calcium ions that are sequestered in the endoplasmic reticulum. The *slow block to polyspermy* is initiated by the release of these calcium ions following sperm and egg fusion (see Figure 42.5 and Figure 15.14).

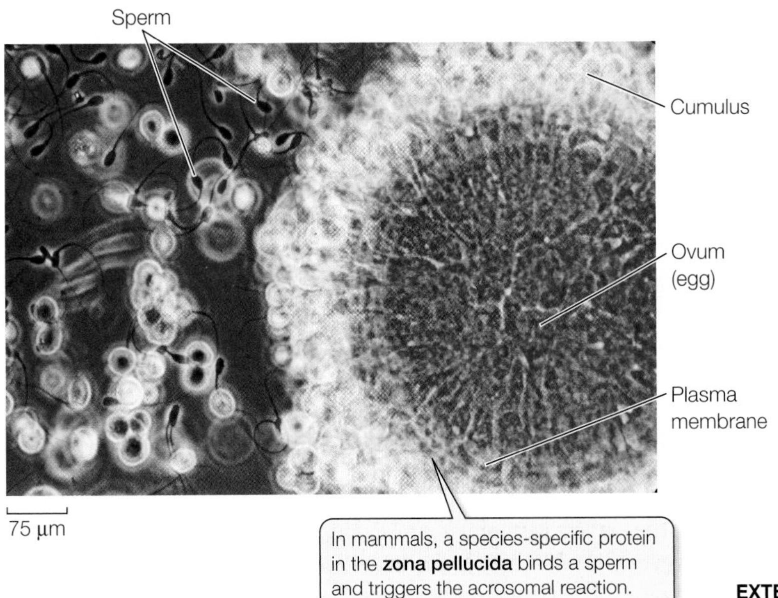

42.6 A Mammalian Egg Is Surrounded by Barriers to Sperm This human egg is protected by the cumulus and zona pellucida, both of which a sperm must penetrate to fertilize the egg. Only one sperm will penetrate the zona pellucida and fuse with the plasma membrane.

EGG ACTIVATION Sperm entry into the sea urchin egg stimulates the release of calcium from the egg's endoplasmic reticulum. The increase in cytosolic calcium causes the egg's cortical granules to fuse with the plasma membrane and release their contents. The cortical granule enzymes break the bonds between the vitelline envelope and the plasma membrane, and other proteins released from the cortical granules attract water into the space between them. As a result, the vitelline envelope rises to form a *fertilization envelope*. Cortical granule enzymes also degrade sperm-binding molecules on the surface of the fertilization envelope and cause it to harden. The hardened fertilization envelope prevents additional sperm from contacting the egg's plasma membrane.

The mechanism of hardening of the fertilization envelope was discovered only recently. One hallmark of fertilization is a rapid increase in oxygen consumption by the fertilized egg, assumed for almost a hundred years to be due to the activation of the various metabolic processes in the fertilized egg. We now know that the big increase in oxygen consumption is due to the rapid production of hydrogen peroxide (H_2O_2) by an enzyme secreted by the cortical granules. This enzyme, called Udx1 for urchin dual oxidase, has two seemingly opposing functions. First, it catalyzes the production of H_2O_2, which provides the oxidizing capacity for hardening of the fertilization envelope; and second, it catalyzes the breakdown of any unused H_2O_2 so it does not enter the zygote and cause damage. Similar mechanisms of rapid oxidation of the materials surrounding the fertilized egg probably occur in most species, including mammals.

In mammals, sperm entry does not cause a rapid change in membrane potential, but it does trigger a release of calcium from the endoplasmic reticulum. As in the sea urchin, the increased calcium causes the cortical granules to fuse with the egg plasma membrane. A fertilization envelope does not form around the mammalian egg, but the cortical granule enzymes destroy the sperm-binding molecules in the zona pellucida. The rise in cytosolic calcium also activates the egg's metabolism and signals it to complete meiosis. The stage is set for the first cell division.

Mating bring eggs and sperm together

As we have just seen, sexual reproduction requires the production of haploid gametes (gametogenesis) and the joining together of those gametes to form a diploid zygote (fertilization). **Mating**, the step between these two processes, gets eggs and sperm close enough together that fertilization can occur. The simplest distinction in mating systems is whether fertilization occurs externally or internally.

EXTERNAL FERTILIZATION In an aquatic environment, animals can bring their gametes together by simply releasing them into the water. This practice is called **external fertilization**. Many simple aquatic animals are not very mobile, but they produce huge numbers of gametes that can travel far from the point of release. A female oyster, for example, will release millions of eggs when she spawns, and the number of sperm produced by a male oyster is astronomical.

Caviar is fish eggs, most of it coming from sturgeon of the Caspian and Black Seas. Much-prized beluga caviar, which can sell for $4,000 a kilogram, comes from beluga sturgeon (*Husa husa*), an evolutionarily ancient fish that can live for 150 years and be 5 meters long. Today, declining population size due to overfishing and pollution means that the average adult beluga is much younger, and far smaller.

But numbers alone do not guarantee that gametes will meet. The reproductive activities of the males and females of a population must be synchronized, since released gametes have a limited life span. Seasonal breeders may use day length, changes in temperature, or changes in weather to time the production and release of their gametes. Social stimulation is also important. Sexual activity by one member of a population can stimulate others to engage in it.

Behavior can play an important role in bringing gametes together even when fertilization is external. Many species travel great distances to congregate with potential mates and release their gametes at the same time in a suitable environment. Salmon are an extreme example, traveling hundreds of miles to spawn in the stream where they hatched.

INTERNAL FERTILIZATION Terrestrial animals cannot simply release their gametes into the environment. Sperm can move only through liquid, and delicate gametes released into air would dry out and die. Terrestrial animals avoid these problems by **internal fertilization**, the release of sperm directly into the female reproductive tract.

Animals have evolved an astonishing diversity of behavioral and anatomical adaptations for internal fertilization. As we saw above, gametogenesis occurs in the gonads, which are the *primary sex organs*. All additional anatomical components of an animal's reproductive system are called *accessory sex organs*. An obvious accessory sex organ in males of many species is the **penis**, which enables the male to deposit sperm in the female's reproductive tract, called in many species a **vagina**. Accessory sex organs include a variety of glands, tubules, ducts, and other structures.

Copulation is the physical joining of male and female accessory sex organs. Transfer of sperm in internal fertilization can also be indirect. Males of many invertebrate species (for example, mites and scorpions) and a few vertebrates (salamanders) deposit *spermatophores*—packets of sperm—in the environment. When a female mite encounters a spermatophore from a potential mate, she straddles it and opens a pair of plates in her abdomen so that the tip of the spermatophore enters her reproductive tract and allows the sperm to enter.

Male squids and spiders play a more active role in spermatophore transfer. The male spider secretes a drop containing sperm onto a bit of web, then uses a special structure on his foreleg to pick up the sperm-containing web and insert it through the female's genital opening. Male squids use one specialized tentacle to pick up a spermatophore and insert it into the female's genital opening.

Most male insects copulate and transfer sperm to the female's vagina through a penis. The **genitalia**—external sex organs—of insects often have species-specific shapes that match in a lock-and-key fashion. This mechanism ensures a tight, secure fit between the mating pair during the prolonged period of sperm transfer. In some insect species in which females mate with more than one male, the males have elaborate structures on their penises that can scoop sperm deposited by other males out of a female's reproductive tract, replacing it with their own.

A single body can function as both male and female

In most species, gametes are produced by individuals that are either male or female. Species that have separate male and female members are called **dioecious** species (from the Greek for "two houses"). In some species, however, a single individual may produce both sperm and eggs. Such species are called **monoecious** ("one house"), or **hermaphroditic**, species.

Almost all invertebrate groups contain some hermaphroditic species. An earthworm is an example of a *simultaneous hermaphrodite*, meaning that it is both male and female at the same time. When two earthworms mate, they exchange sperm, and as a result, the eggs of each are fertilized (see Figure 32.13C). Some vertebrates, such as the anemone fish or clown fish, which live in small groups within large sea anemones, are *sequential hermaphrodites*, meaning that an individual may function as a male or a female at different times in its life (**Figure 42.7**). All anemone fish are born as males. The largest one in a group becomes a functional female. If she is removed from the group, the largest male becomes a female. Also, the second largest anemone fish in the group is the only male in breeding condition.

What is the selective advantage of hermaphroditism? Some simultaneous hermaphrodites, such as parasitic tapeworms, have a low probability of meeting a potential mate. Even though a tapeworm may be large and troublesome for its host, it may be the only tapeworm in the host. Tapeworms can fertilize their own eggs. Most simultaneous hermaphrodites, however, must mate with another individual; but since each member of the population is both male and female, the probability of encountering a possible mate is double what it would be in strictly monoecious species. In some sequential hermaphrodites, all siblings are either male or female at the same time, thus reducing the incidence of inbreeding.

The evolution of vertebrate reproductive systems parallels the move to land

The earliest vertebrates evolved in aquatic environments. The closest living relatives of those earliest vertebrates are modern-day fishes. They remain exclusively aquatic animals, and most practice external fertilization. The most primitive of the fishes, the lampreys and hagfishes, simply release their gametes into the environment. In most fishes, however, mating behaviors bring females and males into close proximity at the time of gamete release. In some sharks and rays, fins have evolved into claspers that hold the male and female together and enable sperm to be transferred directly into the female reproductive tract.

Amphibians were the first vertebrates to live in terrestrial environments. They dealt with the challenge of a dry environment

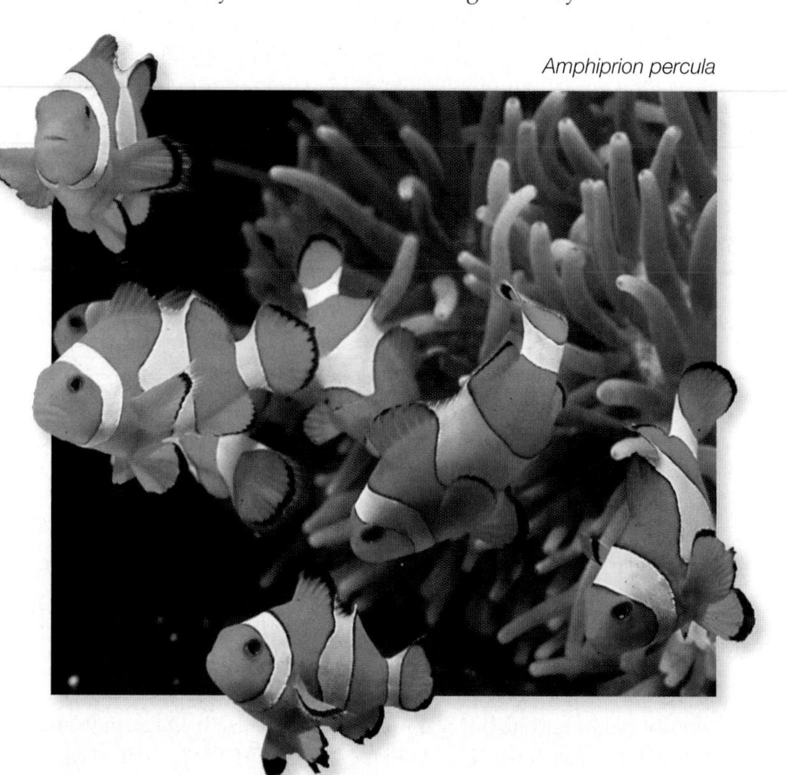

42.7 When Size Determines Sex Anemone fish (also known as "clown fish") live in groups of about a dozen centered on a single sea anemone. All anemone fish are born male, and the largest one in the group becomes a female. Thus any one fish may function first as a male and then as a female.

by returning to water to reproduce, as most amphibians still do today.

Reptiles were the first vertebrate group to solve the problem of reproduction in the terrestrial environment. Their solution, the **amniote egg**, is shared with the birds (see Section 33.4). A good example is the chicken egg, which contains a supply of food (yolk) and water for the developing embryo. A hard shell protects the embryo and impedes water loss while allowing the diffusion of oxygen and carbon dioxide (**Figure 42.8A**). The eggshell creates an obvious problem for fertilization, however: Sperm cannot penetrate the shell, so they must reach the egg before the shell forms. Hence internal fertilization and the evolution of accessory sex organs were necessary for the evolution of the amniote egg.

Male snakes and lizards have paired *hemipenes*, which can be filled with blood and thereby extruded from the male's body. Only one hemipenis is inserted into the female's reproductive tract at a time. It is usually rough or spiny at the end to achieve a secure hold while sperm are transferred down a groove on its surface. Retractor muscles pull the hemipenis back into the male's body when mating is completed. Some evolutionarily ancient bird species have erectile penises that channel sperm along a groove into the female's reproductive tract. Birds with more recent evolutionary origins, however, do not have erectile penises; instead, the male and female simply bring their genital openings close together to transfer sperm. Usually this involves the male standing on the female's back (**Figure 42.8B**).

All mammals practice internal fertilization and, except for the prototherian mammals (see Figure 33.25), they have done away with the shelled egg. The developing mammalian embryo is retained in the female reproductive tract, at least through its early stages; mammalian species vary enormously as to the developmental stage of their offspring at the time of birth.

42.8 The Shelled Egg The shelled egg was a major evolutionary step that allows reptiles and birds to reproduce in the terrestrial environment. (A) A female green sea turtle deposits her eggs in the sand. (B) The shelled egg requires that sperm meet egg before the shell forms, thus terrestrial animals must practice internal fertilization, as these European bee-eaters are doing.

(A) *Chelonia mydas*

(B) *Merops apiaster*

Reproductive systems are distinguished by where the embryo develops

Two patterns of care and nurture of the embryo have evolved in animals: *oviparity* (egg laying) and *viviparity* (live bearing).

- **Oviparous** animals lay eggs in the environment and their embryos develop outside the mother's body. Oviparous terrestrial animals such as insects, reptiles, and birds protect their eggs from desiccation with waterproof membranes or shells.

Oviparity is possible because eggs are stocked with abundant nutrients to supply the needs of the embryo. Some oviparous animals engage in various forms of parental behavior to protect their eggs, but until the eggs hatch, the embryos depend entirely on the nutrients stored in the egg.

- **Viviparous** animals retain the embryo within the mother's body during its early developmental stages. All mammals except the monotremes are viviparous.

Although examples of viviparity exist in all other vertebrate groups except the crocodiles, turtles, and birds (even some sharks retain fertilized eggs in their bodies and give birth to free-living offspring), there is a big difference between viviparity in mammals and in other species. Mammals have a specialized portion of the female reproductive tract, called the **uterus** or *womb*, that holds the embryo and interacts with it to produce a **placenta**, which enables the exchange of nutrients and wastes between the blood of the mother and that of the embryo. Very few nonmammalian species have evolved such a connection between the embryo and the mother.

Most nonmammalian viviparous animals simply retain fertilized eggs in the mother's body until they hatch. These embryos still receive nutrition from stores in the egg, so this reproductive adaptation is called **ovoviviparity**.

42.2 RECAP

Sexual reproduction involves gametogenesis, mating, and fertilization. Fertilization can be external or internal and involves mechanisms for ensuring that only one sperm from the right species enters the egg.

- Can you outline the steps by which a sperm penetrates an egg? See Figure 42.5
- Can you explain how polyspermy is prevented, and do you understand why it is crucial to do so? See p. 902 and Figure 42.5
- Can you describe the reproductive adaptations that made life on land possible? See pp. 904–905

Now that we have seen some of the general aspects of animal reproduction in gametogenesis and fertilization, and have been briefly introduced to the great diversity of mating systems, we will consider the human reproductive systems in detail.

42.3 How Do the Human Male and Female Reproductive Systems Work?

In this section we will describe the structures and functions of the male and female sex organs in mammals—specifically, in human beings—and discuss hormonal regulation of both male and female systems. Our discussion will include the primary sex organs (testes in males and ovaries in females) that produce gametes and serve endocrine functions. It will also include the accessory sex organs—the genitalia, the ducts through which the gametes pass, and the various glands that empty into those ducts. We will also refer to *secondary sexual characteristics*, which are not directly involved in reproduction, but comprise the external differences between males and females.

Male sex organs produce and deliver semen

Semen is the product of the male reproductive system. Besides sperm, semen contains a complex mixture of fluids and molecules that support the sperm and facilitate fertilization. Sperm make up less than 5 percent of the volume of the semen.

The male reproductive organs are diagrammed in **Figure 42.9**. Sperm are produced in the testes, the paired male gonads. The testes of most mammals are located outside the body cavity in a pouch of skin called the **scrotum**.

Why should the testes be located outside the body cavity? The optimal temperature for spermatogenesis in most mammals is slightly lower than the normal body temperature. The scrotum keeps the testes at this optimal temperature. Muscles in the scrotum contract in a cold environment, bringing the testes closer to the warmth of the body; in a hot environment they relax, cooling the testes by suspending them farther from the body.

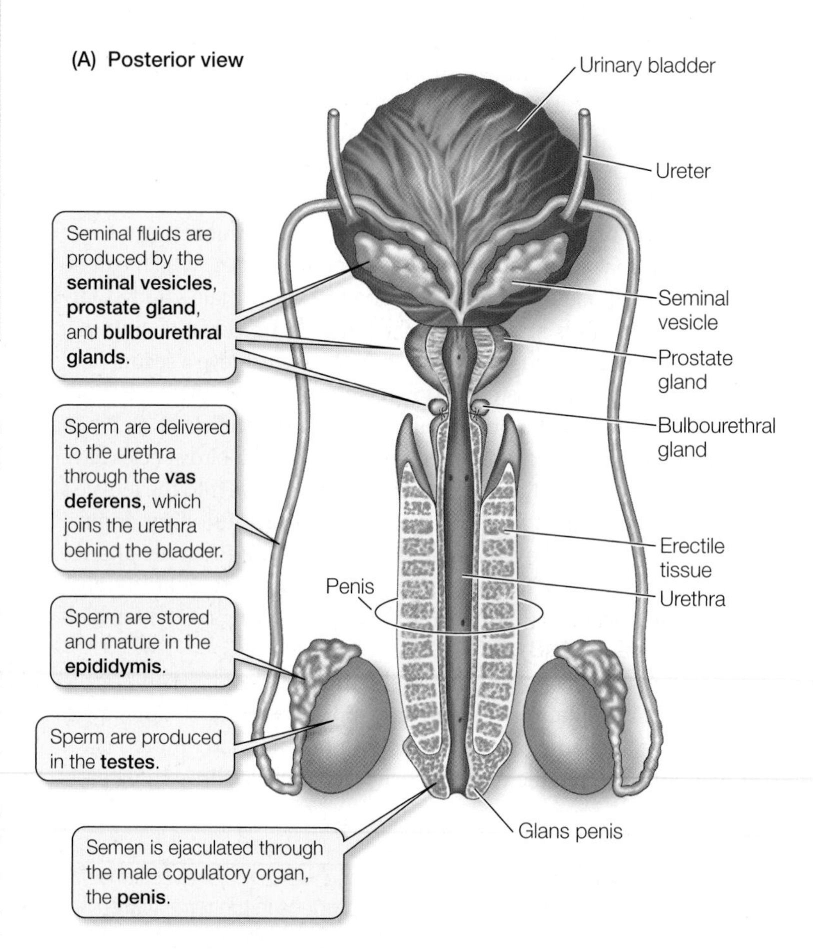

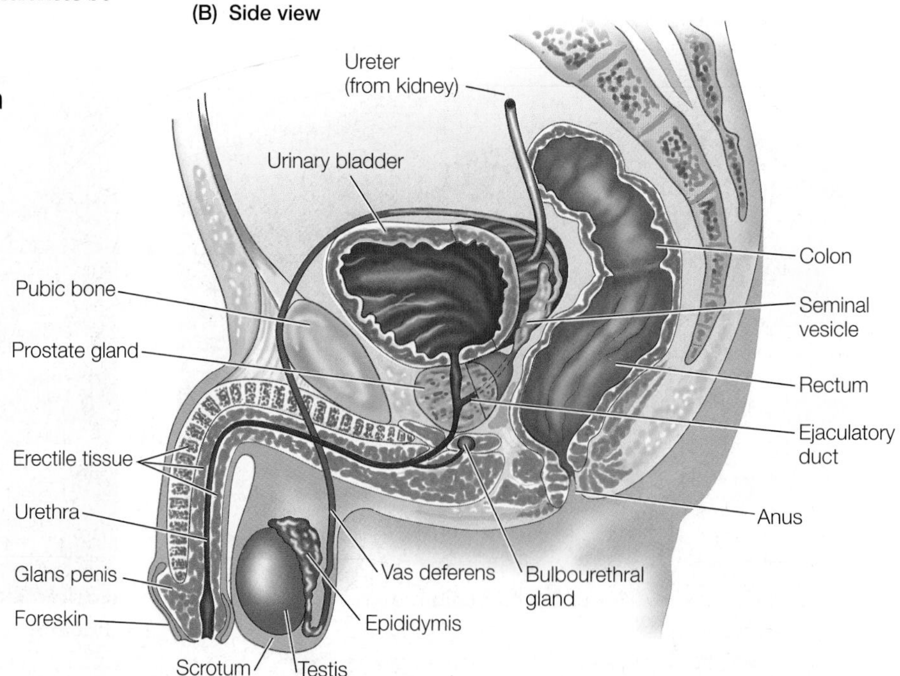

42.9 The Reproductive Tract of the Human Male The male reproductive organs are shown (A) from the rear and (B) in side section.

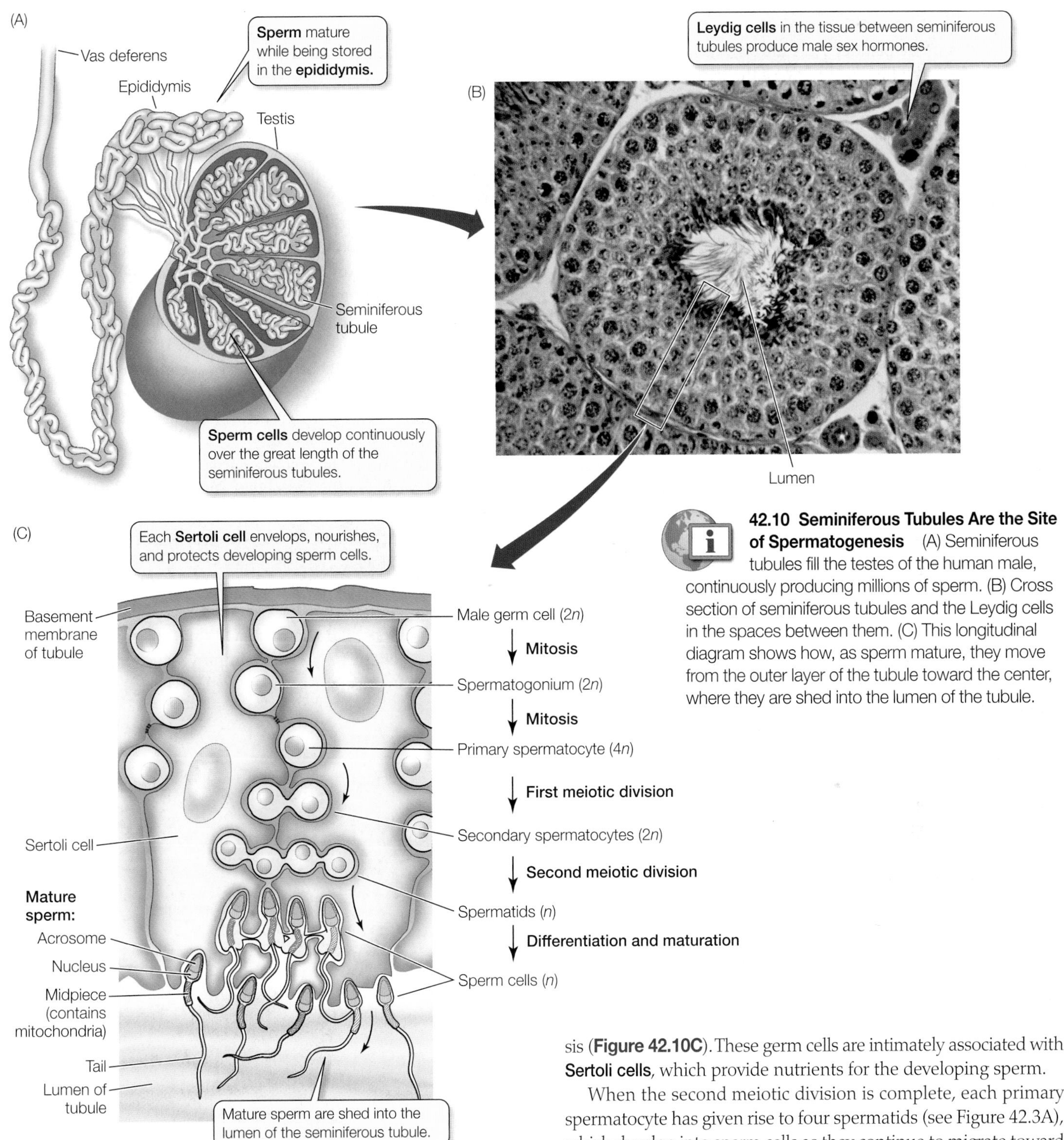

42.10 Seminiferous Tubules Are the Site of Spermatogenesis (A) Seminiferous tubules fill the testes of the human male, continuously producing millions of sperm. (B) Cross section of seminiferous tubules and the Leydig cells in the spaces between them. (C) This longitudinal diagram shows how, as sperm mature, they move from the outer layer of the tubule toward the center, where they are shed into the lumen of the tubule.

Spermatogenesis takes place within the **seminiferous tubules**, which are tightly coiled in each testis (**Figure 42.10A**). Between the seminiferous tubules are clusters of *Leydig cells*, or *interstitial cells*, which produce testosterone (**Figure 4.10B**). Spermatogonia reside in the outer layers of the epithelium that lines the tubules. Moving inward from these outer layers toward the lumen of the tubule, we find germ cells in successive stages of spermatogenesis (**Figure 42.10C**). These germ cells are intimately associated with **Sertoli cells**, which provide nutrients for the developing sperm.

When the second meiotic division is complete, each primary spermatocyte has given rise to four spermatids (see Figure 42.3A), which develop into sperm cells as they continue to migrate toward the lumen of the seminiferous tubule. The nucleus becomes compact and the surrounding cytoplasm is lost. A flagellum, or sperm tail, develops. The mitochondria, which will provide energy for tail motility, become condensed into a midpiece between the head and the tail. An acrosome forms over the nucleus in the head of the sperm. Fully differentiated sperm are shed into the lumen of the seminiferous tubule.

From the seminiferous tubules, sperm move into a storage structure called the **epididymis** (see Figure 42.9), where they mature and become motile. The epididymis connects to the **urethra** via a tube called the **vas deferens** (plural *vasa deferentia*). The urethra originates in the bladder, runs through the penis, and opens to the outside of the body at the tip of the penis. It serves as the common duct for the urinary and reproductive systems. The components of the semen other than sperm come from several accessory glands. About 60 percent of the volume of semen is secreted by the paired **seminal vesicles**, which empty into the vas deferens just before it joins the urethra. Seminal fluid is thick because it contains mucus and fibrinogen, a protein also found in the blood, where it can polymerize to form blood clots. Seminal fluid also contains fructose, an energy source for the sperm.

Men whose work exposes them to high heat (furnace operators and cooks, for example) may experience arrested spermatogenesis and low fertility. This effect may also occur in some men who exercise vigorously or who wear very tight underwear.

The **prostate gland** surrounds the urethra and contributes about 30 percent of the volume of the semen. Prostate fluid is alkaline, so it neutralizes the acidity in the male and female reproductive tracts and makes those environments more hospitable to sperm. The prostate also secretes a clotting enzyme that causes the fibrinogen from the seminal vesicles to convert the semen into a gelatinous mass, facilitating its propulsion into and retention in the upper regions of the female reproductive tract. Another enzyme in the prostate fluid, profibrinolysin, is inactive when secreted, but activated shortly after it enters the female reproductive tract. Active fibrinolysin dissolves the clotted semen and liberates the sperm.

The **bulbourethral glands** produce a small volume of an alkaline, mucoid secretion that helps to neutralize acidity in the urethra and lubricate it to facilitate the passage of semen at the climax of sexual intercourse. Secretions of the bulbourethral glands precede the climax of the sex act and can carry with them residual sperm from prior sexual activity. Therefore, it is possible for pregnancy to occur even if the penis is withdrawn from the female just before climax (a rather ineffective birth control practice known as *coitus interruptus*).

The penis and the scrotum are the male genitalia. The shaft of the penis is covered with normal skin, but the highly sensitive tip, or **glans penis**, is covered with thinner, more sensitive skin that is especially responsive to sexual stimulation. A fold of skin called the *foreskin* covers the glans of the human penis. The procedure known as *circumcision* removes a portion of the foreskin. Considerable controversy exists over this cultural practice. The current position of the American Academy of Pediatrics cites possible medical benefits along with definite associated risks and concludes that scientific evidence does not support circumcision as a routine practice.

Sexual stimulation triggers responses in the nervous system that result in penile **erection**. Nerve endings release a gaseous neurotransmitter, nitric oxide (NO), onto blood vessels leading into the penis. NO stimulates production of the second messenger cGMP (Section 15.3), which causes these vessels to dilate. The increased blood flow that results fills and swells shafts of spongy, erectile tissue located along the length of the penis. The enlargement of these blood-filled cavities compresses the vessels that normally carry blood out of the penis. As a result, the erectile tissue becomes more and more engorged with blood. The penis becomes hard and erect, facilitating its insertion into the female's vagina. Many species of mammals, but not humans, have a bone in the penis, but these species still depend on erectile tissue for copulation.

At the climax of copulation, about 2 to 6 ml of semen is propelled through the vasa deferentia and the urethra in two steps, emission and ejaculation. During **emission** rhythmic contractions of smooth muscles in the vasa deferentia and accessory glands move the semen into the urethra at the base of the penis. **Ejaculation** is caused by contractions of other muscles at the base of the penis surrounding the urethra. The rigidity of the erect penis allows these contractions to force the gelatinous mass of semen through the urethra and out of the penis. The muscle contractions of ejaculation are accompanied by feelings of intense pleasure known as **orgasm**. They are also accompanied by transient increases in heart rate, blood pressure, breathing, and skeletal muscle contractions throughout the body.

After ejaculation, NO release decreases and enzymes break down cGMP, causing the blood vessels flowing into the penis to constrict. The blood pressure in the erectile tissue decreases, relieving the compression of the blood vessels leaving the penis, and the erection declines.

Erectile dysfunction, or *impotence*, is the inability to achieve or sustain an erection. Drugs used to treat erectile dysfunction act by inhibiting the breakdown of cGMP, thus enhancing the effect of NO released in the penis, which improves the ability to achieve and maintain an erection.

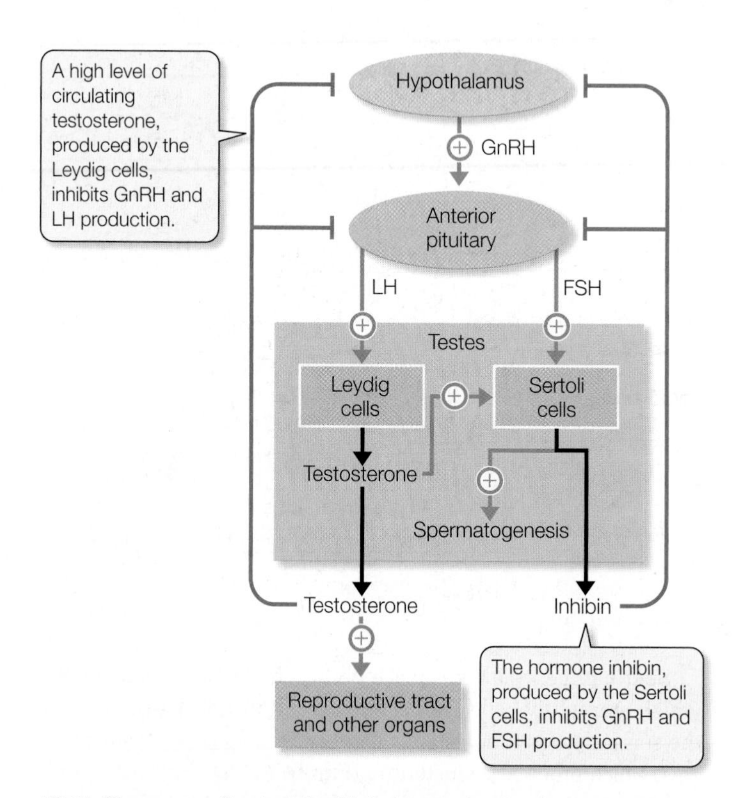

42.11 Hormones Control the Male Reproductive System The male reproductive system is under hormonal control by the hypothalamus and the anterior pituitary.

Male sexual function is controlled by hormones

Spermatogenesis and maintenance of male secondary sexual characteristics depend on testosterone, which is produced by the Leydig cells of the testes. As described in Section 41.3, increased production of testosterone at puberty results from an increased release of gonadotropin-releasing hormone (GnRH) by the hypothalamus, which stimulates anterior pituitary cells to increase their secretion of luteinizing hormone (LH) and follicle-stimulating hormone (FSH) (**Figure 42.11**). Higher levels of LH stimulate the Leydig cells to increase their production and release of testosterone. Testosterone exerts negative feedback on the anterior pituitary and the hypothalamus. At the time of puberty, the sensitivity of the hypothalamus to negative feedback from testosterone declines, and the level of circulating testosterone increases.

Increased testosterone in pubertal males causes the development of secondary sexual characteristics (such as pubic and facial hair, deeper voice, and enlarged genitals) and an increased growth rate. Testosterone also promotes increased muscle mass and maturation of the testes. Continued production of testosterone after puberty is essential for the maintenance of secondary sexual characteristics and the production of sperm.

Spermatogenesis is controlled by the influence of FSH and testosterone on the Sertoli cells in the seminiferous tubules. The Sertoli cells also produce a hormone called *inhibin*, which exerts negative feedback on the anterior pituitary cells that produce and secrete FSH.

Female sex organs produce eggs, receive sperm, and nurture the embryo

When a mammalian egg matures, it is released from the ovary directly into the body cavity. But the egg does not go far. Each ovary is enveloped by the undulating, fringed opening of an **oviduct** (also known as a *Fallopian tube*), which sweeps the egg into that tube (**Figure 42.12**). Fertilization takes place in the oviduct. Whether or not the egg is fertilized, cilia lining the oviduct propel it slowly toward the uterus, a muscular, thick-walled cavity shaped in humans like an upside-down pear. The uterus is where the embryo develops if the egg is fertilized. At the bottom, the uterus narrows into a region called the **cervix**, which leads into the **vagina**.

In humans, two sets of skin folds surround the opening of the vagina and the opening of the urethra, through which urine passes. The inner, more delicate folds are the *labia minora*; the outer, thicker folds are the *labia majora*. At the anterior tip of the labia minora is the *clitoris*, a small bulb of erectile tissue that has the same developmental origins as the penis. The clitoris is highly sensitive and plays an important role in sexual response. The labia minora and the clitoris become engorged with blood in response to sexual stimulation.

The external opening of an infant's vagina is usually, but not always, partly covered by a thin membrane, the *hymen*. Eventually the hymen can be torn by vigorous physical activity or by first sexual intercourse; it can sometimes make first intercourse difficult or painful for the female.

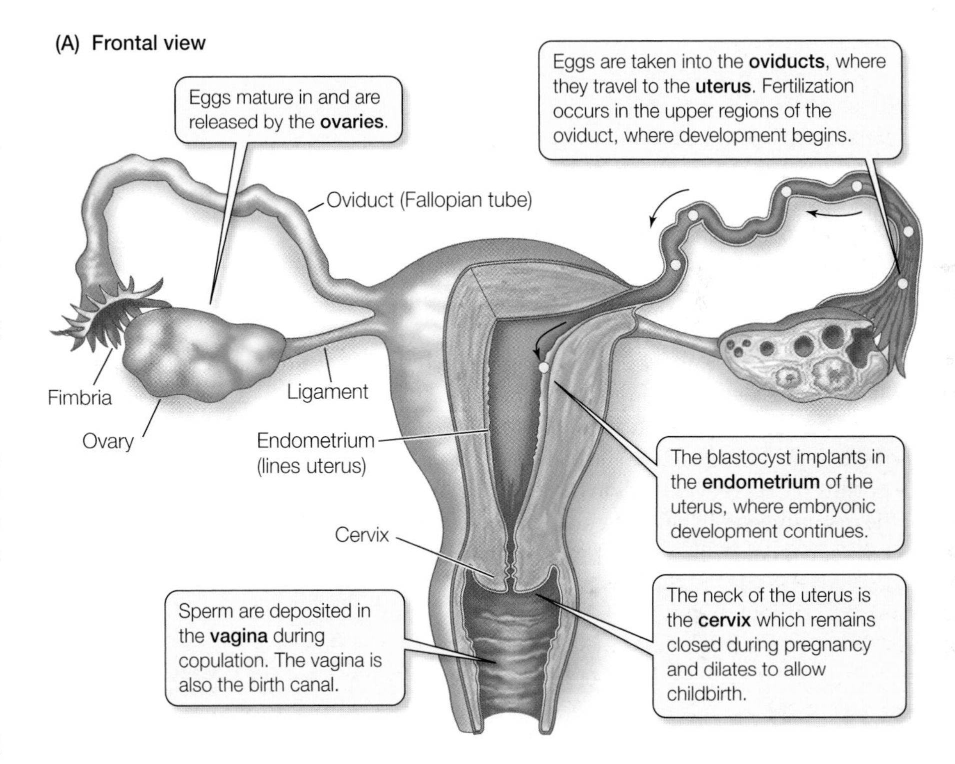

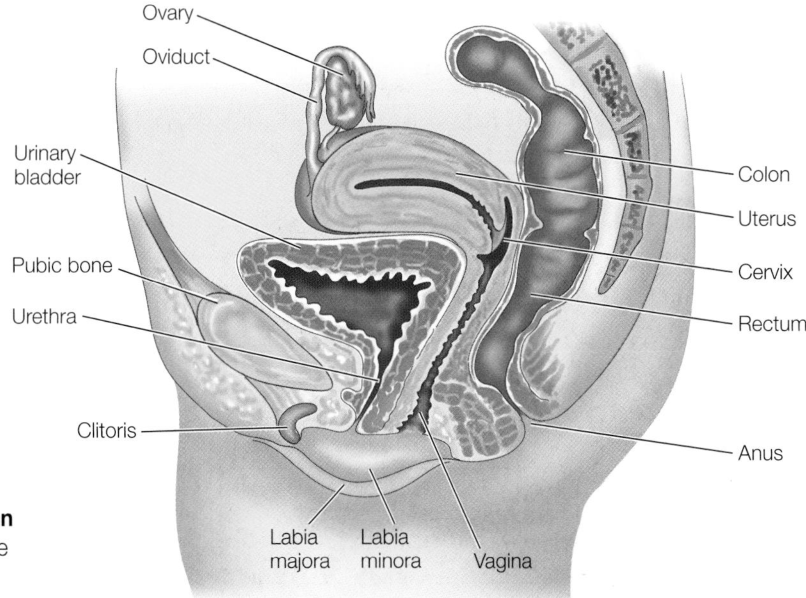

42.12 The Reproductive Tract of the Human Female The female reproductive organs are shown in front and side views.

To fertilize an egg, sperm deposited in the vagina swim and are propelled by contractions of the female reproductive tract through the cervical opening, the uterus, and most of the oviduct. The egg (actually a secondary oocyte) is fertilized in the upper region of the oviduct. Fertilization stimulates the completion of the second meiotic division, after which the haploid nuclei of the sperm and the egg can fuse to produce a diploid zygote nucleus. Still in the oviduct, the zygote undergoes its first few cell divisions to become a **blastocyst**. The blastocyst moves down the oviduct to the uterus, where it attaches itself to the epithelial lining of the uterus, the **endometrium**. Once attached to the endometrium, the blastocyst burrows into it—a process called **implantation**—and interacts with it to form the placenta as we will see in the next chapter. The placenta nurtures the embryo and also produces hormones that help sustain pregnancy.

As the egg matures in the ovary, the endometrium thickens. If a blastocyst does not arrive in the uterus, the endometrium regresses or is sloughed off. Thus the female reproductive cycle actually consists of two linked cycles: an *ovarian cycle* that produces eggs and hormones, and a *uterine cycle* that prepares the endometrium for the arrival of a blastocyst.

The ovarian cycle produces a mature egg

An **ovarian cycle** is about 28 days long in the human female, but it varies considerably among individuals. During the first half of each cycle, at least one primary oocyte matures into a secondary oocyte (egg) and is expelled from the ovary (*ovulation*). During the second half of the cycle, cells in the ovary that were associated with the maturing oocyte develop endocrine functions and then regress if the egg is not fertilized. The progression of these events is shown diagrammatically in **Figure 42.13**.

Some mammals have ovarian cycles longer than 28 days and some shorter. Rats and mice have a 4-day cycle, elephants have a 4-month cycle, and some seasonal breeders have a 1-year cycle. In some species, such as rabbits, ovulation is induced by copulation—a "recipe" for prolific breeding.

A newborn human female has about a million primary oocytes in each ovary. By the time she reaches puberty, she has only about 200,000; the rest have degenerated. During a woman's fertile years, her ovaries will go through about 450 ovarian cycles. During each cycle, a number of oocytes will begin to mature, but usually only one will mature completely and be released; the others degenerate. At around the age of 50, a woman reaches **menopause**—the end of fertility—and may have only a few oocytes left in each ovary. Throughout a woman's life, oocytes are degenerating, and no new ones are produced.

Each primary oocyte in the ovary is surrounded by a layer of ovarian cells. An oocyte and its surrounding cells constitute the functional unit of the ovary, the **follicle**. Between puberty and menopause, six to twelve follicles begin to mature each month. In each follicle, the oocyte enlarges and the surrounding follicular cells proliferate. After about a week, one of these follicles is larger than the rest, and it continues to grow, while the others cease to develop and shrink. In the enlarged follicle, the follicular cells nurture the growing egg, supplying it with nutrients, growth factors, and hormonal stimulation.

In humans, after 2 weeks of follicular growth, **ovulation** occurs: The follicle ruptures and the egg is released. Following ovulation,

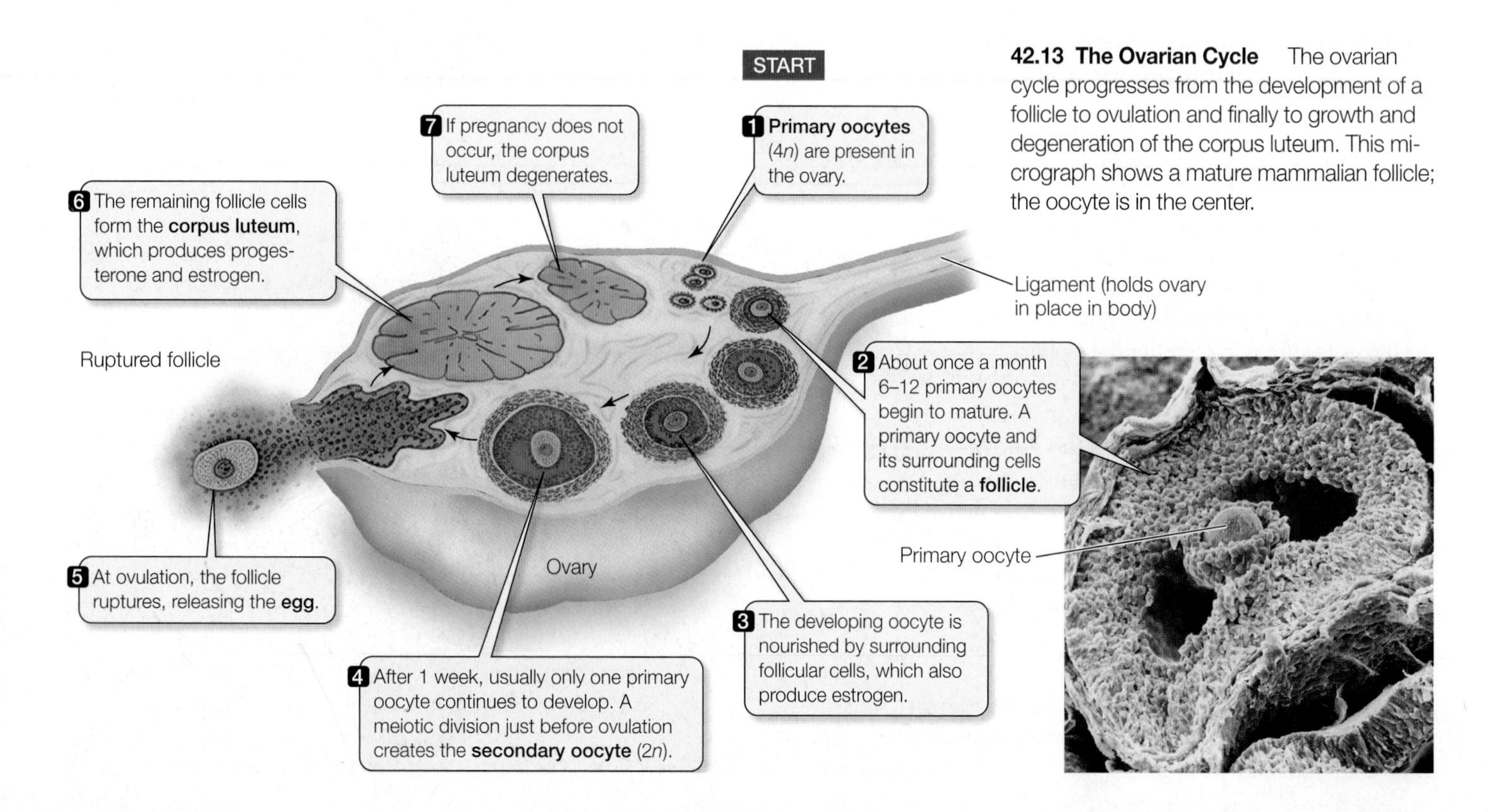

42.13 The Ovarian Cycle The ovarian cycle progresses from the development of a follicle to ovulation and finally to growth and degeneration of the corpus luteum. This micrograph shows a mature mammalian follicle; the oocyte is in the center.

the follicle cells continue to proliferate and form a mass of endocrine tissue about the size of a marble. This structure, which remains in the ovary, is the *corpus luteum* (plural *corpora lutea*). It functions as an endocrine gland, producing estrogen and progesterone for about 2 weeks. It then degenerates unless a blastocyst implants in the endometrium.

The uterine cycle prepares an environment for the fertilized egg

The **uterine cycle** parallels the ovarian cycle and consists of a buildup and then a breakdown of the endometrium (**Figure 42.14**). About 5 days into the ovarian cycle, the endometrium starts to grow in preparation for receiving a blastocyst. The uterus attains its maximum state of preparedness about 5 days after ovulation and remains in that state for another 9 days. If a blastocyst has not arrived by that time, the endometrium begins to break down, and the sloughed-off tissue, including blood, flows from the body through the vagina—the process of **menstruation** (from *menses*, the Latin word for "months").

The uterine cycles of most mammals other than humans do not include menstruation; instead, the uterine lining typically is resorbed. In these species, the most obvious correlate of the ovarian cycle is a state of sexual receptivity called **estrus** around the time of ovulation. You may be aware of the bloody discharge that occurs in dogs at the time of estrus. This discharge is not the same as menstruation—in fact it is exactly the opposite. Bleeding in dogs occurs during the *proliferation* of the uterine lining, which occurs just prior to ovulation. When the female mammal comes into estrus, or "heat," she actively solicits male attention and may be aggressive to other females. Women are unusual among mammals in that they are potentially sexually receptive throughout their ovarian cycles and at all seasons of the year.

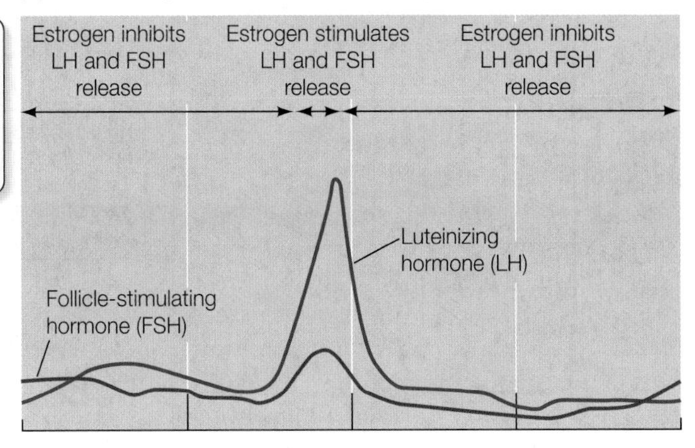

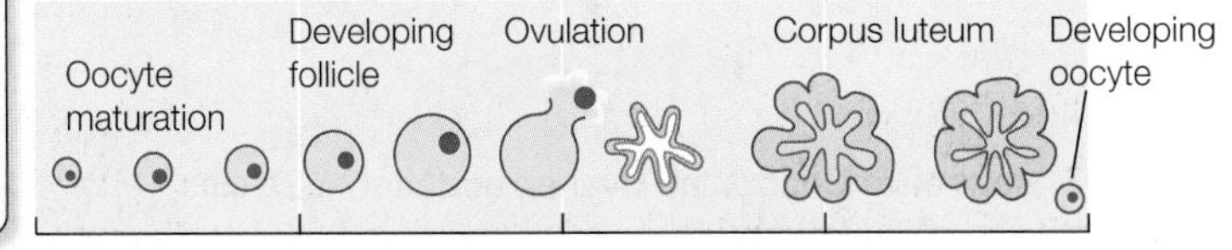

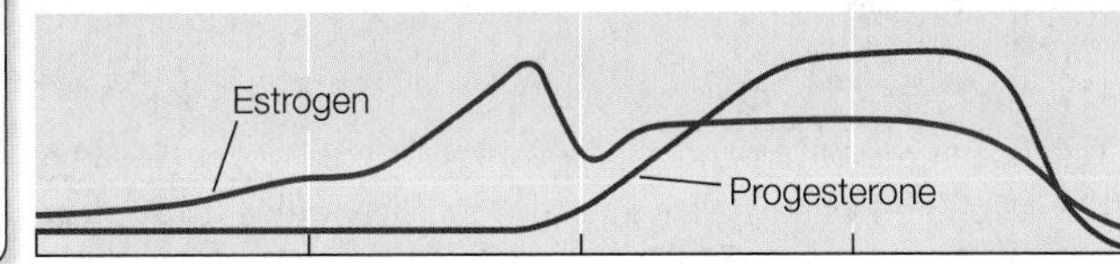

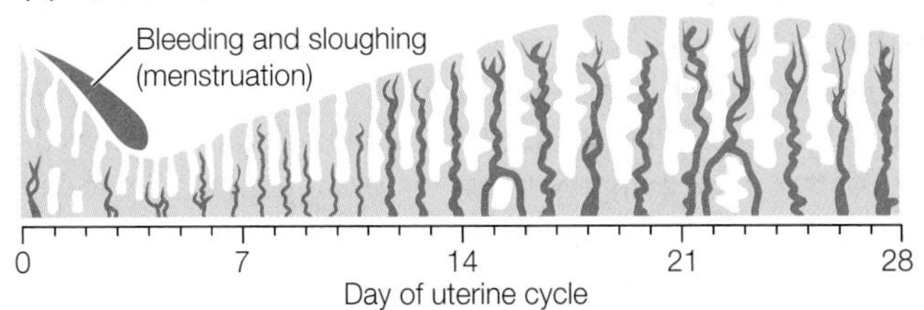

42.14 The Ovarian and Uterine Cycles During a woman's ovarian and uterine cycles, coordinated changes occur in (A) gonadotropin release by the anterior pituitary, (B) the ovary, (C) the release of female sex steroids, and (D) the uterus. The cycles begin with the onset of menstruation; ovulation is at midcycle.

Hormones control and coordinate the ovarian and uterine cycles

The ovarian and uterine cycles are coordinated and timed by the same hormones that initiate sexual maturation. Gonadotropins secreted by the anterior pituitary are the central elements of this control. Before puberty (that is, before about 11 years of age), the secretion of gonadotropins is low and the ovaries are inactive. At puberty, the hypothalamus increases its release of GnRH, stimulating the anterior pituitary to secrete FSH and LH.

In response to FSH and LH, ovarian tissue grows and produces estrogen. The rise in estrogen causes the maturation of the accessory sex organs and the development of female secondary sexual characteristics. Between puberty and menopause, interactions of GnRH, gonadotropins, and sex steroids control the ovarian and uterine cycles.

Menstruation marks the beginning of each uterine and ovarian cycle (see Figure 42.14). A few days before menstruation begins, the anterior pituitary begins to increase its secretion of FSH and LH. In response, some 10 to 20 follicles begin to mature in the ovaries, and these follicles steadily increase their production of estrogen. After about a week, all but one of the follicles wither away.

Estrogen exerts negative feedback control on gonadotropin release by the anterior pituitary during the first 12 days of the ovarian cycle. Then, on about day 12, estrogen exerts positive rather than negative feedback control on the pituitary (**Figure 42.15**).

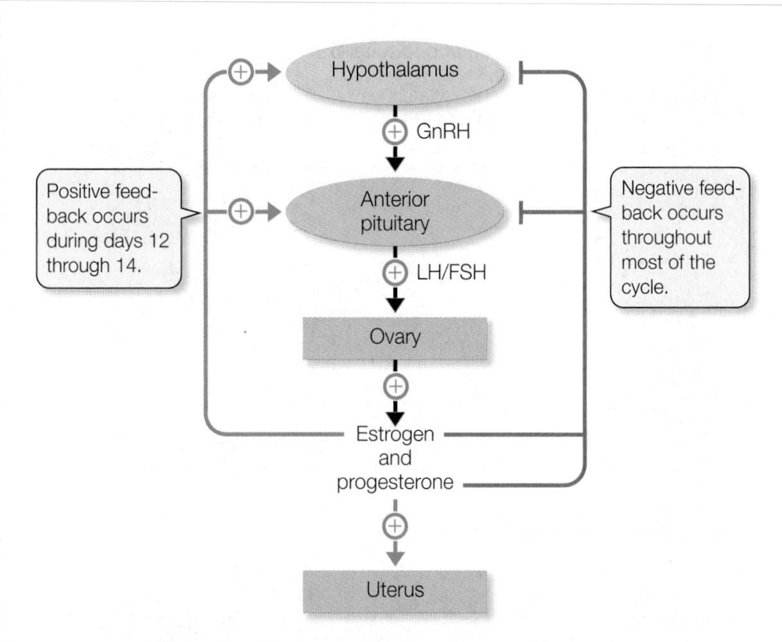

42.15 Hormones Control the Ovarian and Uterine Cycles The ovarian and uterine cycles are under a complex series of positive and negative feedback controls involving several hormones.

As a result, a surge of LH and a lesser surge of FSH occur (see Figure 42.14A). The LH surge triggers the mature follicle to rupture and release its egg, and it stimulates the cells of the ruptured follicle to develop into a corpus luteum.

The corpus luteum becomes an endocrine gland. Estrogen and especially progesterone secreted by the corpus luteum following ovulation are crucial to continued growth and maintenance of the endometrium. In addition, these sex steroids exert negative feedback control on the pituitary, inhibiting gonadotropin release and thus preventing new follicles from beginning to mature.

If the egg is not fertilized, the corpus luteum degenerates on about day 26 of the cycle. Without production of progesterone by the corpus luteum, the endometrium sloughs off, and menstruation occurs. The decrease in circulating steroids also releases the hypothalamus and pituitary from negative feedback control, so GnRH, FSH, and LH all begin to increase. The increase in these hormones induces the next round of follicle development, and the ovarian cycle begins again.

In pregnancy, hormones from the extraembryonic membranes take over

If the egg is fertilized and a blastocyst arrives in the uterus and implants in the endometrium, a new hormone comes into play. A layer of cells covering the blastocyst begins to secrete **human chorionic gonadotropin** (hCG). This gonadotropin, a molecule similar to LH, stimulates the corpus luteum to continue to produce estrogen and progesterone to support the growth and maintenance of the endometrium and thereby prevent menstruation. Because hCG is present only in the blood of pregnant women, the presence of this hormone is the basis for pregnancy testing. Modern pregnancy tests make use of an antibody to detect hCG in urine; thus, they take only minutes and can be done at home. These tests are so sensitive that they are 99 percent accurate and in most cases can detect a pregnancy even before the first missed menstrual period.

Tissues derived from the blastocyst also begin to produce estrogen and progesterone, eventually replacing the corpus luteum as the most important source of these sex steroids. Continued high levels of estrogen and progesterone prevent the pituitary from secreting gonadotropins; thus, the ovarian cycle ceases for the duration of pregnancy. The same mechanism is exploited by birth control pills, which contain synthetic hormones resembling estrogen and progesterone that exert negative feedback control on the hypothalamus and pituitary.

Childbirth is triggered by hormonal and mechanical stimuli

Throughout pregnancy, the muscles of the uterine wall periodically undergo slow, weak, rhythmic contractions called *Braxton-Hicks contractions*. These contractions become gradually stronger during the third trimester of pregnancy and are sometimes called *false labor contractions*. True labor contractions usually mark the beginning of childbirth. Both hormonal and mechanical stimuli contribute to the onset of labor.

Progesterone inhibits and estrogen stimulates contractions of uterine muscle. Toward the end of the third trimester, the estrogen–progesterone ratio shifts in favor of estrogen. The onset of labor is marked by increased oxytocin secretion by the pituitaries of both mother and fetus. Oxytocin is a powerful stimulant of uterine muscle contraction.

Mechanical stimuli come from the stretching of the uterus by the fully grown fetus and the pressure of the fetal head on the cervix. These mechanical stimuli increase the release of oxytocin by the posterior pituitary, which in turn increases the activity of uterine muscle, which causes even more pressure on the cervix. This positive feedback loop converts the weak, slow, rhythmic Braxton-Hicks contractions into stronger labor contractions (**Figure 42.16A**).

In the early stage of labor, the contractions of the uterus are 15–20 minutes apart, and each lasts 45–60 seconds. During this time, hormonal changes and pressure created by the contractions cause the cervix to dilate (expand) until it is large enough to allow the baby to pass through. Gradually the contractions become more frequent and more intense. This stage of labor lasts an average of 12 to 15 hours in a first pregnancy; it is usually 8 hours or less in subsequent ones.

The second stage of labor, called *delivery*, begins when the cervix is fully dilated to a diameter of about 10 centimeters (**Figure 42.16B**). The baby's head moves into the vagina; passage of the fetus through the vagina is assisted by the mother's bearing down ("pushing") with her abdominal and other muscles. Once the head and shoulders of the baby clear the cervix, the rest of its body eases out rapidly, but it is still connected to the placenta by the umbilical cord. Delivery may take as little as a minute, or up to half an hour or more in a first pregnancy.

As soon as the baby clears the birth canal, it can start breathing and become independent of its mother's circulation. The umbilical cord may then be clamped and cut. The segment still attached to the baby dries up and sloughs off in a few days, leaving behind its distinctive signature, the belly button—more properly called the

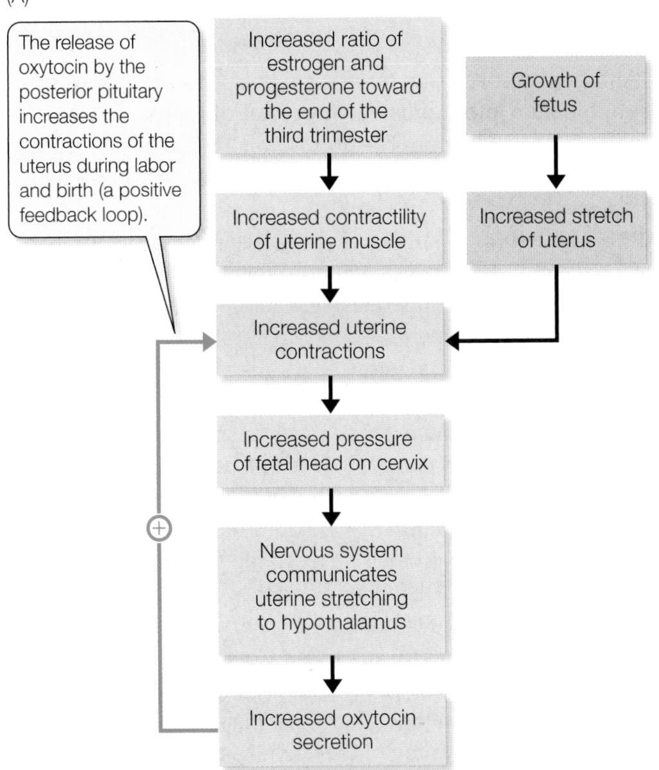

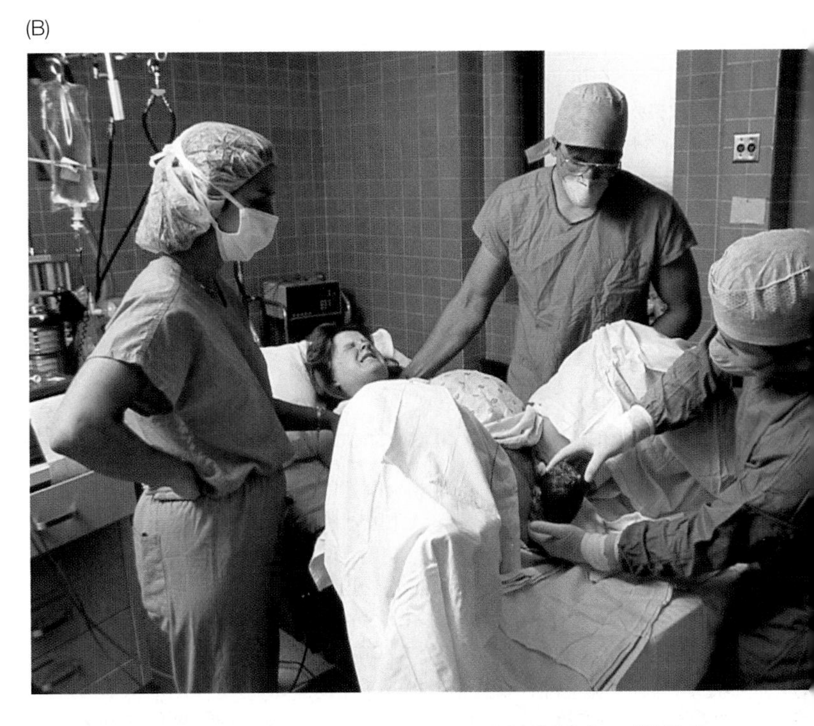

42.16 Control of Uterine Contractions and Childbirth (A) Both mechanical and hormonal signals are involved in stimulating the uterine contractions of labor and delivery. (B) A new person is delivered into the world head first.

umbilicus. The detachment and expulsion of the placenta and fetal membranes takes from a few minutes to an hour, and may be accompanied by uterine contractions. If the baby suckles at the breast immediately following birth, its suckling stimulates additional secretion of oxytocin, which augments uterine contractions that reduce the size of the uterus and help stop bleeding.

42.3 RECAP

The reproductive systems of human males and females produce gametes and hormones, and these functions are controlled by hypothalamic and anterior pituitary hormones. In females the hormonal control of reproductive functions produces linked ovarian and uterine cycles.

- Can you describe the path the human sperm and ovum take in moving from their respective gonads to the point at which fertilization occurs? See Figures 42.9 and 42.12
- Increased production of testosterone at puberty in males stimulates the release of what two hormones of the anterior pituitary? What effect do these hormones have? See p. 909
- Can you explain the events in the ovarian cycle that result in release of a single ovum each month? What events prepare the uterus to receive the egg? See Figures 42.13 and 42.14

Understanding the physiology of human reproduction has led to numerous methods and technologies for controlling it either to prevent unwanted pregnancies or to overcome infertility.

42.4 How Can Fertility Be Controlled and Sexual Health Maintained?

Sexual issues and sexual behavior are dominant aspects of our society, and reproductive technologies have had huge impacts on our sexual and reproductive lives.

Human sexual responses have four phases

The responses of both women and men to sexual stimulation consist of four phases: excitement, plateau, orgasm, and resolution. As sexual *excitement* begins in a woman, her heart rate and blood pressure rise, muscular tension increases, breasts swell, and nipples become erect. Her external genitals, including the sensitive clitoris, swell as they become filled with blood, and the walls of the vagina secrete lubricating fluid that facilitates copulation. In the *plateau* phase, her blood pressure and heart rate rise further, her breathing becomes rapid. The sensitivity once focused in the clitoris spreads over the external genitals, and the clitoris itself becomes even more sensitive. *Orgasm* may last as long as a few minutes, and, unlike men, some women can experience several orgasms in rapid succession. During the *resolution* phase, blood drains from the genitals, and body physiology returns to close to normal.

In the male, the excitement phase is marked by an increase in blood pressure, heart rate, and muscle tension, and penile erection. In the plateau phase, breathing becomes rapid, the diameter of the glans increases, and a clear lubricating fluid from the bulbourethral gland oozes from the penis. Pressure and friction against the nerve endings in the glans and in the skin along the shaft of the penis eventually trigger orgasm. Massive spasms of the muscles in the genital area and contractions in the accessory reproductive organs result in ejaculation.

Within a few minutes after ejaculation, the penis shrinks to its former size, and body physiology returns to resting conditions. The male sexual response includes a *refractory period* immediately after orgasm. During this period, which may last from minutes to hours, a man cannot achieve a full erection or another orgasm, regardless of the intensity of sexual stimulation.

Humans use a variety of methods to control fertility

According to a recent study, almost half of the more than 6 million pregnancies that occur in the United States each year are unintended. The only failure-proof methods of preventing pregnancy are complete abstinence from sexual activity, or the surgical removal of the gonads. Since those approaches are not acceptable to most younger people, they turn to other methods to prevent pregnancy. Many of these methods prevent fertilization or implantation (*conception*) and are therefore referred to as **contraception**.

Some methods of contraception are used by the woman, and others by the man. They vary from blocking gametogenesis to blocking implantation of a blastocyst; they also vary enormously in their effectiveness. **Table 42.1** lists some of the most common methods and their relative failure rates.

For women of college age, a single act of unprotected intercourse in the two days prior to ovulation carries a chance of conception as high as 50 percent.

NONTECHNOLOGICAL APPROACHES An approach to contraception that does not involve physical or pharmacological technologies is periodic abstinence (the "rhythm method"), which attempts to separate sperm and egg in time. The couple avoids sex from day 10 to day 20 of the ovarian cycle, when the woman is most likely to be fertile. The cycle can be tracked by use of a calendar, supplemented by the basal body temperature method, which is based on the observation that a woman's body temperature drops on the day of ovulation and rises sharply on the day after. Changes in the stickiness of the cervical mucus also help identify the day of ovulation.

The "rhythm method" has a high failure rate. Sperm deposited in the female reproductive tract may remain viable for up to 6 days. Similarly, the egg remains viable for 12 to 36 hours after ovulation. These facts, added to individual variation in the timing of ovulation, result in an annual failure rate of between 15 and 35 percent for the rhythm method. In other words, 15 to 35 percent of women using only the rhythm method over the course of 1 year will become pregnant during that time.

Another approach is to try to separate sperm and egg in space through **coitus interruptus**—withdrawal of the penis from the vagina before ejaculation. The annual failure rate of this method (mostly due to lack of willpower) may be as high as 40 percent.

BARRIER METHODS Techniques that place a physical barrier between egg and sperm have been used for centuries. The condom is a sheath made of an impermeable material such as latex that can be fitted over the erect penis. A condom traps semen so that sperm do not enter the vagina. Latex condoms also help prevent the spread of many sexually transmitted diseases. In theory, condom use has a failure rate near zero. However, in practice the annual failure rate is about 15 percent due to leakage because of tearing or poor fit (for example, with the loss of the erection).

Diaphragms and *cervical caps* are dome-shaped pieces of rubber that fit over the woman's cervix to block sperm from entering the uterus. Both the diaphragm and the cervical cap are treated first with jelly or cream containing a *spermicide* (a chemical that kills or incapacitates sperm) and then inserted through the vagina before sexual intercourse. Annual failure rates are about 15 percent, the same as for male condoms. A female condom, which creates an impermeable lining of the vagina, is also available.

Spermicidal foams, jellies, and creams can be used alone by placing them in the vagina with special applicators. Used in this way, they have an annual failure rate of 25 percent or more. *Douching* (flushing the vagina with liquid) after intercourse, despite popular belief, is almost useless as a method of birth control, because sperm can arrive in the oviducts (which cannot be flushed) within 10 minutes after ejaculation.

PREVENTING OVULATION Oral contraceptives, or birth control pills, are the leading method of birth control in the United States and

TABLE 42.1

Methods of Contraception

METHOD	MODE OF ACTION	FAILURE RATE[a]
Unprotected	No form of birth control	85
Douche	Supposedly flushes sperm from vagina	80
Periodic abstinence	Abstinence near time of ovulation	15–35
Coitus interruptus (withdrawal prior to ejaculation)	Prevents sperm from reaching egg	10–40
Vaginal jelly or foam	Kills sperm; blocks sperm movement	3–30
Diaphragm/jelly	Prevents sperm from entering uterus; kills sperm	3–25
Condom	Prevents sperm from entering vagina	3–20
Intrauterine device (IUD)	Prevents implantation of fertilized egg	0.5–6
RU-486	Prevents development of fertilized egg	0–15
Birth control pill	Prevents ovulation	0–3
Vasectomy	Prevents release of sperm	0–0.15
Tubal ligation	Prevents egg from entering uterus	0–0.05

[a]Number of pregnancies per 100 women per year.

are used by almost 12 million women. Used correctly, "the pill" is the most effective method of contraception other than sterilization, with an annual failure rate of less than 1 percent.

Birth control pills work by preventing ovulation. Their mechanisms of action take advantage of the roles of estrogen and progesterone as negative feedback signals to the hypothalamus and the pituitary. The most common pills contain low doses of synthetic estrogens and progesterones (progestins). By keeping the circulating levels of gonadotropins low, these hormones interfere with the maturation of follicles and eggs, suspending the ovarian cycle. The uterine cycle is usually allowed to continue, however, by including in the daily regime pills that contain no hormones every 21 to 23 days.

The negative side effects of oral contraceptives include somewhat increased risk of blood clot formation, heart attack, stroke, and breast cancer. However, these side effects are associated mostly with pills containing higher hormone concentrations than are used in the modern formulations. For birth control pills in use today, the risk is low for non-smokers, especially women under the age of 35. The risk of death from using birth control pills is far less than that associated with a full-term pregnancy.

Long-lasting injectable or implantable steroids are also used to block ovulation through negative feedback effects. Depo-Provera is an injectable progestin that blocks the release of gonadotropins for several months. Another device, called Norplant, consists of thin, flexible tubes filled with progestin. Several of these tubes are inserted under the skin, where they continue to release progestin slowly for years. Another means of controlling blood levels of sex steroids that does not require a daily pill is the transdermal patch.

Another contraceptive device is the plastic *vaginal ring* which is placed around the cervix where it steadily releases estrogen and progestins to prevent ovulation. It is left in place for 3 weeks, removed to allow menstruation, and then a new ring is put in place.

PREVENTING IMPLANTATION A highly effective method of contraception (failure rate 1–7 percent) is the intrauterine device, or IUD, a small piece of plastic or copper that is inserted into the uterus. The IUD works by causing an inflammatory response that includes the release of prostaglandins, which prevent implantation of the fertilized egg.

Another way of interfering with implantation is through the use of "morning-after pills," which deliver high doses of steroids, primarily estrogens. By acting in several ways on the oviducts and the endometrium, this treatment prevents implantation. Morning-after pills can be effective up to several days after sexual intercourse.

The drug RU-486 is not a contraceptive pill, but a *contragestational* pill. It blocks progesterone receptors, thereby interfering with the normal action of progesterone produced by the corpus luteum, which is necessary for the maintenance of the endometrium in early pregnancy. If RU-486 is administered as a morning-after pill, it prevents implantation. However, RU-486 can be effective even if taken at the time of the first missed menstrual period, after implantation has begun. After a few days of treatment with RU-486, the endometrium regresses and sloughs off, along with the embryo, which is in very early stages of development.

STERILIZATION One foolproof method of contraception is sterilization. Male sterilization by **vasectomy** is a simple operation (cutting and tying of the vasa deferentia) that can be performed under a local anesthetic in a doctor's office (**Figure 42.17A**). After this minor surgery, the semen no longer contains sperm. Sperm production continues, but since the sperm cannot move out of the testes, they are destroyed by macrophages. Vasectomy does not affect a man's hormone levels or his sexual responses, and even the amount of semen he ejaculates is essentially unchanged.

In female sterilization, the aim is to prevent the egg from traveling to the uterus and to block sperm from reaching the egg. The most common method is **tubal ligation**: cutting and tying of the oviducts (**Figure 42.17B**). Alternatively, the oviducts may be burned (cauterized) to seal them off. As in the male, these procedures do not alter reproductive hormones or sexual responses. In the United States, female sterilization is second only to use of the pill as a means of limiting fertility.

ABORTION Once a fertilized egg is successfully implanted in the uterus, any termination of the pregnancy is called an **abortion**. A *spontaneous abortion* is the medical term for what most people call a miscarriage. Miscarriages are common early in pregnancy. Most miscarriages are due to a chromosomal abnormality in the fetus or to a breakdown in the process of implantation. Miscarriages often occur before a woman even realizes she is pregnant. Studies indicate that as many as 50 percent of all human conceptions may miscarry before implantation, and of those that implant

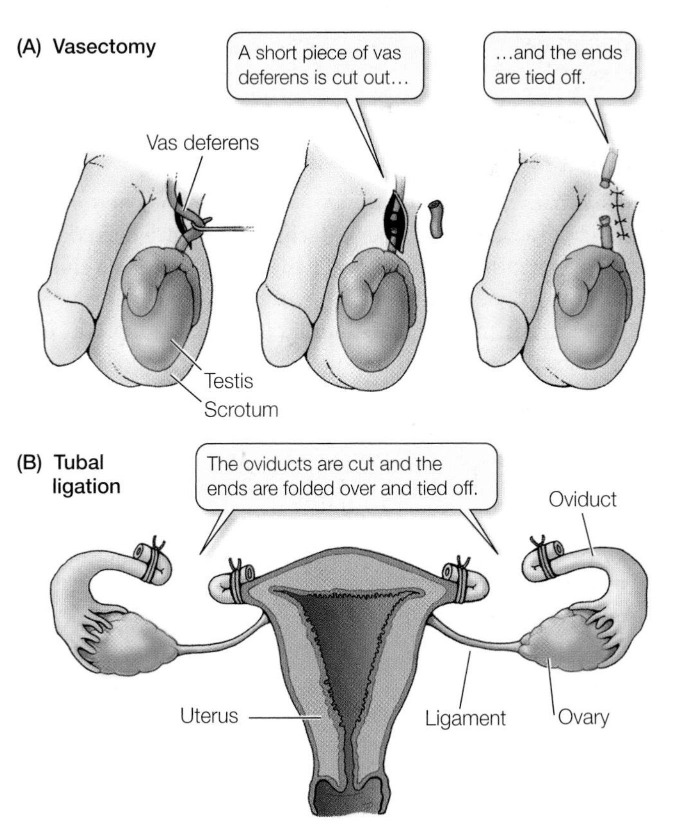

42.17 Sterilization Techniques (A) Vasectomy is the technique for male sterilization. (B) Tubal ligation is the sterilization procedure most commonly performed on human females.

about 25 percent miscarry before the pregnancy can be clinically confirmed. About 10 percent of clinically confirmed pregnancies miscarry.

Abortions that are the result of medical intervention may be performed either for therapeutic purposes or for fertility control. A therapeutic abortion may be necessary to protect the health of the mother, or it may be performed because prenatal testing reveals that the fetus has a severe defect. Of the approximately 3 million unintended pregnancies in the United States each year, almost half are ended by abortion.

In a medical abortion, the cervix is dilated and the endometrium and implanted fetus are removed from the uterus. When performed in the first trimester of a pregnancy, a medical abortion carries less risk of death to the mother than a full-term pregnancy. After the first 12 weeks of pregnancy, the risk rises, but even through the second trimester it is less than that of a full-term pregnancy.

CONTROLLING MALE FERTILITY You may well ask why all the pharmacological approaches to controlling fertility are applied to females. The control of male fertility is a difficult problem. First, spermatogenesis is a continuous rather than a cyclical event, and it is difficult to block a particular step in a continuous process. The ovarian cycle is more vulnerable to manipulation because certain events must happen at certain times and in a certain sequence for ovulation and implantation to occur. Second, the suppression of spermatogenesis must be total to be effective, since technically it takes only a single sperm to fertilize an egg, and normally millions are produced continuously. Such suppression requires powerful and constant chemical intervention, with associated side effects.

Reproductive technologies help solve problems of infertility

There are many reasons why a man and woman may not be able to have children, and a number of technologies have been developed to overcome barriers to conceiving and/or bearing a child.

The simplest treatment available is **artificial insemination**, in which the physician positions sperm in the female's reproductive tract. This technique is useful if the male's sperm count is low, if his sperm lack motility, or if problems in the female's reproductive tract prevent the normal movement of sperm up to and through the oviducts. Artificial insemination is used widely in the production of domesticated animals such as cattle.

More recent advances, called **assisted reproductive technologies**, or **ARTs**, involve procedures that remove unfertilized eggs from the ovary, combine them with sperm outside the body, and then place fertilized eggs or egg–sperm mixtures in the appropriate location in the female's reproductive tract for development to take place.

Approximately 15 percent of all couples in the United States are infertile. The causes are about equally distributed between men and women.

The first successful ART was *in vitro fertilization* (*IVF*). In IVF, the female is treated with hormones that stimulate many follicles in her ovaries to mature. Eggs are collected from these follicles, and sperm are collected from the male. Eggs and sperm are combined in a culture medium outside the body and fertilization takes place. The resulting embryos can be injected into the mother's uterus in the blastocyst stage or kept frozen for implantation later. The first "test-tube baby" resulting from IVF was born in 1978. Since that time, more than 3 million babies have been produced by this ART.

A major cause of failure of IVF is failure of sperm to gain access to the egg plasma membrane (see Figure 42.6). To solve this problem, methods have been developed to inject a sperm cell directly into the cytoplasm of an egg. In *intracytoplasmic sperm injection* (*ICSI*), an egg is held in place by suction applied to a polished glass pipette. A slender, sharp pipette is then used to penetrate the egg and inject a sperm (**Figure 42.18**). This ART was used successfully for the first time in 1992; now thousands of these procedures are performed in U.S. clinics each year, with a success rate of about 25 percent.

IVF, coupled with sensitive techniques of genetic analysis, can eliminate the risk that adults who are carriers of genetic diseases will produce affected children. It is now possible to take a cell from a blastocyst at the 4- or 8-cell stage without damaging its developmental potential. The sampled cell can be subjected to molecular analysis to determine whether it carries the harmful gene. This procedure, called *preimplantation genetic diagnosis* (PGD) makes it possible to determine whether an embryo produced by IVF carries the genetic defect of concern.

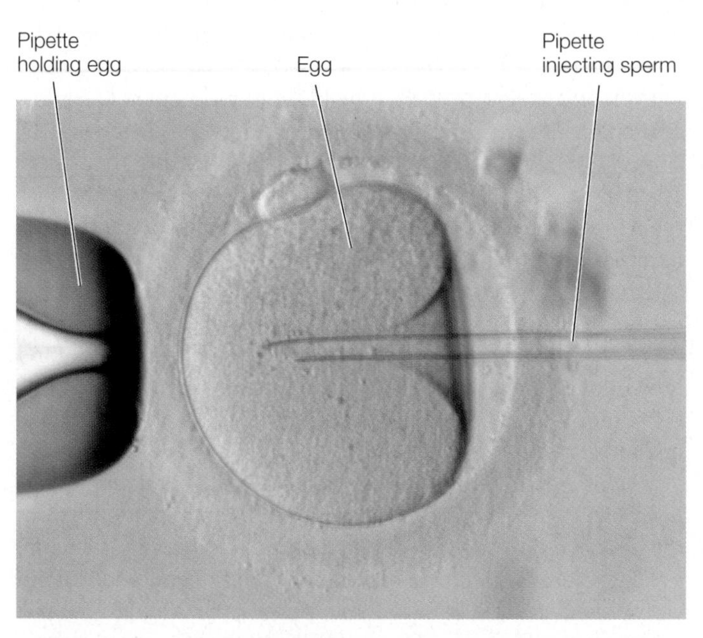

42.18 Intracytoplasmic Sperm Injection In this procedure, sperm are injected directly into a mature egg cell. The fertilized egg is then placed in the female reproductive tract, where it can implant and develop into a fetus.

TABLE 42.2

Some Sexually Transmitted Diseases

DISEASE	INCIDENCE IN UNITED STATES	SYMPTOMS
Syphilis	80,000 new cases/yr	Primary stage (weeks): skin lesion (chancre) at site of infection Secondary stage (months): skin rash and flu-like symptoms; may be followed by a latent period Tertiary stage (years): deterioration of the cardiovascular and central nervous systems; death
Gonorrhea	800,000 new cases/yr	Pus-filled discharge from penis or vagina; burning urination. Infection can also start in throat or rectum
Chlamydia	>4,000,000 new cases/yr	Symptoms similar to gonorrhea, although often there are no obvious symptoms. Can lead to pelvic inflammatory disease in females
Genital herpes	500,000 new cases/yr	Small blisters that can cause itching or burning sensations are accompanied by inflammation and by secondary infections
Genital warts	10% of adults infected	Small growths on genital tissues. Increases risk of cervical cancer in women
Hepatitis B	5–20% of population	Fatigue, fever, nausea, loss of appetite, jaundice, abdominal pain, muscle and joint pain. Can lead to destruction of liver or liver cancer
HIV / AIDS	Approximately 900,000 cases[a]	Failure of the immune system (see Section 18.6)

[a]HIV/AIDS is widespread in other parts of the world, most notably in the southern part of the African continent. The infection is spreading rapidly in Southeast Asia and India. Estimated number of people infected with HIV worldwide in 2006 was 40 million.

Sexual behavior transmits many disease organisms

Disease-causing organisms usually have a very limited ability to survive outside a host organism, which means getting from host to host is a major evolutionary challenge. One of the most intimate types of contact that hosts can have is copulation. It is not surprising, then, that many pathogens have evolved to depend on sexual contact between their hosts as their means of transmission. These organisms are the causes of **sexually transmitted diseases** (commonly referred to as **STDs**), and they include viruses, bacteria, yeasts, and protists. A summary of the most common STDs is presented in **Table 42.2**.

STDs have been with humans since ancient times, and they are one of the most serious public health problems today. Over 10 million new cases of STDs occur each year in the United States, and about two-thirds of these cases occur in people between the ages of 15 and 30. About half of U.S. youth will contract an STD before the age of 25. The only contraceptive device that is effective against the transmission of STDs is the condom.

42.4 RECAP

Controlling fertility and maintaining sexual health are important aspects of human life. Decreasing the probability of pregnancy is achieved through methods that prevent sperm and egg from meeting and from preventing implantation. Pregnancies can be facilitated through medical technology.

- Can you describe the various barrier methods of contraception? See Table 42.1
- Do you understand how implantation of a blastocyst can be prevented? See p. 915

CHAPTER SUMMARY

42.1 How do animals reproduce without sex?

Asexual reproduction produces offspring that are genetically identical to their parent and to one another. A disadvantage of asexual reproduction is that no genetic diversity is produced.

Means of asexual reproduction include **budding**, **regeneration**, and **parthenogenesis**. Review Figures 42.1 and 42.2

42.2 How do animals reproduce sexually?

Sexual reproduction consists of three basic steps: **gametogenesis**, **mating**, and **fertilization**.

Gametogenesis and fertilization are similar in all animals, but mating includes a great variety of anatomical, physiological, and behavioral adaptations.

In sexually reproducing species, genetic diversity is created by crossing over and independent assortment of chromosomes during gametogenesis. Fertilization also contributes to genetic diversity.

Gametogenesis occurs in **testes** and **ovaries**. In **spermatogenesis** (the production of sperm) and **oogenesis** (the production of eggs), the **germ cells** proliferate mitotically, undergo meiosis, and mature into gametes.

Each **primary spermatocyte** can produce four haploid sperm through the two divisions of meiosis. Review Figure 42.3A

Primary oocytes immediately enter prophase of the first meiotic division, and in many species, including humans, their development is arrested at this point. Each **oogonium** produces only one egg. Review Figure 42.3B

Fertilization involves sperm activation, species-specific binding of sperm to egg, the **acrosomal reaction**, digestion of a path through the protective coverings of the egg, and fusion of sperm and egg plasma membranes. Fusion of these two membranes triggers **blocks to polyspermy**, which prevent additional sperm from entering the egg, and in mammals, signal the egg to complete meiosis and begin development. Review Figure 42.5, Web/CD Tutorial 42.1

Fertilization can occur externally, as is common in aquatic species, or internally, as is common in terrestrial species. Internal fertilization usually involves copulation.

Hermaphroditic, or **monoecious**, species have both male and female reproductive systems in the same individual, either sequentially or simultaneously. **Dioecious** species have separate male and female members.

Internal fertilization is necessary for terrestrial species. The shelled egg is an important adaptation to the terrestrial environment.

Animals can be classified as **oviparous** or **viviparous**, depending on whether the early stages of development occur outside or inside the mother's body.

42.3 How do the human male and female reproductive systems work?

Males produce **semen** and deliver it into the female reproductive tract. Semen consists of sperm suspended in seminal fluid, which nourishes them and facilitates fertilization.

Sperm are produced in the **seminiferous tubules** of the testes, mature in the **epididymis**, and are delivered to the **urethra** through the **vas deferens**. Other components of semen are produced in the **seminal vesicles**, **prostate gland**, and **bulbourethral gland**. Review Figures 42.9 and 42.10, Web/CD Activities 42.1 and 42.2

All components of the semen join in the urethra at the base of the penis and are ejaculated through the erect penis by muscle contractions at the culmination of copulation.

Spermatogenesis depends on testosterone secreted by the **Leydig cells** of the testes, which are under the control of LH from the anterior pituitary. Spermatogenesis is also controlled by FSH from the pituitary. Hypothalamic GnRH controls pituitary secretion of LH and FSH. The production of these hormones by the hypothalamus and pituitary is controlled by negative feedback from testosterone and another hormone, inhibin, produced by the **Sertoli cells** of the testes. Review Figure 42.11

Eggs mature in the female's ovaries and are released into the **oviducts**. Sperm deposited in the vagina during copulation move up through the **cervix** and **uterus** into the oviducts. Review Figure 42.12, Web/CD Activity 42.3

Fertilization occurs in the upper regions of the oviducts. The zygote becomes a **blastocyst** as it passes down the oviduct. Upon arrival in the uterus, the blastocyst implants in the **endometrium** and forms a **placenta**.

The maturation and release of eggs constitute an **ovarian cycle**. In humans, this cycle takes about 28 days. Review Figure 42.13

The uterus also undergoes a cycle that prepares it for receipt of a blastocyst. If no blastocyst is implanted, the lining of the uterus sloughs off in the process of **menstruation**. Review Figure 42.14, Web/CD Tutorial 42.2

Both the ovarian and the uterine cycles are under the control of hypothalamic and pituitary hormones, which in turn are under the feedback control of estrogen and progesterone. Review Figure 42.15

Childbirth is initiated by hormonal and mechanical stimuli that increase the contraction of uterine muscle. Review Figure 42.16

42.4 How can fertility be controlled and sexual health maintained?

Human sexual responses consist of four phases: excitement, plateau, orgasm, and resolution. In addition, males have a refractory period during which renewed excitement is not possible.

Methods of contraception include abstention from copulation and the use of technologies that decrease the probability of fertilization. Review Table 42.1

Barrier methods of contraception, such as condoms, diaphragms, and spermicidal substances, kill sperm or block their passage through the female reproductive tract.

Methods to prevent ovulation, such as birth control pills and other hormonal treatments, interfere with the ovarian cycle so that mature, fertile eggs are not produced and released.

Males and females can be sterilized by surgical blockage of the vasa deferentia (**vasectomy**) or oviducts (**tubal ligation**). Review Figure 42.17

Methods to prevent implantation include intrauterine devices, excess doses of steroids, and a progesterone receptor blocker. After implantation, the termination of a pregnancy is called an **abortion**.

Assisted reproductive technologies have been developed to increase fertility.

Many disease-causing organisms are transmitted through sexual behavior. Many **sexually transmitted diseases** are curable if treated early, but can have serious long-term consequences if not treated. Review Table 42.2

SELF-QUIZ

1. A species in which the individual possesses both male and female reproductive systems is termed
   a. dioecious.
   b. parthenogenetic.
   c. hermaphroditic.
   d. monoecious.
   e. ovoviviparous.
2. The major advantage of internal fertilization is that
   a. it ensures paternity.
   b. it permits the fertilization of many gametes.
   c. it reduces the incidence of destructive competitive interactions between the members of a group.
   d. it increases the number of sperm that have access to each egg.
   e. it gives the developing organism a greater degree of protection during the early phases of development.
3. Which statement about oocytes is *true*?
   a. At birth, the human female has produced all the oocytes she will ever produce.
   b. At the onset of puberty, ovarian follicles produce new oocytes in response to hormonal stimulation.
   c. At the onset of menopause, the human female stops producing oocytes.
   d. Oocytes are produced by the human female throughout adolescence.
   e. Oocytes produced by the female are stored in the oviducts.
4. Spermatogenesis and oogenesis differ in that
   a. spermatogenesis produces gametes with greater stores of raw materials than those produced by oogenesis.
   b. spermatocytes remain in prophase of the first meiotic division longer than oocytes.
   c. oogenesis produces four equally functional haploid cells per meiotic event and spermatogenesis does not.
   d. spermatogenesis produces many gametes with meager energy reserves, whereas oogenesis produces relatively few, well-provisioned gametes.
   e. spermatogenesis begins before birth in humans, whereas oogenesis does not start until the onset of puberty.
5. Semen contains all of the following except
   a. fructose.
   b. mucus.
   c. clotting enzymes.
   d. substances to lower the pH of the uterine environment.
   e. substances to increase the contraction of the uterine muscle.
6. During oogenesis in mammals, the second meiotic division occurs
   a. in the formation of the primary oocyte.
   b. in the formation of the secondary oocyte.
   c. before ovulation.
   d. after fertilization.
   e. after implantation.
7. One of the major differences between the sexual responses of human males and females is
   a. the increase in blood pressure in males.
   b. the increase in heart rate in females.
   c. the presence of a refractory period in females.
   d. the presence of a refractory period in males after orgasm.
   e. the increase in muscle tension in males.
8. Which of the following is *true* of sexually transmitted diseases?
   a. They are always caused by viruses or bacteria.
   b. Using contraception will prevent them.
   c. The organisms that cause them have evolved to depend on intimate physical contact between hosts as their means of transmission.
   d. Their transmission has a high probability of failure.
   e. You cannot catch one from someone you love.
9. Contractions of muscles in the uterine wall and in the breasts are stimulated by
   a. progesterone.
   b. estrogen.
   c. prolactin.
   d. oxytocin.
   e. human chorionic gonadotropin.
10. Which method of contraception is *most* likely to fail?
    a. Rhythm method
    b. Birth control pills
    c. Diaphagms
    d. Vasectomy
    e. Condoms

FOR DISCUSSION

1. In the very deep ocean, there are species of fish in which the male is very much smaller than the female and actually lives attached to her body. In terms of the selective pressures that operate on sexual and asexual reproduction and in terms of the deep-sea environment, what factors do you think resulted in the evolution of this extreme sexual dimorphism?
2. What are two main differences between the immediate products of the first and second meiotic divisions in spermatogenesis and oogenesis? Why do these differences exist?
3. At the beginning of each ovarian cycle in humans, six to twelve follicles begin to develop in response to rising levels of FSH, but after a week, only one follicle continues to develop, and the others wither away. Given that follicles produce estrogen, estrogen stimulates follicle cells to produce FSH receptors, and estrogen exerts negative feedback on FSH production in the pituitary, can you explain how one follicle is "selected" to grow?
4. Compare the actions of LH and FSH in the ovaries and testes.
5. Ovarian and uterine events in the month following ovulation differ depending on whether fertilization occurs. Describe the differences and explain their hormonal controls.

FOR INVESTIGATION

No male contraceptive methods exist other than the condom. How would you go about developing a pharmacological method to block sperm production that could lead to a male pill without affecting the maintenance of male secondary sexual characteristics or male sexual behavior?

CHAPTER 43 Animal Development: From Genes to Organisms

Thar she blows!

The whale blows its nose from the top of its head. The spout from the whale's blowhole is exhaled air and water vapor from its nasal passages. It is adaptive for a marine mammal to breathe out of the top of its head because most of its body can remain under water as it breathes. But in most terrestrial mammals, the nose is on the front of the head. How did the whale's nose get to the top of its head? This is an evolutionary question, but the answer is to be found in development—the processes whereby a fertilized egg becomes an adult organism.

The vertebrate body varies enormously among species in form and function, yet its basic structural design does not. For example, the whale flipper, the bat wing, and the human arm all have the same bones. During development, these bones assume different shapes and dimensions to adapt the forelimbs of each species to various functions: swimming, flying, and tool use.

Similarly, all vertebrates have the same bones in their heads, but through development, these bones grow differently, and therefore the skull takes on different shapes in different species. In both whales and humans, the nasal passages are in the nasal bone, which is just above the bones of the upper jaw. In the human, that places the nasal bone just above the jaw on the front of the face. Things are different in the whale. During development, the bones of the whale's upper jaw grow enormously relative to the other bones of its skull. These jaw bones project far forward to form the cavernous mouth. As a result of this differential forward growth of the jaw bone, the nasal bone ends up on the top of the skull, rather than on the front. Thus, the answers to how the whale's nose ends up on the top of its head and how its forelimbs become flippers are found in the processes of development. These processes form and shape the components of the basic vertebrate body plan.

Development begins with the joining of sperm and egg. As described in Section 42.2, in many species the egg will attract a large number of sperm, but only one sperm can supply the genetic material to fertilize the egg. Biochem-

Thar She Blows! The nasal passages of whales such as *Balaenoptera musculus* (the blue whale) are on top of its head because of the extreme growth of its jaw bones during development.

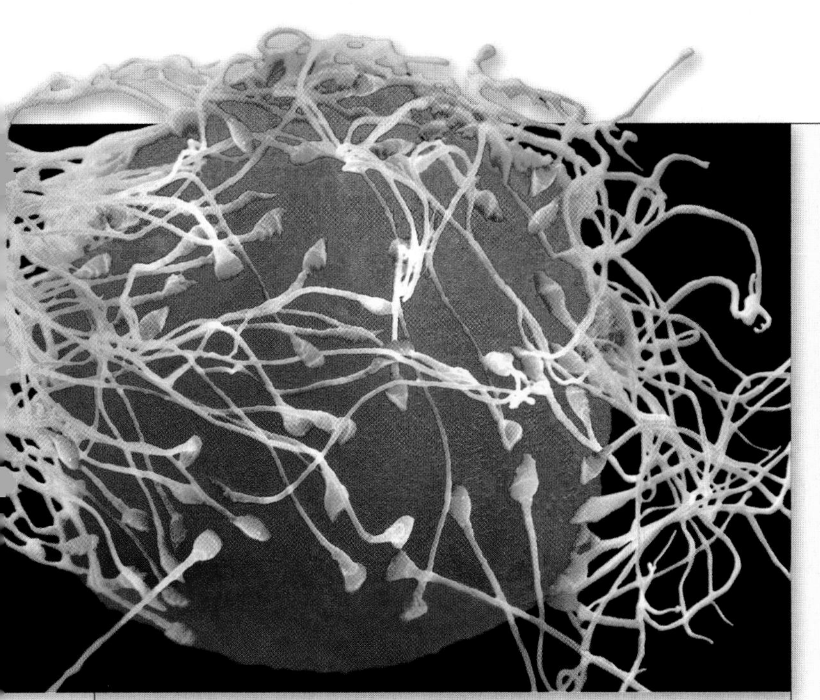

When Sperm Meet Egg Development begins with the fertilization of an egg by a single sperm. Humans are typical of animals in that the egg (artificially colored pink) is much larger than the sperm (gold). The egg cytoplasm is loaded with factors that will direct development and nourish the growing embryo.

ical blocks to polyspermy ensure that the proper genetic complement for the species will be achieved.

The fertilized egg goes through an initial rapid series of cell divisions without growth that subdivides the egg cytoplasm into a mass of smaller undifferentiated cells. Although this mass of cells shows no hints of the eventual body plan, the uneven distribution of molecules in the cytoplasm of the fertilized egg provides positional information that will direct the fates of cells and set up the body plan.

IN THIS CHAPTER we will see how a single cell becomes a multicellular animal through orderly cell movements that create multiple layers and set up cell–cell interactions. The regional and temporal differences in gene expression that control cell differentiation, described in Chapters 19 and 20, lead to the emergence of the body plan of the animal. We will discuss these early developmental steps in four model organisms that have been studied extensively: sea urchins, frogs, chicks, and humans.

CHAPTER OUTLINE

43.1 How Does Fertilization Activate Development?

Two things must be noted at the outset of this discussion. First, in animals that reproduce asexually, development proceeds without fertilization. And second, critical developmental steps have already been taken in the maturation of the egg in all animals. Therefore, the question we really are asking is how the act of fertilization re-starts or activates development in sexually reproducing animals.

In sexually reproducing animals, fertilization does more than just restore a full diploid complement of maternal and paternal genes. The entry of a sperm into an egg sets up blocks to polyspermy, causes ion fluxes, changes pH, stimulates protein synthesis, increases the metabolism of the egg, and initiates the rapid series of cell divisions that produce a multicellular embryo. In many species, the point of entry of the sperm creates an asymmetry in the radially symmetrical egg. This asymmetry enables the emergence of a bilateral body plan from the radial symmetry of the egg. We described the mechanisms of fertilization in Section 42.2. Here we take a closer look at the cellular and molecular interactions of sperm and egg that initiate the first steps of development.

The sperm and the egg make different contributions to the zygote

As shown in the above micrograph, eggs are much larger than sperm. Egg cytoplasm is well stocked with organelles, nutrients, and a variety of molecules including transcription factors and mRNAs. The sperm is little more than a DNA delivery vehicle. Therefore, nearly everything that the embryo needs during its early stages of development comes from the mother. In addition to its haploid nucleus, however, the sperm makes one other important contribution to the zygote in some species: a centriole. The centriole becomes the centrosome of the zygote, which produces the mitotic spindles for subsequent cell divisions (see Figure 9.9).

The entry of the sperm into the egg stimulates rearrangements of the egg cytoplasm that establish the polarity of the embryo. As mentioned in Section 19.4, the nutrients and molecules in the cytoplasm of the zygote are not homogeneously distributed, and therefore will not be divided equally among all daughter cells when cell divisions begin. This unequal distribution of cytoplasmic determinants sets the stage for the signaling cascades that orchestrate the sequential steps of development: determination, differentiation, and morphogenesis. A good example of these events is provided by the frog, an organism in which they have been well studied.

Rearrangements of egg cytoplasm set the stage for determination

The rearrangements of egg cytoplasm in some frog species are easily observed because of pigments in the egg cytoplasm. The nutrient molecules in an unfertilized frog egg are dense, and they are therefore concentrated by gravity in the lower half of the egg, which is called the **vegetal hemisphere**. The haploid nucleus of the egg is located at the opposite end of the egg, in the **animal hemisphere**. The outermost (*cortical*) cytoplasm of the animal hemisphere is heavily pigmented, and the underlying cytoplasm has more diffuse pigmentation. The vegetal hemisphere is not pigmented. How is the cytoplasm rearranged when the egg is fertilized?

The surface of the frog egg has specific sperm-binding sites, but a sperm always enter the egg in the animal hemisphere. When a sperm enters, the cortical cytoplasm rotates toward the site of sperm entry. This rotation reveals a band of diffusely pigmented cytoplasm on the side of the egg opposite the site of sperm entry. This band, called the **gray crescent**, will be the site of important developmental events (**Figure 43.1**).

The cytoplasmic rearrangements that create the gray crescent bring different regions of cytoplasm into contact with each other on opposite sides of the egg. Therefore, bilateral symmetry is imposed on what was a radially symmetrical egg. In addition to the up–down difference of the animal and vegetal hemispheres that defines the anterior–posterior axis of the embryo, the movement of the cytoplasm sets the stage for the creation of the dorsal–ventral (back–belly) body axis. In the frog, the site of sperm entry will become the *ventral* region of the embryo, and the gray crescent will become the *dorsal* region. The region of the gray crescent that borders the animal pole cortical cytoplasm points anteriorly, and the gray crescent region that borders the vegetal pole cytoplasm points posteriorly.

The one non-nuclear organelle that the sperm contributes to the egg—the centriole—initiates cytoplasmic reorganization. The sperm centriole organizes the microtubules in the vegetal hemisphere cytoplasm into a parallel array that guides the movement of the cortical cytoplasm. These microtubules also appear to be directly responsible for movement of specific organelles and proteins, because these organelles and proteins move from the vegetal hemisphere to the gray crescent region even faster than the cortical cytoplasm rotates.

As a result of movement of cytoplasm, proteins, and organelles, the distribution of critical developmental signals changes. A key transcription factor in early development is β-catenin, which is produced from maternal mRNA and is found throughout the cytoplasm of the egg. Also present throughout the egg cytoplasm is a protein kinase called glycogen synthase kinase-3 (GSK-3), which phosphorylates and thereby targets β-catenin for degradation. However, an inhibitor of GSK-3 is segregated in the vegetal cortex of the egg. After sperm entry, this inhibitor is moved along microtubules to the gray crescent, where it prevents the degradation of β-catenin. As a result, the concentration of β-catenin is higher on the dorsal side than on the ventral side of the developing embryo (**Figure 43.2**).

Evidence supports the hypothesis that β-catenin is a key player in the cell–cell signaling cascade that begins the process of cell determination and the formation of the embryo in the region of the gray crescent. But before cell–cell signaling can occur, multiple cells must be in place; let's turn to the early series of cell divisions that transforms the zygote into a multicellular embryo.

Cleavage repackages the cytoplasm

The transformation of the diploid zygote into a mass of cells occurs through a rapid series of cell divisions called **cleavage**. Because the cytoplasm of the zygote is not homogeneous, these first cell divisions result in the differential distribution of nutrients and cytoplasmic determinants among the cells of the early embryo. In most animals, cleavage proceeds with rapid DNA replication and mitosis, but no cell growth and little gene expression. The embryo becomes a solid ball of smaller and smaller cells, called a *morula* (from the Latin word for "mulberry"). Eventually, this ball forms a central fluid-filled cavity called a **blastocoel**, at which point the embryo is called a **blastula**. Its individual cells are called **blastomeres**. The pattern of cleavage in different species influences the form of their blastulas.

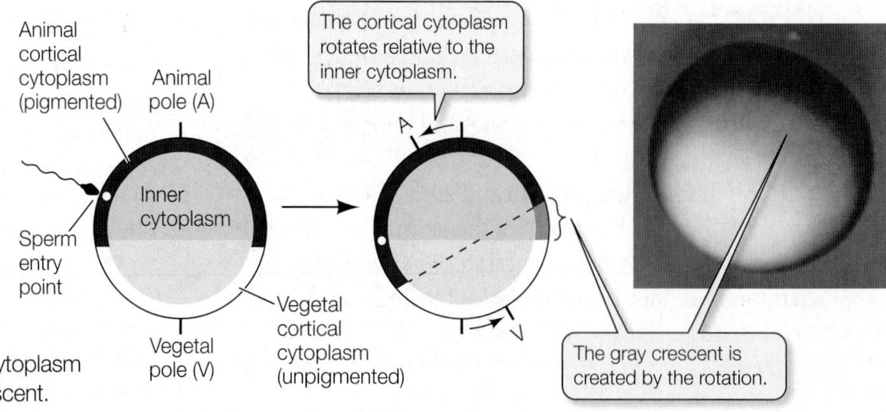

43.1 The Gray Crescent Rearrangement of the cytoplasm of the frog eggs after fertilization creates the gray crescent.

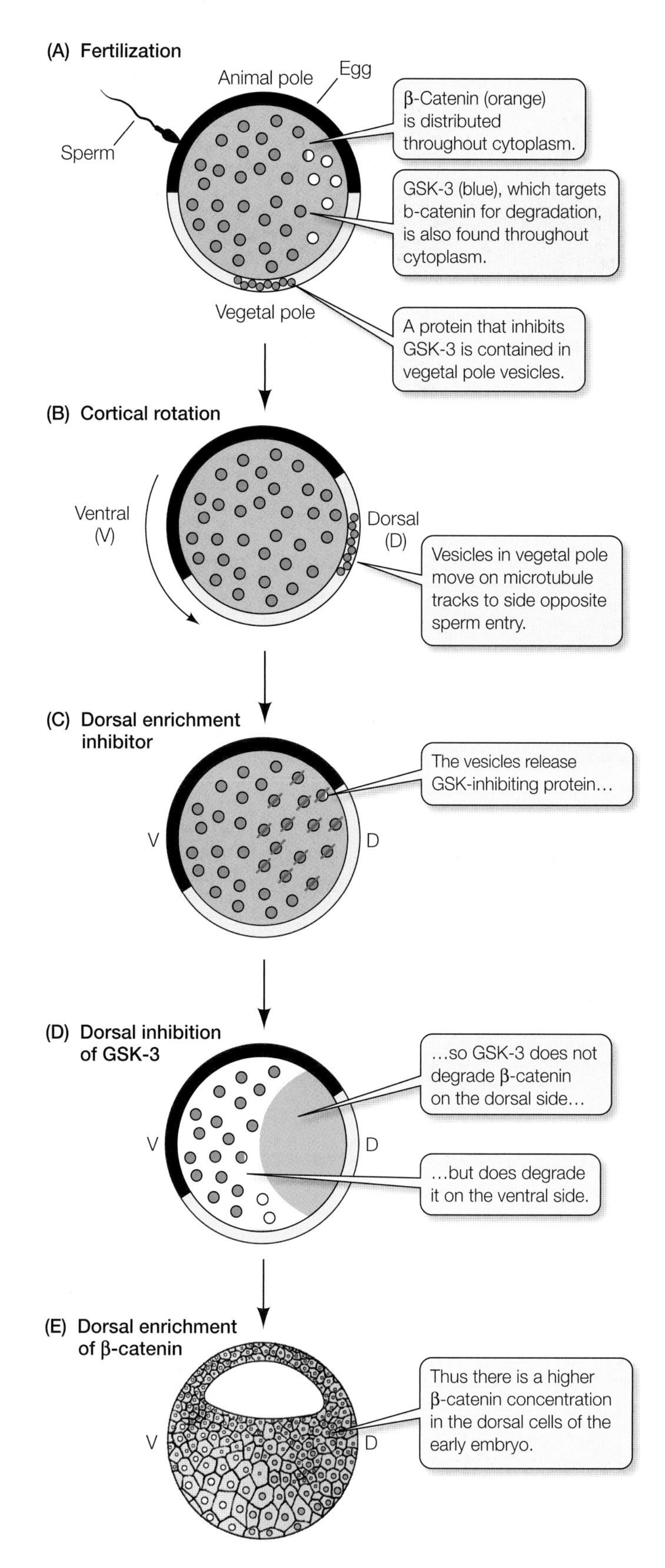

43.2 Cytoplasmic Factors Set Up Signaling Cascades Cytoplasmic movement changes the distributions of critical developmental signals. In the frog zygote, the interaction of the protein kinase GSK-3, its inhibitor, and the protein β-catenin are crucial in specifying the dorsal–ventral (back–belly) axis of the embryo.

- **Complete cleavage** occurs in most eggs that have little **yolk** (stored nutrients). In this pattern, early cleavage furrows divide the egg completely and the daughter cells are of similar size. The sea urchin egg provides an example (**Figure 43.3A**). The frog egg also undergoes complete cleavage, but because the vegetal pole of the frog egg contains more yolk, the division of the cytoplasm is unequal and the daughter cells in the animal hemisphere are smaller than those in the vegetal hemisphere (**Figure 43.3B**).
- **Incomplete cleavage** occurs in many species in which the egg contains a lot of yolk which the cleavage furrows do not penetrate. **Discoidal cleavage** is a type of incomplete cleavage common in fishes, reptiles, and birds, the eggs of which contain a dense mass of yolk. The embryo forms as a disc of cells, called a **blastodisc**, that sits on top of the yolk mass (**Figure 43.3C**).
- **Superficial cleavage** is a variation of incomplete cleavage that occurs in insects such as the fruit fly (*Drosophila*). Early in development, cycles of mitosis occur without cell division, producing a *syncytium*—a single cell with many nuclei. The nuclei eventually migrate to the periphery of the egg, and after several more mitotic cycles, the plasma membrane of the egg grows inward, partitioning the nuclei into individual cells surrounding a core of yolk (**Figure 43.3D**).

The positions of the mitotic spindles during cleavage are not random but are defined by cytoplasmic factors that were produced from the maternal genome and stored in the egg (see Section 19.4). The orientation of the mitotic spindles can determine the planes of cleavage and, therefore, the arrangement of the daughter cells.

In complete cleavage, if the mitotic spindles of successive cell divisions form parallel or perpendicular to the animal–vegetal axis of the zygote, the cleavage pattern is **radial**, as in the sea urchin and the frog. In these organisms, the first two cell divisions are parallel to the animal–vegetal axis and the third is perpendicular to it (see Figures 43.3A,B). **Spiral cleavage** results when the mitotic spindles are at oblique angles to the animal–vegetal axis. In spiral cleavage, each new cell layer is shifted to the left or right, depending on the orientation of the mitotic spindles. Mollusks have spiral cleavage, which is reflected by the coiling pattern of snail shells. If each new layer of cells shifts to the left, the snail shell will coil to the left (sinistral). If each new layer of cells shifts to the right, the snail shell will coil to the right (dextral). Sinistral and dextral coiling snails are mirror images of each other.

Cleavage in mammals is unique

Several features of cleavage in eutherians—placental mammals—are very different from those seen in other animal groups. First, their pattern of cleavage is *rotational*: the first cell division is parallel to the animal–vegetal axis, yielding two blastomeres. The subsequent cell division of those two blastomeres occurs at right angles to each other: one blastomere divides parallel to the animal–vegetal axis, while the other divides perpendicular to it (**Figure 43.4A**).

Cleavage in mammals is very slow; cell divisions are 12–24 hours apart, compared with tens of minutes to a few hours in nonmammalian species. Also, the cell divisions of mammalian blastomeres are not in synchrony with each other. Because the blas-

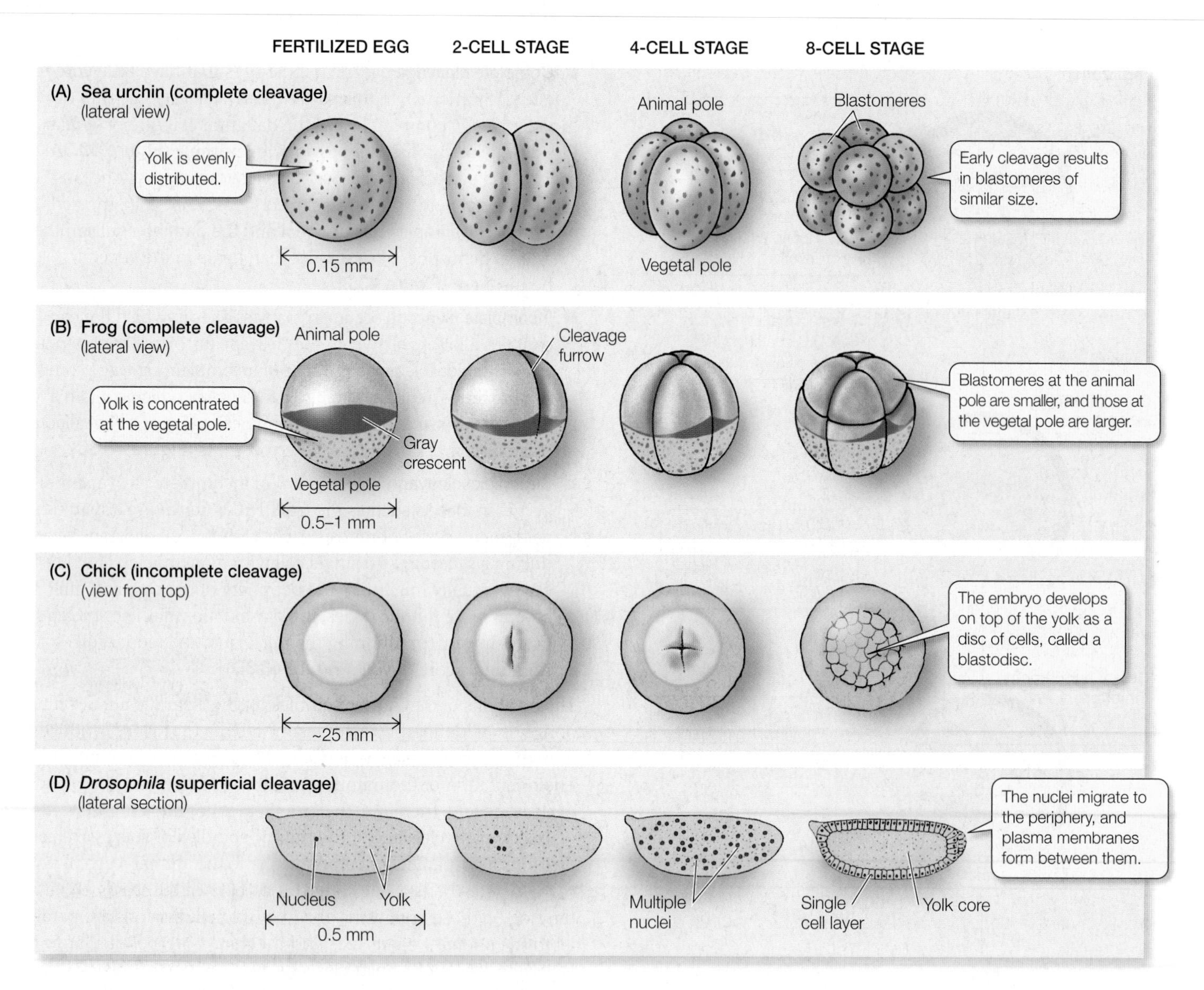

43.3 Patterns of Cleavage in Four Model Organisms Differences in patterns of early embryonic development reflect differences in the way the egg cytoplasm is organized.

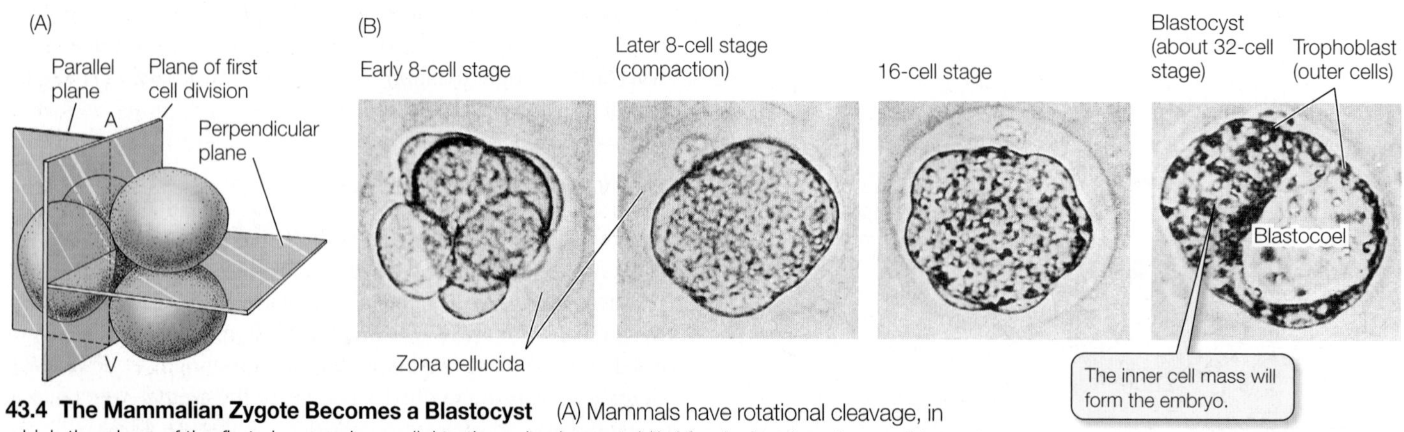

43.4 The Mammalian Zygote Becomes a Blastocyst (A) Mammals have rotational cleavage, in which the plane of the first cleavage is parallel to the animal–vegetal (A, V) axis, but the planes of the second cell division (shown in beige) are at right angles to each other. (B) Starting late in the eight-cell stage, the mammalian embryo undergoes compaction of its cells, resulting in a blastocyst. At the thirty two-cell stage, the blastocyst is a dense mass of cells adjacent to a fluid filled blastocoel and surrounded by trophoblast cells.

tomeres do not undergo mitosis at the same time, the number of cells in the embryo does not increase in the regular (2, 4, 8, 16, 32, etc.) progression typical of other species.

Another unique feature of the slow mammalian cleavage is that the products of genes expressed at this time play roles in cleavage. In animals such as sea urchins and frogs, gene transcription does not occur in the blastomeres, and cleavage is directed exclusively by molecules that were present in the egg before fertilization.

As in other animals that have complete cleavage, the early cell divisions in a mammalian zygote produce a loosely associated ball of cells. However, at about the eight-cell stage, the behavior of the mammalian blastomeres changes. They change shape to maximize their surface contact with one another, form tight junctions, and become a very compact mass of cells (**Figure 43.4B**).

At the transition from the 16- to the 32-cell stage (the fourth division), the cells separate into two groups. The **inner cell mass** will become the embryo, while the surrounding outer cells become an encompassing sac called the **trophoblast**. Trophoblast cells secrete fluid, creating a cavity—the *blastocoel*—with the inner cell mass at one end (final photo in Figure 43.4B). At this stage, the mammalian embryo is called a **blastocyst**, distinguishing it from the blastulas of other animal groups.

Why is mammalian cleavage so different? Remember that placental mammals are *viviparous*: the embryo develops within the uterus of the mother. To support the developing embryo, a connection has to be developed between the circulatory system of the embryo and that of the mother. As we will see later in this chapter, the placenta and the umbilical cord are the structures that provide this connection. Thus, the mammalian blastocyst must produce both the embryo (from the inner cell mass) and its support structures (from the trophoblast).

Fertilization in mammals occurs in the upper reaches of the mother's oviduct, and cleavage occurs as the zygote travels down the oviduct to the uterus. When the blastocyst arrives in the uterus, the trophoblast adheres to the *endometrium* (the lining of the uterus). This event begins the process of **implantation**. In humans, implantation begins on about the sixth day after fertilization. The trophoblast cells of the blastocyst secrete adhesion molecules and enzymes to burrow into the wall of the uterus (**Figure 43.5**). As the blastocyst moves down the oviduct to the uterus, it must not embed itself in the oviduct (Fallopian tube) wall, or the result will be an *ectopic* or *tubal pregnancy*—a very dangerous condition. Early implantation is normally prevented by the zona pellucida, which surrounded the egg and remains around the cleaving ball of cells (see Section 42.2). At about the time the blastocyst reaches the uterus, it hatches from the zona pellucida, and implantation can occur.

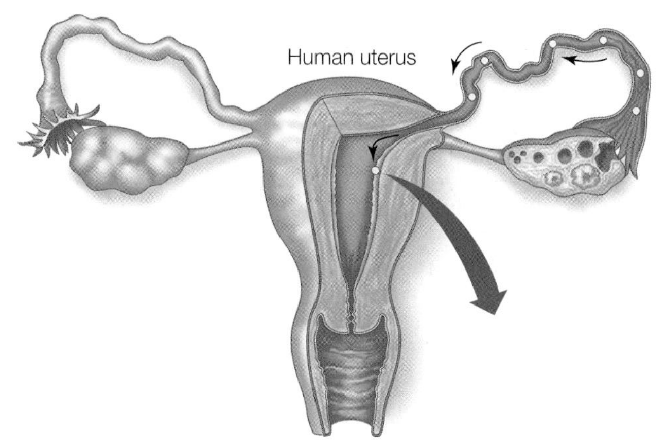

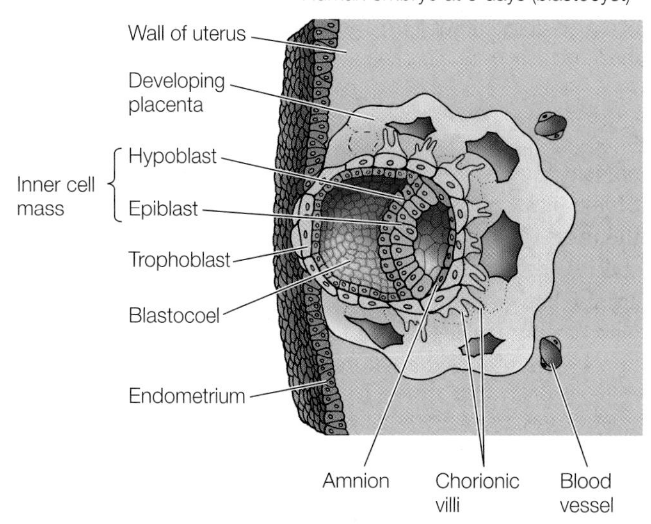

43.5 A Human Blastocyst at Implantation Adhesion molecules and proteolytic enzymes secreted by trophoblast cells allow the blastocyst to burrow into the endometrium. Once implanted within the wall of the uterus, the trophoblast cells send out numerous projections—the chorionic villi—which increase the embryo's area of contact with the mother's bloodstream.

Specific blastomeres generate specific tissues and organs

In all animal species, cleavage results in a repackaging of the egg cytoplasm into a large number of small cells surrounding a central cavity. Little cell differentiation and little if any gene expression occur during cleavage. Nevertheless, cells in different regions of the blastula possess different complements of the nutrients and cytoplasmic determinants that were present in the egg.

The blastocoel prevents cells from different regions of the blastula from coming into contact and interacting, but that will soon change. During the next stage of development, the cells of the blastula will move around and come into new associations with one another, communicate instructions to one another, and begin to differentiate. In many animals, these movements of the blastomeres are so regular and well orchestrated that it is possible to label a specific blastomere with a dye and identify the tissues and organs that form from its progeny. Such labeling experiments produce **fate maps** of the blastula (**Figure 43.6**).

Blastomeres become **determined**—committed to specific fates—at different times in different species. In some species, such as roundworms, the fates of blastomeres are restricted as early as the two-cell stage. If one of these blastomeres is experimentally removed, a particular portion of the embryo will not form. This type of development has been called **mosaic development** because each blastomere seems to contribute a specific set of "tiles" to the final "mosaic" that is the adult animal. In contrast, in **regulative devel-**

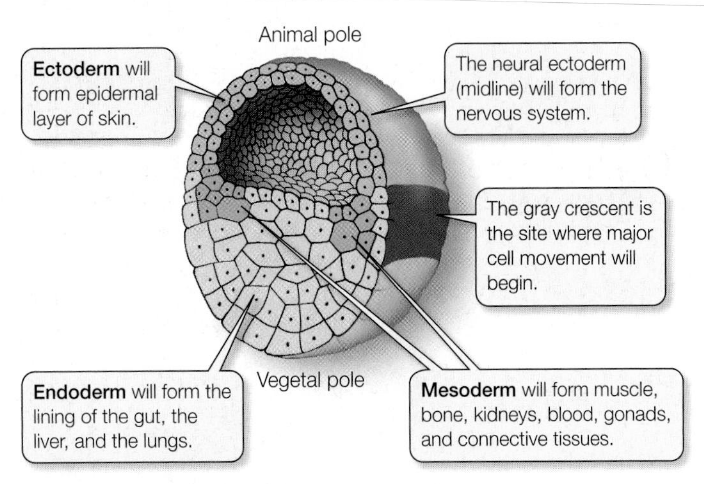

43.6 Fate Map of a Frog Blastula The colors indicate the portions of the blastula that will form the three germ layers and subsequently the frog's tissues and organs.

opment, the loss of some cells during cleavage does not affect the developing embryo, because the remaining cells compensate for the loss. Regulative development is typical of many vertebrate species, including humans. The pluripotent cells of the blastocyst are also known as *embryonic stem cells*, currently the subject of much research, particularly because of their therapeutic potential. (Section 19.2 describes the nature of embryonic stem cells and stem cell research.)

In about 1 out of 50,000 human pregnancies, genetic or environmental factors cause the inner cell mass to split partially. The result is twins that are *conjoined* at some point on their bodies, usually sharing some or most of their organs and/or limbs.

If some blastomeres can change their fate to compensate for the loss of other cells during cleavage and blastula formation, can those cells form an entire embryo? To a certain extent, yes. During cleavage or early blastula formation in mammals, for example, if the blastomeres are physically separated into two groups, both groups can produce complete embryos (**Figure 43.7**). Since the two embryos come from the same zygote, they will be *monozygotic twins*—genetically identical. Nonidentical twins occur when two separate eggs are fertilized by two separate sperm. Thus, while identical twins are always of the same sex, nonidentical twins have a 50 percent chance of being the same sex.

43.1 RECAP

The egg is stocked with nutrients and informational molecules that power and direct the early stages of development. Fertilization activates the egg and stimulates a rearrangement of the cytoplasm, setting up the body axes and positional information that initiate signaling cascades, which control determination and differentiation.

- Can you explain how β-catenin becomes concentrated in only certain blastomeres? See p. 922 and Figure 43.2
- Describe in general terms the difference between complete and incomplete cleavage. See p. 923 and Figure 43.3
- What does a fate map tell us? How are fate maps constructed? See p. 925 and Figure 43.6

We learned in Section 31.1 that the patterns and processes of early development are essential tools in recreating evolutionary history and discerning relationships among the different animal groups, because it is these patterns that result in the diversity of animal body plans. We next describe the events of *gastrulation*, a crucial series of events that initiate the differentiation of embryonic cells and tissues and sets the stage for the emergence of the organs and other structures of the body.

43.2 How Does Gastrulation Generate Multiple Tissue Layers?

The blastula is typically a fluid-filled ball of cells. How does this simple ball of cells become an embryo made up of multiple tissue layers with head and tail ends and dorsal and ventral sides? **Gastrulation** is the process whereby the blastula is transformed by massive movements of cells into an embryo with multiple tissue layers and visible body axes. The resulting spatial relationships between tissues make possible the inductive interactions that trigger differentiation and organ formation (see Figure 19.11).

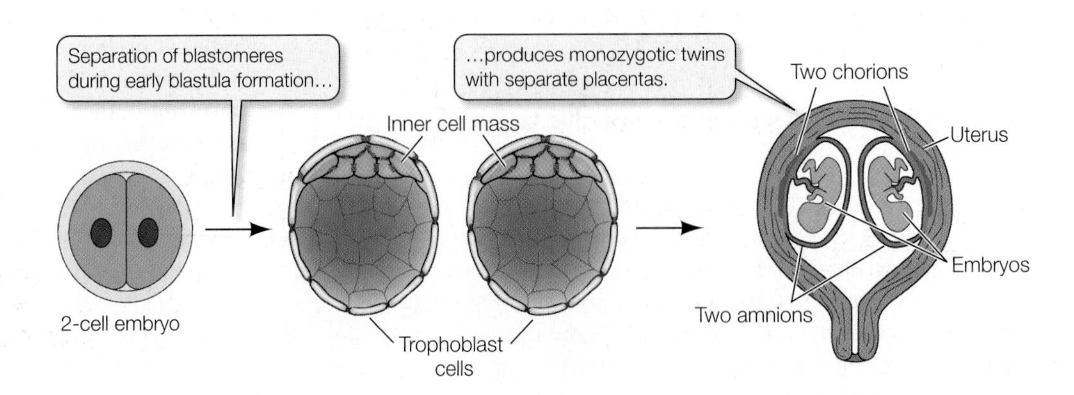

43.7 Twinning in Humans Because humans have regulative development, remaining cells can compensate when cells are lost in early cleavages. Monozygotic (identical) twins can result when cells in the early blastula become physically separated and each group of cells goes on to produce a separate embryo.

During gastrulation, three **germ layers** (also called *cell layers* or *tissue layers*) form (see Figure 43.6):

- The **endoderm** is the innermost germ layer, created as some blastomeres move together as a sheet to the inside of the embryo. The endoderm gives rise to the lining of the digestive tract, respiratory tract, pancreas, and liver.
- The **ectoderm** is the outer germ layer, formed from those cells remaining on the outside of the embryo. The ectoderm gives rise to the nervous system, including the eyes and ears; and to the epidermal layer of the skin and structures derived from skin, such as hair, feathers, nails or claws, sweat glands, and oil glands.
- The **mesoderm**, or middle layer, is made up of cells that migrate between the endoderm and the ectoderm. The mesoderm contributes tissues to many organs, including heart, blood vessels, muscle, and bones.

Some of the most interesting and important challenges in animal development have dealt with two related questions: what directs the cell movements of gastrulation, and what is responsible for the resulting patterns of cell differentiation and organ formation? Scientists have made significant progress in answering both these questions at the molecular level. In the discussion that follows, we will consider the similarities and differences in gastrulation among sea urchins, frogs, reptiles, birds, and mammals. We'll also review some of the exciting discoveries about the mechanisms underlying these phenomena.

Invagination of the vegetal pole characterizes gastrulation in the sea urchin

The sea urchin blastula is a simple, hollow ball of cells that is only one cell thick. The end of the blastula stage is marked by a dramatic slowing of the rate of mitosis, and the beginning of gastrulation is marked by a flattening of the vegetal hemisphere (**Figure 43.8**). Some cells at the vegetal pole bulge into the blastocoel and migrate into the cavity. These cells become *primary mesenchyme* cells—cells of the middle germ layer, the mesoderm. Mesenchymal cells are not organized in tightly packed sheets or tubes like epithelial cells are, and they act more as independent units migrating into and among the other tissue layers.

The flattening at the vegetal pole results from changes in the shape of the individual blastomeres. These cells, which are originally rather cuboidal, become wedge-shaped, with smaller outer edges and larger inner edges. As a result of these shape changes, the vegetal pole bulges inward, or *invaginates*, as if someone were poking a finger into a hollow ball (see Figure 43.8). Some of the cells that invaginate become mesoderm; others become the endoderm and form the primitive gut called the *archenteron*. At the tip of the archenteron more cells enter the blastocoel to form more mesoderm, the *secondary mesenchyme*.

Changes in cell shapes cause the initial invagination of the archenteron but eventually it is pulled by the secondary mesenchyme cells. These cells, attached to the tip of the archenteron, send out extensions called filopodia that adhere to the overlying ectoderm and contract. Where the archenteron eventually makes contact with the ectoderm, the mouth of the animal will form. The opening created by the invagination of the vegetal pole is called the **blastopore**; it will become the anus of the animal.

What mechanisms control the various cell movements of sea urchin gastrulation? The immediate answer is that specific properties of particular blastomeres change. For example, some vegetal cells change shape and bulge into the blastocoel, and these cells become the primary mesenchyme. Once they lose contact with their neighboring cells on the surface of the blastula, they send out filopodia that then move along an extracellular matrix of proteins laid down by the cells lining the blastocoel.

A deeper understanding of gastrulation requires that we discover the molecular mechanisms whereby certain blastomeres develop properties different from those of others. Cleavage systematically divides up the cytoplasm of the egg. The sea urchin blastula at the 64-cell stage is radially symmetrical, but it has *polarity*, as described in Section 19.4. It consists of tiers of cells. As in the frog blastula, the top is the animal pole and the bottom the vegetal pole.

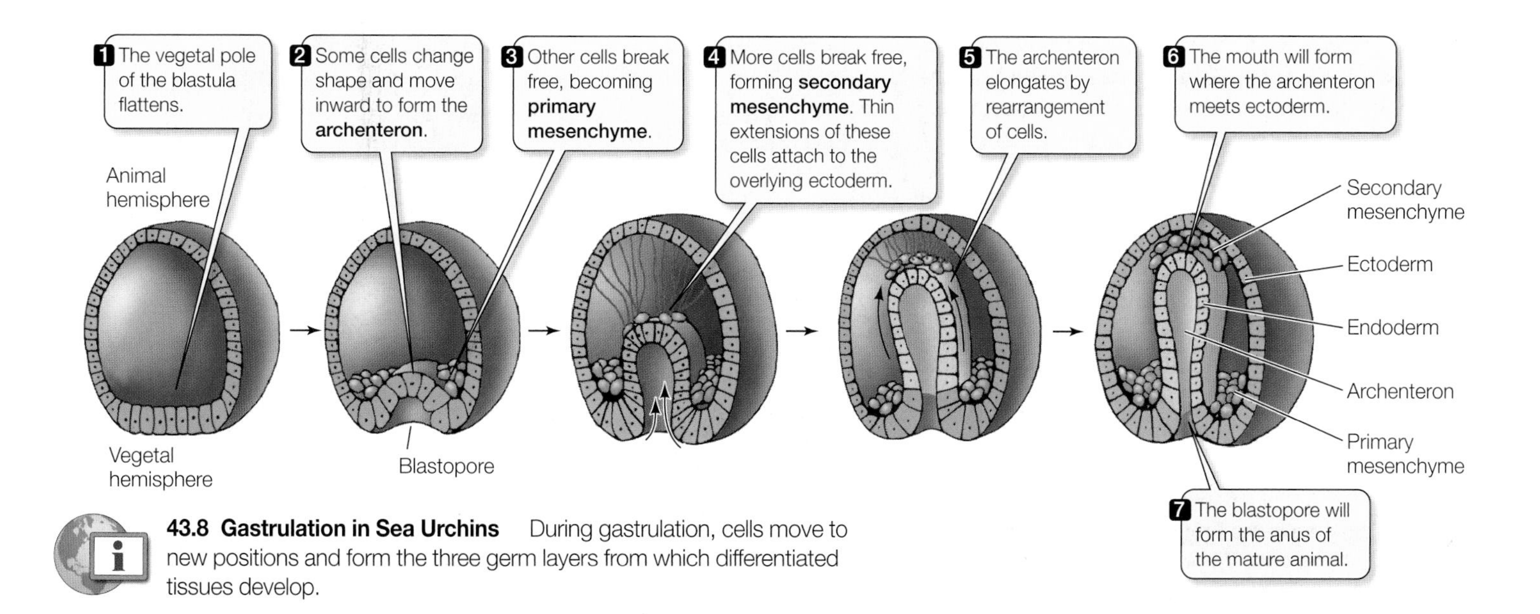

43.8 Gastrulation in Sea Urchins During gastrulation, cells move to new positions and form the three germ layers from which differentiated tissues develop.

If different tiers of blastula cells are separated, they show different developmental potentials; only cells from the vegetal pole are capable of initiating the development of a complete larva (see Figure 19.9). It has been proposed that these differences are due to uneven distribution of various transcriptional regulatory proteins in the egg cytoplasm. As cleavage progresses, these proteins end up in different combinations in different groups of cells. Therefore, specific sets of genes are activated in different cells, determining their different developmental capacities. Let's turn now to gastrulation in the frog, in which a number of key signaling molecules have been identified.

Gastrulation in the frog begins at the gray crescent

Amphibian blastulas have considerable yolk and are more than one cell thick; therefore, gastrulation is more complex in amphibians than in sea urchins. Variation is considerable among different species of amphibians, but in this brief account, we will mix results from studies done on different species to produce a generalized picture of amphibian development.

Amphibian gastrulation begins when certain cells in the gray crescent region change their shapes and cell adhesion properties. These cells bulge inward toward the blastocoel while they remain attached to the outer surface of the blastula by slender necks. Because of their shape, these cells are called *bottle cells*.

The bottle cells mark the spot where the **dorsal lip** of the blastopore will form (**Figure 43.9**). As the bottle cells move inward, the dorsal lip is created, and a sheet of cells moves over it into the blastocoel. This process is called **involution**. One group of involuting cells is the prospective endoderm, and they form the primitive gut, or archenteron. Another group will move between the endoderm and the outermost cells to form the mesoderm. As gastrulation proceeds, cells from the animal hemisphere move toward the site of involution in a process called **epiboly**. The blastopore lip widens and eventually forms a complete circle surrounding a "plug" of yolk-rich cells. As cells continue to move inward through the blastopore, the archenteron grows, gradually displacing the blastocoel.

As gastrulation comes to an end, the amphibian embryo consists of three germ layers: ectoderm on the outside, endoderm on the inside, and mesoderm between. The embryo also has a dorsal–ventral and anterior–posterior organization. Most importantly, however, the fates of specific regions of the endoderm, mesoderm, and ectoderm have been determined. The discovery of the events whereby determination takes place in the amphibian embryo is one of the most exciting stories in animal development.

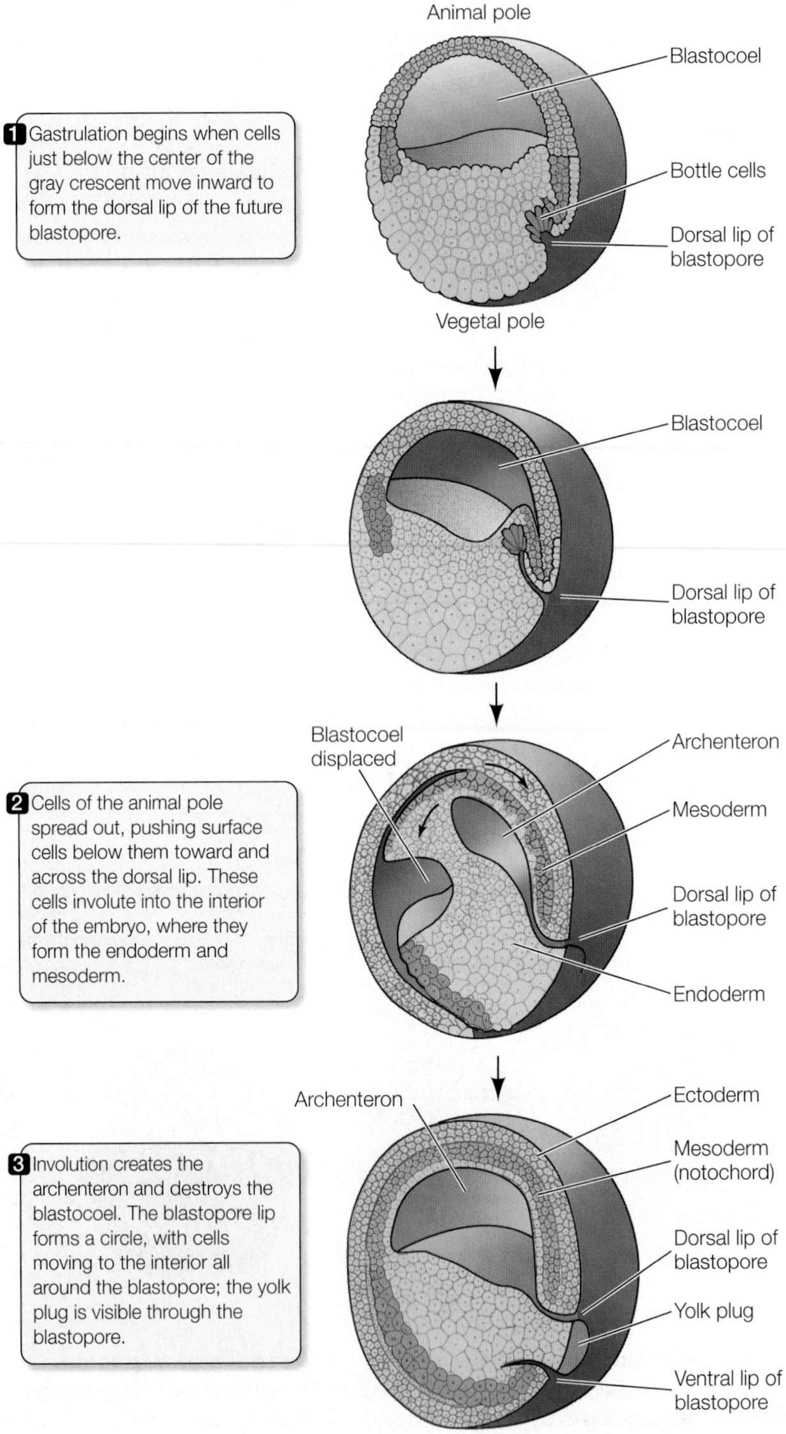

The dorsal lip of the blastopore organizes embryo formation

In the 1920s, the German biologist Hans Spemann was studying the development of salamander eggs. He was interested in finding out whether the nuclei of blastomeres remain capable of directing the development of complete embryos. With great patience

1 Gastrulation begins when cells just below the center of the gray crescent move inward to form the dorsal lip of the future blastopore.

2 Cells of the animal pole spread out, pushing surface cells below them toward and across the dorsal lip. These cells involute into the interior of the embryo, where they form the endoderm and mesoderm.

3 Involution creates the archenteron and destroys the blastocoel. The blastopore lip forms a circle, with cells moving to the interior all around the blastopore; the yolk plug is visible through the blastopore.

43.9 Gastrulation in the Frog Embryo The colors in this diagram are matched to those in the frog fate map (see Figure 43.6).

and dexterity, he formed loops from single human baby hairs to constrict fertilized eggs, effectively dividing them in half.

When Spemann's loops bisected the gray crescent, both halves of the zygote gastrulated and developed into complete embryos (**Figure 43.10, Experiment 1**). But when the gray crescent was on only one side of the constriction, only that half of the zygote developed into a complete embryo. The half lacking gray crescent material became a clump of undifferentiated cells that Spemann called a "belly piece" (**Figure 43.10, Experiment 2**). Spemann hypothesized that cytoplasmic factors unequally distributed in the fertilized egg were necessary for gastrulation and thus for the development of a normal organism.

To test the hypothesis that cells receiving different complements of cytoplasmic factors had different developmental fates, he transplanted pieces of early gastrulas to various locations on other gastrulas. Guided by fate maps (see Figure 43.6), he was able to take a piece of ectoderm he knew would develop into skin and transplant it to a region that normally becomes part of the nervous system, and vice versa.

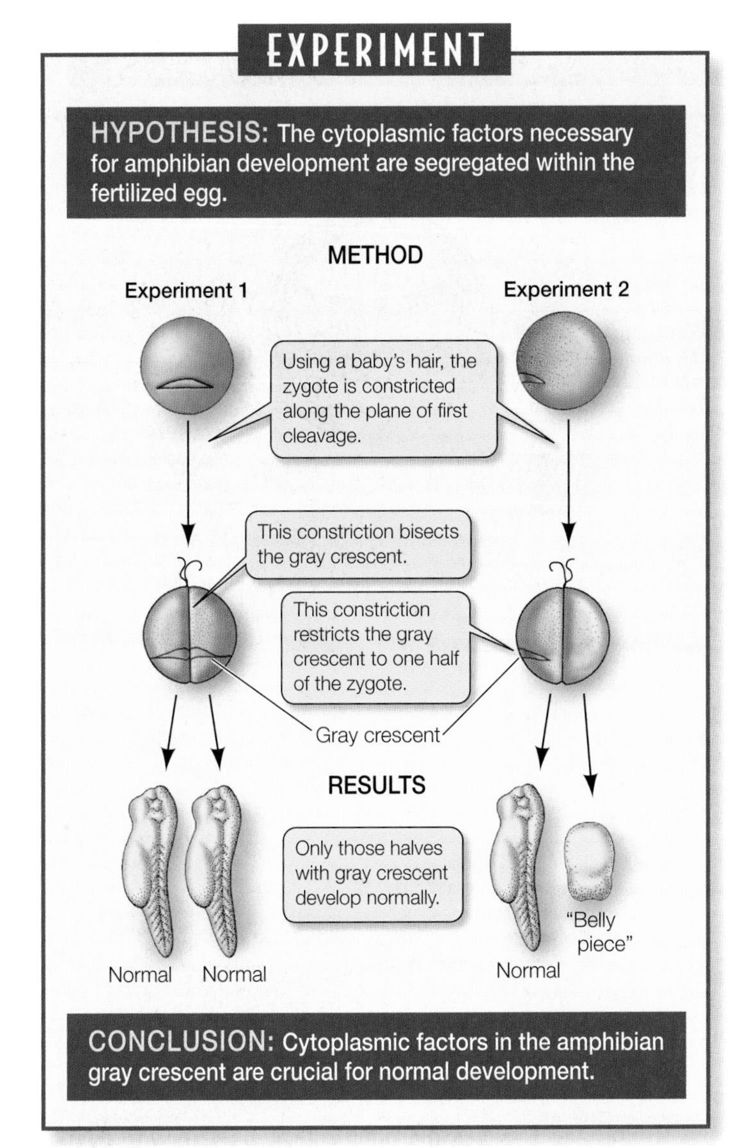

43.10 Spemann's Experiment Spemann's research revealed that gastrulation and subsequent normal development in salamanders depended on cytoplasmic determinants localized in the gray crescent.

When he performed these transplants in early gastrulas—when the blastopore was just beginning to form—the transplanted pieces always developed into tissues that were appropriate for the location where they were placed. Donor-presumptive epidermis (that is, cells destined to become epidermis in their original location) developed into host neural ectoderm (nervous system tissue), and donor-presumptive neural ectoderm developed into host epidermis. Thus, he learned that the fates of the transplanted cells had not been determined before the transplantation (see Figure 19.2). In late gastrulas, however, the same experiment yielded opposite results. Donor-presumptive epidermis produced patches of skin cells in the host nervous system, and donor-presumptive neural ectoderm produced nervous system tissue in the host skin. At some point during gastrulation, the fates of the embryonic cells had become determined.

Spemann's next experiment, done with his student Hilde Mangold, produced momentous results: they transplanted the dorsal lip of the blastopore (**Figure 43.11**). When this small piece of tissue was transplanted into the presumptive belly area of another gastrula, it stimulated a second site of gastrulation—and a second complete embryo formed belly-to-belly with the original embryo! Because the dorsal lip of the blastopore was apparently capable of inducing the host tissue to form an entire embryo, Spemann and Mangold dubbed the dorsal lip tissue the **primary embryonic organizer**, or simply the **organizer**. For more than 80 years, the organizer has been an active area of developmental biology research.

The molecular mechanisms of the organizer involve multiple transcription factors

The primary embryonic organizer has been studied intensively to discover the molecular mechanisms involved in its action. The distribution of the transcription factor β-catenin in the late blastula corresponds to the location of the organizer in the early gastrula, so β-catenin is a candidate for the initiator of organizer activity. To prove that a protein is an inductive signal, it has to be shown that it is both *necessary* and *sufficient* for the proposed effect. In other words, the effect should not occur if the candidate protein is not present (necessity), and the candidate protein should be capable of inducing the effect where it would otherwise not occur (sufficiency).

The criteria of necessity and sufficiency have indeed been satisfied for β-catenin. If β-catenin mRNA transcripts are depleted by injections of antisense RNA into the egg (see Section 16.5), gastrulation does not occur. If β-catenin is experimentally overexpressed in another region of the blastula, it can induce a second axis of embryo formation, as the transplanted dorsal lip did in the Spemann–Mangold experiments. Thus, β-catenin appears to be both necessary and sufficient for the formation of the primary embryonic organizer—but it is only one component of a complex signaling process.

How the presence of β-catenin creates the organizer and how the organizer then induces the beginnings of the body plan involves a complex series of interactions between transcription fac-

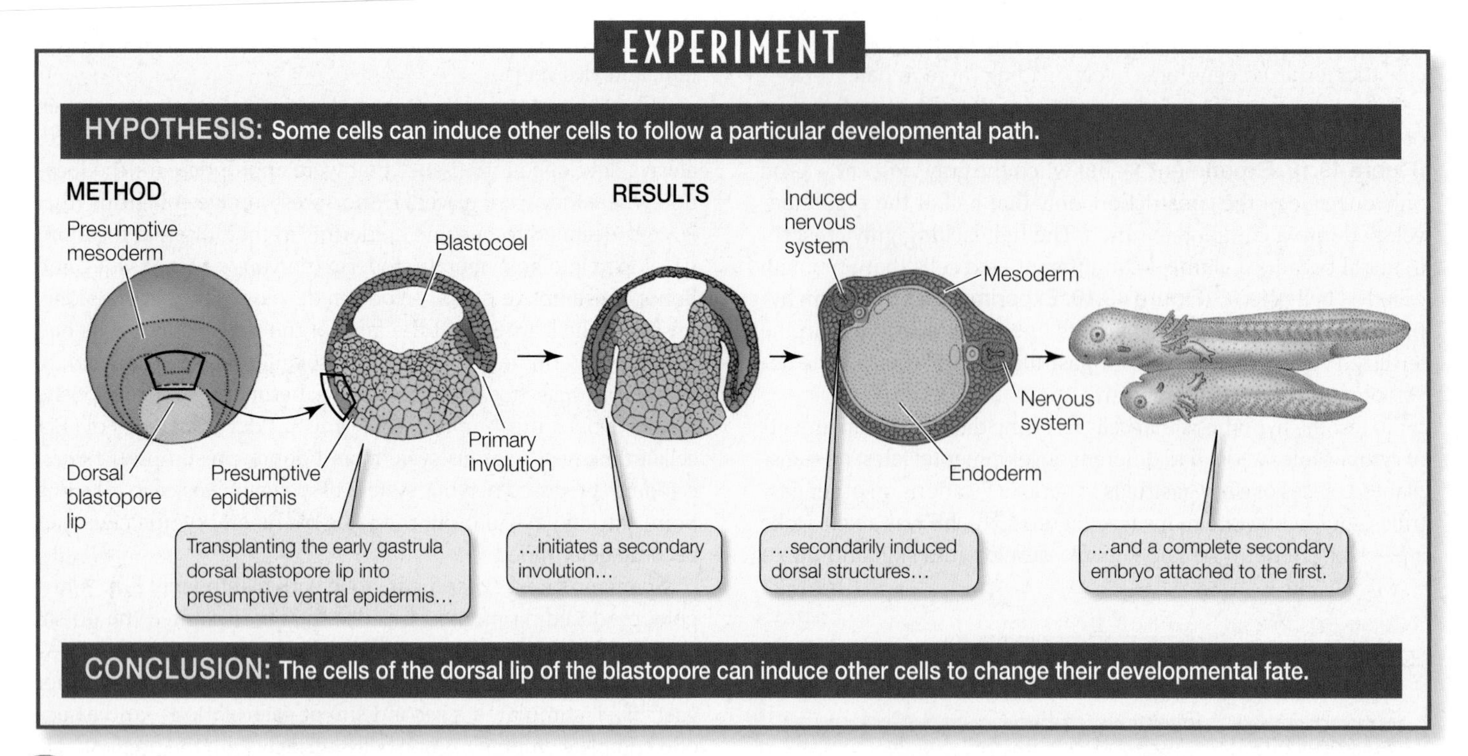

43.11 The Dorsal Lip Induces Embryonic Organization In a famous experiment, Spemann and Mangold transplanted the dorsal lip of the blastopore. The transplanted tissue induced a second site of gastrulation and the formation of a second embryo.

tors and growth factors. What follows is only a portion of this complex and still emerging story. What you should take from this description is not the names of the genes and gene products involved. Rather, we hope you will gain a basic appreciation of how signaling molecules interact to produce different combinations of signals that convey positional and temporal information that guides cells into different paths of determination and differentiation.

Studies of early gastrulas revealed that primary embryonic organizer activity is generated in vegetal cells just below the gray crescent. The concentration of β-catenin is highest here. One critical property of the organizer is expression of the transcription factor Goosecoid. Expression of the *goosecoid* gene depends on two signaling pathways, both of which involve β-catenin.

The first of these pathways involves a *goosecoid*-promoting transcription factor called Siamois. The *siamois* gene is normally repressed by a ubiquitous transcription factor called Tcf-3, but in cells in which β-catenin is present, an interaction between Tcf-3 and β-catenin induces *siamois* expression (**Figure 43.12**). But Siamois protein alone is not sufficient for *goosecoid* expression.

Vegetal cells receive mRNA transcripts from the original egg cytoplasm for proteins in the transforming growth factor-β (TGF-β) superfamily of cell signaling molecules. The signaling pathways activated by one or more proteins from this superfamily interact with the Siamois protein by cooperatively activating the promoter of the *goosecoid* gene and thereby controlling its transcription. Thus it is a particular combination of factors that determine which cells become the primary organizer.

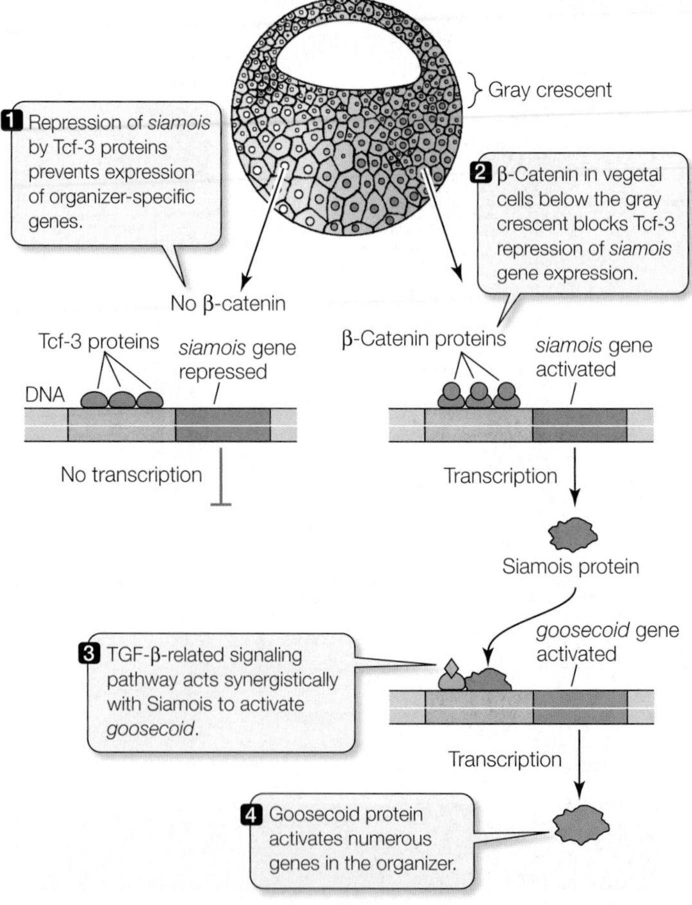

43.12 Molecular Mechanisms of the Primary Embryonic Organizer The organizing potential of the gray crescent depends on the activity of the *goosecoid* gene, which in turn is activated by signaling pathways set up in the vegetal cells below the gray crescent.

The organizer changes its activity as it migrates from the dorsal lip

Organizer cells begin the process of formation of the dorsal lip of the blastopore. Specifically, these cells are at the center of the dorsal lip and involute, moving forward on the midline (i.e., the middle of the anterior–posterior axis) to become mesoderm. Those organizer cells that involute first will move the farthest forward and will induce neighboring cells to participate in making structures of the head. Later organizer cells will induce structures of the trunk, and the last of the organizer cells to move inward from the dorsal lip will induce structures of the tail. How does the capacity of the organizer cells change to enable them to induce head, trunk, or tail structures?

As we learned above, the early organizer cells express the transcription factor Goosecoid, which activates genes encoding soluble signals that influence activity of the cells in contact with the organizer cells in their eventual anterior position. Those neighboring cells produce a number of growth factors. Inhibition of certain of these growth factors is critical for determination of head structures. Under the influence of Goosecoid, the anterior organizer cells produce antagonists to those growth factors. The induction of trunk structures requires inhibition of a different set of growth factors. In organizer cells that involute later than the head organizers, Goosecoid is no longer the dominant transcription factor, and these cells express different growth factor antagonists. The induction of tail structures requires still different activities of the organizer cells that involute last. Thus, the organizer cells express appropriate sets of growth factor antagonists at the right times to achieve different patterns of differentiation on the anterior–posterior axis.

How the activity of the organizer cells changes is a more complex story we will take up later; the main points to grasp are, first, that it is common for mechanisms of development to employ different interactions of transcription factors and growth factors; and, second, the critical interactions are frequently inhibitory.

Reptilian and avian gastrulation is an adaptation to yolky eggs

The eggs of reptiles and birds contain a mass of yolk, and the blastulas of these groups develop as a disc of cells on top of the yolk (see Figure 43.3C). We will use the chicken egg to show how gastrulation proceeds in a flat disc of cells rather than in a ball of cells.

Cleavage in the chick results in a flat, circular layer of cells called a blastodisc (**Figure 43.13**). Between the blastodisc and the yolk mass is a fluid-filled space. Some cells from the blastodisc break free and move into this space. These cells come together to form a continuous layer called the **hypoblast**, which will later contribute to *extraembryonic membranes* that will support and nourish the developing embryo. The overlying cells make up the **epiblast**, from which the embryo proper will form. Thus, the avian blastula is a flattened structure consisting of an upper epiblast and a lower hypoblast, which are joined at the margins of the blastodisc. The blastocoel is the fluid-filled space between the epiblast and hypoblast.

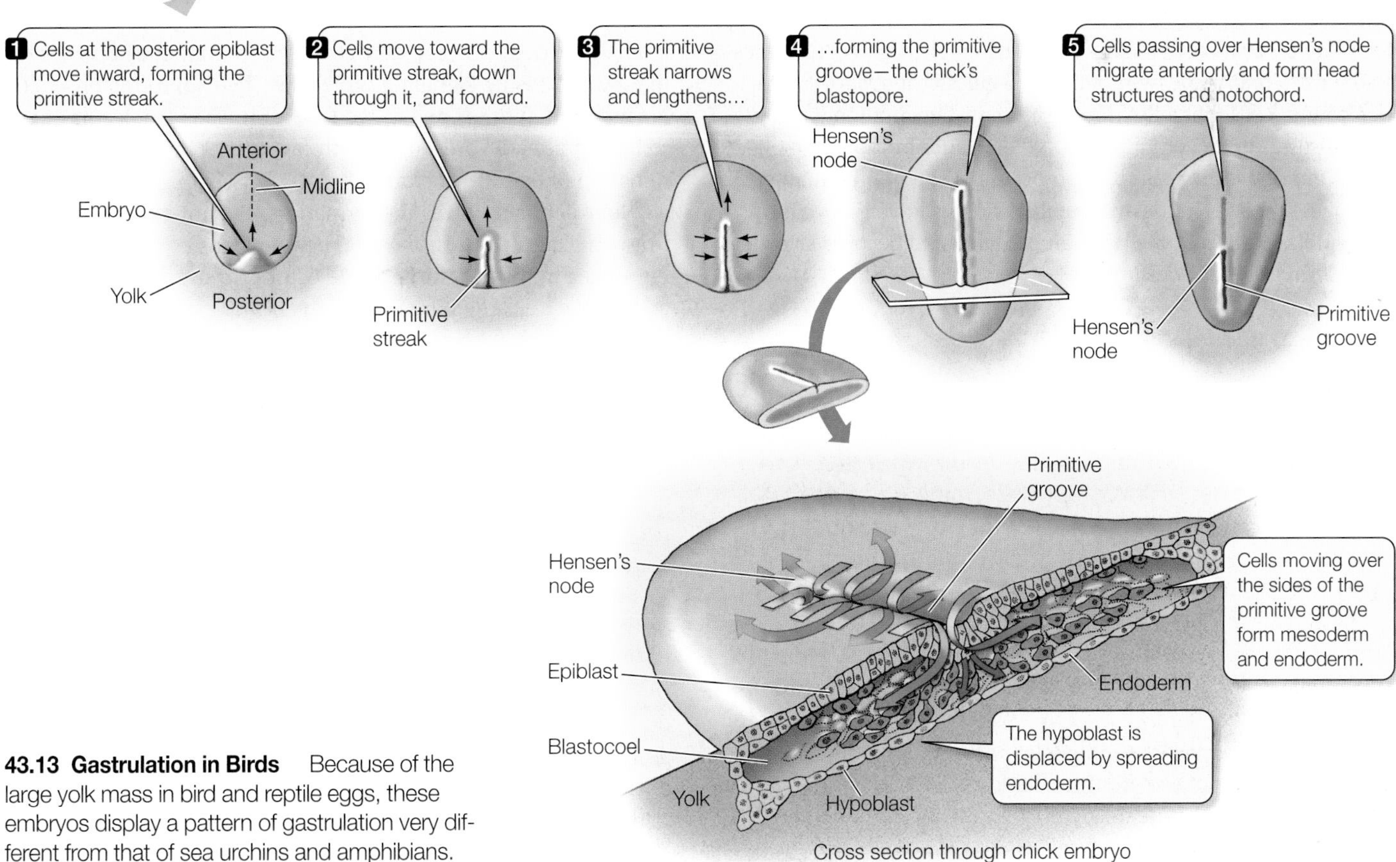

43.13 Gastrulation in Birds Because of the large yolk mass in bird and reptile eggs, these embryos display a pattern of gastrulation very different from that of sea urchins and amphibians.

Gastrulation begins with a thickening in the posterior region of the epiblast caused by the movement of cells toward the midline and then forward along the midline (see Figure 43.13). The result is a midline ridge called the *primitive streak*. A depression called the *primitive groove* forms along the length of the primitive streak. The primitive groove functions as the blastopore, and cells migrate through it into the blastocoel to become endoderm and mesoderm.

In the chick embryo, no archenteron forms, but the endoderm and mesoderm migrate forward to form the gut and other structures. At the anterior end of the primitive groove is a thickening called **Hensen's node**, which is the equivalent of the dorsal lip of the amphibian blastopore. Many signaling molecules that have been identified in the frog organizer are also expressed in Hensen's node. Cells that pass over Hensen's node become determined by the time they reach their final destination, where they differentiate into certain tissues and structures of the head and dorsal midline.

Placental mammals have no yolk but retain the avian–reptilian gastrulation pattern

Both mammals and birds evolved from reptilian ancestors, so it is not surprising that they share patterns of early development, even though the mammalian eggs have no yolk. Earlier we described the development of the mammalian trophoblast and the inner cell mass, which is the equivalent of the avian epiblast.

As in avian development, the inner cell mass splits into an upper layer called the epiblast and a lower layer called the hypoblast, with a fluid-filled cavity between them. The embryo will form from the epiblast, and the hypoblast will contribute to the extraembryonic membranes that will encase the developing embryo and help form the placenta (see Figure 43.5). The epiblast also contributes to the extraembryonic membranes; specifically, it splits off an upper layer of cells that will form the amnion. The amnion will grow to surround the developing embryo as a membranous sac filled with amniotic fluid. Gastrulation occurs in the mammalian epiblast just as it does in the avian epiblast. A primitive groove forms, and epiblast cells migrate through the groove to become layers of endoderm and mesoderm.

43.2 RECAP

The cell movements of gastrulation convert the blastula into an embryo with three tissue layers. New contacts between cells set up inductive signaling interactions that determine cell fates. Dorsal lip tissue is the source of organizer cells that induce development of preliminary head, trunk, and tail structures.

- Compare the cell movements that occur during gastrulation in a sea urchin, a frog, and a bird. See Figures 43.8, 43.9, and 43.13
- Can you explain the molecular basis for the inductive capabilities of the organizer? See pp. 929–930 and Figure 43.12

We have described how the fertilized egg develops into an embryo with three germ layers and how cellular signals trigger different patterns of differentiation. In the next section we describe how organs and organ systems develop in the early embryo.

43.3 How Do Organs and Organ Systems Develop?

Gastrulation produces an embryo with three germ layers that are positioned to influence one another through inductive interactions. During the next phase of development, called **organogenesis**, many organs and organ systems develop simultaneously and in coordination with one another. An early process of organogenesis in chordates that is directly related to gastrulation is neurulation. **Neurulation** is the initiation of the nervous system. We will examine neurulation in the amphibian embryo, but it occurs in a similar fashion in reptiles, birds, and mammals.

The stage is set by the dorsal lip of the blastopore

As we learned in the previous section, one group of cells that passes over the dorsal lip of the blastopore moves anteriorly and becomes the endodermal lining of the digestive tract. The other group of cells that involutes over the dorsal lip will become mesoderm, and these cells that are closest to midline have organizer functions (see Figure 43.9). This mesoderm is called the *chordamesoderm* because it produces a rod of connective tissue called the **notochord**. The notochord gives structural support to the developing embryo; it is eventually replaced by the vertebral column. After gastrulation, the organizing capacity of the chordamesoderm induces the overlying ectoderm to begin forming the nervous system by expressing signaling molecules (one appropriately called Noggin and another one called Chordin). These molecules neutralize an already-present growth factor that inhibits the determination of neural structures.

Neurulation involves the formation of an internal neural tube from an external sheet of cells. The first signs of neurulation are flattening and thickening of the ectoderm overlying the notochord; this thickened area forms the *neural plate* (**Figure 43.14**). The edges of the neural plate that run in an anterior–posterior direction continue to thicken to form ridges or folds. Between these neural folds, a groove forms and deepens as the folds roll over it to converge on the midline. The folds fuse, forming a cylinder, the **neural tube**, and a continuous overlying layer of epidermal ectoderm. The neural tube develops bulges at the anterior end, which become the major divisions of the brain; the rest of the tube becomes the spinal cord.

In humans, failure of the neural tube to develop normally can result in serious birth defects. If the neural folds fail to fuse in a posterior region, the result is a condition known as *spina bifida*. If they fail to fuse at the anterior end, an infant can develop without a forebrain—a condition called *anencephaly*. Although several genetic factors can cause neural tube defects, other factors are environmental, including diet. The incidence of neural tube defects in the United States in the early 1900s was as high as 1 in 300 live births, but it has declined to less than 1 per 1,000 today. A major

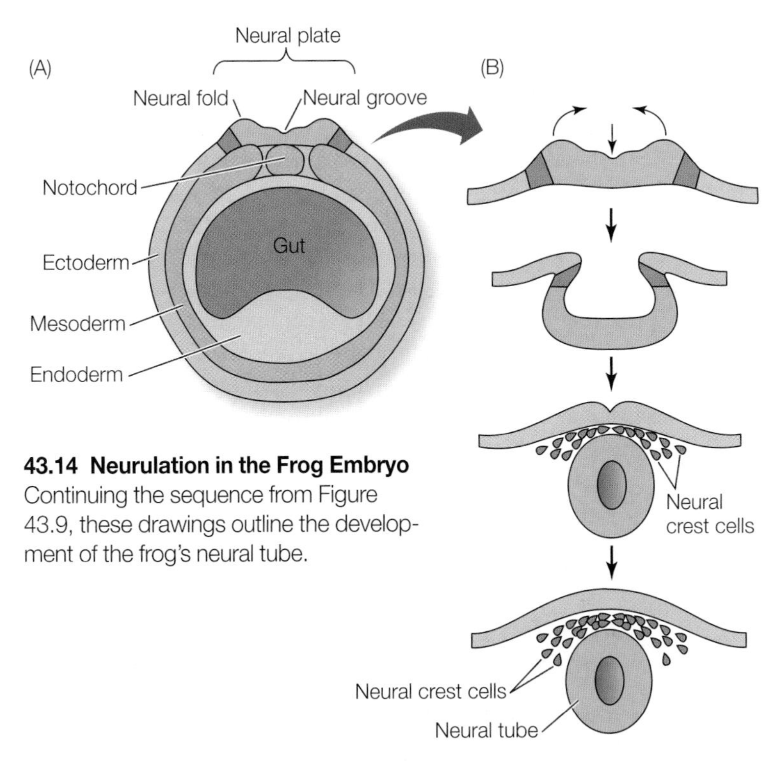

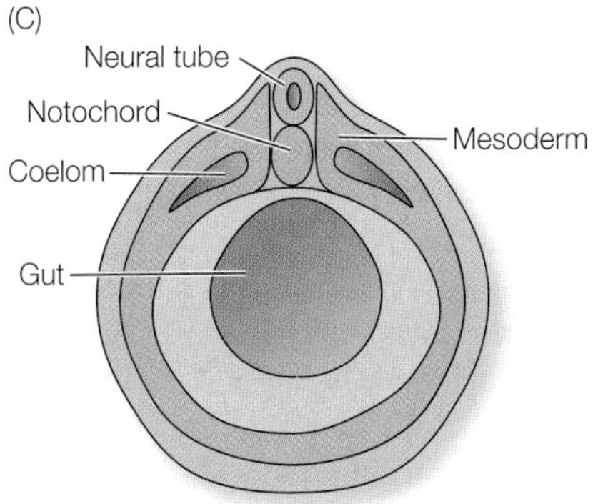

43.14 Neurulation in the Frog Embryo Continuing the sequence from Figure 43.9, these drawings outline the development of the frog's neural tube.

factor in this improvement has been the inclusion of an adequate amount of folic acid (a B vitamin) in the mother's diet.

Body segmentation develops during neurulation

Like the arthropods, vertebrates have a body plan consisting of repeating segments that are modified during development. These segments are most evident as the repeating patterns of vertebrae, ribs, nerves, and muscles along the anterior–posterior axis.

As the neural tube forms, mesodermal tissues gather along the sides of the notochord to form separate, segmented blocks of cells called **somites** (**Figure 43.15**). The somites produce cells that will become the vertebrae, ribs, muscles of the trunk and limbs, and the lower layer of the skin.

The nerves that connect the brain and spinal cord with tissues and organs throughout the body are also arranged segmentally. The somites help guide the organization of these peripheral nerves, but the nerves are not of mesodermal origin. When the neural tube fuses, cells adjacent to the line of closure break loose and migrate inward between the epidermis and the somites and through the somites. These cells, called *neural crest cells,* contribute to a number of structures, including the peripheral nerves, which grow out to the body tissues and back into the spinal cord.

As development progresses, the different segments of the body change. Regions of the spinal cord differ, regions of the vertebral column differ in that some vertebrae grow ribs of various sizes and others do not, forelegs arise in the anterior part of the embryo, and hind legs arise in the posterior region.

Hox genes control development along the anterior–posterior axis

How is mesoderm in the anterior part of a mouse embryo programmed to produce forelegs rather than hind legs? In Section 19.5, we saw how homeotic genes control body segmentation in *Drosophila*. We also learned that all homeotic genes contain a DNA sequence called the *homeobox*. Some of the genes directing gastrulation in the frog are homeobox genes—for example, *goosecoid* and *siamois*. In the mouse, four families of homeotic **Hox genes** control differentiation along the anterior–posterior body axis.

Each mammalian Hox gene family resides on a different chromosome, in clusters of about 10 genes each. Remarkably, the temporal and spatial expression of these genes follows the same pattern as their linear order on their chromosome. That is, the Hox genes closest to the 3′ end of each

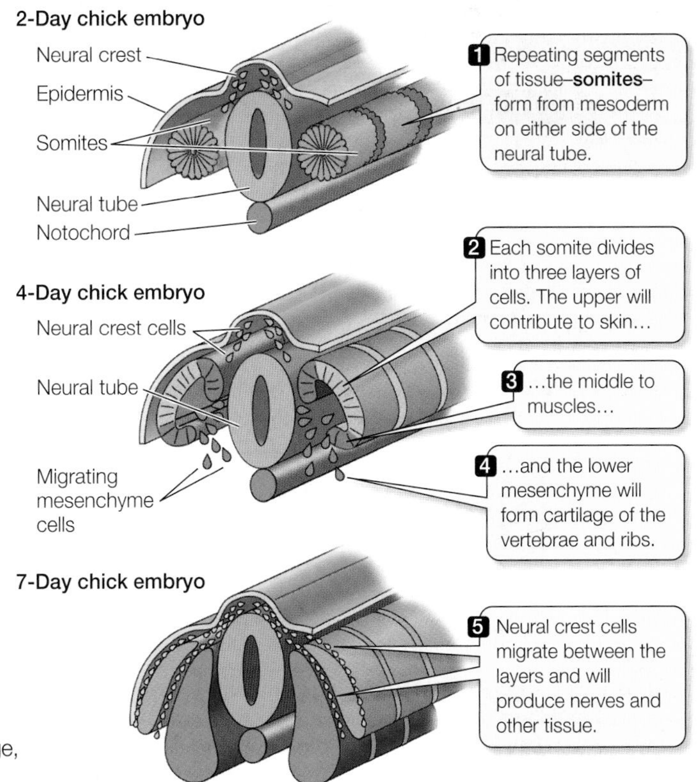

43.15 The Development of Body Segmentation Repeating blocks of tissue called somites form on either side of the neural tube. Muscle, cartilage, bone, and the lower layer of the skin form from the somites.

gene complex are expressed first and in the anterior of the embryo. The Hox genes closer to the 5′ end of the gene complex are expressed later and in a more posterior part of the embryo. As a result, different segments of the embryo receive different combinations of Hox gene products, which serve as transcription factors (**Figure 43.16**; see also Figure 20.1).

Whereas Hox genes give cells information about their position on the anterior–posterior (head–tail) body axis, other genes provide information about their dorsal–ventral (back–belly) position. Tissues in each segment of the body differentiate according to their dorsal–ventral location. In the spinal cord, for example, sensory nerve connections develop in the dorsal region, and motor nerve connections develop in the ventral region. In the somites, dorsal cells develop into skin and muscle and ventral cells develop into cartilage and bone (see Figure 43.15). An example of a gene that provides dorsal–ventral information in vertebrates is *sonic hedgehog*, which is expressed in the mammalian notochord and induces cells in the overlying neural tube to have fates characteristic of ventral spinal cord cells.

The mammalian *sonic hedgehog* gene is homologous to a *Drosophila* gene known simply as *hedgehog*—a gene whose name comes from the distinctive spiny appearance of the back in mutant flies. The mammalian homolog was given the name of the video game character to distinguish it from its insect counterpart.

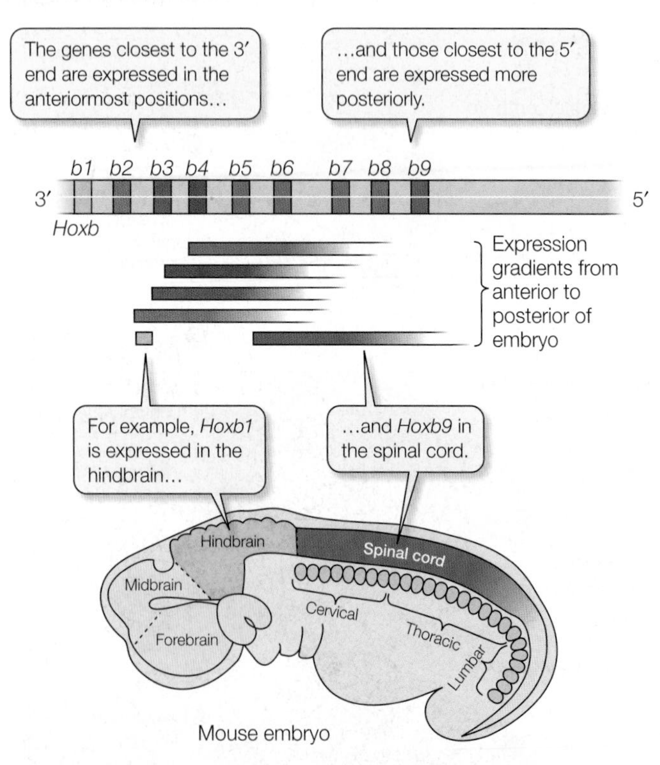

43.16 Hox Genes Control Body Segmentation Hox genes are expressed along the anterior–posterior axis of the embryo in the same order as their arrangement between the 3′ and 5′ ends of the gene complex.

One family of homeobox genes, the *Pax* genes, plays many roles in nervous system and somite development. The role of *Pax6* in eye development was described at the start of Chapter 20. Another gene in the same family, *Pax3*, is expressed in those neural tube cells that will develop into dorsal spinal cord structures. Sonic hedgehog protein represses the expression of the *Pax3* gene, and their interaction is one source of dorsal–ventral information for the differentiation of the spinal cord.

After the development of body segmentation, the formation of organs and organ systems progresses rapidly. The development of an organ involves extensive inductive interactions of the kind we saw in Section 19.4 in the example of the vertebrate eye. These inductive interactions are a current focus of study for developmental biologists.

43.3 RECAP

Gastrulation sets up tissue interactions that initiate organogenesis. Neurulation is initiated by organizer mesoderm that forms the notochord.

- Describe the formation of the neural tube in vertebrates. See p. 932 and Figure 43.14
- Can you explain how somites relate to segmentation of the body axis? See p. 933 and Figure 43.15
- Using information from this chapter and from Chapters 19 and 20, can you explain what Hox genes are and how they instruct patterns of differentiation along the body axis? See Figures 19.19, 20.1, and 43.16

You may be aware that in mammals the circulatory systems of the fetus and mother are separate and that nourishment reaches the fetus through the placenta and the umbilical cord. In the next section we will examine the developmental events that result in the creation of the placenta.

43.4 What Is the Origin of the Placenta?

There is more to a developing reptile, bird, or mammal than the embryo itself. As mentioned earlier, the embryos of these vertebrates are surrounded by several **extraembryonic membranes**, which originate from the embryo but are not part of it. The extraembryonic membranes function in nutrition, gas exchange, and waste removal. In mammals, they interact with tissues of the mother to form the placenta.

Extraembryonic membranes form with contributions from all germ layers

We will use the chick to demonstrate how the extraembryonic membranes form from the germ layers created during gastrulation. In the chick, four membranes form—the *yolk sac*, the *allantoic membrane*, the *amnion*, and the *chorion*. The **yolk sac** is the first to form, and it does so by extension of the endodermal tissue of the hypoblast layer along with some adjacent mesoderm. The yolk sac grows to enclose

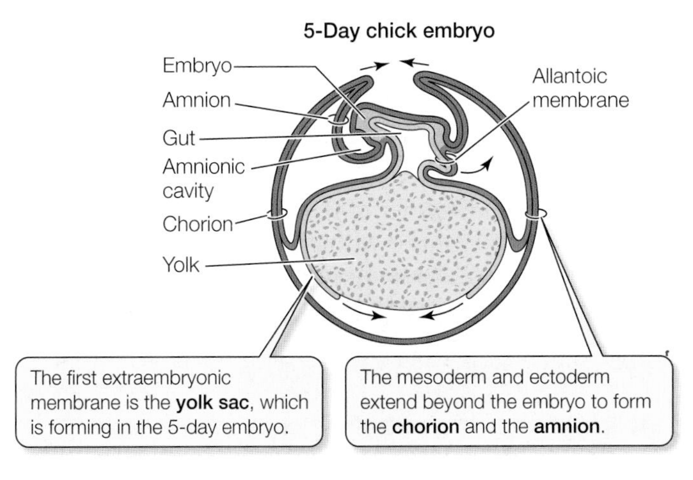

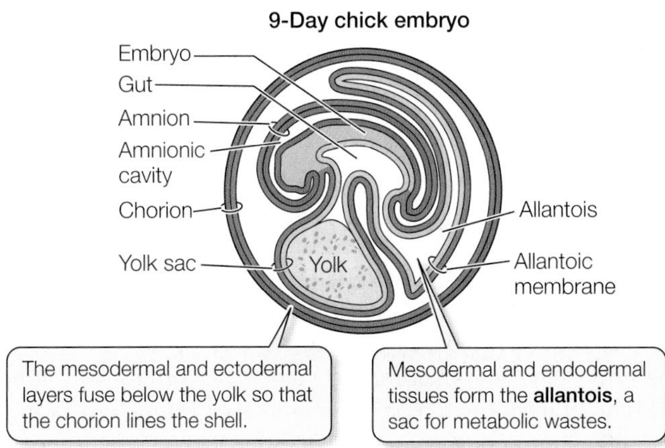

43.17 The Extraembryonic Membranes In birds, reptiles, and mammals, the embryo constructs four extraembryonic membranes. The yolk sac encloses the yolk, and the amnion and chorion enclose the embryo. Fluids secreted by the amnion fill the amniotic cavity, providing an aqueous environment for the embryo. The chorion, along with the allantois, mediates gas exchange between the embryo and its environment. The allantois stores the embryo's waste products.

the entire body of yolk in the egg (**Figure 43.17**). It constricts at the top to create a tube that is continuous with the gut of the embryo. However, yolk does not pass through this tube. Yolk is digested by the cells of the yolk sac, and the nutrients are transported to the embryo through blood vessels that form from mesoderm and line the outer surface of the yolk sac. The **allantoic membrane** is also an outgrowth of the extraembryonic endoderm plus adjacent mesoderm. It forms the *allantois,* a sac for storage of metabolic wastes.

Just as the endoderm and mesoderm of the hypoblast grow out from the embryo to form the yolk sac and the allantoic membrane, ectoderm and mesoderm combine and extend beyond the limits of the embryo to form the other extraembryonic membranes. Two layers of cells extend all along the inside of the eggshell, both over the embryo and below the yolk sac. Where they meet, they fuse, forming two membranes, the inner **amnion** and the outer **chorion**. The amnion surrounds the embryo, forming the amniotic cavity. The amnion secretes fluid into the cavity, providing a protective environment for the embryo. The outer membrane, the chorion, forms a continuous membrane just under the eggshell (see Figure 43.17). It limits water loss from the egg and also works with the enlarged allantoic membrane to exchange respiratory gases between the embryo and the outside world.

Extraembryonic membranes in mammals form the placenta

In mammals, the first extraembryonic membrane to form is the trophoblast, which is already apparent by the fifth cell division (see Figure 43.4). When the blastocyst reaches the uterus and hatches from its encapsulating zona pellucida, the trophoblast cells interact directly with the endometrium. Adhesion molecules expressed on the surfaces of these cells attach them to the uterine wall. By secreting proteolytic enzymes, the trophoblast burrows into the endometrium, beginning the process of implantation (see Figure 43.5). Eventually, the entire trophoblast is within the wall of the uterus. The trophoblast cells then send out numerous projections, or villi, to increase the surface area of contact with maternal blood.

Meanwhile, the hypoblast cells proliferate to form what in the bird would be the yolk sac. But there is no yolk in eggs of placental mammals, so the yolk sac contributes mesodermal tissues that interact with trophoblast tissues to form the chorion. The chorion, along with tissues of the uterine wall, produces the **placenta**, the organ that exchanges nutrients, respiratory gases, and metabolic wastes between the mother and the embryo (**Figure 43.18**).

At the same time the yolk sac is forming from the hypoblast, the epiblast produces the amnion, which grows to enclose the entire embryo in a fluid-filled amniotic cavity. The rupturing of the

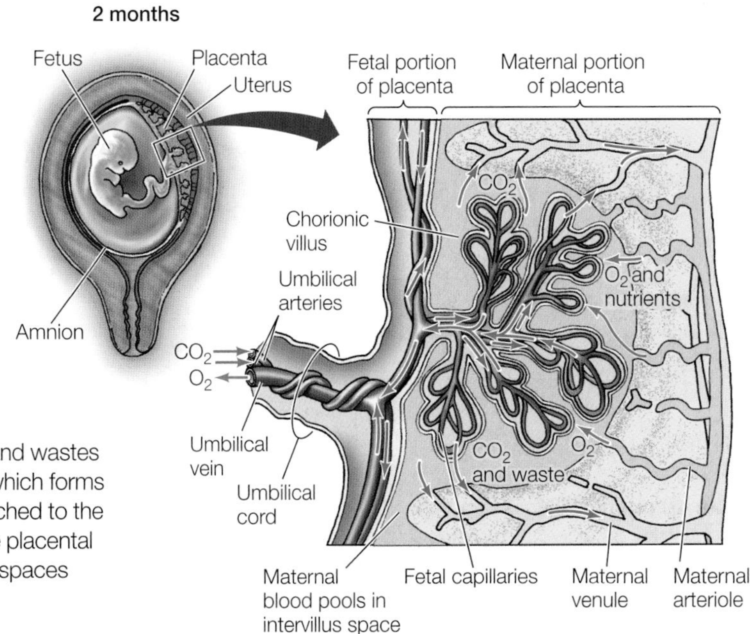

43.18 The Mammalian Placenta In most mammals, nutrients and wastes are exchanged between maternal and fetal blood in the placenta, which forms from the chorion and tissues of the uterine wall. The embryo is attached to the placenta by the umbilical cord. Embryonic blood vessels invade the placental tissue to form fingerlike chorionic villi. Maternal blood flows into the spaces surrounding the villi.

amnion and chorion and the loss of the amniotic fluid (a process called "water breaking") herald the onset of labor in humans.

An allantois also develops in mammals, but its importance depends on how well nitrogenous wastes can be transferred across the placenta. In humans the allantois is minor; in pigs it is important. In humans and other mammals, allantoic tissues contribute to the formation of the umbilical cord, by which the embryo is attached to the chorionic placenta. It is through the blood vessels of the umbilical cord that nutrients and oxygen from the mother reach the developing fetus, and wastes, including carbon dioxide and urea, are removed (see Figure 43.18).

The extraembryonic membranes provide means of detecting genetic diseases

Cells slough off of the developing human embryo and float in the amniotic fluid that bathes it. Later in development, a small volume of the amniotic fluid may be extracted with a needle as the first step of a process called **amniocentesis.** Cells from the fluid can be cultured and used for biochemical and genetic analyses that can reveal the sex of the fetus, as well as genetic markers for diseases such as cystic fibrosis, Tay-Sachs disease, and Down syndrome.

If amniocentesis is performed, it is usually not until after the fourteenth week of pregnancy, and the tests require two weeks to complete. If abnormalities in the fetus are detected, termination of the pregnancy at that stage would put the mother's health at greater risk than would an earlier abortion. Therefore, a newer technique, called **chorionic villus sampling,** is now in common use. In this test, a small sample of the tissue from the surface of the chorion is taken (**Figure 43.19**). This test can be done as early as the eighth week of pregnancy, and the results are available in several days.

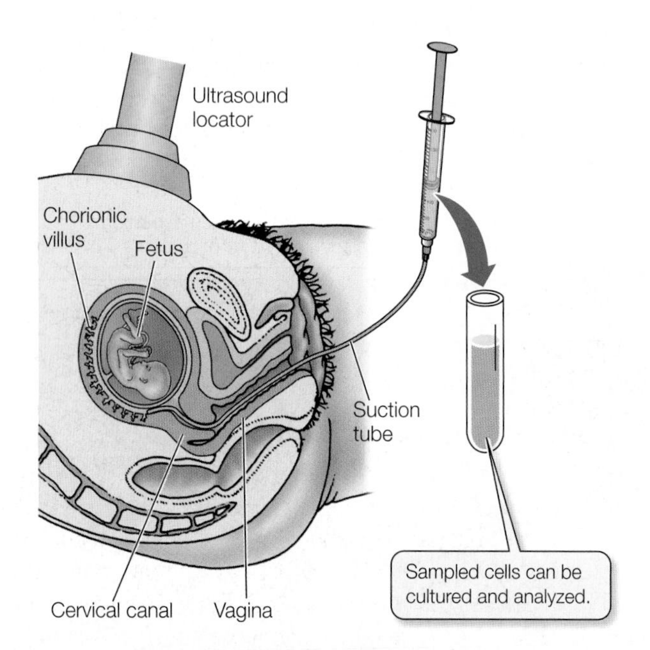

43.19 Chorionic Villus Sampling Information about genetic defects can be obtained from chorionic tissues. The fetus and placenta are imaged by a sonogram to guide a catheter, through which a sample of fetal cells is removed from a chorionic villus for testing.

43.4 RECAP

The extraembryonic membranes of reptiles, birds, and mammals sustain the growing embryo. In reptiles and birds, these membranes surround the embryo within the shelled egg. In mammals the extraembryonic membranes form the placenta, an organ that exchanges nutrients, respiratory gases, and metabolic wastes between the mother and the embryo.

- Can you describe each of the four extraembryonic membranes and their functions in the developing chick egg? See pp. 934–935 and Figure 43.17
- Can you explain the role of the trophoblast in the early development of a mammalian embryo? See p. 935

43.5 What Are the Stages of Human Development?

In humans, **gestation**, or pregnancy, lasts about 266 days, or 9 months. In smaller mammals gestation is shorter—for example, 21 days in mice—and in larger mammals it is longer—for example, 330 days in horses and 600 days in elephants. The events of human gestation can be divided into three periods of roughly 3 months each, called *trimesters*.

The embryo becomes a fetus in the first trimester

Implantation of the human blastocyst begins on about the sixth day after fertilization. After implantation, gastrulation occurs, tissues differentiate, the placenta forms, and organs begin to develop. The heart begins to beat during week 4, and limbs form by week 8 (**Figure 43.20A,B**). By the end of the first trimester, most organs have started to form. The embryo is about 8 centimeters long and weighs about 40 grams (less than 2 ounces); it would fit neatly in a teaspoon. At about this point in time, the human embryo is medically and legally referred to as a **fetus**. (This distinction is not made for other mammals; developing mice, for example, remain embryos until they are born.)

An embryo can be damaged before the mother even knows she is pregnant. A classic and tragic case is that of thalidomide, a drug widely prescribed in Europe in the late 1950s to treat nausea. Women who took this drug in the fourth and fifth week of pregnancy, when the embryo's limbs are beginning to form, gave birth to children with missing or severely malformed arms and legs.

The first trimester is a time of rapid cell division and tissue differentiation. Signal transduction cascades and the resulting branching sequences of developmental processes are in their early stages. Therefore, the first trimester is the period during which the em-

(A) 4 weeks

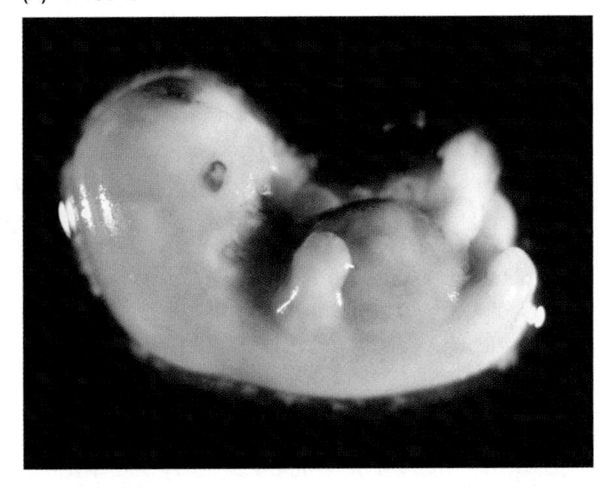

(C) 4 months

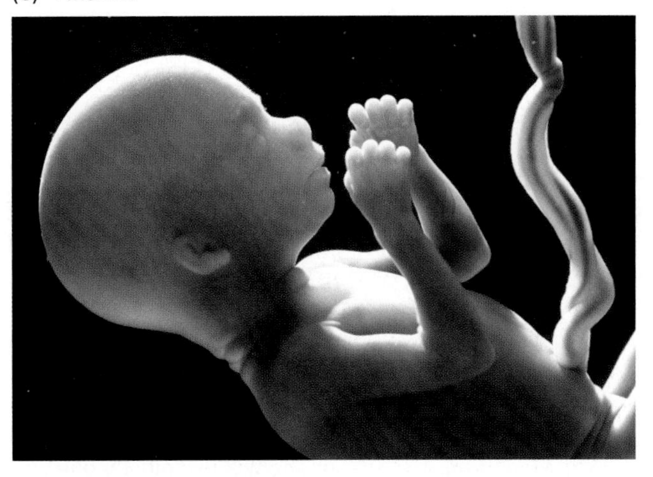

(B) 8 weeks

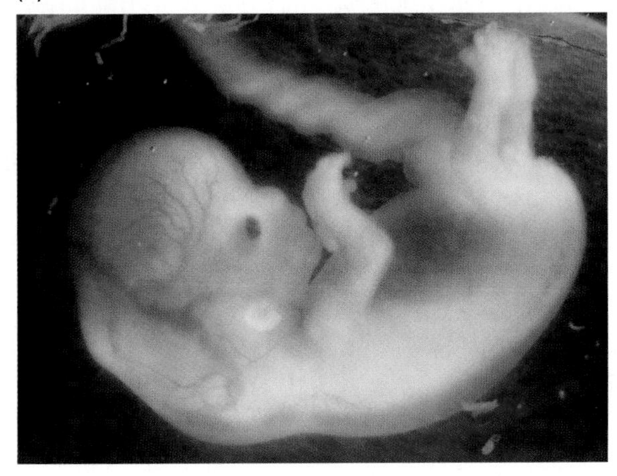

(D) 9 months

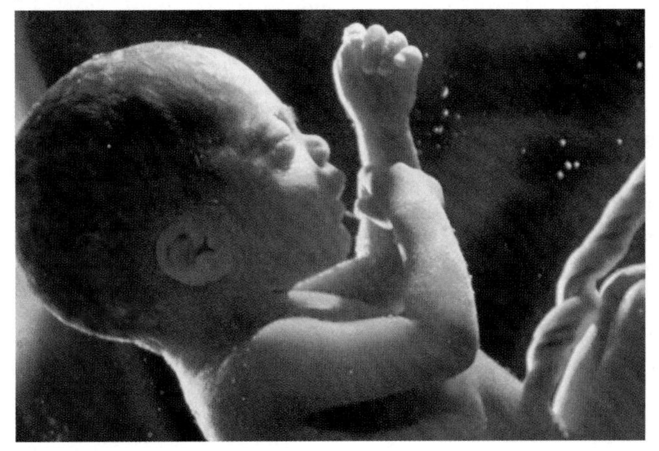

43.20 Stages of Human Development (A) At 4 weeks of gestation, most of the human embryo's organ systems have been formed and the heart is beating. (B) The body structures of this 8-week-old embryo are forming rapidly, and it is visibly a male. The umbilical cord attaches the embryo to the placenta (upper left). (C) At 4 months, the fetus has fully formed limbs with fingers and toes, and moves freely within the amniotic cavity. (D) This fetus is well along in its ninth month. Soon its lungs will be mature enough to trigger the onset of contractions and birth.

bryo is most sensitive to damage from radiation, drugs, chemicals, and pathogens that can cause birth defects.

Hormonal changes cause major and noticeable responses in the mother during the first trimester. Soon after the blastocyst implants itself, it begins to secrete the hormone human chorionic gonadotropin (hCG). Because of this early release, detection of hCG is used as an early test for pregnancy. The presence of hCG stimulates the mother's ovaries to continue producing the hormones estrogen and progesterone, which help maintain pregnancy. These hormonal changes cause pregnancy's well-known symptoms, including morning sickness, mood swings, changes in the senses of taste and smell, and swelling of the breasts.

The fetus grows and matures during the second and third trimesters

During the second trimester the fetus grows rapidly to a weight of about 600 g, and the mother's abdomen enlarges considerably. The limbs of the fetus elongate, and the fingers, toes, and facial features become well formed (**Figure 43.20C**). Eyebrows and fingernails grow. Fetal movements are first felt by the mother early in the second trimester, and they become progressively stronger and more coordinated. By the end of the second trimester, the fetus may suck its thumb. The nervous system undergoes rapid development.

The fetus grows rapidly during the third trimester (**Figure 43.20D**). As the third trimester approaches its end, internal organs mature. The digestive system begins to function, the liver stores glycogen, the kidneys produce urine, and the brain undergoes cycles of sleep and waking. A human infant is born as soon as the last of its critical organs—the lungs—mature. If development were to continue for longer inside the mother's body, the baby's head would grow larger than the birth canal. But if the fetus is born before its lungs mature, the baby cannot breathe on its own.

Although the first-trimester embryo is the most susceptible to adverse effects of drugs, chemicals, and diseases, the potential for serious effects from exposure to environmental factors continues throughout pregnancy. Severe protein malnutrition, alcohol consumption, and cigarette smoking are examples of factors that can result in low birth weight, mental retardation, and other developmental complications.

Developmental changes continue throughout life

Development does not end with birth. Obviously, growth continues until adult size is reached, and even when growth stops, organs of the body continue to repair and renew themselves through cycles of cell replacement by the progeny of undifferentiated stem cells. In humans especially, enormous developmental changes occur in the brain in the years between birth and adolescence. Especially in the early years, there is a great deal of plasticity in the organization of the nervous system as the connections between neurons develop.

For example, if a child is born with its eyes misaligned, a condition known as *strabismus*, he or she will use mostly one eye. The connections to the brain from that eye will become strong, and connections from the other eye will become weak. The child will develop with reduced visual acuity and depth perception. If the eye alignment is corrected in the first 3 years of life, however, the connections between the eyes and the brain will improve, and the child is likely to develop normal vision. If the eye alignment is corrected after 3 years of age, the correct connections between the eyes and the brain are less likely to improve, and visual impairments may persist. Thus, plasticity in the development of the visual system in humans continues for several years after birth, but is gradually lost.

A very exciting area of current research is the role of learning in stimulating the production and differentiation of new neurons in the brains of young and even adult animals (see Section 46.3). The first observations of new neurons were made on song birds, which relearn their songs each year prior to mating. At that time, the regions of their brains involved in vocalization grow and new neurons appear. Subsequently it was shown that mice and rats, when exposed to complex environments offering ample opportunity for exploration and activity, also acquired new neurons in certain parts of their brains. Scientists believe that acquisition of new neurons also occurs in adult humans.

43.5 RECAP

Gestation in humans lasts 9 months and can be divided into three trimesters. By the end of the first trimester, the fetus is very small but most of its organs have begun to form. In the second trimester, limbs elongate and the fetus moves. By the end of the third trimester, most organs have begun to function.

- Do you understand why the first trimester is a time of particular sensitivity for the embryo with respect to environmental risks? See p. 936

CHAPTER SUMMARY

43.1 How does fertilization activate development?

The sperm and the egg contribute differentially to the zygote. The sperm contributes a haploid nucleus and, in some species, a centriole. The egg contributes a haploid nucleus, nutrients, ribosomes, mitochondria, mRNAs, and proteins.

The cytoplasmic contents of the egg are not distributed homogeneously, and they are rearranged after fertilization to set up the major axes of the future embryo. The nutrient molecules are generally found in the **vegetal hemisphere**, whereas the nucleus is found in the **animal hemisphere**. Review Figures 43.1 and 43.2

Cleavage is a period of rapid cell division without cell expansion or gene expression. Cleavage can be complete or incomplete, and the pattern of cell divisions depends on the orientation of the mitotic spindles. The result of cleavage is a ball or mass of cells called a **blastula**. Review Figure 43.3

Cleavage in mammals is unique in that cell divisions are very slow and genes are expressed early in the process. Cleavage results in an inner cell mass that becomes the embryo and an outer cell mass that becomes the **trophoblast**. The mammalian embryo at this stage is called a **blastocyst**. At the time of implantation, the trophoblast secretes molecules that help the blastocyst attach to and penetrate the uterine wall. Review Figures 43.4 and 43.5

A **fate map** can be created by labeling specific blastomeres and observing what tissues and organs are formed by their progeny. Review Figure 43.6

Some species undergo mosaic development, in which the fate of each cell is determined during early divisions. Other species, including vertebrates, undergo regulative development, in which remaining cells can compensate for cells lost in early cleavages.

43.2 How does gastrulation generate multiple tissue layers?

Gastrulation involves massive cell movements that produce three **germ layers** and place cells from various regions of the blastula into new associations with one another. Review Figure 43.8, Web/CD Tutorial 43.1

The initial step of sea urchin and amphibian gastrulation is inward movement of certain blastomeres. The site of inward movement becomes the **blastopore**. Cells that move into the blastula become the **endoderm** and **mesoderm**; cells remaining on the outside become the **ectoderm**. Cytoplasmic factors in the vegetal pole cells are essential to initiate development. Review Figures 43.8 and 43.9

The **dorsal lip** of the amphibian blastopore is a critical site for cell determination. It has been called the primary embryonic **organizer** because it induces determination in cells that pass over it during gastrulation. Review Figures 43.9, 43.10, and 43.11, Web/CD Tutorial 43.2

The protein β-catenin activates a signaling cascade that induces the primary embryonic organizer and sets up the anterior–posterior body axis. Review Figures 43.2 and 43.12

Gastrulation in reptiles and birds differs from that in sea urchins and frogs because the large amount of yolk in their eggs causes the blastula to form a flattened disc of cells. Review Figure 43.13

Mammals have a pattern of gastrulation similar to that of reptiles and birds, although their eggs have no yolk.

CHAPTER SUMMARY

43.3 How do organs and organ systems develop?

Gastrulation is followed by **organogenesis**, the process whereby tissues interact to form organs and organ systems.

In the formation of the vertebrate nervous system, one group of cells that migrates over the blastopore lip is determined to become the **notochord**. The notochord induces the overlying ectoderm to thicken, form parallel ridges, and fold in on itself to form a **neural tube** below the epidermal ectoderm. The nervous system develops from this neural tube. Review Figure 43.14

The notochord and neural crest cells participate in the segmental organization of mesoderm into structures called **somites** along the body axis. Rudimentary organs and organ systems form during these stages. Review Figure 43.15

Four families of **Hox genes** determine the pattern of anterior–posterior differentiation along the body axis in mammals. Other genes, such as *sonic hedgehog*, contribute to dorsal–ventral differentiation. Review Figure 43.16

43.4 What is the origin of the placenta?

The embryos of reptiles, birds, and mammals are protected and nurtured by four **extraembryonic membranes**. In birds and reptiles, the **yolk sac** surrounds the yolk and provides nutrients to the embryo, the **chorion** lines the eggshell and participates in gas exchange, the **amnion** surrounds the embryo and encloses it in an aqueous environment, and the **allantois** stores metabolic wastes. Review Figure 43.17, Web/CD Activity 43.1

In mammals, the chorion and the trophoblast cells interact with the maternal uterus to form a **placenta**, which provides the embryo with nutrients and gas exchange. The amnion encloses the embryo in an aqueous environment. Review Figure 43.18

Samples of amniotic fluid or pieces of the chorion can be analyzed either by **amniocentesis** or **chorionic villus sampling** for genetic analysis that can reveal the presence of genes that can cause birth defects or disease. Review Figure 43.19

43.5 What are the stages of human development?

Human pregnancy, or **gestation**, can be divided into three trimesters. The embryo forms in the first trimester; during this time, it is most vulnerable to environmental factors that can lead to birth defects. During the second and third trimesters the **fetus** grows, the limbs elongate, and the organ systems mature.

Development continues throughout childhood and throughout life.

SELF-QUIZ

1. Fertilization involves all of the following *except*
   a. joining of most cell organelles from sperm and egg.
   b. joining of sperm and egg haploid nuclei.
   c. induction of rearrangements of the egg cytoplasm.
   d. sperm binding to specific sites on the egg surface.
   e. metabolic activation of the egg.
2. Which of the following does *not* occur during cleavage in frogs?
   a. A high rate of mitosis
   b. Reduction in the size of cells
   c. Expression of genes critical for blastula formation
   d. Orientation of cleavage planes at right angles
   e. Unequal division of cytoplasmic determinants
3. How does cleavage in mammals differ from cleavage in frogs?
   a. Slower rate of cell division
   b. Formation of tight junctions
   c. Expression of the embryo's genome
   d. Early separation of cells that will not contribute to the embryo
   e. All of the above
4. Which statement about gastrulation is *true*?
   a. In frogs, gastrulation begins in the vegetal hemisphere.
   b. In sea urchins, gastrulation produces the notochord.
   c. In birds, cells from the surface of the blastodisc move down through the primitive groove to form the hypoblast.
   d. In mammals, gastrulation occurs in the hypoblast.
   e. In sea urchins, gastrulation produces only two germ layers.
5. Which of the following was a conclusion from the experiments of Spemann and Mangold?
   a. Cytoplasmic determinants of development are homogeneously distributed in the amphibian zygote.
   b. In the late blastula, certain regions of cells are determined to form skin or nervous tissue.
   c. The dorsal lip of the blastopore can be isolated and will form a complete embryo.
   d. The dorsal lip of the blastopore can initiate gastrulation.
   e. The dorsal lip of the blastopore gives rise to the neural tube.
6. Which of the following is true of human development?
   a. Most organs begin to form during the second trimester.
   b. Gastrulation takes place in the oviducts.
   c. Genetic diseases can be detected by sampling cells from the chorion.
   d. Implantation occurs through interactions of the zona pellucida with the uterine lining.
   e. Exposure to drugs and chemicals is most likely to cause birth defects when it occurs in the third trimester.
7. Which of the following characterizes neurulation?
   a. The notochord forms a neural tube.
   b. The neural tube is formed from ectoderm.
   c. A neural tube forms around the notochord.
   d. The neural tube forms somites.
   e. In birds, the neural tube forms from the primitive groove.
8. Which statement about trophoblast cells is *true*?
   a. They are capable of producing monozygotic twins.
   b. They are derived from the hypoblast of the blastocyst.
   c. They are endodermal cells.
   d. They secrete proteolytic enzymes.
   e. They prevent the zona pellucida from attaching to the oviduct.
9. Which membrane is part of the embryonic contribution to placenta formation?
   a. Amnion
   b. Chorion
   c. Epiblast
   d. Allantois
   e. Zona pellucida

10. A major factor in the determination and differentiation of tissues along the anterior–posterior axis of the mouse is the
a. differential expression of Hox genes.
b. concentration gradient of β-catenin.
c. differential expression of the *sonic hedgehog* gene.
d. distance of the tissue from the gray crescent.
e. distribution of GSK-3, which degrades β-catenin.

FOR DISCUSSION

1. If you found a protein that was localized to a small group of cells in the frog blastula, how would you determine whether that protein played a role in development? Address the issues of sufficiency and necessity.
2. During gastrulation in birds, the *sonic hedgehog* gene is expressed only on the left side of Hensen's node. What might be the significance of this expression pattern?
3. Much of the early work of describing animal development was done on sea urchins, amphibians, and chicks. Most recent work on the molecular mechanisms of animal development has been done on nematodes, fruit flies, zebrafish, and mice. Why do you think there has been a shift in the animal models used by developmental biologists?
4. If all the mitochondria and mitochondrial DNA in the embryo come from the egg, what implications does this have for using mitochondrial DNA for molecular evolutionary studies?
5. There is currently much controversy over therapeutic cloning as a way of obtaining embryonic stem cells to treat diseases. Given that human development is regulative—in other words, twinning can occur if an early blastocyst is divided into two cell masses—can you think of a way to guarantee a source of isogenic (i.e., identically matching a person's own body) stem cells for an individual without resorting to therapeutic cloning? Assume isolated cells can be preserved indefinitely in a frozen state.

FOR INVESTIGATION

The early gastrula is bilaterally symmetrical. However, the fetus is not bilaterally symmetrical. For example, the top of your heart tilts to the right side of your body and the aorta comes off of the left side of the heart. Your spleen is on the left side of your body. Your large intestine goes from right to left. These asymmetries are set up during gastrulation. How would you investigate the mechanisms involved?

CHAPTER 44 Neurons and Nervous Systems

Fear and survival in the brain

Charles Whitman was a normal and responsible child. He became the youngest Eagle Scout in the country. He was a fine son and husband, and received commendations as a U.S. Marine. But while he was in the service, he began having unexplained fits of anger, among other personality disorders. He was discharged from the Marines and entered the University of Texas. Several times he visited campus doctors and complained about having violent thoughts. Then, on August 1, 1966, after killing his wife and mother, he gathered several high-powered rifles, went to the top of the tall clock tower on the University campus, and barricaded himself inside. From this vantage point, he killed 14 people and wounded 38 others before being shot and killed by Austin police. His autopsy revealed a tumor pressing on his amygdala.

The *amygdala* (Latin for "almond," which describes this structure's shape) is the brain's center for the emotion and memory of fear. When the cells of this structure are activated, your heart beats faster, your breathing becomes rapid and shallow, and your hands get cold and clammy. If you watch a horror movie, your amygdala is activated. If you encounter a threatening face, your amygdala is activated. If you are alone at night and hear an unusual noise, your amygdala is activated. What would life be like without an amygdala? You wouldn't get scared—and *not* being scared could be hazardous to your health.

A rare case of brain damage left a woman without a functional amygdala. When shown pictures of faces registering different emotions, she could not pick out the ones that were threatening or scary. She could not recall every having a frightening experience. In tests where she was administered mild electrical shocks, she developed no anticipatory fear; even though she knew that seeing a red card meant she was about to receive a shock, she never reacted to the red card. Thus, in real life, she would not have the reflex to pull away from a threat.

People with damage to the amygdala frequently have trouble engaging in normal social relationships. They cannot "read" the nature, mood, or intentions of other people by looking at their faces. The presence of pressure on Charles Whitman's amygdala may have been a factor in the emotions that drove him to mass

Fear Factor The fear response—nerves tense, heart racing, cold sweat—kicks in when we encounter a scary animal, person, or situation. Even a scary movie can trigger this primitive and protective reaction.

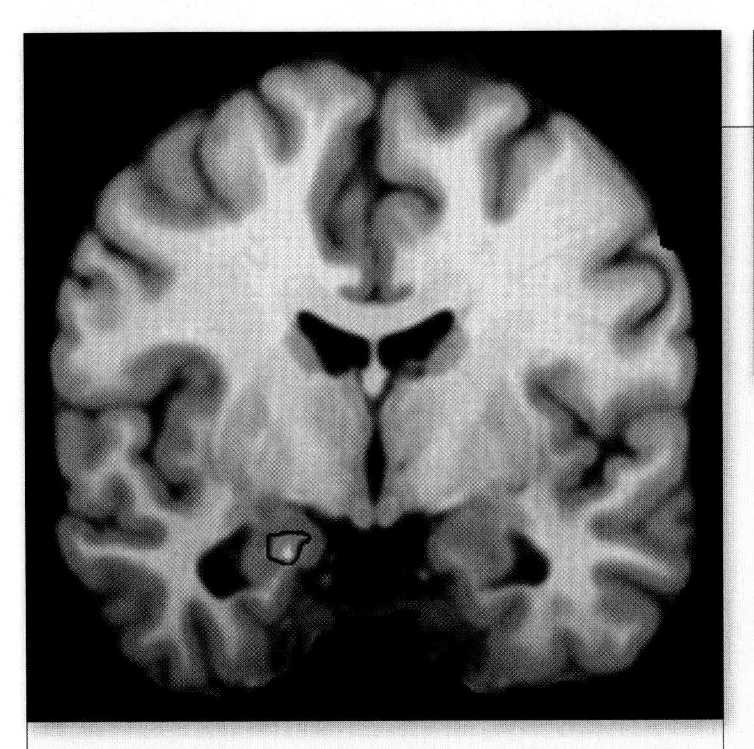

Source of the Fear Response Frightening situations—or even memories of such a situation—activate a group of cells in the amygdala, a structure deep in the brain.

murder; this diagnosis remains a matter for medical speculation.

Our nervous system enables us to experience the world around us and to react to it. But in between sensing and reacting, there is much interpretation based on memory, learning, emotions, and beliefs—all of which are based on the activities of cells in the nervous system. To understand how the eyes see, how the fingers play the piano, or how emotions affect our behavior, we have to understand how cells in different parts of our brains work and interact.

IN THIS CHAPTER we will discuss the general properties of nervous systems. We will begin with a look at the special cells that constitute nervous systems. Next we will see how some of these cells, the neurons, transmit information by generating electrical signals and conducting them from place to place in the body. Finally we will examine the electrical and biochemical mechanisms by which neurons communicate these signals to each other and to other cells in the body.

CHAPTER OUTLINE

44.1 What Cells Are Unique to the Nervous System?

Nervous systems are composed of two unique categories of cells: *nerve cells* or **neurons** and *glial cells* or **glia** (see Figure 40.6). Neurons are *excitable*: they can generate and propagate electrical signals, which are known as nerve impulses, or **action potentials**. Most neurons have long extensions called **axons** that enable them to conduct action potentials over long distances. Glial cells do not conduct action potentials; rather, they support neurons physically, immunologically, and metabolically. A **nerve** (as opposed to a neuron) is a bundle of axons that come from many different neurons.

Nervous systems can process information because their neurons are organized into networks. These networks include three functional categories of cells, which can be thought of as being involved with input, integration, and output. In the first category, **afferent neurons** carry sensory information into the nervous system. That information comes from specialized **sensory neurons** that transduce (convert) various kinds of sensory input into action potentials. Next, **efferent neurons** carry commands to physiological and behavioral *effectors* such as muscles and glands. The third category of cells, called **interneurons**, integrate and store information and facilitate communication between sensors and effectors.

Neuronal networks range in complexity

Simple animals such as cnidarians (e.g., sea anemones) can process information with simple networks of neurons that do little more than provide direct lines of communication from sensory cells to effectors (**Figure 44.1A**). The cnidarian's *nerve net* is most developed around the tentacles and the oral opening, where it facilitates detection of food or danger and causes tentacles to extend or retract. Animals that are more complex and move around the environment to search for food and mates need to process and integrate larger amounts of information. Even animals such as earthworms fit this description, and their increased need for information processing is met by higher numbers of neurons organized into clusters called **ganglia**. Ganglia serving different functions may be distributed around the body, as in the earthworm or the squid (**Figure 41.1B,C**). In animals that are bilaterally symmetrical, ganglia frequently come in pairs, one on each side of the body. Also, as animals increase in complexity, generally one pair of ganglia is larger than the others, and is therefore given the designation of **brain**.

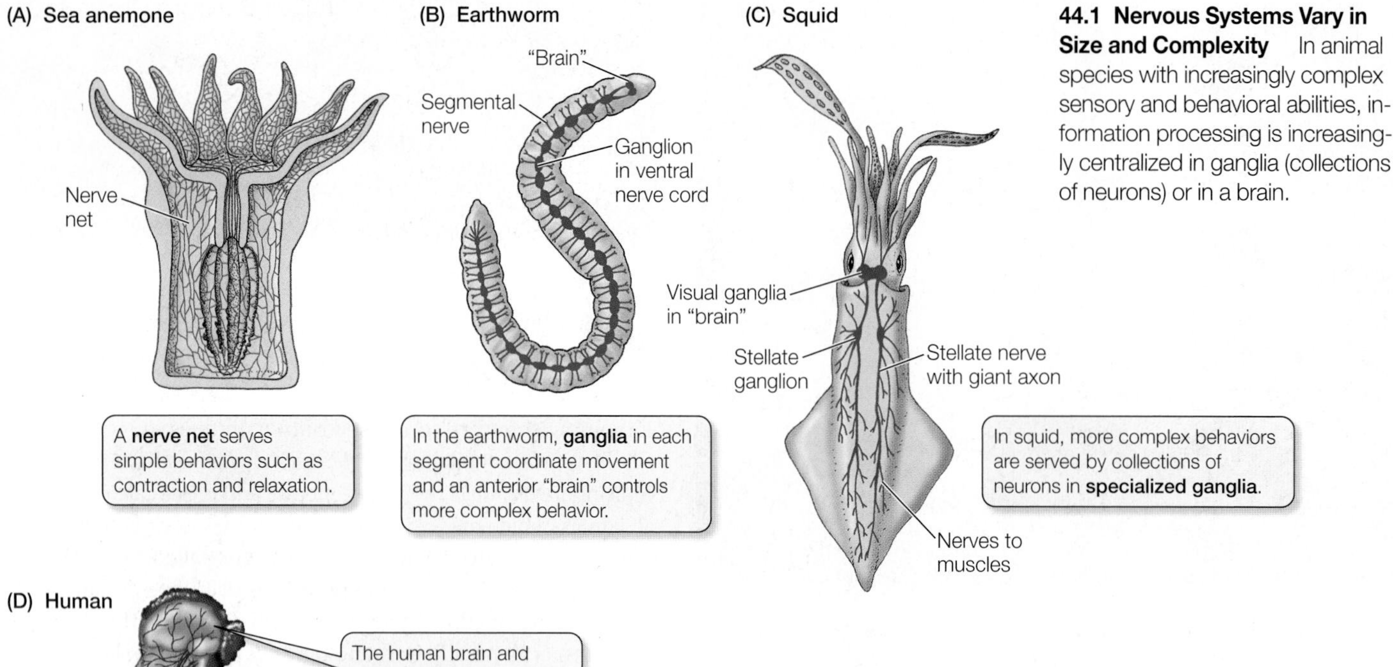

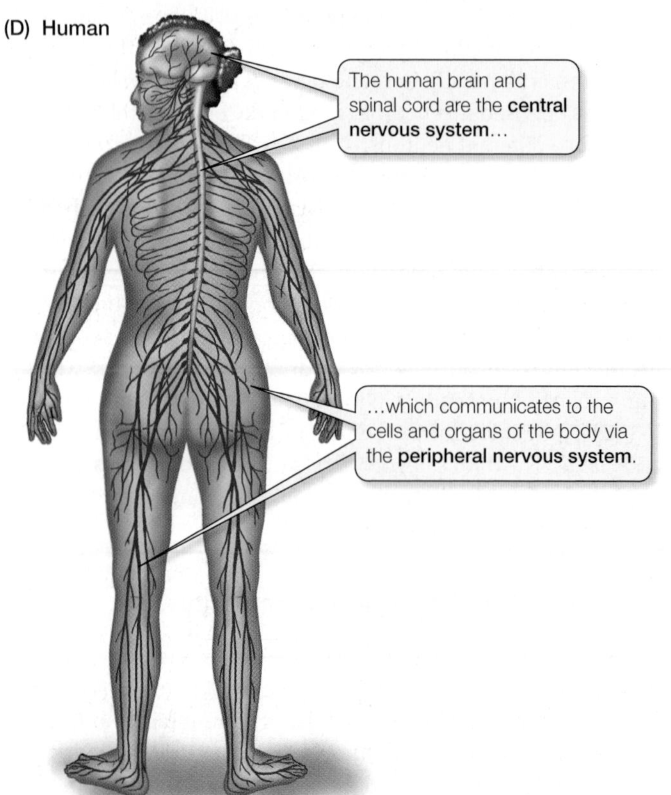

44.1 Nervous Systems Vary in Size and Complexity In animal species with increasingly complex sensory and behavioral abilities, information processing is increasingly centralized in ganglia (collections of neurons) or in a brain.

The small nervous systems of invertebrates can be remarkably complex. Consider the nervous systems of spiders, which have programmed within them the thousands of precise movements necessary to construct a beautiful web without prior experience or opportunities to learn the specific web architecture of their species.

In vertebrates, most cells of the nervous system are found in the brain and the **spinal cord**, the sites of most information processing, storage, and retrieval (**Figure 44.1D**). Therefore, the brain and spinal cord are called the **central nervous system** (**CNS**). Information is transmitted from sensory cells to the CNS and from the CNS to effectors via neurons that extend or reside outside of the brain and the spinal cord; these neurons and their supporting cells are called the **peripheral nervous system** (**PNS**). Vertebrates differ greatly in their behavioral complexity and in their physiological specializations, and their nervous systems reflect this diversity. **Figure 44.2** shows the brains of four vertebrate species of similar body mass drawn to the same scale.

The human nervous system contains an estimated 10^{11} neurons. Information is passed from one neuron to another where they come into close proximity at structures called **synapses**. The cell that sends the message is the **presynaptic neuron**, and the cell that receives it is the **postsynaptic neuron**. A given neuron in the brain can receive information from a thousand or more synapses. Thus the human brain may contain up to 10^{14} synapses, which can be highly plastic—strengthening with use and weakening with disuse. Therein lies the incredible ability of the human brain to process information, to learn, to do complex tasks, to remember, and even to have emotions. This astronomical number of neurons and synapses is divided into thousands of distinct but interacting networks that function in parallel. Before we can understand how even one of these circuits works, we must understand the properties of individual neurons.

Neurons are the functional units of nervous systems

Although nervous systems of different species vary enormously in structure and function, neurons behave similarly in animals as different as squids and humans. Their plasma membranes generate action potentials and conduct these signals from one location on a neuron to the most distant reaches of that cell—a distance that can be more than a meter in a human and many meters in a whale. Moreover, this transmission of action potentials can be rapid—up to 100 meters per second or more—making it possible to sense, process, and act on information very quickly.

Most neurons have four regions—a *cell body, dendrites*, an *axon*, and *axon terminals* (**Figure 44.3A**)—but the variation among different types of neurons is considerable (**Figure 44.3B**). The *cell body* contains the nucleus and most of the cell's organelles. Many pro-

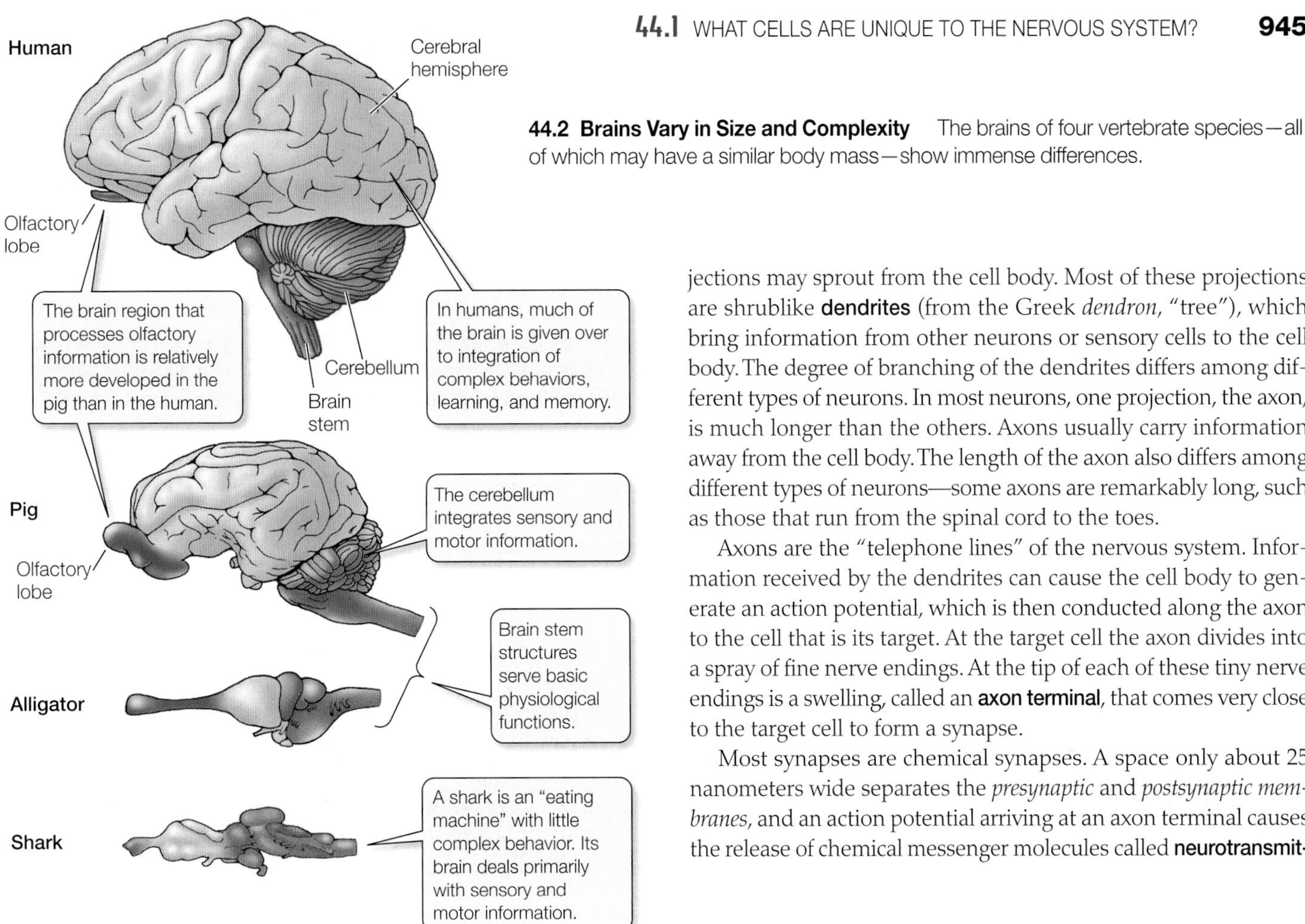

44.2 Brains Vary in Size and Complexity The brains of four vertebrate species—all of which may have a similar body mass—show immense differences.

jections may sprout from the cell body. Most of these projections are shrublike **dendrites** (from the Greek *dendron,* "tree"), which bring information from other neurons or sensory cells to the cell body. The degree of branching of the dendrites differs among different types of neurons. In most neurons, one projection, the axon, is much longer than the others. Axons usually carry information away from the cell body. The length of the axon also differs among different types of neurons—some axons are remarkably long, such as those that run from the spinal cord to the toes.

Axons are the "telephone lines" of the nervous system. Information received by the dendrites can cause the cell body to generate an action potential, which is then conducted along the axon to the cell that is its target. At the target cell the axon divides into a spray of fine nerve endings. At the tip of each of these tiny nerve endings is a swelling, called an **axon terminal**, that comes very close to the target cell to form a synapse.

Most synapses are chemical synapses. A space only about 25 nanometers wide separates the *presynaptic* and *postsynaptic membranes,* and an action potential arriving at an axon terminal causes the release of chemical messenger molecules called **neurotransmit-**

44.3 Neurons (A) A generalized diagram of a neuron. (B) Neurons from different parts of the mammalian nervous system are specifically adapted to their functions.

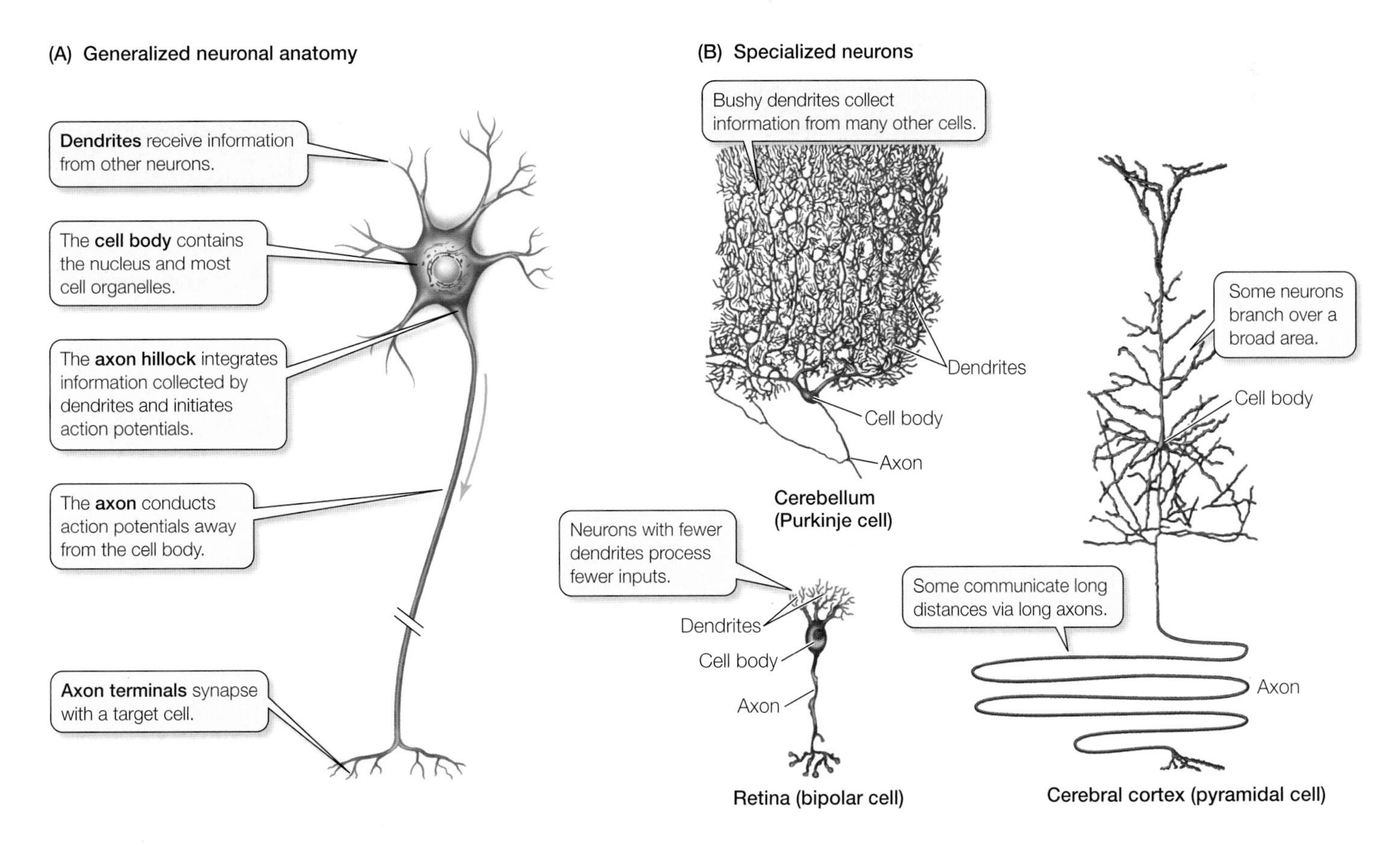

ters from the axon terminal. The released neurotransmitters diffuse across the space and bind to receptors on the plasma membrane of the postsynaptic or target cell. (We will discuss this process of synaptic transmission in more detail later in the chapter.) Integration of information in the nervous system is possible because a neuron can receive information (synaptic inputs) from many sources before producing action potentials that travel down its single axon to target cells. There are also electrical synapses that transmit the action potential from one neuron to the next; we will discuss electrical synapses later.

Glial cells are also important components of nervous systems

Glial cells, or simply *glia*, are another class of nervous system cells. There are many more glial cells than neurons in the human brain. Like neurons, glia come in several forms and have a diversity of functions. They are not excitable and do not transmit electrical signals. Some glia physically support and orient the neurons and help them make the right contacts during embryonic development. Others supply neurons with nutrients, maintain the extracellular environment, consume foreign particles and cellular debris, or insulate axons.

In the CNS some glia called **oligodendrocytes** wrap around the axons of neurons, covering them with concentric layers of insulating plasma membrane. In the PNS, glia called **Schwann cells** perform this function (**Figure 44.4**). **Myelin** is the covering produced by oligodendrocytes and Schwann cells, and it gives many parts of the nervous system a glistening white appearance. Not all axons are myelinated, but those that are can conduct action potentials more rapidly than those axons that are not myelinated. Later in this chapter we will see how the electrical insulation provided by myelin increases the speed with which axons can conduct action potentials.

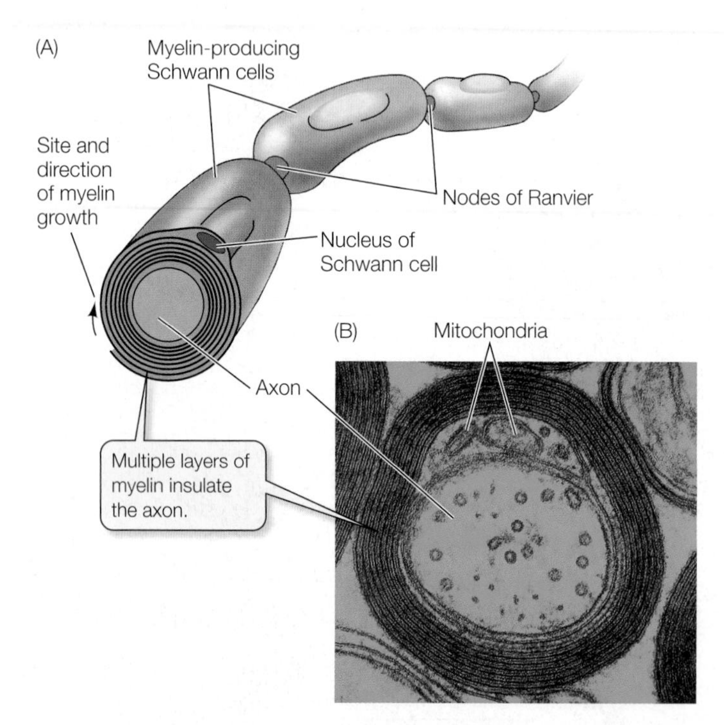

44.4 Wrapping Up an Axon (A) Schwann cells produce layers of myelin, a type of plasma membrane that provides electrical insulation to the axon. At the intervals between Schwann cells—the nodes of Ranvier—the axon is exposed. Action potentials travel along the axon by "jumping" from node to node. (B) A myelinated axon, seen in cross section through an electron microscope.

Glia called **astrocytes** (because they look like stars; see Figure 40.6B) contribute to the **blood–brain barrier**, which protects the brain from toxic chemicals in the blood. Blood vessels throughout the body are very permeable to many chemicals, including toxic ones, which would reach the brain if this special barrier did not exist. Astrocytes help form the blood–brain barrier by surrounding the smallest, most permeable blood vessels in the brain. The barrier is not perfect, however. Since it consists of plasma membranes, it is permeable to fat-soluble substances such as anesthetics and alcohol, which explains why these substances have such rapid and marked effects on the nervous system.

44.1 RECAP

Nervous systems are made up of two unique categories of cells, neurons and glia. Neurons are arranged in circuits. Sensory neurons transduce information from the external or internal environment; interneurons integrate and process information; and efferent neurons carry commands to target tissues of effector organs. Glial cells have a wide variety of supporting roles.

- Comparing nervous systems of anemones, worms, fish, and humans, what trends seem to reflect increasing complexity? See pp. 943–944 and Figures 44.1 and 44.2
- Describe the different parts of neurons and their functions. See pp. 944–945 and Figure 44.3
- What are some types of glial cells and what are their functions? See p. 946

The one feature common to all nervous systems is that they process information in the form of action potentials. In the next section we will focus on how these signals are produced and used by nervous systems.

44.2 How Do Neurons Generate and Conduct Signals?

Action potentials are generated when ion channels in the plasma membranes of neurons open for a short time and permit ions to move across the membrane. The movement of these charged molecules is driven by differences in their concentration gradients and by electrical charge differences on the two sides of the membrane. At rest, the inside of a neuron is electrically negative compared to the outside. Any difference in electric potential across the plasma membrane is a **membrane potential**, which is measured in *millivolts*. When the neuron is resting and not firing action potentials, the membrane potential is called a **resting potential**.

Simple electrical concepts underlie neuronal function

Voltage (electric potential difference) is a force that causes electrically charged particles to move between two points. Voltage is to the flow of electrically charged particles as pressure is to the flow of water. If the negative and the positive poles of a battery are connected by a wire, an electric current will flow through the wire because there is a voltage difference between the two poles of the battery. This flow of electric current can be used to do work, just as a current of water can be used to do work.

In wires, electric current is carried by electrons, but in solutions and across cell membranes, electric current is carried by ions. The major ions that carry electric charges across the plasma membranes of neurons are sodium (Na^+), potassium (K^+), calcium (Ca^{2+}), and chloride (Cl^-). Recall that ions with opposite charges attract one another, and those with like charges repel one another. How do these basic principles of bioelectricity establish the resting potential of the neuronal plasma membrane? And how is the flow of ions through membrane channels turned on and off to generate action potentials? We address these questions next.

Membrane potentials can be measured with electrodes

An *electrode* can be made from a glass pipette with a very sharp tip filled with a solution that conducts electric charges. Using electrodes, we can record electrical events in a cell. If one electrode is placed inside the plasma membrane of an axon, and another electrode is placed just outside of the axon, the difference in voltage can be measured. The typical resting potential of an axon thus measured is a voltage difference usually between –60 and –70 millivolts (mV) (**Figure 44.5**).

The resting potential provides a means for neurons to respond to a stimulus. Because of the voltage difference across the membrane, ions would cross the membrane if they could. Since the inside of the resting cell is negative, for example, positively charged ions such as sodium (Na^+) would enter if they could. Therefore, any chemical or physical stimulus that changes the permeability of the plasma membrane to ions will produce a change in the cell's membrane potential. The most extreme change in membrane potential is the action potential, a sudden and rapid reversal in the voltage across a portion of the plasma membrane. For 1 or 2 milliseconds, positively charged ions flow into the cell, making the inside of the cell *more positive* than the outside. Action potentials are conducted along axons, usually from the cell body to the terminal endings of its axon.

Ion pumps and channels generate membrane potentials

The plasma membranes of neurons, like those of all other cells, are lipid bilayers that are impermeable to ions, but contain many protein molecules that serve as ion channels and ion pumps (see Section 5.3). Ion pumps and channels are responsible for the distribution of charges across the membrane that create resting and action potentials.

Ion pumps require energy to move ions or other molecules against their concentration or electrical gradients. A major ion pump in the plasma membranes of neurons (and all other cells) is the **sodium–potassium pump**, so called because it actively expels Na^+ from inside the cell, exchanging it for K^+ from outside the cell (**Figure 44.6A**). The Na^+–K^+ pump is also known as **sodium–potassium ATPase**, a term emphasizing it as an enzyme complex requir-

44.5 Measuring the Resting Potential If the difference in electric charge across the plasma membrane of an unstimulated neuron, is measured, it is constant (about –60 mV), and is known as the resting potential.

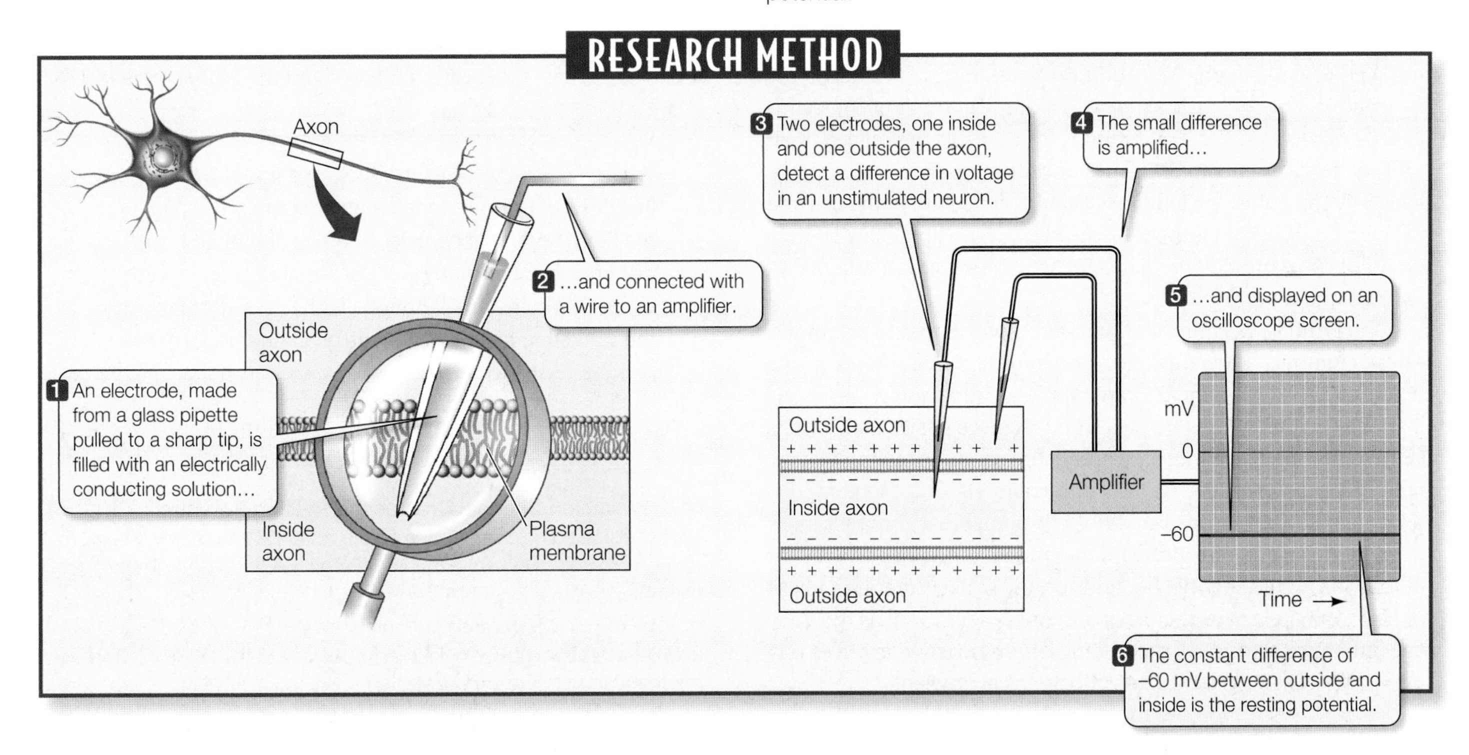

(A) $Na^+ - K^+$ pump (ATPase)

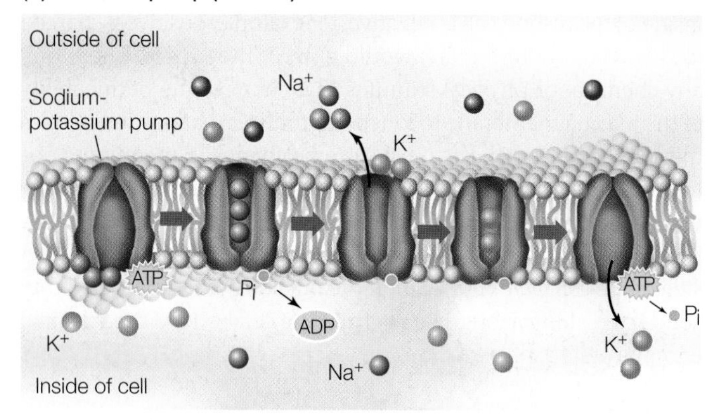

(B) $Na^+ - K^+$ channels

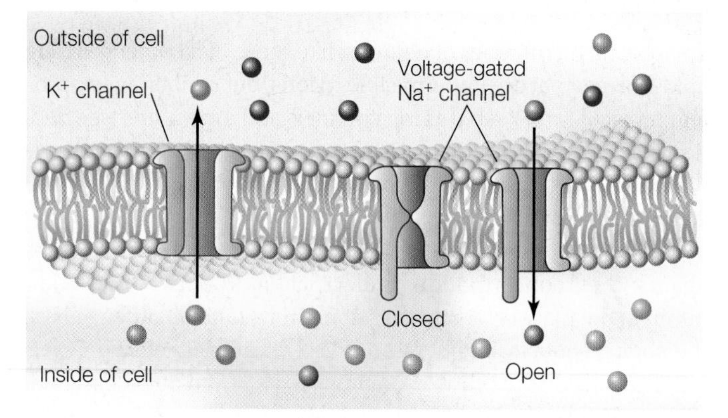

44.6 Ion Pumps and Channels (A) The sodium–potassium pump actively moves K^+ to the inside of a neuron and Na^+ to the outside. (B) Ion channels allow specific ions to diffuse down their concentration gradients; K^+ tends to leave neurons when potassium channels are open, and Na^+ tends to enter neurons when sodium channels are open.

ing ATP to do its work. The sodium–potassium pump keeps the concentration of K^+ inside the cell greater than that of the extracellular fluid, and the concentration of Na^+ inside the cell less than that of the extracellular fluid. The concentration differences established by the pump mean that K^+ would diffuse out of the cell and Na^+ would diffuse in if the ions could cross the lipid bilayer. How do these concentration gradients relate to the electric gradients we learned about above?

Ion channels permit the diffusion of ions across membranes. These channels are water-filled pores formed by proteins in the lipid bilayer (see Section 5.1) and are generally *selective*—they allow some types of ions to pass through more easily than others (**Figure 44.6B**). Thus, there are potassium channels, sodium channels, chloride channels, and calcium channels, and there are different kinds of each. Ions can diffuse through these channels in either direction. The direction and magnitude of the net movement of ions through a channel depends on the concentration gradient of that ion type across the plasma membrane, as well as the voltage difference across that membrane. These two motive forces acting on an ion are termed its **electrochemical gradient**.

Potassium channels are the most common open channels in the plasma membranes of resting (nonstimulated) neurons. As a consequence, resting neurons are more permeable to K^+ than to any other ion. Thus, open potassium channels are largely responsible for the membrane potential. Because the potassium channels make the plasma membrane permeable to K^+, and because the sodium–potassium pump keeps the concentration of K^+ inside the cell much higher than that outside the cell, K^+ tends to diffuse down its electrochemical gradient, out of the cell, through the channels. As these positively charged potassium ions diffuse out of the cell, they leave behind unbalanced negative charges, generating an electric potential across the membrane that tends to pull K^+ back into the cell.

The membrane potential at which the net diffusion of K^+ out of the cell ceases (that is, the point at which K^+ diffusion out due to the concentration gradient is balanced by its movement in due to the negative electric potential) is called the **potassium equilibrium potential**. The value of the potassium equilibrium potential can be calculated from the concentrations of K^+ on the two sides of the membrane using the **Nernst equation** (**Figure 44.7**). This equation, developed in the late 1800s, illustrates that the nature of ion channels in neuronal membranes was hypothesized long before their specific structures and properties were described.

In the late 1940s, A. L. Hodgkin and A. F. Huxley at the University of Cambridge set out to study the electrical properties of axonal membranes. With the techniques available at that time, the necessary measurements could be made only if you had a very large axon to work with. Such an axon exists in nature, in the giant neuron that controls the escape response of squid. Hodgkin and Huxley used electrodes to measure the voltage across the plasma membrane of this large axon, as seen in Figure 44.7, and to pass electric current into it to change its resting potential. They also changed the concentrations of Na^+ and K^+ both inside and outside the squid axon and measured the resulting changes in membrane potential.

On the basis of their many careful experiments, Hodgkin and Huxley developed virtually all of our basic concepts about the electrical properties of neurons, and received the Nobel prize in 1963.

When J. Z. Young called attention to the squid giant axon in 1936, it was a milestone for neuroscience. In a paper for the Royal Society of London, Hodgkin described how "a distinguished neurophysiologist remarked recently at a congress dinner (not, I thought, with the utmost tact), 'It's the squid that really ought to be given the Nobel prize.'"

44.7 Which Ion Channel Creates the Resting Potential? The Nernst equation calculates membrane potential when only one type of ion can cross a membrane that separates solutions with different concentrations of that ion. FURTHER INVESTIGATION: If you were to change the ion concentration in the seawater bathing your squid giant axon preparation, how would investigate what other ion(s) might be contributing to the resting potential?

EXPERIMENT

HYPOTHESIS: The resting potential of neurons is due to permeability of the membrane to potassium ions.

METHOD

1. Measure concentrations of ions inside and outside of a neuron.

To measure the concentration of ions in a neuron, the neuron (and its axon) must be big. Squid have giant neurons that control their escape response. It is possible to sample the cytoplasm of these axons, which are about 1 mm in diameter.

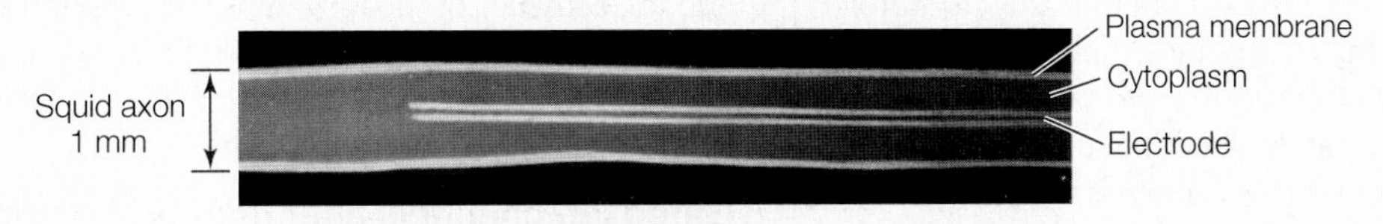

2. Use the Nernst equation to calculate what the membrane potential would be if it were permeable to each of these ions: Na^+, K^+, Ca^{2+}, and Cl^-.

Recall the Nernst equation from Section 5.3. It predicts the membrane potential resulting from membrane permeability to a single type of ion that differs in concentration on the two sides of the membrane. The equation is written

$$E_{ion} = 2.3\frac{RT}{zF}\log\frac{[\text{ion}]_o}{[\text{ion}]_i}$$

where E is the equilibrium (resting) membrane potential (the voltage across the membrane in mV), R is the universal gas constant, T is the absolute temperature, z is the charge on the ion (+1, +1, +2, or −1, respectively, for the ions used here), and F is the Faraday constant. The subscripts o and i indicate the ion concentrations outside and inside the cell, respectively.

At this point you could just "plug and play," but do you understand this equation?

A concentration difference of ions across a membrane creates a *chemical* force that pushes the ions across the membrane; however, the resulting unbalanced *electrical charges* will pull the ions back the other way. At *equilibrium*, the work done moving ions in each direction will be the same.

The *chemical* work pushing the ions will equal 2.3 RT log $[\text{ion}]_o/[\text{ion}]_i$
The *electrical* work pulling the ions will equal zEF. So, at equilibrium:

$$zEF = 2.3\,RT\log\frac{[\text{ion}]_o}{[\text{ion}]_i}$$

Rearranging the equation to solve for E, we get the Nernst equation:

$$E_{ion} = 2.3\frac{RT}{zF}\log\frac{[\text{ion}]_o}{[\text{ion}]_i}$$

We can simplify the equation by picking a temperature—let's use "room temperature," or 20°C—and solving for 2.3 RT/F. At 20°C, 2.3 RT/F equals 58. Thus:

$$E_{ion} = 58/z\log\frac{[\text{ion}]_o}{[\text{ion}]_i}$$

3. Measure the membrane potential across the squid giant axon and compare with calculated values for each ion.

RESULTS

1. Measuring ion concentrations in squid giant axon cytoplasm and in seawater, then solving the Nernst equation for each ion, we find:

Ion	Ion concentration (n*M*) in squid axon	in seawater	Predicted membrane potential (mV)
K^+	400	20	−75
Na^+	50	460	+56
Ca^{2+}	0.5	10	+38
Cl^-	50	560	−60

2. Using the method shown in Figure 44.5, the actual resting membrane potential of a squid giant axon is recorded to be −66 mV.

CONCLUSION: The resting potential of the squid giant axon can be due to permeability to K^+, but there is probably some permeability to another ion as well.

We now know that, in general, the resting potential is less negative than the Nernst equation predicts because resting neurons are also slightly permeable to other ions, such as Na^+ and Cl^-. Another equation, called the Goldman equation, takes all of the ions that can cross the membrane into account and therefore can calculate the membrane potential accurately.

Ion channels and their properties can now be studied directly

Hodgkin and Huxley were working long before the laboratory techniques emerged that enabled the demonstration of ion channels, and so could only hypothesize their properties. With the advent of a technique called **patch clamping**, developed in the 1980s by B. Sakmann and E. Neher, neurobiologists can now record currents caused by the openings and closings of single ion channels.

The patch clamp electrode is a polished glass micropipette filled with an electrically conductive solution that has the same composition as extracellular fluids. The tip of this electrode is then positioned right up against the membrane of a cell. When slight suction is applied to the electrode, it forms a seal with the membrane. Once the seal is formed, any exchanges of ions across the patch of membrane bounded by the seal are exchanges with the fluid in the micropipette, and they can be recorded as electric currents. If a single ion channel or a few ion channels happen to be in that patch of membrane, then the openings and closings of individual channels will be recorded by the pipette-electrode. If the pipette is retracted, it can tear the patched membrane away from the cell, and the activities of the ion channels in the patch can continue to be recorded (**Figure 44.8**). In 1991, Sakmann and Neher received the Nobel prize for their work with the patch clamp technique.

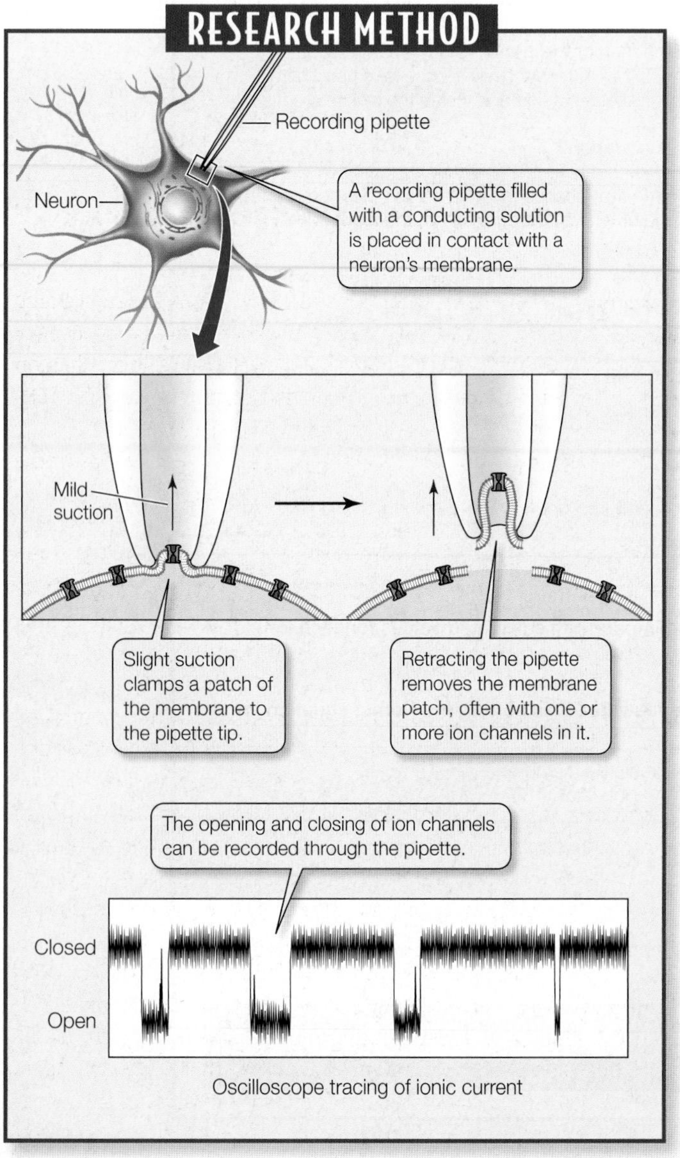

RESEARCH METHOD

Recording pipette

Neuron

A recording pipette filled with a conducting solution is placed in contact with a neuron's membrane.

Mild suction

Slight suction clamps a patch of the membrane to the pipette tip.

Retracting the pipette removes the membrane patch, often with one or more ion channels in it.

The opening and closing of ion channels can be recorded through the pipette.

Closed

Open

Oscilloscope tracing of ionic current

44.8 Patch Clamping The patch clamping technique can record the opening and closing of a single ion channel.

Gated ion channels alter membrane potential

Many ion channels in the plasma membranes of neurons behave as if they contain "gates" that are open under some conditions and closed under other conditions. **Voltage-gated channels** open or close in response to a change in the voltage across the plasma membrane. **Chemically gated channels** open or close depending on the presence or absence of a specific molecule that binds to the channel protein, or to a separate receptor that in turn alters the channel protein. **Mechanically gated channels** open or close in response to mechanical force applied to the plasma membrane. Gated channels play important roles in neuronal function.

Openings and closings of gated channels perturb the resting potential. Imagine what happens, for example, if sodium channels in the plasma membrane open. Na^+ diffuses into the neuron down its electrochemical gradient. As a result of the entry of Na^+, the inside of the cell becomes less negative. When the inside of a neuron becomes less negative (or more positive) in comparison to its resting condition, its plasma membrane is said to be **depolarized** (**Figure 44.9**).

An opposite change in the resting potential occurs if gated K^+ channels open. When K^+ efflux from the neuron increases, the membrane potential becomes even more negative, and the plasma membrane is said to be **hyperpolarized**.

The opening and closing of ion channels, which result in changes in the voltage across the plasma membrane, are the basic mechanisms by which neurons respond to stimuli whether they be electrical, chemical, or mechanical. How does a neuron use a change in its resting membrane potential to process and transmit information?

A change in membrane potential may result from the activity at a synapse. When an action potential traveling along an axon reaches the axon terminal, it causes the release of a chemical neurotransmitter. Neurotransmitter molecules diffuse across the small gap between the presynaptic and postsynaptic membranes and bind to receptors on the postsynaptic membrane. Those receptors may be ion channels or they may control ion channels indirectly. In either case, the neurotransmitter causes the channel to open, ions flow down their electrochemical gradients, and the

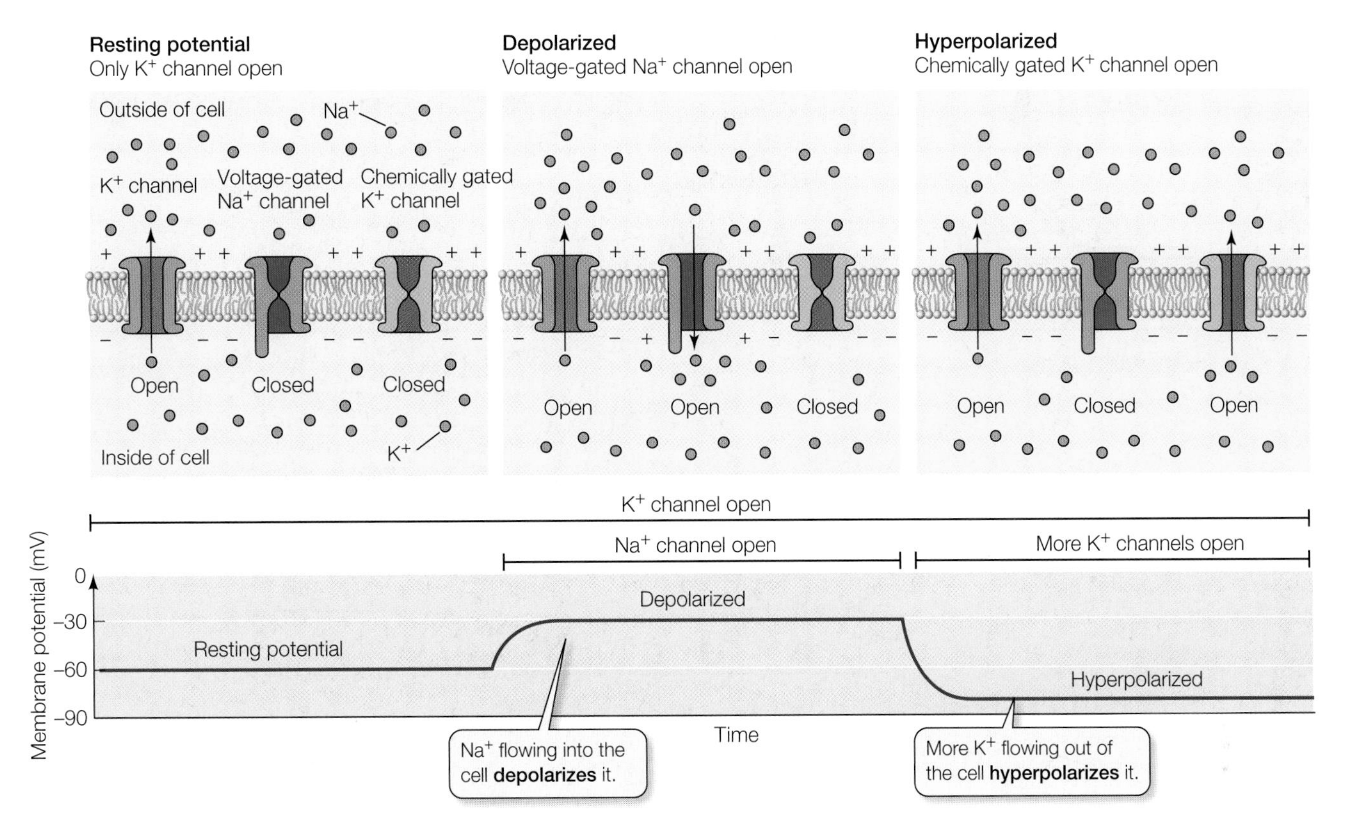

44.9 Membranes Can Be Depolarized or Hyperpolarized The resting potential is produced by open K^+ channels. A shift from the resting potential to a less negative membrane potential, as occurs when Na^+ enters the cell through a gated sodium channel, is called depolarization. Hyperpolarization occurs when the membrane potential becomes more negative, as when additional K^+ leaves the cell through gated K^+ channels.

membrane potential of the postsynaptic neuron changes. How is this local change in membrane potential communicated to other parts of the cell?

A local change in the membrane potential of the postsynaptic neuron causes a flow of ions that spreads the change in membrane potential to adjacent regions of the membrane. For example, when Na^+ enters the postsynaptic neuron through open sodium channels at one location, those positively charged ions are attracted to adjacent areas on the inside of the membrane that are more negative, and thus there is a rapid flow of electric current from the site of the open Na^+ channels. However, this local flow of electric current decays as it spreads and therefore does not spread very far.

Electric currents do not spread far in cells because cell membranes are not completely impermeable to ions. An electric current traveling along a membrane is like water flowing through a leaky hose. The flow of electric current along plasma membranes is useful for transmitting signals over only very short distances. Therefore, axons do not transmit information as a continuous flow of electric current (as telephone wires do). However, the local flow of electric current is an important part of the mechanism by which neurons generate action potentials, the truly long-distance signals.

Sudden changes in Na^+ and K^+ channels generate action potentials

Action potentials are sudden, transient, large changes in membrane potential that are conducted along unmyelinated axons at speeds of up to 2 meters per second, but in myelinated axons the conduction velocity can be 100 meters per second. Think of running the 100-meter dash—the world record is slightly under 10 seconds.

If we place the tips of a pair of electrodes on either side of the plasma membrane of a resting axon and measure the voltage difference, the reading is about –60 mV, as we saw in Figure 44.5. If these electrodes are in place when an action potential travels down the axon, they register a rapid change in membrane potential, from –60 mV to about +50 mV. The membrane potential then rapidly returns to its resting level of –60 mV as the action potential passes (**Figure 44.10**).

Voltage-gated Na^+ and K^+ channels in the plasma membrane of the axon are responsible for action potentials. At the resting potential, most of these channels are closed. Depolarization of the membrane causes them to open. For example, if synaptic input to a neuron is sufficiently strong to cause the plasma membrane of its cell body to depolarize, that depolarization can spread by local current flow to the axon hillock at the base of the axon (see Figure 44.3), where voltage-gated Na^+ channels are concentrated. When the plasma membrane in this area depolarizes, some of these voltage-gated channels open briefly—for less than a millisecond. The Na^+ concentration is much higher outside the axon than inside, and the inside is negatively charged, so when these channels open, Na^+ rushes into the axon. The entering Na^+ depolarizes the membrane even more, causing more Na^+ channels to open—a positive

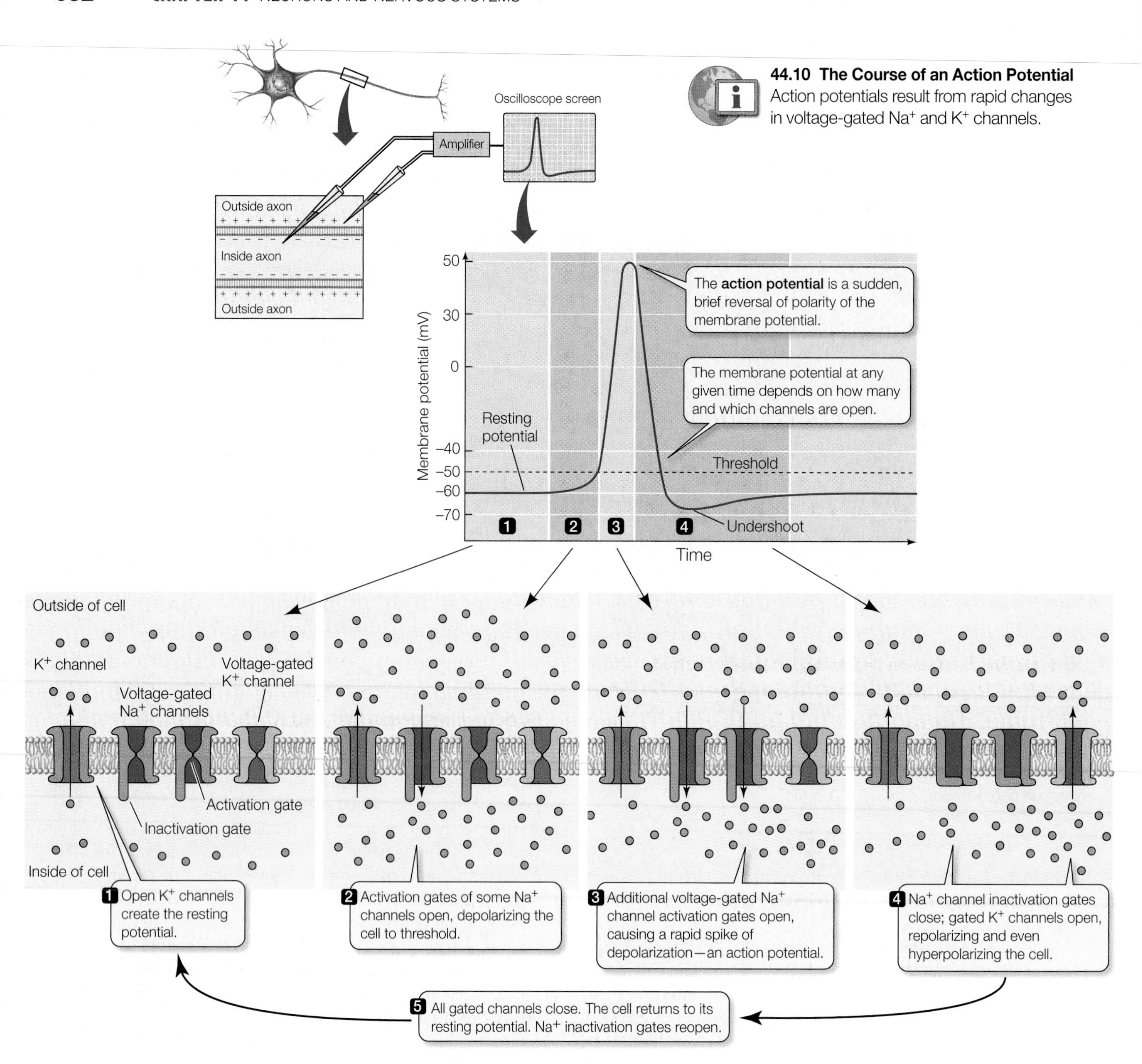

44.10 The Course of an Action Potential Action potentials result from rapid changes in voltage-gated Na^+ and K^+ channels.

feedback effect. When the membrane is depolarized about 5 to 10 mV above the resting potential, a **threshold** is reached at which the depolarizing influx of Na^+ can no longer be offset by the efflux of K^+. At this point, a large number of sodium channels open, and the membrane potential becomes positive—an action potential. The rising phase of the action potential halts abruptly in 1 to 2 milliseconds, and the membrane potential rapidly becomes negative once again.

What causes the axon to return to resting potential? There are two contributing factors: The voltage-gated Na^+ channels close, and voltage-gated K^+ channels may open. The voltage-gated K^+ channels open more slowly than the Na^+ channels and stay open longer, allowing K^+ to carry excess positive charges out of the axon. As a result, the membrane potential returns to a negative value and usually becomes even more negative than the resting potential until all of the voltage-gated K^+ channels close.

Another feature of the voltage-gated Na^+ channels is that once they open and close, they have a **refractory period** of 1 to 2 milliseconds during which they cannot open again. This property can be explained by the channels having two gates, an **activation gate** and an **inactivation gate** (see Figure 44.10). Under resting conditions, the activation gate is closed and the inactivation gate is open. Depolarization of the membrane to the threshold level causes both gates to change state, but the activation gate responds faster. As a result, the channel is open for a brief time between the opening of the activation gate and the closing of the inactivation gate. Inactivation gates remain closed for 1 to 2 milliseconds before they spontaneously open again, thus explaining why the membrane has a refractory period before it can fire another action potential. When the inactivation gate finally opens, the activation gate is closed, and the membrane is poised to respond once again to a depolarizing stimulus by firing another action potential. Another contribution to the refractory period is the duration of the opening of the voltage-gated K^+ channels, as we saw above. The dip in the membrane potential following an action potential is called the *after-hyperpolarization* or *undershoot*.

The difference in the concentration of Na^+ across the plasma membrane and the negative resting potential constitute the "battery" that drives action potentials. How rapidly does the battery run down? It might seem that a substantial number of ions would have to cross the membrane for the membrane potential to change from –60 mV to +50 mV and back to –60 mV again. In fact, only a vanishingly small percentage of the Na^+ concentrated outside the plasma membrane moves through the channels during the passage of an action potential. Thus the effect of a single action potential on the concentration gradients of Na^+ and K^+ is very small, and it is possible in most cases for the sodium–potassium pump to keep the "battery" charged, even when the neuron is generating many action potentials every second.

Action potentials are conducted along axons without loss of signal

Action potentials can travel over long distances with no loss of signal. If we place two pairs of electrodes at two different locations along an axon, we can record an action potential at those two locations as it travels along the axon (**Figure 44.11A**). The magnitude of the action potential does not change between the two recording sites. This constancy is possible because an action potential is an all-or-none, self-regenerating event.

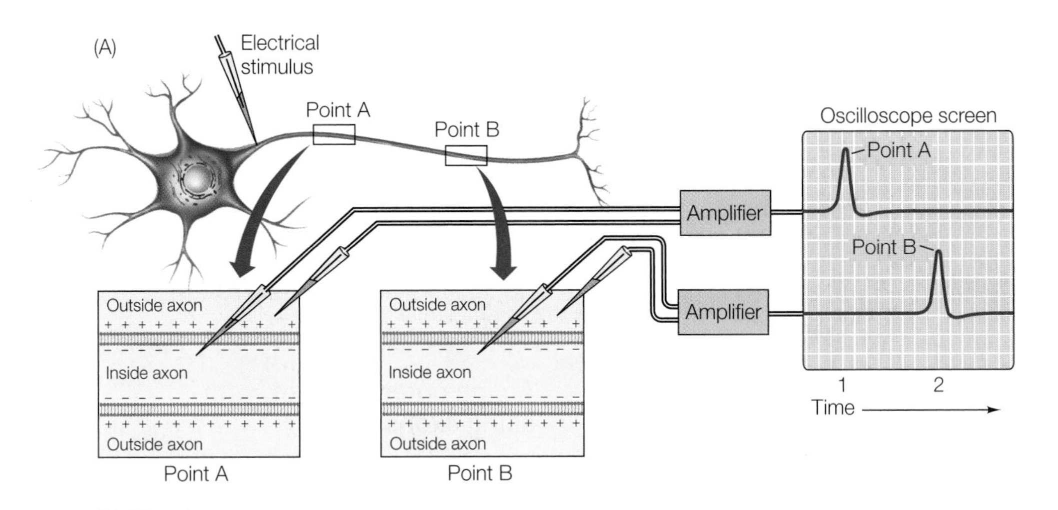

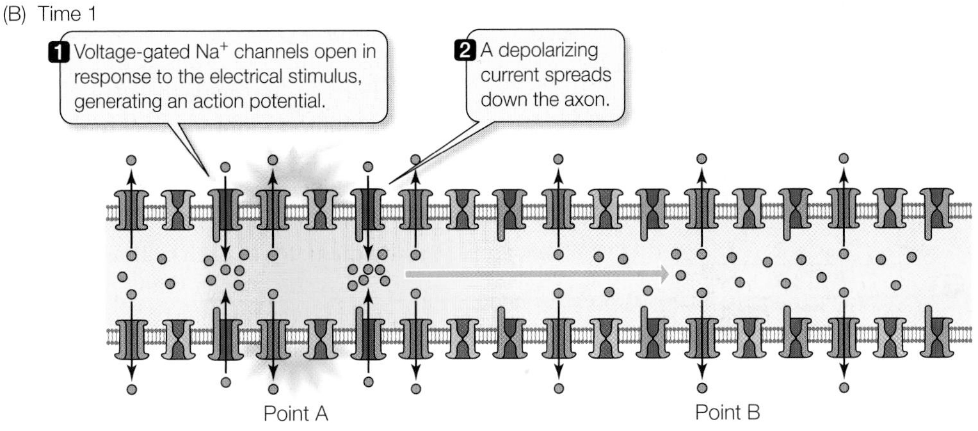

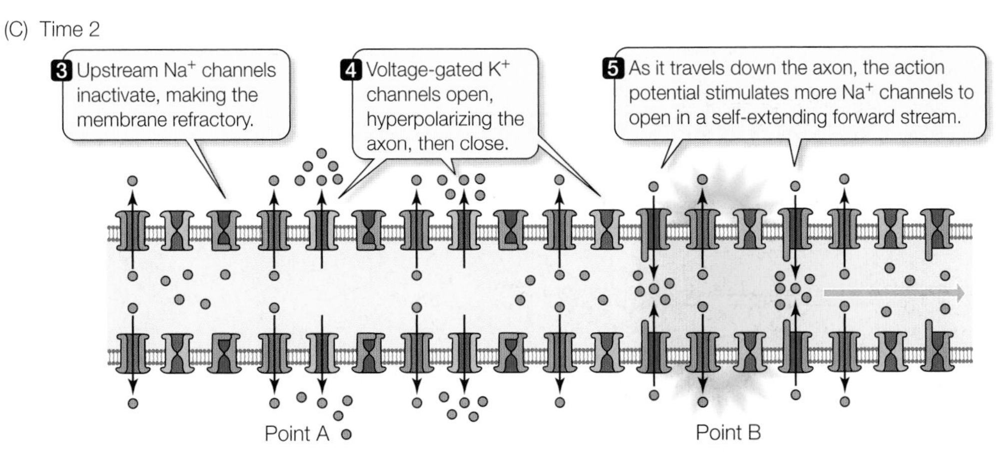

44.11 Action Potentials Travel along Axons (A) There is no loss of signal as an action potential travels along an axon. (B) When an action potential is stimulated in one region of membrane, electric current flows to adjacent areas of membrane and depolarizes them. (C) The advancing wave of depolarization causes more Na^+ channels to open, and the action potential is generated anew in the next section of membrane. Meanwhile, in the region where the action potential has just fired, the Na^+ channels are inactivated and the voltage-gated K^+ channels are still open, rendering this section of the axon refractory. Hence the action potential cannot "back up," but moves continuously forward along the axon, regenerating itself as it goes.

- An action potential is *all-or-none* because of the interaction between the voltage-gated Na^+ channels and the membrane potential. If the membrane is depolarized slightly, some voltage-gated Na^+ channels open. Some sodium ions cross the plasma membrane and depolarize it even more, opening more voltage-gated Na^+ channels, and so on, until the membrane reaches threshold and generates an action potential. This positive feedback mechanism ensures that action potentials always rise to their maximum value.
- An action potential is *self-regenerating* because it spreads by local current flow to adjacent regions of the plasma membrane. The resulting depolarization brings those neighboring areas of membrane to threshold. So when an action potential occurs at one location on an axon, it stimulates the adjacent region of axon to generate an action potential, and so on down the length of the axon.

We can initiate an action potential by using a stimulating electrode to deliver an electric current that depolarizes the membrane enough to reach threshold. We can then observe the changes in membrane potential associated with the passage of that action potential past the recording electrodes (**Figure 44.11B**). The action potential normally is propagated in only one direction, away from the cell body. It cannot reverse itself because the region of membrane it came from is in its refractory period.

Have you ever touched something very hot and felt the pain before you felt the heat? The neurons that carry the sharp pain sensations are myelinated and transmit action potentials about 10 times faster than the unmyelinated neurons that transmit the sensations of hot or cold.

Action potentials do not travel along all axons at the same speed. They travel faster in myelinated than in nonmyelinated axons, and they travel faster in large-diameter axons than in small-diameter axons. Invertebrates do not have myelin, and therefore the rate of conduction in their axons depends mostly on axon diameter. Invertebrate axons that transmit messages involved in escape behavior are very large, like the squid giant axon in Figure 44.7.

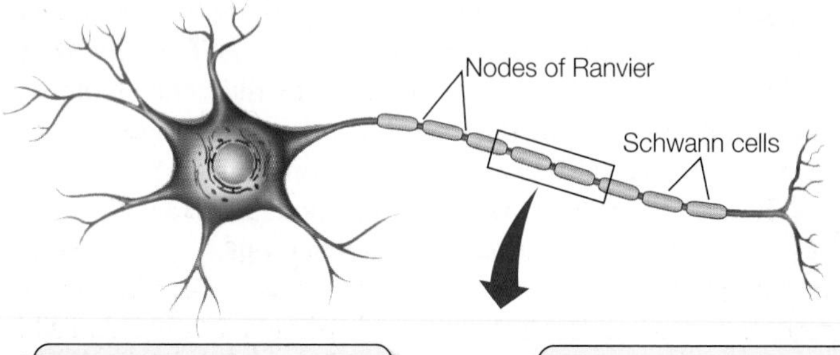

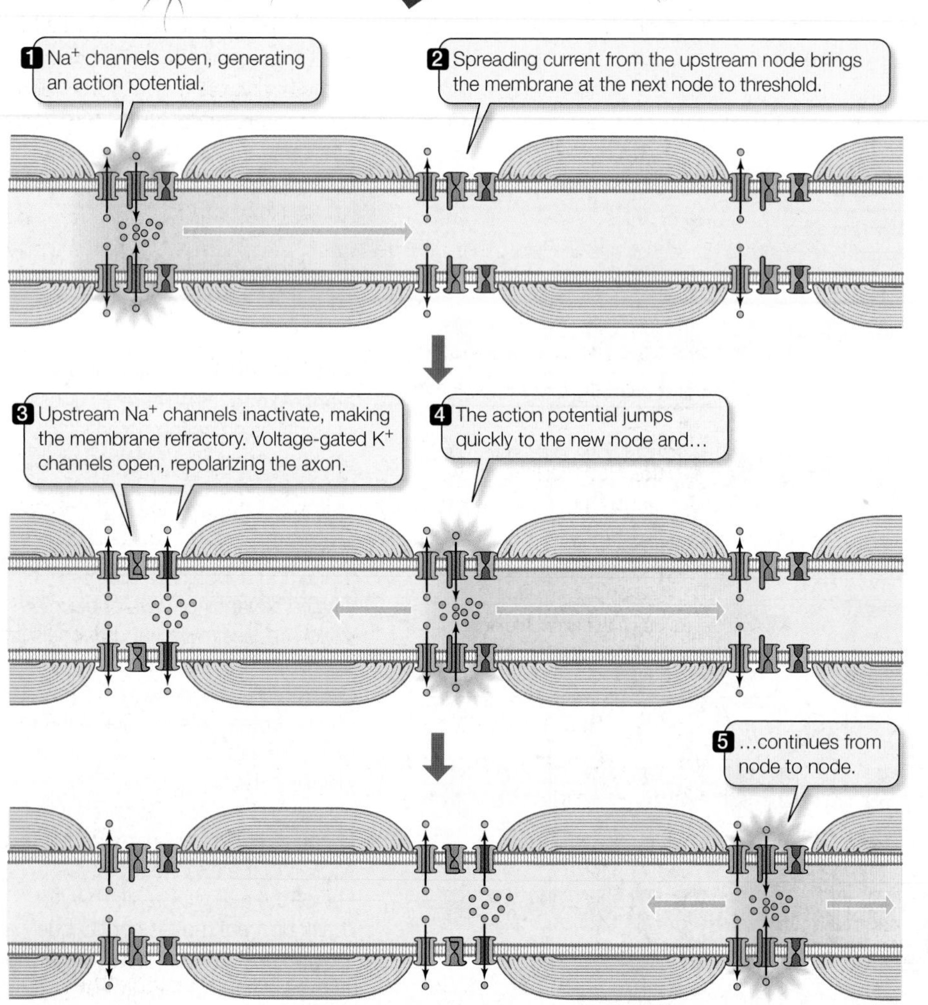

44.12 Saltatory Action Potentials Action potentials appear to jump from node to node in myelinated axons.

Action potentials can jump along axons

In vertebrate nervous systems, increasing the speed of action potentials by increasing the diameter of axons is not feasible because of the huge number of axons in these organisms. Each of our eyes, for example, has about a million axons connecting it to the brain. These axons conduct action potentials at about the same speed as does the squid giant axon—about 20 meters per second—yet the diameter of each of these axons is about 200 times smaller than the diameter of the squid axon. Imagine your optic nerves being 200 times larger than they are. Each would be over half a meter in diameter! Vertebrates have found a different way of increasing conduction velocity of axons, and that adaptation is myelination.

When glial cells wrap themselves around axons, covering them with concentric layers of myelin (see Figure 44.4), they leave regularly spaced gaps, called **nodes of Ranvier**, where the axon is not covered (**Figure 44.12**). The leakage of ions across the regions of the plasma membrane that are wrapped in myelin is reduced, so electric current can spread farther along the inside of a myelinated axon than it can along a nonmyelinated axon. Additionally, voltage-gated ion channels are clustered at the nodes of Ranvier. Thus an axon can fire action potentials only at nodes, and those action potentials cannot be propagated through the adjacent patch of membrane covered with myelin. The positive charges that flow into the axon at the node do, however, flow down the inside of the axon in the form of electric current. When the current reaches the next node, the plasma membrane at that node is depolarized to threshold and fires another an action potential. Action potentials therefore appear to jump from node to node along the axon.

The speed of conduction is increased in these myelin-wrapped axons because electric current flows much faster through the cytoplasm than ion channels can open and close. This form of rapid impulse propagation is called **saltatory conduction** (Latin *saltare*, "to jump").

44.2 RECAP

Neurons generate action potentials—rapid, all-or-none changes in membrane potential that are conducted along axons from the neuron body to the axonal terminals. At axon teriminals, the information is communicated across synapses to target cells.

- How are membrane resting potentials generated? See pp. 947–948 and Figures 44.6 and 44.9
- How are action potentials generated? See pp. 951–953 and Figure 44.10
- How are action potentials propagated along axons? See pp. 953–955 and Figures 44.11 and 44.12

Once we understand how action potentials are generated in and conducted along axons, the question remains as to exactly what happens at the synapse when an action potential gets to the axon terminal. How do synapses convey information from one cell to another?

44.3 How Do Neurons Communicate with Other Cells?

Neurons communicate with each other and with target cells at synapses. The most common type of synapse in the nervous system is the **chemical synapse**—one in which chemical messages from a presynaptic cell induce changes in a postsynaptic cell. In **electrical synapses** the action potential spreads directly from presynaptic to postsynaptic cell.

The neuromuscular junction is a model chemical synapse

Neuromuscular junctions are synapses between motor neurons and the skeletal muscle cells they innervate. They are excellent models for how chemical synaptic transmission works. Like other neurons, a motor neuron has only one axon, but that axon can have many branches, each with axon terminals that form neuromuscular junctions with a muscle cell. At each axon terminal an enlarged knob or button-like structure contains many vesicles filled with the chemical messenger—neurotransmitter molecules. The neurotransmitter used by all vertebrate motor neurons is **acetylcholine (ACh)**. ACh is released by exocytosis when the membrane of a vesicle fuses with the presynaptic membrane.

Where does the neurotransmitter come from? Some neurotransmitters, such as ACh, are synthesized in the axon terminal and packaged in vesicles. The enzymes required for ACh biosynthesis, however, are produced in the cell body of the motor neuron and are transported along microtubules down the axon in membrane-bound packages to the terminals. Other kinds of neurotransmitters, such as peptide neurotransmitters, are produced in the cell body and transported in membrane-bound packages down the axon to the terminals.

The postsynaptic membrane of the neuromuscular junction is a modified part of the muscle cell plasma membrane called a **motor end plate** (**Figure 44.13**). It appears as a depression in the muscle cell membrane, and the terminals of the motor neuron sit in the depression. The space between the presynaptic membrane and the postsynaptic membrane is the **synaptic cleft**, which in chemical synapses is about 20 to 40 nanometers wide. Neurotransmitter released into the cleft by the presynaptic cell diffuses across to the postsynaptic membrane. The membrane of the motor end plate is highly folded. ACh receptors are on the crests of the folds, and voltage-gated cation channels are at the bottoms of the folds and in the surrounding muscle cell membrane.

The arrival of an action potential causes the release of neurotransmitter

Neurotransmitter is released when an action potential arrives at the axon terminal and causes the opening of voltage-gated Ca^{2+} channels in the presynaptic membrane. Because the Ca^{2+} concentration is greater outside the cell than inside, Ca^{2+} enters the axon terminal near the sites of vesicle exocytosis. The increase in Ca^{2+} inside the axon terminal causes the vesicles containing ACh to fuse with the presynaptic membrane and empty their contents into the synaptic cleft.

Two well-known deadly toxins, botulinum toxin and tetanus toxin, act by interfering with the proteins responsible for the fusion of neurotransmitter vesicles and the presynaptic membrane.

In neuromuscular synapses, vesicle fusion and emptying is all-or-none. The vesicle membrane is incorporated into the presy-

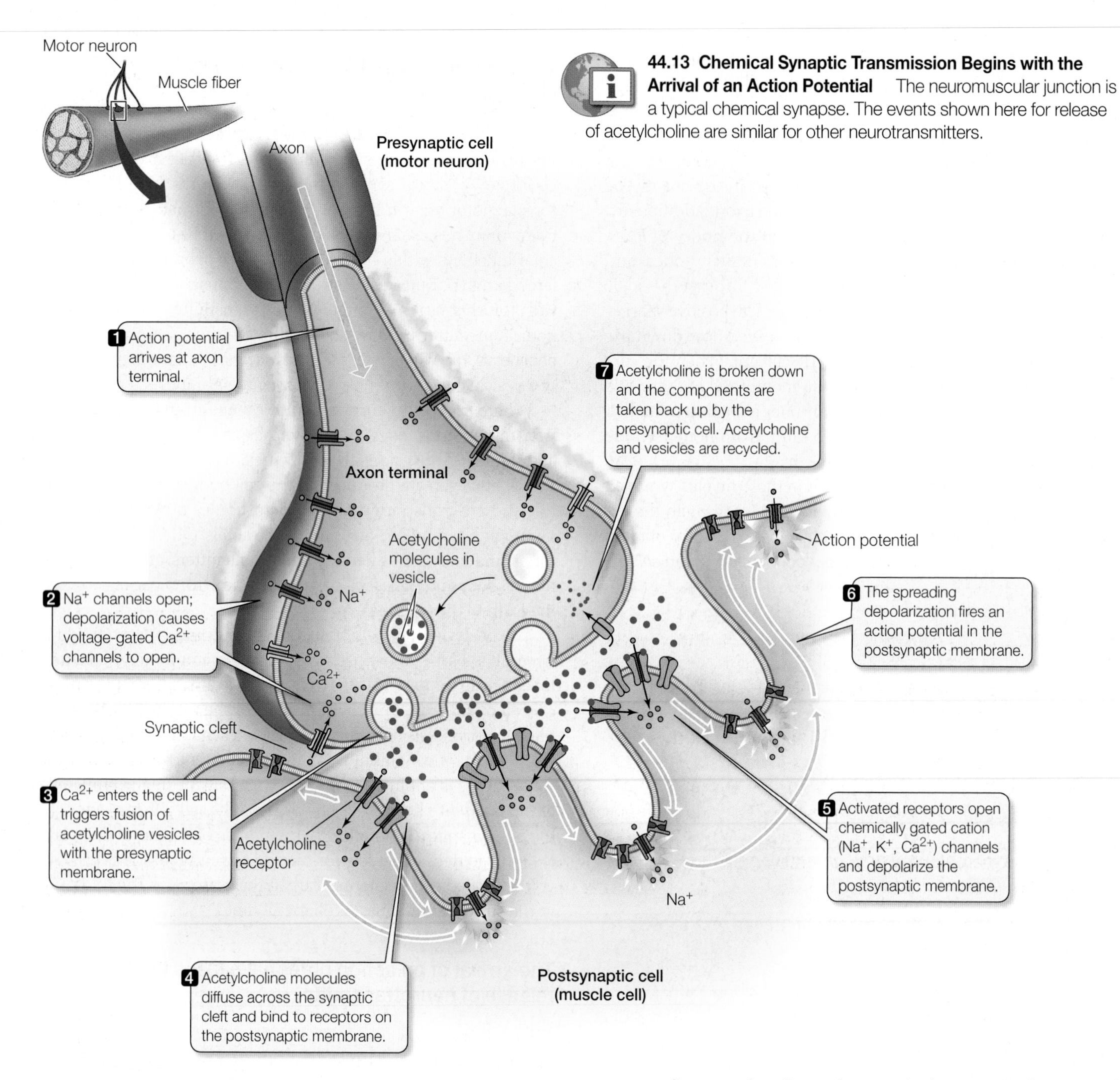

44.13 Chemical Synaptic Transmission Begins with the Arrival of an Action Potential The neuromuscular junction is a typical chemical synapse. The events shown here for release of acetylcholine are similar for other neurotransmitters.

naptic membrane, which actually gets larger as a result—at least until the extra membrane is recycled through endocytosis. The recycled membrane is processed through endosomes to become new vesicles that are then refilled with neurotransmitter.

The postsynaptic membrane responds to neurotransmitter

When ACh is released at a synapse, some of it diffuses across the synaptic cleft and binds to ACh receptors on the postsynaptic membrane (**Figure 44.14**). The ACh receptors are channels that allow both Na^+ and K^+ to flow through, but since the electrochemical gradients favor a net influx of Na^+, the response of the motor end plate to ACh is to depolarize. That depolarization spreads to the depths of the folds of the motor end plate membrane and to surrounding muscle cell membrane, which contain voltage-gated Na^+ channels.

If the axon terminal of a motor neuron releases sufficient amounts of ACh to adequately depolarize a motor end plate, that spreading depolarization will activate the voltage-gated Na^+ channels causing the firing of an action potential. This action potential is then conducted throughout the muscle cell's system of membranes, causing the cell to contract. (We'll learn more about muscle membrane action potentials and the contraction of muscle cells in Section 47.1.)

How much neurotransmitter is enough? Neither a single ACh molecule nor the contents of an entire vesicle (about 10,000 ACh molecules) will bring the plasma membrane of a muscle cell to threshold. However, a single action potential in an axon terminal releases the contents of about 100 vesicles, which is more than enough to fire an action potential in the muscle cell and cause it to contract.

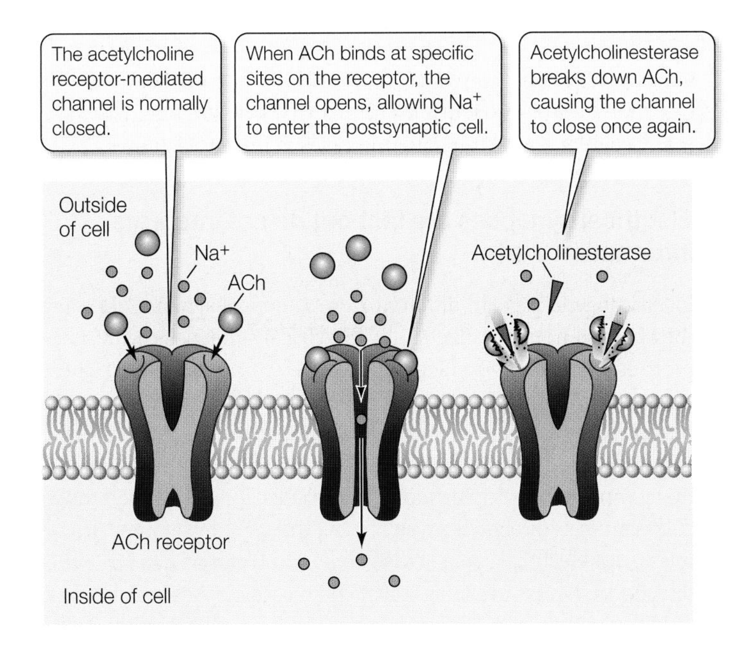

44.14 The Acetylcholine Receptor Is a Chemically Gated Channel The motor end plate contains acetylcholine receptors, which are chemically gated ion channels. When one of these receptors binds ACh, its channel pore opens, and Na^+ ions move into the postsynaptic cell, depolarizing its plasma membrane. Acetylcholinesterase breaks down ACh in the synapse, closing the channel; the breakdown products are then taken up by the presynaptic membrane and resynthesized into more ACh.

Synapses between neurons can be excitatory or inhibitory

In vertebrates, the synapses between motor neurons and muscle cells are always **excitatory**; that is, motor end plates always respond to ACh by depolarizing the postsynaptic membrane. Synapses between neurons, however, are not always excitatory.

Recall that a neuron may have many dendrites. Axon terminals from many other neurons may form synapses with those dendrites and with the cell body. The axon terminals of different presynaptic neurons may store and release different neurotransmitters, and the plasma membrane of the dendrites and cell body of a postsynaptic neuron may have receptors for a variety of neurotransmitters. Thus, at any one time, a postsynaptic neuron may receive a several different chemical messages. If the postsynaptic neuron's response to a neurotransmitter is depolarization, as at the neuromuscular junction, the synapse is excitatory; if its response is hyperpolarization, the synapse is **inhibitory**.

The postsynaptic cell sums excitatory and inhibitory input

What determines when an individual neuron will fire an action potential? Neurons sum excitatory and inhibitory postsynaptic potentials, and when the sum of the potentials surpasses the threshold, the neuron generates an action potential. This summation ability is the major mechanism by which the nervous system integrates information. Each neuron may receive a thousand or more synaptic inputs, but it has only one output: an action potential in a single axon. All of the inputs a neuron receives are reduced to the rate at which that neuron generates action potentials in its axon.

For most neurons, summation takes place in the **axon hillock**, the region of the cell body at the base of the axon (see Figure 44.3). The plasma membrane of the axon hillock is not insulated by glial cells and has many voltage-gated Na^+ channels. Excitatory and inhibitory postsynaptic potentials from synapses anywhere on the dendrites or the cell body may spread to the axon hillock by local current flow. If the resulting combined potential depolarizes the axon hillock to threshold, it fires an action potential. Because postsynaptic potentials decrease in strength as they spread from the site of the synapse, a synapse at the tip of a dendrite has less influence than a synapse on the cell body, near the axon hillock.

Excitatory and inhibitory postsynaptic potentials are summed over space or over time. **Spatial summation** adds up the simultaneous influences of synapses at different sites on the postsynaptic cell (**Figure 44.15A**). **Temporal summation** adds up postsynaptic potentials generated at the same site in a rapid sequence (**Figure 44.15B**).

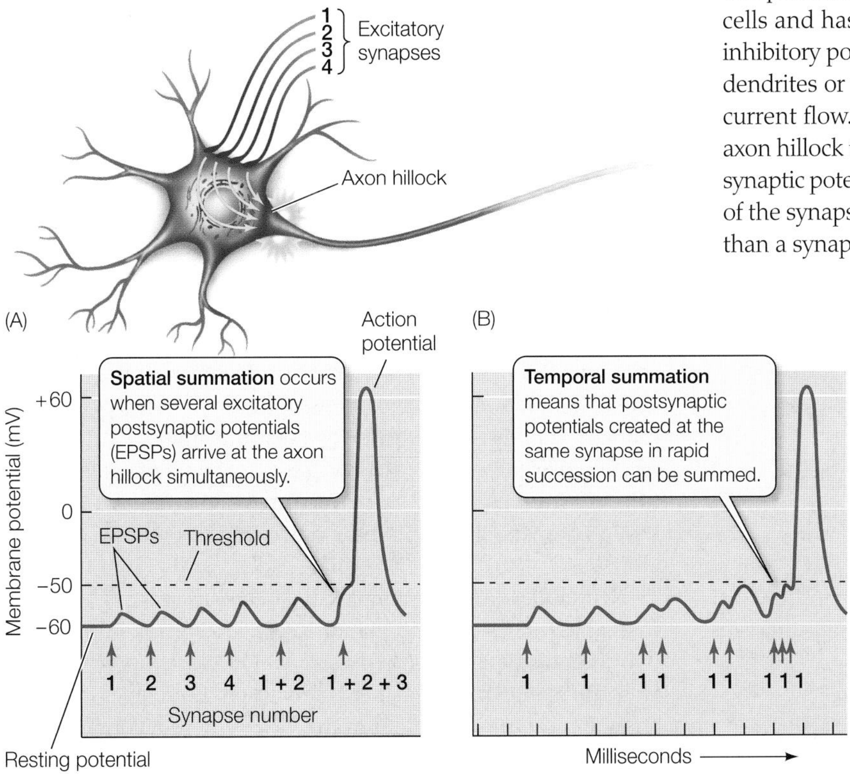

44.15 The Postsynaptic Neuron Sums Information Individual neurons sum excitatory and inhibitory postsynaptic potentials over space (A) and time (B). When the sum of the potentials depolarizes the axon hillock to threshold, the neuron generates an action potential.

Synapses can be fast or slow

Most neurotransmitter receptors induce changes in postsynaptic cells by opening or closing ion channels. How they do so is the basis for grouping receptors into two general categories:

- **Ionotropic receptors** are ion channels themslves. Neurotransmitter binding to an ionotropic receptor causes a direct change in ion movement across the plasma membrane of the postsynaptic cell. These proteins enable fast, short-lived responses.
- **Metabotropic receptors** are not ion channels, but they induce signaling cascades in the postsynaptic cell that secondarily lead to changes in ion channels. Postsynaptic cell responses mediated by metabotropic receptors are generally slower and longer-lived than those induced by ionotropic receptors.

The ACh receptor of the motor end plate is an example of an ionotropic receptor. It consists of five subunits, each of which extends through the plasma membrane. When assembled, the subunits create a central pore that allows ions to pass through (see Figure 44.14). Of several different kinds of subunits, only one kind has the ability to bind ACh. Each functional receptor has two of the ACh-binding subunits and three other subunits.

Metabotropic receptors are also transmembrane proteins, but instead of acting as ion channels, they initiate an intracellular signaling process that can result in the opening or closing of an ion channel. These receptors have seven transmembrane domains, and they are linked to G proteins (**Figure 44.16**). When a neurotransmitter binds to the extracellular domain of a metabotropic receptor, the intracellular domain activates a G protein. In its inactive state, the G protein has three subunits, one of which (the α subunit) is bound to a molecule of GDP. When the receptor binds its neurotransmitter, the GDP is replaced by a GTP molecule, and the α subunit separates from the other two subunits (called β and γ). The subunits move laterally in the membrane until they bind to and open an ion channel or bind to an effector protein that activates a second messenger cascade, which in turn opens an ion channel. G proteins working in just this way are key players in many cellular signal transduction processes (see Section 15.2).

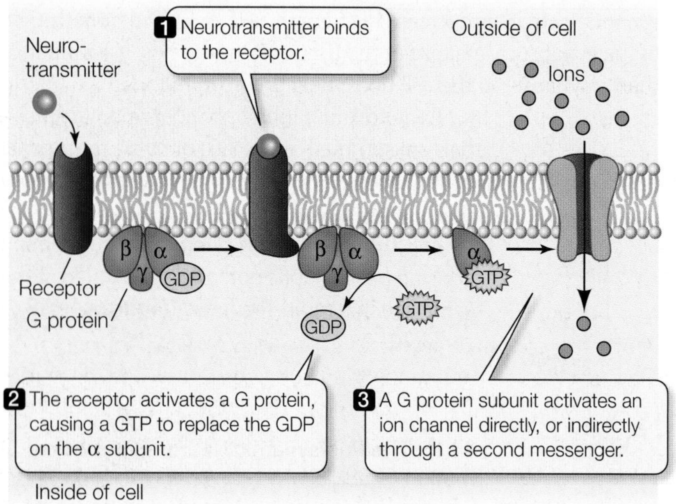

44.16 Metabotropic Receptors Act through G Proteins
Metabotropic receptors activate G proteins, which can influence ion channels directly or through second messengers.

Electrical synapses are fast but do not integrate information well

Electrical synapses are different from chemical synapses because they couple neurons electrically. Electrical synapses contain numerous *gap junctions* (see Figures 5.7C and 15.19) At these synapses, the presynaptic and postsynaptic cell membranes are separated by a space of only 2 to 3 nanometers, and membrane proteins called *connexons* link the two neurons by forming pores that connect the cytoplasm of the two cells. Ions and small molecules can pass directly from cell to cell through these pores. Transmission at electrical synapses is very fast and can proceed in either direction, whereas transmission at chemical synapses is slower and unidirectional.

Electrical synapses are less common in the nervous systems of vertebrates than are chemical synapses for several reasons. First, electrical continuity between neurons does not allow temporal summation of synaptic inputs. Second, an effective electrical synapse requires a large area of contact between the presynaptic and postsynaptic cells. This condition rules out the possibility of thousands of synaptic inputs to a single neuron—which is the norm in complex nervous systems. Third, electrical synapses cannot be inhibitory. Thus, electrical synapses are useful for rapid communication, but they are less useful for processes of integration and learning.

The action of a neurotransmitter depends on the receptor to which it binds

More than 50 neurotransmitters are now recognized, and more will surely be discovered. **Table 44.1** describes some of the best-known neurotransmitters. ACh, as we have seen, is an important neurotransmitter because it is how the nervous system commands muscles to contract. ACh also plays roles in certain synapses between neurons in the CNS, but it accounts for only a small percentage of the total neurotransmitter content of the CNS. The workhorse neurotransmitters of the CNS are simple amino acids: glutamate (excitatory) and glycine and γ-aminobutyrate (GABA, a derivative of glutamate) (inhibitory). Another important group of neurotransmitters in the CNS is the monoamines, which are derivatives of amino acids. They include dopamine and norepinephrine (derivatives of tyrosine) and serotonin (a derivative of tryptophan). Peptides also function as neurotransmitters. An exciting recent discovery revealed that two gases, carbon monoxide and nitric oxide, are used by neurons as intercellular messengers, although the neurons do not have associated receptors (see Figure 15.15).

Neurotransmission is complex in part because each neurotransmitter has multiple receptor types. Acetylcholine, for example, has two receptor types: *nicotinic receptors*, which are ionotropic, and *muscarinic receptors*, which are metabotropic. Both types of acetylcholine receptors are found in the CNS, where nicotinic receptors tend to be excitatory and muscarinic receptors tend to be in-

TABLE 44.1

Some Well-Known Neurotransmitters

NEUROTRANSMITTER	ACTIONS	COMMENTS
Acetylcholine	The neurotransmitter of vertebrate motor neurons and of some neural pathways in the brain	Broken down in the synapse by acetylcholinesterase; blockers of this enzyme are powerful poisons
MONOAMINES		
Norepinephrine	Used in certain neural pathways in the brain. Also found in the peripheral nervous system, where it causes gut muscles to relax and the heart to beat faster	Related to epinephrine and acts at some of the same receptors
Dopamine	A neurotransmitter of the central nervous system	Involved in schizophrenia. Loss of dopamine neurons is the cause of Parkinson's disease
Histamine	A minor neurotransmitter in the brain	Involved in maintaining wakefulness
Serotonin	A neurotransmitter of the central nervous system that is involved in many systems, including pain control, sleep/wake control, and mood	Certain medications that elevate mood and counter anxiety act by inhibiting the reuptake of serotonin (so it remains active longer)
PURINES		
ATP	Co-released with many neurotransmitters	Large family of receptors may shape postsynaptic responses to classical neurotransmitters
Adenosine	Transported across cell membranes; not synaptically released	Not released synaptically; mainly has inhibitory effects on neighboring cells
AMINO ACIDS		
Glutamate	The most common excitatory neurotransmitter in the central nervous system	Some people have reactions to the food additive monosodium glutamate because it can affect the nervous system
Glycine Gamma-aminobutyric acid (GABA)	Common inhibitory neurotransmitters	Drugs called benzodiazepines, used to reduce anxiety and produce sedation, mimic the actions of GABA
PEPTIDES		
Endorphins Enkephalins	Modulation of pain pathways	Receptors are activated by narcotic drugs: opium, morphine, heroin, codeine
Substance P	Used by certain sensory nerves, especially in pain pathways	Released by neurons sensitive to heat and pain
GAS		
Nitric oxide	Widely distributed in the nervous system	Not a classic neurotransmitter, it diffuses across membranes rather than being released synaptically. A means whereby a postsynaptic cell can influence a presynaptic cell

hibitory. Acetylcholine actions can differ outside of the CNS as well. Acetylcholine acting through nicotinic receptors causes the smooth muscle of the gut to depolarize and therefore increases its motility, but acetylcholine acting through muscarinic receptors causes cardiac muscle to hyperpolarize and therefore decreases the contractility of the heart.

We could give many more examples of neurotransmitters that have different effects in different tissues, but the important thing to remember is that the action of a neurotransmitter depends on the receptor to which it binds.

Glutamate receptors may be involved in learning and memory

Glutamate is a neurotransmitter that can bind to a variety of receptors, including both metabotropic and ionotropic receptors. The glutamate receptors are divided into several classes because they can be differentially activated by other chemicals that mimic the action of glutamate. One class of ionotropic glutamate receptors is the *NMDA receptors,* which can be activated by the chemical N-methyl-D-aspartate. Another class of ionotropic glutamate receptors is activated by a different chemical, abbreviated as *AMPA* (which stands for α-amino-3-hydroxy-5-methyl-4-isoxazole propionate).

Glutamate is an excitatory neurotransmitter, so activation of ionotropic glutamate receptors always results in Na^+ entry into the neuron and depolarization. But the *timing* of the response to activation by these different types of receptors differs significantly. The AMPA receptors allow a rapid influx of Na^+ into the postsynaptic cell, while the NMDA receptors allow a slower and longer-lasting influx of Na^+. The NMDA receptors also require that the cell be somewhat depolarized through the action of other receptors before their pores will open and permit Na^+ influx. When they do open, these receptors also allow Ca^{2+} to enter the cell. Calcium ions

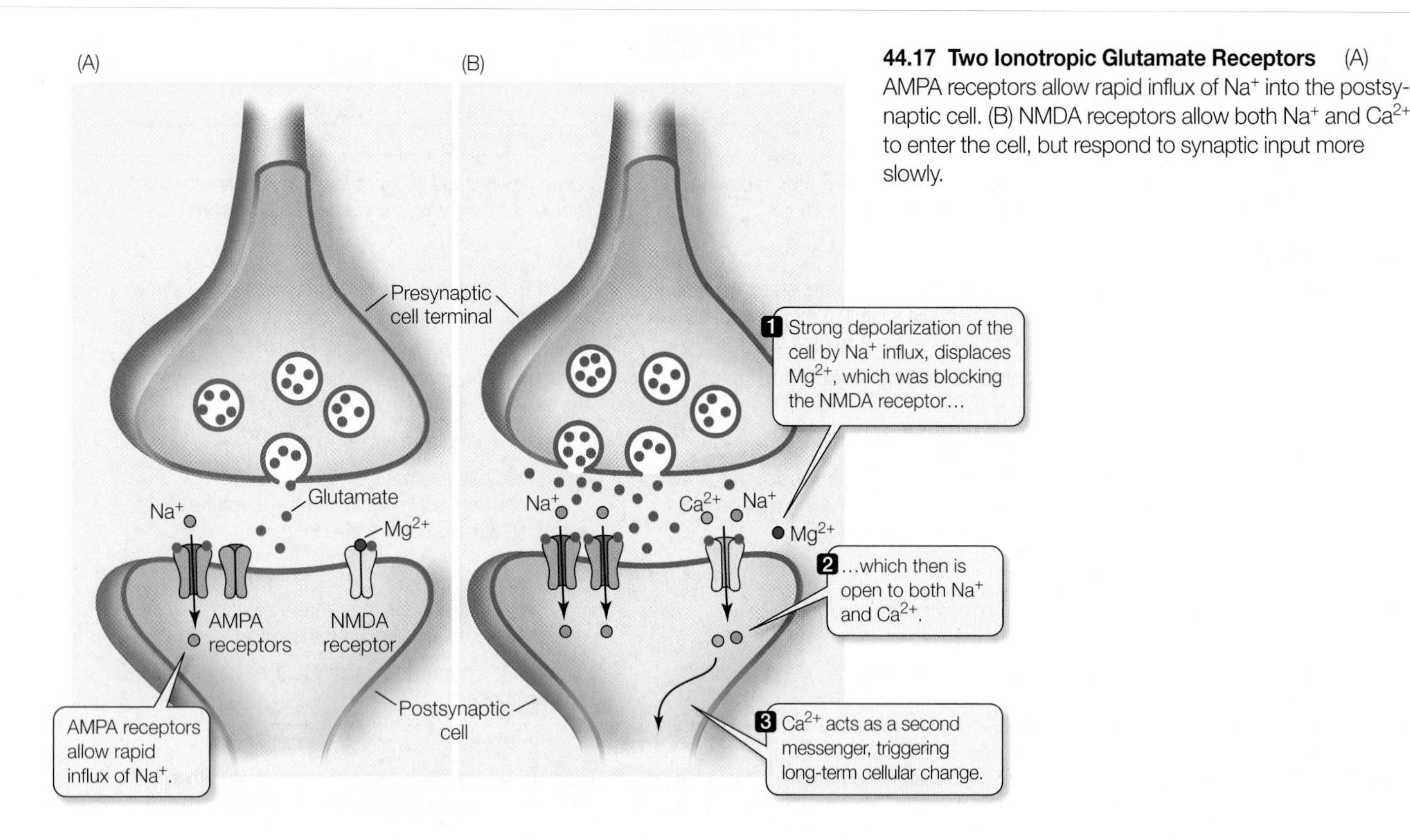

44.17 Two Ionotropic Glutamate Receptors (A) AMPA receptors allow rapid influx of Na^+ into the postsynaptic cell. (B) NMDA receptors allow both Na^+ and Ca^{2+} to enter the cell, but respond to synaptic input more slowly.

act as second messengers in the cell and can trigger a variety of long-term cellular changes.

Figure 44.17 shows how the AMPA and NMDA receptors can work in concert. At resting potential, the NMDA receptor is blocked by a magnesium ion (Mg^{2+}). Strong depolarization of the neuron due to other inputs—such as the activation of AMPA receptors—displaces Mg^{2+} from the NMDA receptors and allows Na^+ and Ca^{2+} to pass through them when they are activated by glutamate. These special properties of the NMDA receptor are probably involved in learning and memory.

Most of the synaptic events we have studied so far happen very quickly. It is therefore a special challenge to understand how the messages carried by action potentials can result in long-term events such as learning and memory. Our understanding of these processes has been greatly affected by study of a phenomenon called **long-term potentiation** (**LTP**), which is studied by neurobiologists working with slices of brain kept alive in dishes of culture medium. Using these brain slice preparations, it is possible to stimulate and record from specific brain regions, and even specific neurons.

In the studies leading to the discovery of LTP, experimenters repeatedly stimulated synaptic inputs to a particular neuron and observed the usual action potential response. When the neuron was stimulated many times in rapid succession, however, they found that the properties of the neuron changed. The magnitude of the postsynaptic response was enhanced, or *potentiated*, and this change lasted for days or weeks.

How does potentiation of a synapse occur? The answer, at least for some areas of the brain, now seems clear. With low levels of stimulation, the glutamate released by presynaptic cells activates only AMPA receptors, and the postsynaptic membrane simply responds with action potentials. With higher levels of stimulation, however, NMDA receptors are also activated, allowing both Na^+ and Ca^{2+} ions to enter the postsynaptic neuron. The Ca^{2+} ions induce long-term changes in the postsynaptic membrane that make it more sensitive to synaptic input (**Figure 44.18**).

Exploiting the LTP system, Joe Tsien and his students and collaborators at Princeton University genetically engineered mice so that their NMDA receptors had a slightly altered structure and were activated for a longer time whenever they bound a molecule of glutamate. These mice learned tasks better, ran mazes faster, and remembered the mazes longer than normal mice. These exciting experiments show that we are on the right track to understanding how the brain achieves learning and memory.

To turn off responses, synapses must be cleared of neurotransmitter

Turning off the action of neurotransmitters is as important as turning it on. If released neurotransmitter molecules simply remained in the synaptic cleft, the postsynaptic membrane would become saturated with neurotransmitter, and receptors would be constantly activated. As a result, the postsynaptic cell would remain hyperpolarized or depolarized and would be unresponsive to short-term changes in the presynaptic cell. The more discrete each separate neuronal signal is, the more information can be processed in a given time. Thus neurotransmitter must be cleared from the synaptic cleft shortly after it is released by the axon terminal.

Neurotransmitter action may be terminated in several ways. First, enzymes may destroy the neurotransmitter. ACh, for example, is rapidly destroyed by the enzyme acetylcholinesterase, which is present in the synaptic cleft in close association with the ACh

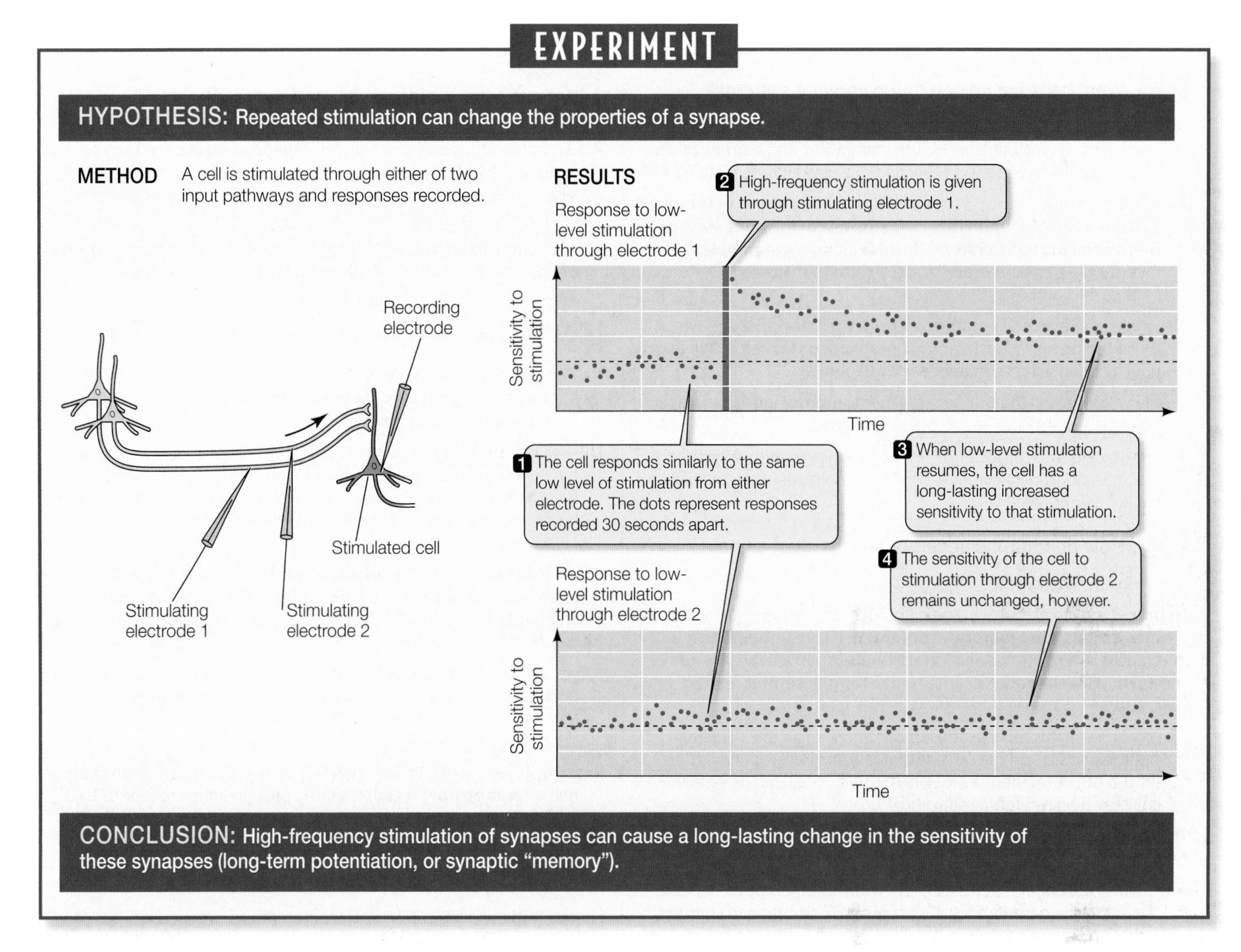

44.18 Repeated Stimulation Can Cause Long-Term Potentiation When a cell receives low-frequency synaptic input, the resulting postsynaptic response remains constant. If, however, that same synaptic pathway is stimulated briefly at a high frequency, the subsequent sensitivity of the postsynaptic cell to the original level of synaptic input is potentiated for a longer period.

receptors on the postsynaptic membrane (see Figure 44.14). Some of the most deadly nerve gases developed for chemical warfare work by inhibiting acetylcholinesterase. As a result, ACh lingers in the synaptic cleft, causing the victim to die of spastic (contracted) muscle paralysis. Some agricultural insecticides, such as malathion, also inhibit acetylcholinesterase and can poison farm workers if used without safety precautions.

Neurotransmitter also may simply diffuse away from the cleft, or be taken up via active transport by nearby cell membranes. Prozac, a drug commonly prescribed to treat depression, slows the reuptake of the neurotransmitter serotonin, thus enhancing its activity at the synapse.

44.3 RECAP

Chemical synapses involve the release of neurotransmitter molecules stored in vesicles in the presynaptic terminal. Action potentials reaching that terminal cause the fusion of vesicles with the presynaptic membrane releasing neurotransmitter that can then bind to receptors on the postsynaptic membrane.

- Can you explain the role of Ca^{2+} channels in synaptic events? See pp. 955–956 and Figure 44.13
- How can some synapses be excitatory and others inhibitory? See p. 957
- How do neurons integrate the input from various synapses? See pp. 957–958 and Figure 44.15

CHAPTER SUMMARY

44.1 What cells are unique to the nervous systems?

Nervous systems include **neurons** and **glial cells**. Neurons are organized in circuits with sensory inputs, integration, and outputs to effectors. Glial cells serve support functions. Review Figures 44.1 and 44.3

In vertebrates, the brain and spinal cord form the **central nervous system**, which communicates with the rest of the body via the **peripheral nervous system**. The CNS increases in complexity from invertebrates to vertebrates and from fish to mammals. Review Figures 44.1 and 44.2

Neurons generally receive information via their **dendrites**, of which there can be many, and transmit information via their single **axons**, which end in axon terminals. Review Figure 44.3

Where neurons and their target cells meet, information is transmitted across specialized junctions called **synapses**.

Schwann cells and **oligodendrocytes** produce **myelin**, which insulates neurons. **Astrocytes** create the **blood–brain barrier**. Review Figure 44.4

44.2 How do neurons generate and conduct signals?

See Web/CD Tutorial 44.1

Neurons have an electric charge difference across their plasma membranes, the **membrane potential**. The membrane potential is created by ion pumps and ion channels. When a neuron is not active, its membrane potential is a **resting potential**. Review Figures 44.5 and 44.6

The **sodium–potassium pump** concentrates K^+ on the inside of a neuron and Na^+ on the outside. Potassium channels allow K^+ to diffuse out of the neuron, leaving behind unbalanced negative charges. Review Figures 44.6 and 44.7

Patch clamping allows the study of single ion channels. Review Figure 44.8

The resting potential is perturbed when ion channels open or close, changing the permeability of the plasma membrane to charged ions. Through this mechanism, the plasma membrane can become **depolarized** or **hyperpolarized**. Review Figure 44.9

An **action potential** is a rapid reversal in charge across a portion of the plasma membrane resulting from the sequential opening and closing of **voltage-gated Na^+ and K^+ channels**. These changes in voltage-gated channels occur when the plasma membrane depolarizes to a **threshold level**. Review Figure 44.10, Web/CD Tutorial 44.2

Action potentials are all-or-none, self-regenerating events. They are conducted down axons because local current flow depolarizes adjacent regions of membrane and brings them to threshold. Review Figure 44.11

In myelinated axons, action potentials appear to jump between **nodes of Ranvier**, patches of axonal plasma membrane that are not covered by myelin. Review Figure 44.12

44.3 How do neurons communicate with other cells?

Neurons communicate with each other and with other cells by transmitting information over **chemical synapses** (with neurotransmitters) or **electrical synapses**.

The **neuromuscular junction** is a well-studied chemical synapse between a motor neuron and a skeletal muscle cell. Its neurotransmitter is **ACh**, which causes a depolarization of the postsynaptic membrane when it binds to its receptor. Review Figure 44.13, Web/CD Tutorial 44.3

When an action potential reaches an **axon terminal**, it causes the release of neurotransmitters, which diffuse across the **synaptic cleft** and bind to receptors on the postsynaptic membrane. Review Figures 44.13 and 44.14

Synapses between neurons can be either excitatory or inhibitory. A postsynaptic neuron integrates information by **summing** excitatory and inhibitory postsynaptic potentials in both space and time. Review Figure 44.15

Ionotropic receptors are ion channels or directly influence ion channels. **Metabotropic receptors** are G protein-linked receptors that influence the postsynaptic cell through various signal transduction pathways and result in the opening or closing of ion channels. The actions of ionotropic synapses are generally faster than those of metabotropic synapses. Review Figure 44.16

There are many different neurotransmitters and even more types of receptors. The action of a neurotransmitter depends on the receptor to which it binds. Review Table 44.1, Web/CD Activity 44.1

With repeated stimulation, a neuron can become more sensitive to its inputs through **LTP**. The properties of the NMDA glutamate receptor appear to explain LTP. Review Figures 44.17 and 44.18

SELF-QUIZ

1. The rising phase of an action potential is due to the
   a. closing of K^+ channels.
   b. opening of chemically gated Na^+ channels.
   c. closing of voltage-gated Ca^{2+} channels.
   d. opening of voltage-gated Na^+ channels.
   e. spread of positive current along the plasma membrane.
2. The resting potential of a neuron is due mostly to
   a. local current spread.
   b. open Na^+ channels.
   c. synaptic summation.
   d. open K^+ channels.
   e. open Cl^- channels.
3. Which statement about synaptic transmission is *not* true?
   a. The synapses between neurons and muscle cells use ACh as their neurotransmitter.
   b. A single vesicle of neurotransmitter cannot cause a muscle cell to contract.
   c. The release of neurotransmitter at the neuromuscular junction causes the motor end plate to fire action potentials.
   d. In vertebrates, the synapses between motor neurons and muscle fibers are always excitatory.
   e. Inhibitory synapses cause the resting potential of the postsynaptic membrane to become more negative.
4. Which statement accurately describes an action potential?
   a. Its magnitude increases along the axon.
   b. Its magnitude decreases along the axon.
   c. All action potentials in a single neuron are of the same magnitude.
   d. During an action potential the membrane potential of a neuron remains constant.
   e. An action potential permanently shifts a neuron's membrane potential away from its resting value.

5. A neuron that has just fired an action potential cannot be immediately restimulated to fire a second action potential. The short interval of time during which restimulation is not possible is called
   a. hyperpolarization.
   b. the resting potential.
   c. depolarization.
   d. repolarization.
   e. the refractory period.
6. The rate of propagation of an action potential depends on
   a. whether or not the axon is myelinated.
   b. the axon's diameter.
   c. whether or not the axon is insulated by glial cells.
   d. the cross-sectional area of the axon.
   e. all of the above.
7. The binding of an inhibitory neurotransmitter to the postsynaptic receptors results in
   a. depolarization of the membrane.
   b. generation of an action potential.
   c. hyperpolarization of the membrane.
   d. increased permeability of the membrane to sodium ions.
   e. increased permeability of the membrane to calcium ions.
8. The difference between slow and fast synapses is
   a. the width of the synaptic cleft.
   b. the size of the synapse.
   c. whether or not the neurotransmitter acts directly on ion channels.
   d. the density of receptors on the postsynaptic membrane.
   e. whether or not the presynaptic vesicles kiss and run.
9. Whether a synapse is excitatory or inhibitory depends on the
   a. type of neurotransmitter.
   b. presynaptic axon terminal.
   c. size of the synapse.
   d. nature of the postsynaptic receptors.
   e. concentration of neurotransmitter in the synaptic space.
10. Which of the following is a likely mechanism for long-term potentiation?
   a. When glutamate binds to postsynaptic AMPA receptors, it activates G proteins that trigger intracellular changes.
   b. When glutamate binds to NMDA receptors, it allows magnesium ions to enter the cell, which initiate intracellular changes.
   c. When sufficient glutamate is released by the presynaptic neuron, it causes an increase in the number of AMPA receptors on the postsynaptic cell.
   d. When sufficient glutamate is released, both AMPA and NMDA receptors are activated, and NMDA receptors allow Ca^{2+} as well as Na^{+} to enter the cell, thus initiating intracellular changes.
   e. When both glutamate and ACh are released together, they create a long-lasting depolarization of the postsynaptic cell.

FOR DISCUSSION

1. The language of the nervous system consists of one "word," the action potential. How can this single message convey a diversity of information, how can that information be quantitative, and how can it be integrated?
2. If you stimulate an axon in the middle, action potentials are conducted in both directions. Yet when an action potential is generated at the axon hillock, it goes only toward the axon terminals and does not backtrack. Explain why action potentials are bidirectional in the first example and unidirectional in the second.
3. The nature of synapses presents various opportunities for plasticity in the nervous system. Discuss at least four synaptic mechanisms that could be altered to change the response of a neuron to a specific input.
4. Benzodiazepines are drugs that act through GABA receptors and open Cl^{-} channels. What effects would you expect these drugs to have?

FOR INVESTIGATION

The patch clamping technique shown in Figure 44.8 produces an "inside-out" patch because the cytoplasmic side of the membrane faces away from the pipette tip and the surface side of the membrane is exposed to the solution in the pipette. Another patch preparation is the "outside-out patch" in which the cytoplasmic side of the membrane is exposed to the solution in the pipette. Describe what kind of experiments you could conduct with the "inside-out" patch, and what kind of experiments you could conduct with the "outside-out" patch.

CHAPTER 45 Sensory Systems

Out of range

A rattlesnake can see to strike and kill a running rodent in complete darkness. How can this be, when "seeing" means using the eyes to detect light, and "complete darkness" means the absence of any light? It is possible because these definitions are based on human capabilities. What we call "light" is actually only a small portion (red, orange, yellow, green, blue, indigo, and violet) of the spectrum of electromagnetic radiation. Other animals see wavelengths we cannot. For example, insects and birds see patterns on flowers that reflect ultraviolet wavelengths invisible to us. Similarly, rattlesnakes can "see" infrared wavelengths that are invisible to us (although at high enough levels of intensity, humans feel these wavelengths as heat).

It is not the snake's eyes that perceive infrared wavelengths. Rattlesnakes and their relatives have *pit organs* containing high densities of infrared-sensitive neurons. The two pits, located between the nostrils and the eyes on each side of the skull, are positioned in such a way that sensory receptor cells in the pits receive directional information. The fields of "view" of the bilateral pits are overlapping and thus convey a three-dimensional perspective. Information from the pit organs goes to the same region of the brain as information from the eyes, so rattlesnakes actually do "see" the world in a range of electromagnetic radiation that is different from the human visual spectrum.

Our definition of silence is as arbitrary as our definition of darkness. "Sound" is actually pressure waves in the environment, and many animals are sensitive to pressure waves with frequencies we cannot hear. Elephants communicate in sound waves that are below human hearing range; such long waves travel great distances, an advantage to large animals that roam over extensive areas. Bats emit incredibly loud, short wavelength sound pulses that are far above our range of hearing, and a flying bat hears echoes of these pulses bouncing off objects in the environment. The pulses are so loud and the echoes are so weak that it is rather like a construction worker trying to overhear a whispered conversation while using a pneumatic drill. Why don't the loud pulses "drown out" the weak echoes for the bat? Small muscles in the bat's ears contract to dampen their hearing sensitivity while the sounds are being emitted, but

Sensing Infrared Radiation The hole to the left of this Aruba rattlesnake's eye is one of its bilateral pit organs. These organs detect infrared radiation from their preferred prey—small rodents—with unerring precision, even in total darkness. The forked tongue also provides positional information, picking up molecular signals which are transmitted to the brain by a specialized organ in the roof of the snake's mouth.

Echolocating around an Obstacle Course A bat's ability to echolocate using sound waves is so precise that, in a totally dark room strung with fine wires, bats can capture small insects while avoiding the wires.

relax in time for the bat to hear the echo—a truly remarkable ability, since the pulses are emitted at rates of 20–80 per second.

Our senses are our windows on the world. "Reality" is what our eyes see, our ears hear, our noses smell, what we touch and taste. But human beings sense only a limited range of the information available. Animals with different ranges of sensitivity process different sources of information and may perceive "reality" quite differently.

IN THIS CHAPTER we will examine the general properties of sensory receptor cells and see how they convert environmental stimuli into the electrochemical signals of nervous systems. We will examine in detail the diversity of cells responsible for our senses and see how they are incorporated into sensory systems that provide the central nervous system with information about the world around and within us. In the course of our study of sensory systems, we will learn about the unusual sensory abilities of many other animals.

CHAPTER OUTLINE

45.1 How Do Sensory Cells Convert Stimuli into Action Potentials?

Sensory cells convert, or *transduce*, physical and chemical stimuli such as light waves, sound waves, touch, pain, and odorant and taste molecules into neuronal signals. These signals are then transmitted to the central nervous system for processing and interpretation.

Sensory receptor proteins act on ion channels

As we learned in Section 44.2, cells use energy to create large gradients of charged ions across their plasma membranes. In this way, cells are like batteries: batteries store potential energy by separating electrical charges between their poles, and cells store potential energy in the ionic gradients across their plasma membranes. If the two poles of a battery are connected by a wire and a switch, current flows through the wire when the switch is closed. Ion channels are like switches. When they open, charged ions can flow down their electrochemical gradient. When the ion channel is selective for one type of ion, the flow of that charged ion creates a change in the electrical potential across the membrane. **Sensory transduction** begins with a membrane receptor protein on a sensory receptor cell that can detect a specific stimulus such as heat, light, chemicals, mechanical force, or electrical fields. The **receptor protein** then directly or indirectly opens or closes ion channels, affecting the resting potential of the cell. This change in resting potential then either causes the sensory cell to fire an action potential or, through release of neurotransmitter, leads an associated neuron one or even two synapses away to fire action potentials.

In Section 44.3 we learned that synaptic receptor proteins are either *ionotropic* or *metabotropic*; the same distinction can be applied to sensory receptor proteins. Ionotropic sensory receptor proteins are either ion channels themselves, or they directly affect the opening of an ion channel. Examples are receptors that respond to physical force (called *mechanoreceptors*), to temperature (*thermoreceptors*), and probably to salty taste. Although *electrosensors* most likely have no receptor protein at all, they are grouped with the ionotropic receptors; the membrane of the sensory cell is sensitive to the voltage across it and releases neurotransmitter in response to slight changes in membrane potential. Metabotropic sensory receptor proteins influence ion channels indirectly, through G proteins and sec-

45.1 Sensory Cell Membrane Receptor Proteins Respond to Stimuli The receptor proteins in mechanoreceptors are ion channels. The activated receptor proteins of metabotropic chemoreceptors and photoreceptors initiate signal transduction cascades that eventually open or close ion channels.

ond messengers. Examples are *chemoreceptors* and *photoreceptors* (**Figure 45.1**).

Sensory transduction involves changes in membrane potentials

In this chapter we will examine several sensory systems. In each case, we can ask the same general question: How do **sensory receptor cells**, also known simply as *receptors* or *sensory cells*, transduce energy from a stimulus into a change in membrane potential? The details differ for different receptors, but those details all fit into one general pattern.

A change in the resting membrane potential of a sensory receptor cell in response to a stimulus is called a **receptor potential**. Receptor potentials can spread by local current flow over short distances, but to travel long distances in the nervous system, they must be converted into action potentials. Receptor potentials produce action potentials in two ways: by generating action potentials within the receptor cell itself, or by causing the release of a neurotransmitter that induces an associated neuron to generate action potentials.

A good example of a sensory receptor cell that can generate action potentials is the stretch receptor of a crayfish (**Figure 45.2**). By placing an electrode in the dendrites or cell body of a crayfish stretch receptor cell, we can record the receptor potentials that result from stretching of the muscle to which the dendrites of the cell are attached. These receptor potentials spread to the base of the cell's axon (the axon hillock), which contains voltage-gated sodium channels. Action potentials generated here travel down

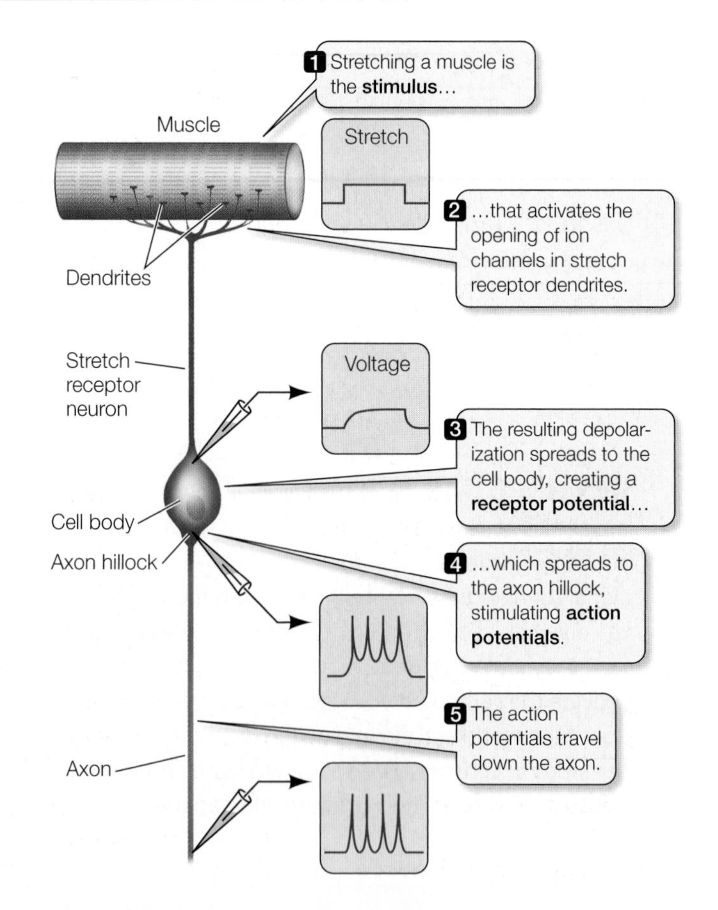

45.2 Stimulating a Sensory Cell Produces a Receptor Potential Signal transduction in the stretch receptor of a crayfish can be investigated by measuring the membrane potential at different places on the stretch receptor neuron while stretching the muscle innervated by that sensory neuron.

the axon to the CNS. The rate at which action potentials are fired by the axon depends on the magnitude of the receptor potential; that, in turn, depends on how much the muscle is stretched.

In a receptor cell that does not fire action potentials, the spreading receptor potential reaches a presynaptic patch of plasma membrane and induces the release of a neurotransmitter. The intensity of the stimulus influences how much neurotransmitter is released. That neurotransmitter binds to receptor proteins on an associated sensory neuron, altering its membrane potential and perhaps causing it to increase or decrease its rate of firing action potentials. In a few cases, this second cell also responds only by changing the rate at which it releases neurotransmitter onto another neuron. Eventually, however, the stimulation of the sensory cell will be coded as a change in firing of action potentials in a sensory circuit.

Sensation depends on which neurons receive action potentials from sensory cells

All the sensory systems process information in the form of action potentials. So how do we perceive different sensations? Sensations such as heat, pressure, pain, light, smell, and sound differ because the messages from different kinds of sensory cells arrive at different places in the CNS. Action potentials arriving in the visual cortex of the brain are interpreted as light, in the auditory cortex as sound, in the olfactory bulb as smell, and so forth.

A small patch of skin on your arm contains some sensory receptor cells that increase their firing rates when the skin is warmed and others that increase their activity when the skin is cooled. Other types of sensory cells in the same patch of skin respond to touch, movement of hairs, irritants such as mosquito bites, and the pain from cuts or burns. These receptor cells transmit their messages through axons that enter the CNS at the spinal cord. The synapses made by those axons in the spinal cord and the subsequent pathways of transmission determine whether the stimulation of the patch of skin on your arm is perceived as warmth, cold, touch, tickle, itch, or pain. So, even though the action potentials carried by all of these sensory axons look the same, the connectivity of each axon is specific for a given sensory modality.

How is the intensity of the stimulus encoded if, as Section 44.2 describes, each action potential is an all-or-none event? Intensity of sensation is coded as the frequency of the action potentials.

Some sensory cells transmit information about internal conditions in the body, but we may not be consciously aware of that information. The brain continuously receives information about body temperature, blood carbon dioxide and oxygen concentrations, arterial pressure, muscle tension, and the positions of the limbs—all of which are important for the maintenance of homeostasis. All sensory cells produce information that the nervous system can use, but that information does not always result in conscious sensation.

Some sensory receptor cells are assembled with other types of cells into *sensory organs*, such as eyes, ears, and noses, that enhance the ability of the sensory cells to collect, filter, and amplify stimuli. We therefore refer to **sensory systems**, which include the sensory cells, the associated structures, and the neuronal networks that process the information.

Many receptors adapt to repeated stimulation

Some sensory cells give gradually diminishing responses to maintained or repeated stimulation. This phenomenon is known as **adaptation**, and it enables an animal to ignore background or unchanging conditions while remaining sensitive to changes or to new information. (Note that this use of the term "adaptation" is different from its application in an evolutionary context.) When you dress, you feel each item of clothing touch your skin, but the sensation of clothes touching your skin is not constantly on your mind throughout the day. You are immediately aware, however, when a seam rips, your shoe comes untied, or someone touches your back.

Animals can discriminate between continuous and changing stimuli partly because some sensory cells adapt; it is also a result of information processing by the CNS. Some sensory cells adapt very little or very slowly; examples are some types of pain receptors and the mechanoreceptors for balance.

In the rest of this chapter we will learn how sensory systems gather and filter stimuli, transduce specific stimuli into action potentials, and transmit action potentials to the CNS.

45.1 RECAP

Sensory receptor cells have receptor proteins that respond to specific stimuli from the external or internal environment by opening or closing ion channels, which results in the generation of action potentials in sensory neurons.

- Can you explain the difference between ionotropic and metabotropic sensory receptor proteins? See p. 965 and Figure 45.1
- How are we able to perceive action potentials—which are all essentially the same—as different sensations? See p. 967

Now that we have a general view of how sensory systems code and process information, we will look in more depth at specific sensory modalities, beginning with chemosensation—the basis of smell and taste.

45.2 How Do Sensory Systems Detect Chemical Stimuli?

A colony of corals responds to a small amount of meat extract in the seawater bathing it by extending bodies and tentacles and searching for food. A solution of a single amino acid can stimulate this response. Conversely, a small amount of sea water from a container in which corals were crushed will stimulate a defensive retraction of the coral polyps. Humans also react strongly to certain chemical stimuli. When we smell freshly baked bread, we salivate and feel hungry, but when we smell diamines from rotting meat we gag and feel nauseated. All animals receive information about chemical stimuli through **chemoreceptors**, which are receptor proteins that bind various ligands and are responsible for smell and taste. Chemoreceptors are also responsible for monitoring aspects of the internal environment such as the level of carbon dioxide in the blood. Information from chemoreceptors can cause powerful physiological behavioral and responses.

Arthropods provide good examples for studying chemoreception

Arthropods use chemical signals to attract mates. These signals, called **pheromones**, demonstrate the sensitivity of chemosensory systems. One of the best-studied examples of this phenomenon is the silkworm moth.

Flies have chemoreceptors on their feet that respond to sugars, amino acids, salts, and even distilled water. A fly tastes a potential food by stepping in it.

To attract a mate, the female silkworm moth (*Bombyx mori*) releases a pheromone called *bombykol* from a gland at the tip of her abdomen. The male silkworm moth has receptors for this molecule on his antennae (**Figure 45.3**). Each feathery antenna carries about 10,000 bombykol-sensitive hairs. A single molecule of bombykol may be sufficient to generate action potentials in the antennal nerve that transmits the signal to the CNS. Because of the male's great sensitivity, the sexual message of a female moth is likely to reach any male within a downwind area stretching over several kilometers. When approximately 200 hairs per second are activated, the male flies upwind in search of the female. Because the rate of firing in the male's sensory nerves is proportional to the bombykol concentration in the air, he can follow the airborne concentration gradient and home in on the signaling female.

Olfaction is the sense of smell

The sense of smell, **olfaction**, depends on chemoreceptors. In vertebrates, the olfactory sensors are neurons embedded in a layer of epithelial tissue at the top of the nasal cavity. These neurons project their axons to the olfactory bulb of the brain, while their dendrites end in olfactory hairs on the surface of the nasal epithelium. A protective layer of mucus covers the epithelium. Molecules from the environment must diffuse through this mucus to reach the receptor proteins on the olfactory hairs. When you have a cold, the amount of mucus in your nose increases, and the epithelium swells. With this in mind, study **Figure 45.4**, and you will easily understand why respiratory infections can cause you to lose your sense of smell. Humans have a fairly sensitive olfactory system, but we are unusual among mammals in that we depend more on vision than on olfaction; a dog has up to 40 million nerve endings per square centimeter of nasal epithelium, many more than we do.

An **odorant** is a molecule that activates an olfactory receptor protein. Odorants bind to receptor proteins on the olfactory cilia of the olfactory neurons. Olfactory receptor proteins are specific for particular odorant molecules. When an odorant molecule binds to its receptor on an olfactory neuron, it activates a G protein. The G protein in turn activates an enzyme that causes an increase of a second messenger (cAMP in vertebrates) in the cytoplasm (see Figures 15.17 and 44.16). The second messenger binds to cation channels in the sensory cell's plasma membrane and opens them, causing an influx of Na^+. The sensory neuron depolarizes to threshold and fires action potentials.

The olfactory world has an enormous number of odors, and accordingly, there are a large number of olfactory receptor proteins. In the 1990s, Linda Buck and Richard Axel discovered in mice a family of about a thousand genes (about 3 percent of the genome) that code for olfactory receptor proteins. Humans have about one-third that number of functional olfactory receptor genes. Each receptor protein that is expressed is found in a limited number of sensory receptor cells in the olfactory epithelium, and each cell expresses just one receptor. Using a combination of patch clamping and molecular techniques, the investigators were able to match specific gene products with the odorants they detect.

45.3 Some Scents Travel Great Distances Mating in silkworm moths of the genus *Bombyx* is coordinated by a pheromone called bombykol.

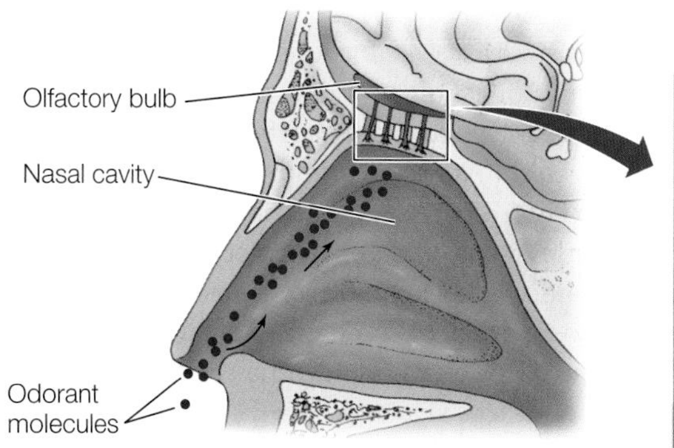

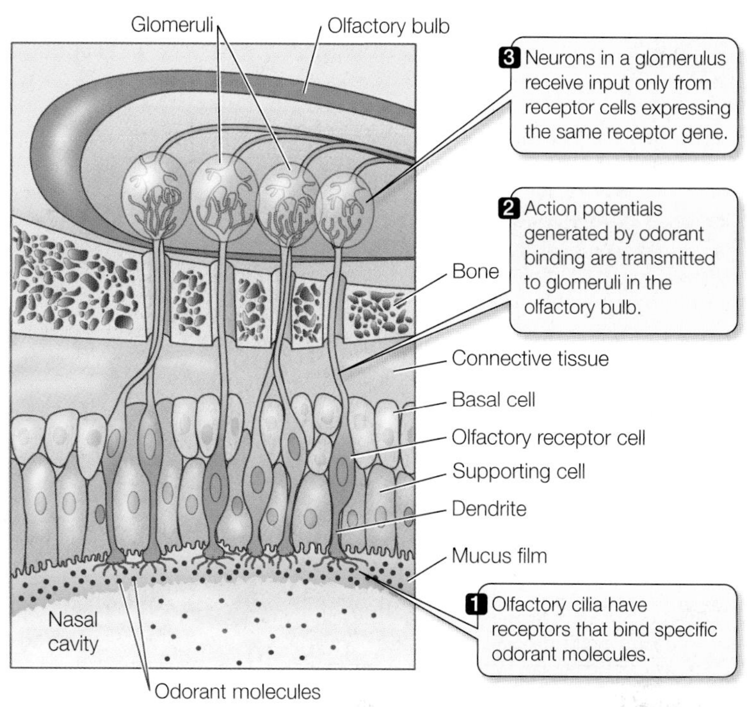

45.4 Olfactory Receptors Communicate Directly with the Brain The receptor cells of the human olfactory system are embedded in epithelial tissues lining the nasal cavity and send their axons to the olfactory bulb of the brain.

Olfactory sensitivity, however, enables discrimination of many more odorants than there are olfactory receptors. An odorant molecule can be quite complex, and different regions of that molecule may bind to different receptor proteins. The next stage of processing of olfactory information is in the olfactory bulb of the brain. In the olfactory bulb, axons from neurons expressing the same receptor protein converge on *glomeruli* (see Figure 45.4), which are clusters of olfactory bulb neurons. Therefore, a complex odorant molecule can activate a unique combination of glomeruli in the olfactory bulb, so an olfactory system with hundreds of different receptor proteins can discriminate an astronomically large number of smells. For their discoveries of the molecular nature of the olfactory system, Buck and Axel received the Nobel prize in 2004.

How does the olfactory receptor cell signal the intensity of a smell? The more odorant molecules that bind to receptors, the greater the frequency of action potentials and the greater the intensity of the perceived smell.

The vomeronasal organ senses pheromones

The **vomeronasal organ** (**VNO**) is a small, paired tubular structure embedded in the nasal epithelium of amphibians, reptiles, and many mammals (maybe even humans). In mammals, the VNO is located on the septum dividing the two nostrils. The VNO has a pore that opens into the nasal cavity. When the animal sniffs, the VNO pulsates and draws a sample of nasal fluid over the chemoreceptors embedded in its walls. The information from these chemoreceptors goes to an accessory olfactory bulb in the brain, and information from there goes to brain regions involved in sexual and other instinctive behaviors.

In 2002, Lawrence Katz and his colleagues at Duke University recorded the neuronal activity in the accessory olfactory bulbs of conscious, behaving mice. To test the hypothesis that the VNO detects pheromones, they placed a mouse attached to recording electrodes with other mice of same or different gender and same or different strain. These neurons were activated when the mouse with the implanted electrodes sniffed another mouse that was placed in the same cage. However, the neurons fired differentially depending on the gender and strain of the strange mouse. These observations support the hypothesis that the VNO is a specialized olfactory organ for pheromones.

In snakes, the VNO opens into the roof of the mouth cavity. Each time the snake's forked tongue darts in and out, the forks fit into the VNO openings and present to the chemoreceptors a sample of molecules from the surrounding air (see the chapter-opening photo). Thus the snake uses its tongue to smell its environment, not to taste it. Why doesn't the snake simply use the flow of air to and from its lungs, as we do, to smell the environment? In reptiles, air flows to and from the lungs slowly (and can even stop entirely for long periods of time), but the tongue can dart in and out many times in a second. It is a quick source of olfactory information.

Gustation is the sense of taste

The sense of taste, or **gustation**, in humans and other vertebrates depends on clusters of chemoreceptor cells called **taste buds**. The taste buds of terrestrial vertebrates are confined to the mouth cavity, but some fishes have taste buds in the skin that enhance their ability to sense their environment. Some fishes living in murky water are very sensitive to small amounts of amino acids in the water around them and can find food without the use of vision. The duck-billed platypus, a prototherian mammal (see Figure 33.24), has similar talents as a result of taste buds on the sensitive skin of its bill.

A human tongue has approximately 10,000 taste buds. The taste buds are embedded in the epithelium of the tongue, and most are found on the papillae of the tongue. (Look at your tongue in a mirror—the papillae make it look fuzzy.) Each papilla has many taste buds—mostly on the sides. The outer surface of a taste bud has a pore that exposes the tips of the sensory receptor cells. Microvilli

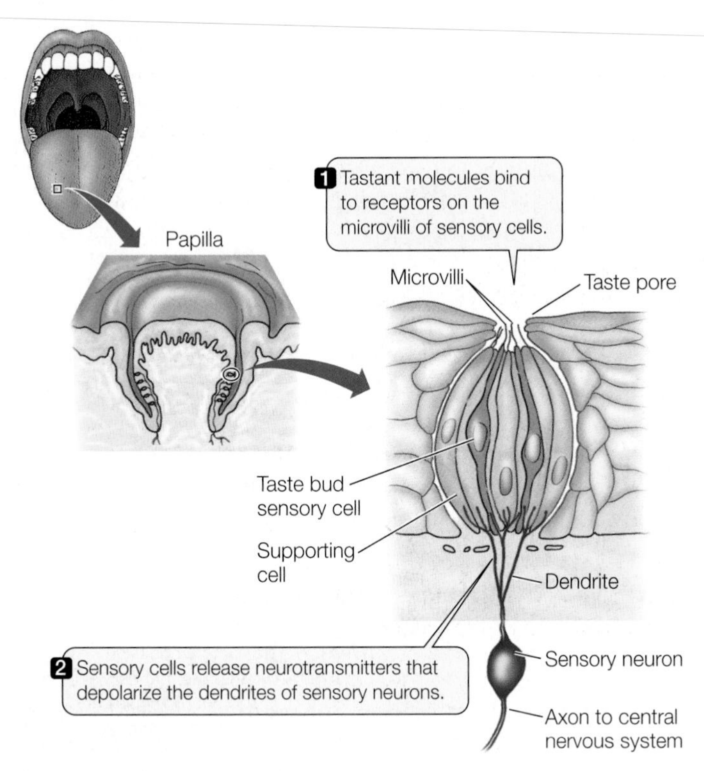

45.5 Taste Buds Are Clusters of Sensory Cells Each taste bud contains a number of sensory cells.

(tiny hairlike projections) increase the surface area of these cells where their tips converge at the pore (**Figure 45.5**). The cells do not generate action potentials, but changes in their membrane potential influence the release of neurotransmitter at their base where they form synapses with sensory neurons.

The tongue does a lot of hard work, so its epithelium, along with cells of its taste buds, are shed and replaced at a rapid rate. Individual taste bud cells last about 10 days before they are replaced, but the sensory neurons associated with them live on, always forming new synapses as new taste buds form.

You may have heard that humans can perceive only four tastes: sweet, salty, sour, and bitter. However, taste buds can distinguish among a variety of sweet-tasting molecules and a variety of bitter-tasting molecules. Recently small families of receptor protein genes for sweet and bitter tastes have been discovered. In addition, we now recognize a fifth taste, called **umami**. Umami is a savory, meaty taste that originates from receptors for amino acids, including monosodium glutamate (MSG). The full complexity of the chemosensitivity that enables us to enjoy the subtle flavors of food comes from the combined activation of gustatory and olfactory receptors, which is why you lose much of your sense of taste when you have a cold.

Gustation begins with receptor proteins in the membranes of the microvilli. The nature of these proteins and the mechanisms by which they depolarize the sensory receptor cell differ for the different basic tastes. Saltiness receptors are ionotropic and simply respond to Na^+ diffusing through open Na^+ channels depolarizing the sensory cell. Sourness receptors are also probably ionotropic. Depolarization of these receptors is similarly due to a direct effect of H^+ ions on Na^+ channels. In contrast, sweetness and bitterness involve families of receptor proteins similar to those involved in olfaction, and they are metabotropic. The bitter taste probably evolved as a protective mechanism enabling animals to detect avoid toxic plant compounds such as quinine, caffeine, and nicotine. Since plants have evolved many such molecules to repel herbivorous predators, a variety of receptors is essential. Similarly a large number of molecules in food could indicate nutritional value, so a variety of receptors is of value. The diversity of sweet receptors helps explain why it has been possible to invent many different artificial sweeteners.

In all cases of taste sensation, changes in the membrane potential of sensory receptor cells cause the cells to release neurotransmitters onto the dendrites of the sensory neurons. Sensory neurons fire action potentials that are conducted to the CNS, where the information is interpreted as specific taste sensations.

45.2 RECAP

Chemoreceptors are the basis of the sensations of olfaction and gustation and the reception of pheromones.

- Why are we able to distinguish so many different smells? Can you see why some people experience more or different odors than others? See pp. 968–969 and Figure 45.4
- Can you describe how different substances in food are transduced into action potentials in taste buds? See pp. 969–970 and Figure 45.5

Chemoreceptors have diverse structures that bind to a tremendous variety of stimulus molecules. Mechanoreceptors, however, need respond only to physical forces. Nevertheless, a considerable diversity of mechanosensory cells and mechanisms has evolved.

45.3 How Do Sensory Systems Detect Mechanical Forces?

Mechanoreceptors are sensory cells that respond to mechanical forces. Physical distortion of a mechanoreceptor's plasma membrane causes ion channels to open, altering the membrane potential of the cell, which in turn leads to the generation of action potentials. The rate of action potentials tells the CNS the strength of the stimulus to the mechanoreceptor. Mechanoreceptor cells are involved in many sensory systems, ranging from interpreting skin sensations to sensing blood pressure.

Many different cells respond to touch and pressure

Objects touching the skin generate varied sensations because skin is packed with diverse mechanoreceptor cells (**Figure 45.6**). The most important tactile receptors found in both hairy and nonhairy skin are *Merkel's discs,* which adapt rather slowly and provide continuous information about things touching the skin. Other mechanoreceptors, called *Meissner's corpuscles,* found primarily in nonhairy skin, are very sensitive, but adapt rapidly, so they provide information

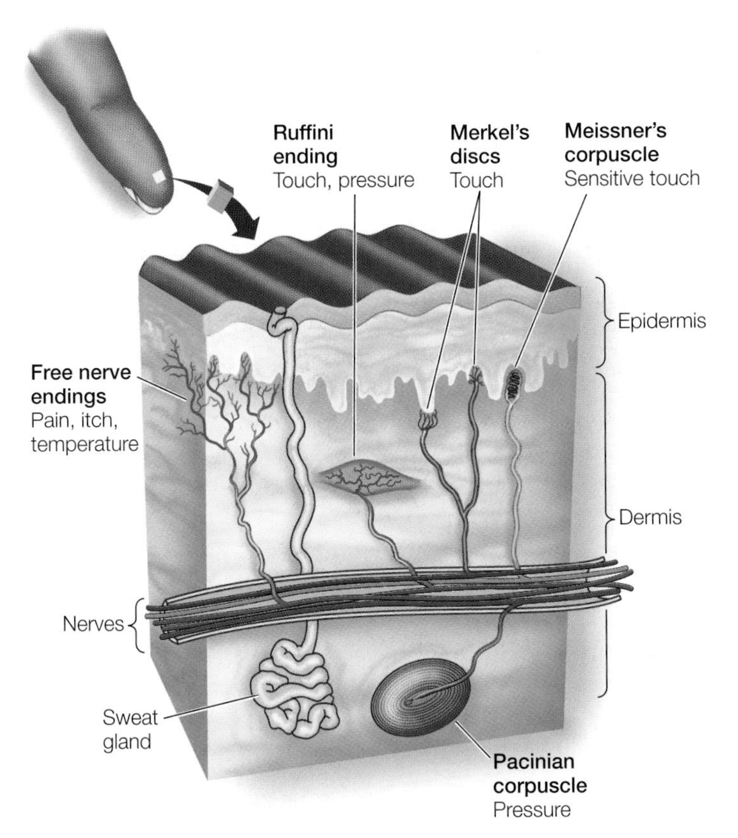

45.6 The Skin Feels Many Sensations Even a very small patch of skin contains a variety of sensory cells.

about changes in things touching the skin. The rapid adaptation of these tactile sensors is why you roll a small object between your fingers, rather than holding it still, to discern its shape and texture. As you roll it, you continue to stimulate Meissner's corpuscles.

Two other kinds of mechanoreceptor cells are found deeper in the skin. *Ruffini endings*, which adapt slowly, are good at providing information about vibrating stimuli of low frequencies. *Pacinian corpuscles*, which adapt rapidly, are good at providing information about vibrating stimuli of higher frequencies. Even deeper in the skin, dendrites of sensory neurons wrap around hair follicles. When the hairs are displaced, those neurons are stimulated.

The density of tactile mechanoreceptor cells varies across the surface of the body. A two-point spatial discrimination test demonstrates this fact. If you lightly touch someone's skin with two toothpicks simultaneously, you can determine how far apart the two stimuli have to be before the person can tell whether he or she is being touched by one or by two toothpicks. On the back, the stimuli have to be rather far apart before they are perceived as two discrete stimuli. The same test applied to the person's lips or fingertips reveals finer spatial discrimination; that is, the person can identify as separate two stimuli that are close together.

Mechanoreceptors are found in muscles, tendons, and ligaments

An animal receives information from mechanoreceptor cells about the position of its limbs and the stresses on its muscles and joints. These mechanoreceptors supply information continuously to the CNS, and this information is essential for postural control and the coordination of movements.

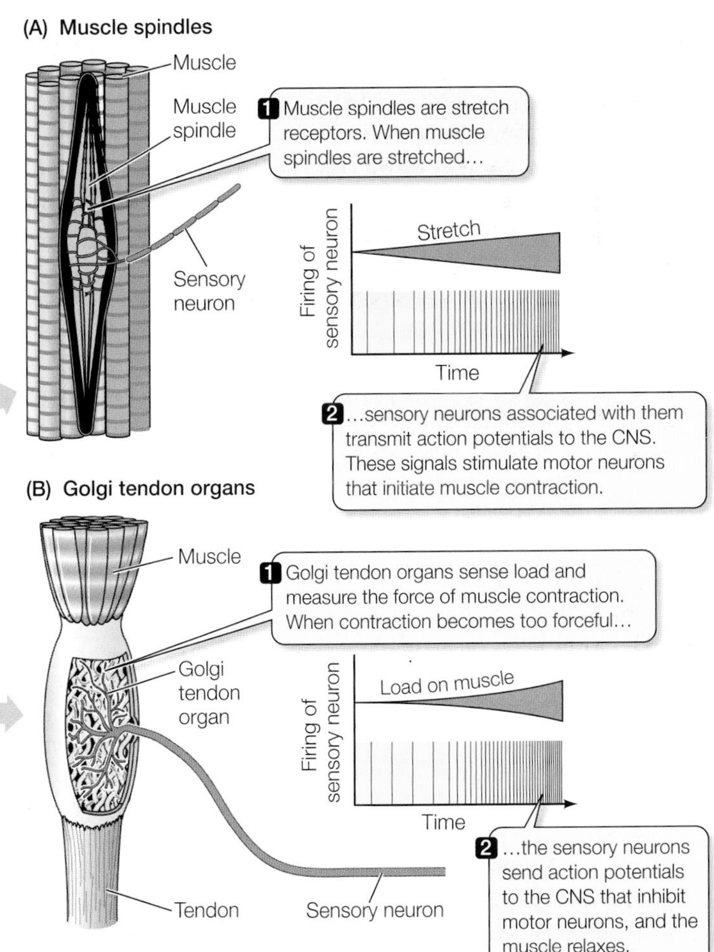

45.7 Stretch Receptors Stretch receptors provide information about the stresses on muscles and joints in an animal's limbs. (A) Signals from muscle spindles to the CNS initiate muscle contraction. (B) Golgi tendon organs in tendons and ligaments inhibit a contraction that becomes too forceful, triggering relaxation and protecting the muscle from tearing.

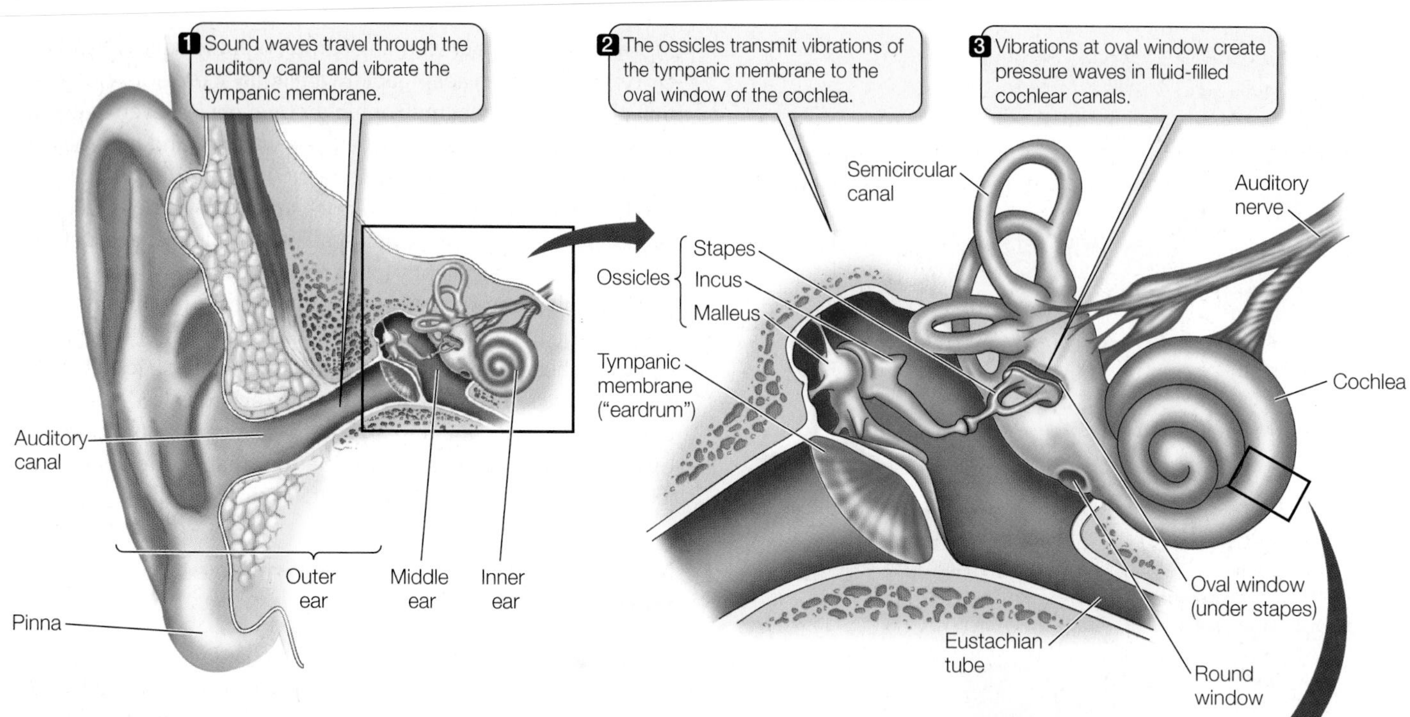

The mechanoreceptor cells found in skeletal muscle are called *muscle spindles*. These are **stretch receptors**, modified muscle cells that are embedded in connective tissue within muscles and innervated by sensory neurons. (Earlier in this chapter, we saw how crayfish stretch receptors transduce physical force into action potentials; see Figure 45.2.) The actions of muscle spindles are similar. Whenever the muscle is stretched, muscle spindles are also stretched, and the neurons transmit action potentials to the central nervous system (**Figure 45.7A**). The CNS uses this information to adjust the strength of contraction of the muscle to match the load put on the muscle. Thus a bartender can hold a beer mug in the same position as he fills it from the tap.

Another type of mechanoreceptor cell, the *Golgi tendon organ*, is found in tendons and ligaments and provides information about the force generated by a contracting muscle. When a contraction becomes too forceful, action potentials from the Golgi tendon organ inhibit the spinal cord motor neurons innervating that muscle, causing it to relax and protecting it from tearing (**Figure 45.7B**).

Auditory systems use hair cells to sense sound waves

The stimuli that animals perceive as sounds are pressure waves. **Auditory systems** use mechanoreceptors to convert pressure waves into receptor potentials. Auditory systems include special structures that gather sound waves, direct them to the sensory organ, and amplify their effect on the mechanoreceptors.

Human hearing provides a good example of an auditory system. The organs of hearing are the ears. The two prominent structures on the sides of our heads are the *pinnae*. The pinna of an ear collects sound waves and directs them into the *auditory canal*, which leads to the actual hearing apparatus in the *middle ear* and the *inner ear* (**Figure 45.8**). If you have ever watched a rabbit, a horse, or a cat change the orientation of its ear pinnae to focus on a particular sound, then you have witnessed the role of pinnae in hearing.

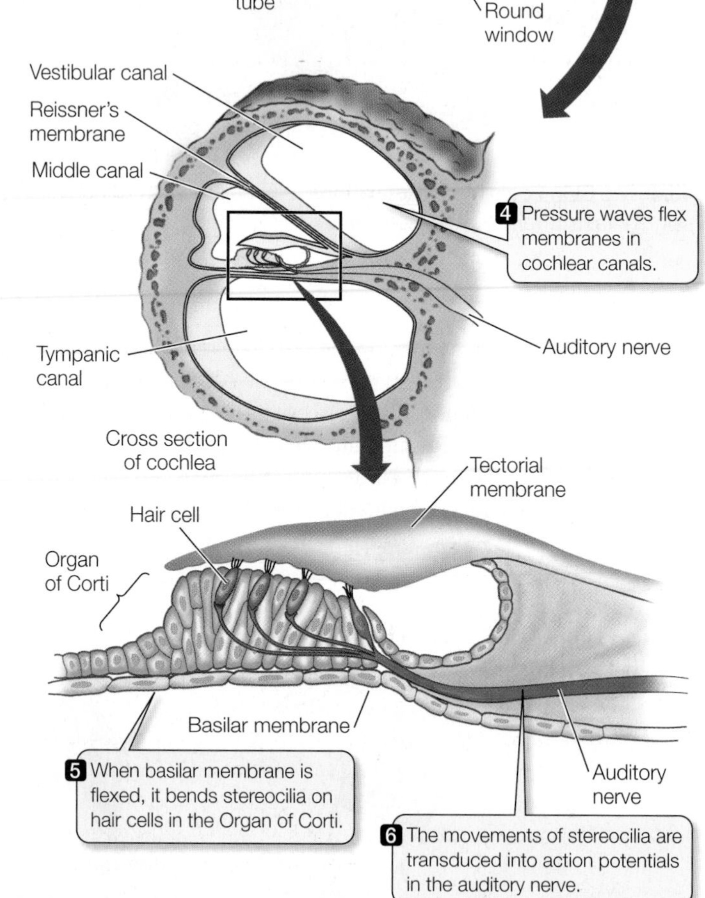

45.8 Structures of the Human Ear The human ear uses hair cells to transduce sound waves into action potentials.

The eardrum, or **tympanic membrane**, covers the end of the auditory canal. The tympanic membrane vibrates in response to pressure waves traveling down the auditory canal. The middle ear, an air-filled cavity, lies on the other side of the tympanic membrane.

The middle ear is open to the throat at the back of the mouth through the *eustachian tube*. Because the eustachian tube is also filled with air, pressure equilibrates between the middle ear and the outside world. When you have a cold or allergy, the tube can become blocked by mucus or by tissue swelling, so you have difficulty "clearing your ears," or equilibrating the pressure in the middle ear with the outside air pressure, which you have to do when ascending or descending in a plane or in a dive using SCUBA gear.

The middle ear contains three delicate bones called the **ossicles**, individually named the *malleus* (Latin, "hammer"), *incus* ("anvil"), and *stapes* ("stirrup"). The ossicles transmit the vibrations of the tympanic membrane to another flexible membrane called the **oval window**. The ossicles act as a lever—like a crow bar—translating a large movement of the tympanic membrane into a smaller movement of the oval window, but a movement of greater force. Also, because the oval window is much smaller than the tympanic membrane, the pressure the stapes transmits to the oval window is more than 20 times greater than the pressure exerted by the sound wave on the tympanic membrane. Behind the oval window lies the fluid-filled inner ear. Movements of the oval window result in pressure changes in the inner ear. These pressure waves in the inner ear are transduced into action potentials.

The inner ear is a long, tapered, coiled chamber called the **cochlea** (from Latin and Greek words for "snail" or "spiral shell"). A cross section of this chamber reveals that it is composed of three parallel canals separated by two membranes: **Reissner's membrane** and the **basilar membrane** (see Figure 45.8). Sitting on the basilar membrane is the **organ of Corti**, the apparatus that transduces pressure waves into action potentials. The organ of Corti contains *hair cells* with *stereocilia* that are in contact with an overhanging, rigid shelf called the *tectorial membrane*. Hair cells do not fire action potentials. However, they form synapses with associated sensory neurons whose axons make up the auditory nerve. When the basilar membrane flexes, the tectorial membrane bends the hair cell stereocilia. When the stereocilia bend, they alter the membrane potential of the hair cell and the rate at which it releases neurotransmitter onto its sensory neuron. As a result, the rates of action potentials traveling to the brain in the auditory nerve change.

What causes the basilar membrane to flex, and how does this mechanism distinguish sounds of different frequencies? In **Figure 45.9**, the cochlea is shown uncoiled to make it easier to understand its structure and function. The upper and lower canals separated by the basilar membrane are joined at the distal end of the cochlea (the end farthest from the oval window), making one continuous canal that turns back on itself. Just as the oval window is a flexible membrane at the beginning of the cochlea, the **round window** is a flexible membrane at the end of the long cochlear canal.

Air is highly compressible, but fluids are not. Therefore, a pressure wave can travel through air without much displacement of the air, but a pressure wave in fluid causes displacement of the fluid. When the stapes pushes on the oval window, the fluid in the upper canal of the cochlea is displaced. If this movement of the oval window occurs slowly, the cochlear fluid pressure wave travels down the upper canal, around the bend, and back through the lower canal. At the end of the lower canal, the displacement pressure is dissipated by the outward bulging of the round window.

If the oval window vibrates in and out rapidly, however, the waves of fluid pressure do not travel all the way to the end of the upper canal and back through the lower canal. Instead, they take a shortcut by crossing the basilar membrane, producing a traveling wave that flexes the basilar membrane. The more rapid the vibration, the greater is the amplitude of the flexion of the basilar membrane closer to the oval and round windows. Slower vibrations cause the largest amplitude flexions of the basilar membrane farther from the oval and round windows. Thus, different pitches of sound flex the basilar membrane at different locations and activate different sets of hair cells. Action potentials stimulated by the mechanoreceptors at different positions along the organ of Corti travel to the brain stem along the auditory nerve.

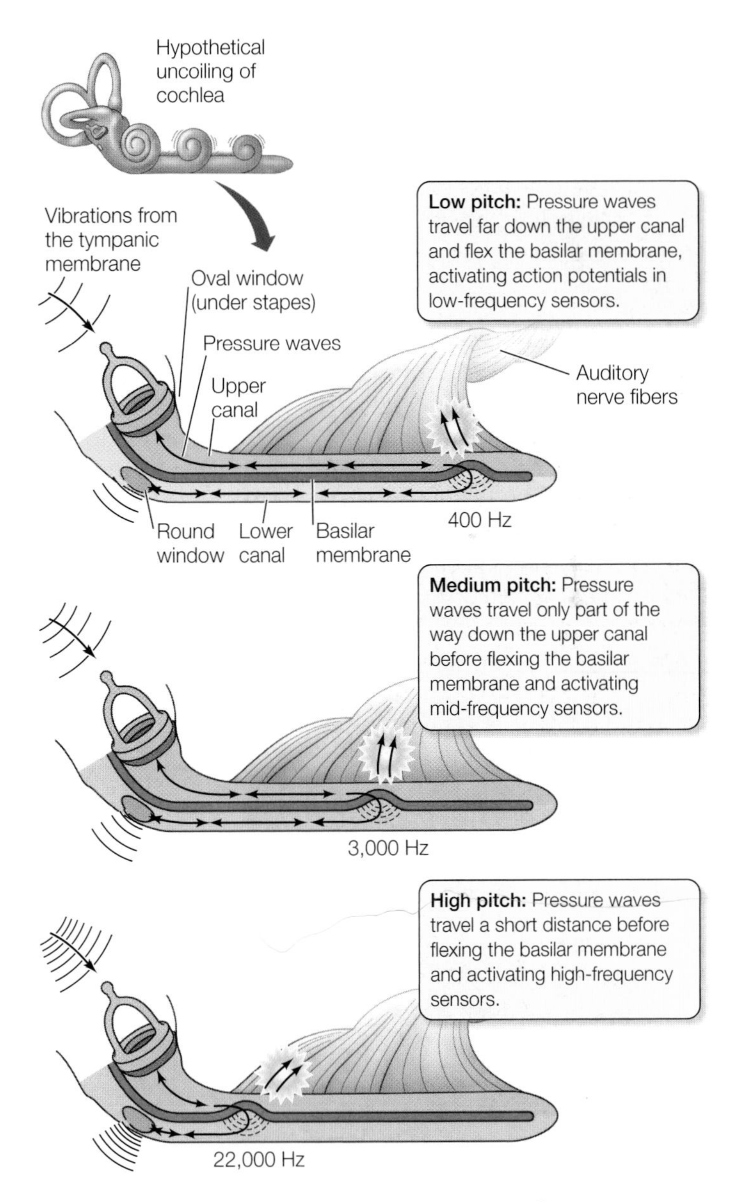

45.9 Sensing Pressure Waves in the Inner Ear Pressure waves of different frequencies flex the basilar membrane at different locations. Information about sound frequency is specified by which hair cells are activated. For simplicity, this representation illustrates the cochlea as uncoiled, and leaves out the middle canal.

Deafness, the loss of the sense of hearing, has two general causes. *Conduction deafness* is caused by the loss of function of the tympanic membrane and/or the ossicles of the middle ear. Repeated infections of the middle ear can cause scarring of the tympanic membrane and stiffening of the connections between the ossicles. The consequence is less efficient conduction of sound waves from the tympanic membrane to the oval window. With increasing age, the ossicles progressively stiffen, resulting in a gradual loss of the ability to hear high-frequency sounds. *Nerve deafness* is caused by damage to the inner ear or the auditory pathways. A common cause of nerve deafness is damage to the hair cells of the delicate organ of Corti by exposure to loud sounds such as jet engines, pneumatic drills, or highly amplified music. This damage is cumulative and irreversible.

Earphones can put you at risk for hearing loss. Although they are small, they can generate high-pressure sound waves close to your tympanic membrane. Consistent exposure to sounds above 85 decibels can damage hearing. Personal stereo earphones can reach 120 decibels, and people commonly use them at 100 decibels (equivalent to being at a rock concert).

Hair cells provide information about displacement

Hair cells are the mechanoreceptors found in organs of hearing and, as we will see, equilibrium. Projecting from the surface of each hair cell is a set of stereocilia, which looks like a set of organ pipes (**Figure 45.10A**). When these stereocilia (which are really microvilli) are bent, they alter ion channels in the hair cell's plasma membrane. When the stereocilia are bent in one direction, open channels close, and the membrane is hyperpolarized (the potential becomes more negative); when they are bent in the opposite direction, closed channels open, and the membrane is depolarized (more positive). When the membrane is adequately depolarized, the hair cell releases a neurotransmitter to the sensory neuron associated with it, and the sensory neuron sends action potentials to the CNS.

How does the bending of the stereocilia open ion channels? The ion channels that are opened are at the ends of the stereocilia. This was discovered by exploring the areas around the stereocilia with microelectrodes and seeing that local currents were created near the tips of the stereocilia when they were bent. Then, careful electron microscopic work revealed minute filaments that connected the tip of each stereocilium to its taller neighbor. It is hypothesized that these filaments are tethers on mechanoreceptor ion channels in the stereocilium plasma membrane, and they act like springs that open the channels. If the taller neighboring stereocilium is bent away, the spring tightens, and the ion channel is opened. If the taller neighbor bends toward its shorter neighbor, the spring is loosened and the channel closes (see **Figure 45.10B**).

Vertebrate organs of equilibrium use hair cells to detect the position of the body with respect to gravity. Within the mammalian inner ear, three **semicircular canals** at angles to one another sense the position and orientation of the head (see Figure 45.8). The **vestibular apparatus** has two chambers that sense the static position of the head as well as linear acceleration produced by movement. The structure and function of these organs are described in **Figure 45.11**.

A very simple use of hair cells to measure displacement can be seen in the **lateral line** sensory system of fishes. The lateral line is a canal just under the surface of the skin that runs down each side of the fish (**Figure 45.12**). Hair cells line the canal and their stereocilia, protected by capsules of gelatinous material called *cupulae* (singular *cupula*), project into the stream of water that flows

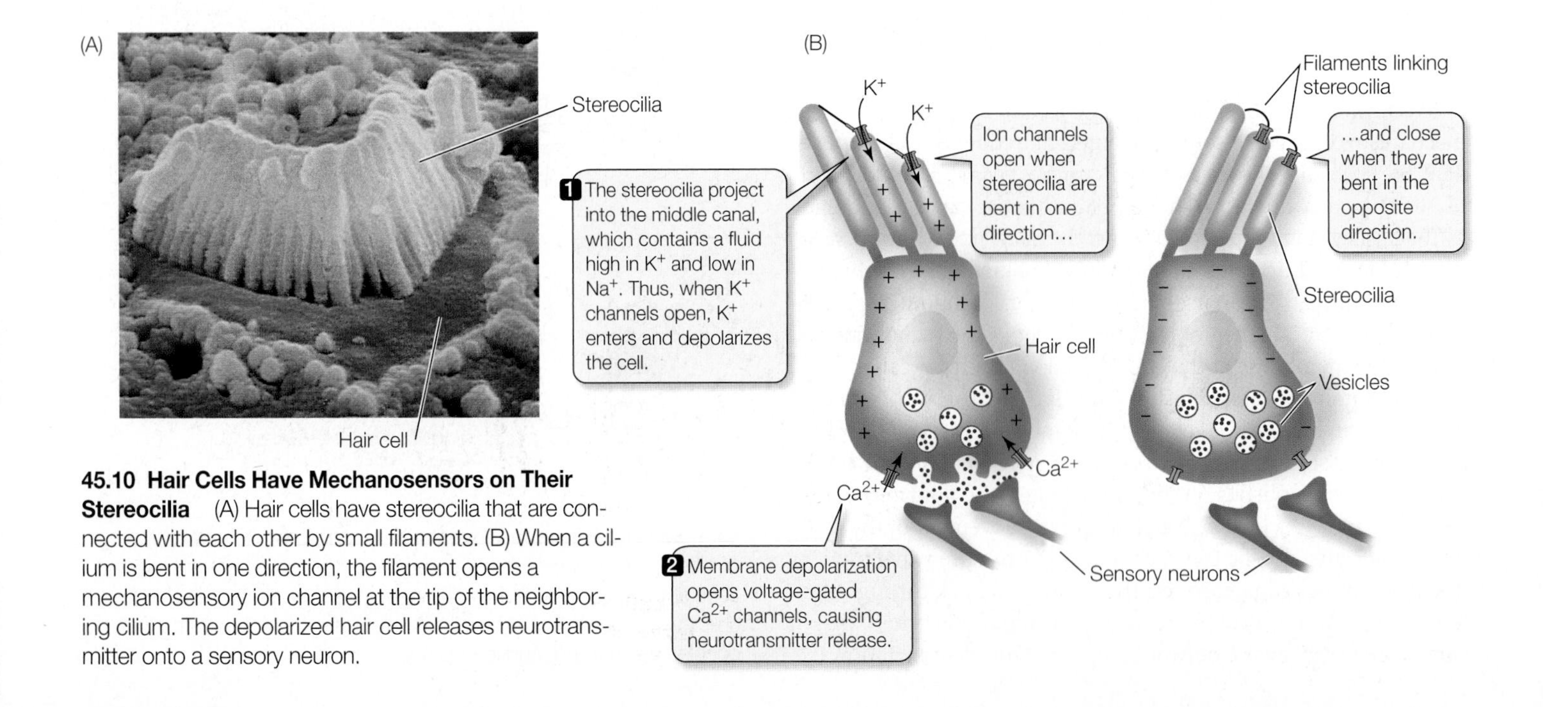

45.10 Hair Cells Have Mechanosensors on Their Stereocilia (A) Hair cells have stereocilia that are connected with each other by small filaments. (B) When a cilium is bent in one direction, the filament opens a mechanosensory ion channel at the tip of the neighboring cilium. The depolarized hair cell releases neurotransmitter onto a sensory neuron.

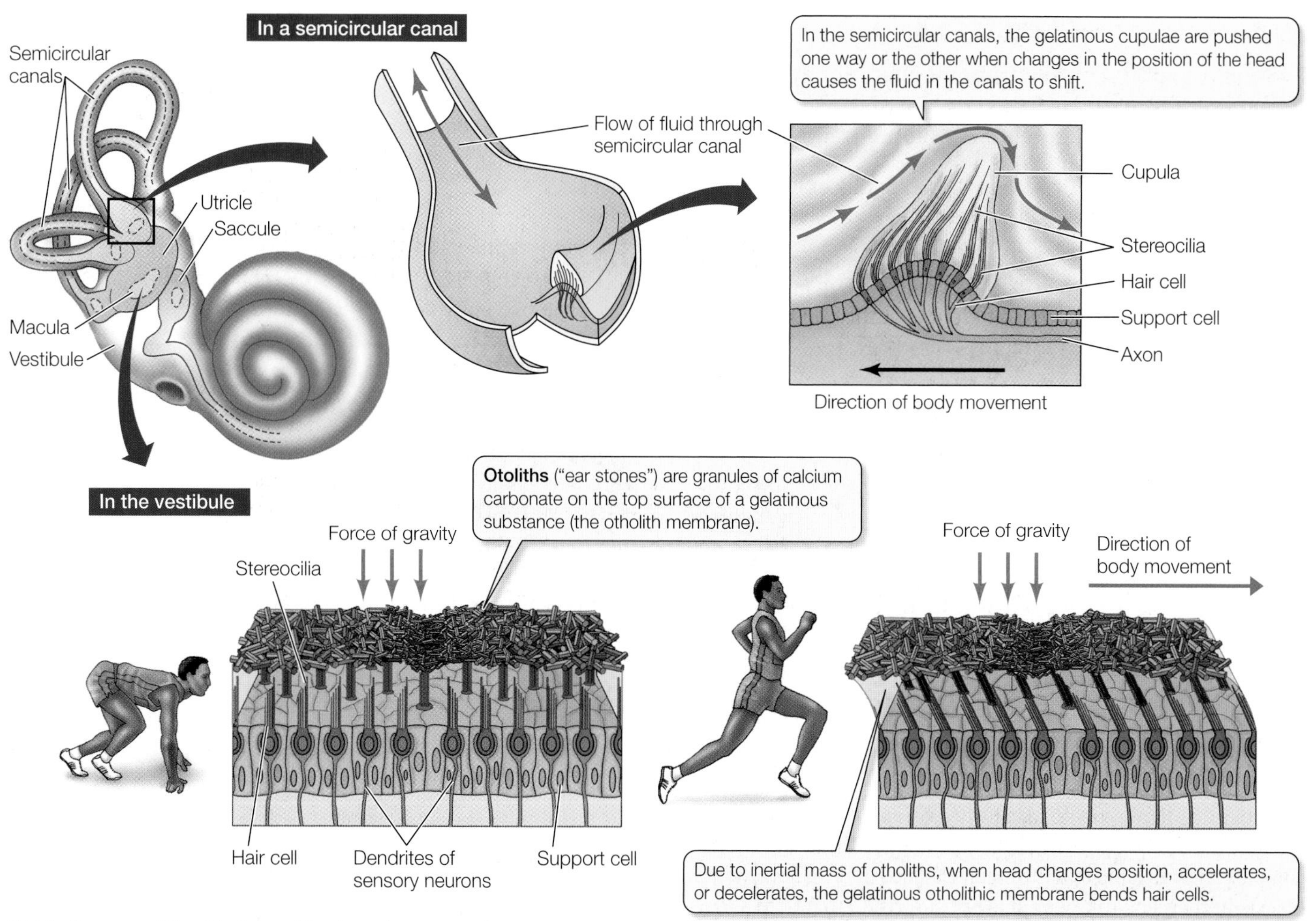

45.11 Organs of Equilibrium The bony inner ear of a human includes organs of equilibrium—three semicircular canals and two vestibular organs—as well as the cochlea, the auditory organ described in Figure 45.9.

through the lateral line canal when the fish moves through the water. Forward movement of the fish puts pressure on the cupulae, causing the stereocilia to bend in the direction that depolarizes the hair cells. Since water is incompressible, disturbances in the water around the fish are translated into pressure waves that can be picked up by the lateral line stereocilia. Thus, the lateral line system provides information about the fish's movements as well as about other moving objects, such as predators or prey. Sound in water also consists of pressure waves, so fish "hear" through their lateral lines.

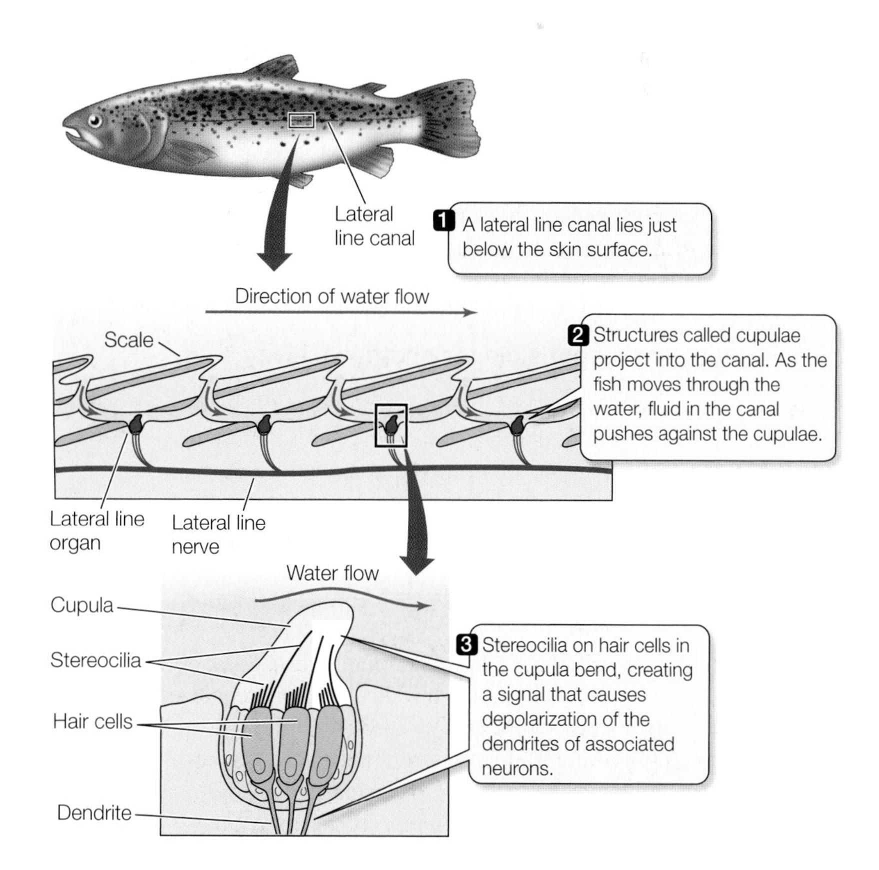

45.12 The Lateral Line Acoustic System Contains Mechanosensors Hair cells in the lateral line of a fish detect movement of the water around the animal, giving the fish information about its own movements and the movements of objects nearby.

45.3 RECAP

Sensations that derive from mechanoreceptors include touch, tickle, pressure, joint position, muscle load, hearing, and equilibrium.

- Describe some of the different mechanoreceptor cells in the skin and their properties. See p. 971 and Figure 45.6
- How do different frequencies of sound result in action potentials being fired in different acoustic neurons? See pp. 972–973 and Figures 45.8 and 45.9
- Can you explain how hair cells transduce force into action potentials? See p. 974 and Figure 45.10

Chemoreception gave us good examples of metabotropic sensory receptors, and mechanoreception has given us good examples of ionotropic sensory receptors. Now we will turn to another example of metabotropic sensory reception, but one in which light energy capture is the stimulus. We will see how light enegy is converted into action potentials.

45.4 How Do Sensory Systems Detect Light?

Sensitivity to light—**photosensitivity**—confers on the simplest animals the ability to orient to the sun and sky and gives more complex animals rapid and extremely detailed information about objects in their environment. It is not surprising that both simple and complex animals can sense and respond to light. What is remarkable is that across the entire range of animal species, evolution has conserved the same basis for photosensitivity: a family of pigments called **rhodopsins**.

In this section we will learn how rhodopsin molecules respond when stimulated by light energy and how that response is transduced into neuronal signals. We will also examine the structures of eyes, the organs that gather light energy and focus it onto **photoreceptor cells**, the metabotropic sensory receptors that transform light energy into action potentials.

Rhodopsins are responsible for photosensitivity

Photosensitivity depends on the ability of rhodopsins to absorb photons of light and to undergo a change in conformation. A rhodopsin molecule consists of a protein, **opsin** (which alone is not photosensitive) and an associated nonprotein light-absorbing group, **11-*cis*-retinal**, cradled in the center of the opsin and bound covalently to it. The entire rhodopsin molecule sits within the plasma membrane of a photoreceptor cell (**Figure 45.13**).

When the 11-*cis*-retinal absorbs a photon of light energy, it changes into a different isomer of retinal, called all-*trans*-retinal. This change puts a strain on the bonds between retinal and opsin, changing the conformation of opsin. This change signals the detection of light. In vertebrate eyes, the retinal and the opsin eventually separate from each other—a process called *bleaching*, which causes the molecule to lose its photosensitivity. A series of enzymatic reactions is then required to return the all-*trans*-retinal to the 11-*cis* isomer, which then recombines with opsin so that it once again becomes the photosensitive pigment rhodopsin.

The amino acid sequence of opsin is similar to that of olfactory receptors, and the retinal binding site is similar to the odorant binding site. Did photoreception evolve as an ability to "smell" light?

How does the conformational change of rhodopsin transduce light into a cellular response? After retinal is converted from the 11-*cis* into the all-*trans* form, its interactions with opsin pass through several unstable intermediate stages. One of these stages triggers a cascade of reactions involving a G protein signaling mechanism that results in the alteration of membrane potential that is the photoreceptor cell's response to light (see Figure 45.13).

To get a better idea of how rhodopsin alters the membrane potential of a photoreceptor cell and how that photoreceptor cell

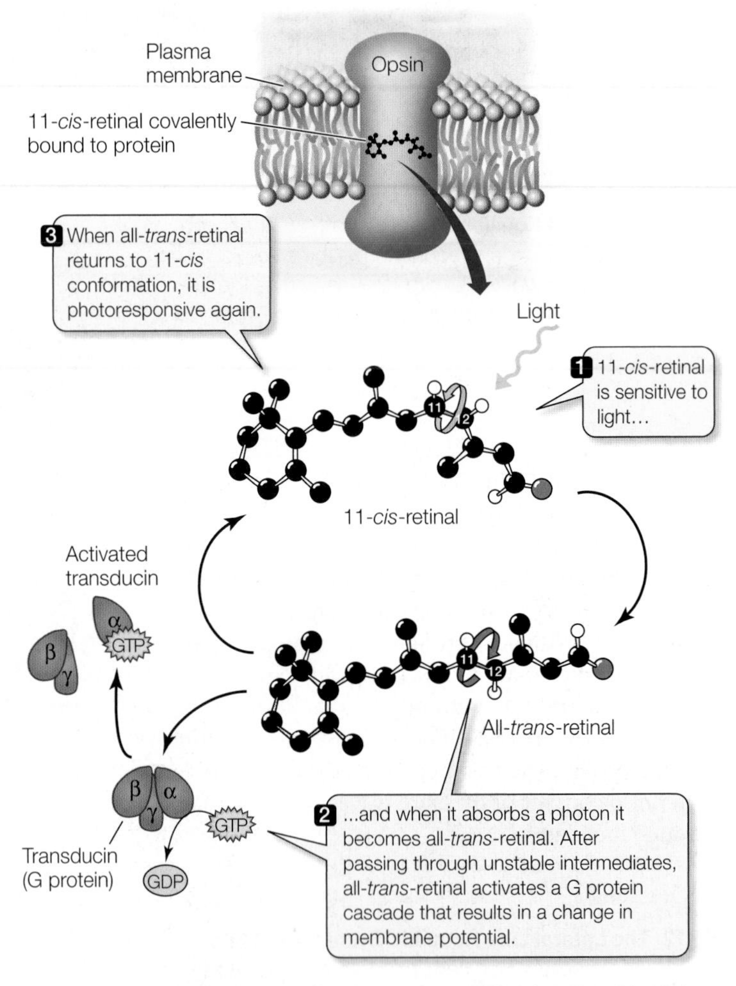

45.13 Light Changes the Conformation of Rhodopsin The light-absorbing molecule 11-*cis*-retinal bonds with the protein opsin to form the pigment rhodopsin, the molecular agent of photosensitivity.

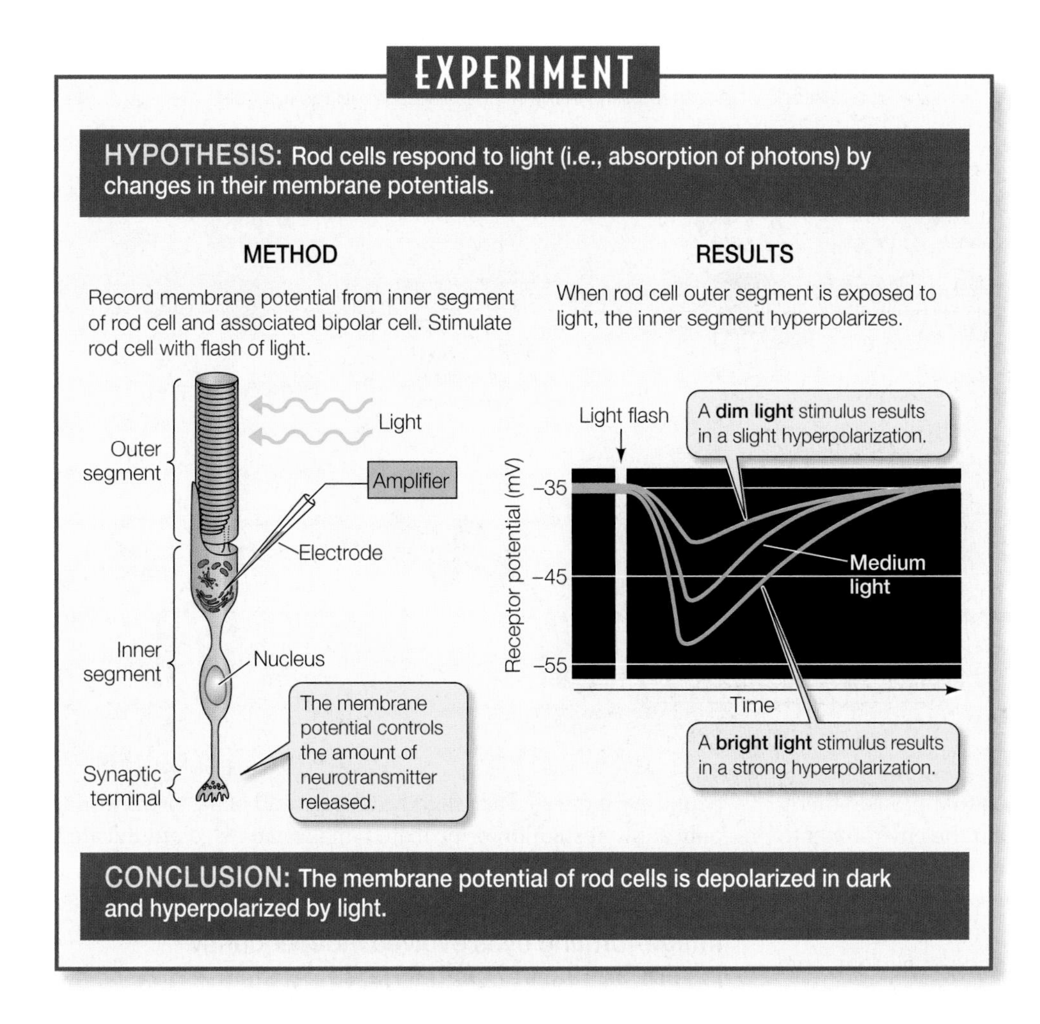

45.14 A Rod Cell Responds to Light The vertebrate rod cell is a neuron modified for photosensitivity. The membranes of a rod cell's discs are densely packed with rhodopsin. The plasma membrane of a rod cell hyperpolarizes—becomes more negative—in response to a flash of light. Rod cells do not fire action potentials. FURTHER RESEARCH: How would you investigate the effect of background illumination on the rod cell's response to light?

signals that it has been stimulated by light, let's look at one type of vertebrate photoreceptor cell, the **rod cell**. The rod cell, named for its shape, is a modified neuron that does not produce action potentials. Rod cells release neurotransmitter from their bases where they form synapses with the next neurons in the visual pathway (**Figure 45.14**). Each rod cell has an outer segment, an inner segment, and a synaptic terminal. The outer segment is highly specialized and contains a stack of disks of plasma membrane densely packed with rhodopsin. The function of the disks is to capture photons of light passing through the rod cell. The inner segment contains the cell nucleus and abundant mitochondria. The synaptic terminal is where the rod cell communicates with other neurons.

To see how a rod cell responds to light, we can penetrate a single rod cell with an electrode and record its membrane potential in the dark and in the light. From what we have learned about other types of sensory receptors, we might expect stimulation of the rod cell by light would make its membrane potential less negative. But the opposite is true. When a rod cell is kept in the dark, it has a relatively depolarized resting potential in comparison with other neurons. In fact, the plasma membrane of the rod cell is almost as permeable to Na^+ as to K^+. In the dark, Na^+ continually enters the outer segment of the cell—the dark current.

When light is flashed on the dark-adapted rod cell, its membrane potential becomes more negative—it hyperpolarizes (see Figure 45.14). The rate of neurotransmitter release changes as membrane potential changes. As the rod cell hyperpolarizes, its release of neurotransmitter decreases.

How does the absorption of light by rhodopsin hyperpolarize the rod cell? When rhodopsin is excited by light, it initiates a cascade of events (**Figure 45.15**). The photoexcited rhodopsin combines with and activates a G protein called *transducin*. Activated transducin in turn activates a phosphodiesterase (PDE), which converts cyclic GMP (cGMP) to GMP. This reaction plays a central role in phototransduction. In the dark, the cGMP in the outer segment binds to cation channels, keeping them open and allowing Na^+ to enter the outer segment. As cGMP is converted to GMP, the channels close, and the cell hyperpolarizes.

This mechanism may seem like a roundabout way of doing business, but its advantage is its enormous amplification ability. Each molecule of photoexcited rhodopsin can activate several hundred transducin molecules, thus activating a large number of PDE molecules. The catalytic capacity of a molecule of PDE is great: It can hydrolyze more than 4,000 molecules of cGMP per second. The bottom line is that a single photon of light can cause a huge number of sodium channels to close.

The dark current in photoreceptors makes evolutionary sense. The ability to detect shadows is a significant selective advantage in avoiding predators.

Invertebrates have a variety of visual systems

Photoreceptors are incorporated into a variety of visual systems, from simple to complex. Flatworms obtain directional information about light from photoreceptor cells that are organized into **eye cups**. The eye cups are paired bilateral structures, each partly shielded from light by a layer of pigmented cells lining the cup. The photoreceptors on the two sides of the animal are unequally stimulated unless the animal is facing directly toward or away from a light source. The flatworm generally uses directional information from the eye cups to move away from light.

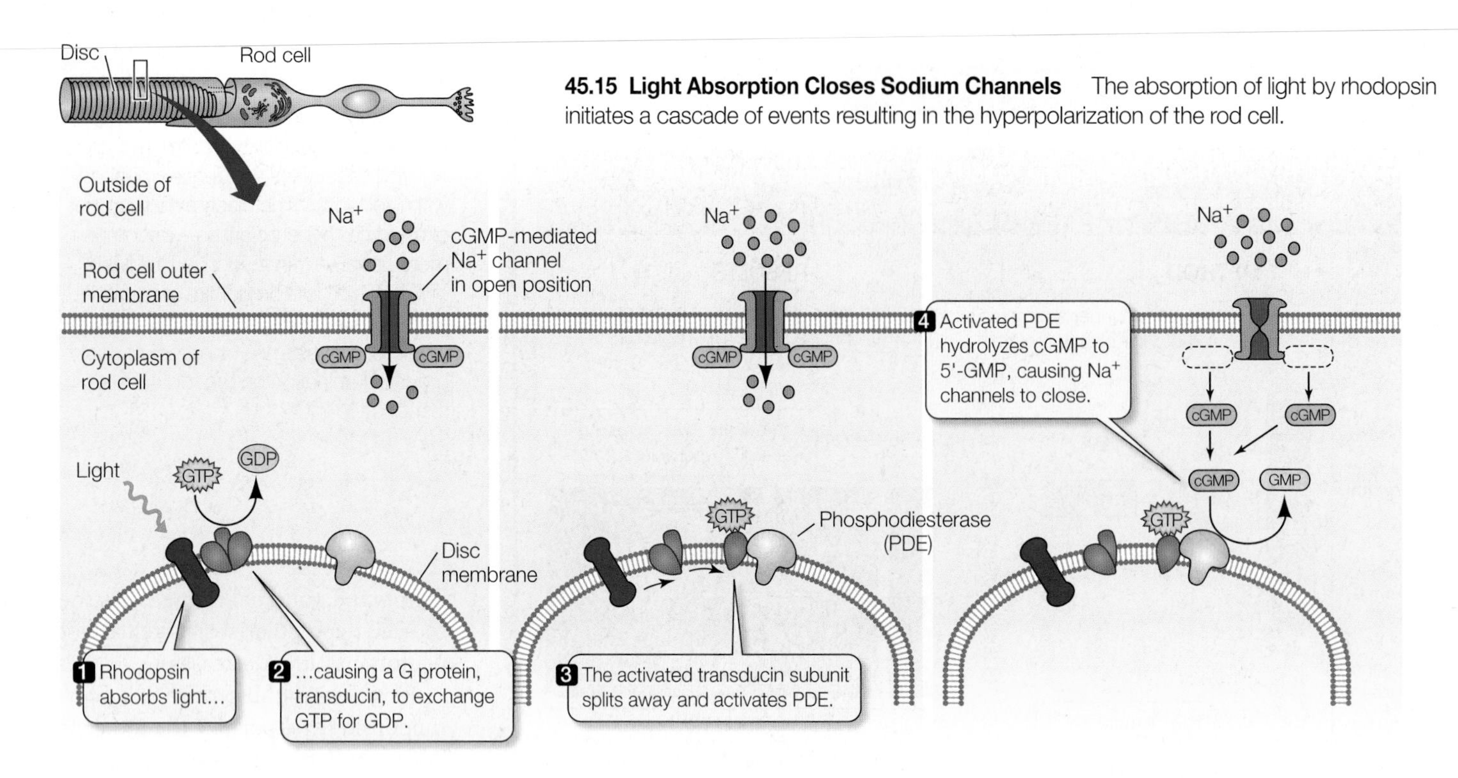

45.15 Light Absorption Closes Sodium Channels The absorption of light by rhodopsin initiates a cascade of events resulting in the hyperpolarization of the rod cell.

Arthropods have evolved **compound eyes** that provide them with information about patterns or images in the environment. These eyes are called compound because each eye consists of many optical units called **ommatidia** (singular *ommatidium*), each with its own narrow-angle lens (**Figure 45.16**). In contrast, a vertebrate eye consists of just one optical unit with a wide-angle lens. The number of ommatidia in a compound eye varies from only a few in some ants, to 800 in fruit flies, to 30,000 in some dragonflies.

Each ommatidium has a lens structure that directs light onto photoreceptor cells. Flies, for example, have eight elongated photoreceptors in each ommatidium. The inner borders of the photoreceptors are covered with microvilli that contain rhodopsin and trap light. Axons from the photoreceptors send the light information to the nervous system. Since each ommatidium of a compound eye is directed at a slightly different part of the visual world, only a low-resolution or pixillated image can be communicated from the compound eye to the CNS.

Image-forming eyes evolved independently in vertebrates and cephalopods

Both vertebrates and cephalopod mollusks have evolved **image-forming eyes**—eyes with exceptional abilities to form detailed images of the visual world. Like cameras, these eyes focus images on an internal surface that is sensitive to light. Considering that they evolved independently of each other, their high degree of similarity is remarkable (**Figure 45.17**).

The vertebrate eye is a spherical, fluid-filled structure bounded by a tough connective tissue layer called the *sclera*. At the front of the eye, the sclera forms the transparent **cornea**, through which light passes to enter the eye. Just inside the cornea is the pigmented **iris**, which gives the eye its color. The function of the iris is to con-

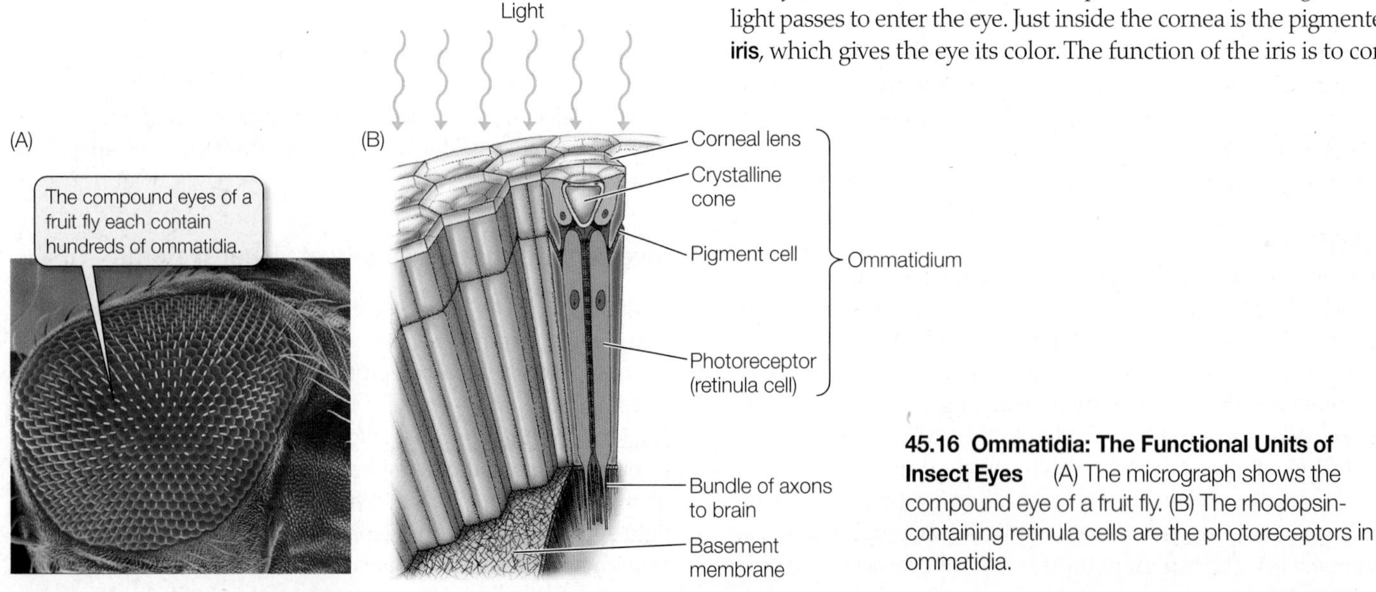

45.16 Ommatidia: The Functional Units of Insect Eyes (A) The micrograph shows the compound eye of a fruit fly. (B) The rhodopsin-containing retinula cells are the photoreceptors in ommatidia.

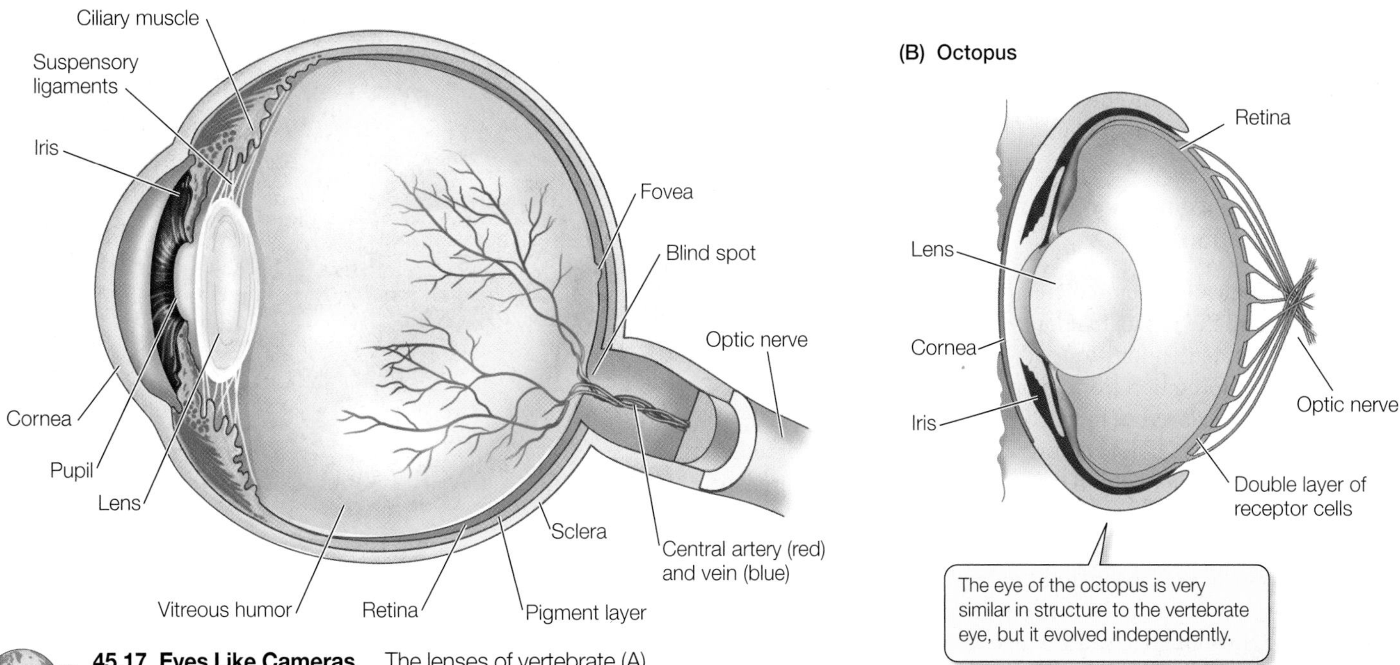

45.17 Eyes Like Cameras The lenses of vertebrate (A) and cephalopod (B) eyes focus images on layers of photoreceptor cells.

trol the amount of light that reaches the photoreceptor cells at the back of the eye, just as the diaphragm of a camera controls the amount of light reaching the film. The central opening of the iris is the **pupil**. The iris is under neuronal control. In bright light, the iris constricts, and the pupil is very small. As light levels fall, the iris relaxes, and the pupil enlarges.

Lenses become less elastic with age, so we lose the ability to focus on objects close at hand without the help of corrective lenses. As a consequence, most people over the age of 45 need the assistance of bifocal lenses or reading glasses.

Behind the iris is the crystalline protein **lens**, which makes fine adjustments in the focus of images falling on the photosensitive layer, the **retina**, at the back of the eye. The cornea and the fluids within the eye are mostly responsible for focusing light on the retina, but the lens allows the eye to *accommodate*—that is, to focus on objects at various locations in the near visual field (**Figure 45.18**). To focus a camera on objects close at hand, you adjust the distance between the lens and the film. Fishes, amphibians, and reptiles accommodate in a similar manner, moving the lenses of their eyes closer to or farther from their retinas. Mammals and birds use a different method: They alter the shape of the lens.

The lens is contained in a connective tissue sheath that tends to keep it in a spherical shape, but it is attached to suspensory ligaments that pull it into a flatter shape. Circular muscles called the *ciliary muscles* counteract the pull of the suspensory ligaments and permit the lens to round up. When the ciliary muscles are at rest, the flatter lens has the correct optical properties to focus dis-

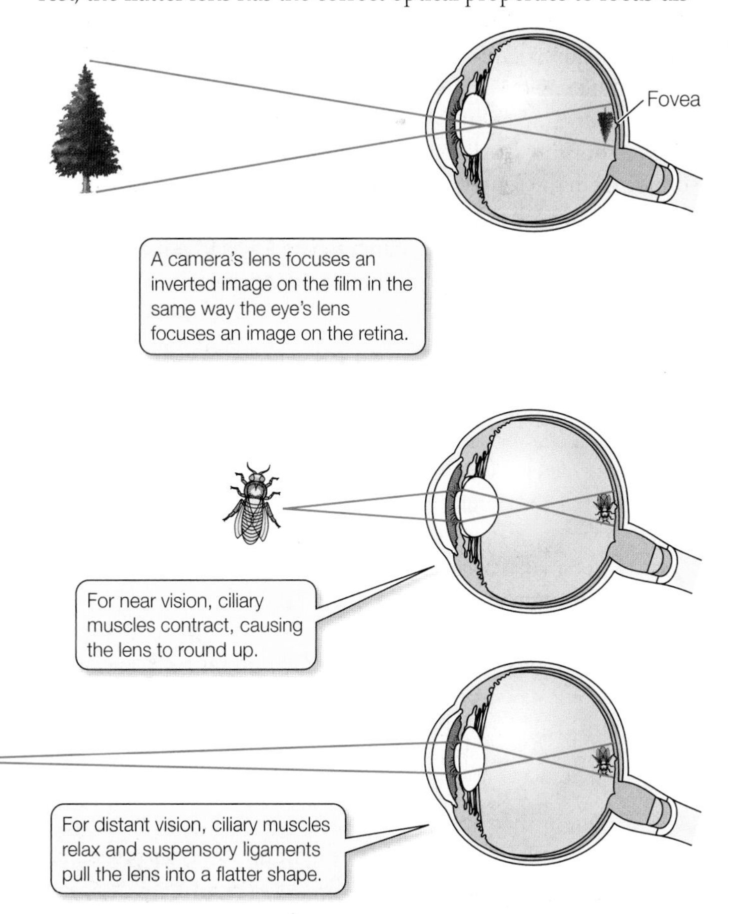

45.18 Staying in Focus Mammals and birds focus their eyes by changing the shape of the lens.

tant images on the retina, but not close images. Contracting the ciliary muscles rounds up the lens, changing its light-bending properties to bring close images into focus.

The vertebrate retina receives and processes visual information

During embryonic development, neuronal tissue grows out from the brain to form the retina. In addition to a layer of photoreceptor cells, the retina includes four additional layers of cells that process visual information from the photoreceptors. Light must pass through all the layers of retinal cells before being captured by rhodopsin. Light that is not captured by rhodopsin is absorbed by a black pigment layer behind the retina. In contrast, nocturnal animals such as deer and raccoons have a white reflective layer behind the retina to maximize the capture of photons. Therefore, the deer in the headlights appears to have bright white eyes. We do not have the white reflective layer in our retinas, but photographic flashes are bright enough to cause a reflection that appears red on photos because of the abundance of blood vessels in the retina.

The pigmented epithelium also plays a role in the renewal of the photoreceptors. New disks are continuously being generated by the inner segments and at the distal ends of the outer segments disks are being shed. The pigmented epithelial cells phagocytose the shed disks. Each outer segment is totally renewed about every two weeks.

THE PHOTORECEPTORS OF THE RETINA Until now we have referred to only one kind of photoreceptor, the rod cell. But there is another major kind of vertebrate photoreceptor, also named for its shape: the **cone cell** (**Figure 45.19**). A human retina has about 5 million cones and about 100 million rods. The density of rods and cones is not the same across the entire retina. In humans, light coming from the center of the visual field falls on the **fovea**, where the density of cone cells is highest. The human fovea has about 160,000 cones per square millimeter. The fovea of a hawk has almost twice the number of photoreceptors per square millimeter, making its vision sharper than ours. In addition, the hawk has two foveas in each eye: one receives light from straight ahead, while the other receives light from below. Thus, while the hawk is flying, it sees both its projected flight path and the ground below (where it might detect a mouse scurrying in the grass).

Because cones have low sensitivity to light, they contribute little to night vision. Night vision depends mostly on rod cells and therefore vision in dim light is mostly in shades of gray and acuity is low. You may have trouble seeing a small object such as a keyhole at night when you are looking straight at it—that is, when its image is falling on your fovea. If you look a little to the side, so that the image falls on a rod-rich area of your retina, you can see the object better. Astronomers looking for faint objects in the sky learned this trick a long time ago.

The human retina has three kinds of cone cells, each containing slightly different opsin molecules, which differ in the wavelengths of light they absorb best. Although the same 11-*cis*-retinal group is the light absorber in all three kinds of cones (see Figure 45.14), its molecular interactions with opsin determine the spectral sensitivity of the rhodopsin molecule as a whole (**Figure 45.20**).

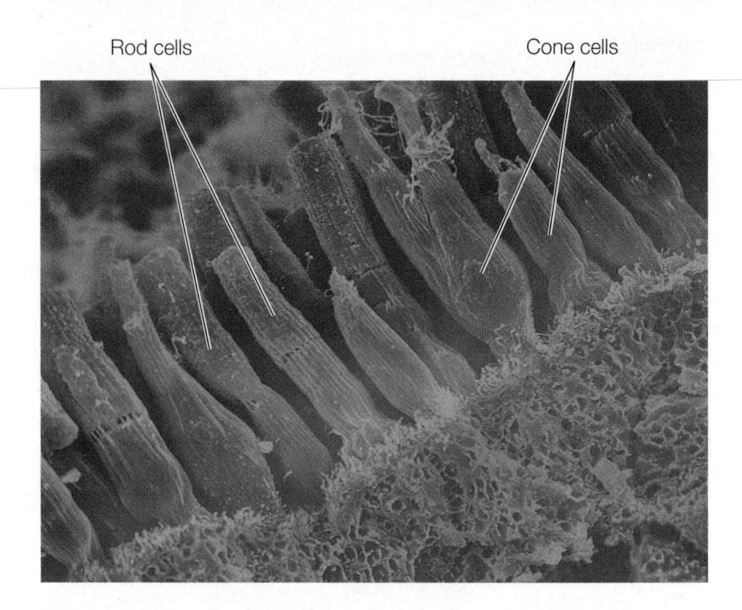

45.19 Rods and Cones This scanning electron micrograph of photoreceptors in the retina of a mud puppy (an amphibian) shows cylindrical rods and tapered cones.

Animals that are nocturnal (such as flying squirrels) have retinas containing a high percentage of rods and may have poor color vision. By contrast, some animals that are active only during the day (such as chipmunks) have mostly cones in their retinas.

Where blood vessels and the optic nerve pass through the back of the eye, there are no photoreceptors, resulting in a blind spot on the retina (see Figure 45.17). You are normally not aware of your blind spot, but you can find it. Stare straight ahead, holding a pencil in your outstretched hand so that the eraser is in the center of your field of vision. While continuing to stare straight ahead, slowly move the pencil to the side until the eraser disappears. When this happens, the light from the eraser is focused directly on your blind spot.

INFORMATION FLOW IN THE RETINA The human retina is organized into five layers of neurons (including the photoreceptor cells) that receive visual information and process it before sending it to the brain (**Figure 45.21**). A first step in understanding how the retina tells the brain what it is seeing is to study how these layers are interconnected and how they influence one another. From our discussion of rod cells above, we know that the photoreceptor cells at the back of the retina hyperpolarize in response to light and do not generate action potentials. The cells at the front of the retina (the cells closest to the lens) are **ganglion cells**. They do fire action potentials, and their axons form the optic nerve that travels to the brain. The layers of cells between the photoreceptors and the ganglion cells process information about the visual field.

The photoreceptors and ganglion cells are connected by *bipolar cells*. Changes in the membrane potential of rods and cones in response to light alter the rates at which the rods and cones release neurotransmitter at their synapses with the bipolar cells. In response to this neurotransmitter, the membrane potentials of the

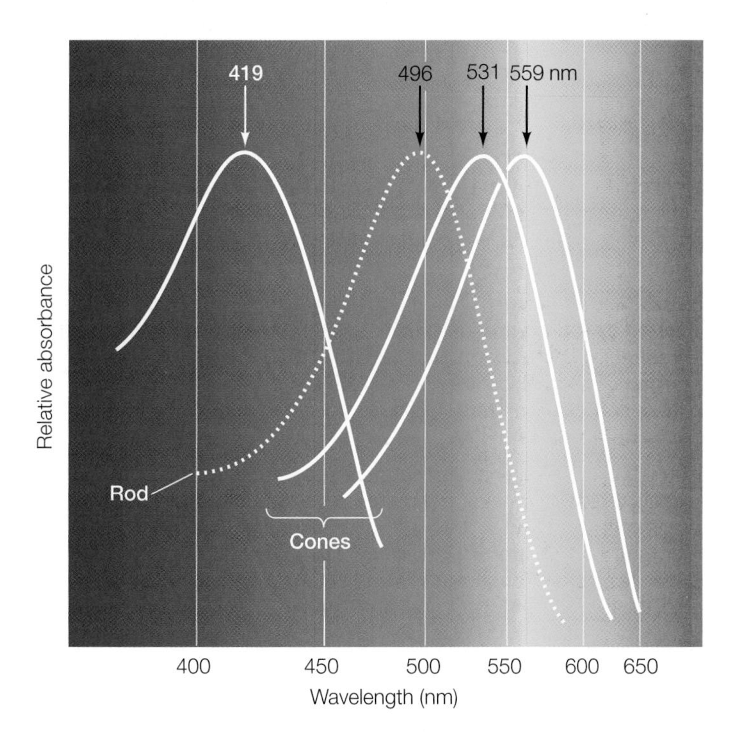

45.20 Absorption Spectra of Cone Cells The three kinds of cone cells contain slightly different opsin molecules, which absorb different wavelengths of light.

bipolar cells change, altering the rate at which they release neurotransmitter onto ganglion cells. The rate of neurotransmitter release from the bipolar cells determines the rate at which the ganglion cells fire action potentials. Thus, the direct flow of information in the retina is from photoreceptor to bipolar cell to ganglion cell. The ganglion cells send the information to the brain.

The other two cell layers, the horizontal cells and the amacrine cells, communicate laterally across the retina. *Horizontal cells* form synapses with neighboring photoreceptors. Thus, light falling on one photoreceptor can influence the sensitivity of its neighbors to light. This lateral flow of information enables the retina to sharpen the perception of contrast between light and dark patterns.

Amacrine cells form local interconnections between bipolar cells and ganglion cells. Some amacrine cell types are highly sensitive to changing illumination or to motion. Others assist in adjusting the sensitivity of the eyes according to the overall level of light falling on the retina. When background light levels change, amacrine cell connections to the ganglion cells help the ganglion cells remain sensitive to temporal changes in stimulation. Thus, even with large changes in background illumination, the eyes are sensitive to smaller, more rapid changes in the pattern of light falling on the retina.

Knowing the path of information in the retina still does not tell us how that information is processed by the brain. What does the eye tell the brain in response to a pattern of light falling on the retina? In Chapter 46 we will learn how the brain reassembles that information into our view of the world.

45.4 RECAP

A family of photopigments called rhodopsins are responsible for light sensitivity in all animals. Receptor cells, including rod and cone cells in humans, transduce the photosensitivity of rhodopsins to light and use it to form images of the environment.

- How does a photon of light change the membrane potential in a rod cell? See p. 977 and Figures 45.14 and 45.15
- What is the mechanism of color vision? See p. 980 and Figure 45.20
- Describe the flow of signals that occurs in the eye in response to light. See pp. 980–981 and Figure 45.21

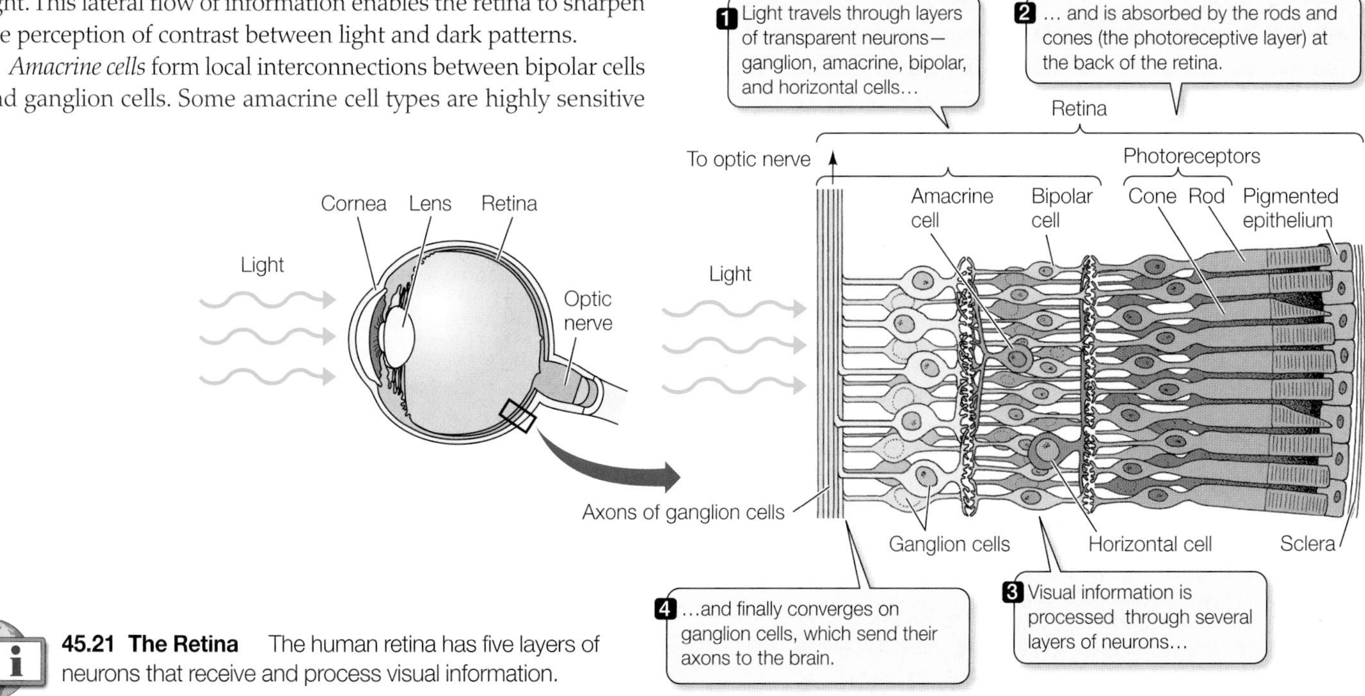

45.21 The Retina The human retina has five layers of neurons that receive and process visual information.

CHAPTER SUMMARY

45.1 How do sensory cells convert stimuli into action potentials?

Sensory receptor cells, also known as sensory cells or simply receptors, transduce information about an animal's external and internal environment into action potentials. Those sensory receptor cells that can fire action potentials are called sensory neurons.

The interpretation of action potentials as particular sensations depends on which neurons in the CNS receive them.

Sensory receptor cells have membrane **receptor proteins** that cause ion channels to open or close, affecting the resting potential of the cell. Metabotropic receptors act through signal transduction pathways to generate **receptor potentials**. Mechanoreceptors are ionotropic sensory receptors that open ion channels physically through forces such as pressure or stretch. Review Figure 45.1

Receptor potentials initiated by a sensory cell can spread to regions of the cell's plasma membrane that generate action potentials, or they can release neurotransmitter in response to changes in membrane potential. Review Figure 45.2

Adaptation enables the nervous system to ignore irrelevant or continuous stimuli while remaining responsive to relevant or new stimuli.

45.2 How do sensory systems detect chemical stimuli?

Chemoreceptors are responsible for smell, taste, and the sensing of **pheromones**.

Mammalian olfactory sensors project directly to the **olfactory bulb** of the brain. Sensors for the same **odorant** project to the same area of the olfactory bulb.

Each olfactory receptor cell expresses one receptor protein that can bind a specific molecule or ion. Binding causes a second messenger to open ion channels, which creates an action potential. Review Figure 45.4

Taste buds in the mouth cavities of vertebrates are responsible for the sense of **gustation**. The five basic tastes are sweet, salt, sour, bitter, and umami. Review Figure 45.5

45.3 How do sensory systems detect mechanical forces?

See Web/CD Tutorial 45.1

The skin contains a variety of ionotropic **mechanoreceptors** that respond to touch and pressure. The density of mechanoreceptors in any skin area determines the sensitivity of that area. Review Figure 45.6

Stretch receptors in muscle spindles and in the Golgi tendon organ found in tendons and ligaments inform the CNS of the positions of and loads on parts of the body. Review Figure 45.7

In mammalian auditory systems, ear pinnae collect and direct sound waves to the **tympanic membrane**, which vibrates in response to sound waves. The movements of the tympanic membrane are amplified through a chain of **ossicles** that conduct the vibrations to the **oval window**. Movements of the oval window create pressure waves in the fluid-filled **cochlea**. Review Figure 45.8, Web/CD Activity 45.1

The **basilar membrane** running down the center of the cochlea is distorted by sound waves at specific locations that depend on their frequency. These distortions cause the bending of hair cells in the **organ of Corti**. Receptor potentials in hair cells cause them to release neurotransmitter, which creates action potentials in the **auditory nerve**. Review Figure 45.9

Hair cells are also mechanoreceptors. The bending of their stereocilia alters receptor proteins and therefore their membrane potentials. Hair cells are found in the auditory organs and organs of equilibrium such as the **lateral line** system of fishes and the **semicircular canals** and **vestibular apparatus** of mammals. Review Figures 45.10, 45.11, and 45.12

45.4 How do sensory systems detect light?

Photosensitivity depends on the absorption of photons of light by **rhodopsin**, a **photoreceptor** molecule that consists of a protein called **opsin** and a light-absorbing prosthetic group called **retinal**. Absorption of light by retinal is the first step in a cascade of intracellular events leading to a change in the membrane potential of the photoreceptor cell. Review Figure 45.13

When excited by light, vertebrate photoreceptor cells hyperpolarize and release less neurotransmitter onto the neurons with which they form synapses. They do not fire action potentials. Review Figures 45.14 and 45.15

Visual systems range from the simple **eye cups** of flatworms, which sense the direction of a light source, to the **compound eyes** of arthropods, which detect shapes and patterns, to the **image-forming eyes** of vertebrates and cephalopods. Review Figures 45.16 and 45.17, Web/CD Activity 45.2

Vertebrate and cephalopod eyes focus detailed images of the visual field onto dense arrays of photoreceptors that transduce the visual image into neuronal signals. Review Figure 45.18

Vertebrates have two types of photoreceptors, **rod cells** and **cone cells**. In humans, the **fovea** contains almost exclusively cone cells, which are responsible for color vision but are not very sensitive in dim light. Color vision arises from three types of cone cells with different spectral absorption properties. Review Figures 45.19 and 45.20

The vertebrate **retina** consists of five layers of neurons lining the back of the eye. The light-absorbing photoreceptor cells are at the back of the retina. Review Figure 45.21, Web/CD Activity 45.3

The axons of the **ganglion cells**, in the innermost layer of the retina, are bundled together in the optic nerve. Between the photoreceptors and the ganglion cells are neurons that process information from the photoreceptors.

SELF-QUIZ

1. Which statement about sensory systems is *not* true?
   a. Sensory transduction involves the conversion (direct or indirect) of a physical or chemical stimulus into changes in membrane potentials.
   b. In general, a stimulus causes a change in the flow of ions across the plasma membrane of a sensory receptor cell.
   c. The term "adaptation" refers to the process by which a sensory system becomes insensitive to a continuing source of stimulation.
   d. The more intense a stimulus, the greater the magnitude of each action potential fired by a sensory neuron.
   e. Sensory adaptation plays a role in the ability of organisms to discriminate between important and unimportant information.

2. The female silkworm moth releases a chemical called bombykol from a gland at the tip of her abdomen. Bombykol is
   a. a sex hormone.
   b. detected by the male only when present in large quantities.
   c. not species-specific.
   d. detected by hairs on the antennae of male silkworm moths.
   e. a chemical basic to the taste process in arthropods.
3. Which statement about olfaction is *not* true?
   a. In general, mammals depend more on vision than on olfaction as their dominant sensory modality.
   b. Olfactory stimuli are recognized by the interaction between odorant molecules and receptor proteins on olfactory hairs.
   c. The more odorant molecules that bind to receptors, the more action potentials are generated.
   d. The greater the number of action potentials generated by an olfactory receptor, the greater the intensity of the perceived smell.
   e. The perception of different smells results from the activation of different combinations of olfactory receptors.
4. The touch receptors located very close to the skin surface
   a. are relatively insensitive to light touch.
   b. adapt very quickly to stimuli.
   c. are uniformly distributed throughout the surface of the body.
   d. are called Pacinian corpuscles.
   e. adapt slowly and only partially to stimuli.
5. The membrane that is most directly responsible for the ability to discriminate different pitches of sound is the
   a. round window.
   b. oval window.
   c. tympanic membrane.
   d. tectorial membrane.
   e. basilar membrane.
6. Which statement is *not* true?
   a. The transmembrane potential of a rod cell becomes more negative when the rod cell is exposed to light.
   b. A photoreceptor releases the most neurotransmitter when in total darkness.
   c. Whereas in vision the intensity of a stimulus is encoded by the degree of hyperpolarization of photoreceptors, in hearing the intensity of a stimulus is encoded by changes in firing rates of sensory neurons.
   d. Stiffening of the ossicles in the middle ear can lead to deafness.
   e. The interaction among hammer (malleus), anvil (incus), and stirrup (stapes) conducts sound waves across the fluid-filled middle ear.
7. In humans, the region of the retina where the central part of the visual field falls is the
   a. central ganglion cell.
   b. fovea.
   c. optic nerve.
   d. cornea.
   e. pupil.
8. The region of the vertebrate eye where the optic nerve passes out of the retina is the
   a. fovea.
   b. iris.
   c. blind spot.
   d. pupil.
   e. visual cortex.
9. Which statement about the cone cells in a human eye is *not* true?
   a. They are responsible for our sharpest vision.
   b. They are responsible for color vision.
   c. They are more sensitive to light than rods are.
   d. They are fewer in number than rods.
   e. They exist in high numbers at the fovea.
10. The color in color vision results from the
    a. ability of each cone cell to absorb all wavelengths of light equally.
    b. lens of the eye acting like a prism and separating the different wavelengths of light.
    c. differential absorption of wavelengths of light by different kinds of rod cells.
    d. three different isomers of opsin in cone cells.
    e. absorption of different wavelengths of light by amacrine and horizontal cells.

FOR DISCUSSION

1. Compare and contrast the functioning of olfactory receptors and photoreceptors. How do these sensory cells enable the CNS to discriminate between an apple and an orange?
2. Amplification of signal is an important feature of sensory systems. Compare mechanisms of amplification in olfactory, visual, and auditory systems.
3. If you were blindfolded and placed in a wheelchair, how would you know if you were being pushed forward or backward?
4. Animals can use visual, olfactory, tactile, and auditory signals to communicate. From what you know about these sensory systems, discuss the relative advantages and disadvantages of these systems for communication.

FOR INVESTIGATION

A certain region of the brain contains neurons that are sensitive to the osmolarity of the blood. You could imagine that these cells are sensitive to the concentration of a particular solute such as NaCl or to tension in their plasma membranes if they take up or lose water osmotically. Using a slice of this brain region in a culture dish, and patch pipettes, how could you investigate the mechanism of signal transduction in these neurons?

CHAPTER 46 The Mammalian Nervous System: Structure and Higher Function

Can our brains be full?

A famous "Far Side" cartoon by Gary Larson shows a classroom in which one student with a noticeably small head has his hand raised, saying, "Mr. Anderson, may I be excused? My brain is full." This lighthearted scene actually suggests a deep question: What is the capacity of the human brain? Is it limited by the organ's size? By the number of synapses? By the number of neurons?

Scientists long believed that we are born with a certain number of neurons, that we steadily lose neurons throughout life, and that we do not grow any new ones. The evidence seemed clear: in the adult brain, we don't see neurons with mitotic structures that would indicate cell division, nor do we see neurons at different stages of maturation. Also, it is hard to imagine how a new neuron, with its many complex connections, could be inserted into an adult brain. Our lifelong ability to learn has been explained solely by the formation and strengthening of synapses.

But one of the most exciting recent discoveries in neurobiology is that new neurons *are* formed in adult avian and mammalian brains, and that the formation of these new neurons seems to be stimulated by experience and learning. The birth of new neurons was first seen when adult rats were injected with radioactively labeled thymidine, which is incorporated into new DNA when cells divide. It came as a huge surprise when thymidine-labeled neurons were discovered in the brains of these animals.

Many reasons were advanced suggesting why these labeled cells could not really be new neurons, and few scientists were ready to give up the old dogma. The debate picked up, however, when Fernando Nottebohm and his colleagues at Rockefeller University showed that new neurons are formed in parts of the bird brain responsible for song at the time of the year when birds are ready to mate and males begin to sing. The researchers further showed that sex hormones and hearing song stimulated the birth of new neurons.

The discovery of neuron generation in bird brains gave new impetus to mammalian studies, and these studies revealed that two structures of the adult mammalian brain can acquire new neurons. One is the olfactory bulb (not surprising, because olfactory bulb neurons extend into the nasal epithelium, which sheds regularly). The other structure is the *hippocampus*, a brain region involved in forming long-term memories. As in birds, experience and learning stimulate neurogenesis in the mammalian hippocampus. Fred Gage and his colleagues at the Salk Institute have recorded action potentials from new hippocampal

New Neurons in an Adult Mouse An adult mouse was injected with a vector carrying a gene for green fluorescent protein (GFP). The gene was taken up and expressed only by dividing cells. Neurons labeled with GFP were later found in the mouse's brain—and these neurons must have arisen *after* injection of the labeled gene.

Birdsong Requires New Neurons A male meadowlark (*Sturnella neglecta*) sings a specific song in order to attract a mate. New neurons are formed in the parts of brain responsible for producing the song, stimulated by sex hormones and hearing the songs of other male meadowlarks.

neurons in mice, showing that they mature into functional neurons with properties identical to those of older neighboring cells.

Many questions arise from these studies. Could neurogenesis be stimulated to repair damage and to counter the effects of aging? Are neurons regularly lost and replaced? Do new neurons facilitate learning? Will you generate new neurons by reading this chapter?

IN THIS CHAPTER we explore the cellular basis of several important subsystems of the human nervous system. The human brain has about 100 billion neurons and probably a thousand times as many synapses, which account for its ability to handle vast amounts of information. But the ability of the brain to serve specific functions—evaluating sensory input, controlling emotions, generating motor output, learning and remembering—arises from the organization of the nervous system into functional subsystems.

CHAPTER OUTLINE

46.1 How Is the Mammalian Nervous System Organized?

We can describe the organization of the mammalian nervous system anatomically or functionally. In anatomical terms, all vertebrate nervous systems consist of three parts: a brain, a spinal cord, and a set of peripheral nerves that reach to all parts of the body. As we learned at the start of Chapter 44, the brain and spinal cord are referred to as the **central nervous system**, or **CNS**, and the cranial and spinal nerves that connect the CNS to all of the tissues of the body are referred to as the **peripheral nervous system** (**PNS**). An additional division of the nervous system exists in the gut—the enteric nervous system, which we will discuss in Chapter 50.

Recall from Section 44.1 that a *neuron* is an electrically excitable cell that communicates via an axon, and a *nerve* is a bundle of axons that carries information about many things simultaneously. Some axons in a nerve may be carrying information to the CNS, while other axons in the same nerve are carrying information from the CNS to the organs of the body. As we will see in this chapter, we can further divide the anatomy of the brain, spinal cord, and PNS into smaller units.

A functional organization of the nervous system is based on flow and type of information

The major avenues of information flow through the nervous system are illustrated in **Figure 46.1**. The **afferent** portion of the peripheral nervous system carries sensory information to the central nervous system. We are conscious of much of this information (for example, light, sound, skin temperature, pain, the position of limbs). We are usually not conscious, however, of the afferent information involved in physiological regulation (for example, blood pressure, deep body temperature, and blood oxygen supply).

The **efferent** portion of the PNS carries information from the CNS to the muscles and glands of the body. Efferent pathways can be divided into a *voluntary* division, which executes our conscious movements, and an *involuntary*, or **autonomic**, division, which controls physiological functions.

In addition to the neuronal information it receives from the PNS, the CNS receives chemical information from hormones circulating in the blood. **Neurohormones** released by neurons into the extracellular fluids of the brain can send chemical information to other neurons in the brain or can leave the brain and enter the circulation. In Chapter 42 we discussed the

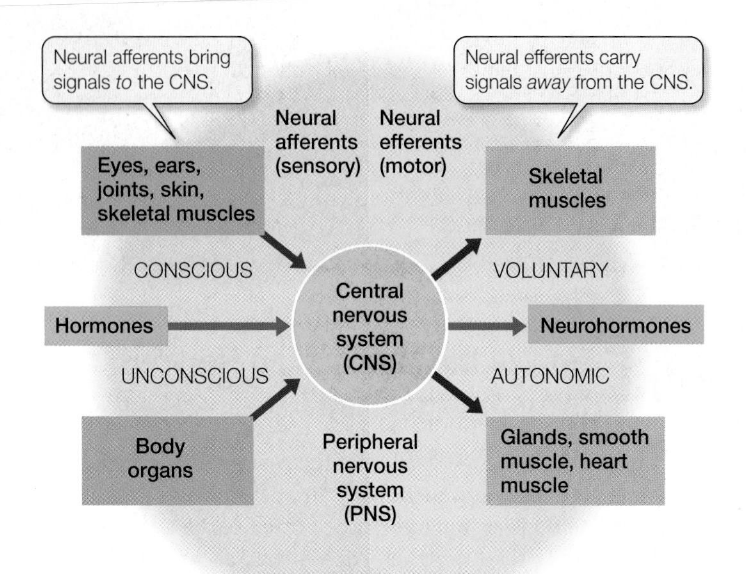

46.1 Organization of the Nervous System The peripheral nervous system (pink, blue) carries information both to and from the central nervous system (gold). The CNS also receives hormonal inputs and produces hormonal outputs (green).

important role of neurohormones (such as GnRH) in the control of the anterior pituitary, and we described how two other neurohormones, oxytocin and ADH, are released into the general circulation from the posterior pituitary.

The vertebrate CNS develops from the embryonic neural tube

Early in the development of a vertebrate embryo, a hollow tube of neural tissue forms (see Section 43.3). This **neural tube** runs the length of the embryo on its dorsal side. At the anterior end of the embryo, the neural tube forms three swellings that become the **hindbrain**, the **midbrain**, and the **forebrain**. The rest of the neural tube becomes the spinal cord (**Figure 46.2**). The cranial and spinal nerves sprout from the neural tube. From these early developmental stages we see the linear axis of information flow in the nervous system. Although the developing brain will fold and become a complex structure, the information flow in the adult nervous system still follows the paths that emerge from the simple linear neural tube.

Each of the three regions of the embryonic brain develops into several structures in the adult brain. From the hindbrain come the **medulla**, the **pons**, and the **cerebellum**. The medulla is continuous with the spinal cord. The pons is in front of the medulla, and the cerebellum is a dorsal outgrowth of the pons. The medulla and pons contain distinct groups of neurons that are involved in the control of physiological functions such as breathing and circulation or basic motor patterns such as swallowing and vomiting. All information traveling between the spinal cord and higher brain areas must pass through the pons and the medulla.

The cerebellum is like the conductor of an orchestra; it receives "copies" of the commands going to the muscles from higher brain areas, and it receives information coming up the spinal cord from

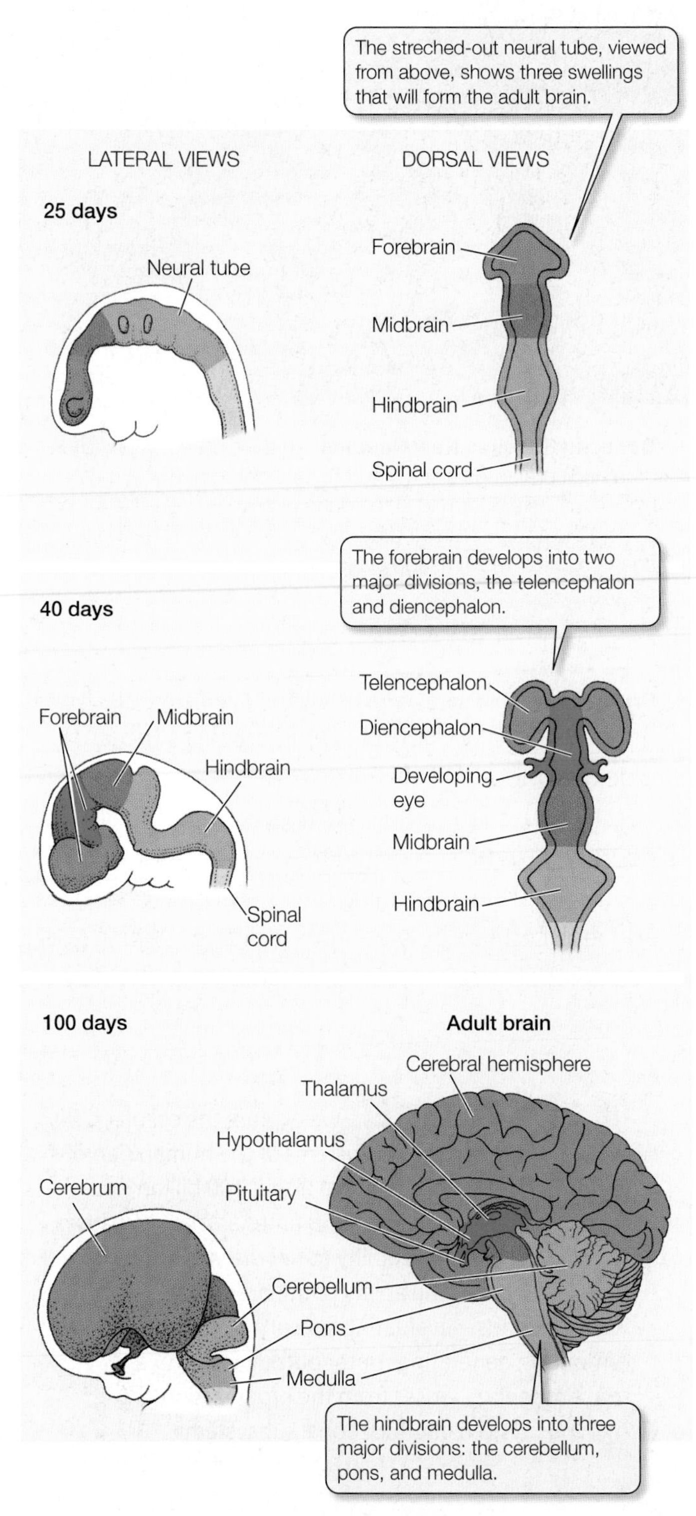

46.2 Development of the Human Nervous System Three swellings at the anterior end of the hollow neural tube in the early vertebrate embryo develop into the regions of the adult brain. The final panel shows an adult human brain cut in half through the midline, a view known as a midsagittal section.

the joints and muscles. It compares the motor "score" with the actual behavior of the muscles and thus refines the motor commands.

From the embryonic midbrain come structures that process aspects of visual and auditory information. In addition, all information traveling between higher brain areas and the spinal cord must pass through the midbrain. The hindbrain and the midbrain are collectively known as the **brain stem**.

The embryonic forebrain develops a central region called the **diencephalon** and a surrounding structure called the **telencephalon**. The diencephalon is the core of the forebrain and consists of an upper structure called the **thalamus** and a lower structure called the **hypothalamus**. The thalamus is the final relay station for sensory information going to the telencephalon, and the hypothalamus regulates many physiological functions (see Section 41.2) and biological drives such as hunger and thirst. The hypothalamus receives a lot of physiological information of which we are not conscious.

The telencephalon consists of two **cerebral hemispheres**, left and right (and is also referred to as the **cerebrum**). In humans, the telencephalon is by far the largest part of the brain and plays major roles in sensory perception, learning, memory, and conscious behavior.

If we compare the classes of vertebrates from fish through amphibians, reptiles, birds, and mammals, the telencephalon increases in size, complexity, and importance—an evolutionary trend called *telencephalization* (see Figure 44.2). The forebrain dominates the nervous systems of mammals, and damage to this region results in severe impairment of sensory, motor, or cognitive functions, and even coma. In contrast, a shark with its telencephalon removed can swim almost normally.

The spinal cord transmits and processes information

The spinal cord conducts information in both directions between the brain and the organs of the body. It also integrates much of the information coming from the PNS, and it responds to that information by issuing motor commands.

A cross section of the spinal cord reveals a central area of gray matter in the shape of a butterfly, surrounded by an area of white matter (**Figure 46.3**). In the nervous system, **gray matter** is rich in neuronal cell bodies, and **white matter** contains axons. The gray matter of the spinal cord contains the cell bodies of the spinal neurons; the white matter contains the axons that conduct information up and down the spinal cord. The white appearance is due to the myelin that wraps most of the axons. Spinal nerves extend from the spinal cord at regular intervals on each side. Each spinal nerve has two roots, one connecting with the *dorsal horn* of the gray matter, and the other connecting with the *ventral horn*. The afferent (sensory) axons in a spinal nerve enter the spinal cord through the *dorsal root*, and the efferent (motor) axons in a spinal nerve leave the spinal cord through the *ventral root*.

The conversion of afferent to efferent information in the spinal cord without participation of the brain is called a **spinal reflex**. The simplest type of spinal reflex involves only two neurons and one synapse and is therefore called a **monosynaptic reflex**. An example is the knee-jerk reflex, which your physician checks with a mallet tap just below your knee. We can diagram the wiring of a monosynaptic reflex by following the flow of information through the spinal cord, as shown in Figure 46.3.

In the case of the knee-jerk reflex, sensory information comes from stretch receptors in the leg muscle that is suddenly stretched when the mallet strikes the tendon that runs over the knee. Each stretch receptor initiates action potentials that are conducted by the axon of a sensory neuron through the dorsal horn of the spinal cord and all the way to the ventral horn. In the ventral horn, the

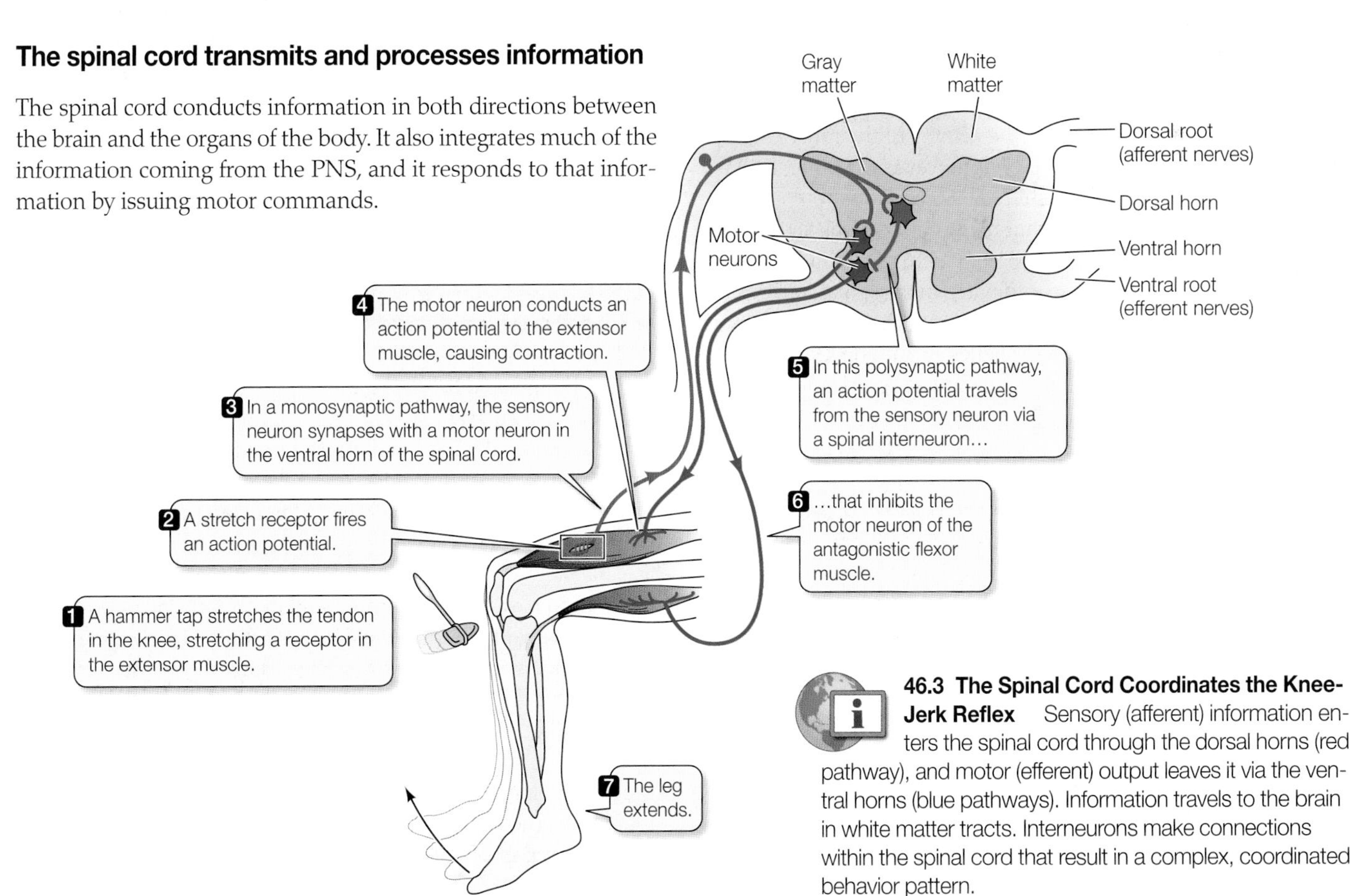

46.3 The Spinal Cord Coordinates the Knee-Jerk Reflex Sensory (afferent) information enters the spinal cord through the dorsal horns (red pathway), and motor (efferent) output leaves it via the ventral horns (blue pathways). Information travels to the brain in white matter tracts. Interneurons make connections within the spinal cord that result in a complex, coordinated behavior pattern.

sensory neuron synapses with motor neurons, causing them to fire action potentials that are then conducted back to the leg extensor muscle, causing it to contract. The function of this simple circuit is to sense an increased load on the limb and to increase the strength of muscle contraction to compensate for the added load and thereby keep muscle length constant.

Most spinal circuits are more complex than this monosynaptic reflex, as we can demonstrate by building on the circuit we have just traced. Limb movement is controlled by *antagonistic* sets of muscles—muscles that work against each other. When one member of an antagonistic set of muscles contracts, it bends, or flexes, the limb; it is therefore called a *flexor*. The antagonist to this muscle straightens, or extends, the limb, and is called an *extensor*. For a limb to move, one muscle of the pair must relax while the other contracts. Thus, sensory input that activates the motor neuron of one muscle also inhibits its antagonist. This coordination is achieved by an **interneuron**, which makes an inhibitory synapse onto the motor neuron of the antagonistic muscle (see Figure 46.3). Thus the reciprocal inhibition of antagonistic muscles involves an interneuron between the sensory cell and the motor neuron of the inhibited muscle, and therefore at least two synapses.

The withdrawal reflex is an example of a polysynaptic spinal reflex that involves many interneurons. When you step on a tack, you immediately pull back the injured foot: the tack stimulates pain receptors in the foot, and the sensory neurons transmit action potentials into the dorsal horn of the spinal cord on the same side of the body. In the dorsal horn, these neurons synapse with interneurons that send information through their axons to the brain resulting in the conscious sensation of pain. But even before the brain is aware of the pain, synapses of the sensory neurons with other interneurons stimulate and inhibit directly a variety of different motor neurons in the spinal cord. Interneurons on the same side of the spinal cord coordinate the activity of the muscles that withdraw the foot and leg. To pull away, however, the other leg has to extend and balance must be shifted. The coordination of these activities involves interneurons that make connections across the spinal cord to motor neurons on the opposite side. Thus a rather complex suite of movements is coordinated in the spinal cord without the brain's participation. Spinal circuits can even generate repetitive motor patterns (such as a shark's swimming movements) even when the individual's telencephalon has been removed.

The reticular system alerts the forebrain

Sensory information ascending the spinal cord to final destinations in the forebrain passes through the brain stem. Many sensory axons give off collateral branches that form synapses with a network of brain stem neurons called the **reticular system**. The reticular system is a highly complex network of axons and dendrites. Within the reticular system are many discrete groups of neurons that share a common characteristic such as the neurotransmitter they produce and release. Such an anatomically distinct group of neurons in the CNS is called a **nucleus** (not to be confused with the nucleus of a single cell).

As axons carrying sensory information ascend through the reticular formation, they make connections with nuclei that are involved in controlling many functions of the body. Information from joints and muscles, for example, is directed to nuclei in the pons and cerebellum that are involved in balance and coordination. Sensory information also goes to reticular formation nuclei that control sleep and wakefulness. High reticular formation activity produces waking; in the absence of such stimulation, sleep occurs. Because of the alerting function of the reticular core of the brain stem, it is called the *reticular activating system*.

If the brain is damaged at midbrain or higher levels, the alerting action of the reticular system on the forebrain can be lost and a person loses the ability to be conscious—they enter a coma. Damage to the brain stem or the spinal cord below the reticular system may cause paralysis but leaves a person with normal patterns of sleep and waking.

The core of the forebrain controls physiological drives, instincts, and emotions

As mentioned earlier, the midbrain connects to the forebrain through the diencephalon, which includes two important structures, the thalamus and the hypothalamus. The thalamus communicates sensory information to the cerebral cortex, and the hypothalamus receives information about physiological conditions in the body and regulates many homeostatic functions. Section 40.4 describes how the hypothalamus is involved in the regulation of body temperature, and Section 41.2 discusses the intimate association between the hypothalamus and the pituitary gland and its control of many homeostatic functions. Axons from neurons in the cerebral cortex that travel down through the brainstem to the spinal cord make up white matter tracts that go around the diencephalon.

The telencephalon is large in birds and mammals in comparison to the other vertebrates, but the regions of the telencephalon that border the diencephalon are the more primitive areas found in all vertebrates. Structures in this phylogenetically older region of the telencephalon form a group of structures collectively known as the **limbic system** (**Figure 46.4**).

The limbic system is responsible for basic physiological drives such as hunger and thirst, instincts, long-term memory formation, and emotions such as fear. Within the limbic system are areas that, when stimulated with small electric currents, can cause intense sensations of pleasure, pain, or rage. If a rat is given the opportunity to stimulate its own pleasure centers by pressing a switch, it will ignore food, water, and even sex, pushing the switch until it is exhausted. Pleasure and pain centers in the limbic system are believed to play roles in learning and in physiological drives.

As we saw in the introduction to Chapter 44, one component of the limbic system—the **amygdala**—is involved in fear and fear memory. If a certain portion of the amygdala is damaged or chemically blocked, an animal cannot learn to be afraid of a stimulus or a situation that would normally induce a strong fear reaction. Moreover, blocking protein synthesis in this part of the limbic system blocks the formation of fear memory.

Another part of the limbic system, the **hippocampus**, is necessary in humans for the transfer of short-term memory to long-term

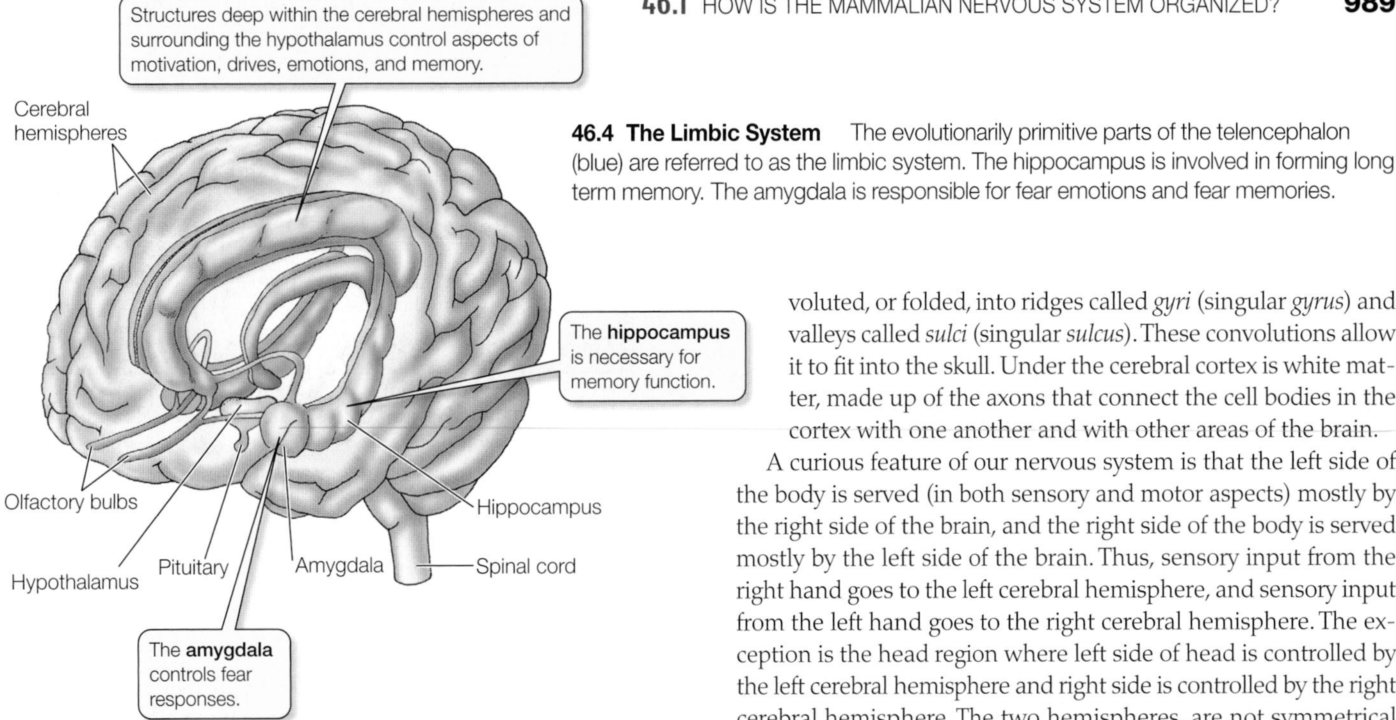

46.4 The Limbic System The evolutionarily primitive parts of the telencephalon (blue) are referred to as the limbic system. The hippocampus is involved in forming long term memory. The amygdala is responsible for fear emotions and fear memories.

memory. If you are told a new telephone number, you may be able to hold it in short-term memory for a few minutes, but within half an hour it is forgotten unless you make a real effort to remember it. The phenomenon of remembering something for more than a few minutes requires its transfer to long-term memory.

Regions of the telencephalon interact to produce consciousness and control behavior

The cerebral hemispheres are the dominant structures in the mammalian brain. In humans, they are so large that they cover all other parts of the brain except the cerebellum (**Figure 46.5A**). A sheet of gray matter called the **cerebral cortex** covers each cerebral hemisphere. It is about 4 mm thick and covers a total surface area over both hemispheres of 1 square meter. The cerebral cortex is convoluted, or folded, into ridges called *gyri* (singular *gyrus*) and valleys called *sulci* (singular *sulcus*). These convolutions allow it to fit into the skull. Under the cerebral cortex is white matter, made up of the axons that connect the cell bodies in the cortex with one another and with other areas of the brain.

A curious feature of our nervous system is that the left side of the body is served (in both sensory and motor aspects) mostly by the right side of the brain, and the right side of the body is served mostly by the left side of the brain. Thus, sensory input from the right hand goes to the left cerebral hemisphere, and sensory input from the left hand goes to the right cerebral hemisphere. The exception is the head region where left side of head is controlled by the left cerebral hemisphere and right side is controlled by the right cerebral hemisphere. The two hemispheres, are not symmetrical with respect to all functions. Language abilities, for example, reside predominantly in the left hemisphere, as we will see below.

If we compare the ratio of brain size to body size for all mammals, humans and dolphins top the list. Humans, however, have the largest amount of association cortex relative to body mass of any mammal (although no measure of size of any area of the human brain correlates with human intelligence).

46.5 The Human Cerebrum (A) Each cerebral hemisphere is divided into four lobes. (B) Different functions are localized in particular areas of the cerebral lobes.

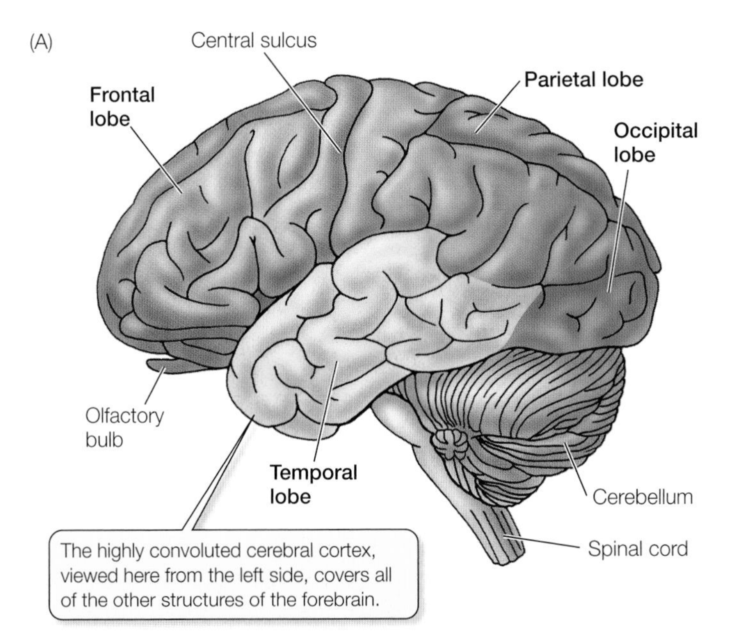

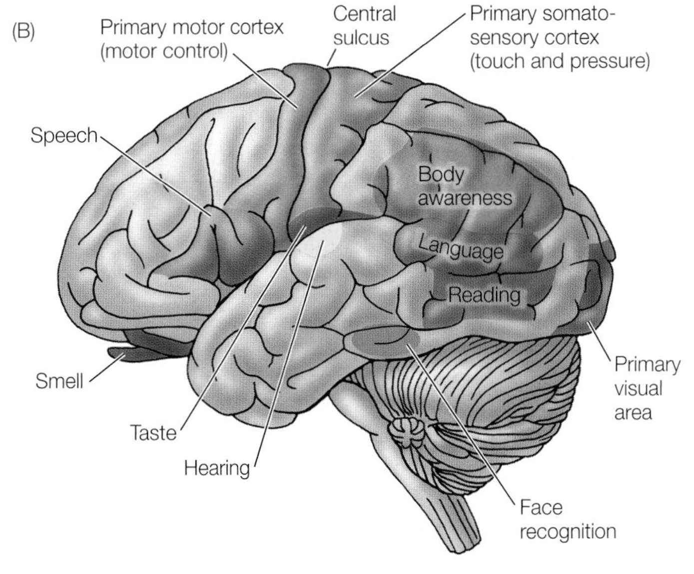

Different regions of the cerebral cortex have specific functions (**Figure 46.5B**). Some of those functions are easily defined, such as receiving and processing sensory information, but most of the cortex is involved in higher-order information processing that is less easy to define. These latter areas are given the general name of **association cortex**, so named because they integrate, or *associate*, information from different sensory modalities and from memory.

To understand the cerebral cortex, it helps to have an anatomical road map. As viewed from the left side, a left cerebral hemisphere looks like a boxing glove for the right hand with the fingers pointing forward, the thumb pointing out, and the wrist at the rear (see Figure 46.5A). The "thumb" area is the **temporal lobe**, the fingers the **frontal lobe**, the back of the hand the **parietal lobe**, and the wrist the **occipital lobe**. A mirror image of this arrangement characterizes the right cerebral hemisphere. Let's look at each lobe of the cerebral cortex separately.

As we explore the functions of the regions of the cerebral cortex and other parts of the brain, you will note frequent mention of persons with damage to their brains. Until recently, the study of such individuals has been the main source of functional information about the human brain, but new imaging technologies such as PET (positron emission tomography) and MRI (magnetic resonance imaging) are providing a wealth of new information and opportunities to study the functioning of the human brain in real time.

THE TEMPORAL LOBE The upper region of the temporal lobe receives and processes auditory information. The association areas of the temporal lobe are involved in the recognition, identification, and naming of objects. Damage to the temporal lobe results in disorders called *agnosias*, in which the individual is aware of a stimulus but cannot identify it.

Damage to a certain area of the temporal lobe results in the inability to recognize faces. Even old acquaintances cannot be identified by facial features, although they may be identified by other attributes such as voice, body features, and characteristic style of walking. Using monkeys, it has been possible to record the activity of neurons in this region that respond selectively to faces in general (**Figure 46.6**). These neurons do not respond to other stimuli in the visual field, and their responsiveness decreases if some of the features of the face are missing or appear in inappropriate locations. Damage to other association areas of the temporal lobe causes deficits in understanding spoken language, although speaking, reading, and writing abilities may be intact.

THE FRONTAL LOBE The frontal and parietal lobes are separated by a deep valley called the *central sulcus*. A strip of the frontal lobe cortex just in front of the central sulcus is called the **primary motor cortex** (see Figure 46.5B). The neurons in this region control muscles in specific parts of the body. The parts of the body have been mapped onto the primary motor cortex largely during neurosurgical procedures. As part of these procedures, electrodes were used to stimulate small areas of cortex. In the area just anterior to the central sulcus, stimulation causes specific muscles to contract. Parts of the body with fine motor control, such as the face and hands, have disproportionate representation (**Figure 46.7**). Stimulation of neurons in the primary motor cortex causes twitches of muscles, not coordinated movements.

The association functions of the frontal lobe are diverse. They are best described as having to do with planning, and they con-

46.6 Neurons in One Region of the Temporal Lobe Respond to Faces The traces represent the firing rate of a neuron in the temporal lobe of a monkey in response to the pictures shown below them.

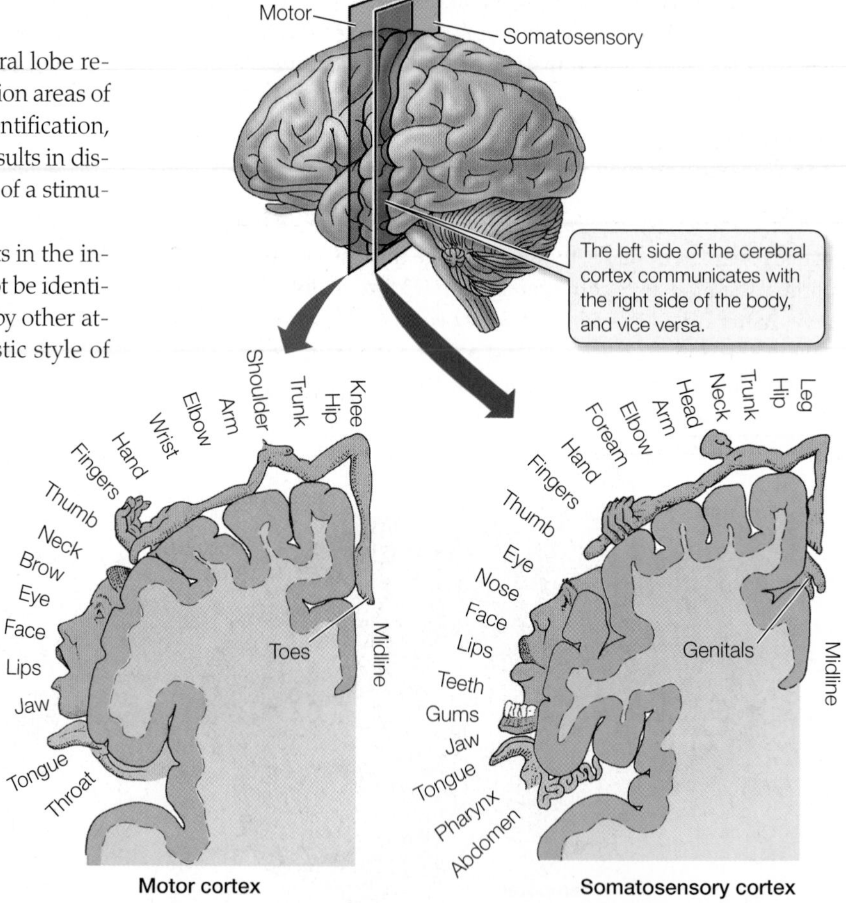

46.7 The Body Is Represented in the Primary Motor Cortex and the Primary Somatosensory Cortex Cross sections through the primary motor and primary somatosensory cortexes can be represented as maps of the human body. Body parts are shown in proportion to the brain area devoted to them.

tribute very significantly to personality. People with frontal lobe damage have drastic alterations of personality because they are unable to perceive themselves in the context of the world around them and cannot plan for future events. A dramatic case of frontal lobe damage is the story of Phineas Gage, who was an industrious, responsible, considerate young railroad construction foreman in 1848. Then a blasting accident shot a meter-long, 3-cm-wide iron tamping rod through his brain. The tamping iron entered Gage's head below his left eye, passed through his frontal lobe, and exited the top of his head (**Figure 46.8**).

Remarkably, Gage survived this terrible accident, but he was a completely different person. He was quarrelsome, bad-tempered, lazy, and irresponsible. He was impatient and obstinate, and he used profane language, which he had never done before. He lost his railroad job and spent the rest of his days as a drifter, earning money by telling his story and exhibiting his scars (and the tamping iron). He died of a seizure disorder in 1860, at the age of 38. If you are in Boston, Massachusetts, you can pay him a visit—his skull, death mask, and the tamping iron are on display in the Warren Anatomical Museum of Harvard Medical School.

THE PARIETAL LOBE The strip of parietal lobe cortex just behind the central sulcus is the **primary somatosensory cortex** (see Figure 46.5B). This area receives touch and pressure information relayed from the body through the thalamus.

The whole body surface can be mapped onto the primary somatosensory cortex (see Figure 46.7). Areas of the body that have a high density of tactile mechanoreceptors and are capable of making fine discriminations in touch (such as the lips and the fingers) have disproportionately large representation. If a very small area of the primary somatosensory cortex is stimulated electrically, the subject reports feeling specific sensations, such as touch, in a localized part of the body.

A major association function of the parietal lobe is attending to complex stimuli. Damage to the right parietal lobe causes a condition called *contralateral neglect syndrome*, in which the individual tends to ignore stimuli from the left side of the body or the left visual field. Such individuals have difficulty performing complex tasks, such as dressing the left side of the body; an afflicted man may not be able to shave the left side of his face. When asked to copy simple drawings, a person who exhibits this syndrome can do well with the right side of the drawing but not the left (**Figure 46.9**). The parietal cortex is not symmetrical with respect to its role in attention, however. Damage to the left parietal cortex does not cause the same degree of neglect of the right side of the body. We will see similar asymmetries in cortical function when we discuss language.

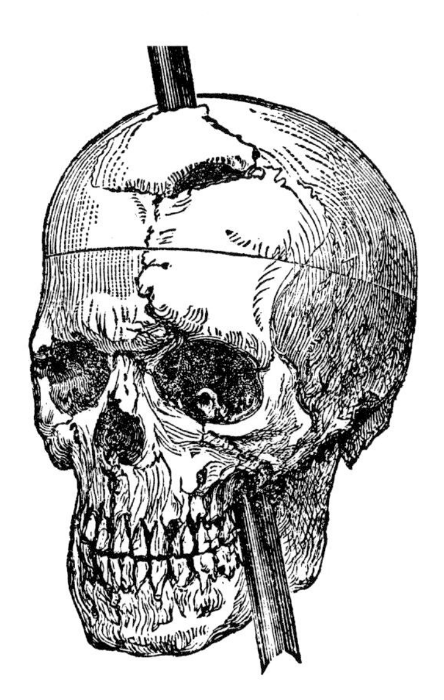

46.8 A Mind-Altering Experience In a nineteenth-century railroad construction accident, an explosion blew a tamping iron through the brain of Phineas Gage. Unbelievably, Gage survived, but his personality was radically changed.

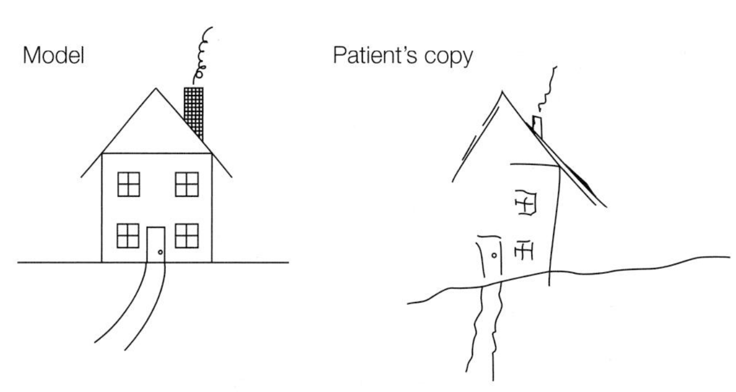

46.9 Contralateral Neglect Syndrome A person with damage to the right parietal association cortex will neglect the left side of a drawing when asked to copy a model.

THE OCCIPITAL LOBE The occipital lobe receives and processes visual information; we'll learn more about that process later in this chapter. The association areas of the occipital cortex are essential for making sense of the visual world and translating visual experience into language. Some deficits resulting from damage to these areas are specific. In one case, a woman with limited damage was unable to see motion. Her vision was intact, but she could see a waterfall only as a still image, and a car approaching only as a series of scenes of a stationary object at different distances.

46.1 RECAP

The central nervous system communicates with the rest of the body through the peripheral nervous system. We are conscious of some sensory information coming into the CNS, but are not conscious of other afferent information used in physiological regulation. Different regions of the brain have specific functions.

- Can you relate the major functional divisions of the nervous system to their origins in the embryonic neural tube? See p. 986–987 and Figure 46.2
- Can you trace the information flow that is involved in a spinal reflex? See p. 987–988 and Figure 46.3
- Can you describe the spatial relations and the functions of the major divisions of the telencephalon? See p. 989–991 and Figure 46.5

Now that we have knowledge of the structure and functions of different regions of the nervous system, we can explore some examples of how information is processed in the neuronal circuitry in some specific regions.

46.2 How Is Information Processed by Neuronal Networks?

We have seen that specific functions are localized in specific parts of the CNS. Therefore, those functions must depend on the neuronal circuits, or networks, in those CNS structures. A major focus of modern neuroscience is to understand how the various functions of the nervous system, ranging from simple reflexes to complex learning and memory, are accomplished by the interactions of neurons in circuits. We now consider two examples of how neuronal networks process information: the autonomic nervous system, which constitutes the efferent pathways of autonomic networks, and the visual system, which processes the information coming from the eyes.

The autonomic nervous system controls involuntary physiological functions

The **autonomic nervous system**, or **ANS**, comprises the output pathways of the CNS that control involuntary functions. The ANS has two divisions, **sympathetic** and **parasympathetic**, that work in opposition to each other in their effects on most organs: one division will cause an increase in an activity and the other a decrease. The sympathetic and parasympathetic divisions of the ANS are easily distinguished by their anatomy, their neurotransmitters, and their actions (**Figure 46.10**).

The best known functions of the ANS are those of the sympathetic division that produce the "fight-or-flight" response: increasing heart rate, blood pressure, and cardiac output and preparing the body for emergencies (see Figure 41.4). In contrast, the parasympathetic division slows the heart and lowers blood pressure, and its actions have been characterized as "rest and digest." It is tempting to think of the sympathetic division as the one that speeds things up and the parasympathetic division as the one that slows things down, but it is not that simple. For example, the sympathetic division slows the digestive system, and the parasympathetic division accelerates it.

Whether sympathetic or parasympathetic, every autonomic efferent pathway begins with a *cholinergic neuron* (one that uses acetylcholine as its neurotransmitter) that has its cell body in the brain stem or spinal cord. These cells are called *preganglionic neurons* because the second neuron in the pathway with which they synapse resides in a collection of neurons outside of the CNS called a *ganglion*. The second neuron is called a *postganglionic neuron* because its axon extends out from the ganglion. The axon of the postganglionic neuron synapses with cells in the target organs.

The postganglionic neurons of the sympathetic division are called *noradrenergic* because they use norepinephrine (also known as noradrenaline) as their neurotransmitter. In contrast, the postganglionic neurons of the parasympathetic division are mostly cholinergic. In organs that receive both sympathetic and parasympathetic input, the target cells respond in opposite ways to norepinephrine and to acetylcholine. A region of the heart called the *pacemaker*, which generates the heartbeat, is an example. Stimulating the sympathetic nerve to the heart or dripping norepinephrine onto the pacemaker region depolarizes the pacemaker cells, increases their firing rate, and causes the heart to beat faster. Stimulating the parasympathetic nerve to the heart or dripping acetylcholine onto the pacemaker region hyperpolarizes the pacemaker cells, decreases their firing rate, and causes the heart to beat more slowly. In contrast, in the digestive tract, norepinephrine hyperpolarizes muscle cells, which slows digestion, and acetylcholine depolarizes muscle cells, which accelerates digestion.

The sympathetic and parasympathetic divisions of the ANS can also be distinguished by anatomy (see Figure 46.10). The preganglionic neurons of the parasympathetic division come from the brain stem and the last segment of the spinal cord (the *sacral* region). The preganglionic neurons of the sympathetic division come from the upper regions of the spinal cord below the neck (the *thoracic* and *lumbar* regions). Most of the ganglia of the sympathetic division are lined up in two chains, one on either side of the spinal cord. The parasympathetic ganglia are close to and sometimes located in the walls of the target organs.

A specialization of the sympathetic division is its innervation of the adrenal gland, which is critical for the "fight or flight" response. The preganglionic sympathetic neuron sends its axon all the way out to the adrenal gland. The medulla (core) of the gland is composed of hormone-secreting cells that are really modified postganglionic sympathetic neurons that have lost their axons, and they secrete their "neurotransmitters" (epinephrine and norepinephrine) into the extracellular fluid (see Section 41.1).

The ANS is an important link between the CNS and many physiological functions. Its control of diverse organs and tissues is crucial to homeostasis. Despite its complexity, work by neurobiologists and physiologists over many decades has made it possible to understand its functions in terms of neuronal properties and circuits. For example, information from pressure receptors in the blood vessels is transmitted to the CNS, where it produces autonomic signals that control the rate of the heartbeat (see Section 49.5).

The first isolation of acetylcholine receptors was possible because of their extremely high concentration in the electric organs of some fish, such as the torpedo ray. Acetylcholine receptors are ion channels, and when many open in a small volume of muscle tissue, they can generate an electric pulse powerful enough to stun a human.

Patterns of light falling on the retina are integrated by the visual cortex

In Section 45.4 we learned how light falling on the retina produces signals that are transmitted through the cellular circuits of the retina resulting in action potentials in the optic nerve. How do these action potentials result in the reconstruction of the visual world in the brain?

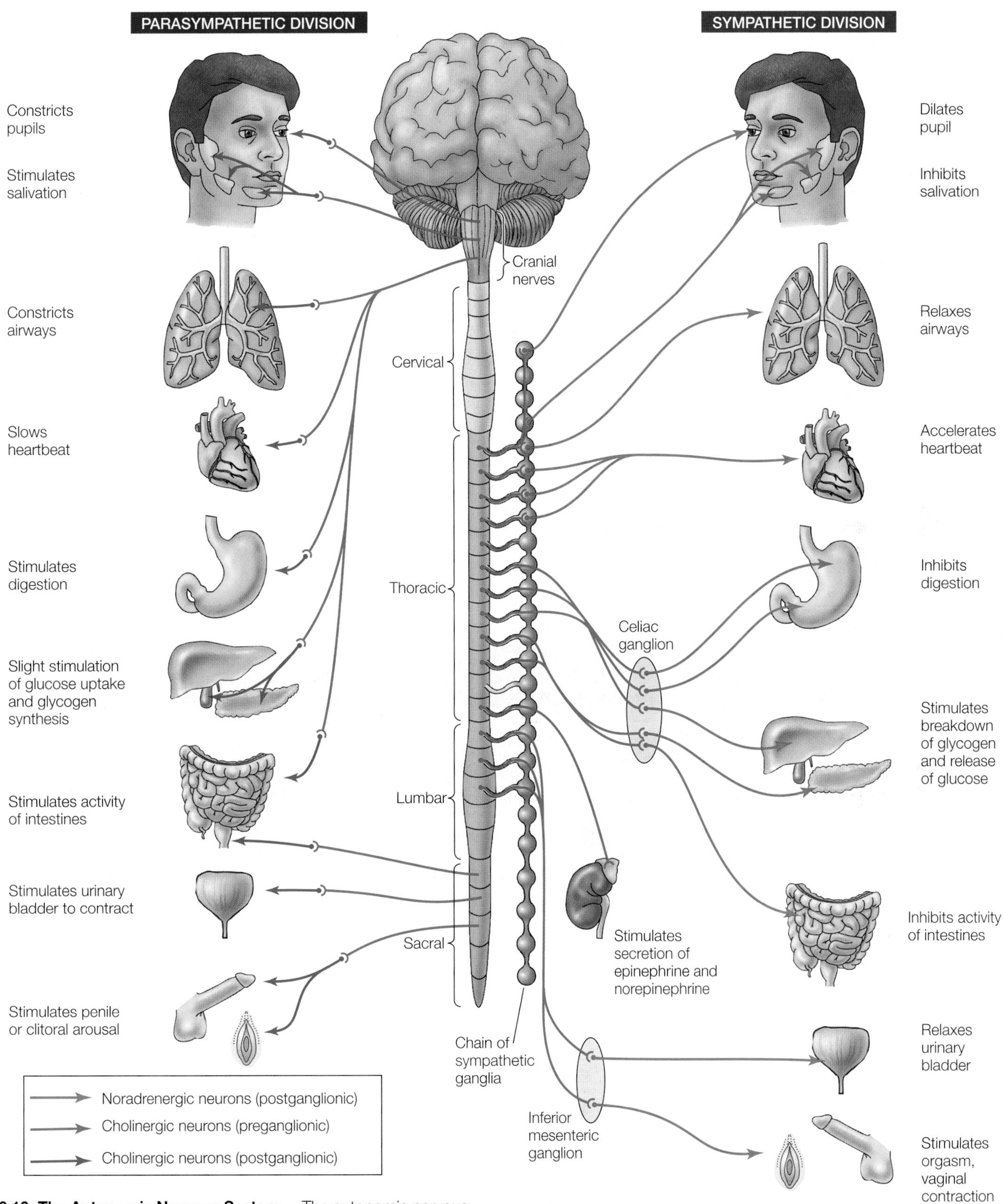

46.10 The Autonomic Nervous System The autonomic nervous system is divided into the sympathetic and parasympathetic divisions. The two divisions which work in opposition to each other in their effects on most organs, with one causing an increase and the other a decrease in activity.

HYPOTHESIS: Retinal ganglion cells are excited or inhibited by light and dark stimuli falling on local areas of the retina.

METHOD

Place electrodes in the optic nerve to record from the axons of ganglion cells while stimulating the retina with different combinations of light and dark stimuli. The stimuli are moved around the retina to find the area of sensitivity for a particular ganglion cell.

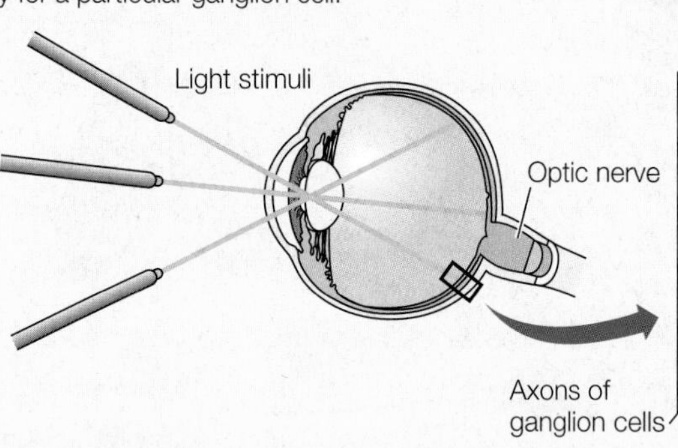

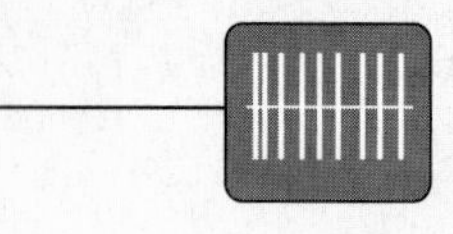

RESULTS

Some ganglion cells are maximally stimulated by light falling on the center of their receptive fields. Others are maximally stimulated by light falling on the surround of their receptive fields.

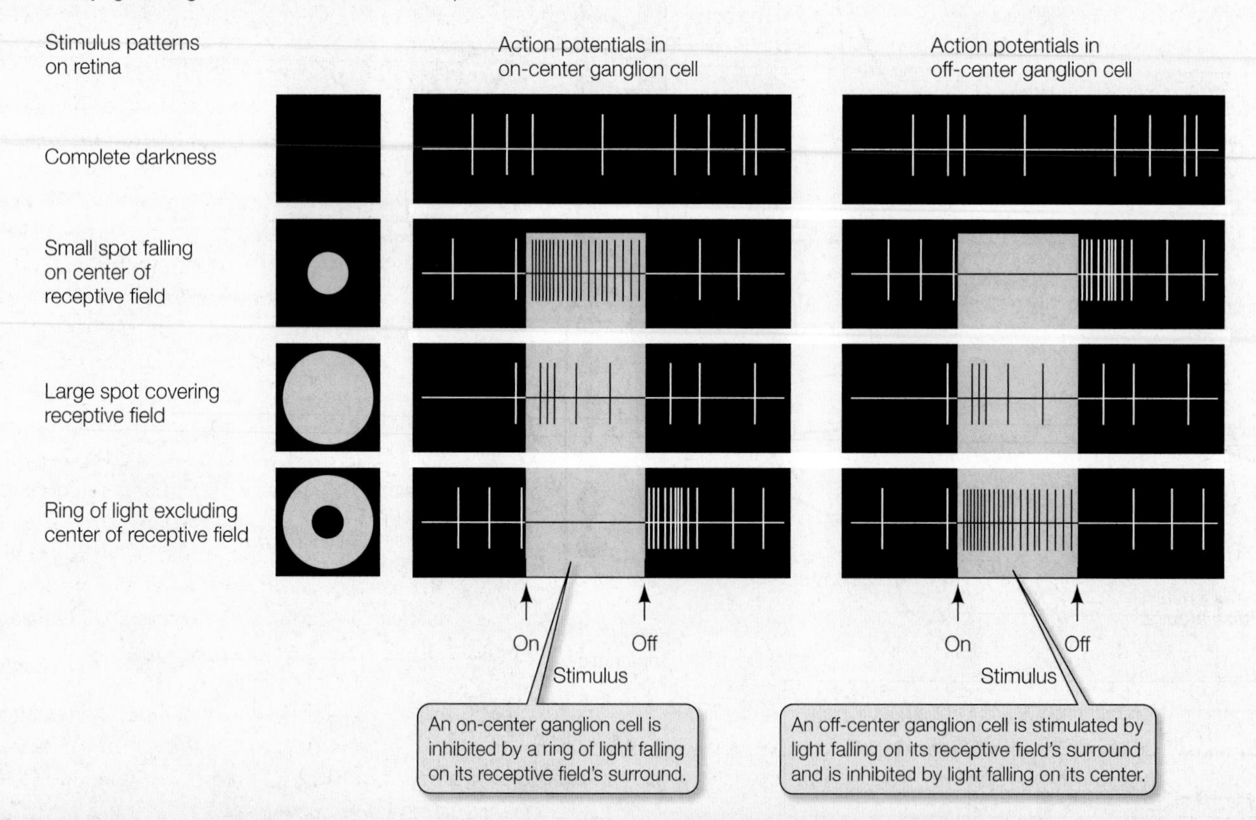

CONCLUSION: Ganglion cells use a center-surround dichotomy to encode patterns of contrast between light and dark.

46.11 What Does the Eye Tell the Brain? When the retina is stimulated with dots and rings of light, individual ganglion cells show different responses.

RETINAL RECEPTIVE FIELDS One aspect of the processing of visual information through the retina is the *convergence of information*. Each human retina contains over 100 million photoreceptors, but only about 1 million *ganglion cells*. It is the axons of the ganglion cells that communicate information from the eyes to the brain. How is the information from so many photoreceptors integrated by the many fewer ganglion cells?

This question was addressed in classic experiments by Stephen Kuffler at Johns Hopkins University in 1953, in which he used electrodes to record the activity in the axons of single ganglion cells while the retina of a cat was stimulated with spots of light (**Figure 46.11**). These experiments were the starting point for our understanding of how the brain assembles information from single cells to create visual images—in other words, of how the brain sees.

Kuffler's studies revealed that each ganglion cell has a well-defined **receptive field** composed of a group of photoreceptor cells that receive light from a small area of the whole visual field. Stimulating these photoreceptors with light activates the ganglion cell, which sends action potentials to the thalamus and on to the visual cortex (the area of the occipital lobe where visual information is processed; see Figure 46.5B). Information from many photoreceptors is therefore communicated to the brain as a single message. However, individual photoreceptors may contribute to the receptive fields of multiple ganglion cells.

The receptive fields of most ganglion cells are circular, but whether a spot of light falling on a receptive field excites or inhibits that ganglion cell depends both on the nature of the receptive field and on where the spot of light falls on the receptive field. Receptive fields have two concentric regions—a *center* and a *surround*—and the field can be either *on-center* or *off-center*. Light falling on an on-center receptive field excites the ganglion cell, and light falling on an off-center receptive field inhibits the ganglion cell. The surround area has the opposite effect: the surround for an on-center receptive field inhibits the ganglion cell and the surround for an off-center field is excitatory. Thus the activity of the ganglion cell reflects how much of the light stimulus is on the center and how much is on the surround of its receptive field (see Figure 46.11).

Center effects are always stronger than surround effects, however. Thus a small dot of light directly on the center of a receptive field has the maximal effect, and a larger light stimulus illuminating the center and parts of the surround has less of an effect. A uniform patch of light falling equally on center and surround has very little effect on the firing rate of the ganglion cell for that receptive field.

How are cells in the retina connected to each other to create receptive fields? The photoreceptors in a receptive field's center are connected to the associated ganglion cell by *bipolar cells*. The photoreceptors in the surround modify the communication between the center photoreceptors and their bipolar cells through the lateral connections of *horizontal cells* (see Figure 45.21). Thus, the receptive field of a ganglion cell results from a pattern of synapses between photoreceptors, horizontal cells, amacrine, and bipolar cells.

Now things can get a bit confusing. Remember that vertebrate photoreceptors are depolarized in the dark and hyperpolarized by light. Thus these cells are releasing neurotransmitter in the dark, whereas light reduces their rate of neurotransmitter release (see Section 45.4). The neurotransmitter of photoreceptors is glutamate which is usually an excitatory neurotransmitter (see Section 44.3). For bipolar cells, however, glutamate can be excitatory or inhibitory depending on the receptors they have. Thus, for bipolar cells of on-center receptive fields, glutamate is inhibitory. In the dark, when the central photoreceptors are releasing the most glutamate, the bipolar cells of on-center receptive fields are hyperpolarized. When light shines on those photoreceptors, they decrease their release of glutamate causing their bipolar cells to depolarize. As a result, these bipolar cells release more neurotransmitter onto their ganglion cells increasing their firing rates. The opposite is true for the bipolar cells of off-center receptive fields as they have glutamate receptors that are excitatory. The general lesson to learn from this seemingly confusing chain of events is that inhibition can be as important as excitation in neural circuits, and sometimes excitation results from inhibition of an inhibitory synapse.

In summary, the neuronal circuitry of the retina results in the generation of signals in the axons of the optic nerve that communicate simple information about the patterns of light and dark falling on different parts of the retina.

RECEPTIVE FIELDS OF CELLS IN THE VISUAL CORTEX We learned how photoreceptors in the retina transduce light into neural signals and how the neural circuits in the retina organize those signals into information about patterns of light and dark that is transmitted to the brain. But once the action potentials in the optic nerve reach their destinations, how does the brain integrate them to construct visual image of the outside world?

The axons of the optic nerves terminate in a region of the thalamus which is a relay station that integrates information from the right and left eyes. From the thalamus the information encoded in the activity of axons in the optic nerves is relayed to the visual cortex in the occipital lobes at the back of the brain. In the 1960s, David Hubel and Torsten Wiesel of Harvard University studied the activity of neurons in the visual cortex by shining spots and bars of light on retinas while recording the activities of single cells in the cortex. They found that neurons in the visual cortex, like retinal ganglion cells, have receptive fields. For their pioneering work in visual neurophysiology, Hubel and Wiesel received the Nobel prize in 1981.

Receptive fields of neurons in the visual cortex, however, differ from the circular receptive fields of retinal ganglion cells. Cortical neurons labeled *simple cells* are maximally stimulated by bars of light with a particular orientation and falling on a small region of the retina. These simple cells probably receive input from several ganglion cells whose circular retinal receptive fields are lined up in a row, creating a bar of sensitivity.

Another class of cortical cells is *complex cells*. These cells responded maximally to bars of light with particular orientations, but the bar of light could fall anywhere on a large area of retina. Thus the receptive field of a complex cell appears to be due to that cell receiving inputs from a number of simple cells that share a certain stimulus orientation but have receptive fields in different locations on the retina (**Figure 46.12**). Some complex cells respond most strongly when the bar of light moves in a particular direction.

The concept that emerges from these experiments is that the brain assembles a mental image of the visual world by analyzing edges in patterns of light falling on the retina. This analysis is conducted in a massively parallel fashion. Each retina sends a million axons to the

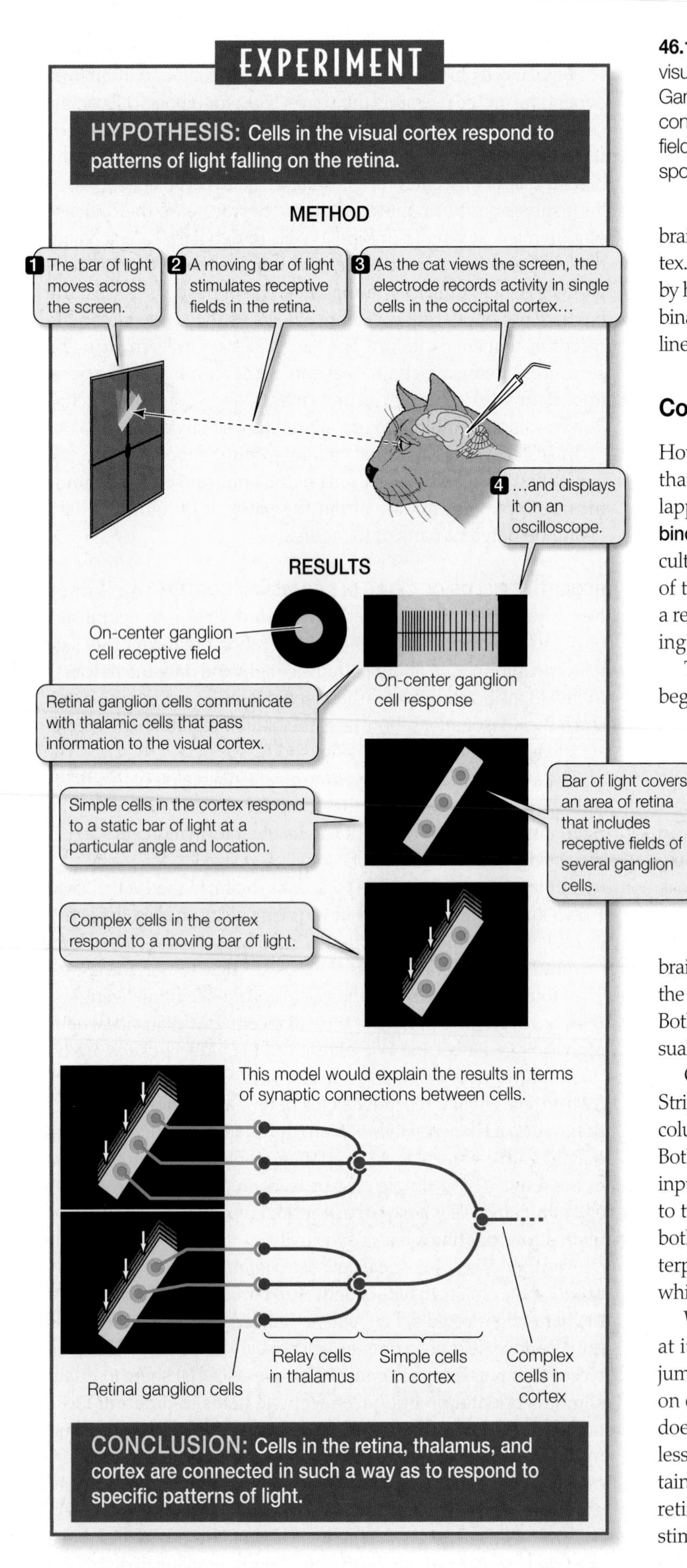

46.12 Receptive Fields of Cells in the Visual Cortex Cells in the visual cortex respond to specific patterns of light falling on the retina. Ganglion cells that transmit information about circular receptive fields converge on the cortex's simple cells, which have linear receptive fields. Simple cells transmit information to complex cells, which respond to linear stimuli falling on different areas of the retina.

brain, but there are *hundreds of millions* of neurons in the visual cortex. The action potentials from one retinal ganglion cell are received by hundreds of cortical neurons, each responsive to a different combination of orientation, position, color, and movement of contrasting lines in the patterns of light and dark falling on the retina.

Cortical cells receive input from both eyes

How do we see objects in three dimensions? The quick answer is that a person's two eyes, located at the front of the head, see overlapping, yet slightly different, visual fields—that is, humans have **binocular vision**. A person who is blind in one eye has great difficulty discriminating distances. Animals whose eyes are on the sides of the head have minimal overlap in their fields of vision and, as a result, poor depth vision; however, they can see predators creeping up from all sides.

The story of how the brain integrates information from two eyes begins with the paths of the optic nerves. From the underside of the brain, the optic nerves from the two eyes appear to join together just under the hypothalamus and then separate again (**Figure 46.13**). The place where they join is called the **optic chiasm**. Axons from the half of each retina closest to your nose cross in the optic chiasm and go to the opposite side of your brain. The axons from the other half of each retina go to the same side of the brain. This means that visual information from the left visual field (everything left of straight ahead) goes to the right side of the brain, as shown in red in Figure 46.13, while visual information from the right visual field goes to the left side of the brain (shown in green). Both eyes transmit information about a specific spot in your right visual field, for example, to the same place in the left visual cortex.

Cells in the visual cortex are organized in stripes and columns. Stripes refer to the organization across the surface of the cortex, and columns refer to the organization through the depth of the cortex. Both stripes and columns alternate according to the source of their input: left eye, right eye, left eye, right eye, and so on. Cells closest to the border between two stripes or columns receive input from both eyes and are therefore called *binocular cells*. Binocular cells interpret distance by measuring the *disparity* between the points at which the same stimulus falls on the two retinas.

What is disparity? Hold your finger out in front of you and look at it, closing one eye and then the other. Your finger appears to jump back and forth, because its image falls on a different position on each retina. Repeat the exercise with an object at a distance. It doesn't appear to jump back and forth as much, because there is less disparity in the positions of the image on the two retinas. Certain binocular cells respond optimally to a stimulus falling on both retinas with a particular disparity. Which set of binocular cells is stimulated depends on how far away the stimulus is.

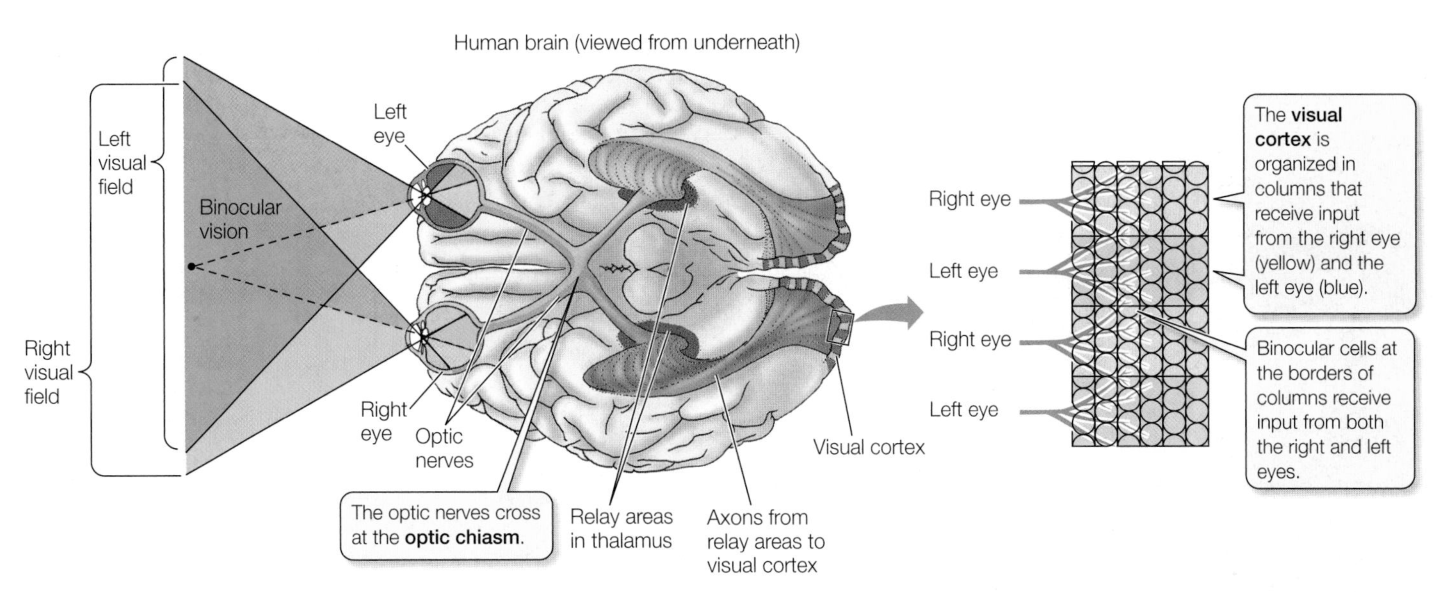

46.13 The Anatomy of Binocular Vision Each eye transmits information to both sides of the brain; however, the right side of the brain processes all information from the left visual field, and the left side of the brain processes all information from the right visual field. The visual cortex sorts visual field information according to whether it comes from the right eye or the left eye.

When we look at something, we can detect its shape, color, depth, and movement. Where does all this information come together? Is there a single cell that fires only when a red sports car drives by? The answer is no. Specific visual experience comes from simultaneous activity in a large collection of cells. In addition, most visual experiences are enhanced by information from the other senses and from memory, which helps explain why about 75 percent of the cerebral cortex is association cortex.

46.2 RECAP

Information in the nervous system is processed by cellular interactions in neuronal networks. The opposing actions of the sympathetic and parasympathetic divisions of the ANS can be understood in terms of neural pathways consisting of just two neurons. Vision involves a more complex interaction of neurons to accomplish the processing of important sensory information.

- Can you describe the anatomical and functional differences between the sympathetic and parasympathetic divisions? See p. 992 and Figure 46.10
- Do you understand the cellular basis for the receptive fields of retinal ganglion cells? See p. 995 and Figure 46.11
- How do cells in the visual cortex get information about the distance of an object? See p. 996 and Figure 46.13

By studying the neural circuitry of the visual system and the autonomic nervous system, you have gained some understanding of how information reaches the central nervous system and how the CNS controls various functions of the body. But what about the higher functions of the mammalian CNS—the complex functions between input and output, such as language, learning, memory, and dreams?

46.3 Can Higher Functions Be Understood in Cellular Terms?

The higher brain functions discussed in the remaining pages of this chapter are undeniably complex. Nevertheless, neuroscientists, using a wide range of techniques, are making considerable progress in understanding some of the cellular and molecular mechanisms involved in those processes. The following discussion presents several complex aspects of brain and behavior that present challenges to neuroscientists: sleep and dreaming, learning and memory, language use, and consciousness.

Sleep and dreaming are reflected in electrical patterns in the cerebral cortex

A dominant feature of behavior is the daily cycle of sleep and waking. All birds and mammals, probably all other vertebrates, and also many invertebrates, sleep. We humans spend one-third of our lives sleeping, yet we do not know why or how. We do know, however, that we need to sleep. Loss of sleep impairs alertness and performance. Many people in our society—certainly most college students—are chronically sleep-deprived. Every day, accidents and serious mistakes that endanger lives can be attributed to impaired alertness due to lack of sleep. Insomnia (difficulty in falling or staying asleep) is one of the most common medical complaints.

(A)

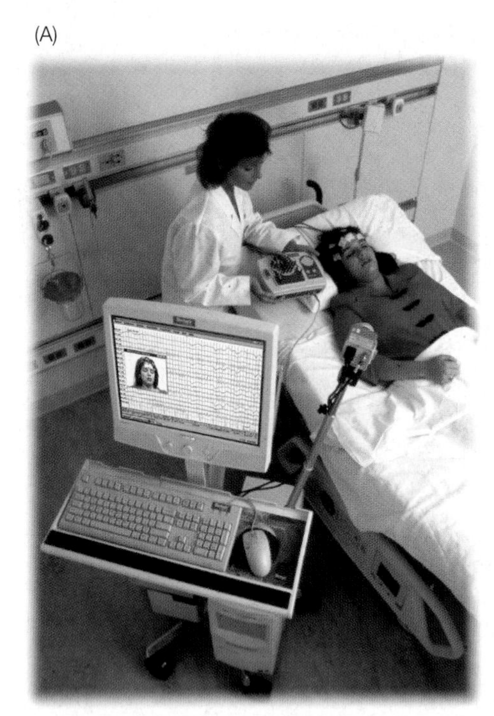

46.14 Patterns of Electrical Activity in the Cerebral Cortex Characterize Stages of Sleep (A) Electrical activity in the cerebral cortex is detected by electrodes placed on the scalp and recorded as changes in voltage between the electrodes through time. (B) The resulting record is an electroencephalogram (EEG). (C) Humans cycle through different stages of sleep throughout the night.

(B)

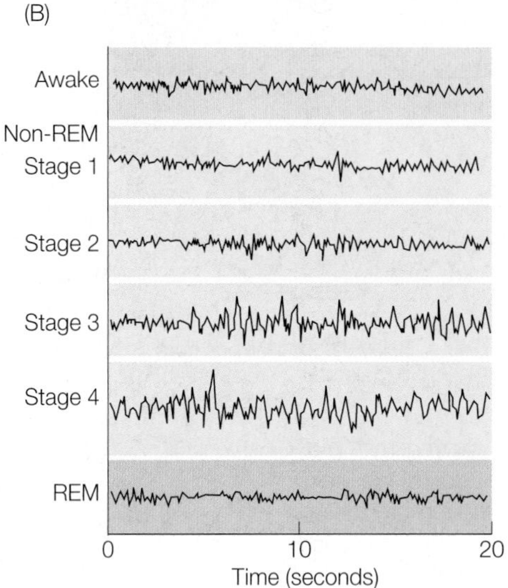

(C)

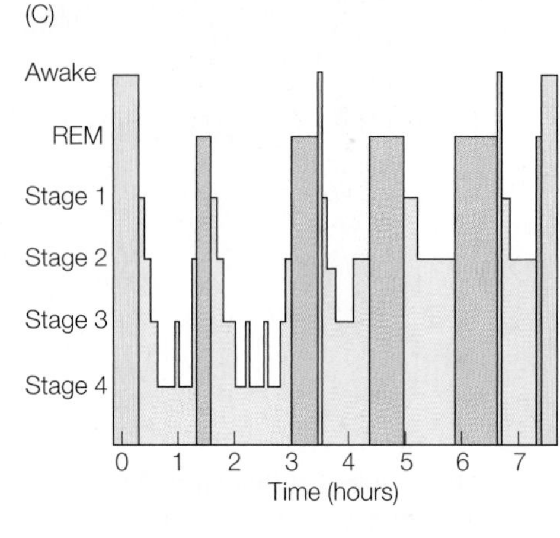

THE ELECTROENCEPHALOGRAM A common tool of sleep researchers is the **electroencephalogram**, or **EEG** (**Figure 46.14A**). Rather than recording the activity of single neurons, the EEG characterizes activity in huge numbers of neurons. EEG electrodes are much larger than the very fine electrodes used to detect single cell activity. EEG electrodes are placed at different locations on the scalp, and changes in the electric potential differences between electrodes are recorded over time. These electric potential differences reflect the electrical activity of the neurons in the brain regions under the electrodes, primarily regions of the cerebral cortex. Usually, the electrical activity of one or more skeletal muscles is also recorded on the chart; this record is called an *electromyogram (EMG)*. Movements of the eyes are recorded as an *electrooculogram (EOG)*.

EEG, EMG, and EOG patterns reveal the transition from being awake to being asleep. They also reveal that there are different states of sleep. In mammals other than humans, two major sleep states are easily distinguished: They are **slow-wave sleep** and **rapid-eye-movement (REM) sleep**. In humans, sleep states are characterized as **non-REM sleep** and **REM sleep**. Of the four stages of human non-REM sleep, only the two deepest stages are considered true slow-wave sleep.

When you fall asleep at night, the first sleep state entered is non-REM sleep, which progresses from stage 1 to stage 4 (**Figure 46.14B**). Stages 3 and 4 are deep, restorative, slow-wave sleep. This first full episode of non-REM sleep is followed by an episode of REM sleep. Throughout the night, you experience four or five cycles of non-REM and REM sleep (**Figure 46.14C**). About 80 percent of your sleep is non-REM sleep.

Vivid dreams and nightmares occur during the 20 percent of sleep that is REM sleep, which gets its name from the jerky movements of the eyeballs that occur during this state. The most remarkable feature of REM sleep is that inhibitory commands from the brain almost completely paralyze the skeletal muscles. Occasional muscle twitches break through the paralysis, as can be seen in a dog that appears to be trying to run in its sleep. The function of muscle paralysis during REM sleep may be to prevent the acting out of dreams. Sleepwalking occurs during non-REM sleep.

CELLULAR CHANGES DURING SLEEP Striking neurophysiological differences distinguish non-REM from REM sleep. Non-REM sleep is characterized by a decrease in the responsiveness of neurons in the thalamus and cerebral cortex. Remember that neurons have a negative resting membrane potential and a threshold for firing action potentials (see Section 44.2). Usually the resting potential is near or slightly below the threshold potential, so the neuron is firing at a low rate. When synaptic input causes the membrane potential to become less negative (depolarized), the cell can exceed threshold and fire action potentials at a higher rate.

During waking, several nuclei in the brain stem reticular formation are continuously active. Axons from neurons in these nuclei extend to the thalamus and throughout the cerebral cortex where they release depolarizing neurotransmitters (acetylcholine, norepinephrine, and serotonin). These broadly distributed neurotransmitters keep the resting potential of the neurons of the thalamus and cortex close to threshold and sensitive to synaptic inputs, thereby maintaining waking. As mentioned earlier, brain damage that cuts off the influence of these brainstem nuclei on the forebrain results in coma.

With the onset of sleep, activity in these brain stem nuclei decreases, and their axon terminals release less neurotransmitter. With the withdrawal of the depolarizing neurotransmitters, the resting potentials of the cells of the thalamus and cortex become more negative (hyperpolarized), and the cells are less sensitive to excitatory synaptic input. Their processing of information is inhibited, and consciousness is lost.

An interesting neuronal event happens as a result of this hyperpolarization: The cells begin to fire action potentials in bursts. The synchronization of these bursts over broad areas of cerebral cortex results in the EEG slow-wave pattern that characterizes non-REM sleep. Studies of neurons of the thalamus and the cortex have shown that their hyperpolarization during non-REM sleep is due to increased opening of K^+ channels, and the bursting is due to Ca^{2+} channels whose inactivation gates close rapidly and require hyperpolarization to be reopened. We can therefore explain the EEG pattern of non-REM sleep in terms of the properties of neurons and ion channels.

At the transition from non-REM to REM sleep, dramatic changes occur. Some of the brain stem nuclei that were inactive during non-REM sleep become active again, causing a general depolarization of cortical neurons. Thus the synchronized bursts of firing cease, and the EEG resembles that of the waking brain. Because the resting potentials of the neurons return to near threshold levels, the cortex can process information, and vivid dreams can occur. However, during non-REM sleep the brain inhibits both afferent (sensory) and efferent (motor) pathways; therefore, the activity in the cortex is unconstrained by its usual sources of information, which is why events in dreams are often quite bizarre.

Knowing the cellular mechanisms of sleep has not led to an understanding of its function, and many questions remain. Why do we have two sleep states with very different neurophysiological characteristics? Why does non-REM sleep always occur first? Why do the two states oscillate on a regular basis through the rest period? We know sleep is essential for life, but we don't know why. One prominent hypothesis is that sleep is necessary for the maintenance and repair of neuronal connections and for the neuronal changes involved in learning and memory—and possibly forgetting. Evidence for such functions is still meager, however.

Some learning and memory can be localized to specific brain areas

Learning is the modification of behavior by experience. *Memory* is the ability of the nervous system to retain what is learned and what is experienced. Even very simple animals can learn and remember, but these two abilities are most highly developed in humans. Consider the amount of information associated with learning a language. The capacity of memory and the rate at which memories can be retrieved are remarkable features of the human nervous system.

LEARNING Learning that leads to long-term memory and modification of behavior must involve long-lasting synaptic changes. An experimental model of how long-term synaptic changes might arise is called **long-term potentiation**, or **LTP** (see Figure 44.18). LTP involves the high-frequency electrical stimulation of certain identifiable circuits that makes these circuits more sensitive to subsequent stimulation. In contrast, continuous, repetitive, low-level stimulation of these hippocampal circuits reduces their responsiveness, a phenomenon that has been called **long-term depression** (**LTD**). LTP and LTD may be fundamental cellular or molecular mechanisms involved in learning and memory.

Several kinds of learning exist. A form of learning that is widespread among animal species is **associative learning**, in which two unrelated stimuli become linked to the same response. The simplest example of associative learning is the **conditioned reflex**, discovered by the Russian physiologist Ivan Pavlov. Pavlov was studying the control of digestive functions in dogs and observed that a dog salivates at the sight or smell of food—a simple autonomic reflex. He discovered that if he rang a bell just before food was presented to the dog, after a few trials the dog would salivate at the sound of the bell, even if no food followed. The salivation reflex was conditioned to be associated with the sound of a bell, a stimulus that normally is unrelated to feeding and digestion.

This simple form of learning has been studied extensively in efforts to understand its underlying neuronal mechanisms. In a series of studies led by Richard Thompson, the eye-blink reflex of a rabbit in response to a puff of air directed at its eyes was conditioned to be associated with a tone stimulus. After conditioning, the rabbit blinked when it heard the tone. A small and specific area of the cerebellum was discovered to be necessary for this conditioned reflex. Thus, it was possible to localize one form of learning to an identifiable set of synapses in the mammalian brain.

MEMORY Attempts to treat human neurological diseases have led to the identification of areas of the brain involved in the formation and recall of memories. Epilepsy is a disorder characterized by uncontrollable increases in neuronal activity in specific parts of the brain. The resulting *seizures*, or "epileptic fits," can endanger the afflicted individual. In the past, serious cases of epilepsy were sometimes treated by destroying the part of the brain from which the surge of activity originated.

To find the right area, the surgery was done under local anesthesia, and different regions of the brain were electrically stimulated with electrodes while the patient reported on the resulting sensations. Stimulation of some regions of the association cortex elicited recall of vivid memories. Such observations were the first evidence that specific areas in the brain are associated with specific memories and that memory can be attributed to properties of neurons and networks of neurons. The destruction of a small area of the brain does not completely erase a memory, however, so it is postulated that memory is a function distributed over many brain regions and can be stimulated via many different routes.

You may recognize several forms of memory from your own experience. There is **immediate memory** for events that are happening now. Immediate memory is almost perfectly photographic, but it lasts only seconds. **Short-term memory** contains less information, but it lasts longer—on the order of 10 to 15 minutes. If you are introduced to a group of new people, you may remember most of their names for 5 or 10 minutes, but you will have forgotten them in an hour or so if you have not repeated them, written them down, or used them in a conversation. Repetition, use, or reinforcement by something that gets your attention (such as the title President) facilitates the transfer of short-term memory to **long-term memory**, which can last for days, months, years, or a lifetime.

Knowledge about neuronal mechanisms for the transfer of short-term memory to long-term memory has come from observations of persons who have lost parts of the limbic system, notably the hippocampus. A famous case is that of a man identified as Henry M., whose hippocampus on both sides of the brain was removed in an effort to control severe epilepsy. After that surgery,

H.M. was not able to transfer information to long-term memory. If someone was introduced to him, had a conversation with him, and then left the room for several minutes, when that person returned, H.M did not know him—it was as if the conversation had never taken place. H.M. retained memories of events that happened before his surgery, but could not remember postsurgery events for more than 10 or 15 minutes.

Memory of people, places, events, and things is called **declarative memory** because you can consciously recall and describe them. Another type of memory, called **procedural memory**, cannot be consciously recalled and described: It is the memory of how to perform a motor task. When you learn to ride a bicycle, ski, or use a computer keyboard, you form procedural memories. Although H.M. was incapable of forming declarative memories, he could form procedural memories. When taught a motor task day after day, he could not recall the lessons of the previous day, yet his performance steadily improved. Thus procedural learning and memory must involve mechanisms different from those used in declarative learning and memory.

Memories can have considerable emotional content. As mentioned earlier, the limbic system plays a major role in controlling emotions. The amygdala is necessary for the emotion of fear and the formation of fear memories. Patients with damage to the amygdala can form declarative memories but do not associate fear reactions with those memories. Normal subjects can be conditioned easily to show anticipatory fear. For example, subjects may be shown cards of different colors, one of which is accompanied by an unpleasant electrical shock, which causes the sympathetic nervous system response of increasing heart rate. Once trained, the heart rate of subjects will increase upon seeing the card even if the shock is not delivered. Patients with damage to the amygdala, however, show the sympathetic response to the shock, but not in response to seeing the appropriate card. They will report that they expect to receive a shock, but do not exhibit a fear response.

Language abilities are localized in the left cerebral hemisphere

No aspect of brain function is as integrally related to human consciousness and intellect as is language. Therefore, studies of the brain mechanisms that underlie the acquisition and use of language are extremely interesting to neuroscientists. A curious observation about language abilities is that they are usually located in only one cerebral hemisphere—which in 97 percent of people is the left hemisphere. This phenomenon is referred to as the *lateralization* of language functions.

Some of the most fascinating research on this subject was conducted by Roger Sperry and his colleagues at the California Institute of Technology; Sperry received a Nobel prize for this work in 1981. The two cerebral hemispheres are connected by a tract of white matter called the *corpus callosum*. In one severe form of epilepsy, bursts of action potentials travel from hemisphere to hemisphere across the corpus callosum. Cutting the tract eliminates the problem, and patients function nearly normally following the surgery. But the "split-brain" subjects of this surgery displayed interesting deficits in language ability. Lacking the connecting tissue between the two hemispheres, knowledge or experience of the right hemisphere could no longer be expressed in language—a function localized in the left hemisphere. Nor could language be used to communicate with the right hemisphere. A patient could respond to verbal instructions to use his right hand (controlled by the left cerebral cortex), but not his left hand (controlled by the right cerebral cortex).

The mechanisms of language in the left hemisphere have been the focus of much research. The experimental subjects are persons who have suffered damage to the left hemisphere and are left with one of many forms of **aphasia**, a deficit in the ability to use or understand words. These studies have identified several language areas in the left hemisphere.

Broca's area, located in the frontal lobe just in front of the primary motor cortex, is essential for speech. Damage to Broca's area results in halting, slow, poorly articulated speech or even complete loss of speech, but the patient can still read and understand language. In the temporal lobe, close to its border with the occipital lobe, is *Wernicke's area*, which is more involved with sensory than with motor aspects of language. Damage to Wernicke's area can cause a person to lose the ability to speak sensibly while retaining the abilities to form the sounds of normal speech and to imitate its cadence. Moreover, such a patient cannot understand spoken or written language. Near Wernicke's area is the *angular gyrus*, which is believed to be essential for integrating spoken and written language.

Normal language ability depends on the flow of information among various areas of the left cerebral cortex. Input from spoken language travels from the auditory cortex to Wernicke's area (**Fig-**

46.15 Language Areas of the Cortex Different regions of the left cerebral cortex participate in the processes of (A) repeating a word that is heard and (B) speaking a written word.

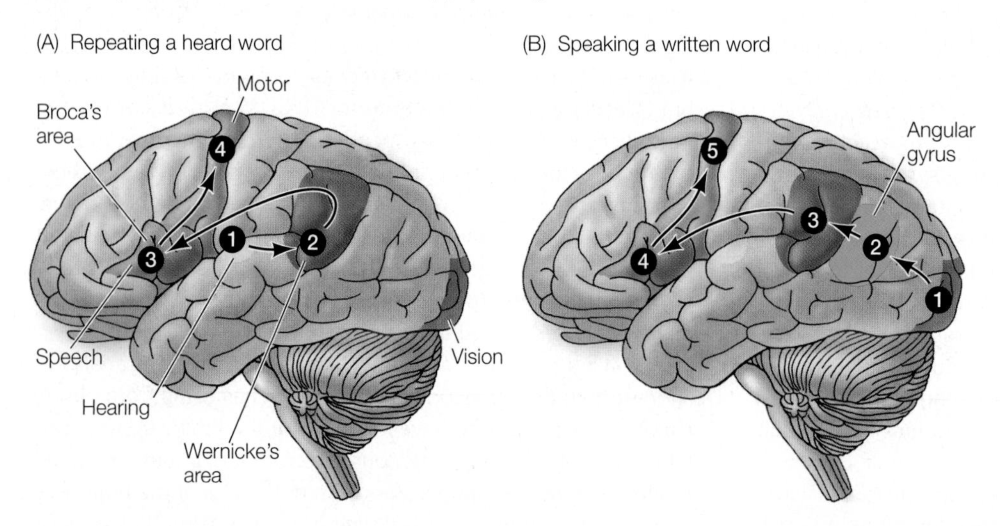

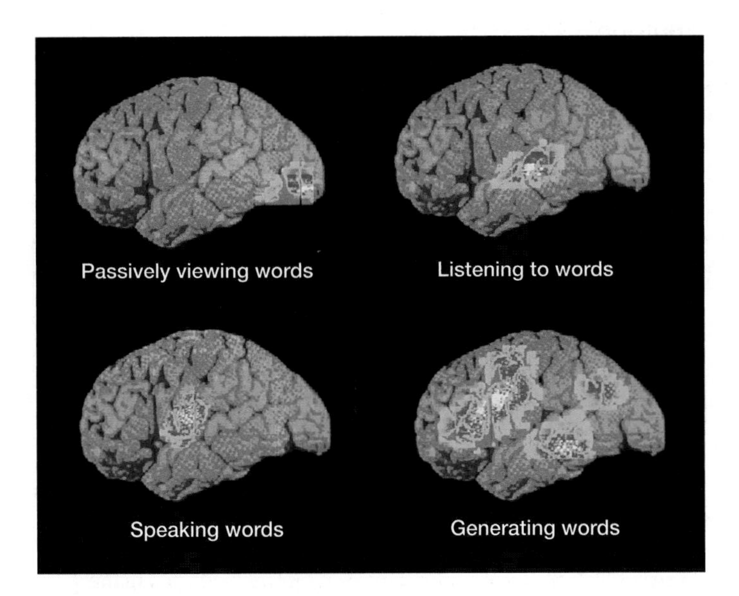

46.16 Imaging Techniques Reveal Active Parts of the Brain Positron emission tomography (PET) scanning reveals the brain regions that are activated by different aspects of language use. Radioactively labeled glucose is given to the subject. Brain areas take up radioactivity in proportion to their metabolic use of glucose. The PET scan visualizes levels of radioactivity in specific brain regions when a particular activity is performed. The red and white areas are the most active.

ure 46.15A). Input from written language, which is happening right now as you read this sentence, travels from the visual cortex to the angular gyrus to Wernicke's area (**Figure 46.15B**). Commands to speak are formulated in Wernicke's area and travel to Broca's area and from there to the primary motor cortex. Damage to any one of those areas or the pathways between them can result in aphasia.

Using modern methods of functional brain imaging, it is possible to see the metabolic activity in different brain areas when the brain is using language (**Figure 46.16**). As noted earlier, most knowledge of regional functions of the human brain have come from studies of patients who have sustained damage—usually injuries from war or accidents. Modern methods of imaging can be used on persons with either damaged or normal brains to study activity in real time; thus, these methods have greatly increased our ability to investigate human brain functions.

What is consciousness?

This chapter has only scratched the surface of our knowledge about the organization and functions of the human brain. Even the most sophisticated new research tools may not allow us to answer the overriding question of "What is consciousness?"

If you see a black dog running across a field, you are conscious that it is a dog, it is black, and it is a Labrador retriever. You may also remember that the dog's name is Sarina, that she belongs to your friend Meera, and that Sarina is 6 years old. From what you have learned in this chapter, imagine how many neurons would be active during this experience: neurons in the visual system, the language areas, and in different regions of association cortex. But is being *conscious* of the dog simply a result of all of these neurons firing at the same time? Your brain is simultaneously processing many other sensory inputs, but you are not necessarily conscious of those inputs. What makes you conscious of the black dog and associated memories and not of other information the brain is processing at the same time?

If we could describe all the neurons and all the synapses involved in the conscious experience of seeing and naming a black dog, and then build a computer with devices that modeled all these neurons and connections, would that computer be conscious? It has been said that the question of consciousness resolves into two types of problems: "easy" and "hard." The easy problems deal with all the cells and circuits that process the information that is involved in conscious experience. The implication of "easy" is that we seem to have the tools to solve these kinds of problems, as complex as they may be. The hard problems involve explaining how properties of cells and networks result in consciousness, and we seem to lack the proper tools or concepts even to begin to solve these problems.

46.3 RECAP

Even complex functions of the nervous system are beginning to be understood in terms of the properties of neurons and neuronal networks.

- Can you describe the differences between slow-wave sleep and REM sleep? See p. 998 and Figure 46.14
- Can you describe the role of the hippocampus in memory and the different types of memories it is involved in? See pp. 999–1000

CHAPTER SUMMARY

46.1 How is the mammalian nervous system organized?

The brain and spinal cord make up the **central nervous system** (**CNS**); the cranial and spinal nerves make up the **peripheral nervous system** (**PNS**).

The nervous system can be modeled conceptually in terms of the direction of information flow and whether we are conscious of the information. The **afferent** component carries information from the PNS to the CNS, and the **efferent** component directs information from the CNS to the peripheral parts of the body. Review Figure 46.1

The vertebrate nervous system develops from a hollow dorsal **neural tube**. The brain forms from three swellings at the anterior end of the neural tube, which become the **hindbrain**, the **midbrain**, and the **forebrain**. Review Figure 46.2

The forebrain develops into the **cerebral hemispheres** (the **telencephalon**) and the underlying **thalamus** and **hypothalamus** (the **diencephalon**). The midbrain and hindbrain develop into the **brain stem** and the **cerebellum**. Review Figure 46.2, Web/CD Tutorial 46.1

The spinal cord communicates information between the brain and the rest of the body. It can issue some commands to the body without input from the brain. Review Figure 46.3

The **reticular system** is a complex network that directs incoming information to appropriate brain stem **nuclei** that control **autonomic functions**, and transmits the information to the forebrain that results in conscious sensation. The reticular system controls the level of arousal of the nervous system, including sleep and wakefulness.

The **limbic system** is an evolutionarily primitive part of the telencephalon that is involved in emotions, physiological drives (such as hunger or thirst), instincts, and memory. Review Figure 46.4

The cerebral hemispheres are the dominant structures of the human brain. Their surfaces are layers of neurons called the **cerebral cortex**. The cerebral hemispheres can be divided into **temporal**, **frontal**, **parietal**, and **occipital lobes**. Many motor functions are localized in parts of the frontal lobe. Information from many sensory receptors projects to a region of the parietal lobe. Visual information projects to the occipital lobe, and auditory information projects to a region of the temporal lobe. Review Figures 46.5, 46.6, and 46.7, Web/CD Activity 46.1

46.2 How is information processed by neuronal networks?

The **autonomic nervous system** (**ANS**) consists of efferent pathways that control the physiological function of organs and organ systems. Its **sympathetic** and **parasympathetic** divisions are characterized by their anatomy, neurotransmitters, and effects on target tissues. Review Figure 46.10

The neuronal network of vision involves patterns of light falling on **receptive fields** in the retina. Receptive fields have a center and a surround, which have opposing effects on ganglion cell firing. Review Figure 46.11, Web/CD Tutorial 46.2

Information from retinal ganglion cells is communicated via the optic nerve to the thalamus and then to the visual cortex. The visual cortex seems to assemble an image of the visual world by analyzing edges of patterns of light. Review Figure 46.12

Binocular vision is possible because information from both eyes is communicated to binocular cells in the visual cortex. These cells interpret distance by measuring the disparity between where the same stimulus falls on the two retinas. Review Figure 46.13

46.3 Can higher functions be understood in cellular terms?

Humans have a daily cycle of sleep and waking. Sleep can be divided into **rapid-eye-movement** (**REM**) **sleep** and **slow-wave** (**non-REM**) **sleep**. Review Figure 46.14

Some learning and memory processes have been localized to specific brain areas. Long-lasting changes in synaptic properties referred to as **long-term potentiation** (**LTP**) and **long-term depression** (**LDP**) may be involved in learning and memory.

Complex memories can be elicited by stimulating small regions of association cortex. Damage to the hippocampus can destroy the ability to form long-term **declarative memories**, but not **procedural memories**.

Language abilities are localized mostly in the left cerebral hemisphere, a phenomenon known as lateralization. Different areas of the left hemisphere—including **Broca's area**, **Wernicke's area**, and the **angular gyrus**—are responsible for different aspects of language. Review Figure 46.15, Web/CD Activity 46.2

See Web/CD Activity 46.3 for a concept review of this chapter.

SELF-QUIZ

1. Which of the following describes the route of sensory information from the foot to the brain?
   a. Ventral horn, spinal cord, medulla, cerebellum, midbrain, thalamus, parietal cortex
   b. Dorsal horn, spinal cord, medulla, pons, midbrain, hypothalamus, frontal cortex
   c. Dorsal horn, spinal cord, medulla, pons, midbrain, thalamus, parietal cortex
   d. Ventral horn, spinal cord, pons, cerebellum, midbrain, thalamus, parietal cortex
   e. Dorsal horn, spinal cord, medulla, pons, midbrain, thalamus, frontal cortex
2. Which statement about the reticular system is *not* true?
   a. Increased activity in the reticular system induces sleep.
   b. The reticular system is located in the brain stem.
   c. Damage to the reticular system in the midbrain can result in coma.
   d. Information from the spinal cord is routed to different nuclei in the reticular system and to the forebrain.
   e. There are groups of neurons called nuclei in the reticular system.
3. Which statement about afferent and efferent pathways is *not* true?
   a. Sympathetic and parasympathetic pathways carry only efferent information.
   b. Visceral afferents carry information about physiological functions of which we are not consciously aware.
   c. The voluntary division of the efferent portion of the peripheral nervous system executes conscious movements.
   d. The cranial nerves and spinal nerves are part of the peripheral nervous system.
   e. Afferent and efferent axons never travel in the same nerve.

4. Which statement about the limbic system is *not* true?
   a. Damage to one structure in the limbic system makes it impossible to form a fear memory.
   b. The limbic system is involved in basic physiological drives, instincts, and emotions.
   c. The limbic system consists of primitive forebrain structures.
   d. In humans, the limbic system is the largest part of the brain.
   e. In humans, a part of the limbic system is necessary for the transfer of short-term memory to long-term memory.
5. Which of the following represents the largest portion of the human cerebral cortex?
   a. The frontal lobes
   b. The primary somatosensory cortex
   c. The temporal cortex
   d. The association cortex
   e. The occipital cortex
6. Which statement about the autonomic nervous system is true?
   a. The sympathetic division is afferent, and the parasympathetic division is efferent.
   b. The transmitter norepinephrine is always excitatory, and acetylcholine is always inhibitory.
   c. Each pathway in the autonomic nervous system includes two neurons, and the neurotransmitter of the first neuron is acetylcholine.
   d. The cell bodies of many sympathetic preganglionic neurons are in the brain stem.
   e. The cell bodies of most parasympathetic postganglionic neurons are in or near the thoracic and lumbar spinal cord.
7. Which statement about cells in the visual cortex is *not* true?
   a. Many cortical cells receive inputs directly from single retinal ganglion cells.
   b. Many cortical cells respond most strongly to bars of light falling at specific locations on the retina.
   c. Some cortical cells respond most strongly to bars of light falling anywhere over large areas of the retina.
   d. Some cortical cells receive inputs from both eyes.
   e. Some cortical cells respond most strongly to an object when it is a certain distance from the eyes.
8. Which of the following characterizes non-REM sleep?
   a. Dreaming
   b. Paralysis of skeletal muscles
   c. EEG slow waves
   d. Rapid and jerky eye movements
   e. It makes up about 20 percent of total sleep time
9. Which conclusion was supported by experiments on split-brain patients?
   a. Language abilities are localized mostly in the left cerebral hemisphere.
   b. Language abilities require both Wernicke's area and Broca's area.
   c. The ability to speak depends on Broca's area.
   d. The ability to read depends on Wernicke's area.
   e. The left hand is served by the left cerebral hemisphere.
10. In the withdrawal reflex,
    a. Action potentials in a pain sensory neuron enter the spinal cord through the ventral root on the same side of the body.
    b. The axon from a pain sensory neuron makes inhibitory synapses with motor neurons on the same side of the spinal cord and inhibitory synapses with motor neurons on the other side of the spinal cord.
    c. Coordinated escape reactions can be initiated before the brain registers the painful sensation.
    d. Axons from the pain sensory neuron make synapses with sympathetic preganglionic neurons in the spinal cord.
    e. Axons from motor neurons on the same side of the spinal cord that receives the input from the pain receptors cross the spinal cord to stimulate muscles on the other side of the body.

FOR DISCUSSION

1. A person receives a stab wound to the left side of his neck. Miraculously, blood vessels are spared. However, following this trauma, his left pupil remains more constricted than his right pupil, and he drools out of the left side of his mouth. How can you explain these symptoms?
2. The stretch receptors in muscles are modified muscle cells, and they have their own motor neurons. What is the function of those motor neurons? To think about this question, remember that the function of the monosynaptic reflex is to adjust muscle tension to a change in load so that the position of the limb does not change.
3. A patient is unable to speak coherently. He can read and write, and he has no obvious loss of muscle function. Where would you expect to find an abnormality if you did brain scans of this patient?
4. How can you investigate how visual images falling on the retina are communicated to the brain? Start with the retina, and imagine how you would continue your investigation in visual areas of the brain.
5. We described the organization of the visual cortex as columns of cells that alternately receive input from the left eye and the right eye. If a young kitten is allowed to see light out of only one eye for a day, more synapses are maintained in the cortical columns receiving input from that eye, while synapses decrease in the intervening columns. This redistribution of synapses does not occur, however, if the kitten is not allowed to sleep. What hypotheses could you propose on the basis of these results?

FOR INVESTIGATION

Figure 46.16 describes technology that enables us to image regional activity in the brain. How would you use this imaging methodology to investigate a higher brain function that is of interest to you? First, formulate your hypothesis. Then design your investigation, keeping in mind that your subject has to lie motionless in a large machine during the imaging. You will have to plan how you will deliver the appropriate stimulus to elicit the response or activity you are interested in. Also, describe what controls you will use to draw clear conclusions about the stimulus-response relationships you expect to observe.

CHAPTER 47 Effectors: How Animals Get Things Done

Champion jumpers

The Olympic record for the woman's long jump is 7.4 meters, set in 1988 by Jackie Joyner-Kersee. Another world record long jump that still stands was set two years earlier by Rosie the Ribeter, who jumped 6.5 meters. Rosie was a frog competing in the Calaveras County Jumping Frog Contest. In some ways, Rosie's jump is more impressive—while Jackie's jump was about 5 times her body length (ie., her height), Rosie's was about *20 times* her body length.

Jackie's jump and Rosie's jump were both powered by skeletal muscle. Muscle tissue is an *effector* that responds to commands from the nervous system. The molecular and cellular mechanisms of muscle contraction are essentially the same in the frog and the human, so why is the frog's jump so much more impressive? The answer involves the concept of *leverage*.

Both frog and human jumping muscles pull on bones that are connected at joints to make levers. A lever makes it possible for the same force to move a large mass a small distance or a small mass a large distance. The ratio of a frog's leg length to its body mass is simply greater than that of a human. Thus the frog's legs are better at moving a small mass a long distance than are the human's legs.

Let's add a flea to our interspecies Olympic competition. The flea can jump over 200 times its body length. This incredible performance is not due to feats of leverage, because no muscle can contract fast enough to explain the take-off velocity of the flea. A different effector mechanism evolved in the flea—a kind of slingshot action. At the base of the flea's jumping legs is an elastic material that is compressed by muscles while the flea is resting. When a trigger mechanism is released, the elastic material recoils and "fires" the flea into the air.

In a contest of jumping endurance, the uncontested champion would be the kangaroo. As a human runs faster, the number of strides and the energy expended per minute increase rapidly. Neither is true for the kangaroo. When moving at speeds from about 5 to 25 kilometers per hour, the kangaroo takes the same number of strides per minute and its metabolic rate does not increase. Why is this so?

A Champion Jumper Jackie Joyner-Kersee set an Olympic record for the women's long jump at the Seoul Olympics in 1988.

Champion Jumpers Relative to their size, many animals have more impressive jumping skills than humans. This Pacific chorus frog (*Pseudacris gegilla*) can leap distances up to 20 times its body length.

In kangaroos, as in frogs and humans, the muscles used to jump are attached to bones by tendons. Like the material at the base of the flea's legs, tendons can be elastic. The kangaroo's tendons stretch when it lands, and their recoil helps power the next jump—similar to the action of a pogo stick. In order to move faster, the kangaroo simply increases the length of its stride, thereby increasing the stretch on its tendons each time it lands and the magnitude of the recoil at the initiation of each jump.

As we saw in Chapter 31, the ability to move is one of the things that distinguishes animals from the other multicellular organisms (i.e., plants and fungi). Our muscles and skeletons—the *musculoskeletal system*—are the *effectors* that produce movement.

IN THIS CHAPTER we will begin by describing muscle structure and the molecular mechanisms of contraction. We then discuss properties of muscles that adapt different types of muscles to different kinds of tasks. To generate specific kinds of movement, muscles must have something to pull on, so we describe the roles of skeletal systems in generating movement. This chapter ends with brief considerations of effector mechanisms other than musculoskeletal ones: glands that secrete, organs that emit light or sound, and structures that produce strong electrical pulses.

CHAPTER OUTLINE

47.1 How Do Muscles Contract?

Most behavior and many physiological actions, such as beating of the heart and moving of food through the digestive tract, depend on muscle contraction. Wherever tissues contract, muscle cells are responsible. There are three types of vertebrate muscle:

- **Skeletal muscle** is responsible for all voluntary movements, such as running or playing a piano. It also generates the unconscious movements of breathing.
- **Cardiac muscle** is responsible for the beating action of the heart.
- **Smooth muscle** creates the movement in many hollow internal organs, such as the gut, bladder, and blood vessels, and it is under the control of the autonomic (involuntary) nervous system.

All three types of muscle use the same contractile mechanism, and we begin our study of effectors by describing the molecular mechanisms underlying this ability to contract. We will use vertebrate skeletal muscle as our primary example here, and discuss its structure in detail. Later we will learn about the differences in cardiac and smooth muscle that adapt them to their particular functions.

Sliding filaments cause skeletal muscle to contract

Skeletal muscle is also called *striated muscle* because of its striped appearance (see Figure 47.7A). Skeletal muscle cells, called **muscle fibers**, are large and have many nuclei. These multinucleate cells form in development through the fusion of many individual embryonic muscle cells called *myoblasts*. A specific muscle such as your biceps (which bends your arm) is composed of hundreds or thousands of muscle fibers bundled together by connective tissue (**Figure 47.1**).

Muscle contraction is due to the interaction between the contractile proteins **actin** and **myosin**. Within muscle cells, actin and myosin molecules are organized into filaments consisting of many molecules. Actin filaments are also called *thin filaments*, and myosin filaments are *thick filaments*. The two kinds of filaments lie parallel to each other. When muscle contraction is triggered, the actin and myosin filaments slide past each other in a telescoping fashion.

What is the relation between a skeletal muscle fiber and the actin and myosin filaments responsible for its contraction? Each

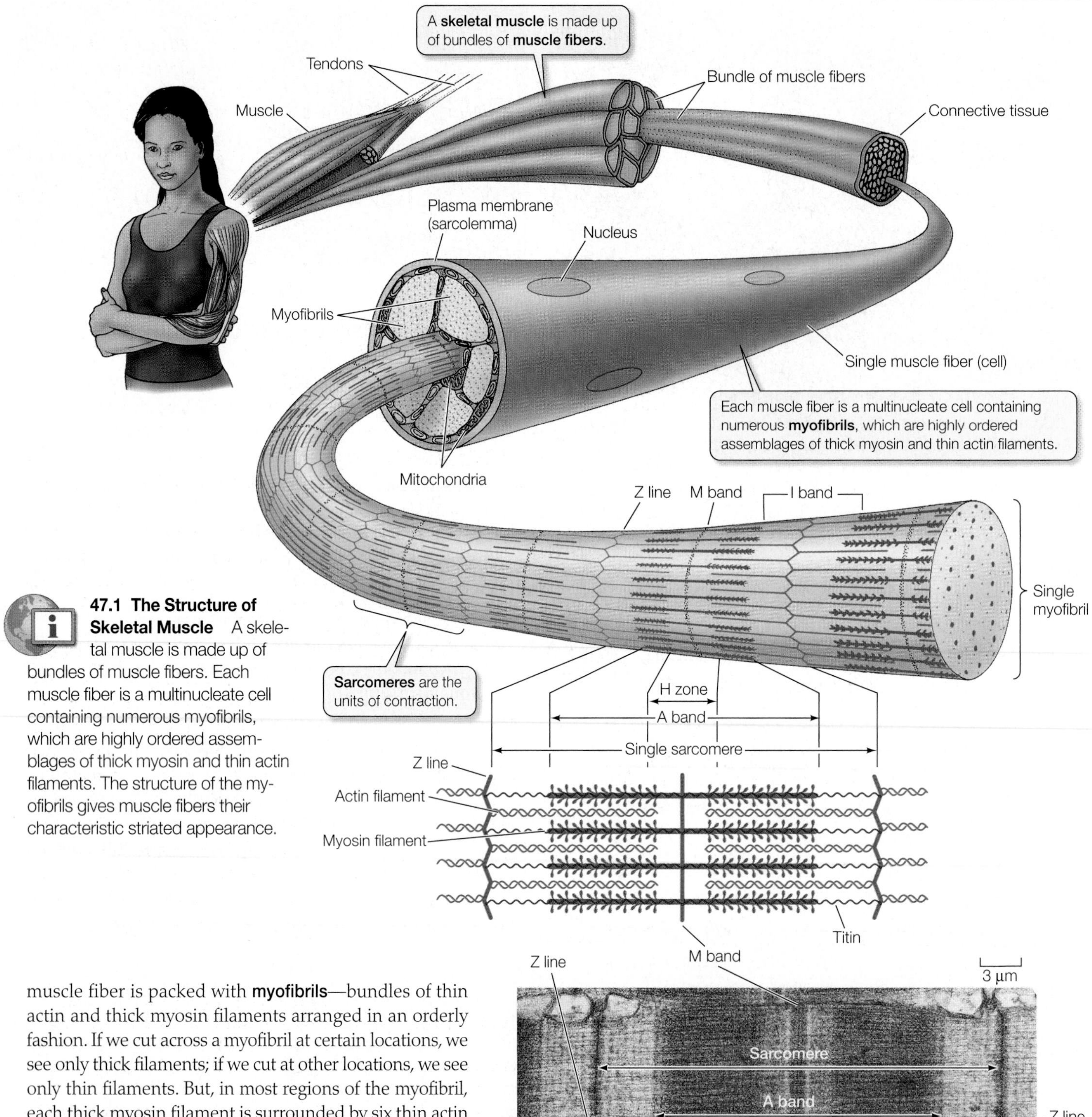

47.1 The Structure of Skeletal Muscle A skeletal muscle is made up of bundles of muscle fibers. Each muscle fiber is a multinucleate cell containing numerous myofibrils, which are highly ordered assemblages of thick myosin and thin actin filaments. The structure of the myofibrils gives muscle fibers their characteristic striated appearance.

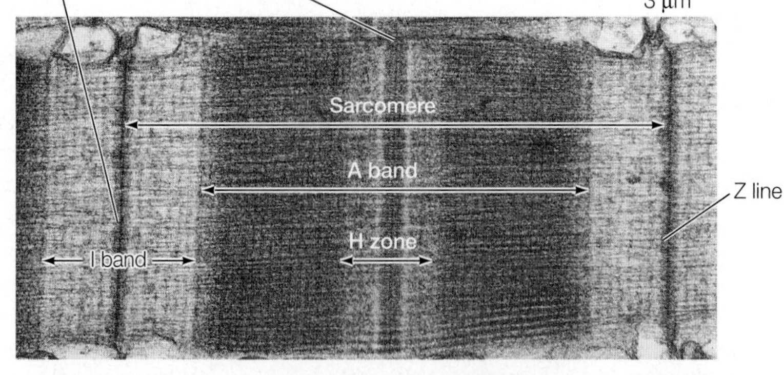

muscle fiber is packed with **myofibrils**—bundles of thin actin and thick myosin filaments arranged in an orderly fashion. If we cut across a myofibril at certain locations, we see only thick filaments; if we cut at other locations, we see only thin filaments. But, in most regions of the myofibril, each thick myosin filament is surrounded by six thin actin filaments, and conversely, each thin actin filament sits within a triangle of three thick myosin filaments.

A longitudinal view of a myofibril reveals why skeletal muscle appears striated. The myofibril consists of repeating units called **sarcomeres**. Each sarcomere is made of overlapping filaments of actin and myosin, which create a distinct banding pattern (see Figure 47.2). The bundles of myosin filaments are held in a centered position within the sarcomere by a protein called **titin**. Titin is the largest protein in the body; it runs the full length of the sarcomere from Z line to Z line. Each titin molecule runs right through a myosin bundle. Between the ends of the myosin bundles and the Z lines, titin molecules are very stretchable, like bungee cords. In a relaxed skeletal muscle, resistance to stretch is mostly due to the elasticity of the titin molecules.

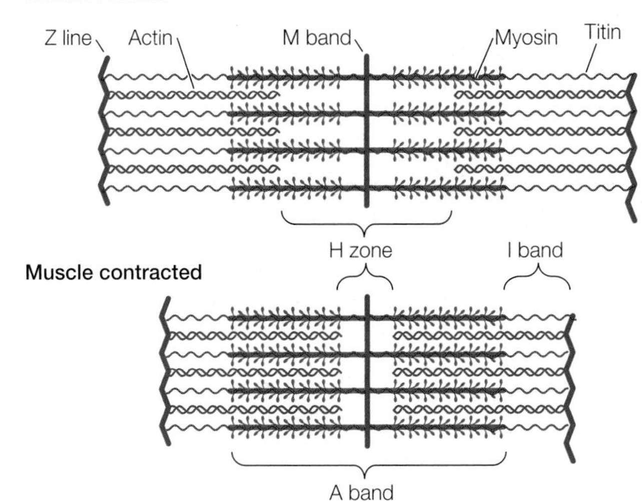

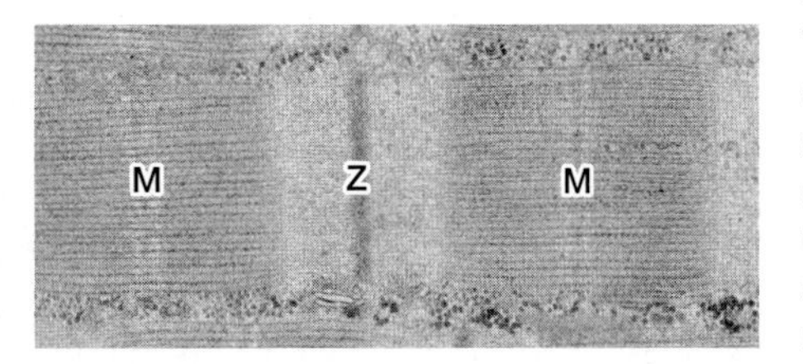

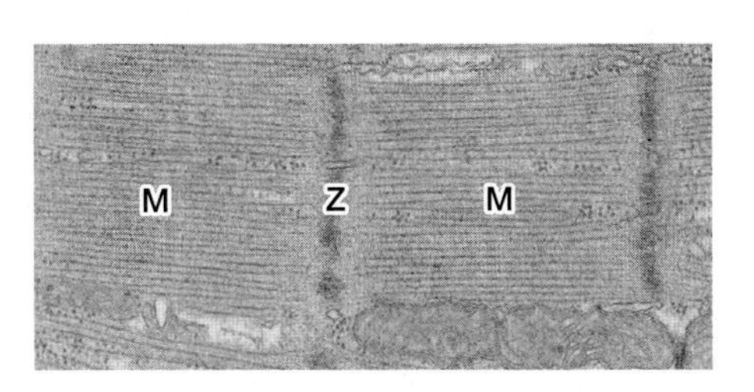

47.2 Sliding Filaments The banding pattern of the sarcomere changes as it shortens. Observations of electron micrographs such as those on the right led to the sliding filament hypothesis of muscle contraction.

Before the molecular nature of the muscle banding pattern was known, the bands were given names that are still used today. Each sarcomere is bounded by *Z lines*, which anchor the thin actin filaments. Centered in the sarcomere is the *A band*, which contains all the myosin filaments. The *H zone* and the *I band*, which appear light, are regions where actin and myosin filaments do not overlap in the relaxed muscle. The dark stripe within the H zone is called the *M band;* it contains proteins that help hold the myosin filaments in their regular arrangement.

As the muscle contracts, the sarcomeres shorten, and the band pattern changes. The H zone and the I band become much narrower, and the Z lines move toward the A band as if the actin filaments were sliding into the region occupied by the myosin filaments (**Figure 47.2**). In the mid 1950s, this observation independently led two teams of British biologists to independently propose the **sliding filament theory** of muscle contraction. It is not uncommon in science for critical breakthroughs to be made simultaneously in different laboratories, but the arm of coincidence was long indeed in this case. The leaders of the two teams were named Hugh Huxley and Andrew Huxley—but they were not related. Working in separate Cambridge University labs, the two groups proposed the sliding filament model at the same time, and both papers were published in the same issue of the journal *Nature.*

Actin–myosin interactions cause filaments to slide

To understand how the sliding filament model explains muscle contraction, we must first examine the structures of actin and myosin (**Figure 47.3**). A myosin molecule consists of two long polypeptide chains coiled together, each ending in a large globular head. A myosin filament is made up of many myosin molecules arranged in parallel, with their heads projecting laterally from one or the other end of the filament. An actin filament consists of two chains of actin monomers twisted together like two strands of pearls in a helix. Twisting around the actin chains is another protein, *tropomyosin*, and attached to the tropomyosin at intervals are molecules of *troponin*. We'll discuss the roles of these last two proteins in the following section.

The myosin heads can bind specific sites on actin and thereby form cross-bridges between the myosin and the actin filaments. Moreover, when a myosin head binds to an actin filament, its conformation changes, and it bends and exerts a force that causes the actin filament to slide 5–10 nanometers relative to the myosin filament. The myosin heads also have ATPase activity; when they are bound to actin, they can bind and hydrolyze ATP. The energy released when this happens changes the conformation of the myosin head, causing it to release the actin and return to its extended position, from which it again can bind to actin. Together, these details explain the cycle of events that cause the actin and myosin filaments to slide past each other and shorten the sarcomere.

47.3 Actin and Myosin Filaments Overlap to Form Myofibrils Myosin filaments are bundles of molecules with globular heads and polypeptide tails; the protein titin holds these filaments centered within the sarcomere. Actin filaments consist of two chains of actin monomers twisted together. They are wrapped by chains of the polypeptide tropomyosin and studded at intervals with another protein, troponin.

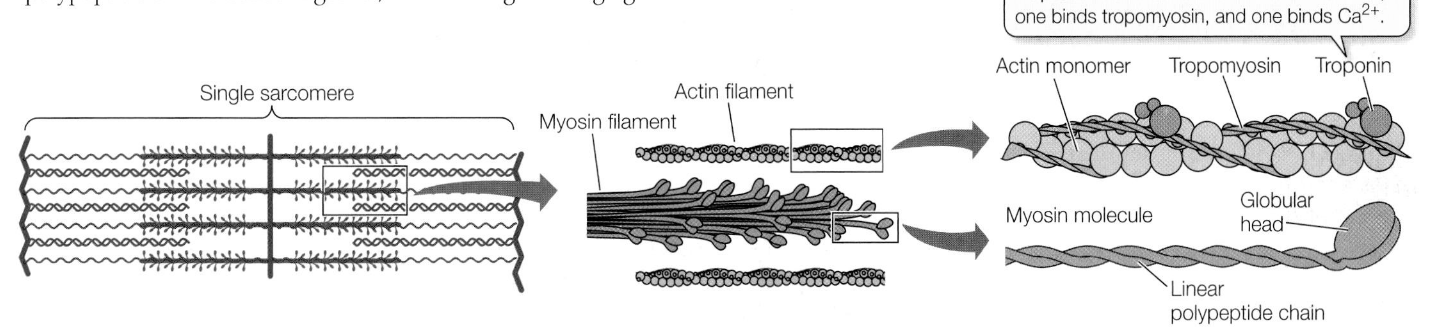

ATP is needed to break the actin–myosin bonds, explaining why muscles stiffen soon after death—*rigor mortis*. ATP production ceases with death, so the actin–myosin bonds cannot be broken, and the muscles stiffen. Eventually, however, the proteins begin to lose their integrity, and the muscles soften. The timing of these events help a coroner estimate the time of death.

As an analogy, think of a person firing an old-fashioned pistol. A finger pulls the pistol hammer back against the force of a spring, cocking the hammer into a high-energy, unstable position. Similarly, ATP hydrolysis converts the potential chemical energy of a phosphate bond into the kinetic energy that "cocks" the myosin head into its unstable, extended position. Pressure on the trigger of a pistol can release the hammer from the cocked position, allowing the stretched spring attached to the hammer to slam it back into its original position. Similarly, the binding of actin destabilizes the myosin head from its extended position, and the spring-like myosin molecule snaps back to its bent configuration, dragging the bound actin filament along with it.

We have been discussing the cycle of contraction in terms of a single myosin head. Remember that each myosin filament has many myosin heads at both ends and is surrounded by six actin filaments; thus the contraction of the sarcomere involves a great many cycles of interaction between actin and myosin molecules. That is why when a single myosin head breaks its contact with actin, the actin filaments do not slip backward.

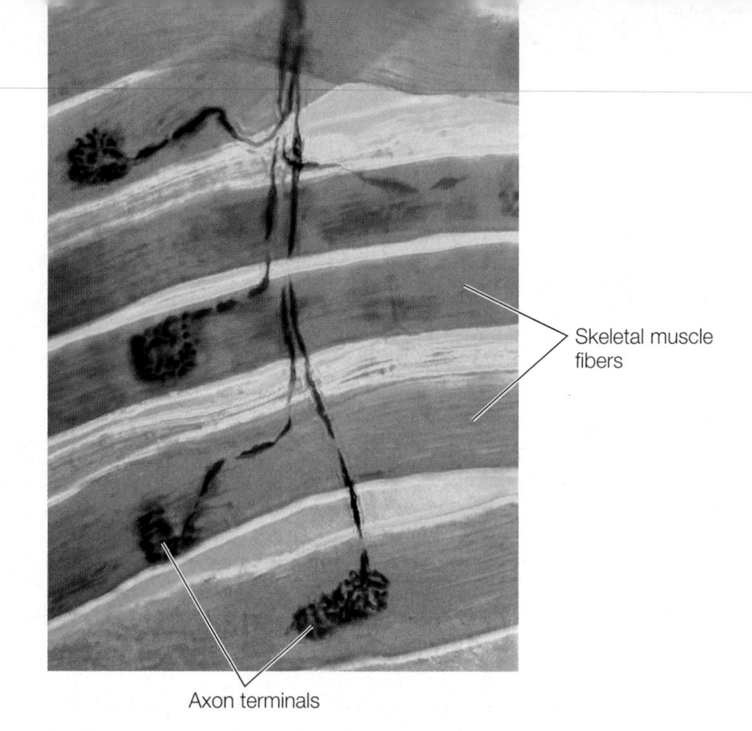

47.4 The Neuromuscular Junction Axon terminals from a single motor neuron innervate multiple skeletal muscle fibers.

Actin–myosin interactions are controlled by calcium ions

Muscle contractions are initiated by action potentials from motor neurons arriving at the *neuromuscular junction* (**Figure 47.4**; also see Figure 44.13). The axon terminals of motor neurons are generally highly branched and can form synapses with hundreds of muscle fibers. All the fibers activated by a single motor neuron constitute a **motor unit** and contract simultaneously in response to action potentials fired by that motor neuron. A muscle can consist of many motor units. Thus, there are two ways to increase the strength of contraction of a muscle—increase the rate of firing in an individual motor neuron or recruit more motor neurons to fire.

Like neurons, muscle cells are *excitable*; that is, their plasma membranes can generate and conduct action potentials. In skeletal muscle fibers (but not smooth or cardiac muscle fibers), all action potentials are initiated by motor neurons. When an action potential arrives at the neuromuscular junction, the neurotransmitter acetylcholine is released from the motor neuron terminals, diffuses across the synaptic cleft, binds to receptors in the postsynaptic membrane, and causes ion channels in the motor end plate to open. Most of the ions that flow through these channels are Na^+, and therefore the motor end plate is depolarized. The depolarization spreads to the surrounding plasma membrane of the muscle fiber, which contains voltage-

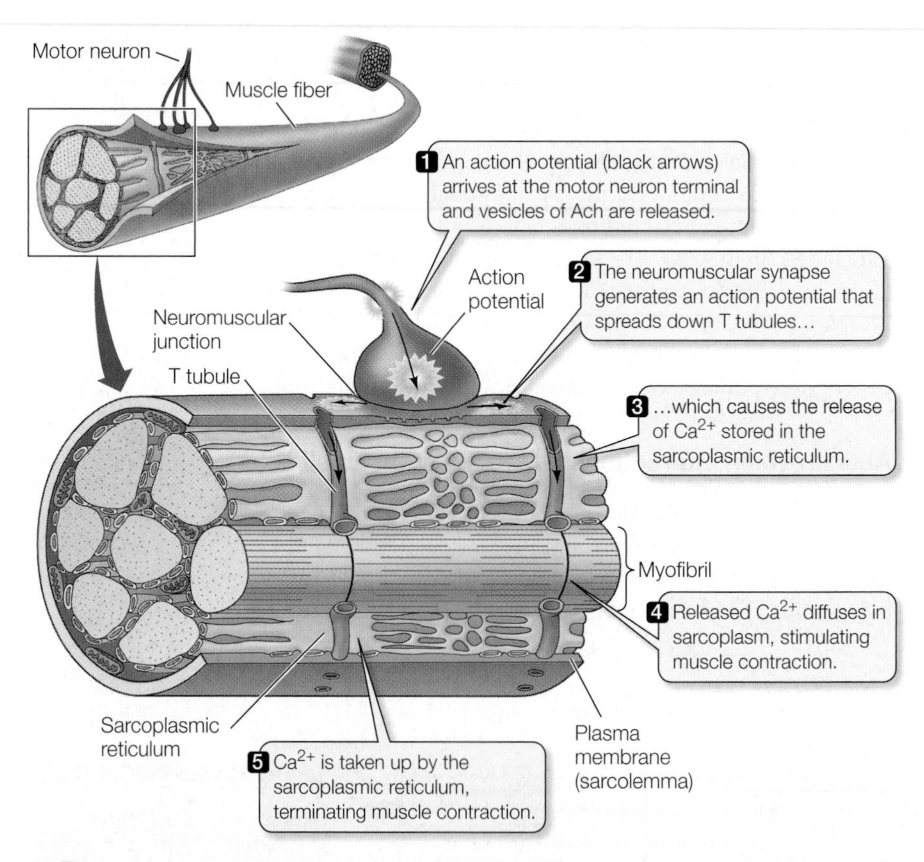

47.5 T Tubules in Action An action potential at the neuromuscular junction spreads throughout the muscle fiber via a network of T tubules, triggering the release of Ca^{2+} from the sarcoplasmic reticulum.

gated sodium channels. When threshold is reached, the plasma membrane fires an action potential that is conducted rapidly to all points on the surface of the muscle fiber.

An action potential in a muscle fiber also travels deep within the cell. The plasma membrane is continuous with a distribution system of tubules that descend into the muscle fiber cytoplasm (also called the **sarcoplasm**) (**Figure 47.5**). The action potential that spreads over the plasma membrane also spreads through this system of transverse tubules, or **T tubules**.

The T tubules come very close to the endoplasmic reticulum of the muscle cell. In muscle cells the ER is called the **sarcoplasmic reticulum**, and it is a closed compartment surrounding every myofibril. Calcium pumps in the sarcoplasmic reticulum cause it to take up Ca^{2+} ions from the sarcoplasm. Therefore, when the muscle fiber is at rest, there is a higher concentration of Ca^{2+} in the sarcoplasmic reticulum and a lower concentration of Ca^{2+} in the sarcoplasm.

Spanning the space between the membranes of the T tubules and the membranes of the sarcoplasmic reticulum are two proteins. One protein, the *dihydropyridine* (DHP) *receptor*, is located in the T tubule membrane; it is voltage-sensitive and changes its conformation when an action potential reaches it. The other protein, the *ryanodine receptor*, is located in the sarcoplasmic reticulum membrane; it is a Ca^{2+} channel. These two proteins are physically connected. When the DHP receptor is activated by an action potential, it changes conformation, causing the ryanodine receptor to allow Ca^{2+} to leave the sarcoplasmic reticulum. Ca^{2+} ions diffuse into the sarcoplasm surrounding the actin and myosin filaments and trigger the interaction of actin and myosin and the sliding of the filaments. How do the Ca^{2+} ions do this?

An actin filament, as we have seen, is a helical arrangement of two strands of actin monomers. Lying in the grooves between the two actin strands is the two-stranded protein **tropomyosin** (**Figure 47.6**; also see Figure 47.3). At regular intervals, the filament also includes a globular protein, **troponin**. The troponin molecule has three subunits: One binds actin, one binds tropomyosin, and one binds Ca^{2+}.

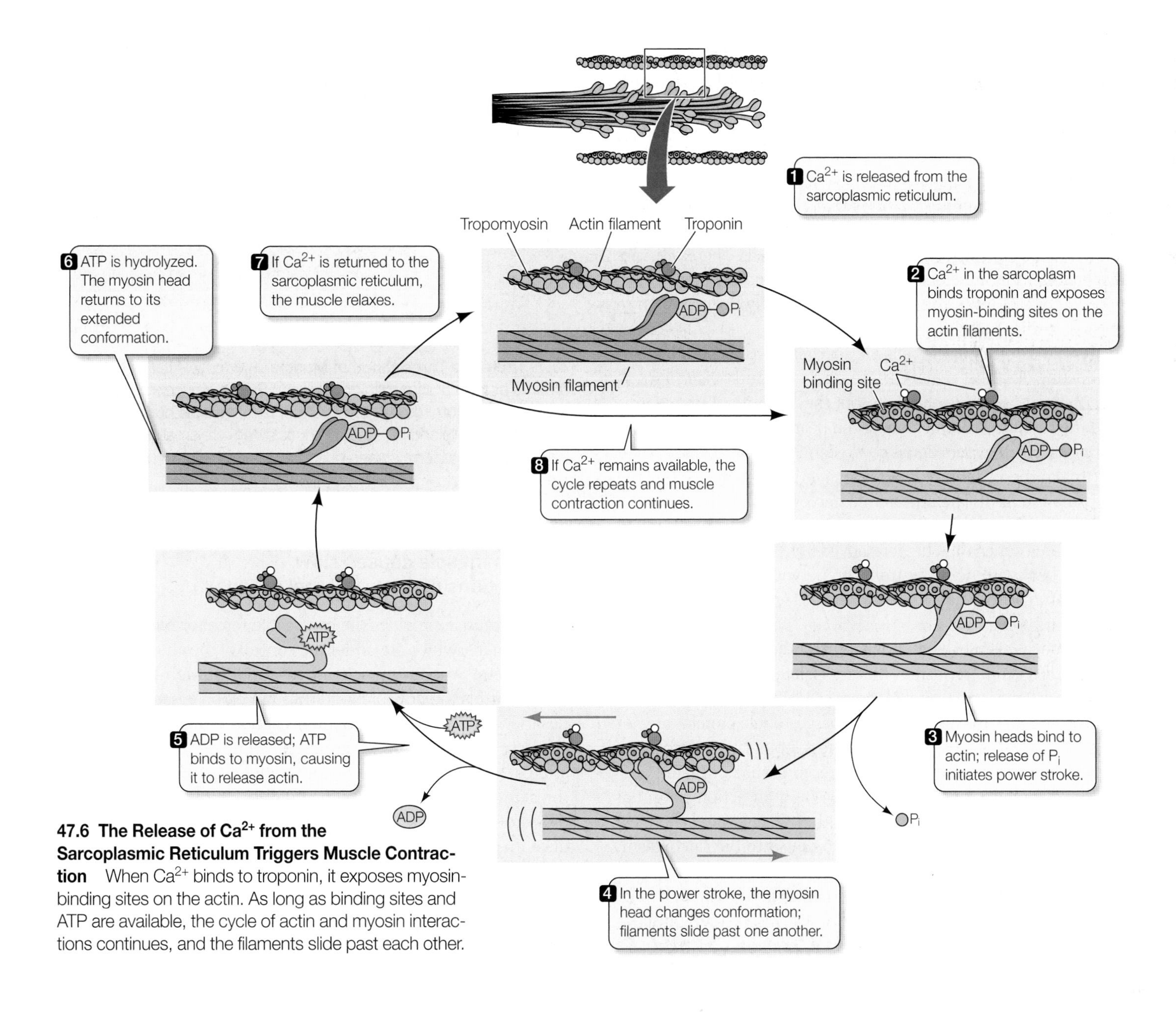

47.6 The Release of Ca^{2+} from the Sarcoplasmic Reticulum Triggers Muscle Contraction When Ca^{2+} binds to troponin, it exposes myosin-binding sites on the actin. As long as binding sites and ATP are available, the cycle of actin and myosin interactions continues, and the filaments slide past each other.

When the muscle is at rest, the tropomyosin strands are positioned so that they block the sites on the actin filament where myosin heads can bind. When Ca^{2+} is released into the sarcoplasm, it binds to troponin, changing its conformation. Because the troponin is bound to the tropomyosin, this conformational change twists the tropomyosin enough to expose the actin–myosin binding sites. Thus the cycle of making and breaking actin–myosin bonds is initiated, the filaments are pulled past each other, and the muscle fiber contracts. When the calcium pumps remove the Ca^{2+} ions from the sarcoplasm, the tropomyosin returns to the position in which it blocks the binding of myosin heads to actin, and the muscle fiber returns to its resting condition. Figure 47.6 summarizes this cycle.

Cardiac muscle causes the heart to beat

Cardiac muscle appears striated as does skeletal muscle because of the regular arrangement of actin and myosin filaments into sarcomeres (**Figure 47.7A,B**). The difference between cardiac and skeletal muscle is that cardiac muscle cells are much smaller and have only one nucleus each. Cardiac muscle cells branch, and the branches of adjoining cells interdigitate into a meshwork that is resistant to tearing. As a result, the heart walls can withstand high pressures while pumping blood without the danger of developing leaks. Adding to the strength of cardiac muscle are *intercalated discs* that provide strong mechanical adhesions between adjacent cells. Gap junctions—protein structures that allow cytoplasmic continuity between cells—in the intercalated discs offer low-resistance pathways for ionic currents to flow between cells. Therefore, cardiac muscle cells are electrically coupled. An action potential initiated at one point in the heart spreads rapidly through a large mass of cardiac muscle.

Certain cardiac muscle cells are specialized for generating and conducting electrical signals. These *pacemaker* and *conducting cells* do not have contractile elements, but they do initiate and coordinate the rhythmic contractions of the heart. (The molecular basis for this pacemaking function is covered in Section 49.3.) Pacemaker cells make the vertebrate heartbeat *myogenic*—generated by the heart muscle itself. The autonomic nervous system modifies the rate of the pacemaker cells, but is not essential for their continued rhythmic function. A heart removed from a vertebrate can continue to beat with no input from the nervous system. The myogenic nature of the heartbeat is a major factor in making heart transplants possible.

The mechanism of excitation–contraction coupling in cardiac muscle cells is different from that in skeletal muscle cells. The T tubules are larger and the voltage-sensitive DHP proteins in the T tubules are actual calcium channels. These T tubule proteins are not physically connected with the ryanodine receptors in the sarcoplasmic reticulum. Instead, the ryanodine receptors are ion-gated Ca^{2+} channels that are sensitive to Ca^{2+}. When an action potential spreads down the T tubules, it causes the voltage-gated channels to open, allowing extracellular Ca^{2+} to flow into the sarcoplasm. This slight rise in sarcoplasmic Ca^{2+} concentration opens the Ca^{2+} channels in the sarcoplasmic reticulum, which in turn causes a huge rise in sarcoplasmic calcium ion concentration, resulting in fiber contraction. This mechanism is called *Ca^{2+}-induced Ca^{2+} release.*

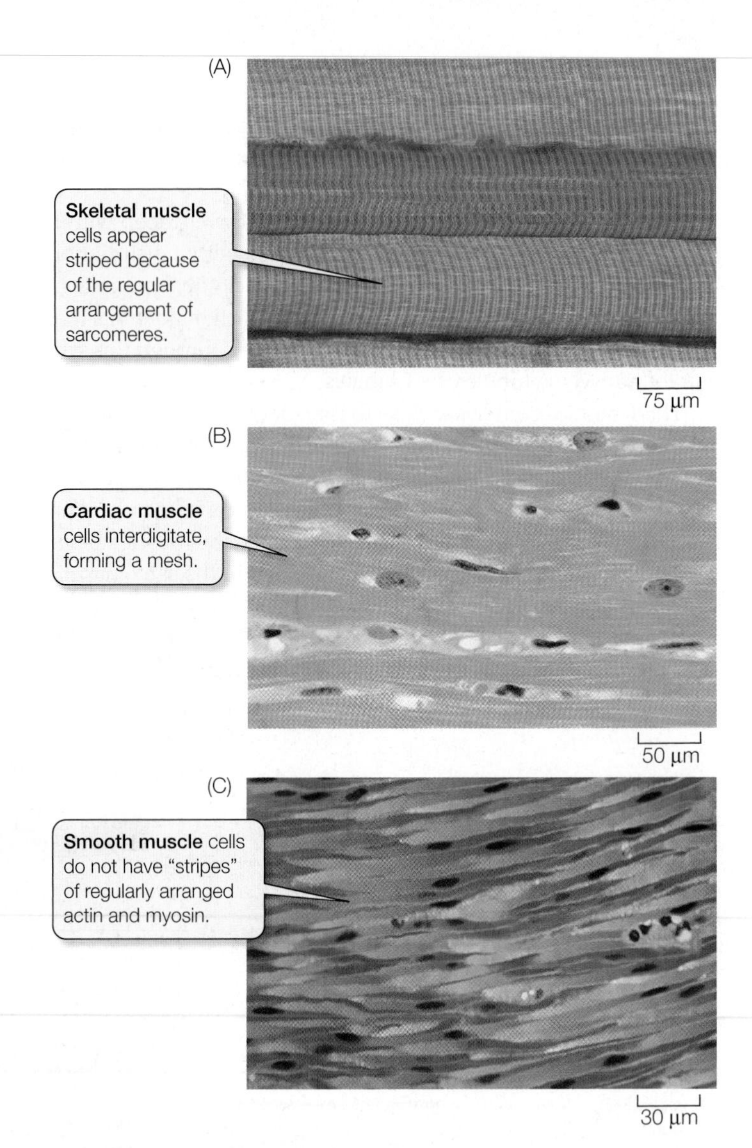

47.7 There are Three Kinds of Muscle All muscle cells contain filaments of actin and myosin proteins. (A) Skeletal muscle cells are very large and form fibers. Each cell contains multiple nuclei. (B) Cardiac muscle cells and (C) smooth muscle cells are smaller, have individual nuclei, and form sheets of contractile tissue in which cells are electrically coupled.

Smooth muscle causes slow contractions of many internal organs

Smooth muscle provides the contractile force for most of our internal organs, which are under the control of the autonomic nervous system. Smooth muscle moves food through the digestive tract, controls the flow of blood through blood vessels, and empties the urinary bladder. Structurally, smooth muscle cells are the simplest muscle cells. They are usually long and spindle-shaped, and each cell has a single nucleus. They are "smooth" because the actin and myosin filaments are not as regularly arranged as they are in skeletal and cardiac muscle, and therefore do not produce the striated appearance (**Figure 47.7C**).

Smooth muscle tissue, such as that from the wall of the digestive tract, has interesting properties. The cells are arranged in sheets, and individual cells in a sheet are in electrical contact with one another through gap junctions, as they are in cardiac muscle. As a result, an action potential generated in the membrane of one smooth muscle cell can spread to all the cells in the sheet of tissue. Thus the cells in the sheet can contract in a coordinated fashion.

The plasma membranes of smooth muscle cells are sensitive to stretch, with important consequences. If the wall of the digestive tract is stretched in one location (as by a mouthful of food passing down the esophagus to the stomach), the membranes of the stretched cells depolarize, reach threshold, and fire action potentials, which cause the cells to contract. Thus, smooth muscle contracts after being stretched, and the harder it is stretched, the stronger it contracts.

The neurotransmitters of the autonomic nervous system can alter the membrane potential of smooth muscle cells as well (**Figure 47.8**). For example, acetylcholine causes smooth muscle cells

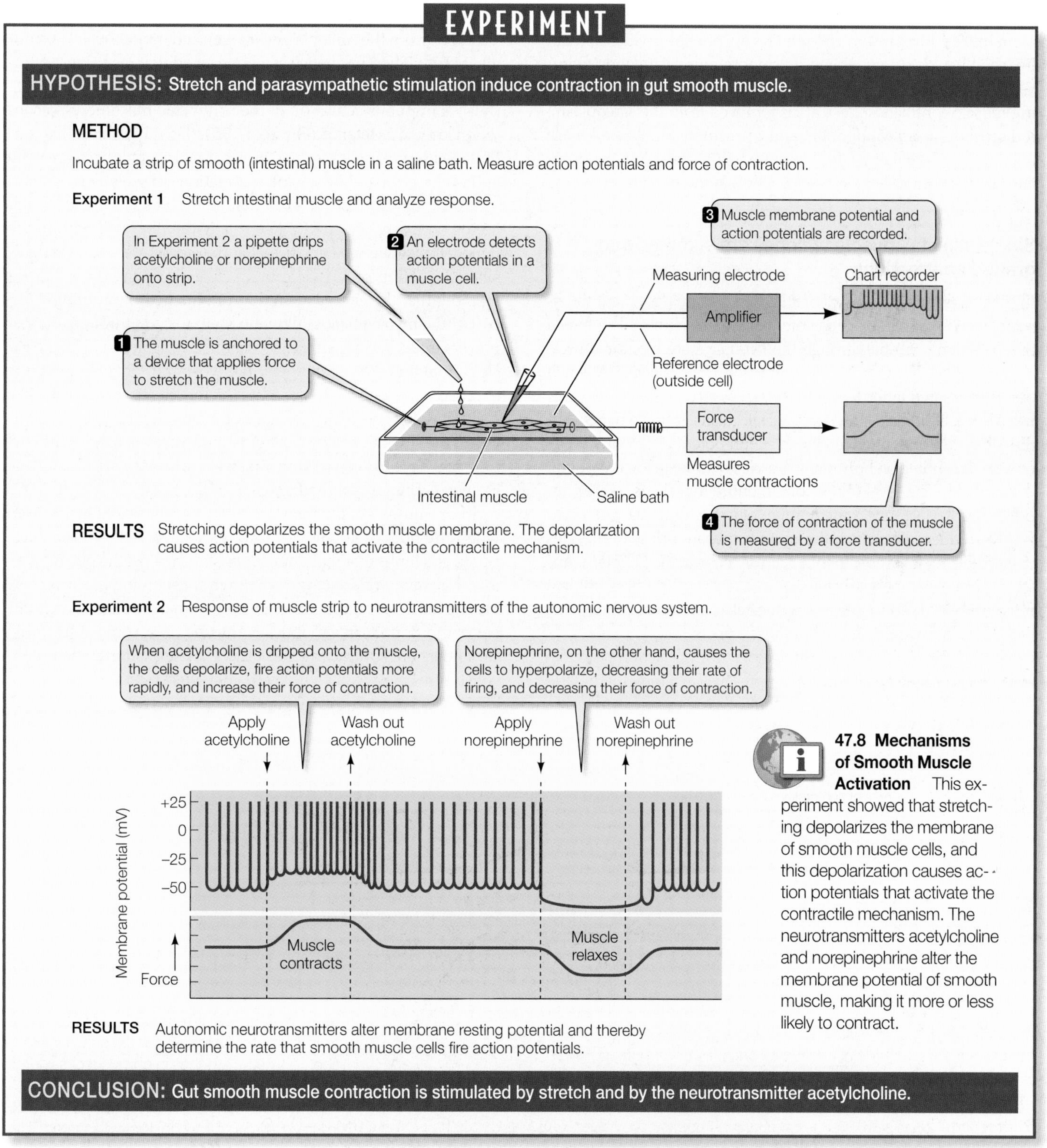

EXPERIMENT

HYPOTHESIS: Stretch and parasympathetic stimulation induce contraction in gut smooth muscle.

METHOD

Incubate a strip of smooth (intestinal) muscle in a saline bath. Measure action potentials and force of contraction.

Experiment 1 Stretch intestinal muscle and analyze response.

In Experiment 2 a pipette drips acetylcholine or norepinephrine onto strip.

2 An electrode detects action potentials in a muscle cell.

3 Muscle membrane potential and action potentials are recorded.

Measuring electrode

Chart recorder

Amplifier

1 The muscle is anchored to a device that applies force to stretch the muscle.

Reference electrode (outside cell)

Force transducer

Measures muscle contractions

Intestinal muscle

Saline bath

4 The force of contraction of the muscle is measured by a force transducer.

RESULTS Stretching depolarizes the smooth muscle membrane. The depolarization causes action potentials that activate the contractile mechanism.

Experiment 2 Response of muscle strip to neurotransmitters of the autonomic nervous system.

When acetylcholine is dripped onto the muscle, the cells depolarize, fire action potentials more rapidly, and increase their force of contraction.

Norepinephrine, on the other hand, causes the cells to hyperpolarize, decreasing their rate of firing, and decreasing their force of contraction.

Apply acetylcholine

Wash out acetylcholine

Apply norepinephrine

Wash out norepinephrine

Membrane potential (mV)

+25

0

–25

–50

Force

Muscle contracts

Muscle relaxes

RESULTS Autonomic neurotransmitters alter membrane resting potential and thereby determine the rate that smooth muscle cells fire action potentials.

CONCLUSION: Gut smooth muscle contraction is stimulated by stretch and by the neurotransmitter acetylcholine.

47.8 Mechanisms of Smooth Muscle Activation This experiment showed that stretching depolarizes the membrane of smooth muscle cells, and this depolarization causes action potentials that activate the contractile mechanism. The neurotransmitters acetylcholine and norepinephrine alter the membrane potential of smooth muscle, making it more or less likely to contract.

of the digestive tract to depolarize; they are thus more likely to fire action potentials and contract. Antagonistically, norepinephrine causes these muscle cells to hyperpolarize, and they are therefore less likely to fire action potentials and contract.

Although smooth muscle cell contraction is not controlled by the troponin–tropomyosin mechanism, calcium still plays a critical role. A Ca^{2+} influx into the sarcoplasm of a smooth muscle cell can be stimulated by action potentials, hormones, or stretching. The Ca^{2+} that enters the sarcoplasm combines with a protein called *calmodulin*. The calmodulin–Ca^{2+} complex activates an enzyme called myosin kinase, which can phosphorylate myosin heads. When the myosin heads in smooth muscle are phosphorylated, they can undergo cycles of binding and releasing actin, causing muscle contraction. As Ca^{2+} is removed from the sarcoplasm, it dissociates from calmodulin, and the activity of myosin kinase falls. An additional enzyme, myosin phosphatase, dephosphorylates the myosin to help stop actin–myosin interactions.

Single skeletal muscle twitches are summed into graded contractions

In skeletal muscle, the arrival of an action potential at a neuromuscular junction causes an action potential in a muscle fiber. The spread of that action potential through the T tubule system of the muscle fiber causes a minimum unit of contraction, called a **twitch**. A twitch can be measured in terms of the *tension*, or force, it generates (**Figure 47.9A**). A single action potential stimulates a single twitch, but the ultimate force generated by an action potential can vary enormously depending on how many muscle fibers are in the motor unit it innervates. The level of tension an entire muscle generates depends on two factors: (1) the number of motor units activated, and (2) the frequency at which the motor units are firing. In muscles responsible for fine movements, such as those of the fingers, a motor neuron may innervate only one or a few muscle fibers, but in a muscle that produces large forces, such as the biceps, a motor neuron innervates a large number of muscle fibers.

At the level of the single muscle fiber, a single action potential stimulates a single twitch. If action potentials reaching the muscle fiber are adequately separated in time, each twitch is a discrete, all-or-none phenomenon. If action potentials are fired more rapidly, however, new twitches are triggered before the myofibrils have had a chance to return to their resting condition. As a result, the twitches sum, and the tension generated by the fiber increases and becomes more sustained. Thus an individual muscle fiber can show a graded response to increased levels of stimulation by its motor neuron.

Twitches sum at high levels of stimulation because the calcium pumps in the sarcoplasmic reticulum are not able to clear the Ca^{2+} ions from the sarcoplasm between action potentials. Eventually a stimulation frequency can be reached that results in continuous presence of Ca^{2+} in the sarcoplasm at high enough levels to cause continuous activation of the contractile machinery—a condition known as **tetanus** (**Figure 47.9B**). (Do not confuse this condition with the disease *tetanus*, which is caused by a bacterial toxin and is characterized by spastic contractions of skeletal muscles.)

How long a muscle fiber can maintain a tetanic contraction depends on its supply of ATP. Eventually the fiber will become fatigued. It may seem paradoxical that the *lack* of ATP causes fatigue, since the action of ATP is to break actin–myosin bonds. But remember that the energy released from the hydrolysis of ATP "re-cocks" the myosin heads, allowing them to cycle through another power stroke. When a muscle is contracting against a load, the cycle of making and breaking actin–myosin bonds must continue to prevent the load from stretching the muscle. The situation is like rowing a boat upstream: You cannot maintain your position relative to the stream bank by just holding the oars out against the current; you have to keep rowing. Likewise, actin–myosin bonds have to keep cycling to maintain tension in the muscle.

Many muscles of the body maintain a low level of tension even when the body is at rest. For example, the muscles of the neck, trunk, and limbs that maintain our posture against the pull of gravity are always working, even when we are standing or sitting still. **Muscle tone** comes from the activity of a small but changing number of motor units in a muscle; at any one time, some of the muscle's fibers are contracting and others are relaxed. Muscle tone is constantly being readjusted by the nervous system.

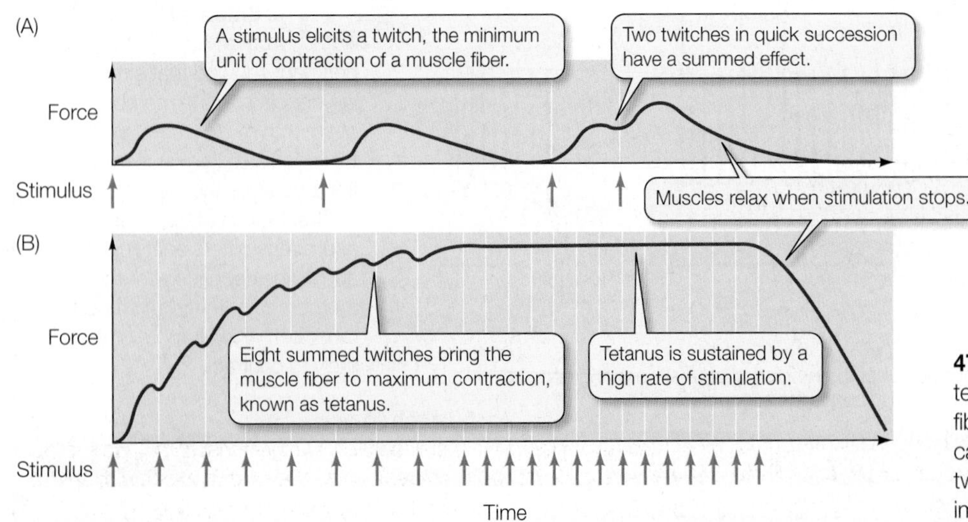

47.9 Twitches and Tetanus (A) Action potentials from a motor neuron cause a muscle fiber to twitch. Twitches in quick succession can be summed. (B) Summation of many twitches can bring the muscle fiber to the maximum level of contraction, known as tetanus.

47.1 RECAP

The most ubiquitous effector is muscle. The contractile ability of muscle derives from interactions between actin and myosin filaments.

- Can you describe how the cellular and subcellular structure of skeletal muscle relate to the sliding filament theory of muscle contraction? See pp. 1007–1008 and Figures 47.1, 47.2, and 47.3
- What is the role of Ca^{2+} in the contractile mechanism of skeletal, cardiac, and smooth muscle? See pp. 1008–1010 and Figures 47.5 and 47.6
- Do you understand the role ATP plays in the actin and myosin interactions that produce contraction? See Figure 47.6

Now that we understand how muscles generate force, we can ask why different muscles have different characteristics, and how individual muscles can change their characteristics with regular use and conditioning.

47.2 What Determines Muscle Strength and Endurance?

The functions that different muscles perform place different demands on them. Some muscles, such as postural muscles, must sustain a load continuously over long periods of time. Other muscles, such as those that control your fingers, generally do not have to sustain long contractions, but they must be able to contract quickly. What differences adapt muscles to specific functions? Can those differences be amplified through conditioning to improve different kinds of human physical performance? A sprinter wants muscles that generate maximum force rapidly, but they do not need to sustain a particular load for a long time. On the other hand, a marathon runner wants muscles with maximum endurance. What properties of muscles determine these functional characteristics?

Muscle fiber types determine endurance and strength

Not all skeletal muscle fibers are alike, and a single muscle contains more than one type of fiber. The two major types of skeletal muscle fibers express different genes for their myosin molecules, and these myosin variants have different rates of ATPase activity. Those with high ATPase activity can recycle their actin–myosin cross-bridges rapidly and are therefore called fast-twitch fibers. Slow-twitch fibers have lower ATPase activity, so they develop tension more slowly but can maintain it longer.

Slow-twitch fibers are also called *oxidative* or *red muscle* because they contain the oxygen-binding protein *myoglobin*, they have many mitochondria, and they are well supplied with blood vessels. These characteristics both increase their capacity for oxidative metabolism and result in their red appearance. The maximum tension a slow-twitch fiber produces is low and develops slowly but is highly resistant to fatigue. Slow-twitch fibers have substantial reserves of fuel (glycogen and fat), so they can maintain steady, prolonged production of ATP as long as oxygen is available. Muscles with high proportions of slow-twitch fibers are good for long-term *aerobic* work (that is, work that requires oxygen). Long-distance runners, swimmers, cyclists, and other athletes whose activities require endurance have leg and arm muscles consisting mostly of slow-twitch fibers (**Figure 47.10**).

Some **fast-twitch fibers** are also called *glycolytic* or *white muscle* because, in comparison to slow-twitch fibers, they have few mitochondria, little or no myoglobin, and fewer blood vessels; thus they look pale. Fast-twitch, glycolytic fibers can develop maximum tension more rapidly than slow-twitch fibers can, and that maximum tension is greater. However, fast-twitch fibers fatigue rapidly. The myosin of these fibers puts the energy of ATP to work very rapidly, but the fibers cannot replenish ATP quickly enough to sustain contraction for a long time. Fast-twitch fibers are especially good for short-term work that requires maximum strength. Weight lifters and sprinters have leg and arm muscles with high proportions of fast-twitch fibers.

There are also fast-twitch fibers that are somewhat oxidative, and therefore intermediate in their properties between slow-twitch and fast glycolytic fibers. These intermediate fibers can become more oxidative with endurance training.

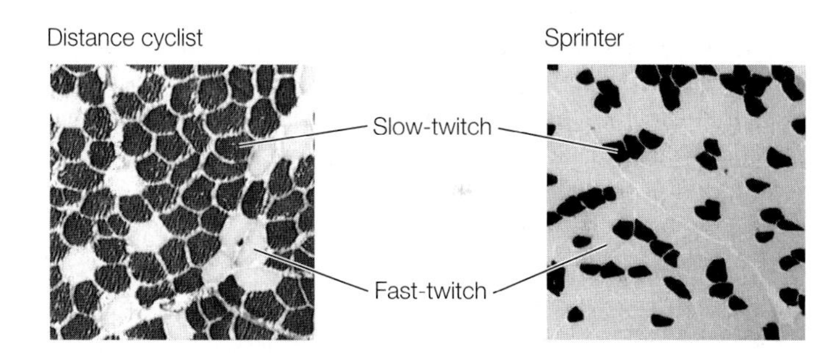

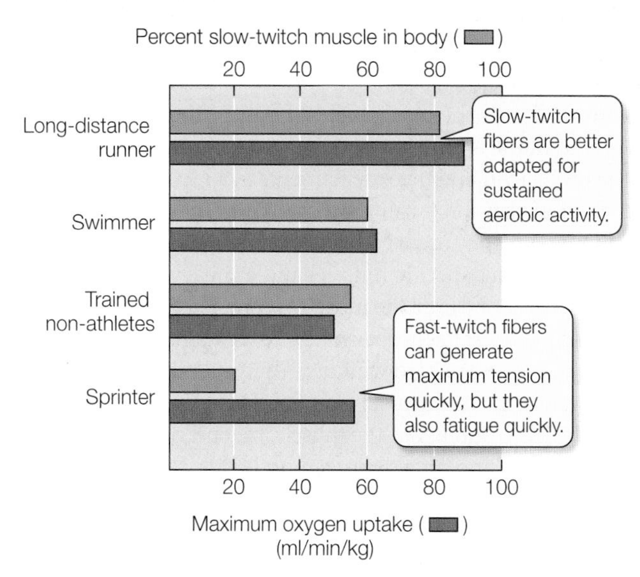

47.10 Slow-and Fast-Twitch Muscle Fibers The skeletal muscles in the micrographs were stained with a reagent that shows slow-twitch fibers as dark; fast-twitch muscle shows up as a light area. Athletes in different sports have different distributions of muscle fiber types.

What determines the proportion of fast- and slow-twitch fibers in your skeletal muscles? The most important factor is genetic heritage, so there is some truth to the statement that champions are born, not made. To a certain extent, you can alter the properties of your muscle fibers through training. But a person born with a high proportion of fast-twitch fibers will never become a champion marathon runner, and one born with a high proportion of slow-twitch fibers will never become a champion sprinter.

A muscle has an optimal length for generating maximum tension

Have you ever done a pull-up? If you have, you know that two parts of this exercise are especially difficult. When you are hanging from the bar with your arms fully extended, it is hard to get the pull-up started; and when your chin is just about to the bar, pulling yourself up the last small distance is difficult. These experiences are explained by the structure of the sarcomere.

When a muscle is stretched and the sarcomeres are lengthened, there is less overlap between the actin and myosin filaments; therefore, fewer cross-bridges can form, and less force can be produced. In fact, if the sarcomeres are stretched too much, actin and myosin do not overlap and no force can be produced. How would a muscle recover from such a difficult situation? The bungee cord-like titin molecules create enough elastic recoil to pull the actin and myosin fibrils back into an overlapping arrangement.

When the muscle is fully contracted, the actin and myosin filaments overlap so much that the myosin bundles are pressed up against the Z lines. Because they have no place to go, additional shortening is difficult. You can see the relationship between the length of a muscle fiber and its ability to develop tension in **Figure 47.11**.

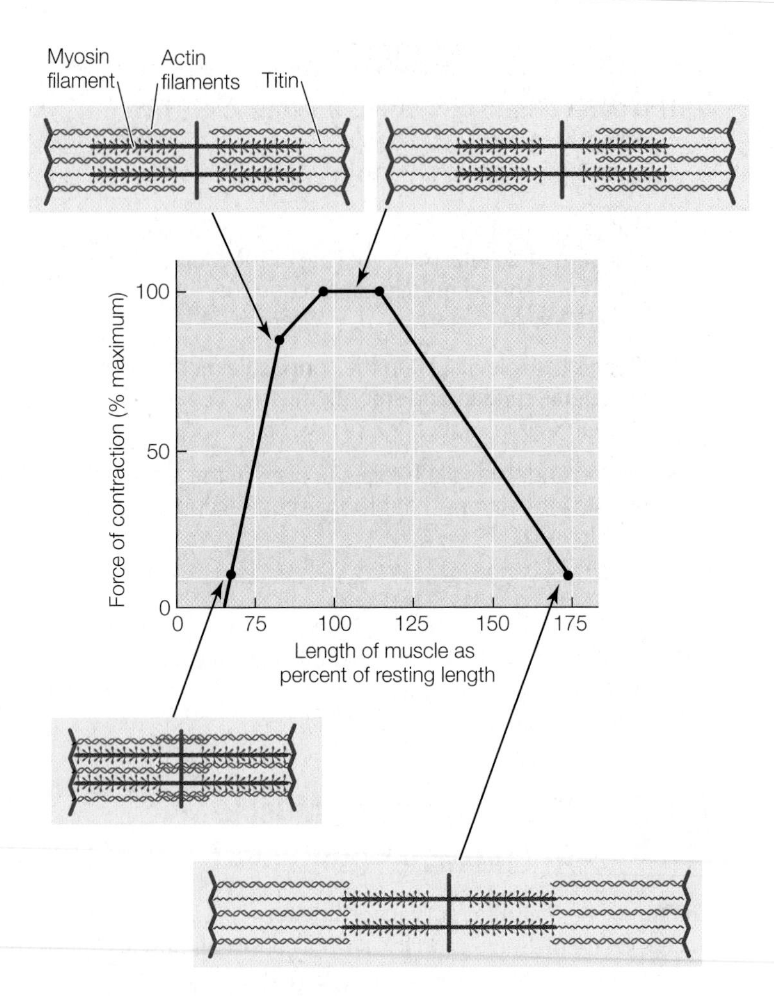

47.11 Strength and Length The amount of force a sarcomere can generate depends on its resting length. When a muscle is stretched, the sarcomeres lengthen, there is less overlap between the actin and myosin filaments, and less force is produced. Overstretched sarcomeres produce no force because there is no overlap between the actin and myosin.

Exercise increases muscle strength and endurance

Different types of exercise produce different physical conditioning responses. In general, anaerobic activities, such as weight lifting, increase strength, and aerobic activities, such as jogging, increase endurance. What is the physiological basis for these differences? Strength is simply a function of the cross-sectional area of muscles: the more actin and myosin filaments in a muscle or a muscle fiber, the more tension it can produce. When athletes undertake strength training, they use weights or exercises such as pull-ups to repeatedly contract specific muscles under heavy loads. Repetitions are usually done until the muscle is completely fatigued. Such stress on a muscle probably does minor tissue damage—hence the soreness the day after a hard workout—but it also induces the formation of new actin and myosin filaments in existing muscle fibers. The muscle fibers, and hence the muscles, get bigger and stronger. In extreme cases, and after serious muscle damage, new muscle fibers can also be produced from stem cells called *satellite cells* in the muscle. In general, however, the major effect of strength training is to produce bigger, rather than more, muscle fibers.

Aerobic exercise has a completely different effect on muscles: it enhances their oxidative capacity. This effect comes from increases in the number of mitochondria, in enzymes involved in energy utilization, and in the density of capillaries that deliver oxygen to the muscle. *Myoglobin* also increases in skeletal muscle cells. **Myoglobin** is an oxygen-binding protein similar to hemoglobin in red blood cells. However, myoglobin has a higher affinity for oxygen than does hemoglobin. Therefore, myoglobin accepts oxygen from the blood, facilitates the diffusion of oxygen throughout the muscle, and provides a store of oxygen for use when oxygen delivery by the blood is insufficient. By increasing the capacity of muscle to use oxygen to produce ATP, aerobic training increases the work load that muscles can sustain through time.

Muscle ATP supply limits performance

Muscles have three systems for obtaining the ATP they need for contraction:

- The *immediate system* uses preformed ATP and creatine phosphate.
- The *glycolytic system* metabolizes carbohydrates to lactate and pyruvate.

- The *oxidative system* metabolizes carbohydrates or fats all the way to H_2O and CO_2.

The capacity of these three systems and the rates at which they can produce ATP determine both work capacity and endurance (**Figure 47.12**).

ATP is stored in muscles in very small amounts. However, muscle fibers also contain a storage compound called *creatine phosphate* (CP). This molecule stores energy in a phosphate bond, which it can transfer to ADP. The total energy available in all the muscles of your body in the form of ATP and CP—the immediate energy system—is only about 10 kilocalories. (A kilocalorie is the amount of energy necessary to raise the temperature of 1 liter of water 1°C). However, the energy from ATP and CP is available immediately, and it enables fast-twitch fibers to generate a lot of force quickly. The immediate system is exhausted in only a few seconds.

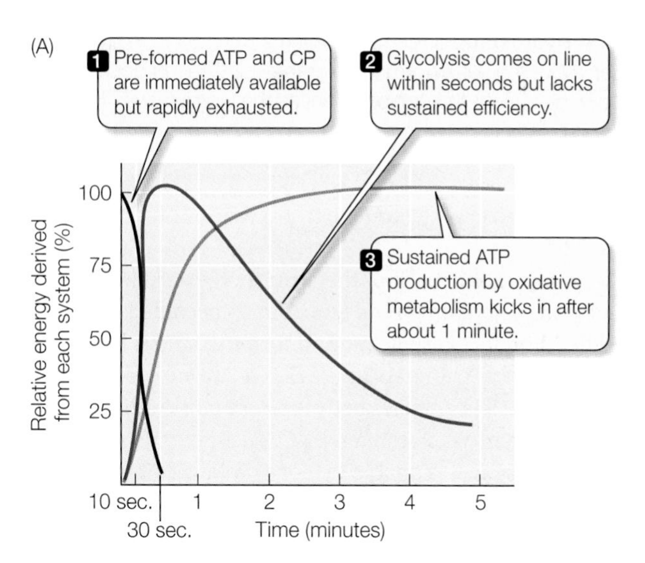

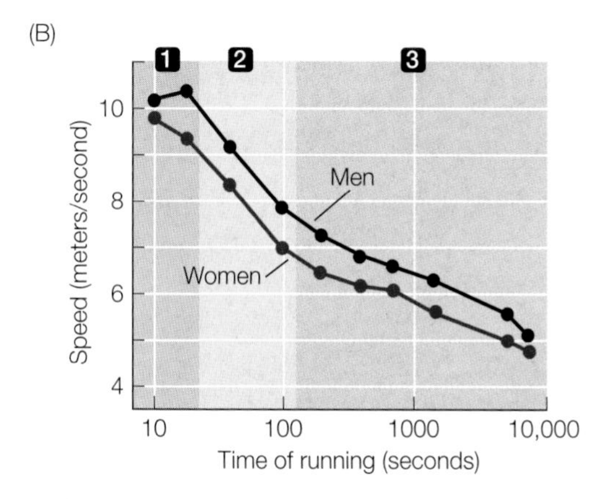

47.12 Supplying Fuel for High Performance (A) Muscles have three systems for obtaining the ATP they need for contraction during exertion such as running. (B) Plotting the time course of world records for running events of different durations, you can see that the performance of world-class athletes corresponds to the time courses of the three energy systems.

The glycolytic system activates within a few seconds to replace the ATP depleted at the onset of muscle activity. The glycolytic enzymes are located in the cytoplasm of the muscle fiber, and therefore the ATP they generate is rapidly available to the myosin filaments. However, as we saw in Chapter 7, glycolysis alone is an inefficient way to produce ATP, and it leads to the accumulation of lactic acid, which slows the process. Thus, the glycolytic system and the immediate system together can provide most of the energy for active muscles for less than a minute (see Figure 47.12).

Oxidative metabolism becomes fully active in about a minute, producing relatively huge amounts of ATP because it can completely metabolize carbohydrates and fats. However, it requires many reactions (see Chapter 7), and it takes place in the mitochondria, so O_2 and substrate must diffuse into the mitochondria, and the formed ATP must diffuse from the mitochondria to the myosin filaments in the muscle. These processes are not instantaneous, so the rate at which oxidative metabolism can make ATP available to do work is slower than the rate at which the other two systems can supply ATP.

The fuel supply available to the muscles influences how long someone can sustain a high level of aerobic exercise. From the circulating blood, muscle receives glucose and free fatty acids, which it can metabolize to generate ATP. At high levels of aerobic exercise, however, most of the fuel used by muscles to produce ATP comes from the reserve of glycogen stored in the muscle itself. Depletion of muscle glycogen results in fatigue. The rate at which muscle glycogen is replenished depends on diet: it is high with a high-carbohydrate diet, low with a high-fat diet, and intermediate with a mixed diet.

After muscle glycogen is depleted by exercise, it is replenished more rapidly by a high-carbohydrate diet than by a diet of fats and protein. This fact is the basis for a practice called "carbo-loading." For 3–5 days, the athlete exercises at a level that depletes muscle glycogen. Then, 2 or 3 days before the event, she tapers down the level of training and eats a diet rich in complex carbohydrates. The result can be glycogen supercompensation, in which the restoration of muscle glycogen stores "overshoots" and reaches above-normal levels.

47.2 RECAP

Depending on the function a muscle serves, it may need to generate maximum force rapidly or sustain activity for a long period. Properties of muscles can adapt them to either of these types of functions.

- Can you describe the differences between slow-twitch and fast-twitch fibers? See p. 1013 and Figure 47.10
- How does exercise influence muscle strength and endurance? See p. 1014
- How do the different sources of ATP influence performance in different types of events? See pp. 1014–1015 and Figure 47.12

47.3 What Roles Do Skeletal Systems Play in Movement?

Muscles can only contract and relax. To create significant movement, they must have something to pull on. In some cases, muscles pull on each other—consider the trunk of the elephant or the arms of an octopus. In most cases, however, **skeletal systems** provide rigid supports against which muscles can pull, creating directed movements. In this section, we'll examine the three types of skeletal systems found in animals: hydrostatic skeletons, exoskeletons, and endoskeletons.

A hydrostatic skeleton consists of fluid in a muscular cavity

The simplest type of skeleton is the hydrostatic skeleton of cnidarians, annelids, and many other soft-bodied invertebrates. As Section 31.2 discusses, a **hydrostatic skeleton** consists of a volume of fluid enclosed in a body cavity surrounded by muscle. When muscles oriented in a certain direction contract, the fluid-filled body cavity bulges out in the opposite direction.

An earthworm uses its hydrostatic skeleton to crawl. The earthworm's body cavity is divided into many separate segments, each of which contains a compartment filled with extracellular fluid. The body wall surrounding each segment has two muscle layers: a circular layer and a longitudinal layer. If the circular muscles in a segment contract, the compartment in that segment narrows and elongates. If the longitudinal muscles in a segment contract, the compartment shortens and bulges outward. Alternating contractions of the earthworm's circular and longitudinal muscles create waves of narrowing and widening, lengthening and shortening, that travel down the body. Bulging, shortened segments serve as anchors as long, narrow segments project forward and longitudinal contractions pull other segments forward. Bristles help the widest parts of the body to hold firm against the substratum (**Figure 47.13**).

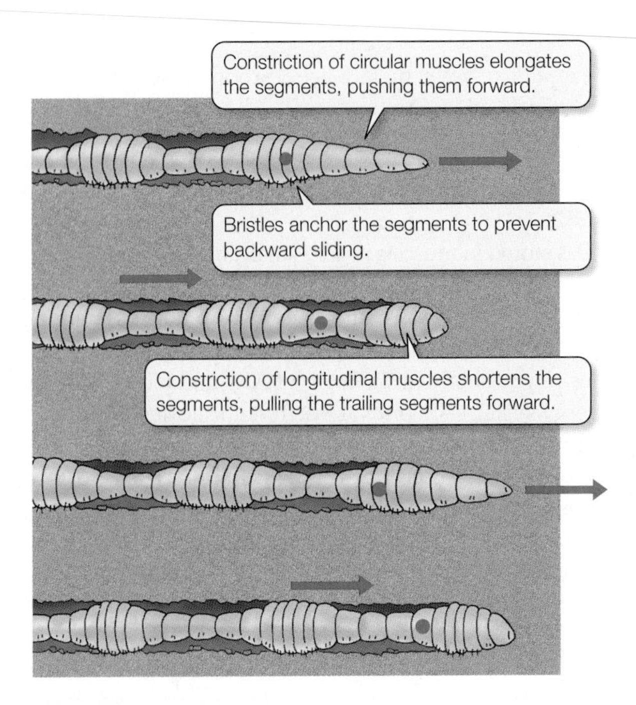

47.13 A Hydrostatic Skeleton Alternating waves of muscle contraction move the earthworm through the soil. The red dot enables you to follow the changes in one segment as the worm moves forward.

Exoskeletons are rigid outer structures

An **exoskeleton** is a hardened outer surface to which muscles can be attached. Contractions of the muscles cause jointed segments of the exoskeleton to move relative to each other. The simplest example of an exoskeleton is the shell of a mollusk. Some marine mollusks, such as clams and snails, have shells composed of protein strengthened by crystals of calcium carbonate (a rock-hard material). These shells can be massive, affording significant protection against predators. The shells of land snails generally lack the hard mineral component and are much lighter. Molluscan shells can grow as the animal grows, and growth rings are usually apparent on the shells (**Figure 47.14**).

The most complex exoskeletons are found among the arthropods. An exoskeleton, or **cuticle**, covers all the outer surfaces of the arthropod's body and all its appendages. It is made up of plates secreted by a layer of cells just below the exoskeleton. The cuticle contains stiffening materials everywhere except at the joints, where flexibility must be retained. Muscles attached to the inner surfaces of the arthropod exoskeleton move its parts around the joints (see Figure 32.4).

The greatest drawback of the arthropod exoskeleton is that it cannot grow. Therefore, if the animal is to become larger, it must *molt*, shedding its exoskeleton and forming a new, larger one. A molting animal is vulnerable because the new exoskeleton takes time to harden. The animal's body is temporarily unprotected, and without a firm exoskeleton against which its muscles can exert maximum tension, it is unable to move rapidly. Soft-shelled crabs, a gourmet delicacy, are crabs caught when they are molting.

47.14 The Clam Shell Is an Exoskeleton The clam is a bivalve, meaning its exoskeleton consists of an upper and lower shell that are hinged. As the clam grows, the shells grow bigger and are marked by growth rings.

Spiders have an unique mechanism of movement. By pumping extracellular fluid into their hollow legs, they cause those legs to extend. Decreasing fluid pressure allows the legs to flex. This is why the legs of dead spiders curl up underneath them.

Vertebrate endoskeletons provide supports for muscles

The **endoskeleton** of vertebrates is an internal scaffolding. Muscles are attached to it and pull against it. Endoskeletons are composed of rodlike, platelike, and tubelike bones connected to one another at a variety of joints that allow a wide range of movements. An advantage of endoskeletons over the exoskeletons of arthropods is that bones within the body can enlarge without the animal shedding its skeleton.

The human skeleton consists of 206 bones, some of which are shown in **Figure 47.15**. It can be divided into an *axial skeleton,* which includes the skull, vertebral column, and ribs, and an *appendicular skeleton,* which includes the pectoral girdle, the pelvic girdle, and the bones of the arms, legs, hands, and feet.

The vertebrate endoskeleton consists of two kinds of connective tissue, *cartilage* and *bone,* which are produced by two kinds of connective tissue cells. *Cartilage cells* produce an extracellular matrix that is a tough, rubbery mixture of polysaccharides and proteins—mainly fibrous collagen. Collagen fibers run in all directions like reinforcing cords through the gel-like matrix and give it the well-known strength and resiliency of "gristle." This matrix, called **cartilage**, is found in parts of the endoskeleton where both stiffness and resiliency are required, such as on the surfaces of joints where bones move against one another. Cartilage is also the supportive tissue in stiff but flexible structures such as the larynx (voice box), the nose, and the ear pinnae. Sharks and rays are called *cartilaginous fishes* because their skeletons are composed entirely of cartilage. In all other vertebrates, cartilage is the principal component of the embryonic skeleton, but during development most of it is gradually replaced by bone.

Bone also contains collagen fibers, but it gets its rigidity and hardness from an extracellular matrix of insoluble calcium phosphate crystals. Bone serves as a reservoir of calcium for the rest of the body and is in dynamic equilibrium with soluble calcium in the extracellular fluids of the body. This equilibrium is under the control of calcitonin and parathyroid hormone (see Figure 41.10). If too much calcium is taken from the skeleton, the bones are seriously weakened. The living cells of bone—called *osteoblasts, osteocytes,* and *osteoclasts*—are responsible for the constant dynamic remodeling of bone (**Figure 47.16**). **Osteoblasts** lay down new matrix material on bone surfaces. These cells gradually become surrounded by matrix and eventually become enclosed within the bone, at which point they cease laying down matrix, but continue to exist within small lacunae (cavities) in the bone. In this state they are called **osteocytes**. Despite the vast amounts of matrix between them, osteocytes remain in contact with one another through long cellular extensions that run through tiny channels in the bone. Communication between osteocytes is important in controlling the activities of the cells that are laying down or removing bone.

The cells that resorb bone are the **osteoclasts**. They are derived from the same cell lineage that produces the white blood cells. Osteoclasts erode bone, forming cavities and tunnels. Osteoblasts follow osteoclasts, depositing new bone. Thus the interplay of osteoblasts and osteoclasts constantly replaces and remodels the bones.

How the activities of the bone cells are coordinated is not understood, but stress placed on bones somehow provides them with information. A remarkable finding in studies of astronauts who spent

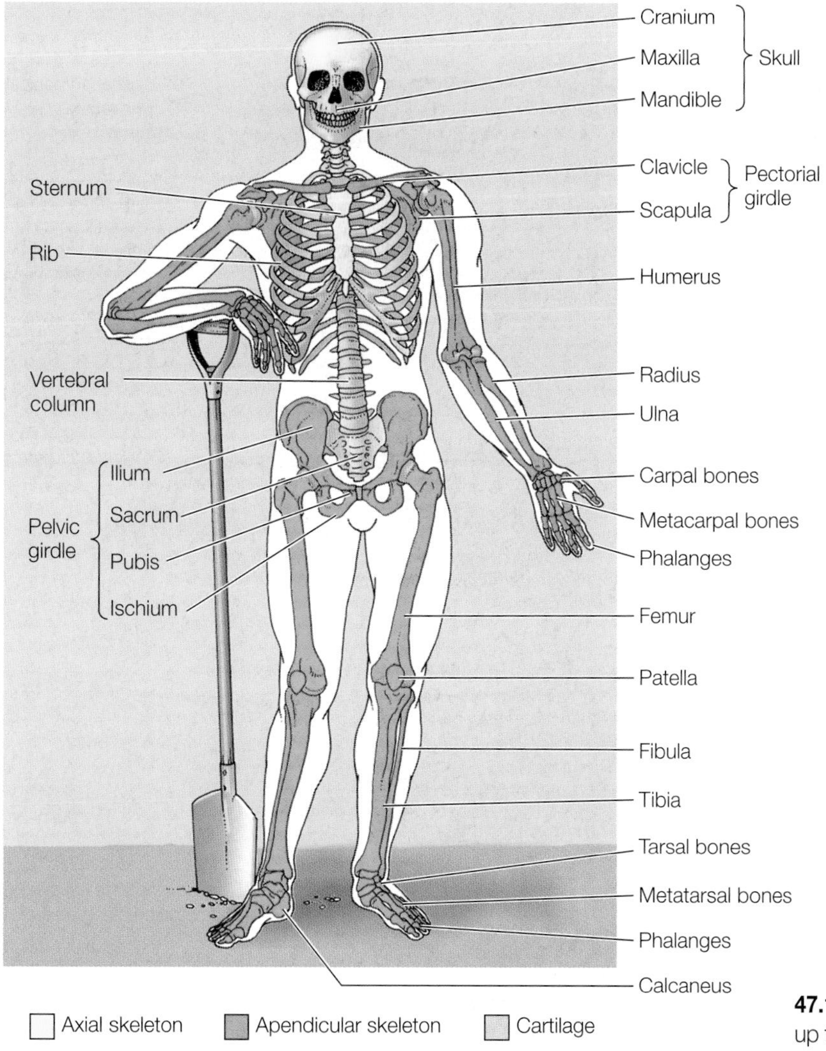

Axial skeleton | Apendicular skeleton | Cartilage

47.15 The Human Endoskeleton Cartilage and bone make up the internal skeleton of a human being.

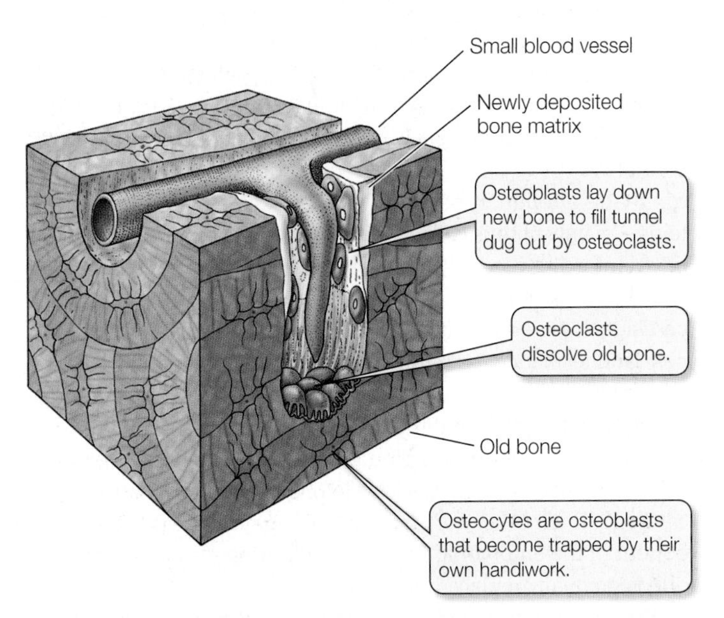

47.16 Renovating Bone Bones are constantly being remodeled by osteoblasts, which lay down bone, and osteoclasts, which resorb bone.

long periods in zero gravity was that their bones decalcified. Conversely, certain bones of athletes thicken during training. Both thickening and thinning of bones are experienced by anyone who has had a leg in a cast for a long time. The bones of the uninjured leg carry the person's weight and thicken, while the bones of the inactive leg in the cast thin.

Because of the positive effects of physical stress on bone deposition, weight-bearing exercise is effective in preventing and treating the loss of bone density (and hence strength) known as *osteoporosis.* Over 25 million people in the United States suffer from this debilitating condition. Although osteoporosis is most commonly a problem for postmenopausal women, it can occur in younger people as a result of malnutrition and sedentary lifestyle.

Bones develop from connective tissues

Bones are divided into two types on the basis of how they develop. **Membranous bone** forms on a scaffold of connective tissue membrane. **Cartilage bone** forms first as a cartilaginous structure resembling the future mature bone, then gradually hardens or *ossifies* to become bone. The outer bones of the skull are membranous bones; the bones of the limbs are cartilage bones.

Cartilage bones can grow throughout the ossification process. The long bones of the legs and arms, for example, ossify first at the centers and later at each end (**Figure 47.17**). Growth can continue until these areas of ossification join. The membranous bones forming the skull cap grow until their edges meet. The soft spot on the top of a baby's head (the fontanelle) is the point at which the skull bones have not yet joined.

The structure of bone may be **compact** (solid and hard) or **cancellous** (having numerous internal cavities that make it appear spongy, although it is rigid). The architecture of a specific bone depends on its position and function, but most bones have both compact and cancellous regions. The shafts of the long bones of the limbs, for example, are cylinders of compact bone surrounding central cavities that contain the bone marrow, where the cellular elements of the blood are made. The ends of the long bones are cancellous (see Figure 47.17). Cancellous bone is lightweight because of its numerous cavities, but it is also strong because its internal meshwork constitutes a support system. It can withstand considerable forces of compression. The rigid, tubelike shaft of compact bone can withstand compression and bending forces. Architects and nature alike use hollow tubes as lightweight structural elements.

Most of the compact bone in mammals is called *Haversian bone* because it is composed of structural units called **Haversian systems** (**Figure 47.18**). Each Haversian system is a set of thin, concentric bony cylinders, between which are the osteocytes in their lacunae. Through the center of each Haversian system runs a narrow canal containing blood vessels and nerves. Adjacent Haversian systems are separated by boundaries called *glue lines*. Haversian bone is resistant to fracturing because cracks tend to stop at glue lines.

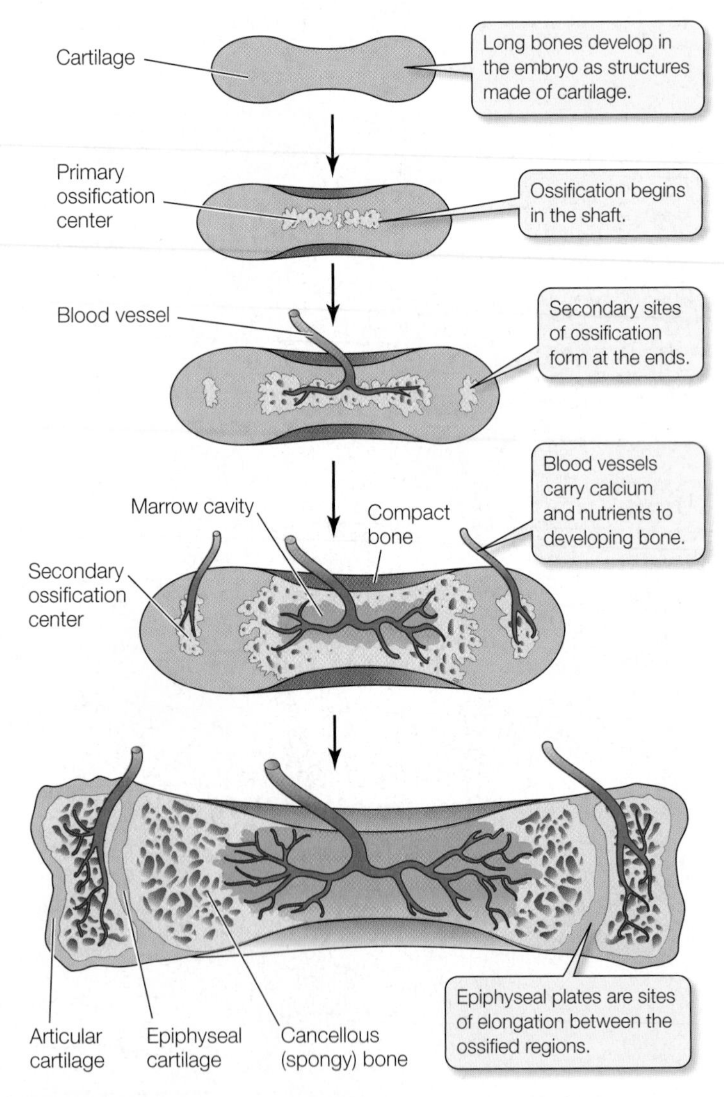

47.17 The Growth of Long Bones In the long bones of human limbs, ossification occurs first at the centers and later at each end.

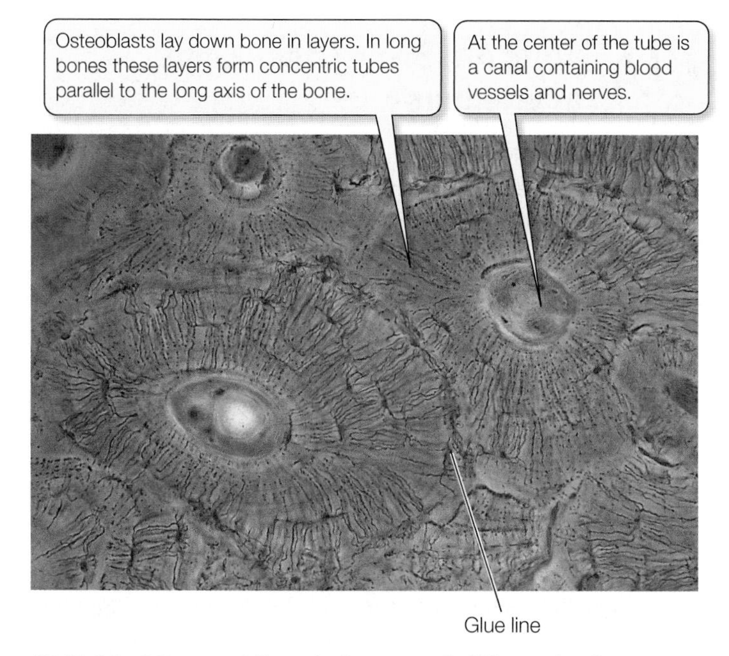

47.18 Most Compact Bone Is Composed of Haversian Systems A micrograph of a section of a long bone shows Haversian systems with their central canals. Glue lines separate Haversian systems.

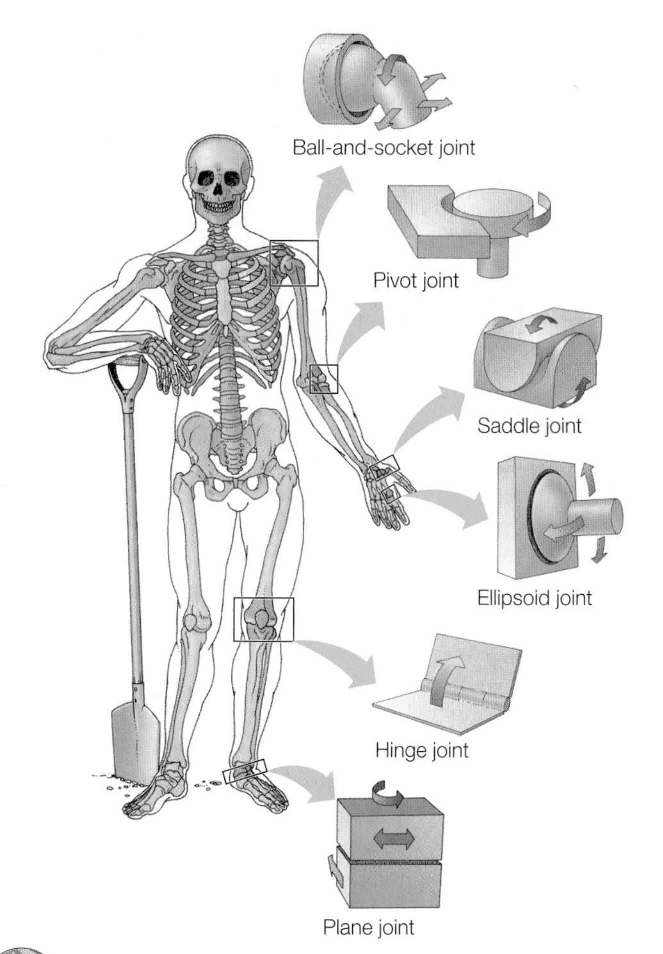

47.19 Types of Joints The designs of joints are similar to mechanical counterparts and enable a variety of movements.

Bones that have a common joint can work as a lever

Muscles and bones work together around **joints**, where two or more bones come together. Different kinds of joints allow motion in different directions (**Figure 47.19**), but muscles can exert force in only one direction. Therefore, muscles create movement around joints by working in antagonistic pairs: When one contracts, the other relaxes. When both contract, the joint becomes rigid. With respect to a particular joint, such as the knee, we can refer to the muscle that bends, or flexes, the joint as the **flexor**, and the muscle that straightens, or extends, the joint as the **extensor**. The bones that meet at the joint are held together by **ligaments**, which are flexible bands of connective tissue. Other straps of connective tissue, called **tendons**, attach the muscles to the bones (**Figure 47.20**). In many kinds of joints, only the tendon spans the joint, sometimes moving over the surfaces of the bones like a rope over a pulley. The tendon of the quadriceps muscle traveling over the knee joint is what is tapped to elicit the knee-jerk reflex (see Figure 46.3).

Bones constitute a system of levers that are moved around joints by the muscles. A lever has a *power arm* and a *load arm* that work around a *fulcrum* (pivot). The length ratio of the two arms determines whether a particular lever can exert a lot of force over a short distance or is better at translating force into large or fast movements. Compare the jaw joint and the knee joint, for example (**Figure 47.21**). The power arm of the jaw is long relative to the load arm, allowing the jaw to apply great force over a small distance. Think of the powerful jaws of carnivores that can easily crack bones. The power arm of the lower leg, on the other hand, is short relative to the load arm, so you can run fast, jump high, and deliver swift kicks.

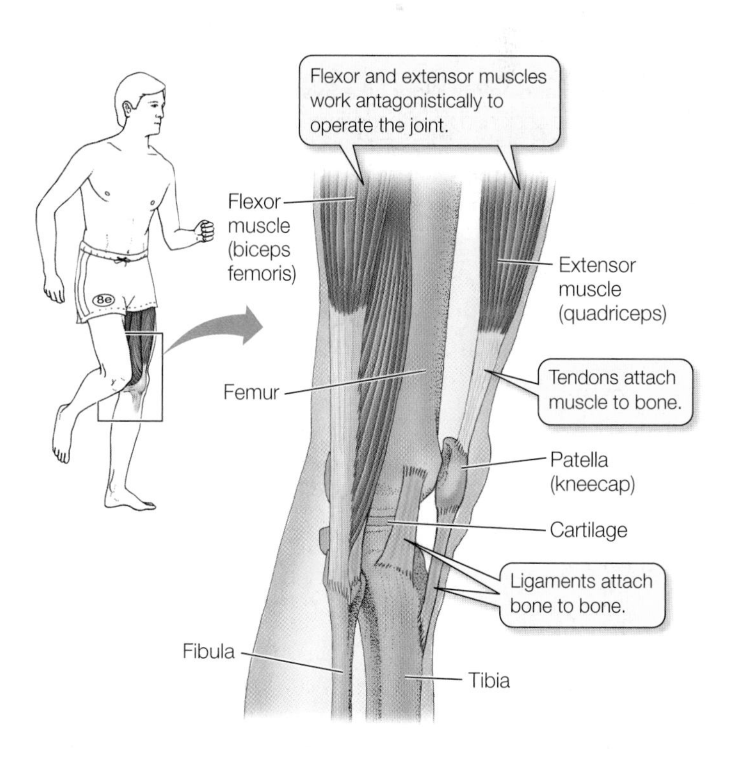

47.20 Joints, Ligaments, and Tendons A side view of the knee shows the interactions of muscle, bone, cartilage, ligaments, and tendons at this crucial and vulnerable human joint.

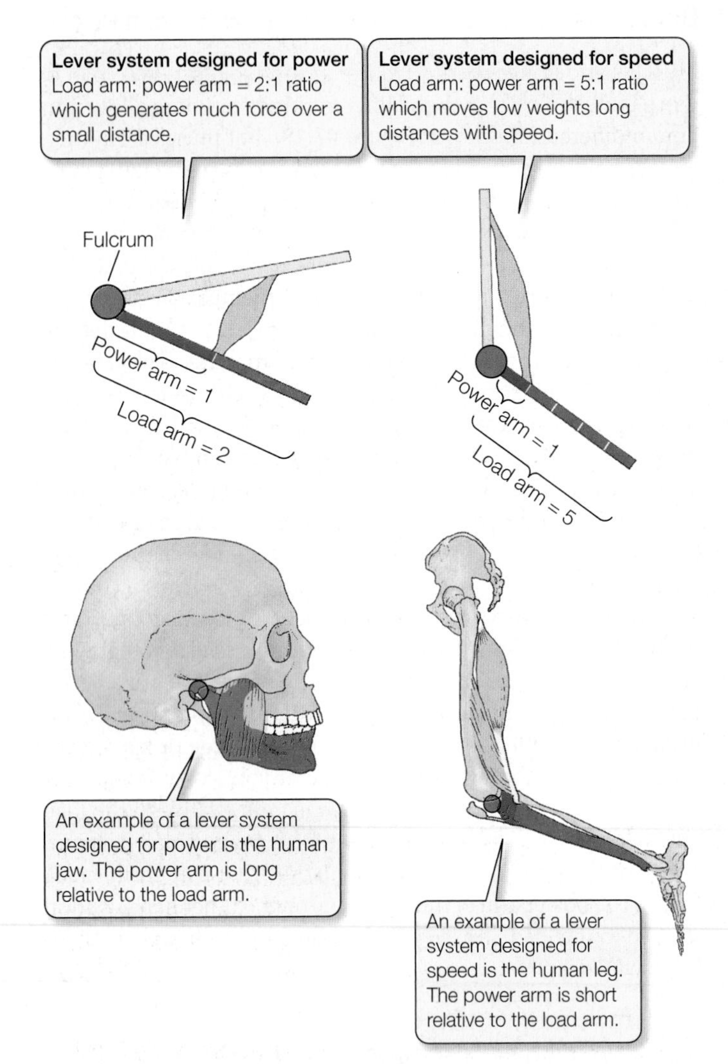

47.21 Bones and Joints Work Like Systems of Levers A lever system can be designed for either power or speed.

47.3 RECAP

Muscles can only contract and relax; to achieve organized movement they have to pull against rigid structures—other muscles, hydrostatic skeletons, exoskeletons, or endoskeletons.

- Can you describe how muscles and a fluid-filled body cavity interact to enable an earthworm to crawl? See p. 1016 and Figure 47.13
- Can you describe the difference between membranous and cartilaginous bone and between compact and cancellous bone? See p. 1018 and Figure 47.17
- In terms of levers can you explain how specific joints can produce maximum force versus maximum speed? See p. 1019 and Figure 47.21

47.4 What Are Some Other Kinds of Effectors?

Musculoskeletal systems, whether they be hydrostatic skeletons, exoskeletons, or endoskeletons, are just one of many effector systems that allow different species of animals to accomplish defense, communication, feeding, and reproduction. For example, in Chapter 31 we learned how cnidarians such as jellyfish use *nematocysts,* an effector that fires miniature "harpoons" to capture prey and repel predators (see Figure 31.10). Several other effector systems are described next, revealing the evolutionary diversity of the mechanisms animals use to get things done.

Chromatophores allow an animal to change its color or pattern

A change in body color is a response that some animals use to camouflage themselves in a particular environment or to communicate with other animals. **Chromatophores** are pigment-containing cells in the skin that can change the color and pattern of the animal. Chromatophores are under neuronal or hormonal control, or both; in most cases, they can effect a change within minutes or even seconds.

Chromatophores enable squids, sole, and flounder, all of which spend much time on the seafloor, as well as the famous chameleons (a group of African lizards; see Figure 52.10) and a few other animals, to blend in with the background on which they are resting and thus escape discovery by predators. Chromatophores with different pigments enable animals to assume different hues or to become mottled to match the background more precisely. In other mollusks, fishes, and lizards, a color change sends a signal to potential mates and territorial rivals of the same species.

There are three principal types of chromatophore cells. The most common type has fixed cell boundaries, within which pigmented granules may be moved about by microfilaments. When the pigment is concentrated in the center of each chromatophore, the animal is pale; the animal turns darker when the pigment is dispersed throughout the cell. Another type of chromatophore is capable of amoeboid motion. These cells can mold themselves into shapes with a minimal surface area, leaving the tissue relatively pale, or they can flatten out to make the tissue appear darker. The third type of chromatophore changes shape as a result of the action of muscle fibers radiating outward from the cell (**Figure 47.22A**). When the muscle fibers are relaxed, the chromatophores are small and compact, and the animal is pale. To darken the animal, the muscle fibers contract and spread the chromatophores over more of the body surface. These chromatophores can change so rapidly that they are used in some species for communication during courtship and aggressive interactions. For example, the cuttlefish, a cephalopod, can signal courtship intentions to a potential mate on one side of its body while signaling aggressive threats to a rival on the other side (**Figure 47.22B**).

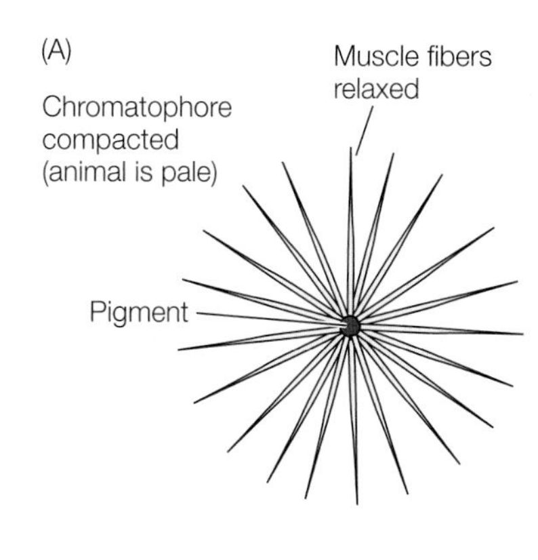

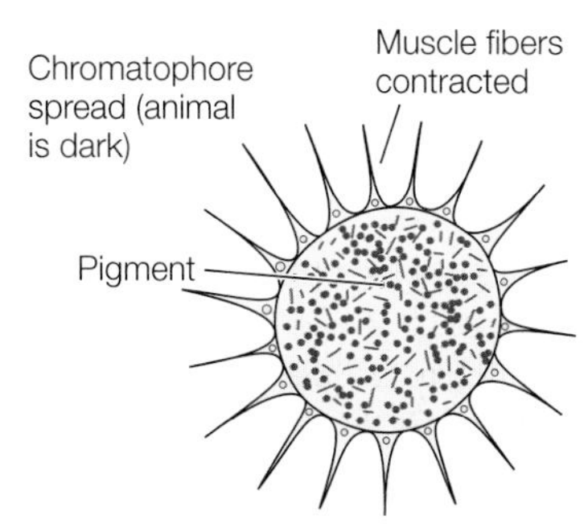

47.22 Chromatophores Help Animals Camouflage Themselves or Communicate (A) Muscle fibers around chromatophores cause the chromatophores to contract. (B) Cuttlefish are cephalopod mollusks that can change color patterns so fast that these changes can be used for rapid communication.

Glands secrete chemicals for defense, communication, or predation

Glands are effector organs that produce and release chemicals. As we learned in Chapter 41, endocrine glands produce hormones for internal signaling and exocrine glands secrete substances into the gut or onto the body surface. In certain animals, some of these secretions are used defensively or to capture prey. Others are *pheromones,* chemical signals released into the environment for communication with other individuals of the same species.

Not all defensive secretions are poisonous; a well-known example is mercaptan, the odoriferous chemical sprayed by skunks as an effective deterrent to most predators.

Certain reptiles, amphibians, mollusks, and fishes have poison glands that are used in capturing prey or defending against predators. Many of the poisons produced by these glands are extremely specific in their modes of action. For example, the poison dendrotoxin, which certain tribes of the Amazonian rainforest use on the tips of their arrows for hunting, comes from the skin of a frog and blocks certain potassium channels. The snake venom α-bungarotoxin inactivates the acetylcholine receptors at the neuromuscular junction. Tetrodotoxin, or TTX (see the opening pages of Chapter 22) blocks voltage-gated sodium channels. Conotoxin, the poison produced by cone snails, blocks calcium channels.

Electric organs generate electricity used for sensing, communication, defense, or attack

Various fishes can generate electricity. These species include the electric eel, the knife fish, the torpedo (a type of ray), and the electric catfish. The electric fields they generate help them sense the environment, communicate, and stun potential predators or prey. The electric organs of these animals evolved from muscles, and they produce electric potentials in the same general way as nerves and muscles do.

Electric organs consist of very large, disc-shaped cells arranged in long rows like stacks of batteries. When these cells discharge simultaneously, the electric organ can generate far more voltage and current than can nerve or muscle tissue. Electric eels, for example, can produce up to 600 volts with an output of approximately 100 watts—enough to light a row of light bulbs or to temporarily stun a person.

Light-emitting organs use enzymes to produce light

Diverse organisms can emit light—**bioluminescence**. Some, such as fireflies, can produce their own light, and others, such as some deep sea fish, use symbiotic luminescent bacteria sequestered in special organs as their light source. The ability to produce light depends on enzymatic oxidation of a substrate. The enzyme is called *luciferase* and the substrate *luciferin*. Bioluminescence serves various functions, including social signaling, attraction of prey, or even avoidance of prey.

47.4 RECAP

Although muscles and the movements they create are ubiquitous effectors in animals, a host of other types of effectors exist, many of which are highly specialized and found in only a limited number of species.

- Can you explain how chromatophores work? See p. 1020 and Figure 47.22
- Can you describe two examples of how effectors are used both for defense and for communication? See p. 1021

CHAPTER SUMMARY

47.1 How do muscles contract?

See Web/CD Tutorial 47.1

Skeletal muscle consists of bundles of **muscle fibers**. Each skeletal muscle fiber is a huge cell containing multiple nuclei.

Skeletal muscles contain numerous **myofibrils**, which are bundles of **actin** and **myosin** filaments. The regular, overlapping arrangement of the actin and myosin filaments into **sarcomeres** gives skeletal muscle its striated appearance. Review Figure 47.1, Web/CD Activity 47.1

The changes in the banding patterns of sarcomeres led to the sliding filament model of muscle contraction. Review Figure 47.2

The molecular mechanism of muscle contraction involves the binding of the globular heads of myosin molecules to actin. Upon binding, the myosin head changes its conformation, causing the two filaments to slide past each other (the **sliding filament** mechanism). Release of the myosin heads from actin and their return to their original conformation requires ATP. Review Figures 47.3 and 47.6

All the fibers activated by a single motor neuron constitute a **motor unit**. When an action potential spreads across the plasma membrane and through the **T tubules**, Ca^{2+} is released from the **sarcoplasmic reticulum**. Review Figure 47.5, Web/CD Activity 47.2

The Ca^{2+} binds to **troponin** and changes its conformation, pulling the **tropomyosin** strands away from the myosin-binding sites on the actin filament. The muscle fiber contracts until the Ca^{2+} is returned to the sarcoplasmic reticulum. Review Figure 47.6

Cardiac muscle cells are striated, uninucleate, branching, and electrically connected by gap junctions, so that action potentials spread rapidly throughout sheets of cardiac muscle and cause coordinated contractions. Some cardiac muscle cells are pacemaker cells that generate and conduct electrical signals.

Smooth muscle provides contractile force for internal organs. Smooth muscle cells respond to stretch and to neurotransmitters from the autonomic nervous system. See Web/CD Tutorial 47.2

In skeletal muscle, a single action potential causes a minimum unit of contraction called a **twitch**. Twitches occurring in rapid succession can be summed, thus increasing the strength of contraction. Maximum sustained tension is known as **tetanus**. Review Figure 47.9

47.2 What determines muscle strength and endurance?

Slow-twitch muscle fibers are adapted for extended, aerobic work; **fast-twitch** fibers are adapted for generating maximum forces for short periods of time. The ratio of slow-twitch to fast-twitch fibers in the muscles of an individual is mostly genetically determined. Review Figure 47.10

The force that a muscle fiber can produce depends on its initial state of extension or contraction. Review Figure 47.11

Anaerobic exercise stimulates the enlargement of muscle fibers through production of new microfilaments. Through aerobic conditioning, muscle fibers can acquire greater oxidative capacity.

Muscle performance depends on ATP supply. Available ATP and creatine phosphate can fuel maximum tension immediately, but are exhausted in seconds. Glycolysis can regenerate ATP rapidly, but is rapidly slowed by accumulation of lactic acid. Oxidative metabolism delivers ATP more slowly, but can continue to do so for a long time. Review Figure 47.12

47.3 What Roles Do Skeletal Systems Play in Movement?

Skeletal systems provide supports against which muscles can pull.

Hydrostatic skeletons are fluid-filled body cavities that can be squeezed by muscles. Review Figure 47.13

Exoskeletons are hardened outer surfaces to which internal muscles are attached.

Endoskeletons are internal systems of rigid rodlike, platelike, and tubelike supports, consisting of **bone** and **cartilage**, to which muscles are attached. Review Figure 47.15

Bone is continually remodeled by **osteoblasts**, which lay down new bone, and **osteoclasts**, which erode bone. Review Figure 47.16

Bones develop from connective tissue membranes (**membranous bone**) or from cartilage (**cartilage bone**) through ossification. Cartilage bone can grow until centers of ossification meet. Review Figure 47.17

Bone can be solid and hard (**compact bone**), or it can contain numerous internal spaces (**cancellous bone**). Most of the compact bone of mammals is composed of **Haversian systems**. Review Figure 47.18

Joints enable muscles to power movements in different directions. Review Figure 47.19, Web/CD Activity 47.3

Tendons connect muscles to bones; **ligaments** connect bones to one another. Review Figure 47.20

Muscles and bones work together around **joints** as systems of levers. Review Figures 47.21

47.4 What Are Some Other Kinds of Effectors?

Effector organs other than muscles include nematocysts, chromatophores, glands, electric organs, and light-emitting organs.

SELF-QUIZ

1. Smooth muscle differs from both cardiac and skeletal muscle in that
   a. it can act as a pacemaker for rhythmic contractions.
   b. contractions of smooth muscle are not due to interactions between neighboring microfilaments.
   c. neighboring cells are electrically connected by gap junctions.
   d. neighboring cells are tightly coupled by intercalated discs.
   e. the membranes of smooth muscle cells are depolarized by stretching.

2. Fast-twitch fibers differ from slow-twitch fibers in that
   a. they are more common in the leg muscles of champion sprinters than marathon runners.
   b. they have more mitochondria.
   c. they fatigue less rapidly.
   d. their abundance is more a product of training than of genetics.
   e. they are more common in postural muscles than in finger muscles.

3. The role of Ca^{2+} in the control of muscle contraction is to
   a. cause depolarization of the T tubule system.
   b. change the conformation of troponin, thus exposing myosin-binding sites.
   c. change the conformation of myosin heads, thus causing microfilaments to slide past each other.
   d. bind to tropomyosin and break actin–myosin cross-bridges.
   e. block the ATP-binding site on myosin heads, enabling muscles to relax.
4. Fifteen minutes into a 10-km run, what is the major energy source of the leg muscles?
   a. Preformed ATP
   b. Glycolysis
   c. Oxidative metabolism
   d. Pyruvate and lactate
   e. High-protein drink consumed right before the race
5. Which statement about skeletal muscle contraction is *not* true?
   a. A single action potential at the neuromuscular junction is sufficient to cause a muscle to twitch.
   b. Once maximum muscle tension is achieved, no ATP is required to maintain that level of tension.
   c. An action potential in the muscle cell activates contraction by releasing Ca^{2+} into the sarcoplasm.
   d. Summation of twitches leads to a graded increase in the tension that can be generated by a single muscle fiber.
   e. The tension generated by a muscle can be varied by controlling how many of its motor units are active.
6. Which statement about the structure of skeletal muscle is *true*?
   a. The light bands of the sarcomere are the regions where actin and myosin filaments overlap.
   b. When a muscle contracts, the A bands of the sarcomere lengthen.
   c. The myosin filaments are anchored in the Z lines.
   d. When a muscle contracts, the H zone of the sarcomere shortens.
   e. The sarcoplasm of the muscle cell is contained within the sarcoplasmic reticulum.
7. The long bones of our arms and legs are strong and can resist both compressional and bending forces because
   a. they are solid rods of compact bone.
   b. their extracellular matrix contains crystals of calcium carbonate.
   c. their extracellular matrix consists mostly of collagen and polysaccharides.
   d. they have a very high density of osteoclasts.
   e. they consist of lightweight cancellous bone with an internal meshwork of supporting elements.
8. If we compare the jaw joint with the knee joint as lever systems,
   a. the jaw joint can apply greater compressional forces.
   b. their ratios of power arm to load arm are about the same.
   c. the knee joint has greater rotational abilities.
   d. the knee joint has a greater ratio of power arm to load arm.
   e. only the jaw is a hinged joint.
9. Which statement about skeletons is *true*?
   a. They can consist of mostly cartilage.
   b. Hydrostatic skeletons cannot be used for locomotion.
   c. An advantage of exoskeletons is that they can continue to grow throughout the life of the animal.
   d. External skeletons must remain flexible, so they never include calcium carbonate crystals, as bones do.
   e. Internal skeletons consist of four different types of bone: compact, cancellous, membranous, and Haversian.
10. Which of the following effectors is *not* used both for avoiding predators and for communication?
    a. Chromatophores
    b. Electric organs
    c. Skeletal muscle
    d. Glandular secretions
    e. All of the above

FOR DISCUSSION

1. You can see from the structure of a sarcomere that it can shorten only by a certain percentage of its resting length. Yet muscles can cause a wide variety of ranges of movement—compare the range of movement of a toe and a leg. What are two adaptive design features of muscles and skeletons that can maximize the ability of a muscle to cause a greater range of movement of an appendage?
2. If athletes train up until a day before competition and then take a day of rest, performance in which types of events will be most affected by what they eat during the rest day?
3. Wombats are powerful digging animals, and kangaroos are powerful jumping animals. How do you think the structures of their legs would compare in terms of their designs as lever systems?
4. Why are ducks better long-distance fliers than chickens?
5. If an adolescent breaks a leg bone close to the ankle joint, after the break heals, that leg may not grow as long as the other one. Why?

FOR INVESTIGATION

For aerobic exercise endurance, the most important limiting factor is muscle glycogen reserve, but the muscle also uses glucose from the blood. How would you design an experiment to test whether or not intake of carbohydrate during exercise will improve endurance, and if it does, what is its effect on the sparing of muscle glycogen?

CHAPTER 48 Gas Exchange in Animals

High fliers

"And so back up the gradual slopes, the wind behind me. A much greater effort this, stopping every few yards with a slight anxiety lest I should not make the distance. As I approached the tents, I was astonished to see a bird, a chough, strutting about on the stones near me. . . . During this day, too, Charles Evans saw what must have been a migration of small grey birds. . . . Neither of us had thought to find any signs of life as high as this."

Sir John Hunt related the above incident in his 1953 book *The Ascent of Everest*, the story of the first expedition to reach the summit of Mount Everest—at 8,850 meters, the highest point on Earth. Hunt's encounter with the chough occurred at the last campsite before the summit attempt, some 8,000 meters up. At that altitude, humans are incapacitated if they do not breathe supplemental oxygen from pressurized bottles. Just before the moment described, Hunt had gone a short distance downhill from his tent without supplemental oxygen.

Humans have a limited capacity to exist when oxygen is in short supply and, to make it worse, we may not even realize when we are in trouble. One example was the infamous flight of the hot air balloon Zenith in 1875, in which three French scientists decided to study the effect of high altitude on humans—themselves. Intending to take physiological measurements on each other during the ascent, the three men loaded the *Zenith* with scientific apparatus and took off. As they reached higher and higher altitudes, the writing in their notebooks became less legible, and finally non-sensical. At no point did they register concern or alarm as they continued to cut away the ballast bags and ascend even higher. Finally all three men went unconscious, and when the balloon finally descended on its own, only one regained consciousness; the other two were dead. What limits our survival at high altitudes, and why didn't these men realize they had reached these limits?

Unlike humans, some birds do indeed function at extreme altitudes. Barhead geese and other birds migrate over the Himalayas at elevations above 9,000 meters. The highest recorded altitude for a bird is for a vulture that collided with an airliner at 11,278 meters. These numbers are even more impressive when you consider that birds in flight consume oxygen at a rate that a well-conditioned human athlete can sustain for only minutes. Birds sustain high rates of

"I Was Astonished To See a Bird" Bar-headed geese (*Anser indicus*) are among those birds that can sustain the high metabolic costs of flight even at the high altitudes of the Himalayas, where oxygen is scarce.

Aquatic Feats Many open-ocean fish such as these barracuda (*Sphyraena genie*) can swim great distances at high speeds, even though the oxygen content of water is only about 5 percent that of the air humans breathe.

CHAPTER OUTLINE

oxygen consumption during very long flights and at extremely high altitudes.

Birds are not the only animals that travel where humans cannot. Fish can swim much faster, farther, and longer than the best human swimmers. Yet the fish are breathing water that has less than 5 percent of the oxygen content of the air humans breathe. The ability of birds and fish to maintain high rates of metabolism even when oxygen is severely limited is explained by adaptations of their respiratory gas exchange systems.

IN THIS CHAPTER we explore adaptations of the respiratory systems of both water and air breathers for exchanging oxygen and carbon dioxide with the environment. We first discuss the physical factors that limit respiratory gas exchange and identify those factors that natural selection has optimized. We then examine the respiratory gas exchange organs of a variety of species and describe the adaptations of the blood for transporting respiratory gases. Finally, we will see how respiratory gas exchange systems are controlled and regulated.

48.1 What Physical Factors Govern Respiratory Gas Exchange?

The **respiratory gases** that animals must exchange are oxygen (O_2) and carbon dioxide (CO_2). Cells need to obtain O_2 from the environment to produce an adequate supply of ATP by cellular respiration (see Chapter 7). CO_2 is an end product of cellular respiration, and it must be removed from the body to prevent toxic effects.

Diffusion is the only means by which respiratory gases are exchanged between the internal body fluids of an animal and the outside medium (air or water). There are no active transport mechanisms to move respiratory gases across biological membranes. Because diffusion is a physical process, knowing the physical factors that influence rates of diffusion helps us understand the diverse adaptations of gas exchange systems. (You might want to review what you learned about the physical nature of diffusion in Section 5.3.) Here we will discuss environmental factors that influence diffusion rates, then consider respiratory system adaptations that facilitate the diffusion of respiratory gases.

Diffusion is driven by concentration differences

Because diffusion results from the random motion of molecules, the net movement of a molecule is always down its concentration gradient. One way biologists express the concentrations of different gases in a mixture is by the *partial pressures* of those gases. First, we have to know what the total pressure is, and we measure that with an instrument called a barometer. There are many types of barometers, but the classical one is a glass tube closed at one end and filled with mercury. This is then inverted over a pool of mercury with the open end of the tube under the surface of the mercury. At sea level, the pressure exerted by the atmosphere will support, and therefore be equal to, a column of mercury in the tube that is about 760 mm high (depending on the weather). Therefore, **barometric pressure** (atmospheric pressure) at sea level is 760 millimeters of mercury (mm Hg).* Because dry air is 20.9 percent O_2, the **partial pressure of oxygen** (P_{O_2}) at sea level is 20.9 percent of 760 mm Hg,

*In S.I. units of pressure, this is 10.1 kilopascals (kPa).

or about 159 mm Hg. That is, the contribution of O_2 to the total air pressure is about 159 mm Hg.

Describing the concentration of respiratory gases in a liquid such as water is a little more complicated because another factor is involved—the solubility of the gas in the liquid. Thus, the actual amount of a gas in a liquid depends on the partial pressure of that gas in the gas phase in contact with the liquid as well as on the solubility of that gas in that liquid. However, the diffusion of the gas between the gaseous phase and the liquid still depends on the partial pressures of the gas in the two phases.

Fick's law applies to all systems of gas exchange

Diffusion is a physical phenomenon that can be described quantitatively with a simple equation called **Fick's law of diffusion**. All environmental variables that limit respiratory gas exchange and all adaptations that maximize respiratory gas exchange are reflected in one or more components of this equation. Fick's law is written as

$$Q = DA\frac{P_1 - P_2}{L}$$

where

- Q is the rate at which a gas such as O_2 diffuses between two locations.
- D is the *diffusion coefficient*, which is a characteristic of the diffusing substance, the medium, and the temperature. (For example, perfume has a higher D than motor oil vapor, and all substances diffuse faster at higher temperatures, as well as diffusing faster in air than in water.)
- A is the cross-sectional area through which the gas is diffusing.
- P_1 and P_2 are the partial pressures of the gas at the two locations.
- L is the path length, or distance, between the two locations.

Therefore, $(P_1 - P_2)/L$ is a *partial pressure gradient*.

Animals can maximize D for respiratory gases by using air rather than water as their gas exchange medium whenever possible; doing so greatly increases Q. All other adaptations for maximizing respiratory gas exchange must influence the surface area (A) for gas exchange or the partial pressure gradient $[(P_1 - P_2)/L]$ across that surface area.

Air is a better respiratory medium than water

Oxygen can be obtained more easily from air than from water for several reasons:

- The O_2 content of air is much higher than the O_2 content of an equal volume of water. The maximum O_2 content of a bubbling stream in equilibrium with air is less than 10 ml of O_2 per liter of water. The O_2 content of the air over the stream is about 200 ml of O_2 per liter of air.
- Oxygen diffuses about 8,000 times more rapidly in air than in water. That is why the O_2 content of a stagnant pond can be zero only a few millimeters below the surface.
- When an animal breathes, it does work to move water or air over its specialized gas exchange surfaces. More energy is required to move water than to move air because water is 800 times more dense than air and about 50 times more viscous.

The slow diffusion of O_2 molecules in water affects air-breathing animals as well as water-breathing ones. Eukaryotic cells carry out cellular respiration in their mitochondria, which are located in the cytoplasm—an aqueous medium. Cells are bathed in extracellular fluid—also an aqueous medium. In addition, all respiratory surfaces must be protected from desiccation by a thin film of fluid through which O_2 must diffuse. The slow rate of O_2 diffusion in water limits the efficiency of O_2 distribution from gas exchange surfaces to the sites of cellular respiration even in air-breathing animals.

Diffusion of O_2 in water is so slow that even animal cells with low rates of metabolism can be no more than a couple of millimeters away from a good source of environmental O_2. Therefore, there are severe size and shape limits on the many species of invertebrates that lack internal systems for distributing O_2. Most of these species are very small, but some have grown larger by evolving a flat, thin body with a large external surface area (**Figure 48.1A**). Still others have very thin bodies that are built around a central cavity through which water circulates (**Figure 48.1B**). A critical factor enabling larger, more complex animal bodies has been the evolution of specialized respiratory systems with large surface areas for enhancing respiratory gas exchange (**Figure 48.1C**).

High temperatures create respiratory problems for aquatic animals

Animals that breathe water are in a double bind when environmental temperatures rise. Most water breathers are *ectotherms*—their body temperatures are closely tied to the temperature of the water around them. As the temperature of the water gets warmer, the ectotherm's body temperature and metabolic rate rise (see Figure 40.9). Thus, water breathers need more O_2 as the water gets warmer. But warm water holds less dissolved gas than cold water does (just think of the gases that escape when you open a warm bottle of soda). In addition, if the animal performs work to move water across its gas exchange surfaces (as fish do), the energy the animal must expend to breathe increases as water temperature rises. Therefore, as water temperature goes up, the water breather must extract more and more O_2 from an environment that is increasingly O_2 deficient, and a lower percentage of that O_2 is available to support activities other than breathing (**Figure 48.2**).

O_2 availability decreases with altitude

Just as a rise in temperature reduces the supply of O_2 available to water breathers, an increase in altitude reduces the O_2 supply for air breathers. At all altitudes, O_2 makes up 20.9 percent of the dry air; however, as you go up in altitude, the total amount of gas per unit volume decreases, as reflected in the barometric pressure. For example, at an altitude of 5,800 m, barometric pressure is only half what it is at sea level, so the P_{O_2} at that altitude is only about 80 mm Hg. At the summit of Mount Everest (8,850 m), P_{O_2} is only about 50 mm Hg—roughly one-third what it is at sea level. Since the movement of O_2 across respiratory gas exchange surfaces and into the body depends on diffusion, its rate of movement depends on the P_{O_2} difference between the air and the body fluids. Therefore, the drastically reduced P_{O_2} in the air at high altitudes constrains O_2 uptake. Because of these constraints, mountain climbers who ven-

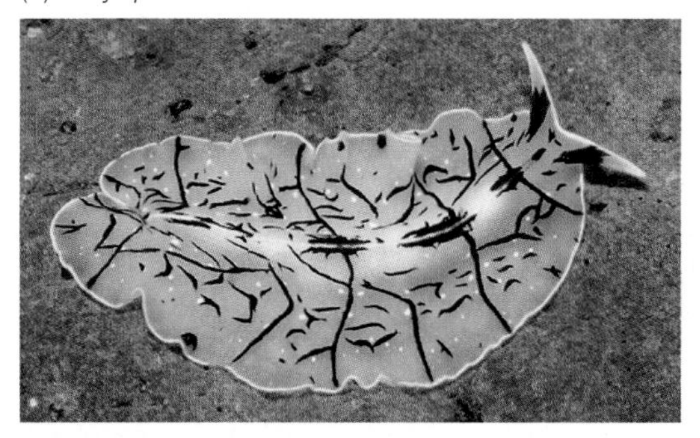

48.1 Keeping in Touch with the Medium (A) No cell in the leaflike body of this marine flatworm is more than a millimeter away from seawater. (B) Sponges have body walls perforated by many channels, which allow water to flow between the outside world and a central cavity. No cell in the sponge is more than a millimeter away from seawater. (C) A feathery fringe of gills on this larval salamander provides a large surface area for gas exchange. Blood circulating through the gills comes into close contact with the respiratory medium.

ture to the heights of Mount Everest usually breathe O_2 from pressurized bottles.

CO_2 is lost by diffusion

Respiratory gas exchange is a two-way process: CO_2 diffuses out of the body as O_2 diffuses in. The direction and rate of diffusion of the respiratory gases across the respiratory exchange surfaces depend on the partial pressure gradients of the gases. The partial pressure gradients of O_2 and CO_2 across these gas exchange surfaces are quite different. The amount of CO_2 in the atmosphere is extremely low (0.03 percent), so for air-breathing animals there is always a large concentration gradient for diffusion of CO_2 from the body to the environment. Whereas the partial pressure gradient for O_2 decreases with altitude, the partial pressure gradient for CO_2 does not. The partial pressure of CO_2 in the atmosphere is close to zero both at sea level and on top of Mount Everest.

In general, getting rid of CO_2 is not a problem for water-breathing animals because CO_2 is much more soluble in water than is O_2. Even in stagnant water, where the P_{CO_2} is higher than in fresh water, the lack of O_2 becomes a problem for the animal long before CO_2 exchange difficulties arise.

48.2 The Double Bind of Water Breathers Fish need more O_2 when the water is warmer, but warm water carries less O_2 than cold water.

48.1 RECAP

Respiratory gases are exchanged by diffusion only. Air is a better respiratory medium than water because there is more O_2 in a given volume of air than in the same volume of water, O_2 diffuses faster in air than in water, and it requires less work to move air over respiratory exchange surfaces than water.

- Can you explain the concept of partial pressures of gases and how they relate to rates of diffusion of O_2 and CO_2 at different altitudes? See pp. 1025–1026
- Why does a rise in water temperature create a double-bind situation for aquatic animals? See p. 1026 and Figure 48.2
- Can you describe how the variables in Fick's law of diffusion relate to respiratory systems?

Now that we know the physical factors that influence the rates of diffusion of respiratory gases between an animal and its environment, let's take a look at some of the adaptations animals have evolved for maximizing their respiratory gas exchange.

48.2 What Adaptations Maximize Respiratory Gas Exchange?

Some common ways the respiratory systems of different organisms maximize the exchange of O_2 and CO_2 with the environment include adaptations for increasing the surface area over which diffusion of gases can occur; for maximizing partial pressure gradients; and for minimizing the diffusion path length through an aqueous medium.

Respiratory organs have large surface areas

Many anatomical adaptations maximize the specialized body surface area (A) over which respiratory gases can diffuse. **External gills** are highly branched and folded extensions of the body surface that provide a large surface area for gas exchange with water (**Figure 48.3A**). External gills are found in larval amphibians and in the larvae of many insect species. Because they consist of thin, delicate tissues, they minimize the length of the path (L) traversed by diffusing molecules of O_2 and CO_2. External gills are vulnerable to damage, however, and are tempting morsels for predators, so in many animals protective body cavities for gills have evolved. Such **internal gills** are found in many mollusks and arthropods, and in all fishes (**Figure 48.3B**).

Air-breathing vertebrates also have surprisingly large surface areas for gas exchange. **Lungs** are internal cavities for respiratory gas exchange with air. Their structure is quite different from that of gills (**Figure 48.3C**). Lungs have a large surface area because they are highly divided, and they are elastic so that they can be inflated with air and deflated.

The most abundant air-breathing invertebrates are insects, which have a respiratory gas exchange system consisting of a network of air-filled tubes called **tracheae** that branch through all tissues of the insect's body (**Figure 48.3D**). The terminal branches of these tubes are so numerous that they have an enormous surface area compared to the external surface area of the insect's body.

Transporting gases to and from the exchange surfaces optimizes partial pressure gradients

Fick's law of diffusion includes variables (other than surface area) that can affect the rate of gas exchange. Partial pressure gradients $[(P_1 - P_2)/L]$ drive diffusion across gas exchange surfaces. These gradients can be maximized in several ways:

- *Minimization of path length*: Very thin tissues in gills and lungs reduce the diffusion path length (L).
- *Ventilation*: Actively moving the respiratory medium over the gas exchange surfaces (breathing) exposes those surfaces regularly to fresh respiratory medium containing maximum O_2 and minimum CO_2 concentrations. Thus, the concentration gradient is maximized.
- *Perfusion*: Circulating blood over the internal side of the exchange surfaces transports CO_2 to those surfaces and O_2 away from those surfaces, thus maximizing the concentration gradients driving diffusion.

An animal's **gas exchange system** is made up of its gas exchange surfaces and the mechanisms it uses to ventilate and perfuse those surfaces. This chapter describes four gas exchange systems. First we'll look at the gas exchange system of insects. Then we will describe fish gills and bird lungs—two remarkably efficient systems. The next section details the structure and functioning of human lungs.

Insects have airways throughout their bodies

The tracheal system that enables insects to exchange respiratory gases extends to all tissues in the insect body. Thus, respiratory gases diffuse through air most of the way to and from every cell. The insect respiratory system communicates with the outside environment through gated openings called *spiracles* in the sides of the abdomen (**Figure 48.4A,B**). The spiracles can open to allow gas exchange, and then close to decrease water loss. Spiracles open into tubes called *tracheae* that branch into even finer tubes, or *tracheoles*, until they end in tiny *air capillaries* (**Figure 48.4C**). In the insect's flight muscles and other highly active tissues, every mitochondrion is close to an air capillary.

Some aquatic insects carry a bubble of air when they dive. As they consume the O_2 in the bubble, the P_{O_2} in the bubble decreases and therefore O_2 diffuses into the bubble from the surrounding water. In this way the bubble serves as a "scuba tank" for the insect.

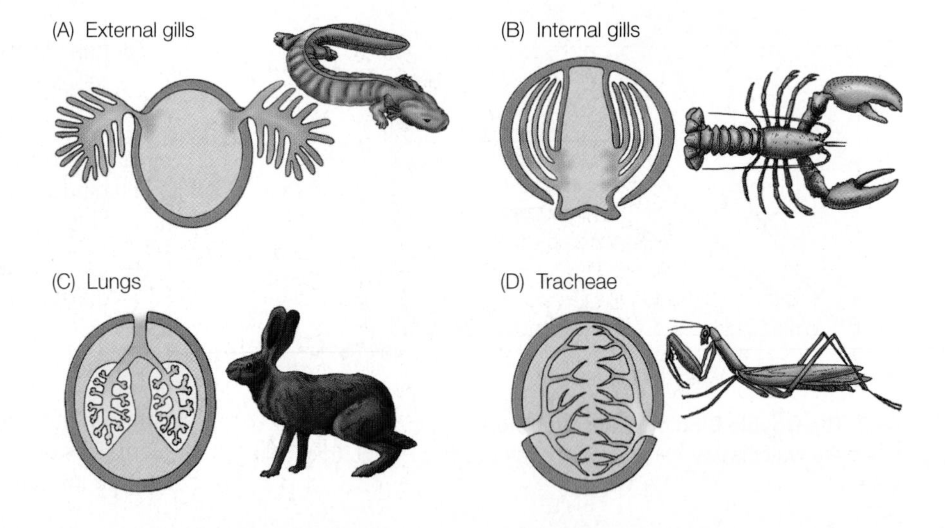

48.3 Gas Exchange Systems Large surface areas (blue in these diagrams) for the diffusion of respiratory gases are common features of animals. Both external (A) and internal (B) gills are adaptations for gas exchange with water. Lungs (C) and tracheae (D) are organs for gas exchange with air.

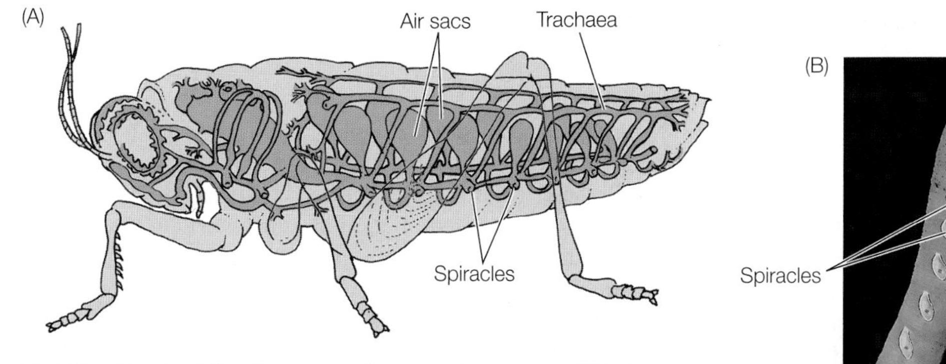

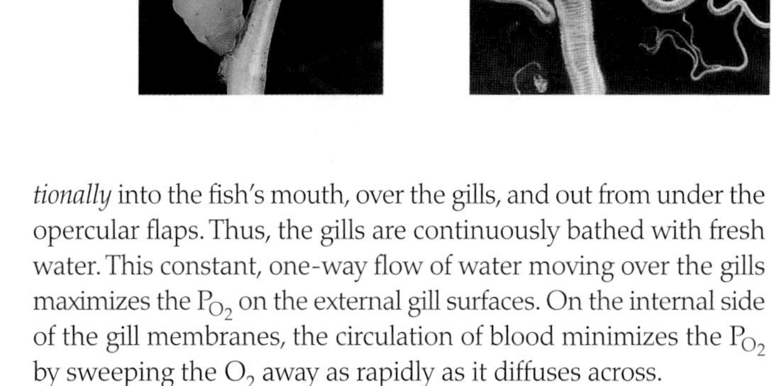

48.4 The Tracheal Gas Exchange System of Insects (A) In insects, respiratory gases diffuse through a system of air tubes (tracheae) that open to the external environment through holes called spiracles. (B) The spiracles of a sphinx moth larva run down its sides. (C) A scanning electron micrograph shows an insect trachea dividing into smaller tracheoles and still finer air capillaries.

Fish gills use countercurrent flow to maximize gas exchange

The internal gills of fish are supported by *gill arches* that lie between the mouth cavity and the protective *opercular flaps* on the sides of the fish just behind the eyes (**Figure 48.5A**). Water flows *unidirectionally* into the fish's mouth, over the gills, and out from under the opercular flaps. Thus, the gills are continuously bathed with fresh water. This constant, one-way flow of water moving over the gills maximizes the P_{O_2} on the external gill surfaces. On the internal side of the gill membranes, the circulation of blood minimizes the P_{O_2} by sweeping the O_2 away as rapidly as it diffuses across.

Gills have an enormous surface area for gas exchange because they are so highly divided. Each gill consists of hundreds of leaf-

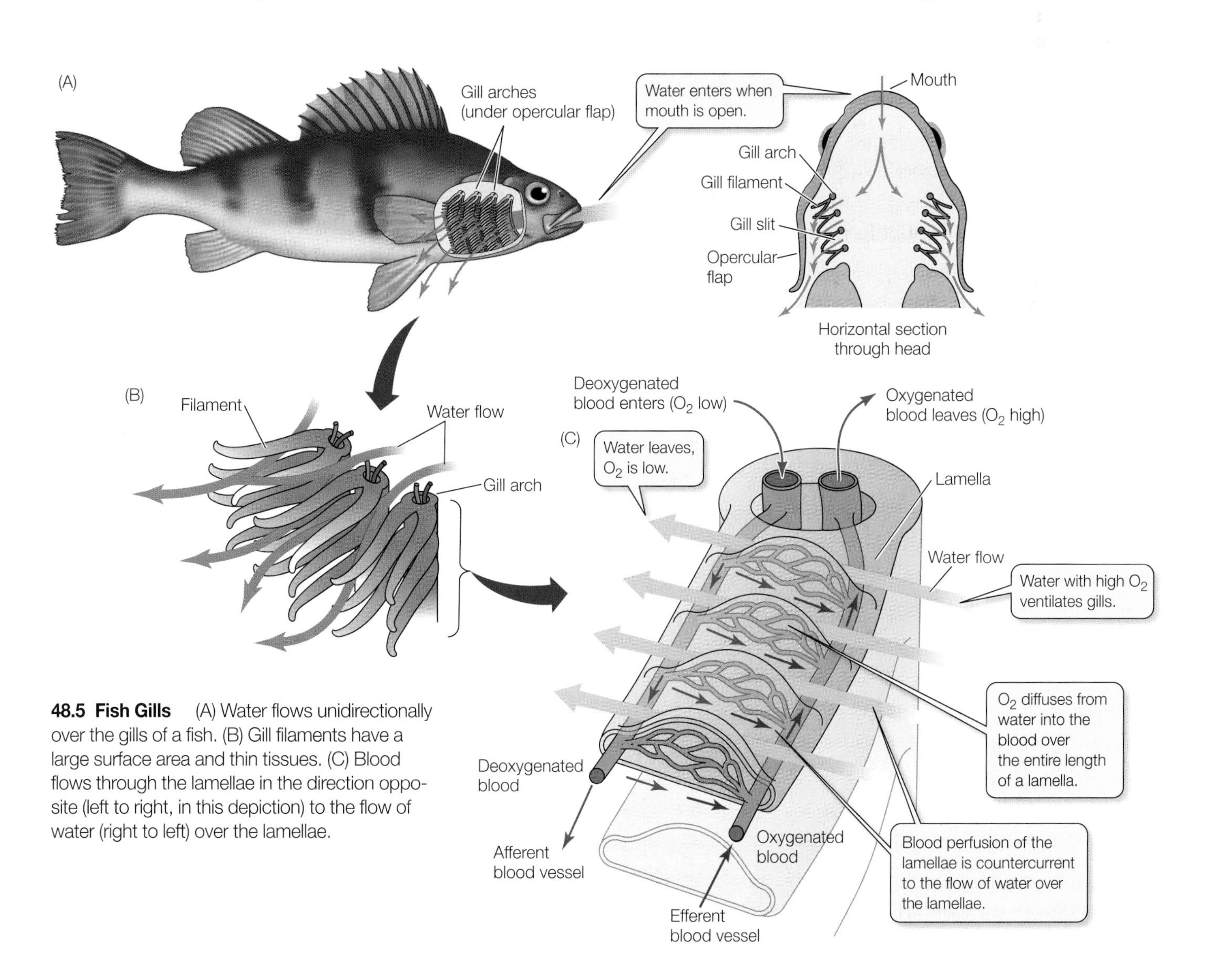

48.5 Fish Gills (A) Water flows unidirectionally over the gills of a fish. (B) Gill filaments have a large surface area and thin tissues. (C) Blood flows through the lamellae in the direction opposite (left to right, in this depiction) to the flow of water (right to left) over the lamellae.

shaped *gill filaments* (**Figure 48.5B**). The upper and lower flat surfaces of each gill filament are covered with rows of evenly spaced folds, or *lamellae*. The lamellae are the actual gas exchange surfaces. Their delicate structure minimizes the path length (L) for diffusion of gases between blood and water. The surfaces of the lamellae consist of highly flattened epithelial cells, so the water and the fish's red blood cells are separated by little more than 1 or 2 μm.

The flow of blood perfusing the inner surfaces of the lamellae, like the flow of water over the gills, is unidirectional. *Afferent* blood vessels bring blood to the gills, while *efferent* blood vessels take blood away from the gills (**Figure 48.5C**). Blood flows through the lamellae in the direction opposite to the flow of water over the lamellae. This **countercurrent flow** optimizes the P_{O_2} gradient between water and blood, making gas exchange more efficient than it would be in a system using concurrent (parallel) flow (**Figure 48.6**).

Some fish, including anchovies, tuna, and certain species of sharks, ventilate their gills by swimming almost constantly with their mouths open. Most fish, however, ventilate their gills by means of a two-pump mechanism. The closing and contracting of the mouth cavity pushes water over the gills, and the expansion of the opercular cavity prior to opening of the opercular flaps pulls water over the gills.

These adaptations allow fish to extract an adequate supply of O_2 from meager environmental sources by maximizing the surface area (A) for diffusion, minimizing the path length (L) for diffusion, and maximizing the P_{O_2} gradient by means of constant, unidirectional, countercurrent flow of blood and water over the opposite sides of their gas exchange surfaces.

Birds use unidirectional ventilation to maximize gas exchange

As we saw at the beginning of this chapter, birds can sustain high levels of activity much longer than mammals can—even at high altitudes where mammals cannot even survive. Yet the lungs of a bird are smaller than the lungs of a similar-sized mammal, and bird lungs expand and contract less during a breathing cycle than do mammalian lungs. Even more puzzling, bird lungs contract during inhalation and expand during exhalation!

The structure of bird lungs allows air to flow unidirectionally through the lungs, rather than bidirectionally through all of the same airways, as it does in mammals. Because mammalian lungs are never completely emptied of air during exhalation, there is always some lung volume that is not ventilated with fresh air. This nonventilated volume is called dead space. Because of continuous/unidirectional airflow, bird lungs have very little dead space, and the fresh incoming air is not mixed with stale air. In this way, a high P_{O_2} gradient is maintained.

In addition to lungs, birds have **air sacs** at several locations in their bodies. The air sacs are interconnected with each other, with the lungs, and with air spaces in some of the bones (**Figure 48.7A**). The air sacs receive inhaled air, but they are not gas exchange surfaces. As in other air-breathing vertebrates, air enters and leaves a bird's gas exchange system through a **trachea** (commonly known as the *windpipe*, and not to be confused with the air-conducting tracheae of insects), which divides into smaller airways called **bronchi** (singular bronchus).

The bronchi divide into tiny tubelike **parabronchi** that run parallel to one another through the lungs (**Figure 48.7B**). Branching off the parabronchi are numerous tiny *air capillaries*. Air flows through the lungs in the parabronchi and diffuses into the air capillaries, which are the gas exchange surfaces. They are so numerous that they provide an enormous surface area for gas exchange. The parabronchi coalesce into larger bronchi that take the air out of the lungs and back to the trachea. Thus the anatomy of the airways of birds allow air to flow unidirectionally and continuously through the lungs.

The puzzle of how birds breathe was solved by an experiment in which small O_2 sensors were placed at different locations in the air sacs and airways of birds. The bird was then exposed to pure O_2 for just a single breath, which made it possible to track that particular inhalation. The experiment demonstrated that a single breath remains in the bird's gas exchange system for two cycles of inhalation and exhalation, and that the air sacs work like bellows; inhalation expands the sacs, and exhalation compresses them to maintain a continuous and unidirectional flow of fresh air through the lungs (**Figure 48.8**).

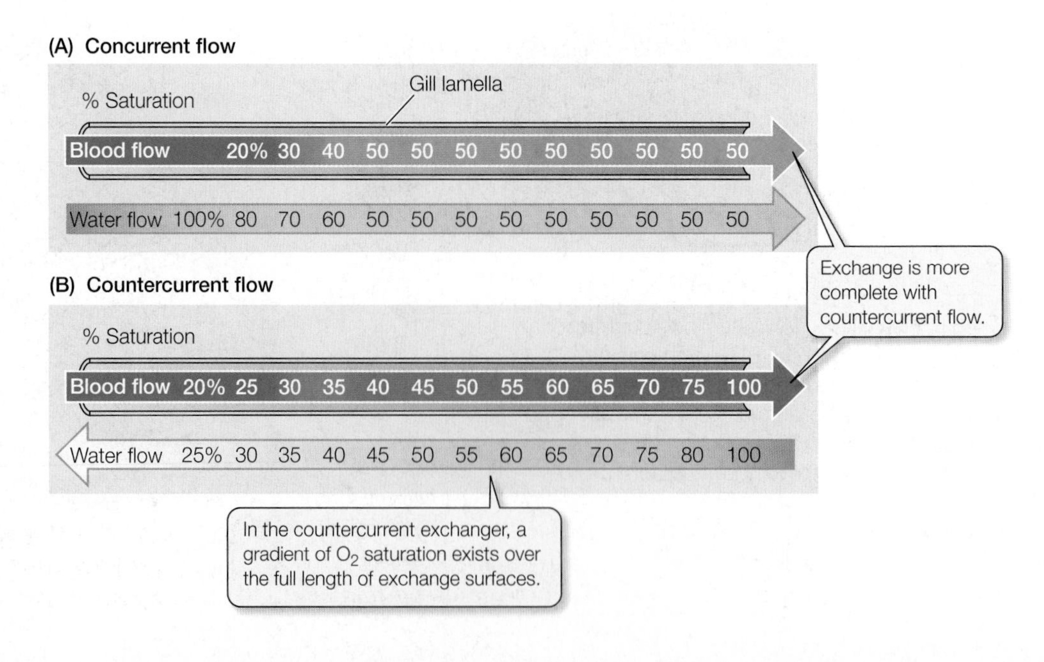

48.6 Countercurrent Exchange Is More Efficient In these models of concurrent and countercurrent gas exchange, the numbers represent the O_2 saturation percentages of blood and water. (A) In a concurrent exchanger, the percentages of saturation of blood and water reaches equilibrium halfway across the exchange surface. (B) A countercurrent exchanger allows more complete gas exchange because the water is always more O_2-saturated than the blood; thus a gradient of O_2 saturation is maintained.

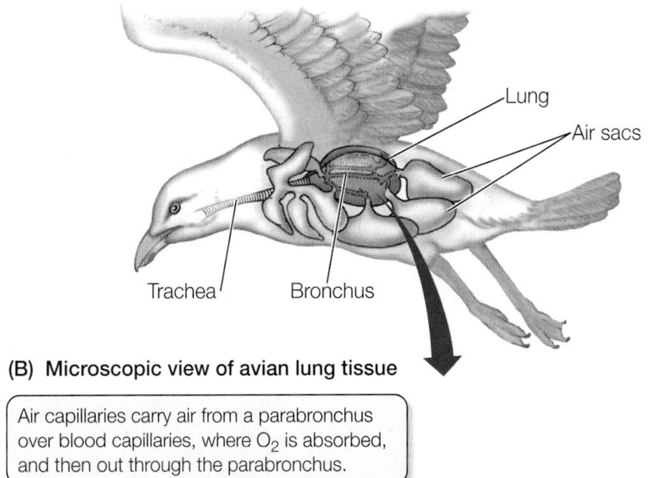

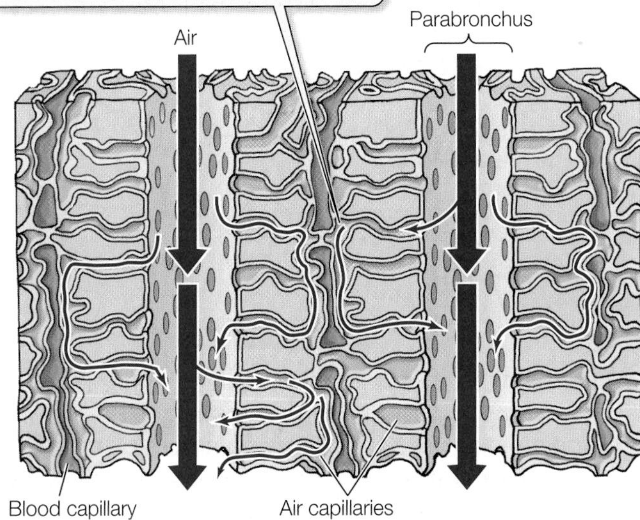

48.7 The Respiratory System of a Bird (A) The air sacs and air spaces in the bones are unique to birds. (B) Air flows through bird lungs unidirectionally in parabronchi. Air capillaries, the site of gas exchange, branch off the parabronchi.

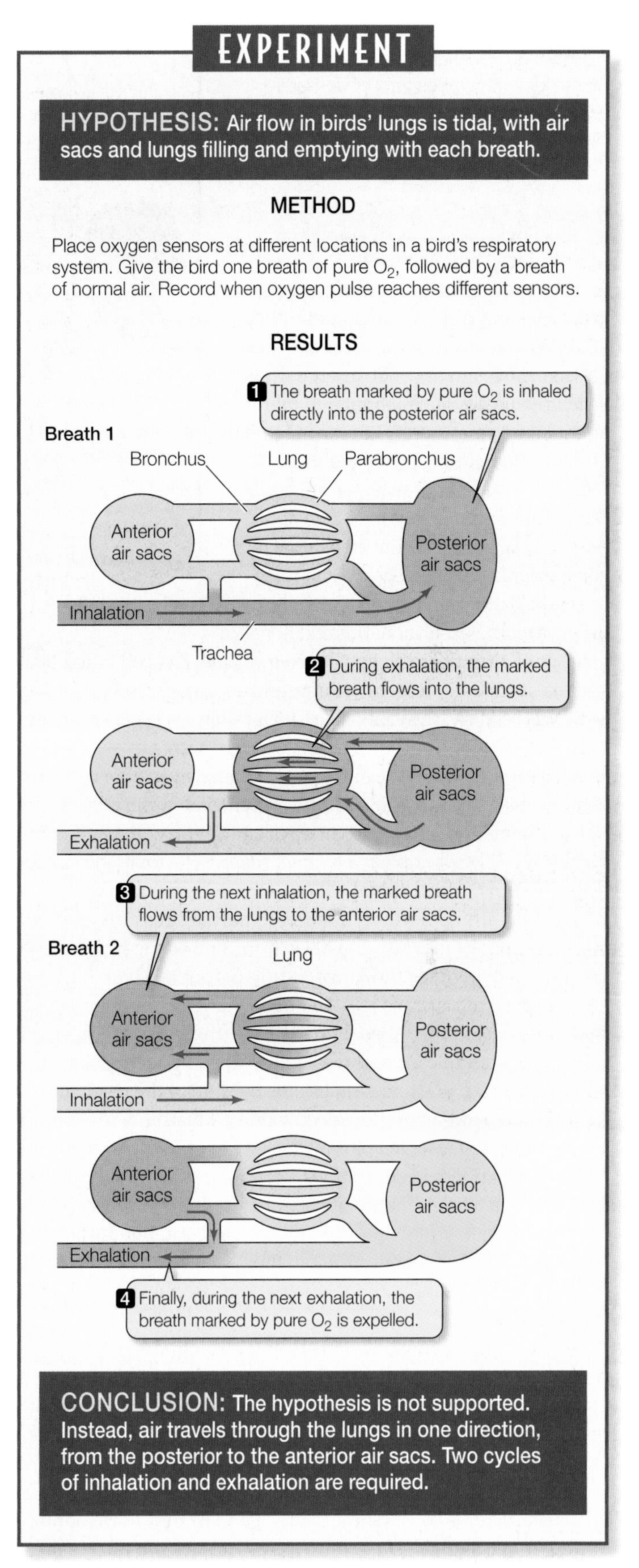

48.8 The Path of Air Flow through Bird Lungs The air a bird takes in with one breath (blue) travels through the lungs in one direction, from the posterior to the anterior air sacs.

The advantages of the bird gas exchange system are similar to those of fish gills. The air sacs keep fresh air flowing unidirectionally and continuously over the gas exchange surfaces. Thus, the bird can supply its gas exchange surfaces with a continuous flow of fresh air that has a P_{O_2} close to that of the ambient air. Even when the P_{O_2} of the ambient air is only slightly above that of the blood, O_2 can diffuse from air to blood.

Tidal ventilation produces dead space that limits gas exchange efficiency

Lungs evolved in the first "air-gulping" vertebrates as outpocketings of the digestive tract. Although their structure has evolved considerably, lungs remain dead-end sacs in all air-breathing vertebrates except birds. Because lungs are dead-end sacs, ventilation cannot be constant and unidirectional, but must be **tidal**: air flows in and exhaled gases flow out by the same route. Since the lungs can never be completely emptied of air, the residual air in the lungs after exhalation represents dead space. We can measure how large that dead space is.

A *spirometer* is a device that measures the volumes of air that a person breathes in or breathes out. A person breathes through a

48.9 Measuring Lung Ventilation
A spirometer measures the volume of air that a person breathes through a mouthpiece. The combined tidal volume, inspiratory reserve volume, and expiratory reserve volume is the vital capacity.

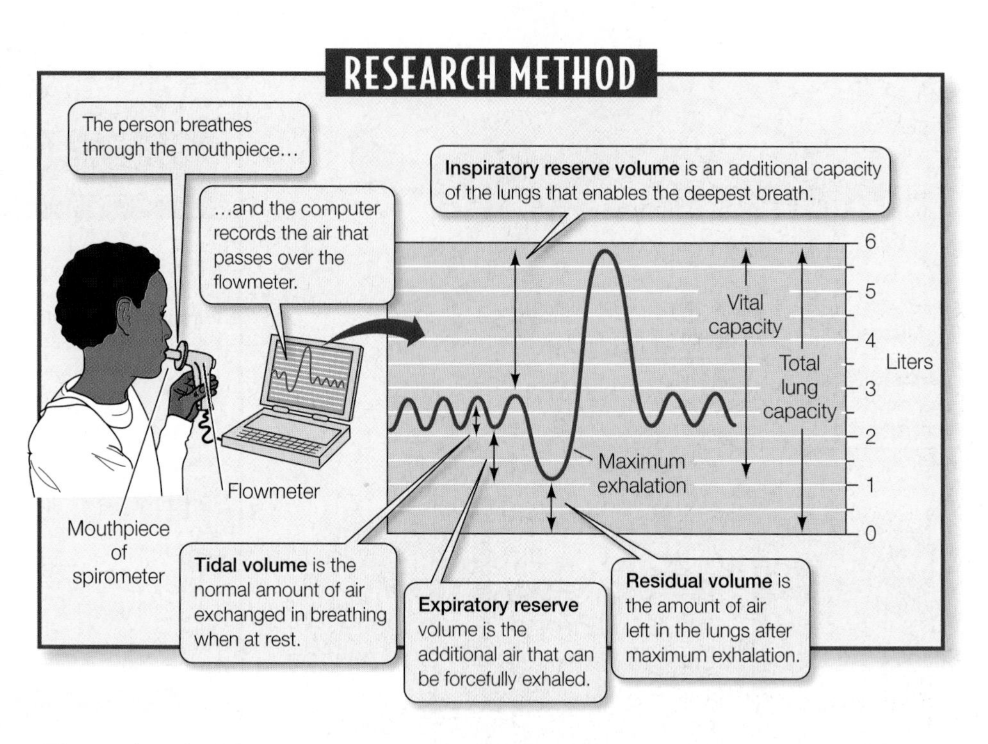

mouthpiece, an electronic flowmeter measures air volumes, and a computer calculates and displays the data (**Figure 48.9**). When we are at rest, the amount of air that moves in and out per breath is called the **tidal volume** (about 500 ml for an average human adult). We can breathe much more deeply and inhale more air than our resting tidal volume; the additional volume of air we can take in above normal tidal volume is our **inspiratory reserve volume**. Conversely, we can forcefully exhale more air than we normally do during a resting exhalation. This additional amount of air is the **expiratory reserve volume**. The combined tidal volume, inspiratory reserve volume, and expiratory reserve volume is the **vital capacity**. The vital capacity of an athlete is generally greater than that of a non-athlete, and vital capacity decreases with age.

Vital capacity is not the entire lung volume. Even after the most extreme exhalation possible, some air remains in the dead space of the lungs. The lungs and airways cannot be collapsed completely; they always contain a *residual volume*. The **total lung capacity** is the sum of the residual volume and the vital capacity.

Tidal breathing severely limits the partial pressure gradient available to drive the diffusion of O_2 from air into the blood. Fresh air is not moving into the lungs during part of the breathing cycle; therefore, the average P_{O_2} of air in the lungs is considerably less than it is outside the lungs. Furthermore, the incoming fresh air mixes with the stale air that was not expelled by the previous exhalation. The volume of this stale air is the sum of the residual volume and, depending on how deeply one is breathing, some or all of the expiratory reserve volume.

The scale in Figure 48.9 shows a typical resting breathing pattern in which a tidal volume of 500 ml of fresh air mixes with over 2,000 ml of stale air before reaching the gas exchange surfaces in our lungs. In this situation, although the P_{O_2} in the ambient air may be 159 mm Hg, the P_{O_2} of the air that reaches the gas exchange surfaces is only about 100 mm Hg.

Considering the mixing of fresh air with air in the residual volume, you can understand why reductions in tidal volume can be a problem—and therefore why patients recovering from surgery are encouraged to breathe deeply, even if it hurts. In addition to reducing the P_{O_2} gradient, tidal breathing makes countercurrent gas exchange impossible. Because air enters and leaves the gas exchange structures by the same route, blood cannot flow countercurrent to the air flow. Offsetting the inefficiencies of tidal breathing, mammalian lungs have some design features to maximize the rate of gas exchange: an enormous surface area and a very short path length for diffusion.

48.2 RECAP

The major adaptations of animals that increase their efficiency of respiratory gas exchange are a large surface area for exchange and maximized concentration gradient across that surface.

- Describe three different ways that the concentration gradient for gas exchange is maximized across fish gills. See pp. 1029–1030 and Figures 48.5 and 48.6
- What respiratory adaptations enable birds to fly at extremely high altitudes? See pp. 1030–1031 and Figures 48.7 and 48.8
- Can you explain why residual volume limits the efficiency of tidal breathing? See pp. 1031–1032

Despite their limitations, mammalian lungs serve the respiratory needs of mammals well. We will use as our example the human respiratory system.

48.3 How Do Human Lungs Work?

In humans, air enters the lungs through the oral cavity or nasal passage, which join together in the *pharynx* (**Figure 48.10A**). Below the pharynx, the esophagus conducts food to the stomach, and the trachea leads to the lungs. At the beginning of this airway is the *larynx*, or voice box, which houses the vocal cords. The larynx is the "Adam's apple" that you can see or feel on the front of your neck. The trachea is about 2 cm in diameter. Its thin walls are prevented from collapsing by C-shaped bands of cartilage as air pressure changes during the breathing cycle. If you run your fingers down the front of your neck just below your larynx, you can feel a couple of these bands of cartilage.

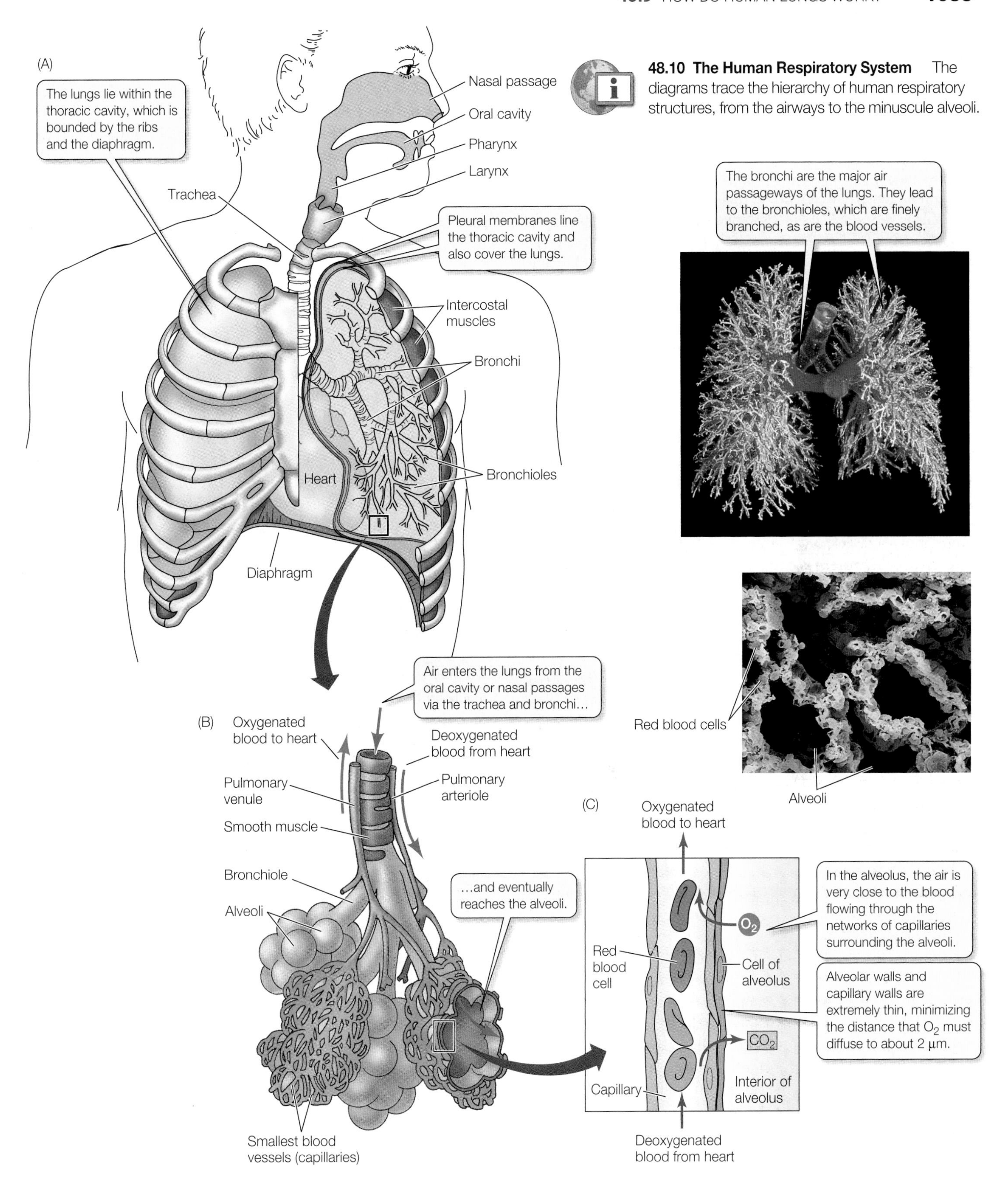

48.10 The Human Respiratory System The diagrams trace the hierarchy of human respiratory structures, from the airways to the minuscule alveoli.

The trachea branches into two **bronchi**, one leading to each lung. The bronchi branch repeatedly to generate a treelike structure of progressively smaller airways extending to all regions of the lungs. After four branchings, the cartilage supports disappear, marking the transition to **bronchioles**. After about 16 branchings, the bronchioles are less than a millimeter in diameter, and tiny, thin-walled air sacs called **alveoli** begin to appear. Alveoli are the sites of gas exchange. Thus, the bronchioles before the appearance of alveoli are called the conducting bronchioles, and those that sprout alveoli are called respiratory bronchioles. After alveoli begin to appear, about six more branchings of the airways occur that then end in clusters of alveoli (**Figure 48.10B**). Because the airways conduct air only to and from the alveoli and do not themselves participate in gas exchange, their volume is dead space.

Human lungs have about 300 million alveoli. Even though each alveolus is very small, their combined surface area for diffusion of respiratory gases is about 70 m^2—about one-fourth the size of a basketball court. Each alveolus is made of very thin cells. Between and surrounding the alveoli are networks of capillaries whose walls are also made up of exceedingly thin cells. Where capillary meets alveolus, very little tissue separates them (**Figure 48.10C**), so the length of the diffusion path between air and blood is less than 2 μm.

Emphysema is a lethal condition in which, over time, inflammation destroys the walls of the alveoli. It is the fourth largest cause of death in the United States. Although genetic factors can influence an individual's susceptibility to emphysema, the principal cause of the disease is smoking.

Respiratory tract secretions aid ventilation

Mammalian lungs produce two secretions that do not directly influence their gas exchange but do affect the process of ventilation: mucus and surfactant.

Many cells lining the airways produce sticky mucus that captures bits of dirt and microorganisms that are inhaled. Other cells lining the airways have cilia whose beating continually sweeps the mucus, with its trapped debris, up toward the pharynx, where it can be swallowed or spit out. This phenomenon, called the *mucus escalator*, can be adversely affected by inhaled pollutants. Smoking one cigarette can immobilize the cilia of the airways for hours. A smoker's cough results from the need to clear the obstructing mucus from the airways when the mucus escalator is out of order. The genetic disease *cystic fibrosis* causes respiratory problems by affecting the respiratory mucus (see Figure 17.3B).

A **surfactant** is a substance that reduces the surface tension of a liquid. **Surface tension** gives the surface of a liquid such as water the properties of an elastic membrane, and it is why certain insects, such as water-striders, can walk on water. As discussed in Section 2.4, surface tension is the result of chemical forces of attraction between water molecules. The attractive forces working on the water molecules at the surface are pulling from below and from the sides, but not from above. This imbalance of forces creates surface tension. The thin film of fluid covering the air-facing surfaces of the alveoli has surface tension that contributes to the elasticity of the lungs. To inflate the lungs, enough force has to be generated to overcome both the elasticity of the lung tissue and the surface tension in the alveoli.

Lung surfactant is a fatty substance that is critical for reducing the work necessary to inflate the lungs. Certain cells in the alveoli release surfactant molecules when they are stretched. If a baby is born more than a month prematurely, however, these cells may not have developed the ability to produce surfactant. Therefore, the baby has great difficulty breathing because an enormous effort is required to stretch the alveoli. A baby with this condition, known as *respiratory distress syndrome*, may die from exhaustion and lack of O_2. Common treatments for premature babies have been to put them on respirators to assist their breathing and to give them hormones to speed lung development. A new approach, however, is to apply surfactant to the lungs via an aerosol.

Lungs are ventilated by pressure changes in the thoracic cavity

Human lungs are suspended in a right and a left **thoracic cavity**. Each thoracic cavity is a closed compartment bounded on the bottom by the domed sheet of muscle called the **diaphragm** (Figure 48.10A). Lining each thoracic cavity and covering the lung in that cavity is a continuous sheet of tissue called the **pleural membrane**. Because the pleural membrane is continuous, it encloses a **pleural space**. There is no real space between the pleural membranes, but there is a thin film of fluid. This fluid lubricates the inner surfaces of the pleural membranes so they can slip and slide against each other during breathing movements. Just as we mentioned above in the explanation of surface tension, there are forces of attraction between the molecules of the fluid in the pleural space. As a result, it is difficult to pull the pleural membranes apart. Think of two wet panes of glass, or two wet microscope slides. You can easily slide them past each other, but it is difficult to separate them. While the inner surfaces of the pleural membranes are "stuck" to each other by surface tension, the outer surfaces of the pleural membranes are attached to the wall of the thoracic cavity and to the surface of the lungs.

Breathing involves changes in the volume of the thoracic cavity (**Figure 48.11**). Because the pleural membranes are attached to the walls of the thoracic cavity, and because the pleural space is a closed compartment, any attempt to increase the volume of the thoracic cavity creates a subatmospheric pressure (which we will refer to as a *negative pressure*) inside the pleural space. Even between breaths, there is normally a slight negative pressure in the pleural space because the rib cage is pulling outward and the elasticity of the lung tissue is pulling inward. This slight negative pressure keeps the alveoli partly inflated even at the end of an exhalation. If the thoracic cavity is punctured—by a knife wound, for example—air leaks into the pleural space, the negative pressure in that closed space is lost, and the lung collapses. If the wound is not sealed, breathing movements pull air into the pleural space rather than into the lung, and there is no ventilation of the alveoli in that lung.

Inhalation is initiated by contraction of the muscular diaphragm (see Figure 48.11). As the domed diaphragm contracts, it pulls down, expanding the thoracic cavity and pulling on the pleural membranes. This pulling on the pleural membranes makes the pressure in the pleural space more negative. The closed pleural space cannot expand, so the pleural membranes pull on the lungs. The lungs are not a closed cavity; they have an airway to the atmosphere, and they can expand in volume. When the diaphragm contracts, air rushes in through the trachea from the outside and the lungs expand. Exhalation begins when the contraction of the

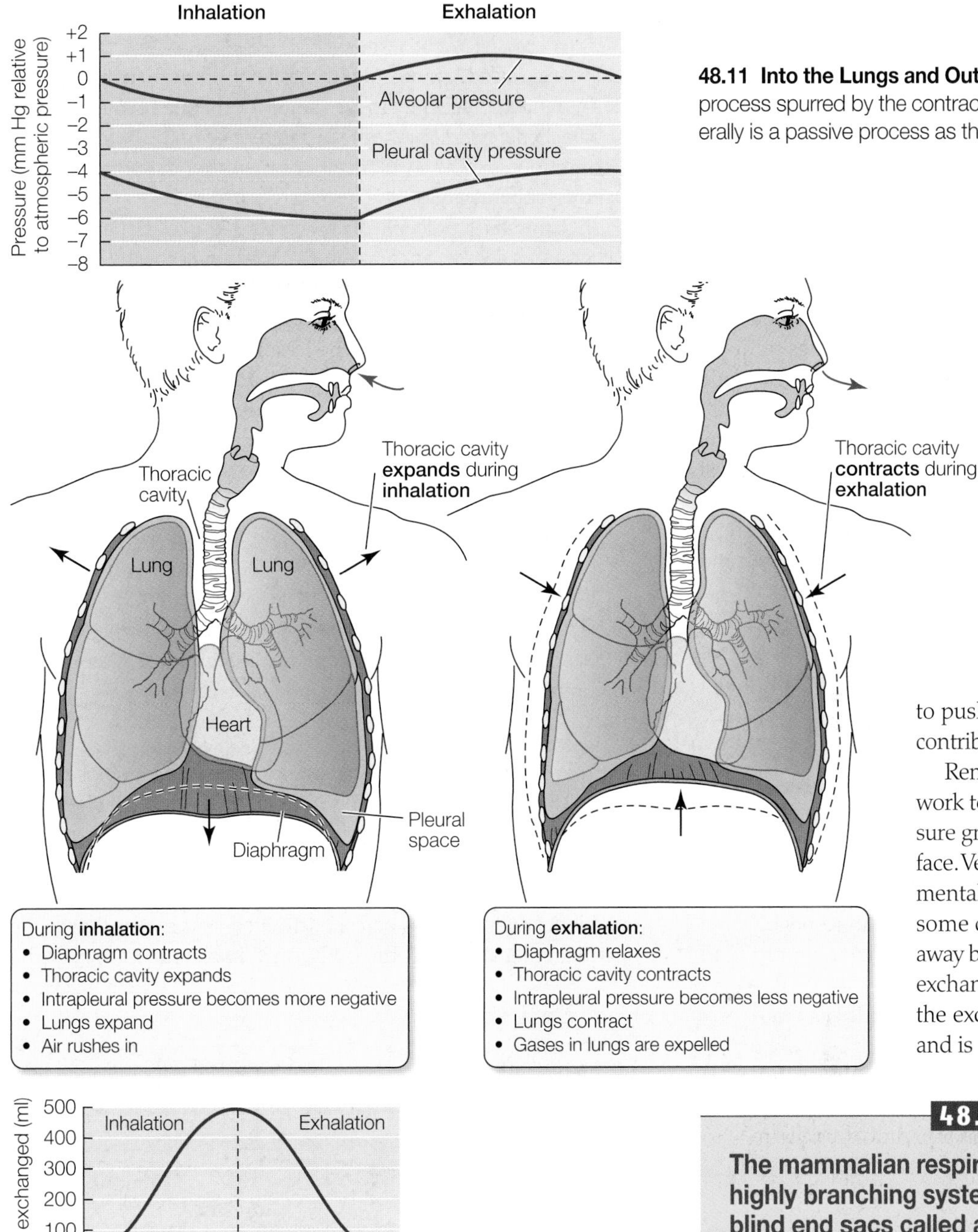

48.11 Into the Lungs and Out Again Inhalation is an active process spurred by the contraction of the diaphragm. Exhalation generally is a passive process as the diaphragm relaxes.

diaphragm ceases. As the diaphragm relaxes, the elastic recoil of the lung tissues pulls the diaphragm up and pushes air out through the airways. When a person is at rest, inhalation is an active process and exhalation is a passive process.

The diaphragm is not the only muscle that can change the volume of the thoracic cavity. Between the ribs are two sets of **intercostal muscles**. The *external intercostal muscles* expand the thoracic cavity by lifting the ribs up and outward. The *internal intercostal muscles* decrease the volume of the thoracic cavity by pulling the ribs down and inward. During strenuous exercise, the external intercostal muscles increase the volume of air inhaled, making use of the inspiratory reserve volume, and the internal intercostal muscles increase the amount of air exhaled, making use of the expiratory reserve volume. The abdominal muscles can also aid in breathing. When they contract, they cause the abdominal contents to push up on the diaphragm and thereby contribute to the expiratory reserve volume.

Remember that ventilation and perfusion work together to maximize the partial pressure gradients across the gas exchange surface. Ventilation delivers O_2 to the environmental side of the exchange surface, where some diffuses into the body and is swept away by perfusion. The reverse is true for the exchange of CO_2. Perfusion delivers CO_2 to the exchange surface, where it diffuses out and is swept away by ventilation.

48.3 RECAP

The mammalian respiratory system consists of a highly branching system of airways that lead to blind end sacs called alveoli which are the gas exchange surfaces. CO_2 and O_2 are exchanged across thin capillary and alveoli walls by diffusion.

- Can you describe the path that a breath of air takes from the nose to the gas exchange surfaces? See p. 1033 and Figure 48.10
- What roles do mucus and surfactant play in maintaining the function of the mammalian respiratory system? See p. 1034
- Can you explain the anatomical and functional relationships between the thoracic cavity, the pleural space, and the lungs? See p. 1034 and Figure 48.11

Now that we have seen how respiratory gases get to and from the environmental side of the gas exchange membranes through ventilation, we can look at how these gases get to and from the internal side of the gas exchange membranes through perfusion.

48.4 How Does Blood Transport Respiratory Gases?

Perfusion of the lungs is one of the functions of the circulatory system. The circulatory system uses a pump (the heart) and a network of vessels to transport extracellular fluids and associated cells (blood) around the body. Circulatory systems are the subject of the next chapter, so here we will discuss only one aspect of perfusion: how the respiratory gases are transported in the blood.

The liquid part of the blood, the *blood plasma*, carries some O_2 in solution, but its ability to transport O_2 is quite limited. The blood plasma of a human can contain in solution only about 0.3 ml of O_2 per 100 ml of plasma, which is inadequate to support even basal metabolism. To increase its O_2 transport capacity, the blood of most animals, vertebrate and invertebrate, also contains molecules that can bind reversibly to O_2 depending on its partial pressure. These molecules pick up or bind O_2 where its partial pressure is high and release it where its partial pressure is lower. There are many O_2 transport molecules in the animal kingdom, but in vertebrates, this role is played by hemoglobin contained in red blood cells. Hemoglobin increases the capacity of blood to transport O_2 by about sixtyfold.

Hemoglobin combines reversibly with oxygen

Red blood cells contain enormous numbers of hemoglobin molecules. **Hemoglobin** is a protein consisting of four polypeptide subunits (see Figure 3.9). Each of these polypeptides surrounds a *heme group*—an iron-containing ring structure that can reversibly bind a molecule of O_2. Thus, each molecule of hemoglobin can bind up to four molecules of O_2.

As O_2 diffuses into the red blood cells, it binds to hemoglobin. Once O_2 is bound, it cannot diffuse back across the red cell plasma membrane. By binding O_2 molecules as they enter the red blood cells, hemoglobin maximizes the partial pressure gradient driving the diffusion of O_2 into the cells. In addition, it enables the red blood cells to carry a large amount of O_2 to the tissues of the body.

The ability of hemoglobin to pick up or release O_2 depends on the P_{O_2} of its environment. When the P_{O_2} of the blood plasma is high, as it usually is in the lung capillaries, each molecule of hemoglobin can carry its maximum load of four molecules of O_2. As the blood circulates through the rest of the body, it encounters lower P_{O_2} values. At these lower P_{O_2} values, the hemoglobin releases some of the O_2 it is carrying (**Figure 48.12**).

The relation between P_{O_2} and the amount of O_2 bound to hemoglobin is not linear, but S-shaped (sigmoidal). The sigmoidal hemoglobin–oxygen binding curve in Figure 48.12 reflects interactions between the four subunits of the hemoglobin molecule. At low P_{O_2} values, only one subunit will bind an O_2 molecule. When it does so, the shape of that subunit changes, causing an alteration in the quaternary structure of the whole hemoglobin molecule. That structural change makes it easier for the other subunits to bind a molecule of O_2; that is, their O_2 *affinity* is increased. Therefore, a smaller increase in P_{O_2} is necessary to get the hemoglobin molecules to bind a second O_2 molecule (that is, to become 50 percent saturated) than was necessary to get them to bind one O_2 molecule (to become 25 percent saturated). This influence of the binding of O_2 by one subunit on the O_2 affinity of the other subunits is called **positive cooperativity**.

Once the third molecule of O_2 is bound, the relationship seems to change, as a larger increase in P_{O_2} is required for the hemoglobin to reach 100 percent saturation. This upper bend of the sigmoid curve is due to a probability phenomenon. The closer we get to having all subunits occupied, the less likely it is that any particular O_2 molecule will find a place to bind. Therefore, it takes a relatively greater P_{O_2} to achieve 100 percent saturation.

Carbon monoxide (CO) from faulty home heating systems can cause unconsciousness and death because CO binds to hemoglobin with a 240-fold higher affinity than O_2. This tightly bound CO prevents hemoglobin from transporting O_2 to the tissues of the body, and the brain is most vulnerable to lack of O_2. Annually more than 5,000 people in the United States die from carbon monoxide poisoning.

The oxygen-binding/dissociation properties of hemoglobin help get O_2 to the tissues that need it most. In the lungs, where the P_{O_2} is about 100 mm Hg, hemoglobin is 100 percent saturated. The P_{O_2} in blood returning to the heart from the body is usually about 40 mm Hg. You can see from Figure 48.12 that at this P_{O_2}, the hemoglobin is still about 75 percent saturated. This means that as the blood circulates around the body, only about one in four of the

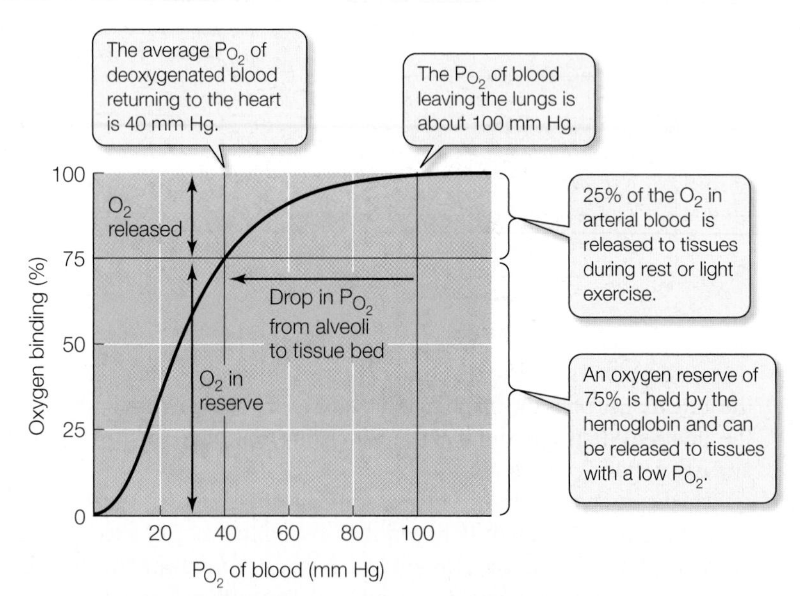

48.12 The Binding of O_2 to Hemoglobin Depends on P_{O_2} Hemoglobin in blood leaving the lungs is 100 percent saturated (four molecules of O_2 are bound to each hemoglobin molecule). Most hemoglobin molecules will drop only one of their four O_2 molecules as they circulate through the body, and are still 75 percent saturated when the blood returns to the lungs. The steep portion of this oxygen binding curve comes into play when tissue P_{O_2} falls below the normal 40 mm Hg, at which point the hemoglobin will "unload" its O_2 reserves.

O_2 molecules it carries is released to the tissues. This system seems inefficient, but it is really quite adaptive, because the hemoglobin keeps 75 percent of its O_2 in reserve to meet peak demands.

When a tissue becomes starved of O_2 and its local P_{O_2} falls below 40 mm Hg, the hemoglobin flowing through that tissue is on the steep portion of its sigmoid binding/dissociation curve. That means that relatively small decreases in P_{O_2} below 40 mm Hg will result in the release of lots of O_2 to the tissue. Thus hemoglobin is very effective in making O_2 available to the tissues precisely when and where it is needed most.

Myoglobin holds an oxygen reserve

Muscle cells have their own oxygen-binding molecule, **myoglobin**. Myoglobin consists of just one polypeptide chain associated with an iron-containing ring structure that can bind one molecule of O_2. Myoglobin has a higher affinity for O_2 than hemoglobin does, so it picks up and holds O_2 at P_{O_2} values at which hemoglobin is releasing its bound O_2 (**Figure 48.13**).

Myoglobin facilitates the diffusion of O_2 in muscle cells and provides a reserve of O_2 for times when metabolic demands are high and blood flow is interrupted. Interruption of blood flow in muscles is common because contracting muscles squeeze blood vessels. When tissue P_{O_2} values are low and hemoglobin can no longer supply more O_2, myoglobin releases its bound O_2. Diving mammals such as seals have high concentrations of myoglobin in their muscles, which is one reason they can stay under water for so long. (We will learn more about adaptations for diving in the next chapter.) Even in nondiving animals, muscles called on for extended periods of work frequently have more myoglobin than muscles that are used for short, intermittent periods, as we saw in the previous chapter.

The affinity of hemoglobin for oxygen is variable

Various factors influence the oxygen-binding/dissociation properties of hemoglobin, thereby influencing O_2 delivery to tissues. In this section we examine three of those factors: the chemical composition of the hemoglobin, pH, and the presence of 2,3-bisphosphoglyceric acid (BPG).

HEMOGLOBIN COMPOSITION There is more than one type of hemoglobin, because the chemical composition of the polypeptide chains that form the hemoglobin molecule varies. The normal hemoglobin of adult humans has two each of two kinds of polypeptide chains—two α-globin chains and two β-globin chains—and the oxygen-binding characteristics shown in Figure 48.12.

Before birth, the human fetus has a different form of hemoglobin, consisting of two α-globin and two γ-globin chains. The functional difference between fetal and adult hemoglobin is that the fetal hemoglobin has a higher affinity for O_2. Therefore, the hemoglobin–oxygen binding/dissociation curve of fetal hemoglobin is shifted to the left in comparison to the curve for adult hemoglobin (see Figure 48.13). You can see from these curves that if both types of hemoglobin are at the same P_{O_2}, the fetal hemoglobin will pick up O_2 more easily than can the maternal hemoglobin. This difference in O_2 affinities facilitates the transfer of O_2 from the mother's blood to the blood of the fetus in the placenta.

Llamas and vicuñas are native to the Andes Mountains of South America. In the natural habitat of these mammals, more than 5,000 m above sea level, the P_{O_2} is below 85 mm Hg, and the P_{O_2} in their lungs is about 50 mm Hg. Thus, the hemoglobins of these animals must pick up O_2 in an environment that has a low P_{O_2}. The hemoglobins of llamas and vicuñas have oxygen binding/dissociation curves to the left of the curves of hemoglobins of most other mammals—in other words, their hemoglobin can become saturated with O_2 at lower P_{O_2} values than those of other mammals.

HEMOGLOBIN AND pH The oxygen-binding properties of hemoglobin are also influenced by physiological conditions. The influence of pH on the function of hemoglobin is known as the **Bohr effect**. As blood passes through metabolically active tissue such as exercising muscle, it picks up acidic metabolites such as lactic acid, fatty acids, and CO_2. As a result, blood pH falls. The excess H^+ binds preferentially to deoxygenated hemoglobin and decreases the affinity of that hemoglobin for O_2. As a result, the oxygen binding/dissociation curve of hemoglobin shifts to the right (see Figure 48.13). This shift means that the hemoglobin will release more O_2 in tissues where pH is low—another way that O_2 is supplied where and when it is most needed.

2,3-BISPHOSPHOGLYCERIC ACID BPG is a metabolite of glycolysis (see Figure 7.5). Mammalian red blood cells have a high concentration of BPG, which serves as an important regulator of hemoglobin function. BPG, like excess H^+, reversibly combines with deoxygenated hemoglobin and lowers its affinity for

Myoglobin
Llama hemoglobin
Human fetal hemoglobin
Human maternal hemoglobin (pH 7.4)
Human hemoglobin (pH 7.2)
Oxygen binding (%)
100
80
60
40
20
20
40
60
80
100
P_{O_2} (mm Hg)

Llama guanicoe

48.13 Oxygen-Binding Adaptations The different hemoglobins and myoglobin have different oxygen-binding properties adapted to different circumstances. The hemoglobin of llamas, for example, is adapted for binding O_2 at high altitudes, where P_{O_2} is low.

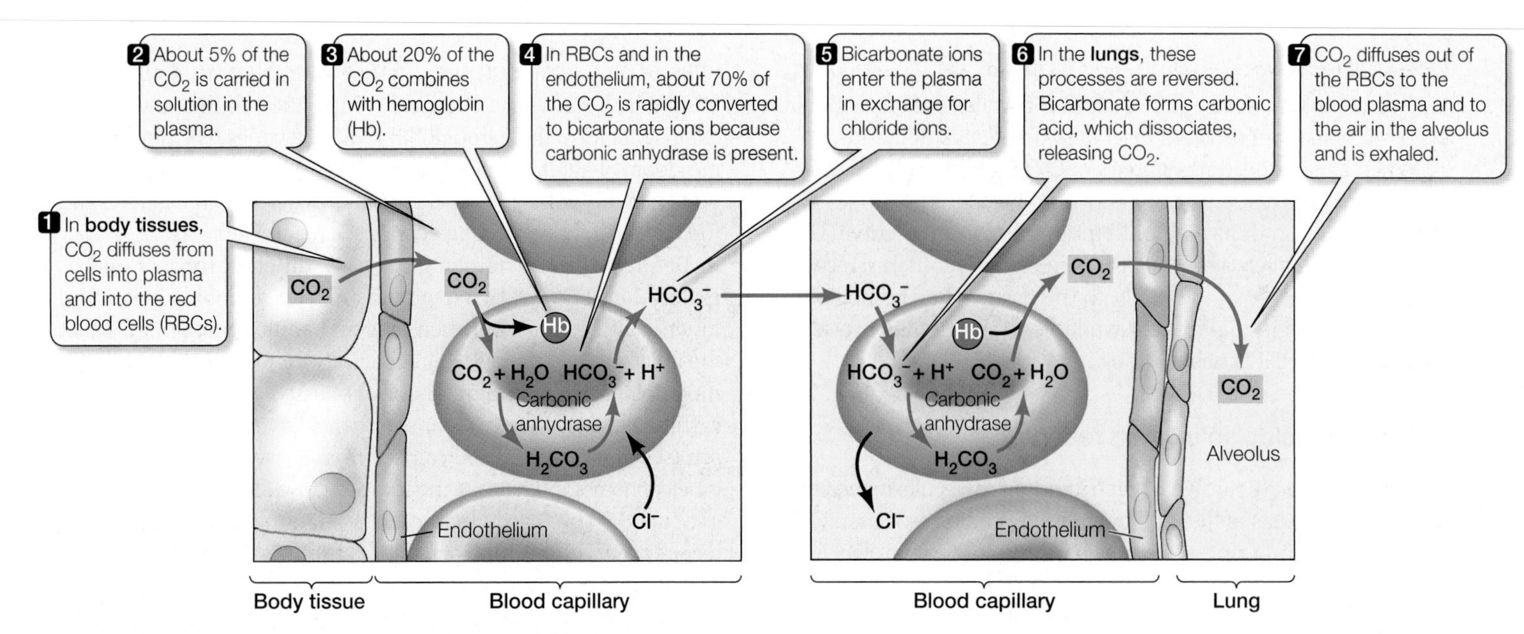

48.14 Carbon Dioxide Is Transported as Bicarbonate Ions Carbonic anhydrase in capillary endothelial cells and in red blood cells facilitates conversion of CO_2 produced by tissues into bicarbonate ions carried by the plasma. In lungs, the process is reversed as CO_2 is exhaled.

O_2. The result is that at any P_{O_2}, hemoglobin releases more of its bound O_2 than it otherwise would. In other words, BPG shifts the oxygen binding/dissociation curve of mammalian hemoglobin to the right. When humans go to high altitudes, or when they cease being sedentary and begin to exercise, the level of BPG in their red blood cells goes up and makes it easier for the hemoglobin to deliver more O_2 to the tissues. The reason that fetal hemoglobin has a left-shifted hemoglobin–oxygen binding/dissociation curve is that the γ-globin chains of fetal hemoglobin have a lower affinity for BPG than do the β-globin chains of adult hemoglobin.

Carbon dioxide is transported as bicarbonate ions in the blood

Delivering O_2 to the tissues is only half of the respiratory function of the blood. The blood also must take CO_2, a metabolic waste product, away from the tissues (**Figure 48.14**). CO_2 is highly soluble and readily diffuses through cell membranes, moving from its site of production in the tissues into the blood, where the partial pressure of carbon dioxide (P_{CO_2}) is lower. However, very little dissolved CO_2 is transported by the blood. Most CO_2 produced by the tissues is transported to the lungs in the form of **bicarbonate ions**, HCO_3^-. CO_2 is converted to HCO_3^-, transported to the lungs, and then converted back to CO_2 in several steps.

When CO_2 dissolves in water, some of it slowly reacts with the water molecules to form carbonic acid (H_2CO_3), some of which then dissociates into a proton (H^+) and a bicarbonate ion (HCO_3^-). This reversible reaction is expressed as follows:

$$CO_2 + H_2O \rightleftharpoons H_2CO_3 \rightleftharpoons H^+ + HCO_3^-$$

In the blood plasma, the reaction between CO_2 and H_2O proceeds slowly. But it is a different story in the endothelial cells of the capillaries and in the red blood cells, where the enzyme *carbonic anhydrase* speeds up the conversion of CO_2 to H_2CO_3. The newly formed carbonic acid dissociates, and the resulting bicarbonate ions enter the plasma in exchange for Cl^- (see Figure 48.14). By converting CO_2 to H_2CO_3, carbonic anhydrase reduces the P_{CO_2} in these cells and in the plasma, facilitating the diffusion of CO_2 from tissue cells to endothelial cells, plasma, and red blood cells. Some CO_2 is also carried in chemical combination with hemoglobin.

In the lungs, the reactions involving CO_2 and bicarbonate ions are reversed. Remember that an enzyme such as carbonic anhydrase only speeds up a reversible reaction; it does not determine its direction. The direction is determined by concentrations of reactants and products (see Section 6.1). Ventilation keeps the CO_2 concentration in the alveoli low, so CO_2 diffuses from the blood plasma into the alveoli, lowering the CO_2 concentration in the blood, which favors the conversion of HCO_3^- into CO_2.

48.4 RECAP

O_2 is transported from the lungs to the tissues of the body in reversible combination with hemoglobin. Each molecule of hemoglobin can reversibly combine with four molecules of O_2; the saturation of the binding sites is a function of the P_{O_2} in the environment of the hemoglobin.

- Can you explain the advantage of having hemoglobin hold on to three molecules of O_2 at the usual P_{O_2} of mixed venous blood? See pp. 1036–1037 and Figure 48.12
- How is the oxygen binding/dissociation curve of hemoglobin influenced by pH? By BPG? By development from fetus to newborn infant? See pp. 1037–1038 and Figure 48.13
- How is CO_2 transported in the blood? See p. 1038 and Figure 48.14

We must breathe every minute of our lives, but we don't usually worry about it, or even think about it very often. In the next section we will examine how the regular breathing cycle is generated and controlled by the central nervous system.

48.5 How is Breathing Regulated?

Breathing is an autonomic function of the central nervous system. The breathing pattern easily adjusts itself around other activities (such as speech and eating), and breathing rates change to match the metabolic demands of our bodies. How is this accomplished?

Breathing is controlled in the brain stem

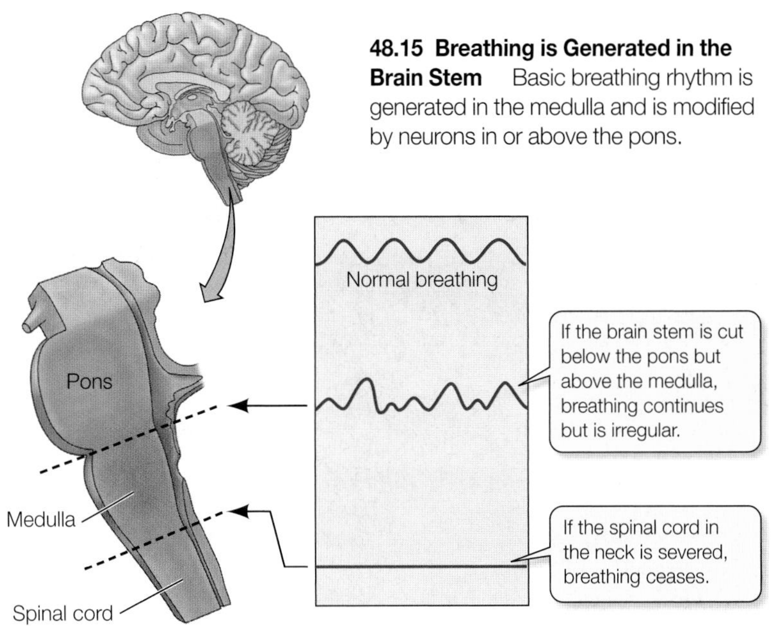

48.15 Breathing is Generated in the Brain Stem Basic breathing rhythm is generated in the medulla and is modified by neurons in or above the pons.

The autonomic nervous system maintains breathing and modifies its depth and frequency to meet the demands of the body for O_2 supply and CO_2 elimination. Breathing ceases if the spinal cord is severed in the neck region, showing that the breathing pattern is generated in the brain. If the brain stem is cut just above the medulla, the segment of the brain stem just above the spinal cord, an irregular breathing pattern remains (**Figure 48.15**).

Groups of respiratory motor neurons within the medulla increase their firing rates just before an inhalation begins. As more and more of these neurons fire—and fire faster and faster—the diaphragm contracts. Suddenly the neurons stop firing, the diaphragm relaxes, and exhalation begins. Exhalation is usually a passive process that depends on the elastic recoil of the lung tissues. When breathing demand is high, however, as during strenuous exercise, motor neurons for the intercostal muscles are recruited, which increases both the inhalation and the exhalation volumes. Brain areas above the medulla modify breathing to accommodate speech, ingestion of food, coughing, and emotional states.

Regulating breathing requires feedback information

When breathing or metabolism changes, it alters the P_{O_2} and the P_{CO_2} in the blood. We should therefore expect the blood levels of one or both of these gases to provide feedback information to the breathing rhythm generator in the medulla. Experiments in which subjects breathe air with different P_{O_2} and P_{CO_2} concentrations lead us to conclude that humans (and other mammals) are remarkably insensitive to falling blood levels of O_2, but very sensitive to increases in the P_{CO_2} of the blood (**Figure 48.16**). This relationship is reversed for water-breathing animals, in which O_2 is the primary feedback stimulus for breathing. Remember that CO_2 is readily lost across gill membranes, so its concentrations in the blood would not be a good index of metabolic needs.

We might ask whether it is an increase in the P_{CO_2} of the blood that stimulates increased breathing when we exercise. To answer this question, researchers in the lab of C. R. Bainton observed dogs running on treadmills at different speeds. As the speed of the treadmill increased, the respiratory gas exchange rate of the dogs increased, but the P_{CO_2} of their blood remained constant. Before concluding that their hypothesis was not supported, and that blood P_{CO_2} does not control breathing rate, the researchers changed their experiment. Instead of increasing the speed of the treadmill, they gradually increased its slope so that the dogs were running at the same speed, but were working harder because they were running uphill (**Figure 48.17**).

In this second experiment, the P_{CO_2} of the blood increased as the slope of the treadmill increased and as the respiratory gas exchange rate increased. The researchers concluded that the P_{CO_2} of the blood is the primary metabolic feedback information for breathing. However, when an animal starts to run or changes its running speed, additional feedback information from receptors in muscles and joints changes the sensitivity of the CO_2 sensors—

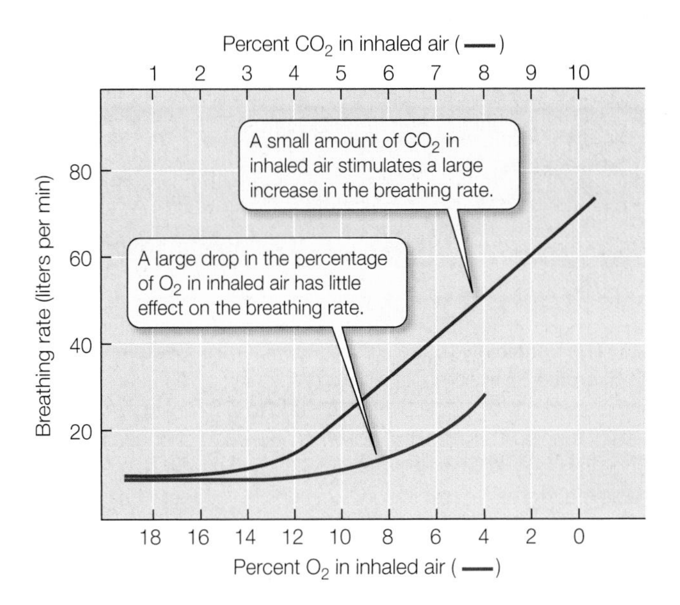

48.16 Carbon Dioxide Affects Breathing Rate Breathing is more sensitive to increased CO_2 content in inhaled air than to its decreased O_2 content.

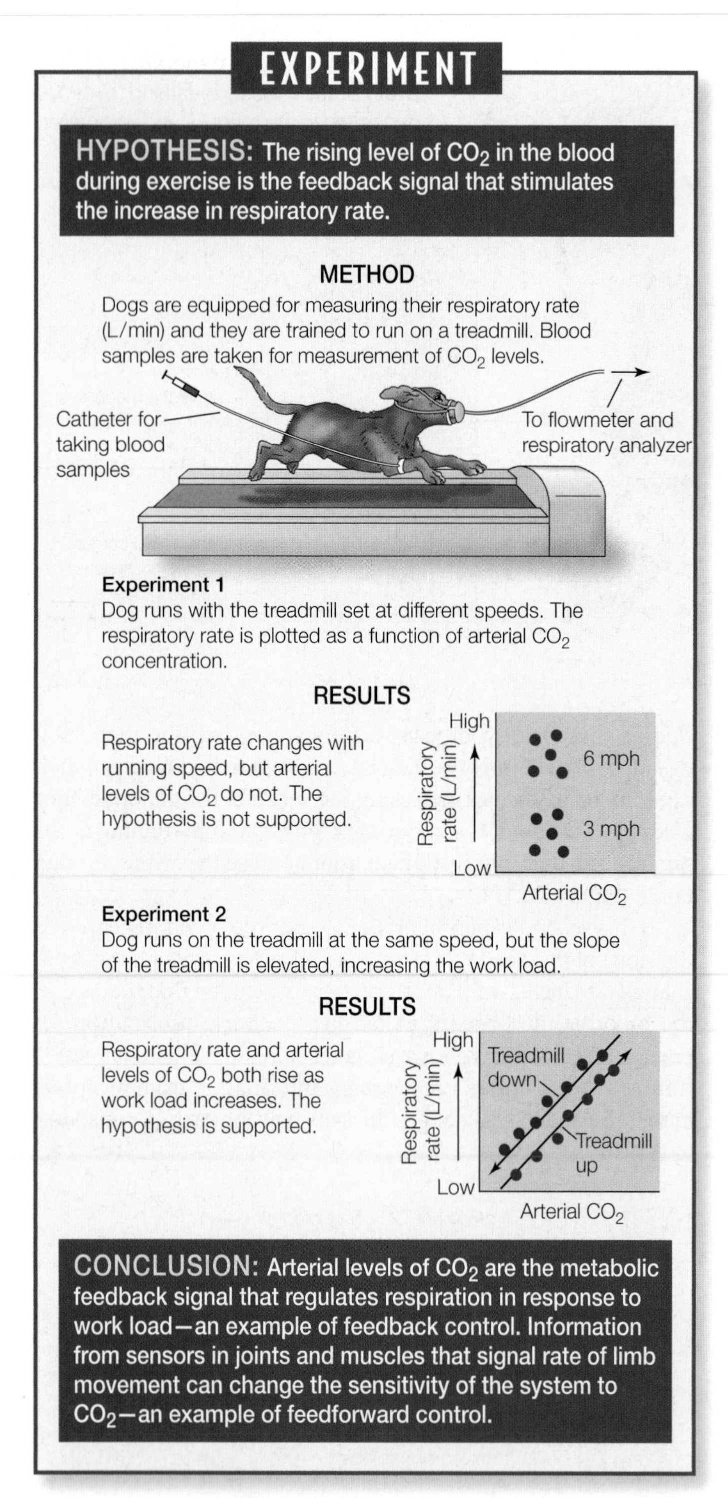

48.17 The Sensitivity of the Respiratory Control System Changes with Exercise Experiments with dogs running on a treadmill show that the sensitivity of the respiratory system to CO_2 changes when speed of running changes, but not when workload changes without a change in speed. FURTHER RESEARCH: Cold exposure is another variable that increases respiration. How would you test whether cold-induced respiration is driven by elevated blood CO_2 levels?

another example of feedforward information. Remember from Section 40.1 that feedforward information can change the sensitivity or the set point of a regulatory system.

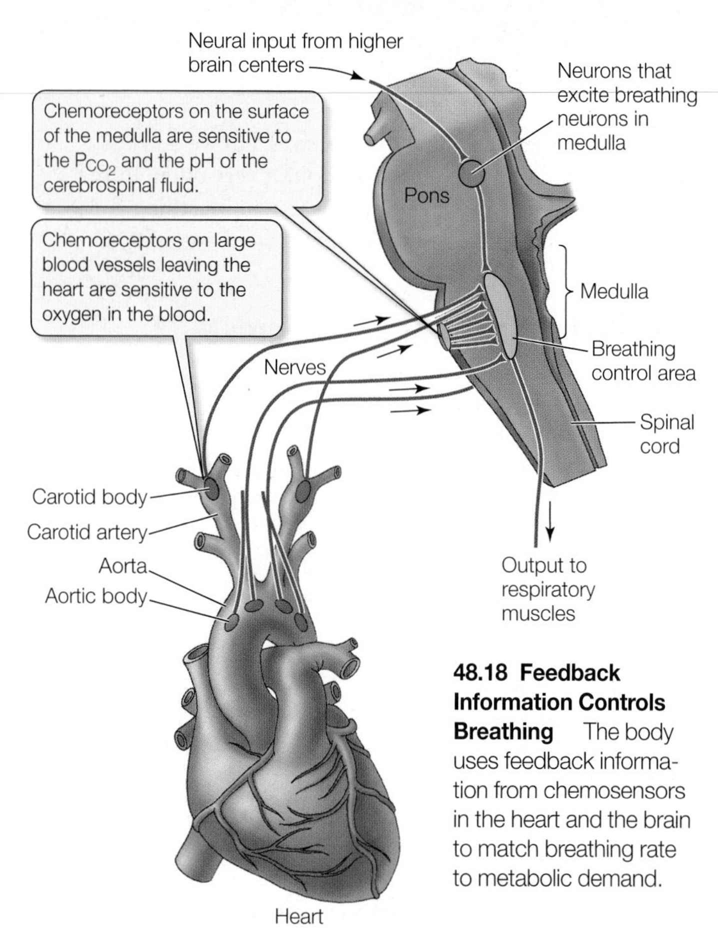

48.18 Feedback Information Controls Breathing The body uses feedback information from chemosensors in the heart and the brain to match breathing rate to metabolic demand.

Where are partial pressures of gases in the blood sensed? The major site of CO_2 sensitivity is an area on the ventral surface of the medulla, not far from the groups of neurons that generate the breathing rhythm. Primary sensitivity to P_{O_2} in the blood resides in nodes of tissue on the large blood vessels leaving the heart: the aorta and the carotid arteries (**Figure 48.18**). These **carotid** and **aortic bodies** receive enormous supplies of blood relative to their small size, and they contain chemoreceptors. If the blood supply to these structures decreases, or if the P_{O_2} of the blood falls dramatically, the chemoreceptors are activated and send nerve impulses to the breathing control center. Although we are not very sensitive to changes in blood P_{O_2}, the carotid and aortic bodies can stimulate increases in breathing during exposure to very high altitudes or when blood volume or blood pressure is very low.

48.5 RECAP

The rhythmic contractions of the respiratory muscles that drive breathing are generated by neurons in the brainstem.

- What is the primary chemical stimulus for controlling the respiratory rate and where is it sensed? See p. 1039 and Figures 48.16 and 48.18
- Do you understand what feedforward information is? Can you give an example of feedforward information in the respiratory control system? See Figure 48.17
- What are the functions of the carotid and aortic bodies? See p. 1040 and Figure 48.18

CHAPTER SUMMARY

48.1 What physical factors govern respiratory gas exchange?

Most cells require a constant supply of oxygen (O_2) and continuous removal of carbon dioxide (CO_2). These **respiratory gases** are exchanged between the body fluids of an animal and its environment by diffusion.

Fick's law of diffusion shows how various physical factors influence the rate of diffusion of gases. Adaptations to maximize respiratory gas exchange influence one or more variables of Fick's law.

In aquatic animals, gas exchange is limited by the low diffusion rate and low amount of O_2 in water. Aquatic animals face a double bind in that the amount of O_2 in water decreases, but their metabolism and the amount of work required to move water over their **gas exchange surfaces** increase, as water temperature rises. Review Figure 48.2

In air, the **partial pressure of oxygen** decreases with altitude.

48.2 What adaptations maximize respiratory gas exchange?

Adaptations to maximize gas exchange include increasing the surface areas for gas exchange, maximizing partial pressure gradients across those exchange surfaces, **ventilating** the outer surface with the respiratory medium, and **perfusing** the inner surface with blood.

Insects distribute air throughout their bodies in a system of **tracheae**, tracheoles, and air capillaries. Review Figure 48.4

The **gills** of fish have large gas exchange surface areas that are ventilated continuously and unidirectionally with water. **Countercurrent blood flow** helps increase the efficiency of gas exchange. Review Figures 48.5 and 48.6

The gas exchange system of birds includes **air sacs** that communicate with the lungs, but are not used for gas exchange. Air flows unidirectionally through bird lungs; gases are exchanged in air capillaries that run between **parabronchi**. Review Figure 48.7

Each breath of air remains in the bird respiratory system for two breathing cycles. The air sacs work as bellows to supply the air capillaries with a continuous unidirectional flow of fresh air. Review Figure 48.8, Web/CD Tutorial 48.1

Breathing in vertebrates other than birds is **tidal** and is therefore less efficient than gas exchange in fish or birds. Although the volume of air exchanged with each breath can vary considerably, the inhaled air is always mixed with stale air. Review Figure 48.9

48.3 How do human lungs work?

See Web/CD Tutorial 48.2

In mammalian lungs, the gas exchange surface area provided by the millions of **alveoli** is enormous, and the diffusion path length between the air and perfusing blood is short. Surface tension in the alveoli would make inflation of the lungs difficult if the alveoli did not produce **surfactant**. Review Figure 48.10, Web/CD Activity 48.1

Inhalation occurs when contractions of the **diaphragm** create subatmospheric pressure in the **thoracic cavities**. Relaxation of the diaphragm increases pressure in the thoracic cavities and causes **exhalation**. Review Figure 48.11

During periods of heavy metabolic demands such as strenuous exercise, the **intercostal muscles**, located between the ribs, increase the volume of air inhaled and exhaled.

48.4 How does blood transport respiratory gases?

See Web/CD Activity 48.2

O_2 is reversibly bound to **hemoglobin** in red blood cells. Each molecule of hemoglobin can carry a maximum of four molecules of O_2. Because of **positive cooperativity**, the affinity of hemoglobin for O_2 depends on the P_{O_2} to which the hemoglobin is exposed. Therefore, hemoglobin picks up O_2 as it flows through respiratory exchange structures and gives up O_2 in metabolically active tissues. Review Figure 48.12

Myoglobin serves as an O_2 reserve in muscle.

There is more than one type of hemoglobin. Fetal hemoglobin has a higher affinity for O_2 than does maternal hemoglobin, allowing fetal blood to pick up O_2 from the maternal blood in the placenta. Review Figure 48.13

Carbon dioxide is transported in the blood principally as bicarbonate ions (HCO_3^-). Review Figure 48.14

48.5 How is breathing regulated?

The breathing rhythm is an autonomic function generated by neurons in the medulla and modulated by higher brain centers. The most important feedback stimulus for breathing is the level of CO_2 in the blood. Review Figures 48.16 and 14.17

The breathing rhythm is sensitive to feedback from chemoreceptors on the ventral surface of the medulla and in the carotid and aortic bodies on the large vessels leaving the heart. Review Figure 48.18

See Web/CD Activity 48.3 for a concept review of this chapter.

SELF-QUIZ

1. Which of the following statements is *not* true?
   a. Respiratory gases are exchanged only by diffusion.
   b. O_2 has a lower rate of diffusion in water than in air.
   c. The O_2 content of water falls as the temperature of water rises.
   d. The amount of O_2 in the atmosphere decreases with increasing altitude.
   e. Birds have evolved active transport mechanisms to augment their respiratory gas exchange.
2. Which statement about the gas exchange system of birds is *not* true?
   a. Respiratory gases are not exchanged in the air sacs.
   b. A bird can achieve more complete exchange of O_2 from air to blood than can the human gas exchange system.
   c. Air passes through birds' lungs in only one direction.
   d. The gas exchange surfaces in bird lungs are the alveoli.
   e. A breath of air remains in the system for two breathing cycles.
3. Which statement about gas exchange in fish is true?
   a. Blood flows over the gas exchange surfaces in a direction opposite to the flow of water.
   b. Gases are exchanged across the gill filaments.
   c. Ventilation of the gills is tidal in fast-swimming fishes.
   d. Less work is needed to ventilate gills in warm water than in cold water.
   e. The path length for diffusion of respiratory gases is determined by the length of the gill filaments.

4. In the human gas exchange system,
   a. the lungs and airways are completely collapsed after a forceful exhalation.
   b. the average P_{O_2} concentration of air inside the lungs is always lower than that in the air outside the lungs.
   c. the P_{O_2} of the blood leaving the lungs is greater than that of the exhaled air.
   d. the amount of air that is moved per breath during normal, at-rest breathing is termed the total lung capacity.
   e. O_2 and CO_2 are actively transported across the alveolar and capillary membranes.
5. Which statement about the human gas exchange system is *not* true?
   a. During inhalation, a subatmospheric pressure exists in the space between the lung and the thoracic wall.
   b. Smoking one cigarette can immobilize the cilia lining the airways for hours.
   c. The respiratory control center in the medulla responds more strongly to changes in arterial O_2 concentration than to changes in arterial CO_2 concentration.
   d. Without surfactant, the work of breathing is greatly increased.
   e. The diaphragm contracts during inhalation and relaxes during exhalation.
6. The hemoglobin of a human fetus
   a. is the same as that of an adult.
   b. has a higher affinity for O_2 than adult hemoglobin has.
   c. has only two protein subunits instead of four.
   d. is supplied by the mother's red blood cells.
   e. has a lower affinity for O_2 than adult hemoglobin has.
7. The amount of O_2 carried by hemoglobin depends on the P_{O_2} in the blood. Hemoglobin in active muscles
   a. becomes saturated with O_2.
   b. takes up only a small amount of O_2.
   c. readily unloads O_2.
   d. tends to decrease the P_{O_2} in the muscle tissues.
   e. is denatured.
8. Most CO_2 in the blood is carried
   a. in the cytoplasm of red blood cells.
   b. dissolved in the plasma.
   c. in the plasma as bicarbonate ions.
   d. bound to plasma proteins.
   e. in red blood cells bound to hemoglobin.
9. Myoglobin
   a. binds O_2 at P_{O_2} values at which hemoglobin is releasing its bound O_2.
   b. has a lower affinity for O_2 than hemoglobin does.
   c. consists of four polypeptide chains, just as hemoglobin does.
   d. provides an immediate source of O_2 for muscle cells at the onset of activity.
   e. can bind four O_2 molecules at once.
10. When the level of CO_2 in the bloodstream *increases,*
   a. the rate of respiration decreases.
   b. the pH of the blood rises.
   c. the respiratory centers become dormant.
   d. the rate of respiration increases.
   e. the blood becomes more alkaline.

FOR DISCUSSION

1. The blood of a certain species of fish that lives in Antarctica has no hemoglobin. What anatomical and behavioral characteristics would you expect to find in this fish, and why is its distribution limited to the waters of Antarctica?
2. Blood banks store whole blood for a much shorter period than they store blood plasma because when blood that has been stored for too long is infused into a patient, it can actually decrease the O_2 availability to the patient's tissues. Why is this so? Explain in terms of the different physiological functions of BPG.
3. From your knowledge of respiratory gas exchange and regulation of respiration, how would you explain the tragic episode of the scientists who ascended in the hot-air balloon *Zenith,* described on the opening page of this chapter? Why do you think the three men continued tossing out ballast to ascend higher and higher, without realizing they were in mortal danger of asphyxiation?
4. In the disease emphysema, the fine structures of alveoli break down, resulting in the formation of larger air cavities in the lungs. Also, the tissue of the lungs becomes less elastic. Explain at least two reasons why patients with emphysema have a low tolerance for exercise.
5. A condition called "the bends" affects scuba divers who surface too quickly after spending an extended period in deep water, where they have been breathing pressurized air. The cause of the bends is tiny bubbles of nitrogen coming out of solution in the blood plasma. Seals spend much more time under water and at deeper depths than scuba divers, yet they do not suffer the bends. Why?

FOR INVESTIGATION

When you suddenly travel to a location at high altitude, you will notice an unusual breathing pattern when you are resting. For a while you stop breathing completely; then suddenly you start breathing rapidly for a short time; then you stop breathing again. This can go on and on in a cyclical pattern called Cheyne-Stokes breathing. It generally goes away after spending a few days at altitude. Can you hypothesize what causes Cheyne-Stokes breathing and design an experiment to test your hypothesis?

CHAPTER 49 Circulatory Systems

You gotta have heart

On April 29, 1993, the Boston Celtics met the Charlotte Hornets in a National Basketball Association playoff game at the Boston Garden. The Celtics captain and star player, 27-year-old Reggie Lewis, had just scored 10 points in 3 minutes when he went up for a rebound and suddenly slumped forward, falling to the floor "as if he'd been shot in the back." He was examined by the team doctor (an orthopedic specialist), who allowed him to return to the game in the second half. But Lewis's legs were "wobbly" and he played only briefly.

Lewis had been experiencing dizzy spells for about a month prior to his collapse. After the playoff incident, he underwent rigorous testing by cardiologists, who diagnosed Lewis as having a dangerous arrhythmia (irregular heartbeat) caused by cardiomyopathy (diseased cardiac muscle). Accepting this diagnosis would mean the end of his professional athletic career, and Lewis sought a second opinion. After another battery of tests, a second medical team felt that Lewis had undergone a transient irregular heartbeat attributable to normal athletic stresses. The condition was deemed treatable, and to have been in part the result of Lewis's enlarged heart—a condition sometimes seen in high-performing athletes. In July of 1993, after an hour spent shooting baskets in a pick-up game, Reggie Lewis collapsed and died of heart failure.

Your heart is a muscular pump that, at rest, beats an average of 72 times per minute. With each beat it circulates about 70 milliliters of blood through the body. Without taking work or exercise into account, that is 300 liters per hour, 7,200 liters per day, 2.6 million liters per year—no time outs. Heart failure accounts for about one-third of the deaths (about 900,000 deaths) each year in the United States, making it the country's leading cause of death. Heart failure is most commonly the result of blockage of the vessels that supply the heart muscle with blood. The risk of such heart failure tends to increase with age. But heart failure is also the leading cause of death among young athletes.

Why does the heart of a young, fit athlete fail? It is usually not due to vessel blockage, but to a mutation that affects the contractile proteins of heart muscle. The heart is good at compensating for these mutant

An Athlete's Heart Boston Celtics basketball star Reggie Lewis died of heart failure at the age of 27. The exact medical situation underlying his heart problems remains clouded in controversy.

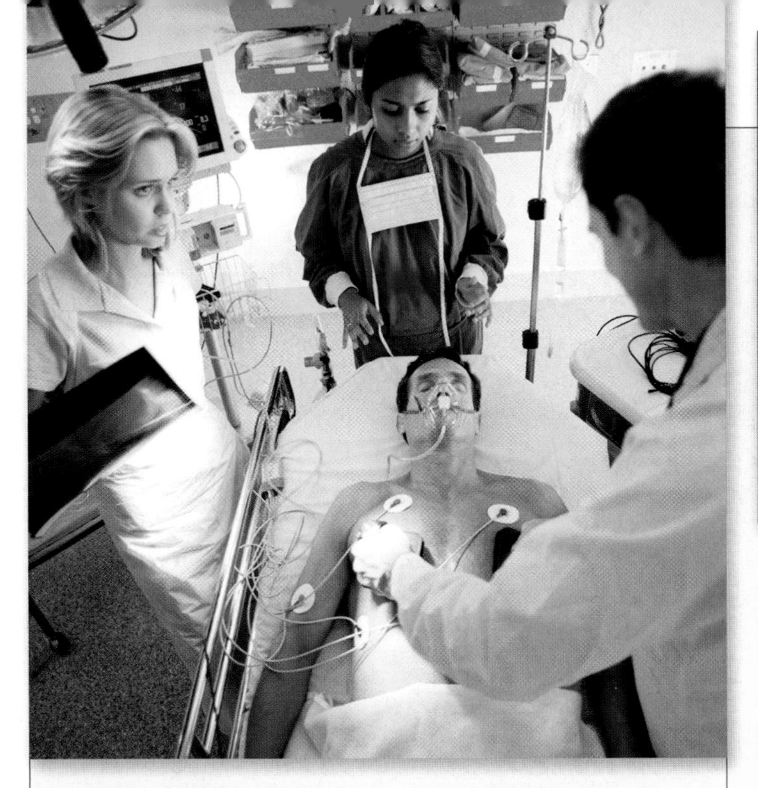

Clear! Hospital emergency rooms deal with heart attacks on a daily basis. Advances in heart surgery and emergency resuscitation techniques have saved many lives, but heart failure remains the number one cause of death in the United States and Europe.

proteins; most people can live their entire lives with the condition and never have a symptom. But with continued heavy exercise, the heart compensates by getting larger. Eventually, the increase in heart size can disrupt the electrical impulses that coordinate the contractions of the heart muscle. When heavy demand is placed on such an enlarged heart, muscle fiber contractions can suddenly become uncoordinated and the heart cannot pump blood. The lack of any prior symptoms is why the condition usually goes undiagnosed in athletes.

IN THIS CHAPTER we will learn about the adaptations of circulatory systems that enable them to match blood supply with demand in a wide variety of species. Taking the human circulatory system as a model, we will explore the mechanics of the beating heart and the characteristics of the vascular system: the arteries, capillaries, and veins. Another component of the circulatory system is the blood, and we will describe the features of this fluid tissue. The chapter ends with a discussion of the hormonal and neuronal regulation of the mammalian circulatory system.

CHAPTER OUTLINE

49.1 Why Do Animals Need a Circulatory System?

A **circulatory system** consists of a muscular pump (the **heart**), a fluid (**blood**), and a series of conduits (**blood vessels**) through which the fluid can be pumped around the body. Heart, blood, and vessels are also known collectively as a **cardiovascular system** (from the Greek *kardia*, "heart," and the Latin *vasculum*, "small vessel"). The function of circulatory systems is to transport things around the body. In preceding chapters we have learned that circulatory systems transport: heat, hormones, respiratory gases, blood cells, platelets, and elements of the immune system. In coming chapters we will add nutrients and waste products to that list. These all seem like important tasks, so why do so many species not have circulatory systems? In this section we will answer that question and examine the general types of circulatory systems found in animals.

Some animals do not have circulatory systems

Single-celled organisms serve all of their needs through direct exchanges with the environment. Such organisms are mostly found in aquatic environments or very moist terrestrial environments. Similarly, for multicellular organisms a circulatory system is unnecessary if all of their cells are close enough to the external environment that nutrients, respiratory gases, and wastes can diffuse between the cells and the environment. Small aquatic invertebrates have structures and body shapes that permit direct exchanges between cells and environment. Many of these animals have flattened, thin body shapes that maximize the amount of surface area that is in contact with the external environment and minimize the diffusion path length for exchanges between the cells and the external environment. The cells of some other aquatic invertebrates are served by highly branched central cavities called *gastrovascular systems* that essentially bring the external environment into the animal. All the cells of a sponge, for example, are in contact with, or very close to, the water that surrounds the animal and circulates through its central cavity. Very small animals without circulatory systems can maintain high levels of metabolism and activity, but bigger animals without circulatory systems such as sponges, coelenterates, and flatworms tend to be inactive, slow, or even sedentary. Larger, active animals need circulatory systems.

Larger and more active animals must support the metabolism of their cells by delivering nutrients to them and taking wastes away from them with circulatory systems. The critical environment for each cell is the extracellular fluid that surrounds it. All of the cell's nutrients—oxygen, fuel, essential molecules—comes from that fluid, and the waste products of the cell go into that fluid. In many animals, the extracellular fluid is continuous with the fluid in the circulatory system. The vessels of these animals empty their fluid directly into the tissues. At other locations the extracellular fluid flows back into the circulatory system to be pumped back out again. These circulatory systems are called *open*. In contrast, *closed* circulatory systems completely contain the circulating fluid, blood, in a continuous system of vessels. However, even in closed circulatory systems, liquid and low-molecular-weight solutes are exchanged between the blood and the extracellular fluids surrounding the cells of the body. In fact, the term *extracellular fluid* refers to both the fluid in the circulatory system and the fluid between the cells of the body. To distinguish the two, we refer to the fluid in the circulatory system as the *blood plasma* and the fluid around the cells as *interstitial fluid*. A normal 70 kilogram person has a total extracellular fluid volume of about 14 liters. A little more than a quarter of it, about 3 liters, is the blood plasma; the rest is interstitial fluid.

Circulatory systems carry materials to and from all regions of the body to maintain the optimum composition of the interstitial fluids, which in turn serve the needs of the cells. Importantly, circulatory systems speed up the delivery of nutrients and deliver them preferentially to the most active tissues.

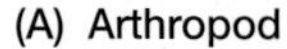

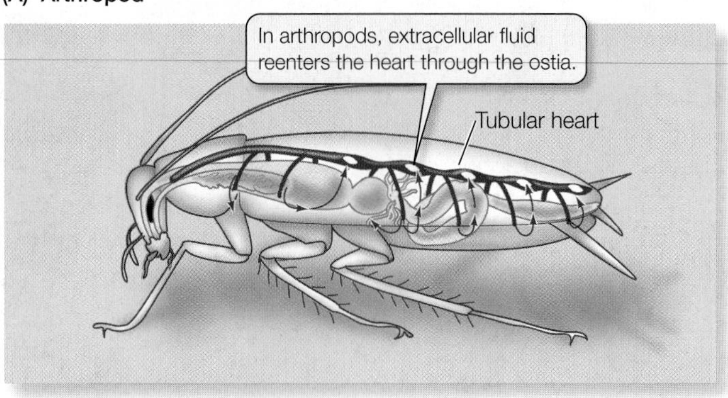

(B) Mollusk

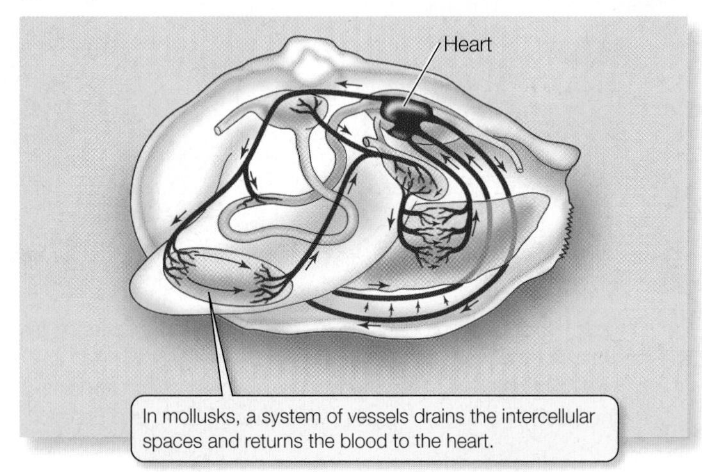

(C) Annelid worm

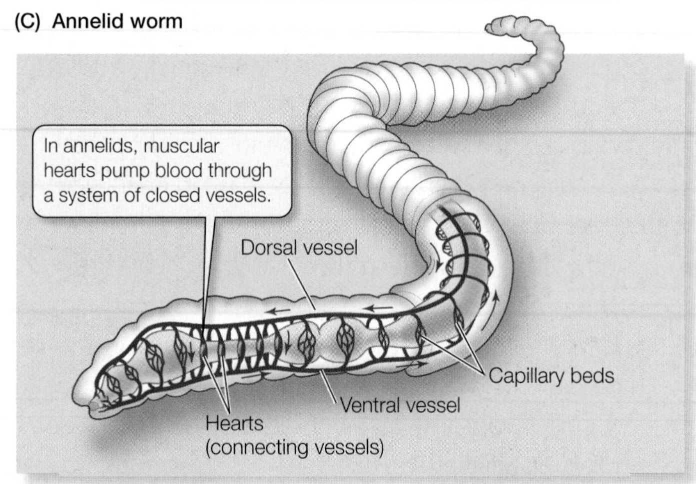

49.1 Circulatory Systems Arthropods, illustrated here by an insect (A) and mollusks such as clams (B) have open circulatory systems. Blood is pumped by a tubular heart and directed to different regions of the body through vessels that open into intercellular spaces. (C) Annelids such as earthworms have closed circulatory systems, in which the cellular and macromolecular elements of the blood are confined in a system of vessels and the blood is pumped through those vessels by one or more muscular hearts.

Open circulatory systems move extracellular fluid

In **open circulatory systems** extracellular fluid squeezes through intercellular spaces as the animal moves. A muscular pump usually assists the distribution of the fluid in these systems. The contractions of this simple heart propel the extracellular fluid through vessels leading to different regions of the body, but the fluid leaves those vessels to trickle through the tissues and eventually return to the heart. Open circulatory systems are found in arthropods, mollusks, and some other invertebrate groups. In the generalized arthropod shown in **Figure 49.1A**, the fluid returns to the heart through valved openings called *ostia*. In mollusks (**Figure 49.1B**), open vessels aid in the return of extracellular fluid to the heart.

Closed circulatory systems circulate blood through a system of blood vessels

In a **closed circulatory system**, a system of vessels keeps circulating blood separate from the interstitial fluid. Blood is pumped through this *vascular system* by one or more muscular hearts, and some components of the blood never leave the vessels. Closed circulatory systems characterize vertebrates, annelids, and some other invertebrate groups.

A simple example of a closed circulatory system is that of the earthworm (**Figure 49.1C**). One large ventral blood vessel carries blood from its anterior end to its posterior end. Smaller vessels branch off and transport the blood to even smaller vessels serving the tissues in each segment of the worm's body. In the smallest vessels, respiratory gases, nutrients, and metabolic wastes diffuse between the blood and the interstitial fluid. The blood then flows from these vessels into larger vessels that lead into one large dorsal vessel, which carries the blood from the posterior to the anterior end of the body. Five pairs of muscular vessels connect the large dorsal and ventral vessels in the anterior end, thus completing the circuit. The dorsal vessel and the five connecting vessels serve as hearts for the earthworm; their contractions keep the blood circulating. The direction of circulation is determined by one-way valves in the dorsal and connecting vessels.

Closed circulatory systems have several advantages over open systems:

- Fluid can flow more rapidly through vessels than through intercellular spaces, and can therefore transport nutrients and wastes to and from tissues more rapidly.
- By changing resistance in the vessels, closed systems can be selective in directing blood to specific tissues.
- Specialized cells and large molecules that aid in the transport of hormones and nutrients can be kept within the vessels, but can drop their cargo in the tissues where it is needed.

It seems logical to accept the premise that in all but very small animals, closed circulatory systems can support higher levels of metabolic activity than open systems can. But we then have to ask: How do highly active insect species achieve high levels of metabolic output with their open circulatory systems? The reason is that insects do not depend on their circulatory systems for respiratory gas exchange. Recall from the last chapter that respiratory gas exchange in insects is through a system of air-filled tubes (see Figure 48.4).

49.1 RECAP

Circulatory systems consist of a pump and an open or closed set of vessels through which is pumped a fluid that transports oxygen, nutrients, wastes, and a variety of other substances.

- Do you understand the difference between extracellular fluid and interstitial fluid? See p. 1046
- Why are most animals with open or with no circulatory systems rather inactive, and why does that not pertain to insects? See p. 1046
- What are some advantages of a closed circulatory system? See pp. 1046–1047

This brief overview of the open and closed systems of several invertebrates introduced some basic concepts about circulatory systems. Now we turn to a more detailed description of the more complex circulatory systems of vertebrates.

49.2 How Have Vertebrate Circulatory Systems Evolved?

Vertebrates have closed circulatory systems and hearts with two or more chambers. When a heart chamber contracts, it squeezes the blood, putting it under pressure. Blood then flows out of the heart and into vessels where pressure is lower. Valves prevent the backflow of blood as the heart cycles between contraction and relaxation.

As we explore the features of the circulatory systems of different classes of vertebrates, a general evolutionary theme will emerge: as circulatory systems become more complex, the blood that flows to the gas exchange organs becomes more completely separated from the blood that flows to the rest of the body. In fish, the phylogenetically oldest vertebrates, blood is pumped from the heart to the gills and then to the tissues of the body and back to the heart. In birds and mammals, the phylogenetically youngest vertebrates, blood is pumped from the heart to the lungs and back to the heart in a **pulmonary circuit**, and then from the heart to the rest of the body and back to the heart in a **systemic circuit**. We will trace the evolution of the separation of the circulation into two circuits.

The closed vascular system of vertebrates begins with vessels called **arteries** that carry blood away from the heart. Arteries give rise to smaller vessels called **arterioles**, which feed blood into **capillary beds. Capillaries** are the tiny, thin-walled vessels where materials are exchanged between the blood and the tissue fluid. Small vessels called **venules** drain capillary beds. The venules join together to form larger vessels called **veins**, which deliver blood back to the heart.

We can trace the evolutionary history of vertebrate circulatory systems by comparing the circulatory systems of fish, lungfish, amphibians, reptiles and crocodilians, and mammals.

Fish have two-chambered hearts

The fish heart has two chambers. An **atrium** (plural *atria*) receives blood from the body and pumps it into a more muscular chamber, the **ventricle**. The ventricle pumps the blood to the gills, where gases are exchanged. Blood leaving the gills collects in a large dorsal artery, the **aorta**, which distributes blood to smaller arteries and arterioles leading to all the organs and tissues of the body. In the tissues, blood flows through beds of tiny capillaries, collects in venules and veins, and eventually returns to the atrium of the heart.

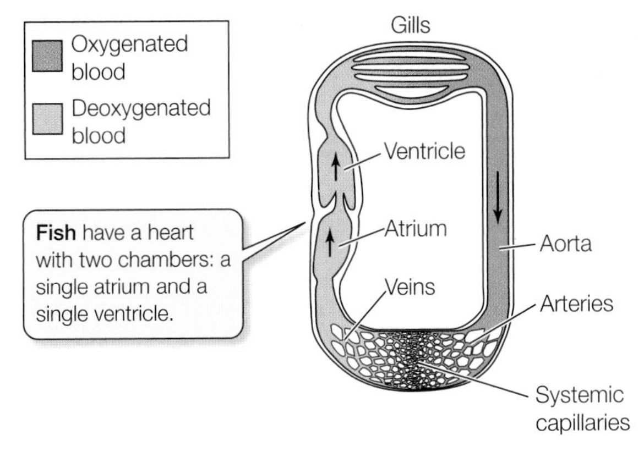

Most of the pressure imparted to the blood by the contraction of the ventricle is dissipated by the high resistance of the narrow spaces in the gill lamellae through which blood flows. As a result, blood leaving the gills and entering the aorta is under low pressure, limiting the maximum capacity of the fish circulatory system to supply the tissues with oxygen and nutrients. Yet this limitation on arterial blood pressure does not seem to hamper the performance of many rapidly swimming species, such as tuna and marlin.

The evolutionary transition from breathing water to breathing air had important consequences for the vertebrate circulatory system. An example of how the system changed to serve a primitive lung can be seen the African lungfish.

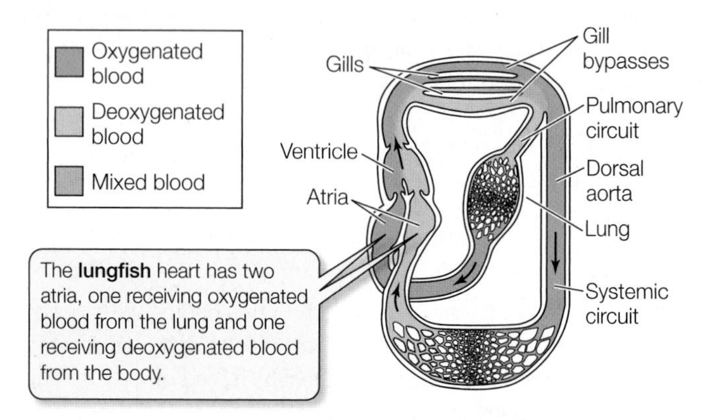

Lungfish are periodically exposed to water with low oxygen content or to situations in which their aquatic environment dries up. Their adaptation for dealing with these conditions is an outpocketing of the gut that serves as a lung. The lung contains many thin-walled blood vessels, so blood flowing through those vessels can pick up oxygen from air gulped into the lung.

How does the lungfish circulatory system take advantage of this new organ? The posterior pair of gill arteries has been modified to carry blood to the lung, and a new vessel carries oxygenated blood from the lung back to the heart. In addition, two anterior gill arches have lost their gills, and their blood vessels deliver blood from the heart directly to the dorsal aorta. Because a few of the gill arches retain gills, the African lungfish can breathe either air or water.

The lungfish heart has adaptations that partially separate the flow of its blood into pulmonary and systemic circuits. Unlike other fish, the lungfish has a partly divided atrium; the left side receives oxygenated blood from the lungs, and the right side receives deoxygenated blood from the other tissues. These two bloodstreams stay mostly separate as they flow through the ventricle and the large vessel leading to the gill arches. As a result, oxygenated blood mostly goes to the anterior gill arteries leading to the dorsal aorta, and the deoxygenated blood mostly goes to the other gill arches that have functional gills as well as to the gill arteries that serve the lung.

We can conclude that the lung of the lungfish evolved as a means of supplementing oxygen uptake from the gills. When the water is fully oxygenated, the lungfish can depend on its gills; but in oxygen-depleted water, the lungfish can augment its oxygen intake by gulping air. This trait in turn resulted in associated modifications of the lungfish vascular system that set the stage for the evolution of separate pulmonary and systemic circulations.

Amphibians have three-chambered hearts

Pulmonary and systemic circulation are partly separated in adult amphibians. A single ventricle pumps blood to the lungs and to the rest of the body. Two atria receive blood returning to the heart. One receives oxygenated blood from the lungs and the other receives deoxygenated blood from the body.

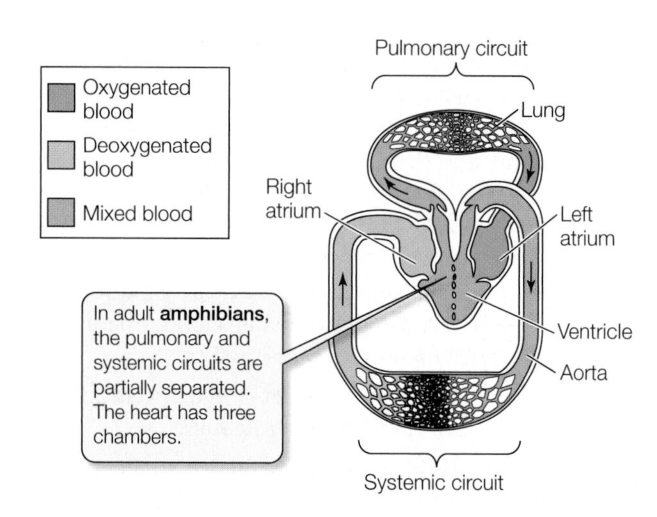

Because both atria deliver blood to the same ventricle, the oxygenated and deoxygenated blood could mix, so that blood going to the tissues would not carry a full load of oxygen. Mixing is limited, however, because anatomical features of the ventricle direct the flow of deoxygenated blood from the right atrium to the pulmonary circuit and the flow of oxygenated blood from the left atrium to the aorta.

Partial separation of pulmonary and systemic circulation has the advantage of allowing blood destined for the tissues to sidestep the high flow resistance of the gas exchange organ. Blood leaving the amphibian heart for the tissues moves directly to the aorta, and hence to the body, at a higher pressure than if it had first flowed through the lungs.

Reptiles have exquisite control of pulmonary and systemic circulation

Turtles, snakes, and lizards are commonly said to have three-chambered hearts because their ventricles are not completely separated into left and right chambers. Crocodilians (crocodiles and alligators), however, have two completely separated ventricles, creating a four-chambered heart. This anatomical description, however, does not reveal the amazing ability of these animals to control the proportion of the blood that is flowing through the lungs and the proportion that is flowing through the rest of the body.

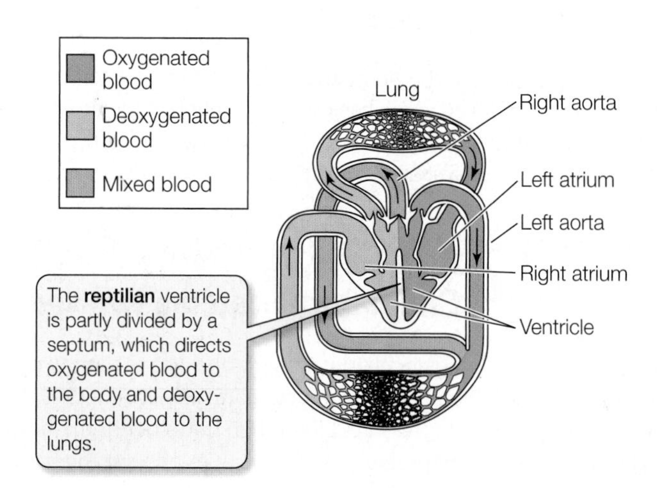

Consider the behavior, ecology, and physiology of reptiles. Many reptiles and crocodilians are active, powerful, fast animals. But their activity is in bursts that are interspersed with long periods of inactivity, during which their metabolic rates are much lower than the resting metabolic rates of birds and mammals. So enormous is the range of metabolic demand in these animals that they do not have to breathe continuously. Some species are accomplished divers and spend long periods under water, where they cannot breathe. When these animals are not breathing, it is a waste of energy for them to pump blood through their lungs. Thus, they have evolved the capability of sending blood to lungs and to the rest of the body when they are breathing, but when they are not breathing, they can bypass the pulmonary circuit and pump all the blood to the body. How do they do this?

Reptiles have two aortas—a left and a right. The left aorta receives oxygenated blood from the left side of the ventricle and carries it to the tissues of the body. The right aorta, however, can receive blood from either the right side or the left side of the ventricle. The two sides of the ventricle are partially divided by a *septum*. When the animal is breathing air, the resistance in the pulmonary circuit is lower than the resistance in the systemic circuit, so blood from the right side of the ventricle tends to flow into the pulmonary artery rather than the right aorta. When the animal is not breathing, pulmonary vessels constrict, resistance in the pulmonary circuit goes up, and therefore blood from the right side of the ventricle tends to flow into the right aorta. As a result, blood from both sides of the ventricle flows through both aortas to the systemic circuit.

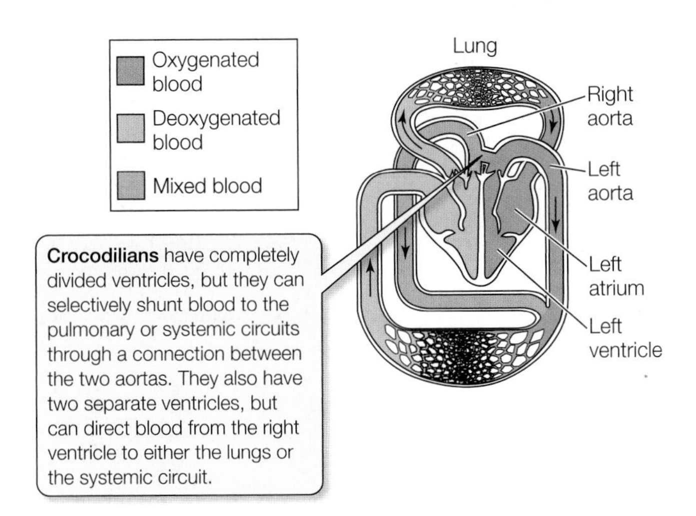

The crocodilians have completely separated ventricles with one aorta originating in each. They can alter the proportion of blood going to their pulmonary and systemic circuits, however, because of a connection between the two aortas just as they leave the heart. When the crocodile or alligator is breathing and resistance in the pulmonary circuit is low, backpressure from the stronger left ventricle prevents blood from the right ventricle from entering the right aorta and it flows into the pulmonary circuit instead. When the animal stops breathing, pulmonary vessels constrict, resistance in the pulmonary circuit rises, and blood from the right ventricle flows into the right aorta.

Birds and mammals have fully separated pulmonary and systemic circuits

The four-chambered hearts of birds and mammals have completely separate pulmonary and systemic circuits. Separate circuits have several advantages for active animals with continuously high metabolic rates:

- Oxygenated and deoxygenated blood cannot mix; therefore, the systemic circuit always receives blood with the highest oxygen content.
- Respiratory gas exchange is maximized because the blood with the lowest oxygen content and highest CO_2 content is sent to the lungs.
- Separate systemic and pulmonary circuits can operate at different pressures.

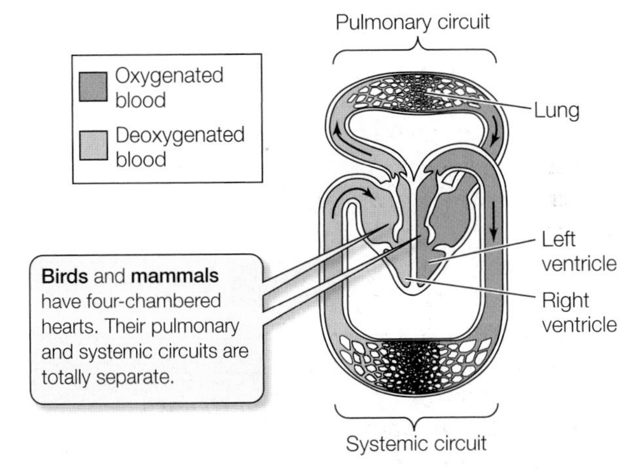

The tissues of birds and mammals have high nutrient demands and thus a very high density of the smallest vessels, the capillaries. Many small vessels present high resistance to the flow of blood. Therefore, high pressure is required in the systemic circuits of birds and mammals. Their pulmonary circuits have fewer capillaries and lower resistance than their systemic circuits, so the pulmonary circuits of birds and mammals can function at lower pressures.

49.2 RECAP

The closed circulatory system of vertebrates has evolved from a two-chambered heart in fish to a four-chambered heart in mammals, birds, and crocodilians.

- Can you explain why fish cannot supply blood to their tissues at high pressure? See p. 1047
- By comparing lungfish and amphibian circulatory systems, can you describe how a three-chambered heart could have evolved? See p. 1048
- Can you name some advantages of a four-chambered heart for mammals and birds? See p. 1049

We now take a closer look at the structure and function of the mammalian heart, using the human heart as our example.

49.3 How Does the Mammalian Heart Function?

The human heart is typical of all mammalian hearts; it has four chambers: two atria and two ventricles (**Figure 49.2**). The atrium and ventricle on the right side of your body are called the right heart. The atrium and ventricle on the left are the left heart. The right heart pumps blood through the pulmonary circuit, and the left heart pumps blood through the systemic circuit.

Valves between the atria and ventricles, the **atrioventricular valves**, prevent backflow of blood into the atria when the ventricles contract. The **pulmonary valve** and the **aortic valve**, positioned between the ventricles and the major arteries, prevent the backflow of blood into the ventricles when they relax.

In this section, we first focus on the flow of blood through the heart and through the body. Next we examine the unique electrical properties of cardiac muscle that result in the heartbeat, and see how the heart's electrical activity can be recorded in an ECG (electrocardiogram).

Blood flows from right heart to lungs to left heart to body

The heart's right atrium receives deoxygenated blood from the **superior** (upper) **vena cava** and the **inferior** (lower) **vena cava** (see Figure 49.2), large veins that collect blood returning to the heart from the upper and lower body, respectively. The veins of the heart itself also drain into the right atrium. From the right atrium, the blood flows through an atrioventricular (AV) valve into the right ventricle. Most of the filling of the ventricle results from passive flow while the heart is relaxed between beats. Just at the end of this period of passive ventricular filling, the atrium contracts and adds a little more blood to the ventricular volume. The right ventricle then contracts, causing the AV valve to close and pumping the blood into the **pulmonary artery** to the lungs.

After gas exchange occurs in the lungs, **pulmonary veins** return oxygenated blood from the lungs to the left atrium, from which

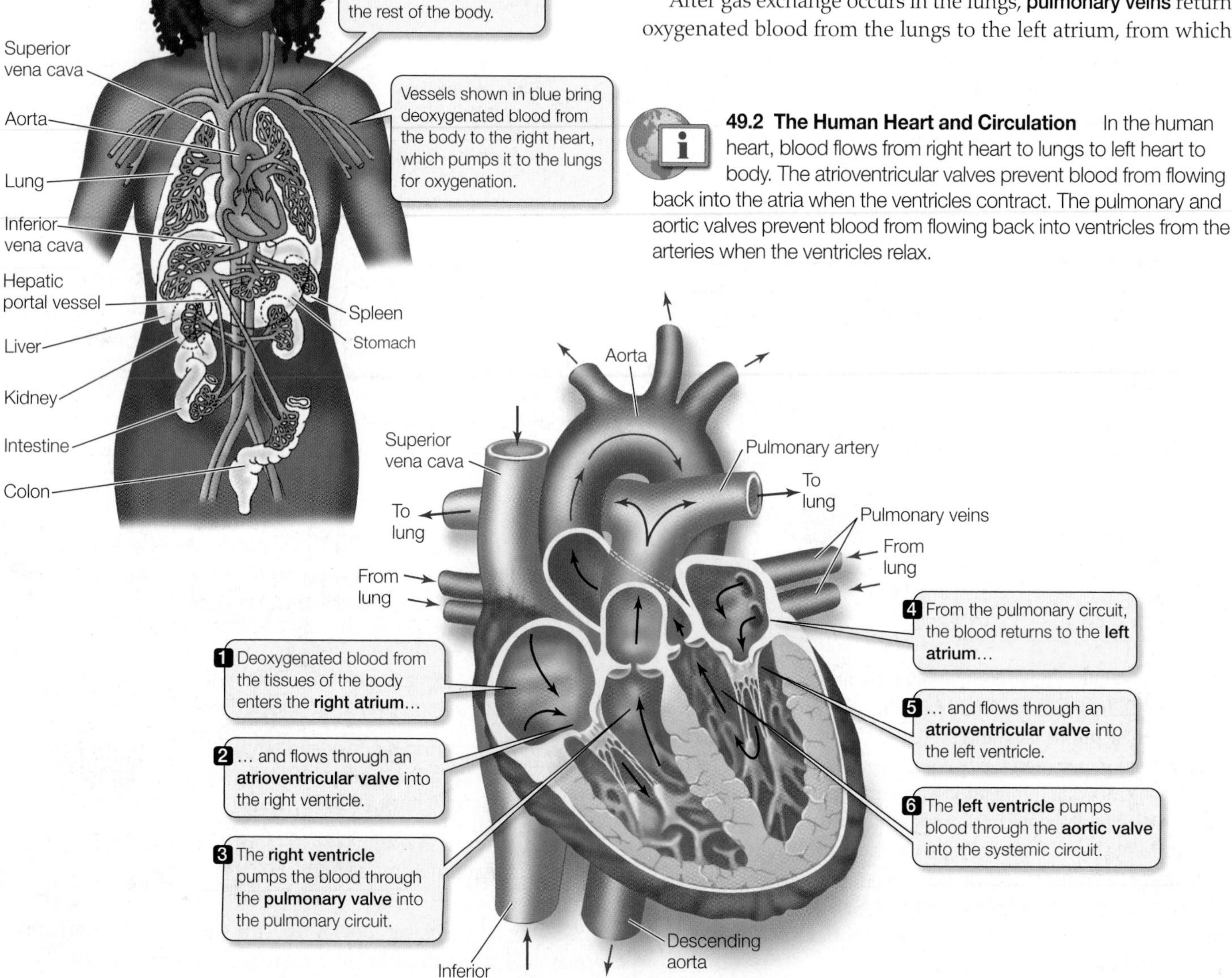

49.2 The Human Heart and Circulation In the human heart, blood flows from right heart to lungs to left heart to body. The atrioventricular valves prevent blood from flowing back into the atria when the ventricles contract. The pulmonary and aortic valves prevent blood from flowing back into ventricles from the arteries when the ventricles relax.

the blood enters the left ventricle through another AV valve. As with the right side of the heart, most left ventricular filling is passive, but the ventricle is topped off by contraction of the atrium just at the end of the period of passive filling.

The walls of the left ventricle are powerful muscles that contract around the blood with a wringing motion starting from the bottom. When pressure in the left ventricle is high enough to push open the aortic valve, the blood rushes into the aorta to begin its circulation throughout the body. In Figure 49.2, observe that the walls at the left ventricle are thicker than those of the right ventricle. The left ventricle has to propel the blood through many more kilometers of blood vessels than does the right ventricle, and must therefore push against more resistance, even though both ventricles pump the same volume of blood.

Both sides of the heart contract at the same time. The contraction of the two atria, followed by the contraction of the two ventricles and then relaxation, is the **cardiac cycle**. The cardiac cycle is divided into two phases: **systole**, when the ventricles contract, and **diastole**, when the ventricles relax (**Figure 49.3**). Just at the end of diastole, the atria contract and top off the volume of blood in the ventricles. The sounds of the cardiac cycle, the "lub-dup" heard through a stethoscope placed on the chest, are created by the heart valves slamming shut. The closing and opening of these valves are simple mechanical events resulting from pressure differences on the two sides of the valves. As the ventricles begin to contract, the pressure in the ventricles rises above the pressure in the atria, so blood starts flowing back into the atria, and the atrioventricular valves close ("lub"). When the ventricles begin to relax, the high pressure in the aorta and pulmonary artery causes blood to start to flow back into the ventricles, and this flow of blood closes the aortic and pulmonary valves ("dup"). Defective valves produce turbulent blood flow and produce the sounds known as *heart murmurs*. For example, if an atrioventricular valve is defective and does not close completely, blood will flow back into the atrium with a "whoosh" sound following the "lub."

The rhythm of the cardiac cycle can be felt in the pulsation of arteries such as the one that supplies blood to your hand. You can

49.3 The Cardiac Cycle The rhythmic contraction (systole) and relaxation (diastole) of the ventricles is called the cardiac cycle. The graphical representation below shows pressure and volume changes during the cardiac cycle for the left ventricle only.

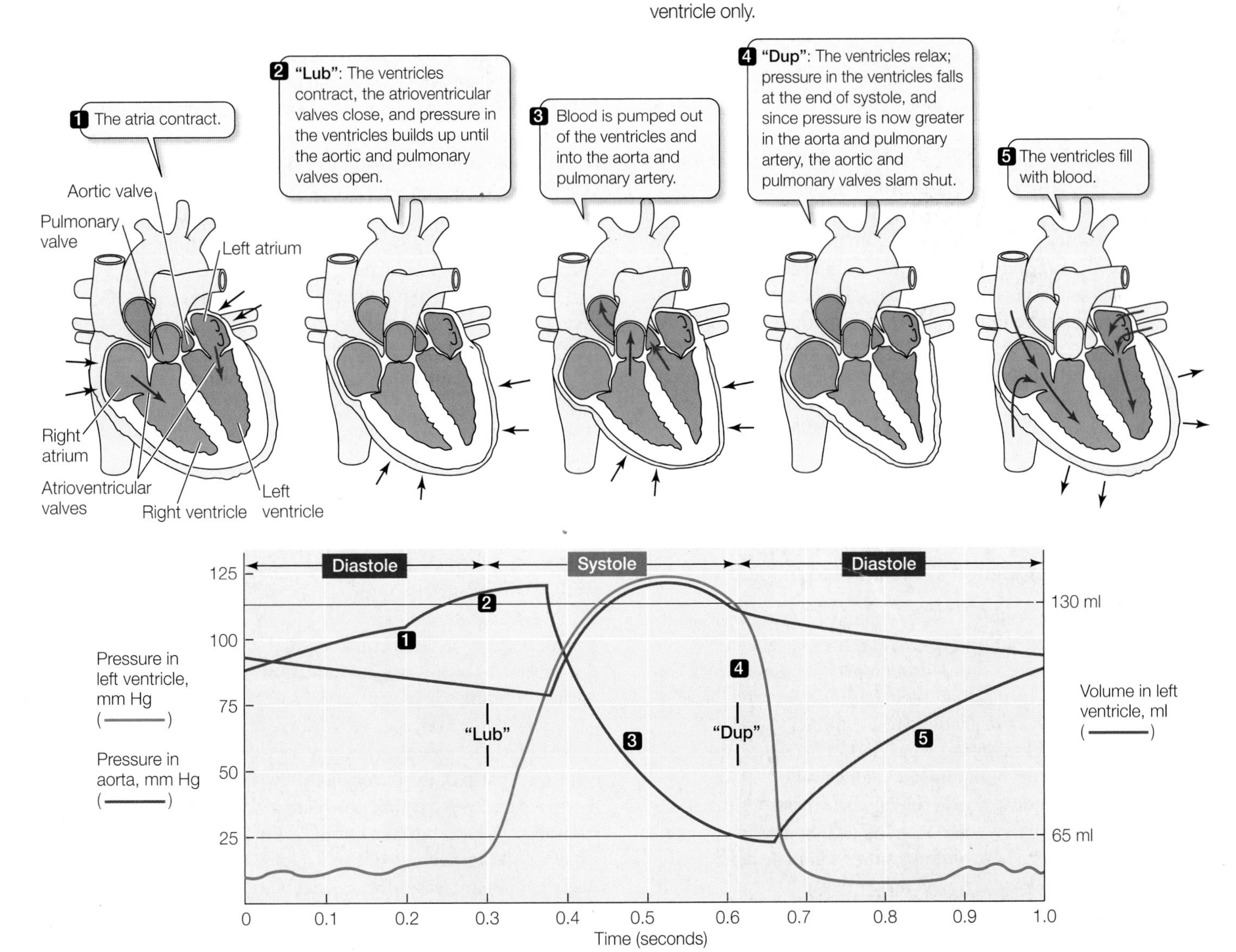

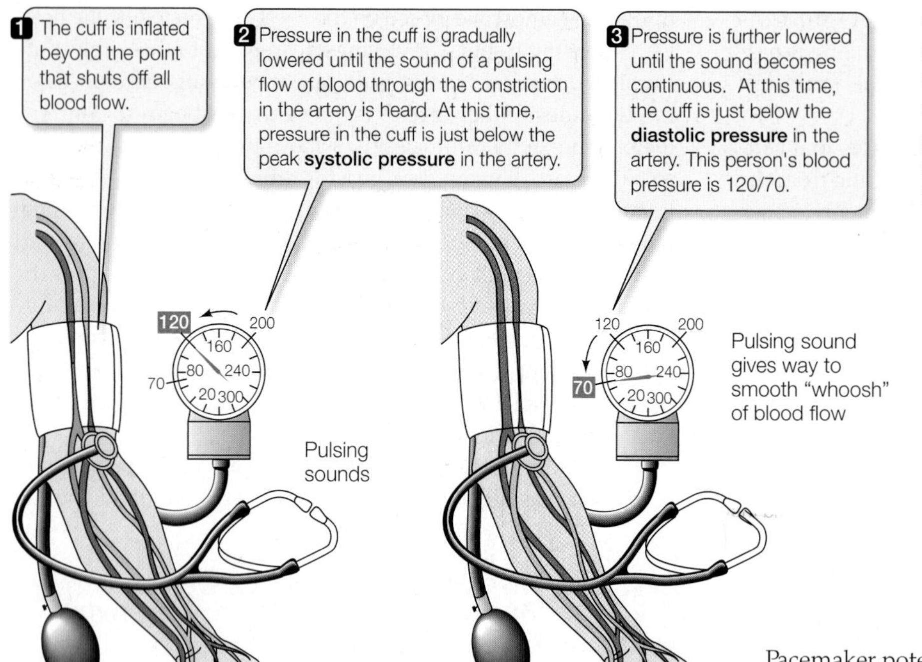

49.4 Measuring Blood Pressure Blood pressure in the major artery of the arm can be measured with a device called a sphygmomanometer, which combines an inflatable cuff and a pressure gauge. A stethoscope is also used to detect sounds created by the blood vessels in response to changes in pressure during the cardiac cycle.

feel your pulse by placing two fingers from one hand lightly over the wrist of the other hand just below the thumb. During systole, the pressure wave created by the contraction of the left ventricle surges through the arteries of your arm, and you can feel this pressure wave as a pulsing of the artery in your wrist.

Blood pressure changes associated with the cardiac cycle can be measured in the large artery in your arm by using an inflatable pressure cuff and a pressure gauge, together called a *sphygmomanometer*, and a stethoscope (**Figure 49.4**). This method measures the minimum pressure necessary to compress an artery so that blood does not flow through it at all (the systolic value) and the minimum pressure that causes intermittent flow through the artery (the diastolic value). A conventional blood pressure reading is expressed as the systolic value placed over the diastolic value. Healthy values for a young adult might be 120 millimeters of mercury (mm Hg) during systole and 70 mm Hg during diastole, or 120/70.

The heartbeat originates in the cardiac muscle

Cardiac muscle (see Section 47.1) has unique adaptations that enable it to function as a pump. First, cardiac muscle cells are in electrical contact with one another through gap junctions, which enable action potentials to spread rapidly from cell to cell. Because a spreading action potential stimulates contraction, large groups of cardiac muscle cells contract in unison. This coordinated contraction is essential for pumping blood effectively.

Second, some cardiac muscle cells are **pacemaker cells** that can initiate action potentials without stimulation from the nervous system. When they fire action potentials, they stimulate neighboring cells to contract. The primary pacemaker of the heart is a nodule of modified cardiac muscle cells, the **sinoatrial node**, located at the junction of the superior vena cava and right atrium. The resting membrane potentials of these cells are less negative than other cardiac muscle cells and they are not stable, but gradually become even less negative until they reach threshold for initiating an action potential. The action potentials of pacemaker cells are very different from those of neurons and other muscle cells. They are slower to rise; they are broader; and they are slower to return to resting potential. These properties of pacemaker cells are due to the ion channels in their membranes (**Figure 49.5**).

Pacemaker potentials involve Na^+, Ca^{2+}, and K^+ channels. Remember from Section 44.2 that when Na^+ or Ca^{2+} channels open, positive charges flow into the cell and the membrane potential becomes less negative. When K^+ channels open, positive charges flow out of the cell and the membrane potential becomes more negative. Because the Na^+ channels of pacemaker cells are more open than are those of other cardiac muscle cells, the pacemaker resting potential is less negative. The action potential of pacemaker cells is not due to voltage-gated Na^+ channels, rather to voltage-gated Ca^{2+} channels. These ion channels open and close more slowly than voltage-gated Na^+ channels, explaining the shape of pacemaker action potentials.

The unstable resting potential of pacemaker cells is due to the behavior of cation channels. As in neurons and other muscle cells, the opening of K^+ channels helps return the cell to its resting potential following an action potential. The resting potential is due to the balance of K^+ ions leaving the cell and Na^+ ions entering the cell, and in pacemaker cells this balance is controlled by cation channels through which both K^+ and Na^+ can pass. These channels open when the cell returns to resting potential following an action potential, and since they are more permeable to Na^+ than to K^+, the resting potential gradually becomes less negative. At a certain point Ca^{2+} channels begin to open, and the membrane potential continues to rise until a threshold is reached that results in a Ca^{2+} action potential.

The nervous system controls the heartbeat (speeds it up or slows it down) by influencing the rate at which the pacemaker cell resting potentials drift upwards. Norepinephrine released onto pacemaker cells by sympathetic nerves increases the permeability of the Na^+/K^+ channels and the Ca^{2+} channels. The result is the resting potential of the pacemaker cells drifts up more rapidly and the interval between action potentials is decreased. Conversely, the parasympathetic neurotransmitter, acetylcholine, has the opposite effect on the rate of rise of the pacemaker cell resting potential. Acetylcholine increases the permeability of K^+ channels and decreases the permeability of Ca^{2+} channels. The result is the interval between pacemaker action potentials shortens.

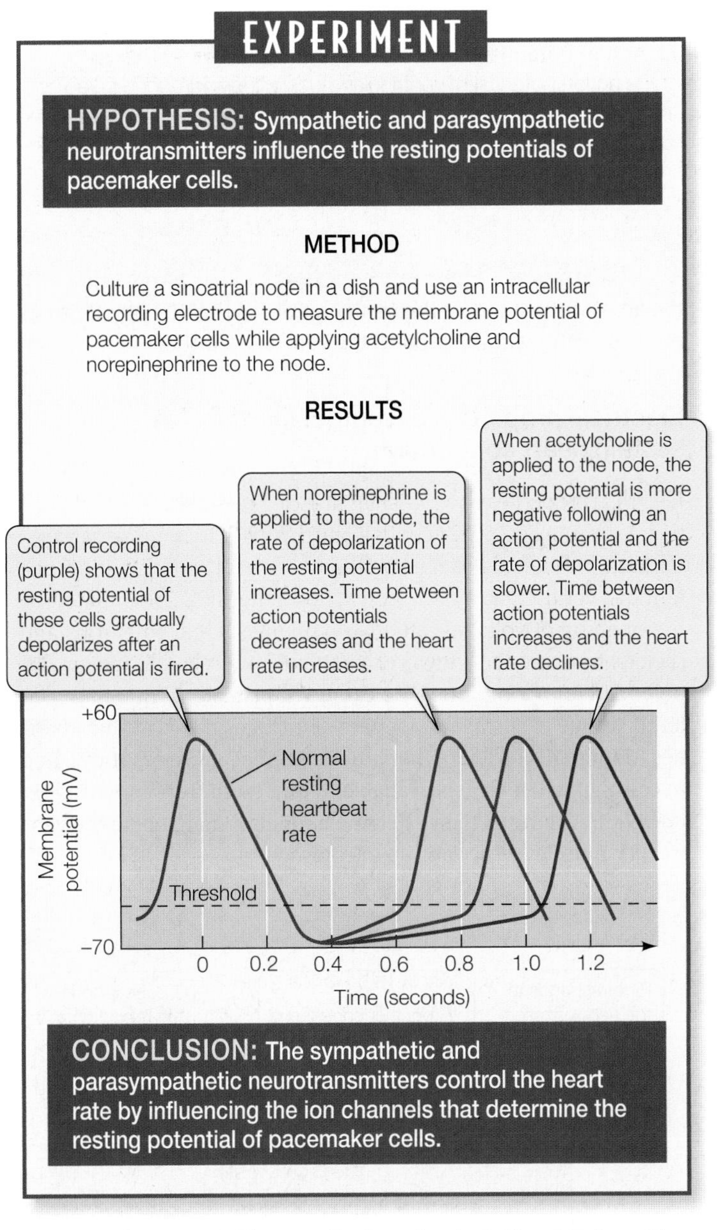

49.5 The Autonomic Nervous System Controls Heart Rate The plasma membranes of pacemaker cells spontaneously depolarize to threshold and fire action potentials. Signals from the two divisions of the autonomic nervous system raise and lower the heart rate by altering the rate at which the cells depolarize. Their effects can be seen in the dynamics of the resting membrane potential.

A conduction system coordinates the contraction of heart muscle

A normal heartbeat begins with an action potential in the sinoatrial node (**Figure 49.6**). This action potential spreads rapidly throughout the electrically coupled cells of the atria, causing them to contract in unison. Since there are no gap junctions between the cells of the atria and those of the ventricles, the action potential does not spread directly to the ventricles. Therefore, the ventricles do not contract in unison with the atria.

The action potential initiated in the atria is regenerated and conducted to and through the ventricular muscle mass by a system of modified, noncontractile cardiac muscle cells. Situated at the junction of the atria and the ventricles is a nodule of modified cardiac muscle cells—the **atrioventricular node**—which is stimulated by the depolarization of the atria. With a slight delay, it generates action potentials that are conducted to the ventricles via the **bundle of His**, which consists of modified cardiac muscle fibers that do not contract. These fibers divide into right and left *bundle branches* that run to the tips of the ventricles and then spread throughout

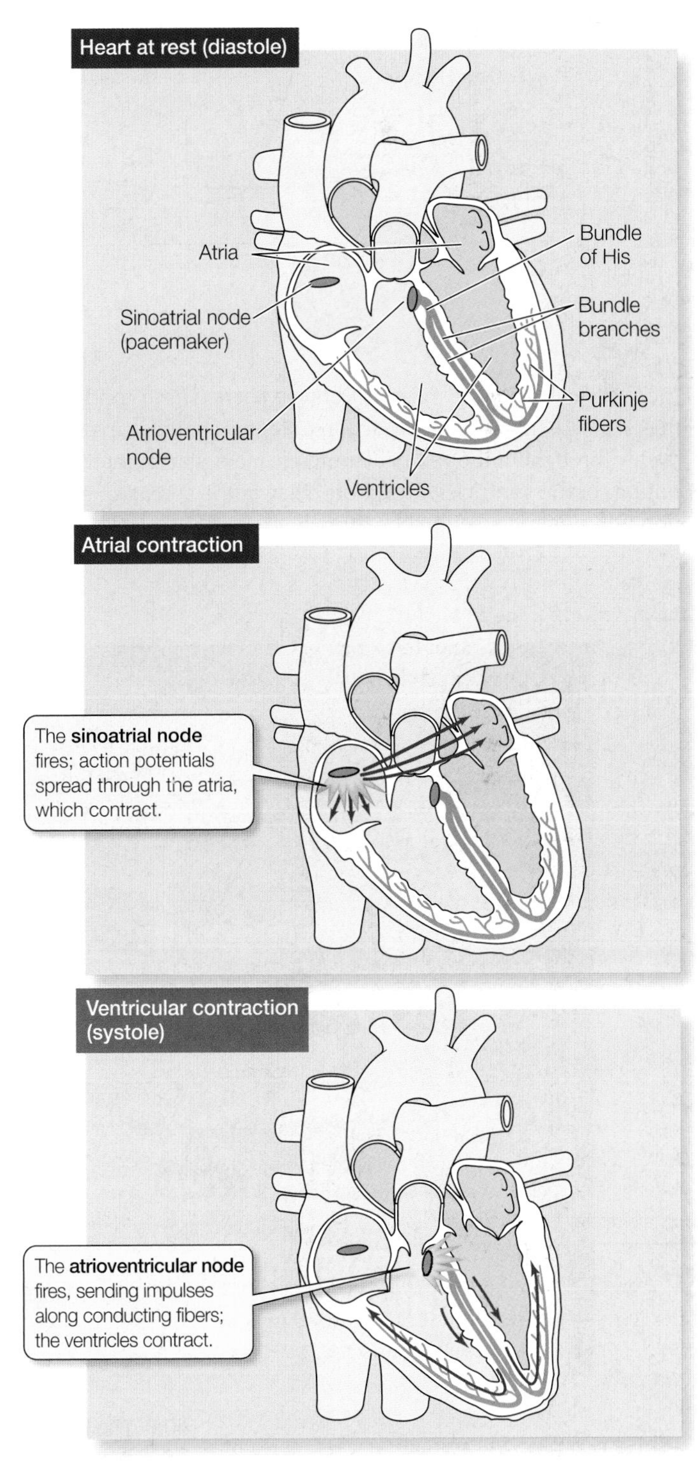

49.6 The Heartbeat Pacemaker cells in the sinoatrial node initiate the heartbeat by firing action potentials.

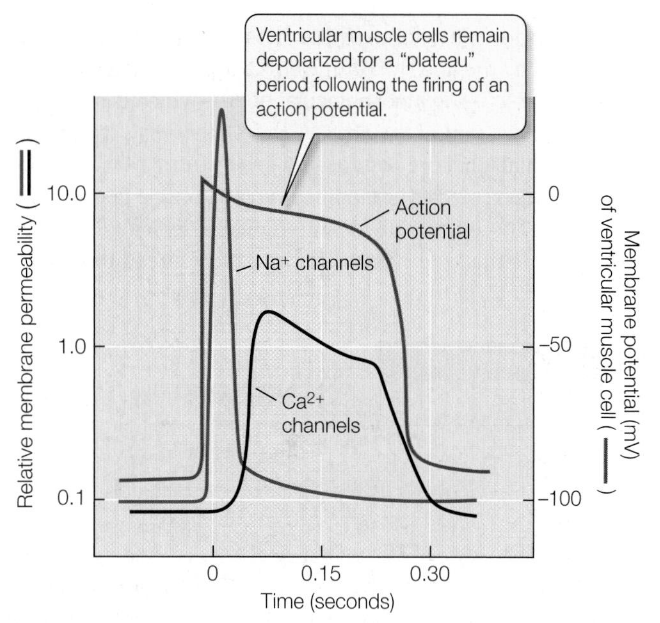

49.7 The Action Potential of Ventricular Muscle Fibers The rising phase of the action potential of ventricular muscle fibers (graphed in red) is due to the opening of voltage-gated Na^+ channels (bluegreen). However, the membrane remains in a depolarized state for a prolonged time because of the opening of voltage-gated Ca^{2+} channels (black).

the ventricular muscle mass as **Purkinje fibers**. These conducting fibers ensure that the cardiac action potential spreads rapidly and evenly throughout the ventricular muscle mass, starting at the very bottom of the ventricles. The short delay in the spread of the action potential imposed by the atrioventricular node ensures that the atria contract before the ventricles do, so that the blood passes progressively from the atria to the ventricles to the arteries.

Electrical properties of ventricular muscles sustain heart contraction

Electrical properties of ventricular muscle fibers allow them to contract for about 300 milliseconds—much longer than those of skeletal muscle fibers. Like neuronal and skeletal muscle action potentials, ventricular muscle cell action potentials are initiated by the opening of voltage-gated Na^+ channels. Unlike neurons and skeletal muscle fibers, however, ventricular muscle cells remain depolarized for a long time. The extended plateau of the action potential is due to sustained opening of voltage-gated calcium channels (**Figure 49.7**). Like other muscle, cardiac muscle is stimulated to contract when Ca^{2+} is available to bind with troponin (see Figure 47.6). As long as their calcium channels remain open, the ventricular muscle cells continue to contract.

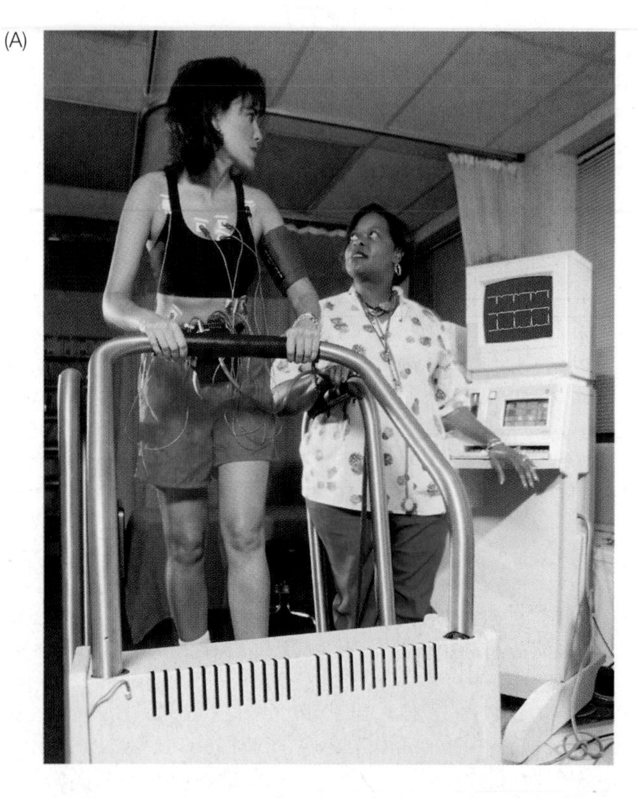

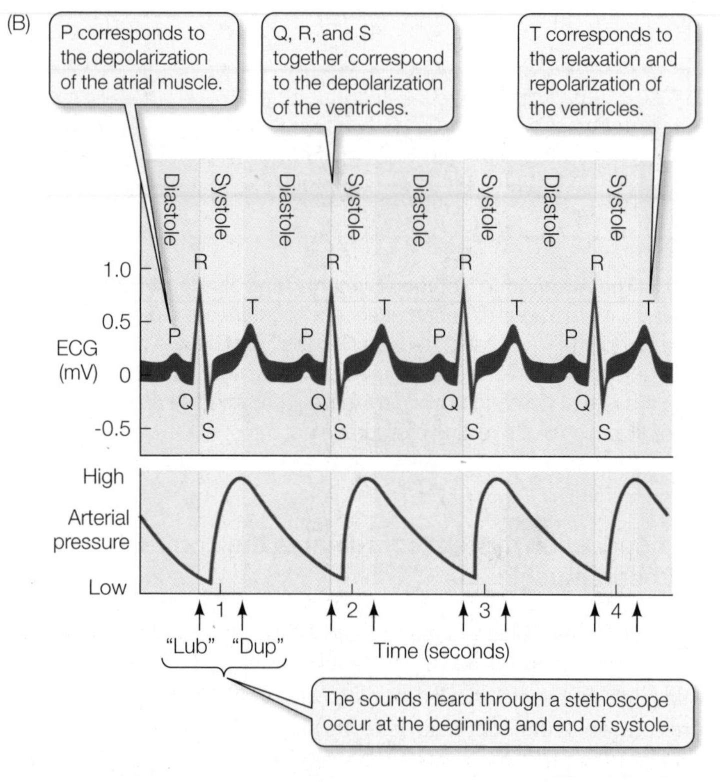

49.8 The Electrocardiogram (A) An electrocardiogram (abbreviated variously as ECG or EKG) is used to monitor heart function. Electrodes attached to the person on the treadmill record an ECG that is amplified and displayed on a monitor. (B) Variations from the normal pattern shown here can be used to diagnose heart problems.

In 1775 a British physician learned that a gypsy woman was successfully treating people for heart problems using an herbal remedy. He discovered that the remedy contained an extract of the purple foxglove plant, *Digitalis purpurea.* This drug, called digitalis and still in use today, increases Ca^{2+} concentration in cardiac muscle cells, thereby strengthening and prolonging systole. The physician, incidentally, became rich; the gypsy woman did not.

The ECG records the electrical activity of the heart

Electrical events in the cardiac muscle during the cardiac cycle can be recorded by electrodes placed on the surface of the body. Such a recording is an **electrocardiogram**, or **ECG** (EKG is also used because German physicians who invented the method used the Greek word for heart and called it the *electrokardiogramm*). The ECG is an important tool for diagnosing heart problems (**Figure 49.8A**).

The action potentials that sweep through the muscles of the atria and the ventricles before they contract are such massive, localized electrical events that they cause electric currents to flow outward from the heart to all parts of the body. Surface electrodes placed at different locations on the body detect those electric currents at different times, and therefore register a voltage difference. The appearance of the ECG depends on the placement of the electrodes. Electrodes placed on the right wrist and left ankle produced the normal ECG shown in **Figure 49.8B**. The wave patterns of the ECG are designated P, Q, R, S, and T, each letter representing a particular event in the cardiac muscle, as shown in the figure.

49.3 RECAP

The mammalian heart is organized into two atria and two ventricles. Modified cardiac muscle tissue in the right atrium functions to spontaneously generate pacemaker action potentials. Other modified cardiac muscle tissue between the atria and ventricles and throughout the ventricles conducts that signal and coordinates the heart contraction.

- Trace the path of blood through both sides of the heart, naming the major blood vessels and heart valves. See p. 1050 and Figure 49.2
- Can you differentiate systole and diastole and describe the events of the cardiac cycle? See p. 1051 and Figure 49.3
- Can you explain how cells of the sinoatrial node generate the heart beat? See pp. 1053–1054 and Figures 49.6 and 49.7

We next consider the composition of the blood and the characteristics of the vessels through which blood circulate around the body, illustrating once again how structure serves function. We will also consider the role of the lymphatic vessels that return interstitial fluid to the blood.

49.4 What Are the Properties of Blood and Blood Vessels?

Blood is a connective tissue. It consists of cells suspended in an extracellular matrix of complex, yet specific, composition. The unusual feature of blood is that the extracellular matrix is a liquid, so blood is a fluid tissue.

The cells of the blood can be separated from the fluid matrix, called **plasma**, by centrifugation (**Figure 49.9**). If a sample of blood is spun in a centrifuge, all the cells move to the bottom of the tube, leaving the clear, straw-colored plasma on top. The *packed-cell volume*, or *hematocrit*, is the percentage of the blood volume made up by cells. Normal hematocrit is about 42 percent for women and 46 percent for men, but these values can vary considerably. They are usually higher, for example, in people who live and work at high altitudes, because the low oxygen concentrations at high altitudes stimulate the production of more red blood cells.

Here we will consider two elements in blood: the *red blood cells* and the *platelets*, which are pinched-off fragments of cells. Anther important class of blood cells—*white blood cells*, or leukocytes—is discussed in Chapter 18.

Red blood cells transport respiratory gases

Most of the cells in the blood are **erythrocytes**, or red blood cells. Mature red blood cells are biconcave, flexible discs packed with hemoglobin. Their function is to transport respiratory gases. Their shape gives them a large surface area for gas exchange, and their flexibility enables them to squeeze through narrow capillaries. There are 5 to 6 million red blood cells per microliter of blood.

Red blood cells, as well as all the other cellular components of blood, are generated by stem cells in the bone marrow, particularly in the ribs, breastbone, pelvis, and vertebrae. Red blood cell production is controlled by a hormone, **erythropoietin**, which is released by cells in the kidney in response to insufficient oxygen—**hypoxia**. Many tissues respond to hypoxia by expressing a transcription factor called *hypoxia-inducible factor 1* (*HIF-1*). When the kidney becomes hypoxic and expresses HIF-1, one of the actions of the transcription factor is to activate the gene encoding erythropoietin.

Under normal conditions, your bone marrow produces about 2 million red blood cells every second. Developing, immature red blood cells divide many times while still in the bone marrow, and during this time they produce hemoglobin. When the hemoglobin content of a red blood cell approaches about 30 percent, the nucleus, endoplasmic reticulum, Golgi apparatus, and mitochondria of the cell begin to break down. This process is almost complete when the new red blood cell squeezes between the endothelial cells of blood vessels in the bone marrow and enters the circulation. Loss of organelles is seen only in mammalian red blood cells. The red blood cells of other vertebrates are nucleated.

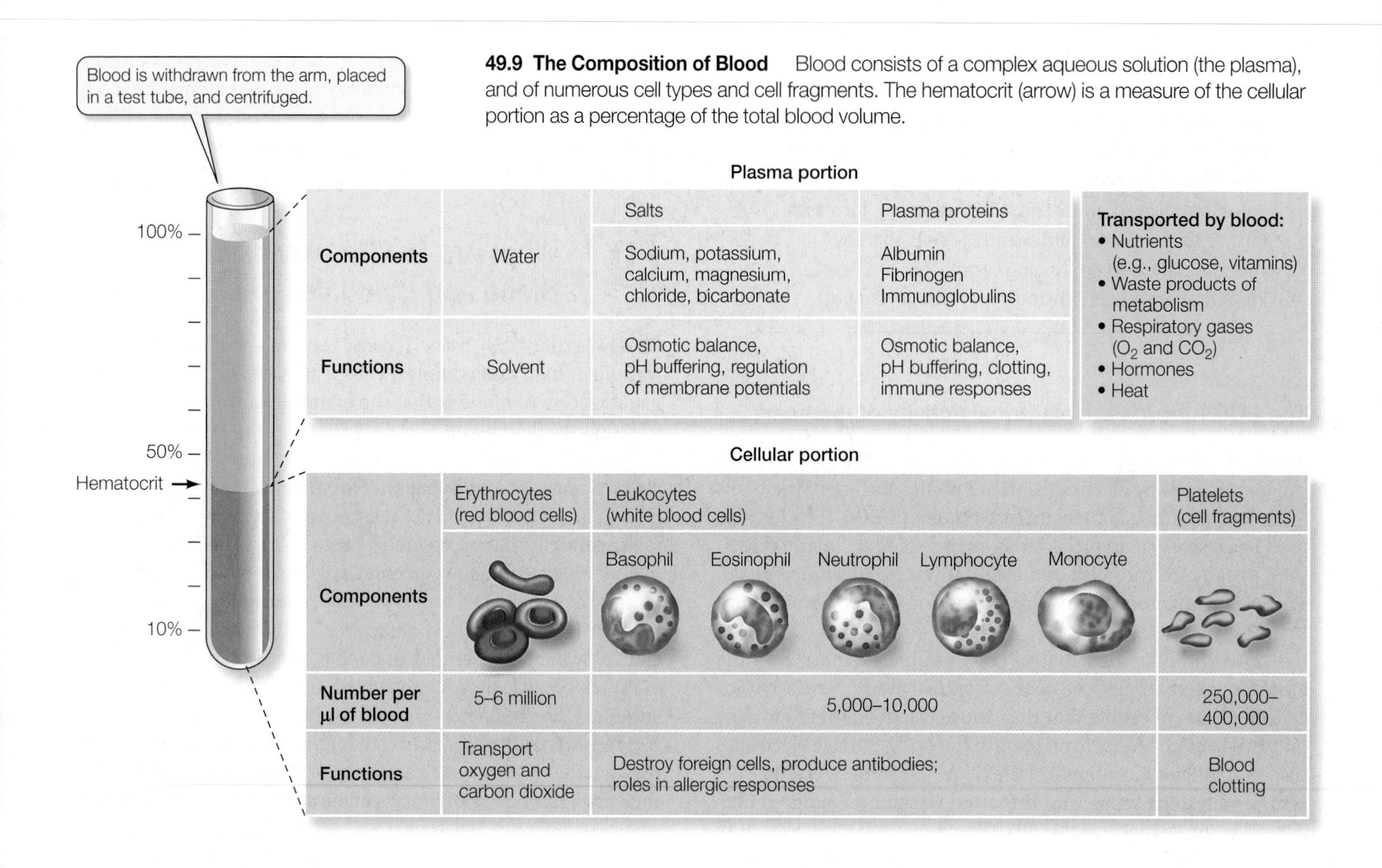

49.9 The Composition of Blood Blood consists of a complex aqueous solution (the plasma), and of numerous cell types and cell fragments. The hematocrit (arrow) is a measure of the cellular portion as a percentage of the total blood volume.

Each red blood cell circulates for about 120 days. As it gets older, its membrane becomes less flexible and more fragile. Therefore, old red blood cells can rupture as they bend to fit through narrow capillaries. One place where they are really squeezed is in the **spleen**, an organ that sits near the stomach in the upper left side of the abdominal cavity. The spleen has many venous cavities, or sinuses, that serve as a reservoir for red blood cells, but to get into the sinuses, the red blood cells must squeeze between spleen cells. When old red blood cells are ruptured by this squeezing, their remnants are taken up and degraded by macrophages.

Platelets are essential for blood clotting

Besides producing erythrocytes and leukocytes, the bone marrow stem cells described in Section 18.1 also produce cells called *megakaryocytes*. Megakaryocytes are large cells that remain in the bone marrow and continually break off cell fragments called **platelets**. A platelet is just a tiny fragment of a cell without cell organelles, but it is packed with enzymes and chemicals necessary for its function: sealing leaks in blood vessels and initiating **blood clotting** (**Figure 49.10**).

Damage to a blood vessel exposes collagen fibers. An encounter with collagen fibers activates a platelet. The platelet swells, becomes irregularly shaped and sticky, and releases chemicals called **clotting factors**, which activate other platelets and initiate the clotting of blood. The sticky platelets also form a plug at the damaged site.

The clotting of blood requires many steps and many clotting factors. The absence of any one of these factors can impair clotting and cause excessive bleeding. Because the liver produces most of the clotting factors, liver diseases such as hepatitis and cirrhosis can result in excessive bleeding. People with hemophilia experience uncontrolled bleeding due to an inherited inability to produce one of the clotting factors.

Blood clotting factors participate in a cascade of chemical reactions that activate other substances circulating in the blood. The cascade begins with cell damage and platelet activation and leads to the conversion of an inactive circulating enzyme, **prothrombin**, to its active form, **thrombin**. Thrombin cleaves molecules of **fibrinogen**, a plasma protein, forming insoluble threads of **fibrin**. The fibrin threads form the meshwork that clots the blood, seals the vessel, and provides a scaffold for the formation of scar tissue (see Figure 49.10).

Plasma is a complex solution

Plasma, the clear, straw-colored liquid portion of the blood, contains dissolved gases, ions, nutrient molecules, proteins, and other molecules, such as hormones and vitamins (see Figure 49.9). Most of the ions are Na^+ and Cl^- (hence the salty taste of blood), but many other ions are also present. The nutrient molecules in plasma include glucose, amino acids, lipids, cholesterol, and lactic acid. The circulating proteins in plasma include the blood clotting factors that we have just mentioned as well as many other proteins.

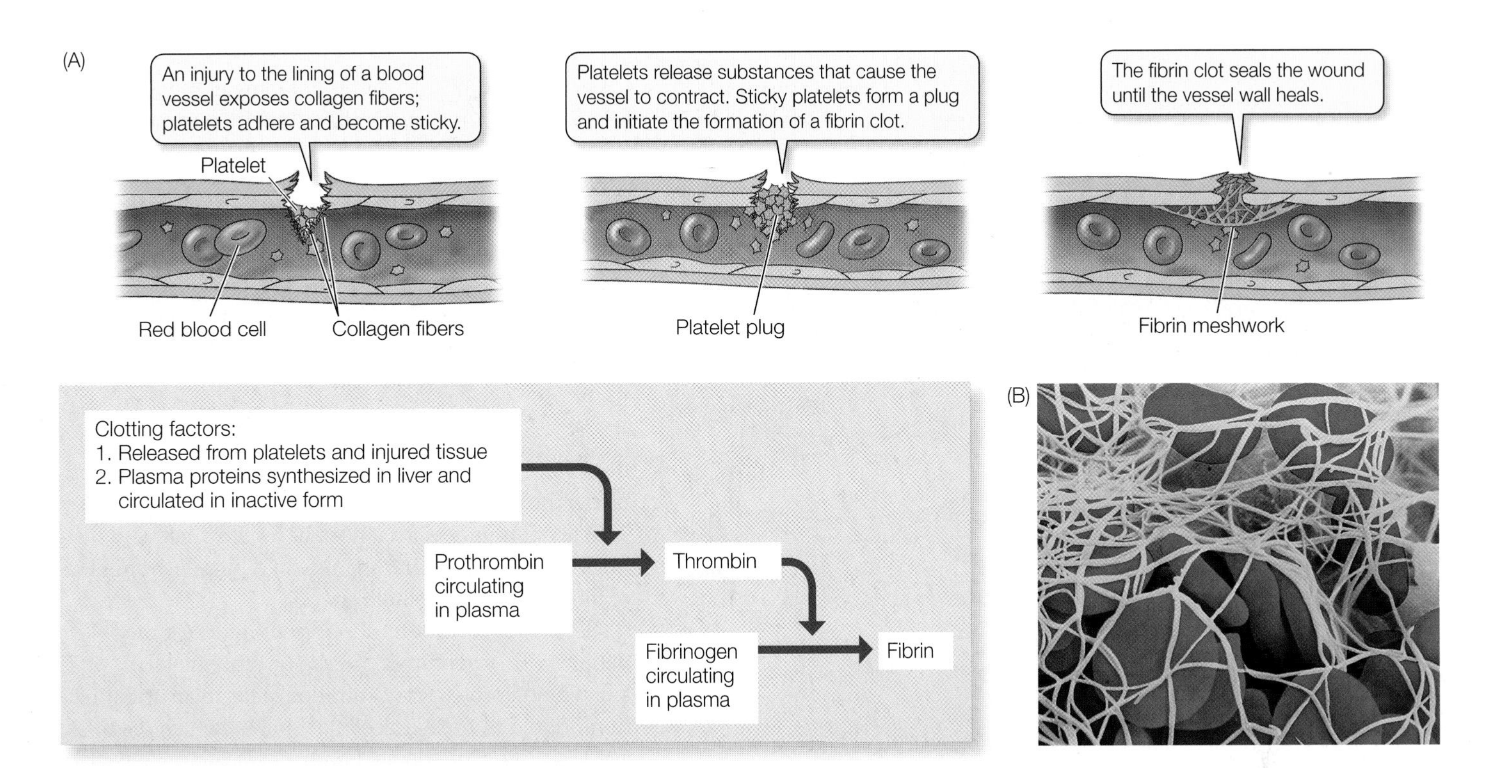

49.10 Blood Clotting (A) Damage to a blood vessel initiates a cascade of events that produces a fibrin meshwork. (B) As the meshwork forms, red blood cells are enmeshed in the fibrin threads, forming a clot, as shown in this color-enhanced electron micrograph.

Plasma is very similar to interstitial fluid in composition, and most of its components move readily between these two fluid compartments of the body. The main difference between the two fluid compartments is the higher concentration of proteins in the plasma.

Blood circulates throughout the body in a system of blood vessels

As we mentioned in Section 49.1, blood circulates through the vertebrate body in a system of closed vessels. Blood leaves the heart in arteries and is distributed throughout the tissues of the body in arterioles, which feed capillary beds. Exchanges of nutrients, wastes, respiratory gases, and hormones occur in the capillaries. Blood leaving capillary beds collects in venules, which empty into veins that conduct the blood back to the heart.

The walls of the large arteries have many extracellular collagen and elastin fibers, which enable them to withstand the high pressures of blood flowing rapidly from the heart (**Figure 49.11A**). These elastic tissues have another important function: they are stretched during systole, and thereby store some of the energy imparted to the blood by the heart. Elastic recoil during diastole returns this energy to the blood by squeezing it and pushing it forward. As a result, even though pressure in the arteries pulsates with the beating of the heart, the flow of blood is smoother than it would be through a system of rigid pipes.

Smooth muscle cells in the walls of the arteries and arterioles allow those vessels to constrict or dilate. When the diameter of the vessels changes, their resistance to blood flow changes as well, and the amount of blood flowing through them changes as a result. Neuronal and hormonal mechanisms control the resistance of the vessels—and therefore the distribution of blood to the different tissues of the body— by influencing the contraction and relaxation of the smooth muscle in the vessel walls. The arteries and arterioles are referred to as the *resistance vessels* because their resistance can vary.

Materials are exchanged in capillary beds by filtration, osmosis, and diffusion

Beds of capillaries lie between arterioles and venules (**Figure 49.11B**). Few cells of the body are more than a couple of cell diameters away from a capillary (notable exceptions include developing oocytes and the cells of the lens and cornea). The cells' needs are served by the exchange of materials between blood and interstitial fluid across the capillary walls. Capillaries have thin, permeable walls, and blood flows through these vessels very slowly, facilitating this exchange.

To anyone who has played with a garden hose, it may seem strange that blood flows through the large arteries rapidly at high pressures, but when it reaches the small capillaries, the pressure and rate of flow decrease (**Figure 49.11C**). When you restrict the diameter of a garden hose by placing your thumb over the opening, the pressure in the hose increases, which in turn increases the velocity of the water spraying out of the hose. This puzzle is solved by two more pieces of information. First, arterioles are highly branched. When flow through one branch is restricted, blood flows into other branches, so pressure does not build up quickly. Second, each arteriole gives rise to a large number of capillaries. Even though each capillary has a diameter so small that red blood cells must pass through in single file, there are so many capillaries that their total cross-sectional area is much greater than that of any

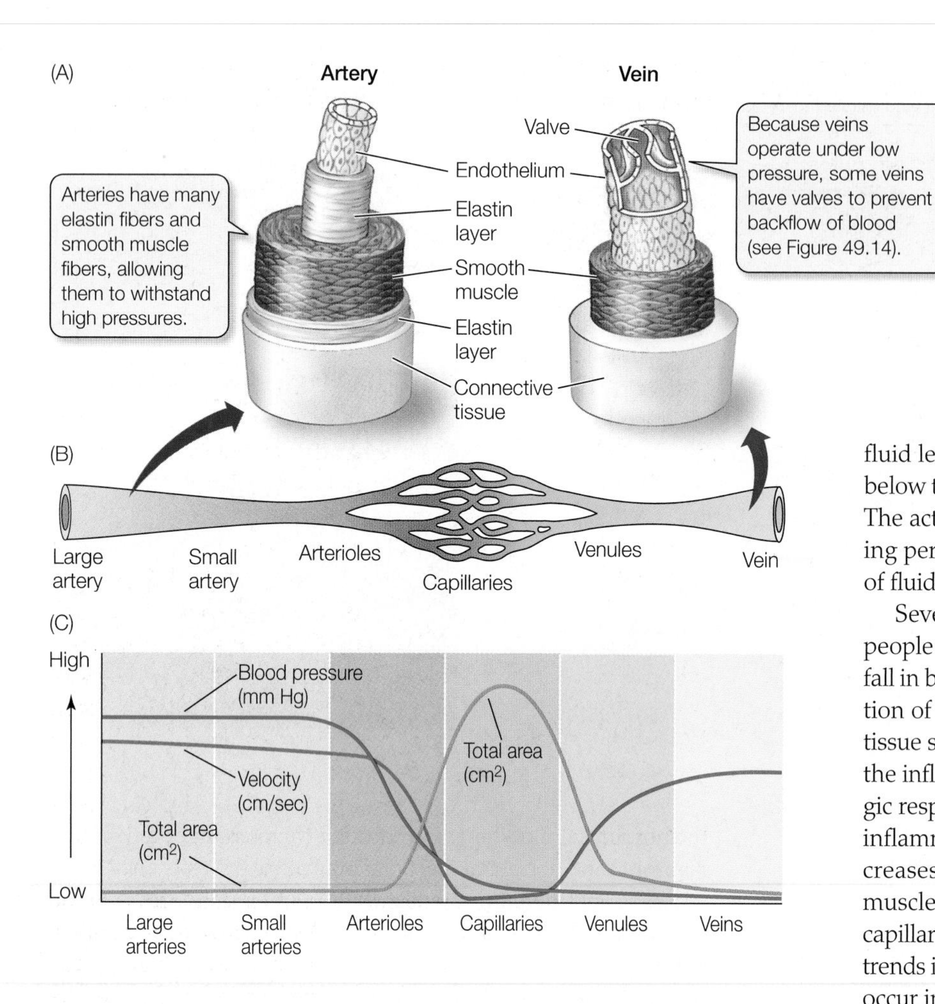

49.11 Anatomy of Blood Vessels (A) The different anatomical characteristics of arteries and veins match their functions. (B) Blood from the arterial system feeds into capillary beds, which exchange nutrients and wastes with the body's cells. The venous system returns the blood to the heart. (C) The area encompassed by each vessel type is graphed along with the pressure and velocity of the blood within them.

other class of vessels. As a result, all of the capillaries together have a much greater capacity for blood than do the arterioles. Returning to our garden hose analogy, if we connected the hose to many junctions leading to small irrigation tubes, the pressure and the flow in each of the irrigation tubes would be quite low.

Capillary walls consist of a single layer of thin endothelial cells (**Figure 49.12**). In most tissues of the body other than the brain, capillaries have tiny holes, or *fenestrations* (Latin *fenestra*, "windows"). Capillaries are permeable to water, some ions, and some small molecules, but not to large molecules such as proteins. Blood pressure therefore squeezes water and some small solutes out of the capillaries and into the surrounding intercellular spaces. Why don't water and small-molecular-weight solutes collect in the intercellular spaces? How is the blood volume maintained if fluid is continuously leaking out of the capillaries?

An answer to this question was put forth more than a hundred years ago by the physiologist E. H. Starling. Starling suggested that water balance in capillary beds is a result of two opposing forces, which have come to be known as **Starling's forces**. One force is blood pressure, which squeezes water and small solutes out of the capillaries, and the other is osmotic pressure created by the large protein molecules that cannot leave the capillaries. Starling called this second force *colloidal osmotic pressure*. He hypothesized that blood pressure is high at the arterial end of a capillary bed and drops steadily as blood flows to the venous end (**Figure 49.13**). The colloidal osmotic pressure, however, is constant along the capillary. As long as the blood pressure is above the osmotic pressure, fluid leaves the capillary, but when blood pressure falls below the osmotic pressure, fluid returns to the capillary. The actual numbers for a normal capillary bed in a resting person suggest that there would be a *slight* net loss of fluid to the intercellular spaces.

Several observations supported Starling's model. In people with severe liver disease or protein starvation, a fall in blood protein concentration leads to an accumulation of fluid in the extracellular spaces, which results in tissue swelling, or **edema**. Edema is also characteristic of the inflammation accompanying tissue damage or allergic responses (see Figure 18.4). *Histamine,* a mediator of inflammation released by certain white blood cells, increases capillary permeability and relaxes the smooth muscles of the arterioles, raising blood pressure in the capillaries, leading to fluid leakage into tissues. Study the trends in Figure 49.13 and you will see that edema should occur in all of these cases.

A few situations are not explained by Starling's hypothesis. During strenuous exercise, the blood pressure in the arterioles serving the muscles rises substantially but does not result in edema. In birds, the blood pressure in arterioles is much higher than in mammals, and the colloidal osmotic pressure is lower. If edema is not a chronic problem in exercising muscles and in birds, what is missing from Starling's model?

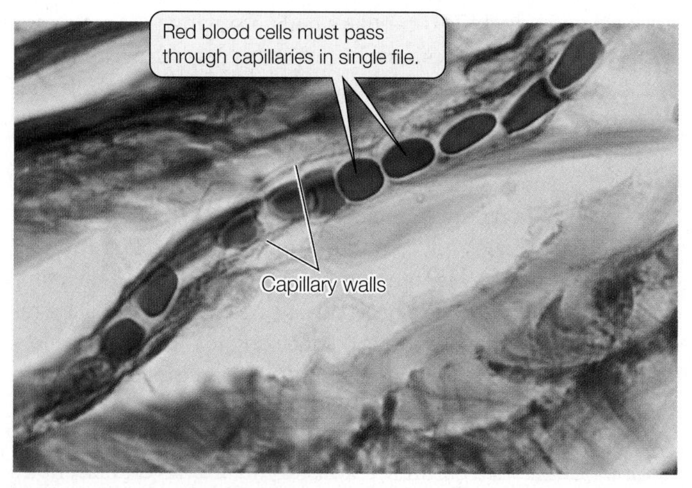

49.12 A Narrow Lane Capillaries have a very small diameter, and blood flows through them slowly.

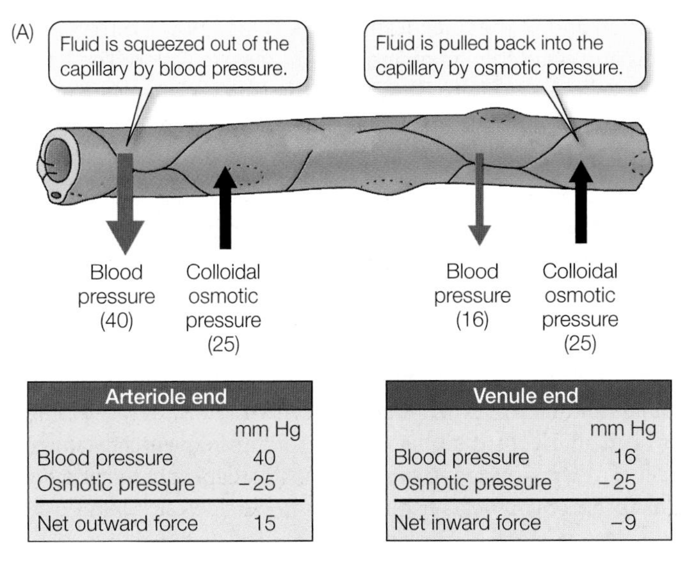

Arteriole end	mm Hg
Blood pressure	40
Osmotic pressure	−25
Net outward force	15

Venule end	mm Hg
Blood pressure	16
Osmotic pressure	−25
Net inward force	−9

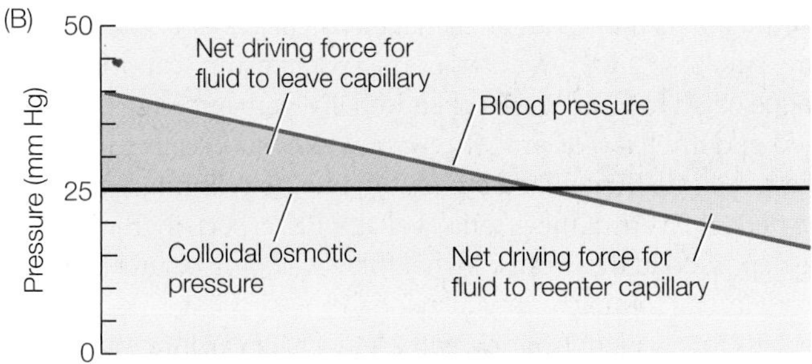

49.13 Starling's Forces Starling's model explains how blood volume is maintained in the capillary beds. (A) When blood pressure is greater than the colloidal osmotic pressure, fluid leaves the capillary; when blood pressure falls below this osmotic pressure, fluid returns to the capillary. (B) The balance of these two forces changes over the capillary bed as blood pressure falls.

Recent research suggests that bicarbonate ions (HCO_3^-) in the blood plasma contribute significantly to the osmotic attraction that draws water back into the capillary. The CO_2 produced by cellular metabolism diffuses into the endothelial cells of the capillaries where it is converted into HCO_3^- and released into the plasma. When the subject is at rest, the increasing HCO_3^- concentration can cause the osmotic pressure of the blood at the venous end to be 30 mm Hg higher than at the arterial end, and during strenuous exercise this difference can be much higher. Thus it appears that CO_2 and HCO_3^- are major factors that pull water back into the capillaries.

Lipid-soluble substances and many small solute molecules can easily pass through capillary walls from an area of higher concentration to one of lower concentration. The capillaries in different tissues, however, are differentially selective as to the sizes of molecules that can pass through them. You can imagine that this is an important issue in the design and delivery of drugs. All capillaries are permeable to O_2, CO_2, glucose, lactate, and small ions such as Na^+ and Cl^-. The capillaries of the brain do not have fenestrations, and therefore not much else can pass through them other than lipid-soluble substances, such as alcohol and anesthetics. This high selectivity of brain capillaries is known as the **blood–brain barrier**. Much less selective capillaries are found in the digestive tract, where nutrients are absorbed, and in the kidneys, where wastes are filtered. Even in the brain there are specific regions where the capillaries are more permeable, enabling the brain to detect non-lipid soluble hormones.

Blood flows back to the heart through veins

The pressure of the blood flowing from capillaries to venules is extremely low, and is insufficient to propel blood back to the heart. The walls of veins are more expandable than the walls of arteries, and blood tends to accumulate in veins. As much as 60 percent of the total blood volume may be in the veins of a resting individual. Because of their high capacity to stretch and store blood, veins are called *capacitance vessels.*

Blood flow through veins that are above the level of the heart is assisted by gravity. Below the level of the heart, however, venous return is against gravity. The most important force propelling blood from these regions is the squeezing of the veins by the contractions of surrounding skeletal muscles. As muscles contract, the vessels are compressed, and blood is squeezed through them. Blood flow may be temporarily obstructed during a prolonged muscle contraction, but when muscles relax, blood is free to move again. One-way valves within the veins of the extremities prevent backflow of blood. Thus, whenever a vein is squeezed, blood is propelled forward toward the heart (**Figure 49.14**).

In a resting person, gravity causes blood accumulation in the veins of the lower body and exerts back pressure on the capillary beds. This back pressure shifts the balance between blood pressure and osmotic pressure, causing increased loss of fluid to the intercellular spaces. That is why your feet swell during a long airline flight.

Varicose veins develop when the veins (especially those of the legs) become so stretched out that the valves can't prevent backflow. People with varicose veins may wear support hose or periodically elevate their legs above heart level to aid venous return.

Because of the one-way valves in the veins of the legs, the contractions of leg muscles act as auxiliary vascular pumps when an animal walks or runs and facilitate the return of blood to the heart from the veins of the lower body. As a greater volume of blood is returned to the heart, the heart contracts more forcefully and its pumping action is enhanced. The heartbeat gets stronger due to a property of cardiac muscle cells described by the **Frank–Starling law**: If the cardiac muscle cells are stretched, as they are when the volume of returning blood increases, they contract more forcefully. The actions of breathing also help return venous blood to the heart. The muscles involved in inhalation create negative pressure that pulls air into the lungs (see Figure 48.11), and this negative pressure also pulls blood toward the chest, increasing venous return to the right atrium. In addition, some of the largest veins closest to the heart contain smooth muscle that contracts at the onset of exercise. Contraction of veins can rapidly increase venous return and

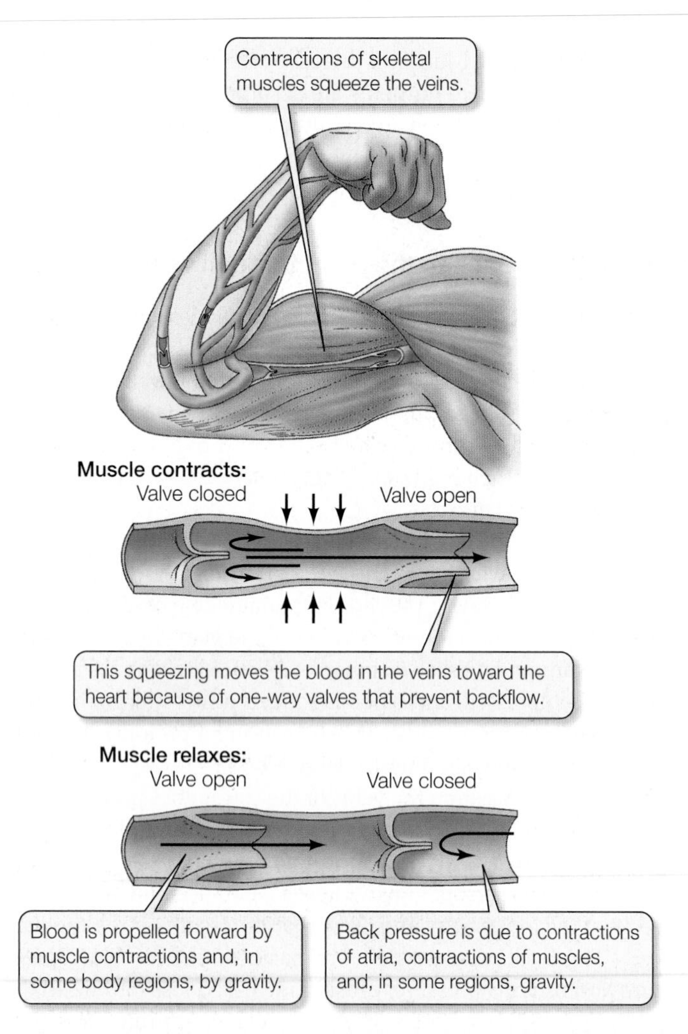

49.14 One-Way Flow Veins have valves that prevent blood from flowing backward, and contractions of skeletal muscle help to move blood toward the heart.

stimulate the heart in accord with the Frank–Starling law, increasing cardiac output.

Lymphatic vessels return interstitial fluid to the blood

The interstitial fluid contains water and small molecules, but no red blood cells, and less protein than found in blood. A separate system of vessels—the **lymphatic system**—returns interstitial fluid to the blood.

Once it enters the lymphatic vessels, the interstitial fluid is called **lymph**. Fine lymphatic capillaries merge into progressively larger vessels and ultimately into two lymphatic vessels—the **thoracic ducts**—that empty into large veins at the base of the neck (see Figure 18.1). The left thoracic duct carries most of the lymph from the lower part of the body and is much larger than the right thoracic duct. Lymph, like blood, is propelled toward the heart by skeletal muscle contractions and breathing movements, and lymphatic vessels, like veins, have one-way valves that keep the lymph flowing toward the thoracic duct.

Mammals and birds have **lymph nodes** along the major lymphatic vessels. Lymph nodes are a major site of lymphocyte production and of the phagocytic action that removes microorganisms and other foreign materials from the circulation. The lymph nodes also act as filters. Particles become trapped there and are digested by phagocytes in the nodes. When you get an infection, the lymph nodes closest to the infection may become swollen and sore. The swelling is due to the accumulation of immune system cells that have been activated to fight the infection.

Vascular disease is a killer

As mentioned at the start of this chapter, cardiovascular disease is responsible for about one-third of all deaths each year in the United States and Europe. The immediate cause of most of these deaths is heart attack or stroke, but those events are frequently the end result of a disease called **atherosclerosis** ("hardening of the arteries") that begins many years before symptoms are detected. Hence atherosclerosis is one of the "silent killer" diseases.

Healthy arteries have a smooth internal lining of endothelial cells (**Figure 49.15A**) that can be damaged by chronic high blood pressure, smoking, a high-fat diet, or microorganisms. Deposits called **plaque** begin to form at sites of endothelial damage. First, the damaged endothelial cells attract certain white blood cells to the site. These cells are then joined by smooth muscle cells migrating from the deeper layers of the arterial wall. Lipids, especially cholesterol, are deposited in these cells, so that the developing plaque becomes fatty. Fibrous connective tissue made by the invading smooth muscle cells in the plaque, along with deposits of calcium, makes the artery wall less elastic—hence, "hardening of the arteries." The growing plaque deposit narrows the artery and causes turbulence in the blood flow. Blood platelets stick to the plaque (see Figure 49.10) and initiate the formation of an intravascular blood clot, a **thrombus**, which can quickly block the artery (**Figure 49.15B**).

The blood supply to the heart muscle flows through the **coronary arteries** which are highly susceptible to atherosclerosis. As these arteries narrow, blood flow to the heart muscle decreases. Chest pain and shortness of breath during mild exertion are symptoms of this condition. A person with atherosclerosis is at high risk of forming a thrombus in a coronary artery. This condition, called **coronary thrombosis**, can totally block the vessel, causing a *heart attack*, or **myocardial infarction**.

A piece of a thrombus that breaks loose, called an **embolus**, is likely to travel to and become lodged in a vessel of smaller diameter, blocking its flow (an **embolism**). Arteries already narrowed by plaque formation are likely places for an embolism. An embolism in an artery in the brain causes the cells fed by that artery to die. This event is a **stroke**. The specific damage resulting from a stroke, such as memory loss, speech impairment, or paralysis, depends on the location of the blocked artery.

Probably the most important determinants of whether you will get atherosclerosis are your genetic predisposition and your age. Environmental risk factors play a large role, however. If you do have a genetic predisposition to atherosclerosis, it is even more important to minimize environmental risk factors. These factors include high-fat and high-cholesterol diets, smoking, and a sedentary lifestyle. Certain untreated medical conditions such as *hypertension* (high blood pressure), obesity, and diabetes are also risk factors for atherosclerosis. Changes in diet and behavior and treatment of

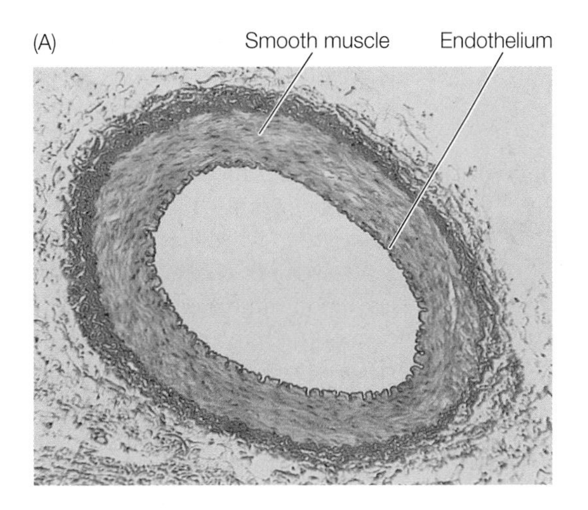

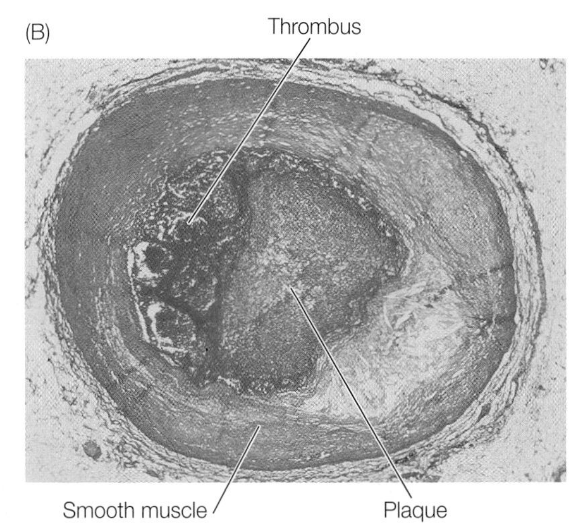

49.15 Atherosclerotic Plaque (A) A healthy, clear artery. (B) An atherosclerotic artery, clogged with plaque and a thrombus.

predisposing medical conditions can prevent and reverse early atherosclerosis and help fend off this silent killer.

49.4 RECAP

Blood is a fluid tissue with cellular components that play roles in transport of respiratory gases, immune system function, and blood clotting. Blood is distributed throughout the body in a system of vessels, and exchanges between the blood and the interstitial fluids occur in the smallest of those vessels, the capillaries.

- Why are arteries called resistance vessels and veins called capacitance vessels? See pp. 1057 and 1059
- What factors control the movement of fluids between the vascular and extravascular spaces? See pp. 1057–1059 and Figure 49.13
- What propels blood from the lower part of the body back to the heart? See pp. 1059–1060 and Figure 49.14

Every tissue in the body requires an adequate supply of oxygen-saturated blood that carries essential nutrients and is relatively free of waste products. But the nervous system cannot monitor and control every capillary bed in the body. We will now discuss how is blood flow is regulated.

49.5 How Is the Circulatory System Controlled and Regulated?

The circulatory system is controlled and regulated by neuronal and hormonal mechanisms at both the local and systemic levels that ensure each tissue is appropriately nourished and maintained. Regulation begins at the local level with each tissue controlling its own blood flow through **autoregulatory mechanisms** that cause the arterioles supplying the tissue to constrict or dilate.

The collective autoregulatory actions of every capillary bed in every tissue in the body influence the pressure and composition of the arterial blood. If many arterioles suddenly dilate, for example, allowing blood to flow through many more capillary beds, arterial blood pressure falls. If all these newly filled capillary beds contribute CO_2 to the blood at one time, the concentration of CO_2 in the blood returning to the heart increases. The nervous and endocrine systems respond to such changes by changing breathing rate, heart rate, and blood distribution to match the metabolic needs of the body.

Autoregulation matches local blood flow to local need

The autoregulatory mechanisms that adjust the flow of blood to a tissue are local but can be influenced by the nervous system and by certain hormones.

The amount of blood that flows through a capillary bed is controlled by the smooth muscle of the arteries and arterioles feeding that bed. The flow of blood in a typical capillary bed is diagrammed in **Figure 49.16**. Blood flows into the bed from an arteriole. Smooth muscle "cuffs," or **precapillary sphincters**, on the arteriole can shut off the supply of blood to the capillary bed. When the precapillary sphincters are relaxed and the arteriole is open, the arterial blood pressure pushes blood into the capillaries.

Autoregulation depends on the sensitivity of the smooth muscle to its local chemical environment. Low O_2 concentrations and high CO_2 concentrations cause the smooth muscle to relax, thus increasing the supply of blood, which brings in more O_2 and carries away CO_2—a response known as **hyperemia**, which means "excess blood." Increases in other by-products of metabolism, such as lactic acid, hydrogen ions, potassium, and adenosine (all of which increase in exercising muscle), promote hyperemia through the same mechanism. Hence, activities that increase the metabolism of a tissue also induce hyperemia in that tissue.

Arterial pressure is controlled and regulated by hormonal and neuronal mechanisms

Control and regulation of the cardiovascular system begins with the local autoregulatory mechanisms we have just described. As more blood flows into the tissues, the central blood pressure falls, and the composition of the blood returning to the heart reflects the exchanges that are occurring in the tissues. Changes in central blood

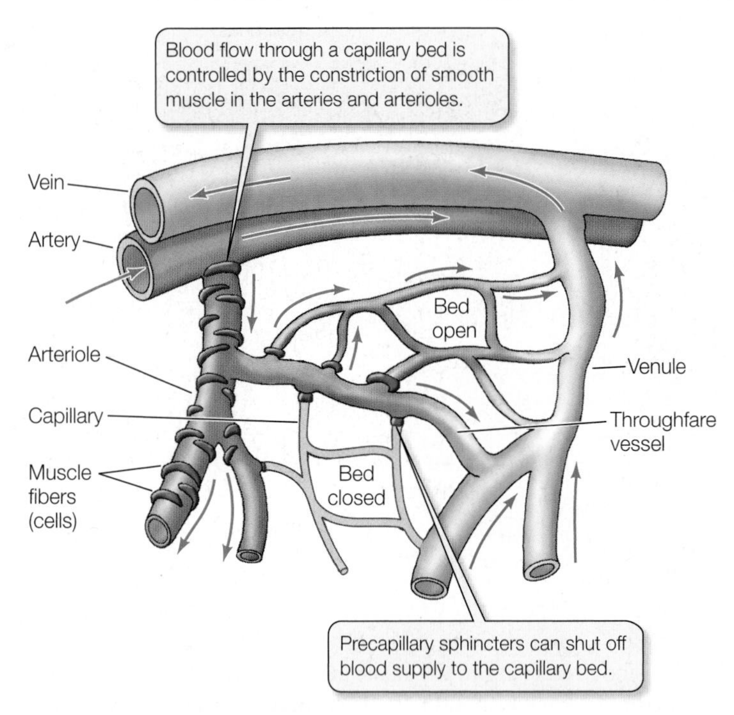

49.16 Local Control of Blood Flow Low O_2 concentrations or high levels of metabolic by-products cause the smooth muscle of the arteries and arterioles to relax, thus increasing the supply of blood to the capillary bed.

pressure and blood composition are sensed, and both endocrine and central nervous system responses are activated to return blood pressure and composition to normal. Thus circulatory functions are matched to the regional and overall needs of the body.

Arteries and arterioles are innervated by the autonomic nervous system, particularly the sympathetic division. The sympathetic postganglionic neurotransmitter norepinephrine causes the smooth muscle cells of arterioles to contract, thus constricting the vessels and reducing blood flow. An exception is found in skeletal muscle, in which specialized sympathetic neurons release acetylcholine, causing the smooth muscle of the arterioles to relax and the vessels to dilate, increasing blood to flow to the muscle.

Hormones also play a role in the regulation of arterial pressure. *Epinephrine*, which has actions similar to those of norepinephrine, is released from the adrenal medulla during massive sympathetic activation stimulated by a fall in arterial pressure or by the activation of the fight-or-flight response to a dangerous threat. *Angiotensin* is produced when blood pressure to the kidneys falls (**Figure 49.17**). These hormones influence arterioles located for the most part in peripheral tissues (extremities) or in tissues whose functions need not be maintained continuously (such as the gut). By reducing blood flow in those arterioles, these hormones increase central blood pressure and blood flow to essential organs such as the heart, brain, and kidneys.

The autonomic nervous system activity that controls heart rate and constriction of blood vessels originates in a cardiovascular control center in the medulla. Many inputs converge on this central integrative network and influence the commands it issues via parasympathetic and sympathetic nerves (**Figure 49.18**). Of special importance is incoming information about changes in blood pressure and composition from stretch receptors and **chemoreceptors** in the walls of the large arteries leading to the brain—the aorta and the carotid arteries. Stretch receptors are called **baroreceptors** because they are sensitive to the blood pressure in the aorta and carotid arteries.

Increased activity in the stretch receptors of the large arteries signals rising blood pressure and inhibits sympathetic nervous system signaling to arteries and arterioles while increasing parasympathetic signaling to the heart's pacemaker. As a result, the heart slows and arterioles in peripheral tissues dilate. If pressure in the large arteries falls, the activity of the stretch receptors decreases, stimulating sympathetic output to the arteries and arterioles while reducing parasympathetic output to the heart's pacemaker. As a result, the heart beats faster, and the arterioles in peripheral tissues constrict. Another hormone that helps to stabilize blood pressure is antidiuretic hormone (ADH, also called vasopressin), which is secreted by the posterior pituitary in response to a fall in the activity of the baroreceptors, signaling a fall in arterial pressure. ADH causes the kidneys to resorb more water and thereby maintain blood volume. Increased activity of the

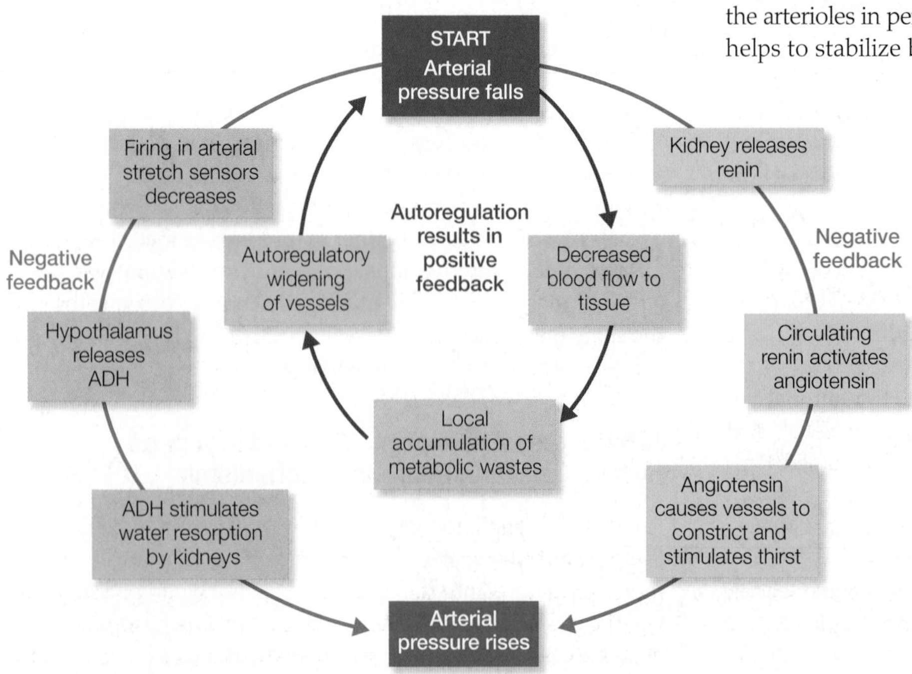

49.17 Control of Blood Pressure through Vascular Resistance A drop in arterial pressure reduces blood flow to tissues, resulting in local accumulation of metabolic wastes. This change in the extracellular environment stimulates autoregulatory opening of the arteries and would lead to a further decrease in central blood pressure if this were not prevented by negative feedback mechanisms, which work by promoting the constriction of arteries in less essential tissues and by maintaining blood volume.

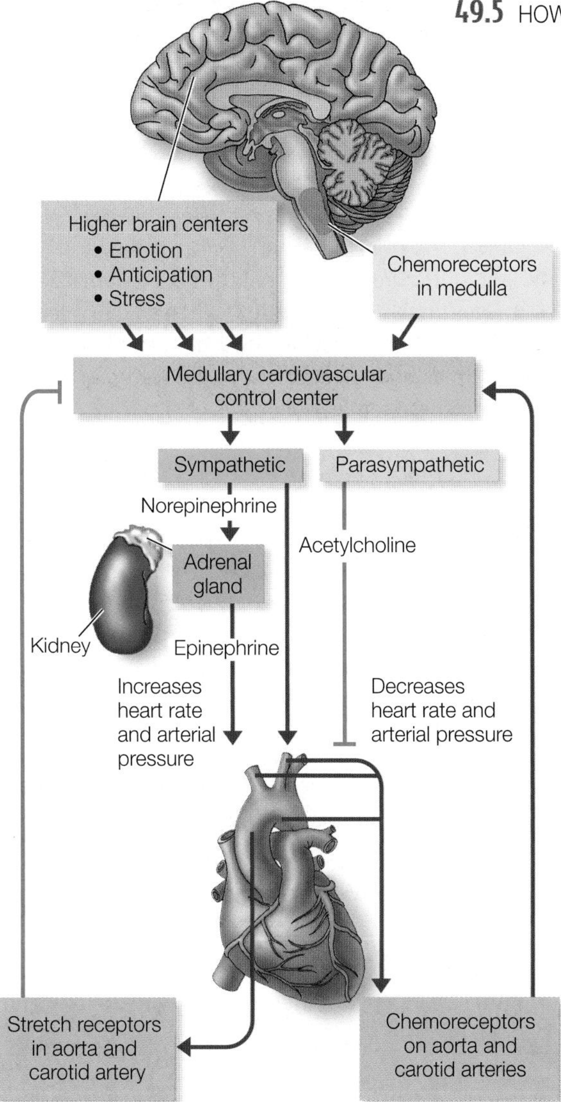

49.18 Regulating Blood Pressure The autonomic nervous system controls heart rate in response to information about blood pressure and blood composition that is integrated by regulatory centers in the medulla.

baroreceptors inhibits the release of ADH, and as a result the kidney excretes more water reducing blood volume and contributing to a fall in arterial pressure (see Figure 49.17).

Other information that causes the cardiovascular control center to increase heart rate and blood pressure comes from chemoreceptors in the medulla, aorta, and the carotid arteries. The medullary chemosensors are activated by increases in arterial CO_2 levels, and the carotid and aortic bodies are activated by falls in arterial O_2 levels. Chemosensors send signals to the regulatory center.

Cardiovascular control in diving mammals conserves oxygen

This chapter and Chapter 48 explained how animals, especially highly active birds and mammals, obtain needed oxygen from the environment and deliver it to their tissues. What happens when environmental oxygen is not available—as when an air-breathing mammal is under water?

EXPERIMENT

HYPOTHESIS: Northern elephant seals spend extended periods of time underwater. A slowed heart rate is part of the physiological reflex that makes extended dives possible.

Mirounga angustirostris

METHOD

While seals are on land for the breeding season, attach electronic tracking and recording devices to individual animals to obtain time and location data, along with readings of heart rate, water pressure (indicates depth of dive), and breathing pattern.

RESULTS

Heart rate (beats/min): 0, 25, 50, 75, 100, 125
Dive depth (m): 0, 200, 400, 600
Time of day: 22:00, 23:00, 24:00, 01:00, 02:00

During many days at sea, seals spent more than 80 percent of their time under water, making repeated dives to depths of 300–600 meters. Each dive lasted an average 20 minutes; surface time averaged less than 3 minutes per dive. Heart rate averaged 110 beats per minute at the surface but only 30 bpm while diving, with heart rate dropping as low as 3–4 bpm during parts of the dive.

CONCLUSION: Northern elephant seals spend more than 80 percent of their time at sea underwater. Reduced heart rate is part of the diving reflex that makes this possible.

49.19 Elephant Seal Diving Ability Northern elephant seals spend months at sea. New field recording technologies reveal the remarkable diving activities of these free-ranging marine mammals.

The diving capabilities of marine mammals are quite remarkable. Several adaptations enable diving seals to spend most of their time underwater, either by increasing the seal's oxygen stores or by limiting its use of oxygen.

Seals have greater oxygen stores in their bodies relative to other mammals because of a greater blood volume, a greater concentration of hemoglobin in that blood, and a greater concentration of myoglobin in their muscles. Yet the body stores of oxygen a seal takes into a dive are only about twice what a human carries into a dive.

The adaptations most critical for sustaining a seal's dive are cardiovascular mechanisms that limit blood flow to non-critical organs. This suite of adaptations is known as the **diving reflex**, which slows the heart (**Figure 49.19**) and constricts the major blood vessels going to all tissues except the critical tissues of the nervous system, heart, and eyes. The seal's central blood pressure remains high, but blood flow to its tissues decreases. This reduced blood flow has two effects: tissues switch to glycolytic (anaerobic) metabolism, and their metabolism drops considerably. While diving, the seal accumulates lactic acid in its muscles, which constitutes an "oxygen debt" to be paid back through elevated metabolism after the dive ends. But the total metabolic "debt" is much less than the metabolism that would have occurred over the same period of time had the seal not dived. The diving reflex causes the seal to be **hypometabolic** (to have a metabolic rate below its basal rate) during the dive. Hypometabolism, increased oxygen stores, and a high capacity for anaerobic metabolism make it possible for the seal to perform its amazing diving feats.

The seal's diving reflex may seem like a unique adaptation, but it provides yet another example of how natural selection shapes biological traits that are widely shared among related species. Humans also have a diving reflex. It is controlled by the vagus nerve and the parasympathetic nervous system. When our faces are submerged, we experience a mild slowing of our heart rate. This reflex probably serves as a protective response during the birth process, when pressure on the umbilical cord can deprive the fetus of maternal oxygen before breathing can begin. There are many cases, in which drowning victims have been submerged in cold water for rather long periods of time, yet have survived with no brain damage. The human diving reflex, along with the rapid cooling of the brain as body heat is lost to the water, is the probable explanation for these remarkable cases of survival.

49.5 RECAP

The delivery of blood to tissues is controlled locally by autoregulatory mechanisms that cause dilation or constriction of arterioles. These local actions are translated into alterations in central blood pressure and composition that are detected by neuronal and hormonal mechanisms, which then mediate corrective cardiovascular adjustments.

- What are the roles of hormones in the regulation of blood pressure? See pp. 1061–1062 and Figure 49.17
- Can you describe the role of stretch receptors and chemoreceptors in regulating blood pressure? See pp. 1062–1063 and Figure 49.18
- What are the major adaptations that enable marine mammals to sustain underwater activity for long periods without breathing? See pp. 1063–1064 and Figure 49.19

CHAPTER SUMMARY

49.1 Why do animals need a circulatory system?

The metabolic needs of the cells of many small animals are met by direct exchange of materials with the external medium. The metabolic needs of the cells of larger animals are met by a circulatory system that transports nutrients, respiratory gases, and metabolic wastes throughout the body.

In **open circulatory systems**, extracellular fluid leaves vessels and percolates through tissues. In **closed circulatory systems**, the blood is contained in a system of vessels. Closed circulatory systems have several advantages, including the ability to selectively direct blood, hormones, and nutrients to specific tissues. Review Figure 49.1

49.2 How has the vertebrate circulatory systems evolved?

See Web/CD Activity 49.1

The circulatory systems of vertebrates consist of a **heart** and a closed system of **vessels** containing blood that is separate from the interstitial fluid. **Arteries** and **arterioles** carry blood from the heart; **capillaries** are the site of exchange between blood and interstitial fluid; **venules** and **veins** carry blood back to the heart.

The vertebrate heart evolved from two chambers in fish to three in amphibians and reptiles and four in crocodilians, mammals, and birds. This evolutionary progression has led to an increasing separation of blood that flows to the gas exchange organs and blood that flows to the rest of the body. Review pages 1047–1049.

The simplest (two-chambered heart) has an **atrium** that receives blood from the body, and a **ventricle** that pumps blood out of the heart. An **aorta** distributes blood to arteries.

In birds and mammals, blood circulates through two circuits: the **pulmonary** which transports blood between the heart and lungs, and the **systemic**, which transports oxygen-rich blood between the heart and tissues.

49.3 How does the mammalian heart function?

The human heart has four chambers. **Valves** in the heart prevent the backflow of blood. Review Figure 49.2, Web/CD Activity 49.2

The **cardiac cycle** has two phases: **systole**, in which the ventricles contract; and **diastole**, in which the ventricles relax. The sequential heart sounds ("lub-dup") are made by the closing of the heart valves. Review Figure 49.3, Web/CD Tutorial 49.1

CHAPTER SUMMARY

Blood pressure can be measured using a sphygmomanometer and a stethoscope. Review Figure 49.4

The autonomic nervous system controls heart rate: Sympathetic activity increases heart rate, and parasympathetic activity decreases it. Norepinephrine increases and acetylcholine decreases the rate of depolarization of the plasma membranes of **pacemaker cells**, affecting heart rate. Review Figure 49.5

The **sinoatrial node** controls the cardiac cycle by initiating a wave of depolarization in the atria, which is conducted to the ventricles through a system consisting of the **atrioventricular node**, the **bundle of His**, and the **Purkinje fibers**. Review Figure 49.6

The sustained contraction of ventricular muscle cells is due to long duration action potentials that are generated by voltage-gated Na^+ and Ca^{2+} channels. Review Figure 49.7

An **electrocardiogram** (**ECG**) records electrical events associated with the contraction and relaxation of the cardiac muscles. Review Figure 49.8

49.4 What are the properties of blood and blood vessels?

Blood can be divided into a **plasma** portion (water, salts, and proteins) and a cellular portion (**erythrocytes** or red blood cells, **platelets**, and **white blood cells**). All of the cellular components are produced from stem cells in the bone marrow. Review Figure 49.9

Erythrocytes transport oxygen. Their production in the bone marrow is stimulated by **erythropoietin**, which is produced in response to **hypoxia** (low oxygen levels) in the tissues.

Platelets, along with circulating proteins, are involved in **blood clotting**, which results in a meshwork of **fibrin threads** that help seal vessels. Review Figure 49.10

Abundant smooth muscle cells allow vessels to change their diameter, altering their resistance and thus blood flow. Arteries and arterioles have many elastic fibers that enable them to withstand high pressures. Review Figure 49.11, Web/CD Activity 49.3

Capillary beds are the site of exchange of materials between blood and tissue fluid.

Starling's forces suggest that blood volume is maintained in the capillary beds by an exchange of fluids driven by both blood pressure and osmotic pressure. Review Figure 49.13

An accumulation of fluid in the extracellular spaces leads to **edema**. Bicarbonate ions in the blood plasma contribute to the osmotic forces that draw water back into capillaries.

The ability of a specific molecule to cross a capillary wall depends on the architecture of the capillary, the type of substance, and the concentration gradient between the blood and the tissue fluid.

Veins have a high capacity for storing blood. Aided by gravity, by contractions of skeletal muscle, and by the actions of breathing, they return blood to the heart. Review Figure 49.14

The **Frank–Starling law** describes forces that increase cardiac output, such as stretch of the cardiac muscles cells due to increased venous return.

The lymphatic system returns the interstitial fluid to the blood.

49.5 How is the circulatory system controlled and regulated?

Blood flow through capillary beds is controlled by local autoregulatory mechanisms, hormones, and the autonomic nervous system. Review Figure 49.16

Blood pressure is controlled in part by the hormones vasopressin and angiotensin, which stimulate contraction of blood vessels. Review Figure 49.17

Heart rate is controlled by the autonomic nervous system, which responds to information about blood pressure and blood composition that is integrated by regulatory centers in the medulla. Review Figure 49.18

Diving mammals conserve blood oxygen stores by slowing the heart rate during dives. Review Figure 49.19

SELF-QUIZ

1. An open circulatory system is characterized by
   a. the absence of a heart.
   b. the absence of blood vessels.
   c. blood with a composition different from that of tissue fluid.
   d. the absence of capillaries.
   e. a higher-pressure circuit through gills than to other organs.
2. Which statement about vertebrate circulatory systems is *not* true?
   a. In fish, oxygenated blood from the gills returns to the heart through the left atrium.
   b. In mammals, deoxygenated blood leaves the heart through the pulmonary artery.
   c. In amphibians, deoxygenated blood enters the heart through the right atrium.
   d. In reptiles, the blood in the pulmonary artery has a lower oxygen content than the blood in the aorta.
   e. In birds, the pressure in the aorta is higher than the pressure in the pulmonary artery.
3. Which statement about the human heart is true?
   a. The walls of the right ventricle are thicker than the walls of the left ventricle.
   b. Blood flowing through atrioventricular valves is always deoxygenated blood.
   c. The second heart sound is due to the closing of the aortic valve.
   d. Blood returns to the heart from the lungs in the vena cava.
   e. During systole, the aortic valve is open and the pulmonary valve is closed.
4. The pacemaker action potentials in the heart
   a. are due to opposing actions of norepinephrine and acetylcholine.
   b. are generated by the bundle of His.
   c. depend on the gap junctions between the cells that make up the atria and those that make up the ventricles.
   d. are due to spontaneous depolarization of the plasma membranes of modified cardiac muscle cells.
   e. result from hyperpolarization of cells in the sinoatrial node.
5. Blood velocity through capillaries is slow because
   a. much blood volume is lost from the capillaries.
   b. the pressure in venules is high.
   c. the total cross-sectional area of capillaries is larger than that of arterioles.
   d. the osmotic pressure in capillaries is very high.
   e. erythrocytes must pass through in single file.

6. How are lymphatic vessels like veins?
   a. Both have nodes where they join together into larger common vessels.
   b. Both carry blood under low pressure.
   c. Both are capacitance vessels.
   d. Both have valves.
   e. Both carry fluids rich in plasma proteins.
7. The production of erythrocytes
   a. ceases if the hematocrit falls below normal.
   b. is stimulated by erythropoietin.
   c. is about equal to the production of white blood cells.
   d. is inhibited by prothrombin.
   e. occurs in bone marrow before birth and in lymph nodes after birth.
8. Which of the following does *not* increase blood flow through a capillary bed?
   a. High concentration of CO_2
   b. High concentration of lactate and hydrogen ions
   c. Histamine
   d. ADH
   e. Increase in arterial pressure
9. Blood clotting
   a. is impaired in patients with hemophilia because they do not produce platelets.
   b. is initiated when platelets release fibrinogen.
   c. involves a cascade of factors produced in the liver.
   d. is initiated by leukocytes forming a meshwork.
   e. requires production of angiotensin.
10. Autoregulation of blood flow to a tissue is due to
   a. sympathetic innervation.
   b. the release of ADH by the hypothalamus.
   c. increased activity of stretch receptors.
   d. chemoreceptors in the aorta and the carotid arteries.
   e. the effect of the local chemical environment on arterioles.

FOR DISCUSSION

1. At the beginning of a race, cardiac output increases immediately before there is any change in blood O_2 or CO_2 concentrations. Explain two factors that contribute to this effect. Include the Frank–Starling law in your answer.
2. Explain how the hearts of crocodilians have the advantages of mammalian hearts during exercise but the efficiency of reptilian hearts during rest.
3. A sudden and massive loss of blood results in a decrease in blood pressure. Describe several mechanisms that help return blood pressure to normal.
4. You can describe the cycle of events in a ventricle of the heart by a graph that plots the pressure in the ventricle on the y axis and the volume of blood in the ventricle on the x axis. What would such a graph look like? Where would the heart sounds be on this graph? How would the graph differ for the left and the right ventricles?
5. If the major arteries become clogged with plaque and become less elastic because of atherosclerosis, the left ventricle must work harder and harder to pump an adequate supply of blood to the body. As a result, the left ventricle can become weakened and begin to fail, even though the right ventricle is healthy. A heart attack primarily affecting the left ventricle can have the same effect. This condition is known as congestive heart failure, and it commonly leads to fatal pulmonary edema. Explain how left ventricular failure can result in pulmonary edema, and why is it said that this condition creates a vicious circle that makes itself worse rapidly.

FOR INVESTIGATION

1. There are strains of rats that are spontaneously hypertensive—they naturally have high blood pressure. Can you formulate a hypothesis about the basis for such inherited hypertension and explain experiments you would do to test your hypothesis?
2. If you were in training to compete in the Tour de France, a grueling 3-week-long bicycle race, a trainer might tell you to "train high, compete low." What do you think is the physiological basis for such advice, and how would you test whether or not that reasoning was correct?

CHAPTER 50 Nutrition, Digestion, and Absorption

An obesity epidemic

For thousands of years, the Pima of southwestern North America were hunters and gatherers who supplemented their diet with subsistence agriculture. Their environment was arid, so they developed sophisticated irrigation systems; even so, they frequently encountered drought and subsequent starvation. Today most individuals of the ethnic Pima population in North America are clinically obese. In fact, as a population, they are one of the heaviest in the world. With obesity come related health problems such as diabetes, high blood pressure, and heart disease. The incidence of diabetes among these native American people rose from an extremely high level of 45 percent of adults in 1965 to a staggering 80 percent in 1999. Moreover, diabetes is occurring in younger individuals than ever before. What could have caused such a radical health change in an entire population? At least two interacting factors are involved: genetics and lifestyle.

Geneticists hypothesize that recurring episodes of starvation produce strong selective pressure for "thrifty genes"—particular alleles of the genes involved in digestion, absorption, and energy storage that result in greater-than-average efficiency in converting food into energy and into energy reserves, such as fat. Thrifty genes would give individuals a strong selective advantage when food is scarce. An example of a "thrifty" phenotype is seen among the Pima. They have a very low resting metabolic rate and convert food into fat readily. As we will see later in this chapter, the hormone insulin facilitates the conversion of dietary sugar into fat tissue. For many Pima, consuming a standard amount of glucose causes their insulin levels to rise three times higher than it does in Americans of European ancestry.

The other factor in the Pima obesity epidemic is an abrupt change in their traditional lifestyle. When food is plentiful and has high caloric content, thrifty genes contribute to obesity by maximizing fat storage. Today the Pima eat a modern Western diet that includes high-fat, high-calorie fast foods. In general, they also engage in less physical activity than their ancestors did.

A comparative study supports the hypotheses that have been put forward to explain the obesity epidemic in the Pima. Another population of Pima live in the Sierra Madre mountains of northern Mexico.

Efficiency Genes The Pima are an example of a population that has probably undergone selection in the past for genes that improve the efficiency of managing the energy obtained from food. With modern diets and lifestyles, these "efficiency genes" can contribute to obesity.

The Great American Lunch High-fat fast foods have become prevalent in much of the developed world. A steady diet of such foods will mean weight gain for most people, and is a major reason for the "obesity epidemic" in the United States.

Genetically they are the same as the Arizona population. However, they live a traditional lifestyle and eat traditional foods. Whereas the Arizona Pima engage in an average of only 2 hours of physical work per week, the Mexico Pima average 23 hours per week. Obesity and diabetes are not prevalent among the Mexican Pima.

High-calorie diet and sedentary lifestyle affect not just the Pima, but contribute to the overall increase in obesity throughout the U.S. population. Researchers are studying the Pima to learn more about the genetics of obesity and related diseases.

IN THIS CHAPTER we will review the nutrients required by organisms for energy, for molecular building blocks, and for specific biochemical functions. We examine diverse adaptations for acquiring, ingesting, and digesting food and absorbing nutrients. Then we will learn how the body regulates its traffic in metabolic fuels, and we return to the question of control of body mass. Lastly, we will explore the issue of natural and artificial toxins in food.

CHAPTER OUTLINE

50.1 What Do Animals Require from Food?

Animals are **heterotrophs**—they derive their nutrition from eating other organisms. In contrast, **autotrophs** (most plants, some bacteria, some archaea, and some protists) can use solar energy or inorganic chemical energy to synthesize all of their components. Directly and indirectly, heterotrophs take advantage of—indeed, depend on—the organic synthesis carried out by autotrophs, and have evolved an enormous diversity of adaptations to exploit this resource (**Figure 50.1**). In this section we will discover how animals use food, be it plants or other animals, to obtain energy and building blocks of complex molecules.

Energy can be measured in calories

Chapters 6 and 7 described how energy in the chemical bonds of food molecules is transferred to the high-energy phosphate bonds of adenosine triphosphate (ATP). ATP provides animals with energy for cellular work; however, each conversion of energy from food molecules to ATP and from ATP to cellular work is inefficient. In fact, most of the energy that was in the food is lost as heat. Even the energy the animal uses is eventually reduced to heat, as molecules that were synthesized are broken down and the energy of movement is dissipated by friction. Therefore, we can talk about the energy requirements of animals and the energy content of food in terms of a measure of heat energy: the calorie.

A **calorie** is the amount of heat necessary to raise the temperature of 1 gram of water 1°C. Since this is a tiny amount of energy compared with the energy requirements of many animals, physiologists commonly use the **kilocalorie** (**kcal**) as a unit of measure (1 kcal = 1,000 calories). Nutritionists also use the kilocalorie as a standard unit of energy, but they traditionally refer to it as the **Calorie** (**Cal**), which is always capitalized to distinguish it from the single calorie. (Scientists are abandoning the calorie as an energy unit as they switch to the International System of Units. In this system, the basic unit of energy is the joule: 1 joule = 0.239 calories.)

The *metabolic rate* of an animal (see Section 40.4) is a measure of the overall energy needs that must be met by the animal's ingestion and digestion of food. The basal metabolic rate

(A) *Ailuropoda melanoleuca*

(B) *Ursus maritimus*

50.1 Heterotrophs Get Energy from Autotrophs (A) Herbivores get their energy directly from autotrophs. Large herbivores such as the giant panda must consume huge amounts of plant matter (this particular animal eats only bamboo plants) to fulfill their nutritional needs. (B) Polar bears are carnivores and ferocious predators. A carnivore's energy is indirectly obtained from autotrophs, since the energy stored in a prey animal was originally obtained from autotrophs.

of a human is about 1,300–1,500 kcal/day for an adult female and 1,600–1,800 kcal/day for an adult male. Physical activity adds to this basal energy requirement. For a person doing sedentary work, about 30 percent of total energy consumption is due to skeletal muscle activity, and for a person doing heavy physical labor, over 95 percent of total caloric expenditure is due to skeletal muscle activity. The components of food that provide energy are fats, carbohydrates, and proteins. Fats yield 9.5 kcal/gram, carbohydrates 4.2 kcal/gram, and proteins about 4.1 kcal/gram. Some equivalencies of food, energy, and energy consumption are shown in **Figure 50.2**.

Energy budgets reveal how animals use their resources

It is possible to quantify the caloric value of any food an animal eats. It is also possible to quantify the caloric cost of anything an animal does. By comparing calories consumed with calories expended, we can construct **energy budgets** that allow ecologists and evolutionists to apply a cost–benefit analysis to behavior.

Consider, for example, territorial aggression: Under what circumstances does it "pay" to fight over a food resource? Such a study was led by Larry Wolf of Syracuse University on African sunbirds, which feed on the nectar of a particular flower. These birds defend feeding territories in some habitats but not others. The in-

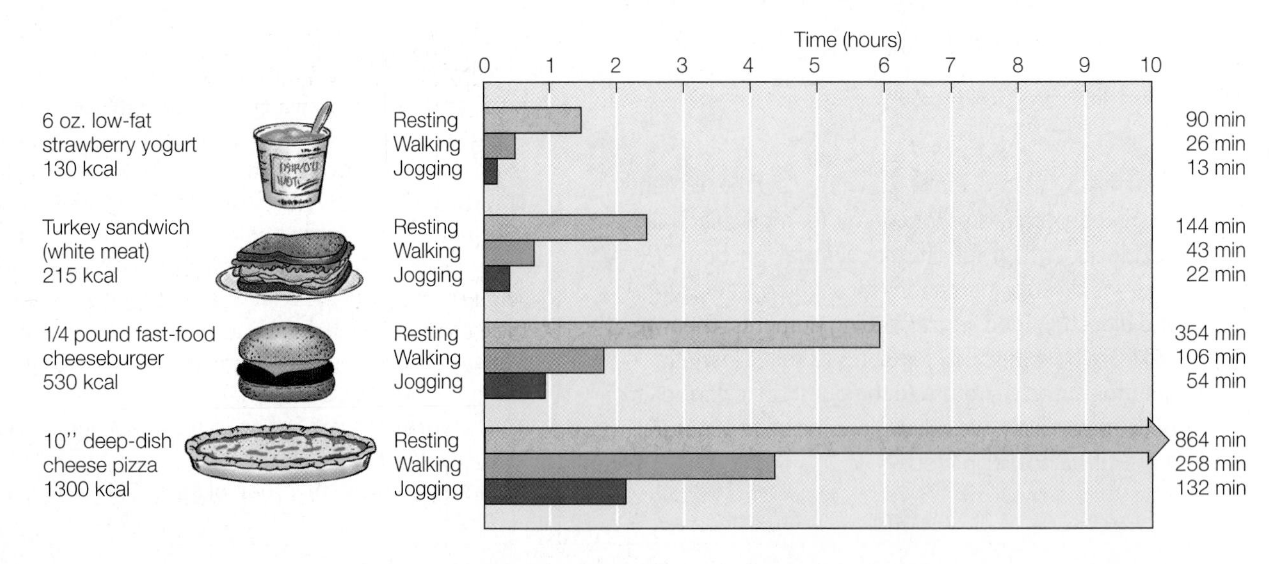

50.2 Food Energy and How We Use It The energy contained in several common food items is shown at the left. The graphs indicate about how long it would take a person with a basal metabolic rate of about 1,800 kcal/day to utilize the equivalent amount of energy while involved in various activities.

vestigators hypothesized that the birds could "afford" to be territorial only if the food resource was rich enough to support the metabolic cost of aggressive behavior. Accordingly, they determined how much nectar a flower produces and how this amount differs with habitat type. They then determined how many calories the birds spent when they were resting, when they were foraging, and when they were aggressively defending a patch of flowers.

According to the investigators' calculations, if a bird had a choice between a patch of flowers that produced nectar at a rate of 1 μl/day per blossom or a patch that produced nectar at 2/μl day per blossom, by selecting the richer patch it could meet its daily caloric requirement with 4 hours of foraging instead of 8 hours of foraging. Since caloric expenditure during foraging is 600 calories/hour greater than when resting, the bird could save 2,400 calories (4 hours × 600 calories/hour) by feeding on the more productive flowers. However, territorial defense costs 2,000 calories more per hour than foraging. Therefore, if a bird has to spend much more than an hour a day chasing intruders away from its rich flower patch, it is better off being nonaggressive and feeding on the less productive flowers. Direct observation and measurement have shown that the actual behavior of the birds tends to agree with predictions based on energy budget calculations.

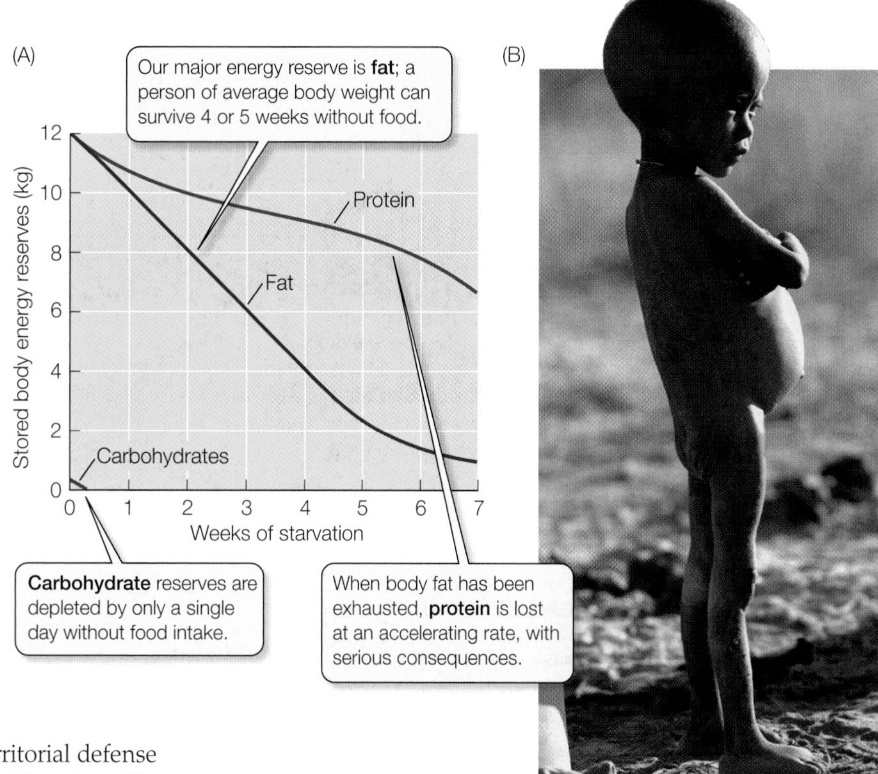

50.3 The Course of Starvation (A) In a person subjected to undernutrition, the energy reserves of the body are eventually depleted. (B) The swollen (due to edema) abdomen, face, hands, and feet of this young Somali girl, as well as her spindly limbs, are symptoms of kwashiorkor. This syndrome results from the body breaking down blood proteins and muscle tissue to fuel metabolism.

Sources of energy can be stored in the body

Although the cells of the body use energy continuously, most animals do not eat continuously. Therefore, animals must store fuel molecules that can be released as needed between meals. Carbohydrates are stored in liver and muscle cells as glycogen, but the total glycogen store represents only about a day's basal energy requirements (1,500–2,000 kcal). Fat is the most important form of stored energy in the bodies of animals. Not only does fat have more energy per gram than glycogen, but it can be stored with little associated water, making it more compact. Migrating birds store energy as fat to fuel their long flights; if they had to store the same amount of energy as glycogen, they would be too heavy to fly! Proteins are not used as energy storage compounds, although body protein can be metabolized as an energy source of last resort.

If an animal takes in too little food to meet its energy requirements, it is **undernourished** and must start metabolizing some of the molecules of its own body. This "self-consumption" begins with the energy storage compounds glycogen and fat. Protein loss is rapid at first, as readily available amino acids are consumed by ongoing protein synthesis. The rate of protein loss lessens as the body maximizes fat utilization and conserves protein (**Figure 50.3A**). Decreased protein synthesis in the liver, however, causes a decrease in blood proteins, resulting in increased loss of fluid from the blood to the interstitial spaces (*edema*; see Section 49.4). Accumulation of fluid in the extremities and abdomen are the classic signs of *kwashiorkor*, a disease caused by chronic protein deficiency (**Figure 50.3B**). When fat reserves are almost gone, protein loss accelerates to the point of damaging the organs of the body, leading eventually to death.

When an animal consistently takes in *more* food than it needs to meet its energy requirements, it is **overnourished**, and the excess nutrients are stored as increased body mass. First, glycogen reserves build up; then additional dietary carbohydrates, fats, and proteins are converted to body fat. In some species, such as hibernators, seasonal overnutrition is an important adaptation for surviving periods when food is not available. In humans, however, overnutrition can be a serious health hazard, increasing the risk of high blood pressure, heart attack, diabetes, and other disorders.

Ironically, at a time when a billion people—one-sixth of the world's population—do not have adequate food, the leading cause of life-threatening undernourishment in some Western nations is a self-imposed starvation syndrome called *anorexia nervosa*. This condition is found mainly among adolescent girls and young women who develop a pathological aversion to real or imagined body fat.

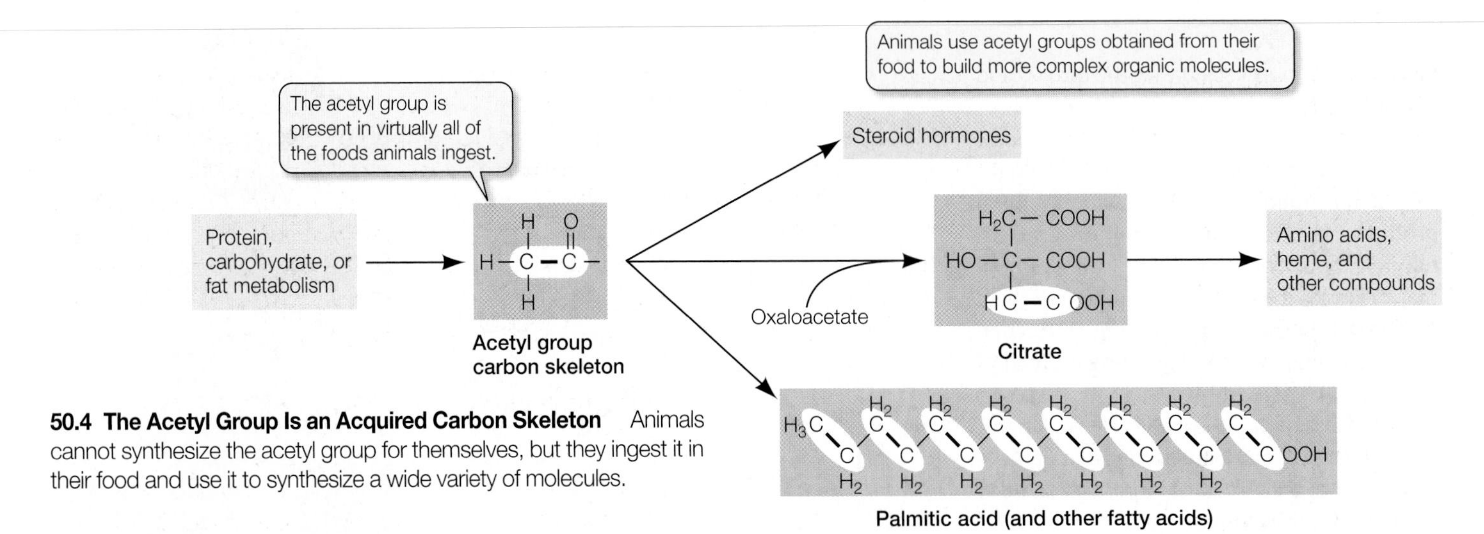

50.4 The Acetyl Group Is an Acquired Carbon Skeleton Animals cannot synthesize the acetyl group for themselves, but they ingest it in their food and use it to synthesize a wide variety of molecules.

Food provides carbon skeletons for biosynthesis

Every animal requires certain basic organic molecules that it cannot synthesize for itself, but needs as building blocks for its own complex organic molecules. The acetyl group (CH_3CO—) is one such required building block supplying the **carbon skeleton** of larger organic molecules (**Figure 50.4**). Animals cannot make acetyl groups from carbon, oxygen, and hydrogen molecules; they must obtain acetyl groups from food. Acetyl groups can be derived from the metabolism of almost any food; they are unlikely ever to be in short supply for an adequately nourished animal. However, some groups supplying carbon skeletons can be deficient in an animal's diet even if its caloric intake is adequate.

Amino acids, the building blocks of proteins, are a good example of carbon skeletons that can be in short supply. Animals can synthesize some of their own amino acids by utilizing carbon skeletons from acetyl or other groups and transferring to them amino groups (—NH_2) derived from other amino acids. Most animals cannot synthesize all the amino acids they need. Each species must obtain certain **essential amino acids** from food. Essential amino acids vary by species. In general, herbivores have fewer essential amino acids than carnivores. If an animal does not take in enough of even one of its essential amino acids, its protein synthesis is impaired.

Humans must obtain eight essential amino acids from their food: isoleucine, leucine, lysine, methionine, phenylalanine, threonine, tryptophan, and valine. All eight are available in milk, eggs, meat, and soybean products, but most plant foods do not contain adequate quantities of all eight. A strict vegetarian diet, therefore, is subject to a risk of protein malnutrition. A complementary dietary mixture of plant foods, however, supplies all eight essential amino acids (**Figure 50.5**). In general, grains (such as rice, wheat, and corn) are complemented by legumes (such as beans and peas). Long before the chemical basis for this complementarity was understood, societies with little access to meat developed complementary diets. Many Central and South American peoples traditionally eat beans with corn, and the native peoples of North America complemented their beans with squash.

Why are dietary proteins completely digested to their constituent amino acids before being used by the body? Wouldn't it be more energy-efficient to reuse some dietary proteins directly? There are several reasons why ingested proteins are not used "as is":

- Macromolecules such as proteins are not readily absorbed by the cells of the gut, but their constituent monomers (such as amino acids) are readily absorbed.
- Protein structure and function are highly species-specific. A protein that functions optimally in one species might not function well in another.
- Foreign proteins entering the body directly from the gut would be recognized as invaders and would be attacked by the immune system.

Humans can synthesize almost all the lipids required by the body using acetyl groups obtained from food (see Figure 50.4), but we must have a dietary source of certain **essential fatty acids**—notably, linoleic acid—that we cannot synthesize. Linoleic acid is an unsaturated fatty acid needed by mammals to synthesize other unsaturated fatty acids, such as arachidonic acid, which is a component of several signaling molecules, including prostaglandins. A deficiency of linoleic acid can lead to problems such as infertility and impaired lactation. Essential fatty acids are also necessary components of membrane phospholipids.

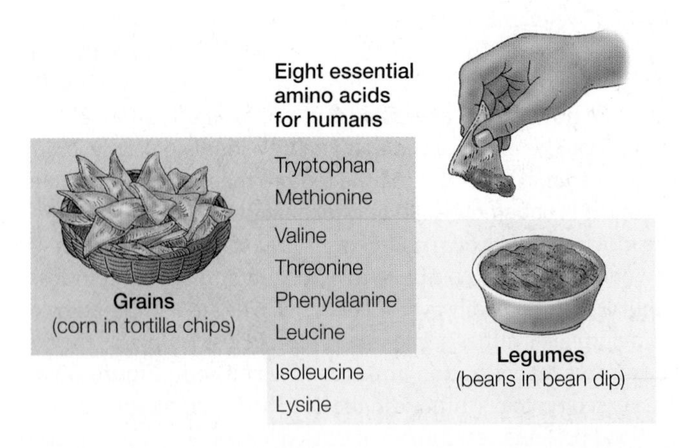

50.5 A Strategy for Vegetarians By combining cereal grains with legumes, a vegetarian can obtain all eight essential amino acids.

Animals need mineral elements for a variety of functions

The principal mineral elements that animals require are listed in **Table 50.1**. Elements required in large amounts are called **macronutrients**; elements required in only tiny amounts are called **micronutrients**. Some micronutrients are required in such minute amounts that deficiencies are never observed, but they are nevertheless essential elements.

Calcium is an example of a macronutrient. It is the fifth most abundant element in the body; a 70-kg person contains about 1.2 kg of calcium. Calcium phosphate is the principal structural material in bones and teeth. Muscle contraction, neuronal function, and many other intracellular functions in animals require calcium ions (Ca^{2+}). The turnover of calcium in the extracellular fluid is high, as bones are constantly being remodeled and calcium is constantly entering and leaving cells. Calcium is lost from the body in urine, sweat, and feces, so it must be replaced regularly. Humans require about 800–1,000 mg of calcium per day in the diet.

Iron is an example of a micronutrient. Iron is found everywhere in the body because it is the oxygen-binding atom in hemoglobin and myoglobin and is a component of enzymes in the electron transport chain. Nevertheless, the total amount of iron in a 70-kg person is only about 4 grams, and since iron is recycled so efficiently in the body and is not lost in the urine, we require only about 15 mg per day in our food. Despite the small amount required, insufficient iron is the most common mineral nutrient deficiency in the world today.

Animals must obtain vitamins from food

Vitamins are another group of essential nutrients. Like essential amino acids and fatty acids, vitamins are carbon compounds that an animal requires for growth and metabolism but cannot synthesize for itself. Most vitamins function as coenzymes or parts of coenzymes and are required in very small amounts compared to the essential amino acids and fatty acids, which have structural roles.

TABLE 50.1

Mineral Elements Required by Animals

ELEMENT	SOURCE IN HUMAN DIET	MAJOR FUNCTIONS
MACRONUTRIENTS		
Calcium (Ca)	Dairy foods, eggs, green leafy vegetables, whole grains, legumes, nuts, meat	Found in bones and teeth; blood clotting; nerve and muscle action; enzyme activation
Chlorine (Cl)	Table salt (NaCl), meat, eggs, vegetables, dairy foods	Water balance; digestion (as HCl); principal negative ion in extracellular fluid
Magnesium (Mg)	Green vegetables, meat, whole grains, nuts, milk, legumes	Required by many enzymes; found in bones and teeth
Phosphorus (P)	Dairy, eggs, meat, whole grains, legumes, nuts	Found in nucleic acids, ATP, and phospholipids; bone formation; buffers; metabolism of sugars
Potassium (K)	Meat, whole grains, fruits, vegetables	Nerve and muscle action; protein synthesis; principal positive ion in cells
Sodium (Na)	Table salt, dairy foods, meat, eggs	Nerve and muscle action; water balance; principal positive ion in extracellular fluid
Sulfur (S)	Meat, eggs, dairy foods, nuts, legumes	Found in proteins and coenzymes; detoxification of harmful substances
MICRONUTRIENTS		
Chromium (Cr)	Meat, dairy, whole grains, legumes, yeast	Glucose metabolism
Cobalt (Co)	Meat, tap water	Found in vitamin B_{12}; formation of red blood cells
Copper (Cu)	Liver, meat, fish, shellfish, legumes, whole grains, nuts	Found in active site of many redox enzymes and electron carriers; production of hemoglobin; bone formation
Fluorine (F)	Most water supplies	Found in teeth; helps prevent decay
Iodine (I)	Fish, shellfish, iodized salt	Found in thyroid hormones
Iron (Fe)	Liver, meat, green vegetables, eggs, whole grains, legumes, nuts	Found in active sites of many redox enzymes and electron carriers, hemoglobin, and myoglobin
Manganese (Mn)	Organ meats, whole grains, legumes, nuts, tea, coffee	Activates many enzymes
Molybdenum (Mo)	Organ meats, dairy, whole grains, green vegetables, legumes	Found in some enzymes
Selenium (Se)	Meat, seafood, whole grains, eggs, milk, garlic	Fat metabolism
Zinc (Zn)	Liver, fish, shellfish, and many other foods	Found in some enzymes and some transcription factors; insulin physiology

The list of vitamins varies from species to species. Most mammals, for example, can make their own ascorbic acid (vitamin C). Primates (including humans) cannot, so for primates, ascorbic acid is a vitamin. If we do not get vitamin C in our food, we develop a disease known as scurvy, characterized by bleeding gums, loss of teeth, subcutaneous hemorrhages, and slow wound healing. Scurvy was a serious and frequently fatal problem for sailors on long voyages until a Scottish physician, James Lind, discovered that the disease could be prevented if the sailors ate fresh greens or fresh fruit. Eventually the British Admiralty made limes standard provisions for its ships, and ever since, British sailors have been called "limeys." The active ingredient in limes was named ascorbic ("without scurvy") acid.

Humans require 13 vitamins, which are listed in **Table 50.2**. They are divided into two groups: water-soluble and fat-soluble. When water-soluble vitamins are ingested in excess of bodily needs, they are simply eliminated in the urine. (This is the fate of much of the vitamin C that people take in excessive doses.) The fat-soluble vitamins, however, can accumulate in body fat, and may build up to toxic levels in the liver if taken in excess.

The fat-soluble vitamin D (cholecalciferol), which is essential for the absorption and metabolism of calcium, is a special case because the body can make it. (Recall from Section 41.3 that vitamin D is by definition a hormone.) Certain lipids present in the human body can be converted into vitamin D by the action of ultraviolet light on the skin. Thus vitamin D must be obtained in the diet only by individuals with inadequate exposure to the sun, such as people who live in cold climates where clothing usually covers most of the body and where the sun may not shine for long periods of time.

The need for vitamin D may have been an important factor in the evolution of skin color. Human races that are adapted to equatorial and low latitudes have dark skin pigmentation as a protection against the damaging effects of ultraviolet radiation. These peoples generally expose extensive areas of skin to the sun on a regular basis, so their skin synthesizes adequate amounts of vitamin D. Most human races that became adapted to higher latitudes lost this dark skin pigmentation. Lighter skin facilitates vitamin D production in the relatively small areas of skin exposed to sunlight during the short days of winter. The Inuit peoples of the Arctic represent an exception to the correlation between latitude and skin pigmentation. These dark-skinned people obtain plenty of vitamin D from the large amounts of meat and fish oils in their diet; they don't require exposure to sunlight to obtain this vitamin.

Nutrient deficiencies result in diseases

The lack of any essential nutrient in the diet produces a state of deficiency called **malnutrition**, and chronic malnutrition leads to a characteristic **deficiency disease**. We have already discussed kwashiorkor (protein deficiency) and scurvy (vitamin C deficiency). A shortage of any of the vitamins results in specific deficiency symptoms (see Table 50.2). Another deficiency disease, *beriberi*, was directly involved in the discovery of vitamins. Beriberi means "extreme weakness." It became prevalent in Asia in the nineteenth century after it became standard practice to mill rice to a high,

TABLE 50.2

Vitamins in the Human Diet

VITAMIN	SOURCE	FUNCTION	DEFICIENCY SYMPTOMS
WATER-SOLUBLE			
B_1 (thiamin)	Liver, legumes, whole grains	Coenzyme in cellular respiration	Beriberi, loss of appetite, fatigue
B_2 (riboflavin)	Dairy, meat, eggs, green leafy vegetables	Coenzyme in FAD	Lesions in corners of mouth, eye irritation, skin disorders
Niacin	Meat, fowl, liver, yeast	Coenzyme in NAD and NADP	Pellagra, skin disorders, diarrhea, mental disorders
B_6 (pyridoxine)	Liver, whole grains, dairy foods	Coenzyme in amino acid metabolism	Anemia, slow growth, skin problems, convulsions
Pantothenic acid	Liver, eggs, yeast	Found in acetyl CoA	Adrenal problems, reproductive problems
Biotin	Liver, yeast, bacteria in gut	Found in coenzymes	Skin problems, loss of hair
B_{12} (cobalamin)	Liver, meat, dairy foods, eggs	Formation of nucleic acids, proteins, and red blood cells	Pernicious anemia
Folic acid	Vegetables, eggs, liver, whole grains	Coenzyme in formation of heme and nucleotides	Anemia
C (ascorbic acid)	Citrus fruits, tomatoes, potatoes	Formation of connective tissues; antioxidant	Scurvy, slow healing, poor bone growth
FAT-SOLUBLE			
A (retinol)	Fruits, vegetables, liver, dairy	Found in visual pigments	Night blindness
D (cholecalciferol)	Fortified milk, fish oils, sunshine	Absorption of calcium and phosphate	Rickets
E (tocopherol)	Meat, dairy foods, whole grains	Muscle maintenance, antioxidant	Anemia
K (menadione)	Intestinal bacteria, liver	Blood clotting	Blood-clotting problems

white polish and discard the hulls present in brown rice. A critical observation was that birds—chickens and pigeons—developed beriberi-like symptoms when fed only polished rice. In 1912, Casimir Funk cured pigeons of beriberi by feeding them the discarded hulls.

At the time of Funk's discovery, all diseases were thought to be either caused by microorganisms or inherited. Funk suggested the radical idea that beriberi and some other diseases are dietary in origin and result from deficiencies in specific substances. Funk coined the term "vitamines" because he mistakenly thought that all these substances vital for life were compounds with amino groups. In 1926, thiamin (vitamin B_1)—the substance lost in the rice milling process—was the first vitamin to be isolated in pure form.

Deficiency diseases can also result from an inability to absorb or process an essential nutrient even if it is present in the diet. Vitamin B_{12} (cobalamin), for example, is present in all foods of animal origin. Since plants neither use nor produce vitamin B_{12}, a strictly vegetarian diet (not supplemented with dairy products or vitamin pills) can lead to a B_{12} deficiency disease called pernicious anemia, characterized by a failure of red blood cells to mature. The most common cause of pernicious anemia, however, is not a lack of vitamin B_{12} in the diet, but an inability to absorb it. Normally, cells in the stomach lining secrete a peptide called intrinsic factor, which binds to vitamin B_{12} and makes it possible for it to be absorbed in the small intestine. Conditions that damage the stomach lining, such as alcoholism or gastritis, can thus lead to pernicious anemia.

Inadequate mineral nutrition can also lead to deficiency diseases. Examples are hypothyroidism and goiter resulting from iodine deficiency (see Section 41.3), and anemia resulting from iron deficiency.

50.1 RECAP

As heterotrophs, animals must obtain the energy and molecular building blocks for biosynthesis from their food. Energy can come from the metabolism of carbohydrates, fats, and proteins. Molecular building blocks include carbon skeletons, vitamins, and minerals.

- Do you understand how metabolic energy measurement can be used to construct energy budgets? See pp. 1070–1071
- Can you give an example of an essential carbon skeleton? A micronutrient? A macronutrient? See pp. 1072–1073, Figure 50.4, and Table 50.1
- Can you explain why fat-soluble vitamins should not be taken in excess? See p. 1074

We have surveyed the essential elements of nutrition in animals. Next, we will look at various methods and adaptations by which animals obtain the food they need, and mechanisms by which the food is processed in the animal's body to extract nutrients.

50.2 How Do Animals Ingest and Digest Food?

Heterotrophic organisms can be classified by how they acquire their nutrition. **Saprobes** (also called *saprotrophs* or *decomposers*) are organisms—mostly protists and fungi—that absorb nutrients from dead organic matter. **Detritivores**, such as earthworms and crabs, actively feed on dead organic material. Animals that feed on living organisms are **predators**: **Herbivores** prey on plants, **carnivores** prey on animals, and **omnivores** prey on both. **Filter feeders**, such as clams and blue whales, prey on small organisms by filtering them from the aquatic environment. **Fluid feeders** include mosquitoes, aphids, leeches, and hummingbirds. The anatomical adaptations that enable a species to exploit a particular source of nutrition are usually quite obvious, but physiological and biochemical adaptations are also important.

The food of herbivores is often low in energy and hard to digest

Most vegetation is coarse and difficult to break down physically, but herbivores must process large amounts of it because its energy content is low. Therefore, herbivores spend a great deal of time feeding. Many have striking adaptations for feeding, such as the trunk (a flexible, gripping nose) of the elephant or the long neck of the giraffe. Many types of grinding, rasping, cutting, and shredding mouthparts have evolved in invertebrates for ingesting plant material, and the teeth of herbivorous vertebrates have been shaped by selection to tear, crush, and grind coarse plant matter. The digestive processes of herbivores can also be quite specialized.

Carnivores must detect, capture, and kill prey

The predatory behaviors of many carnivores are legendary. One need only call to mind the hunting skills of hawks, wolves, or tigers. Carnivores have evolved stealth, speed, power, large jaws, sharp teeth, and strong gripping appendages. Carnivores also have evolved remarkable means of detecting prey. Bats use echolocation, pit vipers sense infrared radiation from the warm bodies of their prey, and certain fishes detect electric fields created in the water by their prey.

Adaptations for killing and ingesting prey are diverse and can be highly specialized. These adaptations are especially important when the prey can inflict damage on the predator. A snake may strike with poisonous fangs, using its venom to immobilize its prey, which may include animals that are very active and have dangerous teeth or claws. To swallow large prey, a snake disengages its lower jaw from its joint with the skull (**Figure 50.6**). The tentacles of jellyfishes, the long, sticky tongues of chameleons, and the webs of spiders are other fascinating examples of adaptations for capturing and immobilizing prey. Some predators digest their prey externally. For example, a spider injects its insect prey with digestive enzymes and then sucks out the liquefied contents, leaving behind the empty exoskeletons frequently seen in old spider webs.

50.6 An Adaptation for Carnivory Snakes such as this corn snake (*Elaphe guttata*) can ingest large prey (in this case, a mouse) by dislocating their jaws.

Vertebrate species have distinctive teeth

Teeth are adapted for the acquisition and initial processing of specific types of foods. Because they are among the hardest structures of the body, an animal's teeth remain in the environment long after it dies. Paleontologists use teeth to identify animals that lived in the distant past and to deduce their feeding behavior.

All mammalian teeth have the same general structure, consisting of three layers (**Figure 50.7A**). An extremely hard material called **enamel**, composed principally of calcium phosphate, covers the crown of the tooth. Both the crown and the root contain a layer of bony material called **dentine**, inside of which is a **pulp cavity** containing blood vessels, nerves, and the cells that produce the dentine.

> Like you, elephants have wisdom teeth (third molars). They also have two more molars and three premolars, but at any one time only one of these huge teeth serves as the main chewing surface. As it wears out, it is replaced by the next one moving forward. The elephant's equivalent of your wisdom teeth do not emerge until the elephant is about 45 years old.

The shapes and organization of mammalian teeth, however, can be very different, since they are adapted to different diets (**Figure 50.7B**). In general, *incisors* are used for cutting, chopping, or gnawing; *canines* are used for stabbing, gripping, and ripping; and *molars* and *premolars* (the cheek teeth) are used for shearing, crushing, and grinding. The highly varied diet of humans is reflected by our multipurpose set of teeth, as is common among omnivores.

Animals digest their food extracellularly

Animals take food into a body cavity that is continuous with the outside environment. They secrete digestive enzymes into that cavity, and the enzymes break down the food into nutrient molecules that can be absorbed by the cells lining the cavity.

The simplest digestive systems are **gastrovascular cavities**, which connect to the outside world through a single opening. Cnidari-

50.7 Mammalian Teeth (A) A mammalian tooth has three layers: enamel, dentine, and a pulp cavity. (B) The teeth of different mammalian species are specialized for different diets. This illustration depicts the teeth of the lower jaw, viewed from above.

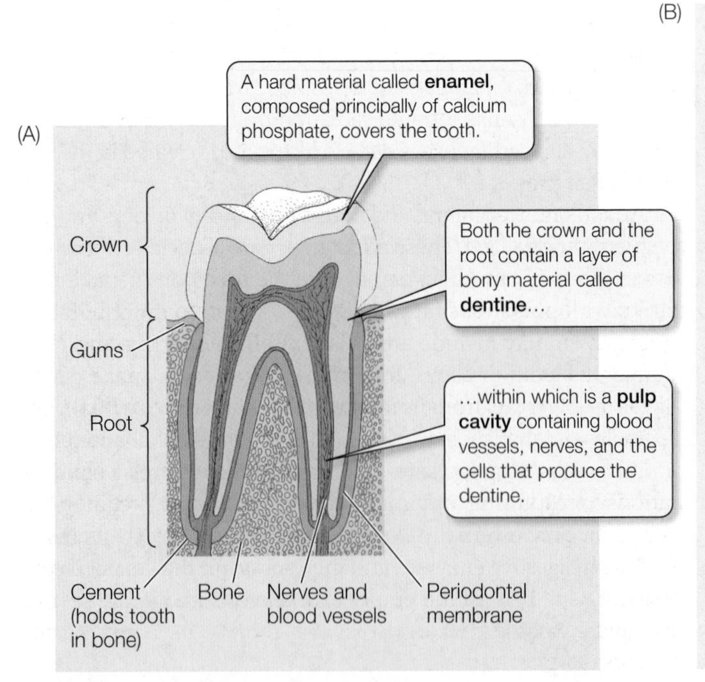

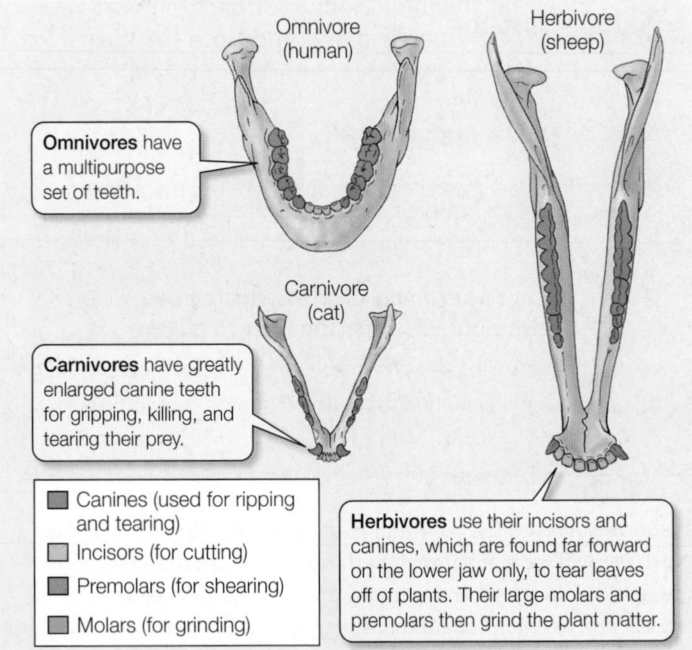

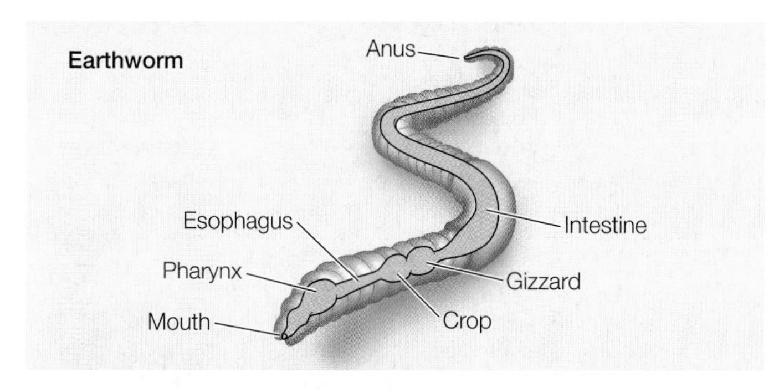

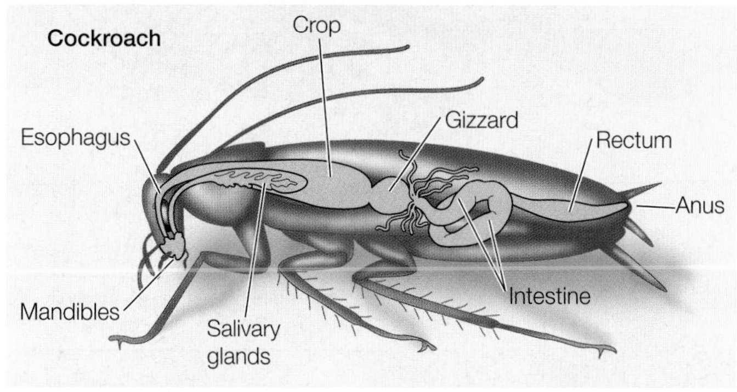

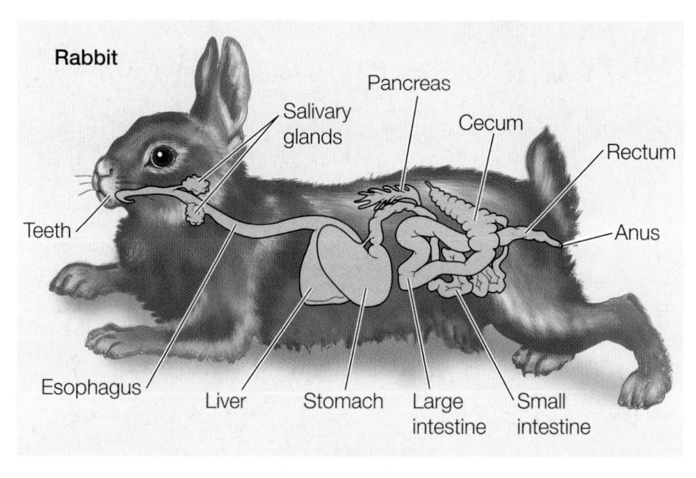

50.8 Compartments for Digestion and Absorption Most animals have tubular guts that begin with a mouth, which takes in food, and end in an anus, which eliminates wastes. Between these two structures are specialized regions for digestion and nutrient absorption; the structures in these regions vary from species to species.

ans, for example, capture prey using their stinging nematocysts (see Figure 31.10) and cram them into their gastrovascular cavities with their tentacles. Enzymes in the gastrovascular cavity partly digest the prey. Cells lining the cavity take in small food particles by endocytosis. The vesicles created by endocytosis then fuse with lysosomes containing digestive enzymes, and intracellular digestion completes the breakdown of the food. Nutrients are released to the cytoplasm as the vesicles break down.

Tubular guts have an opening at each end

The guts of most animals are tubular: A **mouth** takes in food; molecules are digested and absorbed throughout the length of the gut; and solid digestive wastes are eliminated through an **anus**. Different regions in the tubular gut are specialized for particular functions (**Figure 50.8**). These functions must be coordinated so that they occur in the proper sequence and at rates that maximize the efficiency of digestion and absorption of nutrients.

At the anterior end of the gut is the mouth cavity. Food may be physically broken up in the mouth cavity, for example by teeth (in many vertebrates), by the **radula** (in snails), or by **mandibles** (in many arthropods). In most birds food is ground by small stones in an early, muscular, portion of the gut called the **gizzard**. Some animals, such as snakes, simply ingest large chunks of food, with little or no fragmentation. **Stomachs** and **crops** are storage chambers that enable animals to ingest relatively large amounts of food when it is available, then digest it gradually. In these storage chambers, food may be further fragmented and mixed, but digestion may or may not occur there, depending on the species. In any case, food delivered into the next section of the gut, the **midgut** or **intestine**, is in small particles and well mixed.

Most nutrients, water, and ions are absorbed in the midgut. To digest food materials, specialized glands secrete some digestive enzymes into the midgut, and the gut wall itself secretes other enzymes. The **hindgut** recovers water and ions and stores undigested wastes, or **feces**, so that they can be released to the environment at an appropriate time or place. A muscular **rectum** near the anus assists in the expulsion of feces.

Within the hindguts of many species are colonies of endosymbiotic bacteria. These bacteria obtain their nutrition from the food passing through the host's gut while contributing to the digestive processes of the host. Members of the leech genus *Hirudo*, for example, produce no enzymes that can digest the proteins in the blood they suck from vertebrates; instead, a colony of gut bacteria performs this service. The resulting amino acids are subsequently used by both the leech and the bacteria. In many animals, the parts of the gut that absorb nutrients have greater surface areas than would be expected of a simple tube (**Figure 50.9A,B**). In vertebrates, the wall of the gut is richly folded, with the individual folds bearing legions of tiny fingerlike projections called **villi** (**Figure 50.9C**). The cells that line the surfaces of the villi, in turn, have microscopic projections called **microvilli**. The microvilli give the gut an enormous internal surface area for the absorption of nutrients.

Digestive enzymes break down complex food molecules

Protein, carbohydrate, and fat macromolecules are broken down into their simplest monomeric units by hydrolytic enzymes secreted at different locations in the digestive tract. All of these enzymes cleave the chemical bonds of macromolecules through hydrolysis, a reaction that adds a water molecule (see Figure 3.4B). Digestive enzymes are classified according to the substances they hydrolyze: *proteases* break the bonds between adjacent amino acids in proteins; *carbohydrases* hydrolyze carbohydrates; *peptidases*, peptides; *lipases*, fats; and *nucleases*, nucleic acids.

How can an organism produce enzymes that hydrolyze biological macromolecules without digesting itself? Most digestive enzymes are produced in an inactive form, known as a **zymogen**, so that they cannot act on the cells that produce them. When secreted into the gut, a zymogen is generally activated by another enzyme. The cells lining the gut are not digested because they are protected by a covering of mucus.

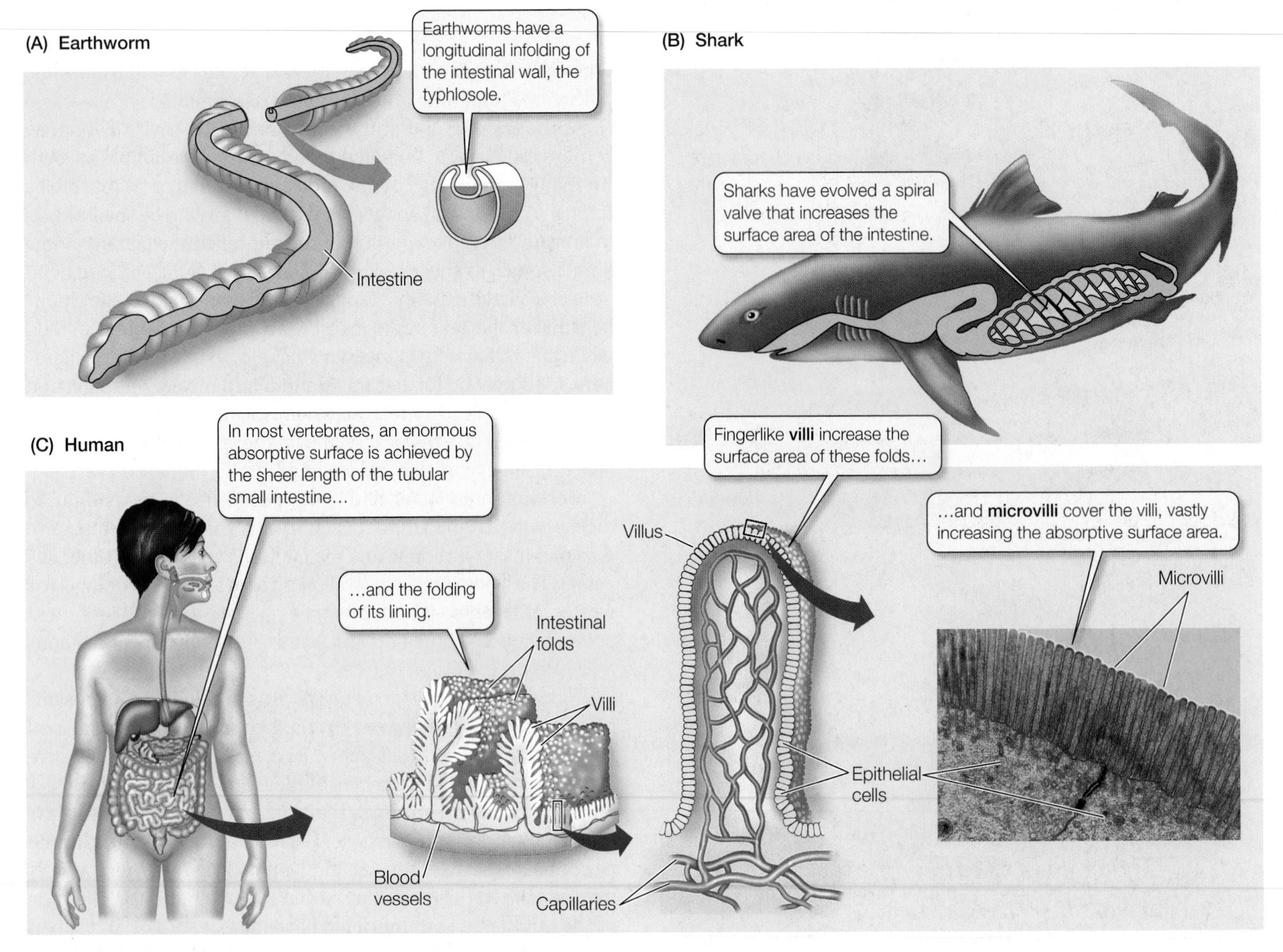

50.9 Greater Intestinal Surface Area Means More Nutrient Absorption The guts of most animals have evolved to maximize their surface area.

50.2 RECAP

Heterotrophs have diverse adaptations for acquiring food. Once captured and/or ingested, food is digested extracellularly by secreted enzymes to release nutrients, which are absorbed into the animal's body, usually via a tubular gut.

- Why do herbivores typically spend a great deal of their time feeding? See p. 1075
- What is the primary purpose of the intestinal microvilli? See p. 1077 and Figure 50.9
- What is a zymogen and why is it important? See p. 1077

Animals have evolved many ways of obtaining and ingesting their food. We have learned that, once ingested, the food may be fragmented and moved into the gut for digestion by hydrolytic enzymes, which release the various nutrients needed by the animal. Next we focus in detail on how those processes occur in vertebrates.

50.3 How Does the Vertebrate Gastrointestinal System Function?

Digestion in vertebrates occurs in the gastrointestinal system, which includes a tubular gut running from mouth to anus and several accessory structures that produce secretions that play important roles in digestion (**Figure 50.10**).

The vertebrate gut consists of concentric tissue layers

The cellular architecture of the vertebrate gut follows a common layered plan throughout its length (**Figure 50.11**). Starting in the cavity, or **lumen**, of the gut, the first layer of cells is the **mucosal epithelium**. These cells have secretory and absorptive functions. Some secrete mucus, which lubricates and protects the walls of the gut; others secrete digestive enzymes or hormones. Mucosal epithelial cells in the stomach secrete hydrochloric acid (HCl). In some regions of the gut, nutrients are absorbed by mucosal epithelial cells. The plasma membranes of these absorptive cells have *microvilli* that increase their surface area (see Figure 50.9C)

At the base of the mucosa are some smooth muscle cells, and just outside the mucosa is the submucosal tissue layer. Here we

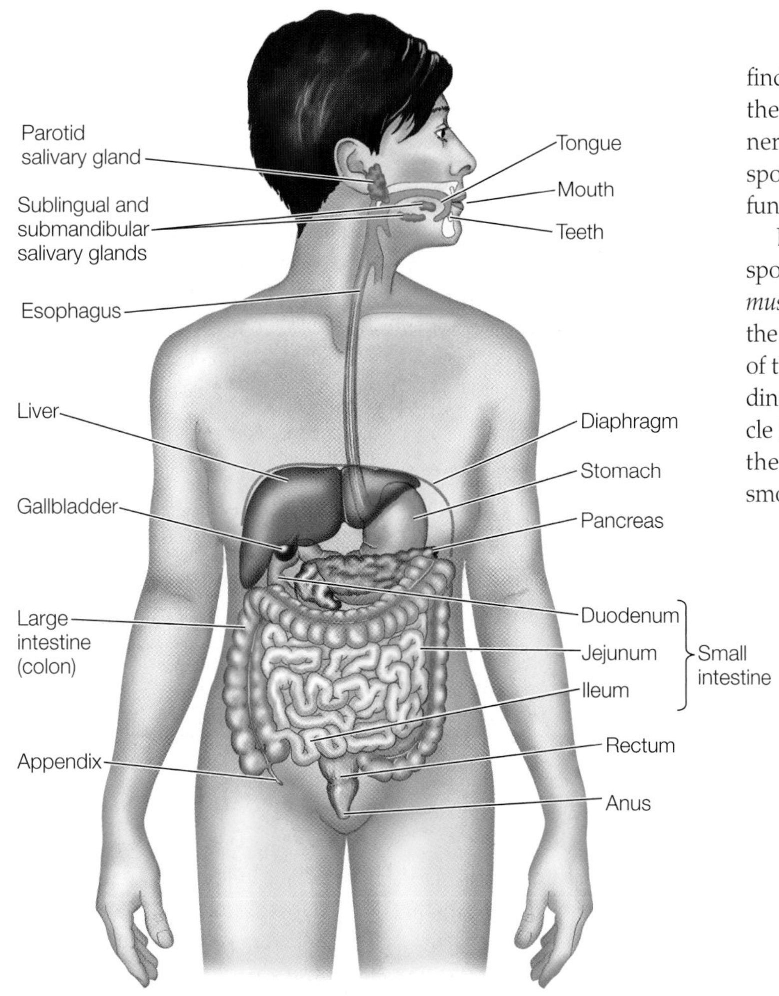

50.10 The Human Digestive System Different compartments within the long tubular gut specialize in digesting food, absorbing nutrients, and storing and expelling wastes. Accessory organs contribute secretions containing enzymes and other molecules.

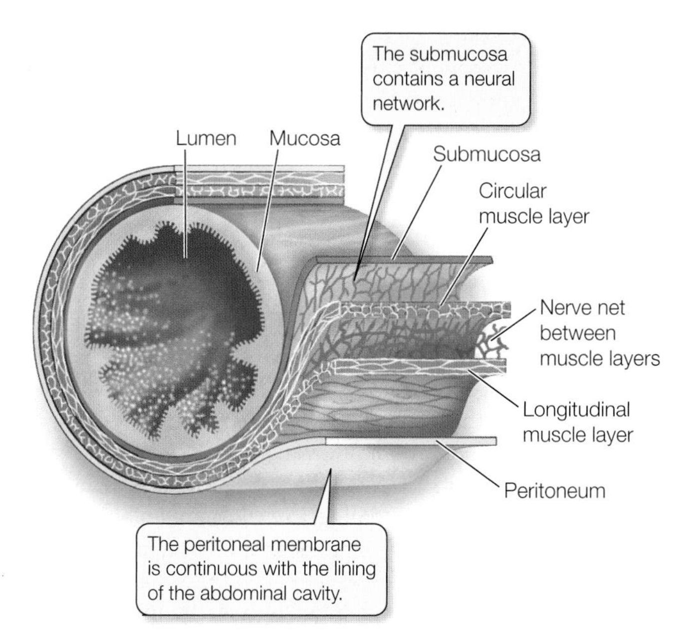

50.11 Tissue Layers of the Vertebrate Gut In all compartments of the gut, the organization of the tissue layers is the same, but specialized adaptations of specific tissues characterize different regions.

find the blood and lymph vessels that carry absorbed nutrients to the rest of the body. The **submucosa** also contains a network of nerves; the neurons in this network have sensory functions (responsible for stomach aches), and they control various secretory functions of the gut.

External to the submucosa are two layers of smooth muscle responsible for the movements of the gut. Innermost is the *circular muscle layer*, with its cells oriented around the gut. Outermost is the *longitudinal muscle layer*, with its cells oriented along the length of the gut. The circular muscles constrict the gut, and the longitudinal muscles shorten it. Between the two layers of smooth muscle is another network of nerves, which controls and coordinates the movements of the gut. The coordinated activity of the two smooth muscle layers moves the gut contents continuously toward the rectum.

A membrane called the **peritoneum** surrounds the gut as it does all of the organs of the abdominal cavity as well as lining the wall of the cavity. The peritoneum includes connective and epithelial tissue, which secrete a fluid that lubricates the organs so they can easily move against each other.

Mechanical activity moves food through the gut and aids digestion

In the mouth cavity, food is chewed and mixed with saliva. Periodically the *tongue* pushes a *bolus* of the chewed food toward the throat. By making contact with the *soft palate* at the back of the mouth cavity, the bolus of food initiates *swallowing*, which is a complex series of reflexes. Swallowing propels the food through the pharynx (where the mouth cavity and the nasal passages join) and into the **esophagus** (the food tube). To prevent the food from entering the trachea (windpipe), the *larynx* (voice box) closes, and a flap of tissue called the *epiglottis* covers the entrance to the larynx (**Figure 50.12**).

Once a bolus of food enters the esophagus, it is moved toward the stomach by waves of muscle contraction called **peristalsis**. The muscle of the upper third of the esophagus is striated (i.e., skeletal muscle) and is controlled by the somatic nervous system; the muscles of the lower two-thirds is smooth muscle, controlled by the autonomic nervous system.

The smooth muscles of the gut contract in response to being stretched. When a bolus of food reaches the smooth-muscle region of the esophagus and stretches it, the muscle responds by contracting, thus pushing the food toward the stomach. Why doesn't the contraction of the esophageal smooth muscle push the food back toward the mouth? The nerve net between the two smooth muscle layers coordinates the muscles so that contraction is always preceded by an *anticipatory wave of relaxation*: When a region of the gut smooth muscle contracts, the smooth muscle just beyond it relaxes, and food is squeezed into that area. The resulting stretch causes that region to constrict and simultaneously the region just beyond it relaxes. In this way peristalsis moves down the

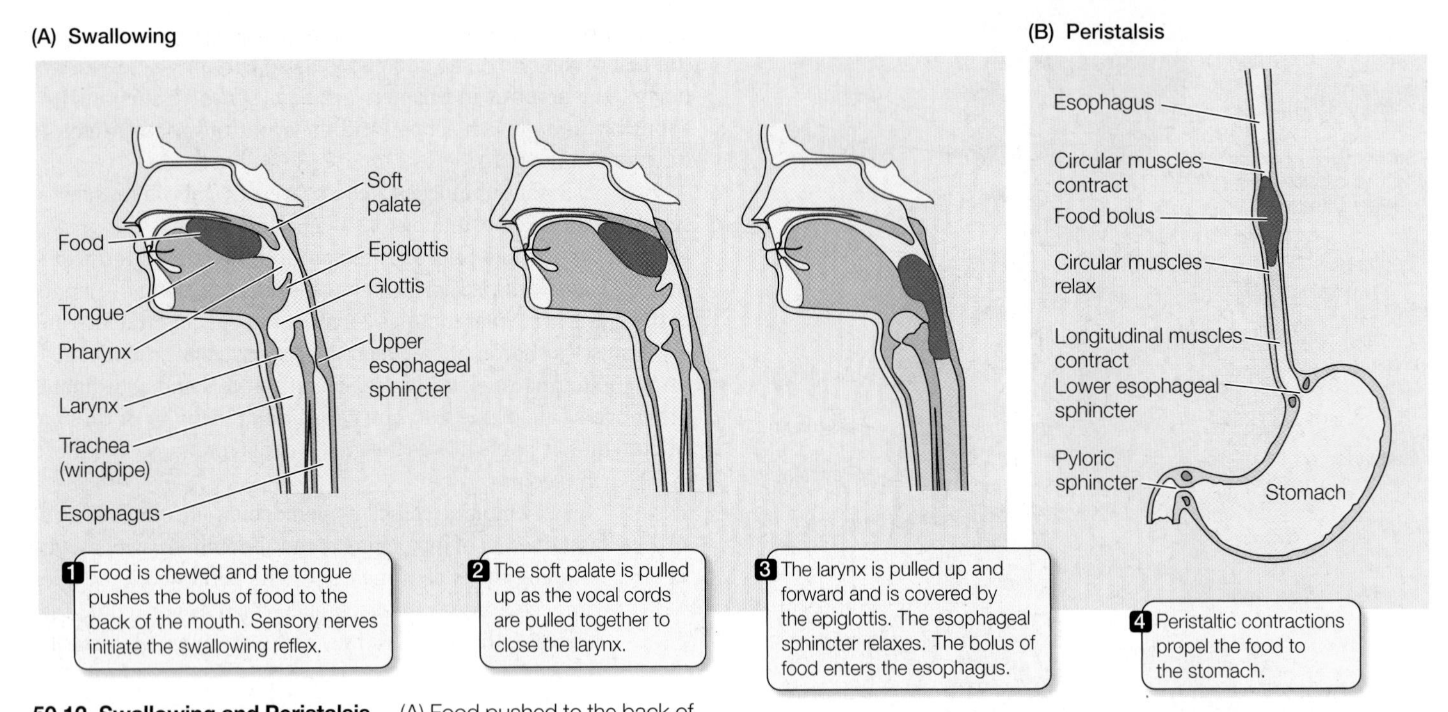

50.12 Swallowing and Peristalsis (A) Food pushed to the back of the mouth triggers the swallowing reflex. (B) Once the food bolus enters the esophagus, peristalsis propels it through the gut.

gut from mouth to anus pushing the contents of the gut in that direction.

The backward movement of food from the stomach into the esophagus is normally prevented by the *lower esophageal sphincter*, a thick ring of circular smooth muscle at the junction of the esophagus and the stomach. This sphincter is normally constricted, but waves of peristalsis cause it to relax enough to let food pass from the esophagus into the stomach. Sphincter muscles are found elsewhere in the digestive tract as well. The *pyloric sphincter* governs the passage of stomach contents into the intestine. Another important sphincter surrounds the anus.

Chemical digestion begins in the mouth and the stomach

The enzyme **amylase** is secreted by the salivary glands and mixed with food as it is chewed. Amylase hydrolyzes the bonds between the glucose monomers that make up starch molecules. The action of amylase is what makes a chewed piece of bread or cracker taste slightly sweet if you hold it in your mouth long enough.

The main role of the stomach is to store food so that digestion can occur slower than ingestion. Therefore, the stomach is very expandable, up to about 1.5 liters after a large meal. To make this expansion possible, the smooth muscle of the stomach is less sensitive to stretching than is the smooth muscle of the esophagus.

The stomach also has secretory functions. Deep infoldings in the walls of the stomach called **gastric pits** are lined with three types of secretory cells (**Figure 50.13A**). One type secretes a proteolytic enzyme that begins the digestion of protein. Another type secretes hydrochloric acid (HCl), which kills most ingested microorganisms. This nasty mix of substances could damage the walls of the stomach, but a third cell type secretes mucus that provides a protective coating of the walls of the gastric pits and the stomach. The mucus contains buffers that maintain the pH at the surface of the gastric mucosa near neutrality, and protease inhibitors that reduce damage from stomach enzymes.

The gastric pit cells that secrete the proteolytic enzyme are called *chief cells*. The enzyme is **pepsin**, and it is secreted as an inactive zymogen called **pepsinogen**. The extremely low pH of the stomach juices activates the conversion of pepsinogen to pepsin by cleaving away a sequence of amino acids that masks the active site of the enzyme. Newly activated pepsin can act on other pepsinogen molecules to activate them. Thus, the activation of pepsin in the stomach is a positive feedback process called **autocatalysis** (**Figure 50.13B**).

The gastric pit cells that secrete the HCl are called *parietal cells*. They can secrete so much HCl—about 2 liters per day—that they can bring the pH of the stomach contents below 1, which is the same as battery acid and 10 times more acidic than pure lemon juice. This means that across their plasma membranes, gastric pits can create a H^+ concentration difference of 3 million-fold. Such a feat of transport is not seen anywhere else in the body. How do they do it? The enzyme carbonic anhydrase in these cells catalyzes the hydration of CO_2 to H_2CO_3, which dissociates into H^+ and HCO_3^-. HCO_3^- is actively exchanged for Cl^- on the blood side of the gastric pits . H^+ is actively exchanged for K^+ from the lumen of the gastric pits (**Figure 50.13C**). However, this K^+ can leak out again down its concentration gradient. Thus, the inward transport of K^+ acts like an endless conveyer belt pushing H^+ out into the stomach lumen. Cl^- also passively leaks out of the gastric lumen side of the parietal cells to maintain electrical neutrality.

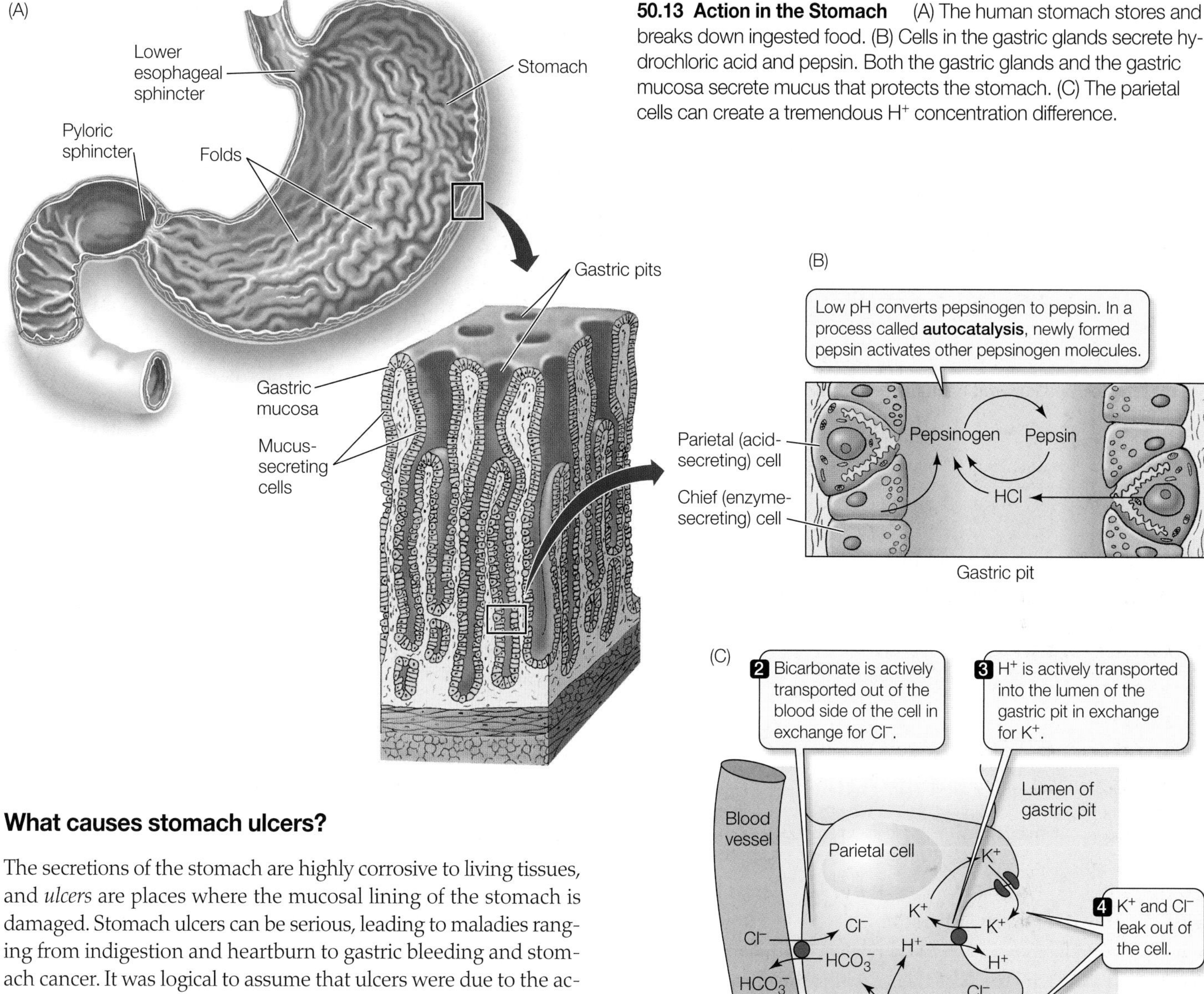

50.13 Action in the Stomach (A) The human stomach stores and breaks down ingested food. (B) Cells in the gastric glands secrete hydrochloric acid and pepsin. Both the gastric glands and the gastric mucosa secrete mucus that protects the stomach. (C) The parietal cells can create a tremendous H^+ concentration difference.

What causes stomach ulcers?

The secretions of the stomach are highly corrosive to living tissues, and *ulcers* are places where the mucosal lining of the stomach is damaged. Stomach ulcers can be serious, leading to maladies ranging from indigestion and heartburn to gastric bleeding and stomach cancer. It was logical to assume that ulcers were due to the actions of the stomach secretions on the stomach mucosa, and therefore that stress and lifestyle issues leading to excess stomach secretions (especially HCl) were the major cause of ulcers. This view led the pharmaceutical industry to develop a plethora of drugs to decrease stomach acid production, which became "billion dollar drugs" because they were prescribed so widely. What seemed like the simplest fact in gastrointestinal medicine, however, was turned on its head by the work of two Australian researchers. In retrospect, their work is a perfect example of the application of Koch's postulates (see Section 26.6) for the proof that a microorganism causes a disease (**Figure 50.14**).

In 1982, Robin Warren, a pathologist, observed an unknown bacterium in biopsies from the stomachs of patients with ulcers. In a study of 100 patients, he and Barry Marshall of the University of Western Australia found that the bacterium was always present in the patients with ulcers. They isolated the bacterium, which they named *Helicobacter pylori*, and grew it in culture. Having thus satisfied Koch's first two postulates, they turned to the last two, as described in Figure 50.14. Their research showed not only that *H. pylori* causes stomach ulcers, but that ulcer patients can be cured with antibiotics.

The medical profession was so certain that no microorganisms could live in the stomach and that stomach acid was the cause of ulcers that at first Warren and Marshall's findings were resisted and even ridiculed. But in 2005 they received the Nobel Prize in Medicine for their important discovery, and antibiotic therapy is now the primary treatment for stomach ulcers worldwide.

Drugs that decrease stomach acid production are still important ulcer medications for several reasons. Once a person has an ulcer, stomach acid exacerbates it. In addition, in many individuals,the lower esophageal sphincter muscle is inadequate to prevent acid from entering the esophagus and causing irritations and

Marshall and Warren set out to satisfy Koch's postulates:

Test 1

The microorganism must be present in every case of the disease.

Results: Biopsies from the stomachs of many patients revealed that the bacterium was always present if the stomach was inflamed or ulcerated.

Test 2

The microorganism must be cultured from a sick host.

Results: The bacterium was isolated from biopsy material and eventually grown in culture media in the laboratory.

Test 3

The isolated and cultured bacteria must be able to induce the disease.

Results: Marshall was examined and found to be free of bacteria and inflammation in his stomach. After drinking a pure culture of the bacterium, he developed stomach inflammation (gastritis).

Test 4

The bacteria must be recoverable from the infected volunteers.

Results: Biopsy of Marshall's stomach 2 weeks after he ingested the bacteria revealed the presence of the bacterium, now christened *Helicobacter pylori*, in the inflamed tissue.

Conclusion

Antibiotic treatment eliminated the bacteria and the inflammation in Marshall. The experiment was repeated on healthy volunteers, and many patients with gastric ulcers were cured with antibiotics. Thus Marshall and Warren demonstrated that the stomach inflammation leading to ulcers is caused by *H. pylori* infections in the stomach.

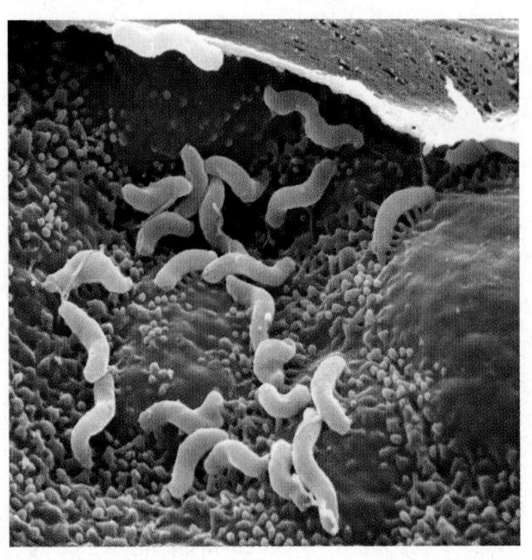

Helicobacter pylori

50.14 Satisfying Koch's Postulates Marshall and Warren showed that ulcers are caused not by the action of stomach acid, but by infection with the bacterium *Helicobacter pylori*.

lesions there. How does *H. pylori* survive in the hostile environment of the stomach? The bacteria live within the mucus layer lining the stomach, and by producing the enzyme urease they convert urea to bicarbonate and ammonium ions, which neutralize the HCl in their local environment.

The stomach gradually releases its contents to the small intestine

Contractions of the smooth muscles in the walls of the stomach churn its contents, thoroughly mixing them with the stomach secretions. The acidic, fluid mixture of gastric juice and partly digested food in the stomach is called **chyme**. A few substances can be absorbed across the stomach wall, including alcohol (hence its rapid effects), aspirin, and caffeine, but even these substances are absorbed in rather small quantities from the stomach.

Contractions of the stomach walls push the chyme toward the bottom end of the stomach. These waves of contractions cause the pyloric sphincter to relax briefly so that little squirts of the chyme can enter the small intestine. In this manner, the human stomach empties itself gradually over a period of approximately 4 hours. This slow introduction of food into the small intestine enables it to work on a little material at a time.

Most chemical digestion occurs in the small intestine

In the **small intestine**, the digestion of carbohydrates and proteins continues, and the digestion of fats and the absorption of nutrients begin. The small intestine takes its name from its diameter; it is in fact a very large organ, about 6 meters long in an adult. Given its length, and because of the folds, villi, and microvilli of its lining, its inner surface area is roughly the size of a tennis court. Across this surface, the small intestine absorbs all the nutrient molecules derived from food.

The small intestine has three sections. The initial section (about 25 cm long) is called the **duodenum** and is the site of most digestion; the **jejunum** and the **ileum** (together about 600 cm) carry out 90 percent of the absorption of nutrients (see Figure 50.10).

Digestion in the small intestine requires many specialized enzymes, as well as several other secretions. Two accessory organs that are not part of the digestive tract—the liver and the pancreas—provide many of these secretions and enzymes.

LIVER The liver synthesizes **bile** from cholesterol. Bile from the liver flows through the *hepatic duct* to the duodenum and through a side branch of the hepatic duct to the **gallbladder** (**Figure 50.15**), where it is stored until it is needed. When fat enters the duodenum, a hormonal signal causes the walls of the gallbladder to contract rhythmically, squeezing bile out of the gallbladder and into the hepatic duct. Below the branch to the gallbladder, the hepatic duct is called the *common bile duct*. Bile flows down the common bile duct to the duodenum.

To understand the role of bile in fat digestion, think of an oil and vinegar salad dressing. The oil, which is hydrophobic, tends to aggregate in large globules. For that reason, many salad dressings include an *emulsifier*—something that prevents oil droplets from aggregating. Mayonnaise, for example, is oil and vinegar with egg yolk added as an emulsifier. Bile contains salts that emulsify fats in the chyme, and thereby greatly enlarge the surface area of the fats exposed to the **lipases**—the enzymes that digest fats. One end of each bile salt molecule is soluble in fat (it is lipophilic and hydrophobic); the other end is soluble in water (it is hydrophilic and lipophobic). Bile molecules bury their lipophilic ends in fat droplets, leaving their hydrophilic ends sticking out. As a result, bile salts prevent the fat droplets from sticking together. The very small fat particles that result are called **micelles** (**Figure 50.16A**).

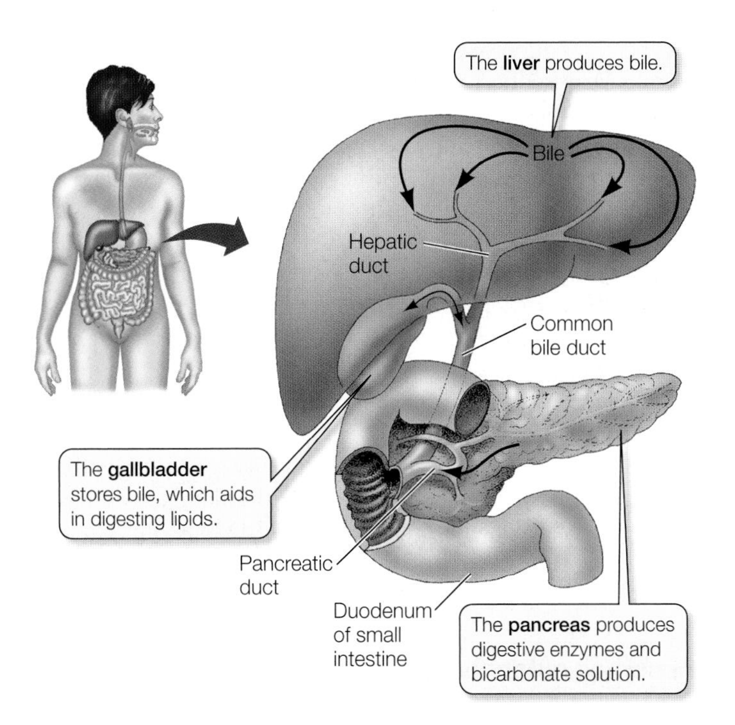

50.15 Ducts of the Gallbladder and Pancreas Bile produced in the liver leaves the liver via the hepatic duct. Branching off this duct is the gallbladder, which stores bile. Below the gallbladder, the hepatic duct is called the common bile duct and is joined by the pancreatic duct before entering the duodenum.

PANCREAS The **pancreas** is a large gland that lies just behind and below the stomach (see Figures 50.10 and 50.15). It is both an endocrine gland (secreting hormones into the tissue fluid; see Section 41.1) and an exocrine gland (secreting other substances through the pancreatic duct to the gut lumen, which is continuous with the outside of the body). The pancreatic duct joins the common bile duct before emptying into the duodenum.

The exocrine tissues of the pancreas produce a host of digestive enzymes, including lipases, amylases, proteases, and nucleases (**Table 50.3**). As in the stomach, some of these enzymes—most notably the proteases—are released as zymogens; otherwise, they would digest the pancreas and its ducts before they ever reached the duodenum. Once in the duodenum, one of these zymogens, **trypsinogen**, is activated by an enzyme called *enterokinase*, which is produced by cells lining the duodenum. This process is similar to the activation of pepsinogen by low pH in the stomach. Active **trypsin** can cleave other trypsinogen molecules to release even more active trypsin. Similarly, trypsin activates other zymogens secreted by the pancreas.

The mixture of zymogens produced by the pancreas can be dangerous if the pancreatic duct is blocked or if the pancreas is injured by infection or physical trauma. A few trypsinogen molecules spontaneously converting to trypsin can initiate a chain reaction of enzyme activity that digests the pancreas in a short time, destroying both its endocrine and exocrine functions.

The pancreas also produces a secretion rich in bicarbonate ions (HCO_3^-). Bicarbonate ions are alkaline (basic) and neutralize the acidic pH of the chyme that enters the duodenum from the stomach. Intestinal enzymes function best at a neutral or slightly alkaline pH.

Nutrients are absorbed in the small intestine

The final step in digesting proteins and carbohydrates and absorbing their components occurs among the microvilli. Mucosal epithelial cells secrete peptidases that cleave polypeptides into absorbable tripeptides, dipeptides, and individual amino acids. These epithelial cells also produce the enzymes maltase, lactase, and sucrase, which cleave the common disaccharides into absorbable monosaccharides—glucose, galactose, and fructose.

Many humans stop producing the enzyme lactase in childhood and thereafter have difficulty digesting lactose (the sugar in milk). Lactose is a disaccharide and cannot be absorbed without being cleaved into its constituents, glucose and galactose. Unabsorbed lactose is metabolized by bacteria in the large intestine, causing intense abdominal cramps, gas, and diarrhea.

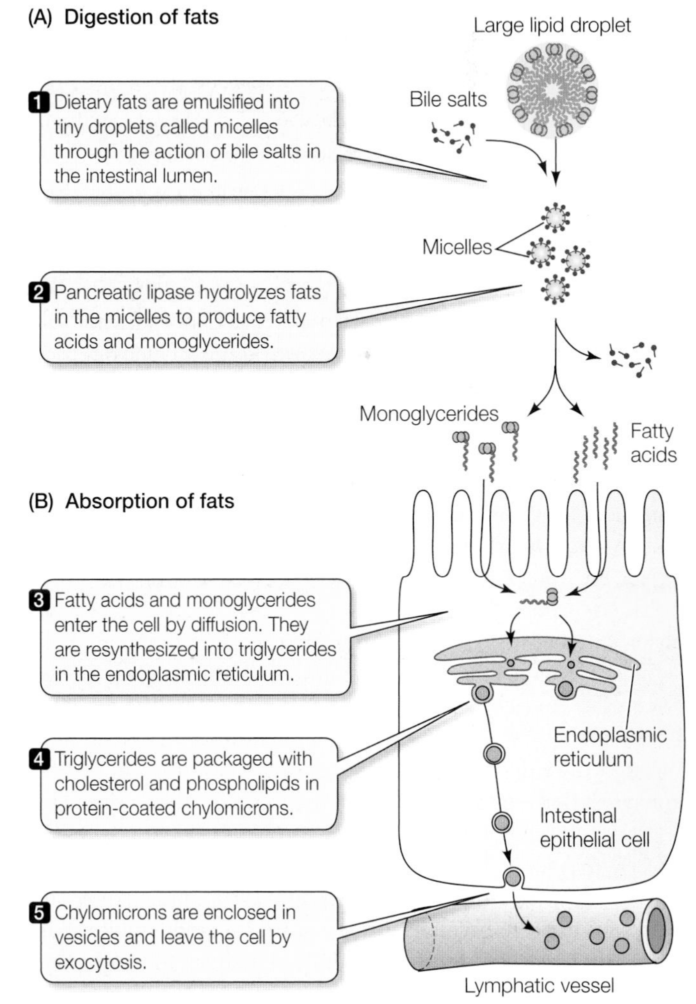

50.16 Digestion and Absorption of Fats (A) Dietary fats are broken up by bile into small micelles that present a large surface area to lipases. (B) The products of fat digestion are absorbed by intestinal mucosal cells, where they are resynthesized into triglycerides and exported to lymphatic vessels.

TABLE 50.3

Major Digestive Enzymes of Humans

SOURCE/ENZYME	ACTION
SALIVARY GLANDS	
Salivary amylase	Starch → Maltose
STOMACH	
Pepsin	Proteins → Peptides; autocatalysis
PANCREAS	
Pancreatic amylase	Starch → Maltose
Lipase	Fats → Fatty acids and glycerol
Nuclease	Nucleic acids → Nucleotides
Trypsin	Proteins → Peptides; zymogen activation
Chymotrypsin	Proteins → Peptides
Carboxypeptidase	Peptides → Shorter peptides and amino acids
SMALL INTESTINE	
Aminopeptidase	Peptides → Shorter peptides and amino acids
Dipeptidase	Dipeptides → Amino acids
Enterokinase	Trypsinogen → Trypsin
Nuclease	Nucleic acids → Nucleotides
Maltase	Maltose → Glucose
Lactase	Lactose → Galactose and glucose
Sucrase	Sucrose → Fructose and glucose

The mechanisms by which cells of the intestinal epithelium absorb nutrients and inorganic ions are diverse and include diffusion, facilitated diffusion, osmosis, active transport, and co-transport. Many inorganic ions such as sodium, calcium, and iron are actively transported by these cells. For example, active Na^+ transporters exist on the basal and lateral sides of the epithelial cells. They maintain a low concentration of Na^+ in those cells so that Na^+ can diffuse in from the chyme in the intestinal lumen. About 30 grams of Na^+ are transported this way every day, and Cl^- follows.

The transport of Na^+ and other ions is also important for water absorption because it creates an osmotic concentration gradient. At least 7–8 liters of water per day moves through the spaces between the epithelial cells in response to this osmotic gradient. Because the water moves through *spaces* between the cells and not through the cells themselves, it can carry with it nutrients that are in solution—a transport mechanism called *solvent drag*.

Many different kinds of transport proteins exist in the epithelial cells. Some, such as the transport protein for fructose, only facilitate diffusion, and that requires a concentration gradient. That works for fructose because once fructose enters the cell it is converted to glucose and the concentration gradient is maintained. A class of transporter proteins known as *sodium co-transporters* exploit the active transport of Na^+, which maintains a low Na^+ concentration in the cells. Co-transporter proteins combine with Na^+ and another molecule, such as glucose, galactose, or an amino acid. As Na^+ is pulled down its concentration gradient into the cell, the "hitchhiking" molecules are carried along with it.

The absorption of the products of fat digestion is relatively simple. Diglycerides, monoglycerides, and fatty acids are lipid-soluble and thus able to pass through the plasma membranes of the microvilli. In the intestinal epithelial cells, these molecules are resynthesized into triglycerides, combined with cholesterol and phospholipids, and coated with protein to form water-soluble **chylomicrons**, which are little particles of fat (**Figure 50.16B**). Rather than entering the blood directly, chylomicrons pass into the lymphatic vessels in the submucosa. They then flow through the lymphatic system and enter the bloodstream through the thoracic ducts at the base of the neck. After a meal rich in fats, chylomicrons can be so abundant in the blood that they give it a milky appearance.

The bile salts that emulsify fats are not absorbed along with the monoglycerides and the fatty acids, but shuttled back and forth between the gut contents and the microvilli. In the ileum, bile salts are actively reabsorbed and returned to the liver via the bloodstream.

Absorbed nutrients go to the liver

All of the blood leaving the digestive tract flows to the liver in the *hepatic portal vein*. This large vein delivers the blood to small spaces called sinusoids between groups of liver cells. These cells absorb the nutrients coming from the digestive tract and either store them or convert them to molecules the body needs. Glucose, sucrose, and fructose are used to synthesize glycogen. Amino acids are used to build proteins. Lipids from the chylomicrons are either stored as triglycerides or used to make lipoproteins.

Water and ions are absorbed in the large intestine

The motility of the small intestine gradually pushes its contents into the *large intestine*, or **colon**. Most of the available nutrients have been removed from the chyme that enters the colon, but it contains a lot of water and inorganic ions.

The colon absorbs water and ions, producing semisolid feces from the chyme it receives from the small intestine. Feces are stored in the rectum of the colon until they are eliminated. Absorption of too much water from the colon can cause *constipation*. The opposite condition, *diarrhea*, results if too little water is absorbed; in this case, water in the colon is excreted with the feces. The excessive diarrhea caused by diseases such as cholera can produce such rapid loss of water and electrolytes that death can occur in hours.

The problem with cellulose

Cellulose is the principal component of the food of herbivores. Most herbivores, however, cannot produce *cellulases*—enzymes that hydrolyze cellulose. Exceptions include silverfish (insects that eat books and stored papers), earthworms, and shipworms. From termites to cattle, herbivores rely on microorganisms in their digestive tracts to digest cellulose.

The stomachs of **ruminants** (cud chewers) such as cattle are large, four-chambered organs that take advantage of their endosymbiotic microorganisms (**Figure 50.17**). The first two chambers, the **rumen** and the **reticulum**, are packed with microorganisms

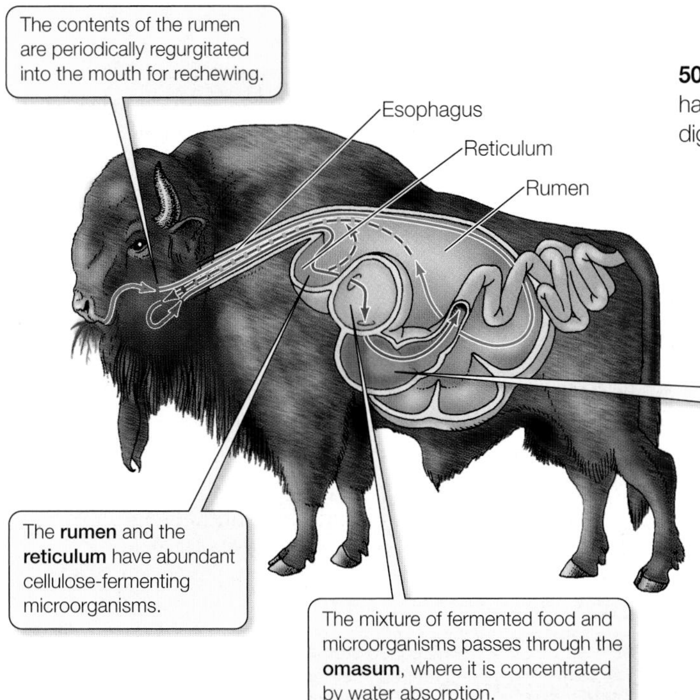

50.17 A Ruminant's Stomach Bison, like their relatives domestic cattle, have a specialized stomach with four compartments that enables them to digest and obtain energy from coarse plant material.

that break down cellulose by fermentation. The ruminant periodically regurgitates the contents of the rumen (the *cud*) into the mouth for re-chewing. When swallowed again, the vegetal fibers present more surface area to the microorganisms. The microorganisms metabolize cellulose and other nutrients to simple fatty acids, which become nutrients for their host.

Enormous numbers of microorganisms leave the rumen along with the partially digested food. This mass is concentrated by water absorption in the **omasum** before it then enters the true stomach, the **abomasum**, where the microorganisms are killed by secreted hydrochloric acid, digested by proteases, and passed on to the small intestine for further digestion and absorption. A cow derives more than 100 grams of protein per day from digestion of its endosymbiotic microorganisms. The rate of multiplication of microorganisms in the rumen offsets their loss, so a well-balanced, mutually beneficial relationship is maintained.

Intestinal bacteria produce gases such as methane and hydrogen sulfide as by-products of their anaerobic metabolism. Beans contain some carbohydrates that humans cannot digest but their intestinal bacteria can. A meal of beans increases gut bacterial metabolic activity and results in the well-earned reputation of beans.

Some mammalian herbivores have a microbial fermentation chamber called a **cecum** extending from the large intestine. An example is the rabbit (see Figure 50.8). Since the cecum empties into the large intestine, absorption of the nutrients produced by the microorganisms is inefficient. Such species frequently produce two kinds of feces – ones that are pure waste and ones that contain cecal material. In a behavior known as **coprophagy**, they re-ingest the cecal feces directly from the anus so they can digest and absorb the nutrients that would otherwise be lost. In humans, the cecum has become the vestigial **appendix**, which serves no digestive function.

50.3 RECAP

The vertebrate gastrointestinal system is a tubular gut that is adapted to ingest food, fragment it, digest it, and absorb nutrients. Peristalsis moves food through the gut. Absorption of nutrients occurs mostly in the small intestine; water and ions are absorbed in the large intestine.

- What digestive functions occur in the mouth and in the stomach? See p. 1080 and Figure 50.13
- How do bile salts assist in the digestion of fats? See pp. 1082–1083 and Figure 50.16
- Can you describe how sodium co-transport drives the absorption of nutrients? See p. 1084

The steps included in ingestion and digestion of food—from fragmentation in the mouth to the digestive processes in the gastrointestinal tract—serve one purpose: to make nutrients in the food available for absorption and ultimately for metabolism. Let's look at how the processes of digestion are controlled and how nutrients are handled by the body once food has been digested.

50.4 How Is the Flow of Nutrients Controlled and Regulated?

The vertebrate gut is an assembly line in reverse—a *dis*assembly line. As with a standard assembly line, the control and coordination of the sequential processes of digestion is critical. Both neuronal and hormonal controls govern these processes. Once the products of digestion are absorbed, their availability to the cells of the body must also be controlled.

You have certainly experienced salivation at the sight or smell of food. That response is an *unconcious reflex*, as is swallowing. Many such autonomic reflexes coordinate activity in different regions of the digestive tract. For example, the introduction of food into the stomach stimulates increased activity in the colon, which can lead to defecation.

The digestive tract is unusual in that it has an intrinsic (its own independent) nervous system. Neuronal messages can travel from one region of the digestive tract to another without being processed by the central nervous system (CNS). One function of the gut's nervous system is the coordination of motility. Of course, this intrinsic nervous system can communicate information to the CNS and receive input from the CNS, but its most important role is to coordinate actions throughout the digestive tract.

Hormones control many digestive functions

Several hormones control the activities of the digestive tract and its accessory organs (**Figure 50.18**). The first hormone discovered came from the duodenum; it was called **secretin** because it causes the pancreas to secrete digestive juices. We now know that secretin is only one of several hormones that control pancreatic secretion; specifically, secretin stimulates the pancreas to secrete a solution rich in bicarbonate ions.

In response to the presence of fats and proteins in the chyme, the mucosa of the small intestine secretes **cholecystokinin**, a hormone that stimulates the gallbladder to release bile and the pancreas to release digestive enzymes. Cholecystokinin and secretin also slow the movements of the stomach, which slows the delivery of chyme into the small intestine.

The stomach secretes a hormone called **gastrin** into the blood. Cells in the lower region of the stomach release gastrin when they are stimulated by the presence of food. Gastrin circulates in the blood until it reaches cells in the upper areas of the stomach wall, where it stimulates the secretions and movements of the stomach. Gastrin release begins to be inhibited when the pH of the stomach contents falls below 3—an example of negative feedback.

Most animals do not eat continuously. When they do eat, food is present in the gut, and nutrients are being absorbed, for a period of time after a meal, called the **absorptive period**. Once the stomach and small intestine are empty, nutrients are no longer being absorbed. During this **postabsorptive period**, the processes of energy metabolism and biosynthesis must run on internal reserves. Nutrient traffic must be controlled so that reserves accumulate during the absorptive period and are used appropriately during the postabsorptive period.

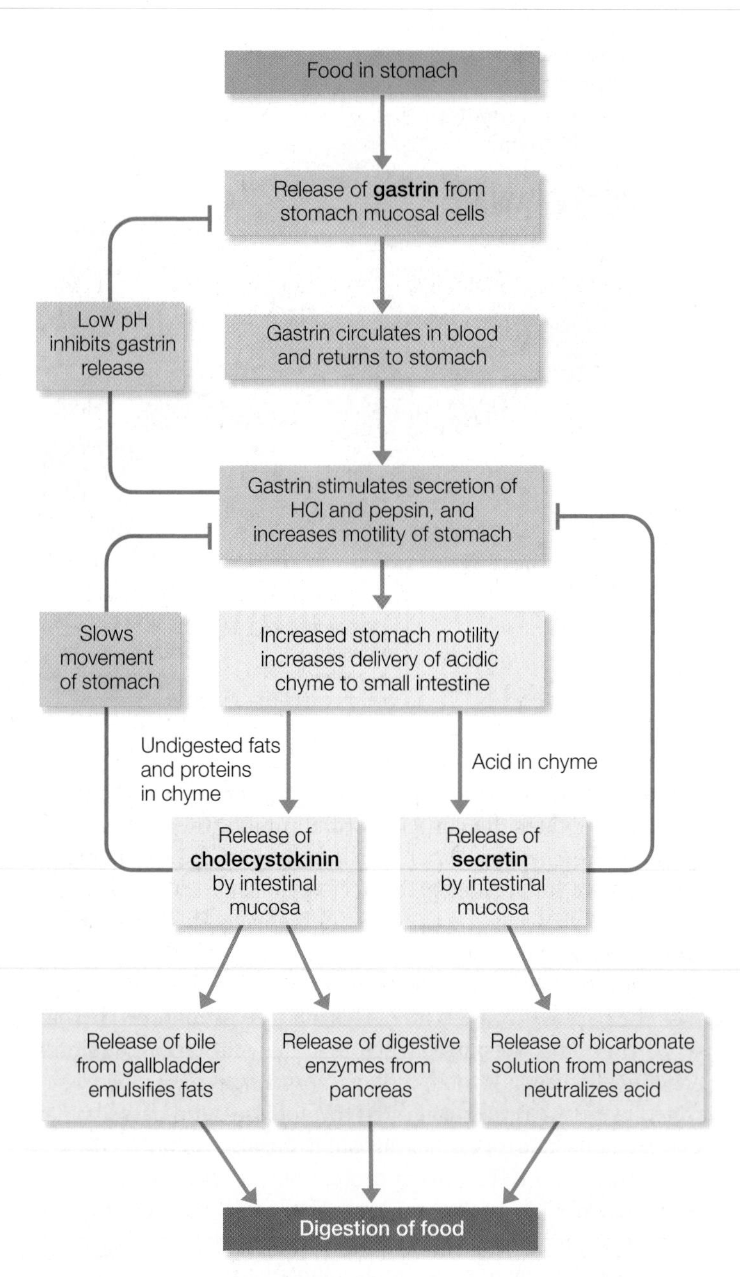

50.18 Hormones Control Digestion The hormones gastrin, cholecystokinin, and secretin are involved in feedback loops that control the sequential processing of food in the digestive tract.

The liver directs the traffic of the molecules that fuel metabolism

When fuel molecules are abundant in the blood, the liver stores them in the form of glycogen and fats. The liver also synthesizes blood plasma proteins from circulating amino acids. When fuel molecule levels in the blood decline, the liver delivers them back into the blood.

The liver has an enormous capacity to interconvert fuel molecules. Liver cells can convert monosaccharides into either glycogen or fats, and vice versa. The liver can also convert certain amino acids and some other molecules, such as pyruvate and lactate, into glucose—this process is termed *gluconeogenesis*. The liver is also the major controller of fat metabolism through its production of *lipoproteins*.

LIPOPROTEINS: THE GOOD, THE BAD, AND THE UGLY In the intestine, bile solves the problem of processing hydrophobic fats in an aqueous medium. The transport of fats in the circulatory system presents the same problem, and lipoproteins are the solution. A **lipoprotein** is a particle made up of a core of hydrophobic fat and cholesterol with a covering of hydrophilic protein that allows it to be suspended in water. The largest lipoprotein particles in the bloodstream are the **chylomicrons** produced by the mucosal cells of the intestine. As the circulation carries chylomicrons through the liver and adipose (fat) tissue, receptors on the capillary walls recognize and bind to the chylomicron protein coats. Lipases begin to hydrolyze the fats, which are then absorbed into liver or fat cells. Thus, the protein coat of the lipoprotein serves as an "address" that directs it to a specific tissue.

Lipoproteins other than chylomicrons are synthesized in the liver. These lipoproteins can be classified according to their density. Fat has a low density (it floats in water) and protein has a high density, so the greater the fat-to-protein ratio in the lipoprotein, the lower its density.

- **High-density lipoproteins (HDLs)** remove cholesterol from tissues and carry it to the liver, where it can be used to synthesize bile. HDL consists of about 50% protein, 35% lipids, and 15% cholesterol.
- **Low-density lipoproteins (LDLs)** transport cholesterol around the body for use in biosynthesis and for storage. LDL consists of about 25% protein, 25% lipids, and 50% cholesterol.
- **Very-low-density lipoproteins (VLDLs)** contain mostly triglyceride fats, which they transport to fat cells in adipose tissues around the body. VLDL consists of about 2% protein, 94% lipids, and 3% cholesterol.

Because of their functions in cholesterol regulation—LDL "adds" it and HDL "removes" it—LDL is sometimes called "bad cholesterol" and HDL "good cholesterol." A high ratio of LDL to HDL in a person's blood is a risk factor for atherosclerotic heart disease. Cigarette smoking lowers HDL levels. Regular exercise increases them.

INSULIN AND GLUCAGON CONTROL FUEL METABOLISM During the absorptive period, blood glucose levels rise as carbohydrates are digested and absorbed. During this time, β cells of the pancreas release insulin, which plays a major role in directing glucose to where it will be used or stored. The actions of insulin vary in different tissues. Glucose enters cells by diffusion facilitated by glucose transporters. However, the glucose transporters in resting skeletal muscle and adipose tissues are normally sequestered in cytoplasmic vesicles until insulin combines with its receptors on the cell surface and triggers the insertion of transporters into the cell membrane. In adipose cells, insulin inhibits lipase and promotes fat synthesis from glucose. In the liver, insulin activates an enzyme that phosphorylates glucose as it enters the cell so it cannot diffuse back out again, enhancing the overall diffusion of glucose into the cells. Insulin also activates the enzymes in liver cells that catalyze the synthesis of glycogen. The many actions of insulin promote the uptake and utilization or storage of glucose by cells all around the body.

During the postabsorptive period, a fall in blood glucose decreases the release of insulin, and the uptake of glucose by most cells is curtailed (**Figure 50.19**). To maintain blood glucose levels, liver cells break down glycogen, which releases glucose into the blood. The liver and the adipose tissues supply fatty acids to the blood, and most cells preferentially use fatty acids as their metabolic fuel. One tissue that does not switch fuel sources during the postabsorptive period, however, is the nervous system.

The cells of the nervous system require a constant supply of glucose, and they can use other fuels to a very limited extent. Most neurons do not require insulin to absorb glucose from the blood, but they do need an adequate glucose concentration gradient to drive the facilitated diffusion of glucose across their plasma membranes. Therefore it is critical that blood glucose levels are maintained during the postabsorptive period. The overall dependence of neuronal tissues on glucose, and their requirement for constant blood glucose levels, are the reasons it is so important for other cells of the body to shift to fat metabolism during the postabsorptive period.

The metabolism of fuel molecules during the postabsorptive period is mostly controlled by the lack of insulin, but if blood glucose falls below a certain level, another pancreatic hormone, **glucagon**, is called into play. Glucagon's effect is opposite that of insulin: it stimulates liver cells to break down glycogen and to carry out gluconeogenesis. Thus, under the influence of glucagon, the liver produces glucose and releases it into the blood.

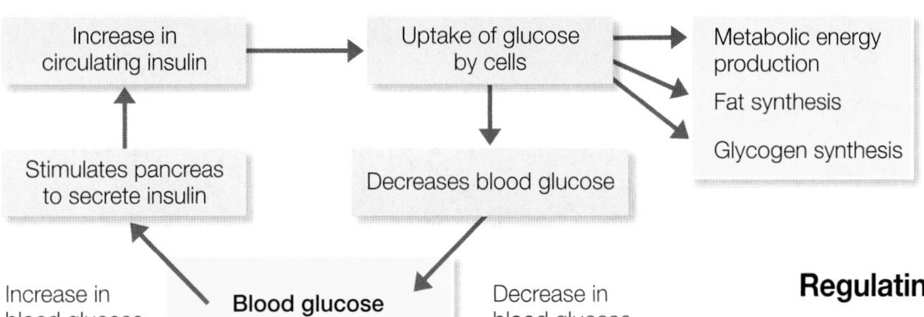

50.19 Regulating Glucose Levels in the Blood Insulin (blue) and glucagon (pink) interactions maintain the homeostasis of circulating glucose.

Regulating food intake is important

Obesity is a major health issue in the United States. People spend billions of dollars on schemes to lose weight, but the problem seems to get worse every year. A simple rule—take in fewer calories than your body burns while eating a balanced diet—should solve the problem, but it doesn't. Why? As we noted at the beginning of this chapter, lifestyle plays a major role in obesity, and genetic and regulatory factors "weigh in" as well.

The amount of food an animal eats is governed by its sensations of hunger and satiety. These sensations are influenced by the hypothalamus (see Section 41.2). If a region in the middle of the hypothalamus of rats, called the ventromedial hypothalamus, is damaged, the animals will increase their food intake and become obese. If a different region, called the lateral hypothalamus, is damaged, rats will decrease their food intake and become thin. In both cases, the rats eventually reach a new equilibrium body weight, which they maintain. Thus, regulation is maintained, but the set point has been changed. Other brain regions have also been implicated in control of hunger and satiety.

In Section 40.1 we learned that regulation involves feedback information and a means of comparing that information with a set point. Some evidence suggests that cells in the hypothalamus and liver are sensitive to glucose and insulin in the blood, with high levels stimulating satiety and low levels stimulating hunger. Even stronger evidence, however, indicates that signals from fat metabolism influence hunger and satiety.

A single-gene mutation in mice, when present in the homozygous condition, results in mice that eat enormous amounts of food and become obese. Geneticists call these mice *ob/ob*, given their double dose of the recessive "obese" allele. Experiments revealed that the wild-type *Ob* allele codes for a protein, which was named **leptin** (Greek *leptos*, "thin"). Leptin is a hormone produced by fat cells and circulates in the blood. Receptors for leptin are found in the regions of the hypothalamus that are involved in control of hunger and satiety. It seems that leptin provides feedback information about the status of the body fat reserves to the brain. When leptin was injected into *ob/ob* mice they ate less and lost body fat. However, a different strain of obese mice, *db/db*, did not respond to leptin injections. Further experiments revealed that *db/db* mice lacked the leptin receptor (**Figure 50.20**).

Could leptin be used to reduce human obesity? In the very few obese people who do not produce the hormone, leptin injections have curbed appetites and facilitated weight loss. Most obese people, however, have higher than normal circulating levels of leptin, so it is likely that their leptin receptors are not completely functional.

Additional feedback signals are most certainly involved in the regulation of food intake. Recently a hormone called **ghrelin** was discovered that is produced and secreted by cells in the stomach. Normally, ghrelin levels rise before a meal and fall after a meal. Fasting causes an increase in ghrelin levels. Ghrelin binding to its receptors in the hypothalamus stimulates appetite. Ghrelin also stimulates cells in the pituitary gland to release growth hormone.

EXPERIMENT

HYPOTHESIS: Strains of mice that become obese lack a satiety factor (or its receptor) in their brains.

METHOD

Using mice from two strains that are genetically obese, surgically join the circulatory systems of normal or wild-type mice with an obese-strain partner so that they share circulating blood (i.e., create parabiotic pairs).

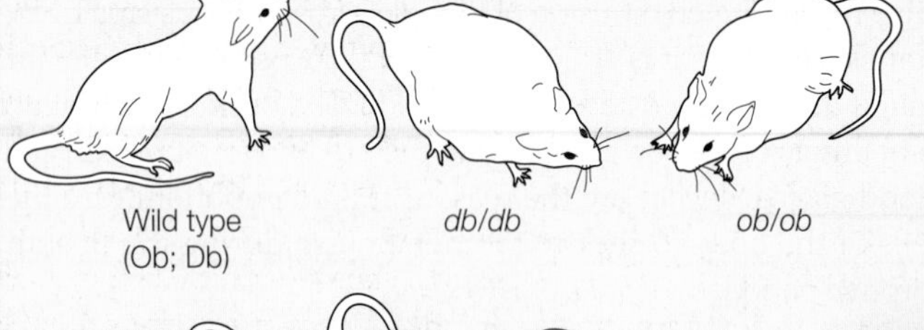

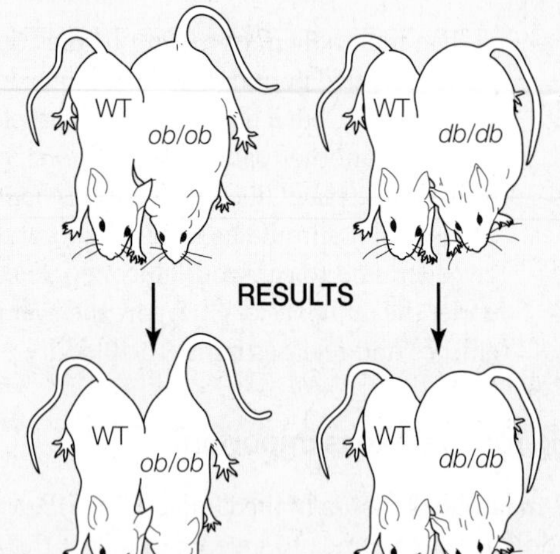

Parabiotic *ob/ob* mice lose fat, but parabiotic *db/db* mice do not. The *ob/ob* mice obtain the satiety factor from the wild-type mouse. *db/db* mice obtain the satiety factor, but are still fat because they lack the receptor.

CONCLUSION: The wild-type *Ob* gene codes for a satiety hormone. This hormone is given the name leptin. The *Db* gene encodes the leptin receptor.

50.20 A Single-Gene Mutation Leads to Obesity in Mice A study of obese and wild-type mice revealed the existence of a satiety factor, leptin, and its receptor. The wild-type *Ob* gene encodes leptin. The recessive allele *ob* is a loss-of-function allele, so *ob/ob* mice do not have leptin; they do not experience satiety and eat so much that they become obese. The wild-type *Db* gene encodes the leptin receptor, so mice homozygous for the loss-of-function allele (*db/db*), even if supplied leptin, cannot use it. The *ob/ob* mice do have the wild-type *Db* gene, so if supplied with leptin they experience satiety and lose fat.

50.4 RECAP

The major controlling factors of gut function are an intrinsic nervous system and the hormones gastrin, secretin, and cholecystokinin. Insulin is the major hormonal controller of fuel metabolism. The hypothalamus controls food intake by generating sensations of hunger and satiety influenced by feedback from blood glucose and the hormones leptin and ghrelin.

- What are the roles of the three different classes of lipoproteins? See pp. 1086–1087
- By what actions does insulin promote uptake and storage of energy during the absorptive period? See p. 1087 and Figure 50.19
- Can you describe the evidence that leptin influences satiety? See p. 1088 and Figure 50.20

The goal of digestion and absorption of food for any animal is nourishment. Not everything an animal ingests, however, is safe; many foods also contain toxins.

50.5 How Do Animals Deal with Ingested Toxins?

Some plant and animal tissues contain more than just nutrients; some can also contain toxic compounds. Plants produce toxic secondary metabolites as defenses against herbivores. One example is the nicotine in tobacco. Animals may use toxins for capturing prey as well as for self-defense, so ingesting certain plant and animal tissues can be dangerous. In addition, human activities add millions of tons of synthetic toxic compounds to the environment every year, and many of these compounds enter the air we breathe, the water we drink, and the food we eat. A new field called **environmental toxicology** addresses the problems of poisons in the environment.

The body cannot metabolize many synthetic toxins

How does the animal body handle synthetic toxins? In many cases, the systems that metabolize natural chemicals can also metabolize synthetic toxins, breaking them apart and eliminating them through the urine. As we saw earlier, everything absorbed from the small intestine is transported first to the liver; liver enzymes called *cytochrome P450*s are largely responsible for the detoxification of absorbed chemicals.

P450s are less specific in their abilities to bind substrates than are most enzymes; thus, each P450 can catalyze reactions with a wide range of compounds, and many P450s exist. Few natural compounds can escape the P450s, even when the body encounters them for the first time.

Some synthetic chemicals, however, fall outside the range of structures that P450s and other enzymes can metabolize. If a synthetic chemical that cannot be metabolized is structurally similar to a hormone, that synthetic chemical may activate the hormone's signaling pathway within target cells. Whereas the natural hormonal signal can be turned off, the synthetic signal cannot be, and control of function is lost.

Some toxins are retained and concentrated

The physical and chemical properties of a toxic compound affect its retention within a biological system. If a compound can dissolve in water, it may be quickly metabolized (and thus detoxified) because it is accessible to the wide variety of enzymes that can break down complex molecules in food. In addition to being degraded or metabolized, many water-soluble compounds can be filtered out of the blood by the kidneys, and therefore do not accumulate in the body. However, some dangerous water-soluble compounds can be incorporated into the body and disrupt normal functions. An example is lead, which can replace iron in blood and calcium in bone.

Lipid-soluble compounds are usually metabolized more slowly than water-soluble compounds, and they are often stored in the body for a long time because they dissolve in adipose tissues. Accumulated lipophilic compounds can reach very high concentrations. Some lipid-soluble toxins, including many pesticides, can **bioaccumulate** in the environment; that is, they can become concentrated in predators that eat contaminated prey. The pesticide load is passed up the food chain from prey to predator, growing increasingly concentrated in the tissues of each consumer. In the top predator, the pesticide may be concentrated thousands or millions of times. Long-lived predators, such as eagles and bears, are particularly at risk for heavy pesticide burdens because they have many years to accumulate them. Bioaccumulated toxins may be responsible for the high rates of cancer and infertility found in some wildlife populations.

One class of toxic synthetic chemicals that has bioaccumulated in animals, including humans, are polychlorinated biphenyls (PCBs). PCBs were produced extensively for use as an insulating fluid in electrical transformers from the 1930s until recently. They are chemically stable, lipophilic, and are now found throughout the environment. They have been shown to bioaccumulate, reaching dangerously high levels in fish from contaminated waters such as the Great Lakes. In communities around the Great Lakes, studies have indicated cognitive impairment in children of mothers with a high body burden of PCBs, probably from eating fish caught in the Great Lakes.

The risks of PCBs are now clear, but it is usually difficult to make a causal connection between a toxin in the environment and specific health effects in a population. Environmental toxicologists must be able to study large populations, use powerful statistical analyses, and do controlled laboratory studies to obtain evidence that will support policy changes to stop and reverse the effects of synthetic environmental toxins.

CHAPTER SUMMARY

50.1 What do animals require from food?

Animals are **heterotrophs** that derive their energy and molecular building blocks, directly or indirectly, from **autotrophs**.

Carbohydrates, fats, and proteins in food supply animals with metabolic energy. A measure of the energy content of food is the **kilocalorie**. Excess caloric intake is stored as glycogen and fat. Review Figure 50.2

For many animals, food provides essential **carbon skeletons** that they cannot synthesize themselves. Review Figure 50.4

Humans require eight **essential amino acids** in the diet. Different animals need mineral elements in different amounts. **Macronutrients** are needed in large quantities. **Micronutrients** are needed in small amounts. Review Figure 50.5 and Table 50.1, Web/CD Activity 50.1

Vitamins are organic molecules that must be obtained in food. Review Table 50.2, Web/CD Activity 50.2

Malnutrition results when any essential nutrient is lacking from the diet. A chronic state of malnutrition causes a **deficiency disease**.

50.2 How do animals ingest and digest food?

Animals can be characterized by how they acquire nutrients: **Saprobes** and **detritivores** depend on dead organic matter, **filter feeders** strain the aquatic environment for small food items, **herbivores** eat plants, and **carnivores** eat animals. Behavioral and anatomical adaptations reflect these feeding strategies. See Web/CD Activity 50.3

Digestion involves the breakdown of complex food molecules into monomers that can be absorbed and utilized by cells. In most animals, digestion takes place in a tubular gut. Review Figure 50.8

Absorptive areas of the gut are characterized by a large surface area produced by extensive folding and numerous **villi** and **microvilli**. Review Figure 50.9

Hydrolytic enzymes break down proteins, carbohydrates, and fats into their monomeric units. To prevent the organism itself from being digested, many of these enzymes are released as inactive **zymogens**, which become activated when secreted into the gut.

50.3 How does the vertebrate gastrointestinal system function?

The vertebrate gut can be divided into several compartments with different functions. Review Figure 50.10, Web/CD Activity 50.4

The cells and tissues of the vertebrate gut are organized in the same way throughout its length. The innermost tissue layer, the **mucosa**, is the secretory and absorptive surface. The **submucosa** contains blood and lymph vessels, and a nerve plexus. External to the submucosa are two **smooth muscle** layers. Between the two muscle layers is another nerve plexus that controls the movements of the gut. Review Figure 50.11

Swallowing is a reflex that pushes the bolus of food into the **esophagus**. **Peristalsis** and other movements of the gut move the bolus down the esophagus and through the entire length of the gut. **Sphincters** block the gut at certain locations, but they relax as a wave of peristalsis approaches. Review Figure 50.12

Digestion begins in the mouth, where **amylase** is secreted with the saliva. Digestion of protein begins in the stomach, where parietal cells secrete HCl and chief cells secrete **pepsinogen**, a zymogen that becomes **pepsin** when activated by low pH and **autocatalysis**. The mucosa also secretes **mucus**, which protects the tissues of the gut. Review Figure 50.13

In the **duodenum**, pancreatic enzymes carry out most of the digestion of food. **Bile** from the **liver** and **gallbladder** emulsify fats into **micelles**. Bicarbonate ions from the pancreas neutralize the pH of the **chyme** entering from the stomach to produce an environment conducive to the actions of pancreatic enzymes such as **trypsin**. Review Figure 50.15 and Table 50.3

Final enzymatic cleavage of polypeptides and disaccharides occurs among the microvilli of the intestinal mucosa. Amino acids, monosaccharides, and inorganic ions are absorbed by the microvilli. Specific transporter proteins are sometimes involved. Sodium co-transport often powers the active transport of nutrients.

Fats broken down by **lipases** are absorbed mostly as monoglycerides and fatty acids and are resynthesized into triglycerides within cells. The triglycerides are combined with cholesterol and phospholipids and coated with protein to form **chylomicrons**, which pass out of the mucosal cells and into lymphatic vessels in the submucosa. Review Figure 50.16, Web/CD Tutorial 50.1

Water and ions are absorbed in the **large intestine** as waste matter is consolidated into **feces**, which is periodically eliminated.

Microorganisms in some compartments of the gut digest materials that their host cannot. Review Figure 50.17

50.4 How is the flow of nutrients controlled and regulated?

Autonomic reflexes coordinate activity of the digestive tract, which has an intrinsic nervous system that can act independently of the CNS.

The actions of the stomach and small intestine are largely controlled by the hormones **gastrin**, **secretin**, and **cholecystokinin**. Review Figure 50.18

The liver plays a central role in directing the traffic of fuel molecules. During the **absorptive period**, the liver takes up and stores fats and carbohydrates, converting monosaccharides to glycogen or fats. The liver also takes up amino acids and uses them to produce blood plasma proteins, and can engage in gluconeogenesis.

Fat and cholesterol are shipped out of the liver as **low-density lipoproteins**. **High-density lipoproteins** act as acceptors of cholesterol and are believed to bring fat and cholesterol back to the liver.

Insulin largely controls fuel metabolism during the absorptive period and promotes glucose uptake as well as glycogen and fat synthesis. During the **postabsorptive** period, lack of insulin blocks the uptake and utilization of glucose by most cells of the body except neurons. If blood glucose levels fall, **glucagon** secretion increases, stimulating the liver to break down glycogen and release glucose to the blood. Review Figure 50.19, Web/CD Tutorial 50.2

Food intake is governed by sensations of hunger and satiety, which are determined by brain mechanisms. Review Figure 50.20

50.5 How do animals deal with ingested toxins?

Toxins in food may come from natural sources, but many come from human activities such as the use of pesticides and the release of pollutants into the environment.

Toxins such as PCBs that accumulate in the bodies of prey are transferred to and further concentrated in the bodies of their predators. This **bioaccumulation** produces high concentrations of toxins in animals high up the food chain.

SELF-QUIZ

1. Most of the metabolic energy needed by a bird for a long-distance migratory flight is stored as
   a. glycogen.
   b. fat.
   c. protein.
   d. carbohydrates.
   e. ATP.
2. Which statement about essential amino acids is true?
   a. They are not found in vegetarian diets.
   b. They are stored by the body until they are needed.
   c. Without them, one is undernourished.
   d. All animals require the same ones.
   e. Humans can acquire all of theirs by eating milk, eggs, and meat.
3. Which statement about vitamins is true?
   a. They are essential inorganic nutrients.
   b. They are required in larger amounts than are essential amino acids.
   c. Many serve as coenzymes.
   d. Vitamin D can be acquired only by eating meat or dairy foods.
   e. When vitamin C is eaten in large quantities, the excess is stored in fat for later use.
4. The digestive enzymes of the small intestine
   a. do not function best at a low pH.
   b. are produced and released in response to circulating secretin.
   c. are produced and released under neuronal control.
   d. are all secreted by the pancreas.
   e. are all activated by an acidic environment.
5. Which statement about nutrient absorption by the intestinal mucosal cells is true?
   a. Carbohydrates are absorbed as disaccharides.
   b. Fats are absorbed as fatty acids and monoglycerides.
   c. Amino acids move across the plasma membrane only by diffusion.
   d. Bile transports fats across the plasma membrane.
   e. Most nutrients are absorbed in the duodenum.
6. Chylomicrons are like the tiny micelles of dietary fat in the lumen of the small intestine in that both
   a. are coated with bile.
   b. are lipid soluble.
   c. travel through the lymphatic system.
   d. contain triglycerides.
   e. are coated with lipoproteins.
7. Microbial fermentation in the gut of a cow
   a. produces fatty acids as a major nutrient for the cow.
   b. occurs in specialized regions of the small intestine.
   c. occurs in the cecum, from which food is regurgitated, chewed again, and swallowed into the true stomach.
   d. produces methane as a major nutrient.
   e. is possible because the stomach wall does not secrete hydrochloric acid.
8. Which of the following is stimulated by cholecystokinin?
   a. Stomach motility
   b. Release of bile
   c. Secretion of hydrochloric acid
   d. Secretion of bicarbonate ions
   e. Secretion of mucus
9. During the absorptive period,
   a. breakdown of glycogen supplies glucose to the blood.
   b. glucagon secretion is high.
   c. the number of circulating lipoproteins is low.
   d. glucose is the major metabolic fuel.
   e. the synthesis of fats and glycogen in muscle is inhibited.
10. During the postabsorptive period,
    a. glucose is the major metabolic fuel.
    b. glucagon stimulates the liver to produce glycogen.
    c. insulin facilitates the uptake of glucose by brain cells.
    d. fatty acids constitute the major metabolic fuel.
    e. liver functions slow down because of low insulin levels.

FOR DISCUSSION

1. Several currently popular diet books recommend high fat and protein intake and low carbohydrate intake as a means of losing body mass. What could the rationale of a high-fat and high-protein diet be, and what health issues should be considered when someone considers going on such a diet?
2. Carnivores generally have more dietary vitamin requirements than herbivores do. Why?
3. It is said that the most important hormonal control of fuel metabolism in the postabsorptive period is the lack of insulin. Explain.
4. Why is obstruction of the common bile duct so serious? Consider in your answer the multiple functions of the pancreas and the way in which digestive enzymes are processed.
5. Trace the history of a fatty acid molecule from a slice of cheese pizza to a plaque on a coronary artery. Into what possible forms and structures might it have been converted as it passed through in the body? Describe a direct and an indirect route it could have taken.

FOR INVESTIGATION

Cystic fibrosis is a genetic disease due to a mutation in the gene for a protein that transports chloride ions out of cells. Heterozygous individuals are normal, but are carriers of the disease. Homozygous individuals have trouble clearing their airways and are subject to respiratory infections. Before the advent of effective respiratory therapy, such individuals succumbed to the disease early in life. How do you think such a gene could have become established in the human population? (A clue might come from our knowledge of the mode of action of the cholera toxin, which overly stimulates chloride transporters in the membranes of intestinal cells, as described on pp. 76–77 of Chapter 5). Could cystic fibrosis, like sickle-cell disease, be a case of heterozygote advantage? How would you investigate that hypothesis?

CHAPTER 51 Salt and Water Balance and Nitrogen Excretion

Blood, sweat, and tears

Blood, sweat, and tears taste salty because they have similar ionic concentrations as the interstitial fluids that bathe the cells of the body. The volume and the composition of the interstitial fluids must remain within certain limits and kept relatively free of wastes. Maintaining homeostasis of the interstitial fluids is the job of the excretory system, and it can be challenging. The nature of the challenge depends on the environment of the animal and its lifestyle. Some animals such as desert insects and small mammals may never experience free water in their environments. They must be able to live their entire lives without drinking. All animals derive water from the metabolism of food, but to make that amount of water do, desert animals must conserve it. Accordingly, they excrete wastes that are extremely concentrated. Insects excrete semi-solid wastes and desert rodents excrete urine that is so concentrated it contains crystals of solute.

Animals that live in fresh water have the opposite problem; water is continuously entering their bodies by osmosis and with the food they eat, so they must constantly bail themselves out by producing copious amounts of dilute urine while they conserve the solutes their bodies need. We will see the adaptations such animals have to maintain salt and water balance, but to show how flexible physiological adaptations to the environment can be, we will discuss an animal that experiences both extreme conditions, and within minutes of each other. We will consider the problems of vampires—not the horror film kind, but the bat kind.

Vampire bats are small tropical mammals that feed on the blood of other mammals, such as cattle. The bat lands on an unsuspecting (usually sleeping) victim, bites into a vein, and drinks blood—a high-protein, liquid food. The bat has only a short time to feed before the victim wakes. To maximize the volume of blood it can ingest, it eliminates water from its food as fast as it can by producing a lot of very dilute urine. The warm trickle down the neck of the victim is not blood!

Once feeding ends, this high rate of water loss must stop. Now the vampire bat is digesting protein and must excrete large amounts of nitrogenous breakdown products while conserving its body water. Within minutes, the excretory system of the vampire bat switches from producing abundant, very dilute urine to producing a tiny

Blood as a Fast Food The vampire bat *Desmodus rotundus* is able to adjust its excretory physiology from water-excreting to water-conserving, depending on whether it is ingesting or digesting its blood meal.

Living without Water Kangaroo rats such as *Dipodomys spectabilis*, a denizen of the Arizona desert, may never see free water during their lifetime. Their dry-adapted excretory systems allow these rodents to derive enough water to survive on from their food.

CHAPTER OUTLINE

amount of highly concentrated urine. The vampire bat rapidly switches from an excretory physiology typical of a mammal living in an environment with abundant fresh water (copious amounts of dilute urine) to an excretory physiology typical of a mammal living in an arid desert (small amounts of concentrated urine).

IN THIS CHAPTER we will discover how animals accomplish the various feats of salt and water balance and waste excretion that adapt them to different environments. We will begin by discussing the challenges presented by those different environments. Then we will explore some invertebrates that illustrate basic mechanisms common to all animal excretory systems. Turning to vertebrates, we will learn about the basic anatomical unit of the excretory system—the nephron—and how it evolved. Finally, we will cover the mechanisms that control and regulate salt and water balance in mammals, giving the vampire bat and other species their remarkable abilities to exploit unusual diets and extreme environments.

51.1 What Roles Do Excretory Organs Play in Maintaining Homeostasis?

Excretory organs control the volume, concentration, and composition of the extracellular fluids of animals. Life evolved in the seas, and seawater is the extracellular environment for the cells of the simplest marine animals. The cells of many small marine animals derive their oxygen directly from the seawater and expel carbon dioxide and nitrogenous wastes into the seawater. Larger and more complex marine animals cannot serve the needs of all of their cells by direct exchanges with seawater, so they maintain an internal environment of extracellular fluids. The salt concentration and composition of the extracellular fluids of most marine invertebrates are similar to that of seawater. The characteristics of extracellular fluids of marine vertebrates and all freshwater and terrestrial animals, in contrast, are considerably different from those of seawater. The concentration of solutes in an animal's extracellular fluid determines the water balance of its cells, and the composition of the extracellular fluid influences the health and functions of the cells it bathes. Recall, for example, the importance of ion concentration gradients between the extracellular fluid and the cytoplasm of nerve and muscle cells (see Sections 44.2 and 47.1).

Water enters or leaves cells by osmosis

The volume of cells depends on whether they take up water from or lose water to the extracellular fluids. The movement of water across cell plasma membranes depends on differences in solute concentration. If the solute concentrations are different on two sides of a membrane permeable to water but not the solutes, the water will flow from the side with the lower concentration to the side with the higher concentration of solute (see Sections 5.3 and 35.1 for more about osmosis). Likewise, if the solute concentration of the extracellular fluid surrounding animal cells is less than that of the cytoplasm, water moves into the cells, causing them to swell and possibly burst. If the solute concentration of the extracellular fluid is greater than that of the cytoplasm, the cells lose water and shrink. Thus the solute concentration of the extracellular fluid affects both the volume and the solute concentration of the cells.

Animal physiologists use the term **osmolarity** in discussing osmosis. The osmolarity of a solution is the number of moles of osmotically active solutes per liter of solvent. Thus, a 1 molar solution of glucose is also a 1 osmolar (1 osmole per liter) solution, but a 1 molar solution of sodium chloride (NaCl) is a 2 osmolar solution, because each NaCl molecule dissociates into two osmotically active ions.

To achieve cellular water balance, animals must maintain the osmolarity of their extracellular fluid within an appropriate range. In addition, they must maintain an appropriate solute composition by saving some substances and eliminating others. To accommodate these needs, most animals have excretory organs.

Excretory organs control extracellular fluid osmolarity by filtration, secretion, and reabsorption

Excretory organs control the osmolarity and the volume of the extracellular fluids (blood and interstitial fluid) by excreting solutes that are present in excess (such as NaCl when we eat lots of salty food) and conserving solutes that are valuable or in short supply (such as glucose and amino acids). In terrestrial organisms, these excretory organs also eliminate the waste products of nitrogen metabolism. The output of the excretory organs is called **urine**.

Certain basic principles apply to all animal excretory organs. In some way, the excretory organ filters extracellular fluid to produce a filtrate that contains no cells or large molecules, such as proteins. The composition of this filtrate is then modified to produce urine. In animals with closed circulatory systems, the blood plasma is filtered across the walls of capillaries. The filtration is driven by blood pressure. Water and small molecular weight solutes cross the capillary wall, but large molecular weight solutes and cells remain in the blood. The filtrate (water and small molecules) then flows through tubules. The cells of the tubules change the composition of the filtrate by active secretion and reabsorption of specific solute molecules. We will see that these three mechanisms—filtration, secretion, and reabsorption—are used in the excretory systems of a wide variety of animals. In all excretory systems *there is no active transport of water*. Water must be moved either by a pressure difference or by a difference in osmolarity.

Animals can be osmoconformers or osmoregulators

Animals that live in marine, freshwater, or terrestrial environments face different salt and water balance problems. In the terrestrial environment salts and water can be scarce and usually must be conserved by excretory systems. In the freshwater environment water is plentiful, but salts are scarce. Freshwater animals have to conserve salts and excrete the water that continuously invades their bodies through osmosis. The osmolarity of ocean water is high—over 1,000 milliosmoles/liter. Most marine invertebrates equilibrate their extracellular fluid osmolarity with the ocean water and are therefore called **osmoconformers**. Other marine animals maintain extracellular fluid osmolarities much lower than seawater and are therefore called **osmoregulators**. All marine vertebrates except for sharks and rays are osmoregulators and maintain their extracellular fluids at about 300 mosm/l like humans do.

There are limits to osmoconformity. Even animals that can osmoconform over a wide range of osmolarities must osmoregulate in extreme environments. For example, no animal could survive if its extracellular fluid had the osmolarity of fresh water, because that would mean there were too few solutes in the extracellular fluid, including nutrients and ions necessary for cell functions. Nor could animals survive with internal osmolarities as high as those that may be reached in an evaporating tide pool. High solute concentrations can cause proteins to denature. A case in point is the brine shrimp *Artemia* (**Figure 51.1**), which can live in environments of almost any osmolarity. *Artemia* are found in huge numbers in the most salty environments known, such as Utah's Great Salt Lake, and in coastal evaporation ponds where salt is concentrated for commercial purposes (see Figure 26.24). The osmolarity of such water reaches 2,500 mosm/l. At these high environmental osmolarities, *Artemia* maintains its tissue fluid osmolarity considerably below that of the environment. Its mechanism of osmoregulation is the active transport of Cl^- from its extracellular fluid out across its gill membranes to the environment. Na^+ ions follow.

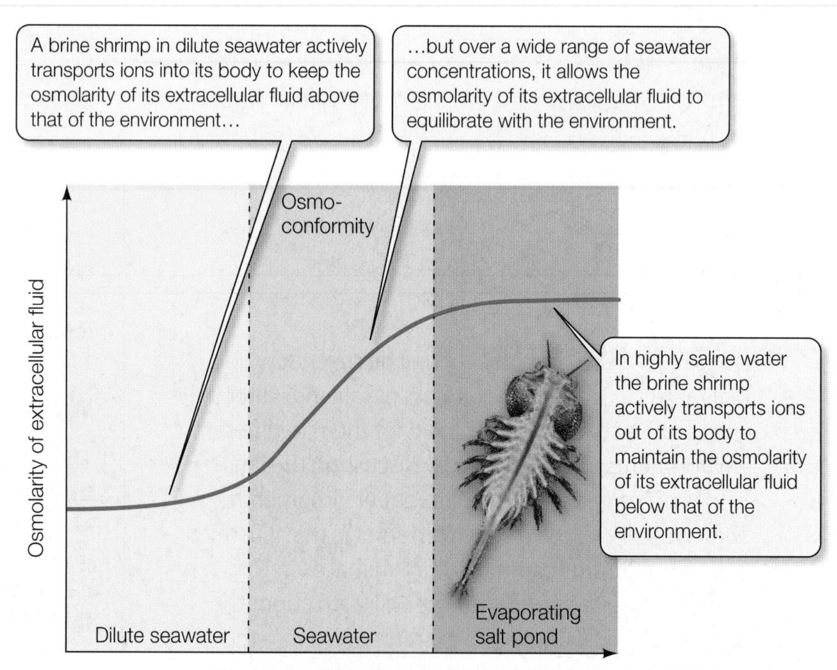

51.1 Environments Can Vary Greatly in Salt Concentration Animals such as the brine shrimp that live at the extremes of environmental osmolarities display flexible osmoregulatory abilities. When they encounter dilute seawater they can regulate their extracellular fluid osmolarity above that of the environment, and when they encounter extreme salt concentrations they can regulate their extracellular fluid osmolarity below that of the environment.

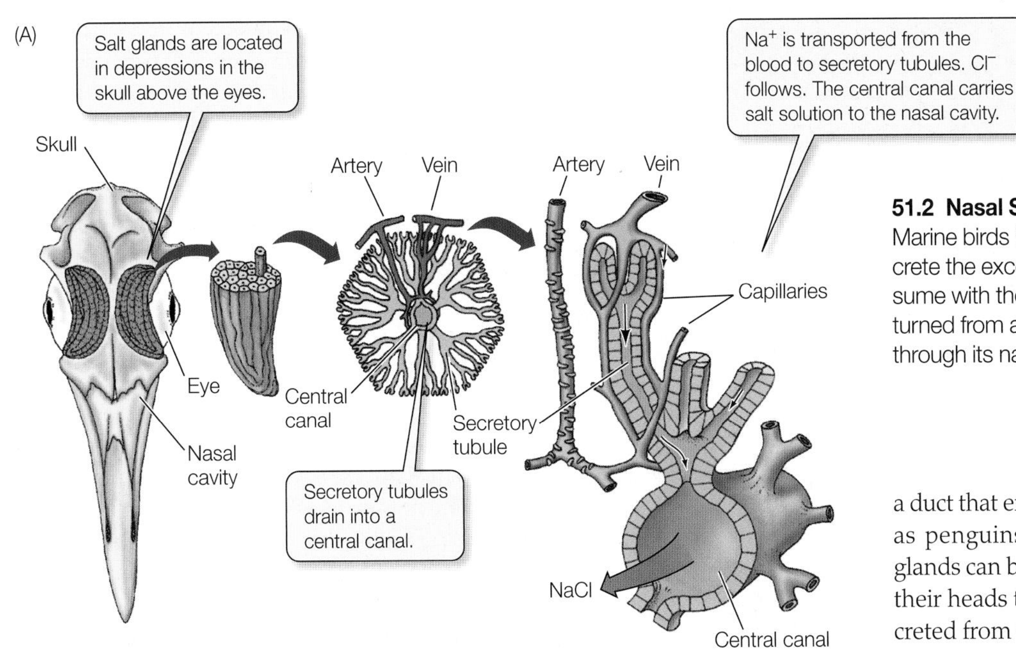

(B) *Macronectes giganteus*

51.2 Nasal Salt Glands Excrete Excess Salt (A) Marine birds have nasal salt glands adapted to excrete the excess salt from the seawater they consume with their food. (B) This giant petrel has returned from a feeding trip at sea and is secreting salt through its nasal salt gland.

Artemia cannot survive in fresh water, but it can live in dilute seawater, in which it maintains the osmolarity of its extracellular fluid above that of the environment. Under these conditions, *Artemia* reverses the direction of Cl^- transport across its gill membranes.

Animals can be ionic conformers or ionic regulators

Osmoconformers can also be ionic conformers, allowing the ionic composition, as well as the osmolarity, of their extracellular fluid to match that of the environment. Most osmoconformers, however, are ionic regulators to some degree; they employ active transport mechanisms to excrete some ions and to maintain other ions in their extracellular fluid at optimal concentrations.

Terrestrial animals obtain their salts from food and regulate the ionic composition of their extracellular fluids by conserving some ions and excreting others. For example, herbivores have to conserve Na^+ because the plants they eat have low concentrations of Na^+. Some terrestrial herbivores travel long distances to naturally occurring salt licks. By contrast, birds that feed on marine animals must excrete the excess of sodium they ingest with their food. Their *nasal salt glands* excrete a concentrated solution of NaCl via a duct that empties into the nasal cavity. Birds, such as penguins and seagulls, that have nasal salt glands can be seen frequently sneezing or shaking their heads to get rid of the very salty droplets excreted from their nasal salt glands (**Figure 51.2**).

51.1 RECAP

Excretory organs control water and salt balance and the excretion of nitrogenous waste products through three mechanisms: filtration of body fluids to form urine, active secretion of substances into the urine, and active reabsorption of substances from the urine.

- Can you explain the two mechanisms used to move water across membranes? See pp. 1093–1094
- Can you describe different salt and water balance problems that animals might encounter in marine, freshwater, and terrestrial environments and some ways they meet those challenges? See p. 1094 and Figures 51.1 and 51.2

51.2 How Do Animals Excrete Toxic Wastes from Nitrogen Metabolism?

In addition to maintaining salt and water balance, animals must eliminate the waste products of metabolism from their extracellular fluids. The end products of the metabolism of carbohydrates and fats are water and carbon dioxide, which are not difficult to eliminate. Proteins and nucleic acids, however, contain nitrogen, so their metabolism produces *nitrogenous wastes* in addition to water and carbon dioxide.

Animals excrete nitrogen in a number of forms

The most common nitrogenous waste is **ammonia** (NH_3). Because it is highly toxic, ammonia must either be excreted continuously to prevent its accumulation, or it must be detoxified by conversion into other molecules (**Figure 51.3**).

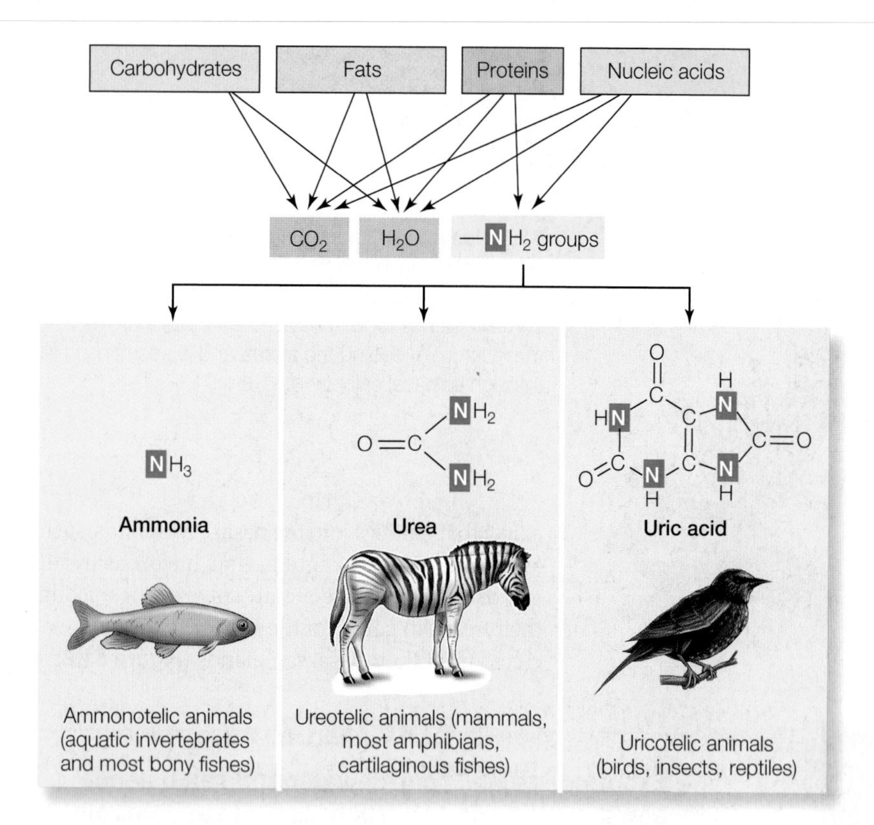

51.3 Waste Products of Metabolism The metabolism of proteins and nucleic acids produces nitrogenous wastes. Most aquatic animals, including most fishes, excrete nitrogenous wastes as ammonia. Most terrestrial animals excrete either urea or uric acid. Urea is more soluble in water and is the major nitrogenous excretory product for mammals, amphibians, and some fishes. Uric acid is not very soluble in water and is the major nitrogenous excretory product for birds, reptiles, and insects.

AMMONIA Ammonia is highly soluble in water and diffuses rapidly, so the continuous excretion of ammonia is relatively simple for aquatic animals. Animals that breathe water continuously lose ammonia from their blood to the environment by diffusion across their gill membranes. Animals that excrete ammonia, such as aquatic invertebrates and bony fishes, are **ammonotelic**.

If ammonia builds up in the extracellular fluids, it becomes toxic at rather low levels and is a dangerous metabolite for terrestrial animals, which cannot continuously excrete ammonia the way aquatic animals do. Therefore, terrestrial (and some aquatic) animals convert ammonia into either **urea** or **uric acid**.

UREA **Ureotelic** animals, such as mammals, amphibians, and cartilaginous fishes (sharks and rays), excrete urea as their principal nitrogenous waste product. Urea is quite soluble in water, but excretion of urea solutions can result in a large loss of water that many terrestrial animals can ill afford. As we will see later in this chapter, mammals have evolved excretory systems that conserve water by producing concentrated urea solutions. The sharks and rays are another story. These marine species keep their extracellular fluids almost isosmotic (same osmotic concentration) to the marine environment by retaining high concentrations of urea.

URIC ACID Animals that conserve water by excreting nitrogenous wastes mostly as uric acid are **uricotelic**. Insects, reptiles, birds, and some amphibians are uricotelic. Uric acid is very insoluble in water, so it can precipitate out of the urine and be excreted as a semisolid (for example, it is the whitish material in bird droppings). A uricotelic animal loses very little water as it disposes of its nitrogenous wastes.

Most species produce more than one nitrogenous waste

Humans are ureotelic, but we also excrete uric acid. The uric acid in human urine comes largely from the metabolism of nucleic acids and caffeine. Humans can also excrete ammonia, which is an important mechanism for regulating the pH of the extracellular fluids. Excreted ammonia buffers the urine and enables the secretion of more hydrogen ions.

Species that live in different habitats at different developmental stages may utilize more than one mechanism of nitrogen excretion. The tadpoles of frogs and toads, for example, excrete ammonia across their gill membranes, but when they develop into adult frogs or toads, they generally excrete urea. Some adult amphibians that live in arid habitats excrete uric acid.

If you wake up at night with a sharp pain in your big toe, you might have the malady of kings—gout. Known for thousands of years, gout was most commonly seen in the overindulgent wealthy. A protein-rich diet coupled with too much alcohol, especially beer, raises blood uric acid levels and causes uric acid crystals to form in joints. (The big toe is especially vulnerable.)

51.2 RECAP

Ammonia is a common toxic waste product of nitrogen metabolism. Aquatic animals excrete ammonia by diffusion into the water. Terrestrial animals detoxify it by conversion to urea or uric acid.

- Why might you expect a species from an arid habitat to use uric acid as its primary nitrogenous waste product? See p. 1096

The challenges animals face with respect to homeostasis of salt and water balance differ with the environment in which they live. Therefore, we can expect to find in animals a variety of adaptations by which they regulate this aspect of their internal environments. However, as we will see in the next section of this chapter, these adaptations are based on a limited number of mechanisms—namely, tubular processing of the extracellular fluids to conserve some solutes and excrete others.

51.3 How Do Invertebrate Excretory Systems Work?

Freshwater and terrestrial invertebrates have a wide variety of adaptations for maintaining salt and water balance and excreting nitrogen. In this section, we will explore three examples of invertebrate excretory systems.

The protonephridia of flatworms excrete water and conserve salts

Many flatworms, such as *Planaria*, live in fresh water. These animals excrete water through an elaborate network of tubules running throughout their bodies. The tubules end in *flame cells*, so called because each cell has a tuft of cilia projecting into the tubule. The beating of the cilia gives the appearance of a flickering flame (**Figure 51.4**). A flame cell and a tubule together form a **protonephridium** (plural protonephridia; from the Greek *proto*, "before," and *nephros*, "kidney").

Extracellular fluid enters the tubules by filtration. The beating of the cilia causes a slight negative pressure within the tubule, and in response to this pressure difference, extracellular fluid flows into the tubule and flows toward the animal's excretory pore. As it flows, the cells of the tubules modify the composition of the fluid. As the modified tubule fluid (urine) leaves the flatworm's body, it is less concentrated than the animal's extracellular fluid, so ions are conserved and water is excreted by the protonephridium.

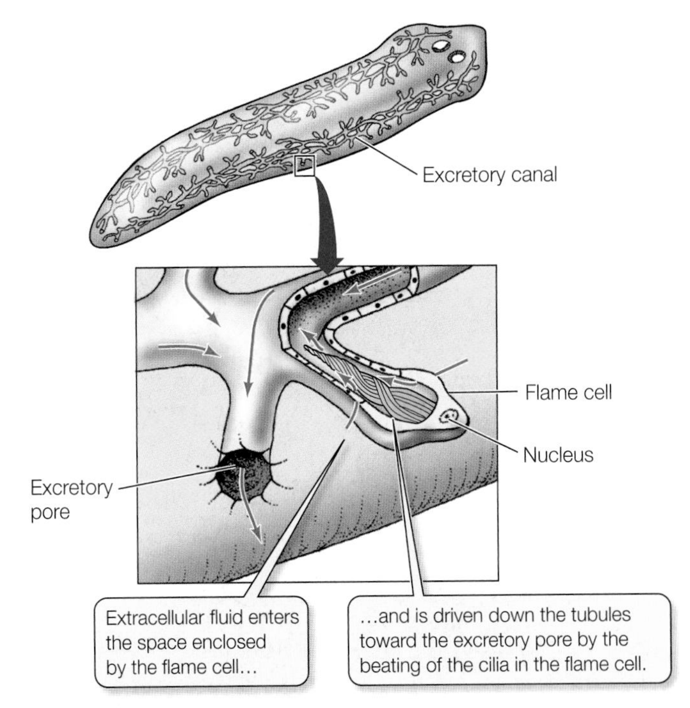

51.4 Protonephridia in Flatworms The protonephridia of the flatworm *Planaria* consist of tubules ending in flame cells. The tubule cells modify the composition of the fluid passing through them.

The metanephridia of annelids process coelomic fluid

Filtration of body fluids and modification of urine by tubules are highly developed processes in annelid worms, such as the earthworm. Recall that annelids are segmented and have a fluid-filled body cavity, called a *coelom*, in each segment (see Figure 32.12). Annelids have a closed circulatory system through which blood is pumped under pressure. The pressure causes the blood to be filtered across the thin, permeable capillary walls into the coelom. The cells and large protein molecules of the blood stay behind in the capillaries, while water and small molecules leave them and enter the coelom. In addition, some waste products, such as ammonia, diffuse directly from the tissues into the coelom. But where does this coelomic fluid go?

Each segment of the earthworm contains a pair of **metanephridia** (singular metanephridium; from the Greek *meta*, "akin to," and *nephros*, "kidney"). Each metanephridium begins in one segment as a ciliated, funnel-like opening in the coelom, called a *nephrostome*, which leads into a tubule in the next segment. The tubule ends in a pore, called the *nephridiopore*, that opens to the outside of the animal (**Figure 51.5**). Coelomic fluid enters the metanephridia through the nephrostomes. As the fluid passes through the tubules, their cells actively reabsorb certain molecules from it and actively secrete other molecules into it. What leaves the animal through the nephridiopores is a dilute urine containing nitrogenous wastes and other solutes.

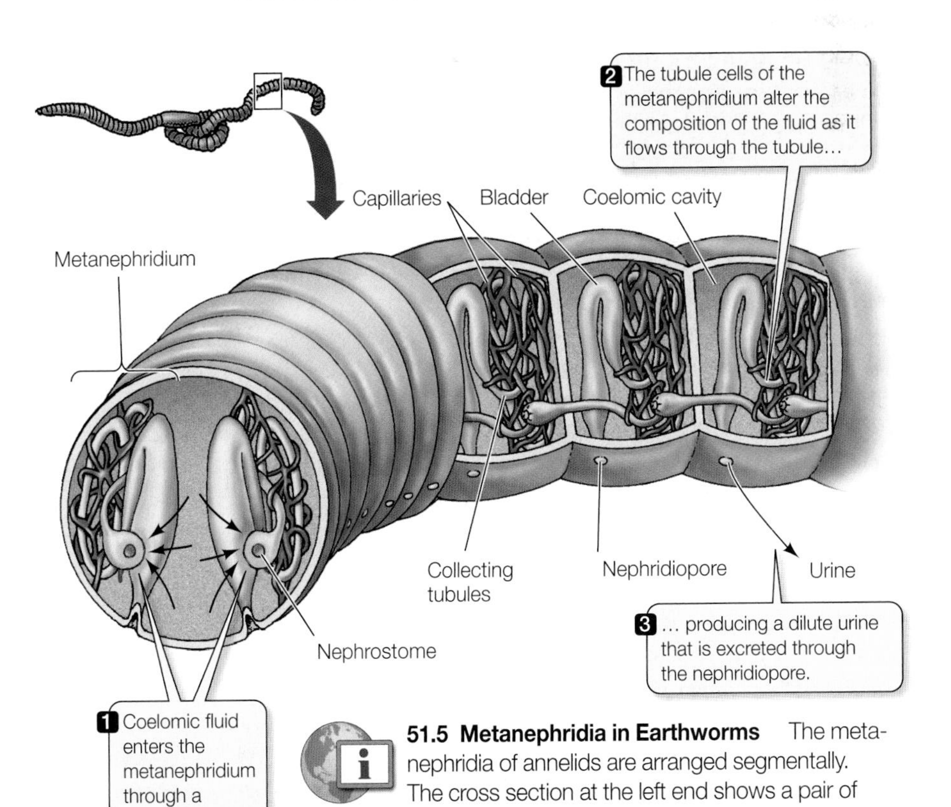

51.5 Metanephridia in Earthworms The metanephridia of annelids are arranged segmentally. The cross section at the left end shows a pair of metanephridia. Three longitudinal sections (right) show only one metanephridium of the two in each segment.

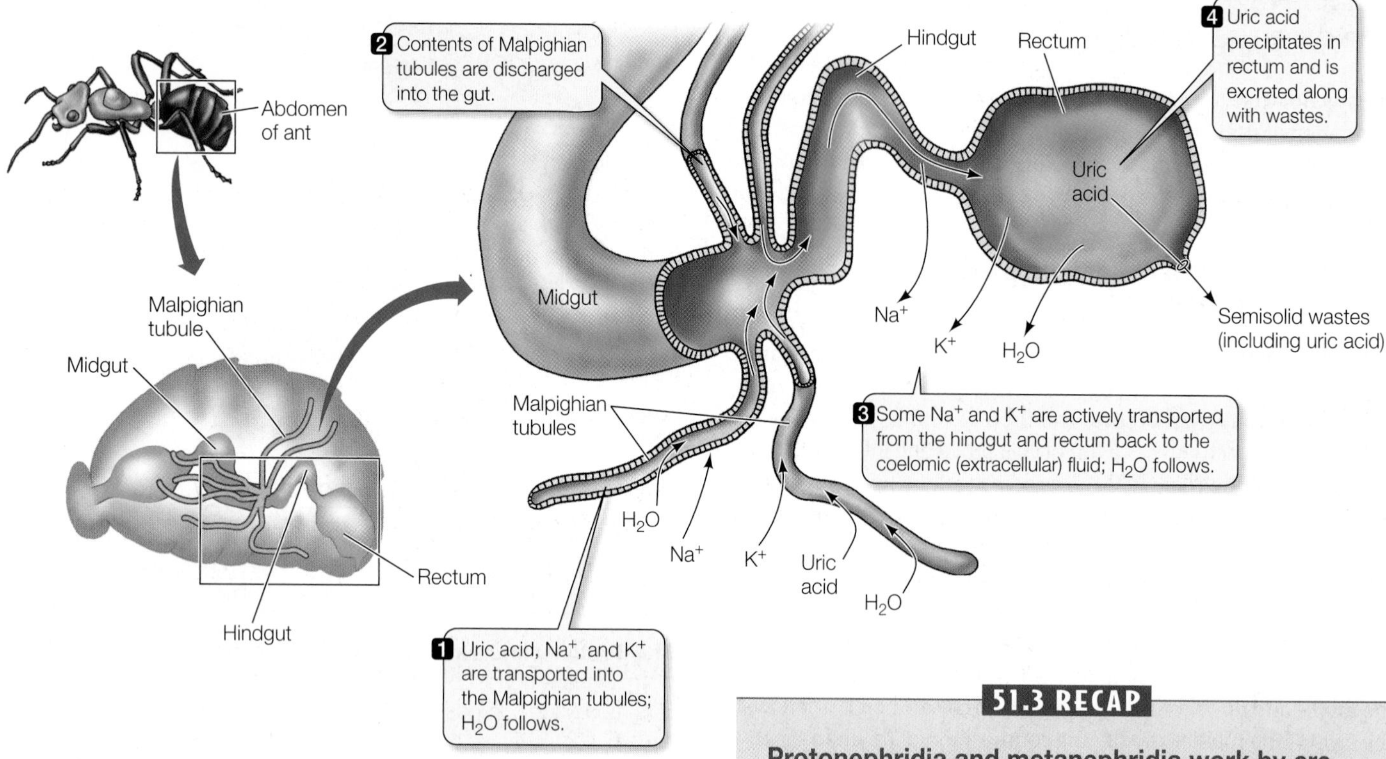

51.6 Malpighian Tubules in Insects The blind, thin-walled Malpighian tubules are attached to the junction of the insect's midgut and hindgut and project into the spaces containing extracellular fluid.

The Malpighian tubules of insects depend on active transport

Insects can excrete nitrogenous wastes with very little loss of water and can therefore live in the driest habitats on Earth. The insect excretory system consists of **Malpighian tubules**. An individual insect has from two to more than a hundred of these blind-ended tubules that open into the gut between the midgut and hindgut (**Figure 51.6**).

Insects have open circulatory systems and therefore cannot use a pressure difference to filter extracellular fluids into the Malpighian tubules. Instead, the cells of the tubules actively transport uric acid, potassium ions, and sodium ions from the extracellular fluid into the tubules. As these solutes are secreted into the tubules, water follows osmotically and flushes the tubule contents toward the gut.

The composition of the tubule fluid changes while it is in the hindgut and rectum. The epithelial cells of the hindgut and rectum actively transport sodium and potassium ions from the gut contents back into the extracellular fluid. This local transport of salts creates an osmotic gradient that pulls water out of the rectal contents. As the uric acid concentration increases, it precipitates and frees more water to be reabsorbed. Remaining in the rectum are crystals of uric acid mixed with other wastes; this semi-dry matter is what the insect eliminates. The Malpighian tubule system is a highly effective mechanism for excreting nitrogenous wastes and some salts without giving up much water.

51.3 RECAP

Protonephridia and metanephridia work by creating a filtrate of the body fluids that is modified by the secretion and reabsorption of specific substances before being excreted. Insect Malpighian tubules actively secrete uric acid and other solutes into closed tubules.

- Do you understand how Malpighian tubules function, and how this system makes it possible for some insect species to survive in the desert? See p. 1098 and Figure 51.6

The survival of animals living in environments with different salt and water combinations depends on their ability to maintain a relatively constant internal environment. Next we will consider how the basic unit of the vertebrate excretory system has evolved to be able to respond to a variety of salt and water balance challenges.

51.4 How Do Vertebrates Maintain Salt and Water Balance?

The main excretory organ of vertebrates is the **kidney**, and the functional unit of the kidney is the **nephron**. The nephron is a structure made up of a long tubule and associated blood vessels that filters the blood and modifies the composition of the filtrate to produce urine. Nephrons can filter large volumes of blood and achieve bulk reabsorption of salts and other valuable molecules such as glucose, making the vertebrate kidney well adapted for the excretion of excess water. If the ancestors of vertebrates evolved in fresh water, as paleontologists propose, the excretory systems of vertebrates would have evolved to excrete excess water. How then have vertebrates adapted to environments where water must be conserved and salts excreted? The answer to this question differs among vertebrate groups. Even among the marine fishes, the excretory adaptations of the bony fishes are different from those of the cartilaginous fishes. The ultimate kidney adaptations for water conservation, however, are found only in birds and mammals.

Marine fishes must conserve water

Marine bony fishes osmoregulate their extracellular fluids to maintain them at one-third to one-half the osmolarity of seawater. Since their only source of water is the sea around them, they must therefore conserve water and excrete excess salts. Marine bony fish cannot produce urine that is more concentrated than their extracellular fluid, so they minimize water loss by producing very little urine. In contrast, fresh water fish produce lots of dilute urine.

If marine bony fish cannot excrete excess salt in their urine, how do they deal with the large salt loads that they ingest with food? Marine bony fish do not absorb from their guts some of the ions they take in, especially divalent ions such as Mg^{2+} or SO_4^{2-}. NaCl, the major salt ingested, is excreted across the gill membranes.

Cartilaginous fishes (sharks and rays) are osmoconformers, but not ionic conformers. They convert nitrogenous wastes to urea and another compound called trimethylamine oxide, and they retain large amounts of these compounds in their extracellular fluids. As a result, their extracellular fluids have an osmolarity close to that of seawater, so they do not lose body water to the environment by osmosis, and may actually gain water. These species have adapted to a concentration of urea in the body fluids that would be toxic to other vertebrates. Sharks and rays still have the problem of excreting the large amounts of salts they take in with their food. They solve this problem by having a gland in the rectum that actively secretes NaCl by a mechanism similar to that of the nasal salt gland of sea birds.

Terrestrial amphibians and reptiles must avoid desiccation

Most amphibians live in or near fresh water, and they stay in humid habitats when they do venture from the water. Like freshwater fishes, most amphibian species produce large amounts of dilute urine and conserve salts. Some amphibians, however, have adapted to habitats that require water conservation.

Amphibians living in very dry terrestrial environments have reduced the permeability of their skin to water. Some secrete a waxy substance over the skin to waterproof it. Several species of frogs that live in arid regions of Australia burrow deep into the ground and remain there during long dry periods. They enter *estivation*, a state of very low metabolic activity and therefore low water turnover. When it rains, the frogs come out of estivation, feed, and reproduce. Their most interesting adaptation is an enormous urinary bladder. Before entering estivation, they fill the bladder with dilute urine, which can amount to one-third of their body weight. This dilute urine serves as a water reservoir that is gradually reabsorbed into the blood during the long period of estivation.

Reptiles occupy habitats ranging from aquatic to extremely hot and dry. In fact, snakes and lizards are among the most prominent members of many desert faunas (including an extensive list of venomous species that inhabit the great deserts of Western Australia).

When Australian aboriginal people need an emergency source of drinking water, they dig up estivating frogs and use the urine in their bladders.

Three major adaptations have freed reptiles from the close association with water that is necessary for most amphibians (see Section 33.4). First, reptiles are *amniotes* that do not need fresh water to reproduce because they employ internal fertilization and lay eggs with shells that retard evaporative water loss. Second, they have a dry, scale-covered epidermis (skin) that retards evaporative water loss. Third, they excrete nitrogenous wastes as uric acid semisolids, losing little water in the process.

Birds and mammals can produce highly concentrated urine

Birds and mammals occupy widely diverse habitats, many of which present special excretory system challenges. The most challenging environments are those in which water is severely limited. Birds and mammals both have adaptations seen in other vertebrate groups that enable them to conserve water. They have skins and surface coverings that impede water loss. Like reptiles, they are amniotes, and thus free from dependence on an aqueous environment for reproduction. Birds have shelled eggs, as reptiles do, whereas mammalian embryos develop internally in the mother (see Section 43.4). Like reptiles, birds produce uric acid as their nitrogenous waste product.

Both mammals and, to a lesser extent birds have a unique ability: they can produce urine that is more concentrated than their blood. They can concentrate their urine because of adaptations of their kidneys that enable a solute concentration gradient to develop that can be used to osmotically reabsorb water from the urine. We will focus on mammals to describe how this important adaptation works, but to do so we first have to understand the structure and function of the vertebrate nephron.

The nephron is the functional unit of the vertebrate kidney

Urine formation in vertebrate nephrons involves three main processes (**Figure 51.7**):

- *Filtration:* The blood is filtered in a dense ball of capillaries called the **glomerulus** (plural glomeruli), which is highly permeable to selected substances but impermeable to large molecules.
- *Tubular reabsorption:* The glomerular filtrate flows into the **renal tubule**. Cells lining the renal tubule modify the glomerular filtrate by reabsorption of specific ions, nutrients, and water, returning these to the blood and leaving behind and concentrating excess ions and waste products such as urea.
- *Tubular secretion:* The glomerular filtrate in the renal tubule is further modified by tubule cells transporting substances into the tubular contents. These are substances that the body needs to excrete.

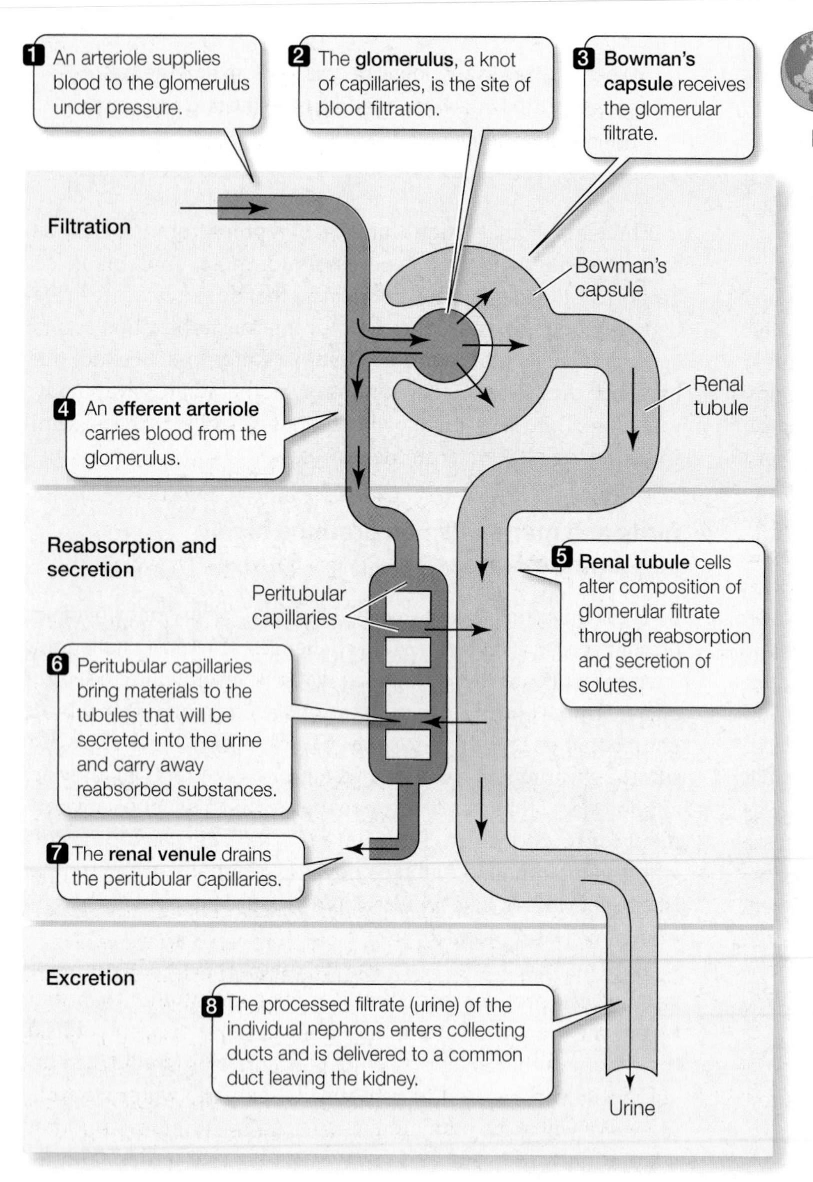

51.7 The Vertebrate Nephron The vertebrate nephron consists of a renal tubule closely associated with two capillary beds, the glomerulus and the peritubular capillaries.

The **peritubular capillaries** are vascular structures closely associated with the renal tubules. These vessels transport substances to and from the renal tubules. By carrying away absorbed substances and delivering substances to be secreted into the urine, the peritubular capillaries work in tandem with the glomerular capillaries.

Blood is filtered in the glomerulus

As seen in Figure 51.7, the vascular arrangement of the nephron is unusual in that the two capillary beds—the glomerulus and the peritubular capillaries—lie in series between the arteriole that supplies them and the venule that drains them. Blood enters the glomerulus through an **afferent arteriole** and exits through an **efferent arteriole** (**Figure 51.8A**). The efferent arteriole gives rise to the peritubular capillaries, which surround the tubule component of the nephron and serve as sites of exchange between the filtrate in the renal tubules and the extracellular fluid.

The tubule component of the nephron—the renal tubule—begins with **Bowman's capsule**, which encloses

51.8 A Tour of the Nephron Scanning electron micrographs illustrate the anatomical basis for blood filtration by the kidneys. (A) In a preparation showing only the blood vessels (tubular tissue has been digested away), the glomeruli appear as balls of capillaries surrounded by the more diffuse peritubular capillaries. (B) Higher magnification of a glomerulus with the tubule cells intact shows the podocytes that wrap around the glomerular capillaries. (C) This cross section of an intact glomerulus shows the tubule cells that form Bowman's capsule.

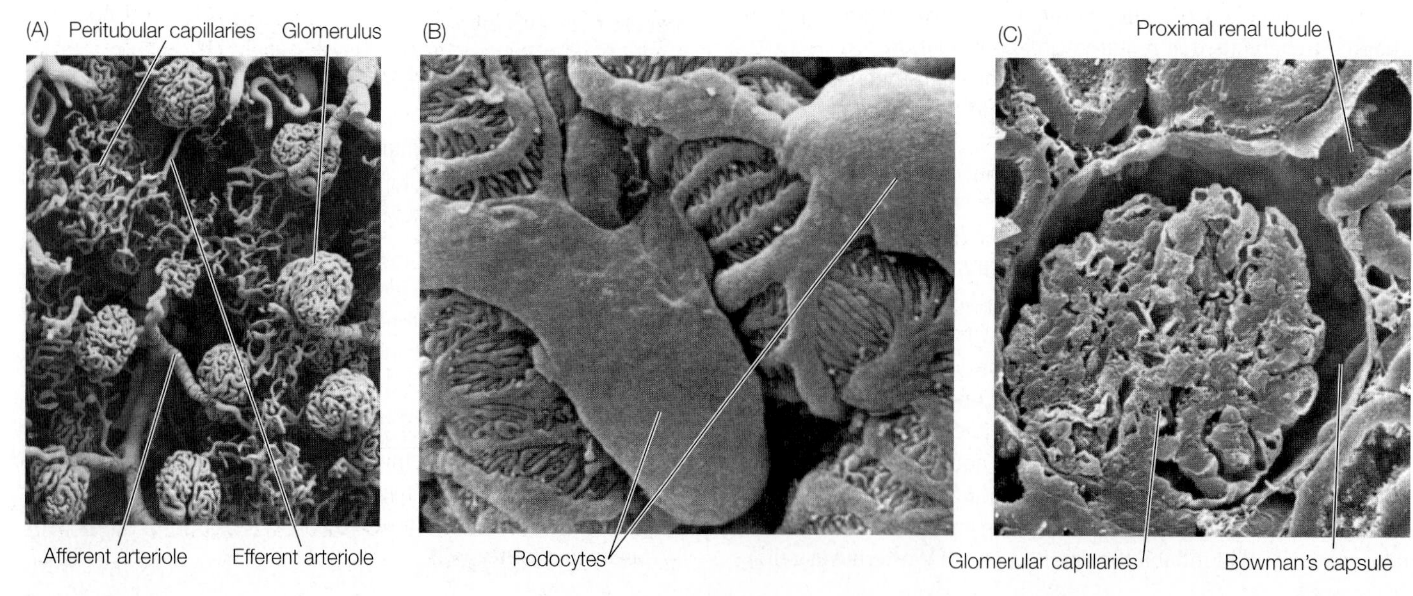

the glomerulus. The glomerulus appears to be pushed into Bowman's capsule much like a fist pushed into an inflated balloon. The cells of the capsule that come into direct contact with the glomerular capillaries are called **podocytes** (**Figure 51.8B**). These highly specialized cells have numerous arm-like extensions, each with hundreds of fine, finger-like processes. The podocytes wrap around the capillaries so that their processes interdigitate and cover the capillaries extensively.

The glomerulus filters the blood to produce a fluid (the renal filtrate) that lacks cells and large molecules. The walls of the capillaries, the basal lamina of the capillary endothelium, and the podocytes of Bowman's capsule all participate in filtration. *Fenestrations* in the walls of the capillaries (see Section 49.4) allow water and many solute molecules, but not red blood cells, to pass through. The meshwork of the basal lamina and the spaces between the processes of the podocytes are even finer and prevent large molecules from leaving the capillaries (**Figure 51.8C**). The arterial pressure of the blood entering the permeable capillaries causes the filtration of water and small molecules in the glomerulus. The glomerular filtration rate is high because blood pressure in the glomerular capillaries is unusually high, and because the capillaries of the glomerulus, along with their covering of podocytes, are more permeable than other capillary beds in the body.

The renal tubules convert glomerular filtrate to urine

The composition of the filtrate that first enters the nephron is similar to that of the blood plasma, with the exception of high-molecular-weight solutes such as proteins. Reabsorption and secretion cause the composition of this fluid to change as it passes down the renal tubule. Cells of the tubule actively reabsorb certain molecules from the tubule fluid (which are returned to the blood flowing through the peritubular capillaries) and actively secrete into the tubule fluid substances delivered by the peritubular capillaries. The urine that eventually exits the body is very different from the original filtrate due to the actions of the renal tubule cells.

51.4 RECAP

The kidney is the major excretory organ of vertebrates. Its functional unit is the nephron, which includes a glomerulus that filters blood and a renal tubule that secretes and reabsorbs solutes, modifying the filtrate to produce urine. The nephron is a mechanism for excreting excess water while conserving valuable solutes.

- Can you explain the difference in osmoregulatory adaptations between marine bony and cartilaginous fishes? See p. 1099
- Can you describe the functional relationships between the glomerular and peritubular capillaries? See pp. 1099–1100 and Figure 51.7
- Can you describe how the blood is filtered in the glomerulus? See p. 1100

The adaptations of mammals and birds that allow them to produce urine that is more concentrated than their extracellular fluids constitute an important step in vertebrate evolution. We now consider the mammalian kidney in more detail.

51.5 How Does the Mammalian Kidney Produce Concentrated Urine?

Mammals and birds have high body temperatures and high metabolic rates, and therefore have the potential for a high rate of water loss. Having an excretory system that minimizes water loss made it possible for these highly active species to occupy arid habitats.

Kidneys produce urine and the bladder stores it

We will use the excretory system of humans as our example of the mammalian excretory system. Humans have two kidneys at the back of the upper region of the abdominal cavity (**Figure 51.9A**). Each kidney filters blood, processes the filtrate into urine, and releases that urine into a duct called the **ureter**. The ureter of each kidney leads to the **urinary bladder**, where the urine is stored until it is excreted through the **urethra**, a short tube that opens to the outside of the body.

Two sphincter muscles surrounding the base of the urethra control the timing of urination. One of these sphincters is a smooth muscle and is controlled by the autonomic nervous system. As the bladder fills, stretch receptors in the walls of the bladder trigger a spinal reflex that relaxes this sphincter. This reflex is the only control of urination in infants. The other sphincter is a skeletal muscle and is controlled by the voluntary nervous system. When the bladder is *very* full, only deliberate conscious effort prevents urination. Infant toilet training involves their learning to control this sphincter.

Nephrons have a regular arrangement in the kidney

The kidney is shaped like a kidney bean; when sliced along its long axis, its key anatomical features are revealed (**Figure 51.9B**). The ureter and the **renal artery** and **renal vein** enter the kidney on its concave (punched-in) side. The ureter extends into the kidney in several branches, the ends of which envelop kidney tissues called **renal pyramids**. The renal pyramids make up the internal core, or **medulla**, of the kidney. The medulla is covered by an outer layer of tissue with a granular appearance, called the **cortex**. In the region between the cortex and the medulla, the renal artery divides into the many arterioles that serve the nephrons. In this same region the renal vein collects the blood from the many venules that drain the peritubular capillaries.

The organization of the nephrons within the kidney is very regular. All of the glomeruli are located in the cortex. The initial segment of a renal tubule is called the **proximal convoluted tubule**—"proximal" because it is the first segment of renal tubule to receive the filtrate from the glomerulus, and "convoluted" because it is twisted (**Figure 51.9C**). All the proximal convoluted tubules are also located in the cortex.

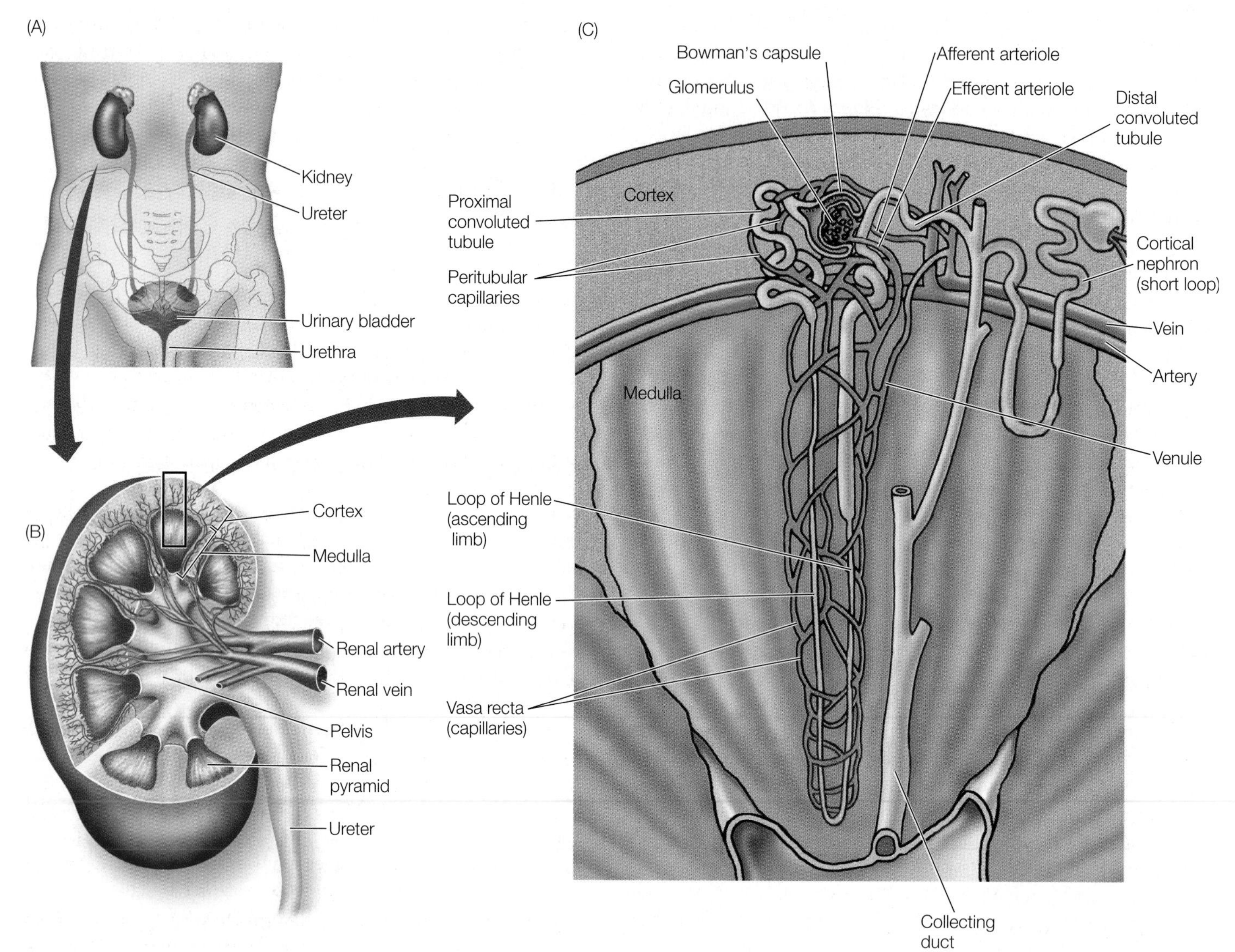

51.9 The Human Excretory System (A) The human kidneys are in the upper dorsal region of the abdominal cavity. (B) The human kidney has a highly organized internal structure that is the basis for its function. (C) The glomeruli and the proximal and distal convoluted tubules are located in the cortex of the kidney, but the loops of Henle run in parallel as straight sections down into the renal medulla and back up to the cortex. Collecting ducts run from the cortex to the inner surface of the medulla, where they open into the ureter.

At the point at which the renal tubule descends into the medulla, it becomes much less convoluted and descends directly down into the medulla. In the medulla the tubule makes a hairpin turn and ascends back to the cortex, forming what is called the **loop of Henle**. Some nephrons have longer loops of Henle than others. Some 20–30 percent of human nephrons that have glomeruli deep in the cortex (i.e., near the border with the medulla) have long loops of Henle that go deep into the medulla. Nephrons that have glomeruli farther up in the cortex generally have short loops of Henle that only descend a short distance into the medulla. As we will see, the long loops are critical to the formation of concentrated urine. The *ascending limb* of the loop of Henle becomes the **distal convoluted tubule** when it reaches the cortex. This section of the renal tubule is called distal because it is farther from the glomerulus. The distal convoluted tubules of many nephrons join a common **collecting duct** in the cortex. The collecting ducts descend back down through the renal pyramid, parallel to and past the tips of the loops of Henle, emptying into the funnel-shaped *pelvis.* Divisions of the pelvis that surround each renal pyramid join together to leave the kidney as the ureter (see Figure 51.9B).

The organization of the blood vessels of the kidney closely parallels the organization of the nephrons (see Figure 51.9C). Smaller and smaller arteries branch from the renal artery and radiate into the cortex. Finally an afferent arteriole carries blood to each glomerulus. Draining each glomerulus is an efferent arteriole that gives rise to the peritubular capillaries, most of which surround the cortical portions of the tubules.

A few peritubular capillaries run into the medulla in parallel with the loops of Henle and the collecting ducts, forming a vascular network called the **vasa recta**. All the peritubular capillaries from a nephron join back together into a venule that joins with venules from other nephrons and eventually leads to the renal vein.

Most of the glomerular filtrate is reabsorbed by the proximal convoluted tubule

Most of the water and solutes filtered out of the glomerulus are reabsorbed and do not appear in the urine. We can reach this conclusion by comparing the rate of filtration by the glomeruli with the rate of urine production. The kidneys receive about 1 liter of blood per minute, or about 1,500 liters of blood per day. How much of this huge volume is filtered out of the glomeruli? The answer is about 12 percent. This is still a large volume—180 liters per day! Since we normally urinate 2–3 liters per day, about 98 percent of the fluid volume that is filtered out of the glomerulus is returned to the blood. Where and how is this enormous fluid volume reabsorbed?

The proximal convoluted tubule (PCT) is responsible for most of the reabsorption of water and solutes from the glomerular filtrate. The cells of this section of the renal tubule have many microvilli that increase their surface area for reabsorption, and they have many mitochondria—an indication that they are metabolically active. PCT cells actively transport Na^+ (with Cl^- following passively) and other solutes, such as glucose and amino acids, out of the tubule fluid. Almost all glucose and amino acid molecules that are filtered from the blood are actively reabsorbed by PCT cells and transported into the extracellular fluid. The active transport of solutes from the tubule into the interstitial fluid causes water to follow osmotically. The water and solutes moved into the interstitial fluid are taken up by the peritubular capillaries and returned to the venous blood.

Despite the large volume of water and solutes reabsorbed by the PCT, the overall osmolarity of the renal filtrate does not change. The fluid that enters the loop of Henle has the same osmolarity as the blood plasma, although its composition is different. How, then, does the kidney produce urine that is more concentrated than the blood plasma?

The loop of Henle creates a concentration gradient in the surrounding tissue

Humans can produce urine that is four times more concentrated than their blood plasma. The vampire bat we encountered at the beginning of this chapter can produce urine that is twenty times more concentrated than its blood plasma. The concentrating ability of the mammalian kidney arises from a **countercurrent multiplier** mechanism made possible by the anatomical arrangement of the loops of Henle. The term "countercurrent" refers to the opposing directions in which the tubule fluid in the descending and ascending limbs flows. The term "multiplier" refers to the ability of this system to create a solute concentration gradient in the renal medulla. The loops of Henle do not themselves produce concentrated urine; rather, they increase the osmolarity of the interstitial fluid in the medulla in a graduated way. The osmolarity of the interstitial fluid of the medulla increases from the top to the bottom of the medulla. More specifically, the interstitial fluid concentration of the medulla increases from about 300 mosm/l (the concentration of blood plasma) at the cortex to about 1,200 mosm/l near the hairpin turn of the loop of Henle (**Figure 51.10**). How do the loops produce this effect?

The segments of the loop of Henle differ anatomically and functionally. Cells of the descending limb and the initial cells of the ascending limb are thin, with no microvilli and few mitochondria. They are not specialized for transport. Partway up the ascending limb, the cells become specialized for active transport. These cells are thick and have many mitochondria. Accordingly, the segments of the loop of Henle are named the *thin descending limb*, the *thin ascending limb*, and the *thick ascending limb*.

The countercurrent multiplier mechanism may be more easily understood by first considering events occurring in the thick ascending limb and their affect on the fluid in the descending limb (see Figure 51.10). The thick ascending limb, with its cells specialized for active transport, actively reabsorbs Na^+ (with Cl^- following passively) from the tubule fluid and moves it into the interstitial fluid. (In the following discussion, it is important to distinguish between the two components of extracellular fluid—the blood plasma and the interstitial fluid.) The thick ascending limb is not permeable to water, so the reabsorption of Na^+ and Cl^- raises the concentration of those solutes in the surrounding interstitial fluid and decreases the concentration of the tubular fluid entering the distal convoluted tubule.

The thin descending limb, in contrast, is highly permeable to water, but not very permeable to Na^+ and Cl^-. Since the local interstitial fluid has been made more concentrated by the Na^+ and Cl^- reabsorbed from the neighboring thick ascending limb, water is withdrawn osmotically from the tubule fluid in the descending limb. Therefore, the fluid in the descending limb becomes more concentrated as it flows toward the hairpin turn at the bottom of the renal medulla.

The thin ascending limb, like the thick ascending limb, is not permeable to water. It is, however, permeable to Na^+ and Cl^-. As the concentrated tubule fluid flows up the thin ascending limb, it is more concentrated than the surrounding interstitial fluid, so Na^+ and Cl^- diffuse out. When the tubule fluid reaches the thick ascending limb, active transport continues to move Na^+ and Cl^- from the tubule fluid to the interstitial fluid.

As a result of the processes described above, the tubule fluid reaching the distal convoluted tubule is less concentrated than the blood plasma, and the solutes that have been left behind in the renal medulla have created a concentration gradient in the interstitial fluid of the medulla. As a result of this countercurrent multiplier mechanism, the interstitial fluid of the renal medulla becomes increasingly concentrated between its border with the cortex and the tips of the renal pyramids.

You may wonder why the blood flow through the medulla does not wash out the concentration gradient established by the loops of Henle. The parallel arrangement of the descending and ascending peritubular capillaries—the vasa recta—in the medulla helps preserve the concentration gradient in the medulla. These capillaries are permeable to salt and water. Therefore, as blood flows down the descending limb of the vasa recta into the increasingly concentrated interstitial fluid of the medulla, it loses water and gains solutes. As the blood flows up from the bottom of the medulla in the ascending limb of the vasa recta, the opposite happens because now the blood is more concentrated than the surrounding interstitial fluid. The dynamics of this countercurrent exchange of salts and water between the blood in the vasa recta and the interstitial fluids result in little change in the composition of the tissue fluid in the medulla, but the excess water that is reabsorbed from the collecting ducts gets returned to the blood.

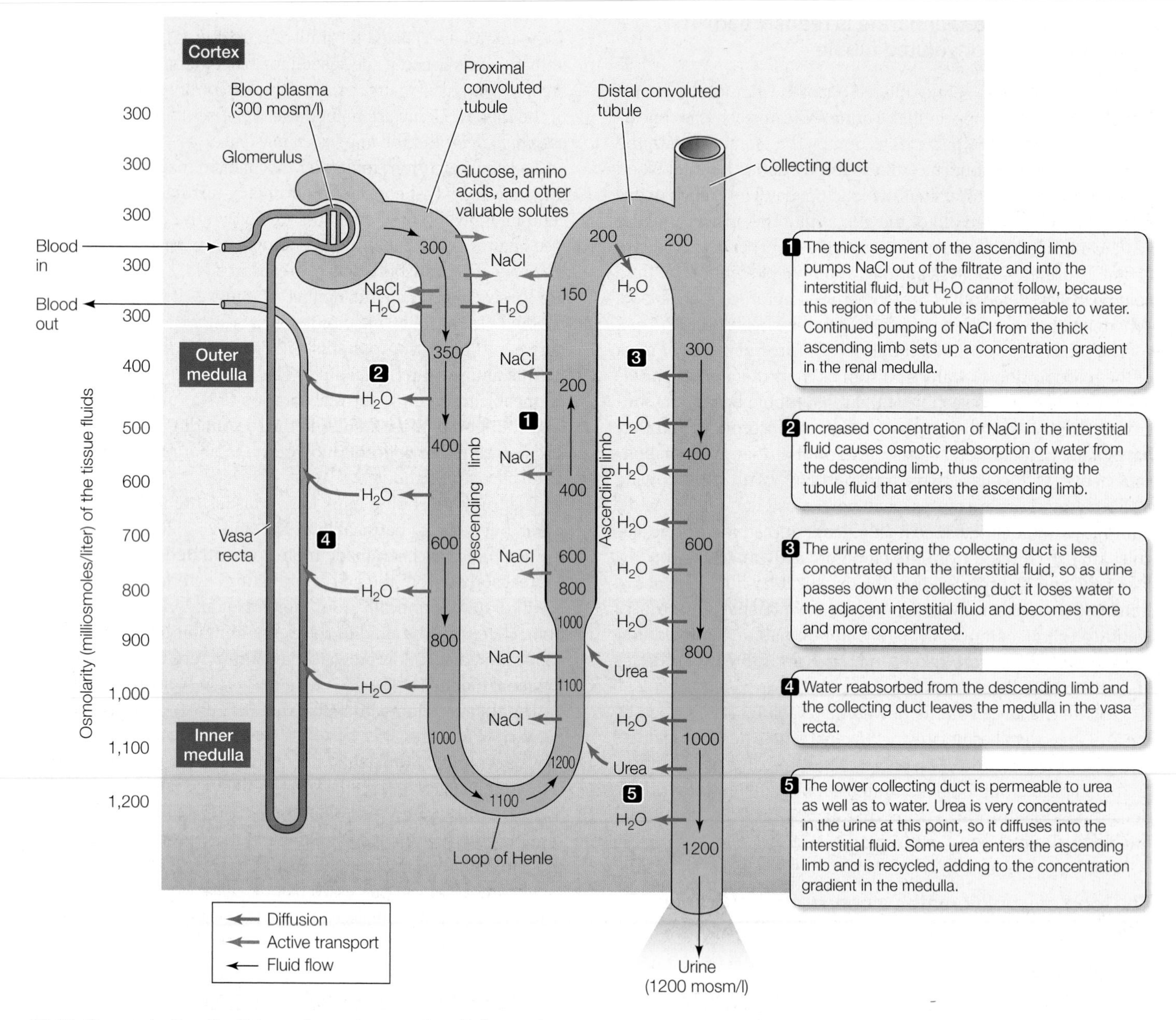

51.10 Concentrating the Urine A countercurrent multiplier mechanism enables the kidney to produce urine that is far more concentrated than mammalian blood plasma.

Water permeability of kidney tubules depends on water channels

We have noted that some tubule regions, such as the PCT, are highly permeable to water while others, such as the thick ascending limb of the loop of Henle, are impermeable to water. How do differences in water permeability in different regions of the nephron arise? Regions of the nephron that are highly permeable to water have greater numbers of *aquaporins*, a class of membrane proteins that form water channels (see Section 5.3). Aquaporins are abundant in kidney PCT cells and in descending limbs of the loops of Henle, but not in the ascending limbs of the loop of Henle. The discovery of aquaporins resulted in Peter Agre of Johns Hopkins University receiving the Nobel Prize in Chemistry in 2003.

Water reabsorption begins in the distal convoluted tubule

The first portion of the distal convoluted tubule has properties similar to the thick ascending limb of the loop of Henle. Na^+ and Cl^- are transported out of the tubule fluid, and water cannot follow. As a result, the tubule fluid becomes even more dilute. The later sections of the distal convoluted tubule, however, can be permeable to water, and water can be osmotically drawn from the tubule into the interstitial fluid. As the tubule fluid flows from the distal tubule to the collecting duct, it can be below or equal to the osmolarity of the blood plasma.

Urine is concentrated in the collecting duct

The tubule fluid entering the collecting duct is at the same solute *concentration* as the blood plasma, but its *composition* is considerably different from that of the plasma. The major solute in the tubular fluid is now urea, since salts were reabsorbed earlier in the nephron. As the tubule fluid flows down the collecting duct, it loses water osmotically to the interstitial fluid, and that water returns to the circulatory system via the vasa recta.

The concentration gradient established in the renal medulla by the loops of Henle creates the osmotic potential that withdraws water from the collecting ducts. The collecting ducts begin in the renal cortex and run through the renal medulla before emptying into the ureter at the tips of the renal pyramids. As the solute concentration of the surrounding interstitial fluid increases, more and more water can be absorbed from the urine in the collecting duct. By the time it reaches the ureter, the urine can become greatly concentrated, with urea as a major solute.

As water is withdrawn from the collecting duct, some urea also leaks out into the medullary interstitial fluid, adding to its osmotic potential. This urea diffuses back into the loop of Henle and is returned to the collecting duct. The recycling of urea in the renal medulla contributes significantly to the ability of the kidney to concentrate the urine in the collecting duct.

Overall, the ability of a mammal to concentrate its urine is determined by the maximum concentration gradient it can establish in its renal medulla. An important adaptation for increasing the concentration gradient is to increase the lengths of the loops of Henle. The desert gerbil, for example, has such extremely long loops of Henle that its renal pyramid (each of its kidneys has only one, in contrast to ours) extends far out of the concave surface of the kidney and into the ureter (**Figure 51.11**). These animals are so effective in conserving water that they can survive just on the water released by the metabolism of their food.

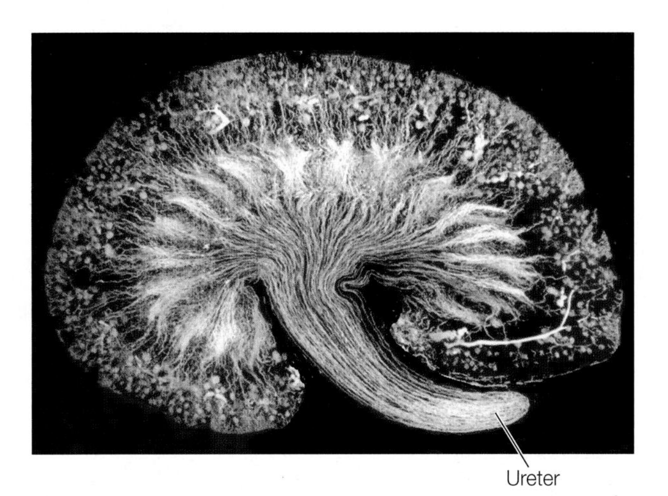

51.11 The Ability to Concentrate The ability of the mammalian kidney to concentrate urine depends on the lengths of its loops of Henle relative to the overall size of the kidney. The kidney of a desert gerbil has a single renal pyramid (the yellow-green area) with loops of Henle so long that the pyramid extends far into the ureter.

Ever wonder how drugs get their names? The prescription drug Lasix® increases water loss by inhibiting sodium reabsorption in the kidney tubules. It is widely used to treat patients with congestive heart failure (a condition that causes them to retain water). The effects of the drug *last* about *six* hours—hence, *Lasix*.

The kidneys help regulate acid–base balance

Besides regulating salt and water balance and excreting nitrogenous wastes, the kidneys have another important role: They regulate the hydrogen ion concentration (the pH) of the blood. Blood pH is a critical variable because it influences the structure, and therefore the function, of proteins.

One way to minimize pH changes in a chemical solution is to add a *buffer*—a substance that can either absorb or release hydrogen ions (Section 2.4). The major buffers in the blood are bicarbonate ions (HCO_3^-; see Figure 48.14) that are formed from the dissociation of carbonic acid, which in turn is formed by the hydration of CO_2 according to the following equilibrium reaction:

$$CO_2 + H_2O \rightleftharpoons H_2CO_3 \rightleftharpoons H^+ + HCO_3^-$$

You can see that if excess hydrogen ions are added to this reaction mixture, the reaction will move to the left and absorb the excess H^+. If hydrogen ions are removed from the reaction mixture, however, the reaction will move to the right and supply more H^+.

The HCO_3^- buffer system is important for controlling the pH of the blood because the reaction can be pushed to the right and pulled to the left physiologically. The lungs control the levels of CO_2 in the blood, thus altering the acid portion of the reaction. CO_2 is considered the acid portion of the reaction because if you add additional CO_2, the reaction shifts to the right, producing more H^+ ions. The kidneys control the base portion of the reaction by removing H^+ from the blood and returning HCO_3^- to the blood. The renal tubules secrete H^+ into the tubule fluid and reabsorb HCO_3^- (**Figure 51.12**). The kidney has other secretory and reabsorption systems that enhances its ability to control blood pH.

Kidney failure is treated with dialysis

Loss of kidney function, or *renal failure*, results in the retention of salts and water (hence high blood pressure), retention of urea (uremic poisoning), and metabolic acids (acidosis). A person who suffers complete renal failure will die within 2 weeks if not treated. A drastic but highly successful treatment is kidney transplant, but it is usually necessary to sustain a patient for considerable time while waiting for a suitable organ to become available. Therefore, artificial kidneys, or renal dialysis machines, are essential modes of treatment.

In a dialysis machine, the patient's blood flows through many small channels made of semipermeable membranes (**Figure 51.13**). A dialysis solution flows on the other side of these membranes, and small molecules can diffuse through these membranes. Molecules and ions diffuse from an area of high concentration to an area of lower concentration, so the composition of the dialysis fluid is crucial. The concentrations of the molecules or ions that need to be conserved must be at the same concentration in the dialysis fluid as they are in the blood. The concentrations of mol-

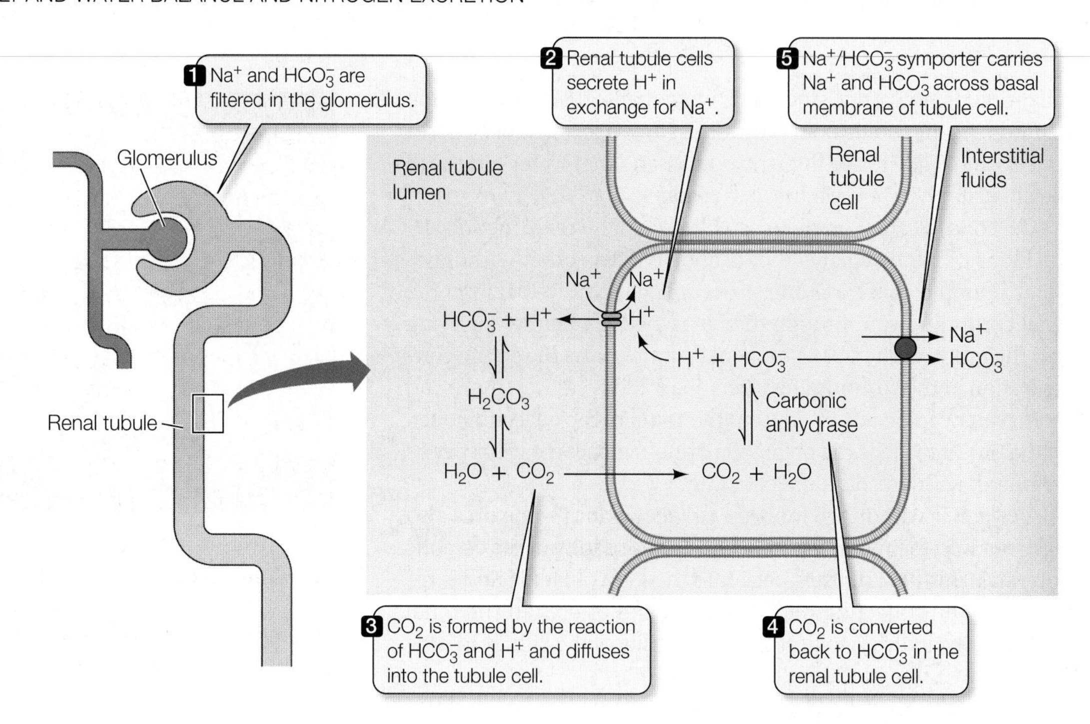

51.12 The Kidney Excretes Acids and Conserves Bases Bicarbonate ions are filtered out at the glomerulus, and renal tubule cells secrete hydrogen ions into the tubule fluid. In the renal tubule, the filtered bicarbonate buffers the secreted hydrogen ions and keeps the urine from becoming too acidic. The CO_2 formed by the reaction of bicarbonate and hydrogen ions is converted back to bicarbonate by the renal tubule cells and transported back into the interstitial fluid.

ecules and ions that need to be removed from the blood are zero in the dialysis fluid. The total osmotic potential of the dialysis fluid must equal that of the plasma.

About 500 ml of the patient's blood is in the dialysis machine at any one time, and the unit processes several hundred milliliters of blood per minute. A patient with no kidney function must be on the dialysis machine for 4–6 hours at least 3 times a week.

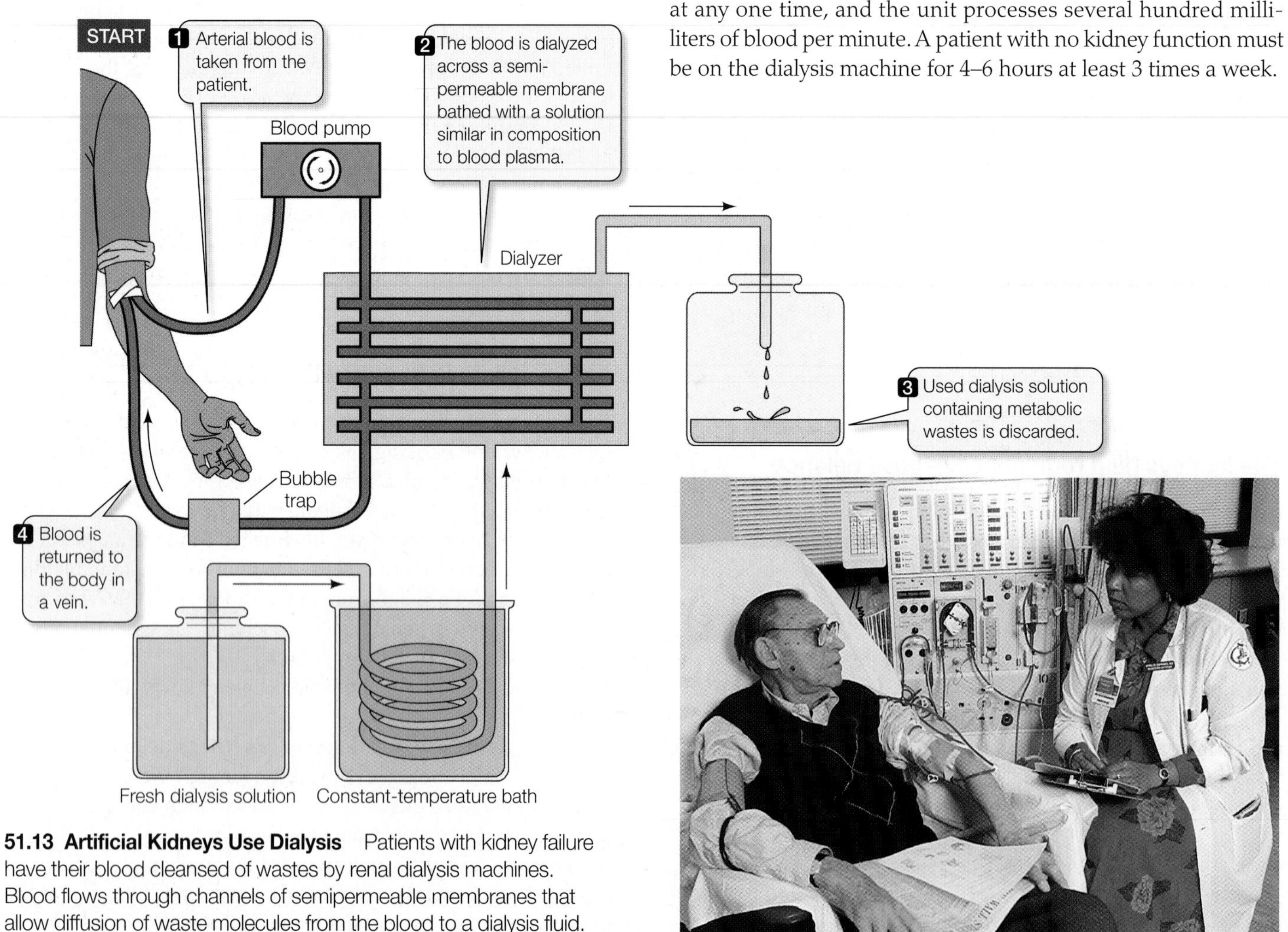

51.13 Artificial Kidneys Use Dialysis Patients with kidney failure have their blood cleansed of wastes by renal dialysis machines. Blood flows through channels of semipermeable membranes that allow diffusion of waste molecules from the blood to a dialysis fluid.

51.5 RECAP

The anatomical organization of the nephrons in the mammalian kidney makes it possible for the kidney to produce a urine more concentrated than the blood. Bulk reabsorption of salts, other valuable solutes, and water takes place in the proximal convoluted tubule. The thick ascending limb of the loop of Henle reabsorbs salt, increasing the osmolarity of the interstitial fluids in the renal medulla. Collecting ducts run through the renal medulla and lose water osmotically to the surrounding interstitial fluids, concentrating the urine.

- What is a countercurrent multiplier mechanism, and how does it work in nephrons to produce concentrated urine? See p. 1103 and Figure 51.10
- How does the kidney contribute to acid–base balance? See p. 1105 and Figure 51.12

The kidney contributes to homeostasis in a number of ways, including maintaining the osmotic concentration and ionic composition of the extracellular fluid and by regulating pH. The kidneys also play a major role in regulating blood pressure.

51.6 What Mechanisms Regulate Kidney Function?

Several regulatory mechanisms act on the kidneys to maintain blood osmolarity and blood pressure. We will discuss these mechanisms separately, but keep in mind that they are always working together.

The kidneys maintain the glomerular filtration rate

If the kidneys stop filtering blood, they cannot accomplish any of their functions. The maintenance of a constant **glomerular filtration rate** (**GFR**) depends on an adequate blood supply to the kidneys at an adequate blood pressure. **Autoregulatory mechanisms** ensure adequate blood supply and blood pressure for kidney function regardless of what is happening elsewhere in the body. The kidney's autoregulatory adjustments compensate for decreases in cardiac output or decreases in blood pressure so that the GFR remains constant.

One autoregulatory mechanism is the dilation (expansion) of the afferent renal arterioles when blood pressure falls. This dilation decreases the resistance in the arterioles and helps maintain blood pressure in the glomerulus. If arteriole dilation does not keep the GFR from falling, the kidney releases an enzyme, **renin**, into the blood. Renin acts on a circulating protein to convert it into an active hormone called **angiotensin**.

Angiotensin has several effects that help restore the GFR to normal. First, angiotensin preferentially constricts the efferent renal arterioles, which elevates blood pressure in the glomerular capillaries. Second, it constricts peripheral blood vessels all over the body—an action that elevates central blood pressure. Third, it stimulates the adrenal cortex to release the hormone **aldosterone**. Aldosterone and angiotensin both stimulate sodium reabsorption by the kidney, thereby making its reabsorption of water more effective. Enhanced water reabsorption helps maintain blood volume and therefore central blood pressure. Finally, angiotensin acts on the brain to stimulate thirst. Increased water intake in response to thirst increases blood volume and blood pressure.

Blood osmolarity and blood pressure are regulated by ADH

Cells in the hypothalamus can stimulate the release from the posterior pituitary of a hormone called **antidiuretic hormone** (**ADH**, also called *vasopressin*) that can act upon water channels to increase the permeability of membranes to water. The higher the circulating levels of ADH, the greater the number of active water channels. Various factors can stimulate or inhibit the release of ADH. Of key importance to kidney function are osmoreceptors that monitor blood osmolarity and stretch receptors that monitor blood pressure (**Figure 51.14**).

Osmoreceptors that detect a rise in blood osmolarity will stimulate the release of ADH. ADH helps regulate blood osmolarity by

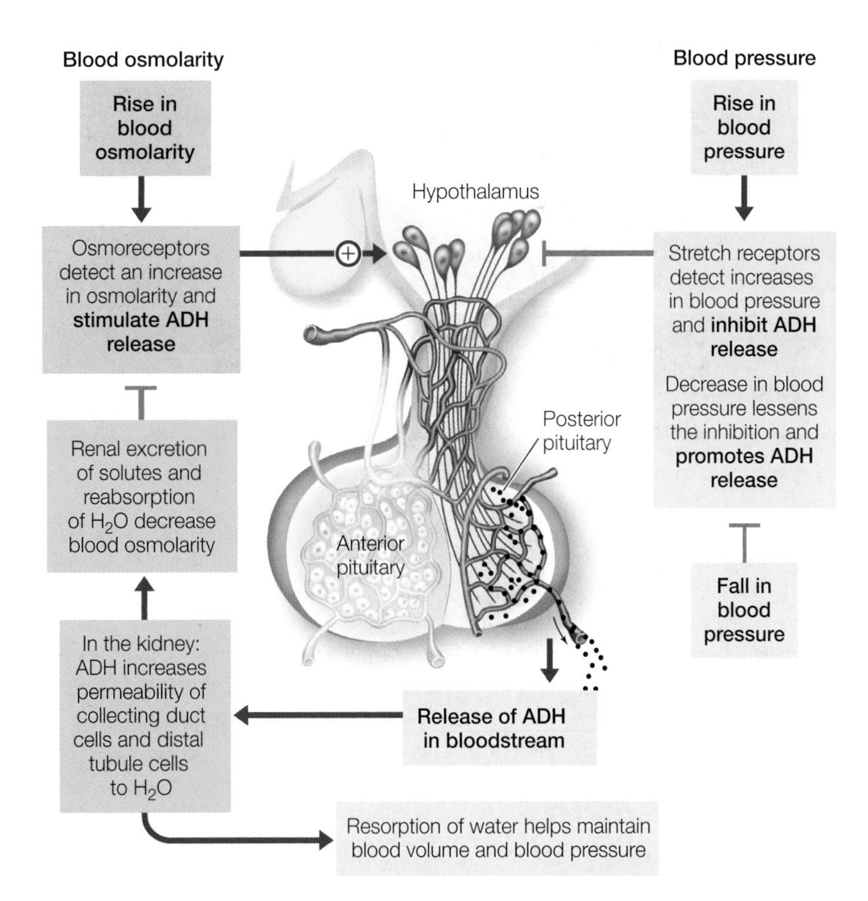

51.14 Antidiuretic Hormone Increases Blood Pressure and Promotes Water Reabsorption ADH is produced by neurons in the hypothalamus and released in the posterior pituitary. The release of ADH is stimulated by hypothalamic osmoreceptors and inhibited by stretch receptors in the great arteries.

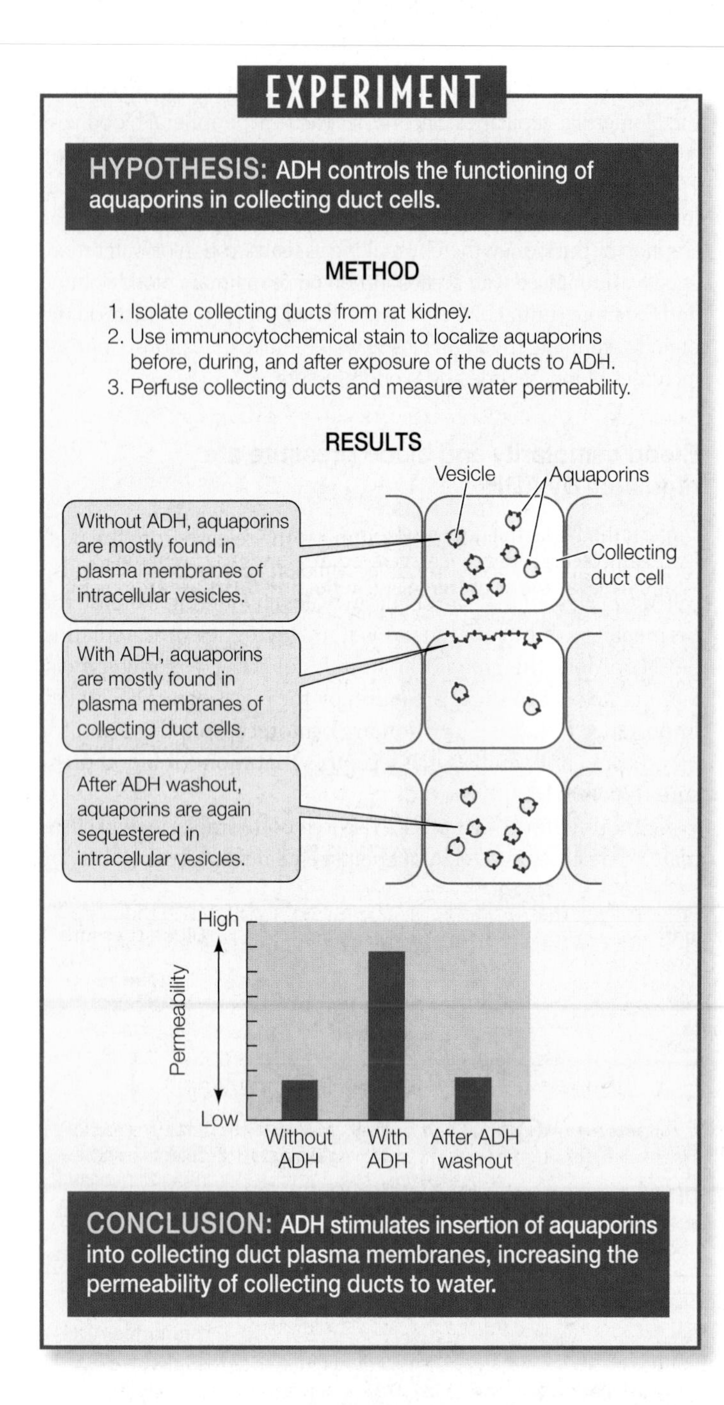

51.15 ADH Induces Insertion of Aquaporins into Collecting Duct Cell Plasma Membranes Aquaporin proteins are ordinarily found in the membranes of intracellular vesicles of collecting duct cells and do not contribute to water permeability. In the presence of ADH, however, these vesicles fuse with the plasma membrane, and the cells become more permeable to water.

controlling water reabsorption. The osmoreceptors also stimulate thirst. The resulting water retention and water intake dilute the blood as they expand blood volume.

Stretch receptors in the walls of the aorta and the carotid arteries (see Figure 49.17) that detect an increase in blood pressure will *inhibit* the release of ADH. With less circulating ADH, less water is reabsorbed, which decreases blood volume and hence acts to lower blood pressure.

If blood pressure falls, as when you lose blood volume through hemorrhage or excessive evaporative water loss, activity of the stretch receptors in the aorta and carotid arteries decreases. Input to the hypothalamus from these receptors inhibits the release of ADH, so when the firing rates of these stretch receptors fall, ADH release increases. More ADH results in more efficient water reabsorption and therefore a protection of blood volume and blood pressure.

Alcohol inhibits ADH release, explaining why excessive beer drinking leads to even more excessive urination and dehydration, which contributes to the symptoms of a hangover.

Given the importance of the collecting ducts in controlling the concentration of the urine and the importance of ADH in regulating collecting duct permeability to water, it was surprising that when the first aquaporins were discovered, they were not found in the collecting ducts of the kidney. It was hypothesized that a different aquaporin was expressed in the epithelial cells of the collecting duct. And, indeed, one was found and was simply named AQP-2. This aquaporin is normally sequestered in the membranes of intracellular vesicles until the collecting duct cells are stimulated by ADH. The immediate effect of circulating ADH is the insertion of AQP-2 into the plasma membranes of the cells of the collecting duct and, consequently, an increase in their permeability to water (**Figure 51.15**). When ADH decreases in the blood, the AQP-2 proteins are again internalized in vesicles and the water permeability of the cells drops.

The heart produces a hormone that influences kidney function

When systemic venous return to the heart increases and the atria of the heart become more stretched, the atrial muscle fibers release a peptide hormone called **atrial natriuretic peptide** (**ANP**). This peptide hormone enters the circulation, and when it reaches the kidney, it decreases the reabsorption of sodium. The result is an increased loss of sodium and water, which has the effect of lowering blood volume and blood pressure.

51.6 RECAP

Glomerular filtration is essential for kidney function, and it is sustained by autoregulatory mechanisms. Sensors that monitor blood pressure and blood osmolarity may stimulate or inhibit the release of hormones that regulate kidney function.

- Can you explain how falling GFR results in an increase in circulating angiotensin and how angiotensin restores GFR? See p. 1107
- Can you explain how falling blood pressure or increasing blood osmolarity results in changes in permeability of the collecting ducts? See Figure 51.14

CHAPTER SUMMARY

51.1 What roles do excretory organs play in maintaining homeostasis?

Excretory organs maintain the osmolarity and the volume of the extracellular fluids and eliminate the waste products of nitrogen metabolism through the processes of filtration, reabsorption, and secretion. **Urine** is the output of excretory organs.

There is no active transport of water, so water must be moved across membranes by a difference in either osmolarities or pressures.

Water enters or leaves cells by osmosis. To achieve cellular water balance, animals must maintain the **osmolarity** of their extracellular fluids within an acceptable range.

Marine animals can be **osmoconformers** or **osmoregulators**. Freshwater animals must be osmoregulators and must continually excrete water and conserve salts. Most animals are ionic regulators to some degree.

Apart from regulating osmolarity of cells and extracellular fluids, most animals must also regulate their ionic composition by conserving some ions and secreting others. Salt glands are adaptations for secretion of NaCl. Review Figure 51.2

51.2 How do animals excrete toxic wastes from nitrogen metabolism?

Aquatic animals can eliminate nitrogenous wastes such as ammonia by diffusion across their gill membranes. Terrestrial animals must detoxify ammonia by converting it to **urea** or **uric acid** before excretion. Review Figure 51.3

Depending on the form in which they excrete their nitrogenous wastes, animals are classified as **ammonotelic**, **ureotelic**, or **uricotelic**.

51.3 How do invertebrate excretory systems work?

The **protonephridia** of flatworms consist of flame cells and excretory tubules. Extracellular fluid is filtered into the tubules, which process the filtrate to produce a dilute urine. Review Figure 51.4

In annelid worms, blood pressure causes filtration of the blood across capillary walls. The filtrate enters the coelomic cavity, where it is taken up by **metanephridia**, which alters the composition of the filtrate by active transport mechanisms. Review Figure 51.5, Web/CD Activity 51.1

The **Malpighian tubules** of insects receive ions and nitrogenous wastes by active transport across the tubule cells. Water follows by osmosis. Ions and water are reabsorbed from the rectum, so the insect excretes semisolid wastes. Review Figure 51.6

51.4 How do vertebrates maintain salt and water balance?

Marine and terrestrial animals conserve water in various ways. Marine bony fishes produce little urine. Cartilaginous fishes retain urea so that the osmolarity of their body fluids remains close to that of seawater. Amphibians remain close to water. Reptiles have scaly skin with low water permeability, lay shelled eggs, and excrete nitrogenous wastes as uric acid.

Only birds and mammals can produce urine more concentrated than their extracellular fluids.

The **nephron**, the functional unit of the vertebrate **kidney**, consists of a **glomerulus**, in which blood is filtered, a **renal tubule**, which processes the filtrate into urine to be excreted, and a system of **peritubular capillaries**, which surround the tubule and serve as site of secretion and reabsorption. Review Figure 51.7, Web/CD Activity 51.2

51.5 How does the mammalian kidney produce concentrated urine?

See Web/CD Tutorial 51.1

The concentrating ability of the mammalian kidney is a function of its anatomy, which is ideal for **countercurrent exchange**. Review Figure 51.9, Web/CD Activity 51.3

The glomeruli and the **proximal** and **distal convoluted tubules** are located in the **cortex** of the kidney. Certain molecules are actively reabsorbed from the glomerular filtrate by the tubule cells, and other molecules are actively secreted. Straight sections of renal tubules called **loops of Henle** and **collecting ducts** are arranged in parallel in the **medulla** of the kidney.

Salts and water are reabsorbed in the proximal convoluted tubule without the renal filtrate becoming more concentrated, although its composition changes.

The loops of Henle create a concentration gradient in the **interstitial** fluid of the renal medulla by a **countercurrent multiplier mechanism**. Urine flowing down the collecting ducts to the ureter is concentrated by the osmotic reabsorption of water caused by the concentration gradient in the surrounding **interstitial fluid**. Review Figure 51.10

Hydrogen ions secreted by the renal tubules are buffered in the urine by bicarbonate and other chemical buffering systems. Review Figure 51.12

Artificial kidneys use dialysis to remove wastes from the blood. Review Figure 51.13

51.6 What mechanisms regulate kidney function?

Kidney function in mammals is controlled by **autoregulatory mechanisms** that maintain a constant high **glomerular filtration rate** even if blood pressure varies.

An important autoregulatory mechanism is the release of **renin** by the kidney when blood pressure falls. Renin activates **angiotensin**, which causes the constriction of efferent glomerular arterioles and peripheral blood vessels, causes the release of **aldosterone** (which enhances water reabsorption), and stimulates thirst.

Changes in blood pressure and osmolarity influence the release of **antidiuretic hormone** (**ADH**), which controls the permeability of the collecting duct to water and therefore the amount of water that is reabsorbed from the urine. ADH stimulates the expression of and controls the intracellular location of aquaporins, which serve as water channels in the membranes of collecting duct cells. Review Figures 51.14 and 51.15

When the volume of blood returning to the heart increases and stretches the atrial walls, **atrial natriuretic peptide** (**ANP**) is released, which causes increased excretion of salt and water.

See Web/CD Activity 51.4 for a review of the major human organ systems.

SELF-QUIZ

1. Which statement is true?
   a. Most marine invertebrates are osmoregulators.
   b. All freshwater invertebrates are osmoconformers.
   c. Marine bony and cartilaginous fishes have similar interstitial osmolarities.
   d. Freshwater fish are ionic regulators.
   e. Marine mammals gain water osmotically.
2. The excretion of nitrogenous wastes
   a. by humans can be in the form of urea and uric acid.
   b. by mammals is never in the form of uric acid.
   c. by marine fishes is mostly in the form of urea.
   d. does not contribute to the osmolarity of the urine.
   e. requires more water if the waste product is the rather insoluble uric acid.
3. How are earthworm metanephridia like mammalian nephrons?
   a. Both process coelomic fluid.
   b. Both take in fluid through a ciliated opening.
   c. Both produce urine more concentrated than the blood.
   d. Both employ tubular secretion and reabsorption to control urine composition.
   e. Both involve a countercurrent multiplier effect.
4. What is the role of renal podocytes?
   a. They prevent red blood cells and large molecules from entering the renal tubules.
   b. They reabsorb most of the glucose that is filtered from the plasma.
   c. They control the glomerular filtration rate by changing the resistance of renal arterioles.
   d. They provide a large surface area for tubular secretion and reabsorption.
   e. They release renin when the glomerular filtration rate falls.
5. Which of the following are *not* found in a renal pyramid?
   a. Collecting ducts
   b. Vasa recta
   c. Peritubular capillaries
   d. Convoluted tubules
   e. Loops of Henle
6. Which part of the nephron is responsible for most of the difference in mammals between the glomerular filtration rate and the urine production rate?
   a. The glomerulus
   b. The proximal convoluted tubule
   c. The loops of Henle
   d. The distal convoluted tubule
   e. The collecting duct
7. For mammals of the same size, what feature of their excretory systems would give them the greatest ability to concentrate their urine?
   a. Higher glomerular filtration rate
   b. Longer convoluted tubules
   c. Increased number of nephrons
   d. More permeable collecting ducts
   e. Longer loops of Henle
8. Which of the following would *not* be a response stimulated by a large drop in blood pressure?
   a. Constriction of afferent renal arterioles
   b. Increased release of renin
   c. Increased release of antidiuretic hormone
   d. Increased thirst
   e. Constriction of efferent renal arterioles
9. Which statement about angiotensin is true?
   a. It is secreted by the kidney when the glomerular filtration rate falls.
   b. It is released by the posterior pituitary when blood pressure falls.
   c. It stimulates thirst.
   d. It increases the permeability of the collecting ducts to water.
   e. It decreases glomerular filtration rate when blood pressure rises.
10. Birds that feed on marine animals ingest a lot of salt, but they excrete most of it by means of
    a. Malpighian tubules.
    b. rectal salt glands.
    c. gill membranes.
    d. concentrated urine.
    e. nasal salt glands.

FOR DISCUSSION

1. Why is it said that the oceans are a physiological desert? For what animals would this apply?
2. Persons with uncontrolled diabetes mellitus can have very high levels of glucose in their blood. Why do such individuals have a high level of urine production?
3. Inulin is a molecule that is filtered out of the glomerulus, but is not secreted or reabsorbed by the renal tubules. If you injected inulin into an animal and after a brief time measured the concentration of inulin in its blood and urine, how could you determine the animal's glomerular filtration rate? Assume that the rate of urine production is 1 milliliter per minute.
4. After you did the inulin experiment to measure glomerular filtration rate, how could you use that information to determine whether another substance is secreted or reabsorbed by the renal tubules? Assume you can measure the concentration of that substance in the blood and in the urine. Urine production is still 1 milliliter per minute.
5. Explain what happens with respect to control and regulation of your salt and water balance when you go to a movie and eat a lot of very salty popcorn.

FOR INVESTIGATION

We mentioned that the GFR is autoregulated, and this regulation is achieved by control over constriction/dilation of the afferent and the efferent renal arterioles. What information could be used in this process? A clue comes from the anatomy of the nephron. When the ascending limb of Henle reaches the cortex, it makes direct contact with the afferent and efferent arterioles of the glomerulus that produced the filtrate flowing in that particular nephron. What aspect of the tubular fluid in the early distal tubule would reflect changes in the volume of filtrate entering the nephron? How might the cells of the early distal tubule communicate with the smooth muscle cells of the arterioles?

Appendix A: The Tree of Life

Phylogeny is the organizing principle of modern biological taxonomy, and a guiding principle of modern phylogeny is *monophyly*: a monophyletic group is considered to be one that contains an ancestral lineage and *all* of its descendants. Any such a group can be extracted from a phylogenetic tree with a single cut. The tree shown here provides a guide to the relationships among the major groups of the extant (living) organisms in the Tree of Life as we have presented them throughout this book. We do include three groups that are not believed to be monophyletic; these are designated with quotation marks.

The position of the branching "splits" indicates the relative branching order of the lineages of life, but the timing of splits in different groups is not drawn on a comparable time scale. In addition, the groups appearing at the branch tips do not necessarily carry equal phylogenetic "weight." For example, the ginkgo [55] is indeed at the apex of its lineage; this gymnosperm group consists of a single living species. In contrast, a phylogeny of the angiosperms [52] would continue on from this point to fill many more trees the size of this one.

The glossary entries that follow are informal descriptions of some major features of the organisms described in Part Six of this book. Each entry gives the group's common name, followed by the formal scientific name of the group (in parentheses). Numbers in square brackets reference the location of the respective groups on the tree.

It is sometimes convenient to use an informal name to refer to a collection of organisms that are not monophyletic but nonetheless all share (or all lack) some common attribute. We call these "convenience terms"; such groups are indicated in these entries by quotation marks, and we do not give them formal scientific names. Examples include "prokaryotes," "protists," and "algae." Note that these groups cannot be removed with a single cut; they represent a collection of distantly related groups that appear in different parts of the tree.

– A –

acorn worms (*Enteropneusta*) Benthic marine hemichordates [109] with an acorn-shaped proboscis, a short collar (neck), and a long trunk.

"algae" A convenience term encompassing various distantly related groups of aquatic, photosynthetic chromalveolates [5] and certain members of the Plantae [8].

alveolates (*Alveolata*) [7] Unicellular eukaryotes with a layer of flattened vesicles (alveoli) supporting the plasma membrane. Major alveolate groups include the dinoflagellates [49], apicomplexans [50], and ciliates [51].

ambulacrarians (*Ambulacraria*) [27] The echinoderms [108] and hemichordates [109].

amniotes (*Amniota*) [34] Mammals, reptiles, and their extinct close relatives. Characterized by many adaptations to terrestrial life, including an amniotic egg (with a unique set of membranes—the amnion, chorion, and allantois), a water-repellant epidermis (with epidermal scales, hair, or feathers), and, in males, a penis that allows internal fertilization.

amoebozoans (*Amoebozoa*) [76] A group of eukaryotes [4] that use lobe-shaped pseudopods for locomotion and to engulf food. Major amoebozoan groups include the loboseans, plasmodial slime molds, and cellular slime molds.

amphibians (*Amphibia*) [118] Tetrapods [33] with glandular skin that lacks epidermal scales, feathers, or hair. Many amphibian species undergo a complete metamorphosis from an aquatic larval form to a terrestrial adult form, although direct development is also common. Major amphibian groups include frogs and toads (anurans), salamanders, and caecilians.

amphipods (*Amphipoda*) Small crustaceans [106] that are abundant in many marine and freshwater habitats. They are important herbivores, scavengers, and micropredators, and are an important food source for many aquatic organisms.

angiosperms (*Anthophyta* or *Magnoliophyta*) [52] The flowering plants. Major angiosperm groups include the monocots, eudicots, and magnoliids.

animals (*Animalia* or *Metazoa*) [19] Multicellular heterotrophic eukaryotes. The majority of animals are bilaterians [21]. Other major groups are the cnidarians [87], ctenophores [86], calcareous sponges [85], demosponges [84], and glass sponges [83]. The closest living relatives of the animals are the choanoflagellates [82].

annelids (*Annelida*) [94] Segmented worms, including earthworms, leeches, and polychaetes. One of the major groups of lophotrochozoans [23].

anthozoans (*Anthozoa*) One of the major groups of cnidarians [87]. Includes the sea anemones, sea pens, and corals.

anurans (*Anura*) Comprising the frogs and toads, this is the largest group of living amphibians [118]. They are tailless, with a shortened vertebral column and elongate hind legs modified for jumping. Many species have an aquatic larval form known as a tadpole.

apicomplexans (*Apicomplexa*) [50] Parasitic alveolates [7] characterized by the possession of an apical complex at some stage in the life cycle.

arachnids (*Arachnida*) Chelicerates [104] with a body divided into two parts: a cephalothorax that bears six pairs of appendages (four pairs of which are usually used as legs) and an abdomen that bears the genital opening. Familiar arachnids include spiders, scorpions, mites and ticks, and harvestmen.

archaeans (*Archaea*) [3] Unicellular organisms lacking a nucleus and lacking peptidoglycan in the cell wall. Once grouped with the bacteria, archaeans possess distinctive membrane lipids.

archosaurs (*Archosauria*) [36] A group of reptiles [35] that includes dinosaurs and crocodilians [123]. Most dinosaur groups became extinct at the end of the Cretaceous; birds [122] are the only surviving dinosaurs.

arrow worms (*Chaetognatha*) [96] Small planktonic or benthic predatory marine worms with fins and a pair of hooked, prey-grasping spines on each side of the head.

arthropods (*Arthropoda*) The largest group of ecdysozoans [24]. Arthropods are characterized by a stiff exoskeleton, segmented bodies, and jointed appendages. Includes the chelicerates [104], myriapods [105], crustaceans [106], and hexapods (insects and their relatives) [107].

ascidians (*Ascidiacea*) "Sea squirts"; the largest group of urochordates [110]. Also known as tunicates, they are sessile (as adults), marine, sac-like filter feeders.

ascomycetes (*Ascomycota*) [78] Fungi that bear the products of meiosis within sacs (asci) if the organism is multicellular. Some are unicellular.

– B –

bacteria (*Eubacteria*) [2] Unicellular organisms lacking a nucleus, possessing distinctive ribosomes and initiator tRNA, and generally containing peptidoglycan in the cell wall. Different bacterial groups are distinguished primarily on nucleotide sequence data.

barnacles (*Cirripedia*) Crustaceans [106] that undergo two metamorphoses—first from a feeding planktonic larva to a nonfeeding swimming larva, and then to a sessile adult that forms a "shell" composed of four to eight plates cemented to a hard substrate.

basidiomycetes (*Basidiomycota*) [77] Fungi [17] that, if multicellular, bear the products of meiosis on club-shaped basidia and possess a long-lasting dikaryotic stage. Some are unicellular.

bilaterians (*Bilateria*) [21] Those animal groups characterized by bilateral symmetry and three distinct tissue types (endoderm, ectoderm, and mesoderm). Includes the protostomes [22] and deuterostomes [26].

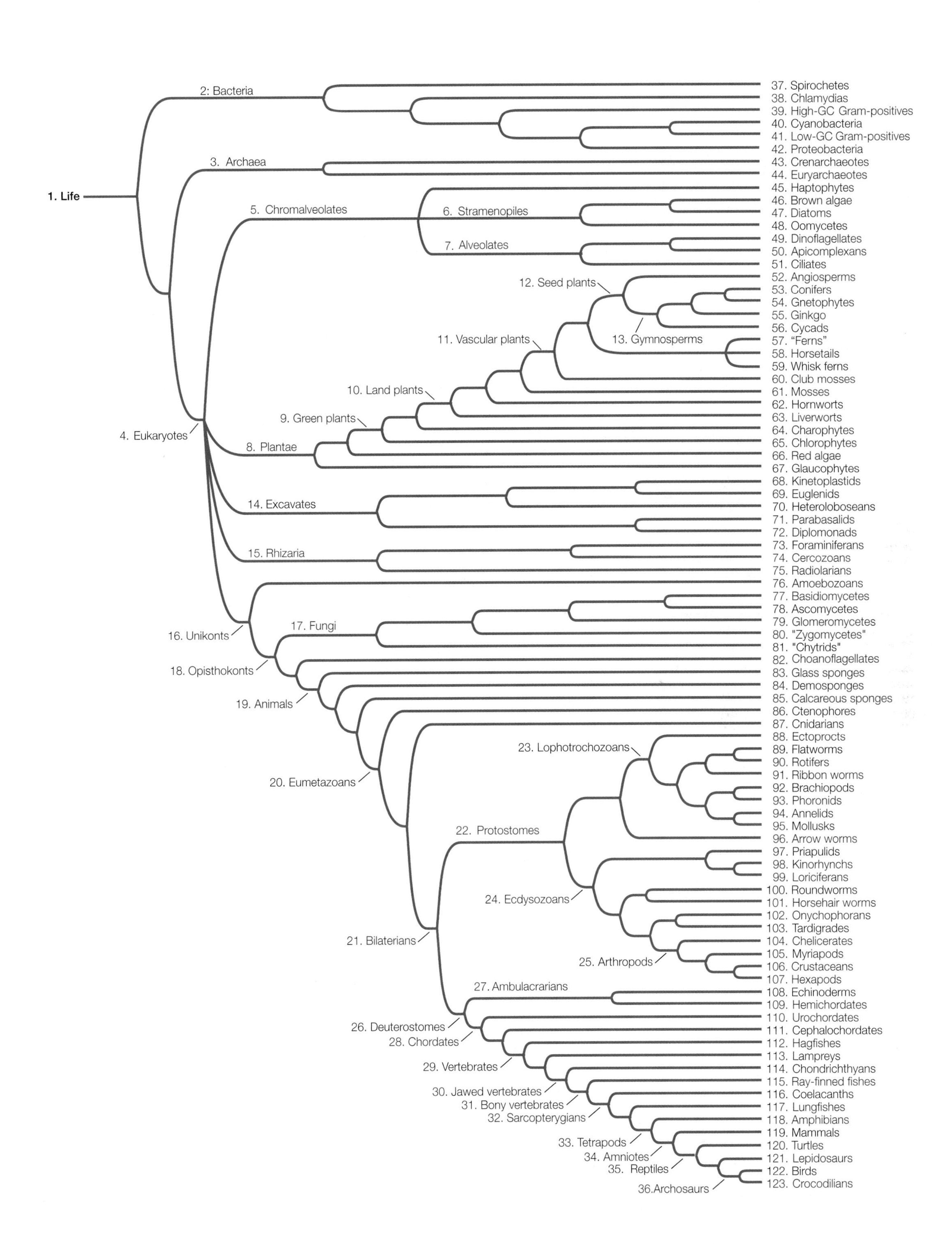
1. Life
2: Bacteria
3. Archaea
4. Eukaryotes
5. Chromalveolates
6. Stramenopiles
7. Alveolates
8. Plantae
9. Green plants
10. Land plants
11. Vascular plants
12. Seed plants
13. Gymnosperms
14. Excavates
15. Rhizaria
16. Unikonts
17. Fungi
18. Opisthokonts
19. Animals
20. Eumetazoans
21. Bilaterians
22. Protostomes
23. Lophotrochozoans
24. Ecdysozoans
25. Arthropods
26. Deuterostomes
27. Ambulacrarians
28. Chordates
29. Vertebrates
30. Jawed vertebrates
31. Bony vertebrates
32. Sarcopterygians
33. Tetrapods
34. Amniotes
35. Reptiles
36.Archosaurs
37. Spirochetes
38. Chlamydias
39. High-GC Gram-positives
40. Cyanobacteria
41. Low-GC Gram-positives
42. Proteobacteria
43. Crenarchaeotes
44. Euryarchaeotes
45. Haptophytes
46. Brown algae
47. Diatoms
48. Oomycetes
49. Dinoflagellates
50. Apicomplexans
51. Ciliates
52. Angiosperms
53. Conifers
54. Gnetophytes
55. Ginkgo
56. Cycads
57. "Ferns"
58. Horsetails
59. Whisk ferns
60. Club mosses
61. Mosses
62. Hornworts
63. Liverworts
64. Charophytes
65. Chlorophytes
66. Red algae
67. Glaucophytes
68. Kinetoplastids
69. Euglenids
70. Heteroloboseans
71. Parabasalids
72. Diplomonads
73. Foraminiferans
74. Cercozoans
75. Radiolarians
76. Amoebozoans
77. Basidiomycetes
78. Ascomycetes
79. Glomeromycetes
80. "Zygomycetes"
81. "Chytrids"
82. Choanoflagellates
83. Glass sponges
84. Demosponges
85. Calcareous sponges
86. Ctenophores
87. Cnidarians
88. Ectoprocts
89. Flatworms
90. Rotifers
91. Ribbon worms
92. Brachiopods
93. Phoronids
94. Annelids
95. Mollusks
96. Arrow worms
97. Priapulids
98. Kinorhynchs
99. Loriciferans
100. Roundworms
101. Horsehair worms
102. Onychophorans
103. Tardigrades
104. Chelicerates
105. Myriapods
106. Crustaceans
107. Hexapods
108. Echinoderms
109. Hemichordates
110. Urochordates
111. Cephalochordates
112. Hagfishes
113. Lampreys
114. Chondrichthyans
115. Ray-finned fishes
116. Coelacanths
117. Lungfishes
118. Amphibians
119. Mammals
120. Turtles
121. Lepidosaurs
122. Birds
123. Crocodilians

birds (*Aves*) [122] Feathered, flying (or secondarily flightless) tetrapods [33].

bivalves (*Bivalvia*) Major mollusk [95]group; clams and mussels. Bivalves typically have two similar hinged shells that are each asymmetrical across the midline.

bony vertebrates (*Osteichthyes*) [31] Vertebrates [29] in which the skeleton is usually ossified to form bone. Includes the ray-finned fishes [115], coelacanths [116], lungfishes [117], and tetrapods [33].

brachiopods (*Brachiopoda*) [92] Lophotrochozoans [23] with two similar hinged shells that are each symmetrical across the midline. Superficially resemble bivalve mollusks, except for the shell symmetry.

brittle stars (*Ophiuroidea*) Echinoderms [108] with five long, whip-like arms radiating from a distinct central disk that contains the reproductive and digestive organs.

brown algae (*Phaeophyta*) [46] Multicellular, almost exclusively marine stramenopiles [6] generally containing the pigment fucoxanthin as well as chlorophylls *a* and *c* in their chloroplasts.

– C –

caecilians (*Gymnophiona*) A group of burrowing or aquatic amphibians [118]. They are elongate, legless, with a short tail (or none at all), reduced eyes covered with skin or bone, and a pair of sensory tentacles on the head.

calcareous sponges (*Calcarea*) [85] Filter-feeding marine sponges with spicules composed of calcium carbonate.

cellular slime molds (*Dictyostelida*) Amoebozoans [76] in which individual amoebas aggregate under stress to form a multicellular pseudoplasmodium.

cephalochordates (*Cephalochordata*) [111] A group of weakly swimming, eel-like benthic marine chordates [28]; also called lancelets. Thought to be the closest living relatives of the vertebrates [29].

cephalopods (*Cephalopoda*) Active, predatory mollusks [95] in which the molluscan foot has been modified into muscular hydrostatic arms or tentacles. Includes octopuses, squids, and nautiluses.

cercozoans (*Cercozoa*) [74] Unicellular eukaryotes [4] that feed by means of threadlike pseudopods. Together with foraminiferans [73] and radiolarians [75], the cercozoans comprise the group *Rhizaria* [15].

charophytes (*Charales*) [64] Multicellular green algae with branching, apical growth and plasmodesmata between adjacent cells. The closest living relatives of the land plants [10], they retain the egg in the parent organism.

chelicerates (*Chelicerata*) [104] A major group of arthropods [25] with pointed appendages (chelicerae) used to grasp food (as opposed to the chewing mandibles of most other arthropods). Includes the arachnids, horseshoe crabs, pycnogonids, and extinct sea scorpions.

chimaeras (*Holocephali*) A group of bottom-dwelling, marine, scaleless chondrichthyan fishes [114] with large, permanent, grinding tooth plates (rather than the replaceable teeth found in other chondrichthyans).

chitons (*Polyplacophora*) Flattened, slow-moving mollusks [95] with a dorsal protective calcareous covering made up of eight articulating plates.

chlamydias (*Chlamydiae*) [38] A group of very small Gram-negative bacteria; they live as intracellular parasites of other organisms.

chlorophytes (*Chlorophyta*) [65] The most abundant and diverse group of green algae, including freshwater, marine, and terrestrial forms; some are unicellular, others colonial, and still others multicellular. Chlorophytes use chlorophylls *a* and *c* in their photosynthesis.

choanoflagellates (*Choanozoa*) [82] Unicellular eukaryotes [4] with a single flagellum surrounded by a collar. Most are sessile, some are colonial. The closest living relatives of the animals [19]

chondrichthyans (*Chondrichthyes*) [114] One of the two main groups of jawed vertebrates [30]; includes sharks, rays, and chimaeras. They have cartilaginous skeletons and paired fins.

chordates (*Chordata*) [28] One of the two major groups of deuterostomes [26], characterized by the presence (at some point in development) of a notochord, a hollow dorsal nerve cord, and a post-anal tail. Includes the urochordates [110], cephalochordates [111], and vertebrates [29].

chromalveolates (*Chromalveolata*) [5] A contested group, said to have arisen from a common ancestor with chloroplasts derived from a red alga and supported by molecular evidence. Major chromalveolate groups include the alveolates [7] and stramenopiles [6].

"chytrids" [81] A convenience term used for a paraphyletic group of mostly aquatic, microscopic fungi [17] with flagellated gametes. Some exhibit alternation of generations.

ciliates (*Ciliophora*) [51] Alveolates [7] with numerous cilia and two types of nuclei (micronuclei and macronuclei).

clitellates (*Clitellata*) Annelids [94] with gonads contained in a swelling (called a clitellum) toward the head of the animal. Includes earthworms (oligochaetes) and leeches.

club mosses (*Lycophyta*) [60] Vascular plants [11] characterized by microphylls.

cnidarians (*Cnidaria*) [87] Aquatic, mostly marine eumetazoans [20] with specialized stinging organelles (nematocysts) used for prey capture and defense, and a blind gastrovascular cavity. The closest living relatives of the bilaterians [21].

coelacanths (*Actinista*) [116] A group of marine sarcopterygians [32] that was diverse from the Middle Devonian to the Cretaceous, but is now known from just two living species. The pectoral and anal fins are on fleshy stalks supported by skeletal elements, so they are also called lobe-finned fishes.

conifers (*Pinophyta* or *Coniferophyta*) [53] Cone-bearing, woody seed plants.

copepods (*Copepoda*) Small, abundant crustaceans [106] found in marine, freshwater, or wet terrestrial habitats. They have a single eye, long antennae, and a body shaped like a teardrop.

craniates (*Craniata*) Some biologist exclude the hagfishes [112] from the vertebrates [29], and use the term craniates to refer to the two groups combined.

crenarchaeotes (*Crenarchaeota*) [43] A major and diverse group of archaeans [3], defined on the basis of rRNA base sequences. Many are extremophiles (inhabit extreme environments), but the group may also be the most abundant archaeans in the marine environment.

crinoids (*Crinoidea*) Echinoderms [108] with a mouth surrounded by feeding arms, and a U-shaped gut with the mouth next to the anus. They attach to the substratum by a stalk or are free-swimming. Crinoids were abundant in the middle and late Paleozoic, but only a few hundred species have survived to the present. Includes the sea lilies and feather stars.

crocodilians (*Crocodylia*) [123] A group of large, predatory, aquatic archosaurs [36]. The closest living relatives of birds [122]. Includes alligators, caimans, crocodiles, and gharials.

crustaceans (*Crustacea*) [106] Major group of marine, freshwater, and terrestrial arthropods [25] with a head, thorax, and abdomen (although the head and thorax may be fused), covered with a thick exoskeleton, and with two-part appendages. Crustaceans undergo metamorphosis from a nauplius larva. Includes decapods, isopods, krill, barnacles, amphipods, copepods, and ostracods.

ctenophores (*Ctenophora*) [86] Radially symmetrical, diploblastic marine animals [19], with a complete gut and eight rows of fused plates of cilia (called ctenes).

cyanobacteria (*Cyanobacteria*) [40] A group of unicellular, colonial, or filamentous bacteria that conduct photosynthesis using chlorophyll *a*.

cycads (*Cycadophyta*) [56] Palmlike gymnosperms with large, compound leaves.

cyclostomes (*Cyclostomata*) This term refers to the possibly monophyletic group of lampreys [113] and hagfishes [112]. Molecular data support this group, but morphological data suggest that lampreys are more closely related to jawed vertebrates [30] than to hagfishes.

– D –

decapods (*Decapoda*) A group of marine, freshwater, and semiterrestrial crustaceans [106] in which five of the eight pairs of thoracic appendages function as legs (the other three pairs, called maxillipeds, function as mouthparts). Includes crabs, lobsters, crayfishes, and shrimps.

demosponges (*Demospongiae*) [84] The largest of the three groups of sponges, accounting for 90 percent of all sponge species. Demosponges have spicules made of silica, spongin fiber (a protein), or both.

deuterostomes (*Deuterostomia*) [26] One of the two major groups of bilaterians [21], in which the mouth forms at the opposite end of the embryo from the blastopore in early development (contrast with protostomes). Includes the ambulacrarians [27] and chordates [28].

diatoms (*Bacillariophyta*) [47] Unicellular, photosynthetic stramenopiles [6] with glassy cell walls in two parts.

dinoflagellates (*Dinoflagellata*) [49] A group of alveolates [7] usually possessing two flagella, one in an equatorial groove and the other in a longitudinal groove; many are photosynthetic.

diplomonads (*Diplomonadida*) [72] A group of eukaryotes [4] lacking mitochondria; most have two nuclei, each with four associated flagella.

– E –

ecdysozoans (*Ecdysozoa*) [24] One of the two major groups of protostomes [22], characterized by periodic molting of their exoskeletons. Roundworms [100] and arthropods [25] are the largest ecdysozoan groups.

echinoderms (*Echinodermata*) [108] A major group of marine deuterostomes [26] with fivefold radial symmetry (at some stage of life) and an endoskeleton made of calcified plates and spines. Includes sea stars, crinoids, sea urchins, sea cucumbers, and brittle stars.

ectoprocts (*Ectoprocta*) [88] A group of marine and freshwater lophotrochozoans [23] that live in colonies attached to substrata. Also known as bryozoans or moss animals.

elasmobranchs (*Elasmobranchii*) The largest group of chondrichthyan fishes [114]. Includes sharks, skates, and rays. In contrast to the other group of living chondrichthyans (the chimaeras), they have replaceable teeth.

eudicots (*Eudicotyledones*) A group of angiosperms [52] with pollen grains possessing three openings. Typically with two cotyledons, net-veined leaves, taproots, and floral organs typically in multiples of four or five.

euglenids (*Euglenida*) [69] Flagellate excavates characterized by a pellicle composed of spiraling strips of protein under the plasma membrane; the mitochondria have disk-shaped cristae. Some are photosynthetic.

eukaryotes (*Eukarya*) [4] Organisms made up of one or more complex cells in which the genetic material is contained in nuclei. Contrast with archaeans [3] and bacteria [2].

eumetazoans (*Eumetazoa*) [20] Those animals [19] characterized by body symmetry, a gut, a nervous system, specialized types of cell junctions, and well-organized tissues in distinct cell layers (although there have been secondary losses of some of these characteristics in some eumetazoans).

euphyllophytes (*Euphyllophyta*) This clade is sister to the club mosses [60] and includes all plants with megaphylls.

euryarchaeotes (*Euryachaeota*) [44] A major group of archaeans [3], diagnosed on the basis of rRNA sequences. Includes many methanogens, extreme halophiles, and thermophiles.

eutherians (*Eutheria*) A group of viviparous mammals [119], eutherians are well developed at birth (contrast to prototherians and marsupials, the other two groups of mammals). Most familiar mammals outside the Australian and South American regions are eutherians (see Table 33.1).

excavates (*Excavata*) [14] Diverse group of unicellular, flagellate eukaryotes, many of which possess a feeding groove; some lack mitochondria.

– F –

"ferns" [57] Vascular plants [11] usually possessing large, frond-like leaves that unfold from a "fiddlehead." Not a monophyletic group, although most fern species are encompassed in a monophyletic clade, the leptosporangiate ferns.

flatworms (*Platyhelminthes*) [89] A group of dorsoventrally flattened and generally elongate soft-bodied lophotrochozoans [23]. May be free-living or parasitic, found in marine, freshwater, or damp terrestrial environments. Major flatworm groups include the tapeworms, flukes, monogeneans, and turbellarians.

flowering plants *See* angiosperms.

flukes (*Trematoda*) A group of wormlike parasitic flatworms [89] with complex life cycles that involve several different host species. May be paraphyletic with respect to tapeworms.

foraminiferans (*Foraminifera*) [73] Amoeboid organisms with fine, branched pseudopods that form a food-trapping net. Most produce external shells of calcium carbonate.

fungi (*Fungi*) [17] Eukaryotic heterotrophs with absorptive nutrition based on extracellular digestion; cell walls contain chitin. Major fungal groups include the "chytrids" [81], "zygomycetes" [80], glomeromycetes [79], ascomycetes [78], and basidiomycetes [77].

– G –

gastropods (*Gastropoda*) The largest group of mollusks [95]. Gastropods possess a well-defined head with two or four sensory tentacles (often terminating in eyes) and a ventral foot. Most species have a single coiled or spiraled shell. Common in marine, freshwater, and terrestrial environments.

ginkgo (*Ginkgophyta*) [55] A gymnosperm [13] group with only one living species. The ginkgo seed is surrounded by a fleshy tissue not derived from an ovary wall and hence not a fruit.

glass sponges (*Hexactinellida*) [83] Sponges with a skeleton composed of four- and/or six-pointed spicules made of silica.

glaucophytes (*Glaucophyta*) [67] Unicellular freshwater algae with chloroplasts containing traces of peptidoglycan, the characteristic cell wall material of bacteria.

glomeromycetes (*Glomeromycota*) [79] A group of fungi [17] that form arbuscular mycorrhizae.

gnathostomes (*Gnathostomata*) *See* jawed vertebrates.

gnetophytes (*Gnetophyta*) [54] A gymnosperm [13] group with three very different lineages; all have wood with vessels, unlike other gymnosperms.

green plants (*Viridiplantae*) [9] Organisms with chlorophylls *a* and *b*, cellulose-containing cell walls, starch as a carbohydrate storage product, and chloroplasts surrounded by two membranes.

gymnosperms (*Gymnospermae*) [13] Seed plants [12] with seeds "naked" (i.e., not enclosed in carpels). Probably monophyletic, but status still in doubt. Includes the conifers [53], gnetophytes [54], ginkgo [55], and cycads [56].

– H –

hagfishes (*Myxini*) [112] Elongate, slimy-skinned vertebrates [29] with three small accessory hearts, a partial cranium, and no stomach or paired fins. *See also* craniata; cyclostomes.

haptophytes (*Haptophyta*) [45] Unicellular, photosynthetic stramenopiles [6] with two slightly unequal, smooth flagella. Abundant as phytoplankton, some form marine algal blooms.

hemichordates (*Hemichordata*) [109] One of the two primary groups of ambulacrarians [27]; marine wormlike organisms with a three-part body plan.

heteroloboseans (*Heterolobosea*) [70] Colorless excavates [14] that can transform among amoeboid, flagellate, and encysted stages.

hexapods (*Hexapoda*) [107] Major group of arthropods [25] characterized by a reduction (from the ancestral arthropod condition) to six walking appendages, and the consolidation of three body segments to form a thorax. Includes insects and their relatives (see Table 32.2).

high-GC Gram-positives (*Actinobacteria*) [39] Gram-positive bacteria with a relatively high G+C/A+T ratio of their DNA, with a filamentous growth habit.

hornworts (*Anthocerophyta*) [62] Nonvascular plants with sporophytes that grow from the base. Cells contain a single large, platelike chloroplast.

horsehair worms (*Nematomorpha*) [101] A group of very thin, elongate, wormlike freshwater ecdysozoans [24]. Largely nonfeeding as adults, they are parasites of insects and crayfish as larvae.

horseshoe crabs (*Xiphosura*) Marine chelicerates [104] with a large outer shell in three parts: a carapace, an abdomen, and a tail-like telson. Only five living species remain, but many additional species are known from fossils.

horsetails (*Sphenophyta* or *Equisetophyta*) [58] Vascular plants [11] with reduced megaphylls in whorls.

hydrozoans (*Hydrozoa*) A group of cnidarians [87]. Most species go through both polyp and mesuda stages, although one stage or the other is eliminated in some species.

– I –

insects (*Insecta*) The largest group within the hexapods [107]. Insects are characterized by exposed mouthparts and one pair of antennae containing a sensory receptor called a Johnston's organ. Most have two pairs of wings as adults. There are more described species of insects than all other groups of life [1] combined, and many species remain to be discovered. The major insect groups are described in Table 32.2.

isopods (*Isopoda*) Crustaceans [106] characterized by a compact head, unstalked compound eyes, and mouthparts consisting of four pairs of appendages. Isopods are abundant and widespread in salt, fresh, and brackish water, although some species (the sow bugs) are terrestrial.

– J –

jawed vertebrates (*Gnathostomata*) [30] A major group of vertebrates [29] with jawed mouths. Includes chondrichthyans [114], ray-finned fishes [115], and sarcopterygians [32].

– K –

kinetoplastids (*Kinetoplastida*) [68] Unicellular, flagellate organisms characterized by the presence in their single mitochondrion of a kinetoplast (a structure containing multiple, circular DNA molecules).

kinorhynchs (*Kinorhyncha*) [98] Small (< 1 mm) marine ecdysozoans [24] with bodies in 13 segments and a retractable proboscis.

korarchaeotes (*Korarchaeota*) A group of archaeans [3] known only by evidence from nucleic acids derived from hot springs. Its phylogenetic relationships within the Archaea are unknown.

krill (*Euphausiacea*) A group of shrimplike marine crustaceans [106] that are important components of the zooplankton.

– L –

lampreys (*Petromyzontiformes*) [113] Elongate, eel-like vertebrates [29] that often have rasping and sucking disks for mouths.

lancelets (*Cephalochordata*) *See* cephalochordates.

land plants (*Embryophyta*) [10] Plants with embryos that develop within protective structures; sporophytes and gametophytes are multicellular. Land plants possess a cuticle. Major groups are the liverworts [63], hornworts [62], mosses [61], and vascular plants [11].

larvaceans (*Larvacea*) Solitary, planktonic urochordates [110] that retain both notochords and nerve cords throughout their lives.

lepidosaurs (*Lepidosauria*) [121] Reptiles [35] with overlapping scales. Includes tuataras and squamates (lizards, snakes, and amphisbaenians).

life (*Life*) [1] The monophyletic group that includes all known living organisms. Characterized by a nucleic-acid based genetic system (DNA or RNA), metabolism, and cellular structure. Some parasitic forms, such as viruses, have secondarily lost some of these features and rely on the cellular environment of their host.

liverworts (*Hepatophyta*) [63] Nonvascular plants lacking stomata; stalk of sporophyte elongates along its entire length.

loboseans (*Lobosea*) A group of unicellular amoebozoans [76]; includes the most familiar amoebas (e.g., *Amoeba proteus*).

"lophophorates" Not a monophyletic group. A convenience term used to describe several groups of lophotrochozoans [23] that have a feeding structure called a lophophore (a circular or U-shaped ridge around the mouth that bears one or two rows of ciliated, hollow tentacles).

lophotrochozoans (*Lophotrochozoa*) [23] One of the two main groups of protostomes [22]. This group is morphologically diverse, and is supported primarily on information from gene sequences. Includes ectoprocts [88], flatworms [89], rotifers [90], ribbon worms [91], brachiopods [92], phoronids [93], annelids [94], and mollusks [95].

loriciferans (*Loricifera*) [99] Small (< 1 mm) ecdysozoans [24] with bodies in four parts, covered with six plates.

low-GC Gram-positives (*Firmicutes*) [41] A diverse group of bacteria [2] with a relatively low G+C/A+T ratio of their DNA, often but not always Gram-positive, some producing endospores.

lungfishes (*Dipnoi*) [117] A group of aquatic sarcopterygians [32] that are the closest living relatives of the tetrapods [33]. They have a modified swim bladder used to absorb oxygen from air, so some species can survive the temporary drying of their habitat.

– M –

magnoliids Major group of angiosperms [52] possessing two cotyledons and pollen grains with a single opening. The group is defined primarily by nucleotide sequence data; it is more closely related to the eudicots and monocots than to three other small angiosperm groups.

mammals (*Mammalia*) [119] A group of tetrapods [33] with hair covering all or part of their skin; females produce milk to feed their developing young. Includes the prototherians, marsupials, and eutherians.

marsupials (*Marsupialia*) Mammals [119] in which the female typically has a marsupium (a pouch for rearing young, which are born at an extremely early stage in development). Includes such familiar mammals as opossums, koalas, and kangaroos.

metazoans (*Metazoa*) *See* animals.

mollusks (*Mollusca*) [95] One of the major groups of lophotrochozoans [23], mollusks have bodies composed of a foot, a mantle (which often secretes a hard, calcareous shell), and a visceral mass. Includes monoplacophorans, chitons, bivalves, gastropods, and cephalopods.

monocots (*Monocotyledones*) Angiosperms [52] characterized by possession of a single cotyledon, usually parallel leaf veins, a fibrous root system, pollen grains with a single opening, and floral organs usually in multiples of three.

monogeneans (*Monogenea*) A group of ectoparasitic flatworms [89].

monoplacophorans (*Monoplacophora*) Mollusks [95] with segmented body parts and a single, thin, flat, rounded, bilateral shell.

mosses (*Bryophyta*) [61] Nonvascular plants with true stomata and erect, "leafy" gametophytes; sporophytes elongate by apical cell division.

multicellular eukaryotes *See* "protists."

myriapods (*Myriapoda*) [105] Arthropods [25] characterized by an elongate, segmented trunk with many legs. Includes centipedes and millipedes.

– N –

nanoarchaeotes (*Nanoarchaeota*) A hypothetical group of extremely small, thermophilic archaeans [3] with a much-reduced genome. The only described example can survive only when attached to a host organism.

nematodes (*Nematoda*) [100] A very large group of elongate, unsegmented ecdysozoans [24] with thick, multilayer cuticles. They are among the most abundant and diverse animals, although most species have not yet been described. Include free-living predators and scavengers, as well as parasites of most species of land plants [10] and animals [19].

neognaths (*Neognathae*) The main group of birds [122], including all living species except the ratites (ostrich, emu, rheas, kiwis, cassowaries) and tinamous (*see* palaeognaths).

– O –

oligochaetes (*Oligochaeta*) An annelid [94] group whose members lack parapodia, eyes, and anterior tentacles, and have few setae. Earthworms are the most familiar oligochaetes.

onychophorans (*Onychophora*) [102] Elongate, segmented ecdysozoans [24] with many pairs of soft, unjointed, claw-bearing legs. Also known as velvet worms.

oomycetes (*Oomycota*) [48] Water molds and relatives; absorptive heterotrophs with nutrient-absorbing, filamentous hyphae.

opisthokonts (*Opisthokonta*) [18] A group of unikonts [16] in which the flagellum on motile cells, if present, is posterior. The opisthokonts include the fungi [17], animals [19], and choanoflagellates [82].

ostracods (*Ostracoda*) Marine and freshwater crustaceans [106] that are laterally compressed and protected by two clam-like calcareous or chitinous shells.

– P –

palaeognaths (*Palaeognathae*) A group of secondarily flightless or weakly flying birds [122]. Includes the flightless ratites (ostrich, emu, rheas, kiwis, cassowaries) and the weakly flying tinamous.

parabasalids (*Parabasalia*) [71] A group of unicellular eukaryotes [4] that lack mitochondria; they possess flagella in clusters near the anterior of the cell.

phoronids (*Phoronida*) [93] A small group of sessile, wormlike marine lophotrochozoans [23] that secrete chitinous tubes and feed using a lophophore.

placoderms (*Placodermi*) An extinct group of jawed vertebrates [30] that lacked teeth. Placoderms were the dominant predators in Devonian oceans.

Plantae [8] The most broadly defined plant group. In most parts of this book, we use the word "plant" as synonymous with "land plant" [10], a more restrictive definition.

plasmodial slime molds (*Myxogastrida*) Amoebozoans [76] that in their feeding stage consist of a coenocyte called a plasmodium.

pogonophorans (*Pogonophora*) Deep-sea annelids [94] that lack a mouth or digestive tract; they feed by taking up dissolved organic matter, facilitated by endosymbiotic bacteria in a specialized organ (the trophosome).

polychaetes (*Polychaeta*) A group of mostly marine annelids [94] with one or more pairs of eyes and one or more pairs of feeding tentacles; parapodia and setae extend from most body segments. May be paraphyletic with respect to the clitellates.

priapulids (*Priapulida*) [97] A small group of cylindrical, unsegmented, wormlike marine ecdysozoans [24] that takes its name from its phallic appearance.

"progymnosperms" Paraphyletic group of extinct vascular plants [11] that flourished from the Devonian through the Mississippian periods. The first truly woody plants, and the first with vascular cambium that produced both secondary xylem and secondary phloem, they reproduced by spores rather than by seeds.

"prokaryotes" Not a monophyletic group; as commonly used, includes the bacteria [2] and archaeans [3]. A term of convenience encompassing all cellular organisms that are not eukaryotes.

proteobacteria (*Proteobacteria*) [42] A large and extremely diverse group of Gram-negative bacteria that includes many pathogens, nitrogen fixers, and photosynthesizers. Includes the alpha, beta, gamma, delta, and epsilon proteobacteria.

"protists" This term of convenience does not describe a monophyletic group but is used to encompass a large number of distinct and distantly related groups of eukaryotes, many but far from all of which are microbial and unicellular. Essentially a "catch-all" term for any eukaryote group not contained within the land plants [10], fungi [17], or animals [19].

protostomes (*Protostomia*) [22] One of the two major groups of bilaterians [21]. In protostomes, the mouth typically forms from the blastopore (if present) in early development (contrast with deuterostomes). The major protostome groups are the lophotrochozoans [23] and ecdysozoans [24].

prototherians (*Prototheria*) A mostly extinct group of mammals [119], common during the Cretaceous and early Cenozoic. The three living species—the echidnas and the duck-billed platypus—are the only extant egg-laying mammals.

pterobranchs (*Pterobranchia*) A small group of sedentary marine hemichordates [109] that live in tubes secreted by the proboscis. They have one to nine pairs of arms, each bearing long tentacles that capture prey and function in gas exchange.

pteridophytes (*Pteridophyta*) A group of vascular plants [11], sister to the seed plants [12], characterized by overtopping and possession of megaphylls. The pteridophytes include the horsetails [58], whisk ferns [59], and "ferns" [57].

pycnogonids (*Pycnogonida*) Treated in this book as a group of chelicerates [104], but sometimes considered an independent group of arthropods [25]. Pycnogonids have reduced bodies and very long, slender legs. Also called sea spiders.

– R –

radiolarians (*Radiolaria*) [75] Amoeboid organisms with needle-like pseudopods supported by microtubules. Most have glassy internal skeletons.

ray-finned fishes (*Actinopterygii*) [115] A highly diverse group of freshwater and marine bony vertebrates [31]. They have reduced swim bladders that often function as hydrostatic organs and fins supported by soft rays (lepidotrichia). Includes most familiar fishes.

red algae (*Rhodophyta*) [66] Mostly multicellular, marine algae characterized by the presence of phycoerythrin in their chloroplasts.

reptiles (*Reptilia*) [35] One of the two major groups of extant amniotes [34], supported on the basis of similar skull structure and gene sequences. The term "reptiles" traditionally excluded the birds [122], but the resulting group is then clearly paraphyletic. As used in this book, the reptiles include turtles [120], lepidosaurs [121], birds [122], and crocodilians [123].

rhizaria (*Rhizaria*) [15] Mostly amoeboid unicellular eukaryotes with pseudopods, many with external or internal shells. Includes the foraminiferans [73], cercozoans [74], and radiolarians [75].

rhyniophytes (*Rhyniophyta*) A group of early vascular plants [11] that appeared in the Silurian and became extinct in the Middle Devonian. Possessed dichotomously branching stems with terminal sporangia but no true leaves or roots.

ribbon worms (*Nemertea*) [91] A group of unsegmented lophotrochozoans [23] with an eversible proboscis used to capture prey. Mostly marine, but some species live in fresh water or on land.

rotifers (*Rotifera*) [90] Tiny (< 0.5 mm) lophotrochozoans [23] with a pseudocoelomic body cavity that functions as a hydrostatic organ and a ciliated feeding organ called the corona that surrounds the head. They live in freshwater and wet terrestrial habitats.

roundworms (*Nematoda*) [100] *See* nematodes.

– S –

salamanders (*Caudata*) A group of amphibians [118] with distinct tails in both larvae and adults and limbs set at right angles to the body.

salps *See* thaliaceans

sarcopterygians (*Sarcopterygii*) [32] One of the two major groups of bony vertebrates [31], characterized by jointed appendages (paired fins or limbs).

scyphozoans (*Scyphozoa*) Marine cnidarians [87] in which the medusa stage dominates the life cycle. Commonly known as jellyfish.

sea cucumbers (*Holothuroidea*) Echinoderms [108] with an elongate, cucumber-shaped body and leathery skin. They are scavengers on the ocean floor.

sea spiders *See* pycnogonids.

sea squirts *See* ascidians.

sea stars (*Asteroidea*) Echinoderms [108] with five (or more) fleshy "arms" radiating from an indistinct central disk. Also called starfishes.

sea urchins (*Echinoidea*) Echinoderms [108] with a test (shell) that is covered in spines. Most are globular in shape, although some groups (such as the sand dollars) are flattened.

"seed ferns" A paraphyletic group of loosely related, extinct seed plants that flourished in the Devonian and Carboniferous. Characterized by large, frond-like leaves that bore seeds.

seed plants (*Spermatophyta*) [12] Heterosporous vascular plants [11] that produce seeds; most produce wood; branching is axillary (not dichotomous). The major seed plant groups are gymnosperms [13] and angiosperms [52].

sow bugs *See* isopods.

spirochetes (*Spirochaetes*) [37] Motile, Gram-negative bacteria with a helically coiled structure and characterized by axial filaments.

"sponges" A term of convenience used for a paraphyletic group of relatively asymmetric, filter-feeding animals that lack a gut or nervous system and generally lack differentiated tissues. (*See* glass sponges [83], demosponges [84], and calcareous sponges [85].)

springtails (*Collembola*) Wingless hexapods [107] with springing structures on the third and fourth segments of their bodies. Springtails are extremely abundant in some environments (especially in soil, leaf litter, and vegetation).

squamates (*Squamata*) The major group of lepidosaurs [121], characterized by the possession of movable quadrate bones (which allow the upper jaw to move independently of the rest of the skull) and hemipenes (a paired set of eversible penises, or penes) in males. Includes the lizards (a paraphyletic group), snakes, and amphisbaenians.

starfish (*Asteroidea*) *See* sea stars

stramenopiles (*Heterokonta* or *Stramenopila*) [6] Organisms having, at some stage in their life cycle, two unequal flagella, the longer possessing rows of tubular hairs. Chloroplasts, when present, surrounded by four membranes. Major stramenopile groups include the brown algae [46], diatoms [47], and oomycetes [48].

– T –

tapeworms (*Cestoda*) Parasitic flatworms [89] that live in the digestive tracts of vertebrates as adults, and usually in various other species of animals as juveniles.

tardigrades (*Tardigrada*) [103] Small (< 0.5 mm) ecdysozoans [24] with fleshy, unjointed legs and no circulatory or gas exchange organs. They live in marine sands, in temporary freshwater pools, and on the water films of plants. Also called water bears.

tetrapods (*Tetrapoda*) [33] The major group of sarcopterygians [32]; includes the amphibians [118] and the amniotes [34]. Named for the presence of four jointed limbs (although limbs have been secondarily reduced or lost completely in several tetrapod groups).

thaliaceans (*Thaliacea*) A group of solitary or colonial planktonic marine urochordates [110]. Also called salps.

therians (*Theria*) Mammals [119] characterized by viviparity (live birth). Includes eutherians and marsupials.

theropods (*Theropoda*) Archosaurs [36] with bipedal stance, hollow bones, a furcula ("wishbone"), elongated metatarsals with three-fingered feet, and a pelvis that points backwards. Includes many well-known extinct dinosaurs (such as *Tyrannosaurus rex*), as well as the living birds [122].

trilobites (*Trilobita*) An extinct group of arthropods [25] related to the chelicerates [104]. Trilobites flourished from the Cambrian through the Permian.

tuataras (*Rhyncocephalia*) A group of lepidosaurs [121] known mostly from fossils; there are just two living tuatara species. The quadrate bone of the upper jaw is fixed firmly to the skull. Sister group of the squamates.

tunicates *See* ascidians.

turbellarians (*Turbellaria*) A group of free-living, generally carnivorous flatworms [89]. Their monophyly is questionable.

turtles (*Testudines*) [120] A group of reptiles [35] with a bony carapace (upper shell) and plastron (lower shell) that encase the body.

– U –

unikonts (*Unikonta*) [16] A group of eukaryotes [4] whose motile cells possess a single flagellum. Major unikont groups include the amoebozoans [76], fungi [17], and animals [19].

urochordates (*Urochordata*) [110] A group of chordates [28] that are mostly saclike filter feeders as adults, with motile larvae stages that resemble a tadpole.

– V –

vascular plants (*Tracheophyta*) [11] Plants with xylem and phloem. Major groups include the club mosses [60] and euphyllophytes.

vertebrates (*Vertebrata*) [29] The largest group of chordates [28], characterized by a rigid endoskeleton supported by the vertebral column and an anterior skull encasing a brain. Includes hagfishes [112], lampreys [113], and the jawed vertebrates [30], although some biologists exclude the hagfishes from this group (see craniates).

– W –

water bears *See* tardigrades.

whisk ferns (*Psilotophyta*) [59] Vascular plants [11] lacking leaves and roots.

– Y –

"yeasts" A convenience term for several distantly related groups of unicellular fungi [17].

– Z –

"zygomycetes" [80] A convenience term for a paraphyletic group of fungi [17] in which hyphae of differing mating types conjugate to form a zygosporangium.

Appendix B: Some Measurements Used in Biology

MEASURES OF	UNIT	EQUIVALENTS	METRIC → ENGLISH CONVERSION
Length	meter (m)	base unit	1 m = 39.37 inches = 3.28 feet
	kilometer (km)	1 km = 1000 (10^3) m	1 km = 0.62 miles
	centimeter (cm)	1 cm = 0.01 (10^{-2}) m	1 cm = 0.39 inches
	millimeter (mm)	1 mm = 0.1 cm = 10^{-3} m	1 mm = 0.039 inches
	micrometer (μm)	1 μm = 0.001 mm = 10^{-6} m	
	nanometer (nm)	1 nm = 0.001 μm = 10^{-9} m	
Area	square meter (m^2)	base unit	1 m^2 = 1.196 square yards
	hectare (ha)	1 ha = 10,000 m^2	1 ha = 2.47 acres
Volume	liter (L)	base unit	1 L = 1.06 quarts
	milliliter (mL)	1 mL = 0.001 L = 10^{-3} L	1 mL = 0.034 fluid ounces
	microliter (μL)	1 μL = 0.001 mL = 10^{-6} L	
Mass	gram (g)	base unit	1 g = 0.035 ounces
	kilogram (kg)	1 kg = 1000 g	1 kg = 2.20 pounds
	metric ton (mt)	1 mt = 1000 kg	1 mt = 2,200 pounds = 1.10 ton
	milligram (mg)	1 mg = 0.001 g = 10^{-3} g	
	microgram (μg)	1 μg = 0.001 mg = 10^{-6} g	
Temperature	degree Celsius (°C)	base unit	°C = (°F – 32)/1.8
			0°C = 32°F (water freezes)
			100°C = 212°F (water boils)
			20°C = 68°F ("room temperature")
			37°C = 98.6°F (human internal body temperature)
	Kelvin (K)*	°C + 273	0 K = –460°F
Energy	joule (J)		1 J ≈ 0.24 calorie = 0.00024 kilocalorie†

*0 K (–273°C) is "absolute zero," a temperature at which molecular oscillations approach 0—that is, the point at which motion all but stops.

†A *calorie* is the amount of heat necessary to raise the temperature of 1 gram of water 1°C. The *kilocalorie*, or nutritionist's calorie, is what we commonly think of as a calorie in terms of food.

Glossary

- A -

abdomen (ab′ duh mun) [L. *abdomin*: belly] In arthropods, the posterior segments of the body; in mammals, the part of the body containing the intestines and most other internal organs, posterior to the thorax.

abiotic (a′ bye ah tick) [Gk. *a*: not + *bios*: life] Nonliving. (Contrast with biotic.)

abscisic acid (ab sighs′ ik) A plant growth substance having growth-inhibiting action. Causes stomata to close.

abscission (ab sizh′ un) [L. *abscissio*: break off] The process by which leaves, petals, and fruits separate from a plant.

absorption (1) Of light: complete retention, without reflection or transmission. (2) Of liquids: soaking up (taking in through pores or cracks).

absorption spectrum A graph of light absorption versus wavelength of light; shows how much light is absorbed at each wavelength.

absorptive period When there is food in the gut and nutrients are being absorbed.

abyssal zone (uh biss′ ul) [Gk. *abyssos*: bottomless] The deep ocean, below the point that light can penetrate.

accessory pigments Pigments that absorb light and transfer energy to chlorophylls for photosynthesis.

acetylcholine A neurotransmitter substance that carries information across vertebrate neuromuscular junctions and some other synapses.

acetylcholinesterase An enzyme that breaks down acetylcholine.

acetyl coenzyme A (acetyl CoA) Compound that reacts with oxaloacetate to produce citrate at the beginning of the citric acid cycle; a key metabolic intermediate in the formation of many compounds.

acid [L. *acidus*: sharp, sour] A substance that can release a proton in solution. (Contrast with base.)

acid precipitation Precipitation that has a lower pH than normal as a result of acid-forming precursors introduced into the atmosphere by human activities.

acidic Having a pH of less than 7.0 (a hydrogen ion concentration greater than 10^{-7} molar).

acoelomate Lacking a coelom.

Acquired Immune Deficiency Syndrome *See* AIDS.

acrosome (a′ krow soam) [Gk. *akros*: highest + *soma*: body] The structure at the forward tip of an animal sperm which is the first to fuse with the egg membrane and enter the egg cell.

ACTH (adrenocorticotropin) A pituitary hormone that stimulates the adrenal cortex.

actin [Gk. *aktis*: ray] One of the two major proteins of muscle; it makes up the thin filaments. Forms the microfilaments found in most eukaryotic cells.

action potential An impulse in a neuron taking the form of a wave of depolarization or hyperpolarization imposed on a polarized cell surface.

action spectrum A graph of a biological process versus light wavelength; shows which wavelengths are involved in the process.

activating enzymes Enzymes that catalyze the addition of amino acids to their appropriate tRNAs. Also called aminoacyl-tRNA synthetases.

activation energy (E_a) The energy barrier that blocks the tendency for a set of chemical substances to react.

active site The region on the surface of an enzyme where the substrate binds, and where catalysis occurs.

active transport The energy-dependent transport of a substance across a biological membrane against a concentration gradient—that is, from a region of low concentration (of that substance) to a region of high concentration. (See also primary active transport, secondary active transport; contrast with facilitated diffusion.)

adaptation (a dap tay′ shun) (1) In evolutionary biology, a particular structure, physiological process, or behavior that makes an organism better able to survive and reproduce. Also, the evolutionary process that leads to the development or persistence of such a trait. (2) In sensory neurophysiology, a sensory cell's loss of sensitivity as a result of repeated stimulation.

adaptive radiation The proliferation of members of a single clade into a variety of different adaptive forms.

adenine (A) (a′ den een) A nitrogen-containing base found in nucleic acids, ATP, NAD, and other compounds.

adenosine triphosphate *See* ATP.

adenylate cyclase Enzyme catalyzing the formation of cyclic AMP (cAMP) from ATP.

adrenal (a dree′ nal) [L. *ad*: toward + *renes*: kidneys] An endocrine gland located near the kidneys of vertebrates, consisting of two glandular parts, the cortex and medulla.

adrenaline *See* epinephrine.

adrenocorticotropin *See* ACTH.

adsorption Binding of a gas or a solute to the surface of a solid.

aerobic (air oh′ bic) [Gk. *aer*: air + *bios:* life] In the presence of oxygen; requiring oxygen.

afferent (af′ ur unt) [L. *ad*: toward + *ferre*: to carry] Carrying to, as in a neuron that carries impulses to the central nervous system, or a blood vessel that carries blood to a structure. (Contrast with efferent.)

AIDS (acquired immune deficiency syndrome) Condition caused by a virus (HIV) in which the body's helper T lymphocytes are reduced, leaving the victim subject to opportunistic diseases.

alcoholic fermentation Breakdown of glucose in cells under anaerobic conditions to produce alcohol.

aldehyde (al′ duh hide) A compound with a —CHO functional group. Many sugars are aldehydes. (Contrast with ketone.)

aldosterone (al dohs′ ter own) A steroid hormone produced in the adrenal cortex of mammals. Promotes secretion of potassium and reabsorption of sodium in the kidney.

allantois (al lan′ to is) A sac-like extraembryonic membrane that contains nitrogen waste from embryo.

allele (a leel′) [Gk. *allos*: other] The alternate forms of a genetic character found at a given locus on a chromosome.

allele frequency The relative proportion of a particular allele in a specific population.

allergy [Ger. *allergie*: altered reaction] An overreaction to amounts of an antigen that do not affect most people; often involves IgE antibodies.

allometric growth A pattern of growth in which some parts of the body of an organism grow faster than others, resulting in a change in body proportions as the organism grows.

allopatric speciation (al′ lo pat′ rick) [Gk. *allos*: other + *patria*: homeland] The formation of two species from one when reproductive isolation occurs because of the interposition of (or crossing of) a physical geographic barrier such as a river. Also called geographic speciation. (Contrast with parapatric speciation, sympatric speciation.)

allopolyploidy The possession of more than two entire chromosomes sets that are derived from a single species.

allostery (al′ lo steer y) [Gk. *allos*: other + *stereos*: structure] Regulation of the activity of a protein by the binding of an effector molecule at a site other than the active site.

alpha (α) helix A prevalent type of secondary protein structure; a right-handed spiral.

alternation of generations The succession of multicellular haploid and diploid phases in some sexually reproducing organisms, notably plants.

alternative splicing A process for generating different mature mRNAs from a single gene by splicing together different sets of exons during RNA processing.

altruism Behavior that harms the individual who performs it but benefits other individuals.

alveolus (al ve′ o lus) (plural: alveoli) [L. *alveus*: cavity] A small, baglike cavity, especially the blind sacs of the lung.

amensalism (a men' sul ism) Interaction in which one animal is harmed and the other is unaffected. (Contrast with commensalism, mutualism.)

amine An organic compound with an amino group. (Compare with amino acid.)

amino acid Organic compounds containing both NH_2 and COOH groups. Proteins are polymers of amino acids.

amino acid replacement A change in a protein sequence in which one amino acid is replaced by another.

ammonotelic (am moan' o teel' ic) [Gk. *telos*: end] Describes an organism in which the final product of breakdown of nitrogen-containing compounds (primarily proteins) is ammonia. (Contrast with ureotelic, uricotelic.)

amnion (am' nee on) The fluid-filled sac in which the embryos of reptiles, birds, and mammals develop.

amniote egg A shelled egg surrounding four extraembryonic membranes and embryo-nourishing yolk. This adaptation allowed animals to colonize the terrestrial environment.

amphipathic (am' fi path' ic) [Gk. *amphi*: both + *pathos*: emotion] Of a molecule, having both hydrophilic and hydrophobic regions.

amylase (am' ill ase) Any of a group of enzymes that digest starch.

anabolic reaction A single reaction that participates in anabolism.

anabolism (an ab' uh liz' em) [Gk. *ana*: upward + *ballein*: to throw] Synthetic reactions of metabolism, in which complex molecules are formed from simpler ones. (Contrast with catabolism.)

anaerobic (an ur row' bic) [Gk. *an*: not + *aer*: air + *bios*: life] Occurring without the use of molecular oxygen, O_2.

anagenesis Evolutionary change in a single lineage over time.

analogy (a nal' o jee) [Gk. *analogia*: resembling] A resemblance in function, and often appearance as well, between two structures that is due to convergent evolution rather than to common ancestry. (Contrast with homology.)

anaphase (an' a phase) [Gk. *ana*: upward progress] The stage in nuclear division at which the first separation of sister chromatids (or, in the first meiotic division, of paired homologs) occurs.

anaphylactic shock A precipitous drop in blood pressure caused by loss of fluid from capillaries because of an increase in their permeability stimulated by an allergic reaction.

ancestral trait The trait originally present in the ancestor of a given group; may be retained or changed in the descendants of that ancestor.

androgens (an' dro jens) The male sex steroids.

aneuploidy (an' you ploy dee) A condition in which one or more chromosomes or pieces of chromosomes are either lacking or present in excess.

angiotensin (an' jee oh ten' sin) A peptide hormone that raises blood pressure by causing peripheral vessels to constrict. Also maintains glomerular filtration by constricting efferent vessels and stimulates thirst and the release of aldosterone.

animal hemisphere The metabolically active upper portion of some animal eggs, zygotes, and embryos; does not contain the dense nutrient yolk. (Contrast with vegetal hemisphere.)

anion (an' eye on) [Gk. *ana*: upward progress] A negatively charged ion. (Contrast with cation.)

anisogamy (an eye sog' a mee) [Gk. *aniso*: unequal + *gamos*: marriage] The existence of two dissimilar gametes (egg and sperm).

annual Referring to a plant whose life cycle is completed in one growing season. (Contrast with biennial, perennial.)

antenna system In photosynthesis, a group of different molecules that cooperate to absorb light energy and transfer it to a reaction center.

anterior pituitary The portion of the vertebrate pituitary gland that derives from gut epithelium and produces tropic hormones.

anther (an' thur) [Gk. *anthos*: flower] A pollen-bearing portion of the stamen of a flower.

antheridium (an' thur id' ee um) [Gk. *antheros*: blooming] The multicellular structure that produces the sperm in nonvascular plants and ferns.

antibody One of the millions of proteins produced by the immune system that specifically binds to a foreign substance and initiates its removal from the body.

anticodon The three nucleotides in transfer RNA that pair with a complementary triplet (a codon) in messenger RNA.

antidiuretic hormone (ADH) A hormone that promotes water reabsorption by the kidney. ADH is produced by neurons in the hypothalamus and released from nerve terminals in the posterior pituitary. Also called vasopressin.

antigen (an' ti jun) Any substance that stimulates the production of an antibody or antibodies in the body of a vertebrate.

antigenic determinant A specific region of an antigen, which is recognized by and binds to a specific antibody.

antiparallel Pertaining to molecular orientation in which a molecule or parts of a molecule have opposing directions.

antipodal cell At one end of the megagametophyte, one of the three cells which eventually degenerate.

antiport A membrane transport process that carries one substance in one direction and another in the opposite direction. (Contrast with symport.)

antisense nucleic acid A single-stranded RNA or DNA complementary to and thus targeted against the mRNA transcribed from a harmful gene such as an oncogene.

anus (a' nus) Opening through which digestive wastes are expelled, located at the posterior end of the gut.

aorta (a or' tah) [Gk. *aorte*: aorta] The main trunk of the arteries leading to the systemic (as opposed to the pulmonary) circulation.

aortic body A nodule of modified muscle tissue on the aorta that is sensitive to the oxygen supply provided by arterial blood.

apex (a' pecks) The tip or highest point of a structure, as the apex of a growing stem or root.

apical (a' pi kul) Pertaining to the apex, or tip, usually in reference to plants.

apical dominance Inhibition by the apical bud of the growth of axillary buds.

apical meristem The meristem at the tip of a shoot or root; responsible for the plant's primary growth.

apomixis (ap oh mix' is) [Gk. *apo*: away from + *mixis*: sexual intercourse] The asexual production of seeds.

apoplast (ap' oh plast) in plants, the continuous meshwork of cell walls and extracellular spaces through which material can pass without crossing a plasma membrane. (Contrast with symplast.)

apoptosis (ap uh toh' sis) A series of genetically programmed events leading to cell death.

aquaporin A transport protein in plant and animal cells through which water passes in osmosis.

aquatic (a kwa' tic) [L. *aqua*: water] Living in water. (Compare with marine, terrestrial.)

aqueous (a' kwee us) Pertaining to water or a watery solution.

archegonium (ar' ke go' nee um) [Gk. *archegonos*: first, foremost] The multicellular structure that produces eggs in nonvascular plants, ferns, and gymnosperms.

archenteron (ark en' ter on) [Gk. *archos*: first + *enteron*: bowel] The earliest primordial animal digestive tract.

area phylogenies Phylogenies in which the names of the taxa are replaced with the names of the places where those taxa live or lived.

arteriosclerosis *See* atherosclerosis.

artery A muscular blood vessel carrying oxygenated blood away from the heart to other parts of the body. (Contrast with vein.)

artficial selection The selection by plant and animal breeders of individuals with certain desirable traits.

ascus (ass' cuss) [Gk. *askos*: bladder] In ascomycete fungi (sac fungi), the club-shaped sporangium within which spores (ascospores) are produced by meiosis.

asexual Without sex.

assortative mating A breeding system in which mates are selected on the basis of a particular trait or group of traits.

atherosclerosis (ath' er oh sklair oh' sis) [Gk. *athero*: gruel, porridge + *skleros*: hard] A disease of the lining of the arteries characterized by fatty, cholesterol-rich deposits in the walls of the arteries. When fibroblasts infiltrate these deposits and calcium precipitates in them, the disease become arteriosclerosis, or "hardening of the arteries."

atmosphere The gaseous mass surrounding our planet. Also a unit of pressure, equal to the normal pressure of air at sea level.

atom [Gk. *atomos*: indivisible] The smallest unit of a chemical element. Consists of a nucleus and one or more electrons.

atomic mass The average mass of an atom of an element; the average depends on the relative amounts of ifferent isotopes of the element on Earth. Also called atomic weight.

atomic number The number of protons in the nucleus of an atom; also equals the number of electrons around the neutral atom. Determines the chemical properties of the atom.

ATP (adenosine triphosphate) An energy-storage compound containing adenine, ribose, and three phosphate groups. When it is formed from ADP, useful energy is stored; when it is broken down (to ADP or AMP), energy is released to drive endergonic reactions.

ATP synthase An integral membrane protein that couples the transport of proteins with the formation of ATP.

atrioventricular node A modified node of cardiac muscle that organizes the action potentials that control contraction of the ventricles.

atrium (a' tree um) [L. *atrium*: central hall] An internal chamber. In the hearts of vertebrates, the thin-walled chamber(s) entered by blood on its way to the ventricle(s). Also, the outer ear.

autocrine Referring to a cell signaling mechanism in which the signal binds to and affects the cell that makes it. An autocrine hormone, for example, is one that influences the cell that releases it. (Compare with endocrine gland;, paracrine)

autoimmune disease A disorder in which the immune system attacks the animal's own antigens.

autonomic nervous system The system that controls such involuntary functions as those of guts and glands. (Compare with central nervous system.)

autopolyploidy The possession of more than two chromosome sets that are derived from more than one species.

autosome Any chromosome (in a eukaryote) other than a sex chromosome.

autotroph (au' tow trow' fik) [Gk. *autos:* self + *trophe*: food] An organism that is capable of living exclusively on inorganic materials, water, and some energy source such as sunlight or chemically reduced matter. (Contrast with heterotroph.)

auxin (awk' sin) [Gk. *auxein*: to grow] In plants, a substance (the most common being indoleacetic acid) that regulates growth and various aspects of development.

auxotroph (awks' o trofe) [Gk. *auxein*: to grow + *trophe*: food] A mutant form of an organism that requires a nutrient or nutrients not usually required by the wild type. (Contrast with prototroph.)

Avogadro's number The number of atoms or molecules in a mole (weighed out in grams) of a substance, calculated to be 6.022×10^{23}.

***Avr* genes (avirulence genes)** Genes in a pathogen that may trigger defenses in plants. *See* gene-for-gene resistance.

axillary bud A bud occurring in the upper angle (axil) between a leaf and stem.

axon [Gk. *axon*: axle] The part of a neuron that conducts action potentials away from the cell body.

axon hillock The junction between an axon and its cell body, where action potentials are generated.

axon terminals The endings of an axon; they form synapses and release neurotransmitter.

axoneme (ax' oh neem) The complex of microtubules and their crossbridges that forms the motile apparatus of a cilium.

- B -

bacillus (bah sil' us) [L: little rod] Any of various rod-shaped bacteria.

bacterial artificial chromosome (BAC) A DNA cloning vector used in bacteria that can carry up to 150,000 base pairs of foreign DNA.

bacteriophage (bak teer' ee o fayj) [Gk. *bakterion*: little rod + *phagein*: to eat] One of a group of viruses that infect bacteria and ultimately cause their disintegration.

bacteroids Nitrogen-fixing organelles that develop from endosymbiotic bacteria.

balanced polymorphism [Gk. *polymorphos*: many forms] The maintenance of more than one form, or the maintenance at a given locus of more than one allele, at frequencies of greater than 1 percent in a population. Often results when heterozygotes are more fit than either homozygote.

bark All tissues outside the vascular cambium of a plant.

baroreceptor [Gk. *baros*: weight] A pressure-sensing cell or organ.

Barr body In female mammals, an inactivated X chromosome.

basal Pertaining to one end—the base—of an axis.

basal body Centriole found at the base of a eukaryotic flagellum or cilium.

basal metabolic rate (BMR) The minimum rate of energy turnover in an awake (but resting) bird or mammal that is not expending energy for thermoregulation.

base (1) A substance tha can accept a hydrogen ion in solution. (Contrast with acid.) (2) In nucleic acids, the purine or pyrimidine that is attached to each sugar in the backbone.

base pairing *See* complementary base pairing.

basic Having a pH greater than 7.0 (i.e., having a hydrogen ion concentration lower than 10^{-7} molar).

basidium (bass id' ee yum) In basidiomycete fungi, the characteristic sporangium in which four spores are formed by meiosis and then borne externally before being shed.

basophils One type of phagocytic white blood cell that releases histamine and may promote T cell development.

Batesian mimicry The convergence in appearance of an edible species (mimic) with an unpalatable species (model).

B cell A type of lymphocyte involved in the humoral immune response of vertebrates. Upon recognizing an antigenic determinant, a B cell develops into a plasma cell, which secretes an antibody. (Contrast with T cell.)

benefit An improvement in survival and reproductive success resulting from performing a behavior or having a trait. (Contrast with cost.)

benign (be nine') Referring to a tumor that grows to a certain size and then stops, uaually with a fibrous capsule surrounding the mass of cells. Benign tumors do not spread (metastasize) to other organs. (Contrast with malignant.)

benthic zone [Gk. *benthos*: bottom] The bottom of the ocean.

beta (β) pleated sheet Type of protein secondary structure; results from hydrogen bonding between polypeptide regions running antiparallel to each other.

biennial Referring to a plant whose life cycle includes vegetative growth in the first year and flowering and senescence in the second year. (Contrast with annual, perennial.)

bilateral symmetry The condition in which only the right and left sides of an organism, divided exactly down the back, are mirror images of each other. (Contrast with biradial symmetry.)

bilayer In membranes, a structure that is two lipid layers in thickness.

bile A secretion of the liver delivered to the small intestine via the common bile duct. In the intestine, bile emulsifies fats.

binary fission Reproduction by cell division of a single-celled organism.

binding domain The region of a receptor molecule where its ligand attaches.

binocular cells Neurons in the visual cortex that respond to input from both retinas; involved in depth perception.

binomial (bye nome' ee al) Consisting of two; for example, the binomial nomenclature of biology in which each species has two names (the genus name followed by the species name).

biodiversity crisis The current high rate of loss of species, caused primarily by human activities.

biofilm A community of microorganisms embedded in a polysaccharide matrix, forming a highly resistant coating on almost any moist surface.

biogeochemical cycle Movement of elements through living organisms and the physical environment.

biogeographic region One of several defined, continental-scale regions of Earth, each of which has a biota distinct from that of the others. (Contrast with biome.)

bioinformatics The use of computers and/or mathematics to analyze complex biological information, such as DNA sequences.

bioluminescence The production of light by biochemical processes in an organism.

biomass The total weight of all the living organisms, or some designated group of living organisms, in a given area.

biome (bye' ome) A major division of the ecological communities of Earth, characterized primarily by distinctive vegetation. A given biogeographic region contains many different biomes.

biosphere (bye' oh sphere) All regions of Earth (terrestrial and aquatic) and Earth's atmosphere in which organisms can live.

biota (bye oh' tah) All of the organisms—animals, plants, fungi, and microorganisms—found in a given area. (Contrast with flora, fauna.)

biotic (bye ah' tick) [Gk. *bios*: life] Alive. (Contrast with abiotic.)

biradial symmetry Radial symmetry modified so that only two planes can divide the animal into similar halves.

blastocoel (blass' toe seal) [Gk. *blastos*: sprout + *koilos*: hollow] The central, hollow cavity of a blastula.

blastocyst (blass' toe cist) An early embryo formed by the first divisions of the fertilized egg (zygote). In mammals, a hollow ball of cells.

blastodisc (blass' toe disk) A disk of cells forming on the surface of a large yolk mass, comparable to a blastula, but occurring in animals such as birds and reptiles, in which the massive yolk restricts cleavage to one side of the egg only.

blastomere A cell produced by the division of a fertilized egg.

blastopore The opening from the archenteron to the exterior of a gastrula.

blastula (blass' chu luh) An early stage in animal embryology; in many species, a hollow sphere of cells surrounding a central cavity, the blastocoel. (Contrast with blastodisc.)

blood–brain barrier A property of the blood vessels of the brain that prevents most chemicals from diffusing from the blood into the brain.

blood plasma (plaz' muh) [Gk. *plassein*: to mold] The liquid portion of blood, in which blood cells and other particulates are suspended.

blue light receptors Pigments in plants that absorbs blue light (400–500 nm). These pigments mediate many plant responses including phototropism, stomatal movements, and expression of some genes.

body cavity Membrane-lined, fluid-filled compartment that lies between the cell layers of many animals.

Bohr effect The fact that low pH decreases the affinity of hemoglobin for oxygen.

bond *See* chemical bond.

bottleneck A stressful period during which only a few individuals of a once large population survive.

Bowman's capsule An elaboration of kidney tubule cells that surrounds a know of capillaries (the glomerulus). Blood is filtered across the walls of these capillaries and the filtrate is collected into Bowman's capsule.

brain stem The portion of the vertebrate brain between the spinal cord and the forebrain.

brassinosteroids Plant steroid hormones that mediate light effects promoting the elongation of stems and pollen tubes.

bronchioles The smallest airways in a vertebrate lung, branching off the bronchi.

bronchus (plural: bronchi) The major airway(s) branching off the trachea into the vertebrate lung.

brown fat Fat tissue in mammals that is specialized to produce heat. It has many mitochondria and capillaries, and a protein that uncouples oxidative phosphorylation.

browser An animal that feeds on the tissues of woody plants.

budding Asexual reproduction in which a more or less complete new organism simply grows from the body of the parent organism and eventually detaches itself.

buffer A substance that can transiently accept or release hydrogen ions and thereby resist changes in pH.

bundle of His Fibers of modified cardiac muscle that conduct action potentials from the atria to the ventricular muscle mass.

bundle sheath cell Part of a tissue that surrounds the veins of plants; contains chloroplasts in C_4 plants.

- C -

C_3 photosynthesis Form of photosynthesis in which 3-phosphoglycerate is the first stable product, and ribulose bisphosphate is the CO_2 receptor.

C_4 photosynthesis Form of photosynthesis in which oxaloacetate is the first stable product, and phosphoenolpyruvate is the CO_2 acceptor. C_4 plants also perform the reactions of C_3 photosynthesis.

calcitonin A hormone produced by the thyroid gland; it lowers blood calcium and promotes bone formation. (Compare with parathyroid hormone.)

calmodulin (cal mod' joo lin) A calcium-binding protein found in all animal and plant cells; mediates many calcium-regulated processes.

calorie [L. *calor*: heat] The amount of heat required to raise the temperature of one gram of water by one degree Celsius (1°C) from 14.5°C to 15.5°C. Calorie spelled with a capital C refers to the kilocalorie (1 kcal = 1,000 cal).

Calvin cycle The stage of photosynthesis in which CO_2 reacts with RuBP to form 3PG, 3PG is reduced to a sugar, and RuBP is regenerated, while other products are released to the rest of the plant. Also known as the Calvin–Benson cycle.

calyx (kay' licks) [Gk. *kalyx*: cup] All of the sepals of a flower, collectively.

CAM *See* crassulacean acid metabolism.

cambium (kam' bee um) [L. *cambiare*: to exchange] A meristem that gives rise to radial rows of cells in stem and root, increasing them in girth; commonly applied to the vascular cambium which produces wood and phloem, and the cork cambium, which produces bark.

Cambrian explosion The rapid diversification of life that took place during the Cambrian period.

cAMP (cyclic AMP) A compound formed from ATP that mediates the effects of numerous animal hormones.

canopy The leaf-bearing part of a tree. Collectively, the aggregate of the leaves and branches of the larger woody plants of an ecological community.

capillaries [L. *capillaris*: hair] Very small tubes, especially the smallest blood-carrying vessels of animals between the termination of the arteries and the beginnings of the veins. Capillaries are the site of exchange of materials between the blood and the interstitial fluid.

capsid The outer shell of a virus that encloses its nucleic acid.

carbohydrates Organic compounds containing carbon, hydrogen, and oxygen in the ratio 1:2:1 (i.e., with the general formula $C_nH_{2n}O_n$). Common examples are sugars, starch, and cellulose.

carboxylase An enzyme that catalyzes the addition of carboxyl groups to a substrate.

carboxylic acid (kar box sill' ik) An organic acid containing the carboxyl group, —COOH, which dissociates to the carboxylate ion, $—COO^-$.

carcinogen (car sin' oh jen) A substance that causes cancer.

cardiac (kar' dee ak) [Gk. *kardia*: heart] Pertaining to the heart and its functions.

cardiac muscle One of the three types of muscle tissue, it makes up the vertebrate heart. Characterized by branching cells with single nuclei and striated (striped) appearance. (Contrast with smooth muscle, striated muscle.)

carnivore [L. *carn*: flesh + *vovare*: to devour] An organism that eats animal tissues. (Contrast with detritivore, herbivore, omnivore.)

carotenoid (ka rah' tuh noid) A yellow, orange, or red lipid pigment commonly found as an accessory pigment in photosynthesis; also found in fungi.

carpel (kar' pel) [Gk. *karpos*: fruit] The organ of the flower that contains one or more ovules.

carrier (1) In facilitated diffusion, a membrane protein that binds a specific molecule and transports it through the membrane. (2) In respiratory and photosynthetic electron transport, a participating substance such as NAD that exists in both oxidized and reduced forms. (3) In genetics, a person heterozygous for a recessive trait.

carrier protein *See* Carrier (1).

cartilage In vertebrates, a tough connective tissue found in joints, the outer ear, and elsewhere. Forms the entire skeleton in some animal groups.

Casparian strip A band of cell wall containing suberin and lignin, found in the endodermis. Restricts the movement of water across the endodermis.

caspase A member of a group of proteases that catalyze cleavage of target proteins and are active in apoptosis.

catabolic reaction A single reaction that participates in catabolism.

catabolism [Gk. *kata*: to break down + *ballein*: to throw] Degradational reactions of metabolism, in which complex molecules are broken down. (Contrast with anabolism.)

catabolite repression In the presence of abundant glucose, the diminished synthesis of catabolic enzymes for other energy sources.

catalyst (cat' a list) [Gk. *kata*: to break down] A chemical substance that accelerates a reaction without itself being consumed in the overall course of the reaction. Catalysts lower the activation energy of a reaction. Enzymes are biological catalysts.

catalytic subunit The polypeptide in an enzyme protein with quaternary structure that contains the active site for the enzyme. (Contrast regulatory subunit.)

cation (cat' eye on) An ion with one or more positive charges. (Contrast with anion.)

caudal [L. *cauda*: tail] Pertaining to the tail, or to the posterior part of the body.

cDNA *See* complementary DNA.

cecum (see' cum) [L. *caecus*: blind] A blind branch off the large intestine. In many nonruminant mammals, the cecum contains a colony of microorganisms that contribute to the digestion of food.

cell adhesion molecules Molecules on animal cell surfaces that affect the selective association of cells during development of the embryo.

cell cycle The stages through which a cell passes between one division and the next. Includes all stages of interphase and mitosis.

cell division The reproduction of a cell to produce two new cells. In eukaryotes, this process involves nuclear division (mitosis) and cytoplasmic division (cytokinesis).

cell junctions Specialized structures associated with the plasma membranes of epithelial cells. Some contribute to cell adhesion, others to intercellular communication.

cell plate A structure that forms at the equator of the spindle in dividing plant cells.

cell recognition Binding of cells to one another mediated by membrane proteins or carbohydrates.

cell wall A relatively rigid structure that encloses cells of plants, fungi, many protists, and most prokaryotes. Gives these cells their shape and limits their expansion in hypotonic media.

cellular immune response Action of the immune system based on the activities of T cells. Directed against parasites, fungi, intracellular viruses, and foreign tissues (grafts). (Contrast with humoral immune system.)

cellular respiration *See* respiration.

cellulose (sell' you lowss) A straight-chain polymer of glucose molecules, used by plants as a structural supporting material.

centimorgan In genetic mapping, a recombinant frequency of 0.01; a map unit.

central dogma The statement that information flows from DNA to RNA to polypeptide (in retroviruses, there is also information flow from RNA to cDNA).

central nervous system That part of the nervous system which is condensed and centrally located, e.g., the brain and spinal cord of vertebrates; the chain of cerebral, thoracic and abdominal ganglia of arthropods. (Compare with autonomic nervous system.)

centrifuge [L. *centrum*: center + *fugere*: to flee] A laboratory device in which a sample is spun around a central axis at high speed. Used to separate suspended materials of different densities.

centriole (sen' tree ole) A paired organelle that helps organize the microtubules in animal and protist cells during nuclear division.

centromere (sen' tro meer) [Gk. *centron*: center + *meros*: part] The region where sister chromatids join.

centrosome (sen' tro soam) The major microtubule organizing center of an animal cell.

cephalization (sef ah luh zay' shun) [Gk. *kephale*: head] The evolutionary trend toward increasing concentration of brain and sensory organs at the anterior end of the animal.

cerebellum (sair uh bell' um) [L.: diminutive of *cerebrum*, brain] The brain region that controls muscular coordination; located at the anterior end of the hindbrain.

cerebral cortex The thin layer of gray matter (neuronal cell bodies) that overlays the cerebrum.

cerebrum (su ree' brum) [L. *cerebrum*: brain] The dorsal anterior portion of the forebrain, making up the largest part of the brain of mammals. In mammals, the chief coordination center of the nervous system; consists of two cerebral hemispheres.

cervix (sir' vix) [L. *cervix*: neck] The opening of the uterus into the vagina.

cGMP (cyclic guanosine monophosphate) An intracellular messenger that is part of signal transmission pathways involving G proteins. (*See* G protein.)

channel protein A membrane protein that forms an aqueous passageway though which specific solutes may pass.

chemical bond An attractive force stably linking two atoms.

chemical equilibrium A state reached in a reversible chemical reaction when the forward and reverse reactions balance each other and there is no net change.

chemical evolution The theory that life originated through the chemical transformation of inanimate substances.

chemical reaction The change in the composition or distribution of atoms of a substance with consequent alterations in properties.

chemiosmosis The formation of ATP in mitochondria and chloroplasts, resulting from a pumping of protons across a membrane (against a gradient of electrical charge and of pH), followed by the return of the protons through a protein channel with ATPase activity.

chemoautotroph *See* chemolithotroph.

chemoheterotroph An organism that must obtain both carbon and energy from organic substances. (Contrast with chemolithotroph, photoautotroph, photoheterotroph.)

chemolithotroph [Gk. *lithos*: stone, rock] An organism that uses carbon dioxide as a carbon source and obtains energy by oxidizing inorganic substances from its environment. (Contrast with chemoheterotroph, photoautotroph, photoheterotroph.)

chemoreceptor A sensory receptor cell or tissue that senses specific substances in its environment.

chemosynthesis Synthesis of food substances, using the oxidation of reduced materials from the environment as a source of energy.

chiasma (kie az' muh) (plural: chiasmata) [Gk. *chiasmata*: cross] An X-shaped connection between paired homologous chromosomes in prophase I of meiosis. A chiasma is the visible manifestation of crossing over between homologous chromosomes.

chitin (kye' tin) [Gk. *kiton*: tunic] The characteristic tough but flexible organic component of the exoskeleton of arthropods, consisting of a complex, nitrogen-containing polysaccharide. Also found in cell walls of fungi.

chlorophyll (klor' o fill) [Gk. *kloros*: green + *phyllon*: leaf] Any of a few green pigments associated with chloroplasts or with certain bacterial membranes; responsible for trapping light energy for photosynthesis.

chloroplast [Gk. *kloros*: green + *plast*: a particle] An organelle bounded by a double membrane containing the enzymes and pigments that perform photosynthesis. Chloroplasts occur only in eukaryotes.

choanocyte (ko' an uh site) The collared, flagellated feeding cells of sponges.

cholecystokinin (ko' luh sis tuh kai' nin) A hormone produced and released by the lining of the duodenum when it is stimulated by undigested fats and proteins. It stimulates the gallbladder to release bile and slows stomach activity.

chorion (kor' ee on) [Gk. *khorion*: afterbirth] The outermost of the membranes protecting mammal, bird, and reptile embryos; in mammals it forms part of the placenta.

chromatid (kro' ma tid) Each of a pair of new sister chromosomes from the time at which the molecular duplication occurs until the time at which the centromeres separate at the anaphase of nuclear division.

chromatin The nucleic acid–protein complex found in eukaryotic chromosomes.

chromatin remodeling Changes in chromatin structure to allow transcription, translation, and chromosome condensation.

chromatophore (krow mat' o for) [Gk. *kroma*: color + *phoreus*: carrier] A pigment-bearing cell that expands or contracts to change the color of the organism.

chromosomal mutation Loss of or changes in position/direction of a DNA segment on a chromosome.

chromosome (krome' o sowm) [Gk. *kroma*: color + *soma*: body] In bacteria and viruses, the DNA molecule that contains most or all of the genetic information of the cell or virus. In eukaryotes, a structure composed of DNA and proteins that bears part of the genetic information of the cell.

chylomicron (ky low my' cron) Particles of lipid coated with protein, produced in the gut from dietary fats and secreted into the extracellular fluids.

chyme (kime) [Gk. kymus, juice] Created in the stomach; a mixture of ingested food with the digestive juices secreted by the salivary glands and the stomach lining.

cilium (sil' ee um) (plural: cilia) [L.: eyelash] Hairlike organelle used for locomotion by many unicellular organisms and for moving water and mucus by many multicellular organisms. Generally shorter than a flagellum.

circadian rhythm (sir kade' ee an) [L. *circa*: approximately + *dies*: day] A rhythm in behavior, growth, or some other activity that recurs about every 24 hours under constant conditions.

circannual rhythm [L. *circa*: approximately + *annus*: year) A rhythm of behavior, growth, or some other activity that recurs on a yearly basis.

citric acid cycle A set of chemical reactions in cellular respiration, in which acetyl CoA is oxidized to carbon dioxide, and hydrogen atoms are stored as NADH and $FADH_2$. Also called the Krebs cycle.

clade [Gk. *klados*: branch] A monophyletic group made up of an ancestor and all of its descendants.

class I MHC molecules These cell surface proteins participate in the cellular immune response directed against virus-infected cells.

class II MHC molecules These cell surface proteins participate in the cell-cell interactions (of helper T cells, macrophages, and B cells) of the humoral immune response.

class switching The process whereby a plasma cell changes the class of immunoglobulin that it synthesizes by changing the DNA region coding for the C segment.

clathrin A fibrous protein on the inner surfaces of animal cell membranes that strengthens coated vesicles and thus participates in receptor-mediated endocytosis.

cleavage First divisions of the fertilized egg of an animal.

climate The average of the atmospheric conditions (temperature, precipitation, wind direction and velocity) found in a region over time.

cline A gradual change in the traits of a species over a geographical gradient.

cloaca (klo ay′ kuh) [L. *cloaca*: sewer] In some invertebrates, the posterior part of the gut; in many vertebrates, a cavity receiving material from the digestive, reproductive, and excretory systems.

clonal anergy Prevention of the synthesis of antibodies against the body's own antigens. When a T cell binds to a self-antigen, it does not receive signals from an antigen-presenting cell; thus the T cell dies (becomes anergic) rather than yielding a clone of active cells.

clonal deletion The inactivation or destruction of lymphocyte clones that would produce immune reactions against the animal's own body.

clonal selection The mechanism by which exposure to antigen results in the activation of selected T- or B-cell clones, resulting in an immune response.

clone [Gk. *klon*: twig, shoot] Genetically identical cells or organisms produced from a common ancestor by asexual means.

cnidocytes (nye′ duh sites) The feeding cells of cnidarians, within which nematocysts are housed.

coacervate (ko as′ er vate) [L. *coacervare*: to heap up] An aggregate of colloidal particles in suspension.

coated vesicle Cytoplasmic vesicle containing distinctive proteins, including clathrin.

coccus (kock′ us) [Gk. *kokkos*: berry, pit] Any of various spherical or spheroidal bacteria.

cochlea (kock′ lee uh) [Gk. *kokhlos* snail] A spiral tube in the inner ear of vertebrates; it contains the sensory cells involved in hearing.

codominance A condition in which two alleles at a locus produce different phenotypic effects and both effects appear in heterozygotes.

codon Three nucleotides in messenger RNA that direct the placement of a particular amino acid into a polypeptide chain. (Contrast with anticodon.)

coelom (see′ lum) [Gk. *koiloma*: cavity] The body cavity of certain animals; the coelom is lined with cells of mesodermal origin.

coelomate Having a coelom.

coenocyte (seen′ a sight) [Gk. *koinos*: common + *kytos*: container] A "cell" enclosed by a single plasma membrane but containing many nuclei.

coenzyme A nonprotein organic molecule that plays a role in catalysis by an enzyme.

cofactor An inorganic ion that is weakly bound to an enzyme and required for its activity.

cohesin Proteins involved in binding chromatids together.

cohesion The tendency of molecules (or any substances) to stick together.

cohort (co′ hort) [L. *cohors*: company of soldiers] A group of similar-aged organisms, considered as it passes through time.

coleoptile A sheath that surrounds and protects the shoot apical meristem and young primary leaves of a grass seedling as they move through the soil.

collagen [Gk. *kolla*: glue] A fibrous protein found extensively in bone and connective tissue.

collecting duct In vertebrates, a tubule that receives urine produced in the nephrons of the kidney and delivers that fluid to the ureter for excretion.

collenchyma (cull eng′ kyma) [Gk. *kolla*: glue + *enchyma*: infusion] A type of plant cell, living at functional maturity, which lends flexible support by virtue of primary cell walls thickened at the corners. (Contrast with parenchyma, sclerenchyma.)

colon [Gk. *kolon*] The large intestine.

common bile duct A single duct that delivers bile from the gallbladder and secretions from the pancreas into the small intestine.

communication A signal from one organism (or cell) that alters the functioning or behavior of another organism (or cell).

community Any ecologically integrated group of species of microorganisms, plants, and animals inhabiting a given area.

companion cell A specialized cell found adjacent to a sieve tube element in flowering plants.

comparative genomics Computer-aided comparison of DNA sequences between different organisms to reveal genes with related functions.

comparative method An experimental design in which two samples or populations exposed to different conditions or treatments are compared to each other.

compensation point The light intensity at which the rates of photosynthesis and of cellular respiration are equal.

competition In ecology, use of the same resource by two or more species, when the resource is present in insufficient supply for the combined needs of the species.

competitive exclusion A result of competition between species for a limiting resource in which one species completely eliminates the other.

competitive inhibitor A nonsubstrate that binds to the active site of an enzyme and thereby inhibits binding of substrate and reaction from part of the environment.

complement system A group of eleven proteins that play a role in some reactions of the immune system. The complement proteins are not immunoglobulins.

complementary base pairing The AT (or AU), TA (or UA), CG, and GC pairing of bases in double-stranded DNA, in transcription, and between tRNA and mRNA.

complementary DNA (cDNA) DNA formed by reverse transcriptase acting with an RNA template; essential intermediate in the reproduction of retroviruses; used as a tool in recombinant DNA technology; lacks introns.

complete metamorphosis A change of state during the life cycle of an organism in which the body is almost completely rebuilt to produce an individual with a very different body form. Characteristic of insects such as butterflies, moths, beetles, ants, wasps, and flies.

compound (1) A substance made up of atoms of more than one element. (2) A structure made up of many units, as the compound eyes of arthropods.

concerted evolution The common evolution of a family of repeated genes, such that changes in one copy of the gene family are replicated in other copies of the gene family.

condensation reaction A reaction in which two molecules become connected by a covalent bond and a molecule of water is released. ($AH + BOH \rightarrow AB + H_2O$.)

condensing A protein complex involved in chromosome condensation during mitosis and meiosis.

conditional mutations Mutations that show characteristic phenotype only under certain environmental conditions such as temperature.

conduction The transfer of heat from one object to another through direct contact. In neurophysiology, the progression of an action potential along an axon.

cone cells (1) In the vertebrate retina, photoreceptor cells that are responsible for color vision. (2) In gymnosperms, reproductive structures consisting of spore-bearing scales inserted on a short axis; the scales are modified branches. (Contrast with strobilus.)

conformation The three-dimensional shape of a protein or other macromolecule.

conidium (ko nid′ ee um) [Gk. *konis*: dust] An asexual fungus spore borne singly or in chains either apically or laterally on a hypha.

conifer (kahn′ e fer) [Gr. *konos*: cone + *phero*: carry] One of the cone-bearing gymnosperms, mostly trees, such as pines and firs.

conjugation (kon ju gay′ shun) [L. *conjugare*: yoke together] The close approximation of two cells during which they exchange genetic material, as in *Paramecium* and other ciliates, or during which DNA passes from one to the other, as in bacteria.

conjugation tube Cytoplasmic connection between two bacterial cells through which DNA passes during conjugation.

connective tissue An animal tissue that connects or surrounds other tissues; its cells are embedded in a collagen-containing matrix.

connexon In a gap junction, a protein channel linking adjacent animal cells.

consensus sequences Short stretches of DNA that appear, with little variation, in many different genes.

conservation biology An applied, normative science that carries out investigations designed to help preserve the diversity of life on Earth.

constant region For a particular class of immunoglobulin molecules, the region with identical amino acid composition.

constitutive enzyme An enzyme that is present in approximately constant amounts in a system, whether its substrates are present or absent. (Contrast with inducible enzyme.)

consumer An organism that eats the tissues of some other organism.

continental drift The gradual movements of the world's continents that have occurred over billions of years.

contraception The act of preventing union of sperm and egg.

controlled experiment An experimental design in which a sample or population is divided into two groups; one group is exposed to a manipulated variable while the other group serves as a nontreated control. The two groups are compared to see if there are changes in a "dependent" variable as a result of the experimental manipulation.

convection The transfer of heat to or from a surface via a moving stream of air or fluid.

convergent evolution The independent evolution of similar features from different ancestral traits.

copulation Reproductive behavior that results in a male depositing sperm in the reproductive tract of a female.

coprophagy Ingesting ones own feces.

corepressor A low-molecular-weight compound that unites with a protein (the repressor) to prevent transcription in a repressible operon.

cork A waterproofing tissue in plants, with suberin-containing cell walls. Produced by a cork cambium.

cornea The clear, transparent tissue that covers the eye and allows light to pass through to the retina.

corolla (ko role' lah) [L. *corolla*: a small crown] All of the petals of a flower, collectively.

coronary (kor' oh nair ee) [L. *corona*: crown] Referring to the blood vessels of the heart.

coronary thrombosis A fibrous clot that blocks a coronary vessel.

corpus luteum (kor' pus loo' tee um) [L.: yellow body] A structure formed from a follicle after ovulation; it produces hormones important to the maintenance of pregnancy.

cortex [L. *cortex*: covering, rind] (1) In plants, the tissue between the epidermis and the vascular tissue of a stem or root. (2) In animals, the outer tissue of certain organs, such as the adrenal cortex and cerebral cortex.

corticosteroids Steroid hormones produced and released by the cortex of the adrenal gland.

cortisol A corticosteroid that mediates stress responses.

cost–benefit analysis An approach for explaining why the traits of organisms evolve as they do; assumes that an organism has a limited amount of time and energy to devote to its activities, and that each activity has costs (e.g., risks, expenditure of energy) as well as benefits (e.g., obtaining food, mating and successful reproduction). (*See also* trade-off.)

cotyledon (kot' ul lee' dun) [Gk. *kotyledon*: hollow space] A "seed leaf." An embryonic organ that stores and digests reserve materials; may expand when seed germinates.

countercurrent exchange An adaptation that promotes maximum exchange of heat or any diffusible substance between two fluids by the fluids flow in opposite directions through parallel tubes close together

countercurrent multiplier The mechanism of increasing the concentration of the interstitial fluid in the mammalian kidney as a result of countercurrent exchange in the loops of Henle and selective permeability and active transport of ions by segments of the loops of Henle.

covalent bond A chemical bond that arises from the sharing of electrons between two atoms. Usually a strong bond.

crassulacean acid metabolism (CAM) A metabolic pathway enabling the plants that possess it to store carbon dioxide at night and then perform photosynthesis during the day with stomata closed.

crista (plural: cristae) A small, shelflike projection of the inner membrane of a mitochondrion; the site of oxidative phosphorylation.

critical night length In the photoperiodic flowering response of short-day plants, the length of night above which flowering occurs and below which the plant remains vegetative. (The reverse applies in the case of long-day plants.)

critical period The age during which some particular type of learning must take place or during which it occurs much more easily than at other times. Typical of song learning among birds.

cross section A section taken perpendicular to the longest axis of a structure. Also called a transverse section.

crossing over The mechanism by which linked markers undergo recombination. In general, the term refers to the reciprocal exchange of corresponding segments between two homologous chromatids.

cryptic [Gk. *kryptos*: hidden] The resemblance of an animal to some part of its environment, which helps it to escape detection by predators.

cryptochromes [Gk. *kryptos*: hidden + *kroma*: color] Photoreceptors mediating some blue-light effects in plants and animals.

culture (1) A laboratory association of organisms under controlled conditions. (2) The collection of knowledge, tools, values, and rules that characterize a human society.

cuticle A waxy layer on the outer surface of a plant or an insect, tending to retard water loss.

cyanobacteria (sigh an' o bacteria) [Gr. *kuanos*: blue] A group of photosynthetic bacteria, formerly referred to as blue-green algae; they use chlorophyll *a* in photosynthesis.

cyclic AMP *See* cAMP.

cyclic electron transport In photosynthetic light reactions, the flow of electrons that produces ATP but no NADPH or O_2.

cyclins Proteins that activate cyclin-dependent kinases, bringing about transitions in the cell cycle.

cyclin-dependent kinase (Cdk) A kinase whose target proteins are involved in transitions in the cell cycle and which is active only when complexed with additional protein subunits, called cyclins.

cytochromes (sy' toe chromes) [Gk. *kytos*: container + *chroma*: color] Iron-containing red proteins, components of the electron-transfer chains in photophosphorylation and respiration.

cytokine A regulatory protein made by immune system cells that affects other target cells in the immune system.

cytokinesis (sy' toe kine ee' sis) [Gk. *kytos*: container + *kinein*: to move] The division of the cytoplasm of a dividing cell. (Compare with mitosis.)

cytokinin (sy' toe kine' in) A member of a class of plant growth substances playing roles in senescence, cell division, and other phenomena.

cytoplasm The contents of the cell, excluding the nucleus.

cytoplasmic determinants In animal development, gene products whose spatial distribution may determine such things as embryonic axes.

cytoplasmic domain The portion of a membrane bound receptor molecule that projects into the cytoplasm.

cytoplasmic segregation The asymmetrical distribution of cytoplasmic determinants in a developing animal embryo.

cytosine (C) (site' oh seen) A nitrogen-containing base found in DNA and RNA.

cytoskeleton The network of microtubules and microfilaments that gives a eukaryotic cell its shape and its capacity to arrange its organelles and to move.

cytosol The fluid portion of the cytoplasm, excluding organelles and other solids.

cytotoxic T cells (T_C) Cells of the cellular immune system that recognize and directly eliminate virus-infected cells. (Compare with helper T cells.)

- D -

DAG *See* diacylglycerol.

daughter chromosomes During mitosis, the separated chromatids from the beginning of anaphase onward.

deciduous [L. *deciduus*: falling off] Refers to a woody plant that sheds it leaves but does not die.

decomposers Organisms that metabolize organic compounds in debris and dead organisms, releasing inorganic material; found among the bacteria, protists, and fungi. *See also* detritivore.

defensin A type of protein made by phagocytes that kills bacteria and enveloped viruses by insertion into their plasma membranes.

degeneracy The situation in which a single amino acid may be represented by any of two or more different codons in messenger RNA. Most of the amino acids can be represented by more than one codon.

delayed hypersensitivity An increased immune reaction against an antigen that does not appear for 1–2 days after exposure. (Contrast with immediate hypersensitivity.)

deletion A mutation resulting from the loss of a continuous segment of a gene or chromosome. Such mutations never revert to wild type. (Contrast with duplication, point mutation.)

deme (deem) [Gk. *demos:* the populace] Any local population of individuals belonging to the same species that interbreed with one another.

demographic processes Events (births, deaths, immigration, and emigration) that change the number of individuals and the age structure of a population.

demographic stochasticity Random variations in the factors influencing the size, density, and distribution of a population.

denaturation Loss of activity of an enzyme or nucleic acid molecule as a result of structural changes induced by heat or other means.

dendrite [Gk. *dendron*: tree] A fiber of a neuron which often cannot carry action potentials. Usually much branched and relatively short compared with the axon, and commonly carries information to the cell body of the neuron.

denitrification Metabolic activity by which nitrate and nitrite ions are reduced to form nitrogen gas; carried on by certain soil bacteria.

density dependence The state in which changes in the severity of action of agents affecting birth and death rates within populations are directly or inversely related to population density.

density independence The state in which the severity of action of agents affecting birth and death rates within a population does not change with the density of the population.

deoxyribonucleic acid *See* DNA.

deoxyribose A five-carbon sugar found in nucleotides and DNA.

depolarization A change in the electric potential across a membrane from a condition in which the inside of the cell is more negative than the outside to a condition in which the inside is less negative, or even positive, with reference to the outside of the cell. (Contrast with hyperpolarization.)

derived trait A trait that differs from the ancestral trait. (Compare with shared derived trait.)

dermal tissue system The outer covering of a plant, consisting of epidermis in the young plant and periderm in a plant with extensive secondary growth. (Contrast with ground tissue system and vascular tissue system.)

desmosome (dez′ mo sowm) [Gk. *desmos*: bond + *soma*: body] An adhering junction between animal cells.

desmotubule A membrane extension connecting the endoplasmic retituclum of two plant cells that traverses the plasmodesma.

determination Process whereby an embryonic cell or group of cells becomes fixed into a predictable developmental pathway.

determined The state of a cell prior to its differentiation but after its developmental fate has been channeled.

detritivore (di try′ ti vore) [L. *detritus*: worn away + *vorare*: to devour] An organism that obtains its energy from the dead bodies and/or waste products of other organisms.

development Progressive change, as in structure or metabolism; in most kinds of organisms, development continues throughout the life of the organism.

developmental plasticity The capacity of an organism to alter its pattern of development in response to environmental conditions.

diacylglycerol (DAG) In hormone action, the second messenger produced by hydrolytic removal of the head group of certain phospholipids.

diaphragm (dye′ uh fram) [Gk. *diaphrassein*: barricade] (1) A sheet of muscle that separates the thoracic and abdominal cavities in mammals; responsible for breathing. (2) A method of birth control in which a sheet of rubber is fitted over the woman's cervix, blocking the entry of sperm.

diastole (dye ass′ toll ee) [Gk. dilation] The portion of the cardiac cycle when the heart muscle relaxes. (Contrast with systole.)

diencephalons The portion of the vertebrate forebrain that becomes the thalamus and hypothalamus.

differential gene expression The hypothesis that, given that all cells contain all genes, what makes one cell type different from another is the difference in transcription and translation of those genes.

differentiation Process whereby originally similar cells follow different developmental pathways. The actual expression of determination.

diffusion Random movement of molecules or other particles, resulting in even distribution of the particles when no barriers are present.

digestion Enzyme-catalyzed process by which large, usually insoluble, molecules (foods) are hydrolyzed to form smaller molecules of soluble substances.

dihybrid cross A mating in which the parents differ with respect to the alleles of two loci of interest.

dikaryon (di care′ ee ahn) [Gk. *di*: two + *karyon*: kernel] A cell or organism carrying two genetically distinguishable nuclei. Common in fungi.

dioecious (die eesh′ us) [Gk.: *di*: two + *oikos*: house] Refers to organisms in which the two sexes are"housed"in two different individuals, so that eggs and sperm are not produced in the same individuals. Examples: humans, fruit flies, date palms. (Contrast with monoecious.)

diploblastic Having two cell layers. (Contrast with triploblastic.)

diploid (dip′ loid) [Gk. *diplos*: double] Having a chromosome complement consisting of two copies (homologs) of each chromosome. Designated $2n$.

diplontic A type of life cycle in which gametes are the only haploid cells and mitosis occurs only in diploid cells. (Contrast with haplontic.)

direct transduction A cell signaling mechanism in which the receptor acts as the effector in the cellular response. (Contrast with indirect transduction.)

directional selection Selection in which phenotypes at one extreme of the population distribution are favored. (Contrast with disruptive selection, stabilizing selection.)

disaccharide A carbohydrate made up of two monosaccharides (simple sugars).

displacement activity Apparently irrelevant behavior performed by an animal under conflict situations, especially when tendencies to attack and escape are closely balanced.

display A behavior that has evolved to influence the actions of other individuals.

disruptive selection Selection in which phenotypes at both extremes of the population distribution are favored. (Contrast with directional selection; stabilizing selection.)

distal Away from the point of attachment or other reference point. (Contrast with proximal.)

disturbance A short-term event that disrupts populations, communities, or ecosystems by changing the environment.

disulfide bridge The covalent bond between twosulfur atoms (–S—S–) linking to molecules or remote parts of the same molecule.

diverticulum (di ver tik′ u lum) [L. *divertere*: turn away] A small cavity or tube that connects to a major cavity or tube.

division A term used by some microbiologists and formerly by botanists, corresponding to the term phylum.

DNA (deoxyribonucleic acid) The fundamental hereditary material of all living organisms. In eukaryotes, stored primarily in the cell nucleus. A nucleic acid using deoxyribose rather than ribose.

DNA chip A small glass or plastic square onto which thousands of single-stranded DNA sequences are fixed. Hybridization of cell-derived RNA or DNA to the target sequences can be performed.

DNA fingerprint An individual's unique pattern of DNA fragments produced by action of restriction endonucleases and separated by electrophoresis.

DNA helicase An enzyme that functions during DNA replication to unwind the double helix.

DNA ligase Enzyme that unites Okazaki fragments of the lagging strand during DNA replication; also mends breaks in DNA strands. It connects pieces of a DNA strand and is used in recombinant DNA technology.

DNA methylation The addition of methyl groups to DNA. Methylation plays a role in regulation of gene expression. Among bacteria, it protects DNA against restriction endonucleases.

DNA polymerase Any of a group of enzymes that catalyze the formation of DNA strands from a DNA template.

DNA sequencing Determining the precise sequence of nucleotides, and thus the sequence of amino acids encoded in a segment DNA.

DNA topoisomerase Enzymes that introduce positive or negative supercoils into the double-stranded DNA of continuous (circular) chromosomes.

docking protein A receptor protein that binds (docks) a ribosome to the membrane of the endoplasmic reticulum by binding the signal sequence attached to a new protein being made at the ribosome.

domain (1) Independent structural elements within proteins that affect the protein's function. Encoded by recognizable nucleotide sequences, a domain often folds separately from the rest of the protein. Similar domains can appear in a variety of different proteins across phylogenetic groups (e.g.,"homeobox domain";"calcium-binding domain.") (2) In phylogenetics, the three monophyletic branches of Life. Members of the three domains (Bacteria, Archaea, and Eukarya) are believed to have been evolving independently of each other for at least a billion years.

dominance In genetics, the ability of one allelic form of a gene to determine the phenotype of a heterozygous individual in which the homologous chromosomes carry both it and a different (recessive) allele. (Contrast with recessive.)

dormancy A condition in which normal activity is suspended, as in some seeds and buds.

dorsal [L. *dorsum*: back] Pertaining to the back or upper surface. (Contrast with ventral.)

dorsal lip In amphibian embryos, the dorsal segment of the blastopore. Also called the"organizer,"this region directs the development of nearby embryonic regions.

double fertilization Virtually unique to angiosperms, a process in which the nuclei of two sperm fertilize one egg. One sperm's nucleus combines with the egg nucleus to produce a zygote, while the other combines with the same egg's two polar nuclei to produce the first cell of the triploid endosperm (the tissue that will nourish the growing plant embryo).

double helix In DNA, the natural, right-handed coil configuration of two complementary, antiparallel strands.

duodenum (do' uh dee' num) The beginning portion of the vertebrate small intestine. (Compare with ileum, jejunum.)

duplication A mutation in which a segment of a chromosome is duplicated, often by the attachment of a segment lost from its homolog. (Contrast with deletion.)

dynein [Gk. *dynamis*: power] A protein that plays a part in the movement of eukaryotic flagella and cilia by means of conformational changes.

- E -

ecdysone (eck die' sone) [Gk. *ek*: out of + *dyo*: to clothe] In insects, a hormone that induces molting.

ecological community The species living together at a particular site.

ecological niche (nitch) [L. *nidus*: nest] The functioning of a species in relation to other species and its physical environment.

ecological succession The sequential replacement of one assemblage of populations by another in a habitat following some disturbance.

ecology [Gk. *oikos*: house + *logos*: study] The scientific study of the interaction of organisms with their living (biotic) and nonliving (abiotic) environment.

ecosystem (eek' oh sis tum) The organisms of a particular habitat, such as a pond or forest, together with the physical environment in which they live.

ectoderm [Gk. *ektos*: outside + *derma*: skin] The outermost of the three embryonic tissue layers first delineated during gastrulation. Gives rise to the skin, sense organs, nervous system, etc.

ectotherm [Gk. *ektos*: outside + *thermos*: heat] An animal unable to control its body temperature. (Contrast with endotherm.)

edema (i dee' mah) [Gk. *oidema*: swelling] Tissue swelling caused by the accumulation of fluid.

edge effect The changes in ecological processes in a community caused by physical and biological factors originating in an adjacent community.

effector Any organ, cell, or organelle that moves the organism through the environment or else alters the environment; for example, muscle, exocrine glands, chromatophores.

effector cell A cell responsible for the effector phase of the immune response.

effector phase Stage of the immune response, when cytotoxic T cells attack virus-infected cells, and helper T cells assist B cells to differentiate into plasma cells.

effector protein In cell signaling, a protein responsible for the cellular reponse to a signal transduction pathway.

efferent [L. *ex*: out + *ferre*: to bear] In physiology, conducting outward or away from an organ or structure. (Contrast with afferent.)

egg In all sexually reproducing organisms, the female gamete; in birds, reptiles, and some other vertebrates, a structure within which early embryonic development occurs. (Compare with amniote egg.)

elasticity The property of returning quickly to a former state after a disturbance.

electrocardiogram (ECG or EKG) A graphic recording of electrical potentials from the heart.

electroencephalogram (EEG) A graphic recording of electrical potentials from the brain.

electromagnetic radiation A self-propagating wave that travels though space and has both electrical and magnetic properties.

electromyogram (EMG) A graphic recording of electrical potentials from muscle.

electron A subatomic particle outside the nucleus carrying a negative charge and very little mass.

electron shell The region surrounding the atomic nucleus at a fixed energy level in which electrons orbit.

electron transport The passage of electrons through a series of proteins with a release of energy which may be captured in a concentration gradient or chemical form such as NADH or ATP.

electronegativity The tendency of an atom to attract electrons when it occurs as part of a compound.

electrostatic Pertaining to the attraction and repulsion of negative and positive charges on atoms due to the number and distribution of electrons.

electrophoresis (e lek' tro fo ree' sis) [L. *electrum*: amber + Gk. *phorein*: to bear] A separation technique in which substances are separated from one another on the basis of their electric charges and molecular weights.

element A substance that cannot be converted to simpler substances by ordinary chemical means.

elongation (1) In molecular biology, the addition of monomers to make a longer RNA or protein during synthesis. (2) Growth of a plant axis or cell primarily in the longitudinal direction.

embolus (em' buh lus) [Gk. *embolos*: inserted object; stopper] A circulating blood clot. Blockage of a blood vessel by an embolus or by a bubble of gas is referred to as an embolism. (Contrast with thrombus.)

embryo [Gk. *en*: within + *bryein*: to grow] A young animal, or young plant sporophyte, while it is still contained within a protective structure such as a seed, egg, or uterus.

embryophyte A photosynthetic organism that develops from a protected embryo; synonymous with"land plant."In this book,"plant,"when unmodified, is synonymous with"embryophyte."

embryo sac In angiosperms, the female gametophyte. Found within the ovule, it consists of eight or fewer cells, membrane bounded, but without cellulose walls between them.

emergent property A property of a complex system that is not exhibited by its individual component parts.

emigration The deliberate and usually oriented departure of an organism from the habitat in which it has been living.

3′ end (3 prime) The end of a DNA or RNA strand that has a free hydroxyl group at the 3′ carbon of the sugar (deoxyribose or ribose).

5′ end (5 prime) The end of a DNA or RNA strand that has a free phosphate group at the 5′ carbon of the sugar (deoxyribose or ribose).

embolism A blockage of a coronary vessel resulting from a circulating blood clot.

endemic (en dem' ik) [Gk. *endemos*: native, dwelling in] Confined to a particular region, thus often having a comparatively restricted distribution.

endergonic reaction A chemical reaction that requires the input of energy in order to proceed. (Contrast with exergonic reaction.)

endocrine gland (en' doh krin) [Gk. *endo*: within + *krinein*: to separate] Any gland, such as the adrenal or pituitary gland of vertebrates, that secretes certain substances, especially hormones, into the body through the blood. The *endocrine system* consists of all cells and glands in the body that produce and release hormones.

endocytosis A process by which liquids or solid particles are taken up by a cell through invagination of the plasma membrane. (Contrast with exocytosis.)

endoderm [Gk. *endo*: within + *derma*: skin] The innermost of the three embryonic tissue layers delineated during gastrulation. Gives rise to the digestive and respiratory tracts and structures associated with them.

endodermis In plants, a specialized cell layer marking the inside of the cortex in roots and some stems. Frequently a barrier to free diffusion of solutes.

endomembrane system Endoplasmic reticulum plus Golgi apparatus; also lysosomes, when present. A system of membranes that exchange material with one another.

endoplasmic reticulum (ER) [Gk. *endo*: within + L. *plasma*: form + L. *reticulum*: net] A system of membranous tubes and flattened sacs found in the cytoplasm of eukaryotes. Exists in two forms: rough ER, studded with ribosomes; and smooth ER, lacking ribosomes.

endorphins, enkephalins Naturally occurring, opiate-like substances in the mammalian brain.

endoskeleton [Gk. *endo*: within + *skleros*: hard] An internal skeleton covered by other, soft body tissues. (Contrast with exoskeleton.)

endosperm [Gk. *endo*: within + *sperma*: seed] A specialized triploid seed tissue found only in angiosperms; contains stored nutrients for the developing embryo.

endosymbiosis [Gk. *endo*: within + *sym*: together + *bios*: life] Two species living together, with one living inside the body (or even the cells) of the other.

endosymbiotic theory The theory that the eukaryotic cell evolved via the engulfing of one prokaryotic cell by another.

endotherm [Gk. *endo*: within + *thermos*: heat] An animal that can control its body temperature by the expenditure of its own metabolic energy. (Contrast with ectotherm.)

end product inhibition A control capacity of some metabolic pathways in which the final product produced inhibits an early enzyme in the pathway.

energetic cost The difference between the energy an animal expends in performing a behavior and the energy it would have expended had it rested.

energy The capacity to do work or move matter against an opposing force. The capacity to accomplish change.

energy budget A quantitative description of all paths of energy exchange between an animal and its environment.

enhancer In eukaryotes, a DNA sequence, lying on either side of the gene it regulates, that stimulates a specific promoter.

enterocoelous development A pattern of development in which the coelum is formed by an outpocketing of the embryonic gut (enteron).

enterokinase (ent uh row kine′ ase) An enzyme secreted by the mucosa of the duodenum. It activates the zymogen trypsinogen to create the active digestive enzyme trypsin.

enthalpy The sum of the internal energy of a system; the product of the volume multiplied by the pressure.

entrainment With respect to circadian rhythms, the process whereby the period is adjusted to match the 24-hour environmental cycle.

entropy (en′ tro pee) [Gk. *tropein*: to change] A measure of the degree of disorder in any system. Spontaneous reactions in a closed system are always accompanied by an increase in disorder and entropy.

environment Whatever surrounds and interacts with a population, organism, or cell. May be external or internal.

environmental carrying capacity (*K*) The number of individuals in a population that the resources of a habitat can support.

enzyme (en′ zime) [Gk. *zyme*: to leaven (as in yeast bread)] A protein, on the surface of which are chemical groups so arranged as to make the enzyme a catalyst for a chemical reaction.

enzyme–substrate complex The complex that forms when an enzyme binds to its substrate(s).

eosinophils Phagocytic white blood cells that attack multicellular parasites once they have been coated with antibodies.

epi- [Gk.: upon, over] A prefix used to designate a structure located on top of another; for example: epidermis, epiphyte.

epiblast The upper or overlying portion of the avian blastula which is joined to the hypoblast at the margins of the blastodisc.

epiboly The movement of cells over the surface of the blastula toward the forming blastopore.

epicotyl (epp′ i kot′ il) [Gk. *epi*: over + *kotyle*: something hollow] That part of a plant embryo or seedling that is above the cotyledons.

epidermis [Gk. *epi*: over + *derma*: skin] In plants and animals, the outermost cell layers. (Only one cell layer thick in plants.)

epididymis (epuh did′ uh mus) [Gk. *epi*: over + *didymos*: testicle] Coiled tubules in the testes that store sperm and conduct sperm from the seminiferous tubules to the vas deferens.

epinephrine (ep i nef′ rin) [Gk. *epi*: over + *nephros*: kidney] The "fight or flight" hormone produced by the medulla of the adrenal gland; it also functions as a neurotransmitter. (Also known as adrenaline.)

epiphyte (ep′ e fyte) [Gk. *epi*: over + *phyton*: plant] A specialized plant that grows on the surface of other plants but does not parasitize them.

epistasis Interaction between genes in which the presence of a particular allele of one gene determines whether another gene will be expressed.

epithelium Sheets of cells that line or cover organs, make up tubules, and cover the surface of the body.

equatorial plate In a cell undergoing mitosis, the region in the middle of a cell where the centromeres will align during metaphase.

equilibrium Any state of balanced opposing forces and no net change.

ER *See* endoplasmic reticulum.

erythrocyte (ur rith′ row site) [Gk. *erythros*: red + *kytos*: container] A red blood cell.

erythropoietin A hormone produced by the kidney in response to lack of oxygen. It stimulates the production of red blood cells.

esophagus (i soff′ i gus) [Gk. *oisophagos*: gullet] That part of the gut between the pharynx and the stomach.

essential element A mineral nutrient element required in order for a seed to develop and complete the plant's life cycle, producing viable new seeds.

essential acids Amino acids or fatty acids that an animal cannot synthesize for itself.

ester linkage A condensation (water-releasing) reaction in which the carboxyl group of a fatty acid reacts with the hydroxyl group of an alcohol. Lipids are formed in this way.

estivation (ess tuh vay′ shun) [L. *aestivalis*: summer] A state of dormancy and hypometabolism that occurs during the summer; usually a means of surviving drought and/or intense heat. (Contrast with hibernation.)

estrogen Any of several steroid sex hormones; produced chiefly by the ovaries in mammals.

estrus (es′ truss) [L. *oestrus*: frenzy] The period of heat, or maximum sexual receptivity, in some female mammals. Ordinarily, the estrus is also the time of release of eggs in the female.

ethology An approach to the study of animal behavior developed in Europe. Emphasizes the causes of the evolution of behavior.

ethylene One of the plant growth hormones, the gas $H_2C{=}CH_2$. Involved in fruit ripening and other growth and developmental responses.

euchromatin Chromatin that is diffuse and non-staining during interphase; may be transcribed. (Contrast with heterochromatin.)

eukaryotes (yew car′ ree oats) [Gk. *eu*: true + *karyon*: kernel or nucleus] Organisms whose cells contain their genetic material inside a nucleus. Includes all life other than the viruses, archaea, and bacteria.

eutrophication (yoo trofe′ ik ay′ shun) [Gk. *eu*: truly + *trephein*: to flourish] The addition of nutrient materials to a body of water, resulting in changes in ecological processes and species composition therein.

evaporation The transition of water from the liquid to the gaseous phase.

evolution Any gradual change. Organic or Darwinian evolution, often referred to as evolution, is any genetic and resulting phenotypic change in organisms from generation to generation. (*See* macroevolution, microevolution; compare with speciation.)

evolutionary agent Any factor that influences the direction and rate of evolutionary change.

evolutionarily conserved Refers to traits that have evolved very slowly and are similar or even identical in individuals of highly divergent groups.

evolutionary radiation The proliferation of species within a single evolutionary lineage.

evolutionary reversal The reappearance of an ancestral trait in a group that had previously acquired a derived trait.

excision repair The removal and damaged DNA and its replacement by the appropriate nucleotides.

excited state The state of an atom or molecule when, after absorbing energy, it has more energy than in its normal, ground state. (Compare with ground state.)

excretion Release of metabolic wastes by an organism.

exergonic reaction A reaction in which free energy is released. (Contrast with endergonic reaction.)

exocrine gland (eks' oh krin) [Gk. *exo*: outside + *krinein*: to separate] Any gland, such as a salivary gland, that secretes to the outside of the body or into the gut. (Contrast with endocrine gland.)

exocytosis A process by which a vesicle within a cell fuses with the plasma membrane and releases its contents to the outside. (Contrast with endocytosis.)

exon A portion of a DNA molecule, in eukaryotes, that codes for part of a polypeptide. (Contrast with intron.)

exoskeleton (eks' oh skel' e ton) [Gk. *exos*: outside + *skleros*: hard] A hard covering on the outside of the body to which muscles are attached. (Contrast with endoskeleton.)

exotoxins Highly toxic proteins released by living, multiplying bacteria.

expanding triplet repeat A three-base-pair sequence in a human gene that is unstable and can be repeated a few to hundreds of times. Often, the more the repeats, the less the activity of the gene involved. Expanding triplet repeats occur in some human diseases such as Huntington's disease and fragile-X syndrome.

experiment A testing process to support or disprove hypotheses and to answer questions. The basis of the scientific method. *See* comparative experiment, controlled experiment.

expiratory reserve volume The amount of air that can be forcefully exhaled beyond the normal tidal expiration. (Compares with inspiratory reserve volume, tidal volume, vital capacity.)

exploitation competition Competition in which individuals reduce the quantities of their shared resources. (Compare with interference competition.)

exponential growth Growth, especially in the number of organisms in a population, which is a geometric function of the size of the growing entity: the larger the entity, the faster it grows. (Contrast with logistic growth.)

expression vector A DNA vector, such as a plasmid, that carries a DNA sequence that includes the adjacent sequences for its expression into mRNA and protein in a host cell.

expressivity The degree to which a genotype is expressed in the phenotype; may be affected by the environment.

extensor A muscle the extends an appendage.

extinction The termination of a lineage of organisms.

extrinsic protein A membrane protein found only on the surface of the membrane. (Contrast with intrinsic protein.)

extracellular matrix In animal tissues, a material of heterogeneous composition surrounding cells and performing many functions including adhesion of cells.

extraembryonic membranes Four membranes that support but are not part of the developing embryos of reptiles, birds, and mammals, defining these groups phylogenetically as amniotes. (*See* amnion, allantois, chorion, and yolk sac.)

- F -

F_1 (first filial generation) The immediate progeny of a parental (P) mating.

F_2 (second filial generation) The immediate progeny of a mating between members of the F_1 generation.

facilitated diffusion Passive movement through a membrane involving a specific carrier protein; does not proceed against a concentration gradient. (Contrast with active transport, diffusion.)

facultative anaerobes (alternatively, facultative aerobes) Prokaryotes that can shift their metabolism between anaerobic and aerobic operations depending on the presence or absence of O_2.

FAD *See* flavin adenine dinucleotide.

fast-twitch fibers Skeletal muscle fibers that can generate high tension rapidly, but fatigue rapidly ("sprinter" fibers). Characterized by an abundance of enzymes of glycolysis.

fat A triglyceride that is solid at room temperature. (Contrast with oil.)

fate In an embryo, the type of cell that a particular cell will become in the adult.

fate map A diagram of the blastula showing which cells (blastomeres) are "fated" to contribute to specific tissues and organs in the mature body.

fatty acid A molecule with a long hydrocarbon tail and a carboxyl group at the other end. Found in many lipids.

fauna (faw' nah) All of the animals found in a given area. (Contrast with flora.)

feces [L. *faeces*: dregs] Waste excreted from the digestive system.

feedback information Information relevant to the rate of a process that can be used by a control system to regulate that process at a particular level.

feedforward information Information that can be used to alter the setpoint of a regulatory process.

fermentation (fur men tay' shun) [L. *fermentum*: yeast] The anaerobic degradation of a substance such as glucose to smaller molecules with the extraction of energy.

ferredoxin A protein containing iron that mediates the transfer of electrons in a number of pathways, including the light reactions of photosynthesis.

fertility factor (F factor) A plasmid that confers on a bacterium the ability to act as a DNA donor during conjugation.

fertilization Union of gametes. Also known as syngamy.

fertilization membrane A membrane surrounding an animal egg which becomes rapidly raised above the egg surface within seconds after fertilization, serving to prevent entry of a second sperm.

fetus Medical and legal term for the latter stages of a developing human embryo, from about the eighth week of pregnancy (the point at which all major organ systems have formed) to the moment of birth.

fiber An elongated, tapering cell of flowering plants, usually with a thick cell wall. Serves a support function.

fibrin A protein that polymerizes to form long threads that provide structure to a blood clot.

fibrinogen A circulating protein that can be stimulated to fall out of solution and provide the structure for a blood clot.

Fick's law of diffusion An equation that describes the factors that determine the rate of diffusion of a molecule from an area of higher concentration to an area of lower concentration.

filter feeder An organism that feeds upon much smaller organisms, that are suspended in water or air, by means of a straining device.

filtration In the excretory physiology of some animals, the process by which the initial urine is formed; water and most solutes are transferred into the excretory tract, while proteins are retained in the blood or hemolymph.

first law of thermodynamics Energy can be neither created nor destroyed.

fission Reproduction of a prokaryote by division of a cell into two comparable progeny cells.

fitness The contribution of a genotype or phenotype to the genetic composition of subsequent generations, relative to the contribution of other genotypes or phenotypes. (*See* also inclusive fitness.)

flagellum (fla jell' um) (plural: flagella) [L. *flagellum*: whip] Long, whiplike appendage that propels cells. Prokaryotic flagella differ sharply from those found in eukaryotes.

flavin adenine dinucleotide (FAD) A coenzyme involved in redox reactions and containing the vitamin riboflavin (B_2).

flexor A muscle that flexes an appendage.

flora (flore' ah) All of the plants found in a given area. (Contrast with fauna.)

floral meristem Meristem that forms the sexual parts of flowering plants (sepals, petals, stamens, and carpels).

florigen A plant hormone involved in the conversion of a vegetative shoot apex to a flower.

flower The total reproductive structure of an angiosperm; its basic parts include the calyx, corolla, stamens, and carpels.

fluid mosaic model A molecular model for the structure of biological membranes consisting of a fluid phospholipid bilayer in which suspended proteins are free to move in the plane of the bilayer.

fluorescence The emission of a photon of visible light by an excited atom or molecule.

follicle [L. *folliculus*: little bag] In female mammals, an immature egg surrounded by nutritive cells.

follicle-stimulating hormone A gonadotropic hormone produced by the anterior pituitary.

food chain A portion of a food web, most commonly a simple sequence of prey species and the predators that consume them.

food vacuole Membrane enclosed structure formed by phagocytosis in which engulfed food particles are digested by the action of lysosomal enzymes.

food web The complete set of food links between species in a community; a diagram indicating which ones are the eaters and which are eaten.

forb Any broad-leaved herbaceous plant. Especially applied to such plants growing in grasslands.

fossil Any recognizable structure originating from an organism, or any impression from such a structure, that has been preserved over geological time.

founder effect Random changes in allele frequencies resulting from establishment of a population by a very small number of individuals.

fovea [L. *fovea*; a small pit] In the vertebrate retina, the area of most distinct vision.

frame-shift mutation A mutation resulting from the addition or deletion of one or two consecutive base pairs in the DNA sequence of a gene, resulting in misreading mRNA during translation and production of a nonfunctional protein. (Compare with missense substitution, nonsense substitution, synonymous substitution.)

Frank–Starling law The stroke volume of the heart increases with increased return of blood to the heart.

free energy That energy which is available for doing useful work, after allowance has been made for the increase or decrease of disorder.

freeze-fracturing Method of tissue preparation for transmission and scanning electron microscopy in which a tissue is frozen and a knife is then used to crack open the tissue; the fracture often occurs in the path of least resistance, within a membrane.

frequency In population genetics, the proportion of all alleles or genotypes in a population composed of a particular allele or genotype.

frequency-dependent selection Selection that changes in intensity with the proportion of individuals in a population having the trait.

fruit In angiosperms, a ripened and mature ovary (or group of ovaries) containing the seeds. Sometimes applied to reproductive structures of other groups of plants.

fruiting body Any structure that bears spores.

functional genomics The assignment of functional roles to genes first identified by sequencing entire genomes.

functional group A characteristic combination of atoms that contribute specific properties when attached to larger molecules.

functional mRNA Eukaryotic mRNA that has been modified after transcription by the removal of introns and the addition of a 5′ cap and a 3′ poly(A) tail.

- G -

G cap A chemically modified GTP added to the 5′ end of mRNA; facilitates binding of mRNA to ribosome and prevents mRNA breakdown.

G_1 phase In the cell cycle, the gap between the end of mitosis and the onset of the S phase.

G_2 phase In the cell cycle, the gap between the S (synthesis) phase and the onset of mitosis.

G protein A membrane protein involved in signal transduction; characterized by binding GDP or GTP.

gametangium (gam uh tan′ gee um) [Gk. *gamos*: marriage + *angeion*: vessel] Any plant or fungal structure within which a gamete is formed.

gamete (gam′ eet) [Gk. *gamete/gametes*: wife, husband] The mature sexual reproductive cell: the egg or the sperm.

gametogenesis (ga meet′ oh jen′ e sis) The specialized series of cellular divisions that leads to the production of sex cells (gametes). (Contrast with oogenesis and spermatogenesis.)

gametophyte (ga meet′ oh fyte) In plants and photosynthetic protists with alternation of generations, the multicellular haploid phase that produces the gametes. (Contrast with sporophyte.)

ganglion (gang′ glee un) [Gk.: tumor] A cluster of neurons that have similar characteristics or function.

gap genes During insect development, the first step of segmentation genes to act organizing the anterior-posterior axis.

gap junction A 2.7-nanometer gap between plasma membranes of two animal cells, spanned by protein channels. Gap junctions allow chemical substances or electrical signals to pass from cell to cell.

gas exchange In animals, the process of taking up oxygen from the environment and releasing carbon dioxide to the environment.

gastrovascular cavity Serving for both digestion (gastro) and circulation (vascular); in particular, the central cavity of the body of jellyfish and other cnidarians.

gastrula (gas′ true luh) [Gk. *gaster*: stomach] An embryo forming the characteristic three cell layers (ectoderm, endoderm, and mesoderm) that give rise to all of the major tissue systems of the adult animal.

gastrulation Development of a blastula into a gastrula.

gated channel A membrane protein that opens and closes in response to binding of specific molecules or to changes in membrane potential. When open, it allows specific ions to move across the membrane.

gene [Gk. *genes*: to produce] A unit of heredity. Used here as the unit of genetic function which carries the information for a single polypeptide or RNA.

gene amplification Creation of multiple copies of a particular gene, allowing the production of large amounts of the RNA transcript (as in rRNA synthesis in oocytes).

gene cloning Formation of a clone of bacteria or yeast cells containing a particular foreign gene.

gene family A set of identical (or once-identical) genes derived from a single parent gene; need not be on the same chromosomes. The vertebrate globin genes constitute a classic example of a gene family.

gene flow Exchange of genes between different species (an extreme case referred to as hybridization) or between different populations of the same species caused by migration followed by breeding.

gene-for-gene resistance A mechanism for resistance to pathogens, in which resistance is triggered by the specific interaction of the products of pathogens' *Avr* genes and plants' *R* genes.

gene frequency *See* allele frequency.

gene library All of the cloned DNA fragments generated by action of a restriction endonuclease on a genome or chromosome.

gene pool All of the different alleles of all of the genes existing in all individual of a population.

gene therapy Treatment of a genetic disease by providing patients with cells containing functioning alleles of the genes that are nonfunctional in their bodies.

generative cell In a pollen tube, a haploid nucleus that undergoes mitosis to produce the two sperm nuclei that participate in double fertilization. (Contrast with tube cell.)

genetic code The set of instructions, in the form of nucleotide triplets, that translate a linear sequence of nucleotides in mRNA into a linear sequence of amino acids in a protein.

genetic drift Changes in gene frequencies from generation to generation as a result of random (chance) processes.

genetic map The positions of genes along a chromosome as revealed by recombination frequencies.

genetic screening The application of medical tests to determine whether an individual carries a specific allele.

genetic stochasticity Random variation in the frequencies of alleles and genotypes in a population over time. (Compare with demographic stochasticity.)

genetics The study of the structure, functioning, and inheritance of genes, the units of hereditary information.

genome (jee′ nome) All the genes in a complete haploid set of chromosomes. (Compare with proteome.)

genomics The study of entire sets of genes and their interactions.

genomic imprinting When a given gene's phenotype is determined by whether that gene is inherited from the male or the female parent.

genotype (jean′ oh type) [Gk. *gen*: to produce + *typos*: impression] An exact description of the genetic constitution of an individual, either with respect to a single trait or with respect to a larger set of traits. (Contrast with phenotype.)

genus (jean′ us) (plural: genera) [Gk. *genos*: stock, kind] A group of related, similar species recognized by taxonomists with a distinct name used in binomial nomenclature.

geotropism *See* gravitropism.

germ cell [L. *germen*: to beget] A reproductive cell or gamete of a multicellular organism. (Contrast with somatic cell.)

germ line mutation A change in the genetic material in a germ cell; such mutations are heritable. (Contrast with somatic mutation.)

germ layers The three embryonic tissue layers formed during gastrulation (ectoderm, mesoderm, endoderm).

germination Sprouting of a seed or spore.

gestation (jes tay′ shun) [L. *gestare*: to bear] The period during which the embryo of a mammal develops within the uterus. Also known as pregnancy.

gibberellin (jib er el′ lin) A class of plant growth substances playing roles in stem elongation, seed germination, flowering of certain plants, etc. Named for the fungus *Gibberella*.

gill An organ for gas exchange in aquatic organisms.

gill arch A skeletal structure that supports gill filaments and the blood vessels that supply them.

gizzard (giz′ erd) [L. *gigeria*: cooked chicken parts] A muscular port of the stomach of birds that grinds up food, sometimes with the aid of fragments of stone.

gland An organ or group of cells that produces and secretes one or more substances.

glans penis Sexually sensitive tissue at the tip of the penis.

glia (glee′ uh) [Gk. *glia*: glue] Cells, found only in the nervous system, that do not conduct action potentials.

glomerular filtration rate (GFR) The rate at which the blood is filtered in the glomeruli of the kidney.

glomerulus (glo mare′ yew lus) [L. *glomus*: ball] Sites in the kidney where blood filtration takes place. Each glomerulus consists of a knot of capillaries served by afferent and efferent arterioles.

glucagon Hormone produced by alpha cells of the pancreatic islets of Langerhans. Glucagon stimulates the liver to break down glycogen and release glucose into the circulation.

glucocorticoids Steroid hormones produced by the adrenal cortex. Secreted in response to ACTH, they inhibit glucose uptake by many tissues in addition to mediating other stress responses.

gluconeogenesis The biochemical synthesis of glucose from other substances, such as amino acids, lactate, and glycerol.

glucose[Gk. *gleukos*: sugar, sweet] The most common monosaccharide; the monomer of the polysaccharides starch, glycogen, and cellulose.

glycerol (gliss′ er ole) A three-carbon alcohol with three hydroxyl groups; a component of phospholipids and triglycerides.

glycogen (gly′ ko jen) An energy storage polysaccharide found in animals and fungi; a branched-chain polymer of glucose, similar to starch.

glycolipid A lipid to which sugars are attached.

glycolysis (gly kol′ li sis) [Gk. *gleukos*: sugar + *lysis*: break apart] The enzymatic breakdown of glucose to pyruvic acid. One of the evolutionarily oldest of the cellular energy-yielding mechanisms.

glycoprotein A protein to which sugars are attached.

glycosidic linkage Bond between carbohydrate (sugar) molecules through an intervening oxygen atom (–O–).

glycosylation The addition of carbohydrates to another type of molecule, such as a protein.

glyoxysome (gly ox′ ee soam) An organelle found in plants, in which stored lipids are converted to carbohydrates.

Golgi apparatus (goal′ jee) A system of concentrically folded membranes found in the cytoplasm of eukaryotic cells; functions in secretion from cell by exocytosis.

gonad (go′ nad) [Gk. *gone*: seed] An organ that produces sex cells in animals: either an ovary (female gonad) or testis (male gonad).

gonadotropin A hormone that stimulates the gonads.

gonadotropin-releasing hormone (GnRH) Hypothalamic hormone that stimulates the anterior pituitary to secrete growth hormone.

Gondwana The large southern land mass that existed from the Cambrian (540 mya) to the Jurassic (138 mya). Present-day remnants are South America, Africa, India, Australia, and Antarctica.

graft Tissue artificially and viably transplanted from one organism to another. In agriculture, refers to the transfer of bud or stem segment from one plant onto another plant as a form of asexual reproduction.

Gram stain A differential purple stain useful in characterizing bacteria. The peptidoglycan-rich cell walls of Gram-positive bacteria stain purple; cell walls of Gram-negative bacteria generally stain orange.

granum (plural: grana) Within a chloroplast, a stack of thylakoids.

gravitropism A directed plant growth response to gravity.

grazer An animal that eats the vegetative tissues of herbaceous plants.

green gland An excretory organ of crustaceans.

greenhouse effect The heating of Earth's atmosphere by gases such as water vapor, carbon dioxide, and methane; such greenhouse gases are transparent to sunlight but opaque to heat; thus sunlight-engendered heat builds up at Earth's surface and cannot be dissipated into the atmosphere.

gross primary production The total energy captured by plants growing in a particular area.

gross primary productivity (GPP) The rate of assimilation of energy by plants growing in a particular area.

ground meristem That part of an apical meristem that gives rise to the ground tissue system of the primary plant body.

ground state The lowest energy state of an atom of molecule. (Compare with excited state.)

ground tissue system Those parts of the plant body not included in the dermal or vascular tissue systems. Ground tissues function in storage, photosynthesis, and support.

group transfer The exchange of atoms between molecules.

growth Irreversible increase in volume (an accurate definition, but at best a dangerous oversimplification).

growth hormone A peptide hormone of the anterior pituitary that stimulates many anabolic processes.

guanine (G) (gwan′ een) A nitrogen-containing base found in DNA, RNA, and GTP.

guard cells In plants, specialized, paired epidermal cells that surround and control the opening of a stoma (pore). *See* stoma.

gut An animal's digestive tract.

guttation The extrusion of liquid water through openings in leaves, caused by root pressure.

gymnosperm (jim′ no sperm) [Gk. *gymnos*: naked + *sperma*: seed] A plant, such as a pine or other conifer, whose seeds do not develop within an ovary (hence, the seeds are"naked").

gyrus The raised or ridged portion of the convoluted surface of the brain. (Contrast with sulcus.)

- H -

habitat The environment in which an organism lives.

habituation (ha bich′ oo ay shun) The simplest form of learning, in which an animal presented with a stimulus without reward or punishment eventually ceases to respond.

hair cell A type of mechanoreceptor in animals. Detects sound waves and other forms of motion in air or water.

half-life The time required for half of a sample of a radioactive isotope to decay to its stable, nonradioactive form, or for a drug or other substance to reach half its initial dosage.

halophyte (hal′ oh fyte) [Gk. *halos*: salt + *phyton*: plant] A plant that grows in a saline (salty) environment.

haploid (hap′ loid) [Gk. *haploeides*: single] Having a chromosome complement consisting of just one copy of each chromosome; designated 1*n* or *n*. (Contrast with diploid.)

haplontic A type of life cycle in which the zygote is the only diploid cell and mitosis occurs only in haploid cells. (Contrast with diplontic.)

Hardy–Weinberg equililbrium The allele frequency at a given locus in a sexually reproducing population that is not being acted on by agents of evolution; the conditions that would result in no evolution in a population.

haustorium (haw stor′ ee um) [L. *haustus*: draw up] A specialized hypha or other structure by which fungi and some parasitic plants draw nutrients from a host plant.

Haversian systems Units of organization in compact bone that reflect the action of intercommunicating osteoblasts.

heat of vaporization The energy that must be supplied to convert a molecule from a liquid to a gas at its boiling point.

heat-shock proteins Chaperone proteins expressed in cells exposed to high or low temperatures or other forms of environmental stress.

helical Shaped like a screw or spring; this shape occurs in DNA and proteins.

helper T cells (T_H) T cells that participate in the activation of B cells and of other T cells; targets of the HIV-I virus, the agent of AIDS. (Contrast with cytotoxic T cells.)

hematocrit (heme at′ o krit) [Gk. *heaema*: blood + *krites*: judge] The proportion of 100 cc of blood that consists of red blood cells.

hemizygous (hem′ ee zie′ gus) [Gk. *hemi*: half + *zygotos*: joined] In a diploid organism, having only one allele for a given trait, typically the case for X-linked genes in male mammals and Z-linked genes in female birds. (Contrast with homozygous, heterozygous.)

hemoglobin (hee′ mo glow bin) [Gk. *heaema*: blood + L. *globus*: globe] Oxygen-transporting protein found in the red blood cells of vertebrates (and found in some invertebrates).

Hensen's node In avian embryos, a structure at the anterior end of the primitive groove; determines the fates of cells passing over it during gastrulation.

hepatic (heh pat′ ik) [Gk. *hepar*: liver] Pertaining to the liver.

hepatic duct Conveys bile from the liver to the gallbladder.

herbivore (ur′ bi vore) [L. *herba*: plant + *vorare*: to devour] An animal that eats plant tissues. (Contrast with carnivore, detritivore, omnivore.)

heritable Able to be inherited; in biology, refers to genetically influenced traits.

hermaphroditism (her maf′ row dite ism) [Gk. Hermes (messenger god) + Aphrodite (goddess of love)] The coexistence of both female and male sex organs in the same organism.

hertz (abbreviated Hz) Cycles per second.

hetero- [Gk.: *heteros*: other, different] A prefix specifying that two or more different conditions are involved; for example, heterotroph, heterozygous.

heterochromatin Chromatin that retains its coiling during interphase; generally not transcribed. (Contrast with euchromatin.)

heterochrony Comparing different species, an alternation in the timing of developmental events, leading to different results in the adult.

heterocyst A large, thick-walled cell in the filaments of certain cyanobacteria; performs nitrogen fixation.

heterogeneous nuclear RNA (hnRNA) The product of transcription of a eukaryotic gene, including transcripts of introns.

heterokaryon In fungi, hypha containing two genetically different nuclei.

heteromorphic (het′ er oh more′ fik) [Gk. *heteros*: different + *morphe*: form] Having a different form or appearance, as two heteromorphic life stages of a plant. (Contrast with isomorphic.)

heterosporous (het′ er os′ por us) Producing two types of spores, one of which gives rise to a female megaspore and the other to a male microspore. (Contrast with homosporous.)

heterosis Situation in which heterozygous genotypes are superior to homozygous genotypes with respect to growth, survival, or fertility. Also called hybrid vigor.

heterotherm An animal that regulates its body temperature at a constant level at some times but not others, such as a hibernator.

heterotroph (het′ er oh trof) [Gk. *heteros*: different + *trophe*: food] An organism that requires preformed organic molecules as food. (Contrast with autotroph.)

heterotypic Referring to adhesion of cells of different types. (Contrast with homotypic.)

heterozygous (het′ er oh zie′ gus) [Gk. *heteros*: different + *zygotos*: joined] Of a diploid organism having different alleles of a given gene on the pair of homologs carrying that gene. (Contrast with homozygous.)

hexose [Gk. *hex*: six] A sugar containing six carbon atoms.

hibernation [L. *hibernum*: winter] The state of inactivity of some animals during winter; marked by a drop in body temperature and metabolic rate.

hierarchical sequencing An approach to DNA sequencing in which markers are mapped and DNA sequences are aligned by matching overlapping sites of known sequence.

highly repetitive DNA Short DNA sequences present in millions of copies in the genome, next to each other (in tandem). In reassociation experiments, denatured highly repetitive DNA reanneals very quickly.

hindbrain The region of the developing vertebrate brain that gives rise to the medulla, pons, and cerebellum.

hippocampus [Gr.: sea horse] A part of the forebrain that takes part in long-term memory formation.

histamine (hiss′ tah meen) A substance released by damaged tissue, or by mast cells in response to allergens. Histamine increases vascular permeability, leading to edema (swelling).

histology [Gk. *histos*: weaving] The study of tissues.

histone Any one of a group of basic proteins forming the core of a nucleosome, the structural unit of a eukaryotic chromosome. (Compare with nucleosome.)

hierarchical sequencing An approach to DNA sequencing in which markers are mapped and DNA sequences are aligned by matching overlapping sites of known sequence.

hnRNA *See* heterogeneous nuclear RNA.

homeobox A 180-base-pair segment of DNA found in certain homeotic genes; regulates the expression of other genes and thus controls large-scale developmental processes.

homeostasis (home′ ee o sta′ sis) [Gk. *homos*: same + *stasis*: position] The maintenance of a steady state, such as a constant temperature or a stable social structure, by means of physiological or behavioral feedback responses.

homeotherm (home′ ee o therm) [Gk. *homos*: same + *thermos*: heat] An animal that maintains a constant body temperature by its own internal heating and cooling mechanisms. (Contrast with heterotherm, poikilotherm.)

homeotic genes (home ee ot′ ic) Genes that determine the developmental fate of entire segments of an animal.

homeotic mutations Mutations in homeotic genes that drastically alter the characteristics of a particular body segment, giving it the characteristics of other segments (as when wings grow from a *Drosophila* thoracic segment that should have produced legs).

homing The ability to return over long distances to a specific site.

homo- [Gk. *homos*: same] Prefix indicating two or more similar conditions, structures, or processes. (Contrast with hetero-.)

homolog (home′ o log′) [Gk. *homos*: same + *logos*: word] In cytogenetics, one of a pair (or larger set) of chromosomes having the same overall genetic composition and sequence. In diploid organisms, each chromosome inherited from one parent is matched by an identical (except for mutational changes) chromosome—its homolog—from the other parent.

homology (ho mol′ o jee) [Gk. *homologia*: of one mind; agreement] A similarity between two or more structures that is due to inheritance from a common ancestor. The structures are said to be homologous, and each is a homolog of the others. (Contrast with analogy.)

homoplasy (home′ uh play zee) [Gk. *homos*: same + *plastikos*: shape, mold] The presence in multiple groups of a trait that is not inherited from the common ancestor of those groups. Can result from convergent evolution, evolutionary reversal, or parallel evolution.

homosporous Producing a single type of spore that gives rise to a single type of gametophyte, bearing both female and male reproductive organs. (Contrast with heterosporous.)

homotypic Referring to adhesion of cells of the same type. (Contrast with heterotypic.)

homozygous (home′ oh zie′ gus) [Gk. *homos*: same + *zygotos*: joined] In a diploid organism, having identical alleles of a given gene on both homologous chromosomes. An individual may be a homozygote with respect to one gene and a heterozygote with respect to another. (Contrast with heterozygous.)

hormone (hore′ mone) [Gk. *hormon*: to excite, stimulate] A substance produced in minute amount at one site in a multicellular organism and transported to another site where it acts on target cells.

host An organism that harbors a parasite or symbiont and provides it with nourishment.

Hox genes Conserved homeotic genes found in vertebrates, *Drosophila*, and other animal groups. Hox genes contain the homeobox domain and specify pattern and axis formation in these animals.

human chorionic gonadotropin (hCG) A hormone secreted by the placenta which sustains the corpus luteum and helps maintain pregnancy.

Human Genome Project An effort to determine the DNA sequence of the entire human genome, understand the structures and functions of the genes, make comparisons with other organisms, and understand the social implications of this information.

humoral immune response The part of the immune system mediated by B cells that produce circulating antibodies active against extracellular bacterial and viral infections.

humus (hew′ muss) The partly decomposed remains of plants and animals on the surface of a soil.

hyaluronidase (high′ uh loo ron′ uh dase) An enzyme that digests proteoglycans. In sperm cells, it digests the coatings surrounding an egg so the sperm can enter.

hybrid (high′ brid) [L. *hybrida*: mongrel] (1) The offspring of genetically dissimilar parents. (2) In molecular biology, a double helix formed of nucleic acids from different sources.

hybridize To combine the genetic material of two distinct species or of two distinguishable populations within a species.

hybrid vigor *See* heterosis.

hybridoma A cell produced by the fusion of an antibody-producing cell with a myeloma cell; it produces monoclonal antibodies.

hybrid zone A narrow zone where two populations interbreed, producing hybrid individuals.

hydrocarbon A compound containing only carbon and hydrogen atoms.

hydrogen bond A weak electrostatic bond which arises from the attraction between the slight positive charge on a hydrogen atom and a slight negative charge on a nearby oxygen or nitrogen atom.

hydrological cycle The movement of water from the oceans to the atmosphere, to the soil, and back to the oceans.

hydrolysis (high drol' uh sis) [Gk. *hydro*: water + *lysis*: break apart] A chemical reaction that breaks a bond by inserting the components of water: AB + H_2O → AH + BOH.

hydrophilic (high dro fill' ik) [Gk. *hydro*: water + *philia*: love] Having an affinity for water. (Contrast with hydrophobic.)

hydrophobic (high dro foe' bik) [Gk. *hydro*: water + *phobia*: fear] Having no affinity for water. Uncharged and nonpolar groups of atoms are hydrophobic; for example, fats and the side chain of the amino acid phenylalanine. (Contrast with hydrophilic.)

hydrostatic pressure Pressure generated by compression of liquid in a confined space. Generated in plants, fungi, and some protists with cell walls by the osmotic uptake of water. Generated in animals with closed circulatory systems by the beating of a heart.

hydrostatic skeleton The incompressible internal liquids of some animals that transfer forces from one part of the body to another when acted upon by the surrounding muscles.

hydroxyl group The —OH group found on alcohols and sugars.

hyper- [Gk. *hyper*: above, over] Prefix indicating above, higher, more.

hyperpolarization A change in the resting potential of a membrane so the inside of a cell becomes more electronegative. (Contrast with depolarization.)

hypersensitive response A defensive response of plants to microbial infection; it results in a "dead spot."

hypertension High blood pressure.

hypertonic Having a greater solute concentration. Said of one solution compared to another. (Contrast with hypotonic, isotonic.)

hypha (high' fuh) (plural: hyphae) [Gk. *hyphe*: web] In the fungi and oomycetes, any single filament.

hypo- [Gk. *hypo*: beneath, under] Prefix indicating underneath, below, less.

hypoblast The lower tissue portion of the avian blastula which is joined to the epiblast at the margins of the blastodisc.

hypocotyl [Gk. *hypo*: beneath + *kotyledon*: hollow space] That part of the embryonic or seedling plant shoot that is below the cotyledons.

hypothalamus The part of the brain lying below the thalamus; it coordinates water balance, reproduction, temperature regulation, and metabolism.

hypothesis A tentative answer to a question, from which testable predictions can be generated. (Contrast with theory.)

hypotonic Having a lesser solute concentration. Said of one solution in comparing it to another. (Contrast with hypertonic, isotonic.)

- I -

ileum The final segment of the small intestine.

imaginal disc [L. *imagos*: image, form] In insect larvae, groups of cells that develop into specific adult organs.

imbibition Water uptake by a seed; first step in germination.

immediate hypersensitivity A rapid, extensive immune reaction against an allergen involving IgE and histamine release. (Contrast with delayed hypersensitivity.)

immune system [L. *immunis*: exempt from] A system in vertebrates that recognizes and attempts to eliminate or neutralize foreign substances (e.g., bacteria, viruses, pollutants).

immunization The deliberate introduction of antigen to bring about an immune response.

immunoassay The use of labeled antibodies to measure the concentration of an antigen in a sample.

immunoglobulins A class of proteins, with a characteristic structure, active as receptors and effectors in the immune system.

immunological memory The capacity to more rapidly and massively respond to a second exposure to an antigen than occurred on first exposure.

immunological tolerance A mechanism by which an animal does not mount an immune response to the antigenic determinants of its own macromolecules.

implantation The process by which the early mammalian embryo becomes attached to and embedded in the lining of the uterus.

imprinting (1) In genetics, the differential modification of a gene depending on whether it is present in a male or a female. (2) In animal behavior, a rapid form of learning in which an animal comes to make a particular response, which is maintained for life, to some object or other organism.

inbreeding Breeding among close relatives.

inclusive fitness The sum of an individual's genetic contribution to subsequent generations both via production of its own offspring and via its influence on the survival of relatives who are not direct descendants.

incomplete dominance Condition in which the heterozygous phenotype is intermediate between the two homozygous phenotypes.

incomplete metamorphosis Insect development in which changes between instars are gradual.

incus (in' kus) [L. *incus*: anvil] The middle of the three bones that conduct movements of the eardrum to the oval window of the inner ear. (*See* malleus, stapes.)

independent assortment During meiosis, the random separation of genes carried on nonhomologous chromosomes. Articulated by Mendel as his second law.

indirect transduction A cell signaling mechanism in which a second messenger mediates the interaction between receptor binding and cellular response. (Contrast with direct transduction.)

individual fitness That component of inclusive fitness resulting from an organism producing its own offspring. (Contrast with kin selection.)

indoleacetic acid *See* auxin.

induced fit A change in enzyme conformation upon binding to substrate with an increase in the rate of catalysis.

induced mutation A mutation resulting from treatment with a chemical or other agent.

inducer (1) In enzyme systems, a small molecule which, when added to a growth medium, causes a large increase in the level of some enzyme. (2) In embryology, a substance that causes a group of target cells to differentiate in a particular way.

inducible enzyme An enzyme that is present in much larger amounts when a particular compound (the inducer) has been added to the system. (Contrast with constitutive enzyme.)

inflammation A nonspecific defense against pathogens; characterized by redness, swelling, pain, and increased temperature.

inflorescence A structure composed of several to many flowers.

inflorescence meristem A meristem that produces floral meristems as well as other small leafy structures (bracts).

inhibitor A substance that binds to the surface of an enzyme and interferes with its action on its substrates.

initial cells In plant meristems, undifferentiated cells that retain the capacity to divide producing both undifferentiated cells (initials) and cells committed to differentiation. (Compare with stem cells.)

initiation In molecular biology, the beginning of transcription or translation.

initiation complex Combination of a ribosomal light subunit, an mRNA molecule, and the tRNA charged with the first amino acid coded for by the mRNA; formed at the onset of translation.

initiation factors Proteins that assist in forming the translation initiation complex at the ribosome.

inner cell mass Derived from the mammalian blastula (bastocyst), the inner cell mass will give rise to the yolk sac (via hypoblast) and embryo (via epiblast).

inositol triphosphate (IP_3) An intracellular second messenger derived from membrane phospholipids.

inspiratory reserve volume The amount of air that can be inhaled above the normal tidal inspiration. (Compares with expiratory reserve volume, tidal volume, vital capacity.)

instar (in' star) An immature stage of an insect between molts.

insulin (in' su lin) [L. *insula*: island] A hormone synthesized in islet cells of the pancreas that promotes the conversion of glucose into the storage material, glycogen.

integral membrane protein A membrane protein embedded in the bilayer of the membrane. (Contrast with peripheral membrane protein.)

integrase An enzyme that integrates retroviral cDNA into the genome of the host cell.

integrated pest management Control of pests by the use of natural predators and parasites in conjunction with sparing use of chemicals; an attempt to limit environmental damage.

integument [L. *integumentum*: covering] A protective surface structure. In gymnosperms and angiosperms, a layer of tissue around the ovule which will become the seed coat.

intercalary meristem A meristematic region in plants which occurs not apically, but between two regions of mature tissue. Intercalary meristems occur in the nodes of grass stems, for example.

intercostal muscles Muscles between the ribs that can augment breathing movements by elevating and suppressing the rib cage.

interference competition Competition in which individuals actively interfere with one another's access to resources. (Compare with exploitation competition.)

interference RNA (RNAi) A mechanism for reducing mRNA translation whereby a double-stranded RNA, made by the cell or synthetically, is processed to a small, single-stranded RNA, and binding of this RNA to a target mRNA results in the latter's breakdown.

interferon A glycoprotein produced by virus-infected animal cells; increases the resistance of neighboring cells to the virus.

interkinesis The period between meiosis I and meiosis II.

interleukins Regulatory proteins, produced by macrophages and lymphocytes, that act upon other lymphocytes and direct their development.

intermediate disturbance hypothesis The hypothesis that explains why species richness is lower in areas with both high and low rates of disturbance than in areas with intermediate rates of disturbance.

intermediate filaments Cytoskeletal component with diameters between the larger microtubules and smaller microfilaments.

internal environment The physical and chemical characteristics of the extracellular fluids of the body.

interneuron A neuron that communicates information between two other neurons.

ionotropic receptors A receptor that that directly alters membrane permeability to a type of ion when it combines with its ligand.

internode The region between two nodes of a plant stem.

interphase The period between successive nuclear divisions during which the chromosomes are diffuse and the nuclear envelope is intact. It is during this period that the cell is most active in transcribing and translating genetic information.

interspecific competition Competition between members of two or more species. (Contrast with intraspecific competition.)

intertropical convergence zone The tropical region where the air rises most strongly; moves north and south with the passage of the sun overhead.

intraspecific competition Competition among members of the same species. (Contrast with interspecific competition.)

intrinsic protein A membrane protein that is embedded in the phospholipid bilayer of the membrane. (Contrast with extrinsic protein.)

intrinsic rate of increase The rate at which a population can grow when its density is low and environmental conditions are highly favorable.

intron A portion of a DNA molecule that, because of RNA splicing, is not involved in coding for part of a polypeptide molecule. (Contrast with exon.)

invagination An infolding of cells during animal embryonic development.

inversion A rare 180° reversal of the order of genes within a segment of a chromosome.

invertebrate A "convenience term" that encompasses any animal that is not a vertebrate—that is, whose nerve cord is not enclosed in a backbone of bony segments.

in vitro [L.: in glass] In a test tube, rather than in a living organism. (Contrast with in vivo.)

in vitro evolution Evolution in the laboratory, as used to produce compounds for industrial and pharmaceutical purposes.

in vivo [L.: in the living state] In a living organism. Many processes that occur in vivo can be reproduced in vitro with the right selection of cellular components. (Contrast with in vitro.)

ion (eye' on) [Gk.: *ion*: wanderer] An atom or group of atoms with electrons added or removed, giving it a negative or positive electrical charge.

ion channel A membrane protein that can let ions diffuse across the membrane. The channel can be ion-selective, and it can be voltage-gated or ligand-gated.

ionic bond An electrostatic attraction between positively and negatively charged ions. Usually a strong bond.

iris (eye' ris) [Gk. *iris*: rainbow] The round, pigmented membrane that surrounds the pupil of the eye and adjusts its aperture to regulate the amount of light entering the eye.

irruption A rapid increase in the density of a population. Often followed by massive emigration.

islets of Langerhans Clusters of hormone-producing cells in the pancreas.

iso- [Gk. *iso*: equal] Prefix used two separate entities that share some element of identity.

isogamous Describes male and female gametes that are morphologically identical.

isolating mechanism Geographical, physiological, ecological, or behavioral mechanisms that lead to a reduction in the frequency of successful matings between individuals in separate populations of a species. Can lead to the eventual evolution of separate species.

isomers Molecules consisting of the same numbers and kinds of atoms, but differing in the bonding patterns by which the atoms are held together.

isomorphic (eye so more' fik) [Gk. *isos*: equal + *morphe*: form] Having the same form or appearance, as when the haploid and diploid life stages of an organism appear identical. (Contrast with heteromorphic.)

isotonic Having the same solute concentration; said of two solutions. (Contrast with hypertonic, hypotonic.)

isotope (eye' so tope) [Gk. *isos*: equal + *topos*: place] Isotopes of a given chemical element have the same number of protons in their nuclei (and thus are in the same position on the periodic table), but differ in the number of neutrons

isozymes Enzymes that have somewhat different amino acid sequences but catalyze the same reaction.

- J -

jasmonates Plant hormones that trigger defenses against pathogens and herbivores.

jejunum (jih jew' num) The middle division of the small intestine, where most absorption of nutrients occurs. (*See* duodenum, ileum.)

joule (jool, or jowl) A unit of energy, equal to 0.24 calories.

juvenile hormone In insects, a hormone maintaining larval growth and preventing maturation or pupation.

- K -

karyogamy The fusion of nuclei of two cells. (Contrast with plasmogamy.)

karyotype The number, forms, and types of chromosomes in a cell.

keratin (ker' a tin) [Gk. keras: horn] A protein which contains sulfur and is part of such hard tissues as horn, nail, and the outermost cells of the skin.

ketone (key' tone) A compound with a C=O group attached to two other groups, neither of which is an H atom. Many sugars are ketones. (Contrast with aldehyde.)

keystone species Species that have a dominant influence on the composition of a community.

kidneys A pair of excretory organs in vertebrates.

kin selection That component of inclusive fitness resulting from helping the survival of relatives containing the same alleles by descent from a common ancestor. (Contrast with individual fitness.)

kinase (kye' nase) An enzyme that transfers a phosphate group from ATP to another molecule. Protein kinases transfer phosphate from ATP to specific proteins, playing important roles in cell regulation.

kinesin Motor protein having the capacity to attach to organelles or vesicles and move them along microtubules of the cytoskeleton.

kinetic energy The energy associated with movement.

kinetochore (kin net' oh core) [Gk. *kinetos*: moving] Specialized structure on a centromere to which microtubules attach.

knockout A molecular genetic method in which a single gene of an organism is permanently inactivated.

Koch's posulates A set of rules for establishing that a particular microorganism causes a particular disease.

Krebs cycle *See* citric acid cycle.

- L -

lactic acid fermentation Fermentation whose end product is lactic acid (lactate).

lagging strand In DNA replication, the daughter strand that is synthesized in discontinuous stretches. (*See* Okazaki fragments.)

lamella (la mell' ah) [L. *lamina*: thin sheet] Layer.

larva (plural: larvae) [L. *lares*: guiding spirits] An immature stage of any invertebrate animal that differs dramatically in appearance from the adult.

larynx (lar' inks) [Gk. *larynx*: voice box] A structure between the pharynx and the trachea that includes the vocal cords.

lateral Pertaining to the side.

lateral gene transfer The transfer of genes from one species to another, common among bacteria and archaea.

lateral meristems The vascular cambium and cork cambium, which give rise to secondary tissue in plants.

laticifers (luh tiss' uh furs) In some plants, elongated cells containing secondary plant products such as latex.

leader sequence A sequence of amino acids at the amino-terminal end of a newly synthesized protein; determines where the protein will be placed in the cell.

leading strand In DNA replication, the daughter strand that is synthesized continuously. (Contrast with lagging strand.)

leaf primordium An outgrowth on the side of the shoot apical meristem that will eventually develop into a leaf.

lenticel (len' ti sill) Spongy region in a plant's periderm, allowing gas exchange.

leukocyte (loo' ko sight) [Gk. *leukos*: clear + *kytos*: container] A white blood cell.

lichen (lie' kun) An organism resulting from the symbiotic association of a true fungus and either a cyanobacterium or a unicellular alga.

life cycle The entire span of the life of an organism from the moment of fertilization (or asexual generation) to the time it reproduces in turn.

life history The stages an individual goes through during its life.

life table A table showing, for a group of equal-aged individuals, the proportion still alive at different times in the future and the number of offspring they produce during each time interval.

ligament A band of connective tissue linking two bones in a joint.

ligand (lig' and) Any molecule that binds to a receptor site of another (usually larger) molecule.

light reactions The initial phase of photosynthesis, in which light energy is converted into chemical energy.

light-independent reactions The phase of photosynthesis in which chemical energy captured in the light reactions is used to drive the reduction of CO_2 to form carbohydrates.

lignin The principal noncarbohydrate component of wood, a polymer that binds together cellulose fibrils in some plant cell walls.

limbic system A group of primitive vertebrate forebrain nuclei that form a network and are involved in emotions, drives, instinctive behaviors, learning, and memory.

limiting resource The required resource whose supply most strongly influences the size of a population.

linkage Association between genetic markers on the same chromosome such that they do not show random assortment and seldom recombine; the closer the markers, the lower the frequency of recombination.

lipase (lip' ase; lye' pase) An enzyme that digests fats.

lipids (lip' ids) [Gk. *lipos*: fat] Substances in a cell which are easily extracted by organic solvents; fats, oils, waxes, steroids, and other large organic molecules, including those which, with proteins, make up the cell membranes. (Compare with phospholipids.)

littoral zone The coastal zone from the upper limits of tidal action down to the depths where the water is thoroughly stirred by wave action.

liver A large digestive gland. In vertebrates, it secretes bile and is involved in the formation of blood.

lobes Regions of the human cerebral hemispheres; includes the temporal, frontal, parietal, and occipital lobes.

locus In genetics, a specific location on a chromosome. May be considered to be synonymous with *gene*.

logistic growth Growth, especially in the size of an organism or in the number of organisms in a population, that slows steadily as the entity approaches its maximum size. (Contrast with exponential growth.)

long-day plant (LDP) A plant that requires long days (actually, short nights) in order to flower.

long-term potentiation (LTP) A long-lasting increase in the sensitivity of a neuron resulting from a period of intense stimulation.

loop of Henle (hen' lee) Long, hairpin loop of the mammalian renal tubule that runs from the cortex down into the medulla, and back to the cortex. Creates a concentration gradient in the interstitial fluids in the medulla.

lophophore A U-shaped fold of the body wall with hollow, ciliated tentacles that encircles the mouth of animals in several different groups. Used for filtering prey from the surrounding water.

lumen (loo' men) [L. *lumen*: light] The open cavity inside any tubular organ or structure, such as the gut or a kidney tubule.

luteinizing hormone A gonadotropin produced by the anterior pituitary. It stimulates the gonads to produce sex hormones.

lymph [L. *lympha*: liquid] A clear, watery fluid that is formed as a filtrate of blood; it contains white blood cells; it collects in a series of special vessels and is returned to the bloodstream.

lymph nodes Specialized tissue regions that act as filters for cells, bacteria and foreign matter.

lymphocyte A major class of white blood cells. Includes T cells, B cells, and other cell types important in the immune response.

lymphoid tissue Tissues of the immune defense system dispersed throughout the body and consisting of: thymus, spleen, bone marrow, lymph nodes, blood, and lymph.

lysis (lie' sis) [Gk. *lysis*: break apart] Bursting of a cell.

lysogenic bacteria Bacteria that harbor a viral chromosome capable of the lysogenic cycle.

lysogenic cycle A form of viral replication in which the virus becomes incorporated into the bacterial chromosome and the host cell is not killed. (Contrast with lytic cycle.)

lysosome (lie' so soam) [Gk. *lysis*: break away + *soma*: body] A membrane-enclosed organelle found in eukaryotic cells (other than plants). Lysosomes contain a mixture of enzymes that can digest most of the macromolecules found in the rest of the cell.

lysozyme (lie' so zyme) An enzyme in saliva, tears, and nasal secretions that attacks bacterial cell walls, as one of the body's nonspecific defense mechanisms.

lytic cycle A form of viral reproduction that lyses the host bacterium releasing the new viruses. (Contrast with lysogenic cycle.)

- M -

M phase The portion of the cell cycle in which mitosis takes place.

macroevolution [Gk. *makros*: large, long] Evolutionary changes occurring over long time spans and usually involving changes in many traits. (Contrast with microevolution.)

macromolecule A giant polymeric molecule. The macromolecules are proteins, polysaccharides, and nucleic acids.

macronutrient A mineral element required by plant tissues in concentrations of at least 1 milligram per gram of their dry matter.

macrophage (mac' roh faj) A type of white blood cell that endocytoses bacteria and other cells.

MADS box A DNA-binding domain in many plant transcription factors that is active in development.

major histocompatibility complex (MHC) A complex of linked genes, with multiple alleles, that control a number of cell surface antigens that identify self and can lead to graft rejection.

malignant Referring to a tumor that can grow indefinitely and/or spread from the original site of growth to other locations in the body. (Contrast with benign.)

malleus (mal' ee us) [L. *malleus*: hammer] The first of the three bones that conduct movements of the eardrum to the oval window of the inner ear. (*See* incus, stapes.)

Malpighian tubule (mal pee' gy un) A type of protonephridium found in insects.

mantle A sheet of specialized tissues that covers most of the viscera of mollusks; provides protection to internal organs and secretes the shell.

mapping In genetics, determining the order of genes on a chromosome and the distances between them.

map unit The distance between two genes, a recombinant frequency of 0.01.

marine [L. *mare*: sea, ocean] Pertaining to or living in the ocean. (Contrast with aquatic, terrestrial.)

marker A gene of identifiable phenotype that indicates the presence on another gene, DNA segment, or chromosome fragment.

mass extinctions Periods of evolutionary history during which rates of extinction were much higher than during intervening times.

mass number The sum of the number of protons and neutrons in an atom's nucleus.

mast cells Typically found in connective tissue, mast cells can be provoked by antigens or inflammation to release histamine.

maternal effect genes These genes code for morphogens that determine the polarity of the egg and larva in the fruit fly *Drosophila melanogaster*.

maternal inheritance Inheritance in which the mother's phenotype is exclusively expressed. Mitochondria and chloroplasts are maternally inherited via egg cytoplasm. Also known as cytoplasmic inheritance.

mating type A particular strain of a species that is incapable of sexual reproduction with another member of the same strain but capable of sexual reproduction with members of other strains of the same species.

maximum likelihood A statistical method of determining which of two or more hypotheses (such as phylogenetic trees) best fit the observed data, given an explicit model of how the data were generated.

mechanoreceptor A cell that is sensitive to physical movement and generates action potentials in response.

medulla (meh dull' luh) (1) The inner, core region of an organ, as in the adrenal medulla (adrenal gland) or the renal medulla (kidneys). (2) The portion of the brain stem that connects to the spinal cord.

megaphyll The generally large leaf of a fern, horsetail, or seed plant, with several to many veins. (Contrast with microphyll.)

megaspore [Gk. *megas*: large + *spora*: to sow] In plants, a haploid spore that produces a female gametophyte.

meiosis (my oh' sis) [Gk. *meiosis*: diminution] Division of a diploid nucleus to produce four haploid daughter cells. The process consists of two successive nuclear divisions with only one cycle of chromosome replication. In *meiosis I,* homologous chromosomes separate but retain their chromatids. The second division *meiosis II,* is similar to mitosis, in which chromatids separate.

melatonin A hormone released by the pineal gland that is involved in photoperiodism and circadian rhythms.

membrane potential The difference in electrical charge between the inside and the outside of a cell, caused by a difference in the distribution of ions.

memory cells Long-lived lymphocytes produced by exposure to antigen. They persist in the body and are able to mount a rapid response to subsequent exposures to the antigen.

Mendelian population A local population of individuals belonging to the same species and exchanging genes with one another.

Mendel's first law *See* segregation.

Mendel's second law *See* independent assortment.

menstrual cycle The monthly sloughing off of the uterine lining if fertilization does not occur in the female. Occurs between puberty and menopause.

meristem [Gk. *meristos*: divided] Plant tissue made up of undifferentiated actively dividing cells.

mesenchyme (mez' en kyme) [Gk. *mesos*: middle + *enchyma*: infusion] Embryonic or unspecialized cells derived from the mesoderm.

mesoderm [Gk. *mesos*: middle + *derma*: skin] The middle of the three embryonic tissue layers first delineated during gastrulation. Gives rise to skeleton, circulatory system, muscles, excretory system, and most of the reproductive system.

mesophyll (mez' uh fill) [Gk. *mesos*: middle + *phyllon*: leaf] Chloroplast-containing, photosynthetic cells in the interior of leaves.

mesosome (mez' uh soam') [Gk. *mesos*: middle + *soma*: body] A localized infolding of the plasma membrane of a bacterium.

messenger RNA (mRNA) A transcript of one of the strands of DNA; carries information (as a sequence of codons) for the synthesis of one or more proteins.

meta- [Gk.: between, along with, beyond] A prefix used in biology to denote a change or a shift to a new form or level; for example, as used in metamorphosis.

metabolic compensation Changes in metabolic properties of an organism that render it less sensitive to temperature changes.

metabolic factor In bacteria, a plasmid that carries genes determining unusual metabolic functions, such as the breakdown of hydrocarbons in oil.

metabolic pathway A series of enzyme-catalyzed reactions so arranged that the product of one reaction is the substrate of the next.

metabolism (meh tab' a lizm) [Gk. *metabole*: to change] The sum total of the chemical reactions that occur in an organism, or some subset of that total (as in respiratory metabolism).

metabotropic receptor A receptor that that indirectly alters membrane permeability to a type of ion when it combines with its ligand.

metamorphosis (met' a mor' fo sis) [Gk. *meta*: between + *morphe*: form, shape] A change occurring between one developmental stage and another, as for example from a tadpole to a frog. (*See* complete metamorphosis, incomplete metamorphosis.)

metaphase (met' a phase) The stage in nuclear division at which the centromeres of the highly supercoiled chromosomes are all lying on a plane (the metaphase plane or plate) perpendicular to a line connecting the division poles.

metapopulation A population divided into subpopulations, among which there are occasional exchanges of individuals.

metastasis (meh tass' tuh sis) The spread of cancer cells from their original site to other parts of the body.

methylation The addition of a methyl group (—CH_3) to a molecule. Extensive methylation of cytosine in DNA is correlated with reduced transcription.

MHC *See* major histocompatibility complex.

micelle a particle of lipid covered with bile salts that is produced in the duodenum and facilitates digestion and absorption of lipids.

microbiology [Gk. *mikros*: small + *bios*: life + *logos*: discourse] The scientific study of microscopic organisms, particularly bacteria, protists, and viruses.

microevolution The small evolutionary changes typically occurring over short time spans; generally involving a small number of traits and minor genetic changes. (Contrast with macroevolution.)

microfilament Minute fibrous structure generally composed of actin found in the cytoplasm of eukaryotic cells. They play a role in the motion of cells.

micronutrient A mineral element required by plant tissues in concentrations of less than 100 micrograms per gram of their dry matter.

microphyll A small leaf with a single vein, found in club mosses and their relatives. (Contrast with megaphyll.)

micropyle (mike' roh pile) [Gk. *mikros*: small + *pylon*: gate] Opening in the integument(s) of a seed plant ovule through which pollen grows to reach the female gametophyte within.

micro RNA A small RNA, typically about 21 bases long, that binds to mRNA to reduce its translation.

microspore [Gk. *mikros*: small + *spora*: to sow] In plants, a haploid spore that produces a male gametophyte.

microtubules Minute tubular structures found in centrioles, spindle apparatus, cilia, flagella, and cytoskeleton of eukaryotic cells. These tubules play roles in the motion and maintenance of shape of eukaryotic cells.

microvilli (singular: microvillus) The projections of epithelial cells, such as the cells lining the small intestine, that increase their surface area.

middle lamella A layer of polysaccharides that separates plant cells; a shared middle lamella lies outside the primary walls of the two cells.

migration The regular, seasonal movements of animals.

mineral An inorganic substance other than water.

mineral nutrients Inorganic ions required by organisms for normal growth and reproduction.

mismatch repair When a single base in DNA is changed into a different base, or the wrong base inserted during DNA replication, there is a mismatch in base pairing with the base on the opposite strand. A repair system removes the incorrect base and inserts the proper one for pairing with the opposite strand.

missense substitution A change in a gene from one nucleotide to another that also results in a change in the amino acid specified by the corresponding codon. (Compare with frame-shift mutation, nonsense substitution, synonymous substitution.)

mitochondrial matrix The fluid interior of the mitochondrion, enclosed by the inner mitochondrial membrane.

mitochondrion (my' toe kon' dree un) [Gk. *mitos*: thread + *chondros*: grain] An organelle in eukaryotic cells that contains the enzymes of the citric acid cycle, the respiratory chain, and oxidative phosphorylation.

mitosis (my toe' sis) [Gk. *mitos*: thread] Nuclear division in eukaryotes leading to the formation of two daughter nuclei, each with a chromosome complement identical to that of the original nucleus.

mitotic center Cellular region that organizes the microtubules for mitosis. In animals a centrosome serves as the mitotic center.

moderately repetitive DNA DNA sequences that appear hundreds to thousands of times in the genome. They include the DNA sequences coding for rRNAs and tRNAs, as well as the DNA at telomeres.

modular organism An organism which grows by producing additional units of body construction (modules) that are very similar to the units of which it is already composed.

mole A quantity of a compound whose weight in grams is numerically equal to its molecular weight expressed in atomic mass units. Avogadro's number of molecules: 6.023×10^{23} molecules.

molecular clock The theory that macromolecules diverge from one another over evolutionary time at a constant rate; this rate may provide insight into the phylogenetic relationships among organisms.

molecular tool kit A set of developmental genes and proteins that is common to most animals and is hypothesized to be responsible for the evoltuion of their differing developmental pathways.

molecular weight The sum of the atomic weights of the atoms in a molecule.

molecule A particle made up of two or more atoms joined by covalent bonds or ionic attractions.

molting The process of shedding part or all of an outer covering, as the shedding of feathers by birds or of the entire exoskeleton by arthropods.

monoclonal antibody Antibody produced in the laboratory from a clone of hybridoma cells, each of which produces the same specific antibody.

monocytes White blood cells that produce macrophages.

monoecious (mo nee' shus) [Gk. *mono*: one + *oikos*: house] Describes organisms in which both sexes are "housed" in a single individual that produces both eggs and sperm. (In some plants, these are found in different flowers within the same plant.) Examples include corn, peas, earthworms, hydras. (Contrast with dioecious.)

monohybrid cross A mating in which the parents differ with respect to the alleles of only one locus of interest.

monomer [Gk. *mono*: one + *meros:* unit] A small molecule, two or more of which can be combined to form oligomers (consisting of a few monomers) or polymers (consisting of many monomers).

monophyletic (mon' oh fih leht' ik) [Gk. *mono*: one + *phylon*: tribe] Referring to a group that consists of an ancestor and all of its descendants. (Compare with paraphyletic, polyphyletic.)

monosaccharide A simple sugar. Oligosaccharides and polysaccharides are made up of monosaccharides.

monosomic Referring to an organism with one less than the normal diploid number of chromosomes.

monosynaptic reflex A neural reflex that begins in a sensory neuron and makes a single synapse before activating a motor neuron.

morphogen A diffusible substances whose concentration gradients determine patterns of development in animals and plants.

morphogenesis (more' fo jen' e sis) [Gk. *morphe*: form + *genesis*: origin] The development of form; the overall consequence of determination, differentiation, and growth.

morphology (more fol' o jee) [Gk. *morphe*: form + *logos*: study, discourse] The scientific study of organic form, including both its development and function.

mosaic development Pattern of animal embryonic development in which each blastomere contributes a specific part of the adult body. (Contrast with regulative development.)

motor end plate The modified area on a muscle cell membrane where a synapse is formed with a motor neuron.

motor neuron A neuron carrying information from the central nervous system to an effector such as a muscle fiber.

motor proteins Specialized proteins that use energy to change shape and move cells or structures within cells. *See* dynein, kinesin.

motor unit A motor neuron and the set of muscle fibers it controls.

mRNA *See* messenger RNA.

mucosa (mew koh' sah) An epithelial membrane containing cells that secrete mucus. The inner cell layers of the digestive and respiratory tracts.

Müllerian mimicry The convergence in appearance of two or more unpalatable species.

multifactorial In medicine, referring to a disease with many interacting causes, both genetic and environmental.

muscle fiber A single muscle cell. In the case of skeletal (striated) muscle, a syncitial, multinucleate cell.

muscle tissue Excitable tissue that can contract due to interactions of actin and myosin. Three types are striated, smooth, and cardiac.

muscle tone The degree of contraction of a muscle.

muscle spindle Modified muscle fibers encased in a connective sheat and functioning as stretch receptors.

mutagen (mute' ah jen) [L. *mutare*: change + Gk. *genesis*: source] Any agent (e.g., chemicals, radiation) that increases the mutation rate.

mutation A detectable, heritable change in the genetic material not caused by recombination.

mutation pressure Evolution (change in gene proportions) by different mutation rates alone (i.e., without the influence of natural selection).

mutualism The type of symbiosis, such as that exhibited by fungi and algae or cyanobacteria in forming lichens, in which both species profit from the association.

mycelium (my seel' ee yum) [Gk. *mykes*: fungus] In the fungi, a mass of hyphae.

mycorrhiza (my' ko rye' za) [Gk. *mykes*: fungus + *rhiza*: root] An association of the root of a plant with the mycelium of a fungus.

myelin (my' a lin) A material forming a sheath around some axons. Formed by Schwann cells that wrap themselves about the axon, myelin insulates the axon electrically and increases the rate of transmission of a nervous impulse.

myocardial infarction A blockage of an artery that carries blood to the heart muscle.

myofibril (my' oh fy' bril) [Gk. *mys*: muscle + L. *fibrilla*: small fiber] A polymeric unit of actin or myosin in a muscle.

myogenic (my oh jen' ik) [Gk. *mys*: muscle + *genesis*: source] Originating in muscle.

myoglobin (my' oh globe' in) [Gk. *mys*: muscle + L. *globus*: sphere] An oxygen-binding molecule found in muscle. Consists of a heme unit and a single globiin chain, and carrys less oxygen than hemoglobin.

myosin One of the two major proteins of muscle, it makes up the thick filaments. (*See* actin.)

- N -

NAD (nicotinamide adenine dinucleotide) A compound found in all living cells, existing in two interconvertible forms: the oxidizing agent NAD^+ and the reducing agent $NADH + H^+$.

NADP (nicotinamide adenine dinucleotide phosphate) A compound similar to NAD, but possessing another phosphate group; plays similar roles but is used by different enzymes.

natural killer cells A nonspecific defensive cell (lymphocyte) that attacks tumor cells and virus infected cells.

natural selection The differential contribution of offspring to the next generation by various genetic types belonging to the same population. The mechanism of evolution proposed by Charles Darwin.

necrosis (nec roh' sis) [Gk. *nekros*: death] Tissue damage resulting from cell death.

negative control The situation in which a regulatory macromolecule (generally a repressor) functions to turn off transcription. In the absence of a regulatory macromolecule, the structural genes are turned on.

negative feedback Information relevant to the rate of a process that can be used by a control system to return the outcome of that process to an optimal level.

nematocyst (ne mat' o sist) [Gk. *nema*: thread + *kystis*: cell] An elaborate, threadlike structure produced by cells of jellyfish and other cnidarians, used chiefly to paralyze and capture prey.

nephridium (nef rid' ee um) [Gk. *nephros*: kidney] An organ which is involved in excretion, and often in water balance, involving a tube that opens to the exterior at one end.

nephron (nef' ron) [Gk. *nephros*: kidney] The functional unit of the kidney, consisting of a structure for receiving a filtrate of blood, and a tubule that absorbs selected parts of the filtrate back into the bloodstream.

nephrostome (nef' ro stome) [Gk. *nephros*: kidney + *stoma*: opening] An opening in a nephridium through which body fluids can enter.

Nernst equation A mathematical statement; calculates the potential across a membrane permeable to a single type of ion that differs in concentration on the two sides of the membrane.

nerve A structure consisting of many neuronal axons and connective tissue.

net primary production Total photosynthesis minus respiration by plants.

neural plate A thickened strip of ectoderm along the dorsal side of the early vertebrate embryo; gives rise to the central nervous system.

neural tube An early stage in the development of the vertebrate nervous system consisting of a hollow tube created by two opposing folds of the dorsal ectoderm along the anterior–posterior body axis.

neurohormone A chemical signal produced and released by neurons; the signal then acts as a hormone.

neuromuscular junction The region where a motor neuron contacts a muscle fiber, creating a synapse.

neuron (noor' on) [Gk. *neuron*: nerve] A nervous system cell that can generate and conduct action potentials along an axon to a synapse with another cell.

neurotransmitter A substance produced in and released by one a neuron (the presynaptic cell) that diffuses across a synapse and excites or inhibits another cell (the postsynaptic cell).

neurula (nure' you la) Embryonic stage during the dorsal nerve cord forms from two ectodermal ridges.

neurulation A stage in vertebrate development during which the nervous system begins to form.

neutral allele An allele that does not alter the functioning of the proteins for which it codes.

neutral theory A view of molecular evolution that postulates that most mutations do not affect the amino acid being coded for, and that such mutations accumulate in a population at rates driven by genetic drift and mutation rates.

neutron (new' tron) One of the three most fundamental particles of matter, with mass approximately 1 amu and no electrical charge.

neutrophils Abundant, short-lived phagocytic leukocytes that attack antibody-coated antigens.

nitrate reduction The process by which nitrate (NO_3^-) is reduced to ammonia (NH_3).

nitric oxide (NO) An unstable molecule (a gas) that serves as a second messenger causing smooth muscle to relax. In the nervous system it operates as a neurotransmitter.

nitrification The oxidation of ammonia to nitrite and nitrate ions, performed by certain soil bacteria.

nitrogenase In nitrogen-fixing organisms, an enzyme complex that mediates the stepwise reduction of atmospheric N_2 to ammonia.

nitrogen fixation Conversion of nitrogen gas to ammonia, which makes nitrogen available to living things. Carried out by certain prokaryotes, some of them free-living and others living within plant roots.

node [L. *nodus*: knob, knot] In plants, a (sometimes enlarged) point on a stem where a leaf is or was attached.

node of Ranvier A gap in the myelin sheath covering an axon; the point where the axonal membrane can fire action potentials.

noncompetitive inhibitor An inhibitor that binds the enzyme at a site other than the active site. (Contrast with competitive inhibitor.)

noncyclic electron transport In photosynthesis, the flow of electrons that forms ATP, NADPH, and O_2.

nondisjunction Failure of sister chromatids to separate in meiosis II or mitosis, or failure of homologous chromosomes to separate in meiosis I. Results in aneuploidy.

nonpolar molecule A molecule whose electric charge is evenly balanced from one end of the molecule to the other.

nonrandom mating The selection by individuals of other individuals of particular genotypes as mates.

non-REM sleep A state of sleep characterized by low muscle tone, but not atonia, quiescence,

nonsense substitution A change in a gene from one nucleotide to another that prematurely terminates a polypeptide. Termination occurs when a codon that specifies an amino acid is changed to one of the codons (UAG, UAA, or UGA) that signal termination of translation. (Compare with frame-shift mutation, missense substitution, synonymous substitution.)

nonspecific defenses Immunologic responses directed against any invading agent without reacting to apecific antigens.

nonsynonymous substitution A change in a gene from one nucleotide to another that changes the amino acid specified by the corresponding codon (i.e., AGC $\rightarrow$ AGA, or serine $\rightarrow$ arginine). (Contrast with synonymous substitution.)

nonvascular plants Those plants lacking well-developed vascular tissue; the liverworts, hornworts, and mosses. (Contrast with vascular plants.)

norepinephrine A neurotransmitter found in the central nervous system and also at the postganglionic nerve endings of the sympathetic nervous system. Also called noradrenaline.

notochord (no' tow kord) [Gk. *notos*: back + *chorde*: string] A flexible rod of gelatinous material serving as a support in the embryos of all chordates and in the adults of tunicates and lancelets.

nuclear envelope The surface, consisting of two layers of membrane, that encloses the nucleus of eukaryotic cells.

nuclear lamina A meshwork of fibers on the inner surface of the nuclear envelope.

nuclear pore complex A protein structure situated in nuclear pores through which RNA and proteins enter and leave the nucleus.

nucleic acid (new klay' ik) A long-chain alternating polymer of deoxyribose or ribose and phosphate groups, with nitrogenous bases—adenine, thymine, uracil, guanine, or cytosine (A, T, U, G, or C)—as side chains. DNA and RNA are nucleic acids.

nucleic acid hybridization A technique in which a single-stranded nucleic acid probe in made that is complementary to, and binds to, a target sequence, either DNA or RNA. The resulting double-stranded molecule is a hybrid.

nucleoid (new' klee oid) The region that harbors the chromosomes of a prokaryotic cell. Unlike the eukaryotic nucleus, it is not bounded by a membrane.

nucleolar organizer (new klee' o lar) A region on a chromosome that is associated with the formation of a new nucleolus following nuclear division. The site of the genes that code for ribosomal RNA.

nucleolus (new klee' oh lus) A small, generally spherical body found within the nucleus of eukaryotic cells. The site of synthesis of ribosomal RNA.

nucleoplasm (new' klee o plazm) The fluid material within the nuclear envelope of a cell, as opposed to the chromosomes, nucleoli, and other particulate constituents.

nucleoside A nucleotide without the phosphate group.

nucleosome A portion of a eukaryotic chromosome, consisting of part of the DNA molecule wrapped around a group of histone molecules, and held together by another type of histone molecule. The chromosome is made up of many nucleosomes.

nucleotide The basic chemical unit in a nucleic acid. A nucleotide in RNA consists of one of four nitrogenous bases linked to ribose, which in turn is linked to phosphate. In DNA, deoxyribose is present instead of ribose.

nucleotide substitution A change of one base pair to another in a DNA sequence.

nucleus (new' klee us) [L. *nux*: kernel or nut] (1) In cells, the centrally located compartment of eukaryotic cells that is bounded by a double membrane and contains the chromosomes. (2) In the brain, an identifiable group of neurons that share common characteristics or functions.

null hypothesis The assertion that an effect proposed by its companion hypothesis does not in fact exist.

nutrient A food substance; or, in the case of mineral nutrients, an inorganic element required for completion of the life cycle of an organism.

- O -

obligate anaerobe An anaerobic prokaryote that cannot survive exposure to O_2.

odorant A molecule that can bind to an olfactory receptor.

oil A triglyceride that is liquid at room temperature. (Contrast with fat.)

Okazaki fragments Newly formed DNA making up the lagging strand in DNA replication. DNA ligase links Okazaki fragments together to give a continuous strand.

olfactory [L. *olfacere*: to smell] Having to do with the sense of smell.

oligomer [Gk.: *oligo*: a few + *meros*: units] A compound molecule of intermediate size, made up of two to a few monomers. (Contrast with monomer, polymer.)

oligosaccharide A polymer containing a small number of monosaccharides.

oligosaccharins Plant hormones, derived from the plant cell wall, that trigger defenses against pathogens.

ommatidium [Gk. *omma*: eye] One of the units which, collected into groups of up to 20,000, make up the compound eye of arthropods.

omnivore [L. *omnis*: everything + *vorare*: to devour] An organism that eats both animal and plant material. (Contrast with carnivore, detritivore, herbivore.)

oncogene [Gk. *onkos*: mass, tumor + *genes*: born] Genes that greatly stimulate cell division, giving rise to tumors.

one-gene, one-polypeptide The principle that each gene codes for a single polypeptide.

oocyte (oh' eh site) [Gk. *oon*: egg + *kytos*: container] The cell that gives rise to eggs in animals.

oogenesis (oh' eh jen e sis) [Gk. *oon*: egg + *genesis*: source] Female gametogenesis, leading to production of the egg.

oogonium (oh' eh go' nee um) In some algae and fungi, a cell in which an egg is produced.

operator The region of an operon that acts as the binding site for the repressor.

operon A genetic unit of transcription, typically consisting of several structural genes that are transcribed together; the operon contains at least two control regions: the promoter and the operator.

opportunity cost The sum of the benefits an animal forfeits by not being able to perform some other behavior during the time when it is performing a given behavior.

opsin (op' sin) [Gk. *opsis*: sight] The protein portion of the visual pigment rhodopsin. (*See* rhodopsin.)

optic chiasm [Gk. *chiasma*: cross] Structure on the lower surface of the vertebrate brain where the two optic nerves come together.

optical isomers Two isomers that are mirror images of each other.

orbital A region in space surrounding the atomic nucleus in which an electron is most likely to be found.

organ [Gk. *organon*: tool] A body part, such as the heart, liver, brain, root, or leaf. Organs are composed of different tissues integrated to perform a distinct function. Organs are in turn often integrated into systems, such as the digestive or reproductive system.

organ identity genes Plant genes that specify the various parts of the flower. *See* homeotic genes.

organ of Corti Structure in the inner ear that transforms mechanical forces produced from pressure waves ("sound waves") into action potentials that are sensed as sound.

organ system An interrelated and integrated group of tissues and organs that work together in a physiological function.

organelles (or gan els') Organized structures found in or on eukaryotic cells. Examples include ribosomes, nuclei, mitochrondria, chloroplasts, cilia, and contractile vacuoles.

organic Pertaining to any aspect of living matter, e.g., to its evolution, structure, or chemistry. The term is also applied to any chemical compound that contains carbon.

organism Any living entity.

organizer Region of an early embryo that directs the development of nearby regions. In amphibian early gastrulas, the dorsal lip of the blastopore is the organizer.

organogenesis The formation of organs and organ systems during development.

origin of replication DNA sequence at which helicase unwinds the DNA double helix and DNA polymerase binds to initiate DNA replication.

orthology (or thol' o jee) A type of homology applied to genes in which the divergence of homologous genes can be traced to speciation events. The genes are said to be *orthologous*, and each is an *ortholog* of the others. (Compare with paralogy)

osmoconformer An aquatic animal that maintains an osmotic concentration of its extracellular fluid that is thet same as that of the external environment.

osmolarity The concentration of osmotically active particles in a solution.

osmoregulation Regulation of the chemical composition of the body fluids of an organism.

osmoreceptor Neuron that converts changes in the solute potential of interstial fluids into action potentials.

osmosis (oz mo' sis) [Gk. *osmos*: to push] The movement of water across a differentially permeable membrane, from one region to another region where the water potential is more negative.

ossicle (oss' ick ul) [L. *os*: bone] The calcified construction unit of echinoderm skeletons.

osteoblasts (oss' tee oh blast) [Gk. *osteon*: bone + *blastos*: sprout] Cells that lay down the protein matrix of bone.

osteoclasts (oss' tee oh clast) [Gk. *osteon*: bone + *klastos*: broken] Cells that dissolve bone.

otolith (oh' tuh lith) [Gk. *otikos*: ear + *lithos*: stone[Structures in the vertebrate vestibular apparatus that mechanically stimulate hair cells when the head moves or changes position.

oval window The flexible membrane that, when moved by the bones of the middle ear, produces pressure waves in the inner ear

ovary (oh' var ee) [L. *ovum*: egg] Any female organ, in plants or animals, that produces an egg.

oviduct [L. *ovum*: egg + *ducere*: to lead] In mammals, the tube serving to transport eggs to the uterus or to outside of the body.

oviparity Reproduction in which eggs are released by the female and development is external to the mother's body. (Contrast with viviparous.)

ovoviviparity Reproduction in which fertilized eggs develop and hatch within the body of the mother but are not attached to the mother by means of a placenta.

ovulation The release of an egg from an ovary.

ovule (oh' vule) In plants, a structure that contains a gametophyte and, within the gametophyte, an egg; when it matures, an ovule becomes a seed.

ovum (oh' vum) [L. *ovum*: egg] The egg; the female sex cell.

oxidation (ox i day' shun) Relative loss of electrons in a chemical reaction; either outright removal to form an ion, or the sharing of electrons with substances having a greater affinity for them, such as oxygen. Most oxidation, including biological ones, are associated with the liberation of energy. (Contrast with reduction.)

oxidative phosphorylation ATP formation in the mitochondrion, associated with flow of electrons through the respiratory chain.

oxidizing agent A substance that can accept electrons from another. The oxidizing agent becomes reduced; its partner becomes oxidized.

oxygenase An enzyme that catalyzes the addition of oxygen to a substrate from O_2.

- P -

P generation Parental generation. The individuals that mate in a genetic cross. Their immediate offspring are the F_1 generation.

pacemaker That part of the heart which undergoes most rapid spontaneous contraction, thus setting the pace for the beat of the entire heart. In mammals, the sinoatrial (SA) node. Also, an artificial device, implanted in the heart, that initiates rhythmic contraction of the organ.

Pacinian corpuscle A modified nerve ending that senses touch and vibration.

pair rule genes Segmentation genes that divide the *Drosophila* larva into two segments each.

pancreas (pan' cree us) A gland located near the stomach of vertebrates that secretes digestive enzymes into the small intestine and releases insulin into the bloodstream.

Pangaea (pan jee' uh) [Gk. *pan*: all, every] The single land mass formed when all the continents came together in the Permian period.

para- [Gk. *para*: akin to, beside] Prefix indicating association in being along side or accessory to.

parabronchi Passages in the lungs of birds through which air flows.

paracrine A substance, such as a hormone, that acts locally, near the site of its secretion. (Compare with autocrine, endocrine gland.)

parallel evolution Repeated evolutionary patterns of change that occur independently in multiple lineages.

paralogy (par al' o jee) A type of homology applied to genes in which the divergence of homologous genes can be traced to gene duplication events. The genes are said to be paralogous, and each is an paralog of the others. (Compare with orthology.)

parapatric speciation [Gk. *para*: along side + *patria*: homeland] Reproductive isolation between subpopulations arising from some non-geographic but physical condition, such as soil nutrient content. (Contrast with allopatric speciation, sympatric speciation.)

paraphyletic (par' a fih leht' ik) [Gk. *para*: beside + *phylon*: tribe] Referring to a group that consists of an ancestor and some (but not all) of its descendants. (Compare with monophyletic, polyphyletic.)

parasite An organism that attacks and consumes parts of an organism much larger than itself. Parasites sometimes, but not always, kill their host.

parasympathetic nervous system A portion of the autonomic (involuntary) nervous system. (Contrast with sympathetic nervous system.)

parathyroids Four glands on the posterior surface of the thyroid that produce and release parathormone.

parathyroid hormone Hormone secreted by the parathyroid glands. Stimulates osteoclast activity and raises blood calcium levels.

parenchyma (pair eng' kyma) A plant tissue composed of relatively unspecialized cells without secondary walls.

parsimony The principle of preferring the simplest among a set of plausible explanations of any phenomenon.

parthenocarpy Formation of fruit from a flower without fertilization.

parthenogenesis (par' then oh jen' e sis) [Gk. parthenos: virgin + genesis: source] The production of an organism from an unfertilized egg.

partial pressure The portion of the barometric pressure of a mixture of gases that is due to one component of that mixture. For example, the partial pressure of oxygen at sea level is 20.9% of barometric pressure.

particulate theory In genetics, the theory that genes are physical entities that retain their identities after fertilization.

passive transport Diffusion across a membrane; may or may not require a channel or carrier protein. (Contrast with active transport.)

patch clamping A technique for isolating a tiny patch of membrane to allow the study of ion movement through a particular channel.

pathogen (path' o jen) [Gk. *pathos*: suffering + *genesis:* source] An organism that causes disease.

pattern formation In animal embryonic development, the organization of differentiated tissues into specific structures such as wings.

pedigree The pattern of transmission of a genetic trait within a family.

penetrance Of a genotype, the proportion of individuals with that genotype who show the expected phenotype.

pentose [Gk. *penta:* five] A sugar containing five carbon atoms.

PEP carboxylase The enzyme that combines carbon dioxide with PEP to form a 4-carbon dicarboxylic acid at the start of C_4 photosynthesis or of crassulacean acid metabolism (CAM).

pepsin [Gk. *pepsis*: digestion] An enzyme in gastric juice that digests protein.

pepsinogen Inactive secretory product that is converted into pepsin by low pH or by enzymatic action.

peptide linkage The bond between amino acids in a protein. Formed between a carboxyl group and amino group (CO—NH^-) with the loss of water molecules.

peptidoglycan The cell wall material of many bacteria, consisting of a single enormous molecule that surrounds the entire cell.

perennial (per ren' ee al) [L. *per*: throughout + *annus*: year] Refers to a plant that survives from year to year. (Contrast with annual, biennial.)

perfect flower A flower with both stamens and carpels, therefore hermaphroditic.

pericycle [Gk. *peri*: around + *kyklos*: ring or circle] In plant roots, tissue just within the endodermis, but outside of the root vascular tissue. Meristematic activity of pericycle cells produces lateral root primordia.

periderm The outer tissue of the secondary plant body, consisting primarily of cork.

period (1) A category in the geological time scale. (2) The duration of a single cycle in a cyclical event, such as a circadian rhythm.

peripheral membrane protein Membrane protein not embedded in the bilayer. (Contrast with integral membrane protein.)

peripheral nervous system Neurons that transmit information to and from the central nervous system and whose cell bodies reside outside the brain or spinal cord.

peristalsis (pair' i stall' sis) [Gk. *peri*: around + *stellein*: place] Wavelike muscular contractions proceeding along a tubular organ, propelling the contents along the tube.

peritoneum The mesodermal lining of the body cavity among coelomate animals.

permease A membrane protein that specifically transports a compound or family of compounds across the membrane.

peroxisome An organelle that houses reactions in which toxic peroxides are formed. The peroxisome isolates these peroxides from the rest of the cell.

petal [Gk. *petalon*: spread out] In an angiosperm flower, a sterile modified leaf, nonphotosynthetic, frequently brightly colored, and often serving to attract pollinating insects.

petiole (pet' ee ole) [L. *petiolus*: small foot] The stalk of a leaf.

pH The negative logarithm of the hydrogen ion concentration; a measure of the acidity of a solution. A solution with pH = 7 is said to be neutral; pH values higher than 7 characterize basic solutions, while acidic solutions have pH values less than 7.

phage (fayj) Short for bacteriophage. A virus that infects bacteria.

phagocyte [Gk. *phagein*: to eat + *kystos*: sac] A white blood cell that ingests microorganisms by endocytosis.

phagocytosis Endocytosis by a cell of another cell or large particle.

pharmacogenomics The relaionship between an individual's genetic makeup and response to drugs.

pharming The use of genetically modified animals to produce medically useful products in their milk.

pharynx [Gk. *pharynx*: throat] The part of the gut between the mouth and the esophagus.

phenotype (fee' no type) [Gk. *phanein*: to show] The observable properties of an individual resulting from both genetic and environmental factors. (Contrast with genotype.)

phenotypic plasticity Refers to the fact that the phenotype of a developing organism is determined by a complex series of processes that are affected by both its genotype and its environment.

pheromone (feer' o mone) [Gk. *pheros*: carry + *hormon*: excite, arouse] A chemical substance used in communication between organisms of the same species.

phloem (flo' um) [Gk. *phloos*: bark] In vascular plants, the tissue that transports sugars and other solutes from sources to sinks. It consists of sieve cells or sieve tubes, fibers, and other specialized cells.

phosphate group The functional group —OPO_3H_2. The transfer of energy from one compound to another is often accomplished by the transfer of a phosphate group.

phosphodiester linkage The connection in a nucleic acid strand, formed by linking two nucleotides.

phospholipids Lipids containing a phosphate group; important constituents of cellular membranes. (*See* lipids.)

phosphorylation The addition of a phosphate group.

photoautotroph An organism that obtains energy from light and carbon from carbon dioxide. (Contrast with chemolithotroph, chemoheterotroph, photoheterotroph.)

photoheterotroph An organism that obtains energy from light but must obtain its carbon from organic compounds. (Contrast with chemolithotroph, chemoheterotroph, photoautotroph.)

photon (foe' ton) [Gk. *photos*: light] A quantum of visible radiation; a "packet" of light energy.

photoperiod (foe' tow peer' ee ud) The duration of a period of light, such as the length of time in a 24-hour cycle in which daylight is present.

photoperiodicity A condition in which physiological and behavioral changes are induced by changes in day length.

photoreceptor (1) A pigment that triggers a physiological response when it absorbs a photon. (2) A sensory receptor cell that senses and responds to light energy.

photorespiration Light-driven uptake of oxygen and release of carbon dioxide, the carbon being derived from the early reactions of photosynthesis.

photosynthesis (foe tow sin' the sis) [literally, "synthesis from light"] Metabolic processes, carried out by green plants, by which visible light is trapped and the energy used to synthesize compounds such as ATP and glucose.

photosystem [Gk. *phos:* light + *systema:* assembly] A light-harvesting complex in the chloroplast thylakoid composed of pigments and proteins.

photosystem I In photosynthesis, the reactions that absorb light at 700 nm, passing electrons to ferrodoxin and thence to NADPH. Rich in chlorophyll *a*.

photosystem II In photosynthesis, the reactions that absorb light at 660 nm, passing electrons to the electron transport chain in the chloroplast. Rich I chlorphyll *b*.

phototropins A class of blue light receptors that mediate phototropism and other plant responses.

phototropism [Gk. *photos*: light + *trope*: turning] A directed plant growth response to light.

phycobilin Photosynthetic pigment that absorbs red, yellow, orange, and green light and is found in cyanobacteria and some red algae.

phylogenetic tree A graphic representation of lines of descent among organisms or their genes.

phylogeny (fy loj' e nee) [Gk. *phylon*: tribe, race + *genesis*: source] The evolutionary history of a particular group of organisms or their genes.

physiology (fiz' ee ol' o jee) [Gk. *physis*: natural form + *logos*: discourse, study] The scientific study of the functions of living organisms and the individual organs, tissues, and cells of which they are composed.

phytoalexins Substances toxic to pathogens, produced by plants in response to fungal or bacterial infection.

phytochrome (fy' tow krome) [Gk. *phyton*: plant + *chroma*: color] A plant pigment regulating a large number of developmental and other phenomena in plants.

pigment A substance that absorbs visible light.

pineal gland A gland located between the cerebral hemispheres that secretes melatonin.

pinocytosis Endocytosis by a cell of liquid containing dissolved substances.

pistil [L. *pistillum*: pestle] The structure of an angiosperm flower within which the ovules are borne. May consist of a single carpel, or of several carpels fused into a single structure. Usually differentiated into ovary, style, and stigma.

pith In plants, relatively unspecialized tissue found within a cylinder of vascular tissue.

pituitary A small gland attached to the base of the brain in vertebrates. Its hormones control the activities of other glands. Also known as the hypophysis.

pits Recessed cavities in the cell walls of a plant vascular element where only the primary wall is present. facilitating the movement of sap between cells.

placenta (pla sen' ta) [Gk. *plax*: flat surface] The organ found in most mammals that provides for the nourishment of the fetus and elimination of the fetal waste products.

placental (pla sen' tal) Pertaining to mammals of the subclass Eutheria, a group characterized by the presence of a placenta; contains the majority of living species of mammals.

plankton Free-floating small organisms inhabiting the surface waters of lakes and oceans. Photosynthetic members of the plankton are referred to as phytoplankton.

plant *See* embryophyte.

planula (plan' yew la) [L. *planum*: flat] The free-swimming, ciliated larva of the cnidarians.

plaque (plack) [Fr.: a metal plate or coin] (1) A circular clearing in a layer (lawn) of bacteria growing on the surface of a nutrient agar gel. (2) An accumulation of prokaryotic organisms on tooth enamel. Acids produced by these microorganisms can cause tooth decay. (3) A region of arterial wall invaded by fibroblasts and fatty deposits (*see* atherosclerosis).

plasma *See* blood plasma.

plasma cell An antibody-secreting cell that developed from a B cell. The effector cell of the humoral immune system.

plasma membrane The membrane that surrounds the cell, regulating the entry and exit of molecules and ions. Every cell has a plasma membrane.

plasmid A DNA molecule distinct from the chromosome(s); that is, an extrachromosomal element. May replicate independently of the chromosome.

plasmodesma (plural: plasmodesmata) [Gk. *plassein*: to mold + *desmos*: band] A cytoplasmic strand connecting two adjacent plant cells.

plasmogamy The fusion of the cytoplasm of two cells. (Contrast with karyogamy.)

plasmolysis (plaz mol' i sis) Shrinking of the cytoplasm and plasma membrane away from the cell wall, resulting from the osmotic outflow of water. Occurs only in cells with rigid cell walls.

plastid Organelle in plants that serves for food manufacture (by photosynthesis) or food storage; bounded by a double membrane.

plastoquinone A mobile electron carrier within the thylakoid membrane of the chloroplast linking photosystems I and II of photosynthesis.

platelet A membrane-bounded body without a nucleus, arising as a fragment of a cell in the bone marrow of mammals. Important to blood-clotting action.

pleiotropy (plee' a tro pee) [Gk. *pleion*: more] The determination of more than one character by a single gene.

pleural membrane [Gk. *pleuras*: rib, side] The membrane lining the outside of the lungs and the walls of the thoracic cavity. Inflammation of these membranes is a condition known as pleurisy.

pluripotent Of a stem cell, having the ability to differentiate into any of a limted number of cell types. (Compare with totipotent.)

podocytes Cells of Bowman's capsule of the nephron that cover the capillaries of the glomerulus, forming filtration slits.

poikilotherm (poy' kill o therm) [Gk. *poikilos*: varied + *thermos*: heat] An animal whose body temperature tends to vary with the surrounding environment. (Contrast with homeotherm, heterotherm.)

point mutation A mutation that results from a small, localized alteration in the chemical structure of a gene; can revert to wild type. (Contrast with deletion.)

polar body A nonfunctional nucleus produced by meiosis, accompanied by very little cytoplasm. The meiosis which produces the mammalian egg produces in addition three polar bodies.

polar molecule A molecule in which the electric charge is not distributed evenly in the covalent bonds.

polar nuclei In flowering plants, the two nuclei in the central cell of the megagametophyte; following fertilization they give rise to the endosperm.

polarity In development, the difference between one end and the other. In chemistry, the property that makes a polar molecule.

pollen [L. *pollin*: fine, powdery flour] In seed plants, the microscopic grains containing the male gametophyte (microgametophyte) and gamete (microspore).

pollination The process of transferring pollen from an anther to the stigma of a pistil in an angiosperm or from a strobilus to an ovule in a gymnosperm.

poly- [Gk. *poly*: many] A prefix denoting multiple entities.

poly(A) tail A long sequence of adenine nucleotides (50–250) added after transcription to the 3′ end of most eukaryotic mRNAs.

polygenes Multiple loci whose alleles increase or decrease a continuously variable phenotypic trait.

polymer [Gk. *poly*: many + *meros*: unit] A large molecule made up of similar or identical subunits called monomers. (Contrast with monomer, oligomer.)

polymerase chain reaction (PCR) An enzymatic technique for the rapid production of millions of copies of a particular stretch of DNA.

polymerization reactions Chemical reactions that generate polymers by linking monomers.

polymorphic Referring to a gene whose most frequent allele in a population is present less than 99% of the time.

polymorphism (pol' lee mor' fiz um) [Gk. *poly*: many + *morphe*: form, shape] In genetics, the coexistence in the same population of two distinct hereditary types based on different alleles.

polyp The sessile asexual stage in the life cycle of most cnidarians.

polypeptide A large molecule made up of many amino acids joined by peptide linkages. Large polypeptides are called proteins.

polyphyletic (pol' lee fih leht' ik) [Gk. *poly*: many + *phylon*: tribe] Referring to a group that consists of multiple distantly related organisms, and does not include the common ancestor of the group. (Compare with monophyletic, paraphyletic.)

polyploidy (pol' lee ploid ee) The possession of more than two entire sets of chromosomes.

polysaccharide A macromolecule composed of many monosaccharides (simple sugars). Common examples are cellulose and starch.

polysome (polyribosome) A complex consisting of a threadlike molecule of messenger RNA and several (or many) ribosomes. The ribosomes move along the mRNA, synthesizing polypeptide chains as they proceed.

polytene (pol' lee teen) [Gk. *poly*: many + *taenia*: ribbon] An adjective describing giant interphase chromosomes, such as those found in the salivary glands of fly larvae. The characteristic pattern of bands and bulges seen on these chromosomes provided a method for preparing detailed chromosome maps of several organisms.

pons [L. *pons*: bridge] Region of the brain stem anterior to the medulla.

population Any group of organisms coexisting at the same time and in the same place and capable of interbreeding with one another.

population bottleneck *See* bottleneck.

population density The number of individuals (or modules) of a population in a unit of area or volume.

population genetics The study of genetic variation and its causes within populations.

population structure The proportions of individuals in a population belonging to different age classes (age structure). Also, the distribution of the population in space.

portal blood vessels Blood vessels that begin and end in capillary beds.

positional cloning A technique for isolating a gene associated with a disease on the basis of its approximate chromosomal location.

positional information Signals by which genes regulate cell functions to locate cells in a tissue during development.

positive control The situation in which a regulatory macromolecule is needed to turn transcription of structural genes on. In its absence, transcription will not occur.

positive cooperativity Occurs when a molecule can bind several ligands and each one that binds alters the conformation of the molecule so that it can bind the next ligand more easily. The binding of four molecules of O_2 by hemoglobin is an example of positive cooperativity.

post [L. *postere*: behind, following after] Prefix denoting something that comes after.

postabsorptive period When there is no food in the gut and no nutrients are being absorbed.

posterior pituitary The portion of the pituitary gland that is derived from neural tissue.

postsynaptic cell The cell whose membranes receive neurotransmitter after its release by another cell (the presynaptic cell) at a synapse.

postzygotic reproductive barriers Barriers to the reproductive process that occur after the union of the nuclei of two gametes. (Contrast with prezygotic reproductive barriers.)

potential energy "Stored" energy not doing work, such as the energy in chemical bonds.

precapillary sphincter A cuff of smooth muscle that can shut off the blood flow to a capillary bed.

pre-mRNA (precursor mRNA) Initial gene transcript before it is modified to produce functional mRNA. Also known as the primary transcript.

predator An organism that kills and eats other organisms.

pressure flow model An effective model for phloem transport in angiosperms. It holds that sieve element transport is driven by an osmotically driven pressure gradient between source and sink.

pressure potential The hydrostatic pressure of an enclosed solution in excess of the surrounding atmospheric pressure. (Contrast with solute potential, water potential.)

presynaptic excitation/inhibition Occurs when a neuron modifies activity at a synapse by releasing a neurotransmitter onto the presynaptic nerve terminal.

prey [L. *praeda*: booty] An organism consumed as an energy source.

prezygotic reproductive barriers Barriers to the reproductive process that occur before the union of the nuclei of two gametes (Contrast with postzygotic reproductive barriers.)

primary active transport Form of active transport in which ATP is hydrolyzed, yielding the energy required to transport ions against their concentration gradients. (Contrast with secondary active transport.)

primary consumer An herbivore; an organism that eats plant tissues.

primary embryonic organizer *See* organizer.

primary growth In plants, growth produced by the apical meristems. (Contrast with secondary growth.)

primary immune response The first response of the immune system to an antigen, involving recognition by lymphocytes and the production of effector cells and memory cells. (Contrast with secondary immune response.)

primary motor cortex The region of the cerebral cortex that contains motor neurons that directly stimulate specific muscle fibers to contract.

primary producer A photosynthetic or chemosynthetic organism that synthesizes complex organic molecules from simple inorganic ones.

primary sex determination Genetic determination of gametic sex, male or female. (Contrast with secondary sex determination.)

primary somatosensory cortex The region of the cerebral cortex that receives input from mechanosensors distributed throughout the body.

primary succession Succession that begins in an area initially devoid of life, such as on recently exposed glacial till or lava flows. (Contrast with secondary succession.)

primary structure The specific sequence of amino acids in a protein.

primary wall Cellulose-rich cell wall layers laid down by a growing plant cell.

primase An enzyme that catalyzes the synthesis of a primer for DNA replication.

primer A short, single-stranded segment of DNA that is the necessary starting material for the synthesis of a new DNA strand, which is synthesized from the 3′ end of the primer.

primitive streak A line running axially along the blastodisc, the site of inward cell migration during formation of the three-layered embryo. Formed in the embryos of birds and fish.

primordium [L. *primordium*: origin] The most rudimentary stage of an organ or other part.

prion An infectious protein that can proliferate by converting other proteins.

pro- [L.: first, before, favoring] A prefix often used in biology to denote a developmental stage that comes first or an evolutionary form that appeared earlier than another. For example, prokaryote, prophase.

probe A segment of single stranded nucleic acid used to identify DNA molecules containing the complementary sequence.

procambium Primary meristem that produces the vascular tissue.

processive Referring to an enzyme that catalyzes many reactions each time it binds to a substrate, as DNA polymerase does during DNA replication.

progesterone [L. *pro*: favoring + *gestare*: to bear] A vertebrate female sex hormone that maintains pregnancy.

prokaryotes (pro kar′ ry otes) [L. *pro*: before + Gk. *karyon*: kernel, nucleus] Organisms whose genetic material is not contained within a nucleus: the bacteria and archaea. Considered an earlier stage in the evolution of life than the eukaryotes.

prometaphase The phase of nuclear division that begins with the disintegration of the nuclear envelope.

promoter The region of an operon that acts as the initial binding site for RNA polymerase.

proofreading The correction of an error in DNA replication just after an incorrectly paired base is added to the growing polynucleotide chain.

prophage (pro′ fayj) The noninfectious units that are linked with the chromosomes of the host bacteria and multiply with them but do not cause dissolution of the cell. Prophage can later enter into the lytic phase to complete the virus life cycle.

prophase (pro′ phase) The first stage of nuclear division, during which chromosomes condense from diffuse, threadlike material to discrete, compact bodies.

prostaglandin Any one of a group of specialized lipids with hormone-like functions. It is not clear that they act at any considerable distance from the site of their production.

prostate gland Glandular tissue that surrounds the male urethra at its junction with the vas deferens; contributes an alkaline fluid to the semen.

prosthetic group Any nonprotein portion of an enzyme.

proteasome In the eukaryotic cytoplasm, a huge protein structure that binds to and digests cellular proteins that have been tagged by ubiquitin.

protein (pro′ teen) [Gk. *protos*: first] One of the most fundamental building substances of living organisms. A long-chain polymer of amino acids with twenty different common side chains. Occurs with its polymer chain extended in fibrous proteins, or coiled into a compact macromolecule in enzymes and other globular proteins.

protein domain *See* domain (1)

protein kinase An enzyme that catalyzes the addition of a phosphate group from ATP to a target protein.

protein kinase cascade aA series of reactions in response to a molecular signal, in which a series of protein kinases activates one another in sequence, amplifying the signal at each step.

proteoglycan A glycoprotein containing a protein core with attached long, linear carbohydrate chains.

proteolysis [protein + Gk. *lysis*: break apart] An enzymatic digestion of a protein or polypeptide.

proteome The total of the different proteins that can be made by an organism. Because of alternate splicing of pre-mRNA, the number of proteins that can be made is usually much larger than the number of protein-coding genes present in the organism's genome.

protobiont [Gk. *protos*: first, before + *bios*: life] Aggregates of abiotically produced molecules that cannot reproduce but do maintain internal chemical environments that differ from their surroundings.

protoderm Primary meristem that gives rise to the plant epidermis.

proton (pro' ton) [Gk. *protos*: first, before] (1) A subatomic particle with a single positive charge. The number of protons in the nucleus of an atom determine its element. (2) A hydrogen ion, H^+.

proton pump An active transport system that uses ATP energy to move hydrogen ions across a membrane generating an electric potential (voltage).

proton motive force A force generated across a membrane expressed in millivolts having two components: a chemical potential (difference in proton concentration) plus an electrical potential due to the electrostatic charge on the proton.

proto-oncogenes The normal alleles of genes possessing oncogenes (cancer-causing genes) as mutant alleles. Proto-oncogenes encode growth factors and receptor proteins.

prototroph (pro' tow trofe') [Gk. *protos*: first + *trophein*: to nourish] The nutritional wild type, or reference form, of an organism. Any deviant form that requires growth nutrients not required by the prototrophic form is said to be a nutritional mutant, or auxotroph.

proximal Near the point of attachment or other reference point. (Contrast with distal.)

pseudocoelom [Gk. *pseudes*: false] A body cavity not surrounded by a peritoneum. Characteristic of nematodes and rotifers.

pseudogene [Gk. *pseudes*: false] A DNA segment that is homologous to a functional gene but is not expressed because of changes to its sequence or changes to its location in the genome.

pseudopod (soo' do pod) [Gk. *pseudes*: false + *podos*: foot] A temporary, soft extension of the cell body that is used in location, attachment to surfaces, or engulfing particles.

pulmonary [L. *pulmo*: lung] Pertaining to the lungs.

punctuated equilibrium An evolutionary pattern in which periods of rapid change are separated by longer periods of little or no change.

Punnett square A method of predicting the results of a genetic cross by arranging the gametes of each parent at the edges of a square.

pupa (pew' pa) [L. *pupa*: doll, puppet] In certain insects (the Holometabola), the encased developmental stage between the larva and the adult.

pupil The opening in the vertebrate eye through which light passes.

purine (pure' een) One of the types of nitrogenous bases. The purines adenine and guanine are found in nucleic acids. (Contrast with pyrimidine.)

Purkinje fibers Specialized heart muscle cells that conduct excitation throughout the ventricular muscle.

pyrimidine (per im' a deen) A type of nitrogenous base. The pyrimidines cytosine, thymine, and uracil are found in nucleic acids.

pyruvate A three-carbon acid; the end product of glycolysis and the raw material for the citric acid cycle.

pyruvate oxidation Conversion of pyruvate to acetyl CoA and CO_2 that occurs in the mitochondrial matrix in the presence of O_2.

-Q-

Q_{10} A value that compares the rate of a biochemical process or reaction over a 10°C range of temperature. A process that is not temperature-sensitive has a Q_{10} of 1; values of 2 or 3 mean the reaction speeds up as temperature increases.

quantum (kwon' tum) [L. *quantus*: how great] An indivisible unit of energy.

quaternary structure The specific three dimensional arrangement of protein subunits.

quiescent center In root meristem, central region where cells do not divide or divide very slowly.

-R-

R factor (resistance factor) A plasmid that contains one or more genes that encode resistance to antibiotics.

***R* genes** Resistance genes that function in plant defenses against bacteria, fungi, and nematodes. *See* gene-for-gene resistance.

R group The distinguishing group of atoms of a particular amino acid.

radial symmetry The condition in which two halves of a body are mirror images of each other regardless of the angle of the cut, providing the cut is made along the center line. Thus, a cylinder cut lengthwise down its center displays this form of symmetry. (Contrast with biradial symmetry.)

radioisotope A radioactive isotope of an element. Examples are carbon-14 (^{14}C) and hydrogen-3, or tritium (^{3}H).

radiometry The use of the regular, known rates of decay of radioisotopes of elements to determine dates of events in the distant past.

rain shadow The relatively dry area on the leeward side of a mountain range.

rapid-eye-movement *See* REM sleep

reactant A chemical substance that enters into a chemical reaction with another substance.

reaction A chemical change in which changes take place in the kind, number, or position of atoms making up a substance.

reaction center A group of electron transfer proteins that receive energy from light-absorbing pigments and convert it to chemical energy by redox reactions.

receptacle The end of a plant stem to which the parts of the flower are attached.

receptive field The area of visual space that activates a particular cell in the visual system.

receptor A site or protein on the outer surface of the plasma membrane or in the cytoplasm to which a specific ligand from another cell binds.

receptor-mediated endocytosis Endocytosis initiated by macromolecular binding to a specific membrane receptor.

receptor potential The change in the resting potential of a sensory cell when it is stimulated.

recessive In genetics, an allele that does not determine phenotype in the presence of a dominant allele. (Contrast with dominance.)

reciprocal crosses A pair of matings in one of which a female of genotype A mates with a male of genotype B and in the other of which a female of genotype B mates with a male of genotype A.

recognition site *See* restriction site.

recombinant An individual, meiotic product, or single chromosome in which genetic materials originally present in two individuals end up in the same haploid complement of genes. The reshuffling of genes can be either by independent segregation, or by crossing over between homologous chromosomes.

recombinant DNA DNA generated in vitro, from more than one source.

recombinant DNA technology The application of restriction endonucleases, plasmids, and transformation to alter and assemble recombinant DNA, with the goal of producing specific proteins.

recombinant frequency The proportion of offspring of a genetic cross that have phenotypes different from the parental phenotypes due to crossing over between linked genes during gamete formation.

reconciliation ecology The practice of making exploited lands more biodiversity-friendly.

rectum The terminal portion of the gut, ending at the anus.

redox reaction A chemical reaction in which one reactant becomes oxidized and the other becomes reduced.

reducing agent A substance that can donate electrons to another substance. The reducing agent becomes oxidized, and its partner becomes reduced.

reduction Gain of electrons by a chemical reactant; any reduction is accompanied by an oxidation. (Contrast with oxidation.)

reflex An automatic action, involving only a few neurons (in vertebrates, often in the spinal cord), in which a motor response swiftly follows a sensory stimulus.

refractory period Of a neuron, the time interval after an action potential, during which another action potential cannot be elicited.

regulative development A pattern of animal embryonic development in which the fates of the first blastomeres are not absolutely fixed. (Contrast with mosaic development.)

regulator sequence A DNA sequence to which the protein product of a regulatory gene binds.

regulatory gene A gene that codes for a protein that controls the transcription of another gene(s).

regulatory subunit The polypeptide in an enzyme protein with quaternary structure that does not contain the active site, but instead binds non-substrate molecules and changes its structure, in turn changing the structure and function of the active site. (Contrast catalytic subunit.)

regulatory system A system that uses feedback information to maintain a physiological function or parameter at an optimal level.

reinforcement The evolution of enhanced reproductive isolation between populations due to natural selection for greater isolation.

releaser A sensory stimulus that triggers the performance of a stereotyped behavior pattern.

releasing hormone One of several hypothalamic hormones that stimulates the secretion of anterior pituitary hormone.

REM sleep A sleep state characterized by vivid dreams, skeletal muscle relaxation, and rapid eye movements.

renal [L. *renes*: kidneys] Relating to the kidneys.

replication Pertaining to the duplication of genetic material.

replication complex The close association of several proteins operating in the replication of DNA.

replication fork A point at which a DNA molecule is replicating. The fork forms by the unwinding of the parent molecule.

replicon A region of DNA controlled by a single origin of replication.

reporter gene A marker gene included in recombinant DNA to indicate the presence of the recombinant DNA in a host cell.

repressible enzyme An enzyme whose synthesis can be decreased or prevented by the presence of a particular compound. A repressible operon often controls the synthesis of such an enzyme.

repressor A protein coded by the regulatory gene. The repressor can bind to a specific operator and prevent transcription of the operon.

reproductive isolating mechanism Any trait that prevents individuals from two different populations from producing fertile hybrids.

reproductive isolation The condition in which a population is not exchanging genes with other populations of the same species.

rescue effect The process by which a few individuals moving among declining subpopulations of a species and reproducing may prevent their extinction.

resolution Of an optical device such as a microscope, the smallest distance between two lines that allows the lines to be seen as separate from one another.

resource Something in the environment required by an organism for its maintenance and growth that is consumed in the process of being used.

respiration (res pi ra' shun) [L. *spirare*: to breathe] (1) Cellular respiration; the catabolic pathways by which electrons are removed from various molecules and passed through intermediate electron carriers to O_2, generating H_2O and releasing energy. (2) Breathing.

respiratory chain The terminal reactions of cellular respiration, in which electrons are passed from NAD or FAD, through a series of intermediate carriers, to molecular oxygen, with the concomitant production of ATP.

resting potential The membrane potential of a living cell at rest. In cells at rest, the interior is negative to the exterior. (Contrast with action potential, electrotonic potential.)

restoration ecology The science and practice of restoring damaged or degraded ecosystems.

restriction digestion Use of restriction enzymes to cleave DNA into fragments in a test tube.

restriction endonuclease *See* restriction enzyme.

restriction enzyme Any one of several enzymes, produced by bacteria, that break foreign DNA molecules at specific sites. Some produce "sticky ends." Extensively used in recombinant DNA technology.

restriction fragment length polymorphism *See* RFLP.

restriction point The specific time during G1 of the cell cycle at which the cell becomes committed to undergo the rest of the cell cycle.

restriction site A specific DNA base sequence recognized and acted on by a restriction endonuclease cutting the DNA.

reticular system A central region of the vertebrate brain stem that includes complex fiber tracts conveying neural signals between the forebrain and the spinal cord, with collateral fibers to a variety of nuclei that are involved in autonomic functions, including arousal from sleep.

retina (rett' in uh) [L. *rete*: net] The light-sensitive layer of cells in the vertebrate or cephalopod eye.

retinal The light-absorbing portion of visual pigment molecules. Derived from β-carotene.

retinoblastoma protein A protein that inhibits an animal cell from passing through the restriction point; inactivation of this protein is necessary for the cell cycle to proceed.

retrovirus An RNA virus that contains reverse transcriptase. Its RNA serves as a template for cDNA production, and the cDNA is integrated into a chromosome of the mammalian host cell.

reversal *See* evolutionary reversal.

reverse transcriptase An enzyme that catalyzes the production of DNA (cDNA), using RNA as a template; essential to the reproduction of retroviruses.

reversible reaction A chemical change that can occur in both the forward and reverse directions.

RFLP (restriction fragment length polymorphism) Coexistence of two or more patterns of restriction fragments (patterns produced by restriction enzymes), as revealed by a probe. The polymorphism reflects a difference in DNA sequence on homologous chromosomes.

rhizoids (rye' zoids) [Gk. *rhiza*: root] Hairlike extensions of cells in mosses, liverworts, and a few vascular plants that serve the same function as roots and root hairs in vascular plants. The term is also applied to branched, rootlike extensions of some fungi and algae.

rhizome (rye' zome) A special underground stem (as opposed to root) that runs horizontally beneath the ground.

rhodopsin A photopigment used in the visual process of transducing photons of light into changes in the membrane potential of photoreceptor cells.

ribonucleic acid *See* RNA.

ribose A five-carbon sugar in nucleotides and RNA.

ribosomal RNA (rRNA) Several species of RNA that are incorporated into the ribosome. Involved in peptide bond formation.

ribosome A small organelle that is the site of protein synthesis.

ribozyme An RNA molecule with catalytic activity.

risk cost The increased chance of being injured or killed as a result of performing a behavior, compared to resting.

RNA (ribonucleic acid) An often single stranded nucleic acid whose nucleotides use ribose rather than deoxyribose and in which the base uracil replaces thymine found in DNA. Serves as genome from some viruses. (*See* rRNA, tRNA, mRMA, and ribozyme.)

RNA editing The alteration of bases on mRNA prior to its translation.

RNA polymerase An enzyme that catalyzes the formation of RNA from a DNA template.

RNA primase A replication complex enzyme that makes the primer strand of DNA needed to initiate DNA replication.

RNA splicing The last stage of RNA processing in eukaryotes, in which the transcripts of introns are excised through the action of small nuclear ribonucleoprotein particles (snRNP).

rod cells Light-sensitive cell in the vertebrate retina; these sensory receptor cells are sensitive in extremely dim light and are responsible for dim light, black and white vision.

root The organ responsible for anchoring the plant in the soil, absorbing water and minerals, and producing certain hormones. Some roots are storage organs.

root cap A thimble-shaped mass of cells, produced by the root apical meristem, that protects the meristem; the organ that perceives the gravitational stimulus in root gravitropism.

root hair A long, thin process from a root epidermal cell that absorbs water and minerals from the soil solution.

rough ER That portion of the endoplasmic reticulum whose outer surface has attached ribosomes. (Contrast with smooth ER.)

rRNA *See* ribosomal RNA.

RT-PCR A technique in which RNA is first converted to cDNA by the use of the enzyme reverse transcriptase, then the cDNA is amplified by the polymerase chain reaction.

rubisco (ribulose bisphosphate carboxylase/ oxygenase) Acronym for the enzyme that combines carbon dioxide or oxygen with ribulose bisphosphate to catalyze the first step of the Calvin-Benson cycle.

rumen (rew' mun) The first division of the ruminant stomach. It stores and initiates bacterial fermentation of food. Food is regurgitated from the rumen for further chewing.

ruminant Herbivorous, cud-chewing mammals such as cows or sheep. The ruminant stomach consists of four compartments.

-S-

S phase In the cell cycle, the stage of interphase during which DNA is replicated. (Contrast with G_1 phase, G_2 phase, M phase.)

saprobe [Gk. *sapros*: rotten] An organism (usually a bacterium or fungus) that obtains its carbon and energy directly from dead organic matter.

sarcomere (sark′ o meer) [Gk. *sark*: flesh + *meros*: unit] The contractile unit of a skeletal muscle.

sarcoplasm The cytoplasm of a muscle cell.

sarcoplasmic reticulum The endoplasmic reticulum of a muscle cell.

saturated fatty acid A fatty acid usually containing from 12 to 18 carbon atoms and no double bonds.

scientific method A means of gaining knowledge about the natural world by making observations, posing hypotheses, and conducting experiments to test those hypotheses.

schizocoelous development [Gk. *schizo*: split + *koiloma*: cavity] Formation of a coelom during embryological development by a splitting of mesodermal masses.

Schwann cell A glial cell that wraps around part of the axon of a peripheral neuron, creating a myelin sheath.

sclereid [Gk. *skleros*: hard] A type of sclerenchyma cell, commonly found in nutshells, that is not elongated.

sclerenchyma (skler eng′ kyma) [Gk. *skleros*: hard + *kymus*: juice] A plant tissue composed of cells with heavily thickened cell walls, dead at functional maturity. The principal types of sclerenchyma cells are fibers and sclereids.

scrotum A sac of skin that encloses the testicles in male mammals.

second law of thermodynamics States that in any real (irreversible) process, there is a decrease in free energy and an increase in entropy.

second messenger A compound, such as cAMP, that is released within a target cell after a hormone (first messenger) has bound to a surface receptor on a cell; the second messenger triggers further reactions within the cell.

secondary active transport Form of active transport which does not use ATP as an energy source; rather, transport is coupled to ion diffusion down a concentration gradient established by primary active transport.**secondary consumer** An organism that eat primary consumers.

secondary growth In plants, growth produced by vascular and cork cambia, contributing to an increase in girth. (Contrast with primary growth.)

secondary immune response A rapid and intense response to a second or subsequent exposure to an antigen, initiated by memory cells. (Contrast with primary immune response.)

secondary metabolite A compound synthesized by a plant that is not needed for basic cellular metabolism. Typically has an antiherbivore or antiparasite function.

secondary sex determination Formation of nongametic features of sex, such as external organs and body hair. (Contrast with primary sex determination.)

secondary structure Of a protein, localized regularities of structure, such as the α helix and the β pleated sheet.

secondary succession Ecological succession after a disturbance that did not eliminate all the organisms originally living on the site. (Contrast with primary succession.)

secondary wall Wall layers laid down by a plant cell that has ceased growing; often impregnated with lignin or suberin.

secretin (si kreet′ in) A peptide hormone secreted by the upper region of the small intestine when acidic chyme is present. Stimulates the pancreatic duct to secrete bicarbonate ions.

section A thin slice, usually for microscopy, as a tangential section or a transverse section.

seed A fertilized, ripened ovule of a gymnosperm or angiosperm. Consists of the embryo, nutritive tissue, and a seed coat.

seed plant Plants in which the embryo is protected and nourished within a seed; the gymnosperms and angiosperms.

seedling A young plant that has grown from a seed (rather than by grafting or by other means.)

segmentation genes In insect larvae, genes that determine the number and polarity of larval segments.

segment polarity genes Genes that determine the boundaries and front-to-back organization of the segments in the *Drosophila* larva.

segregation In genetics, the separation of alleles, or of homologous chromosomes, from each other during meiosis so that each of the haploid daughter nuclei produced by meiosis contains one or the other member of the pair found in the diploid parent cell, but never both. This principle was articulated by Mendel as his first law.

selective permeability Allowing certain substances to pass through while other substances are excluded; a characteristic of membranes.

self incompatability In plants, the rejection of their own pollen; promotes genetic variation and limits inbreeding.

selfish act A behavioral act that benefits its performer but harms the recipients.

semen (see′ men) [L. *semin*: seed] The thick, whitish liquid produced by the male reproductive organ in mammals, containing the sperm.

semiconservative replication The way in which DNA is synthesized. Each of the two partner strands in a double helix acts as a template for a new partner strand. Hence, after replication, each double helix consists of one old and one new strand.

seminiferous tubules The tubules within the testes within which sperm production occurs.

senescence [L. *senescere*: to grow old] Aging; deteriorative changes with aging; the increased probability of dying with increasing age.

sensory receptor cells Cells that are responsive to a particular type of physical or chemical stimulation.

sensory transduction The transformation of environmental stimuli or information into neural signals.

sepal (see′ pul) [L. *sepalum*: covering] One of the outermost structures of the flower, usually protective in function and enclosing the rest of the flower in the bud stage.

septum [L. *saeptum*: wall, fence] (1) A partition or cross-wall appearing in the hyphae of some fungi. (2) The bony structure dividing the nasal passages.

Sertoli cells Cells in the seminiferous tubules that nuture the developing sperm.

sessile (sess′ ul) [L. *sedere*: to sit] Permanently attached; not moving.

set point In a regulatory system, the threshold sensitivity to the feedback stimulus.

sex chromosome In organisms with a chromosomal mechanism of sex determination, one of the chromosomes involved in sex determination.

sex linkage The pattern of inheritance characteristic of genes located on the sex chromosomes of organisms having a chromosomal mechanism for sex determination.

sex pilus (pill′ us) [L. *pilus*: hair] A structure on the cell wall that allows one bacterium to adhere to another prior to conjugation.

sexual reproduction Reproduction involving union of gametes.

sexual selection Selection by one sex of characteristics in individuals of the opposite sex. Also, the favoring of characteristics in one sex as a result of competition among individuals of that sex for mates.

shared derived trait A trait that arose in the ancestor of a phylogenetic group and is present (sometimes in modified form) in all of its members, thus helping define that group. Also called a synapomorphy.

shoot system The aerial parts of a vascular plant, consisting of the leaves, stem(s), and flowers.

short-day plant (SDP) A plant that requires short days (or long nights) in order to flower.

short tandem repeat (STR) An inherited, short (2–5 base pairs), moderately repetitive sequence of DNA.

shotgun sequencing A relatively rapid method of analyzing DNA sequences in which a large DNA molecule is broken up into overlapping fragments, each fragment is sequenced, and computers are used to analyze and realign the fragments.

sieve tube A column of specialized cells found in the phloem, specialized to conduct organic matter from sources (such as photosynthesizing leaves) to sinks (such as roots). Found principally in flowering plants.

sieve tube element A single cell of a sieve tube, containing cytoplasm but relatively few organelles, with highly specialized perforated end walls leading to elements above and below.

signal A chemical (neurotransmitter or hormone) or light message emitted from a cell or cells or organism(s) and received by others to cause some change in function or behavior.

signal recognition particle (SRP) A complex of RNA and protein that recognizes both the signal sequence on a growing polypeptide and receptor protein on the surface of the ER.

signal sequence The sequence of a protein that directs the protein through a particular cellular membrane.

signal transduction pathway The series of biochemical steps whereby a stimulus to a cell (such as a hormone or neurotransmitter binding to a receptor) is translated into a response of the cell.

silencer sequence A sequence of eukaryotic DNA that binds proteins that inhibit the transcription of an associated gene.

silent substitution A change in gene sequence that, due to the redundancy of the genetic code, has no effect on the amino acid produced, and thus no effect on the protein phenotype. Also called a synonymous substitution.

similarity matrix A matrix used to compare the degree of divergence among pairs of objects. For molecular sequences, constructed by summing the number or percentage of nucleotidies or amino acids that are identical in each pair of sequences.

single nucleotide polymorphisms (SNPs) Inherited variations in a single nucleotide base in DNA.

single-strand binding protein In DNA replication, a protein that binds to single strands of DNA after they have been separated from each other, keeping the two strands separate for replication.

sinoatrial node (sigh' no ay' tree al) [L. *sinus*: curve + *atrium*: hall, chamber] The pacemaker of the mammalian heart.

sink In plants, any organ that imports the products of photosynthesis, such as roots, developing fruits, immature leaves. (Contrast with source.)

sinus (sigh' nus) [L. *sinus*: curve, hollow] A cavity in a bone, a tissue space, or an enlargement in a blood vessel.

sister chromatid In the eukaryotic cell, a chromatid resulting from chromosome replication during interphase.

sister groups Two phylogenetic groups that are each other's closest relatives.

skeletal muscle *See* striated muscle.

sliding DNA clamp A protein complex that keeps DNA polymerase bound to DNA during replication.

sliding filament theory A proposed mechanism of muscle contraction based on formation and breaking of crossbridges between actin and myosin filaments, causing them to slide together.

slow-twitch fibers Skeletal muscle fibers that generate tension slowly, but are resistant to fatigue ("marathon" fibers). They have abundant mitochondria, enzymes of aerobic metabolism, myoglobin, and blood supply.

slow-wave sleep A state of mammalian and avian sleep characterized by high amplitude slow waves in the EEG.

small intestine The portion of the gut between the stomach and the colon; consists of the duodenum, the jejunum, and the ileum.

small nuclear ribonucleoprotein particle (snRNP) A complex of an enzyme and a small nuclear RNA molecule, functioning in RNA splicing.

smooth muscle One of three types of muscle tissue. Usually consists of sheets of mononucleated cells innervated by the autonomic nervous system. (Compare with cardiac muscle, striated muscle.)

sodium–potassium pump (Na–K pump) The complex protein in plasma membranes that is responsible for primary active transport; it pumps sodium ions out of the cell and potassium ions into the cell, both against their concentration gradients.

solute A substance that is dissolved in a liquid (solvent) to form a solution.

solute potential A property of any solution, resulting from its solute contents; it may be zero or have a negative value. The more negative the solute potential, the greater the tendency of the solution to take up water through a differentially permeable membrane. (Contrast with pressure potential, water potential.)

solution A liquid (the solvent) and its dissolved solutes.

somatic [Gk. *soma*: body] Pertaining to the body. Somatic cells are cells of the body (as opposed to germ cells).

somatic mutation Permament genetic change in a somatic cell. These mutations affect the individual only; they are not passed on to offspring. (Contrast with germ line mutation)

somite (so' might) One of the segments into which an embryo becomes divided longitudinally, leading to the eventual segmentation of the animal as illustrated by the spinal column, ribs, and associated muscles.

source In plants, an organ exporting photosynthetic products in excess of its own needs. For example, a mature leaf or storage organ. (Contrast with sink.)

spatial summation In the production or inhibition of action potentials in a postsynaptic neuron, the interaction of depolarizations and hyperpolarizations produced by several terminal boutons.

spawning The direct release of sex cells into the water.

speciation (spee' shee ay' shun) The process of splitting one population into two populations that are reproductively isolated from one another.

species (spee' shees) [L. *species*: kind] The basic lower unit of classification, consisting of an ancestor–descendant lineage of populations of closely related and similar organisms. The more narrowly defined "biological species" consists of individuals capable of interbreeding freely with each other but not with members of other species.

species–area relationship The relationship between the sizes of areas and the numbers of species they support.

species diversity A weighted representation of the species of organisms living in a region; large and common species are given greater weight than are small and rare ones. (Contrast with species richness.)

species richness The total number of species living in a region. (Contrast with species diversity.)

specific defenses Defensive reactions of the immune system that are based on antibody reaction with a specific antigen.

specific heat The amount of energy that must be absorbed by a gram of a substance to raise its temperature by one degree centigrade. By convention, water is assigned a specific heat of one.

sperm [Gk. *sperma*: seed] A male gamete (reproductive cell).

spermatogenesis (spur mat' oh jen' e sis) [Gk. *sperma*: seed + *genesis*: source] Male gametogenesis, leading to the production of sperm.

sphincter (sfink' ter) [Gk. *sphinkter*: something that binds tightly] A ring of muscle that can close an orifice, for example at the anus.

spindle apparatus An array of microtubules stretching from pole to pole of a dividing nucleus and playing a role in the movement of chromosomes at nuclear division. Named for its shape.

spiracle (spy' rih kel) [L. *spirare*: to breathe] An opening of the treacheal respiratory system of terrestrial arthorpods.

spleen An organ that serves as a reservoir for venous blood and eliminates old, damaged red blood cells from the circulation.

spliceosome An RNA–protein complex that splices out introns from eukaryotic pre-mRNAs.

splicing Removal of introns and connecting of exons in eukaryotic pre-mRNAs.

spontaneous mutation A genetic change caused by internal cellular mechanisms, such as an error in DNA replication. (Contrast with induced mutation.)

spontaneous reaction A chemical reaction that will proceed on its own, without any outside influence. A spontaneous reaction need not be rapid.

sporangium (spor an' gee um) [Gk. *spora*: seed + *angeion*: vessel or reservoir] In plants and fungi, any specialized stucture within which one or more spores are formed.

spore [Gk. *spora*: seed] Any asexual reproductive cell capable of developing into an adult organism without gametic fusion. In plants, haploid spores develop into gametophytes, diploid spores into sporophytes. In prokaryotes, a resistant cell capable of surviving unfavorable periods.

sporocyte Specialized cells of the diploid sporophyte that will divide by meiosis to produce four haploid spores. Germination of these spores produces the haploid gametophyte.

sporophyte (spor' o fyte) [Gk. *spora*: seed + *phyton*: plant] In plants and protists with alternation of generations, the diploid phase that produces the spores. (Contrast with gametophyte.)

stabilizing selection Selection against the extreme phenotypes in a population, so that the intermediate types are favored. (Contrast with disruptive selection.)

stamen (stay' men) [L. *stamen*: thread] A male (pollen-producing) unit of a flower, usually composed of an anther, which bears the pollen, and a filament, which is a stalk supporting the anther.

starch [O.E. *stearc*: stiff] A polymer of glucose; used by plants to store energy.

start codon The mRNA triplet (AUG) that acts as a signal for the beginning of translation at the ribosome. (Contrast with stop codon.)

stasis [Gk. *stasis*: to stop, stand still] Period during which little or no evolutionary change takes place within a lineage or groups of lineages.

statocyst (stat' oh sist) [Gk. *statos*: stationary + *kystos*: cell] An organ of equilibrium in some invertebrates.

statolith (stat' oh lith) [Gk. *statos*: stationary + *lithos*: stone] A solid object that responds to gravity or movement and stimulates the mechanoreceptors of a statocyst.

stele (steel) [Gk. *stylos*: pillar] The central cylinder of vascular tissue in a plant stem.

stem Plant structure that holds leaves and/or flowers; it is the site for transporting and distributing material throughout the plant.

stem cells In animals, undifferentiated cells that are capable of extensive proliferation. A stem cell generates more stem cells and a large clone of differentiated progeny cells.

steroid Any of numerous lipids based on a 17-carbon atom ring system.

sticky ends On a piece of two-stranded DNA, short, complementary, one-stranded regions produced by the action of a restriction endonuclease. Sticky ends allow the joining of segments of DNA from different sources.

stigma [L. *stigma*: mark, brand] The part of the pistil at the apex of the style that is receptive to pollen, and on which pollen germinates.

stimulus [L. *stimulare*: to goad] Something causing a response; something in the environment detected by a receptor.

stolon [L. *stolon*: branch, sucker] A horizontal stem that forms roots at intervals.

stoma (plural: stomata) [Gk. *stoma*: mouth, opening] Small opening in the plant epidermis that permits gas exchange; bounded by a pair of guard cells whose osmotic status regulates the size of the opening.

stop codon Any of the thre mRNA codons that signal the end of protein translation at the ribosome: UAG, UGA, UAA.

stratosphere The upper part of Earth's atmosphere, above the troposphere; extends from approximately 18 kilometers upward to approximately 50 kilometers above the surface.

stratum (plural strata) [L. *stratos*: layer] A layer of sedimentary rock laid down at a particular time in the past.

striated muscle Contractile tissue characterized by multinucleated cells containing highly ordered arrangements of actin and myosin microfilaments. Also known as skeletal muscle. (Compare with cardiac muscle, smooth muscle.)

strobilus A conelike structure consisting of spore-bearing scales (modified leaves) inserted on an axis. (Contrast with cone.)

stroma The fluid contents of an organelle, such as a chloroplast.

stromatolites Composite, flat-to-domed structures composed of successive mineral layers produced by the action of cyanobacteria in water; ancient ones provide evidence for early life on the earth.

structural gene A gene that encodes the primary structure of a protein.

structural isomers Molecules made up of the same kinds and numbers of atoms, in which the atoms are bonded differently.

style [Gk. *stylos*: pillar or column] In flowering plants, a column of tissue extending from the tip of the ovary, and bearing the stigma or receptive surface for pollen at its apex.

sub- [L. *sub*: under] A prefix often used to designate a structure that lies beneath another or is less than another. For example, subcutaneous (beneath the skin); subspecies.

suberin A waxlike lipid that acts as a barrier to water and solute movement across the Casparian strip of the endodermis. Suberin is the waterproofing element in the cell walls of cork.

submucosa (sub mew koe' sah) The tissue layer just under the epithelial lining of the lumen of the digestive tract.

substrate (sub' strayte) The molecule or molecules on which an enzyme exerts catalytic action.

substrate-level phosphorylation Reaction in which ATP is formed from ADP by the addition of a phosphate group directly from a reactant.

substratum The base material on which a sessile organism lives.

succession In ecology, the gradual, sequential series of changes in species composition of a community following a disturbance.

sulcus [L. *sulcare*: to plow] The valleys or creases between the raised portions of the convoluted surface of the brain. (Contrast with gyrus.)

summation The ability of a neuron to fire action potentials in response to numerous subthreshold postsynaptic potentials arriving simultaneously at differentiated places on the cell, or arriving at the same site in rapid succession.

surface area-to-volume ratio For any cell, organism, or geometrical solid, the ratio of surface area to volume; this is an important factor in setting an upper limit on the size a cell or organism can attain.

surface tension The attractive intermolecular forces at the surface of liquid; especially important in water.

surfactant A substance that decreases the surface tension of a liquid. Lung surfactant, secreted by cells of the alveoli, is mostly phospholipid and decreases the amount of work necessary to inflate the lungs.

survivorship The proportion of individuals in a cohort that is alive at some time in the future.

suspensor In the embryos of seed plants, the stalk of cells that pushes the embryo into the endosperm and is a source of nutrient transport to the embryo.

symbiosis (sim' bee oh' sis) [Gk. *sym*: together + *bios*: living] The living together of two or more species in a prolonged and intimate ecological relationship. (Compare with parasitism, mutualism.)

symmetry Describes an attribute of an animal body in which at least one plane can divide the body into similar, mirror-image halves. (*See* bilateral symmetry, biradial symmetry, radial symmetry.)

sympathetic nervous system A division of the autonomic (involuntary) nervous system. (Contrast with parasympathetic nervous system.)

sympatric speciation (sim pat' rik) [Gk. *sym*: same + *patria*: homeland] Speciation due to reproductive isolation without any physical separation of the subpopulation. (Contrast with allopatric speciation, parapatric speciation.)

symplast The continuous meshwork of the interiors of living cells in the plant body, resulting from the presence of plasmodesmata. (Contrast with apoplast.)

symport A membrane transport process that carries two substances in the same direction across the membrane. (Contrast with antiport.)

synapomorphy *See* shared derived trait.

synapse (sin' aps) [Gk. *syn*: together + *haptein*: to fasten] The narrow gap between the terminal bouton of one neutron and the dendrite or cell body of another.

synapsis (sin ap' sis) The highly specific parallel alignment (pairing) of homologous chromosomes during the first division of meiosis.

synaptic vesicle A membrane-bounded vesicle containing neurotransmitter; the neurotransmitter is produced in and discharged by a presynaptic neuron.

synergids [Gk. *syn*: together + *ergos*: performing work] In flowering plants, the two cells accompanying the egg cell at one end of the megmagametophyte.

syngamy (sing' guh mee) [Gk. *syn*: together + *gamos*: marriage] Union of gametes. Also known as fertilization.

synonymous substitution A change of one nucleotide in a sequence to another when that change does not affect the amino acid specified (i.e., UUA → UUG, both specifying leucine). (Compare with nonsynonymous substitution, missense substitution, nonsense substitution.)

systematics The scientific study of the diversity of organisms, and of their relationships.

systemic circulation The part of the circulatory system serving those parts of the body other than the lungs or gills (which are served by the pulmonary circulation.)

systemic acquired resistance A general resistance to many plant pathogens following infection by a single agent.

systemin The only polypeptide plant hormone; participates in response to tissue damage.

systems biology The study of an organism as an integrated and interacting system of genes, proteins, and biochemical reactions.

systole (sis' tuh lee) [Gk. *systole*: contraction] Contraction of a chamber of the heart, driving blood forward in the circulatory system.

- T -

T cell A type of lymphocyte, involved in the cellular immune response. The final stages of its development occur in the thymus gland. (Contrast with B cell; see also cytotoxic T cell, helper T cell, suppressor T cell.)

T cell receptor A protein on the surface of a T cell that recognizes the antigenic determinant for which the cell is specific.

target cell A cell with the appropriate receptors to bind and respond to a particular hormone or other chemical mediator.

taste bud A structure in the epithelium of the tongue that includes a cluster of chemoreceptors innervated by sensory neurons.

TATA box An eight-base-pair sequence, found about 25 base pairs before the starting point for transcription in many eukaryotic promoters, that binds a transcription factor and thus helps initiate transcription.

taxis (tak' sis) [Gk. *taxis*: arrange, put in order] The movement of an organism or its part directly toward or away from the stimulus. For example, positive phototaxis is movement toward a light source, negative geotaxis is movement away from gravity).

taxon A biological group (typically a species or a clade) that is given a name.

telencephalon The frontmost division of the vertebrate brain; becomes the cerebrum.

telomerase An enzyme that catalyzes the addition of telomeric sequences lost from chromosomes during DNA replication.

telomeres (tee' lo merz) [Gk. *telos*: end + *meros*: units, segments] Repeated DNA sequences at the ends of eukaryotic chromosomes.

telophase (tee' lo phase) [Gk. *telos*: end] The final phase of mitosis or meiosis during which chromosomes became diffuse, nuclear envelopes reform, and nucleoli begin to reappear in the daughter nuclei.

template (1) In biochemistry, a molecule or surface upon which another molecule is synthesized in complementary fashion, as in the replication of DNA. (2) In the brain, a pattern that responds to a normal input but not to incorrect inputs.

template strand In double-stranded DNA, the strand that is transcribed to create an RNA transcript that will be processed into a protein. Also refers to a strand of RNA that is used to create a complementary RNA.

temporal summation [L. *tempus*: time; *summus*: highest amount] In the production or inhibition of action potentials in a postsynaptic neuron, the interaction of depolarizations or hyperpolarizations produced by rapidly repeated stimulation of a single point.

tendon A collagen-containing band of tissue that connects a muscle with a bone.

termination The end of protein synthesis triggered by a stop codon which binds release factor that causes the polypeptide to release from the ribosome.

terminator A sequence at the 3′ end of mRNA that causes the RNA strand to be released from the transcription complex.

terrestrial (ter res' tree al) [L. *terra*: earth] Pertaining to the land. (Contrast with aquatic, marine.)

territory A fixed area from which an animal or group of animals excludes other members of the same (and sometimes other) species by aggressive behavior or displays.

tertiary structure In reference to a protein, the relative locations in three-dimensional space of all the atoms in the molecule. The overall shape of a protein. (Contrast with primary, secondary, and quaternary structures.)

test cross Mating of a dominant-phenotype individual (who may be either heterozygous or homozygous) with a homozygous-recessive individual.

testis (tes' tis) (plural: testes) [L. *testis*: witness] The male gonad; the organ that produces the male sex cells.

testosterone (tes toss' tuhr own) A male sex steroid hormone.

tetanus [Gk. *tetanos*: stretched] (1) A state of sustained maximal muscular contraction caused by rapidly repeated stimulation. (2) In medicine, an often fatal disease ("lockjaw") caused by the bacterium *Clostridium tetani*.

tetrad [Gk. *tettares*: four] During prophase I of meiosis, the association of a pair of homologous chromosomes or four chromatids.

thalamus [Gk. *thalamos*: chamber] A region of the vertebrate forebrain; involved in integration of sensory input.

thallus (thal' us) [Gk. *thallos*: sprout] Any algal body which is not differentiated into root, stem, and leaf.

therapeutic cloning The use of cloning by nuclear transfer to produce an embryo that will provide embryonic stem cells to be used in therapy.

theory [Gk. *theoria*: analysis of facts] A far-reaching explanation of observed facts that is supported by such a wide body of evidence, with no significant contradictory evidence, that it is scientifically accepted as a factual framework. Examples are Newton's theory of gravity and Darwin's theory of evolution. (Contrast with hypothesis.)

thermoneutral zone [Gk. *thermos*: temperature] The range of temperatures over which an endotherm does not have to expend extra energy to thermoregulate.

thermoreceptor A cell or structure that responds to changes in temperature.

thoracic cavity [Gk. *thorax*: breastplate] The portion of the mammalian body cavity bounded by the ribs, shoulders, and diaphragm. Contains the heart and the lungs.

thoracic duct The connection between the lymphatic system and the circulatory system.

thorax [Gk. *thorax*: breastplate] In an insect, the middle region of the body, between the head and abdomen. In mammals, the part of the body between the neck and the diaphragm.

thrombin An enzyme that converts fibrinogen to fibrin, thus triggering the formation of blood clots.

thrombus (throm' bus) [Gk. *thrombos*: clot] A blood clot that forms within a blood vessel and remains attached to the wall of the vessel. (Contrast with embolus.)

thylakoid (thigh la koid) [Gk. *thylakos*: sack or pouch] A flattened sac within a chloroplast. Thylakoid membranes contain all of the chlorophyll in a plant, in addition to the electron carriers of photophosphorylation. Thylakoids stack to form grana.

thymine (T) A nitrogen-containing base found in DNA.

thymus [Gk. *thymos*: warty] A ductless, glandular portion of the lymphoid system, involved in development of the immune system of vertebrates. In humans, the thymus degenerates during puberty.

thyroid [Gk. *thyreos*: door-shaped] A two-lobed gland in vertebrates. Produces the hormone thyroxin.

thyrotropin (thyroid-stimulating hormone, TSH) A hormone of the anterior pituitary that stimulates the thyroid gland to produce and release thyroxin.

thyrotropin-releasing hormone (TRH) A hypothalamic hormone that stimulates anterior pituitary cells to release TSH.

thyroxine The hormone produced by the thyroid gland that controls many metabolic processes.

tidal volume The amount of air that is exchanged during each breath when a person is at rest. (Compare with vital capacity.)

tight junction A junction between epithelial cells, in which there is no gap whatever between the adjacent cells. Materials may pass through a tight junction only by entering the epithelial cells themselves.

tissue A group of similar cells organized into a functional unit; usually integrated with other tissues to form part of an organ.

toll A member of a receptor family that responds to the binding of a molecule from a pathogen by initiating a protein kinase cascade, resulting in the synthesis of defensive proteins.

tonus (toe' nuss) [L. *tonus*: tension] A low level of muscular tension that is maintained even when the body is at rest.

topsoil The uppermost soil layer; contains most of the organic matter of soil, but may be depleted of most mineral nutrients.

totipotent [L. *toto*: whole, entire + *potens*: powerful] Of a cell, possessing all the genetic information and other capacities necessary to form an entire individual. (Compare with pluripotent.)

toxic [L. *toxicum*: poison] Injurious to the tissues of the host organism.

trachea (tray' kee ah) [Gk. *trakhoia*: tube] A tube that carries air to the bronchi of the lungs of vertebrates. When plural (*tracheae*), refers to the major airways of insects.

tracheary element Refers to either or both types of conductive xylem cells: tracheids and vessel elements.

tracheid (tray' kee id) A distinctive conducting and supporting cell found in the xylem of nearly all vascular plants, characterized by tapering ends and walls that are pitted but not perforated. (Contrast with vessel element.)

tracheophytes [Gk. *trakhoia*: tube + *phyton*: plant] *See* vascular plants.

trade-off The relationship between the costs of performing a behavior or other trait and the benefits the individual gains from the trait or behavior. (*See also* cost–benefit analysis.)

trait In genetics, one form of a character: eye color is a character; brown eyes and blue eyes are traits. (Compare with character.)

transcription The synthesis of RNA using one strand of DNA as the template.

transcription factors Proteins that assemble on a eukaryotic chromosome, allowing RNA polymerase II to perform transcription.

transduction (1) Transfer of genes from one bacterium to another, with a bacterial virus acting as the carrier of the genes. (2) In sensory cells, the transformation of a stimulus (e.g., light energy, sound pressure waves, chemical or electrical stimulants) into action potentials.

transfection Uptake, incorporation, and expression of recombinant DNA.

transfer cell A modified parenchyma cell that transports solutes from its cytoplasm into its cell wall, thus moving the solutes from the symplast into the apoplast.

transfer RNA (tRNA) A family of double-stranded RNA molecules. Each tRNA carries a specific amino acid and anticodon that will pair with the complementary codon in mRNA during translation.

transformation Mechanism for transfer of genetic information in bacteria in which pure DNA extracted from bacteria of one genotype is taken in through the cell surface of bacteria of a different genotype and incorporated into the chromosome of the recipient cell.

transforming principle An early term for the as yet unidentified chemical substance responsible for bacterial tranformation.

transgenic organism An organism containing recombinant DNA incorporated into its genetic material.

transition-state species A short-lived, unstable intermediate with high potential energy in a chemical reaction.

translation The synthesis of a protein (polypeptide). Takes place on ribosomes, using the information encoded in messenger RNA.

transmembrane domain The portion of a protein that lies inside the membrane bilayer.

transmembrane protein An integral membrane protein that spans the lipid bilayer.

translocation (1) In genetics, a rare mutational event that moves a portion of a chromosome to a new location, generally on a nonhomologous chromosome. (2) In vascular plants, movement of solutes in the phloem.

transpiration [L. *spirare*: to breathe] The evaporation of water from plant leaves and stem, driven by heat from the sun, and providing the motive force to raise water (plus mineral nutrients) from the roots.

transposable element A segment of DNA that can move to, or give rise to copies at, another locus on the same or a different chromosome.

transposon Mobile DNA segment that can insert into a chromosome and cause genetic change.

triglyceride A simple lipid in which three fatty acids are combined with one molecule of glycerol.

triplet *See* codon.

triplet repeat The occurrence of repeated triplet of bases in a gene, often leading to genetic disease, as does excessive repetition of CGG in the gene responsible for fragile-X syndrome.

triploblastic Having three cell layers. (Contrast with diploblastic.)

trisomic Containing three rather than two members of a chromosome pair.

tRNA *See* transfer RNA.

trophic cascade The progression over successively lower trophic levels of the indirect effects of a predator.

trophic level [Gk *trophes*: nourishment] A group of organisms united by obtaining their energy from the same part of the food web of a biological community.

trophoblast [Gk *trophes*: nourishment + *blastos*: sprout] At the 32-cell stage of mammalian development, the outer group of cells that will become part of the placenta and thus nourish the growing embryo. (Comtrast with inner cell mass.)

trochophore (troke' o fore) [Gk. *trochos*: wheel + *phoreus*: bearer] The free-swimming larva of some annelids and mollusks, distinguished by a wheel-like band of cilia around the middle.

tropic hormones Hormones of the anterior pituitary that control the secretion of hormones by other endocrine glands.

tropism [Gk. *tropos*: to turn] In plants, growth toward or away from a stimulus such as light (phototropism) or gravity (gravitropism).

tropomyosin [troe poe my' oh sin] A protein that, along with actin, constitutes the thin filaments of myofibrils. It controls the interactions of actin and myosin necessary for muscle contraction.

troposphere The lowest atmospheric zone, reaching upward from the Earth's surface approximately 17 km in the tropics and subtropics but only to about 10 km at higher latitudes. The zone in which virtually all the water vapor in the atmosphere is located.

true-breeding A genetic cross in which the same result occurs every time with respect to the trait(s) under consideration, due to homozygous parents.

trypsin A protein-digesting enzyme. Secreted by the pancreas in its inactive form (trypsinogen), it becomes active in the duodenum of the small intestine.

T tubules A system of tubules that runs throughout the cytoplasm of muscle fibers, through which action potentials spread.

tube cell The larger of the two cells in a pollen grain; responsible for growth of the pollen tube. *See* generative cell.

tubulin A protein that polymerizes to form microtubules.

tumor [L. *tumor*: a swollen mass] A disorganized mass of cells, often growing out of control. Malignant tumors spread to other parts of the body.

tumor suppressor genes Genes which, when homozygous mutant, result in cancer. Such genes code for protein products that inhibit cell proliferation.

turgor pressure [L. *turgidus*: swollen] *See* pressure potential.

tympanic membrane [Gk. *tympanum*: drum] The eardrum.

- U -

ubiquinone (yoo bic' kwi known) [L. *ubique*: everywhere] A mobile electron carrier of the mitochondrial respiratory chain. Similar to plastoquinone found in chloroplasts.

ubiquitin A small protein that is covalently linked to other cellular proteins identified for breakdown by the proteosome.

umbilical cord Tissue made up of embryonic membranes and blood vessels that connects the embryo to the placenta in eutherian mammals.

understory The aggregate of smaller plants growing beneath the canopy of dominant plants in a forest.

unicellular (yoon' e sell' yer ler) [L. *unus*: one + *cella*: chamber] Consisting of a single cell, as in a unicellular organism. (Contrast with multicellular.)

uniport [L. *unus*: one + portal: doorway] A membrane transport process that carries a single substance. (Contrast with antiport, symport.)

unsaturated hydrocarbon A compound containing only carbon and hydrogen atoms, with one or more pairs of carbon atoms that are connected by double bonds.

upwelling The surfaceward movement of nutrient-rich, cooler water from deeper layers of the ocean.

uracil (U) A pyrimidine base found in nucleotides of RNA.

urea A compound serving as the main excreted form of nitrogen by many animals, including mammals.

ureotelic Describes an organism in which the final product of the breakdown of nitrogen-containing compounds (primarily proteins) is urea. (Contrast with ammonotelic, uricotelic.)

ureter (your' uh tur) A long duct leading from the vertebrate kidney to the urinary bladder or the cloaca.

urethra (you ree' thra) In most mammals, the canal through which urine is discharged from the bladder and which serves as the genital duct in males.

uric acid A compound that serves as the main excreted form of nitrogen in some animals, particularly those which must conserve water, such as birds, insects, and reptiles.

uricotelic Describes an organism in which the final product of the breakdown of nitrogen-containing compounds (primarily proteins) is uric acid. (Contrast with ammonotelic, ureotelic.)

urine (you' rin) In vertebrates, the fluid waste product containing the toxic nitrogenous by-products of protein and amino acid metabolism.

uterus (yoo' ter us) [L. *utero*: womb] The uterus or womb is a specialized portion of the female reproductive tract in certain mammals. It receives the fertilized egg and nurtures the embryo in its early development.

- V -

vaccination Injection of virus or bacteria or their proteins into the body, to induce immunization. The injected material is usually attenuated (weakened) before injection.

vacuole (vac' yew ole) [Fr.: small vacuum] A liquid-filled, membrane-enclosed compartment in cytoplasm; may function as digestive chambers, storage chambers, waste bins.

vagina (vuh jine' uh) [L.: sheath] In female mammals, the passage leading from the external genital orifice to the uterus; receives the copulatory organ of the male in mating.

van der Waals forces Weak attractions between atoms resulting from the interaction of the electrons of one atom with the nucleus of another. This type of attraction is about one-fourth as strong as a hydrogen bond.

variable regions The part of an immunoglobulin molecule or T-cell receptor that includes the antigen-binding site.

vasa recta Blood vessels that parallel the loops of Henle and the collecting ducts in the renal medulla of the kidney.

vascular (vas' kew lar) [L. *vasculum*: a small vessel] Pertaining to organs and tissues that conduct fluid, such as blood vessels in animals and phloem and xylem in plants.

vascular bundle In vascular plants, a strand of vascular tissue, including conducting cells of xylem and phloem as well as thick-walled fibers.

vascular cambium A lateral meristem giving rise to secondary xylem and phloem.

vascular rays In vascular plants, radially oriented sheets of cells produced by the vascular cambium, carrying materials laterally between the wood and the phloem.

vascular system The conductive system of the plant, consisting primarily of xylem and phloem.

vas deferens The duct that transfers sperm from the epididymis to the urethra.

vasopressin *See* antidiuretic hormone.

vector (1) An agent, such as an insect, that carries a pathogen affecting another species. (2) A plasmid or virus that carries an inserted piece of DNA into a bacterium for cloning purposes in recombinant DNA technology.

vegetal hemisphere The lower portion of some animal eggs, zygotes, and embryos, in which the dense nutrient yolk settles. The vegetal pole refers to the very bottom of the egg or embyro. (Contrast with animal hemisphere.)

vegetative Nonreproductive, nonflowering, or asexual.

vegetative reproduction Asexual reproduction.

vein [L. *vena*: channel] A blood vessel that returns blood to the heart. (Contrast with artery.)

ventral [L. *venter*: belly, womb] Toward or pertaining to the belly or lower side. (Contrast with dorsal.)

ventricle A muscular heart chamber that pumps blood through the lungs or through the body.

vernalization [L. *vernalis*: spring] Events occurring during a required chilling period, leading eventually to flowering.

vertebral column [L. *vertere*: to turn] The jointed, dorsal column that is the primary support structure of vertebrates.

vesicle A membrane enclosed compartment within the cytoplasm.

vessel element In plants, a nonliving water-conducting cell with perforated end walls. (Contrast with tracheid.)

vestibular apparatus (ves tib' yew lar) [L. *vestibulum*: an enclosed passage] Structures associated with the vertebrate ear; these structures sense changes in position or momentum of the head, affecing balance and motor skills.

vestigial (ves tij' ee al) [L. *vestigium*: footprint, track] The remains of body structures that are no longer of adaptive value to the organism and therefore are not maintained by selection.

vicariant distribution A population distribution resulting from the disruption of a formerly continuous range by a vicariant event.

vicariant event (vye care' ee unce) [L. *vicus*: change] The splitting of the range of a taxon by the imposition of some barrier to interchange among its members.

villus (vil' lus) (plural: villi) [L. *villus*: shaggy hair or beard] A hairlike projection from a membrane; for example, from many gut walls.

virion (veer' e on) The virus particle, the minimum unit capable of infecting a cell.

viroid (vye' roid) An infectious agent consisting of a single-stranded RNA molecule with no protein coat; produces diseases in plants.

virulent [L. *virus*: poison, slimy liquid] Causing or capable of causing disease and death.

virus Any of a group of ultramicroscopic particles constructed of nucleic acid and protein (and, sometimes, lipid) that require living cells in order to reproduce. Viruses probably evolved from eukaryotic cells, secondarily losing some cellular attributes.

vital capacity The sum total of the tidal volume and the inspiratory and expiratory reserve volumes.

vitamins [L. *vita*: life] Organic compounds that an organism cannot synthesize, but nevertheless requires in small quantity for normal growth and metabolism.

viviparous (vye vip' uh rus) [L. *vivus*: alive] Reproduction in which fertilization of the egg and development of the embryo occur inside the mother's body. (Contrast with oviparous.)

VNTRs (variable number of tandem repeats) In the human genome, short DNA sequences that are repeated a characteristic number of times in related individuals. Can be used to make a DNA fingerprint.

voltage-gated channels Ion channels in membranes that change conformation and therefore ion conductance when a certain potential difference exists across the membrane in which they are inserted.

- W -

waggle dance The running movement of a working honey bee on the hive, during which the worker traces out a repeated figure eight. The dance contains elements that transmit to other bees the location of the food.

water potential In osmosis, the tendency for a system (a cell or solution) to take up water from pure water through a differentially permeable membrane. Water flows toward the system with a more negative water potential. (Contrast with solute potential, pressure potential.)

wavelength The distance between successive peaks of a wave train, such as electromagnetic radiation.

wild-type Geneticists' term for standard or reference type. Deviants from this standard, even if the deviants are found in the wild, are usually referred to as mutant. (Note that this terminology is not usually applied to human genes.)

wood Secondary xylem tissue.

- X -

xanthophyll (zan' tho fill) [Gk. *xanthos*: yellowish-brown + *phyllon*: leaf] A yellow or orange pigment commonly found as an accessory pigment in photosynthesis, but found elsewhere as well. An oxygen-containing carotenoid.

X-linked A character that is coded for by a gene on the X chromosome; a sex-linked trait.

xerophyte (zee' row fyte) [Gk. *xerox*: dry + *phyton*: plant] A plant adapted to an environment with a limited water supply.

xylem (zy' lum) [Gk. *xylon*: wood] In vascular plants, the tissue that conducts water and minerals; xylem consists, in various plants, of tracheids, vessel elements, fibers, and other highly specialized cells.

- Y -

yolk [M.E. *yolke*: yellow] The stored food material in animal eggs, usually rich in protein and lipid.

yolk sac In reptiles, birds, and mammals, the extraembryonic membrane that forms from the endoderm of the hypoblast; it encloses and digests the yolk.

- Z -

Z-DNA A form of DNA in which the molecule spirals to the left rather than to the right.

zeaxanthin A blue-light receptor involved in the opening of plant stomata.

zona pellucida A jellylike substance that surrounds the mammalian ovum when it is released from the ovary.

zoospore (zoe' o spore) [Gk. *zoon*: animal + *spora*: seed] In algae and fungi, any swimming spore. May be diploid or haploid.

zygote (zye' gote) [Gk. *zygotos*: yoked] The cell created by the union of two gametes, in which the gamete nuclei are also fused. The earliest stage of the diploid generation.

zymogen An inactive precursor of a digestive enzyme secreted into the lumen of the gut, where a protease cleaves it to form the active enzyme.

Answers to Self-Quizzes

Chapter 2
1. b 6. a
2. d 7. d
3. c 8. a
4. c 9. c
5. d 10. b

Chapter 3
1. e 6. a
2. e 7. c
3. c 8. e
4. d 9. a
5. b 10. d

Chapter 4
1. b 6. e
2. d 7. a
3. c 8. d
4. e 9. b
5. a 10. d

Chapter 5
1. e 6. c
2. c 7. c
3. a 8. b
4. d 9. e
5. c 10. c

Chapter 6
1. c 6. e
2. e 7. d
3. b 8. b
4. c 9. d
5. c 10. e

Chapter 7
1. d 6. d
2. d 7. a
3. e 8. b
4. e 9. a
5. c 10. e

Chapter 8
1. c 6. c
2. b 7. c
3. d 8. d
4. b 9. d
5. e 10. b

Chapter 9
1. d 6. d
2. c 7. e
3. b 8. d
4. d 9. c
5. c 10. c

Chapter 10*
1. e 6. d
2. a 7. b
3. d 8. b
4. d 9. b
5. d 10. b

Chapter 11
1. c 6. b
2. a 7. d
3. c 8. d
4. b 9. c
5. e 10. c

Chapter 12
1. c 6. d
2. d 7. b
3. e 8. d
4. b 9. d
5. a 10. a

Chapter 13
1. b 6. d
2. e 7. d
3. a 8. c
4. c 9. b
5. c 10. d

Chapter 14
1. c 6. c
2. c 7. c
3. a 8. b
4. a 9. e
5. c 10. d

Chapter 15
1. d 6. a
2. d 7. e
3. c 8. b
4. c 9. c
5. d 10. a

Chapter 16
1. b 6. b
2. a 7. c
3. a 8. a
4. c 9. e
5. e 10. e

Chapter 17
1. a 6. b
2. c 7. e
3. b 8. d
4. b 9. c
5. d 10. b

Chapter 18
1. a 6. a
2. b 7. d
3. e 8. d
4. e 9. a
5. c 10. d

Chapter 19
1. c 6. c
2. a 7. d
3. a 8. b
4. b 9. a
5. b 10. b

Chapter 20
1. d 6. b
2. e 7. a
3. c 8. c
4. b 9. b
5. c 10. b

Chapter 21
1. d 6. a
2. b 7. c
3. e 8. b
4. c 9. c
5. a 10. e

*Answers to Chapter 10 Genetics Problems

1. Each of the eight boxes in the Punnett squares should contain the genotype *Tt*, regardless of which parent was tall and which dwarf.

2. See Figure 10.4, page 212.

3. The trait is autosomal. Mother *dp dp*, father *Dp dp*. If the trait were sex-linked, all daughters would be wild-type and sons would be *dumpy*.

4. Yellow parent = $s^Y s^b$; offspring 3 yellow (s^Y–): 1 black ($s^b s^b$).
Black parent = $s^b s^b$; offspring all black ($s^b s^b$). Orange parent = $s^O s^b$; offspring 3 orange (s^O–): 1 black ($s^b s^b$). Both s^O and s^Y are dominant to s^b.

5. All females wild-type; all males spotted.

6. F_1 all wild-type, *PpSwsw*; F_2 9:3:3:1 in phenotypes. See Figure 10.7, page 214, for analogous genotypes.

7*a*. Ratio of phenotypes in F_2 is 3:1 (double dominant to double recessive).

7*b*. The F_1 are $Pby\ pB^Y$; they produce just two kinds of gametes (*Pby* and *pBy*). Combine them carefully and see the 1:2:1 phenotypic ratio fall out in the F_2.

7*c*. Pink-blistery.

7*d*. See Figures 9.16 and 9.18 (pages 196–198). Crossing over took place in the F_1 generation.

8. The genotypes are:

PpSwsw
Ppswsw
ppSwsw
ppswsw

Ratio: 1:1:1:1

The phenotypes are:

wild eye, long wing
wild eye, short wing
pink eye, long wing
pink eye, short wing

Ratio: 1:1:1:1

9*a*. 1 black:2 blue:1 splashed white

9*b*. Always cross black with splashed white.

10*a*. $w^+ > w^e > w$

10*b*. Parents $w^e w$ and $w^+ Y$. Progeny $w{+}w^e$, $w{+}w$, $w^e Y$, and wY.

11. All will have normal vision because they inherit dad's wild-type X chromosome, but half of them will be carriers.

12. Agouti parent *AaBb*. Albino offspring *aaBb* and *aabb*; black offspring *Aabb*; agouti offspring *AaBb*.

13. Because the gene is carried on mitochondrial DNA, it is passed through the mother only. Thus if the woman does not have the disease but her husband does, their child will not be affected. On the other hand, if the woman has the disease but her husband does not, their child *will* have the disease.

Chapter 22
1. d
2. e
3. d
4. c
5. d
6. d
7. b
8. e
9. b
10. c

Chapter 23
1. c
2. e
3. d
4. c
5. a
6. d
7. a
8. a
9. c
10. e

Chapter 24
1. a
2. a
3. d
4. a
5. b
6. e
7. e
8. e
9. b
10. e

Chapter 25
1. b
2. e
3. a
4. b
5. e
6. b
7. e
8. a
9. e
10. d

Chapter 26
1. e
2. e
3. b
4. c
5. e
6. b
7. d
8. d
9. c
10. d

Chapter 27
1. a
2. e
3. c
4. d
5. c
6. b
7. d
8. b
9. a
10. d

Chapter 28
1. d
2. c
3. e
4. b
5. b
6. e
7. c
8. b
9. b
10. d

Chapter 29
1. d
2. c
3. d
4. a
5. d
6. c
7. a
8. e
9. c
10. a

Chapter 30
1. b
2. d
3. e
4. c
5. d
6. a
7. e
8. a
9. c
10. c

Chapter 31
1. c
2. d
3. c
4. d
5. c
6. b
7. c
8. d
9. e
10. d

Chapter 32
1. b
2. e
3. c
4. d
5. a
6. e
7. b
8. d
9. d
10. e

Chapter 33
1. b
2. a
3. c
4. c
5. d
6. e
7. a
8. e
9. c
10. b

Chapter 34
1. d
2. b
3. e
4. e
5. a
6. b
7. b
8. c
9. a
10. d

Chapter 35
1. c
2. d
3. b
4. b
5. b
6. d
7. d
8. e
9. e
10. a

Chapter 36
1. d
2. d
3. c
4. a
5. a
6. c
7. e
8. a
9. d
10. e

Chapter 37
1. a
2. e
3. c
4. d
5. b
6. c
7. e
8. c
9. a
10. b

Chapter 38
1. d
2. b
3. e
4. b
5. d
6. e
7. a
8. b
9. c
10. d

Chapter 39
1. e
2. b
3. c
4. c
5. d
6. a
7. b
8. c
9. d
10. a

Chapter 40
1. c
2. c
3. a
4. d
5. b
6. b
7. e
8. a
9. e
10. b

Chapter 41
1. b
2. a
3. b
4. e
5. e
6. b
7. d
8. e
9. c
10. c

Chapter 42
1. c
2. e
3. a
4. d
5. d
6. d
7. d
8. c
9. d
10. a

Chapter 43
1. a
2. c
3. e
4. a
5. d
6. c
7. b
8. b
9. b
10. a

Chapter 44
1. d
2. d
3. c
4. c
5. e
6. e
7. c
8. c
9. d
10. d

Chapter 45
1. d
2. d
3. a
4. b
5. e
6. e
7. b
8. c
9. c
10. d

Chapter 46
1. c
2. a
3. e
4. d
5. d
6. c
7. a
8. c
9. a
10. c

Chapter 47
1. e
2. a
3. b
4. c
5. b
6. d
7. e
8. a
9. a
10. e

Chapter 48
1. e
2. d
3. a
4. b
5. c
6. b
7. c
8. c
9. a
10. d

Chapter 49
1. d
2. a
3. c
4. d
5. c
6. d
7. b
8. d
9. c
10. e

Chapter 50
1. b
2. e
3. c
4. a
5. b
6. d
7. a
8. b
9. d
10. d

Chapter 51
1. d
2. a
3. d
4. a
5. d
6. b
7. e
8. a
9. c
10. e

Chapter 52
1. c
2. a
3. a
4. c
5. a
6. c
7. d
8. c
9. d
10. a

Chapter 53
1. c
2. e
3. c
4. a
5. d
6. d
7. c
8. c
9. d
10. e

Chapter 54
1. c
2. c
3. a
4. d
5. c
6. b
7. d
8. c
9. a
10. d

Chapter 55
1. a
2. a
3. b
4. c
5. d
6. e
7. c
8. a
9. d
10. e

Chapter 56
1. d
2. d
3. e
4. e
5. c
6. b
7. a
8. c
9. d
10. b

Chapter 57
1. b
2. d
3. e
4. e
5. b
6. a
7. d
8. c
9. a
10. c

Illustration Credits

Cover *Amor de Madre* © Max Billder.

Frontispiece © Lynn M. Stone/Naturepl.com.

Part Openers and Back Cover:

Part 1 *Canopy researcher*: © Mark Moffett/ Minden Pictures.

Part 2 *Mitochondrion*: © Bill Loncore/SPL/Photo Researchers, Inc.

Part 3 *T4 bacteriophage*: © Dept. of Microbiology, Biozentrum/SPL/Photo Researchers, Inc.

Part 4 *Dividing cells*: © Steve Gschmeissner/SPL/ Photo Researchers, Inc.

Part 5 *Fossil mesosaur*: © John Cancalosi/AGE Fotostock.

Part 6 *Tiger moth*: © Piotr Naskrecki/Minden Pictures.

Part 7 *Protea flower*: © Gerry Ellis/Minden Pictures.

Part 8 *Great blue heron*: © Larry Jon Friesen.

Part 9 *Zebras and wildebeest*: © Gerry Ellis/ Minden Pictures.

Table of Contents:

p. xix *Leaf*: David McIntyre.

p. xxix *Passionflower*: © Larry Jon Friesen.

p. xxxv *Feather stars*: © André Seale/AGE Fotostock.

Chapter 1 *Frogs*: © Frans Lanting/Minden Pictures. *Pieter Johnson*: © Steven Holt/ stockpix.com. 1.1A: © Eye of Science/SPL/Photo Researchers, Inc. 1.1B: © Dennis Kunkel Microscopy, Inc. 1.1C: © Markus Geisen/NHMPL. 1.1D: © Frans Lanting/Minden Pictures. 1.1E: © Glen Threlfo/Auscape/Minden Pictures. 1.1F: © Piotr Naskrecki/Minden Pictures. 1.1G: © Tui De Roy/Minden Pictures. 1.2A: © Dr. Cecil H. Fox/ Photo Researchers, Inc. 1.2B: From R. Hooke, 1664. *Micrographia*. 1.2C: © Astrid & Hanns-Frieder Michler/Photo Researchers, Inc. 1.2D: © John Durham/SPL/Photo Researchers, Inc. 1.3 *Maple*: © Simon Colmer & Abby Rex/Alamy. 1.3 *Spruce*: David McIntyre. 1.3 *Lily pad*: © Pete Oxford/ Naturepl.com. 1.3 *Cucumber*: David McIntyre. 1.3 *Pitcher plant*: © Nick Garbutt/Naturepl.com. 1.5A: © Bob Bennett/Jupiter Images. 1.5B: © Frans Lanting/Minden Pictures. 1.6 *Organism*: © Sami Sarkis/Painet Inc. 1.6 *Population*: © LogicStock/ Painet Inc. 1.6 *Community*: © Georgie Holland/ AGE Fotostock. 1.6*Biosphere*: Courtesy of NASA. 1.7A: © Frans Lanting/Minden Pictures. 1.7B: © Marguerite Smits Van Oyen/Naturepl.com. 1.8: © Chris Howes/Wild Places Photography/Alamy. 1.10: © Dennis Kunkel Microscopy, Inc. 1.12: Courtesy of Wayne Whippen. 1.13 *Frog (inset)*: © Frans Lanting/Minden Pictures. 1.13, 1.14: After P. T. Johnson et al., 1999. *Science* 284: 802.

Chapter 2 *Enceladus*: Courtesy of the NASA Jet Propulsion Laboratory/Space Science Institute. *Rover*: Courtesy of the NASA Jet Propulsion Laboratory. 2.4: From N. D. Volkow et al., 2001. *Am. J. Psychiatry* 158: 377. 2.14: © Mitsuaki Iwago/ Minden Pictures. 2.15: David McIntyre.

Chapter 3 *Cookies*: David McIntyre. 3.8: Data from PDB 1IVM. T. Obita, T. Ueda, & T. Imoto, 2003. *Cell. Mol. Life Sci.* 60: 176. 3.9A: Data from PDB 2HHB. G. Fermi et al., 1984. *J. Mol. Biol.* 175: 159. 3.16C *left*: © Biophoto Associates/Photo Researchers, Inc. 3.16C *middle*: © Ken Wagner/ Visuals Unlimited. 3.16C *right*: © CNRI/SPL/Photo Researchers, Inc. 3.17 *Cartilage*: © Robert Brons/ Biological Photo Service. 3.17 *Beetle*: © Scott Bauer/ USDA. 3.26: Data from S. Arnott & D. W. Hukins, 1972. *Biochem. Biophys. Res. Commun.* 47(6): 1504. 3.27: Courtesy of NASA.

Chapter 4 *Fossil*: © Stanley M. Awramik/ Biological Photo Service. *Cyanobacterium*: © Dennis Kunkel Microscopy, Inc. *Protobionts*: © Sidney Fox/Science VU/Visuals Unlimited. 4.1: After N. Campbell, 1990. *Biology*, 2nd Ed., Benjamin Cummings. 4.3 *upper row, left*: Courtesy of the IST Cell Bank, Genoa. 4.3 *upper row, center and right, middle row, left*: © Michael W. Davidson, Florida State U. 4.3 *middle row, center*: © Dr. Gopal Murti/SPL/Photo Researchers, Inc. 4.3 *middle row, right*: © Richard J. Green/SPL/Photo Researchers, Inc. 4.3 *bottom row, left*: © Dr. Gopal Murti/Visuals Unlimited. 4.3 *bottom row, center*: © K. R. Porter/ SPL/Photo Researchers, Inc. 4.3 *bottom row, right*: © D. W. Fawcett/Photo Researchers, Inc. 4.4: © J. J. Cardamone Jr. & B. K. Pugashetti/Biological Photo Service. 4.5A: © Dennis Kunkel Microscopy, Inc. 4.5B: Courtesy of David DeRosier, Brandeis U. 4.7 *Mitochondrion*: © K. Porter, D. Fawcett/Visuals Unlimited. 4.7 *Cytoskeleton*: © Don Fawcett, John Heuser/Photo Researchers, Inc. 4.7 *Nucleolus*: © Richard Rodewald/Biological Photo Service. 4.7 *Peroxisome*: © E. H. Newcomb & S. E. Frederick/ Biological Photo Service. 4.7 *Cell wall*: © Biophoto Associates/Photo Researchers, Inc. 4.7 *Ribosome*: From M. Boublik et al., 1990. *The Ribosome*, p. 177. Courtesy of American Society for Microbiology. 4.7 *Centrioles*: © Barry F. King/Biological Photo Service. 4.7 *Plasma membrane*: Courtesy of J. David Robertson, Duke U. Medical Center. 4.7 *Rough ER*: © Don Fawcett/Science Source/Photo Researchers, Inc. 4.7 *Smooth ER*: © Don Fawcett, D. Friend/Science Source/Photo Researchers, Inc. 4.7 *Chloroplast*: © W. P. Wergin, E. H. Newcomb/ Biological Photo Service. 4.7 *Golgi apparatus*: Courtesy of L. Andrew Staehelin, U. Colorado. 4.8 *left*: From U. Aebi et al., 1986. *Nature* 323: 560. © Macmillan Publishers Ltd. 4.8 *upper right*: © D. W. Fawcett/Photo Researchers, Inc. 4.8 *lower right*: Courtesy of Dr. Ron Milligan, Scripps Research Institute. 4.9A: © Barry King, U. California, Davis/Biological Photo Service. 4.9B: © Biophoto Associates/Science Source/Photo Researchers, Inc. 4.10: © Don Fawcett/Visuals Unlimited. 4.11: © B. Bowers/Photo Researchers, Inc. 4.12: © Sanders/Biological Photo Service. 4.13: © K. Porter, D. Fawcett/Visuals Unlimited. 4.14: © W. P. Wergin, E. H. Newcomb/Biological Photo Service. 4.14 *inset*: © W. P. Wergin/Biological Photo Service. 4.15A: © John Durham/SPL/Photo Researchers, Inc. 4.15B: © Paul W. Johnson/ Biological Photo Service. 4.15C: © Gerald & Buff Corsi/Visuals Unlimited. 4.16A: David McIntyre. 4.16A *inset*: © Richard Green/Photo Researchers, Inc. 4.16B: David McIntyre. 4.16B *inset*: Courtesy of R. R. Dute. 4.17: © E. H. Newcomb & S. E. Frederick/Biological Photo Service. 4.18: Courtesy of M. C. Ledbetter, Brookhaven National Laboratory. 4.19: © Michael Abbey/Visuals Unlimited. 4.20: Courtesy of Vic Small, Austrian Academy of Sciences, Salzburg, Austria. 4.21: Courtesy of N. Hirokawa. 4.22A *top*: © SPL/Photo Researchers, Inc. 4.22A *bottom*, 4.22B, C: © W. L. Dentler/Biological Photo Service. 4.23: From N. Pollack et al., 1999. *J. Cell Biol.* 147: 493. Courtesy of R. D. Vale. 4.24: © Biophoto Associates/Photo Researchers, Inc. 4.25 *left*: Courtesy of David Sadava. 4.25 *upper right*: From J. A. Buckwalter & L. Rosenberg, 1983. *Coll. Rel. Res.* 3: 489. Courtesy of L. Rosenberg. 4.25 *lower right*: © J. Gross, Biozentrum/SPL/Photo Researchers, Inc.

Chapter 5 *Vibrio*: © Dennis Kunkel Microscopy, Inc. *Cholera treatment*: Courtesy of the WHO Global Consultation on Child and Adolescent Health and Development. 5.2: After L. Stryer, 1981. *Biochemistry*, 2nd Ed., W. H. Freeman. 5.3: Courtesy of L. Andrew Staehelin, U. Colorado. 5.7A: Courtesy of D. S. Friend, U. California, San Francisco. 5.7B: Courtesy of Darcy E. Kelly, U. Washington. 5.7C: Courtesy of C. Peracchia. 5.17: From M. M. Perry, 1979. *J. Cell Sci.* 39: 26.

Chapter 6 *Enzyme*: Data from PDB 1CW3. L. Ni et al., 1999. *Protein Sci.* 8: 2784. *Champagne*: © Purestock/Alamy. 6.1: Violet Bedell-McIntyre. 6.5B: © Darwin Dale/Photo Researchers, Inc. 6.11A: Data from PDB 1AL6. B. Schwartz et al., 1997. 6.11B: Data from PDB 1BB6. V. B. Vollan et al., 1999. *Acta Crystallogr. D. Biol. Crystallogr.* 55: 60. 6.11C: Data from PDB 1AB9. N. H. Yennawar, H. P. Yennawar, & G. K. Farber, 1994. *Biochemistry* 33: 7326. 6.13: Data from PDB 1A7K. H. Kim & W. G. Hol, 1998. *J Mol. Biol.* 278: 5.

Chapter 7 *Marathoners*: © Chuck Franklin/ Alamy. *Mouse*: © Royalty-Free/Corbis. 7.13: Courtesy of Ephraim Racker.

Chapter 8 *Rainforest*: © Jon Arnold Images/ photolibrary.com. *Meteor*: © Don Davis. 8.1: © Andrew Syred/SPL/Photo Researchers, Inc. 8.12: Courtesy of Lawrence Berkeley National Laboratory. 8.16, 8.18: © E. H. Newcomb & S. E. Frederick/Biological Photo Service. 8.20: © Aflo Foto Agency/Alamy.

Chapter 9 *Cells*: © Dr. Torsten Wittmann/Photo Researchers, Inc. *Henrietta Lacks*: © Obstetrics and Gynaecology/Photo Researchers, Inc. 9.1A: © David M. Phillips/Visuals Unlimited. 9.1B, C: © John D. Cunningham/Visuals Unlimited. 9.2B: © John J. Cardamone Jr./Biological Photo Service. 9.7: Courtesy of G. F. Bahr. 9.8 *inset*: © Biophoto Associates/Science Source/Photo Researchers, Inc. 9.9B: © Conly L. Rieder/Biological Photo Service. 9.10: © Nasser Rusan. 9.12A: © T. E. Schroeder/ Biological Photo Service. 9.12B: © B. A. Palevitz, E. H. Newcomb/Biological Photo Service. 9.13A: © Dr. John Cunningham/Visuals Unlimited. 9.13B:

© Garry T. Cole/Biological Photo Service. 9.14 *left*: © Andrew Syred/SPL/Photo Researchers, Inc. 9.14 *center*: David McIntyre. 9.14 *right*: © Gerry Ellis, DigitalVision/PictureQuest. 9.15: Courtesy of Dr. Thomas Ried and Dr. Evelin Schröck, NIH. 9.16: © C. A. Hasenkampf/Biological Photo Service. 9.17: Courtesy of J. Kezer. 9.21A: © Gopal Murti/ Photo Researchers, Inc.

Chapter 10 *Bris*: © David H. Wells/CORBIS. *Eye test*: © Brand X Pictures/Alamy. 10.1: © the Mendelianum. 10.11: Courtesy the American Netherland Dwarf Rabbit Club. 10.14: Courtesy of Madison, Hannah, and Walnut. 10.15: Courtesy of Pioneer Hi-Bred International, Inc. 10.16: © Carolyn A. McKeone/Photo Researchers, Inc. 10.17: © Peter Morenus/U. of Connecticut. *Bay scallops*: © Barbara J. Miller/Biological Photo Service.

Chapter 11 *T. rex*: © The Natural History Museum, London. *Earrings*: David McIntyre/ Model: Ashley Ying. 11.3: © Biozentrum, U. Basel/ SPL/Photo Researchers, Inc. 11.6 *X-ray crystallograph*: Courtesy of Prof. M. H. F. Wilkins, Dept. of Biophysics, King's College, U. London. 11.6 *Franklin*: © CSHL Archives/Peter Arnold, Inc. 11.8A: © A. Barrington Brown/Photo Researchers, Inc. 11.8B: Data from S. Arnott & D. W. Hukins, 1972. *Biochem. Biophys. Res. Commun.* 47(6): 1504. 11.21C: © Dr. Peter Lansdorp/Visuals Unlimited.

Chapter 12 *Ricinus*: © Alan L. Detrick/Photo Researchers, Inc. *Ribosome*: Data from PDB 1GIX and 1G1Y. M. M. Yusupov et al., 2001. *Science* 292: 883. 12.4: Data from PDB 1I3Q. P. Cramer et al., 2001 *Science* 292: 1863. 12.8: Data from PDB 1EHZ. H. Shi & P. B. Moore, 2000. *RNA* 6: 1091. 12.9B: Data from PDB 1EUQ. L. D. Sherlin et al., 2000. *J. Mol. Biol.* 299: 431. 12.10: Data from PDB 1GIX and 1G1Y. M. M. Yusupov et al., 2001. *Science* 292: 883. 12.14: Courtesy of J. E. Edström and *EMBO J.* 12.18: © Stanley Flegler/Visuals Unlimited.

Chapter 13 *Market*: © Derek Brown/Alamy. *Scientist*: © Tek Image/SPL/Photo Researchers, Inc. 13.1A: © Dept. of Microbiology, Biozentrum/SPL/ Photo Researchers, Inc. 13.1B: © Dennis Kunkel Microscopy, Inc. 13.2A: © Dr. Harold Fisher/ Visuals Unlimited. 13.2B: © E.O.S./Gelderblom/ Photo Researchers, Inc. 13.2C: © BSIP Agency/ Index Stock Imagery/Jupiter Images. 13.7: Courtesy of Roy French. 13.11: Courtesy of L. Caro and R. Curtiss. 13.22: Based on an illustration by Anthony R. Kerlavage, Institute for Genomic Research. *Science* 269: 449 (1995). *For Investigation*: Adapted from J. J. Perry, J. T. Staley, & S. Lory, 2002. *Microbial Life*. Sinauer Associates.

Chapter 14 *Cheetah*: © John Giustina/ ImageState/Jupiter Images. *Panther*: © Lynn M. Stone/Naturepl.com. 14.7: From D. C. Tiemeier et al., 1978. *Cell* 14: 237. 14.18: Courtesy of Murray L. Barr, U. Western Ontario. 14.20: Courtesy of O. L. Miller, Jr.

Chapter 15 *Student*: © Ryan McVay, Photodisc Green/Getty Images. *Ethiopian*: © JTB Photo/ photolibrary.com. 15.2: © Biophoto Associates/ Photo Researchers, Inc. 15.3: From A. M. de Vos, M. Ultsch, & A. A. Kossiakoff, 1992. *Science* 255: 306. 15.14: © Stephen A. Stricker, courtesy of Molecular Probes, Inc.

Chapter 16 *Destruction*: © Suzanne Plunkett/AP Images. *Baby 81*: © Gemunu Amarasinghe/AP Images. 16.2: © Philippe Plailly/Photo Researchers, Inc. 16.5: Bettmann/CORBIS. 16.6: Courtesy of Keith Weller, De Wood, Chris Pooley, & Scott Bauer/USDA ARS. 16.19A: Courtesy of Ingo Potrykus, Swiss Federal Institute of Technology. 16.19B: Joan Gemme. 16.20: Courtesy of Eduardo Blumwald.

Chapter 17 *Champlain*: © North Wind Picture Archives/Alamy. *Family*: © David Sanger Photography/Alamy. 17.6: From C. Harrison et al., 1983. *J. Med. Genet.* 20: 280. 17.11: Courtesy of Harvey Levy and Cecelia Walraven, New England Newborn Screening Program. 17.14: © Dennis Kunkel Microscopy, Inc. 17.19B: © David M. Martin, M.D./SPL/Photo Researchers, Inc. 17.23: From P. H. O'Farrell, 1975. High resolution two-dimensional electrophoresis of proteins. *J. Biol. Chem.* 250: 4007-21. Courtesy of Patrick H. O'Farrell. 17.24: After N. M. Morel et al., 2004. *Mayo Clin. Proc.* 79: 651.

Chapter 18 *Vaccination*: Bettmann/CORBIS. *T cell*: © Dr. Andrejs Liepins/SPL/Photo Researchers, Inc. 18.3: © Dennis Kunkel Microscopy, Inc. 18.8: © Dr. Gopal Murti/SPL/Photo Researchers, Inc. 18.14: © David Phillips/Science Source/Photo Researchers, Inc.

Chapter 19 *Embryo*: © Dr. Yorgas Nikas/Photo Researchers, Inc. *Centrifuge*: Courtesy of Cytori Therapeutics. 19.4: © Roddy Field, the Roslin Institute. 19.5: Courtesy of T. Wakayama and R. Yanagimachi. 19.12: From J. E. Sulston & H. R. Horvitz, 1977. *Dev. Bio.* 56: 100. 19.15B, 19.16 *left*: Courtesy of J. Bowman. 19.16 *right*: Courtesy of Detlef Weigel. 19.17: Courtesy of W. Driever and C. Nüsslein-Vollhard. 19.18B: Courtesy of C. Rushlow and M. Levine. 19.18C: Courtesy of T. Karr. 19.18D: Courtesy of S. Carroll and S. Paddock. 19.20: Courtesy of F. R. Turner, Indiana U.

Chapter 20 *Fly head*: © David Scharf/Photo Researchers, Inc. *Mutant leg*: From G. Halder et al., 1995. *Science* 267: 1788. Courtesy of W. J. Gehring and G. Halder. 20.2A: Courtesy of E. B. Lewis. 20.2B: From H. Le Mouellic et al., 1992. *Cell* 69: 251. Courtesy of H. Le Mouellic, Y. Lallemand, and P. Brûlet. 20.5: Courtesy of J. Hurle and E. Laufer. 20.6: Courtesy of J. Hurle. 20.7 *Cladogram*: After R. Galant & S. Carroll, 2002. *Nature* 415: 910. 20.7 *Beetle*: © Stockbyte/PictureQuest. 20.7 *Centipede*: © Burke/Triolo/Brand X Pictures/PictureQuest. 20.8: Courtesy of S. Carroll and P. Brakefield. 20.9: © Erick Greene. 20.10: Photograph by C. Laforsch & R. Tollrian, courtesy of A. A. Agrawal. 20.11: © Nigel Cattlin, Holt Studios International/Photo Researchers, Inc. 20.13: © Simon D. Pollard/Photo Researchers, Inc. 20.14: Courtesy of Mike Shapiro and David Kingsley.

Chapter 21 *Capybara*: © Frans Lanting/Minden Pictures. *Grand Canyon*: © Robert Fried/Tom Stack & Assoc. 21.4A: David McIntyre. 21.4B: © Robin Smith/photolibrary.com. 21.7: © François Gohier/ The National Audubon Society Collection/Photo Researchers, Inc. 21.8: © Jeff J. Daly/Visuals Unlimited. 21.9 *left*: © Ken Lucas/Visuals Unlimited. 21.9 *right*: © The Natural History Museum, London. 21.10: © Chip Clark. 21.11: © Hans Steur/Visuals Unlimited. 21.12: © Tom McHugh/Field Museum, Chicago/Photo Researchers, Inc. 21.13: Courtesy of Conrad C. Labandeira, Department of Paleobiology, National Museum of Natural History, Smithsonian Institution. 21.14: © Chase Studios, Cedarcreek, MO. 21.16: © The Natural History Museum, London. 21.18: © K. Simons and David Dilcher. 21.20: David McIntyre. 21.22: © John Worrall. 21.23: © Calvin Larsen/Photo Researchers, Inc.

Chapter 22 *Snake*: © Joseph T. Collins/Photo Researchers, Inc. *Newt*: © Robert Clay/Visuals Unlimited. *Pufferfish*: © Georgette Douwma/ Naturepl.com. 22.1A, B: © SPL/Photo Researchers, Inc. 22.2: From W. Levi, 1965. *Encyclopedia of Pigeon Breeds*. T. F. H. Publications, Jersey City, NJ. (A, B: photos by R. L. Kienlen, courtesy of Ralston Purina Company; C, D: photos by Stauber). 22.9A: © S. Maslowski/Visuals Unlimited. 22.9B: © Anthony Cooper/SPL/Photo Researchers, Inc. 22.11, 22.17A: David McIntyre. 22.21A: © Marilyn Kazmers/Dembinsky Photo Assoc. 22.21B: © Paul Osmond/Painet Inc. 22.22 *upper*: © Jeff Foott/ Naturepl.com. 22.22 *lower*: © franzfoto.com/Alamy.

Chapter 23 *Hummingbird*: © Tui De Roy/Minden Pictures. *Swift*: © Kim Taylor/Naturepl.com. 23.1A *left*: © Gary Meszaros/Dembinsky Photo Assoc. 23.1A *right*: © Lior Rubin/Peter Arnold, Inc. 23.1B: © Fi Rust/Painet Inc. 23.8: © Jan Vermeer/Foto Natura/Minden Pictures. 23.9A: © Virginia P. Weinland/Photo Researchers, Inc. 23.9B: © Pablo Galán Cela/AGE Fotostock. 23.10A: © J. S. Sira/ photolibrary.com. 23.10B: © Daniel L. Geiger/ SNAP/Alamy. 23.12: © Boris I. Timofeev/Pensoft. 23.14A: © Tony Tilford/photolibrary.com. 23.14B: © W. Peckover/VIREO. 23.15 *Madia*: © Peter K. Ziminsky/Visuals Unlimited. 23.15 *Argyroxiphium*: © Elizabeth N. Orians. 23.15 *Wilkesia*: © Gerald D. Carr. 23.15 *Dubautia*: © Noble Proctor/The National Audubon Society Collection/Photo Researchers, Inc.

Chapter 24 *Vaccination*: Courtesy of Chris Zahniser, B.S.N., R.N., M.P.H./CDC. *Virus*: © Dr. Tim Baker/Visuals Unlimited. 24.3 *Rice*: data from pdb 1CCR. *Tuna*: data from pdb 5CYT. 24.4: From P. B. Rainey & M. Travisano, 1998. *Nature* 394: 69. © Macmillan Publishers Ltd. 24.7A: © Barrie Britton/Naturepl.com. 24.7B: © M. Graybill/ J. Hodder/Biological Photo Service.

Chapter 25 *HIV*: © James Cavallini/Photo Researchers, Inc. *Chimpanzees*: © John Cancalosi/ Naturepl.com. 25.6A: © Mark Smith/Photo Researchers, Inc. 25.6B: After E. Verheyen et al., 2003. *Science* 300: 325. 25.7: © Larry Jon Friesen. 25.9: © Alexandra Basolo. 25.10: David M. Hillis and Matthew Brauer. 25.11A: © Helen Carr/ Biological Photo Service. 25.11B: © Michael Giannechini/Photo Researchers, Inc. 25.11C: © Skip Moody/Dembinsky Photo Assoc.

Chapter 26 *River*: © Felipe Rodriguez/Alamy. *Salmonella* and *Methanospirillum*: © Kari Lounatmaa/Photo Researchers, Inc. 26.2A: © David Phillips/Photo Researchers, Inc. 26.2B: © R. Kessel & G. Shih/Visuals Unlimited. 26.2C: Courtesy of Janice Carr/NCID/CDC. 26.4: From F. Balagaddé et al., 2005. *Science* 309: 137. Courtesy of Frederick Balagaddé. 26.5A *left*: © David M. Phillips/Visuals Unlimited. 26.5A *right*: Courtesy of Peter Hirsch and Stuart Pankratz. 26.5B *left*: Courtesy of the CDC. 26.5B *right*: Courtesy of Peter Hirsch and Stuart Pankratz. 26.6A: © J. A. Breznak & H. S. Pankratz/Biological Photo Service. 26.6B: © J. Robert Waaland/Biological Photo Service. 26.7: © USDA/Visuals Unlimited. 26.8: © Steven Haddock and Steven Miller. 26.12: Courtesy of David Cox/CDC. 26.13: Courtesy of Randall C. Cutlip. 26.14: © David Phillips/Visuals Unlimited. 26.15A: © Paul W. Johnson/Biological Photo Service. 26.15B: © H. S. Pankratz/Biological Photo Service. 26.15C: © Bill Kamin/Visuals

Unlimited. 26.16: © Dr Kari Lounatmaa/Photo Researchers, Inc. 26.17: © Dr. Gary Gaugler/ Visuals Unlimited. 26.18: © Michael Gabridge/ Visuals Unlimited. 26.20: © Phil Gates/Biological Photo Service. 26.21: From K. Kashefi & D. R. Lovley, 2003. *Science* 301: 934. Courtesy of Kazem Kashefi. 26.23: © Krafft/Hoa-qui/Photo Researchers, Inc. 26.24: © David Sanger Photography/Alamy. 26.25: From H. Huber et al., 2002. *Nature* 417: 63. © Macmillan Publishers Ltd. Courtesy of Karl O. Stetter.

Chapter 27 *Leishmania*: © Dennis Kunkel Microscopy, Inc. *Macrocystis*: © Karen Gowlett-Holmes/photolibrary.com. 27.1: © Steve Gschmeissner/Photo Researchers, Inc. 27.2: © David Patterson, Linda Amaral Zettler, Mike Peglar, & Tom Nerad/micro*scope. 27.3: © London School of Hygiene/SPL/Photo Researchers, Inc. 27.4A: © Bill Bachman/Photo Researchers, Inc. 27.4B: © Markus Geisen/NHMPL. 27.5: © Astrid & Hanns-Frieder Michler/SPL/Photo Researchers, Inc. 27.9: © Mike Abbey/Visuals Unlimited. 27.12A: © Wim van Egmond/Visuals Unlimited. 27.12B: © Biophoto Associates/Photo Researchers, Inc. 27.18: © Dennis Kunkel Microscopy, Inc. 27.19A: © Mike Abbey/Visuals Unlimited. 27.19B: © Dennis Kunkel Microscopy, Inc. 27.19C: © Paul W. Johnson/Biological Photo Service. 27.19D: © M. Abbey/Photo Researchers, Inc. 27.21: © Manfred Kage/Peter Arnold, Inc. 27.22: After A. Ianora et al., 2005. *Nature* 429: 403. 27.23A: © Duncan McEwan/Naturepl.com. 27.23B, C: © Larry Jon Friesen. 27.23D: © J. N. A. Lott/Biological Photo Service. 27.24: © James W. Richardson/Visuals Unlimited. 27.25A: © Larry Jon Friesen. 27.25B: © Milton Rand/Tom Stack & Assoc. 27.26A: © Carolina Biological/Visuals Unlimited. 27.26B: © Larry Jon Friesen. 27.27A: © J. Paulin/Visuals Unlimited. 27.27B: © Dr. David M. Phillips/Visuals Unlimited. 27.29: © Andrew Syred/SPL/Photo Researchers, Inc. 27.30A: © William Bourland/ micro*scope. 27.30B: © David Patterson & Aimlee Laderman/micro*scope. 27.31: © Larry Jon Friesen. 27.32: Courtesy of R. Blanton and M. Grimson.

Chapter 28 *Fossils*: © Sinclair Stammers/SPL/ Photo Researchers, Inc. *Rainforest*: © Photo Resource Hawaii/Alamy. 28.2A: © Ronald Dengler/ Visuals Unlimited. 28.2B: © Larry Mellichamp/ Visuals Unlimited. 28.3: © Brian Enting/Photo Researchers, Inc. 28.5: © J. Robert Waaland/ Biological Photo Service. 28.6: After L. E. Graham et al., 2004. *PNAS* 101: 11025. Micrograph courtesy of Patricia Gensel and Linda Graham. 28.8: © U. Michigan Exhibit Museum. 28.10: Courtesy of the Biology Department Greenhouses, U. Massachusetts, Amherst. 28.11: After C. P. Osborne et al., 2004. *PNAS* 101: 10360. 28.13A, B: David McIntyre. 28.13C: © Harold Taylor/ photolibrary.com. 28.14: © Daniel Vega/AGE Fotostock. 28.15: © Danilo Donadoni/AGE Fotostock. 28.16A: © Ed Reschke/Peter Arnold, Inc. 28.16B: © Carolina Biological/Visuals Unlimited. 28.17A: © J. N. A. Lott/Biological Photo Service. 28.17B: © David Sieren/Visuals Unlimited. 28.18: Courtesy of the Biology Department Greenhouses, U. Massachusetts, Amherst. 28.19 A: © Rod Planck/Dembinsky Photo Assoc. 28.19B: © Nuridsany et Perennou/Photo Researchers, Inc. 28.19C: Courtesy of the Talcott Greenhouse, Mount Holyoke College. 28.20 *inset*: David McIntyre.

Chapter 29 *Masada*: © Eddie Gerald/Alamy. *Coconut*: © Ben Osborne/Naturepl.com. 29.1 *Seed fern*: David McIntyre. 29.1 *Cycad*: © Patricio Robles Gil/Naturepl.com. 29.1 *Magnolia*: © Plantography/ Alamy. 29.4: © NaturalVisions/Alamy. 29.5A: © Patricio Robles Gil/Naturepl.com. 29.5B: © Dave Watts/Naturepl.com. 29.5C: © M. Graybill/J. Hodder/Biological Photo Service. 29.5D: © Frans Lanting/Minden Pictures. 29.7A *left*: David McIntyre. 29.7A *right*: © Stan W. Elems/Visuals Unlimited. 29.7B *left*: David McIntyre. 29.7B *right*: © Dr. John D. Cunningham/Visuals Unlimited. 29.10A: David McIntyre. 29.10B: © Aflo Foto Agency/Naturepl.com. 29.10C: © Richard Shiell/ Dembinsky Photo Assoc. 29.11A: © Plantography/ Alamy. 29.11B: © Thomas Photography LLC/ Alamy. 29.15A: © Inga Spence/Tom Stack & Assoc. 29.15B: © Holt Studios/Photo Researchers, Inc. 29.15C: © Catherine M. Pringle/Biological Photo Service. 29.15D: © blickwinkel/Alamy. 29.17A: Courtesy of Stephen McCabe, U. California, Santa Cruz, and UCSC Arboretum. 29.17B: David McIntyre. 29.17C: © Rob & Ann Simpson/Visuals Unlimited. 29.17D: © R. C. Carpenter/Photo Researchers, Inc. 29.17E: © Geoff Bryant/Photo Researchers, Inc. 29.17F: © José Antonio Jiménez/AGE Fotostock. 29.18A: © Andrew E. Kalnik/Photo Researchers, Inc. 29.18B: © Ed Reschke/Peter Arnold, Inc. 29.18C: © Adam Jones/ Dembinsky Photo Assoc. 29.19A: © Willard Clay/ photolibrary.com. 29.19B: © Adam Jones/ Dembinsky Photo Assoc. 29.19C: © Alan & Linda Detrick/The National Audubon Society Collection/Photo Researchers, Inc. 29.20: © Diaphor La Phototheque/photolibrary.com.

Chapter 30 *Fusarium*: © Dr. Gary Gaugler/ Visuals Unlimited. *Ant*: © L. E. Gilbert/Biological Photo Service. 30.3: © David M. Phillips/Visuals Unlimited. 30.5A: © G. T. Cole/Biological Photo Service. 30.6: © N. Allin & G. L. Barron/Biological Photo Service. 30.7: © Richard Packwood/photo-library.com. 30.8A: © Amy Wynn/Naturepl.com. 30.8B: © Geoff Simpson/Naturepl.com. 30.8C: © Gary Meszaros/Dembinsky Photo Assoc. 30.10A: © R. L. Peterson/Biological Photo Service. 30.10B: © Ken Wagner/Visuals Unlimited. 30.13A: © J. Robert Waaland/Biological Photo Service. 30.13B: © M. F. Brown/Visuals Unlimited. 30.13C: © Dr. John D. Cunningham/Visuals Unlimited. 30.13D: © Biophoto Associates/Photo Researchers, Inc. 30.14 *left*: © William E. Schadel/Biological Photo Service. 30.14 *right*: © Dr. John D. Cunningham/Visuals Unlimited. 30.15: © John Taylor/Visuals Unlimited. 30.16: © G. L. Barron/ Biological Photo Service. 30.17A: © Richard Shiell/Dembinsky Photo Assoc. 30.17B: © Matt Meadows/Peter Arnold, Inc. 30.18: © Andrew Syred/SPL/Photo Researchers, Inc. 30.19A: © Manfred Danegger/Photo Researchers, Inc. 30.19B: © Botanica/photolibrary.com.

Chapter 31 *Fossil*: Photograph by Diane Scott. *Dinosaur*: © Joe Tucciarone/Photo Researchers, Inc. 31.2A: Courtesy of J. B. Morrill. 31.2B: From G. N. Cherr et al., 1992. *Microsc. Res. Tech.* 22: 11. Courtesy of J. B. Morrill. 31.5A: © Jurgen Freund/ Naturepl.com. 31.5B: © Henry W. Robison/Visuals Unlimited. 31.6A: © Brian Parker/Tom Stack & Assoc. 31.6B: © Tui De Roy/Minden Pictures. 31.8: © Colin M. Orians. 31.9A: © T. Kitchin & V. Hurst/Photo Researchers, Inc. 31.9B: © Stockbyte/ PictureQuest. 31.10: Adapted from F. M. Bayerand & H. B. Owre, 1968. *The Free-Living Lower Invertebrates*, Macmillan Publishing Co. 31.11A: © Scott Camazine/Alamy. 31.11B, C: David McIntyre. 31.13A: © Stephen Dalton/Minden Pictures. 31.13B: © Gerald & Buff Corsi/Visuals Unlimited. 31.14A: © Dave Watts/Naturepl.com. 31.14B: © Larry Jon Friesen. 31.16A: © Jurgen Freund/Naturepl.com. 31.16B: © Larry Jon Friesen. 31.16C: © David Wrobel/Visuals Unlimited. 31.17B: © Larry Jon Friesen. 31.18: Adapted from F. M. Bayerand & H. B. Owre, 1968. *The Free-Living Lower Invertebrates*, Macmillan Publishing Co. 31.19A: © Larry Jon Friesen. 31.19B: © Michael Patrick O'Neill/Alamy. 31.19C, D: © Larry Jon Friesen. 31.20A: From J. E. N. Veron, 2000. *Corals of the World*. © J. E. N. Veron. 31.20B: © Jurgen Freund/Naturepl.com. 31.21: Adapted from F. M. Bayerand & H. B. Owre, 1968. *The Free-Living Lower Invertebrates*, Macmillan Publishing Co.

Chapter 32 *Strepsipterans*: Courtesy of Dr. Hans Pohl. *Wasp*: © Paulo de Oliveira/OSF/Jupiter Images. 32.2: © Robert Brons/Biological Photo Service. 32.3A: From D. C. García-Bellido & D. H. Collins, 2004. *Nature* 429: 40. Courtesy of Diego García-Bellido Capdevila. 32.3B: © Piotr Naskrecki/ Minden Pictures. 32.6: © Alexis Rosenfeld/Photo Researchers, Inc. 32.7A: © Ed Robinson/ photolibrary.com. 32.8B: © Robert Brons/Biological Photo Service. 32.9B: © Larry Jon Friesen. 32.10A: © Stan Elems/Visuals Unlimited. 32.11: © David J. Wrobel/Visuals Unlimited. 32.13A: © Jurgen Freund/Naturepl.com. 32.13B: Courtesy of R. R. Hessler, Scripps Institute of Oceanography. 32.13C: © Roger K. Burnard/Biological Photo Service. 32.13D: © Larry Jon Friesen. 32.15A: © Larry Jon Friesen. 32.15B: © Dave Fleetham/ photolibrary.com. 32.15C, D, E: © Larry Jon Friesen. 32.15F: © Orion Press/Jupiter Images. 32.16A: Courtesy of Jen Grenier and Sean Carroll, U. Wisconsin. 32.16B: Courtesy of Graham Budd. 32.16C: Courtesy of Reinhardt Møbjerg Kristensen. 32.17: © R. Calentine/Visuals Unlimited. 32.18B, C: © James Solliday/Biological Photo Service. 32.20A: © Michael Fogden/ photolibrary.com. 32.20B: © Diane R. Nelson/ Visuals Unlimited. 32.21: © Ken Lucas/Visuals Unlimited. 32.22A: © Frans Lanting/Minden Pictures. 32.22B: © Larry Jon Friesen. 32.22C: © Tom Branch/Photo Researchers, Inc. 32.22D: © Norbert Wu/Minden Pictures. 32.25: © Mark Moffett/Minden Pictures. 32.26: © Oxford Scientific Films/Jupiter Images. 32.27A: © Larry Jon Friesen. 32.27B: © Larry Jon Friesen. 32.27C: © Meul/ARCO/Naturepl.com. 32.27D: © Piotr Naskrecki/Minden Pictures. 32.27E: © Colin Milkins/Oxford Scientific Films/photolibrary.com. 32.27F: © Peter J. Bryant/Biological Photo Service. 32.27G: © Larry Jon Friesen. 32.27H: David McIntyre. 32.29A: © John Mitchell/Jupiter Images. 32.29B: © John R. MacGregor/Peter Arnold, Inc. 32.30A: © Norbert Wu/Minden Pictures. 32.30B: © Frans Lanting/Minden Pictures. 32.31A: © Kelly Swift, www.swiftinverts.com. 32.31B: © Larry Jon Friesen. 32.31C: © W. M. Beatty/Visuals Unlimited. 32.31D: Photo by Eric Erbe; colorization by Chris Pooley/USDA ARS.

Chapter 33 *Homo floresiensis*: © Christian Darkin/Alamy. *Tunicates*: © Larry Jon Friesen. 33.2: From S. Bengtson, 2000. Teasing fossils out of shales with cameras and computers. *Palaeontologia Electronica* 3(1). 33.4A: © Larry Jon Friesen. 33.4B: © Hal Beral/Visuals Unlimited. 33.4C: © Randy Morse/Tom Stack & Assoc. 33.4D: © Mark J. Thomas/Dembinsky Photo Assoc. 33.4E: © Peter Scoones/Photo Researchers, Inc. 33.4F: © Larry Jon Friesen. 33.5A: © C. R. Wyttenbach/Biological Photo Service. 33.6: © Robert Brons/Biological Photo Service. 33.7A: © Jurgen Freund/ Naturepl.com. 33.7B: © David Wrobel/Visuals Unlimited. 33.9A: © Brian Parker/Tom Stack &

Assoc. 33.9B: © Reijo Juurinen/Naturbild/Naturepl.com. 33.11B: © Roger Klocek/Visuals Unlimited. 33.12A: © Dave Fleetham/Tom Stack & Assoc. 33.12B: © Kelvin Aitken/AGE Fotostock. 33.12C: © Dave Fleetham/Tom Stack & Assoc. 33.13A: © Tobias Bernhard/Oxford Scientific Films/photolibrary.com. 33.13B: © Larry Jon Friesen. 33.13C: © Dave Fleetham/Visuals Unlimited. 33.13D: © Larry Jon Friesen. 33.14A, B: © Tom McHugh, Steinhart Aquarium/The National Audubon Society Collection/Photo Researchers, Inc. 33.14C: © Ted Daeschler/Academy of Natural Sciences/VIREO. 33.16A: © Ken Lucas/Biological Photo Service. 33.16B: © Michael & Patricia Fogden/Minden Pictures. 33.16C: © Gary Meszaros/Dembinsky Photo Assoc. 33.16D: © Larry Jon Friesen. 33.19A: © Dave B. Fleetham/Tom Stack & Assoc. 33.19B: © C. Alan Morgan/Peter Arnold, Inc. 33.19C: © Dave Watts/Alamy. 33.19D: © Larry Jon Friesen. 33.20A: © Frans Lanting/Minden Pictures. 33.20B: © Gerry Ellis, DigitalVision/PictureQuest. 33.21A: From X. Xu et al., 2003. *Nature* 421: 335. © Macmillan Publishers Ltd. 33.21B: © Tom & Therisa Stack/Painet, Inc. 33.23A: © Roger Wilmhurst/Foto Natura/Minden Pictures. 33.23B: © Andrew D. Sinauer. 33.23C: © Tom Vezo/Minden Pictures. 33.23D: © Skip Moody/Dembinsky Photo Assoc. 33.24A: © Ed Kanze/Dembinsky Photo Assoc. 33.24B: © Dave Watts/Tom Stack & Assoc. 33.25A: © Ingo Arndt/Naturepl.com. 33.25B: © JTB Photo Communications/photolibrary.com. 33.25C: © Jany Sauvanet/Photo Researchers, Inc. 33.26A: © Tim Jackson/OSF/Jupiter Images. 33.26B: © Claude Steelman/OSF/Jupiter Images. 33.26C: © Michael S. Nolan/Tom Stack & Assoc. 33.26D: © Erwin & Peggy Bauer/Tom Stack & Assoc. 33.28: © Pete Oxford/Minden Pictures. 33.29A: © mike lane/Alamy. 33.29B: © Cyril Ruoso/JH Editorial/Minden Pictures. 33.30A: © Frans Lanting/Minden Pictures. 33.30B: © Anup Shah/Dembinsky Photo Assoc. 33.30C: © Stan Osolinsky/Dembinsky Photo Assoc. 33.30D: © Anup Shah/Dembinsky Photo Assoc.

Chapter 34 *Poster*: Courtesy of NJSP Museum. *Wood*: © Werner H. Muller/Peter Arnold, Inc. 34.3: David McIntyre. 34.4A: © Joyce Photographics/Photo Researchers, Inc. 34.4B: © C. K. Lorenz/The National Audubon Society Collection/Photo Researchers, Inc. 34.4C: © Renee Lynn/Photo Researchers, Inc. 34.8: © Biophoto Associates/Photo Researchers, Inc. 34.9A: © Biodisc/Visuals Unlimited. 34.9B: © P. Gates/Biological Photo Service. 34.9C: © Biophoto Associates/Photo Researchers, Inc. 34.9D: © Jack M. Bostrack/Visuals Unlimited. 34.9E: © John D. Cunningham/Visuals Unlimited. 34.9F: © J. Robert Waaland/Biological Photo Service. 34.9G: © Randy Moore/Visuals Unlimited. 34.10: © R. Kessel & G. Shih/Visuals Unlimited. 34.11 *upper*: © Larry Jon Friesen. 34.11 *lower*: © Biodisc/Visuals Unlimited. 34.13: © D. Cavagnaro/Visuals Unlimited. 34.14B: © Microfield Scientific LTD/Photo Researchers, Inc. 34.16A: © Larry Jon Friesen. 34.16B: © Ed Reschke/Peter Arnold, Inc. 34.16C: © Dr. James W. Richardson/Visuals Unlimited. 34.17A *left*: © Ed Reschke/Peter Arnold, Inc. 34.17A *right*: © Biodisc/Visuals Unlimited. 34.17B: © Ed Reschke/Peter Arnold, Inc. 34.19: © John N. A. Lott/Biological Photo Service. 34.20: © Jim Solliday/Biological Photo Service. 34.21: From C. Maton & B. L. Gartner, 2005. *Am. J. Botany* 92: 123. Courtesy of Barbara L. Gartner. 34.22: © Biodisc/Visuals Unlimited. 34.23B: Courtesy of Thomas Eisner, Cornell U. 34.23C: © Larry Jon Friesen.

Chapter 35 *Sunflower*: © Aflo/Naturepl.com. *Sequoia*: © Jeff Foott/Naturepl.com. 35.5: David McIntyre. 35.8: After M. A. Zwieniecki, P. J. Melcher, & N. M. Holbrook, 2001. *Science* 291: 1059. 35.9A: © David M. Phillips/Visuals Unlimited. 35.10: After G. D. Humble & K. Raschke, 1971. *Plant Physiology* 48: 447. 35.12: © M. H. Zimmermann.

Chapter 36 *Family*: Dorothea Lange/Library of Congress, Prints & Photographs Division, FSA/OWI Collection. *Plow*: © Liz Hahn. 36.1: David McIntyre. 36.5: © Kathleen Blanchard/Visuals Unlimited. 36.7: © Hugh Spencer/Photo Researchers, Inc. 36.9 *left*: © E. H. Newcomb & S. R. Tandon/Biological Photo Service. 36.9 *right*: © Dr. Jeremy Burgess/SPL/Photo Researchers, Inc. 36.11A: © J. H. Robinson/The National Audubon Society Collection/Photo Researchers, Inc. 36.11B: © Kim Taylor/Naturepl.com. 36.12: Courtesy of Susan and Edwin McGlew.

Chapter 37 *Rubber ball*: © Luiz Claudio Marigo/Naturepl.com. *Tapping tree*: © J. D. Dallet/AGE Fotostock. 37.2: © Tom J. Ulrich/Visuals Unlimited. 37.3: © Joe Sohm/Pan America/Jupiter Images. 37.5: Courtesy of J. A. D. Zeevaart, Michigan State U. 37.11: © Ed Reschke/Peter Arnold, Inc. 37.13: © Biophoto Associates/Photo Researchers, Inc. 37.17, 37.20: David McIntyre. 37.21: Courtesy of Dr. Eva Huala, Carnegie Institution of Washington.

Chapter 38 *Rafflesia*: © Ingo Arndt/Minden Pictures. *Amorphophallus*: © Deni Bown/OSF/Jupiter Images. 38.2: © RMF/Visuals Unlimited. 38.4: From J. Bowman (ed.), 1994. *Arabiopsis: An Atlas of Morphology and Development.* Springer-Verlag, New York. Photo by S. Craig & A. Chaudhury, Plate 6.2. 38.8: David McIntyre. 38.14: After A. N. Dodd et al., 2005. *Science* 309: 630. 38.15: After M. J. Yanovsky & S. A. Kay, 2002. *Nature* 419: 308. 38.17: © Gerry Ellis/Minden Pictures. 38.18A: © Nigel Cattlin, Holt Studios International/Photo Researchers, Inc. 38.18B: © Maurice Nimmo/SPL/Photo Researchers, Inc.

Chapter 39 *Flood*: © Abir Abdullah/Peter Arnold, Inc. *Salty soil*: © STRDEL/AFP/Getty Images. 39.2: © Holt Studios International Ltd/Alamy. 39.5: After A. Steppuhn et al., 2004. *PLoS Biology* 2: 1074. 39.6: After S. Rasmann et al., 2005. *Nature* 434: 732. 39.8: Courtesy of Thomas Eisner, Cornell U. 39.9: © Adam Jones/Dembinsky Photo Assoc. 39.10: © John N. A. Lott/Biological Photo Service. 39.11 *left*: David McIntyre. 39.11 *right*: © Frans Lanting/Minden Pictures. 39.12: © Simon Fraser/SPL/Photo Researchers, Inc. 39.13: © imagebroker/Alamy. 39.14: © John N. A. Lott/Biological Photo Service. 39.15: © Jurgen Freund/Naturepl.com. 39.17: © Budd Titlow/Visuals Unlimited.

Chapter 40 *Radcliffe*: © Nick Laham/Getty Images. *Soldier*: Courtesy of Major Bryan "Scott" Robison. 40.3A: © Gladden Willis/Visuals Unlimited. 40.3B: © Ed Reschke/Peter Arnold, Inc. 40.4: © Manfred Kage/Peter Arnold, Inc. 40.5: © Ed Reschke/Peter Arnold, Inc. 40.6A: © James Cavallini/Photo Researchers, Inc. 40.6B: © Innerspace Imaging/SPL/Photo Researchers, Inc. 40.10B: © Frans Lanting/Minden Pictures. 40.12: © Gerry Ellis/PictureQuest/Jupiter Images. 40.14: Courtesy of Anton Stabentheiner. 40.17: © Gladden Willis/Visuals Unlimited. 40.18A: © Robert Shantz/Alamy. 40.18B: © Jim Brandenburg/Minden Pictures.

Chapter 41 *Bicyclists*: © Peter Dejong/AP Images. *Body builder*: © Tiziana Fabi/AFP/Getty Images. 41.9A: © Ed Reschke/Peter Arnold, Inc. 41.9C: © Scott Camazine/Photo Researchers, Inc. 41.14: Courtesy of Gerhard Heldmaier, Philipps U.

Chapter 42 *Queen bee*: David McIntyre. *Swarm*: © Stephen Dalton/Minden Pictures. 42.1A: © Biophoto Associates/Photo Researchers, Inc. 42.1B: © Constantinos Petrinos/Naturepl.com. 42.2A: © Patricia J. Wynne. 42.6: © SIU/Peter Arnold, Inc. 42.7: © ullstein - Ibis Bildagentur / Peter Arnold, Inc. 42.8A: © Mitsuaki Iwago/Minden Pictures. 42.8B: © Dave Watts/Naturepl.com. 42.10B: © Ed Reschke/Peter Arnold, Inc. 42.13: © P. Bagavandoss/Photo Researchers, Inc. 42.16B: © S. I. U. School of Med./Photo Researchers, Inc. 42.18: Courtesy of The Institute for Reproductive Medicine and Science of Saint Barnabas, New Jersey.

Chapter 43 *Whale*: © François Gohier/Photo Researchers, Inc. *Egg and sperm*: © Dr. David M. Phillips/Visuals Unlimited. 43.1: Courtesy of Richard Elinson, U. Toronto. 43.4: From J. G. Mulnard, 1967. *Arch. Biol. (Liege)* 78: 107. Courtesy of J. G. Mulnard. 43.20A: © CNRI/SPL/Photo Researchers, Inc. 43.20B: © Dr. G. Moscoso/SPL/Photo Researchers, Inc. 43.20C: © Tissuepix/SPL/Photo Researchers, Inc. 43.20D: © Petit Format/Photo Researchers, Inc.

Chapter 44 *Movie still*: © New Line Cinema/Courtesy of the Everett Collection. *Brain scan*: Courtesy of Dr. Kevin LaBar, Duke U. 44.4B: © C. Raines/Visuals Unlimited. 44.7: From A. L. Hodgkin & R. D. Keynes, 1956. *J. Physiol.* 148: 127.

Chapter 45 *Snake*: © James Gerholdt/Peter Arnold, Inc. *Bat*: © Stephen Dalton/Photo Researchers, Inc. 45.3A: © Hans Pfletschinger/Peter Arnold, Inc. 45.3B: David McIntyre. 45.10A: © P. Motta/Photo Researchers, Inc. 45.16A: © Dennis Kunkel Microscopy, Inc. 45.19: © Omikron/Science Source/Photo Researchers, Inc.

Chapter 46 *Neuron*: From H. van Praag et al., 2002. *Nature* 415: 1030. © Macmillan Publishers Ltd. *Meadowlark*: © Werner Bollmann/AGE Fotostock. 46.8: From J. M. Harlow, 1869. *Recovery from the passage of an iron bar through the head.* Boston: David Clapp & Son. 46.14A: © David Joel Photography, Inc. 46.16: © Wellcome Dept. of Cognitive Neurology/SPL/Photo Researchers, Inc.

Chapter 47 *Joyner-Kersee*: © AFP/Getty Images. *Frog*: © Michael Durham/Minden Pictures. 47.1: © Frank A. Pepe/Biological Photo Service. 47.2: © Tom Deerinck/Visuals Unlimited. 47.4: © Kent Wood/Peter Arnold, Inc. 47.7A: © Manfred Kage/Peter Arnold, Inc. 47.7B: © Innerspace Imaging/SPL/Photo Researchers, Inc. 47.7C: © SPL/Photo Researchers, Inc. 47.10: Courtesy of Jesper L. Andersen. 47.14: From the collection of Andrew Sinauer/photo by David McIntyre. 47.18: © Robert Brons/Biological Photo Service. 47.22B *upper*: © Ken Lucas/Visuals Unlimited. 47.22B *lower*: © Fred McConnaughey/The National Audubon Society Collection/Photo Researchers, Inc.

Chapter 48 *Geese*: © Konrad Wothe/Minden Pictures (geese) and John Warden/PictureQuest (mountains). *Barracudas*: © Fred Bavendam/Minden Pictures. 48.1A: © Larry Jon Friesen. 48.1B: © Norbert Wu/Minden Pictures. 48.1C: © Tom McHugh/Photo Researchers, Inc. 48.4B: © Skip Moody/Dembinsky Photo Assoc. 48.4C:

Courtesy of Thomas Eisner, Cornell U. 48.10A: © SPL/Photo Researchers, Inc. 48.10C: © P. Motta/Photo Researchers, Inc. 48.13: © Berndt Fischer/Jupiter Images. 48.17: After C. R. Bainton, 1972. *J. Appl. Physiol.* 33: 775.

Chapter 49 *Lewis*: © NBAE/Getty Images. *Emergency room*: © BananaStock/Alamy. 49.8A: © Brand X Pictures/Alamy. 49.9: After N. Campbell, 1990. *Biology*, 2nd Ed., Benjamin Cummings. 49.10B: © CNRI/Photo Researchers, Inc. 49.12: © Ed Reschke/Peter Arnold, Inc. 49.15A: © Chuck Brown/Science Source/Photo Researchers, Inc. 49.15B: © Biophoto Associates/Science Source/Photo Researchers, Inc. 49.19: © Doc White/Nature Picture Library.

Chapter 50 *Pima*: © Marilyn "Angel" Wynn/Nativestock.com. *Fast food*: © Matt Bowman/FoodPix/Jupiter Images. 50.1A: © Gerry Ellis, DigitalVision/PictureQuest. 50.1B: © Rinie Van Meurs/Foto Natura/Minden Pictures. 50.3: © AP/Wide World Photos. 50.6: © Ace Stock Limited/Alamy. 50.9: © Dennis Kunkel Microscopy, Inc. 50.14: © Eye of Science/SPL/Photo Researchers, Inc. 50.20: © Science VU/Jackson/Visuals Unlimited.

Chapter 51 *Bat*: © Michael & Patricia Fogden/Minden Pictures. *Kangaroo rat*: © Mary McDonald/Naturepl.com. 51.1 *inset*: © Kim Taylor/Naturepl.com. 51.2B: © Rod Planck/Photo Researchers, Inc. 51.8: From R. G. Kessel & R. H. Kardon, 1979. *Tissues and Organs.* W. H. Freeman, San Francisco. 51.11: From L. Bankir & C. de Rouffignac, 1985. *Am. J. Physiol.* 249: R643-R666. Courtesy of Lise Bankir, INSERM Unit, Hôpital Necker, Paris. 51.13: © Hank Morgan/Photo Researchers, Inc.

Chapter 52 *Katrina*: Courtesy of the NOAA Satellite and Information Service. *Swamp*: © Tim Fitzharris/Minden Pictures. 52.1: © Tim Fitzharris/Minden Pictures. *Tundra, upper*: © Tim Acker/Auscape/Minden Pictures. *Tundra, lower*: © Elizabeth N. Orians. *Boreal, upper*: © Carr Clifton/Minden Pictures. *Boreal, lower*: © Robert Harding Picture Library Ltd/Alamy. *Temperate deciduous*: © Paul W. Johnson/Biological Photo Service. *Temperate grasslands, upper*: © Robert & Jean Pollock/Biological Photo Service. *Temperate grasslands, lower*: © Elizabeth N. Orians. *Cold desert, upper*: © Edward Ely/Biological Photo Service. *Cold desert, lower*: © Robert Harding Picture Library Ltd./photolibrary.com. *Hot desert, left*: © Terry Donnelly/Tom Stack & Assoc. *Hot desert, right*: © Dave Watts/Tom Stack & Assoc. *Chaparral, left*: © Elizabeth N. Orians. *Chaparral, right*: © Larry Jon Friesen. *Thorn forest*: © Frans Lanting/Minden Pictures. *Savanna*: © Nigel Dennis/AGE Fotostock. *Tropical deciduous*: Courtesy of Donald L. Stone. *Tropical evergreen*: © Elizabeth N. Orians. 52.6: © Elizabeth N. Orians. 52.9 *Chameleon*: © Pete Oxford/Naturepl.com. 52.9 *Tenrec*: © Nigel J. Dennis/Photo Researchers, Inc. 52.9 *Fossa*: © Frans Lanting/Minden Pictures. 52.9 *Lemur*: © Wendy Dennis/Dembinsky Photo Associates. 52.9 *Baobab*: © Pete Oxford/Naturepl.com. 52.9 *Pachypodium*: © Michael Leach/Oxford Scientific Films/photolibrary.com. 52.12: Courtesy of E. O. Wilson. 52.13A *left*: © Bill Lea/Dembinsky Photo Associates. 52.13A *right*: © Stephen J. Krasemann/Photo Researchers, Inc. 52.13B *left*: © Frans Lanting/Minden Pictures. 52.13B *center*: © Kenneth W. Fink/Photo Researchers, Inc. 52.13B *right*: © Chris Gomersall/Naturepl.com.

Chapter 53 *Macaques*: © Frans de Waal, Emory U. *Spider*: © Mark Gibson/Jupiter Images. 53.2A: From J. R. Brown et al., 1996. *Cell* 86: 297. Courtesy of Michael Greenberg. 53.4A: © Nina Leen/Time Life Pictures/Getty Images. 53.4B: © Frans Lanting/Minden Pictures. 53.9A: © Anup Shah/Naturepl.com. 53.9B: © Mitsuaki Iwago/Minden Pictures. 53.9C: © Konrad Wothe/Minden Pictures. 53.11: © Frans Lanting/Minden Pictures. 53.13: Courtesy of John Alcock, Arizona State U. 53.16: © Tui De Roy/Minden Pictures. 53.18: © Cyril Ruoso/JH Editorial/Minden Pictures. 53.21: © Nigel Dennis/AGE Fotostock. 53.22: © Piotr Naskrecki/Minden Pictures. 53.23: © Steve & Dave Maslowski/Photo Researchers, Inc. 53.24: Courtesy of John Alcock, Arizona State U. 53.25: © José Fuste Raga/AGE Fotostock.

Chapter 54 *Caterpillars*: Courtesy of John R. Hosking, NSW Agriculture, Australia. *Farmer*: © Thomas Shjarback/Alamy. 54.1A: © Flip Nicklin/Minden Pictures. 54.1B: © PhotoStockFile/Alamy. 54.1C: © Robert McGouey/Alamy. 54.5A: © Frans Lanting/Minden Pictures. 54.5B: © David Nunuk/SPL/Photo Researchers, Inc. 54.6A: © Michael Durham/Minden Pictures. 54.6B: Courtesy of Colin Chapman. 54.7: © Larry Jon Friesen. 54.11: After P. A. Marquet, 2000. *Science* 289: 1487. 54.12: © Ed Reschke/Peter Arnold, Inc. 54.13: © Adam Jones/Photo Researchers, Inc. 54.14: © T. W. Davies/California Academy of Sciences. 54.18: © Kathie Atkinson/OSF/Jupiter Images.

Chapter 55 *Ant on spine*: © Piotr Naskrecki/Minden Pictures. *Ant with Beltian bodies*: © Mark Moffett/Minden Pictures. 55.1: After R. H. Whittaker, 1960. *Ecological Monographs* 30: 279. 55.4: © Alan & Sandy Carey/Photo Researchers, Inc. 55.6: © Dave Watts/Naturepl.com. 55.7: © Lawrence E. Gilbert/Biological Photo Service. 55.8: © Shehzad Noorani/Peter Arnold, Inc. 55.11: © Mitsuaki Iwago/Minden Pictures. 55.12A: © Kim Taylor/Naturepl.com. 55.12B: © Perennou Nuridsany/Photo Researchers, Inc. 55.13D: Courtesy of William W. Dunmire/National Park Service. 55.14A: David McIntyre. 55.15: © Jim Zipp/Photo Researchers, Inc. 55.16: Courtesy of Jim Peaco/National Park Service. 55.18: After M. Begon, J. Harper, & C. Townsend, 1986. *Ecology*. Blackwell Scientific Publications.

Chapter 56 *Crane*: © Mark Moffett/Minden Pictures. *Researchers*: Courtesy of Christian Koerner, U. Basel. 56.1: After M. C. Jacobson et al., 2000. *Introduction to Earth System Science.* Academic Press. 56.9A: © Corbis Images/PictureQuest.

Chapter 57 *Flight*: © Tom Hugh-Jones/Naturepl.com. *Crane suit*: © Mark Payne-Gill/Naturepl.com. 57.1: © Michael Long/NHMPL. 57.2B: © Mark Godfrey/The Nature Conservancy. 57.6: Richard Bierregaard, Courtesy of the Smithsonian Institution, Office of Environmental Awareness. 57.7: © Mr_Jamsey/iStockphoto.com. 57.8A: © Terry Whittaker/Alamy. 57.8B: © Paul Johnson/Naturepl.com. 57.9: © Michael & Patricia Fogden/Minden Pictures. 57.10A: After R. B. Aronson et al., 2000. *Nature* 405: 36. 57.10B: © Fred Bavendam/Minden Pictures. 57.12: Courtesy of WWF-US, GIS Map provided by J. Morrison. 57.14B: © Jim Brandenburg/Minden Pictures. 57.14C: Courtesy of Jesse Achtenberg/U.S. Fish and Wildlife Service. 57.15A: Courtesy of Christopher Baisan and the Laboratory of Tree-Ring Research, U. Arizona, Tucson. 57.15B: © Karen Wattenmaker/Painet, Inc. 57.17: After S. H. Reichard & C. W. Hamilton, 1997. *Conservation Biology* 11: 193. 57.18A: © Elizabeth N. Orians. 57.20: © Tom Vezo/Nature Picture Library.

Index

Numbers in **boldface** indicate a definition of the term. Numbers in *italic* indicate the information will be found in a figure, caption, or table.

- A -

- B -

- C -

- D -

- F -

- I -

- J -

- K -

- L -

- M -

- N -

- U -